Physical Chemistry
Third Edition

Gilbert W. Castellan
University of Maryland

▲ Addison-Wesley Publishing Company
Reading, Massachusetts
Menlo Park, California • London • Amsterdam • Don Mills, Ontario • Sydney

To Joan and our family

Sponsoring Editor: *Robert L. Rogers*
Production Editor: *Margaret Pinette*
Copy Editor: *Jerrold A. Moore*

Text Designer: *Debbie Syrotchen*
Design Coordinator: *Herb Caswell*
Illustrators: *VAP International Communications, Ltd.*
Cover Designer: *Richard Hannus, Hannus Design Associates*
Cover Photograph: *The Image Bank, U. Schiller*
Art Coordinator: *Joseph K. Vetere*

Production Manager: *Herbert Nolan*

The text of this book was composed in Monophoto Times Roman by Composition House Limited.

Reprinted with corrections, November 1983

Copyright © 1983, 1971, 1964 by Addison-Wesley Publishing Company, Inc.

All rights reserved. No part of this publication may be reproduced, stored in a retrieval system, or transmitted, in any form or by any means, electronic, mechanical, photocopying, recording, or otherwise, without the prior written permission of the publisher. Printed in the United States of America. Published simultaneously in Canada. Library of Congress Catalog Card No. 82-74043.

ISBN 0-201-10386-9

BCDEFGHIJ-MA-89876543

Foreword
to the Student

On most campuses the course in physical chemistry has a reputation for difficulty. It is not, nor should it be, the easiest course available; but to keep the matter in perspective it must be said that the IQ of a genius is not necessary for understanding the subject.

The greatest stumbling block that can be erected in the path of learning physical chemistry is the notion that memorizing equations is a sensible way to proceed. Memory should be reserved for the fundamentals and important definitions. Equations are meant to be understood, not to be memorized. In physics and chemistry an equation is not a jumbled mass of symbols, but is a statement of a relation between physical quantities. As you study keep a pencil and scratch paper handy. Play with the final equation from a derivation. If it expresses pressure as a function of temperature, turn it around and express the temperature as a function of pressure. Sketch the functions so that you can "see" the variation. How does the sketch look if one of the parameters is changed? Read physical meaning into the various terms and the algebraic signs which appear in the equation. If a simplifying assumption has been made in the derivation, go back and see what would happen if that assumption were omitted. Apply the derivation to a different special case. Invent problems of your own involving this equation and solve them. Juggle the equation back and forth until you understand its meaning.

In the first parts of the book much space is devoted to the meaning of equations; I hope that I have not been too long-winded about it, but it is important to be able to interpret the mathematical statement in terms of its physical content.

By all means try to keep a good grasp on the fundamental principles that are being applied; memorize *them* and above all *understand* them. Take the time to understand the *methods* that are used to attack a problem.

In Appendix I there is a brief recapitulation of some of the most important mathematical ideas and methods that are used. If any of these things are unfamiliar to you, take the time to review them in a mathematics text. Once the relations between variables have been established, the algebra and calculus are simply mechanical devices, but they should be respected as precision tools.

If problems baffle you, learn the technique of problem solving. The principles contained in G. Polya's book, *How to Solve It*, have helped many of my students.* It is available as a paperback and is well worth studying. Work as many problems as possible. Numerical answers to *all* problems can be found in Appendix VII. Make up your own problems as often as possible. Watching your teacher perform will not make you into an actor; problem solving will. To aid in this, get a good "scientific" calculator (the serious student will want a programmable one with continuous memory) and learn how to use it to the limit of its capability. Reading the instructions will save you hundreds of hours!

Finally, don't be put off by the reputation for difficulty. Many students have enjoyed learning physical chemistry.

---

* G. Polya, *How to Solve It*. Anchor Book No. 93. New York: Doubleday & Co., 1957.

Preface

An introductory course in physical chemistry must expose the fundamental principles that are applicable to all kinds of physicochemical systems. Beyond the exposition of fundamentals, the first course in physical chemistry takes as many directions as there are teachers. I have tried to cover the fundamentals and some applications in depth. The primary aim has been to write a book that the student can, with effort, read and understand; to provide the beginner with a reliable and understandable guide for study in the teacher's absence. I hope that this book is readable enough so that teachers may leave the side issues and the more elementary aspects for assigned reading while they use the lectures to illuminate the more difficult points. Chapters 1, 5, and 6, and most of Chapter 19 contain some general background material and are intended exclusively for reading.

Except where it would needlessly overburden the student, the subject is presented in a mathematically rigorous way. In spite of this, no mathematics beyond the elementary calculus is required. The justification for a rigorous treatment is pedagogical; it makes the subject simpler. The beginner may find it difficult at first to follow a lengthy derivation, but *can* follow it if it is rigorous and logical. Some "simplified" derivations are not difficult to follow, but impossible.

CHANGES IN THIS EDITION

There are several important differences between this edition and the earlier one. I am grateful to Professor James T. Hynes, University of Colorado, who kindly supplied the groups of questions at the end of each chapter. These are an important addition to the book. The questions range in difficulty; some are relatively simple while others challenge the student to take up a line of reasoning from the chapter and apply it beyond the topics that are discussed explicitly. Many new problems have been added; the total is over 750, about twice the number in the second edition. Answers to all the problems are given in Appendix VII. More worked examples are included; these are now set apart from the text, while before they were sometimes hidden in the textual material. A separate solutions manual is in preparation in which representative problems are worked out in detail. Certain sections of the text are marked with a star. The star indicates that the material is either (1) an additional illustration of or a side issue related to the topic under discussion, or (2) a more advanced topic.

In the treatment of thermodynamics, some errors have been corrected, some passages clarified, and a few new topics introduced. The emphasis on the laws of thermodynamics as generalizations from experience is maintained. The chapter on electrochemical cells has been revised and a discussion of electrochemical power sources has been added. The chapter on surface phenomena now includes sections on the BET isotherm and on the properties of very small particles.

The chapters on the quantum mechanics of simple systems have been retained with only minor revisions, while the chapter on the covalent bond has been extended to include a description of molecular energy levels. The basic ideas of group theory are introduced here and illustrated by constructing symmetry-adapted molecular orbitals for simple molecules. There is a new chapter on atomic spectroscopy; the chapter on molecular spectroscopy has been expanded and reorganized.

The treatment of statistical thermodynamics has been extended to include the calculation of equilibrium constants for simple chemical reactions. At the end of the book, new sections on photophysical kinetics, electrochemical kinetics, and a brief chapter on polymers have been added.

TERMINOLOGY AND UNITS

With only a few exceptions I have followed the recommendations of the International Union of Pure and Applied Chemistry (IUPAC) for symbols and terminology. I have retained the traditional name, "advancement of the reaction" for the parameter ξ, rather than "extent of reaction," which is recommended by IUPAC. The connotation in English of the words "advancement" and "advance" when applied to chemical reactions allow a variety of expression that "extent" and its derivatives do not. For thermodynamic work I have retained the sign convention used in the earlier edition. I attempted (unsuccessfully, I thought) to write a clear discussion of the Carnot cycle and its consequences using the alternate sign convention. Then, after examining some other recent books that use the alternate sign convention, I came to the opinion that their discussions of the second law are not distinguished by their clarity. It seems to me that if the subterfuges used in some of these books are needed for clarity, then the game is not worth the candle.

The SI has been used almost exclusively throughout the book. Except for the thermodynamic equations that involve 1 atm or 1 mol/L as standard states (and a few other equations that explicitly involve non-SI units), all the equations in this book have been written in the SI, so that if the values of all the physical quantities are expressed in the correct SI unit, the quantity desired will be obtained in the correct SI unit. The net result is that the calculations of physical chemistry are not just simplified, they are *enormously* simplified. The student no longer has to assemble and store all the mental clutter that was formerly needed to use many of the equations of physical chemistry. One of the great blessings conferred on the student by the SI is that there is only one numerical value of the gas constant, R. The systematic value of R is the only one used and the only one printed in this book. To those who wish to use any other value, I leave the opportunity to muddle the situation and suffer the consequences.

ACKNOWLEDGMENTS

In this third edition my aim has been to preserve the best parts of the earlier editions and to improve the others, hoping all the while for the wisdom to know which is which. I have been aided in this by the following individuals who reviewed either the entire manuscript or major parts of it. My best thanks go to Professors Irving Epstein, Brandeis University; James T. Hynes, University of Colorado; Paul J. Karol, Carnegie-Mellon University; Lawrence Lohr, University of Michigan; Alden C. Mead, University of Minnesota; Earl Mortenson, Cleveland State University. These reviews were thorough and constructive; the final book owes much to them. I am particularly grateful for their willingness to review a manuscript that was not always in a neat and clean form.

My thanks are due to earlier authors in physical chemistry who have shaped my thoughts on various topics. Most particular thanks are due to my first teachers in the subject, Professors Karl F. Herzfeld, Walter J. Moore, and Francis O. Rice. In addition, I am deeply indebted to Professor James A. Beattie for his kind permission to reprint definitions from his book, *Lectures on Elementary Chemical Thermodynamics*. I believe that the influence of this remarkably clear exposition may be noticeable throughout the material on thermodynamics in this book. Chapter 8, the introduction to the second law, is particularly indebted to Professor Beattie's *Lectures*.

I am grateful to all my colleagues at the University of Maryland who have made suggestions, pointed out errors, responded to my questions, and helped in other ways. Particular thanks go to Professors Raj Khanna and Paul Mazzocchi, who supplied laboratory spectra for illustrations; to Professor Robert J. Munn, who wrote the computer program to construct the index; to Professors Isadore Adler and James M. Stewart, who read and commented on the sections dealing with x-ray spectroscopy and x-ray diffraction; and to Professor E. C. Lingafelter, University of Washington, who was kind enough to write detailed comments on the chapter on x-ray diffraction. Thanks to them a number of errors have been corrected and several passages clarified. Donald D. Wagman and David Garvin of the thermochemistry section of The National Bureau of Standards were most helpful and patient in answering my questions and kindly arranged for me to see a copy of NBS Technical Note 270-8 almost before the ink was dry. Professor D. H. Whiffen, The University, Newcastle-upon-Tyne, was most helpful in correspondence on the use of SI units in quantum mechanics.

I wish to express my appreciation to all the teachers, students, and casual readers who have taken the time to write letters with questions, criticisms, and suggestions. The book is much improved as a result of their comments.

I also wish to thank the editors and production staff of Addison-Wesley for their excellent work. Robert L. Rogers, the Senior Science Editor, smoothed my path throughout the preparation of the manuscript, helped with advice, secured timely reviews, and made the necessary editorial decisions promptly and wisely. Margaret Pinette, the Senior Production Editor, resolved all my proofreading complaints and problems, always pleasantly and with good humor. Joseph Vetere, the Art Coordinator, often went the extra mile to fulfill my wishes on the many illustrations in the book. It has been a pleasure to work with all of them.

Finally, to my wife, Joan McDonald Castellan, and our children, Stephen, Bill, David, and Susan, for their constant encouragement and patient endurance, I am grateful in measure beyond words.

G.W.C.

College Park, Md.
October 1982

Contents

8
Introduction to the Second Law of Thermodynamics   153

9
Properties of the Entropy and the Third Law of Thermodynamics   171

10
Spontaneity and Equilibrium   203

11
Systems of Variable Composition; Chemical Equilibrium   221

12
Phase Equilibrium in Simple Systems; The Phase Rule   259

13
Solutions
I. The Ideal Solution and Colligative Properties   277

14
Solutions
II. More Than One Volatile Component; The Ideal Dilute Solution   295

15
Equilibria between Condensed Phases   319

16
Equilibria in Nonideal Systems   347

27
Structure of Solids   681

28
Electronic Structure and Macroscopic Properties   709

29
Structure and Thermodynamic Properties   721

30
Transport Properties   745

31
Electrical Conduction   765

32
Chemical Kinetics
I. Empirical Laws and Mechanism   **799**

33
Chemical Kinetics
II. Theoretical Aspects   **847**

34
Chemical Kinetics
III. Heterogeneous Reactions, Electrolysis, Photochemistry 867

35
Polymers 913

APPENDICES

I
Some Useful Mathematics   A-1

II
Some Fundamentals of Electrostatics   A-11

III
The International System of Units; Le Système International d'Unités; SI   A-17

IV   A-22

1

Some Fundamental
Chemical Concepts

1.1 INTRODUCTION

We begin the study of physical chemistry with a brief statement of a few fundamental ideas and common usages in chemistry. These are very familiar things, but it is worth while recalling them to mind.

1.2 THE KINDS OF MATTER

The various kinds of matter can be separated into two broad divisions: (1) substances and (2) mixtures of substances.

Under a specified set of experimental conditions a *substance* exhibits a definite set of physical and chemical properties that do not depend on the previous history or on the method of preparation of the substance. For example, after appropriate purification, sodium chloride has the same properties whether it has been obtained from a salt mine or prepared in the laboratory by combining sodium hydroxide with hydrochloric acid.

On the other hand, mixtures may vary widely in chemical composition. Consequently their physical and chemical properties vary with composition, and may depend on the manner of preparation. By far the majority of naturally occurring materials are mixtures of substances. For example, a solution of salt in water, a handful of earth, or a splinter of wood are all mixtures.

1.3 THE KINDS OF SUBSTANCES

Substances are of two kinds: elements and compounds. An element cannot be broken down into simpler substances by ordinary chemical methods, but a compound can be. An ordinary chemical method is any method involving an energy of the order of 1000 kJ/mol or less.

For example, the element mercury cannot undergo any *chemical* decomposition of the type $Hg \rightarrow X + Y$, in which X and Y individually have smaller masses than the original mass of mercury. In this definition, both X and Y must have masses at least as large as that of the hydrogen atom, since the reaction $Na \rightarrow Na^+ + e^-$ is a chemical reaction involving an energy of about 500 kJ/mol. In contrast, the compound methane can be decomposed chemically into simpler substances individually less massive than the original methane: $CH_4 \rightarrow C + 2H_2$.

All natural materials can be chemically broken down ultimately into 89 elements. In addition to these, at least 18 other elements have been synthesized using the methods of nuclear physics (methods involving energies of the order of 10^8 kJ/mol or larger). Because of the great difference in the energies involved in chemical methods and nuclear methods, there is no likelihood of confusing the two. The nuclei of atoms endure through chemical reactions; only the outermost electrons of the atoms, the valence electrons, are affected.

Atoms of one element can combine chemically with atoms of another element to form the minute parts of the compound called molecules; for example, four atoms of hydrogen can combine with one atom of carbon to form a molecule of methane, CH_4. Atoms of a single element can also combine with themselves to form molecules of the element, for example, H_2, O_2, Cl_2, P_4, S_8.

1.4 ATOMIC AND MOLAR MASSES

Any atom has a tiny nucleus, diameter $\sim 10^{-14}$ m, in the center of a relatively enormous electron cloud, diameter $\sim 10^{-10}$ m. The negative charge of the electron cloud exactly balances the positive nuclear charge. Each atom, or nuclide, can be described by specifying two numbers, Z and A, where Z, the atomic number, is the number of protons in the nucleus and A, the mass number, is equal to $Z + N$, where N is the number of neutrons in the nucleus. The atoms of different elements are distinguished by having different values of Z. The atoms of a single element all have the same value of Z, but may have different values of A. Atoms with the same Z and different values of A are the *isotopes* of the element. The nuclides described by $Z = 1$, $A = 1$, or 2, or 3 are the three isotopes of hydrogen symbolized by $^1_1H, ^2_1H, ^3_1H$. The three principal isotopes of carbon are $^{12}_6C, ^{13}_6C, ^{14}_6C$.

The isotope of carbon with mass number 12 has been chosen as the defining element for the scale of atomic masses. We define the *atomic mass unit*, symbol u, as exactly 1/12 of the mass of one atom of carbon-12. Then $u = 1.6605655 \times 10^{-27}$ kg. The relative atomic mass of an atom, A_r, is defined by: $A_r = m/u$, where m is the mass of the atom; for example, $A_r(^1_1H) = 1.007825$; $A_r(^{12}_1C) = 12$ (exactly); $A_r(^{16}_8O) = 15.99491$. In any macroscopic sample of an element there may be several different isotopes present in the naturally occurring isotopic mixture. The entry in the table of atomic masses is the *average* of the relative atomic masses of all the atoms in this natural mixture. If x_i is the atom fraction of the particular isotope in the mixture, then the average, $\langle A_r \rangle$, is

$$\langle A_r \rangle = x_1(A_r)_1 + x_2(A_r)_2 + \cdots = \sum_i x_i(A_r)_i. \tag{1.1}$$

■ **EXAMPLE 1.1**   The isotopic composition of naturally occurring nitrogen is 99.63% $^{14}_7N$ for which $(A_r)_{14} = 14.00307$ and 0.37% $^{15}_7N$ for which $(A_r)_{15} = 15.00011$. Then the average relative atomic mass is

$$\langle A_r \rangle = 0.9963(14.00307) + 0.0037(15.00011) = 14.007$$

The variability in isotopic composition of samples of an element from different sources is often the principal origin of the uncertainty in the average relative atomic mass of that element.

The relative molar mass of a molecule can be computed by adding the relative atomic masses of all the atoms in it. By adding the atomic mass of carbon, 12.011, to four times the atomic mass of hydrogen, 4(1.008), the molar mass of CH_4, methane, 16.043, is obtained. This method of computing molar masses assumes that there is no change in mass when the carbon atom combines with four hydrogen atoms to form methane. That is, in the reaction

$$C + 4H \longrightarrow CH_4$$

the total mass on the left, 16.043 units, is equal to the total mass on the right, 16.043 units, if the molar mass of CH_4 is computed by the rule given above.

The question of whether or not mass is conserved in chemical reactions has been the subject of very extensive and very precise experimental investigations, and in no case has any change in mass during a chemical reaction been demonstrated. The law of conservation of mass holds accurately for chemical reactions within the limits of precision of experiments conducted thus far. The expected change in mass accompanying any chemical reaction can be computed from the mass–energy equivalence law of relativity theory. If the energy involved in the chemical reaction is ΔU, and Δm is the associated change in mass, then $\Delta U = (\Delta m)c^2$, where c, the velocity of light, equals 3×10^8 m/s. Computation shows that the change in mass is of the order of 10^{-11} gram per kilojoule of energy involved in a reaction. This change in mass is too small to be detected by contemporary methods. Therefore the law of conservation of mass may be considered exact in all chemical situations.

Note that the terms "atomic mass" and "molar mass" are interchangeable with the traditional terms, "atomic weight" and "molecular weight."

1.5 SYMBOLS; FORMULAS

Over the years a set of symbols for the elements has evolved. Depending on the context, the symbol for an element may stand for several different things: it may merely be an abbreviation of the name of the element; it may symbolize one atom of the element; often it represents 6.022×10^{23} atoms of the element, a *mole*.

The formulas of compounds are interpreted in a variety of ways, but in every instance the formula describes the relative composition of the compound. In substances such as quartz and salt, discrete molecules are not present. Therefore the formulas SiO_2 and NaCl are given only empirical meaning; these formulas describe the relative numbers of atoms of the elements present in the compound and nothing more.

For substances that consist of discrete molecules, their formulas describe the relative numbers of the constituent atoms and the total number of atoms in a molecule; for example, acetylene, C_2H_2; benzene, C_6H_6; sulfur hexafluoride, SF_6.

Structural formulas are used to describe the way atoms are connected within a molecule. Within the limitations imposed by a two-dimensional diagram, they display the geometry of a molecule. Bonding within a molecule is illustrated by using conventional symbols for single and multiple bonds, electron pairs, and positive and negative centers of charge in the molecule. Structural formulas have their greatest utility in representing substances with discrete molecules. As yet, no satisfactory abbreviated way of representing the structural complexity of substances such as quartz and salt has been devised. In using any structural formula, a great deal must be mentally supplied to the diagram.

1.6 THE MOLE

The concept of *amount of substance* is central to chemical measurement. The amount of substance of a system is proportional to the number of elementary entities of that substance present in the system. The elementary entities must be described; they may be atoms, molecules, ions, or specified groups of such particles. The entity itself is a natural unit for measuring the amount of substance; for example, we can describe the amount of substance in a sample of iron by saying that there are 2.0×10^{24} Fe atoms in the sample. The amount of substance in a crystal of NaCl can be described by saying that there are 8.0×10^{20} ion pairs, Na^+Cl^-, in the crystal.

Since any tangible sample of matter contains such an enormous number of atoms or molecules, a unit larger than the entity itself is needed to measure the amount of substance. The SI unit for amount of substance is the *mole*. The mole is defined as the amount of substance in exactly 0.012 kg of carbon-12. One mole of any substance contains the same number of elementary entities as there are carbon atoms in exactly 0.012 kg of carbon-12. This number is the Avogadro constant, $N_A = 6.022045 \times 10^{23}$ mol^{-1}.

1.7 CHEMICAL EQUATIONS

A chemical equation is a shorthand method for describing a chemical transformation. The substances on the left-hand side of the equation are called *reactants*; those on the right-hand side are called *products*. The equation

$$MnO_2 + HCl \longrightarrow MnCl_2 + H_2O + Cl_2$$

expresses the fact that manganese dioxide will react with hydrogen chloride to form manganous chloride, water, and chlorine. As it is written, the equation does little besides record the fact of the reaction and the proper formulas for each substance.

If the equation is balanced,

$$MnO_2 + 4\,HCl \longrightarrow MnCl_2 + 2\,H_2O + Cl_2,$$

it expresses the fact that the number of atoms of a given kind must be the same on both sides of the equation. Most important, *the balanced chemical equation is an expression of the law of conservation of mass*. Chemical equations provide the relationship between the masses of the various reactants and products, which is ordinarily of utmost importance in chemical problems.

1.7.1 Stoichiometry

Consider a system having an initial composition described by a set of mole numbers: $n_1^0, n_2^0, \ldots, n_i^0$. If a reaction occurs, these mole numbers change as the reaction progresses. The mole numbers of the various species do not change independently; the changes are related by the stoichiometric coefficients in the chemical equation. For example, if the reaction of manganese dioxide and hydrogen chloride given above occurs once as written, we say that *one mole of reaction* has occurred. This means that 1 mole of MnO_2 and 4 moles of HCl are consumed and that 1 mole of $MnCl_2$, 2 moles of H_2O and 1 mole of Cl_2 are produced. After ξ moles of reaction occur, the mole numbers of the substances are given by

$$n_{MnO_2} = n_{MnO_2}^0 - \xi; \qquad n_{HCl} = n_{HCl}^0 - 4\xi;$$

$$n_{MnCl_2} = n_{MnCl_2}^0 + \xi; \qquad n_{H_2O} = n_{H_2O}^0 + 2\xi; \qquad n_{Cl_2} = n_{Cl_2}^0 + \xi. \tag{1.1}$$

Since reactants are consumed and products are produced, the algebraic signs appear as shown in Eqs. (1.1).

The variable ξ was first introduced by DeDonder who called it the "degree of advancement" of the reaction. Here we shall call it simply the *advancement* of the reaction. Equations (1.1) show that the composition at any stage of the reaction is described by the initial mole numbers, the stoichiometric coefficients, and the advancement.

We can see how to generalize this description if we rewrite the chemical reaction by moving the reactants to the right side of the equation. It becomes

$$0 = MnCl_2 + 2H_2O + Cl_2 + (-1)MnO_2 + (-4)HCl$$

This form suggests that any chemical reaction can be written in the form

$$0 = \sum_i v_i A_i \tag{1.2}$$

where the A_i represent the chemical formulas of the various species in the reaction, and we agree that the stoichiometric coefficients, v_i, will be given a negative sign for reactants and a positive sign for products. Then we see that each of the mole numbers in Eqs. (1.1) has the form

$$n_i = n_i^0 + v_i \xi \tag{1.3}$$

Equation (1.3) is the general relation between the mole numbers and the advancement of any reaction.

Differentiating, we obtain

$$dn_i = v_i \, d\xi$$

or

$$\frac{dn_i}{v_i} = d\xi \tag{1.4}$$

This equation relates changes of all the mole numbers to the change in the one variable, $d\xi$.

1.7.2 The Advancement Capacity

The value of ξ increases as the reaction advances, reaching a limiting value when one or more of the reactants is exhausted. This limiting value of ξ is the *advancement capacity*, ξ^0, of the reaction mixture. If we divide Eq. (1.3) by $-v_i$, we obtain

$$n_i = (-v_i)\left(\frac{n_i^0}{-v_i} - \xi\right) \tag{1.4}$$

If we define $\xi_i^0 = n_i^0/(-v_i)$, then we have

$$n_i = (-v_i)(\xi_i^0 - \xi) \tag{1.5}$$

The quantity, $n_i^0/(-v_i) = \xi_i^0$, is called the *advancement capacity* of the ith substance. Clearly, if substance i is a reactant, then $-v_i$ is positive; thus the advancement capacities of the reactants are all positive. If the values of ξ_i^0 are all equal, then this common value of $\xi_i^0 = \xi^0$, the *advancement capacity of the mixture*. If the ξ_i^0 are not all equal, then there is at least one smallest value, ξ_j^0. This value identifies the substance j as the limiting reagent, and $\xi_j^0 = \xi^0$, the advancement capacity of the mixture. The value of ξ may not exceed ξ^0, since that would mean that reactant j (and possibly others) would have a negative mole number. Thus, ξ^0 is the greatest value of ξ.

Similarly, if the substance i is a product, then $-v_i$ is negative and $n_i^0/(-v_i) = \xi_i^0$ is negative. Then it is possible, if *none* of the product n_i is zero, for the reaction to move in the reverse direction (ξ is negative). In this case the ξ_i^0 are the advancement capacities of the products. If ξ_k^0 is the least negative of this set, the substance k is the limiting reagent for the reverse reaction and ξ_k^0 is the advancement capacity of the mixture for the reverse reaction. The value of ξ may not be less than ξ_k^0, since this would mean that the product k (and possibly others) would have a negative mole number. Thus $\xi_k^0 = \xi_l$, the least value of ξ. (Note: quite commonly, no products are present at the beginning of the reaction. Then $n_i^0 = 0$ for all the products; the advancement capacity of the reverse reaction is zero and ξ may only have positive values.)

If the reaction goes to completion, then $\xi = \xi^0$, and the final number of moles of the various species is given by

$$n_i(\text{final}) = n_i^0 + v_i \xi^0 = (-v_i)(\xi_i^0 - \xi^0) \tag{1.6}$$

If there were no products present initially, then for the product species, $n_i(\text{final}) = v_i \xi^0$; the number of moles of any product is the advancement capacity of the mixture multiplied by its stoichiometric coefficient.

The utility of this formulation for simple stoichiometric calculations is illustrated by Example 1.2, in which the quantities appropriate to each species are arranged underneath the formula of that species in the chemical equation. Its utility in other applications will be demonstrated in later parts of the book.

■ **EXAMPLE 1.2**   Assume that 0.80 mole of ferric oxide reacts with 1.20 mol of carbon. What amount of each substance is present when the reaction is complete?

Equation:	Fe_2O_3	+	$3C$	$\longrightarrow$	$2Fe$	+	$3CO$
v_i	-1		-3		$+2$		$+3$
n_i^0	0.80		1.20		0		0
$\xi_i^0 = n_i^0/(-v_i)$	0.80		0.40		0		0
Therefore, $\xi^0 = 0.40$							
$n_i = (-v_i)(\xi_i^0 - \xi)$	$0.80 - \xi$		$3(0.40 - \xi)$		2ξ		3ξ
When $\xi = \xi^0 = 0.40$							
$n_i(\text{final}) = (-v_i)(\xi_i^0 - \xi^0)$	0.40		0		0.80		1.20

1.8 THE INTERNATIONAL SYSTEM OF UNITS, SI

In the past, several systems of metric units were commonly used by scientists, each system having its advantages and disadvantages. Recently international agreement was reached on the use of a single set of units for the various physical quantities, as well as on a recommended set of symbols for the units and for the physical quantities themselves. The SI will be used in this book with only a few additions. Because of its importance in defining the standard state of pressure, the atmosphere will be retained as a unit of pressure in addition to the pascal, the SI unit. The litre will be used with the understanding that $1 \, L = 1 \, dm^3$(exactly).

Any system of units depends on the selection of "base units" for the set of physical properties that are chosen as a dimensionally independent set. In Appendix III we give the definitions of the base units, some of the most commonly used derived units, and a list of the prefixes that are used to modify the units. You should become thoroughly familiar with the units, their symbols, and the prefixes because they will be used in the text without explanation.

2

Empirical Properties of Gases

2.1 BOYLE'S LAW; CHARLES'S LAW

Of the three states of aggregation, only the gaseous state allows a comparatively simple quantitative description. For the present we shall restrict this description to the relations among such properties as mass, pressure, volume, and temperature. We shall assume that the system is in equilibrium so that the values of the properties do not change with time, so long as the external constraints on the system are not altered.

A system is in a definite state or condition when all of the properties of the system have definite values, which are determined by the state of the system. Thus the state of the system is described by specifying the values of some or all of its properties. The important question is whether it is necessary to give values of fifty different properties (or twenty or five) to ensure that the state of the system is completely described. The answer depends to a certain extent upon how accurate a description is required. If we were in the habit of measuring the values of properties to twenty significant figures, and thank heaven we are not, then quite a long list of properties would be required. Fortunately, even in experiments of great refinement, only four properties—mass, volume, temperature, and pressure—are ordinarily required.

The *equation of state* of the system is the mathematical relationship between the values of these four properties. Only three of these must be specified to describe the state; the fourth can be calculated from the equation of state, which is obtained from knowledge of the experimental behavior of the system.

The first quantitative measurements of the pressure–volume behavior of gases were made by Robert Boyle in 1662. His data indicated that the volume is inversely proportional to the pressure: $V = C/p$, where p is the pressure, V is the volume, and C is a constant. Figure 2.1 shows V as a function of p. Boyle's law may be written in the form

$$pV = C; \tag{2.1}$$

it applies only to a fixed mass of gas at a constant temperature.

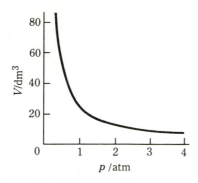

Figure 2.1   Volume as a function of pressure, Boyle's law (T = 25 °C).

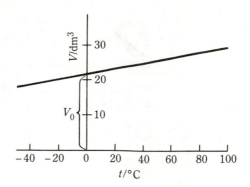

Figure 2.2   Volume as a function of temperature, Charles's law (p = 1 atm).

Later experiments by Charles showed that the constant C is a function of temperature. This is a rough statement of Charles's law.

Gay-Lussac made measurements of the volume of a fixed mass of gas under a fixed pressure and found that the volume was a linear function of the temperature. This is expressed by the equation

$$V = a + bt, \tag{2.2}$$

where t is the temperature, and a and b are constants. A plot of volume as a function of temperature is shown in Fig. 2.2. The intercept on the vertical axis is $a = V_0$, the volume at 0 °C. The slope of the curve is the derivative $b = (\partial V/\partial t)_p$.* Thus Eq. (2.2) can be written in the equivalent form

$$V = V_0 + \left(\frac{\partial V}{\partial t}\right)_p t. \tag{2.3}$$

Charles's experiments showed that for a fixed mass of gas under a constant pressure, the *relative* increase in volume per degree increase in temperature was *the same for all gases* on which he made measurements. At a fixed pressure the increase in volume per degree is $(\partial V/\partial t)_p$; hence, the relative increase in volume per degree at 0 °C is $(1/V_0)(\partial V/\partial t)_p$. This quantity is the *coefficient of thermal expansion* at 0 °C, for which we use the symbol α_0:

$$\alpha_0 = \frac{1}{V_0}\left(\frac{\partial V}{\partial t}\right)_p. \tag{2.4}$$

Then Eq. (2.3) may be written in terms of α_0:

$$V = V_0(1 + \alpha_0 t) = V_0\alpha_0\left(\frac{1}{\alpha_0} + t\right), \tag{2.5}$$

which is convenient because it expresses the volume of the gas in terms of the volume at zero degrees and a constant, α_0, which is the same for all gases and, as it turns out, is very nearly

---

*   The partial derivative is used rather than the ordinary derivative, since the volume depends on the pressure; a and b are constants only if the pressure is constant. The partial derivative $(\partial V/\partial t)_p$ is the rate of change of volume with temperature at constant pressure; this is the slope of the line under the conditions of the experiment.

independent of the pressure at which the measurements are made. If we measure α_0 at various pressures we find that for all gases α_0 approaches the same limiting value at $p = 0$. The form of Eq. (2.5) suggests a transformation of coordinates that should be useful; namely, define a new temperature T in terms of the old temperature through the equation

$$T = \frac{1}{\alpha_0} + t. \tag{2.6}$$

Equation (2.6) defines a new temperature scale, called a *gas scale* of temperature or, more exactly, an ideal gas scale of temperature. The importance of this scale lies in the fact that the limiting value of α_0, and consequently $1/\alpha_0$, has the same value for all gases. On the other hand, α_0 does depend on the scale of temperature used originally for t. If t is in degrees Celsius (symbol: °C), then $1/\alpha_0 = 273.15$ °C. The resulting T-scale is numerically identical to the thermodynamic temperature scale, which we will discuss in detail in Chapter 8. The SI unit of thermodynamic temperature is the kelvin (symbol: K). Temperatures on the thermodynamic scale are frequently called absolute temperatures or kelvin temperatures. According to Eq. (2.6) (see also Appendix III, Sect. A-III-6),

$$T = 273.15 + t. \tag{2.7}$$

Equations (2.5) and (2.6) are combined to yield

$$V = \alpha_0 V_0 T, \tag{2.8}$$

which states that the volume of a gas under a fixed pressure is directly proportional to the thermodynamic temperature.

2.2 MOLAR MASS OF A GAS. AVOGADRO'S LAW; THE IDEAL GAS LAW

So far, two relations between the four variables have been obtained: Boyle's law, Eq. (2.1) (fixed mass, constant temperature), and Gay-Lussac's, or Charles's law, Eq. (2.8) (fixed mass, constant pressure). These two equations may be combined into one general equation by noting that V_0 is the volume at 0 °C, and so is related to the pressure by Boyle's law, $V_0 = C_0/p$, where C_0 is the value of the constant at $t = 0$. Then, Eq. (2.8) becomes

$$V = \frac{C_0 \alpha_0 T}{p} \qquad \text{(fixed mass).} \tag{2.9}$$

The restriction of fixed mass is removed by realizing that if the temperature and pressure are kept constant and the mass of the gas is doubled, the volume will double. This means that the constant C_0 is proportional to the mass of gas; hence, we write $C_0 = Bw$, where B is a constant and w is the mass. Introducing this result into Eq. (2.9), we obtain

$$V = \frac{B\alpha_0 wT}{p}, \tag{2.10}$$

which is the general relation between the four variables V, w, T, and p. Each gas has a different value of the constant B.

For Eq. (2.10) to be useful, we would have to have at hand a table of values for B for all the various gases. To avoid this, B is expressed in terms of a characteristic mass for each gas. Let M denote the mass of gas in the container under a set of standard conditions: T_0, p_0, V_0. If different gases are confined in the standard volume V_0 under the standard temperature and pressure T_0 and p_0, then by Eq. (2.10), for each gas

$$M = \left(\frac{1}{B\alpha_0}\right)\left(\frac{p_0 V_0}{T_0}\right). \tag{2.11}$$

Since the standard conditions are chosen to suit our convenience, the ratio $R = p_0 V_0/T_0$ has a fixed numerical value for any particular choice and has, of course, the same value for all the gases (R is called the *gas constant*). Equation (2.11) may then be written

$$M = \frac{R}{B\alpha_0} \quad \text{or} \quad B = \frac{R}{M\alpha_0}.$$

Using this value for B in Eq. (2.10), we obtain

$$V = \left(\frac{w}{M}\right)\frac{RT}{p}. \tag{2.12}$$

Let the number of characteristic masses of the gas contained in the mass w be $n = w/M$. Then $V = nRT/p$, or

$$pV = nRT. \tag{2.13}$$

Equation (2.13), the *ideal gas law*, has great importance in the study of gases. It does not contain anything that is characteristic of an individual gas, but is a generalization applicable to all gases.

We now inquire about the significance of the characteristic mass M. Avogadro's law says that equal volumes of different gases under the same conditions of temperature and pressure contain equal numbers of molecules; that is, they contain the same amount of substance. We have compared equal volumes, V_0, under the same temperature and pressure, T_0 and p_0, to obtain the characteristic masses of the different gases. According to Avogadro's law these characteristic masses must contain the same number of molecules. If we choose p_0, T_0, and V_0 so that the number is equal to $N_A = 6.022 \times 10^{23}$, then the amount of substance in the characteristic mass is one mole and M is the molar mass. Also, M is N_A times the mass of the individual molecule, m, or

$$M = N_A m. \tag{2.14}$$

In Eq. (2.13) n is the number of moles of the gas present. Since the value of R is directly related to the definition of molar mass, we shall find that the gas constant appears in equations that describe molar properties of solids and liquids, as well as gases.

The mole was originally defined through the kind of procedure described above. First, the normal isotopic mixture of oxygen was arbitrarily assigned a molar mass of exactly 32 g/mol. Then a flask of accurately known volume was filled with oxygen at 0 °C and 1 atm and the mass of oxygen in the flask was measured. Finally, from this measurement the volume required to contain exactly 32 g of oxygen (at 0 °C, 1 atm) was calculated. This is V_0, the standard molar volume. Knowing V_0, we can calculate the molar mass of any other gas from a measurement of the gas density.

The modern value of V_0, based on the carbon-12 definition of the mole, is $V_0 = 22.41383$ L/mol $= 22.41383 \times 10^{-3}$ m³/mol. Since $T_0 = 273.15$ K (exactly), and $p_0 = 1$ atm $= 1.01325 \times 10^5$ Pa (exactly), the value of R is

$$R = \frac{p_0 V_0}{T_0} = \frac{(1.01325 \times 10^5 \text{ Pa})(22.41383 \times 10^{-3} \text{ m}^3/\text{mol})}{298.15 \text{ K}}$$

$$= 8.31441 \text{ Pa m}^3 \text{ K}^{-1} \text{ mol}^{-1} = 8.31441 \text{ J K}^{-1} \text{ mol}^{-1}.$$

For most of our calculations here, the approximate value,

$$R = 8.314 \text{ J K}^{-1} \text{ mol}^{-1},$$

is sufficiently accurate. Note that R has the dimensions: energy kelvin^{-1} mole^{-1}.

2.2.1 Comments on Units

The SI unit of pressure is the pascal (Pa) defined by

$$1 \text{ Pa} = 1 \text{ N/m}^2 = 1 \text{ J/m}^3 = 1 \text{ kg m}^{-1} \text{ s}^{-2}.$$

The common practical units of pressure are the atmosphere (atm), the torr (Torr), and the millimetre of mercury (mmHg). The standard atmosphere is defined by

$$1 \text{ atm} = 1.01325 \times 10^5 \text{ Pa} \qquad \text{(exactly)}.$$

The torr is defined by

$$760 \text{ Torr} = 1 \text{ atm} \qquad \text{(exactly)}.$$

The conventional millimetre of mercury (mmHg) is the pressure exerted by a column exactly 1 mm high of a fluid having a density of exactly 13.5951 g/cm^3 in a location where the acceleration of gravity is exactly 9.80665 m/s^2. The millimetre of mercury is greater than the torr by about 14 parts in 10^8. For our purposes, 1 mmHg = 1 Torr.

The SI unit of volume is the cubic metre. The practical units of volume are the cubic centimetre and the litre, L. The relations are

$$1 \text{ L} = 1 \text{ dm}^3 = 1000 \text{ cm}^3 = 10^{-3} \text{ m}^3 \qquad \text{(all are exact)}.$$

In working problems with the ideal gas law, temperatures are expressed in kelvins, pressures in pascals, and volumes in cubic metres.

■ **EXAMPLE 2.1**   One mole of an ideal gas occupies 12 L at 25 °C. What is the pressure the gas?

The required relation between the data and the unknown is the ideal gas law. Converting to SI we have

$$T = 273.15 + 25 = 298 \text{ K} \qquad \text{and} \qquad V = 12 \text{ L} \times 10^{-3} \text{ m}^3/\text{L} = 0.012 \text{ m}^3.$$

Then

$$p = \frac{nRT}{V} = \frac{1 \text{ mol}(8.314 \text{ J K}^{-1} \text{ mol}^{-1})(298 \text{ K})}{0.012 \text{ m}^3} = 2.06 \times 10^5 \text{ J/m}^3$$

$$= 2.06 \times 10^5 \text{ Pa} = 206 \text{ kPa}.$$

If the pressure is needed in atm, then $p = 206 \text{ kPa}(1 \text{ atm}/101 \text{ kPa}) = 2.04 \text{ atm}$.

■ **EXAMPLE 2.2**   A gas is contained in 50 L under 8 atm pressure at 20 °C. How many moles of gas are in the container?

Changing to SI, $T = 273.15 + 20 = 293 \text{ K}$, $V = 50 \text{ L}(10^{-3} \text{ m}^3/\text{L}) = 0.050 \text{ m}^3$, and

$$p = 8 \text{ atm}(1.013 \times 10^5 \text{ Pa/atm}) = 8(1.013 \times 10^5) \text{ Pa}.$$

Then,

$$n = \frac{pV}{RT} = \frac{8(1.013 \times 10^5 \text{ Pa})(0.050 \text{ m}^3)}{8.314 \text{ J K}^{-1} \text{ mol}^{-1}(293 \text{ K})} = 16.6 \text{ mol}.$$

2.3 THE EQUATION OF STATE; EXTENSIVE AND INTENSIVE PROPERTIES

The ideal gas law, $pV = nRT$, is a relation between the four variables that describe the state of any gas. As such, it is an *equation of state*. The variables in this equation fall into two classes: n and V are extensive variables (extensive properties), while p and T are intensive variables (intensive properties).

The value of any extensive property is obtained by summing the values of that property in every part of the system. Suppose that the system is subdivided into many small parts, as in Fig. 2.3. Then the total volume of the system is obtained by adding together the volumes of each small part. Similarly, the total number of moles (or total mass) in the system is obtained by summing the number of moles in (or mass of) each part. By definition, such properties are extensive. It should be clear that the value obtained is independent of the way in which the system is subdivided.

Intensive properties are not obtained by such a process of summation but are measured at any point in the system, and each has a uniform value throughout a system at equilibrium; for example, T and p.

Extensive variables are proportional to the mass of the system. For the ideal gas, as an example, $n = w/M$, and $V = wRT/Mp$. Both n and V are proportional to the mass of the system. Dividing V by n, we obtain $\overline{V}$, the volume per mole:

$$\overline{V} = \frac{V}{n} = \frac{RT}{p}. \tag{2.15}$$

The ratio of V to n is not proportional to the mass, because in forming the ratio the mass drops out and $\overline{V}$ is an intensive variable. The ratio of any two extensive variables is *always* an intensive variable.

If the ideal gas law is written in the form

$$p\overline{V} = RT, \tag{2.16}$$

it is a relation between *three intensive variables*: pressure, temperature, and molar volume. This is important because we can now discuss the properties of the ideal gas without continually worrying about whether we are dealing with ten moles or ten million moles. It should be clear that no fundamental property of the system depends on the accidental choice of 20 g rather than 100 g of material for study. In the atom bomb project, micro quantities of material were used in preliminary studies, and vast plants were built based on the properties determined on this tiny scale. If fundamental properties depended on the *amount* of substance used, one could imagine the government giving research grants for the study of extremely large systems; enormous buildings might be required, depending upon the ambition of the investigators! For the discussion of principles, the intensive variables are the significant ones. In practical applications such as design of apparatus and engineering, the extensive properties are important as well, because they determine the size of apparatus, the horsepower of an engine, the production capacity of a plant in tons per day, and so forth.

Figure 2.3   Subdivision of the system.

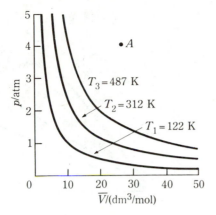

Figure 2.4   Isotherms of the ideal gas.

2.4 PROPERTIES OF THE IDEAL GAS

If arbitrary values are assigned to any two of the three variables p, $\overline{V}$, and T, the value of the third variable can be calculated from the ideal gas law. Hence, any set of two variables is a set of *independent* variables; the remaining variable is a *dependent* variable. The fact that the state of a gas is completely described if the values of any two intensive variables are specified allows a very neat geometric representation of the states of a system.

In Fig. 2.4, p and $\overline{V}$ have been chosen as independent variables. Any point, such as A, determines a pair of values of p and $\overline{V}$; this is sufficient to describe the state of the system. Therefore every point in the p–$\overline{V}$ quadrant (both p and $\overline{V}$ must be positive to make physical sense) describes a different state of the gas. Furthermore, every state of the gas is represented by some point in the p–$\overline{V}$ diagram.

It is frequently useful to pick out all of the points that correspond to a certain restriction on the state of the gas, as, for example, the points that correspond to the same temperature. In Fig. 2.4 the curves labeled T_1, T_2, and T_3 collect all the points that represent states of the ideal gas at the temperatures T_1, T_2, and T_3, respectively. These curves are called *isotherms*. The isotherms of the ideal gas are rectangular hyperbolas determined by the relation

$$p = \frac{RT}{\overline{V}}. \tag{2.17}$$

For each curve, T has a different constant value.

In Fig. 2.5 every point corresponds to a set of values for the coordinates $\overline{V}$ and T; again each point represents a state of the gas, just as in Fig. 2.4. In Fig. 2.5 points corresponding to the same pressure are collected on the lines, which are called *isobars*. The isobars of the ideal gas are described by the equation

$$\overline{V} = \left(\frac{R}{p}\right)T, \tag{2.18}$$

where the pressure is assigned various constant values.

As in the other figures, every point in Fig. 2.6 represents a state of the gas, because it determines values of p and T. The lines of constant molar volume, *isometrics*, are described by the equation

$$p = \left(\frac{R}{\overline{V}}\right)T, \tag{2.19}$$

where $\overline{V}$ is assigned various constant values.

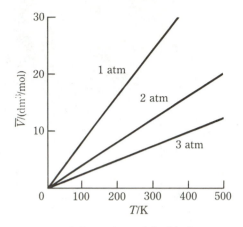

Figure 2.5   Isobars of the ideal gas.

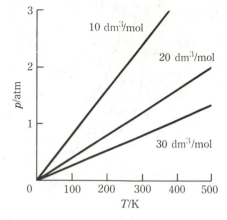

Figure 2.6   Isometrics of the ideal gas.

These diagrams derive their great utility from the fact that all the gaseous, liquid, and solid states of any pure substance can be represented on the same diagram. We will use this idea extensively, particularly in Chapter 12.

A careful examination of Figs. 2.4, 2.5, and 2.6 and of Eqs. (2.17), (2.18), and (2.19) leads to some rather bizarre conclusions about the ideal gas. For example, Fig. 2.5 and Eq. (2.18) say that the volume of an ideal gas confined under a constant pressure is zero at $T = 0$ K. Similarly, Fig. 2.4 and Eq. (2.17) tell us that the volume of the ideal gas kept at a constant temperature approaches zero as the pressure becomes infinitely large. These predictions do not correspond to the observed behavior of real gases at low temperatures and high pressures. As a real gas under a constant pressure is cooled, we observe a decrease in volume, but at some definite temperature the gas liquefies; after liquefaction occurs, not much decrease in the volume is observed as the temperature is lowered. Similarly, isothermal compression of a real gas may produce liquefaction, and thereafter further increase in pressure produces little change in the volume. It is apparent from this that there is good reason for referring to the relation $p\overline{V} = RT$ as the *ideal* gas law. The above discussion shows that we may expect the ideal gas law to fail in predicting the properties of a real gas at low temperatures and at high pressures. Experiment shows that the behavior of all real gases approaches that of the ideal gas as the pressure approaches zero.

In Chapter 3 deviations from the ideal gas law are discussed in detail. For the moment, a few general remarks will suffice on the question of when the ideal gas law may reasonably be used for predicting properties of real gases. In practice, if only a rough approximation is required, the ideal gas law is used without hesitation. This rough approximation is in many cases quite good, within perhaps 5 %. For a rule of such broad scope, the ideal gas law is astonishingly accurate in many practical situations.

The ideal gas law is more accurate the higher the temperature is above the critical temperature of the substance, and the lower the pressure is below the critical pressure* of the substance. In precision work the ideal gas law is never used.

---

*   Above the critical temperature, and above the critical pressure, it is not possible to distinguish liquid and vapor as separate entities; see Sec. 3.5.

2.5 DETERMINATION OF MOLAR MASSES OF GASES AND VOLATILE SUBSTANCES

The ideal gas law is useful in determining the molar masses of volatile substances. For this purpose a bulb of known volume is filled with the gas at a measured pressure and temperature. The mass of the gas in the bulb is measured. These measurements suffice to determine the molar mass of the substance. From Eq. (2.12) we have $pV = (w/M)RT$; then

$$M = \left(\frac{w}{V}\right)\frac{RT}{p} = \left(\frac{\rho}{p}\right)RT, \qquad (2.20)$$

where $\rho = w/V$; ρ is the density. All of the quantities on the right-hand side of Eq. (2.20) are known from the measurements; hence, M can be calculated.

A rough value of the molar mass is usually sufficient to determine the molecular formula of a substance. For example, if chemical analysis of a gas yields an empirical formula $(CH_2)_n$, then the molar mass must be some multiple of 14 g/mol; the possibilities are 28, 42, 56, 70, and so on. If a molar mass determination using Eq. (2.20) yields a value of 54 g/mol, then we may conclude that $n = 4$ and that the material is one of the butenes. The fact that the gas is not strictly ideal does not hinder us in this conclusion at all. In this example the possible values of M are well enough separated so that even if the ideal gas law were wrong by 5%, we would still have no difficulty in assigning the correct molecular formula to the gas. In this example it is unlikely that the ideal gas law would be in error by as much as 2% for a convenient choice of experimental conditions.

Since the determination of molar mass together with chemical analysis establishes the molecular formula of the gaseous substance, the results are of great importance. For example, some very common substances exhibit *dimerization*, a doubling of a simple unit. Table 2.1 lists some of these substances, all of which are solids or liquids at room temperature. Measurements of molar mass must be made at temperatures sufficiently high to vaporize the materials.

The fact that the behavior of a real gas approaches that of the ideal gas as the pressure is lowered is used as a basis for the precise determination of the molar masses of gases. According to Eq. (2.20), the ratio of density to pressure should be independent of pressure: $\rho/p = M/RT$. This is correct for an ideal gas, but if the density of a real gas is measured at one temperature and at several different pressures, the ratio of density to pressure is found to depend slightly on the pressure. At sufficiently low pressures, ρ/p is a linear function of

Table 2.1
Dimerization

Compound	Empirical formula	Molecular formula in the vapor
Aluminum chloride	$AlCl_3$	Al_2Cl_6
Aluminum bromide	$AlBr_3$	Al_2Br_6
Formic acid	$HCOOH$	$(HCOOH)_2$
Acetic acid	CH_3COOH	$(CH_3COOH)_2$
Arsenic trioxide	As_2O_3	As_4O_6
Arsenic pentoxide	As_2O_5	As_4O_{10}
Phosphorus trioxide	P_2O_3	P_4O_6
Phosphorus pentoxide	P_2O_5	P_4O_{10}

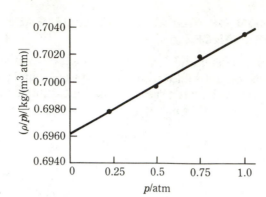

Figure 2.7   Plot of ρ/p versus p for ammonia at 25 °C.

the pressure. The straight line can be extrapolated to yield a value of ρ/p at zero pressure $(\rho/p)_0$, which is appropriate to the ideal gas and can be used in Eq. (2.20) to give a precise value of M:

$$M = \left(\frac{\rho}{p}\right)_0 RT. \tag{2.21}$$

This procedure is illustrated for ammonia at 25°C in Fig. 2.7.

2.6 MIXTURES; COMPOSITION VARIABLES

The state or condition of a mixture of several gases depends not only on the pressure, volume, and temperature, but also on the composition of a mixture. Consequently a method of specifying the composition must be devised. The simplest method would be to state the mole numbers $n_1, n_2, \ldots$ of the several substances in the mixture (the masses would also serve). This method has the disadvantage that the mole numbers are extensive variables. It is preferable to express the composition of a mixture in terms of a set of intensive variables.

It has been shown that the ratio of two extensive variables is an intensive variable. The mole numbers can be converted to intensive variables by dividing each one by some extensive variable. This can be done in several ways.

The volume concentrations are obtained by dividing the amount of each substance by the volume of the mixture.

$$\tilde{c}_i = \frac{n_i}{V} \tag{2.22}$$

The SI unit for volume concentration is mol/m^3. We will reserve the symbol $\tilde{c}_i$ for the volume concentration expressed in mol/m^3. We will use the symbol c_i for the volume concentration in the more commonly used unit, $mol/L = mol/dm^3$, called the *molar concentration* or the *molarity*. Volume concentrations are satisfactory for describing the composition of liquid or solid mixtures because the volume is comparatively insensitive to changes in temperature and pressure. Since the volume of a gas depends markedly on temperature and pressure, volume concentrations are not usually convenient for describing the composition of gas mixtures.

Mole ratios, r_i, are obtained by choosing one of the mole numbers and dividing all the others by that one. Choosing n_1 as the divisor, we have

$$r_i = \frac{n_i}{n_1}. \tag{2.23}$$

A variant of the mole ratio description, the molal concentration m_i, is often used for liquid solutions. Let the solvent be component 1, with a molar mass M_1. The molality of component i is the number of moles of i per unit mass (kg) of solvent. Since the mass of the solvent is $n_1 M_1$, the number of moles of solute per kilogram of solvent is m_i:

$$m_i = \frac{n_i}{n_1 M_1} = \frac{r_i}{M_1}. \tag{2.24}$$

The molality is the mole ratio multiplied by a constant, $1/M_1$. Since the mole ratios and the molalities are completely independent of temperature and pressure, they are preferable to the molar concentrations for the physico-chemical description of mixtures of any kind.

Mole fractions, x_i, are obtained by dividing each of the mole numbers by the total number of moles of all the substances present, $n_t = n_1 + n_2 + \cdots$,

$$x_i = \frac{n_i}{n_t}. \tag{2.25}$$

The sum of the mole fractions of all the substances in a mixture must be unity:

$$x_1 + x_2 + x_3 + \cdots = 1. \tag{2.26}$$

Because of this relation, the composition of the mixture is described when the mole fractions of all but one of the substances are specified; the remaining mole fraction is computed using Eq. (2.26). Like molalities and mole ratios, mole fractions are independent of temperature and pressure, and thus are suitable for describing the composition of any mixture. Gas mixtures are commonly described by the mole fractions, since the pVT relations have a concise and symmetrical form in these terms.

2.7 EQUATIONS OF STATE FOR A GAS MIXTURE; DALTON'S LAW

Experiment shows that for a mixture of gases, the ideal gas law is correct in the form

$$pV = n_t RT, \tag{2.27}$$

where n_t is the total number of moles of all the gases in the volume V. Equation (2.27) and the statement of the mole fractions of all but one of the constituents of the mixture constitute a complete description of the equilibrium state of the system.

It is desirable to relate the properties of complicated systems to those of simpler systems, so we attempt to describe the state of a gas mixture in terms of the states of pure unmixed gases. Consider a mixture of three gases described by the mole numbers n_1, n_2, n_3 in a container of volume V at a temperature T. If $n_t = n_1 + n_2 + n_3$, then the pressure exerted by this mixture is given by

$$p = \frac{n_t RT}{V} \tag{2.28}$$

We define the partial pressure of each gas in the mixture as the pressure the gas would exert if it were alone in the container of volume V at temperature T. Then the partial pressures p_1, p_2, p_3 are given by

$$p_1 = n_1 \frac{RT}{V}, \qquad p_2 = n_2 \frac{RT}{V}, \qquad p_3 = n_3 \frac{RT}{V}. \tag{2.29}$$

Adding these equations, we obtain

$$p_1 + p_2 + p_3 = (n_1 + n_2 + n_3) \frac{RT}{V} = n_t \frac{RT}{V}.$$

Comparison of this equation with Eq. (2.28) shows that

$$p = p_1 + p_2 + p_3. \tag{2.30}$$

This is Dalton's law of partial pressures, which states that at any specified temperature the total pressure exerted by a gas mixture is equal to the sum of the partial pressures of the constituent gases. The first gas is said to exert a partial pressure p_1, the second gas exerts a partial pressure p_2, and so on. Partial pressures are calculated using Eqs. (2.29).

Partial pressures are simply related to the mole fractions of the gases in the mixture. Dividing both sides of the first of Eqs. (2.29) by the total pressure p, we obtain

$$\frac{p_1}{p} = \frac{n_1 RT}{pV}; \tag{2.31}$$

but, by Eq. (2.28), $p = n_t RT/V$. Using this value for p on the right-hand side of Eq. (2.31), we have

$$\frac{p_1}{p} = \frac{n_1}{n_t} = x_1.$$

Thus

$$p_1 = x_1 p, \qquad p_2 = x_2 p, \qquad p_3 = x_3 p.$$

These equations are conveniently abbreviated by writing

$$p_i = x_i p \qquad (i = 1, 2, 3, \ldots), \tag{2.32}$$

where p_i is the partial pressure of the gas that has a mole fraction x_i. Equation (2.32) allows the calculation of the partial pressure of any gas in a mixture from the mole fraction of that gas and the total pressure of the mixture.

Two things should be noted about Eq. (2.32): first, if either molar concentrations or mole ratios had been used, the final result would not be as simple an expression as Eq. (2.32); second, examination of the steps leading to Eq. (2.32) shows that it is not restricted to a mixture of three gases; it is correct for a mixture containing any number of gases.

2.8 THE PARTIAL-PRESSURE CONCEPT

The definition given in Eqs. (2.29) for the partial pressures of the gases in a mixture is a purely mathematical one; we now ask whether or not this mathematical concept of partial pressure has any physical significance. The results of two experiments, illustrated in Figs. 2.8 and 2.9, provide the answer to this question. First consider the experiment shown in Fig. 2.8. A container, Fig. 2.8(a), is partitioned into two compartments of equal volume

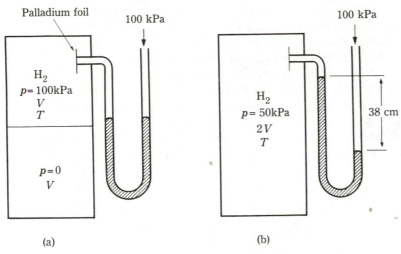

Figure 2.8 (a) Partition in place. (b) Partition removed.

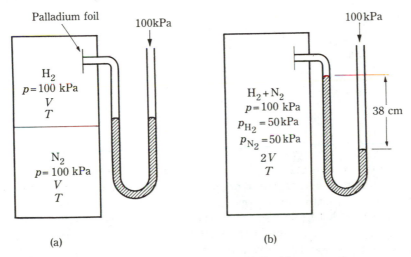

Figure 2.9 (a) Partition in place. (b) Partition removed.

V. The upper compartment contains hydrogen under a pressure of one atmosphere; the lower compartment is evacuated. One arm of a manometer is covered by a thin palladium foil and is connected to the hydrogen-filled compartment. The other arm of the manometer is open to a pressure of 1 atm which is kept constant during the experiment as is the temperature. At the beginning of the experiment, the mercury levels in the two arms of the manometer stand at the same height. This is possible because the palladium membrane is permeable to hydrogen but not to other gases, and so the membrane does not block the entrance of hydrogen to the manometer arm.

The partition is removed, and the hydrogen fills the entire vessel. After a period of time, the mercury levels rest in the final positions shown in Fig. 2.8(b). Since the volume available to the hydrogen has doubled, the pressure in the container has fallen to one-half its original value. (We neglect the volume of the manometer arm in this computation.)

In the second experiment, Fig. 2.9, the lower compartment contains nitrogen (which cannot pass the palladium foil) under 1 atm pressure. At the beginning of the experiment, the mercury levels stand at the same height. The partition is removed and the gases mix throughout the container. After a period of time the levels stand at the positions shown in Fig. 2.9(b). The result of this experiment is *exactly* the same as in the first experiment in which the lower compartment was evacuated. The hydrogen behaves exactly as if the nitrogen were not present. This important result means that the concept of partial pressure has a physical meaning as well as a mathematical one.

The interpretation of each experiment is direct. In the first experiment, the manometer read the total pressure both before and after the partition was removed:

$$p_{\text{initial}} = \frac{n_{\text{H}_2} RT}{V} = 1 \text{ atm},$$

$$p_{\text{final}} = \frac{n_{\text{H}_2} RT}{2V} = \tfrac{1}{2} \text{ atm.}$$

In the second experiment, the manometer read *total pressure* before the membrane was removed, and *partial pressure of hydrogen in the mixture* after removal of the membrane:

$$p_{\text{initial}} = \frac{n_{\text{H}_2} RT}{V} = 1 \text{ atm},$$

$$p_{\text{H}_2 \text{ (final)}} = \frac{n_{\text{H}_2} RT}{2V} = \tfrac{1}{2} \text{ atm},$$

$$p_{\text{N}_2 \text{ (final)}} = \frac{n_{\text{N}_2} RT}{2V} = \tfrac{1}{2} \text{ atm},$$

$$p_{\text{total, final}} = p_{\text{H}_2} + p_{\text{N}_2} = \tfrac{1}{2} + \tfrac{1}{2} = 1 \text{ atm.}$$

Note that the total pressure in the container does not change upon removal of the partition.

It is possible to measure the partial pressure of any gas in a mixture directly if there is a membrane that is permeable to that gas alone; for example, palladium is permeable to hydrogen and certain types of glass are permeable to helium. The fact that at present only a few such membranes are known does not destroy the physical reality of the concept of partial pressure. Later it will be shown that in chemical equilibria involving gases and in physical equilibria such as solubility of gases in liquids and solids, it is the partial pressures of the gases in the mixture that are significant (further confirmation of the physical content of the concept).

2.9 THE BAROMETRIC DISTRIBUTION LAW

In the foregoing discussion of the behavior of ideal gases it has been tacitly assumed that the pressure of the gas has the same value everywhere in the container. Strictly speaking, this assumption is correct only in the absence of force fields. Since all measurements are made on laboratory systems that are always in the presence of a gravitational field, it is important to know what effect is produced by the influence of this field. It may be said that, for gaseous systems of ordinary size, the influence of the gravity field is so slight as to be imperceptible even with extremely refined experimental methods. For a fluid of higher density such as a liquid, the effect is quite pronounced, and the pressure will be different at different vertical positions in a container.

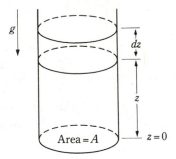

Figure 2.10   Column of fluid in a gravity field.

A column of fluid, Fig. 2.10, having a cross-sectional area A, at a uniform temperature T, is subjected to a gravitational field acting downward to give a particle an acceleration g. The vertical coordinate z is measured upward from ground level where $z = 0$. The pressure at any height z in the column is determined by the total mass of fluid above that height, m. The downward force on this mass is mg; this force divided by the area is the pressure at the height z:

$$p = \frac{mg}{A}.$$ (2.33)

Let the pressure at the height $z + dz$ be $p + dp$; then

$$p + dp = \frac{m'g}{A},$$

where m' is the mass of fluid above the height $z + dz$. But

$$m' + dm = m \qquad \text{or} \qquad m' = m - dm,$$

if dm is the mass of fluid in the slice between z and $z + dz$. Then

$$p + dp = \frac{(m - dm)g}{A} = \frac{mg}{A} - \frac{g\,dm}{A}$$

In view of Eq. (2.33) this becomes

$$dp = -\frac{g\,dm}{A}.$$

If ρ is the density of the fluid, then $dm = \rho A\,dz$; using this in the expression for dp yields

$$dp = -\rho g\,dz.$$ (2.34)

The differential equation, Eq. (2.35), relates the change in pressure, dp, to the density of the fluid, the gravitational acceleration, and the increment in height dz. The negative sign means that if the height increases (dz is $+$), the pressure of the fluid will decrease (dp is $-$). The effect of change in height on the pressure is proportional to the density of the fluid; thus the effect is important for liquids and negligible for gases.

If the density of a fluid is independent of pressure, as is the case for liquids, then Eq. (2.34) may be integrated immediately. Since ρ and g are constants, they are removed from the integral and we obtain

$$\int_{p_0}^{p} dp = -\rho g \int_{0}^{z} dz,$$

which, after integrating, gives

$$p - p_0 = -\rho g z, \tag{2.35}$$

where p_0 is the pressure at the bottom of the column, and p is the pressure at the height z above the bottom of the column. Equation (2.35) is the usual equation for hydrostatic pressure in a liquid.

To apply Eq. (2.34) to a gas, it must be recognized that the density of the gas is a function of the pressure. If the gas is ideal, then from Eq. (2.20), $\rho = Mp/RT$. Using this in Eq. (2.34), we have

$$dp = - \frac{Mgp \, dz}{RT}.$$

Separating variables yields

$$\frac{dp}{p} = - \frac{Mg \, dz}{RT} \tag{2.36}$$

and integrating, we obtain

$$\ln p = - \frac{Mgz}{RT} + C. \tag{2.37}$$

The integration constant C is evaluated in terms of the pressure at ground level; when $z = 0$, $p = p_0$. Using these values in Eq. (2.37), we find that $\ln p_0 = C$. Substituting this value for C and rearranging reduces Eq. (2.37) to

$$\ln \left(\frac{p}{p_0} \right) = - \frac{Mgz}{RT} \tag{2.38}$$

or

$$p = p_0 e^{-Mgz/RT}. \tag{2.39}$$

Since the density is proportional to the pressure, and the number of moles per cubic metre is proportional to the pressure, Eq. (2.39) can be written in two other equivalent forms:

$$\rho = \rho_0 e^{-Mgz/RT} \qquad \text{or} \qquad \tilde{c} = \tilde{c}_0 e^{-Mgz/RT}, \tag{2.40}$$

where ρ and ρ_0 are the densities and $\tilde{c}$ and $\tilde{c}_0$ are the concentrations in mol/m^3 at z and at ground level. Either of the equations (2.39) or (2.40) is called the barometric distribution law or the gravitational distribution law. The equation is a distribution law, because it describes the distribution of the gas in the column. Equation (2.39) relates the pressure at any height z to the height, the temperature of the column, the molecular weight of the gas, and the acceleration produced by the gravity field. Figure 2.11 shows a plot of p/p_0 versus z for nitrogen at three temperatures, according to Eq. (2.39). Figure 2.11 shows that at the higher temperature, the distribution is smoother than at the lower temperature. The variation in pressure with height is less pronounced the higher the temperature; if the temperature were infinite, the pressure would be the same everywhere in the column.

It is advisable to look more closely at this exponential type of distribution law, since it occurs so frequently in physics and physical chemistry in a more general form as the *Boltzmann distribution*. Equation (2.36) is most informative in discussing the exponential distribution; it can be written

$$\frac{-dp}{p} = \frac{Mg \, dz}{RT}, \tag{2.41}$$

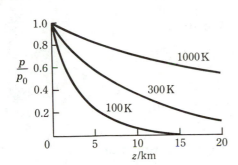

Figure 2.11 Plot of p/p_0 versus z for nitrogen.

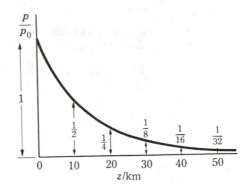

Figure 2.12 The constant relative decrease in pressure with equal increments in height.

which says that the relative decrease* in pressure, $-dp/p$, is a constant, Mg/RT, multiplied by the increase in height, dz. It follows that this relative decrease is the same at all positions in the column; therefore it cannot matter where the origin of z is chosen. For example, suppose that for a certain gas the pressure at ground level is 1 atm and the distribution shows that the pressure decreases to $\frac{1}{2}$ atm at a height of 10 km. Then for this same gas, the pressure at a height $z + 10$ km is one-half the value of the pressure at the height z. Thus at any height, the pressure is one-half the value it has at a height 10 km below. This aspect of the distribution is emphasized in Fig. 2.12.

The argument does not depend on the choice of one-half as the relative value. Suppose that for some gas the pressure at a height of 6.3 km is 0.88 of its value at ground level. Then in another interval of 6.3 km, the pressure will drop again by the factor 0.88. The pressure at $2(6.3) = 12.6$ km will then be $(0.88)(0.88) = 0.774$ of the ground level value (see Problem 2.33).

Another point to note about Eq. (2.41) is that the relative decrease in pressure is proportional to Mg/RT. Consequently, for any particular gas, the relative decrease is less at higher temperatures (see Fig. 2.11). At a specified temperature the relative decrease is larger for a gas having a high molecular weight than for a gas with a low molecular weight.

For a gas mixture in a gravity field, it can be shown that each of the gases obeys the distribution law independently of the others. For each gas

$$p_i = p_{io}e^{-M_igz/RT}, \tag{2.42}$$

where p_i is the partial pressure of the ith gas in the mixture at the height z, p_{io} is the partial pressure of the gas at ground level, and M_i is the molar mass of the gas. The interesting consequence of this law is that the partial pressures of very light gases decrease less rapidly with height than do those of heavier gases. Thus in the earth's atmosphere the percentage composition at very great heights is quite different from that at ground level. At a height of 100 km the light gases such as helium and neon form a higher percentage of the atmosphere than they do at ground level.

Using Eq. (2.42), we can estimate the atmospheric composition at different altitudes. Even though the atmosphere is not isothermal and not in equilibrium, these estimates are not bad.

---

*   Since dp is an increase, $-dp$ is a decrease.

■ **EXAMPLE 2.3**   The partial pressure of argon in the atmosphere is 0.0093 atm. What is the argon pressure at 50 km if the temperature is 20 °C? $g = 9.807$ m/s^2.

In SI, $M_{Ar} = 0.0399$ kg/mol and $z = 50$ km $= 5 \times 10^4$ m. Then

$$\frac{Mgz}{RT} = \frac{(0.0399 \text{ kg/mol})(9.807 \text{ m/s}^2)(5 \times 10^4 \text{ m})}{(8.314 \text{ J/K mol})(293 \text{ K})} = 8.03,$$

and

$$p = p_0 e^{-Mgz/RT} = 0.0093 \text{ atm } e^{-8.03} = 3.0 \times 10^{-6} \text{ atm.}$$

★ 2.9.1 The Distribution of Particles in a Colloidal Solution

The distribution law in Eq. (2.40) not only applies to gases but also describes the dependence of the concentration of colloidal particles or polymer particles suspended in liquid solution on their position in the solution. The total number of moles of substance in the element of volume between z_1 and z_2 is given by dn:

$$dn = \tilde{c} \, dV = \tilde{c}A \, dz \tag{2.43}$$

To obtain the total number of moles, $n(z_1, z_2)$, between any two positions, z_1 and z_2, in the column, we integrate Eq. (2.43) between those positions:

$$n(z_1, z_2) = \int_{z_1}^{z_2} dn = \int_{z_1}^{z_2} \tilde{c}A \, dz. \tag{2.44}$$

The volume enclosed between z_1 and z_2 is

$$V(z_1, z_2) = \int_{z_1}^{z_2} A \, dz.$$

The average concentration, $\langle \tilde{c} \rangle$, in the layer is

$$\langle \tilde{c} \rangle = \frac{n(z_1, z_2)}{V(z_1, z_2)} = \frac{\displaystyle\int_{z_1}^{z_2} \tilde{c}A \, dz}{\displaystyle\int_{z_1}^{z_2} A \, dz}. \tag{2.45}$$

If the column is uniform in cross section, then the area A is constant and we obtain

$$\langle \tilde{c} \rangle = \frac{\displaystyle\int_{z_1}^{z_2} \tilde{c} \, dz}{z_2 - z_1}. \tag{2.46}$$

We use $\tilde{c}$ as a function of z from Eq. (2.40) to evaluate the integral. In this way we can relate the concentration in any part of the container to the total number of moles. Since the distribution of polymer molecules in a solution is determined by the molar mass of the polymer, the difference in concentration between the top and bottom of the solution can be used to measure the molar mass of the polymer.

■ **EXAMPLE 2.4**   Consider a column of air at 20 °C in the earth's gravity field. What fraction of the nitrogen present in the atmosphere lies below an altitude of 20 km?

The number of moles of gas below the height z is given by Eq. (2.44):

$$n(0, z) = \int_0^z dn = \int_0^z \tilde{c}A \, dz = A\tilde{c}_0 \int_0^z e^{-Mgz/RT} \, dz = A\tilde{c}_0 \frac{RT}{Mg} (1 - e^{-Mgz/RT}).$$

The total number of moles is

$$n(0, \infty) = \int_0^\infty dn = A\tilde{c}_0 \frac{RT}{Mg} \lim_{z \to \infty} (1 - e^{-Mgz/RT}) = A\tilde{c}_0 \frac{RT}{Mg}.$$

The fraction lying below z is $n(0, z)/n(0, \infty) = 1 - e^{-Mgz/RT}$. For our case, since for nitrogen $M = 0.0280$ kg/mol, $z = 2 \times 10^4$ m, and $T = 293$ K,

$$\frac{Mgz}{RT} = \frac{(0.0280 \text{ kg/mol})(9.807 \text{ m/s}^2)(2 \times 10^4 \text{ m})}{(8.314 \text{ J/K mol})(293 \text{ K})} = 2.25;$$

then

$$\frac{n(0, 20 \text{ km})}{n(0, \infty)} = 1 - e^{-2.25} = 1 - 0.10 = 0.90.$$

QUESTIONS

2.1 Why are four values of the properties mass, volume, temperature, and pressure insufficient to describe the state of a *nonequilibrium* gas; for example, a turbulent gas?

2.2 Could n in the ideal gas law have been identified as the number of moles without Avogadro's hypothesis?

2.3 According to Dalton's law, what is most of the pressure of the atmosphere (that is, air) due to?

2.4 Why don't all the gas molecules in the atmosphere simply fall to earth?

2.5 The force on an ion of negative charge $-q$ in a constant electric field E in the z direction is $F = -qE$. By analogy to the gravitational case, what is the spatial distribution of such ions immersed in a column of gas and subject to a constant vertical field E? (Ignore the effect of gravity on the ions and on the gas.)

PROBLEMS

Conversion factors:

  Volume: $1 \text{ L} = 1 \text{ dm}^3 = 10^{-3} \text{ m}^3$ (all are exact).

  Pressure: $1 \text{ atm} = 760 \text{ Torr} = 1.01325 \times 10^5 \text{ Pa}$ (all are exact).

2.1 A sealed flask with a capacity of 1 dm^3 contains 5 g of ethane. The flask is so weak that it will burst if the pressure exceeds 1 MPa. At what temperature will the pressure of the gas reach the bursting pressure?

2.2 A large cylinder for storing compressed gases has a volume of about 0.050 m^3. If the gas is stored under a pressure of 15 MPa at 300 K, how many moles of gas are contained in the cylinder? What would be the mass of oxygen in such a cylinder?

2.3 Helium is contained at 30.2 °C in the system illustrated in Fig. 2.13. The leveling bulb L can be raised so as to fill the lower bulb with mercury and force the gas into the upper part of the device. The volume of bulb 1 to the mark b is 100.5 cm^3 and the volume of bulb 2 between the marks a and b is 110.0 cm^3. The pressure exerted by the helium is measured by the difference between the mercury levels in the device and in the evacuated arm of the manometer. When the mercury level is at a, the difference in levels is 20.14 mm. The density of mercury at 30.2 °C is 13.5212 g/cm^3 and the acceleration of gravity is 9.80665 m/s^2. What is the mass of helium in the container?

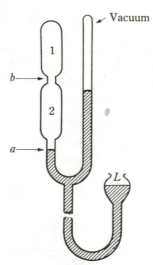

Figure 2.13

2.4 The same type of apparatus is used as in Problem 2.3. In this case the volume v_1 is not known; the volume of bulb 2, v_2, is 110.0 cm^3. When the mercury level is at a the difference in levels is 15.42 mm. When the mercury level is raised to b, the difference in levels is 27.35 mm. The temperature is 30.2 °C. Use the values of the density of mercury and g given in Problem 2.3.

a) What is the mass of helium in the system?
b) What is the volume of bulb 1?

2.5 Suppose that in setting up the scales of atomic masses the standard conditions had been chosen as $p_0 = 1$ atm, $V_0 = 0.03$ m^3 (exactly), and $T_0 = 300$ K (exactly). Compute the "gas-constant," the "Avogadro constant," and the masses of a "mole" of hydrogen atoms and oxygen atoms.

2.6 The coefficient of thermal expansion α is defined by $\alpha = (1/V)(\partial V/\partial T)_p$. Using the equation of state, compute the value of α for an ideal gas.

2.7 The coefficient of compressibility κ is defined by $\kappa = -(1/V)(\partial V/\partial p)_T$. Compute the value of κ for an ideal gas.

2.8 For an ideal gas, express the derivative $(\partial p/\partial T)_V$ in terms of α and κ.

2.9 Consider a gas mixture in a 2 dm^3 flask at 27 °C. For each mixture calculate the partial pressure of each gas, the total pressure, and the composition of the mixture in mole percent. Compare the results of the four calculations.

a) 1 g H_2 and 1 g O_2
b) 1 g N_2 and 1 g O_2
c) 1 g CH_4 and 1 g NH_3
d) 1 g H_2 and 1 g Cl_2

2.10 A sample of air is collected over water at 20 °C. At equilibrium the total pressure of the moist air is 1 atm. The equilibrium vapor pressure of water at 20 °C is 17.54 Torr; the composition of dry air is 78 mole % N_2, 21 mole % O_2, and 1 mole % Ar.

a) Calculate the partial pressures of nitrogen, oxygen, and argon in the wet mixture.
b) Calculate the mole fractions of nitrogen, oxygen, argon, and water in the wet mixture.

2.11 Consider a 20 L sample of moist air at 60 °C under a total pressure of 1 atm in which the partial pressure of water vapor is 0.120 atm. Assume the composition of dry air given in Problem 2.10.

a) What are the mole percentages of each of the gases in the sample?
b) The percent relative humidity is defined as % R.H. = 100 p_w/p_{wo}, where p_w is the partial pressure of water in the sample and p_{wo} is the equilibrium vapor pressure of water at the

temperature in question. At 60 °C, $p_{wo} = 0.197$ atm. What volume must the mixture occupy at 60 °C if the relative humidity is to be 100%?

c) What fraction of the water will be condensed if the total pressure of the mixture is increased isothermally to 200 atm?

2.12 A box contains liquid water in equilibrium with water vapor at 30 °C. The equilibrium vapor pressure of water at 30 °C is 31.82 Torr. If the volume of the box is increased, some of the liquid water evaporates to maintain the equilibrium pressure. There is 0.90 g of water present. What must the volume of the box be if all the liquid is to evaporate? (The volume of the liquid water may be ignored.)

2.13 The total pressure of a mixture of oxygen and hydrogen is 1.00 atm. The mixture is ignited and the water formed is removed. The remaining gas is pure hydrogen and exerts a pressure of 0.40 atm when measured under the same conditions of T and V as the original mixture. What was the original composition of the mixture (mole %)?

2.14 A mixture of nitrogen and water vapor is admitted to a flask that contains a solid drying agent. Immediately after admission, the pressure in the flask is 760 Torr. After standing some hours, the pressure reaches a steady value of 745 Torr.

a) Calculate the composition, in mole percent, of the original mixture.

b) If the experiment is done at 20 °C and the drying agent increases in weight by 0.150 g, what is the volume of the flask? (The volume occupied by the drying agent may be ignored.)

2.15 A mixture of oxygen and hydrogen is analyzed by passing it over hot copper oxide and through a drying tube. Hydrogen reduces the CuO according to the equation

$$CuO + H_2 \longrightarrow Cu + H_2O.$$

Oxygen then reoxidizes the copper formed:

$$Cu + \tfrac{1}{2}O_2 \longrightarrow CuO.$$

100.0 cm^3 of the mixture measured at 25 °C and 750 Torr yields 84.5 cm^3 of dry oxygen measured at 25 °C and 750 Torr after passage over CuO and the drying agent. What is the original composition of the mixture?

2.16 A sample of C_2H_6 is burned in a volume of air sufficient to provide twice the amount of oxygen needed to burn the C_2H_6 completely to CO_2 and H_2O. After the ethane is completely burned, what is the composition (mole fraction) of the gas mixture? Assume that all the water is present as vapor and that air is 78% nitrogen, 21% oxygen, and 1% argon.

2.17 A gas sample is known to be a mixture of ethane and butane. A bulb of 200.0 cm^3 capacity is filled with the gas to a pressure of 100.0 kPa at 20.0 °C. If the weight of gas in the bulb is 0.3846 g, what is the mole percent of butane in the mixture?

2.18 A bulb of 138.2 mL volume contains 0.6946 g of gas at 756.2 Torr and 100.0 °C. What is the molar mass of the gas?

2.19 Consider an isothermal column of an ideal gas at 25 °C. What must the molar mass of this gas be if the pressure is 0.80 of its ground level value at (a) 10 km, (b) 1 km, and (c) 1 m. (d) What kinds of molecules have molar masses of the magnitude in (c)?

2.20 Assuming that air has a mean molar mass of 28.9 g/mol and that the atmosphere is isothermal at 25 °C, compute the barometric pressure at Denver, which is 1600 m above sea level; compute the barometric pressure at the top of Mt. Evans, 4348 m above sea level. The pressure at sea level may be taken as 760 Torr.

2.21 Consider an "ideal potato gas," which has the following properties: it obeys the ideal gas law, and the individual particles have a mass of 100 g, but occupy no volume (that is, they are point masses).

a) At 25 °C compute the height at which the number of potatoes per cubic metre falls to one-millionth of its ground-level value.

b) Recognizing that real potatoes do occupy a volume, is there any correspondence between the result of the calculation in (a) and the observed spatial distribution of potatoes in a paper bag?

2.22 Consider the pressure at a height of 10 km in a column of air, $M = 0.0289$ kg/mol. If the pressure at ground level remains at 1 atm but the temperature changes from 300 K to 320 K, what will be the change in pressure at the 10 km altitude?

2.23 At 300 K a gas mixture in a gravity field exerts a total pressure of 1.00 atm and consists of 0.600 mole fraction of nitrogen, $M = 0.0280$ kg/mol; the remainder is carbon dioxide, $M = 0.0440$ kg/mol.

a) Calculate the partial pressures of N_2 and CO_2, the total pressure, and the mole fraction of N_2 in the mixture at 50 km altitude.

b) Calculate the number of moles of nitrogen between 0 and 50 km altitude in a column having a cross-sectional area of 5 m^2.

2.24 The approximate composition of the atmosphere at sea level is given in the following table.

Gas	Mole percent	Gas	Mole percent
Nitrogen	78.09	Helium	0.0005
Oxygen	20.93	Krypton	0.0001
Argon	0.93	Hydrogen	5×10^{-5}
Carbon dioxide	0.03	Xenon	8×10^{-6}
Neon	0.0018	Ozone	5×10^{-5}

By permission from *Scientific Encyclopedia*, 3d ed. New York: D. Van Nostrand, 1958, p. 34.

Ignoring the last four components, compute the partial pressures of the others, the total pressure, and the composition of the atmosphere in mole percent, at altitudes of 50 and 100 km ($t = 25$ °C).

2.25 A solution of a polymer, $M = 200$ kg/mol, at 27 °C, fills a container to a depth of 10 cm. If the concentration of the polymer at the bottom of the solution is c_0, what is the concentration at the top of the solution?

2.26 At 300 K consider a colloidal solution, $M = 150$ kg/mol, in a gravity field. If the concentration of the colloid is 0.00080 mol/L at the top of the solution and 0.0010 mol/L at the bottom,

a) How deep is the solution?

b) Calculate the average concentration of the colloid in the lowest 0.10 m of the solution.

c) Calculate the number of moles in the lowest 0.10 m of the solution, if the cross-sectional area of the container is 20 cm^2.

2.27 A polymer solution has an average concentration, $\langle \tilde{c} \rangle = 0.100$ mol/m^3, and an average molar mass of 20.0 kg/mol. At 25 °C the solution fills a cylinder that is 50 cm high. What are the concentrations of the polymer at the top and at the bottom of the cylinder?

2.28 At 300 K a polymer solution fills a cylinder to a depth of 0.20 m; the cross-sectional area is 20 cm^2.

a) If the concentration at the top of the solution is 95% of that at the bottom, what is the molar mass of the polymer?

b) Calculate the total mass of polymer in the container, if $\tilde{c}_0 = 0.25$ mol/m^3.

c) Calculate the average concentration of the polymer in the solution.

2.29 A balloon having a capacity of 10,000 m^3 is filled with helium at 20 °C and 1 atm pressure. If the balloon is loaded with 80% of the load that it can lift at ground level, at what height will the balloon come to rest? Assume that the volume of the balloon is constant, the atmosphere isothermal, 20 °C, the molar mass of air is 28.9 g/mol, and the ground level pressure is 1 atm. The mass of the balloon is 1.3×10^6 g.

2.30 When Julius Caesar expired, his last exhalation had a volume of about 500 cm^3. This expelled air was $1 \text{ mol}\%$ argon. Assume that the temperature was 300 K and the ground-level pressure was 1 atm. Assume that the temperature and pressure are uniform over the earth's surface and still have the same values. If Caesar's argon molecules have all remained in the atmosphere and have been completely mixed throughout the atmosphere, how many inhalations, 500 cm^3 each, must we make on average to inhale one of Caesar's argon molecules? The mean radius of the earth is 6.37×10^6 m.

2.31 Show that $x_i = (y_i/M_i)/[(y_1/M_1) + (y_2/M_2) + \cdots]$, in which x_i, y_i, and M_i are the mole fraction, the weight percent, and the molar mass of component i, respectively.

2.32 Express the partial pressures in a mixture of gases (a) in terms of the volume concentrations $\tilde{c}_i$, and (b) in terms of the mole ratios r_i.

2.33 If at a specified height Z the pressure of a gas is p_Z, and that at $z = 0$ is p_0, show that at any height z, $p = p_0 f^{z/Z}$, where $f = p_Z/p_0$.

2.34 Consider an ideal gas with a fixed molar mass and at a specified temperature in a gravity field. If at 5.0 km altitude, the pressure is 0.90 of its ground-level value, what fraction of the ground-level value will the pressure be at 10 km? At 15 km?

2.35 a) Show that if we calculate the total number of molecules of a gas in the atmosphere using the barometric formula, we would get the same result if we assumed that the gas had the ground-level pressure up to a height $z = RT/Mg$ and had zero pressure above that level.

b) Show that the total mass of the earth's atmosphere is given by Ap_0/g, where p_0 is the total ground-level pressure and A is the area of the earth's surface. Note that this result does not depend on the composition of the atmosphere. (Do this problem first by calculating the mass of each constituent, mole fraction $= x_i$, molar mass $= M_i$, and summing. Then by examining the result, do it the easy way.)

c) If the mean radius of the earth is 6.37×10^6 m, and $p_0 = 1$ atm, calculate the mass of the atmosphere.

2.36 Since the gases in the atmosphere are distributed differently according to their molar masses, the *average* percentage of each gas is different from the percentage at ground level. The values, x_i^0, of mole fractions at ground level are given.

a) Derive a relation between the average mole fraction of the gas in the atmosphere and the mole fractions at ground level.

b) If the mole fractions of N_2, O_2, and Ar at ground level are 0.78, 0.21, and 0.01, respectively, compute the average mole fractions of N_2, O_2, and Ar in the atmosphere.

c) Show that the *average mass fraction* of any gas in the atmosphere is equal to its *mole fraction at ground level*.

2.37 Consider a column of gas in a gravity field. Calculate the height Z, determined by the condition that half the mass of the column lies below Z.

2.38 For the dissociation $N_2O_4 \rightleftharpoons 2NO_2$, the equilibrium constant at 25 °C is $K = 0.115$; it is related to the degree of dissociation α and the pressure in atm by $K = 4\alpha^2 p/(1 - \alpha^2)$. If n is the number of moles of N_2O_4 that would be present if no dissociation occurred, calculate V/n at $p = 2$ atm, 1 atm, and 0.5 atm, assuming that the equilibrium mixture behaves ideally. Compare the results with the volumes if dissociation did not occur.

2.39 For the mixture described in Problem 2.38, show that as p approaches zero, the compressibility factor $Z = pV/nRT$ approaches 2 instead of the usual value of unity. Why does this happen?

3

Real Gases

3.1 DEVIATIONS FROM IDEAL BEHAVIOR

Since the ideal gas law does not accurately represent the behavior of real gases, we shall now attempt to formulate more realistic equations of state for gases and explore the implications of these equations.

If measurements of pressure, molar volume, and temperature of a gas do not confirm the relation $p\overline{V} = RT$, within the precision of the measurements, the gas is said to deviate from ideality or to exhibit nonideal behavior. To display the deviations clearly, the ratio of the observed molar volume $\overline{V}$ to the ideal molar volume $\overline{V}_{id}(=RT/p)$ is plotted as a function of pressure at constant temperature. This ratio is called the *compressibility factor Z*. Then,

$$Z \equiv \frac{\overline{V}}{\overline{V}_{id}} = \frac{p\overline{V}}{RT}. \tag{3.1}$$

For the ideal gas, $Z = 1$ and is independent of pressure and temperature. For real gases $Z = Z(T, p)$, a function of both temperature and pressure.

Figure 3.1 shows a plot of Z as a function of pressure at 0 °C for nitrogen, hydrogen, and the ideal gas. For hydrogen, Z is greater than unity (the ideal value) at all pressures. For nitrogen, Z is less than unity in the lower part of the pressure range, but is greater than unity at very high pressures. Note that the pressure range in Fig. 3.1 is very large; near one atmosphere both of these gases behave nearly ideally. Also note that the vertical scale in Fig. 3.1 is much expanded compared to that in Fig. 3.2.

Figure 3.2 shows a plot of Z versus p for several gases at 0 °C. Note that for those gases that are easily liquefied, Z dips sharply below the ideal line in the low-pressure region.

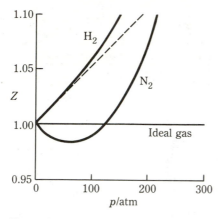

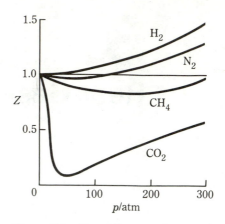

Figure 3.1 Plot of Z versus p for H_2, N_2, and ideal gas at 0 °C.

Figure 3.2 Plot of Z versus p for several gases at 0 °C.

3.2 MODIFYING THE IDEAL GAS EQUATION; THE VAN DER WAALS EQUATION

How can the ideal gas law be modified to yield an equation that will represent the experimental data more accurately? We begin by correcting an obvious defect in the ideal gas law, namely the prediction that under any finite pressure the volume of the gas is zero at the absolute zero of temperature: $\overline{V} = RT/p$. On cooling, real gases liquefy and ultimately solidify; after liquefaction the volume does not change very much. We can arrange the new equation so that it predicts a finite, positive volume for the gas at 0 K by adding a positive constant b to the ideal volume:

$$\overline{V} = b + \frac{RT}{p}. \tag{3.2}$$

According to Eq. (3.2) the molar volume at 0 K is b, and we expect that b will be roughly comparable with the molar volume of the liquid or solid.

Equation (3.2) also predicts that as the pressure becomes infinite the molar volume approaches the limiting value b. This prediction is more in accord with experience than the prediction of the ideal gas law that the molar volume approaches zero at very high pressures.

Now it would be interesting to see how well Eq. (3.2) predicts the curves in Figs. 3.1 and 3.2. Since by definition $Z = p\overline{V}/RT$, multiplication of Eq. (3.2) by p/RT yields

$$Z = 1 + \frac{bp}{RT}. \tag{3.3}$$

Since Eq. (3.3) requires Z to be a linear function of pressure with a positive slope b/RT, it cannot possibly fit the curve for nitrogen in Fig. 3.1, which starts from the origin with a negative slope. However, Eq. (3.3) can represent the behavior of hydrogen. In Fig. 3.1 the dashed line is a plot of Eq. (3.3) fitted at the origin to the curve for hydrogen. In the low-pressure region the dashed line represents the data very well.

We can conclude from Eq. (3.3) that the assumption that the molecules of a gas have finite size is sufficient to explain values of Z greater than unity. Apparently this size effect is the dominating one in producing deviations from ideality in hydrogen at 0 °C. It is also clear that some other effect must produce the deviations from ideality in gases such as nitrogen and methane, since the size effect cannot explain their behavior in the low-pressure range. This other effect must now be sought.

We have already noted that the worst offenders in the matter of having values of Z less than unity are methane and carbon dioxide, which are easily liquefied. Thus we begin to suspect a connection between ease of liquefaction and the compressibility factor, and to ask why a gas liquefies. First of all, energy, the heat of vaporization, must be supplied to take a molecule out of the liquid and put it into the vapor. This energy is required because of the forces of attraction acting between the molecule and its neighbors in the liquid. The force of attraction is strong if the molecules are close together, as they are in a liquid, and very weak if the molecules are far apart, as they are in a gas. The problem is to find an appropriate way of modifying the gas equation to take account of the effect of these weak attractive forces.

The pressure exerted by a gas on the walls of a container acts in an outward direction. Attractive forces between the molecules tend to pull them together, thus diminishing the outward thrust against the wall and reducing the pressure below that exerted by the ideal gas. This reduction in pressure should be proportional to the force of attraction between the molecules of the gas.

Consider two small volume elements v_1 and v_2 in a container of gas (Fig. 3.3). Suppose that each volume element contains one molecule and that the attractive force between the two volume elements is some small value f. If another molecule is added to v_2, keeping one molecule in v_1, the force acting between the two elements should be $2f$; addition of a third molecule to v_2 should increase the force to $3f$, and so on. The force of attraction between the two volume elements is therefore proportional to $\tilde{c}_2$, the concentration of molecules in v_2. If at any point in the argument, the number of molecules in v_2 is kept constant and molecules are added to v_1, then the force should double and triple, etc. The force is therefore proportional to $\tilde{c}_1$, the concentration of molecules in v_1. Thus, the force acting between the two elements can be written as: force $\propto \tilde{c}_1\tilde{c}_2$. Since the concentration in a gas is everywhere the same, $\tilde{c}_1 = \tilde{c}_2 = \tilde{c}$, and so, force $\propto \tilde{c}^2$. But $\tilde{c} = n/V = 1/\overline{V}$; consequently, force $\propto 1/\overline{V}^2$.

We rewrite Eq. (3.2) in the form

$$p = \frac{RT}{\overline{V} - b}. \tag{3.4}$$

Because of the attractive forces between the molecules, the pressure is less than that given by Eq. (3.4) by an amount proportional to $1/\overline{V}^2$, so a term is subtracted from the right-hand side to yield

$$p = \frac{RT}{\overline{V} - b} - \frac{a}{\overline{V}^2}, \tag{3.5}$$

where a is a positive constant roughly proportional to the energy of vaporization of the liquid. Two things should be noted about the introduction of the $a/\overline{V}^2$ term. First, forces

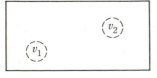

Figure 3.3   Volume elements in a gas.

Table 3.1
van der Waals constants

Gas	$a/\mathrm{Pa\ m^6\ mol^{-2}}$	$b/10^{-6}\ \mathrm{m^3\ mol^{-1}}$
He	0.00345	23.4
H_2	0.0247	26.6
O_2	0.138	31.8
CO_2	0.366	42.9
H_2O	0.580	31.9
Hg	0.820	17.0

Francis Weston Sears, *An Introduction to Thermo-dynamics, the Kinetic Theory of Gases, and Statistical Mechanics.* Reading, Mass.: Addison-Wesley, 1950.

acting on any volume element in the interior of the gas balance out to zero; only those elements of volume near the wall of the container experience an unbalanced force that tends to pull them toward the center. Thus the effect of the attractive forces is felt only at the walls of the vessel. Second, the derivation assumed an effective range of action of the attractive forces of the order of centimetres; in fact the range of these forces is of the order of nanometres. In Chapter 26 we do the derivation without this assumption and obtain the same result.

Equation (3.5) is the *van der Waals equation*, proposed by van der Waals, who was the first to recognize the influence of molecular size and intermolecular forces on the pressure of a gas. These weak forces of attraction are called van der Waals forces. The van der Waals constants, *a* and *b*, for a few gases are given in Table 3.1. The van der Waals equation is frequently written in the equivalent but less instructive forms

$$\left(p + \frac{a}{\overline{V}^2}\right)(\overline{V} - b) = RT \qquad \text{or} \qquad \left(p + \frac{n^2a}{V^2}\right)(V - nb) = nRT, \qquad (3.6)$$

where $V = n\overline{V}$ has been used in the second writing.

3.3 IMPLICATIONS OF THE VAN DER WAALS EQUATION

The van der Waals equation takes two effects into account: first, the effect of molecular size, Eq. (3.2),

$$p = \frac{RT}{(\overline{V} - b)}.$$

Since the denominator in the above equation is smaller than the denominator in the ideal gas equation, the size effect by itself increases the pressure above the ideal value. According to this equation it is the empty space between the molecules, the "free" volume, that follows the ideal gas law. Second, the effect of intermolecular forces, Eq. (3.5),

$$p = \frac{RT}{\overline{V} - b} - \frac{a}{\overline{V}^2},$$

is taken into account. The effect of attractive forces by itself reduces the pressure below the ideal value and is taken into account by subtracting a term from the pressure.

To calculate Z for the van der Waals gas we multiply Eq. (3.5) by $\overline{V}$ and divide by RT; this yields

$$Z = \frac{p\overline{V}}{RT} = \frac{\overline{V}}{\overline{V} - b} - \frac{a}{RT\overline{V}}.$$

The numerator and denominator of the first term on the right-hand side are divided by $\overline{V}$:

$$Z = \frac{1}{1 - b/\overline{V}} - \frac{a}{RT\overline{V}}.$$

At low pressures $b/\overline{V}$ is small compared with unity, so the first term on the right may be developed into a power series in $1/\overline{V}$ by division; thus $1/(1 - b/\overline{V}) = 1 + (b/\overline{V}) + (b/\overline{V})^2 + \cdots$. Using this result in the preceding equation for Z and collecting terms, we have

$$Z = 1 + \left(b - \frac{a}{RT}\right)\frac{1}{\overline{V}} + \left(\frac{b}{\overline{V}}\right)^2 + \left(\frac{b}{\overline{V}}\right)^3 + \cdots, \tag{3.7}$$

which expresses Z as a function of temperature and molar volume. It would be preferable to have Z as a function of temperature and pressure; however, this would entail solving the van der Waals equation for $\overline{V}$ as a function of T and p, then multiplying the result by p/RT to obtain Z as a function of T and p. Since the van der Waals equation is a cubic equation in $\overline{V}$, the solutions are too complicated to be particularly informative. We content ourselves with an approximate expression for $Z(T, p)$ which we obtain from Eq. (3.7) by observing that as $p \to 0$, $(1/\overline{V}) \to 0$, and $Z = 1$. This expansion of Z, correct to the term in p^2, is

$$Z = 1 + \frac{1}{RT}\left(b - \frac{a}{RT}\right)p + \frac{a}{(RT)^3}\left(2b - \frac{a}{RT}\right)p^2 + \cdots. \tag{3.8}$$

The correct coefficient for p could have been obtained by simply replacing $1/\overline{V}$ by the ideal value in Eq. (3.7); however, this would yield incorrect values of the coefficients of the higher powers of pressure. (See Section 3.3.1 for the derivation of Eq. (3.8).)

Equation (3.8) shows that the terms responsible for nonideal behavior vanish not only as the pressure approaches zero but also as the temperature approaches infinity. Thus, as a general rule, real gases are more nearly ideal when the pressure is lower and the temperature is higher.

The second term on the right of Eq. (3.8) should be compared with the second term on the right of Eq. (3.3), which considered only the effect of finite molecular volume. The slope of the Z versus p curve is obtained by differentiating Eq. (3.8) with respect to pressure, keeping the temperature constant:

$$\left(\frac{\partial Z}{\partial p}\right)_T = \frac{1}{RT}\left(b - \frac{a}{RT}\right) + \frac{2a}{(RT)^3}\left(2b - \frac{a}{RT}\right)p + \cdots.$$

At $p = 0$, all of the higher terms drop out and this derivative reduces simply to

$$\left(\frac{\partial Z}{\partial p}\right)_T = \frac{1}{RT}\left(b - \frac{a}{RT}\right), \qquad p = 0, \tag{3.9}$$

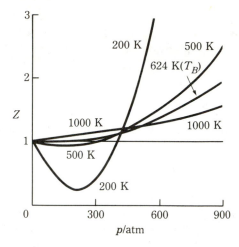

Figure 3.4 Plot of Z versus p for ethylene at several temperatures (T_B = Boyle temperature).

where the derivative is the *initial slope* of the Z versus p curve. If $b > a/RT$, the slope is positive; the size effect dominates the behavior of the gas. On the other hand, if $b < a/RT$, then the initial slope is negative; the effect of the attractive forces dominates the behavior of the gas. Thus the van der Waals equation, which includes both the effects of size and of the intermolecular forces, can interpret either positive or negative slopes of the Z versus p curve. In interpreting Fig. 3.2, we can say that at 0 °C the effect of the attractive forces dominates the behavior of methane and carbon dioxide, while the molecular size effect dominates the behavior of hydrogen.

Having examined the Z versus p curves for several gases at one temperature, we focus attention on the Z versus p curves for a *single* gas at different temperatures. Equation (3.9) shows that if the temperature is low enough, the term a/RT will be larger than b and so the initial slope of Z versus p will be negative. As the temperature increases, a/RT becomes smaller and smaller; if the temperature is high enough, a/RT becomes less than b, and the initial slope of the Z versus p curve becomes positive. Finally, if the temperature is extremely high, Eq. (3.9) shows that the slope of Z versus p must approach zero. This behavior is shown in Fig. 3.4.

At some intermediate temperature T_B, the Boyle temperature, the initial slope must be zero. The condition for this is given by Eq. (3.9) as $b - a/RT_B = 0$. This yields

$$T_B = \frac{a}{Rb}. \tag{3.10}$$

At the Boyle temperature the Z versus p curve is tangent to the curve for the ideal gas at $p = 0$ and rises above the ideal gas curve only very slowly. In Eq. (3.8) the second term drops out at T_B, and the remaining terms are small until the pressure becomes very high. Thus at the Boyle temperature the real gas behaves ideally over a wide range of pressures, because the effects of size and of intermolecular forces roughly compensate. This is also shown in Fig. 3.4. The Boyle temperatures for several different gases are given in Table 3.2.

The data in Table 3.2 make the curves in Fig. 3.2 comprehensible. All of them are drawn at 0 °C. Thus hydrogen is above its Boyle temperature and so always has Z-values greater than unity. The other gases are below their Boyle temperatures and so have Z-values less than unity in the low-pressure range.

Table 3.2
Boyle temperatures for several gases

Gas	He	H_2	N_2	Ar	CH_4	CO_2	C_2H_4	NH_3
T_B/K	23.8	116.4	332	410	506	600	624	995

The van der Waals equation is a distinct improvement over the ideal gas law in that it gives qualitative reasons for the deviations from ideal behavior. This improvement is gained at considerable sacrifice, however. The ideal gas law contains nothing that depends on the individual gas; the constant R is a universal constant. The van der Waals equation contains two constants that are different for every gas. In this sense a different van der Waals equation must be used for each different gas. In Section 3.8 it will be seen that this loss in generality can be remedied for the van der Waals equation and for certain other equations of state.

3.3.1 A Mathematical Trick

As we pointed out for the van der Waals equation, it is impractical to obtain Z as a function of T and p in a straightforward manner. It is necessary to use a mathematical trick to transform Eq. (3.7) to a series in powers of the pressure.

At low pressures we can expand Z as a power series in the pressure.

$$Z = 1 + A_1 p + A_2 p^2 + A_3 p^3 + \cdots,$$

in which the coefficients $A_1, A_2, A_3, \ldots$, are functions of temperature only. To determine these coefficients, we use the definition of Z in Eq. (3.1) to write $(1/\overline{V}) = p/RTZ$. Using this value of $(1/\overline{V})$ in Eq. (3.7) brings it to the form

$$1 + A_1 p + A_2 p^2 + A_3 p^3 + \cdots$$

$$= 1 + \left(b - \frac{a}{RT}\right)\frac{p}{RTZ} + \left(\frac{b}{RT}\right)^2 \frac{p^2}{Z^2} + \left(\frac{b}{RT}\right)^3 \frac{p^3}{Z^3} + \cdots.$$

We subtract 1 from each side of this equation and divide the result by p to obtain

$$A_1 + A_2 p + A_3 p^2 + \cdots$$

$$= \frac{1}{RT}\left(b - \frac{a}{RT}\right)\frac{1}{Z} + \left(\frac{b}{RT}\right)^2 \frac{p}{Z^2} + \left(\frac{b}{RT}\right)^3 \frac{p^2}{Z^3} + \cdots. \qquad (3.11)$$

In the limit of zero pressure, $Z = 1$, and this equation becomes

$$A_1 = \frac{1}{RT}\left(b - \frac{a}{RT}\right)$$

which is the required value of A_1. Using this value of A_1 in Eq. (3.11) brings it to the form,

$$A_1 + A_2 p + A_3 p^2 + \cdots = A_1\left(\frac{1}{Z}\right) + \left(\frac{b}{RT}\right)^2 \frac{p}{Z^2} + \left(\frac{b}{RT}\right)^3 \frac{p^2}{Z^3} + \cdots.$$

We repeat the procedure by subtracting A_1 from both sides of this equation, dividing by p

and taking the limiting value at zero pressure. Note that $(Z - 1)/p = A_1$ at zero pressure. Then,

$$A_2 = \left(\frac{b}{RT}\right)^2 - A_1^2 = \frac{a}{(RT)^3}\left(2b - \frac{a}{RT}\right),$$

which is the required coefficient of p^2 shown in Eq. (3.8). This procedure can be repeated to obtain A_3, A_4, and so on, but the algebra becomes more tedious with each repetition.

3.4 THE ISOTHERMS OF A REAL GAS

If the pressure–volume relations for a real gas are measured at various temperatures, a set of isotherms such as are shown in Fig. 3.5 are obtained. At high temperatures the isotherms look much like those of an ideal gas, while at low temperatures the curves have quite a different appearance. The horizontal portion of the low-temperature curves is particularly striking. Consider a container of gas in a state described by point A in Fig. 3.5. Imagine one wall of the container to be movable (a piston); keeping the temperature at T_1, we slowly push in this wall thus decreasing the volume. As the volume becomes smaller, the pressure rises slowly along the curve until the volume V_2 is reached. Reduction of the volume beyond V_2 produces no change in pressure until V_3 is reached. The small reduction in volume from V_3 to V_4 produces a large increase in pressure from p_e to p'. This is a rather remarkable sequence of events; particularly the decrease in volume over a wide range in which the pressure remains at the constant value p_e.

If we look into the container while all this is going on, we observe that at V_2 the first drops of liquid appear. As the volume goes from V_2 to V_3 more and more liquid forms; the constant pressure p_e is the equilibrium vapor pressure of the liquid at the temperature T_1. At V_3 the last trace of gas disappears. Further reduction of the volume simply compresses the liquid; the pressure rises very steeply, since the liquid is almost incompressible. The steep lines at the left of the diagram are therefore isotherms of the liquid. At a somewhat higher temperature the behavior is qualitatively the same, but the range of volume over which condensation occurs is smaller and the vapor pressure is larger. In going to still higher temperatures, the plateau finally shrinks to a point at a temperature T_c, the critical temperature. As the temperature is increased above T_c, the isotherms approach more and more closely those of the ideal gas; no plateau appears above T_c.

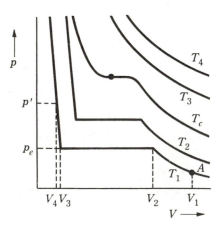

Figure 3.5   Isotherms of a real gas.

3.5 CONTINUITY OF STATES

In Fig. 3.6 the endpoints of the plateaus in Fig. 3.5 have been connected with a dashed line. Just as in any p–V diagram every point in Fig. 3.6 represents a state of the system. From the discussion in the preceding paragraph it can be seen that a point, such as A, on the extreme left of the diagram represents a liquid state of the substance. A point, such as C, on the right side of the diagram represents a gaseous state of the substance. Points under the "dome" formed by the dashed line represent states of the system in which liquid and vapor coexist in equilibrium. It is always possible to make a sharp distinction between states of the system in which one phase is present and states in which two phases* coexist in equilibrium, that is, between those points on and under the "dome" and those outside the "dome." However, it should be noted that there is no dividing line between the liquid states and the gaseous states. The fact that it is not always possible to distinguish between a liquid and a gas is the *principle of continuity of states*.

In Fig. 3.6 points A and C lie on the same isotherm, T_1. Point C clearly represents a gaseous state, and point A clearly represents the liquid obtained by compressing the gas isothermally. However, suppose that we begin at C and increase the temperature of the gas, keeping the volume constant. The pressure rises along the line CD. Having arrived at point D, the pressure is kept constant and the gas is cooled; this decreases the volume along the line DE. Having arrived at point E, the volume is again kept constant and the gas is cooled; this decreases the pressure until the point A is reached. At no time in this series of changes did the state point pass through the two-phase region. Condensation in the usual sense of the term did not occur. Point A could reasonably be said to represent a highly compressed gaseous state of the substance. The statement that point A *clearly* represented a liquid state must be modified. The distinction between liquid and gas is not always clear at all. As this demonstration shows, these two states of matter can be transformed into one another continuously. Whether we refer to states in the region of point A as liquid states or as highly compressed gaseous states depends purely upon which viewpoint happens to be convenient at the moment.

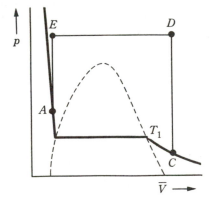

Figure 3.6 Two-phase region and continuity of states.

---

* A *phase* is a region of uniformity in a system. This means a region of uniform chemical composition and uniform physical properties. Thus a system containing liquid and vapor has two regions of uniformity. In the vapor phase, the density is uniform throughout. In the liquid phase, the density is uniform throughout but has a value different from that in the vapor phase.

If the state point of the system lies under the dome, the liquid and gas can be distinguished, since both are present in equilibrium and there is a surface of discontinuity separating them. In the absence of this surface of discontinuity there is no fundamental way of distinguishing between liquid and gas.

3.6 THE ISOTHERMS OF THE VAN DER WAALS EQUATION

Consider the van der Waals equation in the form

$$p = \frac{RT}{\bar{V} - b} - \frac{a}{\bar{V}^2}. \tag{3.12}$$

When $\bar{V}$ is very large this equation approximates the ideal gas law, since $\bar{V}$ is very large compared with b and $a/\bar{V}^2$ is very small compared with the first term. This is true at all temperatures. At high temperatures, the term $a/\bar{V}^2$ can be ignored, since it is small compared with $RT(\bar{V} - b)$. A plot of the isotherms, p versus $\bar{V}$, calculated from the van der Waals equation, is shown in Fig. 3.7. It is apparent from the figure that in the high-volume region the isotherms look much like the isotherms for the ideal gas, as does the isotherm at high temperature T_3.

At lower temperatures and smaller volumes, none of the terms in the equation may be neglected. The result is rather curious. At the temperature T_c the isotherm develops a point of inflection, point E. At still lower temperatures, the isotherms exhibit a maximum and a minimum.

Comparison of the van der Waals isotherms with those of a real gas shows similarity in certain respects. The curve at T_c in Fig. 3.7 resembles the curve at the critical temperature in Fig. 3.5. The curve at T_2 in Fig. 3.7 predicts three values of the volume, $\bar{V}'$, $\bar{V}''$, and $\bar{V}'''$, at the pressure p_e. The corresponding plateau in Fig. 3.5 predicts infinitely many volumes of the system at the pressure p_e. It is worthwhile to realize that even if a very complicated function had been written down, it would not exhibit a plateau such as that in Fig. 3.5. The oscillation of the van der Waals equation in this region is as much as can be expected of a simple continuous function.

The sections AB and DC of the van der Waals curve at T_2 can be realized experimentally. If the volume of a gas at temperature T_2 is gradually reduced, the pressure rises along the isotherm until the point D, at pressure p_e, is reached. At this point con-

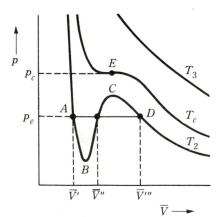

$\bar{V} \longrightarrow$ **Figure 3.7**   Isotherms of the van der Waals gas.

densation should occur; however, it may happen that liquid does not form, so that further reduction in volume produces an increase in pressure along the line DC. In this region (DC) the pressure of the gas exceeds the equilibrium vapor pressure of the liquid, p_e, at the temperature T_2; these points are therefore state points of a supersaturated (or super-cooled) vapor. Similarly, if the volume of a liquid at temperature T_2 is increased, the pressure falls until point A, at pressure p_e, is reached. At this point vapor should form; however, it may happen that vapor does not form, so that further increase in the volume produces a reduction of pressure along the line AB. Along the line AB the liquid exists under pressures that correspond to equilibrium vapor pressures of the liquid at temperatures below T_2. The liquid is at T_2 and so these points are state points of a superheated liquid. The states of the superheated liquid and those of the supercooled vapor are *metastable* states; they are unstable in the sense that slight disturbances are sufficient to cause the system to revert spontaneously into the stable state with the two phases present in equilibrium.

The section BC of the van der Waals isotherm cannot be realized experimentally. In this region the slope of the p–$\overline{V}$ curve is positive; increasing the volume of such a system would increase the pressure, and decreasing the volume would decrease the pressure! States in the region BC are *unstable*; slight disturbances of a system in such states as B to C would produce either explosion or collapse of the system.

3.7 THE CRITICAL STATE

If the van der Waals equation is taken in the form given by Eq. (3.6), the parentheses cleared, and the result multiplied by $\overline{V}^2/p$, it can be arranged in the form

$$\overline{V}^3 - \left(b + \frac{RT}{p}\right)\overline{V}^2 + \frac{a}{p}\,\overline{V} - \frac{ab}{p} = 0. \tag{3.13}$$

Because Eq. (3.13) is a cubic equation, it may have three real roots for certain values of pressure and temperature. In Fig. 3.7 these three roots for T_2 and p_e are the intersections of the horizontal line at p_e with the isotherm at T_2. All three roots lie on the boundary of or within the two-phase region. As we have seen in both Figs. 3.6 and 3.7 the two-phase region narrows and finally closes at the top. This means that there is a certain maximum pressure p_c and a certain maximum temperature T_c at which liquid and vapor can coexist. This condition of temperature and pressure is the critical point and the corresponding volume is the critical volume $\overline{V}_c$. As the two-phase region narrows, the three roots of the van der Waals equation approach one another, since they must lie on the boundary or in the region. At the critical point the three roots are all equal to $\overline{V}_c$. The cubic equation can be written in terms of its roots $\overline{V}'$, $\overline{V}''$, $\overline{V}'''$:

$$(\overline{V} - \overline{V}')(\overline{V} - \overline{V}'')(\overline{V} - \overline{V}''') = 0.$$

At the critical point $\overline{V}' = \overline{V}'' = \overline{V}''' = \overline{V}_c$, so that the equation becomes $(\overline{V} - \overline{V}_c)^3 = 0$. Expanding, we obtain

$$\overline{V}^3 - 3\overline{V}_c\overline{V}^2 + 3\overline{V}_c^2\overline{V} - \overline{V}_c^3 = 0. \tag{3.14}$$

Under these same conditions, Eq. (3.13) becomes

$$\overline{V}^3 - \left(b + \frac{RT_c}{p_c}\right)\overline{V}^2 + \frac{a}{p_c}\,\overline{V} - \frac{ab}{p_c} = 0. \tag{3.15}$$

Equations (3.14) and (3.15) are simply different ways of writing the same equations; thus the coefficients of the individual powers of $\overline{V}$ must be the same in both equations. Setting the coefficients equal, we obtain three equations:*

$$3\overline{V}_c = b + \frac{RT_c}{p_c}, \qquad 3\overline{V}_c^2 = \frac{a}{p_c}, \qquad \overline{V}_c^3 = \frac{ab}{p_c}. \qquad (3.16)$$

Equations (3.16) may be looked at in two ways. First, the set of equations can be solved for $\overline{V}_c$, p_c, and T_c in terms of a, b, and R; thus

$$\overline{V}_c = 3b, \qquad p_c = \frac{a}{27b^2}, \qquad T_c = \frac{8a}{27Rb}. \qquad (3.17)$$

If the values of a and b are known, Eqs. (3.17) can be used to calculate $\overline{V}_c$, p_c, and T_c.

Taking the second point of view, we solve the equations for a, b, and R in terms of p_c, $\overline{V}_c$, and T_c. Then

$$b = \frac{\overline{V}_c}{3}, \qquad a = 3p_c\overline{V}_c^2, \qquad R = \frac{8p_c\overline{V}_c}{3T_c}. \qquad (3.18)$$

Using Eqs. (3.18), we can calculate values of the constants a, b, and R from the critical data. However, the value of R so obtained does not agree well at all with the known value of R, and some difficulty arises.

Since experimentally it is hard to determine $\overline{V}_c$ accurately, it would be better if a and b could be obtained from p_c and T_c only. This is done by taking the third member of Eqs. (3.18) and solving it for $\overline{V}_c$. This yields

$$\overline{V}_c = \frac{3RT_c}{8p_c}.$$

This value of $\overline{V}_c$ is put in the first two of Eqs. (3.18) to yield

$$b = \frac{RT_c}{8p_c}, \qquad a = \frac{27(RT_c)^2}{64p_c}. \qquad (3.19)$$

Using Eqs. (3.19) and the ordinary value of R, we can calculate a and b from p_c and T_c only. This is the more usual procedure. However, to be honest we should compare the

---

* An equivalent method of obtaining these relations is to use the fact that the point of inflection on the p versus $\overline{V}$ curve occurs at the critical point p_c, T_c, $\overline{V}_c$. The conditions for the point of inflection are

$$(\partial p/\partial \overline{V})_T = 0, \qquad (\partial^2 p/\partial \overline{V}^2)_T = 0.$$

From the van der Waals equation,

$$\left(\frac{\partial p}{\partial \overline{V}}\right)_T = \frac{-RT}{(\overline{V} - b)^2} + \frac{2a}{\overline{V}^3},$$

$$\left(\frac{\partial^2 p}{\partial \overline{V}^2}\right)_T = \frac{2RT}{(\overline{V} - b)^3} - \frac{6a}{\overline{V}^4}.$$

Hence, at the critical point,

$$0 = -RT_c/(\overline{V}_c - b)^2 + 2a/\overline{V}_c^3, \qquad 0 = 2RT_c/(\overline{V}_c - b)^3 - 6a/\overline{V}_c^4.$$

These two equations, together with the van der Waals equation itself,

$$p_c = RT_c/(\overline{V}_c - b) - a/\overline{V}_c^2,$$

are equivalent to Eqs. (3.16).

Table 3.3
Critical constants for gases

Gas	p_c/MPa	$\overline{V}_c/10^{-6}\,\text{m}^3$	T_c/K	Gas	p_c/MPa	$\overline{V}_c/10^{-6}\,\text{m}^3$	T_c/K
He	0.229	62	5.25	CO_2	7.40	95	304
H_2	1.30	65	33.2	SO_2	7.8	123	430
N_2	3.40	90	126	H_2O	22.1	57	647
O_2	5.10	75	154	Hg	360	40	1900

Francis Weston Sears, *An Introduction to Thermodynamics, the Kinetic Theory of Gases, and Statistical Mechanics.* Reading, Mass.: Addison-Wesley, 1950.

value, $\overline{V}_c = 3RT_c/8p_c$, with the measured value of $\overline{V}_c$. The result is again very bad. The observed and calculated values of $\overline{V}_c$ disagree by more than can be accounted for by the experimental difficulties.

The whole trouble is that the van der Waals equation is not very accurate near the critical state. This fact, together with the fact that the values of these constants are nearly always calculated (one way or another) from the critical data, means that the van der Waals equation cannot be used for a precise calculation of the gas properties—although it is an improvement over the ideal gas law. The great virtue of the van der Waals equation is that the study of its predictions gives an excellent insight into the behavior of gases and their relation to liquids and the phenomenon of liquefaction. The important thing is that the equation does predict a critical state; it is too bad that it does not describe its properties to six significant figures, but that is of secondary importance. Other equations that are very precise are available. Critical data for a few gases are given in Table 3.3.

3.8 THE LAW OF CORRESPONDING STATES

Using the values of a, b, and R given by Eqs. (3.18), we can write the van der Waals equation in the equivalent form

$$p = \frac{8p_c \overline{V}_c T}{3T_c(\overline{V} - \overline{V}_c/3)} - \frac{3p_c \overline{V}_c^2}{\overline{V}^2},$$

which can be rearranged to the form

$$\frac{p}{p_c} = \frac{8(T/T_c)}{3(\overline{V}/\overline{V}_c) - 1} - \frac{3}{(\overline{V}/\overline{V}_c)^2}. \tag{3.20}$$

Equation (3.20) involves only the ratios p/p_c, T/T_c, and $\overline{V}/\overline{V}_c$. This suggests that these ratios, rather than p, T, and $\overline{V}$, are the significant variables for the characterization of the gas. These ratios are called the *reduced variables* of state, π, τ, and ϕ:

$$\pi = p/p_c, \qquad \tau = T/T_c, \qquad \phi = \overline{V}/\overline{V}_c.$$

Written in terms of these variables, the van der Waals equation becomes

$$\pi = \frac{8\tau}{3\phi - 1} - \frac{3}{\phi^2}. \tag{3.21}$$

The important thing about Eq. (3.21) is that it does not contain any constants that are peculiar to the individual gas; therefore it should be capable of describing all gases.

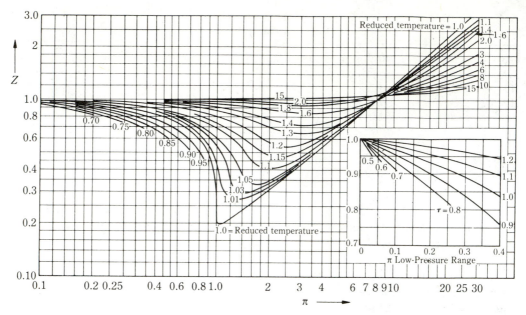

Figure 3.8 The compressibility factor as a function of the reduced pressure and the reduced temperature. (From O. A. Hougen and K. M. Watson, *Chemical Process Principles*, part II. New York: John Wiley and Sons, 1947.)

In this way, the loss in generality entailed in using the van der Waals equation, compared with the ideal gas equation, is regained. Equations, such as Eq. (3.21), which express one of the reduced variables as a function of the other two reduced variables are expressions of the *law of corresponding states*.

Two gases at the same reduced temperature and under the same reduced pressure are in corresponding states. By the law of corresponding states, they should both occupy the same reduced volume. For example, argon at 302 K and under 16 atm pressure, and ethane at 381 K and under 18 atm are in corresponding states, since each has $\tau = 2$ and $\pi = \frac{1}{3}$.

Any equation of state that involves only two constants in addition to R can be written in terms of the reduced variables only. For this reason equations that involved more than two constants were, at one time, frowned upon as contradicting the law of corresponding states. At the same time, hopes were high that an accurate two-constant equation could be devised to represent the experimental data. These hopes have been abandoned; it is now recognized that the experimental data do not support the law of corresponding states as a law of great accuracy over all ranges of pressure and temperature. Although the law is not exact, it has a good deal of importance in engineering practice; in the range of industrial pressures and temperatures, the law often holds with accuracy sufficient for engineering calculations. Plots of Z versus p/p_c at various reduced temperatures are ordinarily used rather than an equation (Fig. 3.8).

3.9 OTHER EQUATIONS OF STATE

The van der Waals equation is only one of many equations which have been proposed over the years to account for the observed pVT data for gases. Several of these equations are listed in Table 3.4, together with the expression for the law of corresponding states for

Table 3.4
Equations of state

The van der Waals equation:

$$p = \frac{RT}{\overline{V} - b} - \frac{a}{\overline{V}^2} \qquad \pi = \frac{8\tau}{3\phi - 1} - \frac{3}{\phi^2} \qquad \frac{RT_c}{p_c \overline{V}_c} = \tfrac{8}{3} = 2.67$$

The Dieterici equation:

$$p = \frac{RTe^{-a/\overline{V}RT}}{\overline{V} - b} \qquad \pi = \frac{\tau e^{2 - 2/\phi\tau}}{2\phi - 1} \qquad \frac{RT_c}{p_c \overline{V}_c} = \tfrac{1}{2}e^2 = 3.69$$

The Berthelot equation:

$$p = \frac{RT}{\overline{V} - b} - \frac{a}{T\overline{V}^2} \qquad \pi = \frac{8}{3\phi - 1} - \frac{3}{\tau\phi^2} \qquad \frac{RT_c}{p_c \overline{V}_c} = \tfrac{8}{3} = 2.67$$

The modified Berthelot equation:

$$p = \frac{RT}{\overline{V}}\left[1 + \frac{9}{128\tau}\left(1 - \frac{6}{\tau^2}\right)\pi\right] \qquad \pi = \frac{128\tau}{9(4\phi - 1)} - \frac{16}{3\tau\phi^2} \qquad \frac{RT_c}{p_c \overline{V}_c} = \tfrac{32}{9} = 3.56$$

General virial equation:

$$p\overline{V} = RT\left(1 + \frac{B}{\overline{V}} + \frac{C}{\overline{V}^2} + \frac{D}{\overline{V}^3} + \cdots\right).$$

$B, C, D, \ldots$ are called the second, third, fourth, $\ldots$ virial coefficients. They are functions of temperature.

Series expansion in terms of pressure:

$$p\overline{V} = RT(1 + B'p + C'p^2 + D'p^3 + \cdots)$$

B', C', and so on, are functions of temperature.

Beattie–Bridgeman equation:

(1) Virial form: $p\overline{V} = RT + \dfrac{\beta}{\overline{V}} + \dfrac{\gamma}{\overline{V}^2} + \dfrac{\delta}{\overline{V}^3}$

(2) Form explicit in the volume: $\overline{V} = \dfrac{RT}{p} + \dfrac{\beta}{RT} + \gamma'p + \delta'p^2 + \cdots$

$$\beta = RT\left(B_0 - \frac{A_0}{RT} - \frac{c}{T^3}\right)$$

$$\gamma = RT\left(-B_0 b + \frac{A_0 a}{RT} - \frac{B_0 c}{T^3}\right) \qquad \gamma' = \frac{1}{RT}\left[\frac{\gamma}{RT} - \left(\frac{\beta}{RT}\right)^2\right]$$

$$\delta = RT\left(\frac{B_0 bc}{T^3}\right) \qquad \delta' = \frac{1}{(RT)^2}\left[\frac{\delta}{RT} - \frac{3\beta\gamma}{(RT)^2} + 2\left(\frac{\beta}{RT}\right)^3\right]$$

the two-constant equations, and the predicted value of the critical ratio $RT_c/p_c \overline{V}_c$. Of these equations, either the Beattie–Bridgeman equation or the virial equation is best suited for precise work. The Beattie–Bridgeman equation involves five constants in addition to R: A_0, a, B_0, b, and c. The values of the Beattie–Bridgeman constants for several gases are given in Table 3.5.

Table 3.5
Constants for the Beattie–Bridgeman equation

Gas	A_0 10^{-3} Pa m^6 mol^{-2}	a 10^{-6} m^3 mol^{-1}	B_0 10^{-6} m^3 mol^{-1}	b 10^{-6} m^3 mol^{-1}	c K^3 m^3 mol^{-1}
He	2.19	59.84	14.00	0.0	0.040
H_2	20.01	-5.06	20.96	-43.59	0.504
O_2	151.09	$+25.62$	46.24	$+4.208$	48.0
CO_2	507.31	71.32	104.76	72.35	660.0
NH_3	242.48	170.31	34.15	191.13	4768.8

Calculated from the values given by Francis Weston Sears, *An Introduction to Thermodynamics, the Kinetic Theory of Gases, and Statistical Mechanics*. Reading, Mass.: Addison-Wesley, 1950.

It is interesting to examine the values of the critical ratio $RT_c/p_c \overline{V}_c$ predicted by the various equations in Table 3.4. The average value of the critical ratio for a large number of nonpolar gases, H_2 and He excepted, is 3.65. Clearly then, the van der Waals equation will be useless at temperatures and pressures near the critical values; see Section 3.6. The Dieterici equation is much better near the critical point; however, it is little used because of the transcendental function involved. Of the two-constant equations, the modified Berthelot equation is most frequently used for estimates of volumes that are better than the ideal gas estimate. The critical temperature and pressure of the gas must be known to use this equation.

Finally, it should be pointed out that all of the equations of state that are proposed for gases are based on the two fundamental ideas first suggested by van der Waals: (1) molecules have size, and (2) forces act between molecules. The more modern equations include the effects of the dependence of the intermolecular forces on the distance of separation of the molecules.

QUESTIONS

3.1 Describe the two types of intermolecular interactions responsible for deviations from ideal gas behavior, and indicate the direction of their effect on the pressure.

3.2 What common phenomena indicate that intermolecular attractions exist between water molecules in the gas phase?

3.3 Does O_2 or H_2O have the higher pressure at the same values of T and $\overline{V}$? (Use Table 3.1 to predict, without calculation.)

3.4 Describe a path between points A and C in Fig. 3.6 along which liquid and gas can be distinguished.

3.5 Give physical arguments explaining why the critical pressure and temperature should increase with increasing van der Waals a values.

PROBLEMS

3.1 A certain gas at 0 °C and 1 atm pressure has $Z = 1.00054$. Estimate the value of b for this gas.

3.2 If $Z = 1.00054$ at 0 °C and 1 atm and the Boyle temperature of the gas is 107 K, estimate the values of a and of b. (Only the first two terms in the expression of Z are needed.)

3.3 The critical constants for water are 374 °C, 22.1 MPa, and 0.0566 L/mol. Calculate values of a, b, and R; using the van der Waals equation, compare the value of R with the correct value and notice the discrepancy. Compute the constants a and b from p_c and T_c only. Using these values and the correct value of R, calculate the critical volume and compare with the correct value.

3.4 Find the relation of the constants a and b of the Berthelot equation to the critical constants.

3.5 Find the relation of the constants a and b of the Dieterici equation to the critical constants. (Note that this cannot be done by setting three roots of the equation equal to one another.)

3.6 The critical temperature of ethane is 32.3 °C, the critical pressure is 48.2 atm. Compute the critical volume using

a) the ideal gas law,
b) the van der Waals equation, realizing that for a van der Waals gas $p_c \overline{V}_c / R T_c = \frac{3}{8}$, and
c) the modified Berthelot equation.
d) Compare the results with the experimental value, 0.139 L/mol.

3.7 The vapor pressure of liquid water at 25 °C is 23.8 Torr and at 100 °C it is 760 Torr. Using the van der Waals equation in one form or another as a guide, show that saturated water vapor behaves more nearly as an ideal gas at 25 °C than it does at 100 °C.

3.8 The compressibility factor for methane is given by $Z = 1 + Bp + Cp^2 + Dp^3$. If p is in atm, the values of the constants are as follows:

T/K	B	C	D
200	-5.74×10^{-3}	6.86×10^{-6}	18.0×10^{-9}
1000	$+0.189 \times 10^{-3}$	0.275×10^{-6}	0.144×10^{-9}

Plot the values of Z as a function of p at these two temperatures in the range from 0 to 1000 atm.

3.9 Using the Beattie–Bridgeman equation, calculate the molar volume of ammonia at 300 °C and 200 atm pressure.

3.10 Compare the molar volume of carbon dioxide at 400 K and 100 atm calculated by the Beattie–Bridgeman equation with that calculated by the van der Waals equation.

3.11 Using the Beattie–Bridgeman equation, calculate the Boyle temperature for O_2 and for CO_2. Compare the values with those calculated from the van der Waals equation.

3.12 At 300 K, for what value of the molar volume will the contribution to the $p\overline{V}$ product of the term in $1/\overline{V}^2$ in the Beattie–Bridgeman equation be equal to that of the term in $1/\overline{V}$ (a) for oxygen? (b) What value of pressure corresponds to this molar volume?

3.13 At low pressures, the Berthelot equation has the form

$$\overline{V} = \frac{RT}{p} + b - \frac{a}{RT^2}$$

in which a and b are constants. Find the expression for α, the coefficient of thermal expansion, as a function of T and p only. Find the expression for the Boyle temperature in terms of a, b, and R.

3.14 Show that $T\alpha = 1 + T(\partial \ln Z/\partial T)_p$, and that $p\kappa = 1 - p(\partial \ln Z/\partial p)_T$.

3.15 If the compressibility factor of a gas is $Z(p, T)$, the equation of state may be written $p\overline{V}/RT = Z$. Show how this affects the equation for the distribution of the gas in a gravity field. From the differential equation for the distribution, show that if Z is greater than unity, the distribution is broader for the real gas than for the ideal gas and that the converse is true if Z is less than unity. If $Z = 1 + Bp$, where B is a function of temperature, integrate the equation and evaluate the constant of integration to obtain the explicit form of the distribution function.

3.16 At high pressures (small volumes), the van der Waals equation, Eq. (3.13), can be rearranged to the form

$$\overline{V} = b + \frac{p}{a}\left(b + \frac{RT}{p}\right)\overline{V}^2 - \left(\frac{p}{a}\right)\overline{V}^3.$$

If the quadratic and cubic terms are dropped, then we obtain $\overline{V}_0 = b$ as a first approximation to the smallest root of the equation. This would represent the volume of the liquid. Using this approximate value of $\overline{V}$ in the higher terms, show that the next approximation for the volume of the liquid is $\overline{V} = b + b^2 RT/a$. From this expression show that the first approximation for the coefficient of thermal expansion of a van der Waals liquid is $\alpha = bR/a$.

3.17 Using the same technique as that used to obtain Eq. (3.8), prove the relation given in Table 3.4 between γ and γ' for the Beattie–Bridgeman equation.

3.18 At what temperature does the slope of the Z versus p curve (at $p = 0$) have a maximum value for the van der Waals gas? What is the value of the maximum slope?

4

The Structure of Gases

4.1 INTRODUCTION

The aim of physics and chemistry is to interpret quantitatively the observed properties of macroscopic systems in terms of the kinds and arrangement of atoms or molecules that make up these systems. We seek an interpretation of behavior in terms of the structure of a system. Having studied the properties of a system, we construct in our mind's eye a model of the system, built of atoms and molecules, and forces of interaction between them. The laws of mechanics and statistics are applied to this model to predict the properties of such an ideal system. If many of the predicted properties are in agreement with the observed properties, the model is a good one. If none or only a few of the predicted properties are in agreement with the observed properties, the model is a poor one. This ideal model of the system may be altered or replaced by different models until the predictions are satisfactory.

Structurally, gases are nature's simplest substances; a simple model and elementary calculation yield results in excellent agreement with experiment. The kinetic theory of gases provides a beautiful and important illustration of the relation of theory to experiment in physics, as well as of the techniques that are commonly used in relating structure to properties.

4.2 KINETIC THEORY OF GASES; FUNDAMENTAL ASSUMPTIONS

The model used in the kinetic theory of gases may be described by three fundamental assumptions about the structure of gases.

1. A gas is composed of a very large number of minute particles (atoms or molecules).
2. In the absence of a force field, these particles move in straight lines. (Newton's first law of motion is obeyed.)
3. These particles interact (that is, collide) with one another only infrequently.

In addition to these assumptions we impose the condition that in any collision the total kinetic energy of the two molecules is the same before and after the collision. This kind of collision is an *elastic collision.*

If the gas consists of a very large number of moving particles, the motion of the particles must be completely random or chaotic. The particles move in all directions with a variety of speeds, some moving quickly, others slowly. If the motion were orderly (let us say that all the particles in a rectangular box were moving in precisely parallel paths), such a condition could not persist. Any slight irregularity in the wall of the box would deflect some particle out of its path; collision of this deflected particle with another particle would deflect the second one, and so on. Clearly, the motion would soon be chaotic.

4.3 CALCULATION OF THE PRESSURE OF A GAS

If a particle collides with a wall and rebounds, a force is exerted on the wall at the moment of collision. This force divided by the area of the wall would be the momentary pressure exerted on the wall by the impact and rebound of the particle. By calculating the force exerted on the wall by the impacts of many molecules, we can evaluate the pressure exerted by the gas.

Consider a rectangular box of length l and cross-sectional area A (Fig. 4.1). In the box there is one particle of mass m traveling with a velocity u_1 in a direction parallel to the length of the box. When the particle hits the right-hand end of the box it is reflected and travels in the opposite direction with a velocity $-u_1$. After a period of time it returns to the right-hand wall, the collision is repeated, and so again and again. If a pressure gauge, sufficiently sensitive to respond to the impact of this single particle, were attached to the wall, the gauge reading as a function of time would be as shown in Fig. 4.2(a). The time interval between the peaks is the time required for the particle to traverse the length of the box and back again, and thus is the distance traveled divided by the speed, $2l/u_1$. If a second particle of the same mass and traveling in a parallel path with a higher velocity is put in the box, the gauge reading will be as shown in Fig. 4.2(b).

In fact a pressure gauge that responds to the impact of individual molecules does not exist. In any laboratory situation, a pressure gauge reads a steady, *average value* of the force per unit area exerted by the impacts of an enormous number of molecules; this is indicated by the dashed line in Fig. 4.2(b).

To compute the average value of the pressure we begin with Newton's second law of motion:

$$F = ma = m\frac{du}{dt} = \frac{d(mu)}{dt}, \tag{4.1}$$

where F is the force acting on the particle of mass m, a is the acceleration, and u is the

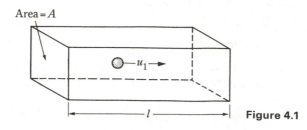

Figure 4.1

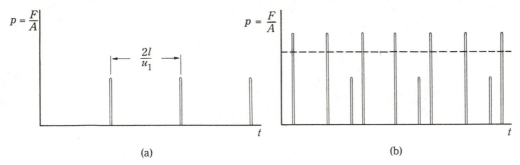

Figure 4.2 Force resulting from a collision of particles with the wall.

velocity of the particle. According to Eq. (4.1) the force acting on the particle is equal to the change of momentum per unit time. The force acting on the wall is equal and opposite in sign to this. For the particle in Fig. 4.1, the momentum before collision is mu_1, while the momentum after collision is $-mu_1$. Then the change in momentum in collision is equal to the difference of the final momentum minus the initial momentum. Thus we have $(-mu_1) - mu_1 = -2mu_1$. The change in momentum in unit time is the change in momentum in one collision multiplied by the number of collisions per second the particle makes with the wall. Since the time between collisions is equal to the time to travel distance $2l$, $t = 2l/u_1$. Then the number of collisions per second is $u_1/2l$. Therefore the change in momentum per second equals $-2mu_1(u_1/2l)$. Thus the force acting on the particle is given by $F = -mu_1^2/l$, and the force acting on the wall by $F_w = +mu_1^2/l$. But the pressure p' is F_w/A; therefore

$$p' = \frac{mu_1^2}{Al} = \frac{mu_1^2}{V}, \tag{4.2}$$

in which $Al = V$, the volume of the box.

Equation (4.2) gives the pressure p', exerted by one particle only; if more particles are added, each traveling parallel to the length of the box with speeds $u_2, u_3, \ldots$, the total force, and so the total pressure p, will be the sum of the forces exerted by each particle:

$$p = \frac{m(u_1^2 + u_2^2 + u_3^2 + \cdots)}{V}. \tag{4.3}$$

The average of the squares of the velocities, $\langle u^2 \rangle$, is defined by

$$\langle u^2 \rangle = \frac{(u_1^2 + u_2^2 + u_3^2 + \cdots)}{N}, \tag{4.4}$$

where N is the number of particles in the box. It is this average of the squares of the velocities that appears in Eq. (4.3). Using Eq. (4.4) in Eq. (4.3), we obtain

$$p = \frac{Nm\langle u^2 \rangle}{V}, \tag{4.5}$$

the final equation for the pressure of a *one-dimensional gas*.* Before using Eq. (4.5), we must examine the derivation to see what effects collisions and the varied directions of motion will have on the result.

---

*   A one-dimensional gas is a gas in which all the molecules are imagined to be moving in one direction (or its reverse) only.

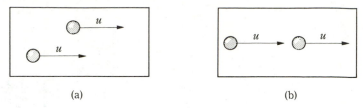

(a)   (b)

Figure 4.3

The effect of collisions is readily determined. It was assumed that all of the particles were traveling in parallel paths. This situation is illustrated for two particles, having the same velocity u, in Fig. 4.3(a). If the two particles travel on the same path, we have the situation shown in Fig. 4.3(b). In this latter case, the molecules collide with one another and each is reflected. One of the molecules never hits the right-hand wall and so cannot transfer momentum to it. However, the other molecule hits the right-hand wall twice as often as in the parallel-path case. Thus the momentum transferred to the wall in a given time does not depend on whether the particles travel on parallel paths or on the same path. We conclude that collisions in the gas do not affect the result in Eq. (4.5). The same is true if the two molecules move with different velocities. An analogy may be helpful: A bucket brigade carries water to a fire; if the brigade consists of two men, the same amount of water will arrive in unit time whether one man relays the bucket to the other at the midpoint between the well and the fire, or both men run the entire distance to the well.

The fact that the molecules are traveling in different directions rather than in the same direction as we originally assumed has an important effect on the result. As a first guess we might say that, on the average, only one-third of the molecules are moving in each of the three directions, so that the factor N in Eq. (4.5) should be replaced by $\frac{1}{3}N$. This alteration gives

$$p = \frac{\frac{1}{3}Nm\langle u^2 \rangle}{V}. \tag{4.6}$$

This simple guess gives the correct result, but the reason is more complex than the one on which the guess was based. To gain a better insight into the effect of directions, Eq. (4.6) will be derived in a different way.

The velocity vector c of the particle can be resolved into one component normal to the wall, u, and two tangential components, v and w. Consider a particle that hits the wall at an arbitrary angle and is reflected (Fig. 4.4). The only component of the velocity that is reversed on collision is the *normal* component u. The tangential component v has the same direction and magnitude before and after the collision. This is true also of the second tangential component w, which is not shown in Fig. 4.4. Since it is only the reversal of the normal component that matters, the change in momentum per collision with the wall is $-2mu$; the number of impacts per second is equal to $u/2l$. Thus Eq. (4.5) should read

$$p = \frac{Nm\langle u^2 \rangle}{V}, \tag{4.7}$$

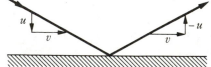

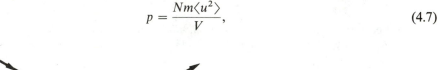

Figure 4.4   Reversal of the normal component of velocity at the wall.

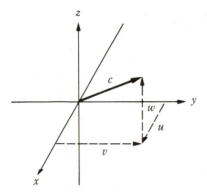

Figure 4.5   Components of the velocity vector.

where $\langle u^2 \rangle$ is the average value of the square of the normal component of the velocity. If the components are taken along the three axes x, y, z, as in Fig. 4.5, then the square of the velocity vector is related to the squares of the components by

$$c^2 = u^2 + v^2 + w^2. \tag{4.8}$$

For any individual molecule, the components of velocity are all different, and so each term on the right-hand side of Eq. (4.8) has a different value. However, if Eq. (4.8) is averaged over all the molecules, we obtain

$$\langle c^2 \rangle = \langle u^2 \rangle + \langle v^2 \rangle + \langle w^2 \rangle \tag{4.9}$$

There is no reason to expect that any one of the three directions is preferred after averaging over all the molecules. Thus we expect that $\langle u^2 \rangle = \langle v^2 \rangle = \langle w^2 \rangle$. Using this result in Eq. (4.9), we obtain

$$\langle u^2 \rangle = \tfrac{1}{3}\langle c^2 \rangle. \tag{4.10}$$

The x-direction is taken as the direction normal to the wall; thus, putting $\langle u^2 \rangle$ from Eq. (4.10) into Eq. (4.7), we obtain the exact equation for the pressure:

$$p = \frac{\tfrac{1}{3}Nm\langle c^2 \rangle}{V}, \tag{4.11}$$

the same as Eq. (4.6) obtained by the guess. Note that in Eq. (4.6) $u = c$, because v and w were zero in the derivation of Eq. (4.6).

Let the kinetic energy of any molecule be $\epsilon = \tfrac{1}{2}mc^2$. If both sides of this equation are averaged over all the molecules, then $\langle \epsilon \rangle = \tfrac{1}{2}m\langle c^2 \rangle$. Using this result in Eq. (4.11), yields $p = \tfrac{2}{3}N\langle \epsilon \rangle/V$, or

$$pV = \tfrac{2}{3}N\langle \epsilon \rangle. \tag{4.12}$$

It is encouraging to note that Eq. (4.12) bears a marked resemblance to the ideal gas law. Consequently, we examine the reason for the form in which the volume appears in Eq. (4.12). If the container in Fig. 4.1 is lengthened slightly, the volume increases by a small amount. If the velocities of the particles are the same, more time is required for a particle to travel between the walls and so it makes fewer collisions per second with the wall, reducing the pressure on the wall. Thus, an increase in volume reduces the pressure simply because there are fewer collisions with the wall in any given time interval.

We now compare Eq. (4.12) with the ideal gas law,

$$pV = nRT.$$

If Eq. (4.12) describes the ideal gas, then it must be that

$$nRT = \tfrac{2}{3}N\langle\epsilon\rangle.$$

Now n and N are related by $n = N/N_A$, where N_A is the Avogadro constant. Thus,

$$RT = \tfrac{2}{3}N_A\langle\epsilon\rangle. \tag{4.13}$$

Let U be the total kinetic energy associated with the random motion of the molecules in one mole of gas. Then $U = N_A\langle\epsilon\rangle$, and

$$U = \tfrac{3}{2}RT. \tag{4.14}$$

Equation (4.14) is one of the most fascinating results of the kinetic theory, for it provides us with an interpretation of temperature. It says that the kinetic energy of the random motion is proportional to the absolute temperature. For this reason, the random or chaotic motion is often called the *thermal motion* of the molecules. At the absolute zero of temperature, this thermal motion ceases entirely. Thus, temperature is a measure of the average kinetic energy of the chaotic motion. It is important to realize that temperature is *not* associated with the kinetic energy of one molecule, but with the *average* kinetic energy of an enormous number of molecules; that is, it is a statistical concept. It is $\langle\epsilon\rangle$ and not ϵ that appears in Eq. (4.13). A system composed of one molecule or even of a few molecules would not have a temperature, properly speaking.

The fact that the ideal gas law does not contain anything that is characteristic of a particular gas implies that at a specified temperature all gases have the same average kinetic energy. Applying Eq. (4.13) to two different gases, we have $\tfrac{3}{2}RT = N_A\langle\epsilon_1\rangle$, and $\tfrac{3}{2}RT = N_A\langle\epsilon_2\rangle$; then $\langle\epsilon_1\rangle = \langle\epsilon_2\rangle$, or $\tfrac{1}{2}m_1\langle c_1^2\rangle = \tfrac{1}{2}m_2\langle c_2^2\rangle$. The root-mean-square speed, c_{rms}, is defined by

$$c_{rms} = \sqrt{\langle c^2\rangle}. \tag{4.15}$$

The ratio of the root-mean-square speeds of two molecules of different masses is equal to the square root of the inverse ratio of the masses:

$$\frac{(c_{rms})_1}{(c_{rms})_2} = \sqrt{\frac{m_2}{m_1}} = \sqrt{\frac{M_2}{M_1}}, \tag{4.16}$$

where $M = N_A m$ is the molar mass. The heavier gas has the smaller rms speed.

The numerical value of the rms velocity of any gas is calculated by combining Eqs. (4.13) and $\langle\epsilon\rangle = \tfrac{1}{2}m\langle c^2\rangle$; thus, $RT = \tfrac{2}{3}N_A\tfrac{1}{2}m\langle c^2\rangle$, or $\langle c^2\rangle = 3RT/M$, and

$$c_{rms} = \sqrt{\frac{3RT}{M}}. \tag{4.17}$$

■ **EXAMPLE 4.1**   If we compare hydrogen, $M_1 = 2$ g/mol, and oxygen, $M_2 = 32$ g/mol, we have

$$(c_{rms})_{H_2} = (c_{rms})_{O_2}\sqrt{\frac{32}{2}} = 4(c_{rms})_{O_2}.$$

At every temperature, hydrogen has an rms speed four times as great as that of oxygen, while their average kinetic energies are the same.

■ **EXAMPLE 4.2** For oxygen at 20 °C, $T = 293$ K; $M = 0.0320$ kg/mol. Then

$$c_{rms} = \sqrt{\frac{3(8.314 \text{ J K}^{-1}\text{ mol}^{-1})(293 \text{ K})}{0.0320 \text{ kg mol}^{-1}}} = \sqrt{22.8 \times 10^4 \text{ m}^2/\text{s}^2} = 478 \text{ m/s} = 1720 \text{ km/hr.}$$

The last figure shows dramatically the magnitude of these molecular speeds.

At room temperature, the usual range of molecular speeds is 300 to 500 m/s. Hydrogen is unusual because of its low mass; its rms speed is about 1900 m/s.

4.4 DALTON'S LAW OF PARTIAL PRESSURES

In a mixture of gases the total pressure is the sum of the forces per unit area produced by the impacts of each kind of molecule on a wall of a container. Each kind of molecule contributes a term of the type in Eq. (4.11) to the pressure. For a mixture of gases we have

$$p = \frac{N_1 m_1 \langle c_1^2 \rangle}{3V} + \frac{N_2 m_2 \langle c_2^2 \rangle}{3V} + \frac{N_3 m_3 \langle c_3^2 \rangle}{3V} + \cdots \quad (4.18)$$

or

$$p = p_1 + p_2 + p_3 + \cdots, \quad (4.19)$$

where $p_1 = N_1 m_1 \langle c_1^2 \rangle / 3V$, $p_2 = N_2 m_2 \langle c_2^2 \rangle / 3V, \ldots$ Dalton's law is thus an immediate consequence of the kinetic theory of gases.

4.5 DISTRIBUTIONS AND DISTRIBUTION FUNCTIONS

The distribution of molecules in a gravity field has been discussed. It was shown that the pressure decreased regularly with increase in height, which implies that the molecules are distributed in such a way that there are fewer per cubic centimetre at the upper levels than at lower levels. The analytical expression that describes this situation is the *distribution function*. A distribution over a space coordinate is a *spatial* distribution. In the kinetic theory of gases it is important to know the velocity distribution, that is, how many molecules have velocities in a given range. The task we undertake in the following sections is to derive the velocity distribution function. Before proceeding to that problem, however, it is necessary to mention a few important ideas about distributions.

First, a distribution is the division of a group of things into classes. If we have a thousand ball bearings and five boxes, and place the ball bearings in the boxes in any particular way we please, the result is a distribution. If we divide the population of the U.S. into classes according to age, the result is an age distribution. Such a distribution shows how many people are between the ages of 0 to 20 years, between 20 to 40 years, 40 to 60 years, and so on. The population could also be divided into classes according to the amount of money in individual savings accounts, according to the amount of money owed to pawnbrokers, or according to any other characteristic. Each of these classifications constitutes a distribution of greater or less importance.

Next, the distribution is used to compute average values. From the distributions mentioned we could compute the average age of persons in the U.S., the average amount per person in savings accounts, and the average amount per person owed to pawnbrokers. For these averages to be reasonably accurate, some attention must be given to the choice of the width of the classification interval. Without going into the details that enter into the choice of the interval width, we can say that it must be small, but not too small. Consider an age distribution: Clearly it is senseless to choose 100 years as the width of the interval; essentially all of any group falls in that one interval and the group would not be divided into classes at all. So the interval width must be smaller. On the other hand, if we choose a very small interval width—for example, one day—then in any small group, say of 10 people, we will find that one person falls in each of ten intervals and zeros fall in all the others. For any large group, the time required just to write down such a detailed distribution would be enormous. Furthermore, if the information were gathered on a different day, the entire distribution would be shifted. Consequently, in constructing a distribution, the interval width chosen must be wide enough to smooth out details of no interest and narrow enough to display significant aspects of the distribution and allow meaningful averages to be calculated.

4.6 THE MAXWELL DISTRIBUTION

In a container of gas, the individual molecules are traveling in various directions with different speeds. We assume that the motions of the molecules are completely random. Then we set the following problem. What is the probability of finding a molecule with a speed between the values c and $c + dc$, regardless of the direction in which the molecule is traveling?

This problem can be broken down into simpler parts; the solution of the problem is achieved by combining the solutions of the simpler problems. Let u, v, and w denote the components of velocity in the x, y, and z directions, respectively. Let dn_u be the number of molecules that have an x component of velocity with a value in the range between u and $u + du$. Then the probability of finding such a molecule is by definition dn_u/N, where N is the number of molecules in the container. If the interval width, du, is small, it is reasonable to expect that doubling the width will double the number of molecules in the interval. Thus dn_u/N is proportional to du. Also the probability dn_u/N will depend on the velocity component u. Thus we write

$$\frac{dn_n}{N} = f(u^2)\, du, \qquad (4.20)$$

where the mathematical form of the function $f(u^2)$ is yet to be determined.*

At this point we must make clear why the function depends on u^2 and not simply on u. Because of the random nature of molecular motion, the probability of finding a molecule with an x component in the range u to $u + du$ must be the same as the probability of finding one with an x component in the range $-u$ to $-(u + du)$. In other words, the molecule has the same chance of going east with a certain speed as it has of going west with the same speed. If the direction mattered, the motion would not be random and the entire mass of gas would have a net velocity in the preferred direction. The required symmetry in the function is assured if we write $f(u^2)$ rather than $f(u)$. In the same way, if the number of molecules having a y component of velocity between v and $v + dv$ is dn_v, the probability of finding a molecule whose y component lies in the range v to $v + dv$ is given by

$$\frac{dn_v}{N} = f(v^2)\, dv, \qquad (4.21)$$

---

*   In writing Eq. (4.20) in this way, we assume implicitly that the probability dn_u/N is not in any way dependent on the values of the y or z components, v and w. This assumption is valid but will not be justified here.

where the function $f(v^2)$ *must have exactly the same form* as the function $f(u^2)$ in Eq. (4.20). These functions must have the same form, since the randomness of the distribution does not allow one direction to be different from another.* With an analogous notation we have for the z component,

$$\frac{dn_w}{N} = f(w^2)\,dw. \tag{4.22}$$

We now ask a more involved question: What is the probability of finding a molecule that has simultaneously an x component in the range u to $u + du$ and a y component in the range v to $v + dv$? Let the number of molecules that satisfy this condition be dn_{uv}; then the probability of finding such a molecule is by definition dn_{uv}/N, the product of the probabilities of finding molecules that satisfy the conditions separately. That is, $dn_{uv}/N = (dn_u/N)(dn_v/N)$, or

$$\frac{dn_{uv}}{N} = f(u^2)f(v^2)\,du\,dv. \tag{4.23}$$

Figure 4.6 illustrates the meaning of Eqs. (4.20), (4.21), and (4.23). The values of u and v for each molecule determine a representative point, marked with a dot, in the u–v coordinate system of Fig. 4.6. The representative points for two different molecules might conceivably coincide; this does not matter. The important thing is that every molecule is so represented. The total number of representative points is N, the total number of molecules in the container. Then the number of molecules having an x component of velocity between the values u and $u + du$ is the number of representative points in the vertical strip at position u and of width du. This number is dn_u, and, by Eq. (4.20), is equal to $Nf(u^2)\,du$. Similarly, the number of representative points in the horizontal strip at the position v and of width dv is the number of molecules having a y component of velocity between v and $v + dv$. The number of molecules that satisfies both conditions simultaneously is the number of representative points in the little rectangle formed by the intersection of the vertical and horizontal strips. By Eq. (4.23) this number of molecules is $dn_{uv} = Nf(u^2)f(v^2)\,du\,dv$. The *density* of representative points at the position (u, v) is the number dn_{uv} divided by the area of the little rectangle $du\,dv$:

$$\text{Point density at } (u, v) = \frac{dn_{uv}}{du\,dv} = Nf(u^2)f(v^2). \tag{4.24}$$

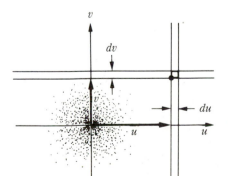

Figure 4.6   Two-dimensional velocity space.

---

*   It is assumed here that there is no force field, such as a gravity field, acting in a particular direction.

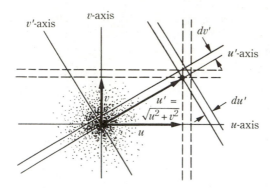

Figure 4.7 Two-dimensional velocity space with different coordinate system.

To derive the form of the function $f(u^2)$, a new set of coordinate axes u' and v' is introduced in the position shown in Fig. 4.7. The velocity ranges in the new coordinate system are du' and dv'. The number of representative points in the area $du'\ dv'$ is given by $dn_{u'v'} = Nf(u'^2)f(v'^2)\ du'\ dv'$. The density of representative points at the position (u', v') is

$$\text{Point density at } (u', v') = \frac{dn_{u'v'}}{du'\ dv'} = Nf(u'^2)f(v'^2). \tag{4.25}$$

However, the position (u', v') is the same as the position (u, v), so the density of representative points must be the same regardless of which coordinate system is used to describe it. From Eqs. (4.24) and (4.25),

$$Nf(u'^2)f(v'^2) = Nf(u^2)f(v^2). \tag{4.26}$$

The position (u, v) in the first coordinate system corresponds to the position $u' = (u^2 + v^2)^{1/2}$, $v' = 0$, in the second coordinate system. Using this relation in Eq. (4.26), we obtain

$$f(u^2 + v^2)f(0) = f(u^2)f(v^2).$$

Since $f(0)$ is a constant, set $f(0) = A$. Then

$$Af(u^2 + v^2) = f(u^2)f(v^2). \tag{4.27}$$

Appendix I shows that the only functions that satisfy Eq. (4.27) are

$$f(u^2) = Ae^{\beta u^2} \quad \text{and} \quad f(u^2) = Ae^{-\beta u^2},$$

where β is a positive constant. The physical situation forces us to choose the negative sign in the exponential; that is

$$f(u^2) = Ae^{-\beta u^2}, \qquad f(v^2) = Ae^{-\beta v^2}. \tag{4.28}$$

Equation (4.20) becomes

$$\frac{dn_u}{N} = Ae^{-\beta u^2}\ du. \tag{4.29}$$

If the positive sign in the exponential had been chosen, Eq. (4.29) would predict that—as the velocity component u becomes infinite—the probability of finding such molecules becomes infinite. This would require infinite kinetic energy for the system and consequently is an impossible case. As it stands, with the negative exponential, Eq. (4.29) predicts that the probability of finding a molecule with infinite x component of velocity is zero; this makes physical sense.

Although the original problem has not been solved, we have made considerable progress. It is well at this point to look back at what has been accomplished. We assumed that the probability of finding a molecule with an x component of velocity in the range u to $u + du$ depended only on the value of u and the width of the range du. This was expressed in Eq. (4.20) as $dn_u/N = f(u^2)\, du$. A rather lengthy argument on the basis of probabilities led us finally to the functional form of $f(u^2) = A \exp(-\beta u^2)$. The important point is the use of the notion of randomness. The argument is almost completely mathematical. Only two specifically physical assumptions are involved: randomness in the motion and a finite value of $f(u^2)$ as $u \to \infty$. The form of the distribution function is completely determined by these two assumptions. The success of the treatment will give us confidence in picturing a gas as a collection of molecules whose motions are completely random. Randomness prompts the use of probability theory. The distribution function $A \exp(-\beta u^2)$ which appears is a famous one in probability theory: it is the Gaussian distribution. This function is the governing rule in any completely random distribution; for example, it expresses the distribution of random errors in all types of experimental measurements.

We are now in a position to solve the original problem, namely to find the distribution of molecular speeds and to evaluate the constants A and β that appear in the distribution function.

The probability dn_{uvw}/N of finding a molecule with velocity components simultaneously in the ranges u to $u + du$, v to $v + dv$, w to $w + dw$ is given by the product of the individual probabilities: $dn_{uvw}/N = (dn_u/N)(dn_v/N)(dn_w/N)$, or

$$\frac{dn_{uvw}}{N} = f(u^2)f(v^2)f(w^2)\, du\, dv\, dw.$$

According to Eqs. (4.28),

$$\frac{dn_{uvw}}{N} = A^3 e^{-\beta(u^2 + v^2 + w^2)}\, du\, dv\, dw. \tag{4.30}$$

A three-dimensional velocity space* is constructed in Fig. 4.8. In this space a molecule is represented by a point determined by the values of the three components of velocity u, v, w. The total number of representative points in the parallelepiped at (u, v, w) is dn_{uvw}. The

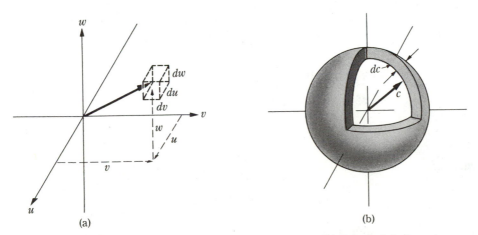

(a)   (b)

Figure 4.8   (a) Three-dimensional velocity space. (b) Spherical shell.

---

*   Figures 4.6 and 4.7 are examples of a two-dimensional velocity space.

density of points in this parallelepiped is

$$\text{Point density at } (u, v, w) = \frac{dn_{uvw}}{du\, dv\, dw} = NA^3 e^{-\beta(u^2 - v^2 + w^2)},\qquad(4.31)$$

where Eq. (4.30) has been used to obtain the last member of Eq. (4.31). Since $c^2 = u^2 + v^2 + w^2$ [see Eq. (4.8) and Fig. 4.5], we have

$$\text{Point density at } (u, v, w) = NA^3 e^{-\beta c^2}.\qquad(4.32)$$

The right-hand side of Eq. (4.32) depends only on the constants N, A, and β and on c^2; consequently, it does not depend in any way on the particular direction of the velocity vector but only on the length of the vector, that is, on the speed. The density of representative points then has the same value everywhere on the sphere of radius c in the velocity space (Fig. 4.8b).

We now pose the question: How many points lie in the shell between spheres of radii c and $c + dc$? This number of points, dn_c, will be equal to the number of molecules having speeds between c and $c + dc$, without regard for the different directions in which the molecules are traveling. The number of points dn_c in the shell is the density of points on the sphere of radius c multiplied by the volume of the shell; that is,

$$dn_c = \text{point density on sphere} \times \text{volume of shell}.\qquad(4.33)$$

The volume of the shell, dV_{shell}, is the difference in volume between the outer and the inner sphere:

$$dV_{\text{shell}} = \frac{4\pi}{3}(c + dc)^3 - \frac{4\pi}{3}c^3 = \frac{4\pi}{3}[3c^2\, dc + 3c(dc)^2 + (dc)^3].$$

The terms on the right which involve $(dc)^2$ and $(dc)^3$ are infinitesimals of higher order that vanish more rapidly than dc in the limit as $dc \to 0$; these higher terms are dropped out and we obtain $dV_{\text{shell}} = 4\pi c^2\, dc$. Using this result and Eq. (4.32) in Eq. (4.33), we have

$$dn_c = 4\pi NA^3 e^{-\beta c^2} c^2\, dc,\qquad(4.34)$$

which relates dn_c, the number of molecules with speeds between c and $c + dc$, to N, c, and dc, and the constants A and β. Equation (4.34) is one form of the Maxwell distribution and is the solution of the problem posed at the beginning of this section. Before we can use Eq. (4.34), the constants A and β must be evaluated.

★4.7 MATHEMATICAL INTERLUDE

In the kinetic theory of gases we deal with integrals of the general type

$$I_n(\beta) = \int_0^\infty x^{2n+1} e^{-\beta x^2}\, dx \qquad (\beta > 0; n > -1).\qquad(4.35)$$

If we make the substitution, $y = \beta x^2$, the integral reduces to the form

$$I_n(\beta) = \tfrac{1}{2}\beta^{-(n+1)} \int_0^\infty y^n e^{-y}\, dy.$$

However, the factorial function, $n!$, is defined by

$$n! = \int_0^\infty y^n e^{-y}\, dy \qquad (n > -1),\qquad(4.36)$$

so that

$$I_n(\beta) = \int_0^\infty x^{2n+1} e^{-\beta x^2} \, dx = \tfrac{1}{2}(n!)\beta^{-(n+1)}. \tag{4.37}$$

The higher-order integrals can be obtained from those of lower order by differentiation; differentiating Eq. (4.37) with respect to β yields

$$\frac{dI_n(\beta)}{d\beta} = -\int_0^\infty x^{2n+3} e^{-\beta x^2} \, dx = -\tfrac{1}{2}[(n+1)!]\beta^{-(n+2)} = -I_{n+1}(\beta),$$

or

$$I_{n+1}(\beta) = -\frac{dI_n(\beta)}{d\beta}. \tag{4.38}$$

Two cases commonly arise.

Case I. $n = 0$ or a positive integer.

In this case we apply Eq. (4.37) directly and no difficulty ensues. The lowest member is

$$I_0(\beta) = \tfrac{1}{2}\beta^{-1}.$$

All other members can be obtained from Eq. (4.37) or by differentiating $I_0(\beta)$ and using Eq. (4.38).

Case II. $n = -\tfrac{1}{2}, \tfrac{1}{2}, \tfrac{3}{2}$ or $n = m - \tfrac{1}{2}$ where $m = 0$ or a positive integer.

In this case we may also use Eq. (4.37) directly, but unless we know the value of the factorial function for half-integral values of the argument we will be in trouble. If $n = m - \tfrac{1}{2}$, the function takes the form

$$I_{m-1/2}(\beta) = \int_0^\infty x^{2m} e^{-\beta x^2} \, dx = \tfrac{1}{2}[(m - \tfrac{1}{2})!]\beta^{-(m+1/2)}. \tag{4.39}$$

When $m = 0$, we have

$$I_{-1/2}(\beta) = \int_0^\infty e^{-\beta x^2} \, dx = \beta^{-1/2} \int_0^\infty e^{-y^2} \, dy = \beta^{-1/2} I_{-1/2}(1), \tag{4.40}$$

where in the second writing, $x = \beta^{-1/2} y$, has been used. Comparing this result with the last member of Eq. (4.39) we find that

$$I_{-1/2}(1) = \int_0^\infty e^{-y^2} \, dy = \tfrac{1}{2}(-\tfrac{1}{2})!. \tag{4.41}$$

The integral, $I_{-1/2}(1)$, cannot be evaluated by elementary methods. We proceed by writing the integral in two ways,

$$I_{-1/2}(1) = \int_0^\infty e^{-x^2} \, dx \quad \text{and} \quad I_{-1/2}(1) = \int_0^\infty e^{-y^2} \, dy,$$

then multiply them together to obtain,

$$I_{-1/2}^2(1) = \int_0^\infty \int_0^\infty e^{-(x^2+y^2)} \, dx \, dy.$$

The integration is over the area of the first quadrant; we change variables to $r^2 = x^2 + y^2$

and replace $dx \, dy$ by the element of area in polar coordinates, $r \, d\phi \, dr$. To cover the first quadrant we integrate ϕ from zero to $\pi/2$ and r from 0 to ∞: the integral becomes

$$I^2_{-1/2}(1) = \int_0^{\pi/2} d\phi \int_0^\infty e^{-r^2} r \, dr = \frac{\pi}{2} \left(\frac{1}{2}\right) \int_0^\infty e^{-r^2} \, d(r^2) = \frac{\pi}{4} \int_0^\infty e^{-y} \, dy.$$

The last integral is equal to $0! = 1$; taking the square root of both sides, we have

$$I_{-1/2}(1) = \tfrac{1}{2}\sqrt{\pi}. \tag{4.42}$$

Comparing Eqs. (4.41) and (4.42), it follows that $(-\tfrac{1}{2})! = \sqrt{\pi}$; now from Eqs. (4.40) and (4.42),

$$I_{-1/2}(\beta) = \tfrac{1}{2}\sqrt{\pi}\,\beta^{-1/2}.$$

By differentiation, and by using Eq. (4.38) we obtain

$$I_{1/2}(\beta) = -\frac{dI_{-1/2}}{d\beta} = \tfrac{1}{2}\sqrt{\pi}(\tfrac{1}{2}\beta^{-3/2})$$

and

$$I_{3/2}(\beta) = -\frac{dI_{1/2}}{d\beta} = \tfrac{1}{2}\sqrt{\pi}(\tfrac{1}{2} \cdot \tfrac{3}{2}\beta^{-5/2}).$$

Repetition of this procedure ultimately yields

$$I_{m-1/2}(\beta) = \int_0^\infty x^{2m} e^{-\beta x^2} \, dx = \tfrac{1}{2}\sqrt{\pi}\,\frac{(2m)!}{2^{2m}m!}\,\beta^{-(m+1/2)}. \tag{4.43}$$

By comparing this result with Eq. (4.39) we obtain the interesting result for half-integral factorials

$$(m - \tfrac{1}{2})! = \sqrt{\pi}\,\frac{(2m)!}{2^{2m}m!}. \tag{4.44}$$

Table 4.1 collects the most commonly used formulas.

Table 4.1
Integrals that occur in the kinetic theory of gases

(1) $\displaystyle\int_{-\infty}^\infty x^{2n} e^{-\beta x^2} \, dx = 2 \int_0^\infty x^{2n} e^{-\beta x^2} \, dx$		(6) $\displaystyle\int_{-\infty}^\infty x^{2n+1} e^{-\beta x^2} \, dx = 0$	
(2) $\displaystyle\int_0^\infty e^{-\beta x^2} \, dx = \tfrac{1}{2}\sqrt{\pi}\,\beta^{-1/2}$		(7) $\displaystyle\int_0^\infty x e^{-\beta x^2} \, dx = \tfrac{1}{2}\beta^{-1}$	
(3) $\displaystyle\int_0^\infty x^2 e^{-\beta x^2} \, dx = \tfrac{1}{2}\sqrt{\pi}\tfrac{1}{2}\beta^{-3/2}$		(8) $\displaystyle\int_0^\infty x^3 e^{-\beta x^2} \, dx = \tfrac{1}{2}\beta^{-2}$	
(4) $\displaystyle\int_0^\infty x^4 e^{-\beta x^2} \, dx = \tfrac{1}{2}\sqrt{\pi}\tfrac{3}{4}\beta^{-5/2}$		(9) $\displaystyle\int_0^\infty x^5 e^{-\beta x^2} \, dx = \beta^{-3}$	
(5) $\displaystyle\int_0^\infty x^{2n} e^{-\beta x^2} \, dx = \tfrac{1}{2}\sqrt{\pi}\,\dfrac{(2n)!\beta^{-(n+1/2)}}{2^{2n}n!}$		(10) $\displaystyle\int_0^\infty x^{2n+1} e^{-\beta x^2} \, dx = \tfrac{1}{2}(n!)\beta^{-(n+1)}$	

★4.7.1 The Error Function

We frequently have occasion to use integrals of the type of Case II above in which the upper limit is not extended to infinity but only to some finite value. These integrals are related to the error function (erf). We define

$$\text{erf}(x) = \frac{2}{\sqrt{\pi}} \int_0^x e^{-u^2} \, du. \qquad (4.45)$$

If the upper limit is extended to $x \to \infty$, the integral is $\frac{1}{2}\sqrt{\pi}$ so that

$$\text{erf}(\infty) = 1.$$

Thus as x varies from zero to infinity, erf (x) varies from zero to unity. If we add the integral from x to ∞ multiplied by $2/\sqrt{\pi}$ to both sides of the equation, we obtain

$$\text{erf}(x) + \frac{2}{\sqrt{\pi}} \int_x^\infty e^{-u^2} \, du = \frac{2}{\sqrt{\pi}} \left[\int_0^x e^{-u^2} \, du + \int_x^\infty e^{-u^2} \, du \right] = \frac{2}{\sqrt{\pi}} \int_0^\infty e^{-u^2} \, du = 1.$$

Therefore

$$\frac{2}{\sqrt{\pi}} \int_x^\infty e^{-u^2} \, du = 1 - \text{erf}(x).$$

We define the co-error function, erfc (x), by

$$\text{erfc}(x) = 1 - \text{erf}(x). \qquad (4.46)$$

Thus

$$\int_x^\infty e^{-u^2} \, du = \frac{\sqrt{\pi}}{2} \text{erfc}(x) \qquad (4.47)$$

Some values of the error function are given in Table 4.2.

Table 4.2
The error function:

$$\text{erf}(x) = \frac{2}{\sqrt{\pi}} \int_0^x e^{-u^2} \, du$$

x	erf(x)	x	erf(x)	x	erf(x)
0.00	0.000	0.80	0.742	1.60	0.976
0.10	0.112	0.90	0.797	1.70	0.984
0.20	0.223	1.00	0.843	1.80	0.989
0.30	0.329	1.10	0.880	1.90	0.993
0.40	0.428	1.20	0.910	2.00	0.995
0.50	0.521	1.30	0.934	2.20	0.998
0.60	0.604	1.40	0.952	2.40	0.9993
0.70	0.678	1.50	0.966	2.50	0.9996

4.8 EVALUATION OF A AND β

The constants A and β are determined by requiring that the distribution yield correct values of the total number of molecules and the average kinetic energy. The total number of molecules is obtained by summing dn_c over all possible values of c between zero and infinity:

$$N = \int_{c=0}^{c=\infty} dn_c. \tag{4.48}$$

The average kinetic energy is calculated by multiplying the kinetic energy, $\frac{1}{2}mc^2$, by the number of molecules, dn_c, which have that kinetic energy, summing over all values of c, and dividing by N, the total number of molecules.

$$\langle \epsilon \rangle = \frac{\int_{c=0}^{c=\infty} \frac{1}{2}mc^2 \, dn_c}{N}. \tag{4.49}$$

Equations (4.48) and (4.49) determine A and β.

Replacing dn_c in Eq. (4.48) by the value given by Eq. (4.34), we have

$$N = \int_0^\infty 4\pi N A^3 e^{-\beta c^2} c^2 \, dc.$$

Dividing through by N and removing constants from under the integral sign yields

$$1 = 4\pi A^3 \int_0^\infty c^2 e^{-\beta c^2} \, dc.$$

From Table 4.1 we have $\int_0^\infty c^2 e^{-\beta c^2} \, dc = \pi^{1/2}/4\beta^{3/2}$. Hence, $1 = 4\pi A^3 \pi^{1/2}/4\beta^{3/2}$. So finally

$$A^3 = \left(\frac{\beta}{\pi}\right)^{3/2}, \tag{4.50}$$

which gives the value of A^3 in terms of β.

In the second condition, Eq. (4.49), we use the value for dn_c from Eq. (4.34):

$$\langle \epsilon \rangle = \frac{\int_0^\infty \frac{1}{2}mc^2 4\pi N A^3 e^{-\beta c^2} c^2 \, dc}{N}.$$

Using Eq. (4.50), we have

$$\langle \epsilon \rangle = 2\pi m \left(\frac{\beta}{\pi}\right)^{3/2} \int_0^\infty c^4 e^{-\beta c^2} \, dc.$$

From Table 4.1, we have $\int_0^\infty c^4 e^{-\beta c^2} \, dc = 3\pi^{1/2}/8\beta^{5/2}$. So $\langle \epsilon \rangle$ becomes $\langle \epsilon \rangle = 3m/4\beta$, and, therefore,

$$\beta = \frac{3m}{4\langle \epsilon \rangle}, \tag{4.51}$$

which expresses β in terms of the average energy per molecule $\langle \epsilon \rangle$. However, Eq. (4.13) relates the average energy per molecule to the temperature:

$$\langle \epsilon \rangle = \frac{3}{2}\left(\frac{R}{N_A}\right)T = \tfrac{3}{2}kT. \tag{4.13a}$$

The gas constant per molecule is the Boltzmann constant, $k = R/N_A = 1.3807 \times 10^{-23}$ J/K. Using this relation in Eq. (4.51) gives β explicitly in terms of m and T.

$$\beta = \frac{m}{2kT}. \tag{4.52}$$

Using Eq. (4.52) in Eq. (4.50), we obtain

$$A^3 = \left(\frac{m}{2\pi kT}\right)^{3/2}, \qquad A = \left(\frac{m}{2\pi kT}\right)^{1/2}. \tag{4.53}$$

Using Eqs. (4.52) and (4.53) for β and A^3 in Eq. (4.34), we obtain the Maxwell distribution in explicit form:

$$dn_c = 4\pi N \left(\frac{m}{2\pi kT}\right)^{3/2} c^2 e^{-mc^2/2kT} \, dc. \tag{4.54}$$

The Maxwell distribution expresses the number of molecules having speeds between c and $c + dc$ in terms of the total number present, the mass of the molecules, the temperature, and the speed. (To simplify computations with the Maxwell distribution, note that the ratio $m/k = M/R$, where M is the molar mass.) The Maxwell distribution is customarily plotted with the function $(1/N)(dn_c/dc)$ as the ordinate and c as the abscissa. The fraction of the molecules in the speed range c to $c + dc$ is dn_c/N; dividing this by dc gives the fraction of the molecules in this speed range per unit width of the interval.

$$\frac{1}{N}\frac{dn_c}{dc} = 4\pi\left(\frac{m}{2\pi kT}\right)^{3/2} c^2 e^{-mc^2/2kT}. \tag{4.55}$$

The plot of the function for nitrogen at two temperatures is shown in Fig. 4.9.

The function shown in Fig. 4.9 is the probability of finding a molecule having a speed between c and $c + dc$, divided by the width dc of the range. Roughly speaking, the ordinate is the probability of finding a molecule with a speed between c and $(c + 1)$ m/s. The curve is parabolic near the origin, since the factor c^2 is dominant in this region, and the exponential function is about equal to unity; at high values of c, the exponential factor dominates the behavior of the function, causing it to decrease rapidly in value. As a consequence of the contrasting behavior of the two factors, the product function has a maximum at a speed c_{mp}. This speed is called the *most probable speed*, since it corresponds to the maximum in the probability curve; c_{mp} can be calculated by differentiating the function on the right of Eq. (4.55) and setting the derivative equal to zero to find the location of the horizontal

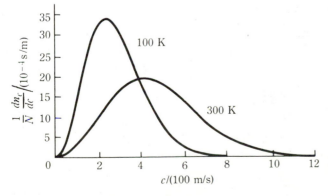

Figure 4.9 Maxwell distribution for nitrogen at two temperatures.

tangents. This procedure yields

$$ce^{-mc^2/2kT}\left(2 - \frac{mc^2}{kT}\right) = 0.$$

The curve has three horizontal tangents: at $c = 0$; at $c \to \infty$, when $\exp(-\frac{1}{2}mc^2/kT) = 0$; and when $2 - mc^2/kT = 0$. This last condition determines c_{mp}.

$$c_{mp} = \sqrt{\frac{2kT}{m}} = \sqrt{\frac{2RT}{M}}. \tag{4.56}$$

Figure 4.9 shows that the chance of finding molecules with very low or very high speeds is nearly zero. The majority of molecules have speeds that cluster around c_{mp} in the middle of the range of c.

Figure 4.9 also shows that an increase in temperature broadens the speed distribution and shifts the maximum to higher values of c. The area under the two curves in Fig. 4.9 must be the same, since it is equal to unity in both cases. This requires the curve to broaden as the temperature is increased. The speed distribution also depends on the mass of the molecule. At the same temperature a heavy gas has a narrower distribution of speeds than a light gas.

The appearance of temperature as the parameter of the distribution yields another interpretation of the, as yet mysterious, notion of temperature. Roughly, the temperature is a measure of the broadness of the speed distribution. If by any means we manage to narrow the distribution, we will discover that the temperature of the system has dropped. At the absolute zero of temperature, the distribution becomes infinitely narrow; all of the molecules have the same kinetic energy, zero.

4.9 CALCULATION OF AVERAGE VALUES USING THE MAXWELL DISTRIBUTION

From the Maxwell distribution, the average value of any quantity that depends on speed can be calculated. If we wish to calculate the average value $\langle g \rangle$ of some function of speed $g(c)$, we multiply the function $g(c)$ by dn_c, the number of molecules that have the speed c; then we sum over all values of c from zero to infinity and divide by the total number of molecules in the gas.

$$\langle g \rangle = \frac{\int_{c=0}^{c=\infty} g(c)\, dn_c}{N}. \tag{4.57}$$

4.9.1 Examples of Average Value Calculations

■ EXAMPLE 4.3   As an example of the use of Eq. (4.57), we can calculate the average kinetic energy of the gas molecules; for this case, $g(c) = \epsilon = \frac{1}{2}mc^2$. Thus Eq. (4.57) becomes

$$\langle \epsilon \rangle = \frac{\int_{c=0}^{c=\infty} \frac{1}{2}mc^2\, dn_c}{N},$$

which is identical with Eq. (4.49). If we put in the value of dn_c and integrate, we would, of course, find that $\langle \epsilon \rangle = \frac{3}{2}kT$, since we used this relation to determine the constant β in the distribution function.

■ **EXAMPLE 4.4**   Another average value of importance is the average speed $\langle c \rangle$. Using Eq. (4.57), we have

$$\langle c \rangle = \frac{\int_{c=0}^{c=\infty} c \, dn_c}{N}.$$

Using the value of dn_c from Eq. (4.54), we obtain

$$\langle c \rangle = 4\pi \left(\frac{m}{2\pi kT} \right)^{3/2} \int_0^\infty c^3 e^{-mc^2/2kT} \, dc.$$

The integral can be obtained from Table 4.1 or can be evaluated by elementary methods through the change in variable: $x = \frac{1}{2}mc^2/kT$. This substitution yields

$$\langle c \rangle = \sqrt{\frac{8kT}{\pi m}} \int_0^\infty x e^{-x} \, dx.$$

But $\int_0^\infty x e^{-x} \, dx = 1$; therefore

$$\langle c \rangle = \sqrt{\frac{8kT}{\pi m}} = \sqrt{\frac{8RT}{\pi M}}. \tag{4.58}$$

It should be noted that the average speed is not equal to the rms speed, $c_{\text{rms}} = (3kT/m)^{1/2}$, but is somewhat smaller. The most probable speed, $c_{\text{mp}} = (2kT/m)^{1/2}$, is smaller yet. The average speed and the rms speed occur most frequently in physico-chemical calculations.

Since the speeds of the molecules are distributed, we can talk about the deviation of the speed of a molecule from the mean value, $\delta = c - \langle c \rangle$. The average deviation from the mean value is zero, of course. However, the square of the deviations from the mean, $\delta^2 = (c - \langle c \rangle)^2$, has an average value different from zero. This quantity gives us a measure of the breadth of the distribution. Calculation of this kind of average value (Problems 4.7 and 4.8) gives us an important insight into the meaning of temperature, particularly in the case of the energy distribution.

★4.10 THE MAXWELL DISTRIBUTION AS AN ENERGY DISTRIBUTION

The speed distribution, Eq. (4.54), can be converted to an energy distribution. The kinetic energy of a molecule is $\epsilon = \frac{1}{2}mc^2$. Then $c = (2/m)^{1/2}\epsilon^{1/2}$. Differentiating, we obtain $dc = (1/2m)^{1/2}\epsilon^{-1/2} \, d\epsilon$. The energy range $d\epsilon$ corresponds to the speed range dc, and so the number of particles dn_c in the speed range corresponds to the number of particles dn_ϵ in the energy range. By replacing c and dc in the velocity distribution by their equivalents according to these equations, we obtain the energy distribution

$$dn_\epsilon = 2\pi N \left(\frac{1}{\pi kT} \right)^{3/2} \epsilon^{1/2} e^{-\epsilon/kT} \, d\epsilon, \tag{4.59}$$

where dn_ϵ is the number of molecules having kinetic energies between ϵ and $\epsilon + d\epsilon$. This form of the distribution function is plotted as a function of ϵ in Fig. 4.10(a). Attention should be given to the distinctly different shape of this curve compared with that of the speed distribution. In particular, the energy distribution has a vertical tangent at the origin

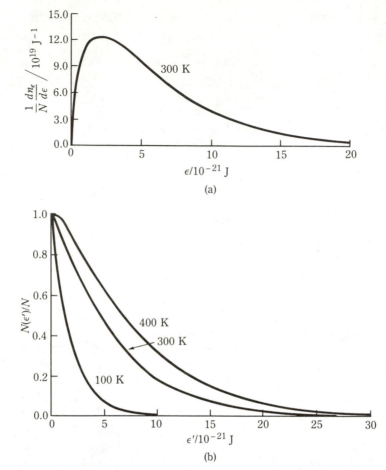

Figure 4.10   (a) Energy distribution at 300 K. (b) Fraction of molecules having energies in excess of ϵ'.

and thus it rises much more quickly than the velocity distribution, which starts with a horizontal tangent. After passing the maximum, the energy distribution falls off more gently than does the velocity distribution. As usual, the distribution is broadened at higher temperatures, a greater proportion of the molecules having higher energies. As before, the areas under the curves for different temperatures must be the same.

It is frequently important to know what fraction of the molecules in a gas have kinetic energies exceeding a specified value ϵ'. This quantity can be calculated from the distribution function. Let $N(\epsilon')$ be the number of molecules with energies greater than ϵ'. Then $N(\epsilon')$ is the sum of the number of molecules in the energy range above ϵ'.

$$N(\epsilon') = \int_{\epsilon'}^{\infty} dn_{\epsilon}. \tag{4.60}$$

The fraction of the molecules with energies above ϵ' is $N(\epsilon')/N$; using the expression in Eq. (4.59) for the integrand in Eq. (4.60), this fraction becomes

$$\frac{N(\epsilon')}{N} = 2\pi \left(\frac{1}{\pi kT}\right)^{3/2} \int_{\epsilon'}^{\infty} \epsilon^{1/2} e^{-\epsilon/kT} \, d\epsilon. \tag{4.61}$$

The substitutions

$$\epsilon = kTx^2, \qquad d\epsilon = kT\, d(x^2), \qquad \epsilon^{1/2} = (kT)^{1/2}x,$$

reduce Eq. (4.61) to

$$\frac{N(\epsilon')}{N} = \frac{2}{\sqrt{\pi}} \int_{\sqrt{\epsilon'/kT}}^{\infty} xe^{-x^2}\, d(x^2) = -\frac{2}{\sqrt{\pi}} \int_{\sqrt{\epsilon'/kT}}^{\infty} x\, d(e^{-x^2}).$$

Integrating by parts, we have

$$\frac{N(\epsilon')}{N} = -\frac{2}{\sqrt{\pi}} \left[xe^{-x^2} \Big|_{\sqrt{\epsilon'/kT}}^{\infty} - \int_{\sqrt{\epsilon'/kT}}^{\infty} e^{-x^2}\, dx \right],$$

$$\frac{N(\epsilon')}{N} = 2\left(\frac{\epsilon'}{\pi kT}\right)^{1/2} e^{-\epsilon'/kT} + \frac{2}{\sqrt{\pi}} \int_{\sqrt{\epsilon'/kT}}^{\infty} e^{-x^2}\, dx. \qquad (4.62)$$

The integral in Eq. (4.62) can be expressed in terms of the co-error function defined in Eq. (4.47).

$$\frac{N(\epsilon')}{N} = 2\left(\frac{\epsilon'}{\pi kT}\right)^{1/2} e^{-\epsilon'/kT} + \text{erfc}\,(\sqrt{\epsilon'/kT}). \qquad (4.63)$$

However, if the energy ϵ' is very much larger than kT, the value of the integral in Eq. (4.62) is approximately zero (since the area under the curve of the integrand is very small from a large value of the lower limit to infinity). In this important case, Eq. (4.62) becomes

$$\frac{N(\epsilon')}{N} = 2\left(\frac{\epsilon'}{\pi kT}\right)^{1/2} e^{-\epsilon'/kT}, \qquad \epsilon' \gg kT. \qquad (4.64)$$

Equation (4.64) has the property that the right-hand side varies quite rapidly with temperature, particularly at low temperatures. Figure 4.10(b) shows the variation of $N(\epsilon')/N$ with ϵ' at three temperatures, calculated from Eq. (4.62). Also, Fig. 4.10(b) shows graphically that the fraction of molecules having energies greater than ϵ' increases markedly with temperature, particularly if ϵ' is in the high-energy range. This property of gases, and of liquids and solids as well, has fundamental significance in connection with the increase in the rates of chemical reactions with temperature. Since only molecules that have more than a certain minimum energy can react chemically, and since the fraction of molecules that have energies exceeding this minimum value increases with temperature according to Eq. (4.62), the rate of a chemical reaction increases with temperature.*

4.11 AVERAGE VALUES OF INDIVIDUAL COMPONENTS; EQUIPARTITION OF ENERGY

It is instructive to compute the average values of the individual components of velocity. For this purpose, the most convenient form of the Maxwell distribution is that in Eq. (4.30). The average value of u is then given by an equation analogous to Eq. (4.57):

$$\langle u \rangle = \frac{\displaystyle\int_{-\infty}^{\infty} \int_{-\infty}^{\infty} \int_{-\infty}^{\infty} u\, dn_{uvw}}{N}.$$

---

* Other conditions being comparable, the rate of chemical reaction depends on temperature through the factor $Ae^{-\epsilon_a/kT}$, where A is a constant and ϵ_a is a characteristic energy. Note the similarity in form to the right-hand side of Eq. (4.64).

The integration is taken over all possible values of all three components; note that any component may have any value from minus infinity to plus infinity. Using dn_{uvw} from Eq. (4.30), we obtain

$$\langle u \rangle = A^3 \int_{-\infty}^{\infty} \int_{-\infty}^{\infty} \int_{-\infty}^{\infty} u e^{-\beta(u^2 + v^2 + w^2)} \, du \, dv \, dw$$

$$= A^3 \int_{-\infty}^{\infty} u e^{-\beta u^2} \, du \int_{-\infty}^{\infty} e^{-\beta v^2} \, dv \int_{-\infty}^{\infty} e^{-\beta w^2} \, dw. \tag{4.65}$$

By Formula (6) in Table 4.1, the first integral on the right-hand side of Eq. (4.65) is zero; thus $\langle u \rangle = 0$. The same result is obtained for the average value of the other components:

$$\langle u \rangle = \langle v \rangle = \langle w \rangle = 0. \tag{4.66}$$

The reason the average value of any individual component must be zero is physically obvious. If the average value of any one of the components had a value other than zero, this would correspond to a net motion of the entire mass of gas in that particular direction; the present discussion applies only to gases at rest.

The distribution function for the x component can be written [see Eqs. (4.20), (4.29), (4.52), (4.53)] as

$$\frac{1}{N} \frac{dn_u}{du} = f(u^2) = \left(\frac{m}{2\pi kT} \right)^{1/2} e^{-mu^2/2kT}, \tag{4.67}$$

which is plotted in Fig. 4.11. It is the symmetry of the function with respect to the origin of u that leads to the vanishing value of $\langle u \rangle$. The interpretation of temperature as a measure of the width of the distribution is clearly illustrated by the two curves in Fig. 4.11. The area under each curve must have the same value, unity. The probability of finding a molecule of velocity u is the same as that of finding one with a velocity $-u$; this was assured in our original choice of the function as one that depended only on u^2.

Although the average value of the velocity component in any one direction is zero, because equal numbers of molecules have components u and $-u$, the average value of kinetic energy associated with a particular component has a positive value. The molecules with velocity component u contribute $\frac{1}{2}mu^2$ to the average and those with component $-u$ contribute $\frac{1}{2}m(-u)^2 = \frac{1}{2}mu^2$. The contributions of particles moving in opposite directions add up in averaging the energy, while in averaging the velocity component the contributions of particles moving in opposite directions exactly cancel one another. To calculate

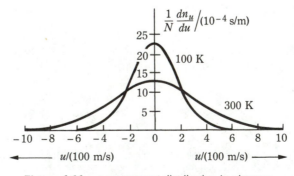

Figure 4.11 x component distribution in nitrogen.

the average value of $\epsilon_x = \frac{1}{2}mu^2$, we use the Maxwell distribution in the same way as before.

$$\langle \epsilon_x \rangle = \frac{\int_{-\infty}^{\infty} \int_{-\infty}^{\infty} \int_{-\infty}^{\infty} \frac{1}{2}mu^2 \, dn_{uvw}}{N}.$$

Using Eq. (4.30), we obtain

$$\langle \epsilon_x \rangle = \frac{1}{2}mA^3 \int_{-\infty}^{\infty} \int_{-\infty}^{\infty} \int_{-\infty}^{\infty} u^2 e^{-\beta(u^2+v^2+w^2)} \, du \, dv \, dw$$

$$= \frac{1}{2}mA^3 \int_{-\infty}^{\infty} u^2 e^{-\beta u^2} \, du \int_{-\infty}^{\infty} e^{-\beta v^2} \, dv \int_{-\infty}^{\infty} e^{-\beta w^2} \, dw.$$

Using Formulas (1) and (2), Table 4.1, we have

$$\int_{-\infty}^{\infty} e^{-\beta v^2} \, dv = \int_{-\infty}^{\infty} e^{-\beta w^2} \, dw = \left(\frac{\pi}{\beta}\right)^{1/2},$$

and, by Formulas (1) and (3), Table 4.1,

$$\int_{-\infty}^{\infty} u^2 e^{-\beta u^2} \, du = \frac{\pi^{1/2}}{2\beta^{3/2}} = \frac{1}{2\beta}\left(\frac{\pi}{\beta}\right)^{1/2}.$$

Introducing these values for the integrals leads to

$$\langle \epsilon_x \rangle = \frac{1}{2}mA^3\left(\frac{1}{2\beta}\right)\left(\frac{\pi}{\beta}\right)^{1/2}\left(\frac{\pi}{\beta}\right)^{1/2}\left(\frac{\pi}{\beta}\right)^{1/2} = \frac{mA^3}{4\beta}\left(\frac{\pi}{\beta}\right)^{3/2}.$$

Using the value of A^3 from Eq. (4.53) and the value of β from Eq. (4.52), we obtain finally

$$\langle \epsilon_x \rangle = \frac{1}{2}kT.$$

The same result can be obtained for $\langle \epsilon_y \rangle$ and $\langle \epsilon_z \rangle$; therefore

$$\langle \epsilon_x \rangle = \langle \epsilon_y \rangle = \langle \epsilon_z \rangle = \frac{1}{2}kT. \tag{4.68}$$

Since the average total kinetic energy is the sum of the three terms, its value is $\frac{3}{2}kT$, the value given by Eq. (4.13a):

$$\langle \epsilon \rangle = \langle \epsilon_x \rangle + \langle \epsilon_y \rangle + \langle \epsilon_z \rangle = \frac{1}{2}kT + \frac{1}{2}kT + \frac{1}{2}kT = \frac{3}{2}kT. \tag{4.69}$$

Equation (4.68) expresses the important law of *equipartition of energy*. It states that the average total energy is equally divided among the three independent components of the motion, which are called *degrees of freedom*. The molecule has three translational degrees of freedom. The equipartition law may be stated in the following way. If the energy of the *individual* molecule can be written in the form of a sum of terms, each of which is proportional to the square of either a velocity component or a coordinate, then each of these square terms contributes $\frac{1}{2}kT$ to the *average* energy. As an example, in a gas the translational energy of the individual molecule is

$$\epsilon = \frac{1}{2}mu^2 + \frac{1}{2}mv^2 + \frac{1}{2}mw^2. \tag{4.70}$$

Because each term is proportional to the square of a velocity component, each contributes $\frac{1}{2}kT$ to the average energy; thus we may write

$$\langle \epsilon \rangle = \frac{1}{2}kT + \frac{1}{2}kT + \frac{1}{2}kT = \frac{3}{2}kT. \tag{4.71}$$

4.12 EQUIPARTITION OF ENERGY AND QUANTIZATION

A mechanical system consisting of N particles is described by specifying three coordinates for each particle, or a total of $3N$ coordinates. Thus there are $3N$ independent components of the motion, or degrees of freedom, in such a system. If the N particles are bound together to form a polyatomic molecule, then the $3N$ coordinates and components of the motion are conveniently chosen as follows.

Translational. Three coordinates describe the position of the center of mass; motion in these coordinates corresponds to translation of the molecule as a whole. The energy stored in this mode of motion is kinetic energy only, $\epsilon_{\text{trans}} = \frac{1}{2}mu^2 + \frac{1}{2}mv^2 + \frac{1}{2}mw^2$. Each of these terms contains the square of a velocity component and therefore, as we have seen previously, each contributes $\frac{1}{2}kT$ to the average energy.

Rotational. Two angles are needed to describe the orientation of a linear molecule in space; three angles are needed for the description of the orientation of a nonlinear molecule. Motion in these coordinates corresponds to rotation about two axes (linear molecule) or three axes (nonlinear molecule) in space. The equation for the energy of rotation has the forms

$$\epsilon_{\text{rot}} = \tfrac{1}{2}I\omega_x^2 + \tfrac{1}{2}I\omega_y^2 \qquad \text{(linear molecule)}$$

$$\epsilon_{\text{rot}} = \tfrac{1}{2}I_x\omega_x^2 + \tfrac{1}{2}I_y\omega_y^2 + \tfrac{1}{2}I_z\omega_z^2 \qquad \text{(nonlinear molecule)}$$

in which ω_x, ω_y, ω_z are angular velocities and I_x, I_y, I_z are moments of inertia about the x-, y-, and z-axes, respectively. (In the linear case, $I_x = I_y = I$.) Since each term in the energy expression is proportional to the square of a velocity component, each term on the average should have its equal share, $\frac{1}{2}kT$, of energy. Thus the average rotational energy of linear molecules is $\frac{2}{2}kT$, while that of nonlinear molecules is $\frac{3}{2}kT$. The rotational modes of a diatomic molecule are illustrated in Fig. 4.12.

Vibrational. There remain $3N - 5$ coordinates for linear molecules and $3N - 6$ coordinates for nonlinear molecules. These coordinates describe the bond distances and bond angles within the molecule. Motion in these coordinates corresponds to the vibrations (stretching or bending) of the molecule. Thus linear molecules have $3N - 5$ vibrational modes; nonlinear molecules have $3N - 6$ vibrational modes. Assuming that the vibrations are harmonic, the energy of each vibrational mode can be written in the form

$$\epsilon_{\text{vib}} = \tfrac{1}{2}\mu\left(\frac{dr}{dt}\right)^2 + \tfrac{1}{2}k(r - r_0)^2,$$

in which μ is an appropriate mass, k is the force constant, r_0 is the equilibrium value of the coordinate r, and dr/dt is the velocity. The first term in this expression is the kinetic energy; the second term is the potential energy. By the equipartition law, the first term should contribute $\frac{1}{2}kT$ to the average energy, since it contains a velocity squared. The second term, since it contains the square of the coordinate $r - r_0$, should also contribute $\frac{1}{2}kT$ to the average energy. Each vibrational mode should contribute $\frac{1}{2}kT + \frac{1}{2}kT = kT$ to the average energy of the system. Thus the average energy of the vibrations should be $(3N - 5)kT$ for linear molecules or $(3N - 6)kT$ for nonlinear molecules. The total average energy per molecule should be

$$\langle \epsilon_t \rangle = \tfrac{3}{2}kT + \tfrac{2}{2}kT + (3N - 5)kT \qquad \text{(linear molecules)}$$

$$\langle \epsilon_t \rangle = \tfrac{3}{2}kT + \tfrac{3}{2}kT + (3N - 6)kT \qquad \text{(nonlinear molecules)}.$$

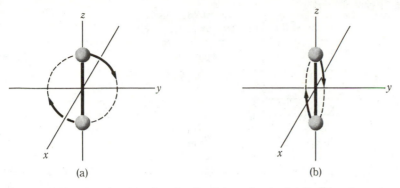

Figure 4.12   Rotational modes of a diatomic molecule. (a) Rotation about the x-axis. (b) Rotation about the y-axis.

If we multiply these values by N_A, the Avogadro number, to convert to average energies per mole, we obtain

Monatomic gases:

$$\bar{U} = \tfrac{3}{2}RT \tag{4.72}$$

Polyatomic gases:

$$\bar{U} = \tfrac{3}{2}RT + \tfrac{2}{2}RT + (3N - 5)RT \qquad \text{(linear)} \tag{4.73}$$

$$\bar{U} = \tfrac{3}{2}RT + \tfrac{3}{2}RT + (3N - 6)RT \qquad \text{(nonlinear)}. \tag{4.74}$$

If heat flows into a gas kept at constant volume, the energy of the gas is increased by the amount of energy transferred by the heat flow. The ratio of the increase in energy to the increase in temperature of the system is the constant volume heat capacity, C_v. Thus, by definition,

$$C_v \equiv \left(\frac{\partial U}{\partial T}\right)_V. \tag{4.75}$$

By differentiating the molar energies with respect to temperature, we obtain the molar heat capacities, $\bar{C}_v$, predicted by the equipartition law.

Monatomic gases:

$$\bar{C}_v = \tfrac{3}{2}R. \tag{4.76}$$

Polyatomic gases:

$$\bar{C}_v = \tfrac{3}{2}R + \tfrac{2}{2}R + (3N - 5)R \qquad \text{(linear)} \tag{4.77}$$

$$\bar{C}_v = \tfrac{3}{2}R + \tfrac{3}{2}R + (3N - 6)R \qquad \text{(nonlinear)}. \tag{4.78}$$

If we examine the values of the heat capacities we find for monatomic gases, $\bar{C}_v/R = 1.5000$, with a high degree of accuracy. This value is independent of temperature over a very wide range.

If we examine the heat capacities of polyatomic gases, Table 4.3, we find two points of disagreement between the data and the equipartition law prediction. The observed heat capacities (1) are always substantially lower than the predicted values, and (2) depend noticeably on temperature. The equipartition principle is a law of classical physics, and these discrepancies were one of the first indications that classical mechanics was not

Table 4.3
Heat capacities of gases at 298.15 K

Monatomic

Species	$\bar{C}_v/R$
He, Ne, Ar, Kr, Xe	1.5000

Diatomic

Species	$\bar{C}_v/R$	Species	$\bar{C}_v/R$
H_2	2.468	F_2	2.78
N_2, HF, HBr, HCl	2.50	Cl_2	3.08
CO	2.505	ICl	3.26
HI	2.51	Br_2	3.33
O_2	2.531	IBr	3.37
NO	2.591	I_2	3.43

Triatomic

Linear	$\bar{C}_v/R$	Nonlinear	$\bar{C}_v/R$
CO_2	3.466	H_2O	3.038
N_2O	3.655	H_2S	3.09
COS	3.99	NO_2	3.56
CS_2	4.490	SO_2	3.79

Tetratomic

Linear	$\bar{C}_v/R$	Nonlinear	$\bar{C}_v/R$
C_2H_2	4.283	H_2CO	3.25
C_2N_2	6.844	NH_3	3.289
		HN_3	4.042
		P_4	7.05

adequate to describe molecular properties. To illustrate this difficulty we choose the case of diatomic molecules, which are of necessity linear. For diatomic molecules, $N = 2$, and we obtain from the equipartition law

$$\frac{\bar{C}_v}{R} = \frac{3}{2} + \frac{2}{2} + 1 = \frac{7}{2} = 3.5.$$

With the exception of H_2, the observed values for diatomic molecules at ordinary temperature fall between 2.5 and 3.5, a number of them being very close to 2.50. Since the translational value 1.5 is observed so accurately in monatomic molecules we suspect the difficulty lies in either the rotational or vibrational motion. When we note that nonlinear molecules have $\bar{C}_v/R > 3.0$ we can narrow the difficulty to the vibrational motion.

The explanation of the observed behavior is that the vibrational motion is quantized. The energy of an oscillator is restricted to certain discrete values and no others. This is in contrast to the classical oscillator, which could have any energy value whatsoever. Now instead of the energies of the various oscillators being distributed continuously over the entire range of energies, the oscillators are distributed in the various quantum states (energy levels). The state of lowest energy is called the *ground state*; the other states are called *excited states*. The permissible values of the energy of a harmonic oscillator are given by the expression

$$\epsilon_s = (s + \tfrac{1}{2})hv \qquad s = 0, 1, 2, \ldots \tag{4.79}$$

in which s, the quantum number, is zero or a positive integer, h is Planck's constant, $h = 6.626 \times 10^{-34}$ J s; and v is the classical frequency of the oscillator, $v = (1/2\pi)\sqrt{\mathcal{k}/\mu}$, in which $\mathcal{k}$ is the force constant and μ is the reduced mass of the oscillator.

The equipartition law depends on the ability of two colliding particles to exchange energy freely between the various modes of motion. This condition is satisfied for both translational and rotational motion because the molecules can accept energy in these modes in any amount, however small, subject only to the dynamic restrictions of conservation of total energy and momentum. But since the vibrational mode is quantized, it can accept only an amount of energy equal to the vibrational quantum, hv. For a molecule such as oxygen this energy quantum is seven times larger than the average energy of translation of the molecules at 25 °C. Thus the collision between two molecules with average kinetic energies cannot raise either molecule to a higher vibrational state because that would require far more energy than they possess. Consequently, essentially all the molecules remain in the ground vibrational state, and the gas does not exhibit a vibrational heat capacity. When the temperature is high enough that the average thermal energy is comparable to the vibrational quantum, hv, the heat capacity approaches the value predicted by the equipartition law. The temperature required depends on the particular vibration.

★4.13 CALCULATION OF VIBRATIONAL HEAT CAPACITY

The distribution of the oscillators is governed by an exponential law (for proof see Section 29.1)

$$n_s = \frac{Ne^{-\epsilon_s/kT}}{Q} \tag{4.80}$$

where n_s is the number of oscillators having the energy ϵ_s. The *partition function*, Q, is determined by the condition that the sum of the number of oscillators in all the energy levels will yield the total number of oscillators, N. That is,

$$\sum_{s=0}^{\infty} n_s = N. \tag{4.81}$$

Hence, substituting from Eq. (4.80) for n_s

$$\sum_{s=0}^{\infty} \frac{Ne^{-\alpha\epsilon_s}}{Q} = N.$$

Thus

$$Q = \sum_{s=0}^{\infty} e^{-\alpha\epsilon_s}, \tag{4.82}$$

where we have set $1/kT = \alpha$ for momentary mathematical convenience.

The average energy is obtained by multiplying the energy of each level by the number in that level, summing over all the levels, and dividing by the total number of molecules.

$$\langle \epsilon \rangle = \frac{\sum\limits_{s=0}^{\infty} \epsilon_s n_s}{N}.$$

Replacing n_s/N by its value from Eq. (4.80) we obtain

$$\langle \epsilon \rangle = \sum_{s=0}^{\infty} \frac{\epsilon_s e^{-\alpha \epsilon_s}}{Q} = \frac{1}{Q} \sum_{s=0}^{\infty} \epsilon_s e^{-\alpha \epsilon_s}.$$

If we differentiate Eq. (4.82) with respect to α, we obtain

$$\frac{dQ}{d\alpha} = -\sum_{s=0}^{\infty} \epsilon_s e^{-\alpha \epsilon_s}. \tag{4.83}$$

Using this result in the expression for $\langle \epsilon \rangle$ reduces it to

$$\langle \epsilon \rangle = -\frac{1}{Q} \frac{dQ}{d\alpha} = -\frac{d \ln Q}{d\alpha}. \tag{4.84}$$

To evaluate Q, we insert $\epsilon_s = (s + \frac{1}{2})hv$ in the expression for Q:

$$Q = \sum_{s=0}^{\infty} e^{-\alpha hv(s+1/2)} = e^{-\alpha hv/2} \sum_{s=0}^{\infty} e^{-\alpha hvs} = e^{-\alpha hv/2} \sum_{s=0}^{\infty} x^s;$$

in the right-hand side we have set $x = e^{-\alpha hv}$. But $\sum_{s=0}^{\infty} x^s = 1 + x + x^2 + \cdots$, and this series is the expansion of $1/(1 - x)$; consequently, we have

$$Q = \frac{e^{-\alpha hv/2}}{1 - x} = \frac{e^{-\alpha hv/2}}{1 - e^{-\alpha hv}}, \quad \text{or} \quad \ln Q = -\tfrac{1}{2}\alpha hv - \ln (1 - e^{-\alpha hv}).$$

Differentiating, we obtain

$$\frac{d \ln Q}{d\alpha} = -\tfrac{1}{2}hv - \frac{hv e^{-\alpha hv}}{1 - e^{-\alpha hv}} = -\tfrac{1}{2}hv - \frac{hv}{e^{\alpha hv} - 1}.$$

Using this expression in Eq. (4.84) for $\langle \epsilon \rangle$, we obtain, after setting $\alpha = 1/kT$,

$$\langle \epsilon \rangle = \tfrac{1}{2}hv + \frac{hv}{e^{hv/kT} - 1}. \tag{4.85}$$

The average energy is made up of the zero point energy $\frac{1}{2}hv$, which is the lowest energy possible for the quantum oscillator, plus a term that depends on temperature.

At very low temperature, $hv/kT \gg 1$, hence $\exp (hv/kT) \gg 1$ so that the second term is very small, and

$$\langle \epsilon \rangle = \tfrac{1}{2}hv \quad \text{and} \quad \bar{C}_v = 0.$$

Effectively, all of the oscillators are in lowest quantum state with $s = 0$.

At very high temperatures such that $hv/kT \ll 1$, we may expand the exponential function: $e^{hv/kT} \approx 1 + hv/kT$; then $e^{hv/kT} - 1 \approx hv/kT$, and we have

$$\langle \epsilon \rangle = \tfrac{1}{2}hv + kT.$$

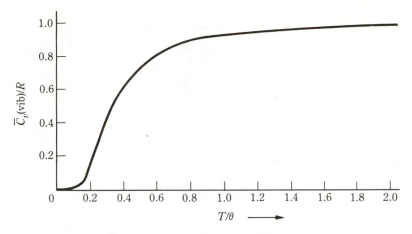

Figure 4.13 Einstein function for $\bar{C}_v/R$ versus T/θ.

For one mole,

$$\bar{U} = N_A \tfrac{1}{2}hv + RT \qquad \text{and} \qquad \bar{C}_v = R.$$

Thus it is only at high temperatures that the vibrational heat capacity attains the classical value, R.

It is customary to define a characteristic temperature $\theta = hv/k$ for each oscillator. Then

$$\langle \epsilon \rangle = \tfrac{1}{2}hv + \frac{k\theta}{e^{\theta/T} - 1}, \tag{4.86}$$

$$\bar{U} = N_A \langle \epsilon \rangle = N_A \tfrac{1}{2}hv + \frac{R\theta}{e^{\theta/T} - 1}, \tag{4.87}$$

and

$$\frac{\bar{C}_v(\text{vib})}{R} = \left(\frac{\theta}{T}\right)^2 \frac{e^{\theta/T}}{(e^{\theta/T} - 1)^2}. \tag{4.88}$$

The function on the right-hand side of Eq. (4.88) is called an Einstein function. The Einstein function is shown as a function of T/θ in Fig. 4.13. Thus for a diatomic molecule we have for the heat capacity

$$\frac{\bar{C}_v}{R} = \frac{5}{2} + \left(\frac{\theta}{T}\right)^2 \frac{e^{\theta/T}}{(e^{\theta/T} - 1)^2}.$$

In the case of polyatomic molecules that have more than one vibration (for example, H_2O, which has three vibrations) there are three distinct frequencies and therefore three distinct characteristic temperatures, so that the heat capacity contains three distinct Einstein functions,

$$\frac{\bar{C}_v}{R} = 3.0 + \left(\frac{\theta_1}{T}\right)^2 \frac{e^{\theta_1/T}}{(e^{\theta_1/T} - 1)^2} + \left(\frac{\theta_2}{T}\right)^2 \frac{e^{\theta_2/T}}{(e^{\theta_2/T} - 1)^2} + \left(\frac{\theta_3}{T}\right)^2 \frac{e^{\theta_3/T}}{(e^{\theta_3/T} - 1)^2}.$$

Values of the characteristic temperatures for a number of molecules are given in Table 4.4.

Table 4.4
Values of θ/K for various gases

H$_2$	6210	Br$_2$	470
N$_2$	3340	I$_2$	310
O$_2$	2230	CO$_2$	$\theta_1 = 1890$
CO	3070		$\theta_2 = 3360$
NO	2690		$\theta_3 = \theta_4 = 954$
HCl	4140		
HBr	3700	H$_2$O	$\theta_1 = 5410$
HI	3200		$\theta_2 = 5250$
Cl$_2$	810		$\theta_3 = 2290$

Terrell L. Hill, *Introduction to Statistical Thermodynamics.*
Reading, Mass.: Addison-Wesley, 1960.

★4.14 THE MAXWELL–BOLTZMANN DISTRIBUTION LAW

Two types of distribution functions have been discussed so far: the spatial distribution of molecules in a gravitational field, the Boltzmann distribution, and the speed distribution in a gas (the Maxwell distribution). These can be written in a combined form, the Maxwell–Boltzmann distribution law.

The barometric formula governs the spatial distribution of the molecules in a gravity field according to the equation

$$\tilde{N} = \tilde{N}_0 e^{-Mgz/RT}, \qquad (4.89)$$

where $\tilde{N}$ and $\tilde{N}_0$ are the numbers of particles per cubic metre at the heights z and zero, respectively. The Boltzmann distribution law governs the spatial distribution of any system in which the particles have a potential energy that depends on the position. For any potential field, the Boltzmann distribution may be written in the form

$$\tilde{N} = \tilde{N}_0 e^{-\epsilon_p/kT}, \qquad (4.90)$$

where ϵ_p is the potential energy of the particle at the point (x, y, z), and $\tilde{N}$ is the number of particles per cubic metre at this position.

For the special case of the gravity field, $\epsilon_p = mgz$. This value of ϵ_p in Eq. (4.90) reduces it to Eq. (4.89), since $m/k = M/R$.

The combined velocity and space distribution is written

$$\frac{d\tilde{N}^*}{\tilde{N}_0} = 4\pi \left(\frac{m}{2\pi kT} \right)^{3/2} c^2 e^{-(mc^2/2 + \epsilon_p)/kT} \, dc, \qquad (4.91)$$

where $d\tilde{N}^*$ is the number of particles per cubic metre at the position (x, y, z) that have speeds in the range c to $c + dc$. Equation (4.91) is the Maxwell–Boltzmann distribution, which is similar to the Maxwell distribution, except that the exponential factor contains the total energy, kinetic plus potential, of the molecule instead of only the kinetic energy.

At any specified position in space, ϵ_p has a definite constant value, and so $\exp(-\epsilon_p/kT)$ is a constant. Then the right-hand side of Eq. (4.91) is simply the Maxwell distribution multiplied by a constant. This means that at any position the distribution of speeds is Maxwellian regardless of the value of the potential energy at that point. For a gas in a gravity field this means that, although there are fewer molecules per cubic metre at 50 km height than at ground level, the fraction of molecules that have speeds in a given range is the same at both levels.

★4.15 EXPERIMENTAL VERIFICATION
OF THE MAXWELL DISTRIBUTION LAW

The amount of indirect evidence for the correctness of the Maxwell distribution law is overwhelming. The relationship of the distribution law to the rate of a chemical reaction has already been mentioned briefly (Section 4.10). We shall see later that the functional form of the experimentally determined temperature dependence of the rate constant agrees with the dependence we expect from the Maxwell distribution. This agreement may be regarded as indirect evidence for the correctness of both the Maxwell distribution and our ideas concerning reaction rates.

Suppose for the sake of argument that speeds were not distributed and that all the molecules moved with the same velocity. Now consider the effect of a gravity field on such a gas. If at ground level all the molecules had the same vertical component of velocity W then all would have a kinetic energy $\frac{1}{2}mW^2$. The maximum height any molecule could reach is that at which all of the ground-level kinetic energy is converted into potential energy; this height H is determined by the equality $mgH = \frac{1}{2}mW^2$, or $H = W^2/2g$. No molecule could reach a height greater than H, and if this situation prevailed, the atmosphere would have a sharp upper boundary. Furthermore, the density of the atmosphere would increase with the height above ground level, since the molecules at the higher levels are moving slowly and thus would spend a larger part of the time at these high levels. None of these predictions is confirmed by observation. The Maxwell distribution, however, says that some molecules have large kinetic energies and so can attain great heights; but the proportion of the molecules with these high energies is small. The Maxwell distribution predicts that the atmospheric density will decrease with increase in height, and that there will be no sharp upper boundary.

A number of direct experimental determinations of the velocity distribution have been made, all of which have verified the Maxwell law within the experimental error. A sketch of the apparatus used in one method is shown in Fig. 4.14. The apparatus is entirely

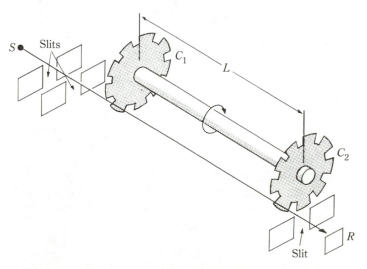

Figure 4.14 Experiment to verify the Maxwell distribution. (Redrawn by permission from K. F. Herzfeld and H. Smallwood, *A Treatise on Physical Chemistry*, H. S. Taylor and S. Glasstone, eds., vol. II, 3d ed. New York: D. Van Nostrand, 1951, p. 37.)

enclosed in a highly evacuated chamber. The molecules escape through a pinhole in the source S, are collimated by the slits, and then pass through one of the openings between the cogs in the cogwheel C_1. The cogwheels C_1 and C_2 are mounted on the same axle, which is rotated rapidly. Only those molecules that have a speed such that they travel the length L in the time required for the cogwheel to be displaced by the width of one opening can get to the detector at R. By changing the velocity of rotation of the cogwheels, molecules having different speeds can be admitted to R. The resemblance of this method to that of Fizeau for measuring the velocity of light should be noted.*

QUESTIONS

4.1 Why are probability laws required to describe gas molecules?

4.2 What is the kinetic theory explanation of the ideal gas law dependence $p \propto V^{-1}$?

4.3 Give a kinetic interpretation of why p for 1 mole of gaseous O_2 molecules is one-half that for 2 moles of gaseous O atoms at a given T and V.

4.4 Why does the Maxwell distribution go to zero at high speeds? (Imagine what such molecules would do to the walls.) At zero speed? (Imagine the fate of a molecule started at rest in the gas.)

4.5 If heavier gas molecules move more slowly than light gas molecules, why is the average kinetic energy independent of the mass?

4.6 Would the average velocity components for gas molecules all vanish for a flowing gas?

PROBLEMS

4.1 Compute the root-mean-square speed, the average speed, and the most probable speed of an oxygen molecule at 300 K and at 500 K. Compare with the values for hydrogen.

4.2 a) Compare the average speed of an oxygen molecule with that of a molecule of carbon tetrachloride at 20 °C.
  b) Compare their average kinetic energies.

4.3 a) Compute the kinetic energy of a mole of a gas at 300 K and 500 K.
  b) Compute the average kinetic energy of a molecule at 300 K.

4.4 Kinetic theory was once criticized on the grounds that it should apply even to potatoes. Compute the average thermal speed at 25 °C of a potato weighing 100 g. Assuming that the earth's gravity field were turned off, how long would it take a potato to traverse 1 m? (After working the problem, compare with the result of Problem 2.21.)

4.5 An oxygen molecule having the average velocity at 300 K is released from the earth's surface to travel upward. If it could move without colliding with other molecules, how high would it go before coming to rest? How high would it go if it had the average kinetic energy at 300 K?

4.6 Suppose that at some initial time all the molecules in a container have the same translational energy, 2.0×10^{-21} J. As time passes, the motion becomes chaotic and the energies finally are distributed in a Maxwellian way.

  a) Compute the final temperature of the system.
  b) What fraction of the molecules finally have energies in the range between 1.98×10^{-21} and 2.02×10^{-21} J? [*Hint*: Since the range of energies in part (b) is small, the differential form of the Maxwell distribution may be used.]

4.7 The quantity $(c - \langle c \rangle)^2 = c^2 - 2c\langle c \rangle + \langle c \rangle^2$ is the square of the deviation of the speed of a molecule from the average speed. Compute the average value of this quantity using the Maxwell

---

*   For a description of several methods of direct determination of the velocity distribution see K. F. Herzfeld and H. Smallwood in *A Treatise on Physical Chemistry*, H. S. Taylor and S. Glasstone, eds., Vol. II, 3rd ed. New York: D. Van Nostrand, 1951, p. 35ff.

distribution, then take the square root of the result to obtain the root-mean-square deviation of the distribution. Note the way in which this last quantity depends on temperature and on the mass of the molecule.

4.8 The quantity $(\epsilon - \langle \epsilon \rangle)^2 = \epsilon^2 - 2\epsilon\langle\epsilon\rangle + \langle\epsilon\rangle^2$ is the square of the deviation of the energy of the molecule from the average energy. Compute the average value of this quantity using the Maxwell distribution. The square root of this quantity is the root-mean-square deviation of the distribution. Note its dependence on temperature and the mass of the molecule.

4.9 The time required for a molecule to travel one metre is $1/c$.

   a) Calculate the average time required for the molecule to travel one metre.
   b) Calculate the root-mean-square deviation of the time from the average time.
   c) What fraction of the molecules require more than the average time to move one metre?

4.10 What fraction of molecules have energies within the range $\langle\epsilon\rangle - \frac{1}{2}kT$ to $\langle\epsilon\rangle + \frac{1}{2}kT$?

4.11 Compute the energy corresponding to the maximum of the energy distribution curve.

4.12 What fraction of the molecules have energies greater than kT? $2kT$? $5kT$? $10kT$?

4.13 What fraction of the molecules have energies between $\langle\epsilon\rangle - \delta\epsilon$ and $\langle\epsilon\rangle + \delta\epsilon$, where $\delta\epsilon$ is the root-mean-square deviation from the average energy?

4.14 What fraction of the molecules have speeds between $\langle c\rangle - \delta c$ and $\langle c\rangle + \delta c$, where δc is the root-mean-square deviation from the average speed?

4.15 The velocity of escape from a planet's surface is given by $v_e = \sqrt{2gR}$. On earth the gravitational acceleration is $g = 9.80$ m/s^2, the earth's radius is $R_E = 6.37 \times 10^6$ m. At 300 K what fraction of

   a) hydrogen molecules have velocities exceeding the escape velocity?
   b) nitrogen molecules have velocities exceeding the escape velocity?

On the moon, $g = 1.67$ m/s^2; the moon's radius is $R_M = 1.74 \times 10^6$ m. Assuming a temperature of 300 K, what fraction of

   c) hydrogen molecules have velocities exceeding the escape velocity?
   d) nitrogen molecules have velocities in excess of the escape velocity?

4.16 What fraction of Cl_2 molecules $(\theta = 810$ K) are in excited vibrational states at 298.15 K? at 500 K? at 700 K?

4.17 The characteristic vibrational temperature for chlorine is 810 K. Calculate the heat capacity of chlorine at 298.15 K; at 500 K; at 700 K.

4.18 The vibrational frequencies in CO_2 are 7.002×10^{13} s^{-1}, 3.939×10^{13} s^{-1}, 1.988×10^{13} s^{-1}, and 1.988×10^{13} s^{-1}. Calculate the corresponding characteristic temperatures and the contributions of each vibration to the heat capacity at 298.15 K.

4.19 The heat capacity of F_2 at 298.15 K is $\bar{C}_v/R = 2.78$. Calculate the characteristic vibrational frequency.

4.20 What is the contribution to $\bar{C}_v(\text{vib})/R$ at $T = 0.1\theta, 0.2\theta, 0.5\theta, \theta, 1.5\theta$?

4.21 The water molecule has three vibrational frequencies: 11.27×10^{13} s^{-1}, 10.94×10^{13} s^{-1}, 4.767×10^{13} s^{-1}. Which of these contributes significantly to the vibrational heat capacity at 298.15 K? Calculate total heat capacity at 298.15 K, 500 K, 1000 K, and 2000 K.

4.22 What fraction of the molecules are in the first excited vibrational state at 300 K, (a) for I_2 with $\theta = 310$ K. (b) for H_2 with $\theta = 6210$ K.

4.23 At 300 K what fraction of CO_2 molecules are in the first, second, and third excited states of the two bending vibrations with $v = 1.988 \times 10^{13}$ s^{-1}?

4.24 What value of T/θ is required if less than half of the molecules are to be in the ground vibrational state? To what temperature would this correspond for I_2 with $\theta = 310$ K?

4.25 As a function of θ/T, sketch the fraction of molecules in

   a) the ground vibrational state.   b) the first excited state.   c) the second excited state.

5

Some Properties of Liquids and Solids

5.1 CONDENSED PHASES

Solids and liquids are referred to collectively as *condensed phases*. This name emphasizes the high density of the liquid or solid as compared with the low density of gases. This difference in density is one of the most striking differences between gases and solids or liquids. The mass of air in a room of moderate size would not exceed two hundred pounds; the mass of liquid required to fill the same room would be some hundreds of tons. Conversely, the volume per mole is very large for gases and very small for liquids and solids. At STP a gas occupies 22,400 cm^3/mole, while the majority of liquids and solids occupy between 10 and 100 cm^3/mole. Under these conditions the molar volume of a gas is 500 to 1000 times larger than that of a liquid or solid.

If the ratio of the gas volume to liquid volume is 1000, then the ratio of the distance between the molecules in the gas to that between those in the liquid is the cube root of the volume ratio, that is, ten. The molecules of the gas are ten times farther apart on the average than are those of the liquid. The distance between molecules in the liquid is roughly equal to the molecular diameter; hence, in the gas the molecules are separated by a distance which is, on the average, ten times their diameter. This large spacing in the gas compared with that in the liquid results in the characteristic properties of the gas and the contrast of these properties with those of the liquid. This comes about simply because of the short-range nature of the intermolecular forces, the van der Waals forces. The effect of these forces decreases very sharply with increase in distance between the molecules and falls to an almost negligible value at distances of four to five times the molecular diameter. If we measure the forces by the magnitude of the term $a/\overline{V}^2$ in the van der Waals equation, then an increase in volume by a factor of 1000 in going from liquid to gas decreases the term by a factor of 10^6. Conversely, in the liquid the effect of the van der Waals forces is a million times larger than it is in the gas.

In gases the volume occupied by the molecules is small compared with the total volume, and the effect of the intermolecular forces is very small. In the first approximation these effects are ignored and any gas is described by the ideal gas law, which is strictly correct only at $p = 0$. This condition implies an infinite separation of the molecules; the intermolecular forces would be exactly zero, and the molecular volume would be completely negligible.

Is it possible to find an equation of state for solids or liquids that has the same generality as the ideal gas law? On the basis of what has been said, the answer must be in the negative. The distances between molecules in liquids and solids are so small, and the effect of the intermolecular forces is correspondingly so large, that the properties of the condensed phases depend on the details of the forces acting between the molecules. Therefore we must expect that the equation of state will be different for each different solid or liquid. If the force law acting between the molecules were a particularly simple one and had the same analytical form for all molecules, we could expect that the law of corresponding states would have universal validity. In fact, the intermolecular forces do not follow such a simple law with precision, so that the law of corresponding states must also be expected to fall short of general applicability. It remains a convenient approximation in many practical situations.

5.2 COEFFICIENTS OF THERMAL EXPANSION AND COMPRESSIBILITY

The dependence of the volume of a solid or liquid on temperature at constant pressure can be expressed by the equation

$$V = V_0(1 + \alpha t),\tag{5.1}$$

where t is the celsius temperature, V_0 is the volume of the solid or liquid at 0 °C, and α is the coefficient of thermal expansion. Equation (5.1) is formally the same as Eq. (2.5), which relates the volume of a gas to the temperature. The important difference between the two equations is that the value of α is approximately the same for all gases, while each liquid or solid has its own particular value of α. Any particular substance has different values of α in the solid and in the liquid state. The value of α is constant over limited ranges of temperature. If the data are to be represented with precision over a wide range of temperature, it it necessary to use an equation with higher powers of t:

$$V = V_0(1 + at + bt^2 + \cdots),\tag{5.2}$$

where a and b are constants. For gases and solids, α is *always* positive, while for liquids α is *usually* positive. There are a few liquids for which α is negative over a small range of temperature. For example, between 0 and 4 °C water has a negative value of α. In this small temperature interval, the specific volume of water decreases as the temperature increases.

In Eq. (5.1), V_0 is a function of pressure. Experimentally, it is found that the relation between volume and pressure is given by

$$V_0 = V_0^0[1 - \kappa(p - 1)],\tag{5.3}$$

where V_0^0 is the volume at 0 °C under one atmosphere pressure, p is the pressure in atmospheres, and κ is the *coefficient of compressibility*, which is a constant for a particular substance over fairly wide ranges of pressure. The value of κ is different for each substance and for the solid and liquid states of the same substance. It is shown in Section 9.2 that the necessary condition for mechanical stability of a substance is that κ must be positive.

According to Eq. (5.3) the volume of a solid or liquid decreases linearly with pressure. This behavior is in sharp contrast to the behavior of gases in which the volume is inversely proportional to the pressure. Furthermore, the values of κ for liquids and solids are extremely small, being of the order of 10^{-6} to 10^{-5} atm^{-1}. If we take $\kappa = 10^{-5}$, then for a pressure of two atmospheres, the volume of the condensed phase is, by Eq. (5.3), $V = V_0^0[1 - 10^{-5}(1)]$. The decrease in volume in going from 1 atm to 2 atm pressure is 0.001%. If a gas were subjected to the same change in pressure, the volume would be halved. Because moderate changes in pressure produce only very tiny changes in the volume of liquids and solids, it is often convenient to consider them to be *incompressible* ($\kappa = 0$) in the first approximation.

The coefficients α and κ are usually given more general definitions than are implied by Eqs. (5.1) and (5.3). The general definitions are

$$\alpha = \frac{1}{V}\left(\frac{\partial V}{\partial T}\right)_p, \qquad \kappa = -\frac{1}{V}\left(\frac{\partial V}{\partial p}\right)_T. \tag{5.4}$$

According to Eq. (5.4), α is the relative increase $(\partial V/V)$ in volume per unit increase in temperature at constant pressure. Similarly, κ is the relative decrease in volume $(-\partial V/V)$ per unit increase in pressure at constant temperature.

If the temperature increment is small, the general definition of α yields the result in Eq. (5.1). Rearranging Eq. (5.4), we have

$$\frac{dV}{V} = \alpha\, dT. \tag{5.5}$$

If the temperature is changed from T_0 to T (corresponding to a change from 0 °C to t °C), then the volume changes from V_0 to V. Integration assuming α is constant yields $\ln(V/V_0) = \alpha(T - T_0)$, or $V = V_0 e^{\alpha(T-T_0)}$. If $\alpha(T - T_0) \ll 1$, we can expand the exponential in series to obtain $V = V_0[1 + \alpha(T - T_0)]$, which is the same as Eq. (5.1) if $T_0 = 273.15$ K. By a similar argument, the definition of κ can be reduced for a small increment in pressure to Eq. (5.3).

Combining Eqs. (5.1) and (5.3) by eliminating V_0 yields an equation of state for the condensed phase:

$$V = V_0^0[1 + \alpha(T - T_0)][1 - \kappa(p - 1)]. \tag{5.6}$$

To use the equation for any particular solid or liquid, the values of α and κ for that substance must be known. Values of α and κ for a few common solids and liquids are given in Table 5.1.

Table 5.1
Coefficients of thermal expansion and compressibility at 20 °C

	Solids					
	Copper	Graphite	Platinum	Quartz	Silver	NaCl
$\alpha/10^{-4}$ K^{-1}	0.492	0.24	0.265	0.15	0.583	1.21
$\kappa/10^{-6}$ atm^{-1}	0.78	3.0	0.38	2.8	1.0	4.2
	Liquids					
	C_6H_6	CCl_4	C_2H_5OH	CH_3OH	H_2O	Hg
$\alpha/10^{-4}$ K^{-1}	12.4	12.4	11.2	12.0	2.07	1.81
$\kappa/10^{-6}$ atm^{-1}	94	103	110	120	45.3	3.85

5.3 HEATS OF FUSION; VAPORIZATION; SUBLIMATION

The absorption or release of heat without any accompanying temperature change is characteristic of a change in the state of aggregation of a substance. The quantity of heat absorbed in the transformation of solid to liquid is the *heat of fusion*. The quantity of heat absorbed in the transformation of liquid to vapor is the *heat of vaporization*. The direct transformation of a solid to vapor is called *sublimation*. The quantity of heat absorbed is the *heat of sublimation*, which is equal to the sum of the heats of fusion and vaporization.

An obvious but important fact about condensed phases is that the intermolecular forces hold the molecules together. The vaporization of a liquid requires the molecules to be pulled apart against the intermolecular forces. The energy required is measured quantitatively by the heat of vaporization. Similarly, energy is required to pull the molecules out of the ordered arrangement in the crystal to the disordered arrangement, usually at a slightly larger distance of separation, existing in the liquid. This energy is measured by the heat of fusion.

Liquids composed of molecules that have comparatively strong forces acting between them have high heats of vaporization, while those composed of weakly interacting molecules have low heats of vaporization. The van der Waals a is a measure of the strength of the attractive forces; we expect the heats of vaporization of substances to fall in the same order as the values of a. This is in fact correct; it can be shown that for a van der Waals fluid the heat of vaporization per mole, Q_{vap}, is equal to a/b.

5.4 VAPOR PRESSURE

If a quantity of a pure liquid is placed in an evacuated container that has a volume greater than that of the liquid, a portion of the liquid will evaporate so as to fill the remaining volume of the container with vapor. Provided that some liquid remains after the equilibrium is established, the pressure of the vapor in the container is a function only of the temperature of the system. The pressure developed is the *vapor pressure* of the liquid, which is a characteristic property of a liquid; it increases rapidly with temperature. The temperature at which the vapor pressure is equal to 1 atm is the *normal boiling point* of the liquid, T_b. Some solids are sufficiently volatile to produce a measurable vapor pressure even at ordinary temperatures; if it should happen that the vapor pressure of the solid reaches 1 atm at a temperature below the melting point of the solid, the solid sublimes. This temperature is called the normal sublimation point, T_s. The boiling point and sublimation point depend upon the pressure imposed upon the substance.

The existence of a vapor pressure and its increase with temperature are consequences of the Maxwell–Boltzmann energy distribution. Even at low temperatures a fraction of the molecules in the liquid have, because of the energy distribution, energies in excess of the cohesive energy of the liquid. As shown in Section 4.10, this fraction increases rapidly with increase in temperature. The result is a rapid increase in the vapor pressure with increase in temperature. The same is true of volatile solids.

The argument implies that at a specified temperature a liquid with a large cohesive energy (that is, a large molar heat of vaporization Q_{vap}) will have a smaller vapor pressure than one with a small cohesive energy. At 20 °C the heat of vaporization of water is 44 kJ/mol, while that of carbon tetrachloride is 32 kJ/mol; correspondingly, the vapor pressures at this temperature are 2.33 kPa for water and 12.13 kPa for carbon tetrachloride.

From the general Boltzmann distribution, the relation between the vapor pressure and heat of vaporization can be made plausible. A system containing liquid and vapor in equilibrium has two regions in which the potential energy of a molecule has different values. The strong effect of the intermolecular forces makes the potential energy low in the liquid; $W = 0$. Comparatively, in the gas the potential energy is high, W. By the Boltzmann law, Eq. (4.90), the number of molecules of gas per cubic metre, $\tilde{N} = A \exp(-W/RT)$, where A is a constant. In the gas, the number of molecules per cubic metre is proportional to the vapor pressure, so we have $p = B \exp(-W/RT)$, where B is another constant. The energy required to take a molecule from the liquid and put it in the vapor is W, the energy of vaporization. As we shall see later, the molar heat of vaporization, Q_{vap}, is related to W by $Q_{vap} = W + RT$. Putting this value of W in the expression for p, we obtain

$$p = p_\infty e^{-Q_{vap}/RT}, \tag{5.7}$$

where p_∞ is also a constant. Equation (5.7) relates the vapor pressure, temperature, and the heat of vaporization; it is one form of the Clausius–Clapeyron equation for which we will give a more rigorous derivation in Section 12.9. The constant p_∞ has the same units as p, and can be evaluated in terms of Q_{vap} and the normal boiling point T_b. At T_b the vapor pressure is 1 atmosphere, so that $1 \text{ atm} = p_\infty e^{-Q_{vap}/RT_b}$. Then

$$p_\infty = (1 \text{ atm})e^{+Q_{vap}/RT_b}. \tag{5.8}$$

The auxiliary Eq. (5.8) suffices to evaluate the constant p_∞.

Taking logarithms, Eq. (5.7) becomes

$$\ln p = -\frac{Q_{vap}}{RT} + \ln p_\infty, \tag{5.9}$$

which is useful for the graphic representation of the variation of vapor pressure with temperature. The function $\ln p$ is plotted against the function $1/T$. Equation (5.9) is then the equation of a straight line, with the slope $-Q_{vap}/R$. (If common logarithms are used, the slope is $-Q_{vap}/2.303R$.) The intercept at $1/T = 0$ is $\ln p_\infty$ (or $\log_{10} p_\infty$). Figure 5.1 is a typical plot of this kind; the vapor-pressure data are for benzene.

A convenient method for determining the heat of vaporization of a liquid is to measure its vapor pressure at several temperatures. After the experimental data are plotted in the manner of Fig. 5.1, the slope of the line is measured and from this the value of Q_{vap} is calculated. If only simple apparatus is used, this method is capable of yielding results of higher accuracy than would a calorimetric determination of Q_{vap} using simple apparatus.

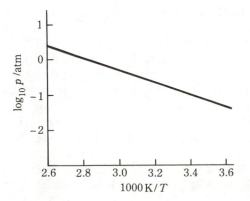

Figure 5.1 Plot of $\log_{10} p$ versus $1/T$ for benzene.

5.5 OTHER PROPERTIES OF LIQUIDS

The viscosity, or more precisely, the coefficient of viscosity, of a liquid measures the resistance to flow under stress. Because the molecules of liquid are very close to one another, a liquid is much more viscous than a gas. The close spacing and the intermolecular forces both contribute to this resistance to flow. Viscosity is discussed in somewhat greater detail in Chapter 30.

A molecule in the bulk of a liquid is attracted by its neighbors about equally, and over a long time interval does not experience an unbalanced force in any particular direction. A molecule in the surface layer of a liquid is attracted by its neighbors, but since it only has neighbors below it in the liquid, it is attracted toward the body of the liquid. Since the molecules on the surface are bound only to the molecules on one side, they do not have as low an energy as do those in the body of the liquid. To move a molecule from the body of the liquid to the surface requires the addition of energy. Since the presence of another molecule in the surface increases the surface area, it follows that energy must be supplied to increase the area of the liquid surface. The energy required to effect an increase of 1 m^2 is called the *surface tension* of the liquid. Surface tension is dealt with in more detail in the chapter on surface properties. For the moment we simply note that the intermolecular forces are responsible for this phenomenon.

5.6 REVIEW OF STRUCTURAL DIFFERENCES BETWEEN SOLIDS, LIQUIDS, AND GASES

We have described the structure of a gas simply in terms of the chaotic motion of molecules (thermal motion), which are separated from one another by distances that are very large compared with their own diameter. The influence of intermolecular forces and finite molecular size is very small and vanishes in the limit of zero pressure.

Since in a liquid the molecules are separated by a distance of the same magnitude as the molecular diameter, the volume occupied by a liquid is about the same as the volume of the molecules themselves. At these close distances the effect of the intermolecular forces is very large, with the result that each molecule has a low potential energy compared with its energy in the gas. The difference in potential energy between gas and liquid is the energy that must be supplied to vaporize the liquid. The motion of the molecules in the liquid is still chaotic, but since the liquid occupies a much smaller volume, there is less randomness in the space distribution of the molecules. The liquid has a very low compressibility simply because there is very little empty space left between the molecules. The liquid is capable of flow under stress because the molecule does have freedom to move anywhere within the volume; it must, however, push other molecules aside to do so, and as a consequence the resistance to flow is greater than for the gas.

The molecules in a solid are locked in a regular pattern; the spatial arrangement is not random as in the gas or liquid, but completely ordered. The solid does not flow under the application of a small stress, as do liquids or gases, but deforms slightly, snapping back when the stress is removed. This highly ordered arrangement is always accompanied by a lower potential energy, so that energy is required to convert the solid to a liquid. The ordered arrangement usually has a somewhat smaller volume (perhaps 5 to 10%) than the liquid volume. The solid has a coefficient of compressibility that is about the same magnitude as that of the liquid.

The distribution of energies in solids and liquids is essentially the same as in the gas and, so long as the temperature is sufficiently high, is described by the Maxwell–Boltzmann distribution function. The motion in gases is characterized by kinetic energy only; in solids and liquids there is a potential energy as well. The motion in solids consists purely of vibration. In liquids, some of the molecules are moving through the liquid while others are momentarily caged by their neighbors and are vibrating in the cage. The motion in the liquid has some of the characteristics of the unhampered motion of molecules in the gas and some of the characteristics of the vibration of molecules in the solid. Overall, the liquid bears a closer resemblance to a solid than to a gas.

QUESTIONS

5.1 Why are liquids, and not gases, used in hydraulic pumps?

5.2 A typical liquid with $\alpha = 10^{-3}\,\text{K}^{-1}$ and $\kappa = 10^{-4}\,\text{atm}^{-1}$ is heated by 10 K. Estimate the external pressure required to keep the density of the liquid constant.

5.3 For most molecular substances, the heat of vaporization is several times larger than the heat of fusion. Explain this on the basis of structure and forces.

5.4 What argument can be given that solid naphthalene (mothballs) has a measureable vapor pressure at room temperature?

5.5 The heat of vaporization of H_2O is about 1.5 times that of CCl_4. Which liquid should have the larger surface tension?

PROBLEMS

5.1 At 25 °C a sealed, rigid container is completely filled with liquid water. If the temperature is raised by 10 °C, what pressure will develop in the container? For water, $\alpha = 2.07 \times 10^{-4}\,\text{K}^{-1}$; $\kappa = 4.50 \times 10^{-5}\,\text{atm}^{-1}$.

5.2 The coefficient of linear expansion is defined by $a = (1/l)(dl/dt)$. If a is very small and has the same value in any direction for a solid, show that the volume expansion coefficient α is approximately equal to $3a$.

5.3 The correction term to the pressure in the van der Waals equation, $a/\overline{V}^2$, has the dimensions of energy per unit volume, J/m^3; therefore $a/\overline{V}$ is an energy per mole. Suppose that the energy per mole of a van der Waals fluid has the form $\overline{U} = f(T) - a/\overline{V}$. At a given temperature find the difference between the energy of water as a gas and the energy of liquid water, assuming that $\overline{V}_{gas} = 24\,\text{dm}^3/\text{mol}$ and $\overline{V}_{liq} = 18\,\text{cm}^3/\text{mol}$. For water, $a = 0.580\,\text{m}^6\,\text{Pa mol}^{-2}$. Compare this difference with the heat of vaporization, 44.016 kJ/mol.

5.4 The heat of vaporization of water is 44.016 kJ/mol. The normal (1 atm) boiling point is 100 °C. Compute the value of the constant p_∞ in Eq. (5.7) and the vapor pressure of water at 25 °C.

5.5 The Clausius–Clapeyron equation relates the equilibrium vapor pressure p to the temperature T. This implies that the liquid boils at the temperature T if it is subjected to a pressure p. Use this idea together with the Boltzmann distribution to derive a relation between the boiling point of a liquid T, the boiling point under 1 atm pressure T_0, and the height above sea level z. Assume that the pressure at sea level is $p_0 = 1$ atm. The temperature of the atmosphere is T_a. If the atmosphere is at 27 °C compute the boiling point of water at 2 km above sea level; $Q_{vap} = 44.016$ kJ/mol; $T_0 = 373$ K.

5.6 If $\alpha = (1/V)(\partial V/\partial T)_p$, show that $\alpha = -(1/\rho)(\partial \rho/\partial T)_p$, where ρ is the density.

5.7 Show that $(d\rho/\rho) = -\alpha\,dT + \kappa\,dp$, where ρ is the density, $\rho = w/V$, where the mass, w, is constant, and V is the volume.

5.8 Since in forming second derivatives of a function of two variables, the order of differentiation does not matter, we have $(\partial^2 V/\partial T\,\partial p) = (\partial^2 V/\partial p\,\partial T)$. Use this relation to show that $(\partial\alpha/\partial p)_T = -(\partial\kappa/\partial T)_p$.

5.9 The following vapor pressure data are available for liquid metallic zinc.

p/mmHg	10	40	100	400
t/°C	593	673	736	844

From an appropriate plot of the data, determine the heat of vaporization of zinc and the normal boiling point.

5.10 From the general definition of α, we find $V = V_0 \exp\left(\int_0^t \alpha \, dt\right)$. If α has the form $\alpha = \alpha_0 + \alpha' t + \frac{1}{2}\alpha'' t^2$ where α_0, α', and α'' are constants, find the relation between α_0, α', and α'' and the constants a, b, and c in the empirical equation

$$V = V_0(1 + at + bt^2 + ct^3).$$

6

The Laws of Thermodynamics: Generalities and the Zeroth Law

6.1 KINDS OF ENERGY AND THE FIRST LAW OF THERMODYNAMICS

Since a physical system may possess energy in a variety of ways, we speak of various kinds of energy.

1. Kinetic energy: energy possessed by a body in virtue of its motion.
2. Potential energy: energy possessed by a body in virtue of its position in a force field; for example, a mass in a gravity field, a charged particle in an electrical field.
3. Thermal energy: energy possessed by a body in virtue of its temperature.
4. Energy possessed by a substance in virtue of its constitution; for example, a compound has "chemical" energy, nuclei have "nuclear" energy.
5. Energy possessed by a body in virtue of its mass; the relativistic mass–energy equivalence.
6. A generator "produces" electrical energy.
7. A motor "produces" mechanical energy.

Many other examples could be mentioned: magnetic energy, strain energy, surface energy, and so forth. The object of thermodynamics is to seek out logically the relations between kinds of energy and their diverse manifestations. The laws of thermodynamics govern the transformation of one kind of energy into another.

In the last two examples a "production" of energy is mentioned. The electrical energy "produced" by the generator did not come from nothing. Some mechanical device, such as a turbine, was needed to run the generator. Mechanical energy disappeared and electrical energy appeared. The quantity of electrical energy "produced" by the generator, plus any friction losses, is exactly equal to the quantity of mechanical energy "lost" by the turbine. Similarly, in the last example, the mechanical energy produced by the motor, plus the friction losses, is exactly equal to the electrical energy supplied to the motor from the power lines. The validity of this conservation law has been established by many most careful and painstaking direct experimental tests and by hundreds of thousands of experiments that confirm it indirectly.

The first law of thermodynamics is the most general statement of this law of conservation of energy; no exception to this law is known. The law of conservation of energy is a generalization from experience and is not derivable from any other principle.

6.2 RESTRICTIONS ON THE CONVERSION OF ENERGY FROM ONE FORM TO ANOTHER

The first law of thermodynamics does not place any restriction on the conversion of energy from one form to another; it simply requires that the total quantity of energy be the same before and after the conversion.

It is always possible to convert any kind of energy into an equal quantity of thermal energy. For example, the output of the generator can be used to operate a toaster immersed in a tub of water. The thermal energy of the water and the toaster is increased by just the amount of electrical energy expended. The electric motor can turn a paddle wheel in the tub of water (as in Joule's experiments), the mechanical energy being converted to an increase in the thermal energy of the water, which is manifested by an increase in the temperature of the water. All kinds of energy can be completely transformed into thermal energy manifested by an increase in temperature of some sample of matter, usually water. The quantity of energy involved can be measured by measuring the temperature rise of a specified mass of water.

Energy may also be classified according to its ability to increase the potential energy of a mass by lifting it against the force of gravity. Only a limited number of the kinds of energy can be completely converted into the lifting of a mass against gravity (for example, the mechanical energy produced by the electric motor). The thermal energy of a steam boiler or the chemical energy of a compound can be only partly converted into the lifting of a mass. The limitations on the conversion of energy from one form to another lead us to the second law of thermodynamics.

6.3 THE SECOND LAW OF THERMODYNAMICS

Imagine the following situation. A hard steel ball is suspended at a height h above a hard steel plate. Upon release the ball travels downward losing its potential energy and simultaneously increasing its velocity and hence its kinetic energy. The ball hits the plate and rebounds. We assume that the collision with the plate is elastic; no energy is lost to the plate in the collision. On the rebound the ball travels upward, gaining in potential energy and losing kinetic energy until it returns to the original height h. At this point the ball has its original energy, mgh, and its original kinetic energy, zero. We can either stop the motion at this point or let the ball repeat the motion as often as we please. The first law of thermodynamics in this case is simply the law of conservation of mechanical energy. The sum of the potential energy and the kinetic energy must be a constant throughout the course of the motion. The first law is not in the least concerned with how much of the energy is potential or how much is kinetic, but requires only that the sum remain constant.

Now imagine a somewhat different situation. The ball is poised above a beaker of water. Upon release, it loses potential energy and gains kinetic energy, then enters the water and comes to rest at the bottom of the beaker. Strictly from the standpoint of mechanics, it seems that some energy has been destroyed, because in the final state the ball has neither potential energy nor kinetic energy, while initially it possessed potential energy. Mechanics makes no prediction of the fate of this energy which has "disappeared." However, careful examination of the system reveals that the temperature of the water is slightly higher after the ball has entered and come to rest than before. The potential energy of the ball has been converted to thermal energy of the ball and the water. The first law of thermodynamics requires that both the ball *and* the beaker of water be included in the system, and that the total of potential energy, kinetic energy, and thermal energy of both ball and water be constant throughout the motion. Using E_b and E_w for the energies of the ball and water, the requirement can be expressed as

$$E_{b(kin)} + E_{w(kin)} + E_{b(pot)} + E_{w(pot)} + E_{b(therm)} + E_{w(therm)} = \text{constant.}$$

As in the case of the ball and the plate, the first law is not concerned with how the constant amount of energy is distributed among the various forms.

There is an important difference between the case of the steel plate and that of the beaker of water. The ball can bounce up and down on the plate for an indefinite period, but it falls only once into the beaker of water. Fortunately, we never observe a ball bearing in a glass of water suddenly leaping out of the glass, leaving the water slightly cooler than it was. It is important to realize, however, that the first law of thermodynamics does not rule out this disconcerting event.

The behavior of the ball and the beaker of water is typical of all real processes in one respect. Every real process has a sequence that we recognize as natural; the opposite sequence is unnatural. We recognize the falling of the ball and its coming to rest in the water as a natural sequence. If the ball were at rest in the beaker and then hopped out of the water, we would admit that this is not a natural sequence of events.

The second law of thermodynamics is concerned with the direction of natural processes. In combination with the first law, it enables us to predict the natural direction of any process and, as a result, to predict the equilibrium situation. To choose a complicated example, if the system consists of a gasoline tank and an engine mounted on wheels, the second law allows us to predict that the natural sequence of events is: consumption of gasoline, the production of carbon dioxide and water, and the forward motion of the whole device. From the second law, the maximum possible efficiency of the conversion of the chemical energy of the gasoline into mechanical energy can be calculated. The second law also predicts that we cannot manufacture gasoline by feeding carbon dioxide and water into the exhaust and pushing this contraption along the highway; not even if we push it *backwards* along the highway!

Obviously, if thermodynamics can predict results of this type, it must have enormous importance. In addition to having far-reaching theoretical consequences, thermodynamics is an immensely practical science. A simpler example of the importance of the second law to the chemist is that it allows the calculation of the equilibrium position of any chemical reaction, and it defines the parameters that characterize the equilibrium (for example, the equilibrium constant).

We shall not deal with the third law of thermodynamics at this point. The principal utility of the third law to the chemist is that it permits the calculation of equilibrium constants from calorimetric data (thermal data) exclusively.

6.4 THE ZEROTH LAW OF THERMODYNAMICS

The law of thermal equilibrium, the zeroth law of thermodynamics, is another important principle. The importance of this law to the temperature concept was not fully realized until after the other parts of thermodynamics had reached a rather advanced state of development; hence the unusual name, zeroth law.

To illustrate the zeroth law we consider two samples of gas.* One sample is confined in a volume V_1, the other in a volume V_2. The pressures are p_1 and p_2, respectively. At the beginning the two systems are isolated from each other and are in complete equilibrium. The volume of each container is fixed, and we imagine that each has a pressure gauge, as shown in Fig. 6.1(a).

The two systems are brought in contact through a wall. Two possibilities exist: when in contact through the wall the systems either influence each other or they do not. If the systems do not influence each other, the wall is an *insulating*, or *adiabatic*, wall; of course, in this situation the pressures of the two systems are not affected by putting the systems in contact. If the systems do influence one another after putting them in contact, we will observe that the readings of the pressure gauges change with time, finally reaching two new values p_1' and p_2' which no longer change with time, Fig. 6.1(b). In this situation the wall is a *thermally conducting* wall; the systems are in *thermal contact*. After the properties of two systems in thermal contact settle down to values that no longer change with time, the two systems are in *thermal equilibrium*. These two systems then have a property in common, the property of being in thermal equilibrium with each other.

Consider three systems A, B, and C arranged as in Fig. 6.2(a). Systems A and B are in thermal contact, and systems B and C are in thermal contact. This composite system is allowed sufficient time to come to thermal equilibrium. Then A is in thermal equilibrium with B, and C is in thermal equilibrium with B. Now we remove A and C from their contact with B and place them in thermal contact with each other (Fig. 6.2b). We then observe that no changes in the properties of A and C occur with time. Therefore A and C are in thermal equilibrium with each other. This experience is summed up in the zeroth law of thermodynamics: Two systems that are both in thermal equilibrium with a third system are in thermal equilibrium with each other.

The temperature concept can be stated precisely by: (1) Systems in thermal equilibrium with each other have the same temperature; and (2) systems not in thermal equilibrium with each other have different temperatures. The zeroth law therefore gives us an operational definition of temperature that does not depend on the physiological sensation of "hotness" or "coldness." This definition is in agreement with the physiological one,

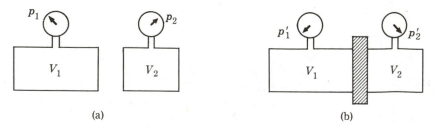

(a)   (b)

Figure 6.1   (a) Systems isolated. (b) Systems in thermal contact.

---

*   The argument does not depend in the least on whether gases, real or ideal, or liquids or solids were chosen.

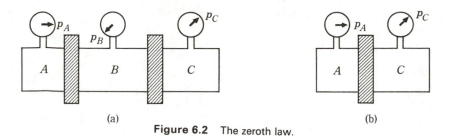

Figure 6.2 The zeroth law.

since two bodies in thermal equilibrium feel the same as far as hotness is concerned. The zeroth law is based on the experience that systems in thermal contact are not in complete equilibrium with one another until they have the same degree of hotness, that is, the same temperature.

6.5 THERMOMETRY

The zeroth law suggests a method for measuring the temperature of any system. We choose a system, the thermometer, having some property y that is conveniently measurable and that varies reasonably rapidly with temperature. The thermometer is allowed to come to thermal equilibrium with a system whose temperature is reproducible (for example, melting ice). The value of y is measured. For the thermometer, suppose that we choose a small quantity of gas confined in a box of constant volume, which is fitted with a pressure gauge. After this thermometer comes to thermal equilibrium with the melting ice, the needle of the pressure gauge will stand in a definite position. This position we can mark with any number we please; let us follow Celsius and mark it zero. The thermometer is next allowed to come to thermal equilibrium with another system having a reproducible temperature: water boiling under 1 atm pressure. The needle stands at some new position, which we can mark with any arbitrary number; again following Celsius we mark the new position with 100. Between the 0 and the 100 we place 99 evenly spaced marks; the dial above 100 and below 0 is divided into intervals of the same width. The thermometer is ready. To measure the temperature of any body, the thermometer is allowed to come to thermal equilibrium with the body; the position of the needle indicates the temperature of the body in degrees. One caution is in order here: the property chosen as the thermometric property must continually increase or decrease in value as the temperature rises in the range of application of the thermometer. The thermometric property may not have a maximum or minimum or stationary value in the temperature range in which the thermometer is to be used.

6.5.1 The Thermometric Equation

The procedure is easily reduced to a formula by which the temperature can be calculated from the measured value of the thermometric property y. Let y_i be the value at the ice point and y_s be the value at the steam point. These points are separated by 100 degrees. Then

$$\frac{dy}{dt} = \frac{y_s - y_i}{100 - 0} = \frac{y_s - y_i}{100}.$$

The right-hand side of this equation is a constant; multiplying through by dt and integrating, we obtain

$$y = \frac{y_s - y_i}{100} t + C, \tag{6.1}$$

where C is an integration constant. But at $t = 0$, $y = y_i$; using these values, Eq. (6.1) becomes $y_i = C$. Using this value for C, Eq. (6.1) reduces to

$$y = \frac{y_s - y_i}{100} t + y_i.$$

Solving for t, we obtain

$$t = \frac{y - y_i}{y_s - y_i} 100, \tag{6.2}$$

which is the thermometric equation. From the measured value of the thermometric property y the temperature on this particular scale can be calculated.

More generally, suppose we choose any two fixed temperatures to which we assign the arbitrary values t_1 and t_2. If y_1 and y_2 are the values of the thermometric property at these temperatures, the thermometric equation, Eq. (6.2), becomes

$$t = t_1 + \frac{y - y_1}{y_2 - y_1} (t_2 - t_1) \tag{6.3}$$

Again, from a measurement of y we can calculate t.

An objection may be raised against this procedure on the grounds that it seems to require that the thermometric property be a linear function of the temperature. The objection is without substance, because we have no way of knowing whether a property is linear with temperature until we have chosen some method of measuring temperature. In fact, the method of operation by its very nature automatically makes the thermometric property a linear function of the temperature measured on that particular scale. This reveals a very real difficulty associated with thermometry. A different scale of temperature is obtained for every different property chosen as the thermometric property. Even with one substance, different scales of temperature will be obtained depending on which property is chosen as the thermometric property. Truly, this is an outrageous turn of events; imagine the consequences if a similar state of affairs existed in the measurement of length: The size of the centimetre would be different depending on whether the metre stick was made of metal or wood or paper.

We can attempt to save the situation by searching for a class of substances all of which have some property that behaves in much the same way with temperature. Gases come to mind. For a given change in temperature, the relative change in pressure at constant volume (or relative change in volume under constant pressure) has nearly the same value for all real gases. The behavior of gases can be generalized in the limit of zero pressure to that of the ideal gas. So we might use an ideal gas in the thermometer and define an *ideal gas scale* of temperature. This procedure is quite useful, as we have seen in Chapter 2. In spite of its utility, the ideal gas scale of temperature does not resolve the difficulty. In the first situation different substances yielded different temperature scales, but at least each of the scales depended upon some property of a *real* substance. The ideal gas scale is a generalization to be sure, but the scale depends on the properties of a *hypothetical* substance!

6.5.2 The Thermodynamic Temperature Scale

Fortunately, there is a way out of this predicament. Using the second law of thermo-dynamics it is possible to establish a temperature scale that is independent of the particular properties of any substance, real or hypothetical. This scale is the *absolute*, or the *thermodynamic*, temperature scale, also called the Kelvin scale after Lord Kelvin, who first demonstrated the possibility of establishing such a scale. By choosing the same size degree, and with the usual definition of the mole of substance, the Kelvin scale and the ideal gas scale become numerically identical. The fact of this identity does not destroy the more fundamental character of the Kelvin scale. We establish this identity because of the convenience of the ideal gas scale compared with other possible scales of temperature.

Having overcome the fundamental difficulties, we use all sorts of thermometers with confidence, requiring only that if the temperatures of two bodies A and B are measured with different thermometers, the thermometers must agree that $t_A > t_B$ or that $t_A = t_B$ or that $t_A < t_B$. The different thermometers need not agree on the numerical value of either t_A or t_B. If it is necessary, the reading of each thermometer can be translated into the temperature in kelvins; then the numerical values must agree.

Originally the ice point on the Kelvin scale was determined by using a constant-volume gas thermometer to measure the pressure and assigning 100 degrees between the ice point and the steam point. The temperature on this centigrade gas scale is given by

$$t = \frac{p - p_i}{p_s - p_i}\,(100),$$

where p is the pressure at t; p_i and p_s are the pressures at the ice point and steam point, respectively. It turns out that the quantity

$$T_0 = \lim_{p_i \to 0} \frac{100 p_i}{p_s - p_i}$$

is a universal constant, independent of the gas in the thermometer. The thermodynamic temperature T is determined by

$$T = \lim_{p_i \to 0} \frac{100 p}{p_s - p_i}.$$

Unfortunately, although the value of T_0 does not depend on the gas, it does depend on how accurately the values of p_i and p_s are measured. As the accuracy of the measurements increased, the value shifted from 273.13 to 273.17. This would not be too troublesome at ordinary temperatures, but for investigators working at very low temperatures it was intolerable. At 1.00 K an uncertainty in the origin of ± 0.01 K would be comparable to an error in the boiling point of water of ± 4 °C.

6.5.3 Current Definition of the Temperature Scale

The current definition of the temperature scale is based on one fixed point, the triple point of water. The absolute temperature of that point is defined arbitrarily as 273.16 K *exactly*. (The triple point of water is that temperature at which pure liquid water is in equilibrium with ice and water vapor.) This definition fixes the size of the *kelvin*, the "degree" on the thermodynamic scale. The size of the Celsius degree is defined to be equal to one kelvin exactly and the origin of the Celsius scale of temperature is *defined* as 273.15 K *exactly*.

This point is very close to the ice point; $t = +0.0002$ °C. Similarly, 100 °C is very close to the steam point but is not exactly at the steam point. The difference is much too small to cause any concern.

QUESTIONS

6.1 A pendulum swinging in a vacuum will continue indefinitely, but will come to rest if immersed in air. How do the first and second laws apply to these situations?

6.2 What is the thermometric property employed in ordinary mercury thermometers?

PROBLEMS

Conversion factor:

   1 watt = 1 joule/second

6.1 An electric motor requires 1 kilowatt-hour to run for a specified period of time. In this same period it produces 3200 kilojoules of mechanical work. How much energy is dissipated in friction and in the windings of the motor?

6.2 A ball bearing having a mass of 10 g falls through a distance of 1 metre and comes to rest. How much energy is dissipated as thermal energy?

6.3 A bullet, mass = 30 g, leaves the muzzle of a rifle with a velocity of 900 m/s. How much energy is dissipated in bringing the bullet to rest?

6.4 One proposal in the so-called synthetic fuels program is to gasefy coal *in situ* by forcing steam into the underground coal seam, thus converting the coal into CO and H_2 by the water–gas reaction:

$$C + H_2O \longrightarrow CO + H_2.$$

To make this reaction go, 175.30 kJ of energy must be supplied for each mole of carbon consumed. This energy is obtained by setting the coal on fire in air or oxygen, then introducing steam into the gas stream. The reaction

$$C + O_2 \longrightarrow CO_2$$

supplies 393.51 kJ for each mole of carbon burned. When the exit mixture is finally used as fuel, 282.98 kJ/mol CO and 285.83 kJ/mol H_2 are recovered.

a) What fraction of the coal must be burned to CO_2 to drive the water–gas reaction?
b) Adjust for the coal burned to drive the process (assuming there are no losses). How much more energy is obtained from the combustion of the CO and H_2 than would have been obtained if the coal had been burned directly?

6.5 a) Suppose that we use the equilibrium vapor pressure of water as a thermometric property in constructing a scale of temperature, t'. In terms of the Celsius temperature, t, the vapor pressure is (to the nearest mmHg)

$t/$°C	0	25	50	75	100
$p/$mmHg	5	24	93	289	760

If the fixed points, ice point and steam point, are separated by 100° on the t' scale, what will be the temperatures t' corresponding to $t = 0, 25, 50, 75,$ and 100 °C? Plot t' versus t.

b) The vapor pressures of benzene and water in terms of the Celsius temperature have the following values:

$t/°C$	7.6	26.1	60.6	80.1
$p\,(C_6H_6)/mmHg$	40	100	400	760
$p\,(H_2O)/mmHg$	8	25	154	356

Plot the vapor pressure of benzene as a function of t', the temperature on the water vapor pressure scale.

6.6 The length of a metal rod is given in terms of the Celsius temperature t by

$$l = l_0(1 + at + bt^2),$$

where a and b are constants. A temperature scale, t', is defined in terms of the length of the metal rod, taking 100° between the ice point and the steam point. Find the relation between t' and t.

6.7 With the present scale of absolute temperature, T, the zero of the Celsius scale is defined as 273.15 K exactly. Suppose we were to define an absolute scale, T', such that the zero of the Celsius scale was at 300 K', exactly. If the boiling point of water on the Celsius scale is 100 °C, what would be the boiling point of water on the T' scale?

7

Energy and the First Law of Thermodynamics; Thermochemistry

7.1 THERMODYNAMIC TERMS: DEFINITIONS

In beginning the study of thermodynamics it is important to understand the precise thermodynamic sense of the terms that are employed. The following definitions have been given succinct expression by J. A. Beattie.*

> *System, Boundary, Surroundings.* A thermodynamic *system* is that part of the physical universe the properties of which are under investigation. . . .
> The system is confined to a definite place in space by the *boundary* which separates it from the rest of the universe, the *surroundings*. . . .
> A system is *isolated* when the boundary prevents any interaction with the surroundings. An isolated system produces no observable effect or disturbance in its surroundings. . . .
> A system is called *open* when mass passes across the boundary, *closed* when no mass passes the boundary. . . .
> *Properties of a System.* The properties of a system are those physical attributes that are perceived by the senses, or are made perceptible by certain experimental methods of investigation. Properties fall into two classes: (1) non-measurable, as the kinds of substances composing a system and the states of aggregation of its parts; and (2) measurable, as pressure and volume, to which a numerical value can be assigned by a direct or indirect comparison with a standard.
> *State of a System.* A system is in a definite state when each of its properties has a definite value. We must know, from an experimental study of a system or from experience with similar systems, what properties must be taken into consideration in order that the state of a system be defined with sufficient precision for the purpose at hand. . . .
> *Change in State, Path, Cycle, Process.* Let a system undergo a change in its state from a specified initial to a specified final state.
> The *change in state* is completely defined when the initial and the final states are specified.

---

* J. A. Beattie, *Lectures on Elementary Chemical Thermodynamics*. Printed by permission from the author.

The *path* of the change in state is defined by giving the initial state, the sequence of inter-mediate states arranged in the order traversed by the system, and the final state.

A *process* is the method of operation by means of which a change in state is effected. The description of a process consists in stating some or all of the following: (1) the boundary; (2) the change in state, the path, or the effects produced in the system during each stage of the process; and (3) the effects produced in the surroundings during each stage of the process.

Suppose that a system having undergone a change in state returns to its initial state. The path of this cyclical transformation is called a *cycle*, and the process by means of which the transformation is effected is called a cyclical process.

State Variable, ... A state variable is one that has a definite value when the state of a system is specified. ...

You should not be misled by the simplicity and clarity of these definitions. The meanings, while apparently "obvious," are precise. These definitions should be thoroughly understood so that when one of the terms appears, it will be immediately recognized as one that has a precise meaning. In the illustrations that follow, these mental questions should be posed: What is the system? Where is the boundary? What is the initial state? What is the final state? What is the path of the transformation? Asking such questions, and other pertinent ones, will help a great deal in clarifying the discussion and is absolutely indispensable before you begin to work any problem.

A system ordinarily must be in a container so that *usually* the boundary is located at the inner surface of the container. As we have seen in Chapter 2, the state of a system is described by giving the values of a sufficient number of state variables; in the case of pure substances, two intensive variables such as T and p are ordinarily sufficient.

7.2 WORK AND HEAT

The concepts of work and of heat are of fundamental importance in thermodynamics, and their definitions must be thoroughly understood; the use of the term *work* in thermo-dynamics is much more restricted than its use in physics generally, and the use of the term *heat* is quite different from the everyday meaning of the word. Again, the definitions are those given by J. A. Beattie.*

Work. In thermodynamics work is defined as any quantity that flows across the boundary of a system during a change in its state and is completely convertible into the lifting of a weight in the surroundings.

Several things should be noted in this definition of work.

1. Work appears only at the boundary of a system.
2. Work appears only *during* a change in state.
3. Work is manifested by an effect in the *surroundings*.
4. The quantity of work is equal to mgh, where m is the mass lifted, g is the acceleration due to gravity, h is the height through which the weight has been raised.
5. Work is an algebraic quantity; it is positive if the mass is lifted (h is $+$), in which case we say that work has been produced in the surroundings or has flowed *to* the surroundings; it is negative if the mass is lowered (h is $-$), in which case we say that work has been destroyed in the surroundings or has flowed *from* the surroundings.†

---

*   J. A. Beattie, *op cit.*

†   Parts of this paragraph follow Beattie's discussion closely. By permission from the author.

Heat. We explain the attainment of thermal equilibrium of two systems by asserting that a quantity of heat Q has flowed from the system of higher temperature to the system of lower temperature.

In thermodynamics heat is defined as a quantity that flows across the boundary of a system during a change in its state in virtue of a difference in temperature between the system and its surroundings and flows from a point of higher to a point of lower temperature.*

Again several things must be emphasized.

1. Heat appears only at the boundary of the system.
2. Heat appears only *during* a change in state.
3. Heat is manifested by an effect in the *surroundings*.
4. The quantity of heat is proportional to the mass of water in the surroundings that is increased by one degree in temperature starting at a specified temperature under a specified pressure. (We must agree to use one particular thermometer.)
5. Heat is an algebraic quantity; it is positive if a mass of water in the surroundings is cooled, in which case we say that heat has flowed *from* the surroundings; it is negative if a mass of water in the surroundings is warmed, in which case we say that heat has flowed *to* the surroundings.†

In these definitions of work and heat, it is of utmost importance that the judgment as to whether or not a heat flow or a work flow has occurred in a transformation is based on observation of *effects produced in the surroundings*, not upon what happens within the system. The following example clarifies this point, as well as the distinction between work and heat.

Consider a system consisting of 10 g of liquid water contained in an open beaker under constant pressure of 1 atm. Initially the water is at 25 °C, so that we describe the initial state by $p = 1$ atm, $t = 25$ °C. The system is now immersed in, let us say, 100 g of water at a high temperature, 90 °C. The system is kept in contact with this 100 g of water until the temperature of the 100 g has fallen to 89 °C, whereupon the system is removed. We say that 100 units of heat has flowed from the surroundings, since the 100 g of water in the surroundings dropped 1 °C in temperature. The final state of the system is described by $p = 1$ atm, $t = 35$ °C.

Now consider the same system, 10 g of water, $p = 1$ atm, $t = 25$ °C, and immerse a stirring paddle driven by a falling mass (Fig. 7.1). By properly adjusting the mass of the falling mass and the height h through which it falls, the experiment can be arranged so that after the mass falls once, the temperature of the system rises to 35 °C. Then the final state is $p = 1$ atm, $t = 35$ °C. In this experiment the *change in state* of the system is exactly the same as in the previous experiment. There was no heat flow, but there was a flow of work. A mass is lower in the surroundings.

If we turned our backs on the experimenter while the change in state had been effected, but had *observed the system* before and after the change in state, we could deduce nothing whatsoever about the heat flow or work flow involved. We could conclude only that the temperature of the system was higher afterward than before; as we shall see later, this implies that the *energy* of the system increased. On the other hand, if we observed the surroundings before and after, we would find cooler bodies of water and/or masses at

---

*   J. A. Beattie, *op cit.*

†   Parts of this paragraph follow Beattie's discussion closely. By permission from the author.

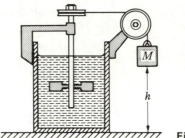

Figure 7.1

lower elevations. From these observations on the surroundings, we could immediately deduce the quantities of heat and work that flowed in the transformation.*

It should be clear that the fact that a system is hotter, that is, has a higher temperature, after some transformation does not mean that it has more "heat"; it could equally well have more "work." The system has neither "heat" nor "work"; this use of these terms is to be avoided at all costs. This usage reflects confusion between the concepts of heat and temperature.

The experiment in Fig. 7.1 is Joule's classic experiment on "the mechanical equivalent of heat." This experiment together with earlier ones of Rumford were instrumental in demolishing the caloric theory of heat and establishing that "heat" is equivalent in a certain sense to ordinary mechanical energy. Even today this experiment is described in the words "work is converted into 'heat'." In the modern definition of the word, there is no "heat" involved in the Joule experiment. Today Joule's observation is described by saying that the destruction of work in the surroundings produces an increase in temperature of the system. Or, less rigidly, work in the surroundings is converted into thermal energy of the system.

The two experiments, immersion of the system in hot water and rotating a paddle in the same system, involved the same *change in state* but different heat and work effects. The quantities of heat and work that flow depend on the process and therefore on the *path* connecting the initial and final states. Heat and work are called *path functions*.

7.3 EXPANSION WORK

If a system alters its volume against an opposing pressure, a work effect is produced in the surroundings. This expansion work appears in most practical situations. The system is a quantity of a gas contained in a cylinder fitted with a piston D (Fig. 7.2a). The piston is assumed to have no mass and to move without friction. The cylinder is immersed in a thermostat so that the temperature of the system is constant throughout the change in state. Unless a specific statement to the contrary is made, in all of these experiments with cylinders it is understood that the space above the piston is evacuated so that no air pressure is pushing down on the piston.

In the initial state the piston D is held against a set of stops S by the pressure of the gas. A second set of stops S' is provided to arrest the piston after the first set is pulled out. The initial state of the system is described by T, p_1, V_1. We place a small mass M on the piston; this mass must be small enough so that when the stops S are pulled out, the piston

---

*   The work of expansion accompanying the temperature increase is negligibly small and has been ignored to avoid obscuring the argument.

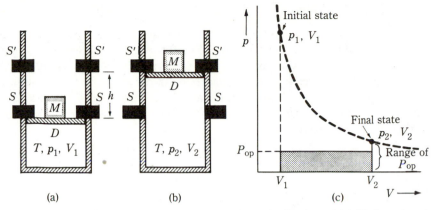

Figure 7.2   Single-stage expansion. (a) Initial state. (b) Final state. (c) Work produced in a single-stage expansion, $W = P_{op}(V_2 - V_1)$.

will rise and be forced against the stops S'. The final state of the system is T, p_2, V_2 (Fig. 7.2b). The boundary is the inner walls of the cylinder and the piston; in the change the boundary expands to enclose a larger volume V_2. Work is produced in this transformation, since a mass M *in the surroundings*, has been lifted a vertical distance h against the force of gravity Mg. The quantity of work produced is

$$W = Mgh. \tag{7.1}$$

If the area of the piston is A, then the downward pressure acting on the piston is $Mg/A = P_{op}$, the pressure which *opposes* the motion of the piston. Thus $Mg = P_{op}A$. Using this value in Eq. (7.1), we obtain

$$W = P_{op}Ah.$$

However, the product Ah is simply the additional volume enclosed by the boundary in the change of state. Thus, $Ah = V_2 - V_1 = \Delta V$, and we have*

$$W = P_{op}(V_2 - V_1). \tag{7.2}$$

The work produced in the change in state, Eq. (7.2), is represented graphically by the shaded area in the p–V diagram of Fig. 7.2(c). The dashed curve is the isotherm of the gas, on which the initial and final states have been indicated. It is evident that M can have any arbitrary value from zero to some definite upper limit and still permit the piston to rise to the stops S'. It follows that P_{op} can have any value in the range $0 \le P_{op} \le p_2$, and so the quantity of work produced may have any value between zero and some upper limit. *Work is a function of the path.* It must be kept in mind that P_{op} is arbitrary and is not related to the pressure of the system.

The sign of W is determined by the sign of ΔV, since $P_{op} = Mg/A$ is always positive. In expansion, $\Delta V = +$, and $W = +$; the mass rises. In compression, $\Delta V = -$, $W = -$; the mass falls.

---

*   Differences between the values of a state function in the final and initial states occur so frequently in thermodynamics that a special short-hand notation is used. The Greek capital delta, Δ, is prefixed to the symbol of the state function. The symbol ΔV is read "delta vee" or "the increase in volume" or "the difference in volume." The symbol Δ always signifies a *difference* of two values, which is *always* taken in the order, final value minus initial value.

7.3.1 Two-Stage Expansion

As it stands, Eq. (7.2) is correct only if P_{op} is constant throughout the change in state. It is easy to imagine more complicated ways of performing the expansion. Suppose that a large mass were placed on the piston during the first part of the expansion from V_1 to some intermediate volume V'; then a smaller mass replaced the large one in the expansion from V' to V_2. In such a two-stage expansion, we apply Eq. (7.2) to each stage of the expansion, using different values of P_{op} in each stage. Then the total work produced is the sum of the amounts produced in each stage:

$$W = W_{\text{first stage}} + W_{\text{second stage}} = P'_{op}(V' - V_1) + P''_{op}(V_2 - V').$$

The quantity of work produced in the two-stage expansion is represented by the shaded areas in Fig. 7.3 for the special case $P''_{op} = p_2$.

Comparison of Figs. 7.2(c) and 7.3 shows that for the same change in state the two-stage expansion produces more work than the single-state expansion could possibly produce. If the heats had been measured, we would also have found different quantities of heat associated with the two paths.

7.3.2 Multistage Expansion

In a multistage expansion the work produced is the sum of the small amounts of work produced in each stage. If P_{op} is constant as the volume increases by an infinitesimal amount dV, then the small quantity of work dW is given by

$$dW = P_{op}\,dV. \tag{7.3}$$

The total work produced in the expansion from V_1 to V_2 is the integral

$$W = \int_1^2 dW = \int_{V_1}^{V_2} P_{op}\,dV, \tag{7.4}$$

which is the general expression for the work of expansion of any system. Once P_{op} is known as a function of the volume, the integral is evaluated by the usual methods.

Observe that the differential dW does not integrate in the ordinary way. The integral of an ordinary differential dx between limits yields a finite difference, Δx,

$$\int_{x_1}^{x_2} dx = x_2 - x_1 = \Delta x,$$

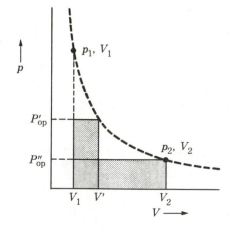

Figure 7.3 Work produced in a two-stage expansion, $W = P'_{op}(V' - V_1) + P''_{op}(V_2 - V')$.

but the integral of dW is the sum of small quantities of work produced along each element of the path,

$$\int_1^2 dW = W,$$

where W is the total amount of work produced. This explains the use of d instead of the ordinary d. The differential dW is an *inexact differential*, dx is an *exact differential*. More about that later.

7.4 WORK OF COMPRESSION

The work destroyed in compression is computed using the same equation that is used to compute the work produced in expansion. In compression the final volume is less than the initial volume, so in every stage ΔV is negative; therefore the total work destroyed is negative. The sign is automatically taken care of by the integration process if the volume of the final state is the upper limit and the volume of the initial state the lower limit in the integral of Eq. (7.4). However, in comparing work of compression with work of expansion, more than a sign change is involved; to compress the gas we need larger masses on the piston than those that were lifted in the expansion. Thus more work is destroyed in the compression of a gas than is produced in the corresponding expansion. The single-stage compression of a gas illustrates this point.

The system is the same as before—a gas, kept at a constant temperature T—but now the initial state is the expanded state T, p_2, V_2, while the final state is the compressed state T, p_1, V_1. The positions of the stops are arranged so that the piston rests on top of them. Figure 7.4(a and b) shows that if the gas is to be compressed to the final volume V_1 in one stage, we must choose a mass large enough to produce an opposing pressure P_{op} which is *at least* as great as the final pressure p_1. The mass may be larger than this but not smaller. If we choose the mass M to be equivalent to $P_{op} = p_1$, then the work destroyed is equal to the area of the shaded rectangle in Fig. 7.4(c) with, of course, a negative sign:

$$W = P_{op}(V_1 - V_2).$$

The work destroyed in the single-stage compression is very much greater than the work

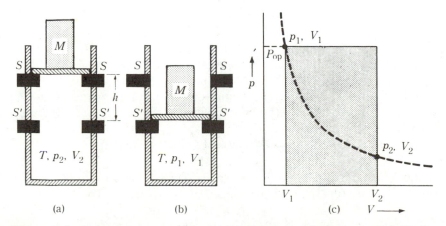

Figure 7.4 Single-stage compression. (a) Initial state. (b) Final state. (c) Work destroyed in a single-stage compression, $W = P_{op}(V_1 - V_2)$.

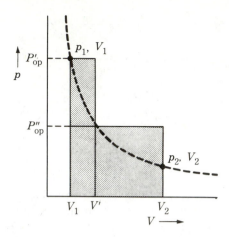

Figure 7.5 Work destroyed in a two-stage compression, $W = P''_{op}(V' - V_2) + P'_{op}(V_1 - V')$.

produced in the single-stage expansion (Fig. 7.2c). We could destroy any greater amount of work in this compression by using larger masses.

If the compression is done in two stages, compressing first with a lighter mass to an intermediate volume and then with the heavier mass to the final volume, less work is destroyed; the work destroyed is the area of the shaded rectangles in Fig. 7.5.

7.5 MAXIMUM AND MINIMUM QUANTITIES OF WORK

In the two-stage expansion more work was produced than in the single-stage expansion. It seems reasonable that if the expansion were done in many stages using a large mass in the beginning and making it smaller as the expansion proceeded, even more work should be produced. This is correct, but there is a limitation to the procedure. The masses that we use must not be so large as to compress the system instead of allowing it to expand. By doing the expansion in a progressively larger number of stages, the work produced can be increased up to a definite maximum value.* Correspondingly, the work destroyed in the two-stage compression is less than that destroyed in the single-stage compression. In a multistage compression, even less work is destroyed.

The expansion work is given by

$$W = \int_{V_i}^{V_f} P_{op} \, dV.$$

For the integral to have a maximum value, P_{op} must have the largest possible value at each stage of the process. But if the gas is to expand, P_{op} must be less than the pressure p of the gas. Therefore, to obtain the maximum work, at each stage we adjust the opposing pressure to $P_{op} = p - dp$, a value just infinitesimally less than the pressure of the gas. Then

$$W_m = \int_{V_i}^{V_f} (p - dp) \, dV = \int_{V_i}^{V_f} (p \, dV - dp \, dV),$$

where V_i and V_f are the initial and final volumes. The second term in the integral is an infinitesimal of higher order than the first and so has a limit of zero. Thus for the maximum

---

*   This is true only if the temperature is constant along the path of the change in state. If the temperature is allowed to vary along the path, there is no upper limit on the work produced.

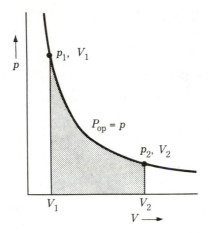

Figure 7.6 W_{max} or W_{min}.

work in expansion

$$W_m = \int_{V_i}^{V_f} p \, dV. \tag{7.5}$$

Similarly, we find the minimum work required for compression by setting the value of P_{op} at each stage just infinitesimally greater than the pressure of the gas; $P_{op} = p + dp$. The argument will obviously yield Eq. (7.5) for the minimum work required for compression if V_i and V_f are the initial and final volumes in the compression. Equation (7.5) is, of course, general and not restricted to gases.

For the ideal gas, the maximum quantity of work produced in the expansion or the minimum destroyed in the compression is equal to the shaded area under the isotherm in Fig. 7.6. For the ideal gas the maximum or minimum work in an isothermal change in state is easily evaluated, since $p = nRT/V$. Using this value for the pressure in Eq. (7.5), we obtain

$$W_{max, min} = \int_{V_i}^{V_f} \frac{nRT}{V} \, dV = nRT \int_{V_i}^{V_f} \frac{dV}{V} = nRT \ln \frac{V_f}{V_i}. \tag{7.6}$$

Under the conditions described, n and T are constant throughout the change and so can be removed from under the integral sign. Note that in expansion $V_f > V_i$, so the logarithm of the ratio is positive; in compression, $V_f < V_i$, the ratio is less than unity so the logarithm is negative. In this way the sign of W takes care of itself.

7.6 REVERSIBLE AND IRREVERSIBLE TRANSFORMATIONS

Consider the same system as before: a quantity of gas confined in a cylinder at a constant temperature T. We expand the gas from the state T, p_1, V_1 to the state T, p_2, V_2 and then we compress the gas to the original state. The gas has been subjected to a *cyclic* transformation returning at the end to its initial state. Suppose that we perform this cycle by two different processes and calculate the net work effect W_{cy} for each process.

Process I. Single-stage expansion with $P_{op} = p_2$; then single-stage compression with $P_{op} = p_1$.

The work produced in the expansion is, by Eq. (7.4),

$$W_{exp} = p_2(V_2 - V_1),$$

while the work produced in the compression is

$$W_{comb} = p_1(V_1 - V_2).$$

The net work effect in the cycle is the sum of these two:

$$W_{cy} = p_2(V_2 - V_1) + p_1(V_1 - V_2) = (p_2 - p_1)(V_2 - V_1).$$

Since $V_2 - V_1$ is positive, and $p_2 - p_1$ is negative, W_{cy} is negative. Net work has been destroyed in this cycle. The system has been restored to its initial state, but the surroundings have not been restored; masses are lower in the surroundings after the cycle.

Process II. The limiting multistage expansion with $P_{op} = p$; then the limiting multistage compression with $P_{op} = p$.

By Eq. (7.5), the work produced in expansion is

$$W_{exp} = \int_{V_1}^{V_2} p\,dV,$$

while the work produced in compression is, by Eq. (7.5),

$$W_{comb} = \int_{V_2}^{V_1} p\,dV.$$

The net work effect in the cycle is

$$W_{cy} = \int_{V_1}^{V_2} p\,dV + \int_{V_2}^{V_1} p\,dV = \int_{V_1}^{V_2} p\,dV - \int_{V_1}^{V_2} p\,dV = 0.$$

(The change in sign of the second integral is effected by interchanging the limits of integration.) If the transformation is conducted by this second method, the system is restored to its initial state, and *the surroundings are also restored to their initial condition*, since no net work effect is produced.

Suppose that a system undergoes a change in state through a specified sequence of intermediate states and then is restored to its original state by traversing the same sequence of states in reverse order. Then if the surroundings are also restored to their original state, the transformation in either direction is *reversible*. The corresponding process is a reversible process. If the surroundings are not restored to their original state after the cycle, the transformation and the process are *irreversible*.

Clearly, the second process just described is a reversible process, while the first is irreversible. There is another important characteristic of reversible and irreversible processes. In the irreversible process just described, a single mass is placed on the piston, the stops are released, and the piston shoots up and settles in the final position. As this occurs the internal equilibrium of the gas is completely upset, convection currents are set up, and the temperature fluctuates. A finite length of time is required for the gas to equilibrate under the new set of conditions. A similar situation prevails in the irreversible compression. This behavior contrasts with the reversible expansion in which at each stage the opposing pressure differs only infinitesimally from the equilibrium pressure in the system, and the volume increases only infinitesimally. In the reversible process the internal equilibrium of the gas is disturbed only infinitesimally and in the limit not at all. Therefore, at any stage in a reversible transformation, the system does not depart from equilibrium by more than an infinitesimal amount.

Obviously, we cannot actually conduct a transformation reversibly. An infinite length of time would be required if the volume increment in each stage were truly infinitesimal. Reversible processes therefore are not real processes, but *ideal ones. Real processes are always irreversible.* With patience and skill the goal of reversibility can be very closely approached, but not attained. Reversible processes are important because the work effects associated with them represent maximum or minimum values. Thus limits are set on the ability of a specified transformation to produce work; in actuality we will get less, but we must not expect to get more.

In the isothermal cycle described above, the net work produced in the irreversible cycle was negative, that is, net work was destroyed. This is a fundamental characteristic of every *irreversible* and therefore every *real* isothermal cyclic transformation. If any system is kept at a constant temperature and subjected to a cyclic transformation by irreversible processes (real processes), a net amount of work is destroyed in the surroundings. This is in fact a statement of the second law of thermodynamics. The greatest work effect will be produced in a reversible isothermal cycle, and this, as we have seen, is $W_{cy} = 0$. Therefore we cannot expect to get a positive amount of work in the surroundings from the cyclic transformation of a system kept at a constant temperature.

Examination of the arguments presented above shows that the general conclusions reached do not depend on the fact that the system chosen for illustration consisted of a gas; the conclusions are valid regardless of how the system is constituted. Therefore to calculate the expansion work produced in the transformation of any system whatsoever we use Eq. (7.4), and to calculate the work produced in the reversible transformation, we set $P_{op} = p$ and use Eq. (7.5).

By appropriate modification of the argument, the general conclusions reached could be shown to be correct for any kind of work: electrical work, work done against a magnetic field, and so on. To calculate the quantities of these other kinds of work we would not, of course, use the integral of pressure over volume, but rather the integral of the appropriate force over the corresponding displacement.

7.7 ENERGY AND THE FIRST LAW OF THERMODYNAMICS

The work produced in a cyclic transformation is the sum of the small quantities of work ${\mathchar'26\mkern-9mu d}W$ produced at each stage of the cycle. Similarly, the heat withdrawn from the surroundings in a cyclic transformation is the sum of the small quantities of heat ${\mathchar'26\mkern-9mu d}Q$ withdrawn at each stage of the cycle. These sums are symbolized by the *cyclic integrals* of ${\mathchar'26\mkern-9mu d}W$ and ${\mathchar'26\mkern-9mu d}Q$:

$$W_{cy} = \oint {\mathchar'26\mkern-9mu d}W, \qquad Q_{cy} = \oint {\mathchar'26\mkern-9mu d}Q.$$

In general, W_{cy} and Q_{cy} are not zero; this is characteristic of path functions.

In contrast, note that if we sum the differential of any *state property* of the system over any cycle the total difference, the cyclic integral, must be zero. Since in any cycle the system returns at the end to its initial state, the total difference in value of any state property must be zero. Conversely, if we find a differential quantity dy such that

$$\oint dy = 0 \qquad \text{(all cycles)}, \tag{7.7}$$

then dy is the differential of some property of the state of the system. This is a purely mathematical theorem, stated here in physical language. Using this theorem and the first law of thermodynamics, we discover the existence of a property of the state of the system, the *energy*.

The first law of thermodynamics is a statement of the following universal experience: *If a system is subjected to any cyclic transformation, the work produced in the surroundings is equal to the heat withdrawn from the surroundings.* In mathematical terms, the first law states that

$$\oint dW = \oint dQ \quad \text{(all cycles)}. \tag{7.8}$$

The system suffers no net change in the cycle, but the condition of the surroundings changes. If masses in the surroundings are higher after the cycle than before, then some bodies in the surroundings must be colder. If masses are lower, then some bodies must be hotter.

Rearranging Eq. (7.8), we have

$$\oint (dQ - dW) = 0 \quad \text{(all cycles)}. \tag{7.9}$$

But if Eq. (7.9) is true, the mathematical theorem requires that the quantity under the integral sign must be the differential of some property of the state of the system. This property of the state is called the *energy*, U, of the system; the differential is dU, defined by

$$dU \equiv dQ - dW; \tag{7.10}$$

then, of course,

$$\oint dU = 0 \quad \text{(all cycles)}. \tag{7.11}$$

Thus from the first law, which relates the heat and work effects observed in the surroundings in a cyclic transformation, we deduce the existence of a property of the state of the system, the energy. Equation (7.10) is an equivalent statement of the first law.

Equation (7.10) shows that when small amounts of heat and work dQ and dW appear at the boundary, the energy of the system suffers a change dU. For a finite change in state, we integrate Eq. (7.10):

$$\int_i^f dU = \int_i^f dQ - \int_i^f dW,$$

$$\Delta U = Q - W, \tag{7.12}$$

where $\Delta U = U_{\text{final}} - U_{\text{initial}}$. Note that only a difference in energy dU or ΔU has been defined, so we can calculate the difference in energy in a change in state, but we cannot assign an *absolute* value to the energy of the system in any particular state.

We can show that energy is conserved in any change in state. Consider an arbitrary transformation in a system A; then

$$\Delta U_{\text{A}} = Q - W$$

where Q and W are the heat and work effects that are manifested in the immediate surroundings by temperature changes of bodies and altitude changes of masses. It is possible to choose a boundary that encloses both system A and its immediate surroundings, and such that no effect resulting from the transformation in A is observable outside this boundary. This boundary separates a new composite system—made up of the original system, A, and of M, the immediate surroundings—from the rest of the universe. Since no heat or work effects are observed outside this composite system, it follows that the energy change of the composite system is zero.

$$\Delta U_{A+M} = 0$$

But the change in energy of the composite system is the sum of the changes in energy of the subsystems, A and M. Thus

$$\Delta U_{A+M} = \Delta U_A + \Delta U_M = 0 \qquad \text{or} \qquad \Delta U_A = -\Delta U_M$$

This equation states that, in any transformation, any increase in energy of system A is exactly balanced by an equal decrease in energy of the surroundings.

It follows that

$$U_A(\text{final}) - U_A(\text{initial}) + U_M(\text{final}) - U_M(\text{initial}) = 0,$$

or

$$U_A(\text{final}) + U_M(\text{final}) = U_A(\text{initial}) + U_M(\text{initial}),$$

which says that the energy of the composite system is constant.

If we imagine the universe to be composed of myriads of such composite systems, in each of which $\Delta U = 0$, then in the aggregate it must also be that $\Delta U = 0$. Thus we have the famous statement of the first law of thermodynamics by Clausius: "The energy of the universe is a constant."

7.8 PROPERTIES OF THE ENERGY

For a specified change in state, the increase in energy ΔU of the system depends only on the initial and final states of the system and not upon the path connecting those states. Both Q and W depend upon the path, but their difference $Q - W = \Delta U$ is independent of the path. This is equivalent to the statement that dQ and dW are inexact differentials, while dU is an exact differential.

The energy is an *extensive* state property of the system; under the same conditions of T and p, 10 mol of the substance composing the system has ten times the energy of 1 mol. The energy per mole is an intensive state property of the system.

Energy is conserved in all transformations. A perpetual motion machine of the first kind is a machine that by its action creates energy by some transformation of a system. The first law of thermodynamics asserts that it is impossible to construct such a machine; not that people have not tried! No one has ever succeeded, but there have been some famous frauds in this field.

7.9 MATHEMATICAL INTERLUDE; EXACT AND INEXACT DIFFERENTIALS

An exact differential integrates to a finite difference, $\int_1^2 dU = U_2 - U_1$, which is independent of the path of integration. An inexact differential integrates to a total quantity, $\int_1^2 dQ = Q$, which depends on the path of integration. The cyclic integral of an exact differential is zero for any cycle, Eq. (7.7). The cyclic integral of an inexact differential is usually not zero.

Note that the symbolism ΔQ and ΔW is *meaningless*. If ΔW meant anything, it would mean $W_2 - W_1$; but the system in either the initial state or the final state does not have any work W_1 or W_2, nor does it have any heat Q_1 or Q_2. Work and heat appear *during a change* in state; they are not properties of the state, but properties of the path.

Properties of the state of a system, such as T, p, V, U, have differentials that are exact. Differentials of properties of the path, such as Q and W, are inexact. For more properties of exact and inexact differentials see Section 9.6.

7.10 CHANGES IN ENERGY IN RELATION TO CHANGES IN PROPERTIES OF THE SYSTEM

Using the first law in the form

$$\Delta U = Q - W,$$

we can calculate ΔU for the change in state from the measured values of Q and W, the effects in the surroundings. However, a change in state of the system implies changes in properties of the system, such as T and V. These properties of the system are readily measurable in the initial and final states, and it is useful to relate the change in energy of the system to, let us say, changes in its temperature and volume. It is to this problem that we now direct our attention.

Choosing a system of fixed mass, we can describe the state by T and V. Then $U = U(T, V)$, and the change in energy dU is related to the changes in temperature dT and in volume dV through the total differential expression

$$dU = \left(\frac{\partial U}{\partial T}\right)_V dT + \left(\frac{\partial U}{\partial V}\right)_T dV. \tag{7.13}$$

The differential of any state property, any exact differential, can be written in the form of Eq. (7.13). (See Appendix I.) Expressions of this sort are used so often that it is essential to understand their physical and mathematical meaning. Equation (7.13) states that if the temperature of the system increases by an amount dT and the volume increases by an amount dV, then the total increase in energy dU is the sum of two contributions: the first term, $(\partial U/\partial T)_V \, dT$, is the increase in energy resulting from the temperature increase alone; the second term, $(\partial U/\partial V)_T \, dV$, is the increase in energy resulting from the volume increase alone. The first term is the rate of increase of energy with temperature at constant volume, $(\partial U/\partial T)_V$, multiplied by the increase in temperature dT. The second term is interpreted in an analogous way. Each time an expression of this kind appears, the effort should be made to give this interpretation to each term until it becomes a habit. The habit of reading a physical meaning into an equation will help enormously in clarifying the derivations that follow.

Since energy is an important property of the system, the partial derivatives $(\partial U/\partial T)_V$ and $(\partial U/\partial V)_T$ are also important properties of the system. These derivatives tell us the rate of change of energy with temperature at constant volume, or with volume at constant temperature. If the values of these derivatives are known, we can integrate Eq. (7.13) and obtain the change in energy from the change of temperature and volume of the system. Therefore we must express these derivatives in terms of measurable quantities.

We begin by combining Eqs. (7.10) and (7.13) to obtain

$$dQ - P_{op} \, dV = \left(\frac{\partial U}{\partial T}\right)_V dT + \left(\frac{\partial U}{\partial V}\right)_T dV, \tag{7.14}$$

where $P_{op} \, dV$ has replaced dW, and work other than expansion work has been ignored. (If other kinds of work must be included, we set $dW = P_{op} \, dV + dW_a$, where dW_a represents the small amounts of other kinds of work.) Next we apply Eq. (7.14) to various changes in state.

7.11 CHANGES IN STATE AT CONSTANT VOLUME

If the volume of a system is constant in the change in state, then $dV = 0$, and the first law, Eq. (7.10), becomes

$$dU = dQ_V, \tag{7.15}$$

where the subscript indicates the restriction to constant volume. But at constant volume, Eq. (7.14) becomes

$$dQ_V = \left(\frac{\partial U}{\partial T}\right)_V dT, \tag{7.16}$$

which relates the heat withdrawn from the surroundings, dQ_V, to the increase in temperature dT of the system at constant volume. Both dQ_V and dT are easily measurable; the ratio, dQ_V/dT, of the heat withdrawn from the surroundings to the temperature increase of the system is C_v, the heat capacity of the system at constant volume. Thus, dividing Eq. (7.16) by dT, we obtain

$$C_v \equiv \frac{dQ_V}{dT} = \left(\frac{\partial U}{\partial T}\right)_V. \tag{7.17}$$

Either member of Eq. (7.17) is an equivalent definition of C_v. The important point about Eq. (7.17) is that it identifies the partial derivative $(\partial U/\partial T)_V$ with an easily measurable quantity C_v. Using C_v for the derivative in Eq. (7.13), and since $dV = 0$, we obtain

$$dU = C_v\, dT \qquad \text{(infinitesimal change)}, \tag{7.18}$$

or, integrating, we have

$$\Delta U = \int_{T_1}^{T_2} C_v\, dT \qquad \text{(finite change)}. \tag{7.19}$$

Using Eq. (7.19) we can calculate ΔU exclusively from properties of the system. Integrating Eq. (7.15), we obtain the additional relation

$$\Delta U = Q_V \qquad \text{(finite change)}. \tag{7.20}$$

Both Eqs. (7.19) and (7.20) express the energy change in a transformation at constant volume in terms of measurable quantities. These equations apply to any system: solids, liquids, gases, mixtures, old razor blades, and so on.

Note in Eq. (7.20) that ΔU and Q_V have the same sign. According to the convention for Q, if heat flows *from* the surroundings, $Q_V > 0$, and so $\Delta U > 0$; the energy of the system increases. If heat flows *to* the surroundings, both Q_V and ΔU are negative; the energy of the system decreases. Furthermore, since C_v is always positive, Eq. (7.18) shows that if the temperature increases, $dT > 0$, the energy of the system increases; conversely, a decrease in temperature, $dT < 0$, means a decrease in the energy of the system, $\Delta U < 0$. For a system maintained at a constant volume, the temperature is a direct reflection of the energy of the system.

Since the energy of the system is an extensive state property, the heat capacity is also. The heat capacity per mole $\bar{C}$, an intensive property, is the quantity found in tables of data. If the heat capacity of the system is a constant in the range of temperature of interest, then Eq. (7.19) reduces to the special form

$$\Delta U = C_v\, \Delta T. \tag{7.21}$$

This equation is quite useful, particularly if the temperature range ΔT is not very large. Over short ranges of temperature the heat capacity of most substances does not change very much.

Although Eqs. (7.19) and (7.20) are completely general for a constant-volume process, a practical difficulty arises if the system consists entirely of solids or liquids. If a liquid or a solid is confined in a container of fixed volume and the temperature is increased by a small amount, the pressure rises to a high value because of the very small compressibility of the liquid. Any ordinary container will be deformed and increase in volume or it will burst. From the experimental standpoint, constant volume processes are practical only for those systems which are, at least partly, gaseous.

■ **EXAMPLE 7.1**   Calculate the ΔU and Q_V for the transformation of 1 mol of helium at constant volume from 25 °C to 45 °C; $\bar{C}_v = \frac{3}{2}R$.

At constant volume

$$\Delta U = \int_{T_1}^{T_2} C_v \, dT = \tfrac{3}{2}R \int_{T_1}^{T_2} dT = \tfrac{3}{2}R \, \Delta T = \tfrac{3}{2}R(20\,\mathrm{K})$$

$$Q_V = \Delta U = \tfrac{3}{2}(8.314\ \mathrm{J/K\ mol})(20\ \mathrm{K}) = 250\ \mathrm{J/mol}.$$

7.12 MEASUREMENT OF $(\partial U/\partial V)_T$; JOULE'S EXPERIMENT

The identification of the differential coefficient $(\partial U/\partial V)_T$ with readily measurable quantities is not so easily managed. For gases it can be done, in principle at least, by an experiment devised by Joule. Two containers A and B are connected through a stopcock. In the initial state, A is filled with a gas at a pressure p, while B is evacuated. The apparatus is immersed in a large vat of water and is allowed to equilibrate with the water at the temperature T, which is read on the thermometer (Fig. 7.7). The water is stirred vigorously to hasten the attainment of thermal equilibrium. The stopcock is opened and the gas expands to fill the containers A and B uniformly. After allowing time for the system to come to thermal equilibrium with the water in the vat, the temperature of the water is read again. Joule observed no temperature difference in the water before and after opening the stopcock.

The interpretation of this experiment is as follows. To begin with, no work is produced in the surroundings. The boundary, which is initially along the interior walls of vessel A, moves in such a way that it always encloses the entire mass of gas; the boundary therefore expands against zero opposing pressure so no work is produced. This is called a *free expansion* of the gas. Setting $đW = 0$, we see that the first law becomes $dU = đQ$.

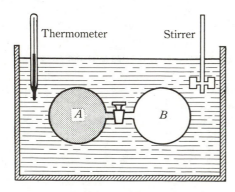

Figure 7.7   Joule expansion experiment.

Since the temperature of the surroundings (the water) is unchanged, it follows that $dQ = 0$. Hence, $dU = 0$. Since the system and the water are in thermal equilibrium, the temperature of the system is also unchanged; $dT = 0$. In this situation, Eq. (7.13) becomes

$$dU = \left(\frac{\partial U}{\partial V}\right)_T dV = 0.$$

Since $dV \neq 0$, it follows that

$$\left(\frac{\partial U}{\partial V}\right)_T = 0. \tag{7.22}$$

If the derivative of energy with respect to volume is zero, the energy is independent of the volume. This means that the energy of the gas is a function only of temperature. This rule of behavior is *Joule's law*, which may be expressed either by Eq. (7.22) or by $U = U(T)$.

Later experiments, notably the Joule–Thomson experiment, have shown that Joule's law is not precisely correct for real gases. In Joule's apparatus the large heat capacity of the vat of water and the small heat capacity of the gas reduced the magnitude of the effect below the limits of observation. For real gases, the derivative $(\partial U/\partial V)_T$ is a very small quantity, usually positive. The ideal gas obeys Joule's law exactly.

Until we have the equations from the second law of thermodynamics, the problem of identifying the derivative $(\partial U/\partial V)_T$ with readily measurable quantities is a clumsy procedure at best. The Joule experiment, which does not work very well with gases, is completely unsuitable for liquids and solids. A fortunate circumstance intervenes to simplify matters for liquids and solids. Very great pressures are required to effect even a small change in volume of a liquid or solid kept at a constant temperature. The energy change accompanying an isothermal change in volume of a liquid or solid is, by integrating Eq. (7.13) with $dT = 0$,

$$\Delta U = \int_{V_1}^{V_2} \left(\frac{\partial U}{\partial V}\right)_T dV.$$

The initial and final volumes V_1 and V_2 are so nearly equal that the derivative is constant over this small range of volume; removing it from under the integral sign and integrating dV, the equation becomes

$$\Delta U = \left(\frac{\partial U}{\partial V}\right)_T \Delta V. \tag{7.23}$$

Even though for liquids and solids the value of the derivative is very large, the value of ΔV is so small that the product in Eq. (7.23) is very nearly zero. Consequently, to a good approximation the energy of all substances can be considered to be a function of temperature only. The statement is precisely true only for the ideal gas. To avoid errors in derivations the derivative will be carried along. Having identified $(\partial U/\partial T)_V$ with C_v, we shall, from now on, write the total derivative of U, Eq. (7.13), in the form

$$dU = C_v\, dT + \left(\frac{\partial U}{\partial V}\right)_T dV. \tag{7.24}$$

7.13 CHANGES IN STATE AT CONSTANT PRESSURE

In laboratory practice most changes in state are carried out under a constant atmospheric pressure, which is equal to the pressure of the system. The change in state at constant pressure can be envisioned by confining the system to a cylinder closed by a weighted

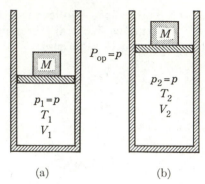

Figure 7.8 Change in state at constant pressure. (a) Initial state. (b) Final state.

(a)   (b)

piston that floats freely (Fig. 7.8), rather than being held in some position by a set of stops. Since the piston floats freely, its equilibrium position is determined by the balance of the opposing pressure developed by the mass M and the pressure in the system. No matter what we do to the system, the piston will move until the condition $p = P_{op}$ is fulfilled. The pressure p in the system may be adjusted to any constant value by appropriately adjusting the mass M. Under ordinary laboratory conditions the mass of the column of air above the system floats on top of the system and maintains the pressure at the constant value p.

Since $P_{op} = p$, for a change in state at constant pressure the first-law statement becomes

$$dU = dQ_p - p\,dV. \tag{7.25}$$

Since p is constant, this integrates at once to yield

$$\int_1^2 dU = \int_1^2 dQ_p - \int_{V_1}^{V_2} p\,dV,$$

$$U_2 - U_1 = Q_p - p(V_2 - V_1).$$

Rearranging, we obtain

$$(U_2 + pV_2) - (U_1 + pV_1) = Q_p. \tag{7.26}$$

Since $p_1 = p_2 = p$, in Eq. (7.26), the first p can be replaced by p_2, the second by p_1:

$$(U_2 + p_2 V_2) - (U_1 + p_1 V_1) = Q_p. \tag{7.27}$$

Since the pressure and the volume of the system depend only on the state, the product pV depends only on the state of the system. The function $U + pV$, being a combination of state variables, is itself a state variable H. We define

$$H \equiv U + pV; \tag{7.28}$$

H is called the *enthalpy* of the system,* an extensive state property.

Using the definition of H, we can rewrite Eq. (7.27) as $H_2 - H_1 = Q_p$, or

$$\Delta H = Q_p, \tag{7.29}$$

which shows that in a constant pressure process the heat withdrawn from the surroundings

---

* It is worthwhile noting that the appearance of the product pV in the definition of enthalpy results from the algebraic form for the expansion work; it has nothing to do with the presence of the pV product in the ideal gas law!

is equal to the increase in enthalpy of the system. Ordinarily, heat effects are measured at constant pressure; therefore these heat effects indicate changes in enthalpy of the system, not changes in its energy. To compute the change in energy in a constant pressure process, Eq. (7.26) is written as

$$\Delta U + p\Delta V = Q_p. \tag{7.30}$$

Knowing Q_p and the change in volume ΔV, we can calculate the value of ΔU.

Equation (7.29) finds immediate application to the vaporization of a liquid under a constant pressure and at a constant temperature. The heat withdrawn from the surroundings is the heat of vaporization Q_{vap}. Since the transformation is done at constant pressure, $Q_{vap} = \Delta H_{vap}$. Similarly, the heat of fusion of a solid is the enthalpy increase in fusion: $Q_{fus} = \Delta H_{fus}$.

For an infinitesimal change in state of a system, Eq. (7.29) takes the form

$$dH = dQ_p. \tag{7.31}$$

Since H is a state function, dH is an exact differential; choosing T and p as convenient variables for H, we can write the total differential as

$$dH = \left(\frac{\partial H}{\partial T}\right)_p dT + \left(\frac{\partial H}{\partial p}\right)_T dp. \tag{7.32}$$

For a transformation at constant pressure, $dp = 0$, and Eq. (7.32) becomes $dH = (\partial H/\partial T)_p \, dT$. Combining this with Eq. (7.31) yields

$$dQ_p = \left(\frac{\partial H}{\partial T}\right)_p dT,$$

which relates the heat withdrawn from the surroundings to the temperature increment of the system. The ratio, dQ_p/dT, is C_p, the heat capacity of the system at constant pressure. Hence, we have

$$C_p \equiv \frac{dQ_p}{dT} = \left(\frac{\partial H}{\partial T}\right)_p, \tag{7.33}$$

which identifies the important partial derivative $(\partial H/\partial T)_p$ with the measurable quantity C_p. From this point on, the total differential in Eq. (7.32) will be written in the form

$$dH = C_p \, dT + \left(\frac{\partial H}{\partial p}\right)_T dp. \tag{7.34}$$

For any constant pressure transformation, since $dp = 0$, Eq. (7.34) reduces to

$$dH = C_p \, dT, \tag{7.35}$$

or for a finite change in state from T_1 to T_2,

$$\Delta H = \int_{T_1}^{T_2} C_p \, dT. \tag{7.36}$$

If C_p is constant in the temperature range of interest, Eq. (7.36) becomes

$$\Delta H = C_p \, \Delta T. \tag{7.37}$$

The equations in this section are quite general and are applicable to any transformation at constant pressure of any system of fixed mass, provided no phase changes or chemical reactions occur.

■ **EXAMPLE 7.2**   For silver, $\bar{C}_p/(\text{J/K mol}) = 23.43 + 0.00628T$. Calculate ΔH if 3 mol of silver are raised from 25 °C to the melting point, 961 °C, under 1 atm pressure.

At constant p for 1 mol, $\Delta H = \int_{T_1}^{T_2} C_p \, dT = \int_{T_1}^{T_2} (23.43 + 0.00628T) \, dT.$

$\Delta H = 23.43(T_2 - T_1) + \frac{1}{2}(0.00628)(T_2^2 - T_1^2) \text{ J/mol}.$

Since $T_1 = 273.15 \text{ K} + 25 \text{ K} = 298.15 \text{ K}$ and $T_2 = 273.15 \text{ K} + 961 \text{ K} = 1234.15 \text{ K}$, $T_2 - T_1 = 936 \text{ K}$.

$\Delta H = 23.43(936) + \frac{1}{2}(0.00628)(1234^2 - 298^2) = 21\,930 + 4500 = 26\,430 \text{ J/mol}.$

For 3 mol, $\Delta H = 3 \text{ mol}(26\,430 \text{ J/mol}) = 79\,290 \text{ J}.$

7.14 THE RELATION BETWEEN C_p AND C_v

For a specified change in state of a system that has a definite temperature change dT associated with it, the heat withdrawn from the surroundings may have different values, since it depends upon the path of the change in state. Therefore it is not surprising that a system has more than one value of heat capacity. In fact, the heat capacity of a system may have any value from minus infinity to plus infinity. Only two values, C_p and C_v, have major importance, however. Since they are not equal, it is important to find the relation between them.

We attack this problem by calculating the heat withdrawn at constant pressure using Eq. (7.14) in the form

$$dQ = C_v \, dT + \left(\frac{\partial U}{\partial V}\right)_T dV + P_{\text{op}} \, dV.$$

For a change at constant pressure with $P_{\text{op}} = p$, this equation becomes

$$dQ_p = C_v \, dT + \left[p + \left(\frac{\partial U}{\partial V}\right)_T\right] dV.$$

Since $C_p = dQ_p/dT$, we divide by dT and obtain

$$C_p = C_v + \left[p + \left(\frac{\partial U}{\partial V}\right)_T\right]\left(\frac{\partial V}{\partial T}\right)_p, \tag{7.38}$$

which is the required relation between C_p and C_v. It is usually written in the form

$$C_p - C_v = \left[p + \left(\frac{\partial U}{\partial V}\right)_T\right]\left(\frac{\partial V}{\partial T}\right)_p. \tag{7.39}$$

This equation is a general relation between C_p and C_v. It will be shown later that the quantity on the right-hand side is always positive; thus C_p is always larger than C_v for any substance. The excess of C_p over C_v is made up of a sum of two terms. The first term,

$$p\left(\frac{\partial V}{\partial T}\right)_p,$$

is the work produced, $p \, dV$, per unit increase in temperature in the constant pressure process. The second term,

$$\left(\frac{\partial U}{\partial V}\right)_T\left(\frac{\partial V}{\partial T}\right)_p,$$

is the energy required to pull the molecules farther apart against the attractive inter-molecular forces.

If a gas expands, the average distance between the molecules increases. A small amount of energy must be supplied to the gas to pull the molecules to this greater separation against the attractive forces; the energy required per unit increase in volume is given by the derivative $(\partial U/\partial V)_T$. In a constant volume process, no work is produced and the average distance between the molecules remains the same. Therefore the heat capacity is small; all of the heat withdrawn goes into the chaotic motion and is reflected by a temperature increase. In a constant pressure process, the system expands against the resisting pressure and produces work in the surroundings; the heat withdrawn from the surroundings is divided into three portions. The first portion produces work in the surroundings; the second portion provides the energy necessary to separate the molecules farther; the third portion goes into increasing the energy of the chaotic motion. Only this last portion is reflected by a temperature increase. To produce a temperature increment of one degree, more heat must be withdrawn in the constant pressure process than is withdrawn in the constant volume process. Thus C_p is greater than C_v.

Another useful quantity is the heat capacity ratio, γ, defined by

$$\gamma \equiv \frac{C_p}{C_v}. \tag{7.40}$$

From what has been said, it is clear that γ is always greater than unity.

The heat capacity difference for the ideal gas has a particularly simple form because $(\partial U/\partial V)_T = 0$ (Joule's law). Then Eq. (7.39) is

$$C_p - C_v = p\left(\frac{\partial V}{\partial T}\right)_p. \tag{7.41}$$

If we are speaking of molar heat capacities, the volume in the derivative is the molar volume; from the equation of state, $\overline{V} = RT/p$. Differentiating with respect to temperature, keeping the pressure constant, yields $(\partial \overline{V}/\partial T)_p = R/p$. Putting this value in Eq. (7.41) reduces it to the simple result

$$\overline{C}_p - \overline{C}_v = R. \tag{7.42}$$

Although Eq. (7.42) is precisely correct only for the ideal gas, it is a useful approximation for real gases.

The heat capacity difference for liquids and solids is usually rather small, and except in work requiring great accuracy it is sufficient to take

$$C_p = C_v, \tag{7.43}$$

although there are some notable exceptions to this rule. The physical reason for the approximate equality of C_p and C_v is plain enough. The thermal expansion coefficients of liquids and solids are very small, so that the volume change on increasing the temperature by one degree is very small; correspondingly, the work produced by the expansion is small and little energy is required for the small increase in the spacing of the molecules. Almost all of the heat withdrawn from the surroundings goes into increasing the energy of the chaotic motion, and so is reflected in a temperature increase which is nearly as large as that in a constant volume process. For the reasons mentioned at the end of Section 7.11, it is impractical to measure the C_v of a liquid or solid directly; C_p is readily measurable. The tabulated values of heat capacities of liquids and solids are values of C_p.

7.15 THE MEASUREMENT OF $(\partial H/\partial p)_T$; JOULE–THOMSON EXPERIMENT

The identification of the partial derivative $(\partial H/\partial p)_T$ with quantities that are readily accessible to experiment is beset with the same difficulties we experienced with $(\partial U/\partial V)_T$ in Section 7.12. These two derivatives are related. By differentiating the definition $H = U + pV$, we obtain

$$dH = dU + p\,dV + V\,dp.$$

Introducing the values of dH and dU from Eqs. (7.24) and (7.34), we have

$$C_p\,dT + \left(\frac{\partial H}{\partial p}\right)_T dp = C_v\,dT + \left[\left(\frac{\partial U}{\partial V}\right)_T + p\right]dV + V\,dp. \qquad (7.44)$$

Restricting this formidable equation to constant temperature, $dT = 0$, and dividing by dp, we can simplify it to

$$\left(\frac{\partial H}{\partial p}\right)_T = \left[p + \left(\frac{\partial U}{\partial V}\right)_T\right]\left(\frac{\partial V}{\partial p}\right)_T + V, \qquad (7.45)$$

which is at best a clumsy equation.

For liquids and solids the first term on the right-hand side of Eq. (7.45) is ordinarily very much smaller than the second term, so that a good approximation is

$$\left(\frac{\partial H}{\partial p}\right)_T = V \qquad \text{(solids and liquids).} \qquad (7.46)$$

Since the molar volume of liquids and solids is very small, the variation of the enthalpy with pressure can be ignored—unless the change in pressure is enormous.

For the ideal gas,

$$\left(\frac{\partial H}{\partial p}\right)_T = 0. \qquad (7.47)$$

This result is most easily obtained from the definition $H = U + pV$. For the ideal gas, $p\overline{V} = RT$, so that

$$\overline{H} = \overline{U} + RT. \qquad (7.48)$$

Since the energy of the ideal gas is a function of temperature only, by Eq. (7.48) the enthalpy is a function of temperature only, and is independent of pressure. The result in Eq. (7.47) could also be obtained from Eq. (7.45) and Joule's law.

The derivative $(\partial H/\partial p)_T$ is very small for real gases, but can be measured. The Joule experiment, in which the gas expanded freely, failed to show a measurable difference in temperature between the initial and final states. Later, Joule and Thomson performed a different experiment, the Joule–Thomson experiment (Fig. 7.9).

A steady flow of gas passes through an insulated pipe in the direction of the arrows; at position A there is an obstruction, which may be a porous disc or a diaphragm with a small hole in it or, as in the original experiment, a silk handkerchief. Because of the obstruction there is a drop in pressure, measured by the gauges M and M', in passing from the left to the right side. Any drop in temperature is measured by the thermometers t and t'. The boundary of the system moves with the gas, always enclosing the same mass. Consider the passage of one mole of gas through the obstruction. The volume on the left decreases by the molar volume $\overline{V}_1$; since the gas is pushed by the gas behind it, which

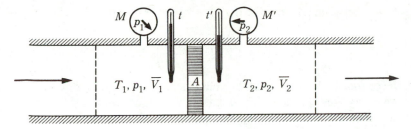

Figure 7.9   The Joule–Thomson experiment.

exerts a pressure p_1, the work produced is

$$W_{\text{left}} = \int_{\overline{V}_1}^{0} p_1 \, dV$$

The volume on the right increases by the molar volume $\overline{V}_2$; the gas coming through must push the gas ahead of it, which exerts an opposing pressure, p_2. The work produced is

$$W_{\text{right}} = \int_{0}^{\overline{V}_2} p_2 \, dV$$

The net work produced is the sum of these two amounts

$$W = \int_{\overline{V}_1}^{0} p_1 \, dV + \int_{0}^{\overline{V}_2} p_2 \, dV = p_1(-\overline{V}_1) + p_2 \overline{V}_2 = p_2 \overline{V}_2 - p_1 \overline{V}_1$$

Since the pipe is insulated, $Q = 0$, and we have the first-law statement

$$\overline{U}_2 - \overline{U}_1 = Q - W = -(p_2 \overline{V}_2 - p_1 \overline{V}_1).$$

Rearranging, we have

$$\overline{U}_2 + p_2 \overline{V}_2 = \overline{U}_1 + p_1 \overline{V}_1, \qquad \overline{H}_2 = \overline{H}_1.$$

The enthalpy of the gas is a constant in the Joule–Thomson expansion. The measured decrease in temperature $-\Delta T$ and the measured decrease in pressure $-\Delta p$ are combined in the ratio

$$\left(\frac{-\Delta T}{-\Delta p}\right)_H = \left(\frac{\Delta T}{\Delta p}\right)_H.$$

The Joule–Thomson coefficient μ_{JT} is defined as the limiting value of this ratio as Δp approaches zero:

$$\mu_{JT} = \left(\frac{\partial T}{\partial p}\right)_H. \tag{7.49}$$

The drop in temperature (Joule–Thomson effect) is easily measurable in this experiment, particularly if the pressure difference is large. A noisy but dramatic demonstration of this effect can be made by partially opening the valve on a tank of compressed nitrogen; after a few minutes the valve is cold enough to form a coating of snow by condensing moisture from the air. (This should not be done with hydrogen or oxygen because of the explosion or fire hazard!) If the tank of gas is nearly full, the driving pressure is about 150 atm and the exit pressure is 1 atm. With this pressure drop, the temperature drop is quite large.

The relation between μ_{JT} and the derivative $(\partial H/\partial p)_T$ is simple. The total differential of H,

$$dH = C_p\, dT + \left(\frac{\partial H}{\partial p}\right)_T dp,$$

expresses the change in H in terms of the changes in T and in p. It is possible to change T and p in such a way that H remains unchanged if we impose the condition $dH = 0$. Under this condition the relation becomes

$$0 = C_p\, dT + \left(\frac{\partial H}{\partial p}\right)_T dp.$$

Dividing by dp, we obtain

$$0 = C_p \left(\frac{\partial T}{\partial p}\right)_H + \left(\frac{\partial H}{\partial p}\right)_T.$$

Using the definition of μ_{JT} and rearranging, we have

$$\left(\frac{\partial H}{\partial p}\right)_T = -C_p \mu_{JT}. \tag{7.50}$$

Thus, if we measure C_p and μ_{JT}, the value of $(\partial H/\partial p)_T$ can be calculated from Eq. (7.50). Note that by combining Eqs. (7.50) and (7.45), a value of $(\partial U/\partial V)_T$ can be obtained in terms of measurable quantities.

The Joule–Thomson coefficient is positive at and below room temperature for all gases, except hydrogen and helium, that have negative Joule–Thomson coefficients. These two gases will be hotter after undergoing this kind of expansion. Every gas has a characteristic temperature above which the Joule–Thomson coefficient is negative, the Joule–Thomson inversion temperature. The inversion temperature for hydrogen is about -80 °C: below this temperature hydrogen will cool in a Joule–Thomson expansion. The inversion temperatures of most gases are very much higher than room temperature.

The Joule–Thomson effect can be used as the basis for a refrigerating device. The cooled gas on the low-pressure side is passed back over the high-pressure line to reduce the temperature of the gas before it is expanded; repetition of this can reduce the temperature on the high-pressure side to quite low values. If the temperature is low enough, then on expansion the temperature falls below the boiling point of the substance and drops of liquid are produced. This procedure is used in the Linde method for producing liquid air. The ordinary household refrigerator has a high- and a low-pressure side separated by an expansion valve, but the cooling results from the evaporation of a liquid refrigerant on the low-pressure side; the refrigerant is liquefied by compression on the high-pressure side.

7.16 ADIABATIC CHANGES IN STATE

If no heat flows during a change in state, $dQ = 0$, and the change in state is *adiabatic*. Experimentally we approximate this condition by wrapping the system in a layer of insulating material or by using a vacuum bottle to contain the system. For an adiabatic change in state, since $dQ = 0$, the first law statement is

$$dU = -dW, \tag{7.51}$$

or, for a finite change in state,

$$\Delta U = -W. \tag{7.52}$$

Turning Eq. (7.52) around, we find that $W = -\Delta U$, which means that the work produced, W, is at the expense of a decrease in energy of the system, $-\Delta U$. A decrease in energy in a system is evidenced almost entirely by a decrease in temperature of the system; hence, if work is produced in an adiabatic change in state, the temperature of the system falls. If work is destroyed in the adiabatic change, W is $-$, and ΔU is $+$; the work destroyed increases the energy and the temperature of the system.

If only pressure–volume work is involved, Eq. (7.51) becomes

$$dU = -P_{op}\, dV, \tag{7.53}$$

from which it is clear that in expansion dV is $+$ and dU is $-$; the energy decreases and so does the temperature. If the system is compressed adiabatically, dV is $-$, and dU is $+$; the energy and the temperature both increase.

7.16.1 Special Case: Adiabatic Changes in State in the Ideal Gas

Because of Joule's law we have for the ideal gas $dU = C_v\, dT$. Using this relation in Eq. (7.53), we obtain

$$C_v\, dT = -P_{op}\, dV, \tag{7.54}$$

which shows immediately that dT and dV have opposite signs. The drop in temperature is proportional to P_{op}, and for a specified increase in volume will have a maximum value when P_{op} has its maximum value; that is, when $P_{op} = p$. Consequently, for a fixed change in volume, the *reversible adiabatic* expansion will produce the greatest drop in temperature; conversely, a reversible adiabatic compression between two specified volumes produces the least increase in temperature.

For a reversible adiabatic change in state of the ideal gas, $P_{op} = p$, and Eq. (7.54) becomes

$$C_v\, dT = -p\, dV. \tag{7.55}$$

To integrate this equation, C_v and p must be expressed as functions of the variables of integration T and V. Since U is a function only of temperature, C_v is also a function only of temperature; from the ideal gas law, $p = nRT/V$. Equation (7.55) becomes

$$C_v\, dT = -nRT\, \frac{dV}{V}.$$

Dividing by T to separate variables, and using $\bar{C}_v = C_v/n$, we have

$$\bar{C}_v\, \frac{dT}{T} = -R\, \frac{dV}{V}.$$

Describing the initial state by T_1, V_1, the final state by T_2, V_2, and integrating, we have

$$\int_{T_1}^{T_2} \bar{C}_v\, \frac{dT}{T} = -R \int_{V_1}^{V_2} \frac{dV}{V}.$$

If $\bar{C}_v$ is a constant, it can be removed from the integral. Integration yields

$$\bar{C}_v \ln\left(\frac{T_2}{T_1}\right) = -R \ln \frac{V_2}{V_1}. \tag{7.56}$$

Since $R = \bar{C}_p - \bar{C}_v$, then $R/\bar{C}_v = (\bar{C}_p/\bar{C}_v) - 1 = \gamma - 1$. This value of $R/\bar{C}_v$ reduces

Eq. (7.56) to

$$\ln \left(\frac{T_2}{T_1} \right) = -(\gamma - 1) \ln \left(\frac{V_2}{V_1} \right),$$

which can be written

$$\frac{T_2}{T_1} = \left(\frac{V_1}{V_2} \right)^{\gamma - 1}$$

or

$$T_1 V_1^{\gamma - 1} = T_2 V_2^{\gamma - 1}. \tag{7.57}$$

Using the ideal gas law, we can transform this equation to the equivalent forms

$$T_1^\gamma p_1^{1-\gamma} = T_2^\gamma p_2^{1-\gamma}, \tag{7.58}$$

$$p_1 V_1^\gamma = p_2 V_2^\gamma. \tag{7.59}$$

Equation (7.59), for example, says that any two states of an ideal gas that can be connected by a reversible adiabatic process fulfill the condition that pV^γ = constant. Equations (7.57) and (7.58) can be given analogous interpretations. Although these equations are rather specialized, occasional use will be made of them.

7.17 A NOTE ABOUT PROBLEM WORKING

So far we have more than fifty equations. Working a problem would be a terrible task if we had to search through such a bewildering array of equations in the hope of quickly finding the right one. Thus only the fundamental equations should be used in application to any problem. The conditions set in the problem immediately limit these fundamental equations to simple forms from which it should be clear how to calculate the "unknowns" in the problem. So far we have only seven fundamental equations:

1. The formula for expansion work: $dW = P_{\text{op}} \, dV$.
2. The definition of energy: $dU = dQ - dW$.
3. The definition of enthalpy: $H = U + pV$.
4. The definition of the heat capacities:

$$C_v = \frac{dQ_V}{dT} = \left(\frac{\partial U}{\partial T} \right)_V, \qquad C_p = \frac{dQ_p}{dT} = \left(\frac{\partial H}{\partial T} \right)_p.$$

5. The purely mathematical consequences:

$$dU = C_v \, dT + \left(\frac{\partial U}{\partial V} \right)_T dV, \qquad dH = C_p \, dT + \left(\frac{\partial H}{\partial p} \right)_T dp.$$

Of course, it is essential to understand the meaning of these equations and the meaning of such terms as isothermal, adiabatic, and reversible. These terms have definite mathematical consequences to the equations. For problems involving the ideal gas, the equation of state, the mathematical consequences of Joule's law, and the relation between the heat capacities should be known. The equations that solve each problem must be derived from these few fundamental equations. Other methods of attack, such as attempting to memorize as many equations as possible, produce only panic, paralysis, and paranoia.

■ **EXAMPLE 7.3**   An ideal gas, $\bar{C}_v = \frac{3}{2}R$, is expanded adiabatically against a constant pressure of 1 atm until it doubles in volume. If the initial temperature is 25 °C, and the initial pressure is 5 atm, calculate T_2; then calculate Q, W, ΔU, and ΔH per mole of gas for the transformation.

Data: Initial state, T_1, p_1, V_1.   Final state, T_2, p_2, $2V_1$.

Moles of gas $= n$ (not given): $P_{op} = 1$ atm.

First law: $dU = dQ - P_{op}\, dV$.

Conditions: Adiabatic; therefore $dQ = 0$, and $Q = 0$. Ideal gas, therefore, $dU = C_v\, dT$. These reduce the first law to $C_v\, dT = -P_{op}\, dV$.

Since both C_v and P_{op} are constant, the first law integrates to

$$C_v \int_{T_1}^{T_2} dT = -P_{op} \int_{V_1}^{V_2} dV \qquad \text{or} \qquad C_v(T_2 - T_1) = -P_{op}(V_2 - V_1).$$

Then

$$C_v = n\bar{C}_v = n\tfrac{5}{2}R; \qquad V_2 - V_1 = 2V_1 - V_1 = V_1 = \frac{nRT_1}{p_1}.$$

The first law becomes

$$n\tfrac{5}{2}R(T_2 - T_1) = -\frac{P_{op}\, nRT_1}{p_1}$$

Solving for T_2, we find

$$T_2 = T_1\left(1 - \frac{2P_{op}}{5p_1}\right) = 298\ \text{K}\left[1 - \frac{2}{5}\left(\frac{1\ \text{atm}}{5\ \text{atm}}\right)\right] = 274\ \text{K}.$$

Substituting for T_2,

$$\Delta U = \bar{C}_v(T_2 - T_1) = \tfrac{5}{2}R(274\ \text{K} - 298\ \text{K}) = \tfrac{5}{2}(8.314\ \text{J/K mol})(-24\ \text{K})$$

$$= -500\ \text{J/mol}.$$

Then $W = -\Delta U = 500$ J/mol.

$$\Delta H = \int_{T_1}^{T_2} \bar{C}_p\, dT = (\bar{C}_v + R)(T_2 - T_1)$$

$$= (\tfrac{5}{2}R + R)(-24\ \text{K}) = -\tfrac{7}{2}(8.314\ \text{J/K mol})(24\ \text{K})$$

$$= -700\ \text{J/mol}$$

Note 1: Since the gas is ideal, we set $\bar{C}_p = \bar{C}_v + R$.

Note 2: We do not need the value of n to calculate T_2. Since we were not given n, we can only calculate the value of W, ΔU and ΔH *per mole* of the gas.

7.18 APPLICATION OF THE FIRST LAW OF THERMODYNAMICS TO CHEMICAL REACTIONS. THE HEAT OF REACTION

If a chemical reaction takes place in a system, the temperature of the system immediately after the reaction generally is different from the temperature immediately before the reaction. To restore the system to its initial temperature, heat must flow either to or from the surroundings. If the system is hotter after the reaction than before, heat must flow *to* the surroundings to restore the system to the initial temperature. In this event the reaction is *exothermic*; by the convention for heat flow, the heat of the reaction is negative. If the system is colder after the reaction than before, heat must flow *from* the surroundings to restore the system to the initial temperature. In this event the reaction is *endothermic*, and the heat of the reaction is positive. The *heat of a reaction* is the heat withdrawn from the surroundings in the transformation of reactants at T and p to products at the *same* T and p.

In the laboratory the majority of chemical reactions are performed under a constant pressure; therefore the heat withdrawn from the surroundings is equal to the change in enthalpy of the system. To avoid mixing the enthalpy change associated with the chemical reaction and that associated with a temperature or pressure change in the system, the initial and final states of the system must have the same temperature and pressure.

For example, in the reaction

$$Fe_2O_3(s) + 3\,H_2(g) \longrightarrow 2\,Fe(s) + 3\,H_2O(l),$$

the initial and final states are:

Initial state	Final state
T, p	T, p
1 mole solid Fe_2O_3	2 moles solid Fe
3 moles gaseous H_2	3 moles liquid H_2O

Since the state of aggregation of each substance must be specified, the letters s, l, and g appear in parentheses after the formulas of the substances. Suppose that we think of the change in state as occurring in two distinct steps. In the first step, reactants at T and p are transformed adiabatically to products at T' and p.

Step 1.

$$\underbrace{Fe_2O_3(s) + 3\,H_2(g)}_{T,\,p} \longrightarrow \underbrace{2\,Fe(s) + 3\,H_2O(l).}_{T',\,p}$$

At constant pressure, $\Delta H = Q_p$; but, since this first step is adiabatic, $(Q_p)_1 = 0$ and $\Delta H_1 = 0$. In the second step, the system is placed in a heat reservoir at the initial temperature T. Heat flows into or out of the reservoir as the products of the reaction come to the initial temperature.

Step 2.

$$\underbrace{2\,Fe(s) + 3\,H_2O(l)}_{T',\,p} \longrightarrow \underbrace{2\,Fe(s) + 3\,H_2O(l),}_{T,\,p}$$

for which $\Delta H_2 = Q_p$. The sum of the two steps is the overall change in state

$$Fe_2O_3(s) + 3\,H_2(g) \longrightarrow 2\,Fe(s) + 3\,H_2O(l)$$

and the ΔH for the overall reaction is the sum of the enthalpy changes in the two steps: $\Delta H = \Delta H_1 + \Delta H_2 = 0 + Q_p$,

$$\Delta H = Q_p, \tag{7.60}$$

where Q_p is the heat of the reaction, the increase in enthalpy of the system resulting from the chemical reaction.

The increase in enthalpy in a chemical reaction can be viewed in a different way. At a specified temperature and pressure, the molar enthalpy $\bar{H}$ of each substance has a definite value. For any reaction, we can write

$$\Delta H = H_{\text{final}} - H_{\text{initial}}. \tag{7.61}$$

But the enthalpy of the initial or the final state is the sum of the enthalpies of the substances present initially or finally. Therefore, for the example,

$$H_{\text{final}} = 2\bar{H}(\text{Fe, s}) + 3\bar{H}(\text{H}_2\text{O, l}),$$

$$H_{\text{initial}} = \bar{H}(\text{Fe}_2\text{O}_3, \text{s}) + 3\bar{H}(\text{H}_2, \text{g}),$$

and Eq. (7.61) becomes

$$\Delta H = [2\bar{H}(\text{Fe, s}) + 3\bar{H}(\text{H}_2\text{O, l})] - [\bar{H}(\text{Fe}_2\text{O}_3, \text{s}) + 3\bar{H}(\text{H}_2, \text{g})]. \tag{7.62}$$

It seems reasonable that measuring ΔH could lead ultimately to the evaluation of the four molar enthalpies in Eq. (7.62). However, there are four "unknowns" and only one equation. We could measure the heats of several different reactions, but this would introduce more "unknowns." We deal with this difficulty in the next two sections.

7.19 THE FORMATION REACTION

We can simplify the result in Eq. (7.62) by considering the formation reaction of a compound. The formation reaction of a compound has *one mole* of the compound and *nothing else* on the product side; only elements in their stable states of aggregation appear on the reactant side. The increase in enthalpy in such a reaction is the *heat of formation*, or *enthalpy of formation*, of the compound, ΔH_f. The following reactions are examples of formation reactions.

$$\text{H}_2(\text{g}) + \tfrac{1}{2}\text{O}_2(\text{g}) \longrightarrow \text{H}_2\text{O(l)}$$

$$2\,\text{Fe(s)} + \tfrac{3}{2}\text{O}_2(\text{g}) \longrightarrow \text{Fe}_2\text{O}_3(\text{s})$$

$$\tfrac{1}{2}\text{H}_2(\text{g}) + \tfrac{1}{2}\text{Br}_2(\text{l}) \longrightarrow \text{HBr(g)}$$

$$\tfrac{1}{2}\text{N}_2(\text{g}) + 2\,\text{H}_2(\text{g}) + \tfrac{1}{2}\text{Cl}_2(\text{g}) \longrightarrow \text{NH}_4\text{Cl(s)}$$

If the ΔH for these reactions is written in terms of the molar enthalpies of the substances, we obtain, using the first two as examples,

$$\Delta H_f(\text{H}_2\text{O, l}) = \bar{H}(\text{H}_2\text{O, l}) - \bar{H}(\text{H}_2, \text{g}) - \tfrac{1}{2}\bar{H}(\text{O}_2, \text{g})$$

$$\Delta H(\text{Fe}_2\text{O}_3, \text{s}) = \bar{H}(\text{Fe}_2\text{O}_3, \text{s}) - 2\bar{H}(\text{Fe, s}) - \tfrac{3}{2}\bar{H}(\text{O}_2, \text{g})$$

Solving for the molar enthalpy of the compound in each example, we have

$$\bar{H}(\text{H}_2\text{O, l}) = \bar{H}(\text{H}_2, \text{g}) + \tfrac{1}{2}\bar{H}(\text{O}_2, \text{g}) + \Delta H_f(\text{H}_2\text{O, l})$$

$$\bar{H}(\text{Fe}_2\text{O}_3, \text{s}) = 2\bar{H}(\text{Fe, s}) + \tfrac{3}{2}\bar{H}(\text{O}_2, \text{g}) + \Delta H_f(\text{Fe}_2\text{O}_3, \text{s}) \tag{7.63}$$

These equations show that the molar enthalpy of a compound is equal to the total enthalpy of the elements that compose the compound plus the enthalpy of formation of the compound. Thus we can write for any compound,

$$\bar{H}(\text{compound}) = \Sigma H(\text{elements}) + \Delta H_f(\text{compound}), \tag{7.64}$$

in which ΣH(elements) is the total enthalpy of the elements (in their stable states of aggregation) in the compound.

Next we insert the values of $\bar{H}(H_2O, l)$ and $\bar{H}(Fe_2O_3, s)$ given by Eq. (7.63) into Eq. (7.62); this yields

$$\Delta H = 2\bar{H}(Fe, s) + 3[\bar{H}(H_2, g) + \tfrac{1}{2}\bar{H}(O_2, g) + \Delta H_f(H_2O, l)]$$
$$- [2\bar{H}(Fe, s) + \tfrac{3}{2}\bar{H}(O_2, g) + \Delta H_f(Fe_2O_3, s)] - 3\bar{H}(H_2, g)$$

Collecting like terms, this becomes

$$\Delta H = 3\Delta H_f(H_2O, l) - \Delta H_f(Fe_2O_3, s) \qquad (7.65)$$

Equation (7.65) states that the change in enthalpy of the reaction depends only on the heats of formation of the *compounds* in the reaction. The change in enthalpy is independent of the enthalpies of the elements in their stable states of aggregation.

A moment's reflection on Eq. (7.64) tells us that this independence of the values of the enthalpies of the elements must be correct for all chemical reactions. If, in the expression for the ΔH of a reaction, we replace the molar enthalpy of every compound by the expression in Eq. (7.64), then it is clear that the sum of the enthalpies of the elements composing the reactants must be equal to the sum of the enthalpies of the elements composing the products. The balanced chemical equation requires this. Therefore the enthalpies of the elements must drop out of the expression. We are left only with the proper combination of the enthalpies of formation of the compounds. This conclusion is correct at every temperature and pressure.

The enthalpy of formation of a compound at 1 atm pressure is the *standard* enthalpy of formation, ΔH_f°. Values of ΔH_f° at 25 °C are tabulated in Appendix V, Table A-V, for a number of compounds.

■ **EXAMPLE 7.4**   Using the values of ΔH_f° given in Table A-V, calculate the heat of the reaction

$$Fe_2O_3(s) + 3H_2(g) \longrightarrow 2Fe(s) + 3H_2O(l),$$

From Table A-V we find

$$\Delta H_f^\circ(H_2O, l) = -285.830 \text{ kJ/mol}; \qquad \Delta H_f^\circ(Fe_2O_3, s) = -824.2 \text{ kJ/mol}.$$

Then

$$\Delta H = 3(-285.830 \text{ kJ/mol}) - 1(-824.2 \text{ kJ/mol}) = (-857.5 + 824.2) \text{ kJ/mol}$$
$$= -33.3 \text{ kJ/mol}.$$

The negative sign indicates that the reaction is exothermic. Note that the stoichiometric coefficients in these expressions are *pure numbers*. The unit for ΔH is kJ/mol. This means per *mole of reaction*. Once we balance the chemical equation in a particular way, as above, this defines the mole of reaction. If we had balanced the equation differently, as, for example,

$$\tfrac{1}{2}Fe_2O_3(s) + \tfrac{3}{2}H_2(g) \longrightarrow Fe(s) + \tfrac{3}{2}H_2O(l)$$

then this amount of reaction would be one mole of reaction and ΔH would be

$$\Delta H = \tfrac{3}{2}(-285.830 \text{ kJ/mole}) - \tfrac{1}{2}(-824.2 \text{ kJ/mol}) = -428.7 + 412.1 = -16.6 \text{ kJ/mol}.$$

7.20 CONVENTIONAL VALUES OF MOLAR ENTHALPIES

The molar enthalpy $\bar{H}$ of any substance is a function of T and p; $\bar{H} = \bar{H}(T, p)$. Choosing $p = 1$ atm as the *standard* pressure, we define the *standard* molar enthalpy $\bar{H}^\circ$ of a substance by

$$\bar{H}^\circ = \bar{H}(T, 1 \text{ atm}). \tag{7.66}$$

From this it is clear that $\bar{H}^\circ$ is a function only of temperature. The degree superscript on any thermodynamic quantity indicates the value of that quantity at the standard pressure. (Because the dependence of enthalpy on pressure is very slight—compare with Section 7.15—we will often use standard enthalpies at pressures other than one atm; the error will not be serious unless the pressure is very large, for example, 1000 atm.)

As we showed in Section 7.19, the enthalpy change in any chemical reaction does not depend on the numerical values of the enthalpies of the elements that compose the compound. Because this is so we may assign any arbitrary, convenient values to the molar enthalpies of the elements in their stable states of aggregation at a selected temperature and pressure. Clearly, if we chose the required values randomly from the numbers in a telephone directory this could introduce a good deal of unnecessary numerical clutter into our work. Since the numbers do not matter, they can all be the same; if they can all be the same, they all might as well be zero and eliminate the clutter entirely.

The enthalpy of every *element* in its stable state of aggregation at 1 atm pressure and at 298.15 K is assigned the value zero. For example, at 1 atm and 298.15 K the stable state of aggregation of bromine is the liquid state. Hence, liquid bromine, gaseous hydrogen, solid zinc, solid (rhombic) sulfur, and solid (graphite) carbon all have $\bar{H}^\circ_{298.15} = 0$. (We will write H_{298} as an abbreviation for $H_{298.15}$.)

For elementary solids that exist in more than one crystalline form, the modification that is stable at 25 °C and 1 atm is assigned $\bar{H}^\circ = 0$; for example, the zero assignment goes to rhombic sulfur rather than to monoclinic sulfur, and to graphite rather than to diamond. In cases in which more than one molecular species exists (for example, oxygen atoms, O; diatomic oxygen, O_2; and ozone, O_3) the zero enthalpy value is assigned to the most stable form at 25 °C and 1 atm pressure; for oxygen, $\bar{H}^\circ_{298}(O_2, g) = 0$. Once the value of the standard enthalpy of the elements at 298.15 K has been assigned, the value at any other temperature can be calculated. Since at constant pressure, $d\bar{H}^\circ = \bar{C}^\circ_p \, dT$, then

$$\int_{298}^{T} d\bar{H}^\circ = \int_{298}^{T} \bar{C}^\circ_p \, dT, \qquad \bar{H}^\circ_T - \bar{H}^\circ_{298} = \int_{298}^{T} \bar{C}^\circ_p \, dT,$$

$$\bar{H}^\circ_T = \bar{H}^\circ_{298} + \int_{298}^{T} \bar{C}^\circ_p \, dT, \tag{7.67}$$

which is correct for both elements and compounds; for elements, the first term on the right-hand side is zero.

Given the definition of the formation reaction, if we introduce the conventional assignment, $\bar{H}^\circ(\text{elements}) = 0$, into the expression for the heat of formation, Eq. (7.63) or Eq. (7.64), we find that for any compound

$$\bar{H}^\circ = \Delta H^\circ_f. \tag{7.68}$$

The standard heat of formation ΔH°_f is the conventional molar enthalpy of the compound relative to the elements that compose it. Accordingly, if the heats of formation ΔH°_f of all the compounds in a chemical reaction are known, the heat of the reaction can be calculated from equations formulated in the manner of Eq. (7.62).

7.21 THE DETERMINATION OF HEATS OF FORMATION

In some cases it is possible to determine the heat of formation of a compound directly by carrying out the formation reaction in a calorimeter and measuring the heat effect produced. Two important examples are

$$C(graphite) + O_2(g) \longrightarrow CO_2(g), \quad \Delta H_f^\circ = -393.51 \text{ kJ/mol}$$

$$H_2(g) + \tfrac{1}{2}O_2(g) \longrightarrow H_2O(l), \quad \Delta H_f^\circ = -285.830 \text{ kJ/mol}.$$

These reactions can be conducted easily in a calorimeter; the reactions go to completion, and conditions can easily be arranged so that only one product is formed. Because of the importance of these two reactions, the values have been determined quite accurately.

The majority of formation reactions are unsuitable for calorimetric measurement; these heats of formation must be determined by indirect methods. For example,

$$C(graphite) + 2H_2(g) \longrightarrow CH_4(g).$$

This reaction has three strikes against it as far as its use in calorimetry is concerned. The combination of graphite with hydrogen does not occur readily; if we did manage to get these materials to react in a calorimeter, the product would not be pure methane, but an exceedingly complex mixture of hydrocarbons. Even if we succeeded in analyzing the product mixture, the result of such an experiment would be impossible to interpret.

There is one method that is generally applicable if the compound burns easily to form definite products. The heat of formation of a compound can be calculated from the measured value of the heat of combustion of the compound. The combustion reaction has one mole of the substance to be burned on the reactant side, with as much oxygen as is necessary to burn the substance completely; organic compounds containing only carbon, hydrogen, and oxygen are burned to gaseous carbon dioxide and liquid water.

For example, the combustion reaction for methane is

$$CH_4(g) + 2O_2(g) \longrightarrow CO_2(g) + 2H_2O(l).$$

The measured heat of combustion is $\Delta H_{comb}^\circ = -890.36$ kJ/mol. In terms of the enthalpies of the individual substances,

$$\Delta H_{comb}^\circ = \bar{H}^\circ(CO_2, g) + 2\bar{H}^\circ(H_2O, l) - \bar{H}^\circ(CH_4, g).$$

Solving this equation for $\bar{H}^\circ(CH_4, g)$,

$$\bar{H}^\circ(CH_4, g) = \bar{H}^\circ(CO_2, g) + 2\bar{H}^\circ(H_2O, l) - \Delta H_{comb}^\circ. \tag{7.69}$$

The molar enthalpies of CO_2 and H_2O are known to a high accuracy; from this knowledge and the measured value of the heat of combustion, the molar enthalpy of methane (the heat of formation) can be calculated by using Eq. (7.69).

$$\bar{H}^\circ(CH_4, g) = -393.51 + 2(-285.83) - (-890.36)$$
$$= -965.17 + 890.36 = -74.81 \text{ kJ/mol}.$$

The measurement of the heat of combustion is used to determine the heats of formation of all organic compounds that contain only carbon, hydrogen, and oxygen. These compounds burn completely to carbon dioxide and water in the calorimeter. The combustion method is used also for organic compounds containing sulfur and nitrogen; however, in these cases the reaction products are not so definite. The sulfur may end up as sulfurous acid or sulfuric acid; the nitrogen may end up in the elementary form or as a mixture of oxy-acids. In these cases considerable skill and ingenuity are required in the determination of the conditions for the reaction and in the analysis of the reaction products. The accuracy of the values obtained for this latter class of compounds is very much less than that obtained for the compounds containing only carbon, hydrogen, and oxygen.

The problem of determining the heat of formation of any compound resolves into that of finding some chemical reaction involving the compound which is suitable for calorimetric measurement, then measuring the heat of this reaction. If the heats of formation of all of the other substances involved in this reaction are known, then the problem is solved. If the heat of formation of one of the other substances in the reaction is not known, then we must find a calorimetric reaction for that substance, and so on.

Devising a series of reactions from which an accurate value of the heat of formation of a particular compound can be obtained can be a challenging problem. A calorimetric reaction must take place quickly (that is, be completed within a few minutes at most), with as few side reactions as possible and preferably none at all. Very few chemical reactions take place without concomitant side reactions, but their effects can be minimized by controlling the reaction conditions so as to favor the main reaction as much as possible. The final product mixture must be carefully analyzed, and the thermal effect of the side reactions must be subtracted from the measured value. Precision calorimetry is demanding work.

7.22 SEQUENCES OF REACTIONS; HESS'S LAW

The change in state of a system produced by a specified chemical reaction is definite. The corresponding enthalpy change is definite, since the enthalpy is a function of the state. Thus, if we transform a specified set of reactants to a specified set of products by more than one sequence of reactions, the total enthalpy change must be the same for every sequence. This rule, which is a consequence of the first law of thermodynamics, was originally known as Hess's law of constant heat summation. Suppose that we compare two different methods of synthesizing sodium chloride from sodium and chlorine.

Method 1.

$$Na(s) + H_2O(l) \longrightarrow NaOH(s) + \tfrac{1}{2}H_2(g), \quad \Delta H = -139.78 \text{ kJ/mol}$$
$$\tfrac{1}{2}H_2(g) + \tfrac{1}{2}Cl_2(g) \longrightarrow HCl(g), \quad \Delta H = -\ 92.31 \text{ kJ/mol}$$
$$HCl(g) + NaOH(s) \longrightarrow NaCl(s) + H_2O(l), \quad \Delta H = -179.06 \text{ kJ/mol}$$

Net change: $Na(s) + \tfrac{1}{2}Cl_2(g) \longrightarrow NaCl(s), \qquad \Delta H_{net} = -411.15 \text{ kJ/mol}$

Method 2.

$$\tfrac{1}{2}H_2(g) + \tfrac{1}{2}Cl_2(g) \longrightarrow HCl(g), \quad \Delta H = -\ 92.31 \text{ kJ/mol}$$
$$Na(s) + HCl(g) \longrightarrow NaCl(s) + \tfrac{1}{2}H_2(g), \quad \Delta H = -318.84 \text{ kJ/mol}$$

Net change: $Na(s) + \tfrac{1}{2}Cl_2(g) \longrightarrow NaCl(s), \qquad \Delta H_{net} = -411.15 \text{ kJ/mol}$

The net chemical change is obtained by adding together all the reactions in the sequence; the net enthalpy change is obtained by adding together all the enthalpy changes in the sequence. The net enthalpy change must be the same for every sequence which has the same net chemical change. Any number of reactions can be added or subtracted to yield the desired chemical reaction; the enthalpy changes of the reactions are added or subtracted algebraically in the corresponding way.

If a certain chemical reaction is combined in a sequence with the reverse of the same reaction, there is no net chemical effect, and $\Delta H = 0$ for the combination. It follows immediately that the ΔH of the reverse reaction is equal in magnitude but opposite in sign to that of the forward reaction.

The utility of this property of sequences, which is really nothing more than the fact that the enthalpy change in a system is independent of the path, is illustrated by the sequence

1) $\qquad\qquad C(graphite) + \frac{1}{2}O_2(g) \longrightarrow CO(g), \qquad \Delta H_1,$

2) $\qquad\qquad\quad CO(g) + \frac{1}{2}O_2(g) \longrightarrow CO_2(g), \qquad \Delta H_2.$

The net change in the sequence is

3) $\qquad\qquad C(graphite) + O_2(g) \longrightarrow CO_2(g), \qquad \Delta H_3.$

Therefore $\Delta H_3 = \Delta H_1 + \Delta H_2$. In this particular instance, ΔH_2 and ΔH_3 are readily measurable in the calorimeter, while ΔH_1 is not. Since the value of ΔH_1 can be computed from the other two values, there is no need to measure it.

Similarly, by subtracting reaction (2) from reaction (1) we obtain

4) $\qquad C(graphite) + CO_2(g) \longrightarrow 2CO(g), \qquad \Delta H_4 = \Delta H_1 - \Delta H_2,$

and the heat of this reaction can also be obtained from the measured values.

★ 7.23 HEATS OF SOLUTION AND DILUTION

The *heat of solution* is the enthalpy change associated with the addition of a specified amount of solute to a specified amount of solvent at constant temperature and pressure. For convenience we shall use water as the solvent in the illustrations, but the argument can be applied to any solvent with slight modification. The change in state is represented by

$$X + nAq \longrightarrow X \cdot nAq, \qquad \Delta H_S.$$

One mole of solute X is added to n moles of water. The water is given the symbol Aq in this equation; it is convenient to assign a conventional enthalpy of zero to the water in these solution reactions.

Consider the examples

$$HCl(g) + 10\,Aq \longrightarrow HCl \cdot 10\,Aq, \qquad \Delta H_1 = -69.01 \text{ kJ/mol}$$
$$HCl(g) + 25\,Aq \longrightarrow HCl \cdot 25\,Aq, \qquad \Delta H_2 = -72.03 \text{ kJ/mol}$$
$$HCl(g) + 40\,Aq \longrightarrow HCl \cdot 40\,Aq, \qquad \Delta H_3 = -72.79 \text{ kJ/mol}$$
$$HCl(g) + 200\,Aq \longrightarrow HCl \cdot 200\,Aq, \qquad \Delta H_4 = -73.96 \text{ kJ/mol}$$
$$HCl(g) + \infty\,Aq \longrightarrow HCl \cdot \infty\,Aq, \qquad \Delta H_5 = -74.85 \text{ kJ/mol}$$

The values of ΔH show the general dependence of the heat of solution on the amount of solvent. As more and more solvent is used, the heat of solution approaches a limiting value, the value in the "infinitely dilute" solution; for HCl this limiting value is given by ΔH_5.

If we subtract the first equation from the second in the above set, we obtain

$$\text{HCl} \cdot 10\,\text{Aq} + 15\,\text{Aq} \longrightarrow \text{HCl} \cdot 25\,\text{Aq}, \qquad \Delta H = \Delta H_2 - \Delta H_1 = -3.02 \text{ kJ/mol}.$$

This value of ΔH is a heat of dilution: the heat withdrawn from the surroundings when additional solvent is added to a solution. The heat of dilution of a solution is dependent on the original concentration of the solution and on the amount of solvent added.

The heat of formation of a solution is the enthalpy associated with the reaction (using hydrochloric acid as an example):

$$\tfrac{1}{2}\text{H}_2(\text{g}) + \tfrac{1}{2}\text{Cl}_2(\text{g}) + n\,\text{Aq} \longrightarrow \text{HCl} \cdot n\,\text{Aq}, \qquad \Delta H_f^\circ,$$

where the solvent Aq is counted as having zero enthalpy.

The heat of solution defined above is the *integral* heat of solution. This distinguishes it from the *differential* heat of solution, which is defined in Section 11.24.

7.24 HEATS OF REACTION AT CONSTANT VOLUME

If any of the reactants or products of the calorimetric reaction are gaseous, it is necessary to conduct the reaction in a sealed bomb. Under this condition the system is initially and finally in a constant volume rather than being under a constant pressure. The measured heat of reaction at constant volume is equal to an energy increment, rather than to an enthalpy increment:

$$Q_V = \Delta U \tag{7.70}$$

The corresponding change in state is

$$\text{R}(T, V, p) \longrightarrow \text{P}(T, V, p'),$$

where $\text{R}(T, V, p)$ represents the reactants in the initial condition T, V, p; and $\text{P}(T, V, p')$ represents the products in the final condition T, V, p'. The temperature and volume remain constant, but the pressure may change from p to p' in the transformation.

To relate the ΔU in Eq. (7.70) to the corresponding ΔH, we apply the defining equation for H to the initial and final states:

$$H_{\text{final}} = U_{\text{final}} + p'V, \qquad H_{\text{initial}} = U_{\text{initial}} + pV.$$

Subtracting the second equation from the first, we have

$$\Delta H = \Delta U + (p' - p)V. \tag{7.71}$$

The initial and final pressures in the bomb are determined by the number of moles of gases present initially and finally; assuming that the gases behave ideally, we have

$$p = \frac{n_\text{R} RT}{V}, \qquad p' = \frac{n_\text{P} RT}{V},$$

where n_R and n_P are the total number of moles of *gaseous* reactants and *gaseous* products in the reaction. Equation (7.71) becomes

$$\Delta H = \Delta U + (n_\text{P} - n_\text{R})RT,$$
$$\Delta H = \Delta U + \Delta n RT. \tag{7.72}$$

Strictly speaking, the ΔH in Eq. (7.72) is the ΔH for the constant-volume transformation. To convert it to the appropriate value for the constant-pressure transformation, we must add to it the enthalpy change of the process:

$$P(T, V, p') \quad\longrightarrow\quad P(T, V', p).$$

For this change in pressure at constant temperature, the enthalpy change is very nearly zero (Section 7.15) and exactly zero if only ideal gases are involved. Thus for practical purposes the ΔH in Eq. (7.72) is equal to the ΔH for the constant-pressure process, while the ΔU refers to the constant-volume transformation. To a good approximation, Eq. (7.72) can be interpreted as

$$Q_p = Q_V + \Delta n R T. \tag{7.73}$$

It is through Eqs. (7.72) or (7.73) that measurements in the bomb calorimeter, $Q_V = \Delta U$, are converted to values of $Q_p = \Delta H$. In precision measurements it may be necessary to include the effect of gas imperfection or of the change in enthalpy of the products with pressure; this would depend on the conditions employed in the experiment.

■ **EXAMPLE 7.5**   Consider the combustion of benzoic acid in the bomb calorimeter:

$$C_6H_5COOH(s) + \tfrac{15}{2}O_2(g) \quad\longrightarrow\quad 7CO_2(g) + 3H_2O(l).$$

In this reaction, $n_P = 7$, while $n_R = \tfrac{15}{2}$. Thus $\Delta n = 7 - \tfrac{15}{2} = -\tfrac{1}{2}$, $T = 298.15$ K; we have

$$Q_p = Q_V - \tfrac{1}{2}(8.3144 \text{ J/K mol})(298.15 \text{ K}), \qquad Q_p = Q_V - 1239 \text{ J/mol}.$$

Note that only the number of moles of *gases* are counted in computing Δn.

7.25 DEPENDENCE OF THE HEAT OF REACTION ON TEMPERATURE

If we know the value of $\Delta H°$ for a reaction at a particular temperature, let us say at 25 °C, then we can calculate the heat of reaction at any other temperature if the heat capacities of all the substances taking part in the reaction are known. The $\Delta H°$ of any reaction is

$$\Delta H° = H°(\text{products}) - H°(\text{reactants}).$$

To find the dependence of this quantity on temperature we differentiate with respect to temperature:

$$\frac{d\,\Delta H°}{dT} = \frac{dH°}{dT}(\text{products}) - \frac{dH°}{dT}(\text{reactants}).$$

But, by definition, $dH°/dT = C_p°$. Hence,

$$\frac{d\,\Delta H°}{dT} = C_p°(\text{products}) - C_p°(\text{reactants})$$

$$\frac{d\,\Delta H°}{dT} = \Delta C_p°. \tag{7.74}$$

Note that since $H°$ and $\Delta H°$ are functions only of temperature (Section 7.20), these are ordinary derivatives rather than partial derivatives.

The value of ΔC_p° is calculated from the individual heat capacities in the same way as ΔH° is calculated from the individual values of the molar enthalpies. We multiply the molar heat capacity of each product by the number of moles of that product involved in the reaction; the sum of these quantities for every product yields the heat capacity of the products. A similar procedure yields the heat capacity of the reactants. The difference in value of the heat capacity of products less that of reactants is ΔC_p.

Writing Eq. (7.74) in differential form, we have

$$d\,\Delta H^\circ = \Delta C_p^\circ\,dT.$$

Integrating between a fixed temperature T_0 and any other temperature T, we obtain

$$\int_{T_0}^{T} d\,\Delta H^\circ = \int_{T_0}^{T} \Delta C_p^\circ\,dT.$$

The integral of the differential on the left is simply ΔH°, which, when evaluated between the limits, becomes

$$\Delta H_T^\circ - \Delta H_{T_0}^\circ = \int_{T_0}^{T} \Delta C_p^\circ\,dT.$$

Rearranging, we have

$$\Delta H_T^\circ = \Delta H_{T_0}^\circ + \int_{T_0}^{T} \Delta C_p^\circ\,dT. \tag{7.75}$$

Knowing the value of the enthalpy increase at the fixed temperature T_0, we can calculate the value at any other temperature T, using Eq. (7.75). If any of the substances change their state of aggregation in the temperature interval, the corresponding enthalpy change must be included.

If the range of temperature covered in the integration of Eq. (7.75) is short, the heat capacities of all the substances involved may be considered constant. If the temperature interval is very large, the heat capacities must be taken as functions of temperature. For many substances this function has the form

$$C_p = a + bT + cT^2 + dT^3 + \cdots, \tag{7.76}$$

where $a, b, c, d, \ldots$ are constants for a specified material. In Table 7.1 values of these constants are given for a number of substances in multiples of R, the gas constant.

■ **EXAMPLE 7.6** Calculate the ΔH° at 85 °C for the reaction

$$\mathrm{Fe_2O_3(s) + 3\,H_2(g) \longrightarrow 2\,Fe(s) + 3\,H_2O(l)}.$$

The data are: $\Delta H_{298}^\circ = -33.29$ kJ/mol; and

Substance	$Fe_2O_3(s)$	$Fe(s)$	$H_2O(l)$	$H_2(g)$
$\bar{C}_p^\circ/(\text{J/K mol})$	103.8	25.1	75.3	28.8

First compute ΔC_p°.

$$\Delta C_p^\circ = 2\bar{C}_p^\circ(\mathrm{Fe, s}) + 3\bar{C}_p^\circ(\mathrm{H_2O, l}) - [\bar{C}_p^\circ(\mathrm{Fe_2O_3, s}) + 3\bar{C}_p^\circ(\mathrm{H_2, g})]$$

$$= 2(25.1) + 3(75.3) - [103.8 + 3(28.8)] = 85.9 \text{ J/K mol}.$$

Table 7.1

Heat capacity of gases as a function of temperature

$\bar{C}_p/R = a + bT + cT^2 + dT^3$

Range: 300 K to 1500 K

	a	$b/10^{-3}\,\text{K}^{-1}$	$c/10^{-7}\,\text{K}^{-2}$	$d/10^{-9}\,\text{K}^{-3}$
H_2	3.4958	$-\ 0.1006$	2.419	
O_2	3.0673	$+\ 1.6371$	$-\ 5.118$	
Cl_2	3.8122	1.2200	$-\ 4.856$	
Br_2	4.2385	0.4901	$-\ 1.789$	
N_2	3.2454	0.7108	$-\ 0.406$	
CO	3.1916	0.9241	$-\ 1.410$	
HCl	3.3876	0.2176	$+\ 1.860$	
HBr	3.3100	0.4805	0.796	
NO	3.5326	$-\ 0.186$	12.81	-0.547
CO_2	3.205	$+\ 5.083$	$-\ 17.13$	
H_2O	3.633	1.195	$+\ 1.34$	
NH_3	3.114	3.969	$-\ 3.66$	
H_2S	3.213	2.870	$-\ 6.09$	
SO_2	3.093	6.967	$-\ 45.81$	$+1.035$
CH_4	1.701	9.080	$-\ 21.64$	
C_2H_6	1.131	19.224	$-\ 55.60$	
C_2H_4	1.424	14.393	$-\ 43.91$	
C_2H_2	3.689	6.352	$-\ 19.57$	
C_3H_8	1.213	28.782	$-\ 88.23$	
C_3H_6	1.637	22.703	$-\ 69.14$	
C_3H_4	3.187	15.595	$-\ 47.59$	
C_6H_6	-0.206	39.061	-133.00	
$C_6H_5CH_3$	$+0.290$	47.048	-157.14	
C(graphite)	-0.637	7.049	$-\ 51.99$	1.384

Calculated from the compilations of H. M. Spencer and J. L. Justice, *J. Am. Chem. Soc.*, **56**:2311 (1934); H. M. Spencer and G. N. Flanagan, *J. Am. Chem. Soc.*, **64**:2511 (1942); H. M. Spencer, *Ind. Eng. Chem.*, **40**:2152 (1948).

Since 85 °C = 358 K, we have

$$\Delta H^\circ_{358} = \Delta H^\circ_{298} + \int_{298}^{358} 85.9\ dT$$

$$= -33.29\ \text{kJ/mol} + 85.9(358 - 298)\ \text{J/mol}$$

$$= -33.29\ \text{kJ/mol} + 85.9(60)\ \text{J/mol} = -33.29\ \text{kJ/mol} + 5150\ \text{J/mol}$$

$$= -33.29\ \text{kJ/mol} + 5.15\ \text{kJ/mol} = -28.14\ \text{kJ/mol}.$$

Note that care must be taken to express both terms in kilojoules or both in joules before adding them together!

■ **EXAMPLE 7.7**   Compute the heat of reaction at 1000 °C = 1273 K for

$$\tfrac{1}{2}H_2(g) + \tfrac{1}{2}Cl_2(g) \longrightarrow HCl(g)$$

Given $\Delta H^\circ_{298} = -92.312$ kJ/mol and the data for C_p from Table 7.1:

$$\bar{C}^\circ_p(H_2)/R = 3.4958 - 0.1006(10^{-3})T + 2.419(10^{-7})T^2$$

$$\bar{C}^\circ_p(Cl_2)/R = 3.8122 + 1.2200(10^{-3})T - 4.856(10^{-7})T^2$$

$$\bar{C}^\circ_p(HCl)/R = 3.3876 + 0.2176(10^{-3})T + 1.860(10^{-7})T^2$$

We begin by computing $\Delta C^\circ_p/R$ for the integral in Eq. (7.75). We arrange the work in columns.

$$\begin{aligned}
\Delta C^\circ_p/R = \quad & 3.3876 + 0.2176(10^{-3})T + 1.860(10^{-7})T^2 \\
& -\tfrac{1}{2}[3.4958 - 0.1006(10^{-3})T + 2.419(10^{-7})T^2] \\
& -\tfrac{1}{2}[3.8122 + 1.2200(10^{-3})T - 4.856(10^{-7})T^2] \\
\Delta C^\circ_p = R[&-0.2664 - 0.3421(10^{-3})T + 3.079(10^{-7})T^2]
\end{aligned}$$

$$\begin{aligned}
\int_{298}^{1273} \Delta C^\circ_p \, dT = R\bigg[&-0.2664 \int_{298}^{1273} dT - 0.3421(10^{-3}) \int_{298}^{1273} T \, dT \\
& + 3.079(10^{-7}) \int_{298}^{1273} T^2 \, dT \bigg]
\end{aligned}$$

$$\begin{aligned}
= R[&-0.2664(1273 - 298) - \tfrac{1}{2}(0.3421)(10^{-3})(1273^2 - 298^2) \\
& + \tfrac{1}{3}(3.079)(10^{-7})(1273^3 - 298^3)]
\end{aligned}$$

$$= R(-259.7 - 262.0 + 209.0) = (8.3144 \text{ J/K mol})(-312.7 \text{ K})$$

$$= -2.600 \text{ kJ/mol}$$

$$\Delta H^\circ_{1273} = \Delta H^\circ_{298} + \int_{298}^{1273} \Delta C^\circ_p \, dT = -92.312 \text{ kJ/mol} - 2.600 \text{ kJ/mol}$$

$$= -94.912 \text{ kJ/mol}.$$

Note that the heat capacities of *all* the substances taking part in the reaction must be included; the elements cannot be omitted as they were in calculating enthalpy differences.

7.26 BOND ENTHALPIES

If we consider the atomization of a gaseous diatomic molecule,

$$O_2(g) \longrightarrow 2O(g) \qquad \Delta H^\circ_{298} = 498.34 \text{ kJ/mol},$$

the value 498.34 kJ is called the bond enthalpy of the oxygen molecule.

Similarly we can write

$$H_2O(g) \longrightarrow 2H(g) + O(g) \qquad \Delta H^\circ_{298} = 926.98 \text{ kJ/mol}$$

and call $\tfrac{1}{2}(926.98) = 463.49$ kJ/mole the average bond enthalpy of the O—H bond in water. So long as we deal only with molecules in which the bonds are equivalent, such as H_2O, NH_3, CH_4, the procedure is straightforward.

On the other hand, in dealing with a molecule such as H_2O_2 in which two different bonds exist, some assumption must be introduced. The assumption is usually made that

Table 7.2
Heats of formation of gaseous atoms at 25 °C

Atom	$\Delta H_f/$(kJ/mol)	Atom	$\Delta H_f/$(kJ/mol)	Atom	$\Delta H_f/$(kJ/mol)
O	249.17	Br	111.86	N	472.68
H	217.997	I	106.762	P	316.5
F	79.39	S	276.98	C	716.67
Cl	121.302	Se	202.4	Si	450.

the average OH bond in H_2O_2 is the same as that in water. The enthalpy of atomization of H_2O_2 is

$$H_2O_2(g) \longrightarrow 2H(g) + 2O(g) \qquad \Delta H^\circ_{298} = 1070.6 \text{ kJ/mol.}$$

If we subtract the enthalpy of two OH bonds in water we obtain: $1070.6 - 927.0 = 143.6$ kJ/mol as the strength of the O—O single bond. Clearly the method does not warrant keeping the fraction, and we would say that the oxygen–oxygen single bond has a strength of about 144 kJ/mol.

The heats of formation of the atoms must be known before we can compute the bond strength. Some of these values are given in Table 7.2.

★ 7.26.1   Bond Energies

If we wish to know the bond energy, assuming all species behave as ideal gases, we can use the relation

$$\Delta U = \Delta H - \Delta nRT.$$

In the case of the oxygen molecule, $\Delta n = 1$, so that

$$\Delta U = 498.34 \text{ kJ/mol} - (1)(8.3144 \text{ J/K mol})(298.15 \text{ K})(10^{-3} \text{ kJ/J})$$
$$= 498.34 \text{ kJ/mol} - 2.48 \text{ kJ/mol} = 495.86 \text{ kJ/mol.}$$

This is the average energy that must be supplied to break one mole of bonds in the oxygen molecule at 25 °C. At this temperature some of the molecules will be in excited rotational and vibrational states; these molecules will require somewhat less energy to break the bonds than will one in its ground state. At 0 K, all the molecules are in the ground state and thus all require the same energy to break the bond. If we correct the value of ΔU to 0 K, we obtain the bond energy. The relation is

$$\Delta U_{298} = \Delta U_0 + \int_0^{298} \Delta C_v \, dT.$$

Since $\bar{C}_v(O, g) = \frac{3}{2}R$ and $\bar{C}_v(O_2, g) = \frac{5}{2}R$, and these values are independent of temperature, this yields $\Delta C_v = 2(\frac{3}{2}R) - \frac{5}{2}R = \frac{1}{2}R$. Then

$$\Delta U_0 = \Delta U_{298} - \frac{1}{2}R \int_0^{298} dT = 495.86 \text{ kJ/mol} - \frac{1}{2}(8.314 \text{ J/K mol})(298.15 \text{ K})(10^{-3} \text{ kJ/J})$$

$$= 495.86 \text{ kJ/mol} - 1.24 \text{ kJ/mol} = 494.62 \text{ kJ/mol}$$

This is the bond energy of the oxygen–oxygen double bond. For any molecule for which the data are available the calculation is straightforward, as above. Note that the difference between the bond enthalpy at 25 °C, ΔH_{298}°, and the bond energy, ΔU_0, is only 3.72 kJ in almost 500 kJ. This is only 0.7 %. The differences are usually of this relative order, so that often we do not bother with them.

★ 7.27 CALORIMETRIC MEASUREMENTS

It is worthwhile to describe how the heat of a reaction is calculated from the quantities that are actually measured in a calorimetric experiment. It is not possible in a brief space to describe all the types of calorimeters or all of the variations and refinements of technique that are necessary in individual cases and in precision work. A highly idealized situation will be described to illustrate the methods involved.

The situation is simplest if the calorimeter is an *adiabatic* calorimeter. In the laboratory, this device is quite elaborate; on paper we shall simply say that the vessel containing the system is perfectly insulated so that no heat flows into or out of the system. Under constant pressure, the first law for any transformation within the calorimeter is

$$\Delta H = Q_p = 0. \tag{7.77}$$

The change in state can be represented by

$$K(T_1) + R(T_1) \longrightarrow K(T_2) + P(T_2) \quad (p = \text{constant}),$$

where K symbolizes the calorimeter, R the reactants, and P the products. Since the system is insulated, the final temperature T_2 differs from the initial temperature T_1; both temperatures are measured as accurately as possible with a sensitive thermometer.

The change in state can be supposed to occur in two steps:

1)
$$R(T_1) \longrightarrow P(T_1), \qquad \Delta H_{T_1},$$

2)
$$K(T_1) + R(T_1) \longrightarrow K(T_2) + P(T_2), \qquad \Delta H_2.$$

By Eq. (7.77) the overall $\Delta H = 0$, so that $\Delta H_{T_1} + \Delta H_2 = 0$, or $\Delta H_{T_1} = -\Delta H_2$. The second step is simply a temperature change of the calorimeter and the reaction products, so

$$\Delta H_2 = \int_{T_1}^{T_2} [C_p(K) + C_p(P)] \, dT,$$

and we obtain for the heat of the reaction at T_1

$$\Delta H_{T_1} = -\int_{T_1}^{T_2} [C_p(K) + C_p(P)] \, dT. \tag{7.78}$$

If the heat capacities of the calorimeter and the products of the reaction are known, the heat of reaction at T_1 can be calculated from the measured temperatures T_1 and T_2.

If the required heat capacities are not known, the value of ΔH_2 can be measured as follows. Cool the calorimeter and the products to the initial temperature T_1. (This assumes that T_2 is greater than T_1.) The calorimeter and the products are then taken from T_1 to T_2 by allowing an electrical current to flow in a resistor immersed in the calorimeter; the change in enthalpy in this step is equal to ΔH_2. This can be related to the electrical work expended in the resistance wire, which can be measured quite accurately, being the product of the current, the potential difference across the resistance, and the time.

If we include electrical work, dW_{el}, at constant pressure, the first law becomes

$$dU = dQ - p \, dV - dW_{\text{el}}. \tag{7.79}$$

Differentiating $H = U + pV$ under the constant pressure condition, we get $dH = dU + p\,dV$. Adding this equation to Eq. (7.79) yields

$$dH = dQ - dW_{el}. \tag{7.80}$$

For an adiabatic process, $dQ = 0$, and Eq. (7.80) integrates to

$$\Delta H = -W_{el}. \tag{7.81}$$

Applying Eq. (7.81) to the electrical method of carrying the products and calorimeter from the initial to the final temperature, we have $\Delta H_2 = -W_{el}$, and so, since $\Delta H_{T_1} + \Delta H_2 = 0$, we obtain

$$\Delta H_{T_1} = W_{el}. \tag{7.82}$$

Since work was destroyed in the surroundings, W_{el}, and therefore ΔH_{T_1}, are negative. The reaction is exothermic, a result of the assumption that T_2 is greater than T_1. For endothermic reactions the procedure is modified in an obvious way.

An alternative scheme can be imagined for the steps in the reaction:

3) $\qquad\qquad K(T_1) + R(T_1) \longrightarrow K(T_2) + R(T_2), \qquad \Delta H_3,$

4) $\qquad\qquad\qquad\qquad R(T_2) \longrightarrow P(T_2), \qquad \Delta H_{T_2}.$

Again, the overall $\Delta H = 0$, so that $\Delta H_3 + \Delta H_{T_2} = 0$, or

$$\Delta H_{T_2} = -\Delta H_3 = -\int_{T_1}^{T_2} [C_p(K) + C_p(R)]\,dT. \tag{7.83}$$

If the heat capacities of the calorimeter and the reactants are known, the heat of reaction at T_2 can be calculated from Eq. (7.83).

If we deal with a bomb calorimeter so that the volume is constant, rather than the pressure, the argument is unchanged. In all of the equations ΔH's are simply replaced by ΔU's, and C_p's by C_v's.

QUESTIONS

7.1 What is the difference between energy and heat? Between energy and work? Between heat and work?

7.2 Some texts define work W' as positive when a weight is *lowered* in the surroundings; that is, when the surroundings do work on the system. How can the first law be expressed in terms of Q and W'? (Justify the sign in front of W'.)

7.3 The deviation of the work performed in a real expansion of a gas from that of reversible expansion can be shown to be of order $\mathcal{U}/\langle u \rangle$. Here $\langle u \rangle$ is the average molecular speed and the $\mathcal{U}$ is the speed of the piston. What piston speed is required for a 10% deviation from the reversible work formula?

7.4 Why is the enthalpy a useful quantity?

7.5 For a constant pressure process, $\Delta H = Q_p$. Does it follow that Q_p is a state function? Why or why not?

7.6 What is the molecular interpretation of the dependence of the thermodynamic energy on the volume?

7.7 What is the connection between Hess's law and the fact that enthalpy is a state function?

7.8 Why does C_p exceed C_v for an ideal gas? Give a molecular explanation.

7.9 Why are heat capacity integrals required for accurate $\Delta H°$ calculations?

7.10 ΔU for most chemical reactions is in the range of 200–400 kJ/mol. At the 10% level, is there any difference between ΔH and ΔU?

PROBLEMS

Before working these problems, read Section 7.17.

7.1 One mole of an ideal gas is subjected to several changes in state. $\bar{C}_v = 12.47$ J/K mol. What will be the change in temperature in each change?

a) 512 J of heat flows out; 134 J of work are destroyed.
b) 500 J of heat flows in; 500 J of work are produced.
c) No heat flows in; 126 J of work are destroyed.

7.2 In a given change in state, 44 J of work are destroyed and the internal energy increases by 170 J. If the temperature of the system rises by 10 K, what is the heat capacity of the system?

7.3 Three moles of an ideal gas expand isothermally against a constant opposing pressure of 100 kPa from 20 dm³ to 60 dm³. Compute Q, W, ΔU, and ΔH.

7.4 a) Three moles of an ideal gas at 27 °C expand isothermally and reversibly from 20 dm³ to 60 dm³. Compute Q, W, ΔU, and ΔH.

b) Compute Q, W, ΔU, and ΔH if the same gas at 27 °C is compressed isothermally and reversibly from 60 dm³ to 20 dm³.

7.5 Three moles of an ideal gas are compressed isothermally from 60 L to 20 L using a constant pressure of 5 atm. Calculate Q, W, ΔU, and ΔH.

7.6 Develop an equation for the work produced in the isothermal, reversible expansion from $\bar{V}_1$ to $\bar{V}_2$ of a gas with the equation of state

$$p\bar{V} = RT + (bRT - a)\left(\frac{1}{\bar{V}}\right)$$

7.7 One mole of a van der Waals gas at 300 K expands isothermally and reversibly from 20 dm³ to 60 dm³ ($a = 0.556$ m⁶ Pa mol⁻²; $b = 0.064$ dm³/mol). For the van der Waals gas, $(\partial U/\partial V)_T = a/\bar{V}^2$. Calculate W, Q, ΔU, and ΔH for the transformation.

7.8 One mole of the ideal gas is confined under a constant pressure, $P_{op} = p = 200$ kPa. The temperature is changed from 100 °C to 25 °C. $\bar{C}_v = \frac{3}{2}R$. Calculate W, Q, ΔU, and ΔH.

7.9 One mole of an ideal gas, $\bar{C}_v = 20.8$ J/K mol, is transformed at constant volume from 0 °C to 75 °C. Calculate Q, W, ΔU, and ΔH.

7.10 Calculate ΔH and ΔU for the transformation of one mole of an ideal gas from 27 °C and 1 atm to 327 °C and 17 atm. $\bar{C}_p = 20.9 + 0.042\,T$ J/K mol.

7.11 If an ideal gas undergoes a reversible polytropic expansion, the relation $pV^n = C$ holds; C and n are constants, $n > 1$.

a) Calculate W for such an expansion if one mole of gas expands from V_1 to V_2 and if $T_1 = 300$ K, $T_2 = 200$ K, and $n = 2$.
b) If $\bar{C}_v = \frac{3}{2}R$, calculate Q, ΔU, and ΔH.

7.12 At 25 °C the coefficient of thermal expansion of water is $\alpha = 2.07 \times 10^{-4}$ K⁻¹ and the density is 0.9970 g/cm³. If the temperature of 200 g of water is raised from 25 °C to 50 °C under a constant pressure of 101 kPa,

a) calculate W.
b) Given $\bar{C}_p/(\text{J/K mol}) = 75.30$, calculate Q, ΔH, and ΔU.

7.13 One mole of an ideal gas is compressed adiabatically in a single stage with a constant opposing pressure equal to 1.00 MPa. Initially the gas is at 27 °C and 0.100 MPa pressure; the final pressure is 1.00 MPa. Calculate the final temperature of the gas, Q, W, ΔU, and ΔH. Do this for two cases:

    Case 1. Monatomic gas, $\bar{C}_v = \frac{3}{2}R$.

    Case 2. Diatomic gas, $\bar{C}_v = \frac{5}{2}R$.

How would the various quantities be affected if n moles were used instead of one mole?

7.14 One mole of an ideal gas at 27 °C and 0.100 MPa is compressed adiabatically and reversibly to a final pressure of 1.00 MPa. Compute the final temperature, Q, W, ΔU, and ΔH for the same two cases as in Problem 7.13.

7.15 One mole of an ideal gas at 27 °C and 1.00 MPa pressure is expanded adiabatically to a final pressure of 0.100 MPa against a constant opposing pressure of 0.100 MPa. Calculate the final temperature, Q, W, ΔU, and ΔH for the two cases, $\bar{C}_v = \frac{3}{2}R$ and $\bar{C}_v = \frac{5}{2}R$.

7.16 One mole of an ideal gas at 27 °C and 1.0 MPa pressure is expanded adiabatically and reversibly until the pressure is 0.100 MPa. Calculate the final temperature, Q, W, ΔU, and ΔH for the two cases, $\bar{C}_v = \frac{3}{2}R$ and $\bar{C}_v = \frac{5}{2}R$.

7.17 In an adiabatic expansion of one mole of an ideal gas from an initial temperature of 25 °C, the work produced is 1200 J. If $\bar{C}_v = \frac{3}{2}R$, calculate the final temperature, Q, W, ΔU, and ΔH.

7.18 If one mole of an ideal gas, $\bar{C}_v = \frac{5}{2}R$, is expanded adiabatically until the temperature drops from 20 °C to 10 °C, calculate Q, W, ΔU, and ΔH.

7.19 An automobile tire contains air at 320 kPa total pressure and 20 °C. The valve stem is removed and the air is allowed to expand adiabatically against the constant external pressure of 100 kPa until the pressure is the same inside and out. The molar heat capacity of air is $\bar{C}_v = \frac{5}{2}R$; the air may be considered an ideal gas. Calculate the final temperature of the gas in the tire, Q, W, ΔU, and ΔH per mole of gas in the tire.

7.20 A bottle at 21.0 °C contains an ideal gas under a pressure of 126.4 kPa. The rubber stopper closing the bottle is removed. The gas expands adiabatically against the constant pressure of the atmosphere, 101.9 kPa. Obviously, some gas is expelled from the bottle. When the pressure in the bottle is equal to 101.9 kPa the stopper is quickly replaced. The gas, which cooled in the adiabatic expansion, now slowly warms up until its temperature is again 21.0 °C. What is the final pressure in the bottle?

    a) If the gas is monatomic, $\bar{C}_v/R = \frac{3}{2}$.

    b) If it is diatomic, $\bar{C}_v/R = \frac{5}{2}$.

7.21 The method described in Problem 7.20 is that of Clément-Désormes for determining γ, the heat capacity ratio. In an experiment, a gas is confined initially under $p_1 = 151.2$ kPa pressure. The ambient pressure, $p_2 = 100.8$ kPa, and the final pressure after temperature equilibration is $p_3 = 116.3$ kPa. Calculate γ for this gas. Assume the gas is ideal.

7.22 When one mole of an ideal gas, $\bar{C}_v = \frac{5}{2}R$, is compressed adiabatically, the temperature rises from 20 °C to 50 °C. Calculate Q, W, ΔU, and ΔH.

7.23 One mole of an ideal gas having $\bar{C}_v = \frac{5}{2}R$ and initially at 25 °C and 100 kPa, is compressed adiabatically using a constant pressure equal to the final pressure until the temperature of the gas reaches 325 °C. Calculate the final pressure, Q, W, ΔU, and ΔH for this transformation.

7.24 One mole of an ideal gas, $\bar{C}_v = \frac{3}{2}R$, initially at 20 °C and 1.0 MPa pressure undergoes a two-stage transformation. For each stage and for the overall change calculate Q, W, ΔU, and ΔH.

    a) Stage I: Isothermal, reversible expansion to double the initial volume.

    b) Stage II: Beginning at the final state of Stage I, keeping the volume constant, the temperature is raised to 80 °C.

7.25 One mole of an ideal gas, $\bar{C}_v = \frac{5}{2}R$, is subjected to two successive changes in state.

    a) From 25 °C and 100 kPa pressure, the gas is expanded isothermally against a constant pressure of 20 kPa to twice the initial volume.

    b) After undergoing the change in (a) the gas is cooled at constant volume from 25 °C to −25 °C. Calculate Q, W, ΔU, and ΔH for the changes in (a) and (b) and for the overall change (a) + (b).

7.26 a) An ideal gas undergoes a single-stage expansion against a constant opposing pressure from T, p_1, V_1, to T, p_2, V_2. What is the largest mass M that can be lifted through a height h in this expansion?

    b) The system in (a) is restored to its initial state by a single-stage compression. What is the smallest mass M' that must fall through the height h to restore the system?

    c) What is the net mass lowered through height h in the cyclic transformation in (a) and (b)?

    d) If $h = 10$ cm, $p_1 = 1.0$ MPa, $p_2 = 0.50$ MPa, $T = 300$ K, and $n = 1$ mol, calculate the values of the masses in (a), (b), and (c).

7.27 One mole of an ideal gas is expanded from T, p_1, V_1 to T, p_2, V_2 in two stages:

	Opposing pressure	Volume change
First stage	P' (constant)	V_1 to V'
Second stage	p_2 (constant)	V' to V_2

We specify that the point P', V' lies on the isotherm at the temperature T.

    a) Formulate the expression for the work produced in this expansion in terms of T, p_1, p_2, and P'.

    b) For what value of P' does the work in this two-stage expansion have a maximum value?

    c) What is the maximum value of the work produced?

7.28 The heat capacity of solid lead oxide, PbO, is given by:

$$\bar{C}_p/(\text{J/K mol}) = 44.35 + 1.67 \times 10^{-3}\, T.$$

Calculate the change in enthalpy of PbO if it is cooled at constant pressure from 500 K to 300 K.

7.29 From the value of $\bar{C}_p$ given in Table 7.1 for oxygen, calculate Q, W, ΔU, and ΔH per mole of oxygen for the changes in state:

    a) $p = $ constant, 100 °C to 300 °C;

    b) $V = $ constant, 100 °C to 300 °C.

7.30 The Joule–Thomson coefficient for a van der Waals gas is given by $\mu_{JT} = [(2a/RT) - b]/\bar{C}_p$. Calculate the value of ΔH for the isothermal, 300 K, compression of 1 mole of nitrogen from 1 to 500 atm: $a = 0.136$ m^6 Pa mol^{-2}; $b = 0.0391$ dm^3/mol.

7.31 The boiling point of nitrogen is −196 °C, and $\bar{C}_p = \frac{7}{2}R$. The van der Waals constants and μ_{JT} are given in Problem 7.30. What must the initial pressure be if nitrogen in a single-stage Joule–Thomson expansion is to drop in temperature from 25 °C to the boiling point? (The final pressure is to be 1 atm.)

7.32 Repeat the calculation in Problem 7.31 for ammonia; b.p. = −34 °C, $\bar{C}_p = 35.6$ J/K mol, $a = 0.423$ m^6 Pa/mol^2, $b = 0.037$ dm^3/mol.

7.33 It can be shown that for a van der Waals gas $(\partial U/\partial V)_T = a/\bar{V}^2$. One mole of a van der Waals gas at 20 °C expands adiabatically and reversibly from 20.0 dm^3 to 60.0 dm^3; $\bar{C}_v = 4.79R$; $a = 0.556$ m^6 Pa mol^{-2}, $b = 64 \times 10^{-6}$ m^3/mol. Calculate Q, W, ΔU, and ΔH.

7.34 If one mole of a van der Waals gas, for which it may be shown that $(\partial U/\partial V)_T = a/\bar{V}^2$, expands isothermally from a volume equal to b, the liquid volume, to a volume of 20.0 L, calculate ΔU for the transformation; $a = 0.136$ m^6 Pa mol^{-2}; $b = 0.0391$ dm^3/mol.

7.35 From the data in Table A-V compute the values of ΔH_{298}° for the following reactions.

a) $2O_3(g) \to 3O_2(g)$.
b) $H_2S(g) + \frac{3}{2}O_2(g) \to H_2O(l) + SO_2(g)$.
c) $TiO_2(s) + 2Cl_2(g) \to TiCl_4(l) + O_2(g)$.
d) $C(\text{graphite}) + CO_2(g) \to 2CO(g)$.
e) $CO(g) + 2H_2(g) \to CH_3OH(l)$.
f) $Fe_2O_3(s) + 2Al(s) \to Al_2O_3(s) + 2Fe(s)$.
g) $NaOH(s) + HCl(g) \to NaCl(s) + H_2O(l)$.
h) $CaC_2(s) + 2H_2O(l) \to Ca(OH)_2(s) + C_2H_2(g)$.
i) $CaCO_3(s) \to CaO(s) + CO_2(g)$.

7.36 Assuming the gases are ideal, calculate ΔU_{298}° for each of the reactions in Problem 7.35.

7.37 At 25 °C and 1 atm pressure, the data are:

Substance	$H_2(g)$	C(graphite)	$C_6H_6(l)$	$C_2H_2(g)$,
$\Delta H_{\text{combustion}}^{\circ}$/(kJ/mol)	-285.83	-393.51	-3267.62	-1299.58.

a) Calculate the ΔH° of formation of liquid benzene.
b) Calculate ΔH° for the reaction $3C_2H_2(g) \to C_6H_6(l)$.

7.38 For the following reactions at 25 °C

		ΔH°/(kJ/mol)
$CaC_2(s) + 2H_2O(l) \longrightarrow$	$Ca(OH)_2(s) + C_2H_2(g)$,	-127.9;
$Ca(s) + \frac{1}{2}O_2(g) \longrightarrow$	$CaO(s)$,	-635.1;
$CaO(s) + H_2O(l) \longrightarrow$	$Ca(OH)_2(s)$,	-65.2.

The heat of combustion of graphite is -393.51 kJ/mol, and that of $C_2H_2(g)$ is -1299.58 kJ/mol. Calculate the heat of formation of $CaC_2(s)$ at 25 °C.

7.39 A sample of sucrose, $C_{12}H_{22}O_{11}$, weighing 0.1265 g is burned in a bomb calorimeter. After the reaction is over, it is found that to produce an equal temperature increment electrically, 2082.3 joules must be expended.

a) Calculate the heat of combustion of sucrose.
b) From the heat of combustion and appropriate data in Table A-V calculate the heat of formation of sucrose.
c) If the temperature increment in the experiment is 1.743 °C, what is the heat capacity of the calorimeter and contents?

7.40 If 3.0539 g of liquid ethyl alcohol, C_2H_5OH, is burned completely at 25 °C in a bomb calorimeter, the heat evolved is 90.447 kJ.

a) Calculate the molar ΔH° of combustion for ethyl alcohol at 25 °C.
b) If the ΔH_f° of $CO_2(g)$ and $H_2O(l)$ are -393.51 kJ/mol and -285.83 kJ/mol, calculate the ΔH_f° of ethyl alcohol.

7.41 From the data at 25 °C:

$Fe_2O_3(s) + 3C(\text{graphite}) \longrightarrow$	$2Fe(s) + 3CO(g)$,	$\Delta H^{\circ} =$	492.6 kJ/mol;
$FeO(s) + C(\text{graphite}) \longrightarrow$	$Fe(s) + CO(g)$,	$\Delta H^{\circ} =$	155.8 kJ/mol;
$C(\text{graphite} + O_2(g)) \longrightarrow$	$CO_2(g)$,	$\Delta H^{\circ} =$	-393.51 kJ/mol;
$CO(g) + \frac{1}{2}O_2(g) \longrightarrow$	$CO_2(g)$,	$\Delta H^{\circ} =$	-282.98 kJ/mol.

Compute the standard heat of formation of $FeO(s)$ and of $Fe_2O_3(s)$.

7.42 From the data at 25 °C:

$O_2(g) \longrightarrow$	$2O(g)$,	$\Delta H^{\circ} = 498.34$ kJ/mol;
$Fe(s) \longrightarrow$	$Fe(g)$,	$\Delta H^{\circ} = 416.3$ kJ/mol.

and $\Delta H_f^{\circ}(\text{FeO, s}) = -272$ kJ/mol.

a) Compute the $\Delta H°$ at 25 °C for the reaction

$$\text{Fe(g)} + \text{O(g)} \longrightarrow \text{FeO(s)}.$$

b) Assuming that the gases are ideal, calculate $\Delta U°$ for this reaction. (The negative of this quantity, $+933$ kJ/mol, is the cohesive energy of the crystal.)

7.43 At 25 °C, the following enthalpies of formation are given:

Compound	$SO_2(g)$	$H_2O(l)$
$\Delta H_f°/(\text{kJ/mol})$	-296.81	-285.83

For the reactions at 25 °C:

$$2\,\text{H}_2\text{S(g)} + \text{Fe(s)} \longrightarrow \text{FeS}_2(s) + 2\,\text{H}_2(g), \qquad \Delta H° = -137.0 \text{ kJ/mol};$$
$$\text{H}_2\text{S(g)} + \tfrac{3}{2}\text{O}_2(g) \longrightarrow \text{H}_2\text{O(l)} + \text{SO}_2(g) \qquad \Delta H° = -562.0 \text{ kJ/mol}.$$

Calculate the heat of formation of $H_2S(g)$ and of $FeS_2(s)$.

7.44 At 25 °C:

Substance	Fe(s)	$FeS_2(s)$	$Fe_2O_3(s)$	S(rhombic)	$SO_2(g)$
$\Delta H_f°/(\text{kJ/mol})$			-824.2		-296.81
$\bar{C}_p/R$	3.02	7.48		2.72	

For the reaction:

$$2\,\text{FeS}_2(s) + \tfrac{11}{2}\text{O}_2(g) \longrightarrow \text{Fe}_2\text{O}_3(s) + 4\,\text{SO}_2(g), \qquad \Delta H° = -1655 \text{ kJ/mol}.$$

Calculate $\Delta H_f°$ of $FeS_2(s)$ at 300 °C.

7.45 a) From the data in Table A-V compute the heat of vaporization of water at 25 °C.

b) Compute the work produced in the vaporization of one mole of water at 25 °C under a constant pressure of 1 atm.

c) Compute the ΔU of vaporization of water at 25 °C.

d) The values of $\bar{C}_p/(\text{J/K mol})$ are: water vapor, 33.577; liquid water, 75.291. Calculate the heat of vaporization at 100 °C.

7.46 At 1000 K, from the data:

$$\text{N}_2(g) + 3\,\text{H}_2(g) \longrightarrow 2\,\text{NH}_3(g), \qquad \Delta H° = -123.77 \text{ kJ/mol};$$

Substance	N_2	H_2	NH_3
$\bar{C}_p/R$	3.502	3.466	4.217

Calculate the heat of formation of NH_3 at 300 K.

7.47 For the reaction:

$$\text{C(graphite)} + \text{H}_2\text{O(g)} \longrightarrow \text{CO(g)} + \text{H}_2(g), \qquad \Delta H_{298}° = 131.28 \text{ kJ/mol}.$$

The values of $\bar{C}_p/(\text{J/K mol})$ are: graphite, 8.53; $H_2O(g)$, 33.58; CO(g), 29.12; and $H_2(g)$, 28.82. Calculate the value of $\Delta H°$ at 125 °C.

7.48 From the data in Tables A-V and 7.1 calculate the $\Delta H_{1000}°$ for the reaction

$$2\,\text{C(graphite)} + \text{O}_2(g) \longrightarrow 2\,\text{CO(g)}.$$

7.49 From the values of $\bar{C}_p$ given in Table 7.1, and from the data:

$$\tfrac{1}{2}\text{H}_2(g) + \tfrac{1}{2}\text{Br}_2(l) \longrightarrow \text{HBr(g)}, \qquad \Delta H_{298}° = -36.38 \text{ kJ/mol};$$
$$\text{Br}_2(l) \longrightarrow \text{Br}_2(g), \qquad \Delta H_{298}° = 30.91 \text{ kJ/mol}.$$

Calculate the $\Delta H_{1000}°$ for the reaction:

$$\tfrac{1}{2}\text{H}_2(g) + \tfrac{1}{2}\text{Br}_2(g) \longrightarrow \text{HBr(g)}.$$

7.50 Using the data in Appendix V and Table 7.1 calculate the $\Delta H_{298}°$ and $\Delta H_{1000}°$ for the reaction:

$$\text{C}_2\text{H}_2(g) + \tfrac{5}{2}\text{O}_2(g) \longrightarrow 2\,\text{CO}_2(g) + \text{H}_2\text{O(g)}.$$

7.51 The data are:

$$CH_3COOH(l) + 2O_2(g) \longrightarrow 2CO_2(g) + 2H_2O(l), \qquad \Delta H^\circ_{298} = -871.5 \text{ kJ/mol};$$
$$H_2O(l) \longrightarrow H_2O(g), \qquad \Delta H^\circ_{373.15} = 40.656 \text{ kJ/mol};$$
$$CH_3COOH(l) \longrightarrow CH_3COOH(g), \qquad \Delta H^\circ_{391.4} = 24.4 \text{ kJ/mol}.$$

Substance	$CH_3COOH(l)$	$O_2(g)$	$CO_2(g)$	$H_2O(l)$	$H_2O(g)$
$\bar{C}_p/R$	14.9	3.53	4.46	9.055	4.038

Calculate $\Delta H^\circ_{391.4}$ for the reaction:

$$CH_3COOH(g) + 2O_2(g) \longrightarrow 2CO_2(g) + 2H_2O(g).$$

7.52 Given the data at 25 °C:

Compound	$TiO_2(s)$	$Cl_2(g)$	$C(graphite)$	$CO(g)$	$TiCl_4(l)$
$\Delta H^\circ_f/(kJ/mol)$	−945			−110.5	
$\bar{C}^\circ_p/(J/K \text{ mol})$	55.06	33.91	8.53	29.12	145.2

For the reaction:

$$TiO_2(s) + 2C(graphite) + 2Cl_2(g) \longrightarrow 2CO(g) + TiCl_4(l), \qquad \Delta H^\circ_{298} = -80 \text{ kJ/mol}$$

a) Calculate ΔH° for this reaction at 135.8 °C, the boiling point of $TiCl_4$.
b) Calculate ΔH°_f for $TiCl_4(l)$ at 25 °C.

7.53 From the heats of solution at 25 °C:

$$HCl(g) + 100 \text{ Aq} \longrightarrow HCl \cdot 100 \text{ Aq}, \qquad \Delta H^\circ = -73.61 \text{ kJ/mol};$$
$$NaOH(s) + 100 \text{ Aq} \longrightarrow NaOH \cdot 100 \text{ Aq}, \qquad \Delta H^\circ = -44.04 \text{ kJ/mol};$$
$$NaCl(s) + 200 \text{ Aq} \longrightarrow NaCl \cdot 200 \text{ Aq}, \qquad \Delta H^\circ = +4.23 \text{ kJ/mol};$$

and the heats of formation of $HCl(g)$, $NaOH(s)$, $NaCl(s)$, and $H_2O(l)$ in Table A-V calculate ΔH° for the reaction

$$HCl \cdot 100 \text{ Aq} + NaOH \cdot 100 \text{ Aq} \longrightarrow NaCl \cdot 200 Aq + H_2O(l).$$

7.54 From the heats of formation at 25 °C:

Solution	$H_2SO_4 \cdot 600 Aq$	$KOH \cdot 200 Aq$	$KHSO_4 \cdot 800 Aq$	$K_2SO_4 \cdot 1000 Aq$
$\Delta H^\circ/(kJ/mol)$	−890.98	−481.74	−1148.8	−1412.98

calculate the ΔH° for the reactions:

$$H_2SO_4 \cdot 600 Aq + KOH \cdot 200 Aq \longrightarrow KHSO_4 \cdot 800 Aq + H_2O(l).$$
$$KHSO_4 \cdot 800 Aq + KOH \cdot 200 Aq \longrightarrow K_2SO_4 \cdot 1000 Aq + H_2O(l).$$

Use Table A-V for the heat of formation of $H_2O(l)$.

7.55 From the heats of formation at 25 °C:

Solution	$\Delta H^\circ/(kJ/mol)$	Solution	$\Delta H^\circ/(kJ/mol)$
$H_2SO_4(l)$	−813.99	$H_2SO_4 \cdot 10 Aq$	−880.53
$H_2SO_4 \cdot 1 Aq$	−841.79	$H_2SO_4 \cdot 20 Aq$	−884.92
$H_2SO_4 \cdot 2 Aq$	−855.44	$H_2SO_4 \cdot 100 Aq$	−887.64
$H_2SO_4 \cdot 4 Aq$	−867.88	$H_2SO_4 \cdot \infty Aq$	−909.27

Calculate the heat of solution of sulfuric acid for the various solutions and plot ΔH_S against the mole fraction of water in each solution.

7.56 From the data at 25 °C:

$$\tfrac{1}{2}H_2(g) + \tfrac{1}{2}O_2(g) \longrightarrow OH(g), \qquad \Delta H° = \quad 38.95 \text{ kJ/mol};$$
$$H_2(g) + \tfrac{1}{2}O_2(g) \longrightarrow H_2O(g), \qquad \Delta H° = -241.814 \text{ kJ/mol};$$
$$H_2(g) \longrightarrow 2H(g), \qquad \Delta H° = \quad 435.994 \text{ kJ/mol};$$
$$O_2(g) \longrightarrow 2O(g), \qquad \Delta H° = \quad 498.34 \text{ kJ/mol}.$$

Compute $\Delta H°$ for

a) $OH(g) \to H(g) + O(g)$,
b) $H_2O(g) \to 2H(g) + O(g)$,
c) $H_2O(g) \to H(g) + OH(g)$.
d) Assuming the gases are ideal, compute the values of $\Delta U°$ for reactions (a), (b), and (c).

Note: The energy change for (a) is called the bond energy of the OH radical; one-half the energy change in (b) is the average OH bond energy in H_2O. The energy change for (c) is the bond-dissociation energy of the OH bond in H_2O.

7.57 Given the data from Table A-V and the heats of formation at 25 °C of the gaseous compounds:

Compound	SiF_4	$SiCl_4$	CF_4	NF_3	OF_2	HF
$\Delta H_f°/(\text{kJ/mol})$	− 1614.9	−657.0	−925	− 125	−22	−271

calculate the single bond energies: Si—F; Si—Cl; C—F; N—F; O—F; H—F.

7.58 Given the data in Table A-V calculate the bond enthalpy of

a) the C—H bond in CH_4;
b) the C—C single bond in C_2H_6;
c) the C=C double bond in C_2H_4;
d) the C≡C triple bond in C_2H_2.

7.59 Using the data in Table A-V, calculate the average bond enthalpy of the oxygen–oxygen bond in ozone.

7.60 The adiabatic flame temperature is the final temperature reached by the system if one mole of the substance is burned adiabatically under the specified conditions. Using the values of $\bar{C}_p$ derived from the $\bar{C}_v$'s in Table 4.3 and data from Table A-V, calculate the adiabatic flame temperature of hydrogen burned in (a) oxygen, (b) air. (c) Assume that for water vapor, $\bar{C}_p/R = 4.0 + f(\theta_1/T) + f(\theta_2/T) + f(\theta_3/T)$ where $f(\theta/T)$ is the Einstein function, Eq. (4.88), and the values of $\theta_1, \theta_2, \theta_3$ are in Table 4.4. Calculate the adiabatic flame temperature in oxygen using this expression for $\bar{C}_p$ and compare the result with (a).

7.61 The heat of combustion of glycogen is about 476 kJ/mol of carbon. Assume that the average rate of heat loss by an adult male is 150 watts. If we were to assume that all of this came from the oxidation of glycogen, how many units of glycogen (1 mol carbon per unit) must be oxidized per day to provide for this heat loss?

7.62 Consider a classroom that is roughly 5 m × 10 m × 3 m. Initially, $t = 20$ °C and $p = 1$ atm. There are 50 people in the class, each losing energy to the room at the average rate of 150 watts. Assume that the walls, ceiling, floor, and furniture are perfectly insulating and do not absorb any heat. How long will the physical chemistry examination last if the professor has foolishly agreed to dismiss the class when the air temperature in the room reaches body temperature, 37 °C? For air, $\bar{C}_p = \tfrac{7}{2}R$. Loss of air to the outside as the temperature rises may be neglected.

7.63 Estimate the enthalpy change for liquid water, $\bar{V} = 18.0$ cm³/mol, if the pressure is increased by 10 atm at constant temperature. Compare this with the enthalpy change produced by a 10 °C increase in temperature at constant pressure; $\bar{C}_p = 75.3$ J/K mol.

7.64 Calculate the final temperature of the system if 20 g of ice at $-5\,°C$ is added to 100 g of liquid water at 21 °C in a Dewar flask (an insulated flask); for the transformation, $H_2O(s) \rightarrow H_2O(l)$; $\Delta H° = 6009$ J/mol.

$$\bar{C}_p(H_2O, s)/(J/K\;mol) = 37.7, \qquad \bar{C}_p(H_2O, l)/(J/K\;mol) = 75.3.$$

7.65 From the equipartition principle and the first law, calculate γ for an ideal gas that is (a) monatomic, (b) diatomic, and (c) nonlinear triatomic; (d) compare the values predicted by the equipartition principle with the values in Table 4.3 for (a) Ar, (b) N_2 and I_2, (c) H_2O; (e) what is the limiting value of γ, assuming equipartition, as the number of atoms in the molecule becomes very large?

7.66 Using Eq. (7.45) and Joule's law show for the ideal gas that $(\partial H/\partial p)_T = 0$.

7.67 From the ideal gas law and Eq. (7.57), derive Eqs. (7.58) and (7.59).

7.68 By applying Eq. (7.44) to a constant volume transformation, show that

$$C_p - C_v = [V - (\partial H/\partial p)_T](\partial p/\partial T)_v.$$

8

Introduction to the Second Law of Thermodynamics

8.1 GENERAL REMARKS

In Chapter 6 we mentioned the fact that all real changes have a direction which we consider natural. The transformation in the opposite sense would be unnatural; it would be unreal. In nature, rivers run from the mountains to the sea, never in the opposite way. A tree blossoms, bears fruit, and later sheds its leaves. The thought of dry leaves rising, attaching themselves to the tree, and later shrinking into buds is grotesque. An isolated metal rod initially hot at one end and cold at the other comes to a uniform temperature; such a metal rod initially at a uniform temperature never develops a hot and a cold end spontaneously.

Yet the first law of thermodynamics tells us nothing of this preference of one direction over the opposite one. The first law requires only that the energy of the universe remain the same before and after the change takes place. In the changes described above, the energy of the universe is not one whit altered; the transformation may go in either direction and satisfy the first law.

It would be helpful if a system possessed one or more properties that always change in one direction when the system undergoes a natural change, and change in the opposite direction if we imagine the system to undergo an "unnatural change." Fortunately, there exists such a property of a system, the entropy, as well as several others derived from it. To prepare a foundation for the mathematical definition of the entropy, we must divert our attention briefly to the characteristics of cyclic transformations. Having done that, we will return to chemical systems and the chemical implications of the second law.

8.2 THE CARNOT CYCLE

In 1824 a French engineer, Sadi Carnot, investigated the principles governing the transformation of thermal energy, "heat," into mechanical energy, work. He based his discussion on a cyclical transformation of a system that is now called the Carnot cycle. The

Carnot cycle consists of four reversible steps, and therefore is a reversible cycle. A system is subjected consecutively to the *reversible* changes in state:

Step 1. Isothermal expansion.   *Step 2.* Adiabatic expansion.

Step 3. Isothermal compression.   *Step 4.* Adiabatic compression.

Since the mass of the system is fixed, the state can be described by any two of the three variables T, p, V. A system of this sort that produces only heat and work effects in the surroundings is called a *heat engine*. A *heat reservoir* is a system that has the same temperature everywhere within it; this temperature is unaffected by the transfer of any desired quantity of heat into or out of the reservoir.

Imagine that the material composing the system, the "working" substance, is confined in a cylinder fitted with a piston. In Step 1, the cylinder is immersed in a heat reservoir at a temperature T_1, and is expanded isothermally from the initial volume V_1 to a volume V_2. The cylinder is now taken out of the reservoir, insulated, and in Step 2 is expanded adiabatically from V_2 to V_3; in this step the temperature of the system drops from T_1 to a lower temperature T_2. The insulation is removed and the cylinder is placed in a heat reservoir at T_2. In Step 3 the system is compressed isothermally from V_3 to V_4. The cylinder is removed from the reservoir and insulated again. In Step 4 the system is compressed adiabatically from V_4 to V_1, the original volume. In this adiabatic compression, the temperature rises from T_2 to T_1, the original temperature. Thus, as it must be in a cycle, the system is restored to its initial state.

The initial and final states and the application of the first law to each step in the Carnot cycle are described in Table 8.1. For the cycle, $\Delta U = 0 = Q_{cy} - W_{cy}$, or

$$W_{cy} = Q_{cy}. \tag{8.1}$$

Summing the first law statements for the four steps yields

$$W_{cy} = W_1 + W_2 + W_3 + W_4, \tag{8.2}$$

$$Q_{cy} = Q_1 + Q_2. \tag{8.3}$$

Combining Eqs. (8.1) and (8.3), we have

$$W_{cy} = Q_1 + Q_2. \tag{8.4}$$

(Note that the subscripts on the Q's have been chosen to correspond with those on the T's.) If W_{cy} is positive, then work has been produced at the expense of the thermal energy of the surroundings. The system suffers no net change in the cycle.

Table 8.1

Step	Initial state	Final state	First-law statement
1	T_1, p_1, V_1	T_1, p_2, V_2	$\Delta U_1 = Q_1 - W_1$
2	T_1, p_2, V_2	T_2, p_3, V_3	$\Delta U_2 = -W_2$
3	T_2, p_3, V_3	T_2, p_4, V_4	$\Delta U_3 = Q_2 - W_3$
4	T_2, p_4, V_4	T_1, p_1, V_1	$\Delta U_4 = -W_4$

8.3 THE SECOND LAW OF THERMODYNAMICS

The important thing about Eq. (8.4) is that W_{cy} is the sum of *two* terms, each associated with a *different* temperature. We might imagine a complicated cyclic process involving many heat reservoirs at different temperatures; for such a case

$$W_{cy} = Q_1 + Q_2 + Q_3 + Q_4 + \cdots,$$

where Q_1 is the heat withdrawn from the reservoir at T_1, Q_2 is the heat withdrawn from the reservoir at T_2, and so on. Some of the Q's will have positive signs, some will have negative signs; the net work effect in the cycle is the algebraic sum of all the values of Q.

It is possible to devise a cyclic process so that W_{cy} is positive; that is, such that after the cycle, masses are truly higher in the surroundings. It can be done in complicated ways using reservoirs at many different temperatures, or it can be done using only two reservoirs at different temperatures, as in the Carnot cycle. However, experience has shown that it is not possible to build such an engine using only one heat reservoir (compare with Section 7.6). Thus, if

$$W_{cy} = Q_1,$$

where Q_1 is the heat withdrawn from a single heat reservoir at a uniform temperature, then W_{cy} must be negative or, at best, zero; that is,

$$W_{cy} \le 0.$$

This experience is embodied in the second law of thermodynamics. *It is impossible for a system operating in a cycle and connected to a single heat reservoir to produce a positive amount of work in the surroundings.* This statement is equivalent to that proposed by Kelvin in about 1850.

8.4 CHARACTERISTICS OF A REVERSIBLE CYCLE

According to the second law, the simplest cyclic process capable of producing a positive amount of work in the surroundings must involve at least two heat reservoirs at different temperatures. The Carnot engine operates in such a cycle; because of its simplicity it has come to be the prototype of cyclic heat engines. An important property of the Carnot cycle is the fact that it is reversible. In a cyclic transformation, reversibility requires, after the complete cycle has been traversed once in the forward sense and once in the opposite sense, that the surroundings be restored to their original condition. This means that the reservoirs and masses must be restored to their initial condition, which can only be accomplished if reversing the cycle reverses the sign of W and of Q_1 and Q_2, individually. The magnitudes of W and of the individual values of Q are not changed by running a reversible engine backwards; only the signs are changed. Hence for a reversible engine we have:

Forward cycle: W_{cy}, Q_1, Q_2, $W_{cy} = Q_1 + Q_2$;
Reverse cycle: $-W_{cy}, -Q_1, -Q_2,$ $-W_{cy} = -Q_1 + (-Q_2)$.

8.5 A PERPETUAL-MOTION MACHINE OF THE SECOND KIND

The Carnot engine with its two heat reservoirs is usually represented schematically by a drawing such as that in Fig. 8.1. The work W produced in the surroundings by the reversible engine E_r is indicated by the arrow directed away from the system. The quantities

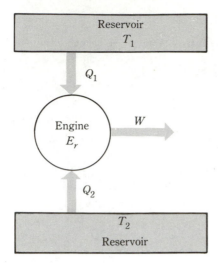

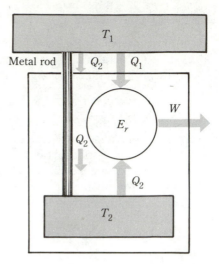

Figure 8.1 Schematic representation of the Carnot engine.

Figure 8.2 An impossible engine.

of heat Q_1 and Q_2 withdrawn from the reservoirs are indicated by arrows directed toward the system. In all of the discussions that follow, we choose T_1 as the higher temperature.

The second law has the immediate consequence that Q_1 and Q_2 cannot have the same algebraic sign. We prove this by making the contrary assumption. Assume that both Q_1 and Q_2 are positive; then W, being the sum of Q_1 and Q_2, is also positive. If Q_2 is positive, then heat flows out of the reservoir at T_2 as indicated by the arrow in Fig. 8.1. Suppose that we restore this quantity of heat Q_2 to the reservoir at T_2 by connecting the two reservoirs with a metal rod, so that heat may flow directly from the high-temperature reservoir to the low-temperature reservoir (Fig. 8.2). By making the rod in the proper shape and size, matters can be arranged so that in the time needed for the engine to traverse one cycle in which it extracts Q_2 from the reservoir, an equal quantity of heat Q_2 flows to the reservoir through the rod. Therefore after a cycle the reservoir at T_2 is restored to its initial state; thus the engine and the reservoir at T_2 form a composite cyclic engine, enclosed by the light line in Fig. 8.2. This composite cyclic engine is connected to a single heat reservoir at T_1 and produces a positive quantity of work. The second law asserts that such an engine is an impossibility. Our assumption that Q_1 and Q_2 are both positive has led to a contradiction of the second law. If we assume that Q_1 and Q_2 are both negative, then W is negative. We reverse the engine; then Q_1 and Q_2 and W are all positive, and the proof goes as before. We conclude that Q_1 and Q_2 must differ in sign; otherwise we could build this impossible engine.

Suppose that we install the impossible engine shown in Fig. 8.2 in the living room. The room itself can serve as the heat reservoir. We set the machine in motion. (Note that we do not have to plug it in!) The machine is now extracting heat from the room and producing mechanical work. Anyone with an ounce of frugality would use this work to run an electrical generator. As we gaily pen a note to the local power company stating that we no longer need their services, we note that the room is getting chilly. Air conditioning! Very nice in the summer, but definitely a nuisance in the winter. In winter we can put the machine outdoors. The heat is then extracted from the atmosphere; that engine will run for a long time before the atmosphere cools by so much as one degree; meanwhile we aren't paying any light bills. The delightful thing about this engine is that of course the atmosphere will never go permanently cold on us. As we use the stored electrical energy, it is returned to the universe mainly in the form of "heat." Unfortunately, this wonderful machine is not available on the market. The truth of the matter is that experience has shown that it is not possible to build such a machine. It is a perpetual-motion machine of the second kind.

8.6 THE EFFICIENCY OF HEAT ENGINES

Experience shows that if a heat engine operates between two temperature reservoirs so that a positive amount of work is produced, then Q_1, the heat withdrawn from the high-temperature reservoir, is positive, and Q_2, the heat withdrawn from the low-temperature reservoir, is negative. The negative value of Q_2 means that the heat flows to the low-temperature reservoir. In producing work the engine extracts an amount of heat Q_1 from the high-temperature reservoir and rejects an amount $-Q_2$ to the low-temperature reservoir. The arrow between reservoir T_2 and the engine in Figs. 8.1 and 8.2 thus seems deceptive. However, we will retain the direction of the arrow and keep in mind that in every case one of the Q's is negative. This preserves our original sign convention for Q: positive when it flows from the surroundings, and the signs will care for themselves without our juggling them back and forth.

The efficiency, ϵ, of a heat engine is defined as the ratio of the work produced to the quantity of heat extracted from the high-temperature reservoir:

$$\epsilon \equiv \frac{W}{Q_1}. \tag{8.5}$$

But, since $W = Q_1 + Q_2$,

$$= 1 + \frac{Q_2}{Q_1}. \tag{8.6}$$

Since Q_1 and Q_2 differ in sign, the second term in Eq. (8.6) is negative; consequently, the efficiency is less than unity. The efficiency is the fraction of the heat withdrawn from the high-temperature reservoir that is converted into work in the cyclic process.

8.7 ANOTHER IMPOSSIBLE ENGINE

We consider two engines E_r and E' both operating in a cycle between the same two heat reservoirs. Is it possible that the efficiencies of these two engines are different? The engines may be designed differently and may use different working substances. Let E_r be a reversible engine, and E' any engine at all, reversible or not. The reservoirs are at T_1 and T_2; $T_1 > T_2$. For the engine E_r we may write

$$W = Q_1 + Q_2, \qquad \text{(forward cycle)};$$
$$-W = -Q_1 + (-Q_2), \qquad \text{(reverse cycle)}.$$

For the engine E'

$$W' = Q_1' + Q_2', \qquad \text{(forward cycle)}.$$

Suppose that we run engine E_r in its reverse cycle and couple it to engine E' running in its forward cycle. This gives us a composite cyclic engine that produces heat and work effects that are simply the sum of the individual effects of the appropriate cycles:

$$-W + W' = -Q_1 + (-Q_2) + Q_1' + Q_2'. \tag{8.7}$$

By making the engine E_r the proper size, matters are arranged so that the composite engine produces no work effect in the surroundings; that is, we adjust E_r until $-W + W' = 0$, or

$$W = W'. \tag{8.8}$$

Equation (8.7) can then be rearranged to the form

$$Q_1 - Q_1' = -(Q_2 - Q_2'). \tag{8.9}$$

We now examine these heat effects in the reservoirs under the assumption that the efficiency of E' is greater than that of E_r, that is,

$$\epsilon' > \epsilon.$$

By the definition of the efficiency, this implies that

$$\frac{W'}{Q_1'} > \frac{W}{Q_1}.$$

Since by Eq. (8.8), $W = W'$, the inequality becomes

$$\frac{1}{Q_1'} > \frac{1}{Q_1},$$

which is equivalent to $Q_1 > Q_1'$, or

$$Q_1' - Q_1 < 0, \qquad \text{(a negative quantity)}.$$

The heat withdrawn from the reservoir at T_1 is Q_1' by E' running forward, and $-Q_1$ by E_r running in reverse. The total heat withdrawn from T_1 is the sum of these two, $Q_1' - Q_1$, and this, by our argument, has a negative value. If the heat withdrawn from the reservoir is negative, then the heat actually flows *to* that reservoir. Thus this engine pours heat into the reservoir at T_1. The heat withdrawn from the reservoir at T_2 is $Q_2' - Q_2$. Our argument, together with Eq. (8.9), would show that this quantity of heat is positive. Heat is extracted from the reservoir at T_2. The various quantities are tabulated in Table 8.2. The quantities for the composite engine are the sums of the quantities for the separate engines, the sum of the preceding two columns in Table 8.2.

The right-hand column shows that the composite engine takes a negative quantity of heat, $Q_1' - Q_1$, out of the reservoir at T_1. Consequently, the engine puts a positive quantity

Table 8.2

	E_r forward	E_r reverse	E' forward	Composite engine E_r(reverse) + E'(forward)
Work produced	W	$-W$	W	0
Heat out of T_1	Q_1	$-Q_1$	Q_1'	$Q_1' - Q_1 = -$
Heat out of T_2	Q_2	$-Q_2$	Q_2'	$Q_2' - Q_2 = +$
First law	$W = Q_1 + Q_2$	$-W = -Q_1 - Q_2$	$W = Q_1' + Q_2'$	$0 = (Q_1' - Q_1) + (Q_2' - Q_2)$

of heat into the high-temperature reservoir and extracts an equal quantity of heat from the low-temperature reservoir. The remarkable aspect of this engine is that it produces no work, nor does it require work to operate it.

Again in our imagination we install this engine in the living room. We pour a bucket of hot water into one end and a bucket of cold water into the other end, then set the machine in motion. It commences to pump heat from the cold end to the hot end. Before long the water in the hot end is boiling while that in the cold end is freezing. If the designer has been provident enough to make the cold end in the shape of an insulated box, we can keep the beer in that end and boil the coffee on the hot end. Any thrifty homemaker would be delighted with this gadget. What a kitchen appliance: a combination stove–refrigerator! And again, no bill from the power company. Experience shows that it is not possible to build this engine; this is another example of perpetual motion of the second kind.

The argument that led to this impossible engine was based only on the first law and an assumption. The assumption that the efficiency of E' is greater than E_r is in error. We therefore conclude that the efficiency of any engine E' must be less than or equal to the efficiency of a reversible engine E_r, both engines operating between the same two temperature reservoirs:

$$\epsilon' \le \epsilon. \tag{8.10}$$

The relation in Eq. (8.10) is another important consequence of the second law. The engine E' is any engine whatsoever; the engine E_r is any *reversible* engine. Consider two *reversible* engines, with efficiencies ϵ_1 and ϵ_2. Since the second one is reversible, the efficiency of the first must be less than or equal to that of the second, by Eq. (8.10):

$$\epsilon_1 \le \epsilon_2. \tag{8.11}$$

But the first engine is reversible; therefore by Eq. (8.10) the efficiency of the second must be less than or equal to that of the first:

$$\epsilon_2 \le \epsilon_1. \tag{8.12}$$

The only way that both (8.11) and (8.12) can be satisfied simultaneously is if

$$\epsilon_1 = \epsilon_2. \tag{8.13}$$

Equation (8.13), a consequence of the second law, means that *all reversible engines operating between the same two temperature reservoirs have the same efficiency.*

According to Eq. (8.13) the efficiency does not depend on the engine, so it cannot depend on the design of the engine or on the working substance used in the engine. Only the reservoirs are left; the only specifications placed on the reservoirs were the temperatures. Hence, the efficiency is a function only of the temperatures of the reservoirs:

$$\epsilon = f(T_1, T_2). \tag{8.14}$$

Since from Eq. (8.6) $\epsilon = 1 + Q_2/Q_1$, the ratio Q_2/Q_1 must be a function only of the temperatures:

$$\frac{Q_2}{Q_1} = g(T_1, T_2). \tag{8.15}$$

From the concept of reversibility it follows that an irreversible engine will produce heat and work effects in the surroundings that are different from those produced by a reversible engine. Therefore the efficiency of the irreversible engine is different from that of a reversible one; the efficiency cannot be greater, so it must be less.

8.8 THE THERMODYNAMIC TEMPERATURE SCALE

For a reversible engine, both the efficiency and the ratio Q_2/Q_1 can be calculated directly from the measurable quantities of work and heat flowing to the surroundings. Therefore we have measurable properties that depend on temperatures only and are independent of the properties of any special kind of substance. Consequently, it is possible to establish a scale of temperature independent of the properties of any individual substance. This overcomes the difficulty associated with empirical scales of temperature described in Section 6.5. This scale is the absolute, or the thermodynamic, temperature scale.

We operate a reversible heat engine in the following way. The low-temperature reservoir is at some fixed low temperature t_0. The t_0 is the temperature on any empirical scale. The heat withdrawn from this reservoir is Q_0. If we run the engine with the high-temperature reservoir at t, an amount of heat Q will flow from this reservoir and a positive amount of work will be produced. Keeping t_0 and Q_0 constant, we increase the temperature of the other reservoir to some higher temperature t'. Experimentally we find that more heat Q' is withdrawn from the reservoir at t'. Thus the heat withdrawn from the high-temperature reservoir increases with increase in temperature. For this reason we choose the heat withdrawn from the high-temperature reservoir as the thermometric property. We can define the thermodynamic temperature θ by

$$Q = a\theta, \tag{8.16}$$

where a is a constant and Q is the heat withdrawn from the reservoir. Writing Eq. (8.15) in the notation for this situation, it becomes $Q_0/Q = g(t, t_0)$. From this it is clear that if Q_0 and t_0 are constant, then Q is a function of t only. In Eq. (8.16) we have arbitrarily chosen Q as a simple and reasonable function of the absolute temperature.

The work produced in the cycle is $W = Q + Q_0$, which, using Eq. (8.16), becomes

$$W = a\theta + Q_0. \tag{8.17}$$

Now if the high-temperature reservoir is cooled until it reaches θ_0, the temperature of the cold reservoir, then the cycle becomes an isothermal cycle, and no work can be produced. Since it is a reversible cycle, $W = 0$, and so $0 = a\theta_0 + Q_0$; hence, $Q_0 = -a\theta_0$. Then Eq. (8.17) becomes

$$W = a(\theta - \theta_0). \tag{8.18}$$

For the efficiency we obtain

$$\epsilon = \frac{\theta - \theta_0}{\theta}. \tag{8.19}$$

Since there is nothing special about the temperature of the cold reservoir, except that $\theta > \theta_0$, Eqs. (8.18) and (8.19) apply to any reversible heat engine operating between any two thermodynamic temperatures θ and θ_0. Equation (8.18) shows that the work produced in a reversible heat engine is directly proportional to the difference in temperatures on the thermodynamic scale, while the efficiency is equal to the ratio of the difference in temperature to the temperature of the hot reservoir. The Carnot formula, Eq. (8.19), which relates the efficiency of a reversible engine to the temperatures of the reservoirs is probably the most celebrated formula in all of thermodynamics.

Lord Kelvin was the first to define the thermodynamic temperature scale, named in his honor, from the properties of reversible engines. If we choose the same size of the degree for both the Kelvin scale and the ideal gas scale, and adjust the proportionality constant a in Eq. (8.16) to conform to the ordinary definition of one mole of an ideal gas, then the ideal gas scale and the Kelvin scale become *numerically* identical. However, the Kelvin scale is the fundamental one. From now on we will use T for the thermodynamic temperature, $\theta = T$, except where the use of θ can supply needed emphasis.

Once one value of the thermodynamic temperature has been assigned a positive value, all other temperatures must be positive; otherwise in some circumstances the Q's for the two reservoirs would have the same sign, resulting, as we have seen, in perpetual motion.

8.9 RETROSPECTION

From the characteristics of a particularly simple kind of heat engine, the Carnot engine, and from universal experience that certain kinds of engine cannot be constructed, we concluded that all reversible heat engines operating between the same two heat reservoirs have the same efficiency, which depends only on the temperatures of the reservoirs. Thus it was possible to establish the thermodynamic scale of temperature, which is independent of the properties of any individual substance, and to relate the efficiency of the engine to the temperatures on this scale:

$$\epsilon = \frac{\theta_1 - \theta_2}{\theta_1} = \frac{T_1 - T_2}{T_1},$$

where $\theta_1 = T_1$ is the temperature of the hot reservoir.

The second law has been stated in the sense that it is impossible for an engine operating in a cycle and connected to a reservoir at only one temperature to produce a positive amount of work in the surroundings. This is equivalent to the Kelvin–Planck statement of the second law. The possibility of another kind of engine is also denied. It is impossible for an engine operating in a cycle to have as its *only* effect the transfer of a quantity of heat from a reservoir of low temperature to a reservoir at a higher temperature. This is the content of the Clausius statement of the second law. Both engines are perpetual motion machines of the second kind. If it were possible to build one of them, the other could be built. (The proof of equivalence is left as an exercise, Problem 8.1.) The Kelvin–Planck statement and the Clausius statement of the second law of thermodynamics are, of course, completely equivalent.

In this study of thermodynamic engines, our goal has been to arrive at the definition of some state property the variation of which in a given change in state would yield a clue as to whether the change in state was a real or natural change. We are at the brink of that definition, but we will first look at the Carnot cycle using an ideal gas as the working substance and also describe the operation of the Carnot refrigerator.

8.10 CARNOT CYCLE WITH AN IDEAL GAS

If an ideal gas is used as the working substance in a Carnot engine, the application of the first law to each of the steps in the cycle can be written as in Table 8.3. The values of W_1 and W_3, which are quantities of work produced in an isothermal reversible expansion of an ideal gas, are obtained from Eq. (7.6). The values of ΔU are computed by integrating the equation $dU = C_v \, dT$. The total work produced in the cycle is the sum of the individual quantities.

$$W = RT_1 \ln\left(\frac{V_2}{V_1}\right) - \int_{T_1}^{T_2} C_v \, dT + RT_2 \ln\left(\frac{V_4}{V_3}\right) - \int_{T_2}^{T_1} C_v \, dT.$$

The two integrals sum to zero, as can be shown by interchanging the limits and thus

Table 8.3

Step	General case	Ideal gas
1	$\Delta U_1 = Q_1 - W_1$	$0 = Q_1 - RT_1 \ln (V_2/V_1)$
2	$\Delta U_2 = -W_2$	$\int_{T_1}^{T_2} C_v \, dT = -W_2$
3	$\Delta U_3 = Q_2 - W_3$	$0 = Q_2 - RT_2 \ln (V_4/V_3)$
4	$\Delta U_4 = -W_4$	$\int_{T_2}^{T_1} C_v \, dT = -W_4$

changing the sign of either of them. Hence

$$W = RT_1 \ln \left(\frac{V_2}{V_1}\right) - RT_2 \ln \left(\frac{V_3}{V_4}\right), \tag{8.20}$$

where the sign of the second term has been changed by inverting the argument of the logarithm.

Equation (8.20) can be simplified if we realize that the volumes V_2 and V_3 are connected by an adiabatic reversible transformation; the same is true for V_4 and V_1. By Eq. (7.57),

$$T_1 V_2^{\gamma-1} = T_2 V_3^{\gamma-1}, \qquad T_1 V_1^{\gamma-1} = T_2 V_4^{\gamma-1}.$$

By dividing the first equation by the second, we obtain

$$\left(\frac{V_2}{V_1}\right)^{\gamma-1} = \left(\frac{V_3}{V_4}\right)^{\gamma-1} \qquad \text{or} \qquad \frac{V_2}{V_1} = \frac{V_3}{V_4}.$$

Putting this result in Eq. (8.20), we obtain

$$W = R(T_1 - T_2) \ln \left(\frac{V_2}{V_1}\right). \tag{8.21}$$

From the equation for the first step in the cycle, we have

$$Q_1 = RT_1 \ln \left(\frac{V_2}{V_1}\right),$$

and the efficiency is given by

$$\epsilon = \frac{W}{Q_1} = \frac{T_1 - T_2}{T_1} = 1 - \frac{T_2}{T_1}. \tag{8.22}$$

Equation (8.21) shows that the total work produced depends on the difference in temperature between the two reservoirs [compare to Eq. (8.18)] and the volume ratio V_2/V_1 (the compression ratio). The efficiency is a function only of the two temperatures [compare to Eq. (8.19)]. It is apparent from Eq. (8.22) that if the efficiency is to be unity, either the cold reservoir must be at $T_2 = 0$ or the hot reservoir must have T_1 equal to infinity. Neither situation is physically realizable.

8.11 THE CARNOT REFRIGERATOR

If a reversible heat engine operates so as to produce a positive amount of work in the surroundings, then a positive amount of heat is extracted from the hot reservoir and heat

Table 8.4

Cycle	Q_1	Q_2	W
Forward	+	−	+
Reverse	−	+	−

is rejected to the cold reservoir. Suppose we call this the forward cycle of the engine. If the engine is reversed, the signs of all the quantities of heat and work are reversed. Work is destroyed, $W < 0$; heat is withdrawn from the cold reservoir and rejected to the hot reservoir. In this reverse cycle, by destroying work, heat is pumped from a cold reservoir to a hot reservoir; the machine is a refrigerator. Note that the refrigerator is quite different from our impossible engine, which pumped heat from a cold end to a hot end of the machine. The impossible engine did not destroy work in the process, as a proper refrigerator would. The signs of the quantities of work and heat in the two modes of operation are shown in Table 8.4 (T_1 is the higher temperature).

The *coefficient of performance*, η, of a refrigerator is the ratio of the heat extracted from the low-temperature reservoir to the work destroyed:

$$\eta \equiv \frac{Q_2}{-W} = \frac{Q_2}{-(Q_1 + Q_2)}, \tag{8.23}$$

since $W = Q_1 + Q_2$. Also, since $(Q_2/Q_1) = -(T_2/T_1)$, we obtain

$$\eta = \frac{T_2}{T_1 - T_2}. \tag{8.24}$$

The coefficient of performance is the heat extracted from the cold box for each unit of work expended. From Eq. (8.24) it is apparent that as T_2, the temperature inside the cold box, becomes smaller, the coefficient of performance drops off very rapidly; this happens because the numerator in Eq. (8.24) decreases and the denominator increases. The amount of work that must be expended to maintain a cold temperature against a specified heat leak into the box goes up very rapidly as the temperature of the box goes down.

8.12 THE HEAT PUMP

Suppose we run the Carnot engine in reverse, as a refrigerator, but instead of having the interior of the refrigerator serve as the cold reservoir we use the outdoors as the cold reservoir and the interior of the house as the hot reservoir. Then the refrigerator pumps heat, Q_2, from outdoors and rejects heat, $-Q_1$, into the house. The coefficient of performance of the heat pump, η_{hp}, is the amount of heat pumped into the high temperature reservoir, $-Q_1$, per unit of work destroyed, $-W$.

$$\eta_{hp} \equiv \frac{-Q_1}{-W} = \frac{Q_1}{W} = \frac{Q_1}{Q_1 + Q_2}. \tag{8.25}$$

Since $Q_2/Q_1 = -T_2/T_1$,

$$\eta_{hp} = \frac{T_1}{T_1 - T_2}. \tag{8.26}$$

This remarkable formula is best illustrated by an example. Suppose that the exterior temperature is 5 °C and the interior is 20 °C. Then if $-W = 1$ kJ the quantity of heat pumped into the house is

$$-Q_1 = \frac{T_1}{T_1 - T_2}(-W) = \frac{293 \text{ K}}{15 \text{ K}}(1 \text{ kJ}) = 20 \text{ kJ}.$$

This means that if we compare a house using electric resistance heating to one using a heat pump, the expenditure of 1 kJ in resistance heating yields 1 kJ of heat to the house, while the expenditure of 1 kJ in a heat pump yields 20 kJ of heat to the house. The advantage of the heat pump over resistance heating is apparent even though the coefficients of performance of real machines are substantially below the theoretical maximum given by the second law. With the given temperatures, the coefficients of performance of real machines range from 2 to 3 (still good multiplication factors). However, when the exterior temperature drops below 5 °C the heat pump runs into trouble. Under the usual heating demand, it is difficult to supply cold air at a sufficient rate to keep the cold coil at the ambient temperature. The coil temperature drops and the performance ratio decreases, as shown by Eq. (8.26).

If we try to assess the relative economy of a heat pump versus burning fossil fuel directly, we must bear in mind that, if the electrical energy to run the heat pump comes from fossil fuel, the power plant is subject to the Carnot limitation. The overall efficiency of a modern steam power plant is about 35 percent. Thus, just to break even on fossil fuel consumption, the heat pump coefficient of performance must be at least $1/0.35 = 2.9$.

8.13 DEFINITION OF ENTROPY

Just as the first law led to the definition of the energy, so also the second law leads to a definition of a state property of the system, the entropy. It is characteristic of a state property that the sum of the changes of that property in a cycle is zero. For example, the sum of changes in energy of a system in a cycle is given by $\oint dU = 0$. We now ask whether the second law defines some new property whose changes sum to zero in a cycle.

We begin by comparing two expressions for the efficiency of a simple reversible heat engine that operates between the two reservoirs at the thermodynamic temperatures θ_1 and θ_2. We have seen that

$$\epsilon = 1 + \frac{Q_2}{Q_1} \quad \text{and} \quad \epsilon = 1 - \frac{\theta_2}{\theta_1}.$$

Subtracting these two expressions yields the result

$$\frac{Q_2}{Q_1} + \frac{\theta_2}{\theta_1} = 0,$$

which can be rearranged to the form

$$\frac{Q_1}{\theta_1} + \frac{Q_2}{\theta_2} = 0. \tag{8.27}$$

The left-hand side of Eq. (8.27) is simply the sum over the cycle of the quantity Q/θ. It could be written as the cyclic integral of the differential quantity dQ/θ:

$$\oint \frac{dQ}{\theta} = 0 \qquad \text{(reversible cycles)}. \tag{8.28}$$

Since the sum over the cycle of the quantity dQ/θ is zero, this quantity is the differential of some property of state; this property is called the *entropy* of the system and is given the symbol S. The defining equation for the entropy is then

$$dS \equiv \frac{dQ_{\text{rev}}}{T}, \tag{8.29}$$

where the subscript "rev" has been used to indicate the restriction to reversible cycles. The symbol θ for the thermodynamic temperature has been replaced by the more usual symbol T. Note that while dQ_{rev} is not the differential of a state property, dQ_{rev}/T is; dQ_{rev}/T is an *exact* differential.

8.14 GENERAL PROOF

We have shown that dQ_{rev}/T has a cyclic integral equal to zero only for cycles that involve only two temperatures. The result can be generalized to any cycle.

Consider a Carnot engine. Then in a cycle

$$W = \oint dQ, \tag{8.30}$$

and we have shown for the Carnot engine that

$$\oint \frac{dQ}{T} = 0. \tag{8.31}$$

(By the definition of the Carnot cycle, the Q is a reversible Q.) Consider another engine E'. Then in a cycle, by the first law,

$$W' = \oint dQ'; \tag{8.32}$$

but let us assume that for this engine,

$$\oint \frac{dQ'}{T} > 0. \tag{8.33}$$

This second engine may execute as complicated a cycle as we please; it may have many temperature reservoirs; it may use any working substance.

The two engines are coupled together to make a composite cyclic engine. The work produced by the composite engine in its cycle is $W_c = W + W'$, which, by Eqs. (8.30) and (8.32), is equal to

$$W_c = \oint (dQ + dQ') = \oint dQ_c \tag{8.34}$$

where $dQ_c = dQ + dQ'$.

If we add Eqs. (8.31) and (8.33), we obtain

$$\oint \frac{(dQ + dQ')}{T} > 0,$$

$$\oint \frac{dQ_c}{T} > 0. \tag{8.35}$$

We now adjust the direction of operation and the size of the Carnot engine so that the composite engine produces no work; the work required to operate E' is supplied by the Carnot engine, or vice versa. Then, $W_c = 0$, and Eq. (8.34) becomes

$$\oint dQ_c = 0. \tag{8.36}$$

Under what condition will the relations Eqs. (8.35) and (8.36) be compatible?

Because each of the cyclic integrals can be considered as a sum of terms, we write Eqs. (8.36) and (8.35) in the forms

$$Q_1 + Q_2 + Q_3 + Q_4 + \cdots = 0, \tag{8.37}$$

and

$$\frac{Q_1}{T_1} + \frac{Q_2}{T_2} + \frac{Q_3}{T_3} + \frac{Q_4}{T_4} + \cdots > 0. \tag{8.38}$$

The sum on the left-hand side of Eq. (8.37) consists of a number of terms, some positive and some negative. But the positive ones just balance the negative ones, and the sum is zero. We have to find numbers (temperatures) such that by dividing each term in Eq. (8.37) by a proper number we can obtain a sum in which the positive terms predominate, and thus fulfill the requirement of the inequality (8.38). We can make the positive terms predominate if we divide the positive terms in Eq. (8.37) by small numbers and the negative terms by larger numbers. However, this means that we are associating positive values of Q with low temperatures and negative values with high temperatures. This implies that heat is extracted from reservoirs at low temperatures and rejected to reservoirs at higher temperatures in the operation of the composite engine. The composite engine is consequently an impossible engine, and our assumption, Eq. (8.33), must be incorrect. It follows that for any engine E',

$$\oint \frac{dQ'}{T} \leq 0. \tag{8.39}$$

We distinguish two cases:

Case I. The engine E' is reversible.

We have excluded the possibility expressed by (8.33). If we assume that for E'

$$\oint \frac{dQ'}{T} < 0,$$

then we can reverse this engine, which changes all the signs but not the magnitudes of the Q's. Then we have

$$\oint \frac{dQ'}{T} > 0,$$

and the proof is the same as before. This forces us to the conclusion that for any system

$$\oint \frac{dQ_{\text{rev}}}{T} = 0 \qquad \text{(all reversible cycles)}. \tag{8.40}$$

Therefore every system has a state property S, the entropy, such that

$$dS = \frac{dQ_{\text{rev}}}{T}. \tag{8.41}$$

The study of the properties of the entropy will be undertaken in the next chapter.

Case II. The engine E' is not reversible.

For any engine we have only the possibilities expressed by (8.39). We have shown that the equality holds for the reversible engine. Since the heat and work effects associated with an irreversible cycle are different from those associated with a reversible cycle, this implies that the value of $\oint dQ/T$ for an irreversible cycle is different from the value, zero, associated with the reversible cycle. We have shown that for any engine the value cannot be greater than zero; consequently, it must be less than zero. Therefore for irreversible cycles we must have

$$\oint \frac{dQ}{T} < 0 \qquad \text{(all irreversible cycles).} \qquad (8.42)$$

8.15 THE CLAUSIUS INEQUALITY

Consider the following cycle: A system is transformed irreversibly from state 1 to state 2, then restored reversibly from state 2 to state 1. The cyclic integral is

$$\oint \frac{dQ}{T} = \int_1^2 \frac{dQ_{\text{irr}}}{T} + \int_2^1 \frac{dQ_{\text{rev}}}{T} < 0,$$

and it is less than zero, by (8.42), since it is an irreversible cycle. Using the definition of dS, this relation becomes

$$\int_1^2 \frac{dQ_{\text{irr}}}{T} + \int_2^1 dS < 0.$$

The limits can be interchanged on the second integral (but not on the first!) by changing the sign. Thus we have

$$\int_1^2 \frac{dQ_{\text{irr}}}{T} - \int_1^2 dS < 0,$$

or, by rearranging, we have

$$\int_1^2 dS > \int_1^2 \frac{dQ_{\text{irr}}}{T}. \qquad (8.43)$$

If the change in state from state 1 to state 2 is an infinitesimal one, we have

$$dS > \frac{dQ_{\text{irr}}}{T}. \qquad (8.44)$$

This is the Clausius inequality, which is a fundamental requirement for a real transformation. The inequality (8.44) enables us to decide whether or not some proposed transformation will occur in nature. We will not ordinarily use (8.44) just as it stands but will manipulate it to express the inequality in terms of properties of the state of a system, rather than in terms of a path property such as dQ_{irr}.

The Clausius inequality can be applied directly to changes in an isolated system. For any change in state in an isolated system, $dQ_{\text{irr}} = 0$. The inequality then becomes

$$dS > 0. \qquad (8.45)$$

The requirement for a real transformation in an isolated system is that dS be positive; the entropy must increase. Any natural change occurring within an isolated system is attended by an increase in entropy of the system. The entropy of an isolated system continues to increase so long as changes occur within it. When the changes cease, the system is in equilibrium and the entropy has reached a maximum value. Therefore the condition of equilibrium *in an isolated system* is that the entropy have a maximum value.

These, then, are also fundamental properties of the entropy: (1) the entropy of an isolated system is increased by any natural change which occurs within it; and (2) the entropy of an isolated system has a maximum value at equilibrium. Changes in a non-isolated system produce effects in the system and in the immediate surroundings. The system and its immediate surroundings constitute a composite isolated system in which the entropy increases as natural changes occur within it. Thus in the universe the entropy increases continually as natural changes occur within it.

Clausius expressed the two laws of thermodynamics in the famous aphorism: "The energy of the universe is constant; the entropy strives to reach a maximum."

8.16 CONCLUSION

By what may seem a rather long route, the existence of a property of a system—the entropy—has been demonstrated. The existence of this property is a consequence of the second law of thermodynamics. The zeroth law defined the temperature of a system; the first law, the energy; and the second law, the entropy. Our interest in the second law stems from the fact that this law has something to say about the natural direction of a transformation. It denies the possibility of constructing a machine that causes heat to flow from a cold to a hot reservoir without any other effect. In the same way, the second law can identify the natural direction of a chemical reaction. In some situations the second law declares that neither direction of the chemical reaction is natural; the reaction must then be at equilibrium. The application of the second law to chemical reactions is the most fruitful approach to the subject of chemical equilibrium. Fortunately, this application is easy and is done without interminable combinations of cyclic engines.

QUESTIONS

8.1 Using the considerations of Section 7.6, how can the Kelvin statement $W_{cy} \leq 0$ of Section 8.3 be amplified to (a) $W_{cy} = 0$ in a reversible cycle and (b) $W_{cy} < 0$ in an irreversible cycle?

8.2 Would the Carnot engine efficiency be increased more by (a) increasing T_1 at fixed T_2 or (b) decreasing T_2 at fixed T_1? Explain.

8.3 How can $\int dQ_{rev}/T$ vanish when integrated around a cycle while the cyclical integral of dQ_{rev} remains finite?

8.4 Verify Eq. (8.43) [with Eq. (8.41)] by (a) evaluating $\int dQ_{irr}/T$ for the irreversible Joule expansion of an ideal gas from volume V_1 to volume V_2 (Fig. 7.7); and (b) evaluating $\int dQ_{rev}/T$ for the isothermal reversible expansion of the gas between the same volumes.

PROBLEMS

Conversion factors:

$$1 \text{ watt} = 1 \text{ joule per second } (1 \text{ W} = 1 \text{ J/s})$$

$$1 \text{ horsepower} = 746 \text{ watts } (1 \text{ hp} = 746 \text{ W})$$

8.1 a) Consider the impossible engine that is connected to only one heat reservoir and produces net work in the surroundings. Couple this impossible engine to an ordinary Carnot engine in such a way that the composite engine is the "stove–refrigerator."

b) Couple the "stove–refrigerator" to an ordinary Carnot engine in such a way that the composite engine produces work in an isothermal cycle.

8.2 What is the maximum possible efficiency of a heat engine that has a hot reservoir of water boiling under pressure at 125 °C and a cold reservoir at 25 °C?

8.3 The Chalk Point, Maryland, generating station is a modern steam generating plant supplying electrical power to the Washington, D.C., and surrounding Maryland areas. Units One and Two have a gross generating capacity of 710 MW. The steam pressure is 3600 lbs/in^2 = 25 MPa and the superheater outlet temperature is 540 °C (1000 °F). The condensate temperature is at 30 °C (86 °F).

a) What is the Carnot efficiency of the engine?

b) If the efficiency of the boiler is 91.2%; the overall efficiency of the turbine, which includes the Carnot efficiency and its mechanical efficiency, is 46.7%; and the efficiency of the generator is 98.4%, what is the overall efficiency of the unit? (Note: Another 5% of the total must be subtracted to account for other plant losses.)

c) One of the coal burning units produces 355 MW. How many metric tons (1 metric ton = 1 Mg) of coal/hr are required to fuel this unit at its peak output if the heat of combustion of the coal is 29.0 MJ/kg?

d) How much heat per minute is rejected to the 30 °C reservoir in the operation of the unit in (c)?

e) If 250,000 gallons/minute of water pass through the condenser, what is the temperature rise of the water? C_p = 4.18 J/K g; 1 gallon = 3.79 litres; density = 1.0 kg/L.

(Data courtesy of William Herrmann, the Potomac Electric Power Company.)

8.4 a) Liquid helium boils at about 4 K, and liquid hydrogen boils at about 20 K. What is the efficiency of a reversible engine operating between heat reservoirs at these temperatures?

b) If we wanted the same efficiency as in (a) for an engine with a cold reservoir at ordinary temperature, 300 K, what must the temperature of the hot reservoir be?

8.5 The solar energy flux is about 4 J/cm^2 min. In a nonfocusing collector the temperature can reach a value of about 90 °C. If we operate a heat engine using the collector as the heat source and a low temperature reservoir at 25 °C, calculate the area of collector needed if the heat engine is to produce 1 horsepower. Assume that the engine operates at maximum efficiency.

8.6 A refrigerator is operated by a $\frac{1}{4}$-hp motor. If the interior of the box is to be maintained at −20 °C against a maximum exterior temperature of 35 °C, what is the maximum heat leak (in watts) into the box that can be tolerated if the motor runs continuously? Assume that the coefficient of performance is 75% of the value for a reversible engine.

8.7 Suppose an electrical motor supplies the work to operate a Carnot refrigerator. If the heat leak into the box is 1200 J/s and the interior of the box is to be maintained at −10 °C while the exterior is at 30 °C, what size motor (in horsepower) must be used if the motor runs continuously? Assume that the efficiencies involved have their largest possible values.

8.8 Suppose an electrical motor supplies the work to operate a Carnot refrigerator. The interior of the refrigerator is at 0 °C. Liquid water is taken in at 0 °C and converted to ice at 0 °C. To convert 1 g of ice to 1 g liquid, ΔH_{fus} = 334 J/g are required. If the temperature outside the box is 20 °C, what mass of ice can be produced in one minute by a $\frac{1}{4}$-hp motor running continuously? Assume that the refrigerator is perfectly insulated and that the efficiencies involved have their largest possible values.

8.9 Under 1 atm pressure, helium boils at 4.216 K. The heat of vaporization is 84 J/mol. What size motor (in horsepower) is needed to run a refrigerator that must condense 2 mol of gaseous helium at 4.216 K to liquid at 4.216 K in one minute? Assume that the ambient temperature is 300 K and that the coefficient of performance of the refrigerator is 50% of the maximum possible.

8.10 A 0.1 horsepower motor is used to run a Carnot refrigerator. If the motor runs continuously, what will be the temperature reached inside the box if the heat leak into the box is 500 J/s and the outside temperature is 20 °C? Assume that the machine performs with maximum efficiency.

8.11 If a heat pump is to provide a temperature of 21 °C inside the house from an exterior reservoir at 1 °C, calculate the maximum value for the coefficient of performance. If the cold end of the heat pump is made as a solar collector, what must the area of the collector be if the 1 °C temperature is maintained while pumping 2 kJ/s into the house as heat? Assume that the solar flux is 40 kJ m^{-2} min^{-1}.

8.12 If a fossil fuel power plant operating between 540 °C and 50 °C provides the electrical power to run a heat pump that works between 25 °C and 5 °C, what is the amount of heat pumped into the house per unit amount of heat extracted from the power plant boiler?

  a) Assume that the efficiencies are equal to the theoretical maximum values.
  b) Assume that the power plant efficiency is 70% of maximum and that the coefficient of performance of the heat pump is 10% of maximum.
  c) If a furnace can use 80% of the energy in fossil fuel to heat the house, would it be more economical in terms of overall fossil fuel consumption to use a heat pump or a furnace? Do the calculations for cases (a) and (b).

8.13 A 23 600 BTU/hr air-conditioning unit has an energy efficiency ratio (EER) of 7.5. The EER is defined as the number of BTU/hr extracted from the room divided by the power consumption of the unit in watts (1 BTU = 1.055 kJ).

  a) What is the actual coefficient of performance of this refrigerator?
  b) If the outside temperature is 32 °C and the inside temperature is 22 °C, what percent of the theoretical maximum value is the coefficient of performance?

8.14 The standard temperatures for evaluating the performance of heat pumps for high temperatures are 70 °F for the inside temperature and 47 °F for the outside temperature. For low-temperature heating the standard temperatures are 70 °F and 17 °F. Calculate the theoretical coefficient of performance for the heat pump under both these conditions. The values achieved by commercial machines range from 1.0[*sic*] to 2.4 for low-temperature heating and from 1.7 to 3.2 for high-temperature heating.

8.15 The standard conditions for evaluating air conditioners are 80 °F interior temperature and 95 °F exterior temperature. Calculate the theoretical coefficient of performance under these conditions. What value of EER does this coefficient of performance translate to? (EER is defined in Problem 8.13.) *Note:* The EER values for commercial machines range from 4.35 to 12.80.

8.16 a) Suppose we choose the efficiency of a reversible engine as the thermometric property for a thermodynamic temperature scale. Let the cold reservoir have a fixed temperature. Measure the efficiency of the engine with the hot reservoir at the ice point, 0 degrees, and with the hot reservoir at the steam point, 100 degrees. What is the relation between temperatures, t, on this scale and the usual thermodynamic temperatures T?

  b) Suppose the hot reservoir has a fixed temperature and we define the temperature scale by measuring efficiency with the cold reservoir at the steam point and at the ice point. Find the relation between t and T for this case. (Choose 100 degrees between the ice point and the steam point.)

8.17 Consider the following cycle using 1 mol of an ideal gas, initially at 25 °C and 1 atm pressure.

Step 1. Isothermal expansion against zero pressure to double the volume (Joule expansion).
Step 2. Isothermal, reversible compression from $\frac{1}{2}$ atm to 1 atm.

  a) Calculate the value of $\oint dQ/T$; note that the sign conforms with Eq. (8.42).
  b) Calculate ΔS for step 2.
  c) Realizing that for the cycle, $\Delta S_{cycle} = 0$, find ΔS for step 1.
  d) Show that ΔS for step 1 is *not* equal to the Q for step 1 divided by T.

9

Properties of the Entropy and the Third Law of Thermodynamics

9.1 THE PROPERTIES OF ENTROPY

Each year the question, "What is entropy?" echoes plaintively in physical chemistry classrooms. The questioner rarely regards the answer given as a satisfactory one. The question springs from a strange feeling most people have that entropy is something they can see or feel or put in a bottle, if only they could squint at the system from the proper angle. The difficulty arises for two reasons. First, it must be admitted that entropy is a more impalpable thing than a quantity of heat or work. Second, the question itself is vague (unintentionally, of course). Sleepless nights can be saved if, at least for the present, we simply ignore the vague question, "What is entropy?" and consider precise questions and statements about entropy. How does the entropy change with temperature under constant pressure? How does the entropy change with volume at constant temperature? If we know how the entropy behaves in various circumstances, we will know a great deal about what it "is." Later, the entropy will be related to "randomness" in a spatial or energy distribution of the constituent particles. However, this relation to "randomness" depends on the assumption of a structural model for a system, while the purely thermodynamic definition is independent of any structural model and, in fact, does not require such a model. The entropy is defined by the differential equation

$$dS = \frac{dQ_{rev}}{T},$$
(9.1)

from which it follows that the entropy is a *single-valued*, *extensive* state property of the system. The differential dS is an *exact* differential. For a finite change in state from state 1 to state 2, we have from Eq. (9.1)

$$\Delta S = S_2 - S_1 = \int_1^2 \frac{dQ_{rev}}{T}.$$
(9.2)

Since the values of S_2 and S_1 depend only on the states 1 and 2, it does not matter in the least whether the change in state is *effected* by a reversible process or an irreversible process; ΔS is the same regardless. However, if we use Eq. (9.2) to *calculate* ΔS, we must use the heat withdrawn along any *reversible* path connecting the two states.

9.2 CONDITIONS OF THERMAL AND MECHANICAL STABILITY OF A SYSTEM

Before beginning a detailed discussion of the properties of the entropy, two facts must be established. The first is that the heat capacity at constant volume C_v is always positive for a pure substance in a single state of aggregation; the second is that the coefficient of compressibility κ is always positive for such a substance. Although each of these statements is capable of elegant mathematical proof from the second law, a simple physical argument will be convincing enough for our purposes.

Suppose that for the system specified, C_v is negative and that the system is kept at constant volume. If a warm draft strikes the system, an amount of heat, $dQ_V = +$, flows from the surroundings; by definition, $dQ_V = C_v \, dT$. Since dQ_V is positive, and by supposition C_v is negative, dT would have to be negative to fulfill this relation. Thus the flow of heat into this system lowers its temperature, which causes more heat to flow in, and the system cools even more. Ultimately, the system would get very cold for no reason but that an accidental draft struck it. By the same argument, an accidental cold draft would result in the system getting extremely hot. It would be too distressing to have objects in a room glowing red hot and freezing up just because of drafts. Therefore C_v must be positive to ensure the thermal stability of a system against chance variations in external temperature.

The coefficient of compressibility has been defined, Eq. (5.4), as

$$\kappa = - \frac{1}{V} \left(\frac{\partial V}{\partial p} \right)_T ;$$

(9.3)

thus at constant temperature $dp = -(dV/V\kappa)$. Suppose that at constant temperature the system is accidentally pushed in a little bit, dV is then negative. If κ is negative, dp must be negative to fulfill the relation. The pressure in the system goes down, which allows the external pressure to push the system in a little more, which lowers the pressure further. The system would collapse. If the volume of the system were accidentally increased, the system would explode. We conclude that κ must be positive if the system is to be mechanically stable against accidental variations in its volume.

9.3 ENTROPY CHANGES IN ISOTHERMAL TRANSFORMATIONS

For any isothermal change in state, T, being constant, can be removed from the integral in Eq. (9.2), which then reduces immediately to

$$\Delta S = \frac{Q_{\text{rev}}}{T}.$$

(9.4)

The entropy change for the transformation can be calculated by evaluating the quantities of heat required to conduct the change in state reversibly.

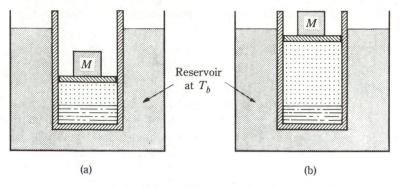

(a)   (b)

Figure 9.1   Reversible vaporization of a liquid.

Equation (9.4) is used to calculate the entropy change associated with a change in state of aggregation at the equilibrium temperature. Consider a liquid in equilibrium with its vapor under a pressure of 1 atm. The temperature is the equilibrium temperature, the normal boiling point of the liquid. Imagine that the system is confined in a cylinder by a floating piston carrying a mass equivalent to the 1 atm pressure (Fig. 9.1a). The cylinder is immersed in a temperature reservoir at the equilibrium temperature T_b. If the temperature of the reservoir is raised infinitesimally, a small quantity of heat flows from the reservoir to the system, some liquid vaporizes, and the mass M rises (Fig. 9.1b). If the temperature of the reservoir is lowered infinitesimally, the same quantity of heat flows back to the reservoir. The vapor formed originally condenses, and the mass falls to its original position. Both the system and the reservoir are restored to their initial condition in this small cycle, and the transformation is reversible; the quantity of heat required is a Q_{rev}. The pressure is constant, so that $Q_p = \Delta H$; hence for the vaporization of a liquid at the boiling point, Eq. (9.4) becomes

$$\Delta S_{vap} = \frac{\Delta H_{vap}}{T_b}. \tag{9.5}$$

By the same argument, the entropy of fusion at the melting point is given by

$$\Delta S_{fus} = \frac{\Delta H_{fus}}{T_m}, \tag{9.6}$$

where ΔH_{fus} is the heat of fusion at the melting point T_m. For any change of phase at the equilibrium temperature T_e, the entropy of transition is given by

$$\Delta S = \frac{\Delta H}{T_e}, \tag{9.7}$$

where ΔH is the heat of transition at T_e.

9.3.1 Trouton's Rule

For many liquids, the entropy of vaporization at the normal boiling point has approximately the same value:

$$\Delta S_{vap} \approx 90 \text{ J/K mol}. \tag{9.8}$$

Equation (9.8) is Trouton's rule. It follows immediately that for liquids which obey this rule,

$$\Delta H_{vap} \approx (90 \text{ J/K mol})T_b \qquad (9.9)$$

which is useful for obtaining an approximate value of the heat of vaporization of a liquid from a knowledge of its boiling point.

Trouton's rule fails for associated liquids such as water, alcohols, and amines. It also fails for substances with boiling points of 150 K or below. Hildebrand's rule, which we describe later, includes these low-boiling substances, but not associated liquids.

There is no equally general rule for entropies of fusion at the melting point. For most substances the entropy of fusion is much less than the entropy of vaporization, lying usually in the range from R to $4R$. If the particles composing the substances are atoms, such as in the metals, the entropy of fusion is about equal to R. If the molecule composing the substance is quite large (a long chain hydrocarbon, for example) the entropy of fusion may be as high as $15R$.

9.4 MATHEMATICAL INTERLUDE. MORE PROPERTIES OF EXACT DIFFERENTIALS. THE CYCLIC RULE

The total differential of a function of two variables $f(x, y)$ is written in the form

$$df = \frac{\partial f}{\partial x} dx + \frac{\partial f}{\partial y} dy. \qquad (9.10)$$

Since the differential coefficients $(\partial f/\partial x)$ and $(\partial f/\partial y)$ are functions of x and y, we may write

$$M(x, y) = \frac{\partial f}{\partial x}, \qquad N(x, y) = \frac{\partial f}{\partial y}, \qquad (9.11)$$

and Eq. (9.10) becomes

$$df = M(x, y) \, dx + N(x, y) \, dy. \qquad (9.12)$$

If we form second derivatives of the function $f(x, y)$, there are several possibilities: $(\partial f/\partial x)$ can be differentiated with respect to either x or y, and the same is true of $(\partial f/\partial y)$. We get

$$\frac{\partial^2 f}{\partial x^2}, \qquad \frac{\partial^2 f}{\partial y \, \partial x}, \qquad \frac{\partial^2 f}{\partial x \, \partial y}, \qquad \frac{\partial^2 f}{\partial y^2}.$$

Of these four, only three are distinct. It can be shown that for a function of several variables, the order of differentiation with respect to two variables such as x and y does not matter and the mixed derivatives are equal; that is,

$$\frac{\partial^2 f}{\partial y \, \partial x} = \frac{\partial^2 f}{\partial x \, \partial y}. \qquad (9.13)$$

Differentiating the first of Eqs. (9.11) with respect to y, and the second with respect to x, we obtain

$$\frac{\partial M}{\partial y} = \frac{\partial^2 f}{\partial y \, \partial x}, \qquad \frac{\partial N}{\partial x} = \frac{\partial^2 f}{\partial x \, \partial y}.$$

These two equations in the light of Eq. (9.13) yield

$$\frac{\partial M}{\partial y} = \frac{\partial N}{\partial x}. \tag{9.14}$$

The derivatives in Eq. (9.14) are sometimes called "cross-derivatives" because of their relation to the total differential, Eq. (9.12):

$$df = M\,dx + N\,dy.$$

(In all the above equations, the subscript on the derivatives denoting constancy of x or y has been dropped to simplify the writing.)

■ **EXAMPLE 9.1**   If we write the first law equation as $dU = dQ_{\text{rev}} - p\,dV$, and then, using Eq. (9.1), set $dQ_{\text{rev}} = T\,dS$, the first law becomes

$$dU = T\,dS - p\,dV.$$

Applying the cross-derivative rule in Eq. (9.14), we obtain

$$\left(\frac{\partial T}{\partial V}\right)_S = -\left(\frac{\partial p}{\partial S}\right)_V. \tag{9.15}$$

Equation (9.15) is one of an important group of equations called the Maxwell relations; its meaning will be discussed later along with that of the other members of the group. The equality of the cross-derivatives is used frequently in later arguments.

The rule in Eq. (9.14) follows from the fact that the differential expression $M\,dx + N\,dy$ is the total differential of some function $f(x, y)$; that is, $M\,dx + N\,dy$ is an exact differential expression. The converse is also true. For example, suppose that we have an expression of the form

$$R(x, y)\,dx + Q(x, y)\,dy. \tag{9.16}$$

This is an exact differential expression *if and only if*

$$\frac{\partial R}{\partial y} = \frac{\partial Q}{\partial x}. \tag{9.17}$$

If Eq. (9.17) is satisfied, then there exists some function of x and y, $g(x, y)$, for which

$$dg = R\,dx + Q\,dy.$$

If Eq. (9.17) is not satisfied, then no such function $g(x, y)$ exists, and the differential expression (9.16) is an inexact differential.

9.4.1 The Cyclic Rule

Another useful relation between partial derivatives is the cyclic rule. The total differential of a function $z(x, y)$ is written

$$dz = \left(\frac{\partial z}{\partial x}\right)_y dx + \left(\frac{\partial z}{\partial y}\right)_x dy. \tag{9.18}$$

We now restrict Eq. (9.18) to those variations of x and y that leave the value of z unchanged; $dz = 0$:

$$0 = \left(\frac{\partial z}{\partial x}\right)_y (\partial x)_z + \left(\frac{\partial z}{\partial y}\right)_x (\partial y)_z.$$

Dividing by $(\partial y)_z$, we have

$$0 = \left(\frac{\partial z}{\partial x}\right)_y \left(\frac{\partial x}{\partial y}\right)_z + \left(\frac{\partial z}{\partial y}\right)_x.$$

Multiplying by the reciprocal of the second term, $(\partial y/\partial z)_x$, we obtain

$$0 = \left(\frac{\partial z}{\partial x}\right)_y \left(\frac{\partial x}{\partial y}\right)_z \left(\frac{\partial y}{\partial z}\right)_x + 1.$$

A slight rearrangement brings this to

$$\left(\frac{\partial x}{\partial y}\right)_z \left(\frac{\partial y}{\partial z}\right)_x \left(\frac{\partial z}{\partial x}\right)_y = -1, \qquad (9.19)$$

which is the cyclic rule. The variables x, y, z in the numerators are related to y, z, x in the denominators and to the subscripts z, x, y by a cyclic permutation. If any three variables are connected by a functional relationship, then the three partial derivatives satisfy a relation of the type of Eq. (9.19). Since in many thermodynamic situations, the variables of state are functions of two other variables, Eq. (9.19) has frequent application. The lovely part of an equation such as Eq. (9.19) is that we do not have to memorize it. Write the three variables down in any order, x, y, z, then underneath them write the variables again in any order so that the vertical columns do not match; there are only two possibilities:

$$xyz, \qquad xyz,$$

$$yzx, \qquad zxy.$$

The first row yields the numerators of the derivatives, the second row the denominators; the subscripts are easily obtained, since in any derivative the same symbol does not occur twice. From the diagrams we write

$$\left(\frac{\partial x}{\partial y}\right)_z \left(\frac{\partial y}{\partial z}\right)_x \left(\frac{\partial z}{\partial x}\right)_y = -1 \quad \text{and} \quad \left(\frac{\partial x}{\partial z}\right)_y \left(\frac{\partial y}{\partial x}\right)_z \left(\frac{\partial z}{\partial y}\right)_x = -1.$$

The first expression is Eq. (9.19); the second is the reciprocal of Eq. (9.19). Since the reciprocal of -1 is also -1, it is almost impossible to write this equation incorrectly.

9.4.2 An Application of the Cyclic Rule

Suppose that the three variables are pressure, temperature, and volume. We write the cyclic rule using the variables p, T, V:

$$\left(\frac{\partial p}{\partial T}\right)_V \left(\frac{\partial T}{\partial V}\right)_p \left(\frac{\partial V}{\partial p}\right)_T = -1.$$

From the definitions of the coefficient of thermal expansion and the coefficient of compressibility, we have

$$\left(\frac{\partial V}{\partial T}\right)_p = V\alpha \quad \text{and} \quad \left(\frac{\partial V}{\partial p}\right)_T = -V\kappa.$$

Using the definitions of α and κ, the cyclic rule becomes

$$\left(\frac{\partial p}{\partial T}\right)_V \frac{1}{V\alpha}(-V\kappa) = -1,$$

so that

$$\left(\frac{\partial p}{\partial T}\right)_V = \frac{\alpha}{\kappa}. \tag{9.20}$$

With the cross-derivative rule and the cyclic relation at our disposal, we are ready to manipulate the equations of thermodynamics into useful forms.

EXERCISES

1. For each of the following functions, calculate $\partial f/\partial x$, $\partial f/\partial y$, and verify that the mixed second derivatives are equal. (a) $x^2 + y^2$; (b) xy; (c) $x^2 y^3 + 2x^3 y^2 - 5x^5 + xy^4$; (d) x/y; (e) $\sin xy^2$.
2. Test each of the expressions to decide which are exact differentials. (a) $2dx - 3dy$; (b) $y\,dx + x\,dy$; (c) $y\,dx - x\,dy$; (d) $3x^2 y\,dx + x^3\,dy$; (e) $y^2\,dx + x^2\,dy$.
3. If $z = xy^3$, calculate $(\partial y/\partial x)_z$. (a) Directly, by solving for y in terms of z and x, then differentiating; (b) by using the cyclic rule.

9.5 ENTROPY CHANGES IN RELATION TO CHANGES IN THE STATE VARIABLES

The defining equation for the entropy,

$$dS = \frac{dQ_{\text{rev}}}{T}, \tag{9.21}$$

relates the change in entropy to an effect, dQ_{rev}, in the surroundings. It would be useful to transform this equation so as to relate the change in entropy to changes in value of state properties of the system. This is quite easily done.

If only pressure–volume work is done, then in a reversible transformation, we have $P_{\text{op}} = p$, the pressure of the system, so that the first law becomes

$$dQ_{\text{rev}} = dU + p\,dV. \tag{9.22}$$

Dividing Eq. (9.22) by T and using the definition of dS, we obtain

$$dS = \frac{1}{T}\,dU + \frac{p}{T}\,dV, \tag{9.23}$$

which relates the change in entropy dS to changes in energy and volume, dU and dV, and to the pressure and temperature of the system. Equation (9.23), a combination of the first and second laws of thermodynamics, is the fundamental equation of thermodynamics;

all our discussions of the equilibrium properties of a system will begin from this equation or equations directly related to it.

For the present, it is sufficient to state that both of the differential coefficients $1/T$ and p/T are always positive. According to Eq. (9.23) there are two independent ways of varying the entropy of a system: by varying the energy or the volume. Note carefully that if the volume is constant ($dV = 0$), an increase in energy (dU is $+$) implies an increase in entropy. Also, if the energy is constant ($dU = 0$), an increase in volume (dV is $+$) implies an increase in entropy. This behavior is a fundamental characteristic of the entropy. At constant volume, the entropy goes up as the energy goes up. At constant energy, the entropy goes up as the volume goes up.

In the laboratory we do not ordinarily exercise control of the energy of the system directly. Since we can conveniently control the temperature and volume, or the temperature and pressure, it is useful to transform Eq. (9.23) to the more convenient sets of variables, T and V, or T and p.

9.6   ENTROPY AS A FUNCTION OF TEMPERATURE AND VOLUME

Considering the entropy as a function of T and V, we have $S = S(T, V)$; the total differential is written as

$$dS = \left(\frac{\partial S}{\partial T}\right)_V dT + \left(\frac{\partial S}{\partial V}\right)_T dV. \tag{9.24}$$

Equation (9.23) can be brought into the form of Eq. (9.24) if we express dU in terms of dT and dV. In these variables,

$$dU = C_v dT + \left(\frac{\partial U}{\partial V}\right)_T dV. \tag{9.25}$$

Using this value of dU in Eq. (9.12), we have

$$dS = \frac{C_v}{T} dT + \frac{1}{T}\left[p + \left(\frac{\partial U}{\partial V}\right)_T\right] dV. \tag{9.26}$$

Since Eq. (9.26) expresses the change in entropy in terms of changes in T and V, it must be identical with Eq. (9.24), which does the same thing. In view of this identity, we may write

$$\left(\frac{\partial S}{\partial T}\right)_V = \frac{C_v}{T}, \tag{9.27}$$

and

$$\left(\frac{\partial S}{\partial V}\right)_T = \frac{1}{T}\left[p + \left(\frac{\partial U}{\partial V}\right)_T\right]. \tag{9.28}$$

Since C_v/T is always positive (Section 9.2), Eq. (9.27) expresses the important fact that at constant volume the entropy increases with increase in temperature. Note that the dependence of entropy on temperature is simple, the differential coefficient being the appropriate heat capacity divided by the temperature. For a finite change in temperature at constant volume

$$\Delta S = \int_{T_1}^{T_2} \frac{C_v}{T} dT. \tag{9.29}$$

■ **EXAMPLE 9.2**   One mole of argon is heated at a constant volume from 300 K to 500 K; $\overline{C}_v = \frac{3}{2}R$. Compute the change in entropy for this change in state.

$$\Delta S = \int_{300}^{500} \frac{\frac{3}{2}R}{T} \, dT = \frac{3}{2}R \ln \frac{500 \text{ K}}{300 \text{ K}} = 0.766R = 0.766(8.314 \text{ J/K mol}) = 6.37 \text{ J/K mol}.$$

Note that if 2 mol were used, C_v would be doubled and so the entropy change would be doubled.

In contrast to the simplicity of the temperature dependence, the volume dependence *at constant temperature* given by Eq. (9.28) is quite complicated. Remember that the volume dependence *at constant energy*, Eq. (9.23), was very simple. We can obtain a simpler expression for the isothermal volume dependence of the entropy by the following device. We differentiate Eq. (9.27) with respect to volume, keeping temperature constant; this yields

$$\frac{\partial^2 S}{\partial V \, \partial T} = \frac{1}{T} \frac{\partial C_v}{\partial V} = \frac{1}{T} \frac{\partial^2 U}{\partial V \, \partial T}.$$

In the right-hand side we have replaced C_v by $(\partial U/\partial T)_V$. Similarly, we differentiate Eq. (9.28) with respect to temperature keeping volume constant, to obtain

$$\frac{\partial^2 S}{\partial T \, \partial V} = \frac{1}{T} \left[\left(\frac{\partial p}{\partial T} \right)_V + \frac{\partial^2 U}{\partial T \, \partial V} \right] - \frac{1}{T^2} \left[p + \left(\frac{\partial U}{\partial V} \right)_T \right].$$

However, since S is a function of T and V (dS is an exact differential) the mixed second derivatives must be equal; hence we have

$$\frac{\partial^2 S}{\partial V \, \partial T} = \frac{\partial^2 S}{\partial T \, \partial V},$$

or

$$\frac{1}{T} \left(\frac{\partial^2 U}{\partial V \, \partial T} \right) = \frac{1}{T} \left(\frac{\partial p}{\partial T} \right)_V + \frac{1}{T} \left(\frac{\partial^2 U}{\partial T \, \partial V} \right) - \frac{1}{T^2} \left[p + \left(\frac{\partial U}{\partial V} \right)_T \right].$$

Now the same consideration applies to U; the mixed second derivatives are equal. This reduces the preceding equation to

$$p + \left(\frac{\partial U}{\partial V} \right)_T = T \left(\frac{\partial p}{\partial T} \right)_V. \tag{9.30}$$

Comparing Eqs. (9.30) and (9.28) we obtain

$$\left(\frac{\partial S}{\partial V} \right)_T = \left(\frac{\partial p}{\partial T} \right)_V. \tag{9.31}$$

Equation (9.31) is a relatively simple expression for the isothermal volume dependence of the entropy in terms of a derivative, $(\partial p/\partial T)_V$, which is readily measurable for any system. From Eq. (9.20), the cyclic rule, we have $(\partial p/\partial T)_V = \alpha/\kappa$. Using this result, we obtain

$$\left(\frac{\partial S}{\partial V} \right)_T = \frac{\alpha}{\kappa}. \tag{9.32}$$

Since κ is positive, the sign of this derivative depends on the sign of α; for the vast majority of substances the volume increases with temperature so that α is positive. According to

Eq. (9.32) then, for the majority of substances the entropy will increase with increase in volume. Water between 0 °C and 4 °C has a negative value of α and so is an exception to the rule.

The equations written in this section are applicable to any substance. Thus for any substance we can write the total differential of the entropy in terms of T and V in the form

$$dS = \frac{C_v}{T}\, dT + \frac{\alpha}{\kappa}\, dV. \tag{9.33}$$

Except for gases, the dependence of entropy on volume at constant temperature is negligibly small in most practical situations.

9.7 ENTROPY AS A FUNCTION OF TEMPERATURE AND PRESSURE

If the entropy is considered as a function of temperature and pressure, $S = S(T, p)$, the total differential is written

$$dS = \left(\frac{\partial S}{\partial T}\right)_p dT + \left(\frac{\partial S}{\partial p}\right)_T dp. \tag{9.34}$$

To bring Eq. (9.23) into this form, we introduce the relation between energy and enthalpy in the form $U = H - pV$; differentiating yields

$$dU = dH - p\, dV - V\, dp.$$

Using this value for dU in Eq. (9.23), we have

$$dS = \frac{1}{T}\, dH - \frac{V}{T}\, dp, \tag{9.35}$$

which is another version of the fundamental equation (9.23); it relates dS to changes in enthalpy and pressure. We can express dH in terms of dT and dp, as we have seen before:

$$dH = C_p\, dT + \left(\frac{\partial H}{\partial p}\right)_T dp. \tag{9.36}$$

Using this value of dH in Eq. (9.35), we obtain

$$dS = \frac{C_p}{T}\, dT + \frac{1}{T}\left[\left(\frac{\partial H}{\partial p}\right)_T - V\right] dp. \tag{9.37}$$

Since Eqs. (9.34) and (9.37) both express dS in terms of dT and dp, they must be identical. Comparison of the two equations shows that

$$\left(\frac{\partial S}{\partial T}\right)_p = \frac{C_p}{T}, \tag{9.38}$$

and

$$\left(\frac{\partial S}{\partial p}\right)_T = \frac{1}{T}\left[\left(\frac{\partial H}{\partial p}\right)_T - V\right]. \tag{9.39}$$

For any substance, the ratio C_p/T is always positive. Therefore, Eq. (9.38) states that at constant pressure the entropy always increases with temperature. Here again, the dependence of entropy on temperature is simple, the derivative being the ratio of the appropriate heat capacity to the temperature.

In Eq. (9.39) we have a rather messy expression for the pressure dependence of the entropy at constant temperature. To simplify matters, we again form the mixed second derivatives and set them equal. Differentiation of Eq. (9.38) with respect to pressure at constant temperature yields

$$\frac{\partial^2 S}{\partial p \, \partial T} = \frac{1}{T}\left(\frac{\partial C_p}{\partial p}\right)_T = \frac{1}{T}\frac{\partial^2 H}{\partial p \, \partial T}.$$

To obtain right-hand equality we have set $C_p = (\partial H/\partial T)_p$. Similarly, differentiation of Eq. (9.39) with respect to temperature yields

$$\frac{\partial^2 S}{\partial T \, \partial p} = \frac{1}{T}\left[\frac{\partial^2 H}{\partial T \, \partial p} - \left(\frac{\partial V}{\partial T}\right)_p\right] - \frac{1}{T^2}\left[\left(\frac{\partial H}{\partial p}\right)_T - V\right].$$

Setting the mixed derivatives equal yields

$$\frac{1}{T}\frac{\partial^2 H}{\partial p \, \partial T} = \frac{1}{T}\frac{\partial^2 H}{\partial T \, \partial p} - \frac{1}{T}\left(\frac{\partial V}{\partial T}\right)_p - \frac{1}{T^2}\left[\left(\frac{\partial H}{\partial p}\right)_T - V\right].$$

Since the mixed second derivatives of H are also equal this equation reduces to

$$\left(\frac{\partial H}{\partial p}\right)_T - V = -T\left(\frac{\partial V}{\partial T}\right)_p. \tag{9.40}$$

Combining this result with Eq. (9.39) we have

$$\left(\frac{\partial S}{\partial p}\right)_T = -\left(\frac{\partial V}{\partial T}\right)_p = -V\alpha. \tag{9.41}$$

To obtain the right-hand equality the definition of α has been used. In Eq. (9.41) we have an expression for the isothermal pressure dependence of the entropy in terms of the quantities V and α which are easily measurable for any system. The entropy can be written in terms of the temperature and pressure in the form

$$dS = \frac{C_p}{T}\,dT - V\alpha\,dp. \tag{9.42}$$

9.7.1 Change in Entropy of a Liquid with Pressure

For solids, $\alpha \approx 10^{-4}\,\mathrm{K}^{-1}$ or less, while for liquids $\alpha \approx 10^{-3}\,\mathrm{K}^{-1}$ or less. Suppose a liquid has a molar volume of $100\ \mathrm{cm}^3/\mathrm{mol} = 10^{-4}\,\mathrm{m}^3/\mathrm{mol}$. What is the entropy change if the pressure is increased by $1\ \mathrm{atm} = 10^5\ \mathrm{Pa}$ at constant temperature?

Since the temperature is constant, we set $dT = 0$ in Eq. (9.42), and obtain $dS = -V\alpha\,dp$. Since V and α are constants, they can be removed from the integral; thus,

$$\Delta S = -\int_{p_1}^{p_2} V\alpha\,dp = -V\alpha\,\Delta p = -(10^{-4}\ \mathrm{m}^3/\mathrm{mol})(10^{-3}\ \mathrm{K}^{-1})(10^5\ \mathrm{Pa})$$

$$= -0.01\ \mathrm{J/K\ mol}.$$

To produce a decrease in entropy of $1\ \mathrm{J/K}$ a pressure of at least 100 atm must be applied to the liquid. Since the variation of entropy of a liquid or a solid with pressure is so small, we will usually ignore it completely. If the pressure on a gas were increased from 1 atm to 2 atm, the corresponding change in entropy would be $\Delta S = -5.76$ J/K mol; the decrease is large simply because the volume has decreased greatly. We cannot ignore the entropy change of a gas accompanying a change in pressure.

9.8 THE TEMPERATURE DEPENDENCE OF THE ENTROPY

Attention has been directed to the simplicity of the dependence of entropy on temperature both at constant volume and constant pressure. This simplicity results from the fundamental definition of the entropy. If the state of the system is described in terms of the temperature and any other independent variable x, then the heat capacity of the system in a reversible transformation at constant x is by definition $C_x = (dQ_{rev})_x/dT$. Combining this equation with the definition of dS, we obtain at constant x

$$dS = \frac{C_x}{T}\,dT \quad \text{or} \quad \left(\frac{\partial S}{\partial T}\right)_x = \frac{C_x}{T}. \tag{9.43}$$

Thus, under any constraint, the dependence of the entropy on temperature is simple; the differential coefficient is always the appropriate heat capacity divided by the temperature. In the majority of practical applications, x is either V or p. Thus we may take as equivalent definitions of the heat capacities

$$C_v = T\left(\frac{\partial S}{\partial T}\right)_V \quad \text{or} \quad C_p = T\left(\frac{\partial S}{\partial T}\right)_p. \tag{9.44}$$

■ **EXAMPLE 9.3**   One mole of solid gold is raised from 25 °C to 100 °C at constant pressure. $\bar{C}_p/(\text{J/K mol}) = 23.7 + 0.00519T$. Calculate ΔS for the transformation.

$$\Delta S = \int_{T_1}^{T_2} \frac{C_p}{T}\,dT = \int_{298.15}^{373.15} \frac{(23.7 + 0.00519T)}{T}\,dT$$

$$= 23.7 \int_{298.15}^{373.15} \frac{dT}{T} + 0.00519 \int_{298.15}^{373.15} dT$$

$$= 23.7 \ln \frac{373.15}{298.15} + 0.00519(373.15 - 298.15) = 5.318 + 0.389 = 5.71 \text{ J/K mol}.$$

9.9 ENTROPY CHANGES IN THE IDEAL GAS

The relations derived in the preceding sections are applicable to any system. They have a particularly simple form when applied to the ideal gas, which is the result of the fact that in the ideal gas the energy and the temperature are equivalent variables: $dU = C_v\,dT$. Using this value of dU in Eq. (9.23), we obtain

$$dS = \frac{C_v}{T}\,dT + \frac{p}{T}\,dV. \tag{9.45}$$

The same result could be obtained by using Joule's law, $(\partial U/\partial V)_T = 0$, in Eq. (9.26). To use Eq. (9.45), all of the quantities must be expressed as functions of the two variables T

and V. Hence, we replace the pressure by $p = nRT/V$; and the equation becomes

$$dS = \frac{C_v}{T} dT + \frac{nR}{V} dV. \tag{9.46}$$

By comparing Eq. (9.46) with (9.24), we see that

$$\left(\frac{\partial S}{\partial V}\right)_T = \frac{nR}{V}. \tag{9.47}$$

This derivative is always positive; in an isothermal transformation, the entropy of the ideal gas increases with increase in volume. The rate of increase is less at large volumes, since V appears in the denominator.

For a finite change in state, we integrate Eq. (9.46) to

$$\Delta S = \int_{T_1}^{T_2} \frac{C_v}{T} dT + nR \int_{V_1}^{V_2} \frac{dV}{V}.$$

If C_v is a constant, this integrates directly to

$$\Delta S = C_v \ln \left(\frac{T_2}{T_1}\right) + nR \ln \left(\frac{V_2}{V_1}\right). \tag{9.48}$$

The entropy of the ideal gas is expressed as a function of T and p by using the property of the ideal gas, $dH = C_p \, dT$, in Eq. (9.35) which reduces to

$$dS = \frac{C_p}{T} dT - \frac{V}{T} dp.$$

To express everything in terms of T and p, we use $V = nRT/p$, so that

$$dS = \frac{C_p}{T} dT - \frac{nR}{p} dp. \tag{9.49}$$

Comparing Eq. (9.49) to Eq. (9.34), we have

$$\left(\frac{\partial S}{\partial p}\right)_T = -\frac{nR}{p}, \tag{9.50}$$

which shows that the entropy decreases with isothermal increase in pressure, a result that would be expected from the volume dependence of the entropy. For a finite change in state, Eq. (9.49) integrates to

$$\Delta S = C_p \ln \left(\frac{T_2}{T_1}\right) - nR \ln \left(\frac{p_2}{p_1}\right), \tag{9.51}$$

where C_p has been taken as a constant in the integration.

■ **EXAMPLE 9.4**   One mole of an ideal gas, $\overline{C}_p = \frac{5}{2}R$, initially at 20 °C and 1 atm pressure, is transformed to 50 °C and 8 atm pressure. Calculate ΔS. Using Eq. (9.51), with $T_1 = 293.15$ K and $T_2 = 323.15$ K, we have

$$\Delta S = \frac{5}{2}R \ln \frac{323.15 \text{ K}}{293.15 \text{ K}} - R \ln \frac{8 \text{ atm}}{1 \text{ atm}} = \frac{5}{2}R(0.0974) - 2.079R$$

$$= -1.836R = -1.836(8.314 \text{ J/K mol}) = -15.26 \text{ J/K mol}.$$

Note that in this example, as well as in the earlier ones, it is essential to express the temperature in kelvins. Note also that in the second part of the problem where *only a pressure ratio* is involved, we may use any unit of pressure so long as it is related to the pascal by a multiplicative constant. In forming the ratio, the conversion factor will disappear, and thus need not have been introduced in the first place.

9.9.1   Standard State for the Entropy of an Ideal Gas

For a change in state at constant temperature, Eq. (9.50) can be written

$$d\bar{S} = -\frac{R}{p} dp.$$

Suppose that we integrate this equation from $p = 1$ atm to any pressure p. Then

$$\bar{S} - \bar{S}^\circ = -R \ln \left(\frac{p}{1 \text{ atm}}\right), \tag{9.52}$$

where $\bar{S}^\circ$ is the value of the molar entropy under 1 atm pressure; it is the standard entropy at the temperature in question.

To calculate a numerical value of the logarithm on the right-hand side of Eq. (9.52), it is essential that the pressure be expressed in atmospheres. Then the ratio ($p/1$ atm) will be a pure number, and the operation of taking the logarithm is possible. (Note that it is not possible to take the logarithm of five oranges.) It is customary to abbreviate Eq. (9.52) to the simple form

$$\bar{S} - \bar{S}^\circ = -R \ln p. \tag{9.53}$$

It must be clearly understood that in Eq. (9.53) the value of p is a pure number, the number obtained by dividing the pressure in atm by 1 atm.

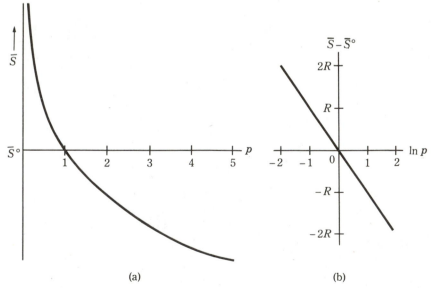

(a)   (b)

Figure 9.2   (a) Entropy of the ideal gas as a function of pressure. (b) Entropy of the ideal gas versus ln p.

The quantity $\bar{S} - \bar{S}^\circ$ is the molar entropy at the pressure p relative to that at 1 atm pressure. A plot of $\bar{S} - \bar{S}^\circ$ for the ideal gas is shown as a function of pressure in Fig. 9.2(a). The rate of decrease of the entropy with pressure is rapid at low pressures and becomes less rapid at higher pressures. There is an evident advantage in using a plot of $\bar{S} - \bar{S}^\circ$ against ln p in this situation (Fig. 9.2b). The plot is linear and a wider range of pressures can be represented on a scale of reasonable length.

9.10 THE THIRD LAW OF THERMODYNAMICS

Consider the constant-pressure transformation of a solid from the absolute zero of temperature to some temperature T below its melting point:

$$\text{Solid } (0 \text{ K, } p) \to \text{Solid } (T, p).$$

The entropy change is given by Eq. (9.38),

$$\Delta S = S_T - S_0 = \int_0^T \frac{C_p}{T} \, dT,$$

$$S_T = S_0 + \int_0^T \frac{C_p}{T} \, dT. \tag{9.54}$$

Since C_p is positive, the integral in Eq. (9.54) is positive; thus the entropy can only increase with temperature. Thus at 0 K the entropy has its smallest possible algebraic value S_0; the entropy at any higher temperature is greater than S_0. In 1913, M. Planck suggested that the value of S_0 is zero for every pure, perfectly crystalline substance. This is the third law of thermodynamics: *The entropy of a pure, perfectly crystalline substance is zero at the absolute zero of temperature.*

When we apply the third law of thermodynamics to Eq. (9.54), it reduces to

$$S_T = \int_0^T \frac{C_p}{T} \, dT, \tag{9.55}$$

where S_T is called the third-law entropy, or simply the entropy, of the solid at temperature T and pressure p. If the pressure is 1 atm, then the entropy is also a standard entropy S_T°. Table 9.1 is a selection of entropy values for a number of different types of substances.

Since a change in the state of aggregation (melting or vaporization) involves an increase in entropy, this contribution must be included in the computation of the entropy of a liquid or of a gas. For the standard entropy of a liquid above the melting point of the substance, we have

$$S_T^\circ = \int_0^{T_m} \frac{C_p^\circ(\text{s})}{T} \, dT + \frac{\Delta H_{\text{fus}}^\circ}{T_m} + \int_{T_m}^T \frac{C_p^\circ(\text{l})}{T} \, dT. \tag{9.56}$$

Similarly, for a gas above the boiling point of the substance

$$S_T^\circ = \int_0^{T_m} \frac{C_p^\circ(\text{s})}{T} \, dT + \frac{\Delta H_{\text{fus}}^\circ}{T_m} + \int_{T_m}^{T_b} \frac{C_p^\circ(\text{l})}{T} \, dT + \frac{\Delta H_{\text{vap}}^\circ}{T_b} + \int_{T_b}^T \frac{C_p^\circ(\text{g})}{T} \, dT. \tag{9.57}$$

If the solid undergoes any transition between one crystalline modification and another, the entropy of transition at the equilibrium temperature must be included also. To calculate the entropy, the heat capacity of the substance in its various states of aggregation must be measured accurately over the range of temperature from absolute zero to

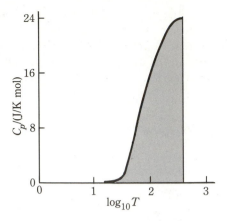

Figure 9.3   Plot of C_p versus $\log_{10} T$.

the temperature of interest. The values of the heats of transition and the transition temperatures must also be measured. All of these measurements can be made calorimetrically.

Measurements of the heat capacity of some solids have been made at temperatures as low as a few hundredths of a degree above the absolute zero. However, this is unusual. Ordinarily, measurements of heat capacity are made down to a low temperature T', which frequently lies in the range from 10 to 15 K. At such low temperatures, the heat capacity of solids follows the Debye "T-cubed" law accurately; that is

$$C_v = aT^3, \tag{9.58}$$

where a is a constant for each substance. At these temperatures C_p and C_v are indistinguishable, so the Debye law is used to evaluate the integral of C_p/T over the interval from 0 K to the lowest temperature of measurement T'. The constant a is determined from the value of $C_p(=C_v)$ measured at T'. From the Debye law, $a = (C_p)_{T'}/T'^3$.

In the range of temperature above T', the integral

$$\int_{T'}^{T} \frac{C_p}{T} \, dT = \int_{T'}^{T} C_p \, d(\ln T) = 2.303 \int_{T'}^{T} C_p \, d(\log_{10} T)$$

is evaluated graphically by plotting either C_p/T versus T, or C_p versus $\log_{10} T$. The area under the curve is the value of the integral. Figure 9.3 shows the plot of $\bar{C}_p$ versus $\log_{10} T$ for a solid from 12 K to 298 K. The total area under the curve when multiplied by 2.303 yields a value of $\bar{S}^\circ_{298} = 32.6$ J/K mol.

In conclusion, we should note that the first statement of the third law of thermodynamics was made by Nernst in 1906, *the Nernst heat theorem*, which states that in any chemical reaction involving only pure, crystalline solids the change in entropy is zero at 0 K. This form is less restrictive than the statement of Planck.

The third law of thermodynamics lacks the generality of the other laws, since it applies only to a special class of substances, namely pure, crystalline substances, and not to all substances. In spite of this restriction the third law is extremely useful. The reasons for exceptions to the law can be better understood after we have discussed the statistical interpretation of the entropy; the entire matter of exceptions to the third law will be deferred until then.

The following general comments may be made about the entropy values that appear in Table 9.1.

Table 9.1
Standard entropies at 298.15 K

Substance	$S^\circ_{298.15}/R$	Substance	$S^\circ_{298.15}/R$
Solids		**Liquids**	
Single unit, simple		Hg	9.129
C (diamond)	0.286	Br_2	18.3068
Si	2.262	H_2O	8.4131
Sn (white)	6.156	$TiCl_4$	30.35
Pb	7.79	CH_3OH	15.2
Cu	3.987	C_2H_5OH	19.3
Fe	3.28		
Al	3.410	**Gases**	
Ca	5.00	*Monatomic*	
Na	6.170	He	15.1591
K	7.779	Ne	17.5856
Single unit, complex		Ar	18.6101
I_2	13.968	Kr	19.7213
P_4	19.77	Xe	20.3951
S_8(rhombic)	30.842	*Diatomic*	
C (graphite)	0.690	H_2	15.7041
Two unit, simple		HF	20.8872
SnO	6.876	HCl	22.4653
PbS	11.0	HBr	23.8844
HgO(red)	8.449	HI	24.8340
AgCl	11.57	Cl_2	26.8167
FeO(wustite)	6.91	O_2	24.6604
MgO	3.241	N_2	23.0325
CaO	4.58	NO	25.336
NaCl	8.68	CO	23.7607
KCl	9.93	*Triatomic*	
KBr	11.53	H_2O	22.6984
KI	12.79	O_3	28.72
Two unit, complex		NO_2	28.86
FeS_2(pyrite)	6.37	N_2O	26.43
NH_4Cl	11.4	CO_2	25.6996
$CaCO_3$(calcite)	11.2	*Tetratomic*	
$NaNO_3$	14.01	SO_3	30.87
$KClO_3$	17.2	NH_3	23.173
Three units, simple		P_4	33.66
SiO_2(α-quartz)	4.987	PCl_3	37.49
Cu_2O	11.20	C_2H_2	24.15
Ag_2O	14.6	*Pentatomic*	
Na_2O	9.03	CH_4	22.389
Five units, simple		SiH_4	24.60
Fe_2O_3	10.51	SiF_4	33.995

Calculated from values in NBS Technical Notes 270–3 through 270–8. U.S. Government Printing Office, 1968–81; and in No. 28 CODATA Recommended Values for Thermodynamics 1977. (April 1978) International Council of Scientific Unions.

1. Entropies of gases are larger than those of liquids, which are larger than those of solids. This is a consequence of Eq. (9.57).

2. The entropy of gases increases logarithmically with the mass; this is illustrated by the monatomic gases, or the series of diatomics, HF, HCl, HBr, HI.

3. Comparing gases having the same mass—Ne, HF, H_2O—we see the effect of the rotational heat capacity. Two degrees of rotational freedom add $3.302R = 27.45$ J/K mol in passing from Ne to HF; one additional rotation in H_2O compared to HF adds $1.811R = 15.06$ J/K mol. Similarly, H_2O and NH_3 have nearly the same entropy. (Both have 3 rotational degrees of freedom.) For molecules with the same mass and the same heat capacities but different shapes, the more symmetrical molecule has the lower entropy; clear-cut examples are few, but compare N_2 to CO and NH_3 to CH_4.

4. In the case of solids consisting structurally of a single simple unit, the heat capacity is exclusively vibrational. A tightly bound solid (high cohesive energy) has high characteristic frequencies (in the sense of Section ★4.13), hence a lower heat capacity and a low entropy; for example, diamond has very high cohesive energy, very low entropy; silicon has lower cohesive energy (also lower vibrational frequencies due to higher mass), hence a higher entropy.

5. Solids made up of two, three, . . . , simple units have entropies that are roughly two, three, . . . , times greater than those composed of one simple unit. The entropy per particle is roughly the same throughout.

6. Where there is a single complex unit, van der Waals forces (very low cohesive forces) bind the solid. The entropy is correspondingly high. Note that the masses are quite large in the examples given in the table.

7. Where complex units occur in the crystal, the entropy is correspondingly greater since the heat capacity is greater due to the additional degrees of freedom associated with these units.

9.11 ENTROPY CHANGES IN CHEMICAL REACTIONS

The standard entropy change in a chemical reaction is computed from tabulated data in much the same way as the standard change in enthalpy. However, there is one important difference: The standard entropy of elements is *not* assigned a conventional value of zero. The characteristic value of the entropy of each element at 25 °C and 1 atm pressure is known from the third law. As an example, in the reaction

$$Fe_2O_3(s) + 3H_2(g) \longrightarrow 2Fe(s) + 3H_2O(l),$$

the standard entropy change is given by

$$\Delta S° = S°_{(final)} - S°_{(initial)}. \tag{9.59}$$

Then

$$\Delta S° = 2\bar{S}°(Fe, s) + 3\bar{S}°(H_2O, l) - \bar{S}°(Fe_2O_3, s) - 3\bar{S}°(H_2, g) \tag{9.60}$$

From the values in Table 9.1, we find for this reaction at 25 °C

$$\Delta S° = R[2(3.28) + 3(8.4131) - 10.51 - 3(15.7041)]$$

$$= -25.82R = -25.82(8.314 \text{ J/K mol}) = -214.7 \text{ J/K mol}.$$

Since the entropy of gases is much larger than the entropy of condensed phases, there is a large decrease in entropy in this reaction; a gas, hydrogen, is consumed to form condensed materials. Conversely, in reactions in which a gas is formed at the expense of condensed materials, the entropy will increase markedly.

$$Cu_2O(s) + C(s) \longrightarrow 2Cu(s) + CO(g) \qquad \Delta S_{298}^\circ = +158 \text{ J/K mol.}$$

From the value of ΔS° for a reaction at any particular temperature T_0, the value at any other temperature is easily obtained by applying Eq. (9.38):

$$\Delta S^\circ = S^\circ(\text{products}) - S^\circ(\text{reactants}).$$

Differentiating this equation with respect to temperature at constant pressure, we have

$$\left(\frac{\partial \, \Delta S^\circ}{\partial T}\right)_p = \left(\frac{\partial S^\circ(\text{products})}{\partial T}\right)_p - \left(\frac{\partial S^\circ(\text{reactants})}{\partial T}\right)_p$$

$$= \frac{C_p^\circ(\text{products})}{T} - \frac{C_p^\circ(\text{reactants})}{T} = \frac{\Delta C_p^\circ}{T}. \qquad (9.61)$$

Writing Eq. (9.61) in differential form and integrating between the reference temperature T_0 and any other temperature T, we obtain

$$\int_{T_0}^{T} d(\Delta S^\circ) = \int_{T_0}^{T} \frac{\Delta C_p^\circ}{T} \, dT;$$

$$\Delta S_T^\circ = \Delta S_{T_0}^\circ + \int_{T_0}^{T} \frac{\Delta C_p^\circ}{T} \, dT, \qquad (9.62)$$

which is applicable to any chemical reaction so long as none of the reactants or products undergoes a change in its state of aggregation in the temperature interval T_0 to T.

9.12 ENTROPY AND PROBABILITY

The entropy of a system in a definite state can be related to what is called the probability of that state of the system. To make this relation, or even to define what is meant by the probability of the state, it is necessary to have some structural model of the system. In contrast, the definition of the entropy from the second law does not require a structural model; the definition does not depend in the least on whether we imagine that the system is composed of atoms and molecules or that it is built with waste paper and baseball bats. For simplicity we will suppose that the system is composed of a very large number of small particles, or molecules.

Imagine the following situation. A large room is sealed and completely evacuated. In one corner of the room there is a small box that confines a gas under atmospheric pressure. The sides of the box are now taken away so that the molecules of gas are free to move into the room. After a period of time we observe that the gas is distributed uniformly throughout the room. At the time the box was opened each gas molecule had a definite position and velocity, if we take a classical view of the matter. At some instant after the gas has filled the room, the position and velocity of each molecule have values that are related in a complicated way to the values of the positions and velocities of all the molecules at the instant the box was opened. At the later time, imagine that each velocity component of every molecule is exactly reversed. Then the molecules will just reverse their original motion, and after a period of time the gas will collect itself in the corner of the room where it was originally sealed in the box.

The strange thing is that there is no reason to suppose that the one particular motion, which led to the uniform filling of the room, is any more probable than the same motion reversed, which leads to the collection of the gas in one corner of the room. If this is so, why is it that we never observe the air in a room collecting in one particular portion of the room? The fact that we never observe some motions of a system, which are inherently just as probable as those we do observe, is called the *Boltzmann paradox*.

This paradox is resolved in the following way. It is true that any exactly specified motion of the molecules has the same probability as any other exactly specified motion. But it is also true that of all the possible exactly specified motions of a group of molecules, the total number of these motions that lead to the uniform filling of the available space is enormously greater than the number of these motions that lead to the occupation of only a small part of the available space. And so, although each individual motion of the system has the same probability, the probability of observing the available space filled uniformly is proportional to the total number of motions that would result in this observation; consequently, the probability of observing the uniform filling is overwhelmingly large compared with the probability of any other observation.

It is difficult to imagine the detailed motion of even one particle, much less that of many particles. Fortunately, for the calculation we do not have to deal with the motions of the particles, but only with the number of ways of distributing the particles in a given volume. A simple illustration suffices to show how the probability of the uniform distribution compares with that of the nonuniform one.

Suppose we have a set of four cells each of which can contain one ball. The set of four cells is then divided in half; each half has two cells, as in Fig. 9.4(a). We place two balls in the cells; the arrangements in Fig. 9.4(b) are possible ($\bigcirc$ indicates an empty cell, $\otimes$ indicates an occupied cell). Of these six arrangements, four correspond to uniform filling; that is, one ball in each half of the box. The probability of uniform filling is therefore $\frac{4}{6} = \frac{2}{3}$, while the probability of finding both balls on one side of the box is $\frac{2}{6} = \frac{1}{3}$. The probability of any *particular* arrangement is $\frac{1}{6}$. But four particular arrangements lead to uniform filling; only two particular arrangements lead to nonuniform filling.

Suppose that there are eight cells and two balls; then the total number of arrangements is 28. Of the 28 arrangements, 16 of them correspond to one ball in each half of the box. The probability of the uniform distribution is therefore $\frac{16}{28} = \frac{4}{7}$. It is easy to show that, as the number of cells increases without limit, the probability of finding one ball in one half of the box and the other in the other half of the box approaches the value $\frac{1}{2}$.

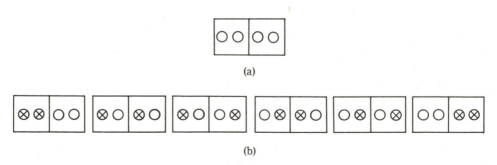

(a)

(b)

Figure 9.4

At this point it seems reasonable to ask what all this has to do with entropy. The entropy of a system in a specified state can be defined in terms of the number of possible arrangements of the particles composing the system that are consonant with the state of the system. Each such possible arrangement is called a *complexion* of the system. Following Boltzmann, we define the entropy by the equation

$$S = k \ln \Omega, \tag{9.63}$$

where k is the Boltzmann constant, $k = R/N_A$, and Ω is the number of complexions of the system that are consonant with the specified state of the system. Since the probability of a specified state of a system is proportional to the number of complexions which make up that state, it is clear from Eq. (9.63) that the entropy depends on the logarithm of the probability of the state.

Suppose we calculate the entropy for two situations in the foregoing example.

Situation 1. The two balls are confined to the left half of the box. There is only one arrangement (complexion) that produces this situation; hence, $\Omega = 1$, and

$$S_1 = k \ln (1) = 0.$$

The entropy of this state is zero.

Situation 2. The two balls may be anywhere in the box. As we have seen, there are six complexions corresponding to this situation; hence, $\Omega = 6$, and

$$S_2 = k \ln (6).$$

The entropy increase associated with the expansion of the system from 2 cells to 4 cells is then

$$\Delta S = S_2 - S_1 = k \ln 6 \qquad \text{for 2 balls}$$
$$= \tfrac{1}{2}k \ln 6 \qquad \text{for 1 ball.}$$

This result is readily generalized to apply to a box having N cells. How many arrangements are possible for two balls in N cells? There are N choices for the placement of the first ball; for each choice of cell for the first ball there are $N - 1$ choices for the second ball. The total number of arrangements of 2 balls in N cells is apparently $N(N - 1)$. However, since we cannot distinguish between ball 1 in position x, ball 2 in position y, and the arrangement ball 1 in y, ball 2 in x, this number must be divided by 2 to obtain the number of distinct arrangements; hence,

$$\Omega_1 = \frac{N(N - 1)}{2}.$$

The entropy of this system is, by Eq. (9.63),

$$S_1 = k \ln \left[\tfrac{1}{2}N(N - 1)\right].$$

If we increase the number of cells available to N', then $\Omega_2 = \tfrac{1}{2}N'(N' - 1)$, and

$$S_2 = k \ln \left[\tfrac{1}{2}N'(N' - 1)\right].$$

The increase in entropy associated with increasing the number of cells from N to N' is

$$\Delta S = S_2 - S_1 = k \ln \left[\frac{N'(N' - 1)}{N(N - 1)}\right]. \tag{9.64}$$

If $N' = 4$ and $N = 2$, this yields the result obtained originally for the expansion from 2 to 4 cells.

A more instructive application of Eq. (9.64) is obtained if we suppose that both N and N' are very large, so large that $N - 1$ can be replaced by N and $N' - 1$ by N'. Then Eq. (9.64) becomes

$$\Delta S = k \ln \left(\frac{N'}{N}\right)^2 = 2k \ln \left(\frac{N'}{N}\right). \tag{9.65}$$

If we ask to what physical situation this random placing of balls in cells might be applied, the ideal gas comes to mind. In the ideal gas the position of a molecule at any time is a result of pure chance. The proximity of the other molecules does not affect the chance of the molecule being where it is. If we apply Eq. (9.65) to an ideal gas, the balls become molecules and the number of cells is proportional to the volume occupied by the gas; thus, $N'/N = V'/V$, and Eq. (9.65) becomes

$$\Delta S \text{ (two molecules)} = 2k \ln \left(\frac{V'}{V}\right), \qquad \Delta S \text{ (one molecule)} = k \ln \left(\frac{V'}{V}\right).$$

Since $N_A k = R$, the gas constant, for one mole, we have

$$\Delta S \text{ (one mole)} = R \ln \left(\frac{V'}{V}\right), \tag{9.66}$$

which is identical to the second term of Eq. (9.48), the expression for the increase in entropy accompanying the isothermal expansion of one mole of an ideal gas from volume V to volume V'.

From the standpoint of this structural and statistical definition of entropy, iso-thermal expansion of a gas increases the entropy because there are more ways of arranging a given number of molecules in a large volume than in a small volume. Since the prob-ability of a given state is proportional to the number of ways of arranging the molecules in that state, the gas confined in a large volume is in a more probable state than if it is confined in a small volume. If we assume that the equilibrium state of the gas is the state of highest probability, then it is understandable why the gas in a room never collects in a small corner. The gas achieves its most probable state by occupying as much volume as is available to it. The equilibrium state has the maximum probability consistent with the constraints on the system and so has a maximum entropy.

9.13 GENERAL FORM FOR OMEGA

To calculate the number of arrangements of three particles in N cells, we proceed in the same way as before. There are N choices for placing the first particle, $N - 1$ choices for the second, and $N - 2$ choices for the third. This would seem to make a total of $N(N - 1)(N - 2)$ arrangements; but again we cannot distinguish between arrangements that are only permutations of the three particles between the cells x, y, z. There are 3! such permutations: $xyz, xzy, yxz, yzx, zxy, zyx$. Hence, for three particles in N cells the number of complexions is

$$\Omega = \frac{N(N - 1)(N - 2)}{3!}. \tag{9.67}$$

Again if the number of cells N is much larger than the number of particles, this reduces for three particles to

$$\Omega = \frac{N^3}{3!}.$$

From this approximate form we can immediately jump to the conclusion that for N_a particles, if N is much larger than N_a, then, approximately,

$$\Omega = \frac{N^{N_a}}{N_a!}. \tag{9.68}$$

On the other hand, if we need the exact form for Ω, Eq. (9.67) can be generalized for N_a particles to

$$\Omega = \frac{N(N-1)(N-2)(N-3)\cdots(N-N_a+1)}{N_a!}.$$

If we multiply this last equation by $(N-N_a)!$ in both numerator and denominator, it reduces to

$$\Omega = \frac{N!}{N_a!(N-N_a)!}. \tag{9.69}$$

The entropy attending the expansion from N to N' cells is easily calculated using Eq. (9.68). For N cells,

$$S = k[\ln N^{N_a} - \ln (N!)],$$

while for N' cells,

$$S' = k[\ln N'^{N_a} - \ln (N_a!)].$$

The value of ΔS is

$$S = S' - S = N_a k \ln \left(\frac{N'}{N}\right).$$

As before, we take the ratio $N'/N = V'/V$; then if $N_a = N_A$, the equation becomes

$$\Delta S = R \ln \left(\frac{V'}{V}\right),$$

which is identical to Eq. (9.66).

9.14 THE ENERGY DISTRIBUTION

It is rather easy to make the translation in concept from arrangements of balls in cells to the physical arrangement of molecules in small elements of volume. By arranging molecules in the elements of volume we obtain a space distribution of the molecules. The problem in the space distribution was simplified considerably by the implicit assumption that there is at most one molecule in a given volume element.

The problem of translating arrangements of balls in cells to an energy distribution is only slightly more difficult. We assume that any molecule can have an energy value between zero and infinity. We partition this entire range of energy into small compartments of width $d\epsilon$; the compartments are labeled, beginning with the one of lowest energy,

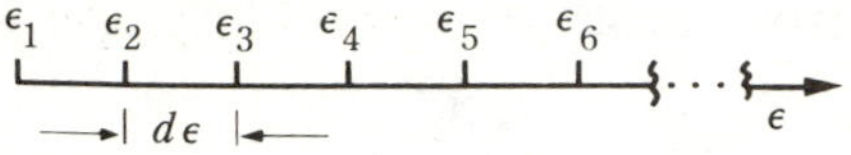

Figure 9.5 Division of the energy range into compartments.

by ϵ_1, ϵ_2, $_3$, ..., as in Fig. 9.5. The energy distribution is described by specifying the number of molecules n_1 having energies lying in the first compartment, the number n_2 in the second compartment, and so on.

Consider a collection of N molecules for which the energy distribution is described by the numbers $n_1, n_2, n_3, n_4, n_5, \ldots$. In how many ways can this particular distribution be achieved? We begin by supposing that there are three molecules in ϵ_1; there are N ways of choosing the first molecule, $(N - 1)$ of choosing the second, and $(N - 2)$ ways of choosing the third. Thus there appear to be $N(N - 1)(N - 2)$ ways of selecting three molecules from N molecules. However, the order of choice does not matter; the same distribution is obtained with molecules 1, 2, and 3 whether they are chosen in the order 123, 132, 213, 231, 312, or 321. We must divide the total number of ways of choosing by 3! to get the number of distinguishable ways of choosing;

$$\frac{N(N - 1)(N - 2)}{3!}.$$

Suppose that there are two molecules in the second compartment; these must be chosen from the $N - 3$ molecules remaining; the first may be chosen in $N - 3$ ways, the second in $N - 4$ ways. Again the order does not matter, so we divide by 2!. The two molecules in the second compartment can be chosen in

$$\frac{(N - 3)(N - 4)}{2!}$$

different ways. The total number of ways of choosing three molecules in the first compartment and two molecules in the second compartment is the product of these expressions:

$$\frac{N(N - 1)(N - 2)(N - 3)(N - 4)}{3! \, 2!}.$$

We then find how many ways there are of choosing the number of molecules in compartment three from the remaining $N - 5$ molecules, and so on. Repetition of this procedure yields the final result for Ω, the total number of ways of placing n_1 molecules in compartment 1, n_2 molecules in compartment 2, ... :

$$\Omega = \frac{N!}{n_1! \, n_2! \, n_3! \, n_4! \ldots}. \tag{9.70}$$

The value of Ω, the number of complexions for a particular distribution, given by Eq. (9.70) seems rather forbidding. However, we do not need to do very much with it to get the information we need. As usual, the entropy resulting from the distribution of molecules over a range of energies is related to the number of complexions by $S = k \ln \Omega$. If Ω is very large, the entropy will be large. It is clear from Eq. (9.70) that the smaller the populations of the compartments, $n_1, n_2, n_3, \ldots$, the larger will be the value of Ω. For example, if every compartment either was empty or contained only 1 molecule, all the factors in the denominator would be either 0! or 1!; the denominator would then be unity and $\Omega = N!$. This would be the largest possible value of Ω, and would correspond to the largest possible value of the entropy. Note that in this situation the molecules are spread out very widely over the energy range; thus a broad energy distribution means a high entropy.

In contrast, consider the situation where all the molecules but one are crowded into the first level; then

$$\Omega = \frac{N!}{(N-1)! \; 1! \; 0! \; 0! \; \cdots} = N.$$

If N is large, then N is very much smaller than $N!$; the entropy in this case is very much smaller than that for the broad distribution.

To achieve a high entropy, the molecules will therefore try to spread out into as broad an energy distribution as possible, just as gas molecules fill as much space as is available. The spatial distribution is limited by the walls of the container. The energy distribution is subject to an analogous limitation. In a specified state, a system has a fixed value of its total energy; from the distribution this value is

$$U = n_1 \epsilon_1 + n_2 \epsilon_2 + n_3 \epsilon_3 + n_4 {}_4 + \cdots.$$

It is clear that the system may not have many molecules in the high-energy compartments; if it did, the distribution would yield a value of energy above the fixed value in the particular state. This restriction limits the number of complexions of a system quite severely. The value of Ω nonetheless reaches a maximum consistent with the restriction that the energy must sum to the fixed value U. The molecules spread themselves over as broad a range of energy as is consistent with the fixed total energy of the system.

If the energy of the system is increased, the distribution can be broader; the number of complexions and the entropy of the system goes up. This is a statistical interpretation of the fact illustrated by the fundamental equation (9.12):

$$dS = \frac{1}{T} dU + \frac{p}{T} dV,$$

from which we obtain the differential coefficient

$$\left(\frac{\partial S}{\partial U} \right)_V = \frac{1}{T}.$$

We noted in Section 9.5 that this coefficient was always positive. For the present we simply note the agreement in the sign of this coefficient with the statistical argument that increase in energy increases the number of complexions and the entropy.

The two fundamental ways of varying the entropy of a system expressed by the fundamental equation are interpreted as the two ways of achieving a broader distribution. By increasing the volume, the spatial distribution broadens; by increasing the energy, the energy distribution broadens. The broader distribution is the more probable one, since it can be made up in a greater number of ways.

It is easy now to understand why the entropies of liquids and solids are nearly unchanged by a change in pressure. The volume of condensed materials is altered so little by a change in pressure that the breadth of the spatial distribution remains about the same. The entropy therefore remains at very nearly the same value.

We can also understand the phenomena in the adiabatic reversible expansion of a gas; in such an expansion, $dQ_{rev} = 0$, so that $dS = 0$. Since the volume goes up, the distribution over space broadens, and this part of the entropy increases. If the total entropy change is to be zero, the distribution over energies must get narrower; this corresponds to a decrease in energy that is reflected in a decrease in the temperature of the gas. The work produced in such an adiabatic expansion of a gas is produced at the expense of the decrease in energy of the system.

In Chapter 4 the Maxwell distribution of kinetic energies in a gas was discussed in detail. There we found that the average energy was given by $\frac{3}{2}RT$. Thus an increase in temperature corresponds to an increase in the energy of the gas; it should also correspond to a broadening of the energy distribution. This broadening of the energy distribution with increase in temperature was emphasized at that time.

From what has been said, it seems reasonable to expect the direction of natural changes to correspond to the direction that increases the probability of the system. Thus in natural transformations we might expect the entropy of the system to increase. This is not quite true. In a natural change both the system and the surroundings are involved. Therefore, in any natural change, the universe *must* reach a state of higher probability and thus of higher entropy. In a natural transformation, the entropy of the system may decrease if there is an increase in entropy in the surroundings that more than compensates for the decrease in the system. The entropy change in a transformation is a powerful clue to the natural direction of the transformation.

9.15 ENTROPY OF MIXING AND EXCEPTIONS TO THE THIRD LAW OF THERMODYNAMICS

The third law of thermodynamics is applicable only to those substances that attain a completely ordered configuration at the absolute zero of temperature. In a pure crystal, for example, the atoms are located in an exact pattern of lattice sites. If we calculate the number of complexions of N atoms arranged on N sites, we find that although there are $N!$ ways of arranging the atoms, since the atoms are identical, these arrangements differ only in the order of choosing the atoms. Since the arrangements are not distinguishable, we must divide by $N!$, and we obtain $\Omega = 1$ for the perfectly ordered crystal. The entropy is therefore

$$S = k \ln (1) = 0.$$

Suppose that we arrange different kinds of atoms A and B on the N sites of the crystal. If N_a is the number of A atoms, and N_b is the number of B atoms, then $N_a + N_b = N$, the total number of sites. In how many distinguishable ways can we select N_a sites for the A atoms and N_b sites for the B atoms? This number is given by Eq. (9.70):

$$\Omega = \frac{N!}{N_a!\, N_b!}. \tag{9.71}$$

The entropy of the mixed crystal is given by

$$S = k \ln \frac{N!}{N_a! N_b!}. \tag{9.72}$$

To evaluate this expression we take advantage of the Stirling approximation: When N is very large, then

$$\ln N! = N \ln N - N. \tag{9.73}$$

The expression for the entropy becomes

$$S = k(N \ln N - N - N_a \ln N_a + N_a - N_b \ln N_b + N_b).$$

Since $N = N_a + N_b$, this becomes

$$S = -k(N_a \ln N_a + N_b \ln N_b - N \ln N).$$

But, $N_a = x_a N$, and $N_b = x_b N$, where x_a is the mole fraction of A and x_b is the mole fraction of B. The expression for the entropy reduces to

$$S_{\text{mix}} = -Nk(x_a \ln x_a + x_b \ln x_b). \tag{9.74}$$

Since the terms in the parentheses in Eq. (9.74) are negative (the logarithm of a fraction is negative), the entropy of the mixed crystal is positive. If we imagine the mixed crystal to be formed from a pure crystal of A and a pure crystal of B, then for the mixing process

$$\text{pure } A + \text{pure } B \quad \longrightarrow \quad \text{mixed crystal.}$$

The entropy change is

$$\Delta S_{\text{mix}} = S \text{ (mixed crystal)} - S \text{ (pure } A) - S \text{ (pure } B).$$

The entropies of the pure crystals are zero, so the ΔS of mixing is simply

$$\Delta S_{\text{mix}} = -Nk(x_a \ln x_a + x_b \ln x_b), \tag{9.75}$$

and is a positive quantity.

Since any impure crystal has at least the entropy of mixing at the absolute zero, its entropy cannot be zero; such a substance does not follow the third law of thermodynamics. Some substances that are chemically pure do not fulfill the requirement that the crystal be perfectly ordered at the absolute zero of temperature. Carbon monoxide, CO, and nitric oxide, NO, are classic examples. In the crystals of CO and NO, some molecules are oriented differently than others. In a perfect crystal of CO, all the molecules should be lined up with the oxygen pointing north and the carbon pointing south, for example. In the actual crystal, the two ends of the molecule are oriented randomly; it is as if two kinds of carbon monoxide were mixed, half and half. The molar entropy of mixing would be

$$\Delta S = -N_A k(\tfrac{1}{2} \ln \tfrac{1}{2} + \tfrac{1}{2} \ln \tfrac{1}{2}) = N_A k \ln 2$$
$$= R \ln 2 = 0.693R = 5.76 \text{ J/K mol.}$$

The actual value for the residual entropy of crystalline carbon monoxide is $0.55R = 4.6$ J/K mol; the mixing is apparently not quite half and half. In the case of NO, the residual entropy is $0.33R = 2.8$ J/K mol, which is about one-half of 5.76 J/K mol; this has been explained by the observation that the molecules in the crystal of NO are dimers, $(NO)_2$. Thus one mole of NO contains only $\tfrac{1}{2}N_A$ double molecules; this reduces the residual entropy by a factor of two.

In ice, a residual entropy remains at the absolute zero because of randomness in the hydrogen bonding of the water molecules in the crystal. The magnitude of residual entropy has been computed and is in agreement with that observed.

It has been found that crystalline hydrogen has a residual entropy of $0.750R = 6.23$ J/K mol at the absolute zero of temperature. This entropy is not the result of disorder in the crystal, but of a distribution over several quantum states. Ordinary hydrogen is a mixture of ortho- and para-hydrogen, which have different values of the total nuclear spin angular momentum. As a consequence of this difference, the rotational energy of ortho-hydrogen at low temperatures does not approach zero as does that of para-hydrogen, but achieves a finite value. Ortho-hydrogen can be in any one of nine states, all having the same energy, while para-hydrogen exists in a single state. As a result of the mixing of the two kinds of hydrogen and the distribution of the ortho-hydrogen in nine different energy states, the system has a randomness and therefore a residual entropy. Pure para-hydrogen, since it exists in a single state at low temperature, would have no residual entropy and would follow the third law. Pure ortho-hydrogen would be distributed over nine states at absolute zero and would have a residual entropy.

From what has been said it is clear that glassy or amorphous substances will have a random arrangement of constituent particles and so will possess a residual entropy at absolute zero. The third law is therefore restricted to pure crystalline substances. A final restriction should be made in the application of the third law: The substance must be in a single quantum state. This last requirement would take care of the difficulty that arises in the case of hydrogen.

QUESTIONS

9.1 Under what special circumstances does $\Delta S = \Delta H/T$?

9.2 Green's theorem in the plane (see any calculus book) states that

$$\oint df = \int_A dx\, dy \left[\frac{\partial^2 f}{\partial x\, \partial y} - \frac{\partial^2 f}{\partial y\, \partial x} \right].$$

In words, the integral of the differential of a function $f(x, y)$ around a cyclical path equals the integral, over the enclosed area A, of the difference of the mixed derivatives shown. Use this theorem to argue that Eq. (9.13) holds when f is a thermodynamic state function.

9.3 The negative value of α for water between 0 °C and 4 °C is attributed to the breakdown of some hydrogen-bonded structure on passing from the solid to the liquid. How does this idea enable us to rationalize the variation of S with V for water in this temperature range?

9.4 Explain the direction of the differences of the standard entropies for the members of each of the pairs (a) C (diamond) and C (graphite); (b) Ar and F_2; (c) NH_3 and PCl_3.

9.5 Why is the third law useful?

9.6 Compare and contrast the entropy changes for (a) reversible isothermal and (b) reversible adiabatic compressions of an ideal gas. Discuss in terms of distributions in space and energy.

PROBLEMS

9.1 The temperature of 1 mol of an ideal gas is increased from 100 K to 300 K; $C_v = \frac{3}{2}R$.

    a) Calculate ΔS if the volume is constant.
    b) Calculate ΔS if the pressure is constant.
    c) What would ΔS be if 3 mol were used instead of 1 mol?

9.2 One mole of gaseous hydrogen is heated at constant pressure from 300 K to 1500 K.

    a) Calculate the entropy change for this transformation using the heat capacity data in Table 7.1.
    b) The standard third-law entropy of hydrogen at 300 K is 130.592 J/K mol. What is the entropy of hydrogen at 1500 K?

9.3 A monatomic solid has a heat capacity, $\bar{C}_p = 3.1\,R$. Calculate the increase in entropy of one mole of this solid if the temperature is increased from 300 K to 500 K at constant pressure.

9.4 For aluminum, $\bar{C}_p/(\text{J/K mol}) = 20.67 + 12.38 \times 10^{-3}T$.

    a) What is ΔS if one mole of aluminum is heated from 25 °C to 200 °C?
    b) If $\bar{S}^\circ_{298} = 28.35$ J/K mol, what is the entropy of aluminum at 200 °C?

9.5 Given the heat capacity of aluminum in Problem 9.4, calculate the mean heat capacity of aluminum in the range 300 K to 400 K.

9.6 At its boiling point, 35 °C, the heat of vaporization of MoF_6 is 25.1 kJ/mol. Calculate ΔS°_{vap}.

9.7 a) At its transition temperature, 95.4 °C, the heat of transition from rhombic to monoclinic sulfur is 0.38 kJ/mol. Calculate the entropy of transition.

b) At its melting point, 119 °C, the heat of fusion of monoclinic sulfur is 1.23 kJ/mol. Calculate the entropy of fusion.

c) The values given in (a) and (b) are for one mole of S, that is, for 32 g; however, in crystalline and liquid sulfur the molecule is S_8. Convert the values in parts (a) and (b) to those appropriate to S_8. (The converted values are more representative of the usual magnitudes of entropies of fusion and transition.)

9.8 a) What is the entropy change if one mole of water is warmed from 0 °C to 100 °C under constant pressure; $\bar{C}_p = 75.291$ J/K mol.

b) The melting point is 0 °C and the heat of fusion is 6.0095 kJ/mol. The boiling point is 100 °C and the heat of vaporization is 40.6563 kJ/mol. Calculate ΔS for the transformation

$$\text{Ice (0 °C, 1 atm)} \rightarrow \text{steam (100 °C, 1 atm).}$$

9.9 At 25 °C and 1 atm, the entropy of liquid water is 69.950 J/K mol. Calculate the entropy of water vapor at 200 °C and 0.5 atm. The data are: $\bar{C}_p(l)/(\text{J/K mol}) = 75.291$; $\bar{C}_p(g)/(\text{J/K mol}) = 33.577$; $\Delta H^\circ_{vap} = 40.6563$ kJ/mol at the boiling point, 100 °C. Water vapor may be assumed to be an ideal gas.

9.10 The standard entropy of lead at 25 °C is $\bar{S}^\circ_{298} = 64.80$ J/K mol. The heat capacity of solid lead is: $\bar{C}_p(s)/(\text{J/K mol}) = 22.13 + 0.01172T + 0.96 \times 10^5 T^{-2}$. The melting point is 327.4 °C and the heat of fusion is 4770 J/mol. The heat capacity of liquid lead is $\bar{C}_p(l)/(\text{J/K mol}) = 32.51 - 0.00301T$.

a) Calculate the standard entropy of liquid lead at 500 °C.

b) Calculate the ΔH in changing solid lead from 25 °C to liquid lead at 500 °C.

9.11 From the data for graphite: $\bar{S}^\circ_{298} = 5.74$ J/K mol, and $\bar{C}_p/(\text{J/K mol}) = -5.293 + 58.609 \times 10^{-3}T - 432.24 \times 10^{-7}T^2 + 11.510 \times 10^{-9}T^3$, calculate the molar entropy of graphite at 1500 K.

9.12 Between 0 °C and 100 °C liquid mercury has $\bar{C}_p/(\text{J/K mol}) = 30.093 - 4.944 \times 10^{-3}T$. If one mole of mercury is raised from 0 °C to 100 °C at constant pressure, calculate ΔH and ΔS.

9.13 One mole of an ideal gas is expanded isothermally to twice its initial volume.

a) Calculate ΔS.

b) What would ΔS be if 5 mol were used instead of 1 mol?

9.14 One mole of carbon monoxide is transformed from 25 °C and 5 atm to 125 °C and 2 atm. If $\bar{C}_p/R = 3.1916 + 0.9241 \times 10^{-3}T - 1.410 \times 10^{-7}T^2$, calculate ΔS. Assume the gas is ideal.

9.15 One mole of an ideal gas, $\bar{C}_v = \frac{3}{2}R$, is transformed from 0 °C and 2 atm to −40 °C and 0.4 atm. Calculate ΔS for this change in state.

9.16 One mole of an ideal gas, initially at 25 °C and 1 atm is transformed to 40 °C and 0.5 atm. In the transformation 300 J of work are produced in the surroundings. If $\bar{C}_v = \frac{3}{2}R$, calculate Q, ΔU, ΔH, and ΔS.

9.17 One mole of a van der Waals gas at 27 °C expands isothermally and reversibly from 0.020 m³ to 0.060 m³. For the van der Waals gas, $(\partial U/\partial V)_T = a/\bar{V}^2$; $a = 0.556$ Pa m⁶/mol²; $b = 64 \times 10^{-6}$ m³/mol. Calculate Q, W, ΔU, ΔH, and ΔS for the transformation.

9.18 Consider one mole of an ideal gas, $\bar{C}_v = \frac{3}{2}R$, in the initial state: 300 K, 1 atm. For each transformation, (a) through (g), calculate Q, W, ΔU, ΔH, and ΔS; compare ΔS to Q/T.

a) At constant volume, the gas is heated to 400 K.

b) At constant pressure, 1 atm, the gas is heated to 400 K.

c) The gas is expanded isothermally and reversibly until the pressure drops to $\frac{1}{2}$ atm.

d) The gas is expanded isothermally against a constant external pressure equal to $\frac{1}{2}$ atm until the gas pressure reaches $\frac{1}{2}$ atm.

e) The gas is expanded isothermally against zero opposing pressure (Joule expansion) until the pressure of the gas is $\frac{1}{2}$ atm.

f) The gas is expanded adiabatically against a constant pressure of $\frac{1}{2}$ atm until the final pressure is $\frac{1}{2}$ atm.

g) The gas is expanded adiabatically and reversibly until the final pressure is $\frac{1}{2}$ atm.

9.19 For metallic zinc the values of $\bar{C}_p$ as a function of temperature are given. Calculate $\bar{S}°$ for zinc at 100 K.

T/K	$\bar{C}_p/(J/K \text{ mol})$	T/K	$\bar{C}_p/(J/K \text{ mol})$	T/K	$\bar{C}_p/(J/K \text{ mol})$
1	0.000720	10	0.1636	50	11.175
2	0.001828	15	0.720	60	13.598
3	0.003791	20	1.699	70	15.426
4	0.00720	25	3.205	80	16.866
6	0.01895	30	4.966	90	18.108
8	0.0628	40	8.171	100	19.154

9.20 Fit the data between 0 and 4 K in Problem 9.19 to the curve: $\bar{C}_p = \gamma T + aT^3$. The first term is a contribution of the electron gas in the metal to the heat capacity. *Hint:* To find the constants γ and a, rearrange to: $\bar{C}_p/T = \gamma + aT^2$, and either plot $\bar{C}_p/T$ *versus* T^2 or do a least squares fit. (See Appendix I, Section A-I-7.)

9.21 Silica, SiO_2, has a heat capacity given by

$$\bar{C}_p(\alpha\text{-quartz, s})/(J/K \text{ mol}) = 46.94 + 34.31 \times 10^{-3}T - 11.30 \times 10^5 T^{-2}.$$

The coefficient of thermal expansion is $0.3530 \times 10^{-4} \text{ K}^{-1}$. The molar volume is $22.6 \text{ cm}^3/\text{mol}$. If the initial state is 25 °C and 1 atm and the final state is 225 °C and 1000 atm, calculate ΔS for one mole of silica.

9.22 For liquid water at 25 °C, $\alpha = 2.07 \times 10^{-4} \text{ K}^{-1}$; the density may be taken as 1.00 g/cm^3. One mole of liquid water is compressed isothermally, 25 °C, from 1 atm to 1000 atm. Calculate ΔS,

a) supposing that water is incompressible; that is, $\kappa = 0$.

b) supposing that $\kappa = 4.53 \times 10^{-5} \text{ atm}^{-1}$.

9.23 For copper, at 25 °C, $\alpha = 0.492 \times 10^{-4} \text{ K}^{-1}$ and $\kappa = 0.78 \times 10^{-6} \text{ atm}^{-1}$; the density is 8.92 g/cm^3. Calculate ΔS for the isothermal compression of one mole of copper from 1 atm to 1000 atm for the same two conditions as in Problem 9.22.

9.24 In the limit, $T = 0$ K, it is known empirically that the value of the coefficient of thermal expansion of solids approaches zero as a limit. Show that, as a consequence, the entropy is independent of pressure at 0 K so that no specification of pressure is necessary in the third-law statement.

9.25 Consider the expression:

$$dS = \frac{C_p}{T} dT - V\alpha \, dp$$

Suppose that water has $\bar{V} = 18 \text{ cm}^3/\text{mol}$, $\bar{C}_p = 75.3 \text{ J/K mol}$ and $\alpha = 2.07 \times 10^{-4} \text{ K}^{-1}$. Compute the decrease in temperature that occurs if water at 25 °C and 1000 atm pressure is brought reversibly and adiabatically to 1 atm pressure. Assume $\kappa = 0$.

9.26 Show that $(\partial\alpha/\partial p)_T = -(\partial\kappa/\partial T)_p$.

9.27 In an insulated flask (Dewar flask) 20 g of ice at -5 °C are added to 30 g of water at $+25$ °C. If the heat capacities are $C_p(H_2O, l) = 4.18 \text{ J/K g}$, and $C_p(H_2O, s) = 2.09 \text{ J/K g}$, what is the final state of the system? (The pressure is constant.) $\Delta H_{\text{fusion}} = 334 \text{ J/g}$. Calculate ΔS and ΔH for the transformation.

9.28 How many grams of water at 25 °C must be added to a Dewar flask containing 20 g of ice at −5 °C to satisfy the conditions in (a) through (d)? Compute the entropy change in each case.

a) The final temperature is −2 °C; all the water freezes.
b) The final temperature is 0 °C; half the water freezes.
c) The final temperature is 0 °C; half the ice melts.
d) The final temperature is 10 °C; all the ice melts.

Predict the sign of ΔS in each case before doing the calculation. (Use the data in Problem 9.27.)

9.29 Twenty grams of steam at 120 °C and 300 g of liquid water at 25 °C are brought together in an insulated flask. The pressure remains at 1 atm throughout. If $C_p(H_2O, 1) = 4.18$ J/K g, $C_p(H_2O, g) = 1.86$ J/K g, and $\Delta H_{vap} = 2257$ J/g at 100°C,

a) what is the final temperature of the system and which phase or phases are present?
b) Calculate ΔS for the transformation.

9.30 An ingot of copper with a mass of 1 kg and an average heat capacity of 0.39 J/K g is at a temperature of 500 °C.

a) If the ingot is quenched in water, what mass of water at 25 °C must be used so that the final state of the system consists of liquid water, steam, and solid copper at 100 °C, half the water having been converted to steam. The heat capacity of water is 4.18 J/K g and the heat of vaporization is 2257 J/g.
b) What is ΔS in this transformation?

9.31 Sketch the possible indistinguishable arrangements of

a) two balls in six cells;
b) four balls in six cells.
c) What is the probability of the uniform distribution in each case?

9.32 Suppose that three indistinguishable molecules are distributed among three energy levels. The energies of the levels are: 0, 1, 2 units.

a) How many complexions are possible if there is no restriction on the energy of the three molecules?
b) How many complexions are possible if the total energy of the three molecules is fixed at one unit?
c) Find the number of complexions if the total energy is two units, and calculate the increase in entropy accompanying the energy increase from one to two units.

9.33 Suppose we have N distinguishable balls that are to be distributed in N_c cells.

a) How many complexions are there if we do not care whether there is more than one ball in the cell?
b) How many complexions correspond to distributions with no more than one ball per cell?
c) Using the results in (a) and (b), calculate the probability that in a group of 23 people no two will have the same birthday.

9.34 Pure ortho-hydrogen can exist in any of nine quantum states at absolute zero. Calculate the entropy of this mixture of nine "kinds" of ortho-hydrogen; each has a mole fraction of $\frac{1}{9}$.

9.35 The entropy of a binary mixture relative to its pure components is given by Eq. (9.74). Since $x_a + x_b = 1$, write the entropy of the mixture in terms of x_a or x_b only, and show that the entropy is a maximum when $x_a = x_b = \frac{1}{2}$. Calculate S_{mix} values for $x_a = 0, 0.2, 0.4, 0.5, 0.6, 0.8,$ and 1. Plot S_{mix} as a function of x_a.

10

Spontaneity and Equilibrium

10.1 THE GENERAL CONDITIONS FOR EQUILIBRIUM AND FOR SPONTANEITY

Our aim now is to find out what characteristics distinguish irreversible (real) transformations from reversible (ideal) transformations. We begin by asking what relation exists between the entropy change in a transformation and the irreversible heat flow that accompanies it. At every stage of a reversible transformation, the system departs from equilibrium only infinitesimally. The system is transformed, yet remains effectively at equilibrium throughout a reversible change in state. The condition for reversibility is therefore a condition of equilibrium; from the defining equation for dS, the condition of reversibility is that

$$T\,dS = dQ_{rev}. \tag{10.1}$$

Therefore Eq. (10.1) is the condition of equilibrium.

The condition placed on an irreversible change in state is the Clausius inequality, (8.44), which we write in the form

$$T\,dS > dQ. \tag{10.2}$$

Irreversible changes are real changes or natural changes or spontaneous changes. We shall refer to changes in the natural direction as spontaneous changes, and the inequality (10.2) as the condition of spontaneity. The two relations Eq. (10.1) and (10.2) can be combined into

$$T\,dS \geq dQ, \tag{10.3}$$

where it is understood that the equality sign implies a reversible value of dQ.

By using the first law in the form $dQ = dU + dW$, the relation in (10.3) can be written

$$T \, dS \geq dU + dW,$$

or

$$-dU - dW + T \, dS \geq 0. \tag{10.4}$$

The work includes all kinds; $dW = P_{op} \, dV + dW_a$. This value for dW brings relation (10.4) to the form

$$-dU - P_{op} \, dV - dW_a + T \, dS \geq 0. \tag{10.5}$$

Both relations (10.4) and (10.5) express the condition of equilibrium ($=$) and of spontaneity ($>$) for a transformation in terms of changes in properties of the system dU, dV, dS, and the amount of work dW or dW_a associated with the transformation.

10.2 CONDITIONS FOR EQUILIBRIUM AND SPONTANEITY UNDER CONSTRAINTS

Under the combinations of restraints usually imposed in the laboratory, relations (10.4) and (10.5) can be expressed in simple and convenient terms. We consider each set of constraints separately.

10.2.1 Transformations in an Isolated System

For an isolated system, $dU = 0$, $dW = 0$, $dQ = 0$; thus relation (10.4) becomes directly

$$dS \geq 0. \tag{10.6}$$

This requirement for an isolated system was discussed in detail in Section 8.14, where it was shown that in an isolated system the entropy can only increase and reaches a maximum at equilibrium.

From relation (10.6) it follows that an isolated system at equilibrium must have the same temperature in all its parts. Assume that an isolated system is subdivided into two parts, α and β. If a positive quantity of heat, dQ_{rev}, passes reversibly from region α to region β, we have

$$dS_\alpha = \frac{-dQ_{rev}}{T_\alpha} \quad \text{and} \quad dS_\beta = \frac{dQ_{rev}}{T_\beta}.$$

The total change in entropy is

$$dS = dS_\alpha + dS_\beta = \left(\frac{1}{T_\beta} - \frac{1}{T_\alpha} \right) dQ_{rev}.$$

If this flow of heat is to occur spontaneously, then by relation (10.6) $dS > 0$. Since dQ_{rev} is positive, this means that

$$\frac{1}{T_\beta} - \frac{1}{T_\alpha} > 0 \quad \text{or} \quad T_\alpha > T_\beta.$$

It follows that heat flows spontaneously from the region of higher temperature, α, to that of lower temperature, β. Furthermore, at equilibrium $dS = 0$. This requires that

$$T_\alpha = T_\beta$$

This is the condition of thermal equilibrium; a system in equilibrium must have the same temperature in all its parts.

10.2.2 Transformations at Constant Temperature

If a system undergoes an isothermal change in state, then $T dS = d(TS)$, and the relation (10.4) can be written

$$-dU + d(TS) \geq đW,$$

$$-d(U - TS) \geq đW. \tag{10.7}$$

The combination of variables $U - TS$ appears so frequently that it is given a special symbol, A. By definition,

$$A \equiv U - TS. \tag{10.8}$$

Being a combination of other functions of the state, A is a function of the state of the system; A is called the *Helmholtz energy* of the system.* The relation (10.7) reduces to the form

$$-dA \geq đW, \tag{10.9}$$

or, by integrating,

$$-\Delta A \geq W. \tag{10.10}$$

The significance of A is given by relation (10.10); the work produced in an isothermal transformation is less than or equal to the decrease in the Helmholtz energy. The equality sign applies to the reversible transformation, so the maximum work obtainable in an isothermal change in state is equal to the decrease in the Helmholtz energy. This maximum work includes all the kinds of work produced in the transformation.

10.2.3 Transformations at Constant Temperature and Under Constant Pressure

The system is confined under a constant pressure, $P_{op} = p$, the equilibrium pressure of the system. Since p is a constant, $p\,dV = d(pV)$. The temperature is constant so that $T\,dS = d(TS)$. The relation (10.5) becomes

$$-[dU + d(pV) - d(TS)] \geq đW_a,$$

$$-d(U + pV - TS) \geq đW_a. \tag{10.11}$$

The combination of variables $U + pV - TS$ occurs frequently and is given a special symbol, G. By definition,

$$G \equiv U + pV - TS = H - TS = A + pV. \tag{10.12}$$

Being a composite of properties of the state of a system, G is a property of the state; G is called the *Gibbs energy* of the system. More commonly, G is called the *free energy* of the system.†

Using Eq. (10.12), relation (10.11) becomes

$$-dG \geq đW_a, \tag{10.13}$$

---

* In the past, A has been known by a number of names: work function, maximum work function, Helmholtz function, Helmholtz free energy, and simply free energy. The IUPAC agreement is to use the symbol A and call it the *Helmholtz energy*.

† In the past, G has been known as: Gibbs function, Gibbs free energy, and simply free energy. The IUPAC agreement is to use G for the symbol and *Gibbs energy* as the name. In using tables of thermodynamic data, you should be aware that many of them will use the symbol F for the Gibbs energy. Unfortunately, in the past F has also been used as a symbol for A. In using any table of data it is best to make sure just what the symbols stand for.

or, by integrating,

$$-\Delta G \geq W_a. \tag{10.14}$$

Fixing our attention on the equality sign in (10.14), we have

$$-\Delta G = W_{a,\,\text{rev}}, \tag{10.15}$$

which reveals an important property of the Gibbs energy; the decrease in Gibbs energy $(-\Delta G)$ associated with a change in state at constant T and p is equal to the maximum work $W_{a,\text{rev}}$ over and above expansion work, which is obtainable in the transformation. By (10.14), in any real transformation the work obtained over and above expansion work is less than the decrease in Gibbs energy that accompanies the change in state at constant T and p.

If the work W_a is to show up in the laboratory, the transformation must be conducted in a device that enables the work to be produced; the most usual chemical example of such a device is the electrochemical cell. If granulated zinc is dropped into a solution of copper sulfate, metallic copper precipitates and the zinc dissolves according to the reaction

$$\text{Zn} + \text{Cu}^{2+} \rightarrow \text{Cu} + \text{Zn}^{2+}.$$

Obviously, the only work produced in this mode of performing the reaction is expansion work, and there is very little of that. On the other hand, this same chemical reaction can be carried out in such a way as to produce a quantity of electrical work $W_a = W_{\text{el}}$. In the Daniell cell shown in Fig. 17.1, a zinc electrode is immersed in a solution of zinc sulfate and a copper electrode is immersed in a solution of copper sulfate; the solutions are in electrical contact through a porous partition that prevents the solutions from mixing. The Daniell cell can produce electrical work W_{el}, which is related to the decrease in Gibbs energy, $-\Delta G$, of the chemical reaction by relation (10.14). If the cell operates reversibly, then the electrical work produced is equal to the decrease in Gibbs energy. The performance of the electrochemical cell is discussed in detail in Chapter 17.

Any spontaneous transformation *may* be harnessed to do some kind of work in addition to expansion work, but it need not necessarily be so harnessed. For the present, our interest is in those transformations that are not harnessed to do special kinds of work; for these cases, $dW_a = 0$, and the condition of equilibrium and spontaneity for a transformation at constant p and T, relation (10.14), becomes

$$-dG \geq 0, \tag{10.16}$$

or, for a finite change,

$$-\Delta G \geq 0. \tag{10.17}$$

Both relations (10.16) and (10.17) require the Gibbs energy to decrease in any real transformation at constant T and p; if the Gibbs energy decreases, ΔG is negative and $-\Delta G$ is positive. Spontaneous changes can continue to occur in such a system as long as the Gibbs energy of the system can decrease, that is, *until the Gibbs energy of the system reaches a minimum value*. The system at equilibrium has a minimum value of the Gibbs energy; this equilibrium condition is expressed by the equality sign in relation (10.16): $dG = 0$, the usual mathematical condition for a minimum value.

Of the several criteria for equilibrium and spontaneity, we shall have the most use for the one involving dG or ΔG, simply because most chemical reactions and phase transformations are subject to the conditions, constant T and p. If we know how to compute the change in Gibbs energy for any transformation, the algebraic sign of ΔG tells us whether this transformation can occur in the direction in which we imagine it. There are three possibilities:

1. ΔG is $-$; the transformation can occur spontaneously, or naturally;

2. $\Delta G = 0$; the system is at equilibrium with respect to this transformation; or

3. ΔG is $+$; the natural direction is opposite to the direction we have envisioned for the transformation (the transformation is nonspontaneous).

The third case is best illustrated by an example. Suppose we ask whether water can flow uphill. The transformation can be written

$$H_2O(\text{low level}) \quad \longrightarrow \quad H_2O(\text{high level}) \qquad (T \text{ and } p \text{ constant}).$$

The value of ΔG for this transformation is calculated and found to be positive. We conclude that the direction of this transformation as it is written is not the natural direction, and that the natural, spontaneous direction is opposite to the way we have written it. In the absence of artificial restraints, water at a high level will flow to a lower level; the ΔG for water flowing downhill is equal in magnitude but opposite in sign to that for water flowing uphill. Transformations with positive values for ΔG include such outlandish things as water flowing uphill, a ball bearing jumping out of a glass of water, an automobile manufacturing gasoline from water and carbon dioxide as it is pushed down the street.

10.3 RECOLLECTION

By comparing real transformations with reversible ones we arrived at the Clausius inequality, $dS > dQ/T$, which gives us a criterion for a real, or spontaneous, transformation. By algebraic manipulation this criterion was given simple expression in terms of the entropy change, or changes in value of two new functions A and G. By examining the algebraic sign of ΔS, ΔA, or ΔG for the transformation in question, we can decide whether it can occur spontaneously or not. At the same time we obtain the condition of equilibrium for the tranformation. These conditions of spontaneity and equilibrium are summarized in Table 10.1; we shall make the greatest use of those on the last line, since the constraints $W_a = 0$, T and p constant are those most frequently used in the laboratory.

Table 10.1

Constraint	Condition for spontaneity		Equilibrium condition	
None	$-(dU + p\,dV - T\,dS) - dW_a = +$		$-(dU + p\,dV - T\,dS) - dW_a = 0$	
	Infinitesimal change	Finite change	Infinitesimal change	Finite change
Isolated system	$dS = +$	$\Delta S = +$	$dS = 0$	$\Delta S = 0$
T constant	$dA + dW = -$	$\Delta A + W = -$	$dA + dW = 0$	$\Delta A + W = 0$
T, p constant	$dG + dW_a = -$	$\Delta G + W_a = -$	$dG + dW_a = 0$	$\Delta G + W_a = 0$
$W_a = 0; T, V$ constant	$dA = -$	$\Delta A = -$	$dA = 0$	$\Delta A = 0$
$W_a = 0; T, p$ constant	$dG = -$	$\Delta G = -$	$dG = 0$	$\Delta G = 0$

The word "spontaneous" applied to changes in state in a thermodynamic sense must not be given too broad a meaning. It means only that the change in state is *possible*. Thermodynamics cannot provide any information about how much time is required for the change in state. For example, thermodynamics predicts that at 25 °C and 1 atm pressure the reaction between hydrogen and oxygen to form water is a spontaneous reaction. However, in the absence of a catalyst or an initiating event, such as a spark, they do not react to form water in any measurable length of time. The length of time required for a spontaneous transformation to come to equilibrium is a proper subject for kinetics, not thermodynamics. Thermodynamics tells us what *can* happen; kinetics tells us whether it will take a thousand years or a millionth of a second. Once we know that a certain reaction *can* happen, it may be worth our while to search for a catalyst that will shorten the time interval required for the reaction to reach equilibrium. It is futile to seek a catalyst for a reaction that is thermodynamically impossible.

What can be done about those transformations that have ΔG positive and are thermodynamically impossible, or nonspontaneous? Human nature being what it is, it does not submit lightly to the judgment that a certain change is "impossible." The "impossible" flow of water uphill can be made "possible," not through the agency of a catalyst that is unchanged in the transformation, but by *coupling* the nonspontaneous uphill flow of a certain mass of water with the spontaneous downhill flow of a greater mass of some substance. A weight cannot by itself jump three feet up from the floor, but if it is coupled through a pulley to a heavier weight that falls three feet, it will jump up. The composite change, light weight up, heavier weight down, is accompanied by a decrease in Gibbs energy and thus is a "possible" one. As we shall see later, coupling one change in state with another can be turned to great advantage in dealing with chemical reactions.

10.4 DRIVING FORCES FOR NATURAL CHANGES

In a natural change at constant temperature and pressure, ΔG must be negative. By definition, $G = H - TS$, so that at constant temperature

$$\Delta G = \Delta H - T\Delta S. \tag{10.18}$$

Two contributions to the value of ΔG can be distinguished in Eq. (10.18): an energetic one, ΔH, and an entropic one, $T\Delta S$. From Eq. (10.18) it is clear that to make ΔG negative, it is best if ΔH is negative (exothermic transformation) and ΔS is positive. In a natural change, the system attempts to achieve the lowest enthalpy (roughly, the lowest energy) and the highest entropy. It is also clear from Eq. (10.18) that a system can tolerate a decrease in entropy in the change in state that makes the second term positive, if the first term is negative enough to over-balance the positive second term. Similarly, an increase in enthalpy, ΔH positive, can be tolerated if ΔS is sufficiently positive so that the second term over-balances the first. In such instances the compromise between low enthalpy and high entropy is reached in such a way as to minimize the Gibbs energy at equilibrium. The majority of common chemical reactions are exothermic in their natural direction, often so highly exothermic that the term $T\Delta S$ has little influence in determining the equilibrium position. In the case of reactions that are endothermic in their natural direction, the term $T\Delta S$ is all-important in determining the equilibrium position.

10.5 THE FUNDAMENTAL EQUATIONS OF THERMODYNAMICS

In addition to the mechanical properties p and V, a system has three fundamental properties T, U, and S, defined by the laws of thermodynamics, and three composite properties

H, A, and G, which are important. We are now in a position to develop the important differential equations that relate these properties to one another.

For the present we restrict the discussion to systems that produce only expansion work so that $dW_a = 0$. With this restriction, the general condition of equilibrium is

$$dU = T\,dS - p\,dV. \tag{10.19}$$

This combination of the first and second laws of thermodynamics is the fundamental equation of thermodynamics. Using the definitions of the composite functions,

$$H = U + pV, \qquad A = U - TS, \qquad G = U + pV - TS,$$

and differentiating each one, we obtain

$$dH = dU + p\,dV + V\,dp,$$
$$dA = dU - T\,dS - S\,dT,$$
$$dG = dU + p\,dV + V\,dp - T\,dS - S\,dT.$$

In each of these three equations, dU is replaced by its value from Eq. (10.19); after collecting terms, the equations become [Eq. (10.19) is repeated first]

$$dU = T\,dS - p\,dV, \tag{10.19}$$

$$dH = T\,dS + V\,dp, \tag{10.20}$$

$$dA = -S\,dT - p\,dV, \tag{10.21}$$

$$dG = -S\,dT + V\,dp. \tag{10.22}$$

These four equations are sometimes called the four fundamental equations of thermodynamics; in fact, they are simply four different ways of looking at the one fundamental equation, Eq. (10.19).

Equation (10.19) relates the change in energy to the changes in entropy and volume. Equation (10.20) relates the change in enthalpy to changes in entropy and pressure. Equation (10.21) relates the change in the Helmholtz energy dA to changes in temperature and volume. Equation (10.22) relates the change in Gibbs energy to changes in temperature and pressure. Because of the simplicity of these equations, S and V are called the "natural" variables for the energy; S and p are the natural variables for the enthalpy; T and V are the natural variables for the Helmholtz energy; and T and p are the natural variables for the Gibbs energy.

Since each of the expressions on the right-hand side of these equations is an exact differential expression, it follows that the cross-derivatives are equal. From this we immediately obtain the four Maxwell relations:

$$\left(\frac{\partial T}{\partial V}\right)_S = -\left(\frac{\partial p}{\partial S}\right)_V; \tag{10.23}$$

$$\left(\frac{\partial T}{\partial p}\right)_S = \left(\frac{\partial V}{\partial S}\right)_p; \tag{10.24}$$

$$\left(\frac{\partial S}{\partial V}\right)_T = \left(\frac{\partial p}{\partial T}\right)_V; \tag{10.25}$$

$$-\left(\frac{\partial S}{\partial p}\right)_T = \left(\frac{\partial V}{\partial T}\right)_p. \tag{10.26}$$

The first two of these equations relate to changes in state at constant entropy, that is, adiabatic, reversible changes in state. The derivative $(\partial T / \partial V)_S$ represents the rate of change of temperature with volume in a reversible adiabatic transformation. We shall not be much concerned with Eqs. (10.23) and (10.24).

Equations (10.25) and (10.26) are of great importance because they relate the iso-thermal volume dependence of the entropy and the isothermal pressure dependence of the entropy to easily measured quantities. We obtained these relations earlier, Eqs. (9.31) and (9.41), by utilizing the fact that dS is an exact differential. Here we obtain them with much less algebraic labor from the facts that dA and dG are exact differentials. The two deriva-tions are clearly equivalent since A and G are functions of the state only if S is a function of the state.

10.6 THE THERMODYNAMIC EQUATION OF STATE

The equations of state discussed so far, the ideal gas law, the van der Waals equation, and others, were relations between p, V, and T obtained from empirical data on the behavior of gases or from speculation about the effects of molecular size and attractive forces on the behavior of the gas. The equation of state for a liquid or solid was simply expressed in terms of the experimentally determined coefficients of thermal expansion and com-pressibility. These relations applied to systems at equilibrium, but there is a more general condition of equilibrium. The second law of thermodynamics requires the relation, Eq. (10.19),

$$dU = T\,dS - p\,dV$$

as an equilibrium condition. From this we should be able to derive an equation of state for any system. Let the changes in U, S, and V of Eq. (10.19) be changes at constant T:

$$(\partial U)_T = T(\partial S)_T - p(\partial V)_T.$$

Dividing by $(\partial V)_T$, we have

$$\left(\frac{\partial U}{\partial V}\right)_T = T\left(\frac{\partial S}{\partial V}\right)_T - p, \tag{10.27}$$

where, from the writing of the derivatives, U and S are considered to be functions of T and V. Therefore the partial derivatives in Eq. (10.27) are functions of T and V. This equation relates the pressure to functions of T and V; it is an equation of state. Using the value for $(\partial S / \partial V)_T$ from Eq. (10.25) and rearranging, Eq. (10.27) becomes

$$p = T\left(\frac{\partial p}{\partial T}\right)_V - \left(\frac{\partial U}{\partial V}\right)_T, \tag{10.28}$$

which is perhaps a neater form for the equation.

By restricting the second fundamental equation, Eq. (10.20), to constant temperature and dividing by $(\partial p)_T$ we obtain

$$\left(\frac{\partial H}{\partial p}\right)_T = T\left(\frac{\partial S}{\partial p}\right)_T + V. \tag{10.29}$$

Using Eq. (10.26) and rearranging, this equation becomes

$$V = T\left(\frac{\partial V}{\partial T}\right)_p + \left(\frac{\partial H}{\partial p}\right)_T, \tag{10.30}$$

which is a general equation of state expressing the volume as a function of temperature and pressure. These thermodynamic equations of state are applicable to any substance whatsoever. Eqs. (10.28) and (10.30) were obtained earlier, Eqs. (9.30) and (9.40), but were not discussed at that point.

10.6.1 Applications of the Thermodynamic Equation of State

If we knew the value of either $(\partial U/\partial V)_T$ or $(\partial H/\partial p)_T$ for a substance, we would know its equation of state immediately from Eqs. (10.28) or (10.30). More commonly we do not know the values of these derivatives, so we arrange Eq. (10.28) in the form

$$\left(\frac{\partial U}{\partial V}\right)_T = T\left(\frac{\partial p}{\partial T}\right)_V - p. \tag{10.31}$$

From the empirical equation of state, the right-hand side of Eq. (10.31) can be evaluated to yield a value of the derivative $(\partial U/\partial V)_T$. For example, for the ideal gas, $p = nRT/V$, so $(\partial p/\partial T)_V = nR/V$. Using these values in Eq. (10.31), we obtain $(\partial U/\partial V)_T = nRT/V - p = p - p = 0$. We have used this result, Joule's law, before; this demonstration proves its validity for the ideal gas.

Since, from Eq. (9.23), $(\partial p/\partial T)_V = \alpha/\kappa$, Eq. (10.31) is often written in the form

$$\left(\frac{\partial U}{\partial V}\right)_T = T\frac{\alpha}{\kappa} - p = \frac{\alpha T - \kappa p}{\kappa}, \tag{10.32}$$

and Eq. (10.30) in the form

$$\left(\frac{\partial H}{\partial p}\right)_T = V(1 - \alpha T). \tag{10.33}$$

It is now possible, using Eqs. (10.32) and (10.33), to write the total differentials of U and H in a form containing only quantities that are easily measurable:

$$dU = C_v\,dT + \frac{(\alpha T - \kappa p)}{\kappa}\,dV, \tag{10.34}$$

$$dH = C_p\,dT + V(1 - \alpha T)\,dp. \tag{10.35}$$

These equations together with the two equations for dS, Eqs. (9.33), and (9.42), are helpful in deriving others.

Using Eq. (10.32), we can obtain a simple expression for $C_p - C_v$. From Eq. (7.39), we have

$$C_p - C_v = \left[p + \left(\frac{\partial U}{\partial V}\right)_T\right]V\alpha.$$

Using the value of $(\partial U/\partial V)_T$ from Eq. (10.32), we obtain

$$C_p - C_v = \frac{TV\alpha^2}{\kappa}, \tag{10.36}$$

which permits the evaluation of $C_p - C_v$ from quantities that are readily measurable for any substance. Since T, V, κ, and α^2 must all be positive, C_p is always greater than C_v.

For the Joule–Thomson coefficient we have, from Eq. (7.50),

$$C_p\mu_{JT} = -\left(\frac{\partial H}{\partial p}\right)_T.$$

Using Eq. (10.33), we obtain for μ_{JT},

$$C_p \mu_{JT} = V(\alpha T - 1). \tag{10.37}$$

Thus, if we know C_p, V, and α for the gas, we can calculate μ_{JT}.

These quantities are much more easily measured than is μ_{JT} itself. At the Joule–Thomson inversion temperature, μ_{JT} changes sign; that is, $\mu_{JT} = 0$; using this condition in Eq. (10.37), we find at the inversion temperature, $T_{inv}\alpha - 1 = 0$.

10.7 THE PROPERTIES OF A

The properties of the Helmholtz energy A are expressed by the fundamental equation (10.21),

$$dA = -S\,dT - p\,dV.$$

This equation views A as a function of T and V, and we have the identical equation

$$dA = \left(\frac{\partial A}{\partial T}\right)_V dT + \left(\frac{\partial A}{\partial V}\right)_T dV.$$

Comparing these two equations shows that

$$\left(\frac{\partial A}{\partial T}\right)_V = -S, \tag{10.38}$$

and

$$\left(\frac{\partial A}{\partial V}\right)_T = -p. \tag{10.39}$$

Since the entropy of any substance is positive, Eq. (10.38) shows that the Helmholtz energy of any substance decreases (minus sign) with an increase in temperature. The rate of this decrease is greater the greater the entropy of the substance. For gases, which have large entropies, the rate of decrease of A with temperature is greater than for liquids and solids, which have comparatively small entropies.

Similarly, the minus sign in Eq. (10.39) shows that an increase in volume decreases the Helmholtz energy; the rate of decrease is greater the higher the pressure.

10.7.1 The Condition for Mechanical Equilibrium

Consider a system at constant temperature and constant total volume that is subdivided into two regions, α and β. Suppose that region α expands reversibly by an amount, dV_α, while region β contracts by an equal amount, $dV_\beta = -dV_\alpha$, since the total volume must remain constant. Then, by Eq. (10.39), we have

$$dA_\alpha = -p_\alpha\,dV_\alpha \qquad \text{and} \qquad dA_\beta = -p_\beta\,dV_\beta.$$

The total change in A is

$$dA = dA_\alpha + dA_\beta = -p_\alpha\,dV_\alpha - p_\beta\,dV_\beta = (p_\beta - p_\alpha)\,dV_\alpha.$$

Since no work is produced, $đW = 0$, and Eq. (10.9) requires $dA < 0$ if the transformation is to be spontaneous. Since dV_α is positive, this means that $p_\alpha > p_\beta$. The high-pressure region expands at the expense of the low-pressure region. The equilibrium requirement is that $dA = 0$; that is,

$$p_\alpha = p_\beta.$$

This is the condition for mechanical equilibrium; namely, that the pressure have the same value in all parts of the system.

10.8 THE PROPERTIES OF *G*

The fundamental equation (10.22),

$$dG = -S\,dT + V\,dp,$$

views the Gibbs energy as a function of temperature and pressure; the equivalent expression is therefore

$$dG = \left(\frac{\partial G}{\partial T}\right)_p dT + \left(\frac{\partial G}{\partial p}\right)_T dp. \tag{10.40}$$

Comparing these two equations shows that

$$\left(\frac{\partial G}{\partial T}\right)_p = -S, \tag{10.41}$$

and

$$\left(\frac{\partial G}{\partial p}\right)_T = V. \tag{10.42}$$

Because of the importance of the Gibbs energy, Eqs. (10.41) and (10.42) contain two of the most important pieces of information in thermodynamics. Again, since the entropy of any substance is positive, the minus sign in Eq. (10.41) shows that increase in temperature decreases the Gibbs energy if the pressure is constant. The rate of decrease is greater for gases, which have large entropies, than for liquids or solids, which have small entropies. Because *V* is always positive, an increase in pressure increases the Gibbs energy at constant temperature, as shown by Eq. (10.42). The larger the volume of the system the greater is the increase in Gibbs energy for a given increase in pressure. The comparatively large volume of a gas implies that the Gibbs energy of a gas increases much more rapidly with pressure than would that of a liquid or a solid.

The Gibbs energy of any pure material is conveniently expressed by integrating Eq. (10.42) at constant temperature from the standard pressure, $p^\circ = 1$ atm, to any other pressure p:

$$\int_{p^\circ}^p dG = \int_{p^\circ}^p V\,dp, \qquad G - G^\circ = \int_{p^\circ}^p V\,dp,$$

or

$$G = G^\circ(T) + \int_{p^\circ}^p V\,dp, \tag{10.43}$$

where $G^\circ(T)$ is the Gibbs energy of the substance under 1 atm pressure, the *standard* Gibbs energy, which is a function of temperature.

If the substance in question is either a liquid or a solid, the volume is nearly independent of the pressure and can be removed from under the integral sign; then

$$G(T, p) = G^\circ(T) + V(p - p^\circ) \qquad \text{(liquids and solids)}. \tag{10.44}$$

Since the volume of liquids and solids is small, unless the pressure is enormous, the second term on the right of Eq. (10.44) is negligibly small; ordinarily for condensed phases we will write simply

$$G = G^\circ(T) \tag{10.45}$$

and ignore the pressure dependence of G.

The volume of gases is very much larger than that of liquids or solids and depends greatly on pressure; applying Eq. (10.43) to the ideal gas, it becomes

$$G = G^\circ(T) + \int_{p^\circ}^{p} \frac{nRT}{p}\, dp, \qquad \frac{G}{n} = \frac{G^\circ(T)}{n} + RT \ln\left(\frac{p\,\text{atm}}{1\,\text{atm}}\right).$$

It is customary to use a special symbol, μ, for the Gibbs energy per mole; we define

$$\mu = \frac{G}{n}. \tag{10.46}$$

Thus for the molar Gibbs energy of the ideal gas, we have

$$\mu = \mu^\circ(T) + RT \ln p. \tag{10.47}$$

As in Section 9.11, the symbol p in Eq. (10.47) represents a pure number, the number which when multiplied by 1 atm yields the value of the pressure in atmospheres.

The logarithmic term in Eq. (10.47) is quite large in most circumstances and cannot be ignored. From this equation it is clear that at a specified temperature, the pressure describes the Gibbs energy of the ideal gas; the higher the pressure the greater the Gibbs energy (Fig. 10.1).

It is worth emphasizing that if we know the functional form of $G(T, p)$, then we can obtain all other thermodynamic functions by differentiation, using Eqs. (10.41) and (10.42), and combining with definitions. (See Problem 10.29.)

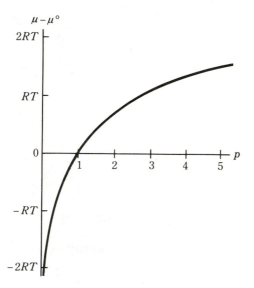

Figure 10.1 Gibbs energy of ideal gas as a function of pressure.

10.9 THE GIBBS ENERGY OF REAL GASES

The functional form of Eq. (10.47) is particularly simple and convenient. It would be helpful if the molar Gibbs energy of real gases could be expressed in the same mathematical form. We therefore "invent" a function of the state that will express the molar Gibbs energy of a real gas by the equation

$$\mu = \mu^\circ(T) + RT \ln f. \tag{10.48}$$

The function f is called the *fugacity* of the gas. The fugacity measures the Gibbs energy of a real gas in the same way as the pressure measures the Gibbs energy of an ideal gas.

An invented function such as the fugacity has little use unless it can be related to measurable properties of the gas. Dividing the fundamental equation (10.22) by n, the number of moles of gas, and restricting to constant temperature, $dT = 0$, we obtain for the real gas $d\mu = \overline{V}\,dp$, while for the ideal gas $d\mu^{\mathrm{id}} = \overline{V}^{\mathrm{id}}\,dp$, where $\overline{V}$ and $\overline{V}^{\mathrm{id}}$ are the molar volumes of the real and ideal gases, respectively. Subtracting these two equations, we obtain $d(\mu - \mu^{\mathrm{id}}) = (\overline{V} - \overline{V}^{\mathrm{id}})\,dp$.

Integrating between the limits p^* and p yields

$$(\mu - \mu^{\mathrm{id}}) - (\mu^* - \mu^{*\mathrm{id}}) = \int_{p^*}^{p} (\overline{V} - \overline{V}^{\mathrm{id}})\,dp.$$

We now let $p^* \to 0$. The properties of any real gas approach their ideal values as the pressure of the gas approaches zero. Therefore, as $p^* \to 0$, $\mu^* \to \mu^{*\mathrm{id}}$. The equation becomes

$$\mu - \mu^{\mathrm{id}} = \int_{0}^{p} (\overline{V} - \overline{V}^{\mathrm{id}})\,dp. \tag{10.49}$$

But by Eq. (10.47), $\mu^{\mathrm{id}} = \mu^\circ(T) + RT \ln p$, and by the definition of f, Eq. (10.48), $\mu = \mu^\circ(T) + RT \ln f$. Using these values for μ and μ^{id}, Eq. (10.49) becomes

$$RT(\ln f - \ln p) = \int_{0}^{p} (\overline{V} - \overline{V}^{\mathrm{id}})\,dp;$$

$$\ln f = \ln p + \frac{1}{RT} \int_{0}^{p} (\overline{V} - \overline{V}^{\mathrm{id}})\,dp. \tag{10.50}$$

The integral in Eq. (10.50) can be evaluated graphically; knowing $\overline{V}$ as a function of pressure, we plot the quantity $(\overline{V} - \overline{V}^{\mathrm{id}})/RT$ as a function of pressure. The area under the curve from $p = 0$ to p is the value of the second term on the right of Eq. (10.50). Or, if $\overline{V}$ can be expressed as a function of pressure by an equation of state, the integral can be evaluated analytically, since $\overline{V}^{\mathrm{id}} = RT/p$. The integral can be expressed neatly in terms of the compressibility factor Z; by definition, $\overline{V} = Z\overline{V}^{\mathrm{id}}$. Using this value for $\overline{V}$, and $\overline{V}^{\mathrm{id}} = RT/p$, in the integral of Eq. (10.50), it reduces to

$$\ln f = \ln p + \int_{0}^{p} \frac{(Z - 1)}{p}\,dp. \tag{10.51}$$

The integral in Eq. (10.51) is evaluated graphically by plotting $(Z - 1)/p$ against p and measuring the area under the curve. For gases below their Boyle temperatures, $Z - 1$ is negative at moderate pressures, and the fugacity, by Eq. (10.51), will be less than the pressure. For gases above their Boyle temperatures, the fugacity is greater than the pressure.

The Gibbs energy of gases will usually be discussed as though the gases were ideal, and Eq. (10.47) will be used. The algebra will be exactly the same for real gases; we need only replace the pressure in the final equations by the fugacity, keeping in mind that the fugacity depends on temperature as well as pressure.

10.10 TEMPERATURE DEPENDENCE OF THE GIBBS ENERGY

The dependence of the Gibbs energy on temperature is expressed in several different ways for convenience in different problems. Rewriting Eq. (10.41), we have

$$\left(\frac{\partial G}{\partial T}\right)_p = -S. \tag{10.52}$$

From the definition $G = H - TS$, we obtain $-S = (G - H)/T$, and Eq. (10.52) becomes

$$\left(\frac{\partial G}{\partial T}\right)_p = \frac{G - H}{T}, \tag{10.53}$$

a form that is sometimes useful.

Frequently it is important to know how the function G/T depends on temperature. By the ordinary rule of differentiation, we obtain

$$\left(\frac{\partial (G/T)}{\partial T}\right)_p = \frac{1}{T}\left(\frac{\partial G}{\partial T}\right)_p - \frac{1}{T^2}G.$$

Using Eq. (10.52), this becomes

$$\left(\frac{\partial (G/T)}{\partial T}\right)_p = -\frac{TS + G}{T^2},$$

which reduces to

$$\left(\frac{\partial (G/T)}{\partial T}\right)_p = -\frac{H}{T^2}, \tag{10.54}$$

the Gibbs–Helmholtz equation, which we use frequently.

Since $d(1/T) = -(1/T^2)\,dT$, we can replace ∂T in the derivative in Eq. (10.54) by $-T^2\,\partial(1/T)$; this reduces it to

$$\left(\frac{\partial (G/T)}{\partial (1/T)}\right)_p = H, \tag{10.55}$$

which is another frequently used relation.

Any of Eqs. (10.52), (10.53), (10.54), (10.55) are simply different versions of the fundamental equation, Eq. (10.52). We will refer to them as the first, second, third, and fourth forms of the Gibbs–Helmholtz equation.

QUESTIONS

10.1 For what sort of experimental conditions is (a) A or (b) G the appropriate indicator of spontaneity?

10.2 The second law states that the entropy of the universe (system and surroundings) increases in a spontaneous process: $\Delta S_{syst} + \Delta S_{surr} \geq 0$. Argue that, at constant T and p, ΔS_{surr} is related to the system enthalpy change by $\Delta S_{surr} = -\Delta H_{syst}/T$. Then argue that Eq. (10.17) follows, where G is the system Gibbs energy.

10.3 Discuss the meaning of the term "spontaneous" in thermodynamics.

10.4 Construct a ΔH and ΔS table, including the four possibilities associated with the two possible signs for each of ΔH and ΔS. Discuss the resulting sign of ΔG and the spontaneity of the process.

10.5 The endothermic process of forming a solution of salt (NaCl) and water is spontaneous at room temperature. Explain how this is possible in terms of the higher entropy of the ions in solution compared to that of ions in the solid.

10.6 Is the increase of μ with increasing p for an ideal gas an enthalpy or an entropy effect?

10.7 Explain why Eqs. (10.17) and (10.47) do *not* imply that an ideal gas at constant temperature will spontaneously reduce its own pressure.

PROBLEMS

10.1 Using the van der Waals equation with the thermodynamic equation of state, evaluate $(\partial U/\partial V)_T$ for the van der Waals gas.

10.2 By integrating the total differential dU for a van der Waals gas, show that if C_v is a constant, $U = U' + C_v T - na/\bar{V}$, where U' is a constant of integration. (The answer to Problem 10.1 is needed for this problem.)

10.3 Calculate ΔU for the isothermal expansion of one mole of a van der Waals gas from 20 dm³/mol to 80 dm³/mol; if $a = 0.141$ m⁶ Pa mol⁻² (nitrogen) and if $a = 3.19$ m⁶ Pa mol⁻² (heptane).

10.4 a) Find the value of $(\partial S/\partial V)_T$ for the van der Waals gas.
b) Derive an expression for the change in entropy for the isothermal expansion of one mole of the van der Waals gas from V_1 to V_2.
c) Compare the result in (b) with the expression for the ideal gas. For the same increase in volume, will the entropy increase be greater for the van der Waals gas or for the ideal gas?

10.5 Evaluate the derivative, $(\partial U/\partial V)_T$, for the Berthelot equation and the Dieterici equation.

10.6 a) Write the thermodynamic equation of state for a substance that follows Joule's law.
b) By integrating the differential equation obtained in (a), show that at constant volume the pressure is proportional to the absolute temperature for such a substance.

10.7 As a first approximation, the compressibility factor of the van der Waals gas is given by

$$\frac{p\bar{V}}{RT} = 1 + \left(b - \frac{a}{RT} \right) \frac{p}{RT}.$$

From this expression and the thermodynamic equation of state show that $(\partial \bar{H}/\partial p)_T = b - (2a/RT)$.

10.8 Using the expression in Problem 10.7 for the compressibility factor, show that for the van der Waals gas

$$\left(\frac{\partial \bar{S}}{\partial p} \right)_T = - \left[\frac{R}{p} + \frac{Ra}{(RT)^2} \right].$$

10.9 Using the results in Problems 10.7 and 10.8, calculate ΔH and ΔS for an isothermal increase in pressure of CO_2 from 0.100 MPa to 10.0 MPa, assuming van der Waals behavior; $a = 0.366$ m⁶ Pa mol⁻², $b = 42.9 \times 10^{-6}$ m³/mol.
a) At 300 K.
b) At 400 K.
c) Compare with the ideal gas values.

10.10 At 700 K calculate ΔH and ΔS for the compression of ammonia from 0.1013 MPa to 50.00 MPa, using the Beattie–Bridgeman equation and the constants in Table 3.5.

10.11 Show that for a real gas

$$\bar{C}_p \mu_{JT} = \frac{RT^2}{p}\left(\frac{\partial Z}{\partial T}\right)_p ,$$

where μ_{JT} is the Joule–Thomson coefficient, and $Z = p\bar{V}/RT$ is the compressibility factor of the gas. Compare to Eq. (7.50).

10.12 Using the value of Z for the van der Waals gas given in Problem 10.7, calculate the value of μ_{JT}. Show that μ_{JT} changes sign at the inversion temperature, $T_{inv} = 2a/Rb$.

10.13 a) Show that Eq. (10.31) can be written in the form

$$\left(\frac{\partial U}{\partial V}\right)_T = T^2\left[\frac{\partial(p/T)}{\partial T}\right]_V = -\left[\frac{\partial(p/T)}{\partial(1/T)}\right]_V$$

b) Show that Eq. (10.30) can be written in the form

$$\left(\frac{\partial H}{\partial p}\right)_T = -T^2\left[\frac{\partial(V/T)}{\partial T}\right]_p = \left[\frac{\partial(V/T)}{\partial(1/T)}\right]_p$$

10.14 At 25 °C calculate the value of ΔA for an isothermal expansion of one mole of the ideal gas from 10 litres to 40 litres.

10.15 By integrating Eq. (10.39) derive an expression for the Helmholtz energy of

a) the ideal gas;
b) the van der Waals gas. (Don't forget the "constant" of integration!)

10.16 Calculate ΔG for the isothermal (300 K) expansion of an ideal gas from 5000 kPa to 200 kPa.

10.17 Using the form given in Problem 10.7 for the van der Waals equation, derive an expression for ΔG if one mole of gas is compressed isothermally from 1 atm to a pressure p.

10.18 Calculate ΔG for the isothermal expansion of the van der Waals gas at 300 K from 5000 kPa to 200 kPa. Compare with the result in Problem 10.16 for O_2 for which $a = 0.138$ m^6 Pa mol^{-2} and $b = 31.8 \times 10^{-6}$ m^3/mol.

10.19 At 300 K one mole of a substance is subjected to an isothermal increase in pressure from 100 kPa to 1000 kPa. Calculate ΔG for each substance in (a) through (d) and compare the numerical values.

a) Ideal gas.
b) Liquid water; $\bar{V} = 18$ cm^3/mol.
c) Copper; $\bar{V} = 7.1$ cm^3/mol.
d) Sodium chloride; $\bar{V} = 27$ cm^3/mol.

10.20 Using the van der Waals equation in the form given in Problem 10.7, derive the expression for the fugacity of the van der Waals gas.

10.21 From the definition of the fugacity and the Gibbs–Helmholtz equation, show that the molar enthalpy, $\bar{H}$, of a real gas is related to the molar enthalpy of the ideal gas, $\bar{H}°$, by

$$\bar{H} = \bar{H}° - RT^2\left(\frac{\partial \ln f}{\partial T}\right)_p$$

and that the molar entropy, $\bar{S}$, is related to the standard molar entropy of the ideal gas $\bar{S}°$ by

$$\bar{S} = \bar{S}° - R\left[\ln f + T\left(\frac{\partial \ln f}{\partial T}\right)_p\right].$$

Show also from the differential equation for dG that $\bar{V} = RT(\partial \ln f/\partial p)_T$.

10.22 Combining the results of Problems 10.20 and 10.21 show that the enthalpy of the van der Waals gas is

$$\bar{H} = \bar{H}° + \left(b - \frac{2a}{RT}\right)p.$$

10.23 From the purely mathematical properties of the exact differential

$$dU = C_v \, dT + \left(\frac{\partial U}{\partial V}\right)_T dV,$$

show that if $(\partial U/\partial V)_T$ is a function only of volume, then C_v is a function only of temperature.

10.24 By taking the reciprocal of both sides of Eq. (10.23) we obtain $(\partial S/\partial p)_V = -(\partial V/\partial T)_S$. Using this equation and the cyclic relation between V, T, and S, show that $(\partial S/\partial p)_V = \kappa C_v/\alpha T$.

10.25 Given $dU = C_v \, dT + [(T\alpha - p\kappa)/\kappa] \, dV$, show that $dU = [C_v + (TV\alpha^2/\kappa) - pV\alpha] \, dT + V(p\kappa - T\alpha) \, dp$. *Hint:* Expand dV in terms of dT and dp.

10.26 Using the result in Problem 10.25 and the data for carbon tetrachloride at 20 °C: $\alpha = 12.4 \times 10^{-4} \, \mathrm{K}^{-1}$; $\kappa = 103 \times 10^{-6} \, \mathrm{atm}^{-1}$; density $= 1.5942 \, \mathrm{g/cm^3}$ and $M = 153.8 \, \mathrm{g/mol}$, show that near 1 atm pressure, $(\partial U/\partial p)_T \approx -VT\alpha$. Calculate the change in molar energy per atm at 20 °C.

10.27 Using the approximate value of the compressibility factor given in Problem 10.7, show that for the van der Waals gas:

a) $\bar{C}_p - \bar{C}_v = R + 2ap/RT^2$
b) $(\partial \bar{U}/\partial p)_T = -a/RT$. [*Hint:* Refer to Problem 10.25.]
c) $(\partial \bar{U}/\partial T)_p = \bar{C}_v + ap/RT^2$.

10.28 Knowing that $dS = (C_p/T) \, dT - V\alpha \, dp$, show that

a) $(\partial S/\partial p)_V = \kappa C_v/T\alpha$.
b) $(\partial S/\partial V)_p = C_p/TV\alpha$.
c) $-(1/V)(\partial V/\partial p)_S = \kappa/\gamma$, where $\gamma \equiv C_p/C_v$.

10.29 By using the fundamental differential equations and the definitions of the functions, determine the functional form of $\bar{S}$, $\bar{V}$, $\bar{H}$, $\bar{U}$ for

a) the ideal gas, given that $\mu = \mu°(T) + RT \ln p$.
b) the van der Waals gas, given that

$$\mu = \mu°(T) + RT \ln p + (b - a/RT)p.$$

10.30 Show that if $Z = 1 + B(T)p$, then $f = pe^{Z-1}$; and that this implies that at low to moderate pressures $f \approx pZ$, and that $p^2 = fp_{\text{ideal}}$. (This last relation states that the pressure is the geometric mean of the ideal pressure and the fugacity.)

11

Systems of Variable Composition; Chemical Equilibrium

11.1 THE FUNDAMENTAL EQUATION

In our study so far we have assumed implicitly that the system is composed of a pure substance or, if it was composed of a mixture, that the composition of the mixture was unaltered in the change of state. As a chemical reaction proceeds, the composition of the system and the thermodynamic properties change. Consequently, we must introduce the dependence on composition into the thermodynamic equations. We do this first only for the Gibbs energy G, since it is the most immediately useful.

For a pure substance or for a mixture of fixed composition the fundamental equation for the Gibbs energy is

$$dG = -S\,dT + V\,dp. \tag{11.1}$$

If the mole numbers, $n_1, n_2, \ldots$, of the substances present vary, then $G = G(T, p, n_1, n_2, \ldots)$, and the total differential is

$$dG = \left(\frac{\partial G}{\partial T}\right)_{p, n_i} dT + \left(\frac{\partial G}{\partial p}\right)_{T, n_i} dp + \left(\frac{\partial G}{\partial n_1}\right)_{T, p, n_j} dn_1 + \left(\frac{\partial G}{\partial n_2}\right)_{T, p, n_j} dn_2 + \cdots, \tag{11.2}$$

where the subscript n_i on the partial derivative means that *all* the mole numbers are constant in the differentiation, and the subscript n_j on the partial derivative means that all the mole numbers except the one in that derivative are constant in the differentiation. For example, $(\partial G/\partial n_2)_{T, p, n_j}$ means that T, p, and all the mole numbers except n_2 are constant in the differentiation.

If the system does not suffer any change in composition, then

$$dn_1 = 0, \qquad dn_2 = 0,$$

and so on, and Eq. (11.2) reduces to

$$dG = \left(\frac{\partial G}{\partial T}\right)_{p,\,n_i} dT + \left(\frac{\partial G}{\partial p}\right)_{T,\,n_i} dp. \tag{11.3}$$

Comparison of Eq. (11.3) with Eq. (11.1), shows that

$$\left(\frac{\partial G}{\partial T}\right)_{p,\,n_i} = -S \quad \text{and} \quad \left(\frac{\partial G}{\partial p}\right)_{T,\,n_i} = V. \tag{11.4a, b}$$

To simplify writing, we define

$$\mu_i = \left(\frac{\partial G}{\partial n_i}\right)_{T,\,p,\,n_j}. \tag{11.5}$$

In view of Eqs. (11.4) and (11.5), the total differential of G in Eq. (11.2) becomes

$$dG = -S\,dT + V\,dp + \mu_1\,dn_1 + \mu_2\,dn_2 + \cdots. \tag{11.6}$$

Equation (11.6) relates the change in Gibbs energy to changes in the temperature, pressure, and the mole numbers; it is usually written in the more compact form

$$dG = -S\,dT + V\,dp + \sum_i \mu_i\,dn_i, \tag{11.7}$$

where the sum includes all the constituents of the mixture.

11.2 THE PROPERTIES OF μ_i

If a small amount of substance i, dn_i moles, is added to a system, keeping T, p, and all the other mole numbers constant, then the increase in Gibbs energy is given by Eq. (11.7), which reduces to $dG = \mu_i\,dn_i$. The increase in Gibbs energy *per mole* of the substance added is

$$\left(\frac{\partial G}{\partial n_i}\right)_{T,\,p,\,n_j} = \mu_i.$$

This equation expresses the immediate significance of μ_i, and is simply the content of the definition of μ_i in Eq. (11.5). For any substance i in a mixture, the value of μ_i is the increase in Gibbs energy that attends the addition of an infinitesimal number of moles of that substance to the mixture *per mole* of the substance added. (The amount added is restricted to an infinitesimal quantity so that the composition of the mixture, and therefore the value of μ_i, does not change.)

An alternative approach involves an extremely large system, let us say a roomful of a water solution of sugar. If one mole of water is added to such a large system, the composition of the system remains the same for all practical purposes, and therefore the μ_{H_2O} of the water is constant. The increase in Gibbs energy attending the addition of one mole of water to the roomful of solution is the value of μ_{H_2O} in the solution.

Since μ_i is the derivative of one extensive variable by another, it is an *intensive* property of the system and must have the same value everywhere within a system at equilibrium.

Suppose that μ_i had different values, μ_i^A and μ_i^B, in two regions of the system, A and B. Then keeping T, p, and all the other mole numbers constant, suppose that we transfer dn_i moles of i from region A to region B. For the increase in Gibbs energy in the two regions, we have from Eq. (11.7), $dG^A = \mu_i^A(-dn_i)$, and $dG^B = \mu_i^B \, dn_i$, since $+dn_i$ moles go into B and $-dn_i$ moles go into A. The total change in Gibbs energy of the system is the sum $dG = dG^A + dG^B$, or

$$dG = (\mu_i^B - \mu_i^A) \, dn_i.$$

Now if μ_i^B is less than μ_i^A, then dG is negative, and this transfer of matter decreases the Gibbs energy of the system; the transfer therefore occurs spontaneously. Thus, substance i flows spontaneously from a region of high μ_i to a region of low μ_i; this flow continues until the value of μ_i is uniform throughout the system, that is, until the system is in equilibrium. The fact that μ_i must have the same value throughout the system is an important equilibrium condition, which we will use again and again.

The property μ_i is called the *chemical potential* of the substance i. Matter flows spontaneously from a region of high chemical potential to a region of low chemical potential just as electric current flows spontaneously from a region of high electrical potential to one of lower electrical potential, or as mass flows spontaneously from a position of high gravitational potential to one of low gravitational potential. Another name frequently given to μ_i is the *escaping tendency* of i. If the chemical potential of a component in a system is high, that component has a large escaping tendency, while if the chemical potential is low, the component has a small escaping tendency.

11.3 THE GIBBS ENERGY OF A MIXTURE

The fact that the μ_i are intensive properties implies that they can depend only on other intensive properties such as temperature, pressure, and intensive composition variables such as the mole ratios, or the mole fractions. Since the μ_i depend on the mole numbers only through intensive composition variables, an important relation is easily derived.

Consider the following transformation:

	Initial State T, p			*Final State* T, p		
Substances	1	2	3	1	2	3
Mole numbers	0	0	0	n_1	n_2	n_3
Gibbs energy	$G = 0$			G		

We achieve this transformation by considering a large quantity of a mixture of uniform composition, in equilibrium at constant temperature and constant pressure. Imagine a small, closed mathematical surface such as a sphere that lies completely in the interior of this mixture and forms the boundary that encloses our thermodynamic system. We denote the Gibbs energy of this system by G^* and the number of moles of the ith species in the system by n_i^*. We now ask by how much the Gibbs energy of the system increases if we enlarge this mathematical surface so that it encloses a greater quantity of the mixture. We may imagine that the final boundary enlarges and deforms in such a way as to enclose any desired amount of mixture in a vessel of any shape. Let the Gibbs energy of the enlarged system be G and the mole numbers be n_i. We obtain this change in

Gibbs energy by integrating Eq. (11.7) at constant T and p; that is,

$$\int_{G^*}^{G} dG = \sum_i \mu_i \int_{n_i^*}^{n_i} dn_i;$$

$$G - G^* = \sum_i \mu_i(n_i - n_i^*). \tag{11.8}$$

The μ_i were taken out of the integrals because, as we have shown above, each μ_i must have the same value everywhere throughout a system at equilibrium. Now we allow our initial small boundary to shrink to the limit of enclosing zero volume; then $n_i^* = 0$, and $G^* = 0$. This reduces Eq. (11.8) to

$$G = \sum_i n_i \mu_i. \tag{11.9}$$

The addition rule in Eq. (11.9) is a very important property of chemical potentials. Knowing the chemical potential and the number of moles of each constituent of a mixture, we can compute, using Eq. (11.9), the total Gibbs energy, G, of the mixture at the specified temperature and pressure. If the system contains only one substance, then Eq. (11.9) reduces to $G = n\mu$, or

$$\mu = \frac{G}{n}. \tag{11.10}$$

By Eq. (11.10), the μ of a pure substance is simply the *molar Gibbs energy*; for this reason the symbol μ was introduced for molar Gibbs energy in Section 10.8. In mixtures, μ_i is the *partial molar Gibbs energy* of the substance i.

11.4 THE CHEMICAL POTENTIAL OF A PURE IDEAL GAS

The chemical potential of a pure ideal gas is given explicitly by Eq. (10.47):

$$\mu = \mu^\circ(T) + RT \ln p. \tag{11.11}$$

This equation shows that at a given temperature the pressure is a measure of the chemical potential of the gas. If inequalities in pressure exist in a container of a gas, then matter will flow from the high-pressure regions (high chemical potential) to those of lower pressure (lower chemical potential) until the pressure is equalized throughout the vessel. The equilibrium condition, equality of the chemical potential everywhere, requires that the pressure be uniform throughout the vessel. For nonideal gases it is the fugacity that must be uniform throughout the vessel; however, since the fugacity is a function of temperature and pressure, at a given temperature equal values of fugacity imply equal values of pressure.

11.5 CHEMICAL POTENTIAL OF AN IDEAL GAS IN A MIXTURE OF IDEAL GASES

Consider the system shown in Fig. 11.1. The right-hand compartment contains a mixture of hydrogen under a partial pressure p_{H_2} and nitrogen under a partial pressure p_{N_2}, the total pressure being $p = p_{H_2} + p_{N_2}$. The mixture is separated from the left-hand side by a palladium membrane. Since hydrogen can pass freely through the membrane, the left-hand side contains pure hydrogen. When equilibrium is attained, the pressure of the pure hydrogen on the left-hand side is equal by definition to the partial pressure of

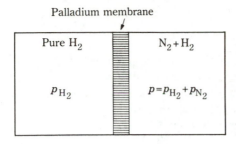

Figure 11.1   Chemical potential of a gas in a mixture.

the hydrogen in the mixture (see Section 2.8). The equilibrium condition requires that the chemical potential of the hydrogen must have the same value in both sides of the vessel:

$$\mu_{H_2(pure)} = \mu_{H_2(mix)}.$$

The chemical potential of pure hydrogen under a pressure p_{H_2} is, by Eq. (11.11),

$$\mu_{H_2(pure)} = \mu^{\circ}_{H_2}(T) + RT \ln p_{H_2}.$$

Therefore in the mixture it must be that

$$\mu_{H_2(mix)} = \mu^{\circ}_{H_2}(T) + RT \ln p_{H_2}.$$

This equation shows that the chemical potential of hydrogen in a mixture is a logarithmic function of the *partial pressure* of hydrogen in the mixture. By repeating the argument using a mixture of any number of ideal gases, and a membrane* permeable only to substance i, it may be shown that the chemical potential of substance i in the mixture is given by

$$\mu_i = \mu^{\circ}_i(T) + RT \ln p_i, \tag{11.12}$$

where p_i is the partial pressure of substance i in the mixture. The $\mu^{\circ}_i(T)$ has the same significance as for a pure gas; it is the chemical potential of the pure gas under 1 atm pressure at the temperature T.

By using the relation $p_i = x_i p$, where x_i is the mole fraction of substance i in the mixture and p is the total pressure, for p_i in Eq. (11.12), and expanding the logarithm, we obtain

$$\mu_i = \mu^{\circ}_i(T) + RT \ln p + RT \ln x_i. \tag{11.13}$$

By Eq. (11.11), the first two terms in Eq. (11.13) are the μ for pure i under the pressure p, so Eq. (11.13) reduces to

$$\mu_i = \mu_{i(pure)}(T, p) + RT \ln x_i. \tag{11.14}$$

Since x_i is a fraction and its logarithm is negative, Eq. (11.14) shows that the chemical potential of any gas in a mixture is always less than the chemical potential of the pure gas under the same total pressure. If a pure gas under a pressure p is placed in contact with a mixture under the same total pressure, the pure gas will spontaneously flow into the mixture. This is the thermodynamic interpretation of the fact that gases, and for that matter liquids and solids as well, diffuse into one another.

The form of Eq. (11.14) suggests a generalization. Suppose we define an *ideal mixture*, or *ideal solution*, in any state of aggregation (solid, liquid, or gaseous) as one in which

---

*   The fact that such membranes are known for only a few gases does not impair the argument.

the chemical potential of every species is given by the expression

$$\mu_i = \mu_i^\circ(T, p) + RT \ln x_i. \tag{11.14a}$$

In Eq. (11.14a) we interpret $\mu_i^\circ(T, p)$ as the chemical potential of the *pure* species i in the *same state of aggregation as the mixture*; that is, in a liquid mixture, $\mu_i^\circ(T, p)$ is the chemical potential, or molar Gibbs energy, of *pure liquid i* at temperature T and pressure p, and x_i is the mole fraction of i in the liquid mixture. We will introduce particular empirical evidence to justify this generalization in Chapter 13.

11.6 GIBBS ENERGY AND ENTROPY OF MIXING

Since the formation of a mixture from pure constituents always occurs spontaneously, this process must be attended by a decrease in Gibbs energy. Our object now is to calculate the Gibbs energy of mixing. The initial state is shown in Fig. 11.2(a). Each of the compartments contains a pure substance under a pressure p. The partitions separating the substances are pulled out and the final state, shown in Fig. 11.2(b), is the mixture under the same pressure p. The temperature is the same initially and finally. For the pure substances, the Gibbs energies are

$$G_1 = n_1 \mu_1^\circ, \qquad G_2 = n_2 \mu_2^\circ, \qquad G_3 = n_3 \mu_3^\circ.$$

The Gibbs energy of the initial state is simply the sum

$$G_{\text{initial}} = G_1 + G_2 + G_3 = n_1 \mu_1^\circ + n_2 \mu_2^\circ + n_3 \mu_3^\circ = \sum_i n_i \mu_i^\circ.$$

The Gibbs energy in the final state is given by the addition rule, Eq. (11.9):

$$G_{\text{final}} = n_1 \mu_1 + n_2 \mu_2 + n_3 \mu_3 = \sum_i n_i \mu_i.$$

The Gibbs energy of mixing, $\Delta G_{\text{mix}} = G_{\text{final}} - G_{\text{initial}}$, on inserting the values of G_{final} and G_{initial}, becomes

$$\Delta G_{\text{mix}} = n_1(\mu_1 - \mu_1^\circ) + n_2(\mu_2 - \mu_2^\circ) + n_3(\mu_3 - \mu_3^\circ) = \sum_i n_i(\mu_i - \mu_i^\circ).$$

Using the value of $\mu_i - \mu_i^\circ$ from Eq. (11.14a), we obtain

$$\Delta G_{\text{mix}} = RT(n_1 \ln x_1 + n_2 \ln x_2 + n_3 \ln x_3) = RT \sum_i n_i \ln x_i,$$

which can be put in a slightly more convenient form by the substitution $n_i = x_i n$, where

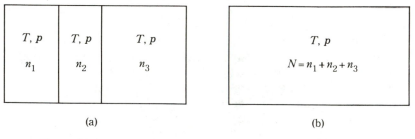

(a)   (b)

Figure 11.2 Free energy of mixing. (a) Initial state. (b) Final state.

n is the total number of moles in the mixture, and x_i is the mole fraction of i. Then

$$\Delta G_{mix} = nRT(x_1 \ln x_1 + x_2 \ln x_2 + x_3 \ln x_3), \qquad (11.15)$$

which is the final expression for the Gibbs energy of mixing in terms of the mole fractions of the constituents of the mixture. Every term on the right-hand side is negative, and so the sum is always negative. From the derivation, it can be seen that in forming an ideal mixture of any number of species the Gibbs energy of mixing will be

$$\Delta G_{mix} = nRT \sum_i x_i \ln x_i. \qquad (11.16)$$

If there are only two substances in the mixture, then if $x_1 = x$, $x_2 = 1 - x$, Eq. (11.16) becomes

$$\Delta G_{mix} = nRT[x \ln x + (1 - x) \ln (1 - x)]. \qquad (11.17)$$

A plot of the function in Eq. (11.17) is shown in Fig. 11.3. The curve is symmetrical about $x = \frac{1}{2}$. The greatest decrease in Gibbs energy on mixing is associated with the formation of the mixture having equal numbers of moles of the two constituents. In a ternary system, the greatest decrease in free energy on mixing occurs if the mole fraction of each substance is equal to $\frac{1}{3}$, and so on.

Differentiation of $\Delta G_{mix} = G_{final} - G_{initial}$, with respect to temperature, yields ΔS_{mix} directly, through Eq. (11.4a):

$$\left(\frac{\partial \Delta G_{mix}}{\partial T}\right)_{p, n_i} = \left(\frac{\partial G_{final}}{\partial T}\right)_{p, n_i} - \left(\frac{\partial G_{initial}}{\partial T}\right)_{p, n_i} = -(S_{final} - S_{initial});$$

$$\left(\frac{\partial \Delta G_{mix}}{\partial T}\right)_{p, n_i} = -\Delta S_{mix}. \qquad (11.18)$$

Differentiating both sides of Eq. (11.16) with respect to temperature, we have

$$\left(\frac{\partial \Delta G_{mix}}{\partial T}\right)_{p, n_i} = nR \sum_i x_i \ln x_i,$$

so that Eq. (11.18) becomes

$$\Delta S_{mix} = -nR \sum_i x_i \ln x_i. \qquad (11.19)$$

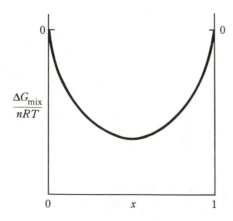

Figure 11.3 $\Delta G_{mix}/nRT$ for a binary ideal mixture.

The functional form of the entropy of mixing is the same as for the Gibbs energy of mixing, except that T does not appear as a factor and a minus sign occurs in the expression for the entropy of mixing. The minus sign means that the entropy of mixing is always positive, while the Gibbs energy of mixing is always negative. The positive entropy of mixing corresponds to the increase in randomness that occurs in mixing the molecules of several kinds. The expression for the entropy of mixing in Eq. (11.19) should be compared to that in Eq. (9.75), which was obtained from the statistical argument. Note that N in Eq. (9.75) is the number of molecules, whereas in Eq. (11.19) n is the number of moles; therefore different constants, R and k, appear in the two equations.

A plot of the entropy of mixing of a binary mixture according to the equation

$$\Delta S_{\text{mix}} = -nR[x \ln x + (1 - x) \ln (1 - x)] \tag{11.20}$$

is shown in Fig. 11.4. The entropy of mixing has a maximum value when $x = \frac{1}{2}$. Using $x = \frac{1}{2}$ in Eq. (11.20), we obtain for the entropy of mixing per mole of mixture

$$\Delta S_{\text{mix}}/n = -R(\tfrac{1}{2} \ln \tfrac{1}{2} + \tfrac{1}{2} \ln \tfrac{1}{2}) = -R \ln \tfrac{1}{2} = +0.693R = 5.76 \text{ J/K mol.}$$

In a mixture containing only two substances, the entropy of mixing per mole of the final mixture varies between 0 and 5.76 J/K, depending on the composition.

The heat of mixing can be calculated by the equation

$$\Delta G_{\text{mix}} = \Delta H_{\text{mix}} - T\Delta S_{\text{mix}}, \tag{11.21}$$

using the values of the Gibbs energy and entropy of mixing from Eqs. (11.16) and (11.19). This reduces Eq. (11.21) to

$$nRT \sum_i x_i \ln x_i = \Delta H_{\text{mix}} + nRT \sum_i x_i \ln x_i,$$

which becomes

$$\Delta H_{\text{mix}} = 0. \tag{11.22}$$

There is no heat effect associated with the formation of an ideal mixture.

Using the result that $\Delta H_{\text{mix}} = 0$, Eq. (11.21) becomes

$$-\Delta G_{\text{mix}} = T\Delta S_{\text{mix}}. \tag{11.23}$$

Equation (11.23) shows that the driving force, $-\Delta G_{\text{mix}}$, that produces the mixing is entirely an entropy effect. The mixed state is a more random state, and therefore is a more

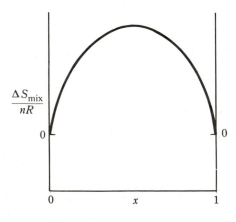

Figure 11.4 $\Delta S_{\text{mix}}/nR$ for a binary ideal mixture.

probable state. If the value of 5.76 J/K mol is used for the entropy of mixing, then at $T = 300$ K, $\Delta G_{mix} = -(300$ K$) (5.76$ J/K mol$) = -1730$ J/mol. Thus the Gibbs energy of mixing of an ideal binary mixture ranges from 0 to -1730 J/mol. Since -1730 J/mol is not large, in nonideal mixtures for which the heat of mixing is not zero, the heat of mixing must either be negative or *only slightly positive* if the substances are to mix spontaneously. If the heat of mixing is more positive than about 1300 to 1600 J/mol of mixture, then ΔG_{mix} is positive, and the liquids are not miscible but remain in two distinct layers.

The volume of mixing is obtained by differentiating the Gibbs energy of mixing with respect to pressure, the temperature and composition being constant,

$$\Delta V_{mix} = \left(\frac{\partial \, \Delta G_{mix}}{\partial p}\right)_{T, \, n_i}.$$

However, inspection of Eq. (11.16) shows that the Gibbs energy of mixing is independent of pressure, so the derivative is zero; hence,

$$\Delta V_{mix} = 0. \tag{11.24}$$

Ideal mixtures are formed without any volume change.

11.7 CHEMICAL EQUILIBRIUM IN A MIXTURE

Consider a closed system at a constant temperature and under a constant total pressure. The system consists of a mixture of several chemical species that can react according to the equation

$$0 = \sum_i v_i A_i \tag{11.25}$$

where the A_i represent the chemical formulas of the substances, while the v_i represent the stoichiometric coefficients. This is the notation used in Sec. 1.7.1 for chemical reactions. It is understood that the v_i are negative for reactants and positive for products.

We now inquire whether the Gibbs energy of the mixture will increase or decrease if the reaction advances in the direction indicated by the arrow. If the Gibbs energy decreases as the reaction advances, then the reaction goes spontaneously in the direction of the arrow; the advance of the reaction and the decrease in Gibbs energy continue until the Gibbs energy of the system reaches a minimum value. When the Gibbs energy of the system is a minimum, the reaction is at equilibrium. If the Gibbs energy of the system increases as the reaction advances in the direction of the arrow, then the reaction will go spontaneously, with a decrease in Gibbs energy, in the opposite direction; again the mixture will reach a minimum value of Gibbs energy at the equilibrium position.

Since T and p are constant, as the reaction advances the change in Gibbs energy of the system is given by Eq. (11.7), which becomes

$$dG = \sum_i \mu_i \, dn_i \tag{11.26}$$

where the changes in the mole numbers, dn_i, are those resulting from the chemical reaction. These changes are not independent because the substances react in the stoichiometric ratios. Let the reaction advance by ξ moles, where ξ is the advancement of the reaction; then the number of moles of each of the substances present is

$$n_i = n_i^0 + v_i \xi \tag{11.27}$$

where the n_i^0 are the numbers of moles of the substances present before the reaction advanced by ξ moles. Since the n_i^0 are constant, by differentiating Eq. (11.27) we obtain

$$dn_i = v_i \, d\xi \tag{11.28}$$

Using Eq. (11.28) in Eq. (11.26), we obtain

$$dG = \left(\sum_i v_i \mu_i \right) d\xi$$

which becomes

$$\left(\frac{\partial G}{\partial \xi} \right)_{T, p} = \sum_i v_i \mu_i \tag{11.29}$$

The derivative, $(\partial G/\partial \xi)_{T, p}$, is the rate of increase of the Gibbs energy of the mixture with the advancement ξ of the reaction. If this derivative is negative, the Gibbs energy of the mixture decreases as the reaction progresses in the direction indicated by the arrow, which implies that the reaction is spontaneous. If this derivative is positive, progress of the reaction in the forward direction would lead to an increase in Gibbs energy of the system; since this is not possible, the reverse reaction would go spontaneously. If $(\partial G/\partial \xi)_{T, p}$ is zero, the Gibbs energy has a minimum value and the reaction is at equilibrium. The equilibrium condition for the chemical reaction is then

$$\left(\frac{\partial G}{\partial \xi} \right)_{T, p, \text{eq}} = 0, \tag{11.30}$$

and

$$\left(\sum_i v_i \mu_i \right)_{\text{eq}} = 0 \tag{11.31}$$

The derivative in Eq. (11.29) has the form of an increase of Gibbs energy, ΔG, since it is the sum of the Gibbs energies of the products of the reaction less the sum of the Gibbs energies of the reactants. Consequently we will write ΔG for $(\partial G/\partial \xi)_{T, p}$ and call ΔG the *reaction Gibbs energy*. From the above derivation it is clear that for any chemical reaction

$$\Delta G = \sum_i v_i \mu_i \tag{11.32}$$

The equilibrium condition for *any chemical reaction* is

$$\Delta G = \left(\sum v_i \mu_i \right)_{\text{eq}} = 0 \tag{11.33}$$

The subscript eq is placed on the quantities in Eqs. (11.31) and (11.33) to emphasize the fact that at equilibrium the values of the μ's are related in the special way indicated by these equations. Since each μ_i is $\mu_i(T, p, n_1^0, n_2^0, \ldots, \xi)$ the equilibrium condition determines ξ_e as a function of T, p, and the specified values of the initial mole numbers.

11.8 THE GENERAL BEHAVIOR OF G AS A FUNCTION OF ξ

Figure 11.5a shows the general behavior of G as a function of ξ in a homogeneous system. The advancement, ξ, has a limited range of variation between a least value, ξ_1, and a greatest value, ξ_g. At ξ_1, one or more of the products has been exhausted, while at ξ_g one or more of the reactants has been exhausted. At some intermediate value, ξ_e, G

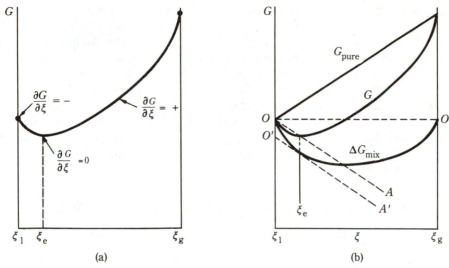

Figure 11.5 Gibbs energy as a function of the advancement.

passes through a minimum. The value ξ_e is the equilibrium value of the advancement. To the left of the minimum, $\partial G/\partial \xi$ is negative, indicating spontaneity in the forward direction, while to the right of the minimum, $\partial G/\partial \xi$ is positive, indicating spontaneity in the reverse direction. Note that even though in the case illustrated the products have an intrinsically higher Gibbs energy than the reactants, the reaction does form some products. This is a consequence of the contribution of the Gibbs energy of mixing.

At any composition the Gibbs energy of the mixture has the form

$$G = \sum_i n_i \mu_i.$$

If we add and subtract $\mu_i^\circ(T, p)$, the chemical potential of the *pure* species i in each term of the sum, we obtain

$$G = \sum_i n_i(\mu_i^\circ + \mu_i - \mu_i^\circ) = \sum_i n_i \mu_i^\circ(T, p) + \sum_i n_i(\mu_i - \mu_i^\circ).$$

The first sum is the total Gibbs energy of the pure gases separately, G_{pure}; the last sum is the Gibbs energy of mixing, ΔG_{mix}. The Gibbs energy of the system is given by

$$G = G_{\text{pure}} + \Delta G_{\text{mix}}. \tag{11.34}$$

The plot of G_{pure}, ΔG_{mix}, and G as a function of the advancement is shown in Fig. 11.5b. Since G_{pure} depends on ξ only through the n_i, each of which is a linear function of ξ, we see that G_{pure} is a linear function of ξ. The minimum in G occurs at the point where ΔG_{mix} decreases as rapidly as G_{pure} increases; by differentiating,

$$\left(\frac{\partial G}{\partial \xi}\right)_{T,p} = \left(\frac{\partial G_{\text{pure}}}{\partial \xi}\right)_{T,p} + \left(\frac{\partial \Delta G_{\text{mix}}}{\partial \xi}\right)_{T,p}.$$

At equilibrium

$$\left(\frac{\partial G_{\text{pure}}}{\partial \xi}\right)_{\text{eq}} = -\left(\frac{\partial \Delta G_{\text{mix}}}{\partial \xi}\right)_{\text{eq}}.$$

This condition can be established geometrically by reflecting the line for G_{pure} in the horizontal line OO, to yield the line OA; the point of tangency of the line $O'A'$, parallel to OA, with the curve for ΔG_{mix} yields the value of the advancement at equilibrium. Equation (11.34) is correct for any equilibrium in a homogeneous system. Equation (11.34) is, in fact, formally correct for any equilibrium, but unless at least one phase is a mixture, the term G_{mix}, will be zero and only the first term, G_{pure}, will appear.

Equation (11.34) shows that a system approaches the equilibrium state of minimum Gibbs energy by forming substances of intrinsically lower Gibbs energy; this makes G_{pure} small. It also lowers its Gibbs energy by mixing the reactants and products. A compromise is reached between a pure material having a low intrinsic Gibbs energy and the highly mixed state.

11.9 CHEMICAL EQUILIBRIUM IN A MIXTURE OF IDEAL GASES

It has been shown, Eq. (11.12), that the μ of an ideal gas in a gas mixture is given by

$$\mu_i = \mu_i^\circ + RT \ln p_i, \tag{11.35}$$

where p_i is the partial pressure of the gas in the mixture. We use this value of μ_i in Eq. (11.29) to compute the ΔG for the reaction.

$$\alpha A + \beta B \longrightarrow \gamma C + \delta D$$

where A, B, C, and D represent the chemical formulas of the substances, while α, β, γ, and δ represent the stoichiometric coefficients. Then

$$\Delta G = \gamma \mu_C^\circ + \gamma RT \ln p_C + \delta \mu_D^\circ + \delta RT \ln p_D - \alpha \mu_A^\circ - \alpha RT \ln p_A - \beta \mu_B^\circ - \beta RT \ln p_B,$$
$$= \gamma \mu_C^\circ + \delta \mu_D^\circ - (\alpha \mu_A^\circ + \beta \mu_B^\circ) + RT[\gamma \ln p_C + \delta \ln p_D - (\alpha \ln p_A + \beta \ln p_B)].$$

Let

$$\Delta G^\circ = \gamma \mu_C^\circ + \delta \mu_D^\circ - (\alpha \mu_A^\circ + \beta \mu_B^\circ); \tag{11.36}$$

ΔG° is the *standard* reaction Gibbs energy. Then, combining the logarithmic terms,

$$\Delta G = \Delta G^\circ + RT \ln \frac{p_C^\gamma p_D^\delta}{p_A^\alpha p_B^\beta}. \tag{11.37}$$

The argument of the logarithm is called the *proper quotient of pressures*; the numerator is the product of partial pressures of the chemical products each raised to the power of its stoichiometric coefficient, and the denominator is the product of the partial pressures of the reactants, each raised to the power of its stoichiometric coefficient. Ordinarily the quotient is abbreviated by the symbol Q_p:

$$Q_p = \frac{p_C^\gamma p_D^\delta}{p_A^\alpha p_B^\beta}. \tag{11.38}$$

This reduces Eq. (11.37) to

$$\Delta G = \Delta G^\circ + RT \ln Q_p. \tag{11.39}$$

The sign of ΔG is determined by the sign and magnitude of $\ln Q_p$, since at a given temperature ΔG° is a constant characteristic of the reaction. If, for example, we compose the mixture so that the partial pressures of the reactants are very large, while those of the products are very small, then Q_p will have a small fractional value, and $\ln Q_p$ will be a large negative number. This in turn will make ΔG more negative and increase the tendency for products to form.

At equilibrium, $\Delta G = 0$, and Eq. (11.37) becomes

$$0 = \Delta G^\circ + RT \ln \frac{(p_C)_e^\gamma (p_D)_e^\delta}{(p_A)_e^\alpha (p_B)_e^\beta}, \tag{11.40}$$

where the subscript e indicates that these are *equilibrium* partial pressures. The quotient of *equilibrium* partial pressures, is the pressure equilibrium constant K_p:

$$K_p = \frac{(p_C)_e^\gamma (p_D)_e^\delta}{(p_A)_e^\alpha (p_B)_e^\beta}. \tag{11.41}$$

Using the more general notation, we put the value of μ_i from Eq. (11.35) in Eq. (11.29) to obtain

$$\Delta G = \left(\frac{\partial G}{\partial \xi} \right)_{T,p} = \sum_i v_i (\mu_i^\circ + RT \ln p_i),$$

which can be written,

$$\Delta G = \sum_i v_i \mu_i^\circ + RT \sum_i v_i \ln p_i.$$

But

$$\sum_i v_i \mu_i^\circ = \Delta G^\circ, \tag{11.36a}$$

the change in the standard reaction Gibbs energy, and $v_i \ln p_i = \ln p_i^{v_i}$; thus the equation becomes

$$\Delta G = \Delta G^\circ + RT \sum_i \ln p_i^{v_i}. \tag{11.37a}$$

But a sum of logarithms is the logarithm of a product:

$$\ln p_1^{v_1} + \ln p_2^{v_2} + \ln p_3^{v_3} + \cdots = \ln (p_1^{v_1} p_2^{v_2} p_3^{v_3} \cdots).$$

This continued product,

$$\prod_i p_i^{v_i} = p_1^{v_1} p_2^{v_2} p_3^{v_3} \cdots,$$

is called the proper quotient of pressures, Q_p.

$$Q_p \equiv \prod_i p_i^{v_i} \tag{11.38a}$$

Note that since the v_i for the reactants are negative, we have for the reaction in question

$$v_1 = -\alpha, \qquad v_2 = -\beta, \qquad v_3 = \gamma, \qquad v_4 = \delta,$$

and

$$Q_p = p_A^{-\alpha} p_B^{-\beta} p_C^\gamma p_D^\delta = \frac{p_B^\gamma p_D^\delta}{p_D^\alpha p_C^\beta} \tag{11.38b}$$

Correspondingly, K_p can be written as

$$K_p \equiv \prod_i (p_i)_e^{v_i} \tag{11.41a}$$

Equation (11.40) becomes

$$\Delta G^\circ = -RT \ln K_p. \tag{11.42}$$

The quantity $\Delta G°$ is a combination of $\mu°$'s, each of which is a function only of temperature; therefore $\Delta G°$ is a function only of temperature, and so K_p is a function only of temperature. From a measurement of the equilibrium constant of the reaction $\Delta G°$ can be calculated using Eq. (11.42). This is the way in which the value of $\Delta G°$ for any reaction is obtained.

■ **EXAMPLE 11.1**   For the reaction

$$\tfrac{1}{2}N_2(g) + \tfrac{3}{2}H_2(g) \rightleftharpoons NH_3(g),$$

the equilibrium constant is 6.59×10^{-3} at $450\,°C$. Compute the standard reaction Gibbs energy at $450\,°C$.

Solution.

$$\Delta G° = -(8.314 \text{ J/K mol})(723 \text{ K}) \ln (6.59 \times 10^{-3})$$
$$= -(6010 \text{ J/mol})(-5.02) = +30\,200 \text{ J/mol.}$$

Since this is the formation reaction for ammonia, it follows that $30\,200$ J/mol is the standard Gibbs energy of formation of ammonia at $450\,°C$.

11.10 CHEMICAL EQUILIBRIUM IN A MIXTURE OF REAL GASES

If the corresponding algebra were carried out for real gases using Eq. (10.48), the equation equivalent to Eq. (11.41) is

$$K_f = \frac{(f_C)_e^{\gamma}(f_D)_e^{\delta}}{(f_A)_e^{\alpha}(f_B)_e^{\beta}}, \tag{11.43}$$

and, corresponding to Eq. (11.42),

$$\Delta G° = -RT \ln K_f. \tag{11.44}$$

For real gases, it is K_f rather than K_p that is a function of temperature only.

11.11 THE EQUILIBRIUM CONSTANTS, K_x AND K_c

It is sometimes advantageous to express the equilibrium constant for gaseous systems in terms of either mole fractions, x_i, or concentrations, $\tilde{c}_i$, rather than partial pressures. The partial pressure, p_i, the mole fraction, and the total pressure, p, are related by $p_i = x_i p$. Using this relation for each of the partial pressures in the equilibrium constant, we obtain from Eq. (11.41)

$$K_p = \frac{(p_C)_e^{\gamma}(p_D)_e^{\delta}}{(p_A)_e^{\alpha}(p_B)_e^{\beta}} = \frac{(x_C p)_e^{\gamma}(x_D p)_e^{\delta}}{(x_A p)_e^{\alpha}(x_B p)_e^{\beta}} = \frac{(x_C)_e^{\gamma}(x_D)_e^{\delta}}{(x_A)_e^{\alpha}(x_B)_e^{\beta}} p^{\gamma + \delta - \alpha - \beta}.$$

The mole fraction equilibrium constant is defined by

$$K_x = \frac{(x_C)_e^{\gamma}(x_D)_e^{\delta}}{(x_A)_e^{\alpha}(x_B)_e^{\beta}}. \tag{11.45}$$

Then

$$K_p = K_x p^{\Delta\nu}, \tag{11.46}$$

where $\Delta\nu = \Sigma\nu_i$ is the sum of stoichiometric coefficients on the right-hand side of the chemical equation minus the sum of the coefficients on the left-hand side. Rearranging Eq. (11.46), we obtain $K_x = K_p p^{-\Delta\nu}$. Since K_p is independent of pressure, K_x will depend on pressure unless $\Delta\nu$ is zero.

Keep in mind that in K_p the p_i are pure numbers—abbreviations for the ratio $p_i/(1 \text{ atm})$ —which we will write as p_i/p°; see the discussion of Eqs. (9.52), (9.53), and (10.47). It follows that the pressure in Eq. (11.46) is also a pure number; it is an abbreviation for $p/p^\circ = p/(1 \text{ atm})$.

In a similar way, since the partial pressure of a gas is given by $p_i = n_i RT/V$ and the concentration is $\tilde{c}_i = n_i/V$, we obtain $p_i = \tilde{c}_i RT$. Introducing the standard pressure explicitly, we have

$$\frac{p_i}{p^\circ} = \frac{\tilde{c}_i RT}{p^\circ}.$$

Before we put this in K_p it is useful to have $\tilde{c}_i$ in a dimensionless ratio, so we multiply and divide by a standard concentration, $\tilde{c}^\circ$. Then we have

$$\frac{p_i}{p^\circ} = \left(\frac{\tilde{c}_i}{\tilde{c}^\circ}\right)\left(\frac{\tilde{c}^\circ RT}{p^\circ}\right) \tag{11.47}$$

Since we have a ratio of concentrations, it follows that

$$\frac{\tilde{c}_i}{\tilde{c}^\circ} = \frac{c_i}{c^\circ}$$

where the c_i and c° are concentrations in mol/L, whereas the $\tilde{c}_i$ and $\tilde{c}^\circ$ are the corresponding concentrations in mol/m³, the SI unit of concentration. As before, we will abbreviate p_i/p° as p_i and $c_i/c^\circ = c_i/(1 \text{ mol/L})$ as c_i; then we have

$$p_i = c_i \left(\frac{\tilde{c}^\circ RT}{p^\circ}\right) \tag{11.48}$$

in which p_i and c_i are to be understood as the pure numbers equal to the ratios $p_i/(1 \text{ atm})$ and $c_i/(1 \text{ mol/L})$. If we insert these values of p_i in K_p by the same argument that we used to obtain Eq. (11.46), we find

$$K_p = K_c \left(\frac{\tilde{c}^\circ RT}{p^\circ}\right)^{\Delta \nu} \tag{11.49}$$

where K_c is a quotient of equilibrium concentrations; K_c is a function of temperature only.

Since the standard concentration was $c^\circ = 1 \text{ mol/L}$, the corresponding value of $\tilde{c}^\circ = 10^3 \text{ mol/m}^3$; thus

$$\frac{\tilde{c}^\circ RT}{p^\circ} = \frac{(10^3 \text{ mol/m}^3)(8.31441 \text{ J/K mol})T}{101\,325 \text{ Pa}} = 0.0820568 \; T/\text{K},$$

and we have

$$K_p = K_c \left(\frac{RT}{101.325 \text{ J/mol}}\right)^{\Delta \nu} = K_c(0.0820568 \; T/\text{K})^{\Delta \nu} \tag{11.50}$$

Note that the quantity in the parentheses is dimensionless, as are K_p and K_c.

11.12 STANDARD GIBBS ENERGIES OF FORMATION

Having obtained values of ΔG° from measurements of equilibrium constants, it is possible to calculate conventional values of the standard molar Gibbs energy μ° of individual compounds. Just as in the case of the standard enthalpies of substances, we are at liberty to assign a value of zero to the Gibbs energy of the elements in their stable state of aggregation at 25 °C and 1 atm pressure. For example, at 25 °C

$$\mu°(H_2, g) = 0, \qquad \mu°(Br_2, l) = 0, \qquad \mu°(S, \text{rhombic}) = 0.$$

For the formation reaction of a compound such as CO, we have

$$C(\text{graphite}) + \tfrac{1}{2}O_2(g) \longrightarrow CO(g),$$

$$\Delta G_f° = \mu°(CO, g) - [\mu°(C, \text{graphite}) + \tfrac{1}{2}\mu°(O_2, g)].$$

Since $\mu°(C, \text{graphite}) = 0$ and $\mu°(O_2, g) = 0$ by convention, we have

$$\Delta G_f° = \mu°(CO, g) \tag{11.51}$$

Consequently, the standard Gibbs energy of formation of any compound is equal to the conventional standard molar Gibbs energy of that compound. Some values of the standard Gibbs energy of formation at 25 °C are given in Table A-V.

It is always possible to relate the composition of an equilibrium mixture to the equilibrium value of the advancement, ξ_e, the initial mole numbers, $n_i°$, and the stoichiometric coefficients, v_i. Two examples will be discussed.

■ **EXAMPLE 11.2**   The dissociation of dinitrogen tetroxide.

$$N_2O_4(g) \rightleftharpoons 2NO_2(g)$$

This equilibrium can be easily studied in the laboratory through a measurement of the vapor density of the equilibrium mixture. In the following formulation the various quantities are listed in columns under the formulas of the compounds in the balanced chemical equation. Let $n°$ be the initial number of moles of N_2O_4, ξ_e the equilibrium advancement, and α_e the fraction dissociated at equilibrium $\alpha_e = \xi_e/n°$.

	$N_2O_4(g)$	$\rightleftharpoons 2NO_2(g)$
Stoichiometric coefficient	-1	$+2$
Initial mole numbers, $n_i°$	$n°$	0
Equilibrium mole numbers, n_i	$n° - \xi_e$	$0 + 2\xi_e$
Total number of moles, $n = n° + \xi_e$		
Mole fractions, x_i	$\dfrac{n° - \xi_e}{n° + \xi_e}$	$\dfrac{2\xi_e}{n° + \xi_e}$
or, since $\alpha_e = \xi_e/n°$, the x_i are	$\dfrac{1 - \alpha_e}{1 + \alpha_e}$	$\dfrac{2\alpha_e}{1 + \alpha_e}$
Partial pressures, $p_i = x_i p$	$\left(\dfrac{1 - \alpha_e}{1 + \alpha_e}\right)p$	$\left(\dfrac{2\alpha_e}{1 + \alpha_e}\right)p$

Using these values of the partial pressures, we obtain

$$K_p = \frac{p_{NO_2}^2}{p_{N_2O_4}} = \frac{\left(\dfrac{2\alpha_e}{1 + \alpha_e}p\right)^2}{\dfrac{1 - \alpha_e}{1 + \alpha_e}} = \frac{4\alpha_e^2 p}{1 - \alpha_e^2} \tag{11.52}$$

By the ideal gas law, $pV = nRT$, where $n = (1 + \alpha_e)n°$. Thus $pV = n°(1 + \alpha_e)RT$. But $n° = w/M$, where w is the mass of gas in the volume V and M is the molar mass of N_2O_4. Thus, if we know p, T, V, and w we can calculate α_e and then, using Eq. (11.52), we can obtain K_p.

A measurement of α_e at any pressure p suffices to determine K_p. From K_p, $\Delta G°$ can be calculated. The dependence of α_e on the pressure can be obtained explicitly by solving Eq. (11.52) for α_e:

$$\alpha_e = \sqrt{\frac{K_p}{K_p + 4p}}.$$

It is clear that as $p \to 0$, $\alpha_e \to 1$, while as $p \to \infty$, $\alpha_e \to 0$. This is what would be expected from the LeChatelier principle. At moderately high pressures, $K_p \ll 4p$ and $\alpha_e = \frac{1}{2}K_p^{1/2}/p^{1/2}$, approximately.

■ **EXAMPLE 11.3**   The ammonia synthesis.

Suppose we mix one mole of N_2 with 3 moles of H_2 (the stoichiometric ratio) and consider the equilibrium:

	$N_2(g)$	$+ 3H_2(g)$	$\rightleftharpoons 2NH_3(g)$
Stoichiometric coefficients	-1	-3	2
Initial mole numbers, n_i^0	1	3	0
Equilibrium mole numbers, n_i	$1 - \xi$	$3 - 3\xi$	2ξ
Total number of moles, $n = 4 - 2\xi$			
Mole fractions, x_i	$\dfrac{1 - \xi}{2(2 - \xi)}$	$\dfrac{3(1 - \xi)}{2(2 - \xi)}$	$\dfrac{2\xi}{2(2 - \xi)}$
Partial pressures, $p_i = x_i p$	$\dfrac{1 - \xi}{2(2 - \xi)}p$	$\dfrac{3(1 - \xi)}{2(2 - \xi)}p$	$\dfrac{2\xi p}{2(2 - \xi)}$

We note immediately that $p_{H_2} = 3p_{N_2}$; using these values in K_p, we get

$$K_p = \frac{p_{NH_3}^2}{p_{N_2}p_{H_2}^3} = \frac{p_{NH_3}^2}{p_{N_2}(3p_{N_2})^3} = \frac{p_{NH_3}^2}{3^3 p_{N_2}^4}.$$

Taking the square root, we have

$$\frac{p_{NH_3}}{p_{N_2}^2} = 3^{3/2}K_p^{1/2}$$

or, using the partial pressures from the table,

$$3^{3/2}K_p^{1/2} = \frac{\dfrac{2\xi p}{2(2 - \xi)}}{\left[\dfrac{(1 - \xi)p}{2(2 - \xi)}\right]^2} = 2^2\,\frac{\xi(2 - \xi)}{(1 - \xi)^2 p}.$$

Analysis of the mixture yields the value of x_{NH_3} from which we can obtain the value of ξ at equilibrium. From the experimental value of ξ we can calculate K_p, and from that, $\Delta G°$. We can also formulate the expression in terms of p_{NH_3} and the total pressure. Since

$p = p_{N_2} + p_{H_2} + p_{NH_3}$ and $p_{H_2} = 3p_{N_2}$, then $p = 4p_{N_2} + p_{NH_3}$ or $p_{N_2} = \frac{1}{4}(p - p_{NH_3})$. Then

$$\frac{p_{NH_3}}{(p - p_{NH_3})^2} = \frac{3^{3/2}K_p^{1/2}}{16}.$$

From this relation, the partial pressure of NH_3 can be calculated at any total pressure. If the conversion to NH_3 is low, then $p - p_{NH_3} \approx p$, and $p_{NH_3} = 0.325K_p^{1/2}p^2$, so that the partial pressure of ammonia is approximately proportional to the square of the pressure. If the reactants are not mixed originally in the stoichiometric ratio, the expression is more complex.

A measurement of the equilibrium partial pressure of NH_3 at a given temperature and pressure yields a value of $\Delta G°$ for this reaction, which is twice the conventional standard molar Gibbs energy of NH_3 at this temperature.

Note that we have suppressed the subscripts on ξ_e and $(p_{NH_3})_e$ to avoid a cumbersome notation. We will usually omit the subscript except when it is needed to avoid confusion. It is to be understood that *all* the quantities in the equilibrium constant are equilibrium values.

11.13 THE TEMPERATURE DEPENDENCE OF THE EQUILIBRIUM CONSTANT

The equilibrium constant can be written as

$$\ln K_p = -\frac{\Delta G°}{RT}. \tag{11.53}$$

Differentiating, we obtain

$$\frac{d\ln K_p}{dT} = -\frac{1}{R}\frac{d(\Delta G°/T)}{dT}. \tag{11.54}$$

Dividing Eq. (11.36a) by T, we obtain

$$\frac{\Delta G°}{T} = \sum_i v_i\left(\frac{\mu_i°}{T}\right).$$

Differentiating, we have

$$\frac{d(\Delta G°/T)}{dT} = \sum_i v_i\frac{d(\mu_i°/T)}{dT} \tag{11.55}$$

where the $\mu_i°$ are standard molar Gibbs energies of pure substances. Using molar values in the Gibbs–Helmholtz equation, Eq. (10.54), we have $d(\mu_i°/T)/dT = -\overline{H}_i°/T^2$. This relation reduces Eq. (11.55) to

$$\frac{d(\Delta G°/T)}{dT} = -\frac{1}{T^2}\sum_i v_i\overline{H}_i° = -\frac{\Delta H°}{T^2}, \tag{11.56}$$

since the summation is the standard enthalpy increase for the reaction, $\Delta H°$. Equation (11.56) reduces Eq. (11.54) to

$$\frac{d\ln K_p}{dT} = \frac{\Delta H°}{RT^2}, \quad \text{or} \quad \frac{d\log_{10}K_p}{dT} = \frac{\Delta H°}{2.303\,RT^2}. \tag{11.57}$$

Equation (11.57) is also called the Gibbs–Helmholtz equation.

If the reaction is exothermic, $\Delta H°$ is negative, and the equilibrium constant decreases with increase in temperature. If the reaction is endothermic, $\Delta H°$ is positive; then K_p increases with increase in temperature. Since an increase in the equilibrium constant implies an increase in the yield of products, Eq. (11.57) is the mathematical expression of one aspect of the LeChatelier principle.

Equation (11.57) can be expressed readily in a form convenient for plotting:

$$d \ln K_p = \frac{\Delta H°}{R} \frac{dT}{T^2} = -\frac{\Delta H°}{R} d\left(\frac{1}{T}\right),$$

$$\frac{d \ln K_p}{d(1/T)} = -\frac{\Delta H°}{R}, \qquad \frac{d \log_{10} K_p}{d(1/T)} = -\frac{\Delta H°}{2.303\,R}. \tag{11.58}$$

Equation (11.58) shows that a plot of $\ln K_p$ versus $1/T$ has a slope equal to $-\Delta H°/R$. Since $\Delta H°$ is almost constant, at least over moderate ranges of temperature, the plot is often linear.

If K_p is measured at several temperatures and the data plotted as $\ln K_p$ versus $1/T$, the slope of the line yields a value of $\Delta H°$ for the reaction through Eq. (11.58). Consequently, it is possible to determine heats of reaction by measuring equilibrium constants over a range of temperature. The values of the heats of reaction obtained by this method are usually not so precise as those obtained by precision calorimetric methods. However, the equilibrium method can be used for reactions that are not suited to direct calorimetric measurement. Later we will find that certain equilibrium constants can be calculated from calorimetrically measured quantities only.

Having obtained values of $\Delta G°$ at several temperatures and a value of $\Delta H°$ from the plot of Eq. (11.58), we can calculate the values of $\Delta S°$ at each temperature from the equation

$$\Delta G° = \Delta H° - T \Delta S°. \tag{11.59}$$

The equilibrium constant can be written as an explicit function of temperature by integrating Eq. (11.57). Suppose that at some temperature T_0, the value of the equilibrium constant is $(K_p)_0$ and at any other temperature T the value is K_p:

$$\int_{\ln (K_p)_0}^{\ln K_p} d(\ln K_p) = \int_{T_0}^{T} \frac{\Delta H°}{RT^2}\, dT, \qquad \ln K_p - \ln (K_p)_0 = \int_{T_0}^{T} \frac{\Delta H°}{RT^2}\, dT,$$

$$\ln K_p = \ln (K_p)_0 + \int_{T_0}^{T} \frac{\Delta H°}{RT^2}\, dT. \tag{11.60}$$

If $\Delta H°$ is a constant, then by integrating, we have

$$\ln K_p = \ln (K_p)_0 - \frac{\Delta H°}{R} \left(\frac{1}{T} - \frac{1}{T_0}\right). \tag{11.61}$$

From the knowledge of ΔH_0 and a value of $(K_p)_0$ at any temperature T_0, we can calculate K_p at any other temperature.

If, in Eq. (11.53), we set $\Delta G° = \Delta H° - T \Delta S°$, we obtain

$$\ln K_p = -\frac{\Delta H°}{RT} + \frac{\Delta S°}{R} \tag{11.61a}$$

This relation is always true. But if $\Delta H°$ is constant, then $\Delta S°$ must also be constant, and this equation is equivalent to Eq. (11.61). (Note that constancy of $\Delta H°$ implies that $\Delta C_p° = 0$; but if $\Delta C_p° = 0$, then $\Delta S°$ must also be constant.)

If $\Delta H°$ is not a constant, it can ordinarily be expressed (see Section 7.24) as a power series in T:

$$\Delta H° = \Delta H_0° + A'T + B'T^2 + C'T^3 + \cdots.$$

Using this value for $\Delta H°$ in Eq. (11.60) and integrating, we obtain

$$\ln K_p = \ln (K_p)_0 - \frac{\Delta H_0°}{R}\left(\frac{1}{T} - \frac{1}{T_0}\right) + \frac{A'}{R}\ln\left(\frac{T}{T_0}\right) + \frac{B'}{R}(T - T_0)$$

$$+ \frac{C'}{2R}(T^2 - T_0^2) + \cdots, \tag{11.62}$$

which has the general functional form

$$\ln K_p = \frac{A}{T} + B + C\ln T + DT + ET^2 + \cdots, \tag{11.63}$$

in which A, B, C, D, and E are constants. Equations having the general form of Eq. (11.63) are often used to calculate an equilibrium constant at 25 °C (so that it can be tabulated) from a measurement at some other (usually higher) temperature. To evaluate the constants, the values of $\Delta H°$ and the heat capacities of all the reactants and products must be known.

11.14 EQUILIBRIA BETWEEN IDEAL GASES AND PURE CONDENSED PHASES

If the substances participating in the chemical equilibrium are in more than one phase, the equilibrium is *heterogeneous*. If the substances are all present in a single phase, the equilibrium is *homogeneous*. We have dealt so far only with homogeneous equilibria in gases. If, in addition to gases, a chemical reaction involves one or more *pure* liquids or solids, the expression for the equilibrium constant is slightly different.

11.14.1 The Limestone Decomposition

Consider the reaction

$$CaCO_3(s) \rightleftharpoons CaO(s) + CO_2(g).$$

The equilibrium condition is

$$[\mu(CaO, s) + \mu(CO_2, g) - \mu(CaCO_3, s)]_{eq} = 0.$$

For each gas present, e.g., CO_2, $[\mu(CO_2, g)]_{eq} = \mu°(CO_2, g) + RT\ln (p_{CO_2})_e$. While for the pure solids (and for pure liquids if they appear), because of the insensitivity of the Gibbs energy of a condensed phase to change in pressure, we have

$$\mu(CaCO_3, s) = \mu°(CaCO_3, s), \qquad \mu(CaO, s) = \mu°(CaO, s).$$

The equilibrium condition becomes

$$0 = \mu°(CaO, s) + \mu°(CO_2, g) - \mu°(CaCO_3, s) + RT\ln (p_{CO_2})_e,$$

$$0 = \Delta G° + RT\ln (p_{CO_2})_e. \tag{11.64}$$

In this case, the equilibrium constant is simply

$$K_p = (p_{CO_2})_e.$$

The equilibrium constant contains only the pressure of the gas; however, the $\Delta G°$ contains the standard Gibbs energies of *all* the reactants and products.

From the data in Table A-V, we find (at 25 °C)

Substance	$CaCO_3(s)$	$CaO(s)$	$CO_2(g)$
$\mu°/(kJ/mol)$	-1128.8	-604.0	-394.36
$\Delta H_f°/(kJ/mol)$	-1206.9	-635.09	-393.51

Then for the reaction

$$\Delta G° = -604.0 - 394.4 - (-1128.8) = 130.4 \text{ kJ/mol},$$

and

$$\Delta H° = -635.1 - 393.5 - (-1206.9) = 178.3 \text{ kJ/mol}.$$

The equilibrium pressure is calculated from Eq. (11.64).

$$\ln (p_{CO_2})_e = -\frac{130\,400 \text{ J/mol}}{(8.314 \text{ J/K mol})(298.15 \text{ K})} = -52.60;$$

$$(p_{CO_2})_e = 1.43 \times 10^{-23} \text{ atm} \qquad \text{(at 298 K)}.$$

Suppose we want the value at another temperature, 1100 K. We use Eq. (11.61):

$$\ln (p_{CO_2})_{1100} = \ln (p_{CO_2})_{298} - \frac{\Delta H°}{R}\left(\frac{1}{T} - \frac{1}{T_0}\right)$$

$$= -52.60 - \frac{178\,300 \text{ J/mol}}{8.314 \text{ J/K mol}}\left(\frac{1}{1100 \text{ K}} - \frac{1}{298.15 \text{ K}}\right) = 0.17;$$

$$(p_{CO_2})_{1100} = 0.84 \text{ atm}.$$

11.14.2 The Decomposition of Mercuric Oxide

Consider the reaction

$$HgO(s) \rightleftharpoons Hg(l) + \tfrac{1}{2}O_2(g).$$

The equilibrium constant is $K_p = (p_{O_2})_e^{1/2}$. Also

$$\Delta G° = \mu°(Hg, l) + \tfrac{1}{2}\mu°(O_2, g) - \mu°(HgO, s) = -\mu°(HgO, s) = 58.56 \text{ kJ/mol}.$$

Then

$$\ln (p_{O_2})_e = -\frac{58\,560 \text{ J/mol}}{(8.314 \text{ J/K mol})(298.15 \text{ K})} = -23.62;$$

$$(p_{O_2})_e = 5.50 \times 10^{-11} \text{ atm}.$$

11.14.3 Vaporization Equilibria

An important example of equilibrium between ideal gases and pure condensed phases is the equilibrium between a pure liquid and its vapor:

$$A(l) \; \rightleftharpoons \; A(g).$$

Let p be the equilibrium vapor pressure. Then

$$K_p = p \quad \text{and} \quad \Delta G^\circ = \mu^\circ(g) - \mu^\circ(l).$$

Using the Gibbs–Helmholtz equation, Eq. (11.57), we have

$$\frac{d \ln p}{dT} = \frac{\Delta H^\circ_{\text{vap}}}{RT^2}, \tag{11.65}$$

which is the Clausius–Clapeyron equation; it relates the temperature dependence of the vapor pressure of a liquid to the heat of vaporization. A similar expression holds for the sublimation of a solid. Consider the reaction

$$A(s) \; \rightleftharpoons \; A(g); \quad K_p = p, \quad \text{and} \quad \Delta G^\circ = \mu^\circ(g) - \mu^\circ(s),$$

where p is the equilibrium vapor pressure of the solid. By the same argument as above

$$\frac{d \ln p}{dT} = \frac{\Delta H^\circ_{\text{sub}}}{RT^2}, \tag{11.66}$$

where $\Delta H^\circ_{\text{sub}}$ is the heat of sublimation of the solid. In either case, a plot of $\ln p$ versus $1/T$ has a slope equal to $-\Delta H^\circ/R$ and is nearly linear.

★11.15 THE LECHATELIER PRINCIPLE

It is fairly easy to show how a change in temperature or pressure affects the equilibrium value of the advancement ξ_e of a reaction. We need only to determine the sign of the derivatives $(\partial \xi_e / \partial T)_p$ and $(\partial \xi_e / \partial p)_T$. We begin by writing the identity

$$\left(\frac{\partial G}{\partial \xi} \right)_{T,p} = \Delta G. \tag{11.67}$$

Since $(\partial G / \partial \xi)_{T,p}$ is itself a function of T, p, and ξ we may write the total differential expression,

$$d\left(\frac{\partial G}{\partial \xi} \right) = \frac{\partial}{\partial T}\left(\frac{\partial G}{\partial \xi} \right) dT + \frac{\partial}{\partial p}\left(\frac{\partial G}{\partial \xi} \right) dp + \frac{\partial}{\partial \xi}\left(\frac{\partial G}{\partial \xi} \right) d\xi. \tag{11.68}$$

Using Eq. (11.67) and setting $(\partial^2 G / \partial \xi^2) = G''$, Eq. (11.68) becomes

$$d\left(\frac{\partial G}{\partial \xi} \right) = \frac{\partial \Delta G}{\partial T} dT + \frac{\partial \Delta G}{\partial p} dp + G'' \, d\xi.$$

From the fundamental equation, $(\partial \Delta G / \partial T) = -\Delta S$ and $(\partial \Delta G / \partial p) = \Delta V$, in which ΔS is the entropy change and ΔV is the volume change for the reaction. Thus

$$d\left(\frac{\partial G}{\partial \xi} \right) = -\Delta S \, dT + \Delta V \, dp + G'' \, d\xi.$$

If we insist that these variations in temperature, pressure, and advancement occur while keeping the reaction at equilibrium, then $\partial G/\partial \xi = 0$ and hence $d(\partial G/\partial \xi) = 0$. At equilibrium, $\Delta S = \Delta H/T$, so the equation becomes

$$0 = -\left(\frac{\Delta H}{T}\right)(dT)_{eq} + \Delta V(dp)_{eq} + G''_e(d\xi_e). \tag{11.69}$$

At equilibrium G is a minimum; therefore G''_e must be positive.

At constant pressure, $dp = 0$, and Eq. (11.69) becomes

$$\left(\frac{\partial \xi_e}{\partial T}\right)_p = \frac{\Delta H}{TG''_e}. \tag{11.70}$$

At constant temperature, $dT = 0$, and Eq. (11.69) becomes

$$\left(\frac{\partial \xi_e}{\partial p}\right)_T = -\frac{\Delta V}{G''_e}. \tag{11.71}$$

Equations (11.70) and (11.71) are quantitative statements of the principle of LeChatelier: They describe the dependence of the advancement of the reaction at equilibrium on temperature and on pressure. Since G''_e is positive, the sign of $(\partial \xi_e/\partial T)_p$ depends on the sign of ΔH. If ΔH is $+$, an endothermic reaction, then $(\partial \xi_e/\partial T)_p$ is $+$, and an increase in temperature increases the advancement at equilibrium. For an exothermic reaction, ΔH is $-$, so $(\partial \xi_e/\partial T)_p$ is $-$; increase in temperature will decrease the equilibrium advancement of the reaction.

Similarly, the sign of $(\partial \xi_e/\partial p)_T$ depends on ΔV. If ΔV is $-$, the product volume is less than the reactant volume and $(\partial \xi_e/\partial p)_T$ is positive; increase in pressure increases the equilibrium advancement. Conversely, if ΔV is $+$, then $(\partial \xi_e/\partial p)_T$ is $-$; increase in pressure decreases the equilibrium advancement.

The net effect of these relations is that an increase in pressure shifts the equilibrium to the low-volume side of the reaction while a decrease in pressure shifts the equilibrium to the high-volume side. Similarly an increment in temperature shifts the equilibrium to the high-enthalpy side, while a decrease in temperature shifts it to the low-enthalpy side.

We may state the principle of LeChatelier in the following way. If the external constraints under which an equilibrium is established are changed, the equilibrium will shift in such a way as to moderate the effect of the change.

For example, if the volume of a nonreactive system is decreased by a specified amount, the pressure rises correspondingly. In a reactive system, the equilibrium shifts to the low-volume side (if $\Delta V \neq 0$), so the pressure increment is less than in the nonreactive case. The response of the system is moderated by the shift in equilibrium position. This implies that the compressibility of a reactive system is much greater than that of a nonreactive one (see Problem 11.39).

Similarly, if we extract a fixed quantity of heat from a nonreactive system, the temperature decreases by a definite amount. In a reactive system, withdrawing the same amount of heat will not produce as large a decrease in temperature because the equilibrium shifts to the low-enthalpy side (if $\Delta H \neq 0$). This implies that the heat capacity of a reactive system is much larger than that of a nonreactive one (see Problem 11.40). This is useful if the system can be used as a heat-transfer or heat-storage medium.

It must be noted here that there are certain types of systems that do not obey the LeChatelier principle in all circumstances (for example, open systems). A very general validity has been claimed for the LeChatelier principle. However, if the principle does have such broad application, the statement of the principle must be very much more complex than that given here or in other elementary discussions.

★11.16 EQUILIBRIUM CONSTANTS FROM CALORIMETRIC MEASUREMENTS. THE THIRD LAW IN ITS HISTORICAL CONTEXT

Using the Gibbs–Helmholtz equation, we can calculate the equilibrium constant of a reaction at any temperature T from a knowledge of the equilibrium constant at one temperature T_0 and the ΔH° of the reaction. For convenience we rewrite Eq. (11.60):

$$\ln K_p = \ln (K_p)_0 + \int_{T_0}^{T} \frac{\Delta H^\circ}{RT^2} \, dT.$$

The ΔH° for any reaction and its temperature dependence can be determined by purely thermal (that is, calorimetric) measurements. Thus, according to Eq. (11.60), a measurement of the equilibrium constant at only *one* temperature together with the thermal measurements of ΔH° and ΔC_p suffice to determine the value of K_p at any other temperature.

The question naturally arises whether or not it is possible to calculate the equilibrium constant exclusively from quantities that have been determined calorimetrically. In view of the relation $\Delta G^\circ = -RT \ln K_p$, the equilibrium constant can be calculated if ΔG° is known. At any temperature T, by definition,

$$\Delta G^\circ = \Delta H^\circ - T \Delta S^\circ. \tag{11.72}$$

Since ΔH° can be obtained from thermal measurements, the problem resolves into the question of whether or not ΔS° can be obtained solely from thermal measurements.

For any single substance

$$S_T^\circ = S_0^\circ + S_{0 \to T}^\circ, \tag{11.73}$$

where S_T° is the entropy of the substance at temperature T; S_0°, the entropy at 0 K, and $S_{0 \to T}^\circ$ is the entropy increase if the substance is taken from 0 K to the temperature T. The $S_{0 \to T}^\circ$ can be measured calorimetrically. For a chemical reaction, using Eq. (11.73) for each substance

$$\Delta S^\circ = \Delta S_0^\circ + \Delta S_{0 \to T}^\circ.$$

Putting this result into Eq. (11.72), we obtain

$$\Delta G^\circ = \Delta H^\circ - T \Delta S_0^\circ - T \Delta S_{0 \to T}^\circ.$$

Therefore

$$\ln K = \frac{\Delta S_0^\circ}{R} + \frac{\Delta S_{0 \to T}^\circ}{R} - \frac{\Delta H^\circ}{RT}. \tag{11.74}$$

Since the last two terms in Eq. (11.74) can be calculated from heat capacities and heats of reaction, the only unknown quantity is ΔS_0°, the change in entropy of the reaction at 0 K. In 1906, Nernst suggested that for all chemical reactions involving pure crystalline solids, ΔS_0° is zero at the absolute zero; the *Nernst heat theorem*. In 1913, Planck suggested that the reason that ΔS_0° is zero is that the entropy of each individual substance taking part in such a reaction is zero. It is clear that Planck's statement includes the Nernst theorem.

However, either one is sufficient for the solution of the problem of determining the equilibrium constant from thermal measurements. Setting $\Delta S_0^\circ = 0$ in Eq. (11.74), we obtain

$$\ln K = \frac{\Delta S^\circ}{R} - \frac{\Delta H^\circ}{RT}, \tag{11.75}$$

where ΔS° is the difference, at temperature T, in the third-law entropies of the substances involved in the reaction. Thus it is possible to calculate equilibrium constants from calorimetric data exclusively, provided that every substance in the reaction follows the third law.

Nernst based the heat theorem on evidence from several chemical reactions. The data showed that, at least for those reactions, ΔG° approached ΔH° as the temperature decreased; from Eq. (11.72)

$$\Delta G^\circ - \Delta H^\circ = -T \Delta S^\circ.$$

If ΔG° and ΔH° approach each other in value, it follows that the product $T \Delta S^\circ \to 0$ as the temperature decreases. This could be because T is getting smaller; however, the result was observed when the value of T was still of the order of 250 K. This strongly suggests that $\Delta S^\circ \to 0$ as $T \to 0$, which is the Nernst heat theorem.

The validity of the third law is tested by comparing the change in entropy of a reaction computed from the third-law entropies with the entropy change computed from equilibrium measurements. Discrepancies appear whenever one of the substances in the reaction does not follow the third law. A few of these exceptions to the third law were described in Section 9.17.

★ 11.17 CHEMICAL REACTIONS AND THE ENTROPY OF THE UNIVERSE

A chemical reaction proceeds from some arbitrary initial state to the equilibrium state. If the initial state has the properties T, p, G_1, H_1, and S_1, and the equilibrium state has the properties T, p, G_e, H_e, S_e, then the Gibbs energy change in the reaction is $\Delta G = G_e - G_1$; the enthalpy change is $\Delta H = H_e - H_1$, and the entropy change of the *system* is $\Delta S = S_e - S_1$. Since the temperature is constant, we have

$$\Delta G = \Delta H - T \Delta S,$$

and since the pressure is constant, $Q_p = \Delta H$. The heat that flows to the surroundings is $Q_s = -Q_p = -\Delta H$. If we suppose that Q_s is transferred reversibly to the immediate surroundings at temperature T, then the entropy increase of the surroundings is $\Delta S_s = Q_s/T = -\Delta H/T$; or $\Delta H = -T \Delta S_s$. In view of this relation we have

$$\Delta G = -T(\Delta S_s + \Delta S).$$

The sum of the entropy changes in the system and the immediate surroundings is the entropy change in the universe; we have the relation

$$\Delta G = -T \Delta S_{\text{universe}}.$$

In this equation we see the equivalence of the two criteria for spontaneity: the Gibbs energy decrease of the system and the increase in entropy of the universe. If $\Delta S_{\text{universe}}$ is positive, then ΔG is negative. Note that it is *not* necessary for spontaneity that the entropy *of the system* increase and in many spontaneous reactions the entropy of the system decreases; for example, $Na + \frac{1}{2}Cl_2 \rightarrow NaCl$. The entropy *of the universe* must increase in any spontaneous transformation.

★11.18 COUPLED REACTIONS

It often happens that a reaction which would be useful to produce a desirable product has a positive value of ΔG. For example, the reaction

$$TiO_2(s) + 2\,Cl_2(g) \longrightarrow TiCl_4(l) + O_2(g), \qquad \Delta G^\circ_{298} = +152.3 \text{ kJ/mol},$$

would be highly desirable for producing titanium tetrachloride from the common ore TiO_2. The high positive value of ΔG° indicates that at equilibrium only traces of $TiCl_4$ and O_2 are present. Increasing the temperature will improve the yield $TiCl_4$ but not enough to make the reaction useful. However, if this reaction is coupled with another reaction that involves a ΔG more negative than -152.3 kJ/mol, then the composite reaction can go spontaneously. If we are to pull the first reaction along, the second reaction must consume one of the products; since $TiCl_4$ is the desired product, the second reaction must consume oxygen. A likely prospect for the second reaction is

$$C(s) + O_2(g) \longrightarrow CO_2(g), \qquad \Delta G^\circ_{298} = -394.36 \text{ kJ/mol}.$$

The reaction scheme is

coupled reactions $\begin{cases} TiO_2(s) + 2\,Cl_2(g) \longrightarrow TiCl_4(l) + O_2(g), & \Delta G^\circ_{298} = +152.3 \text{ kJ/mol}, \\ \quad\ C(s) + O_2(g) \longrightarrow CO_2(g), & \Delta G^\circ_{298} = -394.4 \text{ kJ/mol}, \end{cases}$

and the overall reaction is

$$C(s) + TiO_2(s) + 2\,Cl_2(g) \longrightarrow TiCl_4(l) + CO_2(g), \qquad \Delta G^\circ_{298} = -242.1 \text{ kJ/mol}.$$

Since the overall reaction has a highly negative ΔG°, it is spontaneous. As a general rule metal oxides cannot be converted to chlorides by simple replacement; in the presence or carbon, the chlorination proceeds easily.

Coupled reactions have great importance in biological systems. Vital functions in an organism often depend on reactions which by themselves involve a positive ΔG; these reactions are coupled with the metabolic reactions, which have highly negative values of ΔG. As a trivial example, the lifting of a weight by Mr. Universe is a nonspontaneous event involving an increase in Gibbs energy. The weight goes up only because that event is coupled with the metabolic processes in the body that involve decreases in Gibbs energy sufficient to more than compensate for the increase associated with the lifting of the weight.

11.19 DEPENDENCE OF THE OTHER THERMODYNAMIC FUNCTIONS ON COMPOSITION

Having established the relation between the Gibbs energy and the composition, we can readily obtain the relation of the other functions to the composition. Considering the fundamental equation, Eq. (11.7),

$$dG = -S\,dT + V\,dp + \sum_i \mu_i\,dn_i.$$

We write the definitions of the other functions in terms of G:

$$U = G - pV + TS,$$

$$H = G + TS,$$

$$A = G - pV.$$

Differentiating each of these definitions, we have

$$dU = dG - p\,dV - V\,dp + T\,dS + S\,dT,$$

$$dH = dG + T\,dS + S\,dT,$$

$$dA = dG - p\,dV - V\,dp.$$

Replacing dG by its value in Eq. (11.7), we obtain

$$dU = T\,dS - p\,dV + \sum_i \mu_i\,dn_i, \tag{11.76}$$

$$dH = T\,dS + V\,dp + \sum_i \mu_i\,dn_i, \tag{11.77}$$

$$dA = -S\,dT - p\,dV + \sum_i \mu_i\,dn_i, \tag{11.78}$$

$$dG = -S\,dT + V\,dp + \sum_i \mu_i\,dn_i. \tag{11.79}$$

Equations (11.76), (11.77), (11.78), and (11.79) are the fundamental equations for systems of variable composition, and they imply that μ_i may be interpreted in four different ways:

$$\mu_i = \left(\frac{\partial U}{\partial n_i}\right)_{S,V,n_j} = \left(\frac{\partial H}{\partial n_i}\right)_{S,p,n_j} = \left(\frac{\partial A}{\partial n_i}\right)_{T,V,n_j} = \left(\frac{\partial G}{\partial n_i}\right)_{T,p,n_j} \tag{11.80}$$

The last equality in Eq. (11.80), namely

$$\mu_i = \left(\frac{\partial G}{\partial n_i}\right)_{T,p,n_j}, \tag{11.81}$$

is the one we have used previously.

11.20 PARTIAL MOLAR QUANTITIES AND ADDITIVITY RULES

Any extensive property of a mixture can be considered as a function of T, p, n_1, n_2, $\ldots$. Therefore, corresponding to any extensive property U, V, S, H, A, G, there are partial molar properties, $\bar{U}_i$, $\bar{V}_i$, $\bar{S}_i$, $\bar{H}_i$, $\bar{A}_i$, $\bar{G}_i$. The partial molar quantities are defined by

$$\bar{U}_i = \left(\frac{\partial U}{\partial n_i}\right)_{T,p,n_j}, \qquad \bar{H}_i = \left(\frac{\partial H}{\partial n_i}\right)_{T,p,n_j}, \qquad \bar{S}_i = \left(\frac{\partial S}{\partial n_i}\right)_{T,p,n_j}, \tag{11.82}$$

$$\bar{V}_i = \left(\frac{\partial V}{\partial n_i}\right)_{T,p,n_j}, \qquad \bar{A}_i = \left(\frac{\partial A}{\partial n_i}\right)_{T,p,n_j}, \qquad \bar{G}_i = \mu_i = \left(\frac{\partial G}{\partial n_i}\right)_{T,p,n_j}.$$

If we differentiate the defining equations for H, A, and G with respect to n_i, keeping T, p, n_j constant, and use the definitions in Eqs. (11.82), we obtain

$$\bar{H}_i = \bar{U}_i + p\bar{V}_i, \qquad \bar{A}_i = \bar{U}_i - T\bar{S}_i, \qquad \mu_i = \bar{H}_i - T\bar{S}_i. \tag{11.83}$$

Equations (11.83) show that the partial molar quantities are related to each other in the same way as the total quantities. (The use of μ_i rather than $\bar{G}_i$ for the partial molar Gibbs energy is customary.)

The total differential of any extensive property then takes a form analogous to Eq. (11.7). Choosing S, V, and H as examples,

$$dS = \left(\frac{\partial S}{\partial T}\right)_{p,\,n_i} dT + \left(\frac{\partial S}{\partial p}\right)_{T,\,n_i} dp + \sum_i \bar{S}_i\, dn_i; \tag{11.84}$$

$$dV = \left(\frac{\partial V}{\partial T}\right)_{p,\,n_i} dT + \left(\frac{\partial V}{\partial p}\right)_{T,\,n_i} dp + \sum_i \bar{V}_i\, dn_i; \tag{11.85}$$

$$dH = \left(\frac{\partial H}{\partial T}\right)_{p,\,n_i} dT + \left(\frac{\partial H}{\partial p}\right)_{T,\,n_i} dp + \sum_i \bar{H}_i\, dn_i. \tag{11.86}$$

Since $\bar{S}_i$, $\bar{V}_i$, and $\bar{H}_i$ are *intensive* properties they must have the same value everywhere in a system at equilibrium. Consequently, we could use precisely the same argument that was used for G in Section 11.3 to arrive at the additivity rules, namely,

$$S = \sum_i n_i \bar{S}_i, \qquad V = \sum_i n_i \bar{V}_i, \qquad H = \sum_i n_i \bar{H}_i. \tag{11.87}$$

However, by proceeding differently we gain some additional insights.

The Gibbs energy of a mixture is given by Eq. (11.9), $G = \sum_i n_i \mu_i$. If we differentiate this with respect to temperature (p and n_i are constant), we obtain

$$\left(\frac{\partial G}{\partial T}\right)_{p,\,n_i} = \sum_i n_i \left(\frac{\partial u_i}{\partial T}\right)_{p,\,n_i}. \tag{11.88}$$

By Eq. (11.79), the derivative on the left of Eq. (11.88) is equal to $-S$. The derivative on the right is evaluated by differentiating Eq. (11.81) with respect to T (suppressing subscripts to simplify writing):

$$\left(\frac{\partial \mu_i}{\partial T}\right)_{p,\,n_i} = \frac{\partial}{\partial T}\left(\frac{\partial G}{\partial n_i}\right) = \frac{\partial}{\partial n_i}\left(\frac{\partial G}{\partial T}\right) = -\left(\frac{\partial S}{\partial n_i}\right)_{T,\,p,\,n_j} = -\bar{S}_i.$$

The second equality is correct since the order of differentiation does not matter (Section 9.6); the third since $\partial G/\partial T = -S$. This reduces Eq. (11.88) to

$$S = \sum_i n_i \bar{S}_i, \tag{11.89}$$

which is the additivity rule for the entropy.

By differentiating Eq. (11.9) with respect to p, keeping T and n_i constant, we obtain

$$\left(\frac{\partial G}{\partial p}\right)_{T,\,n_i} = \sum_i n_i \left(\frac{\partial \mu_i}{\partial p}\right)_{T,\,n_i}. \tag{11.90}$$

Differentiating Eq. (11.81) with respect to p, we obtain

$$\left(\frac{\partial \mu_i}{\partial p}\right)_{T,\,n_i} = \frac{\partial}{\partial p}\left(\frac{\partial G}{\partial n_i}\right) = \frac{\partial}{\partial n_i}\left(\frac{\partial G}{\partial p}\right) = \left(\frac{\partial V}{\partial n_i}\right)_{T,\,p,\,n_j} = \bar{V}_i,$$

since $(\partial G/\partial p)_{T,\,n_i} = V$. Equation (11.90) then reduces to

$$V = \sum_i n_i \bar{V}_i, \tag{11.91}$$

which is the additivity rule for the volume. The other additivity rules can be established from these by taking the appropriate equation from the set (11.83). For example, multiply the last equation in the set by n_i and sum:

$$\sum_i n_i \mu_i = \sum_i n_i \bar{H}_i - T \sum_i n_i \bar{S}_i.$$

In view of Eqs. (11.9) and (11.89) this becomes

$$G = \sum_i n_i \bar{H}_i - TS,$$

but, by definition, $G = H - TS$; therefore

$$H = \sum_i n_i \bar{H}_i. \tag{11.92}$$

In the same way, the additivity rules for U and A can be derived.

Any extensive property J of a system follows the additivity rule

$$J = \sum_i n_i \bar{J}_i, \tag{11.93}$$

where $\bar{J}_i$ is the partial molar quantity

$$\bar{J}_i = \left(\frac{\partial J}{\partial n_i} \right)_{T, p, n_j}. \tag{11.94}$$

This is true also for the total number of moles, $N = \sum_i n_i$, or the total mass, $M = \sum_i n_i M_i$. The partial molar mole numbers are all equal to unity. The partial molar mass of a substance is its molar mass.

11.21 THE GIBBS–DUHEM EQUATION

An additional relation between the μ_i can be obtained by differentiating Eq. (11.9):

$$dG = \sum_i (n_i \, d\mu_i + \mu_i \, dn_i),$$

but, by the fundamental equation,

$$dG = -S \, dT + V \, dp + \sum_i \mu_i \, dn_i.$$

Subtracting, the two equations yield

$$\sum_i n_i \, d\mu_i = -S \, dT + V \, dp, \tag{11.95}$$

which is the Gibbs–Duhem equation. An important special case arises if the temperature and pressure are constant and only variations in composition occur; Eq. (11.95) becomes

$$\sum_i n_i \, d\mu_i = 0 \qquad (T, p \text{ constant}). \tag{11.96}$$

Equation (11.96) shows that if the composition varies, the chemical potentials do not change independently but in a related way. For example, in a system of two constituents, Eq. (11.96), becomes

$$n_1 \, d\mu_1 + n_2 \, d\mu_2 = 0 \qquad (T, p \text{ constant}).$$

Rearranging, we have

$$d\mu_2 = -\left(\frac{n_1}{n_2}\right) d\mu_1.$$ (11.97)

If a given variation in composition produces a change $d\mu_1$ in the chemical potential of the first component, then the concomitant change in the chemical potential of the second component $d\mu_2$ is given by Eq. (11.97).

By a similar argument it can be shown that the variations with composition of any of the partial molar quantities are related by the equation

$$\sum_i n_i \, d\bar{J}_i = 0 \qquad (T, p \text{ constant}),$$ (11.98)

where $\bar{J}_i$ is any partial molar quantity.

11.22 PARTIAL MOLAR QUANTITIES IN MIXTURES OF IDEAL GASES

The various partial molar quantities for the ideal gas are obtained from μ_i. From Eq. (11.13),

$$\mu_i = \mu_i^\circ(T) + RT \ln p + RT \ln x_i = \mu_{i(\text{pure})} + RT \ln x_i.$$

Differentiating, we have

$$\left(\frac{\partial \mu_i}{\partial T}\right)_{p, n_i} = \left(\frac{\partial \mu_i^\circ}{\partial T}\right)_{p, n_i} + R \ln p + R \ln x_i.$$

But $(\partial \mu_i / \partial T)_{p, n_i} = -\bar{S}_i$, so that

$$\bar{S}_i = \bar{S}_i^\circ - R \ln p - R \ln x_i = \bar{S}_{i(\text{pure})} - R \ln x_i.$$ (11.99)

Similarly, differentiation of μ_i with respect to pressure, keeping T and all n_i constant, yields

$$\left(\frac{\partial \mu_i}{\partial p}\right)_{T, n_i} = \frac{RT}{p}.$$

Since $(\partial \mu_i / \partial p)_{T, n_i} = \bar{V}_i$, we obtain

$$\bar{V}_i = \frac{RT}{p}.$$ (11.100)

For an ideal gas mixture we have $V = nRT/p$, where n is the total number of moles of all the gases in the mixture. Therefore

$$\bar{V}_i = \frac{V}{n},$$ (11.101)

which shows that in a mixture of ideal gases, the partial molar volume is simply the average molar volume, and that the partial molar volume of all the gases in the mixture has the same value.

From Eqs. (11.13), (11.83), (11.99), and (11.100) it is easy to show that $\bar{H}_i = \mu_i^\circ + T\bar{S}_i^\circ = \bar{H}_i^\circ$, and that $\bar{U}_i = \bar{H}_i^\circ - RT = \bar{U}_i^\circ$.

★11.23 DIFFERENTIAL HEAT OF SOLUTION

If dn moles of pure solid i, with molar enthalpy $\overline{H}_i^\circ$, are added at constant T and p to a solution in which the partial molar enthalpy is $\overline{H}_i$, then the heat absorbed is $dq = dH = (\overline{H}_i - \overline{H}_i^\circ)\, dn$. (The system contains both solid and solution.) The *differential heat of solution* is defined as dq/dn:

$$\frac{dq}{dn} = \overline{H}_i - \overline{H}_i^\circ. \tag{11.102}$$

The differential heat of solution is a more generally useful quantity than the integral heat of solution defined in Section 7.22.

QUESTIONS

11.1 What is the importance of the chemical potential? What is its interpretation?

11.2 How can the quantity $-\partial G/\partial \xi$ be viewed as a "driving force" towards chemical equilibrium. Discuss.

11.3 Sketch G versus ξ for a reaction for which $\Delta G^\circ < 0$. What are the roles of both ΔG° and the mixing Gibbs energy in determining the equilibrium position?

11.4 What is the distinction between K_p and Q_p for a gas phase reaction?

11.5 If initially $Q_p < K_p$ for a reaction system, what is the sign of the slope $\Delta G = \partial G/\partial \xi$? What subsequently happens to the pressures of the species in the system? Answer the same questions for $Q_p > K_p$.

11.6 Sketch a G versus ξ plot for the "reaction" $A(l) \rightleftharpoons A(g)$ for three different external pressures: P_{ext} less than, equal to, and greater than $\exp\left[-\Delta G^\circ/RT\right]$. ($\xi$ = the fraction of A in the gaseous state.) What does the equilibrium condition $\partial G/\partial \xi = 0$ give for the equilibrium vapor pressure in terms of P_{ext}?

11.7 What is the connection between the temperature effects on equilibrium described by Eqs. (11.58) and (11.70)?

11.8 Apply the LeChatelier principle, Eq. (11.71), to predict the effect of pressure on the gas phase equilibria (a) $N_2 + 3H_2 \rightleftharpoons 2NH_3$; (b) $N_2O_4 \rightleftharpoons 2NO_2$.

11.9 What is the practical value of the Nernst heat theorem in calculating equilibrium constants?

11.10 What is the origin of the increased entropy of the universe in a reaction for which $\Delta H^\circ \ll 0$ and $\Delta S^\circ < 0$?

PROBLEMS

In all of the following problems, the gases are assumed to be ideal.

11.1 Plot the value of $(\mu - \mu^\circ)/RT$ for an ideal gas as a function of pressure.

11.2 The conventional standard Gibbs energy of ammonia at 25 °C is -16.5 kJ/mol. Calculate the value of the molar Gibbs energy at $\frac{1}{2}$, 2, 10, and 100 atm.

11.3 Consider two pure gases A and B, each at 25 °C and 1 atm pressure. Calculate the Gibbs energy relative to the unmixed gases of

a) a mixture of 10 mol of A and 10 mol of B;

b) a mixture of 10 mol of A and 20 mol of B.

c) Calculate the change in Gibbs energy if 10 mol of B are added to the mixture of 10 mol of A with 10 mol of B.

11.4 a) Calculate the entropy of mixing 3 mol of hydrogen with 1 mol of nitrogen.

b) Calculate the Gibbs energy of mixing at 25 °C.

c) At 25 °C, calculate the Gibbs energy of mixing $1 - \xi$ mol of nitrogen, $3(1 - \xi)$ mol of hydrogen, and 2ξ mol of ammonia as a function of ξ. Plot the values from $\xi = 0$ to $\xi = 1$ at intervals of 0.2.

d) If $\Delta G_f^\circ(NH_3) = -16.5$ kJ/mol at 25 °C, calculate the Gibbs energy of the mixture for values of $\xi = 0$ to $\xi = 1$ at intervals of 0.2. Plot G versus ξ if the initial state is the mixture of 1 mol N_2 and 3 mol H_2. Compare the result with Figure 11.5.

e) Calculate G for ξ_e at $p = 1$ atm.

11.5 Four moles of nitrogen, n mol of hydrogen and $(8 - n)$ mol of oxygen are mixed at $T = 300$ K and $p = 1$ atm.

a) Write the expression for ΔG_{mix}/mol of mixture.

b) Calculate the value of n for which ΔG_{mix}/mol has a minimum.

c) Calculate the value of ΔG_{mix}/mol of the mixture at the minimum.

11.6 Show that in an ideal ternary mixture, the minimum Gibbs energy is obtained if $x_1 = x_2 = x_3 = \frac{1}{3}$.

11.7 Consider the reaction

$$H_2(g) + I_2(g) \longrightarrow 2\,HI(g).$$

a) If there are 1 mol of H_2, 1 mol of I_2, and 0 mol of HI present before the reaction advances, express the Gibbs energy of the reaction mixture in terms of the advancement ξ.

b) What form would the expression for G have if the iodine were present as the solid?

11.8 At 500 K, we have the data

Substance	ΔH_{500}°/(kJ/mol)	S_{500}°/(J/K mol)
HI(g)	32.41	221.63
$H_2(g)$	5.88	145.64
$I_2(g)$	69.75	279.94

One mole of H_2 and one mole of I_2 are placed in a vessel at 500 K. At this temperature only gases are present and the equilibrium

$$H_2(g) + I_2(g) \rightleftharpoons 2\,HI(g)$$

is established. Calculate K_p at 500 K and the mole fraction of HI present at 500 K and 1 atm. What would the mole fraction of HI be at 500 K and 10 atm?

11.9 a) Equimolar amounts of H_2 and CO are mixed. Using data from Table A-V calculate the equilibrium mole fraction of formaldehyde, HCHO(g), at 25 °C as a function of the total pressure; evaluate this mole fraction for a total pressure of 1 atm and for 10 atm.

b) If one mole of HCHO(g) is placed in a vessel, calculate the degree of dissociation into $H_2(g)$ and CO(g) at 25 °C for a total pressure of 1 atm and 10 atm.

c) Calculate K_x at 10 atm and K_c for the synthesis of HCHO.

11.10 For ozone at 25 °C, $\Delta G_f^\circ = 163.2$ kJ/mol.

a) At 25 °C, compute the equilibrium constant K_p for the reaction

$$3\,O_2(g) \rightleftharpoons 2\,O_3(g)$$

b) Assuming that the advancement at equilibrium, ξ_e, is very much less than unity, show that $\xi_e = \frac{3}{2}\sqrt{pK_p}$. (Let the original number of moles of O_2 be three, and of O_3 be zero.)

c) Calculate K_x at 5 atm and K_c.

11.11 Consider the equilibrium

$$2\,NO(g) + Cl_2(g) \rightleftharpoons 2\,NOCl(g).$$

At 25 °C for NOCl(g), $\Delta G_f^{\circ} = 66.07$ kJ/mol; for NO(g), $\Delta G_f^{\circ} = 86.57$ kJ/mol. If NO and Cl_2 are mixed in the molar ratio 2:1, show that $x_{NO} = (2/pK_p)^{1/3}$ and $x_{NOCl} = 1 - \frac{3}{2}(2/pK_p)^{1/3}$ at equilibrium. (Assume that $x_{NOCl} \approx 1$.) Note how each one of these quantities depends on pressure. Evaluate x_{NO} at 1 atm and at 10 atm.

11.12 Consider the dissociation of nitrogen tetroxide: $N_2O_4(g) \rightleftharpoons 2\,NO_2(g)$ at 25 °C. Suppose 1 mol of N_2O_4 is confined in a vessel under 1 atm pressure. Using data from Table A-V,

a) calculate the degree of dissociation.
b) If 5 mol of argon are introduced and the mixture confined under 1 atm total pressure, what is the degree of dissociation?
c) The system comes to equilibrium as in (a). If the volume of the vessel is then kept constant and 5 mol of argon are introduced, what will be the degree of dissociation?

11.13 From the data in Table A-V compute K_p for the reaction $H_2(g) + S(rhombic) \rightleftharpoons H_2S(g)$ at 25 °C. What is the mole fraction of H_2 present in the gas phase at equilibrium?

11.14 Consider the following equilibrium at 25 °C:

$$PCl_5(g) \rightleftharpoons PCl_3(g) + Cl_2(g).$$

a) From the data in Table A-V compute ΔG° and ΔH° at 25 °C.
b) Calculate the value of K_p at 600 K.
c) At 600 K calculate the degree of dissociation at 1 atm and at 5 atm total pressure.

11.15 At 25 °C the data are

Compound	$\Delta G_f^{\circ}/(kJ/mol)$	$\Delta H_f^{\circ}/(kJ/mol)$
$C_2H_4(g)$	68.1	52.3
$C_2H_2(g)$	209.2	226.7

a) Calculate K_p at 25 °C for the reaction

$$C_2H_4(g) \rightleftharpoons C_2H_2(g) + H_2(g).$$

b) What must the value of K_p be if 25 percent of the C_2H_4 is dissociated into C_2H_2 and H_2 at a total pressure of 1 atm?
c) At what temperature will K_p have the value determined in (b)?

11.16 At 25 °C, for the reaction

$$Br_2(g) \rightleftharpoons 2\,Br(g),$$

we have $\Delta G^{\circ} = 161.67$ kJ/mol, and $\Delta H^{\circ} = 192.81$ kJ/mol.

a) Compute the mole fraction of bromine atoms present at equilibrium at 25 °C and $p = 1$ atm.
b) At what temperature will the system contain 10 mol percent bromine atoms in equilibrium with bromine vapor at $p = 1$ atm.

11.17 For the reaction

$$H_2(g) + I_2(g) \rightleftharpoons 2\,HI(g),$$

$K_p = 50.0$ at 448 °C and 66.9 at 350 °C. Calculate ΔH° for this reaction.

11.18 At 600 K the degree of dissociation of $PCl_5(g)$ according to the reaction

$$PCl_5(g) \rightleftharpoons PCl_3(g) + Cl_2(g)$$

is 0.920 under 5 atm pressure.

a) What is the degree of dissociation under 1 atm pressure?
b) If the degree of dissociation at 520 K is 0.80 at 1 atm pressure, what are ΔH°, ΔG°, and ΔS° at 520 K?

11.19 At 800 K, 2 mol of NO are mixed with 1 mol of O_2. The reaction

$$2\,NO(g) + O_2(g) \;\rightleftharpoons\; 2\,NO_2(g)$$

comes to equilibrium under a total pressure of 1 atm. Analysis of the system shows that 0.71 mol of oxygen are present at equilibrium.
a) Calculate the equilibrium constant for the reaction.
b) Calculate $\Delta G°$ for the reaction at 800 K.

11.20 Consider the equilibrium

$$C_2H_6(g) \;\rightleftharpoons\; C_2H_4(g) + H_2(g).$$

At 1000 K and 1 atm pressure, C_2H_6 is introduced into a vessel. At equilibrium, the mixture consists of 26 mol percent H_2, 26 mol percent C_2H_4 and 48 mol percent C_2H_6.
a) Calculate K_p at 1000 K.
b) If $\Delta H° = 137.0\ kJ/mol$, calculate the value of K_p at 298.15 K.
c) Calculate $\Delta G°$ for this reaction at 298.15 K.

11.21 Consider the equilibrium

$$NO_2(g) \;\rightleftharpoons\; NO(g) + \tfrac{1}{2}O_2(g).$$

One mole of NO_2 is placed in a vessel and allowed to come to equilibrium at a total pressure of 1 atm. Analysis shows that

T	700 K	800 K
p_{NO}/p_{NO_2}	0.872	2.50

a) Calculate K_p at 700 K and 800 K.
b) Calculate $\Delta G°$ and $\Delta H°$.

11.22 Consider the equilibrium

$$CO(g) + H_2O(g) \;\rightleftharpoons\; CO_2(g) + H_2(g).$$

a) At 1000 K the composition of a sample of the equilibrium mixture is

Substance	CO_2	H_2	CO	H_2O
mol %	27.1	27.1	22.9	22.9

Calculate K_p and $\Delta G°$ at 1000 K.
b) Given the answer to part (a) and the data

Substance	$CO_2(g)$	$H_2(g)$	$CO(g)$	$H_2O(g)$
$\Delta H_f°/(kJ/mol)$	-393.51	0	-110.52	-241.81

Calculate $\Delta G°$ for this reaction at 298.15 K.

11.23 Nitrogen trioxide dissociates according to the equation

$$N_2O_3(g) \;\rightleftharpoons\; NO_2(g) + NO(g).$$

At 25 °C and 1 atm total pressure the degree of dissociation is 0.30. Calculate $\Delta G°$ for this reaction at 25 °C.

11.24 Consider the synthesis of formaldehyde:

$$CO(g) + H_2(g) \;\rightleftharpoons\; CH_2O(g).$$

At 25 °C, $\Delta G° = 24.\ kJ/mol$ and $\Delta H° = -7\ kJ/mol$. For $CH_2O(g)$ we have: $\bar{C}_p/R = 2.263 + 7.021(10^{-3})T - 1.877(10^{-6})T^2$. The heat capacities of $H_2(g)$ and $CO(g)$ are given in Table 7.1.

a) Calculate the value of K_p at 1000 K assuming $\Delta H°$ is independent of temperature.

b) Calculate the value of K_p at 1000 K taking into account the variation of $\Delta H°$ with temperature, and compare the result with that in (a).

c) At 1000 K compare the value of K_x at 1 atm pressure with that at 5 atm pressure.

11.25 At 25 °C for the reaction

$$H_2O(l) \longrightarrow H_2O(g),$$

$\Delta H° = 44.016$ kJ/mol. If $\bar{C}_p(l) = 75.29$ J/K mol and $\bar{C}_p(g) = 33.58$ J/K mol, calculate $\Delta H°$ for this reaction at 100 °C.

11.26 Liquid bromine boils at 58.2 °C; the vapor pressure at 9.3 °C is 100 Torr. Calculate the standard Gibbs energy of $Br_2(g)$ at 25 °C.

11.27 Consider the reaction

$$FeO(s) + CO(g) \rightleftharpoons Fe(s) + CO_2(g)$$

for which we have

$t/°C$	600	1000
K_p	0.900	0.396

a) Calculate $\Delta H°$, $\Delta G°$, and $\Delta S°$ for the reaction at 600 °C.

b) Calculate the mole fraction of CO_2 in the gas phase at 600 °C.

11.28 If the reaction

$$Fe_2N(s) + \tfrac{3}{2}H_2(g) \rightleftharpoons 2\,Fe(s) + NH_3(g)$$

comes to equilibrium at a total pressure of 1 atm, analysis of the gas shows that at 700 K and 800 K $p_{NH_3}/p_{H_2} = 2.165$ and 1.083, respectively, if only H_2 was present initially with an excess of Fe_2N. Calculate

a) K_p at 700 K and 800 K.

b) $\Delta H°$ and $\Delta S°$.

c) $\Delta G°$ at 298.15 K.

11.29 From the data in Table A-V find the values of $\Delta G°$ and $\Delta H°$ for the reactions

$$MCO_3(s) \rightleftharpoons MO(s) + CO_2(g); \quad (M = Mg, Ca, Sr, Ba).$$

Under the rash assumption that $\Delta H°$ for these reactions does not depend on temperature, calculate the temperatures at which the equilibrium pressure of CO_2 in these carbonate–oxide systems reaches 1 atm. (This is the decomposition temperature of the carbonate.)

11.30 Solid white phosphorus has a conventional standard Gibbs energy of zero at 25 °C. The melting point is 44.2 °C and $\Delta H°_{fus} = 2510$ J/mol P_4. The vapor pressure of white phosphorus has the values

$t/°C$	76.6	128.0	197.3
$p/$Torr	1	10	100

a) Calculate $\Delta H°_{vap}$ of liquid phosphorus.

b) Calculate the boiling point of the liquid.

c) Calculate the vapor pressure at the melting point.

d) Assuming that solid, liquid, and gaseous phosphorus are in equilibrium at the melting point, calculate the vapor pressure of *solid* white phosphorus at 25 °C.

e) Calculate the standard Gibbs energy of *gaseous* phosphorus at 25 °C.

11.31 For the reaction at 25 °C

$$Zn(s) + Cl_2(g) \rightleftharpoons ZnCl_2(s),$$

$\Delta G° = -369.43$ kJ/mol and $\Delta H° = -415.05$ kJ/mol. Sketch $\Delta G°$ as a function of temperature in the range from 298 K to 1500 K for this reaction under the condition that all the substances are in their stable states of aggregation at every temperature. The data are (T_m = melting point; T_b = boiling point):

	T_m/K	$\Delta H_{\text{fus}}/(\text{kJ/mol})$	T_b/K	$\Delta H_{\text{vap}}/(\text{kJ/mol})$
Zn	692.7	7.385	1180	114.77
$ZnCl_2$	548	23.0	1029	129.3

11.32 For the reaction

$$Hg(l) + \tfrac{1}{2}O_2(g) \;\rightleftharpoons\; HgO(s),$$

$$\Delta G^\circ/(\text{J/mol}) = -91\,044 + 1.54T \ln T - 10.33(10^{-3})T^2 - \frac{0.42 \times 10^5}{T} + 103.81T$$

a) What is the vapor pressure of oxygen over liquid mercury and solid HgO at 600 K?
b) Express $\ln K_p$, ΔH°, and ΔS° as functions of temperature.

11.33 Consider the reaction

$$Ag_2O(s) \;\rightleftharpoons\; 2Ag(s) + \tfrac{1}{2}O_2(g),$$

for which $\Delta G^\circ/(\text{J/mol}) = 32\,384 + 17.32T \log_{10} T - 116.48T$.

a) At what temperature will the equilibrium pressure of oxygen be 1 atm?
b) Express $\log_{10} K_p$, ΔH°, and ΔS° as functions of temperature.

11.34 The values of ΔG° and ΔH° at 25 °C for the reactions

$$C(\text{graphite}) + \tfrac{1}{2}O_2(g) \;\rightleftharpoons\; CO(g) \quad \text{and} \quad CO(g) + \tfrac{1}{2}O_2(g) \;\rightleftharpoons\; CO_2(g)$$

can be obtained from the data in Table A-V.

a) Assuming that the values of ΔH° do not vary with temperature, compute the composition (mole percent) of the gas in equilibrium with solid graphite at 600 K and 1000 K if the total pressure is 1 atm. Qualitatively, how would the composition change if the pressure were increased?
b) Using the heat capacity data in Table 7.1, compute the composition at 600 K and 1000 K (1 atm) and compare the results with those in (a).
c) Using the equilibrium constants from (b), compute the composition at 1000 K and 10 atm pressure.

11.35 At 25 °C, the data for the various isomers of C_5H_{10} in the gas phase are

Substance	$\Delta H_f^\circ/(\text{kJ/mol})$	$\Delta G_f^\circ/(\text{kJ/mol})$	$\log_{10} K_f$
A = 1-pentene	− 20.920	78.605	− 13.7704
B = cis-2-pentene	− 28.075	71.852	− 12.5874
C = trans-2-pentene	− 31.757	69.350	− 12.1495
D = 2-methyl-1-butene	− 36.317	64.890	− 11.3680
E = 3-methyl-1-butene	− 28.953	74.785	− 13.1017
F = 2-methyl-2-butene	− 42.551	59.693	− 10.4572
G = cyclopentane	− 77.24	38.62	− 6.7643

Consider the equilibria

$$A \;\rightleftharpoons\; B \;\rightleftharpoons\; C \;\rightleftharpoons\; D \;\rightleftharpoons\; E \;\rightleftharpoons\; F \;\rightleftharpoons\; G,$$

which might be established using a suitable catalyst.

a) Calculate the mole ratios (A/G), (B/G), $\dots$, (F/G) present at equilibrium at 25 °C.
b) Do these ratios depend on the total pressure?

c) Calculate the mole percent of the various species in the equilibrium mixture.
d) Calculate the composition of the equilibrium mixture at 500 K.

11.36 The following data are given at 25 °C.

Compound	CuO(s)	Cu$_2$O(s)	Cu(s)	O$_2$(g)
ΔH_f°/(kJ/mol)	−157	−169	—	—
ΔG_f°/(kJ/mol)	−130	−146	—	—
C_p°/(J/K mol)	42.3	63.6	24.4	29.4

a) Calculate the equilibrium pressure of oxygen over copper and cupric oxide at 900 K and at 1200 K; that is, the equilibrium constant for the reaction $2\,\text{CuO(s)} \rightleftharpoons 2\,\text{Cu(s)} + \text{O}_2\text{(g)}$.
b) Calculate the equilibrium pressure of oxygen over Cu$_2$O and Cu at 900 K and 1200 K.
c) At what temperature and pressure do Cu, CuO, Cu$_2$O, and O$_2$ coexist in equilibrium?

11.37 The standard state of zero Gibbs energy for phosphorus is solid white phosphorus, P$_4$(s). At 25 °C,

$$P_4(s) \rightleftharpoons P_4(g), \qquad \Delta H^\circ = 58.9 \text{ kJ/mol}, \qquad \Delta G^\circ = 24.5 \text{ kJ/mol};$$
$$\tfrac{1}{4}P_4(s) \rightleftharpoons P(g), \qquad \Delta H^\circ = 316.5 \text{ kJ/mol}, \qquad \Delta G^\circ = 280.1 \text{ kJ/mol};$$
$$\tfrac{1}{2}P_4(s) \rightleftharpoons P_2(g), \qquad \Delta H^\circ = 144.0 \text{ kJ/mol}, \qquad \Delta G^\circ = 103.5 \text{ kJ/mol}.$$

a) The P$_4$ molecule consists of four phosphorus atoms at the corners of a tetrahedron. Calculate the bond strength of the P—P bond in the tetrahedral molecule. Calculate the bond strength in P$_2$.
b) Calculate the mole fractions of P, P$_2$, and P$_4$ in the vapor at 900 K and 1200 K, and 1 atm total pressure.

11.38 In a gravity field the chemical potential of a species is increased by the potential energy required to raise one mole of the material from ground level to the height z. Then $\mu_i(T, p, z) = \mu_i(T, p) + M_i g z$, in which $\mu_i(T, p)$ is the value of μ_i at ground level, M_i is the molar mass, and g is the gravitational acceleration.

a) Show that if we require the chemical potential to be the same everywhere in an isothermal column of an ideal gas, this form of the chemical potential yields the barometric distribution law, $p_i = p_{io}\exp\,(-M_i g z/RT)$.
b) Show that the condition of chemical equilibrium is independent of the presence or absence of a gravity field.
c) Derive expressions for the entropy and enthalpy as functions of z. (*Hint:* Write the differential of μ_i in terms of dT, dp, and dz.)

11.39 The degree of dissociation, α, of N$_2$O$_4$ is a function of the pressure. Show that if the mixture remains in equilibrium as the pressure is changed, the apparent compressibility $(-1/V)(\partial V/\partial p)_T = (1/p)[1 + \tfrac{1}{2}\alpha_e(1 - \alpha_e)]$. Show that the quantity in brackets has a maximum value at $p = \tfrac{3}{4}K_p$.

11.40 One mole of N$_2$O$_4$ is placed in a vessel. When the equilibrium

$$\text{N}_2\text{O}_4\text{(g)} \rightleftharpoons 2\,\text{NO}_2$$

is established, the enthalpy of the equilibrium mixture is

$$H = (1 - \xi_e)\overline{H}(\text{N}_2\text{O}_4) + 2\xi_e\overline{H}(\text{NO}_2)$$

If the mixture remains in equilibrium as the temperature is raised,
a) show that the heat capacity is given by
$$C_p/R = \overline{C}_p(\text{N}_2\text{O}_4)/R + \xi_e\Delta C_p/R + \tfrac{1}{2}\xi_e(1 - \xi_e^2)(\Delta H^\circ/RT)^2;$$
b) show that the last term has a maximum value when $\xi_e = \tfrac{1}{3}\sqrt{3}$;
c) plot C_p/R versus T from 200 K to 500 K at $p = 1$ atm using

$\overline{C}_p(\text{N}_2\text{O}_4)/R = 9.29$ and $\overline{C}_p(\text{NO}_2)/R = 4.47$; $\Delta H_{298}^\circ = 57.20$ kJ/mol; $\Delta G_{298}^\circ = 4.77$ kJ/mol.

11.41 Consider the equilibrium

$$TiO_2(s) + 2Cl_2(g) \rightleftharpoons TiCl_4(l) + O_2(g).$$

$\Delta H_{vap}(TiCl_4) = 35.1$ kJ/mol at 409 K, the normal boiling point of $TiCl_4$. At 298.15 K

Substance	$TiO_2(s)$	$TiCl_4(l)$
ΔH_f°/(kJ/mol)	-945	-804
ΔG_f°/(kJ/mol)	-890	-737

a) Calculate K_p for the reaction at 500 K and 1000 K under 1 atm pressure.
b) Using data from Table A-V for the reaction

$$C(graphite) + O_2(g) \rightleftharpoons CO_2(g),$$

calculate the value of K_p for the reaction

$$C(graphite) + TiO_2(s) + 2Cl_2(g) \rightleftharpoons TiCl_4(g) + CO_2(g)$$

at 500 K and 1000 K.
c) If one mol of TiO_2 and 2 mol of Cl_2 (and 1 mol C when needed) are placed in the vessel, calculate the fraction of TiO_2 converted to $TiCl_4$ at 500 K and 1000 K if the total pressure is 1 atm. Do this for the reactions in (a) and (b). Compare the yield in (a) to that in (b).

11.42 Consider the two equilibria,

$$A_2 \rightleftharpoons 2A \tag{1}$$
$$AB \rightleftharpoons A + B, \tag{2}$$

and assume that the ΔG° and therefore K_p is the same for both. Show that the equilibrium value of ξ_2 is greater than the equilibrium value of ξ_1. What is the physical reason for this result?

11.43 An athlete in the weight room lifts a 50 kg mass through a vertical distance of 2.0 m; $g = 9.8$ m/s^2. The mass is allowed to fall through the 2.0 m distance while coupled to an electrical generator. The electrical generator produces an equal amount of electrical work, which is used to produce aluminum by the Hall electrolytic process.

$$Al_2O_3(sln) + 3C(graphite) \longrightarrow 2Al(l) + 3CO(g).$$

$\Delta G^\circ = 593$ kJ/mol. How many times must the athlete lift the 50 kg mass to provide sufficient Gibbs energy to produce one soft drink can (≈ 27 g). *Note:* This is the energy for the electrolysis. It is estimated that the total energy expenditure required to produce aluminum from the ore is about three times this amount.

Phase Equilibrium in Simple Systems; The Phase Rule

12.1 THE EQUILIBRIUM CONDITION

For a system in equilibrium the chemical potential of each constituent must be the same everywhere in the system. If there are several phases present, the chemical potential of each substance must have the same value in every phase in which that substance appears.

For a system of one component, $\mu = G/n$; dividing the fundamental equation by n, we obtain

$$d\mu = -\bar{S}\,dT + \bar{V}\,dp, \tag{12.1}$$

where $\bar{S}$ and $\bar{V}$ are the molar entropy and molar volume. Then

$$\left(\frac{\partial \mu}{\partial T}\right)_p = -\bar{S} \quad \text{and} \quad \left(\frac{\partial \mu}{\partial p}\right)_T = \bar{V}. \tag{12.2a, b}$$

The derivatives in Eqs. (12.2a, b) are the slopes of the curves μ versus T and μ versus p, respectively.

12.2 STABILITY OF THE PHASES OF A PURE SUBSTANCE

By the third law of thermodynamics, the entropy of a substance is always positive. This fact combined with Eq. (12.2a) shows that $(\partial \mu/\partial T)_p$ is always negative. Consequently, the plot of μ versus T at constant pressure is a curve with a negative slope.

For the three phases of a single substance we have

$$\left(\frac{\partial \mu_{\text{solid}}}{\partial T}\right)_p = -\bar{S}_{\text{solid}} \quad \left(\frac{\partial \mu_{\text{liq}}}{\partial T}\right)_p = -\bar{S}_{\text{liq}} \quad \left(\frac{\partial \mu_{\text{gas}}}{\partial T}\right)_p = -\bar{S}_{\text{gas}}. \tag{12.3}$$

At any temperature, $\bar{S}_{gas} \gg \bar{S}_{liq} > \bar{S}_{solid}$. The entropy of the solid is small so that in Fig. 12.1 the μ versus T curve for the solid, curve S, has a slight negative slope. The μ versus T curve for the liquid has a slope which is slightly more negative than that of the solid, curve L. The entropy of the gas is very much larger than that of the liquid, so the slope of curve G has a large negative value. The curves have been drawn as straight lines; they should be slightly concave downward. However, this refinement does not affect the argument.

The thermodynamic conditions for equilibrium between phases at constant pressure are immediately apparent in Fig. 12.1. Solid and liquid coexist in equilibrium when $\mu_{solid} = \mu_{liq}$; that is, at the intersection point of curves S and L. The corresponding temperature is T_m, the melting point. Similarly, liquid and gas coexist in equilibrium at the temperature T_b, the intersection point of curves L and G at which $\mu_{liq} = \mu_{gas}$.

The temperature axis is divided into three intervals. Below T_m the solid has the lowest chemical potential. Between T_m and T_b the liquid has the lowest chemical potential. Above T_b the gas has the lowest chemical potential. *The phase with the lowest value of the chemical potential is the stable phase.* If liquid were present in a system at a temperature below T_m, Fig. 12.2, the chemical potential of the liquid would have the value μ_a while the solid has the value μ_b. Thus, liquid could freeze spontaneously at this temperature, since freezing will decrease the Gibbs energy. At a temperature above T_m the situation is reversed: the μ of the solid is greater than that of the liquid and the solid melts spontaneously to decrease the Gibbs energy of the system. At T_m the chemical potentials of solid and liquid are equal, so neither phase is preferred; they coexist in equilibrium. The situation is much the same near T_b. Just below T_b liquid is stable, while just above T_b the gas is the stable phase.

The diagram illustrates the familiar sequence of phases observed if a solid is heated under constant pressure. At low temperatures the system is completely solid; at a definite temperature T_m the liquid forms; the liquid is stable until it vaporizes at a temperature T_b. This sequence of phases is a consequence of the sequence of entropy values, and so is an immediate consequence of the fact that heat is absorbed in the transformation from solid to liquid, and from liquid to gas.

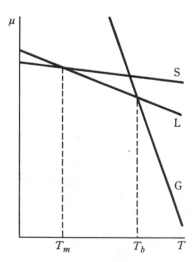

Figure 12.1 μ versus T at constant pressure.

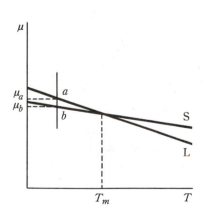

Figure 12.2 μ versus T at constant pressure.

12.3 PRESSURE DEPENDENCE OF μ VERSUS *T* CURVES

At this point it is natural to ask what happens to the curves if the pressure is changed. This question is answered using Eq. (12.2b) in the form $d\mu = \bar{V}\, dp$. If the pressure is decreased, dp is negative, $\bar{V}$ is positive; hence $d\mu$ is negative, and the chemical potential decreases in proportion to the volume of the phase. Since the molar volumes of the liquid and solid are very small, the value of μ is decreased only slightly; for the solid from a to a', for the liquid from b to b' (Fig. 12.3a). The volume of the gas is roughly 1000 times larger than that of the solid or liquid, so the μ of the gas decreases greatly; from c to c'. The curves at the lower pressure are shown as dashed lines parallel to the original lines in Fig. 12.3(b). (The figure has been drawn for the case $\bar{V}_{\text{liq}} > \bar{V}_{\text{solid}}$.) Figure 12.3(b) shows that both equilibrium temperatures (both intersection points) have shifted; the shift in the melting point is small, while the shift in the boiling point is relatively large. The melting point shift has been exaggerated for emphasis; it is actually very small. The decrease in boiling point of a liquid with decrease in pressure is neatly illustrated. At the lower pressure the range of stability of the liquid is noticeably decreased. If the pressure is reduced to a sufficiently low value, the boiling point of the liquid may even fall below the melting point of the solid (Fig. 12.4). Then there is no temperature at which the liquid is stable; the solid sublimes. At the temperature T_s, the solid and vapor coexist in equilibrium. The temperature T_s is the sublimation temperature of the solid. It is very dependent on the pressure.

Clearly there is some pressure at which the three curves intersect at the same temperature. This temperature and pressure define the *triple point*; all three phases coexist in equilibrium at the triple point.

Whether or not a particular material will sublime under reduced pressure rather than melt depends entirely on the individual properties of the substance. Water, for example, sublimes at pressures below 611 Pa. The higher the melting point, and the smaller the difference between the melting point and boiling point at 1 atm pressure, the higher will be the pressure below which sublimation is observed. The pressure (in atm) below which sublimation is observed can be estimated for substances obeying Trouton's rule by the formula

$$\ln p = -10.8\left(\frac{T_b - T_m}{T_m}\right). \tag{12.4}$$

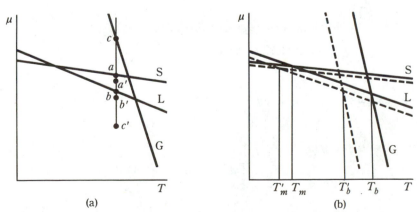

(a)   (b)

Figure 12.3 Effect of pressure on melting and boiling points. Solid line indicates high pressure; dashed line low pressure.

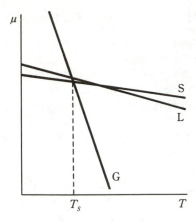

Figure 12.4 μ versus T for a substance that sublimes.

12.4 THE CLAPEYRON EQUATION

The condition for equilibrium between two phases, α and β, of a pure substance is

$$\mu_\alpha(T, p) = \mu_\beta(T, p). \tag{12.5}$$

If the analytical forms of the functions μ_α and μ_β were known, it would be possible, in principle at least, to solve Eq. (12.5) for

$$T = f(p) \qquad \text{or} \qquad p = g(T). \tag{12.6a, b}$$

Equation (12.6a) expresses the fact, illustrated in Fig. 12.3(b), that the equilibrium temperature depends on the pressure.

   In the absence of this detailed knowledge of the functions μ_α and μ_β, it is possible nonetheless to obtain a value for the derivative of the temperature with respect to pressure. Consider the equilibrium between two phases α and β under a pressure p; the equilibrium temperature is T. Then, at T and p, we have

$$\mu_\alpha(T, p) = \mu_\beta(T, p). \tag{12.7}$$

If the pressure is changed to a value $p + dp$, the equilibrium temperature will change to $T + dT$, and the value of each μ will change to $\mu + d\mu$. Hence at $T + dT, p + dp$ the equilibrium condition is

$$\mu_\alpha(T, p) + d\mu_\alpha = \mu_\beta(T, p) + d\mu_\beta. \tag{12.8}$$

Subtracting Eq. (12.7) from Eq. (12.8), we obtain

$$d\mu_\alpha = d\mu_\beta. \tag{12.9}$$

We write $d\mu$ explicitly in terms of dp and dT using the fundamental equation, Eq. (12.1):

$$d\mu_\alpha = -\bar{S}_\alpha \, dT + \bar{V}_\alpha \, dp \qquad d\mu_\beta = -\bar{S}_\beta \, dT + \bar{V}_\beta \, dp. \tag{12.10}$$

Using Eqs. (12.10) in Eq. (12.9), we get

$$-\bar{S}_\alpha \, dT + \bar{V}_\alpha \, dp = -\bar{S}_\beta \, dT + \bar{V}_\beta \, dp.$$

Rearranging, we have

$$(\bar{S}_\beta - \bar{S}_\alpha) \, dT = (\bar{V}_\beta - \bar{V}_\alpha) \, dp. \tag{12.11}$$

If the transformation is written $\alpha \rightarrow \beta$, then $\Delta S = \bar{S}_\beta - \bar{S}_\alpha$, and $\Delta V = \bar{V}_\beta - \bar{V}_\alpha$, and

Eq. (12.11) becomes

$$\frac{dT}{dp} = \frac{\Delta V}{\Delta S} \quad \text{or} \quad \frac{dp}{dT} = \frac{\Delta S}{\Delta V}. \qquad (12.12a, b)$$

Either of Eqs. (12.12) is called the Clapeyron equation.

The Clapeyron equation is fundamental to any discussion of the equilibrium between two phases of a pure substance. Note that the left-hand side is an ordinary derivative and not a partial derivative. The reason for this should be apparent from Eqs. (12.6).

Figure 12.3(b) shows that the equilibrium temperatures depend on the pressure, since the intersection points depend on pressure. The Clapeyron equation expresses the quantitative dependence of the equilibrium temperature on pressure, Eq. (12.12a), or the variation in the equilibrium pressure with temperature, Eq. (12.12b). Using this equation, we can plot the equilibrium pressure versus temperature schematically for any phase transformation.

12.4.1 The Solid–Liquid Equilibrium

Applying the Clapeyron equation to the transformation solid → liquid, we have

$$\Delta S = \bar{S}_{\text{liq}} - \bar{S}_{\text{solid}} = \Delta S_{\text{fus}} \qquad \Delta V = \bar{V}_{\text{liq}} - \bar{V}_{\text{solid}} = \Delta V_{\text{fus}}.$$

At the equilibrium temperature, the transformation is reversible; hence $\Delta S_{\text{fus}} = \Delta H_{\text{fus}}/T$. The transformation from solid to liquid always entails an absorption of heat, (ΔH_{fus} is $+$); hence

$$\Delta S_{\text{fus}} \quad \text{is} + \quad \text{(all substances)}.$$

The quantity ΔV_{fus} may be positive or negative, depending on whether the density of the solid is greater or less than that of the liquid; therefore

$$\Delta V_{\text{fus}} \quad \text{is} + \quad \text{(most substances)};$$

$$\Delta V_{\text{fus}} \quad \text{is} - \quad \text{(a few substances, such as } H_2O).$$

The ordinary magnitudes of these quantities are

$$\Delta S_{\text{fus}} = 8 \text{ to } 25 \text{ J/(K mol)} \qquad \Delta V_{\text{fus}} = \pm(1 \text{ to } 10) \text{ cm}^3/\text{mole}.$$

If, for illustration, we choose: $\Delta S_{\text{fus}} = 16$ J/(K mol) and $\Delta V_{\text{fus}} = \pm 4$ cm^3/mol, then for the solid–liquid equilibrium line,

$$\frac{dp}{dT} = \frac{16 \text{ J/(K mol)}}{\pm 4(10^{-6}) \text{ m}^3/\text{mol}} = \pm 4(10^6) \text{ Pa/K} = \pm 40 \text{ atm/K}.$$

Inverting, we obtain $dT/dp = \pm 0.02$ K/atm. This value shows that a change in pressure of 1 atm alters the melting point by a few hundredths of a kelvin. In a plot of pressure as a function of temperature, the slope is given by Eq. (12.12b); (40 atm/K in the example); this slope is large and the curve is nearly vertical. The case dp/dT is $+$ is shown in Fig. 12.5(a); over a moderate range of pressure the curve is linear.

The line in Fig. 12.5(a) is the locus of all points (T, p) at which the solid and liquid can coexist in equilibrium. Points that lie to the left of the line correspond to temperatures below the melting point; these points are conditions (T, p) under which only the solid is stable. Points immediately to the right of the line correspond to temperatures above the melting point; hence these points are conditions (T, p) under which the liquid is stable.

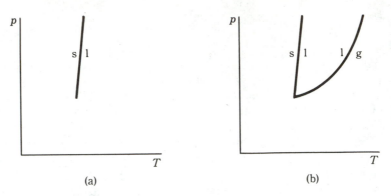

Figure 12.5   Equilibrium lines. (a) Solid–liquid. (b) Liquid–vapor.

12.4.2 The Liquid–Gas Equilibrium

Application of the Clapeyron equation to the transformation liquid → gas yields

$$\Delta S = \bar{S}_{\text{gas}} - \bar{S}_{\text{liq}} = \frac{\Delta H_{\text{vap}}}{T} \quad \text{is } + \qquad \text{(all substances)},$$

$$\Delta V = \bar{V}_{\text{gas}} - \bar{V}_{\text{liq}} \quad \text{is } + \qquad \text{(all substances)},$$

and, consequently,

$$\frac{dp}{dT} = \frac{\Delta S}{\Delta V} \quad \text{is } + \qquad \text{(all substances)}.$$

The liquid–gas equilibrium line always has a positive slope. At ordinary T and p the magnitudes are

$$\Delta S \approx +90 \text{ J/K mol} \qquad \Delta V \approx +20{,}000 \text{ cm}^3 = 0.02 \text{ m}^3.$$

However, ΔV depends strongly on T and p because $\bar{V}_{\text{gas}}$ depends strongly on T and p. The slope of the liquid–gas curve is small compared with that of the solid–liquid curve:

$$\left(\frac{dp}{dT} \right)_{\text{liq, gas}} \approx \frac{90 \text{ J/K mol}}{0.02 \text{ m}^3/\text{mol}} = 4000 \text{ Pa/K} = 0.04 \text{ atm/K}.$$

Figure 12.5(b) shows the l–g curve as well as the s–l curve. In Fig. 12.5(b), curve l–g is the locus of all points (T, p) at which liquid and gas coexist in equilibrium. Points just to the left of l–g are below the boiling point and so are conditions under which the liquid is stable. Points to the right of l–g are conditions under which the gas is stable.

   The intersection of curves s–l and l–g corresponds to a temperature and pressure at which solid, liquid, and gas all coexist in equilibrium. The values of T and p at this point are determined by the conditions

$$\mu_{\text{solid}}(T, p) = \mu_{\text{liq}}(T, p) \qquad \text{and} \qquad \mu_{\text{liq}}(T, p) = \mu_{\text{gas}}(T, p). \tag{12.13}$$

Equations (12.13) can, in principle at least, be solved for definite numerical values of T and p. That is,

$$T = T_t \qquad p = p_t, \tag{12.14}$$

where T_t and p_t are the triple-point temperature and pressure. There is only one such triple point at which a specific set of three phases (for example, solid–liquid–gas) can coexist in equilibrium.

12.4.3 The Solid–Gas Equilibrium

For the transformation solid $\rightarrow$ gas, we have

$$\Delta S = \bar{S}_{gas} - \bar{S}_{solid} = \frac{\Delta H_{sub}}{T} \quad \text{is} \; + \quad \text{(all substances)},$$

$$\Delta V = \bar{V}_{gas} - \bar{V}_{solid} \quad \text{is} \; + \quad \text{(all substances)},$$

and the Clapeyron equation is

$$\left(\frac{dp}{dT}\right)_{s-g} = \frac{\Delta S}{\Delta V} \quad \text{is} \; + \quad \text{(all substances)}.$$

The slope of the s–g curve is steeper at the triple point than the slope of the l–g curve. Since $\Delta H_{sub} = \Delta H_{fus} + \Delta H_{vap}$, then

$$\left(\frac{dp}{dT}\right)_{l-g} = \frac{\Delta H_{vap}}{T \, \Delta V} \quad \text{and} \quad \left(\frac{dp}{dT}\right)_{s-g} = \frac{\Delta H_{sub}}{T \, \Delta V}.$$

The ΔV's in the two equations are very nearly equal. Since ΔH_{sub} is greater than ΔH_{vap}, the slope of the s–g curve in Fig. 12.6 is steeper than that of the l–g curve.

Points on the s–g curve are those sets of temperatures and pressures at which solid coexists in equilibrium with vapor. Points to the left of the line lie below the sublimation temperature, and so correspond to conditions under which the solid is stable. Those points to the right of the s–g curve are points above the sublimation temperature, and so are conditions under which the gas is the stable phase. The s–g curve must intersect the others at the triple point because of the conditions expressed by Eqs. (12.13).

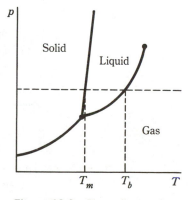

Figure 12.6 Phase diagram for a simple substance.

12.5 THE PHASE DIAGRAM

Examination of Fig. 12.6 at a constant pressure, indicated by the dashed horizontal line, shows the melting point and boiling point of the substance as the intersections of the horizontal line with the s–l and l–g curves. These intersection points correspond to the intersections of the μ–T curves in Fig. 12.1. At temperatures below T_m, the solid is stable; at the points between T_m and T_b the liquid is stable, while above T_b the gas is stable. Illustrations such as Fig. 12.6 convey more information than those such as 12.1 and 12.3(b). Figure 12.6 is called a *phase diagram*, or an *equilibrium diagram*.

The phase diagram shows at a glance the properties of the substance; melting point, boiling point, transition points, triple points. Every point on the phase diagram represents a state of the system, since it describes values of T and p.

The lines on the phase diagram divide it into regions, labeled *solid*, *liquid*, and *gas*. If the point that describes the system falls in the solid region, the substance exists as a solid. If the point falls in the liquid region, the substance exists as a liquid. If the point falls on a line such as l–g, the substance exists as liquid and vapor in equilibrium.

The l–g curve has a definite upper limit at the critical pressure and temperature, since it is not possible to distinguish between liquid and gas above this pressure and temperature.

12.5.1 The Phase Diagram for Carbon Dioxide

The phase diagram for carbon dioxide is shown schematically in Fig. 12.7. The solid–liquid line slopes slightly to the right, since $\overline{V}_{\text{liq}} > \overline{V}_{\text{solid}}$. Note that liquid CO_2 is not stable at pressures below 5 atm. For this reason "dry ice" is dry under ordinary atmospheric pressure. When carbon dioxide is confined to a cylinder under pressure at 25 °C, the diagram shows that if the pressure reaches 67 atm, liquid CO_2 will form. Commercial cylinders of CO_2 commonly contain liquid and gas in equilibrium; the pressure in the cylinder is about 67 atm at 25 °C.

12.5.2 The Phase Diagram for Water

Figure 12.8 is the phase diagram for water under moderate pressure. The solid–liquid line leans slightly to the left because $\overline{V}_{\text{liq}} < \overline{V}_{\text{solid}}$. The triple point is at 0.01 °C and 611 Pa. The normal freezing point of water is at 0.0002 °C. An increase in pressure decreases the melting

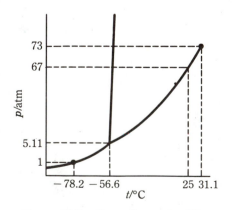

Figure 12.7 Phase diagram for CO_2.

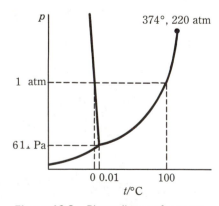

Figure 12.8 Phase diagram for water.

point of water. This lower melting point under the pressure exerted by the weight of the skater through the knife edge of the skate blade is part of the reason that ice skating is possible. This effect together with the heat developed by friction combine to produce a lubricating layer of liquid water between the ice and the blade. In this connection, it is interesting to note that if the temperature is too low, the skating is not good.

If water is studied under very high pressures, several crystalline modifications of ice are observed. The equilibrium diagram is shown in Fig. 12.9. Ice I is ordinary ice; ices II, III, V, VI, VII are modifications that are stable at higher pressures. The range of pressure is so large in Fig. 12.9 that the s–g and l–g curves lie only slightly above the horizontal axis; they are not shown in the figure. It is remarkable that under very high pressures, melting ice is quite hot! Ice VII melts at about 100 °C under a pressure of 25 000 atm.

12.5.3 The Phase Diagram for Sulfur

Figure 12.10 shows two phase diagrams for sulfur. The stable form of sulfur at ordinary temperatures and under 1 atm pressure is rhombic sulfur, which, if heated slowly, trans-

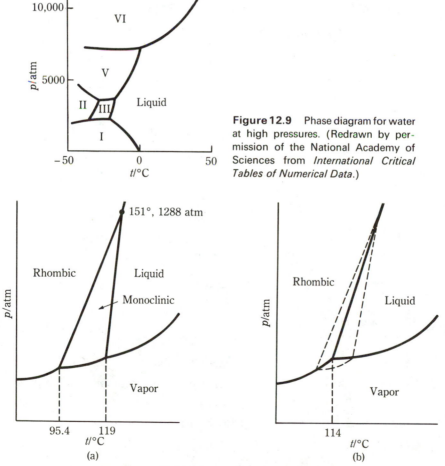

Figure 12.9 Phase diagram for water at high pressures. (Redrawn by permission of the National Academy of Sciences from *International Critical Tables of Numerical Data*.)

Figure 12.10 Phase diagram for sulfur.

forms to solid monoclinic sulfur at 95.4 °C (see Fig. 12.10a). Above 95.4 °C monoclinic sulfur is stable, until 119 °C is reached; monoclinic sulfur melts at 119 °C. Liquid sulfur is stable up to the boiling point, 444.6 °C. The transformation of one crystalline modification to another is often very slow and, if rhombic sulfur is heated quickly to 114 °C, it melts. This melting point of rhombic sulfur is shown as a function of pressure in Fig. 12.10(b). The equilibrium S(rhombic) $\rightleftharpoons$ S(l) is an example of a *metastable* equilibrium, since the line lies in the region of stability of monoclinic sulfur, shown by dashed lines in Fig. 12.10(b). In this region the reactions

$$S(rh) \longrightarrow S(mono) \quad \text{and} \quad S(liq) \longrightarrow S(mono)$$

both can occur with a decrease in Gibbs energy.

In Fig. 12.10(a) there are three triple points. The equilibrium conditions are

$$\text{at } 95.4 \text{ °C}; \qquad \mu_{rh} = \mu_{mono} = \mu_{gas},$$

$$\text{at } 119 \text{ °C}: \qquad \mu_{mono} = \mu_{liq} = \mu_{gas},$$

$$\text{at } 151 \text{ °C}: \qquad \mu_{rh} = \mu_{mono} = \mu_{liq}.$$

12.6 THE INTEGRATION OF THE CLAPEYRON EQUATION

12.6.1 Solid–Liquid Equilibrium

The Clapeyron equation is

$$\frac{dp}{dT} = \frac{\Delta S_{fus}}{\Delta V_{fus}}.$$

Then

$$\int_{p_1}^{p_2} dp = \int_{T_m}^{T'_m} \frac{\Delta H_{fus}}{\Delta V_{fus}} \frac{dT}{T}.$$

If ΔH_{fus} and ΔV_{fus} are nearly independent of T and p, the equation integrates to

$$p_2 - p_1 = \frac{\Delta H_{fus}}{\Delta V_{fus}} \ln \frac{T'_m}{T_m}, \tag{12.15}$$

where T'_m is the melting point under p_2; T_m is the melting point under p_1. Since $T'_m - T_m$ is usually quite small, the logarithm can be expanded to

$$\ln \left(\frac{T'_m}{T_m} \right) = \ln \left(\frac{T_m + T'_m - T_m}{T_m} \right) = \ln \left(1 + \frac{T'_m - T_m}{T_m} \right) \approx \frac{T'_m - T_m}{T_m};$$

then Eq. (12.15) becomes

$$\Delta p = \frac{\Delta H_{fus}}{\Delta V_{fus}} \frac{\Delta T}{T_m}, \tag{12.16}$$

where ΔT is the increase in melting point corresponding to the increase in pressure Δp.

12.6.2 Condensed-Phase–Gas Equilibrium

For the equilibrium of a condensed phase, either solid or liquid, with vapor, we have

$$\frac{dp}{dT} = \frac{\Delta S}{\Delta V} = \frac{\Delta H}{T(\overline{V}_g - \overline{V}_c)},$$

where ΔH is either the molar heat of vaporization of the liquid or the molar heat of sublimation of the solid, and $\overline{V}_c$ is the molar volume of the solid or liquid. In most circumstances, $\overline{V}_g - \overline{V}_c \approx \overline{V}_g$, and this, assuming that the gas is ideal, is equal to RT/p. Then the equation becomes

$$\frac{d \ln p}{dT} = \frac{\Delta H}{RT^2}, \tag{12.17}$$

which is the Clausius–Clapeyron equation, relating the vapor pressure of the liquid (solid) to the heat of vaporization (sublimation) and the temperature. Integrating between limits, under the additional assumption that ΔH is independent of temperature yields

$$\int_{p_0}^{p} d \ln p = \int_{T_0}^{T} \frac{\Delta H}{RT^2} \, dT,$$

$$\ln \frac{p}{p_0} = -\frac{\Delta H}{R}\left(\frac{1}{T} - \frac{1}{T_0}\right) = -\frac{\Delta H}{RT} + \frac{\Delta H}{RT_0}, \tag{12.18}$$

where p_0 is the vapor pressure at T_0, and p is the vapor pressure at T. (In Section 5.4, this equation was derived in a different way.) If $p_0 = 1$ atm, then T_0 is the normal boiling point of the liquid (normal sublimation point of the solid). Then

$$\ln p = \frac{\Delta H}{RT_0} - \frac{\Delta H}{RT}, \qquad \log_{10} p = \frac{\Delta H}{2.303 RT_0} - \frac{\Delta H}{2.303 RT}. \tag{12.19}$$

According to Eq. (12.19), if $\ln p$ or $\log_{10} p$ is plotted against $1/T$, a linear curve is obtained with a slope equal to $-\Delta H/R$ or $-\Delta H/2.303R$. The intercept at $1/T = 0$ yields a value of $\Delta H/RT_0$. Thus, from the slope and intercept, both ΔH and T_0 can be calculated. Heats of vaporization and sublimation are often determined through the measurement of the vapor pressure of the substance as a function of temperature. Figure 12.11 shows a plot of $\log_{10} p$ versus $1/T$ for water; Fig. 12.12 is the same plot for solid CO_2 (dry ice).

Compilations of data on vapor pressure frequently use an equation of the form $\log_{10} p = A + B/T$, and tabulate values of A and B for various substances. This equation has the same functional form as Eq. (12.19).

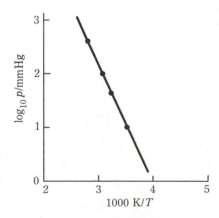

Figure 12.11 $\log_{10} p/\text{mmHg}$ versus $1/T$ for water.

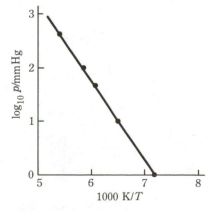

Figure 12.12 $\log_{10} p/\text{mmHg}$ versus for $1/T$ solid CO_2.

For substances that obey Trouton's rule, Eq. (12.19) takes a particularly simple form, which is useful for estimating the vapor pressure of a substance at any temperature T from a knowledge of the boiling point only (Problem 12.11).

12.7 EFFECT OF PRESSURE ON THE VAPOR PRESSURE

In the preceding discussion of the liquid–vapor equilibrium it was implicitly assumed that the two phases were under the same pressure p. If by some means it is possible to keep the liquid under a pressure P and the vapor under the vapor pressure p, then the vapor pressure depends on P. Suppose that the liquid is confined in the container shown in Fig. 12.13. In the space above the liquid, the vapor is confined together with a foreign gas that is insoluble in the liquid. The vapor pressure p plus the pressure of the foreign gas is P, the total pressure exerted on the liquid. As usual, the equilibrium condition is

$$\mu_{\text{vap}}(T, p) = \mu_{\text{liq}}(T, P). \tag{12.20}$$

At constant temperature this equation implies that $p = f(P)$. To discover the functionality, Eq. (12.20) is differentiated with respect to P, keeping T constant:

$$\left(\frac{\partial \mu_{\text{vap}}}{\partial p}\right)_T \left(\frac{\partial p}{\partial P}\right)_T = \left(\frac{\partial \mu_{\text{liq}}}{\partial P}\right)_T.$$

Using the fundamental equation, Eq. (12.2b), this becomes

$$\overline{V}_{\text{vap}}\left(\frac{\partial p}{\partial P}\right)_T = \overline{V}_{\text{liq}} \qquad \text{or} \qquad \left(\frac{\partial p}{\partial P}\right)_T = \frac{\overline{V}_{\text{liq}}}{\overline{V}_{\text{vap}}}. \tag{12.21}$$

The Gibbs equation, Eq. (12.21), shows that the vapor pressure increases with the total pressure on the liquid; the rate of increase is very small since $\overline{V}_{\text{liq}}$ is very much less than $\overline{V}_{\text{vap}}$. If the vapor behaves ideally, Eq. (12.21) can be written

$$\frac{RT}{p} \, dp = \overline{V}_{\text{liq}} \, dP, \qquad RT \int_{p_0}^{p} \frac{dp}{p} = \overline{V}_{\text{liq}} \int_{p_0}^{P} dP,$$

where p is the vapor pressure under a pressure P, p_0 is the vapor pressure when liquid and vapor are under the same pressure p_0, the orthobaric pressure. Thus

$$RT \ln \left(\frac{p}{p_0}\right) = \overline{V}_{\text{liq}}(P - p_0). \tag{12.22}$$

We will use Eqs. (12.21) and (12.22) in discussing the osmotic pressure of a solution.

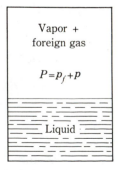

Vapor + foreign gas

$P = p_f + p$

Liquid

Figure 12.13

12.8 THE PHASE RULE

The coexistence of two phases in equilibrium implies the condition

$$\mu_\alpha(T, p) = \mu_\beta(T, p), \qquad (12.23)$$

which means that the two intensive variables ordinarily needed to describe the state of a system are no longer independent, but are related. Because of this relation, only one intensive variable, either temperature or pressure, is needed to describe the state of the system. The system has one *degree of freedom*, or is *univariant*, whereas if only one phase is present, two variables are needed to describe the state, and the system has two degrees of freedom, or is *bivariant*. If three phases are present, two relations exist between T and p:

$$\mu_\alpha(T, p) = \mu_\beta(T, p) \qquad \mu_\alpha(T, p) = \mu_\gamma(T, p). \qquad (12.24)$$

These two relations determine T and p completely. No other information is necessary for the description of the state of the system. Such a system is *invariant*; it has no degrees of freedom. Table 12.1 shows the relation between the number of degrees of freedom and the number of phases present for a one-component system. The table suggests a rule relating the number of degrees of freedom, F, to the number of phases, P, present.

$$F = 3 - P, \qquad (12.25)$$

which is the *phase rule* for a one-component* system.

It would be helpful to have a simple rule by which we can decide how many independent variables are required for the description of the system. Particularly in the study of systems in which many components and many phases are present, any simplification of the problem is welcome.

We begin by finding the total conceivable number of intensive variables that would be needed to describe the state of the system containing C components and P phases. These are listed in Table 12.2. Each equation that connects these variables implies that one

Table 12.1

Number of phases present	1	2	3
Degrees of freedom	2	1	0

Table 12.2

Kind of variable	Total number of variables
Temperature and pressure	2
Composition variables (in each phase the mole fraction of each component must be specified; thus, C mole fractions are required to describe one phase; PC are needed to describe P phases)	PC
Total number of variables	$PC + 2$

---

*   The term "component" is defined in Section 12.9.

Table 12.3

Kind of equation	Total number of equations
In each phase there is a relation between the mole fractions: $$x_1 + x_2 + \cdots + x_C = 1.$$ For P phases, there are P equations	P
The equilibrium conditions: For each component there exists a set of equations $$\mu_i^\alpha = \mu_i^\beta = \mu_i^\gamma = \cdots = \mu_i^P.$$ There are $P - 1$ equations in the set. Since there are C components, there are $C(P - 1)$ equations.	$C(P - 1)$
Total number of equations	$P + C(P - 1)$

variable is dependent rather than independent. So we must find the total number of equations connecting the variables. These are listed in Table 12.3.

The number of independent variables, F, is obtained by subtracting the total number of equations from the total number of variables:

$$F = PC + 2 - P - C(P - 1),$$
$$F = C - P + 2. \tag{12.26}$$

Equation (12.26) is the phase rule of J. Willard Gibbs. The best way to remember the phase rule is by realizing that increasing the number of components increases the number of variables, therefore C enters with a positive sign. Increasing the number of phases increases the number of equilibrium conditions and the number of equations, thus eliminating some of the variables; therefore P enters with a negative sign.

In a one-component system, $C = 1$, so $F = 3 - P$. This result is, of course, the same as Eq. (12.25) obtained by inspection of Table 12.1. Equation (12.25) shows that the greatest number of phases that can coexist in equilibrium in a one-component system is three. In the sulfur system, for example, it is not possible for rhombic, monoclinic, liquid, and gaseous sulfur to coexist in equilibrium with one another. Such a quadruple equilibrium would imply three independent conditions on two variables, which is an impossiblity.

For a system of only one component it is possible to derive, as was done in Table 12.1, the consequences of the phase rule quite easily. The equilibria are readily represented by lines and their intersections in a two-dimensional diagram of the type we have used in this chapter. It hardly seems necessary to have the phase rule for such a situation. However, if the system has two components, then three variables are required and the phase diagram consists of surfaces and their intersections in three dimensions. If three components are present, surfaces in a fourdimensional space are required. Visualization of the entire situation is difficult in three dimensions, impossible for four or more dimensions. Yet the phase rule, with exquisite simplicity, expresses the limitations that are placed on the intersections of the surfaces in these multidimensional spaces. For this reason, the Gibbs phase rule is counted among the truly great generalizations of physical science.

12.9 THE PROBLEM OF COMPONENTS

The number of components in a system is defined as the least number of *chemically independent* species that is required to describe the composition of every phase in the system. At face value, the definition seems simple enough, and in ordinary practice it is simple. A number of examples will show up the joker in the deck, that little phrase, "chemically independent."

■ **EXAMPLE 12.1**   The system contains the *species* PCl_5, PCl_3, Cl_2. There are *three species* present but only *two components*, because the equilibrium

$$PCl_5 \rightleftharpoons PCl_3 + Cl_2$$

is established in this system. One can alter the number of moles of any two of these chemical individuals arbitrarily; the alteration in the number of moles of the third species is then fixed by the equilibrium condition, $K_x = x_{PCl_3} x_{Cl_2}/x_{PCl_5}$. Consequently, any two of these species are chemically independent; the third is not. There are only two components.

■ **EXAMPLE 12.2**   Liquid water presumably contains an enormous number of chemical species: H_2O, $(H_2O)_2$, $(H_2O)_3$, ..., $(H_2O)_n$. Yet there is only one component, because, as far as is known, all of the equilibria

$$H_2O + H_2O \rightleftharpoons (H_2O)_2,$$

$$H_2O + (H_2O)_2 \rightleftharpoons (H_2O)_3,$$

$$\vdots$$

$$H_2O + (H_2O)_{n-1} \rightleftharpoons (H_2O)_n$$

are established in the system; thus, if there are n species, there are $n - 1$ equilibria connecting them, and so only one species is chemically independent. There is only one component, and we may choose the simplest species, H_2O, as that component.

■ **EXAMPLE 12.3**   In the system water–ethyl alcohol, two species are present. No known equilibrium connects them at ordinary temperature; thus there are two components also.

■ **EXAMPLE 12.4**   In the system $CaCO_3$–CaO–CO_2, there are three species present; also, there are three distinct phases: solid $CaCO_3$, solid CaO, and gaseous CO_2. Because the equilibrium $CaCO_3 \rightleftharpoons CaO + CO_2$ is established, there are only two components. These are most simply chosen as CaO and CO_2; the composition of the phase $CaCO_3$ is then described as one mole of component CO_2 plus one mole of component CaO. If $CaCO_3$ and CO_2 were chosen as components, the composition of CaO would be described as one mole of $CaCO_3$ minus one mole of CO_2.

There is still another point to be made concerning the number of components. Our criterion is the establishment of a chemical equilibrium in a system; the existence of such an equilibrium reduces the number of components. There are instances where this test is not very clear-cut. Take the example of water, ethylene, and ethyl alcohol; at high temperatures several equilibria are established in this system; we consider only one, $C_2H_5OH \rightleftharpoons C_2H_4 + H_2O$. The question arises as to the temperature at which the system shifts from a three-component system, which it surely is at room temperature, to the two-component system that it is at high temperature. The answer lies in how long it takes us to make successive measurements on the system! If we measure a certain property of the system at a series of pressures, and if the time required to make the measurements is very short compared with the time required for the equilibrium to shift under the change in pressure, the system is effectively a three-component system; the equilibrium may as well not be there at all. On the other hand, if the equilibrium shifts very quickly under the change in pressure, in a very short time compared with the time we need to make the measurement, then the fact of the equilibrium matters very much, and the system is indeed a two-component system.

Liquid water is a good example of both types of behavior. The equilibria between the various polymers of water shift very rapidly, within 10^{-11} s at most. Ordinary measurements require much longer times, so the system is effectively a one-component system. In contrast to this behavior, the system H_2, O_2, H_2O, is a three-component system. The equilibrium that could reduce the number of components is $H_2 + \frac{1}{2}O_2 \rightleftharpoons H_2O$. In the absence of a catalyst, eons are required for this equilibrium to shift from one position to another. For practical purposes the equilibrium is not established.

It is clear that an accurate assignment of the number of components in a system presupposes some experimental knowledge of the system. This is an unavoidable pitfall in the use of the phase rule. Failure to realize that an unsuspected equilibrium has been established in a system sometimes leads an investigator to rediscover, the hard way, the second law of thermodynamics.

QUESTIONS

12.1 Illustrate by a μ versus T graph how the fact that ΔS_{fus} and ΔS_{sub} are always positive guarantees that the solid phase is the most stable at low temperature.

12.2 How do the liquid and gas phase lines at $T = T_b$ in Fig. 12.3(b) illustrate the LeChatelier principle, Eq. (11.71)?

12.3 In the winter, lakes that have frozen surfaces remain liquid at their bottoms (this allows survival of many species!). How do you explain this in terms of Fig. 12.8?

12.4 Removal of water from a mixture by "freeze drying" involves cooling below 0 °C, reduction of pressure below the triple point, and subsequent warming. How do you explain this in terms of Fig. 12.8?

12.5 How do the two phase diagrams for sulfur illustrate the "problem of components" for the phase rule?

PROBLEMS

12.1 Dry ice has a vapor pressure of 1 atm at -72.2 °C and 2 atm at -69.1 °C. Calculate the ΔH of sublimation for dry ice.

12.2 The vapor pressure of liquid bromine at 9.3 °C is 100 Torr. If the heat of vaporization is 30 910 J/mol, calculate the boiling point of bromine.

12.3 The vapor pressure of diethyl ether is 100 Torr at -11.5 °C and 400 Torr at 17.9 °C. Calculate

    a) the heat of vaporization;
    b) the normal boiling point and the boiling point in Denver where the barometric pressure is 620 Torr;
    c) the entropy of vaporization at the boiling point;
    d) $\Delta G°$ of vaporization at 25 °C.

12.4 The heat of vaporization of water is 40 670 J/mol at the normal boiling point, 100 °C. The barometric pressure in Denver is about 620 Torr.

    a) What is the boiling point of water in Denver?
    b) What is the boiling point under 3 atm pressure?

12.5 At 25 °C, $\Delta G_f°(H_2O, g) = -228.589$ kJ/mol and $\Delta G_f°(H_2O, l) = -237.178$ kJ/mol. What is the vapor pressure of water at 298.15 K?

12.6 The vapor pressures of liquid sodium are

$t/°C$	439	549	701
$p/$Torr	1	10	100

By plotting these data appropriately, determine the boiling point, the heat of vaporization, and the entropy of vaporization at the boiling point for liquid sodium.

12.7 Naphthalene, $C_{10}H_8$, melts at 80.0 °C. If the vapor pressure of the liquid is 10 Torr at 85.8 °C and 40 Torr at 119.3 °C, and that of the solid is 1 Torr at 52.6 °C, calculate

a) the ΔH_{vap} of the liquid, the boiling point, and ΔS_{vap} at T_b;
b) the vapor pressure at the melting point.
c) Assuming that the melting-point and triple-point temperatures are the same, calculate ΔH_{sub} of the solid and ΔH_{fus}.
d) What must the temperature be if the vapor pressure of the solid is to be less than 10^{-5} Torr?

12.8 Iodine boils at 183.0 °C; the vapor pressure of the liquid at 116.5 °C is 100 Torr. If $\Delta H°_{fus} = 15.65$ kJ/mol and the vapor pressure of the solid is 1 Torr at 38.7 °C, calculate

a) the triple point temperature and pressure;
b) $\Delta H°_{vap}$, and $\Delta S°_{vap}$;
c) $\Delta G°_f$ (I_2, g) at 298.15 K.

12.9 For ammonia we have

$t/°C$	4.7	25.7	50.1	78.9
$p/$atm	5	10	20	40

Plot or do a least squares fit of the data to $\ln p$ versus $1/T$, to obtain ΔH_{vap}, and the normal boiling point.

12.10 a) By combining the barometric distribution with the Clausius–Clapeyron equation, derive an equation relating the boiling point of a liquid to the temperature of the atmosphere, T_a, and the altitude, h. In (b) and (c) assume $t_a = 20$ °C.

b) For water, $t_b = 100$ °C at 1 atm, and $\Delta H_{vap} = 40.670$ kJ/mol. What is the boiling point on top of Mt. Evans, $h = 14\,260$ ft?

c) For diethyl ether, $t_b = 34.6$ °C at 1 atm, and $\Delta H_{vap} = 29.86$ kJ/mol. What is the boiling point on top of Mt. Evans?

12.11 a) From the boiling point T_b of a liquid and the assumption that the liquid follows Trouton's rule, calculate the value of the vapor pressure at any temperature T.

b) The boiling point of diethyl ether is 34.6 °C. Calculate the vapor pressure at 25 °C.

12.12 For sulfur, $\Delta S°_{vap} = 14.6$ J/K per mole S, and for phosphorus, $\Delta S°_{vap} = 22.5$ J/K per mole P. The molecular formulas of these substances are S_8 and P_4. Show that if the correct molecular formulas are used, the entropies of vaporization have more normal values.

12.13 Derive Eq. (12.4).

12.14 If the vapor is an ideal gas, there is a simple relation between the vapor pressure p and the concentration $\tilde{c}$ (mol/m³) in the vapor. Consider a liquid in equilibrium with its vapor. Derive the expression for the temperature dependence of $\tilde{c}$ in such a system.

12.15 Assuming that the vapor is ideal and that ΔH_{vap} is independent of temperature, calculate

a) The molar concentration of the vapor at the boiling point T_b of the liquid.
b) The Hildebrand temperature, T_H, is that temperature at which the vapor concentration is $(1/22.414)$ mol/L. Using the result in Problem 12.14, find the expression for T_H in terms of ΔH_{vap} and T_b.
c) The Hildebrand entropy, $\Delta S_H = \Delta H_{vap}/T_H$, is very nearly constant for many normal liquids. If $\Delta S_H = 92.5$ J/K mol, use the result in (b) to compute values of T_b for various values of T_H. Plot T_H as a function of T_b. (Choose values of $T_H = 50, 100, 200, 300, 400$ K to compute T_b.)

d) For the following liquids compute ΔS_H and the Trouton entropy, $\Delta S_T = \Delta H_{vap}/T_b$. Note that ΔS_H is more constant than ΔS_T (Hildebrand's rule).

Liquid	$\Delta H_{vap}/(\text{kJ/mol})$	T_b/K
Argon	6.519	87.29
Oxygen	6.820	90.19
Methane	8.180	111.67
Krypton	9.029	119.93
Xenon	12.640	165.1
Carbon disulfide	26.78	319.41

12.16 The density of diamond is 3.52 g/cm^3 and that of graphite is 2.25 g/cm^3. At 25 °C the Gibbs energy of formation of diamond from graphite is 2.900 kJ/mol. At 25 °C what pressure must be applied to bring diamond and graphite into equilibrium?

12.17 At 1 atm pressure, ice melts at 273.15 K. $\Delta H_{fus} = 6.009 \text{ kJ/mol}$, density of ice $= 0.92 \text{ g/cm}^3$, density of liquid $= 1.00 \text{ g/cm}^3$.

a) What is the melting point of ice under 50 atm pressure?

b) The blade of an ice skate is ground to a knife edge on each side of the skate. If the width of the knife edge is 0.001 in, and the length of the skate in contact with the ice is 3 in, calculate the pressure exerted on the ice by a 150 lb man.

c) What is the melting point of ice under this pressure?

12.18 At 25 °C we have for rhombic sulfur: $\Delta G_f^\circ = 0$, $S^\circ = 31.88 \pm 0.17 \text{ J/K mol}$; and for monoclinic sulfur: $\Delta G_f^\circ = 63 \text{ J/mol}$, $S^\circ = 32.55 \pm 0.25 \text{ J/K mol}$. Assuming that the entropies do not vary with temperature, sketch the value of μ versus T for the two forms of sulfur. From the data determine the equilibrium temperature for the transformation of rhombic sulfur to monoclinic sulfur. Compare this temperature with the experimental value, 95.4 °C, noting the uncertainties in the values of S°.

12.19 The transition

$$\text{Sn(s, gray)} \rightleftharpoons \text{Sn(s, white)}$$

is in equilibrium at 18 °C and 1 atm pressure. If $\Delta S = 8.8 \text{ J/K mol}$ for the transition at 18 °C and if the densities are 5.75 g/cm^3 for gray tin and 7.28 g/cm^3 for white tin, calculate the transition temperature under 100 atm pressure.

12.20 For the transition, rhombic sulfur $\rightarrow$ monoclinic sulfur, the value of ΔS is positive. The transition temperature increases with increase in pressure. Which is denser, the rhombic or the monoclinic form? Prove your answer mathematically.

12.21 Liquid water under an air pressure of 1 atm at 25 °C has a larger vapor pressure than it would in the absence of air pressure. Calculate the increase in vapor pressure produced by the pressure of the atmosphere on the water. The density of water $= 1 \text{ g/cm}^3$; the vapor pressure (in the absence of the air pressure) $= 3167.2 \text{ Pa}$.

13

Solutions

I. The Ideal Solution and Colligative Properties

13.1 KINDS OF SOLUTIONS

A solution is a homogeneous mixture of chemical species dispersed on a molecular scale. By this definition, a solution is a single phase. A solution may be gaseous, liquid, or solid. *Binary* solutions are composed of two constituents, *ternary* solutions three, *quaternary* four. The constituent present in the greatest amount is ordinarily called the *solvent*, while those constituents—one or more—present in relatively small amounts are called the *solutes*. The distinction between solvent and solute is an arbitrary one. If it is convenient, the constituent present in relatively small amount may be chosen as the solvent. We shall employ the words *solvent* and *solute* in the ordinary way, realizing that nothing fundamental distinguishes them. Examples of kinds of solution are listed in Table 13.1.

Gas mixtures have been discussed in some detail in Chapter 11. The discussion in this chapter and in Chapter 14 is devoted to liquid solutions. Solid solutions are dealt with as they occur in connection with other topics.

Table 13.1

Gaseous solutions	Mixtures of gases or vapors
Liquid solutions	Solids, liquids, or gases, dissolved in liquids
Solid solutions	
Gases dissolved in solids	H_2 in palladium, N_2 in titanium
Liquids dissolved in solids	Mercury in gold
Solids dissolved in solids	Copper in gold, zinc in copper (brasses), alloys of many kinds

13.2 DEFINITION OF THE IDEAL SOLUTION

The ideal gas law is an important example of a *limiting* law. As the pressure approaches zero, the behavior of any real gas approaches that of the ideal gas as a limit. Thus all real gases behave ideally at zero pressure, and for practical purposes they are ideal at low finite pressures. From this generalization of experimental behavior, the ideal gas is defined as one that behaves ideally at any pressure.

We arrive at a similar limiting law from observing the behavior of solutions. For simplicity, we consider a solution composed of a volatile solvent and one or more involatile solutes, and examine the equilibrium between the solution and the vapor. If a pure liquid is placed in a container that is initially evacuated, the liquid evaporates until the space above the liquid is filled with vapor. The temperature of the system is kept constant. At equilibrium, the pressure established in the vapor is $p°$, the vapor pressure of the pure liquid (Fig. 13.1a). If an involatile material is dissolved in the liquid, the equilibrium vapor pressure p over the solution is observed to be less than over the pure liquid (Fig. 13.1b).

Since the solute is involatile, the vapor consists of pure solvent. As more involatile material is added, the pressure in the vapor phase decreases. A schematic plot of the vapor pressure of the solvent against the mole fraction of the involatile solute in the solution, x_2, is shown by the solid line in Fig. 13.2. At $x_2 = 0$, $p = p°$; as x_2 increases, p decreases. The important feature of Fig. 13.2 is that the vapor pressure of the dilute solution (x_2 near zero) approaches the dashed line connecting $p°$ and zero. Depending on the particular combination of solvent and solute, the experimental vapor-pressure curve at higher concentrations of solute may fall below the dashed line, as in Fig. 13.2, or above it, or even lie exactly on it. However, for all solutions the experimental curve is tangent to the dashed line at $x_2 = 0$, and approaches the dashed line very closely as the solution becomes more and more dilute. The equation of the ideal line (the dashed line) is

$$p = p° - p°x_2 = p°(1 - x_2).$$

If x is the mole fraction of solvent in the solution, then $x + x_2 = 1$, and the equation

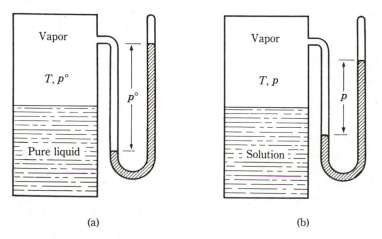

(a)   (b)

Figure 13.1   Vapor pressure lowering by an involatile solute.

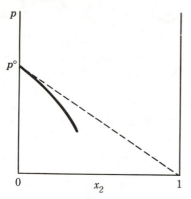

Figure 13.2   Vapor pressure as a function of x_2.

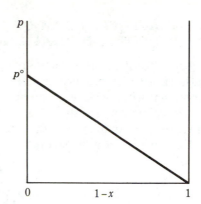

Figure 13.3   Raoult's law for the solvent.

becomes

$$p = xp°, \tag{13.1}$$

which is *Raoult's law*. It states that the vapor pressure of the solvent over a solution is equal to the vapor pressure of the pure solvent multiplied by the mole fraction of the solvent in the solution.

Raoult's law is another example of a limiting law. Real solutions follow Raoult's law more closely as the solution becomes more dilute. The *ideal solution* is defined as one that follows Raoult's law over the entire range of concentrations. The vapor pressure of the solvent over an ideal solution of an involatile solute is shown in Fig. 13.3. All real solutions behave ideally as the concentration of the solutes approaches zero.

From Eq. (13.1) the *vapor pressure lowering, $p° - p$,* can be calculated:

$$p° - p = p° - xp° = (1 - x)p°,$$
$$p° - p = x_2 p°. \tag{13.2}$$

The vapor pressure lowering is proportional to the mole fraction of the *solute*. If several solutes, 2, 3, ..., are present, then it is still true that $p = xp°$; but, in this case, $1 - x = x_2 + x_3 + \cdots$ and

$$p° - p = (x_2 + x_3 + \cdots)p°. \tag{13.3}$$

In a solution containing several involatile solutes, the vapor pressure lowering depends on the sum of the mole fractions of the various solutes. Note particularly that it does not depend on the kinds of solutes present, except that they be involatile. The vapor pressure depends only on the relative numbers of solute molecules.

In a gas mixture, the ratio of the partial pressure of the water vapor to the vapor pressure of pure water at the same temperature is called the *relative humidity*. When multiplied by 100, it is the *percent relative humidity*. Thus

$$\text{R.H.} = \frac{p}{p°} \quad \text{and} \quad \%\text{R.H.} = \frac{p}{p°}(100).$$

Over an aqueous solution that obeys Raoult's law, the relative humidity is equal to the mole fraction of water in the solution.

13.3   ANALYTICAL FORM OF THE CHEMICAL POTENTIAL IN IDEAL LIQUID SOLUTIONS

As a generalization of the behavior of real solutions the ideal solution follows Raoult's law over the entire range of concentration. Taking this definition of an ideal liquid solution and combining it with the general equilibrium condition leads to the analytical expression of the chemical potential of the solvent in an ideal solution. If the solution is in equilibrium with vapor, the requirement of the second law is that the chemical potential of the solvent have the same value in the solution as in the vapor, or

$$\mu_{\mathrm{liq}} = \mu_{\mathrm{vap}}, \tag{13.4}$$

where μ_{liq} is the chemical potential of the solvent in the liquid phase, μ_{vap} the chemical potential of the solvent in the vapor. Since the vapor is pure solvent under a pressure p, the expression for μ_{vap} is given by Eq. (10.47); assuming that the vapor is an ideal gas $\mu_{\mathrm{vap}} = \mu_{\mathrm{vap}}^{\circ} + RT \ln p$. Then Eq. (13.4) becomes

$$\mu_{\mathrm{liq}} = \mu_{\mathrm{vap}}^{\circ} + RT \ln p.$$

Using Raoult's law, $p = xp^{\circ}$, in this equation and expanding the logarithm, we obtain

$$\mu_{\mathrm{liq}} = \mu_{\mathrm{vap}}^{\circ} + RT \ln p^{\circ} + RT \ln x.$$

If pure solvent were in equilibrium with vapor, the pressure would be p°; the equilibrium condition is

$$\mu_{\mathrm{liq}}^{\circ} = \mu_{\mathrm{vap}}^{\circ} + RT \ln p^{\circ},$$

where $\mu_{\mathrm{liq}}^{\circ}$ signifies the chemical potential of the pure liquid solvent. Subtracting this equation from the preceding one, we obtain

$$\mu_{\mathrm{liq}} - \mu_{\mathrm{liq}}^{\circ} = RT \ln x.$$

In this equation, nothing pertaining to the vapor phase appears; omitting the subscript liq, the equation becomes

$$\mu = \mu^{\circ} + RT \ln x. \tag{13.5}$$

The significance of the symbols in Eq. (13.5) must be clearly understood: μ is the chemical potential of the solvent in the solution, μ° is the chemical potential of the pure liquid solvent, a function of T and p, and x is the mole fraction of solvent in the solution. This equation is the result we suggested in Section 11.5, as a generalization from the form obtained for the μ of an ideal gas in a mixture.

13.4   CHEMICAL POTENTIAL OF THE SOLUTE IN A BINARY IDEAL SOLUTION; APPLICATION OF THE GIBBS–DUHEM EQUATION

The Gibbs–Duhem equation can be used to calculate the chemical potential of the solute from that of the solvent in a binary ideal system. The Gibbs–Duhem equation, Eq. (11.96), for a binary system (T, p constant) is

$$n \, d\mu + n_2 \, d\mu_2 = 0. \tag{13.6}$$

The symbols without subscripts in Eq. (13.6) refer to the solvent; those with the subscript 2 refer to the solute. From Eq. (13.6), $d\mu_2 = -(n/n_2)\,d\mu$; or, since $n/n_2 = x/x_2$, we have

$$d\mu_2 = -\frac{x}{x_2}\,d\mu.$$

Differentiating Eq. (13.5) keeping T and p constant, we obtain for the solvent $d\mu = (RT/x)\,dx$, so that $d\mu_2$ becomes

$$d\mu_2 = -RT\,\frac{dx}{x_2}.$$

However, $x + x_2 = 1$, so that $dx + dx_2 = 0$, or $dx = -dx_2$. Then $d\mu_2$ becomes

$$d\mu_2 = RT\,\frac{dx_2}{x_2}.$$

Integrating, we have

$$\mu_2 = RT\ln x_2 + C, \tag{13.7}$$

where C is the constant of integration; since T and p are kept constant throughout this manipulation, C can be a function of T and p and still be a constant for this integration. If the value of x_2 in the liquid is increased until it is unity, the liquid becomes pure *liquid* solute, and μ_2 must be μ_2°, the chemical potential of pure *liquid* solute. So if $x_2 = 1, \mu_2 = \mu_2^\circ$. Using these values in Eq. (13.7), we find $\mu_2^\circ = C$, and Eq. (13.7) becomes

$$\mu_2 = \mu_2^\circ + RT\ln x_2. \tag{13.8}$$

Equation (13.8) relates the chemical potential of the solute to the mole fraction of the solute in the solute. This expression is analogous to Eq. (13.5), and the symbols have corresponding significances. Since the μ for the solute has the same form as the μ for the solvent, the solute behaves ideally. This implies that in the vapor over the solution the partial pressure of the solute is given by Raoult's law:

$$p_2 = x_2 p_2^\circ. \tag{13.9}$$

If the solute is involatile, p_2° is immeasurably small and Eq. (13.9) cannot be proved experimentally; thus it has academic interest only.

13.5 COLLIGATIVE PROPERTIES

Since the second term in Eq. (13.5) is negative, the chemical potential of the solvent in solution is less than the chemical potential of the pure solvent by an amount $-RT\ln x$. Several related properties of the solution have their origin in this lower value of the chemical potential. These properties are: (1) the vapor pressure lowering, discussed in Section 13.2; (2) the freezing-point depression; (3) the boiling-point elevation; and (4) the osmotic pressure. Since these properties are all bound together through their common origin, they are called *colligative properties* (colligative: from Latin: *co-*, together, *ligare*, to bind). All of these properties have a common characteristic: They do not depend on the nature of the solute present but only on the number of solute molecules relative to the total number of molecules present.

The μ versus T diagram displays the freezing-point depression and the boiling-point elevation clearly. In Fig. 13.4(a) the solid lines refer to the pure solvent. Since the solute is

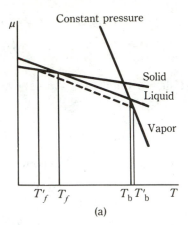

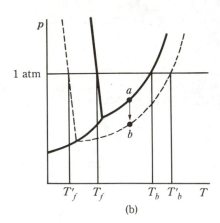

Figure 13.4 Colligative properties.

involatile, it does not appear in the gas phase, so the curve for the gas is the same as for the pure gas. If we assume that the solid contains only the solvent, then the curve for the solid is unchanged. However, because the liquid contains a solute, the μ of the solvent is lowered at each temperature by an amount $-RT \ln x$. The dashed curve in Fig. 13.4(a) is the curve for the solvent in an ideal solution. The diagram shows directly that the intersection points with the curves for the solid and the gas have shifted. The new intersection points are the freezing point, T'_f, and the boiling point, T'_b, of the solution. It is apparent that the boiling point of the solution is higher than that of the pure solvent (boiling-point elevation), while the freezing point of the solution is lower (freezing-point depression). From the figure it is obvious that the change in the freezing point is greater than the change in the boiling point for a solution of the same concentration.

The freezing-point depression and boiling-point elevation can be illustrated on the ordinary phase diagram of the solvent, shown for water by the solid curves in Fig. 13.4(b). If an involatile material is added to the liquid solvent, then the vapor pressure is lowered at every temperature as, for example, from point a to point b. The vapor-pressure curve for the solution is shown by the dotted line. The dashed line shows the new freezing point as a function of pressure. At 1 atm pressure, the freezing points and boiling points are given by the intersections of the solid and dashed lines with the horizontal line at 1 atm pressure. This diagram also shows that a given concentration of solute produces a greater effect on the freezing point than on the boiling point.

The freezing point and boiling point of a solution depend on the equilibrium of the solvent in the solution with pure solid solvent or pure solvent vapor. The remaining possible equilibrium is that between solvent in solution and pure liquid solvent. This equilibrium can be established by increasing the pressure on the solution sufficiently to raise the μ of the solvent in solution to the value of the μ of the pure solvent. The additional pressure on the solution that is required to establish the equality of the μ of the solvent both in the solution and in the pure solvent is called the *osmotic pressure* of the solution.

13.6 THE FREEZING-POINT DEPRESSION

Consider a solution that is in equilibrium with pure solid solvent. The equilibrium condition requires that

$$\mu(T, p, x) = \mu_{\text{solid}}(T, p), \tag{13.10}$$

where $\mu(T, p, x)$ is the chemical potential of the solvent in the solution, $\mu_{\text{solid}}(T, p)$ is the chemical potential of the pure solid. Since the solid is *pure*, μ_{solid} does not depend on any composition variable. In Eq. (13.10), T is the equilibrium temperature, the freezing point of the solution; from the form of Eq. (13.10), T is some function of pressure and x, the mole fraction of solvent in the solution. If the pressure is constant, then T is a function only of x.

If the solution is ideal, then $\mu(T, p, x)$ in the solution is given by Eq. (13.5), so that Eq. (13.10) becomes

$$\mu^\circ(T, p) + RT \ln x = \mu_{\text{solid}}(T, p).$$

Rearrangement yields

$$\ln x = -\frac{\mu^\circ(T, p) - \mu_{\text{solid}}(T, p)}{RT}. \tag{13.11}$$

Since μ° is the chemical potential of the pure liquid, $\mu^\circ(T, p) - \mu_{\text{solid}}(T, p) = \Delta G_{\text{fus}}$, where ΔG_{fus} is the molar Gibbs energy of fusion of the pure solvent at the temperature T. Equation (13.11) becomes

$$\ln x = -\frac{\Delta G_{\text{fus}}}{RT}. \tag{13.12}$$

To discover how T depends on x, we evaluate $(\partial T/\partial x)_p$. Differentiating Eq. (13.12) with respect to x, p being constant, we obtain

$$\frac{1}{x} = -\frac{1}{R}\left[\frac{\partial(\Delta G_{\text{fus}}/T)}{\partial T}\right]_p\left(\frac{\partial T}{\partial x}\right)_p.$$

Using the Gibbs–Helmholtz equation, Eq. (10.54), $[\partial(\Delta G/T)/\partial T]_p = -\Delta H/T^2$, we obtain

$$\frac{1}{x} = \frac{\Delta H_{\text{fus}}}{RT^2}\left(\frac{\partial T}{\partial x}\right)_p. \tag{13.13}$$

In Eq. (13.13), ΔH_{fus} is the heat of fusion of the *pure* solvent at the temperature T. The procedure is now reversed and we write Eq. (13.13) in differential form and integrate:

$$\int_1^x \frac{dx}{x} = \int_{T_0}^T \frac{\Delta H_{\text{fus}}}{RT^2}\, dT. \tag{13.14}$$

The lower limit $x = 1$ corresponds to pure solvent having a freezing point T_0. The upper limit x corresponds to a solution that has a freezing point T. The first integral can be evaluated immediately; the second integration is possible if ΔH_{fus} is known as a function of temperature. For simplicity we assume that ΔH_{fus} is a constant in the temperature range from T_0 to T; then Eq. (13.14) becomes

$$\ln x = -\frac{\Delta H_{\text{fus}}}{R}\left(\frac{1}{T} - \frac{1}{T_0}\right). \tag{13.15}$$

This equation can be solved for the freezing point T, or rather more conveniently for $1/T$,

$$\frac{1}{T} = \frac{1}{T_0} - \frac{R \ln x}{\Delta H_{\text{fus}}}, \tag{13.16}$$

which relates the freezing point of an ideal solution to the freezing point of the pure solvent, T_0, the heat of fusion of the solvent, and the mole fraction of the solvent in the solution, x.

The relation between freezing point and composition of a solution can be simplified considerably if the solution is dilute. We begin by expressing the freezing-point depression $-dT$ in terms of the total molality of the solutes present, m, where $m = m_2 + m_3 + \cdots$. Let n and M be the number of moles and molar mass of the solvent, respectively; then the mass of solvent is nM. Then $m_2 = n_2/nM$; $m_3 = n_3/nM$; ...; or $n_2 = nMm_2$; $n_3 = nMm_3$; ... The mole fraction of the solvent is given by

$$x = \frac{n}{n + n_2 + n_3 + \cdots} = \frac{n}{n + nM(m_2 + m_3 + \cdots)}$$

$$x = \frac{1}{1 + Mm} \tag{13.17}$$

Taking logarithms and differentiating, we obtain $\ln x = -\ln(1 + Mm)$, and

$$d \ln x = -\frac{M\,dm}{1 + Mm}. \tag{13.18}$$

Equation (13.13) can be written

$$dT = \frac{RT^2}{\Delta H_{\text{fus}}}\,d \ln x.$$

Replacing $d \ln x$ by the value in Eq. (13.18), we obtain

$$dT = -\frac{MRT^2}{\Delta H_{\text{fus}}} \frac{dm}{(1 + Mm)}. \tag{13.19}$$

If the solution is very dilute in all solutes, then m approaches zero and T approaches T_0, and Eq. (13.19) becomes

$$-\left(\frac{\partial T}{\partial m}\right)_{p,\, m=0} = \frac{MRT_0^2}{\Delta H_{\text{fus}}} = K_f. \tag{13.20}$$

The subscript, $m = 0$, designates the limiting value of the derivative, and K_f is the freezing-point depression constant. The freezing-point depression $\theta_f = T_0 - T$, $d\theta_f = -dT$, so for dilute solutions we have

$$\left(\frac{\partial \theta_f}{\partial m}\right)_{p,\, m=0} = K_f, \tag{13.21}$$

which integrates immediately, if m is small, to

$$\theta_f = K_f m. \tag{13.22}$$

The constant K_f depends only on the properties of the pure solvent. For water, $M = 0.0180152$ kg/mol, $T_0 = 273.15$ K, and $\Delta H_{\text{fus}} = 6009.5$ J/mol. Thus

$$K_f = \frac{(0.0180152 \text{ kg/mol})(8.31441 \text{ J/K mol})(273.15 \text{ K})^2}{6009.5 \text{ J/mol}} = 1.8597 \text{ K kg/mol}.$$

Equation (13.22) provides a simple relation between the freezing-point depression and the molal concentration of solute in a dilute ideal solution, which is often used to determine the molar mass of a dissolved solute. If w_2 kg of a solute of unknown molar mass, M_2, are dissolved in w kg of solvent, then the molality of solute is $m = w_2/wM_2$. Using this value

Table 13.2
Freezing-point depression constants

Compound	$M/(\text{kg/mol})$	$t_m/^\circ\text{C}$	$K_f/(\text{K kg/mol})$
Water	0.0180	0	1.86
Acetic acid	0.0600	16.6	3.57
Benzene	0.0781	5.45	5.07
Dioxane	0.0881	11.7	4.71
Naphthalene	0.1283	80.1	6.98
p-dichlorobenzene	0.1470	52.7	7.11
Camphor	0.1522	178.4	37.7
p-dibromobenzene	0.2359	86	12.5

for m in Eq. (13.22) and solving for M_2 yields

$$M_2 = \frac{K_f w_2}{\theta_f w}.$$

The measured values of θ_f, w_2, and w, together with a knowledge of K_f of the solvent, suffice to determine M_2. It is clear that for a given value of m, the larger the value of K_f, the greater will be θ_f. This increases the ease and accuracy of the measurement of θ_f; consequently, it is desirable to choose a solvent having a large value of K_f. By examining Eq. (13.20) we can decide what sorts of compounds will have large values of K_f. First of all, we replace ΔH_{fus} by $T_0 \Delta S_{\text{fus}}$; this reduces Eq. (13.20) to

$$K_f = \frac{RMT_0}{\Delta S_{\text{fus}}}, \tag{13.23}$$

which shows that K_f increases as the product MT_0 increases. Since T_0 increases as M increases, K_f increases rapidly as the molar mass of the substance increases. The increase is not very uniform, simply because ΔS_{fus} may vary a good deal, particularly when M is very large. Table 13.2 illustrates the behavior of K_f with increasing M. Because of variations in the value of ΔS_{fus}, marked exceptions occur; the general trend is apparent, however.

★13.7 SOLUBILITY

The equilibrium between solid solvent and solution was considered in Section 13.6. The same equilibrium may be considered from a different point of view. The word "solvent" as we have seen is ambiguous. Suppose we consider the equilibrium between solute in solution and pure solid solute. In this condition the solution is *saturated* with respect to the solute. The equilibrium condition is that the μ of the solute must be the same everywhere, that is

$$\mu_2(T, p, x_2) = \mu_{2(\text{solid})}(T, p), \tag{13.24}$$

where x_2 is the mole fraction of solute in the saturated solution, and therefore is the *solubility* of the solute expressed as a mole fraction. If the solution is ideal, then

$$\mu_2^\circ(T, p) + RT \ln x_2 = \mu_{2(\text{solid})}(T, p),$$

where $\mu_2^\circ(T, p)$ is the chemical potential of the pure *liquid* solute. The argument then

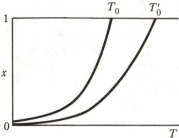

Figure 13.5   Ideal solubility versus T.

proceeds in exactly the same way as for the freezing-point depression; the symbols refer to the solute, however. The equation corresponding to Eq. (13.15) is

$$\ln x_2 = -\frac{\Delta H_{fus}}{R}\left(\frac{1}{T} - \frac{1}{T_0}\right);\tag{13.25}$$

ΔH_{fus} is the heat of fusion of pure solute, T_0 the freezing point of pure solute. Using $\Delta H_{fus} = T_0\,\Delta S_{fus}$ in Eq. (13.25), we obtain

$$\ln x_2 = \frac{\Delta S_{fus}}{R}\left(1 - \frac{T_0}{T}\right).\tag{13.26}$$

Either Eq. (13.25) or Eq. (13.26) is an expression of the *ideal law of solubility*. According to this law, the solubility of a substance is the same in all solvents with which it forms an ideal solution. The solubility of a substance in an ideal solution depends on the properties of that substance only. Low melting point T_0 and low heat of fusion both favor enhanced solubility. Figure 13.5 shows the variation of the solubility, x, as a function of temperature for two substances with the same entropy of fusion but different melting points.

The use of Eq. (13.25) can be illustrated by the solubility of naphthalene. The melting point is 80.0 °C; the heat of fusion is 19 080 J/mol. Using these data we find from Eq. (13.25) that the ideal solubility $x = 0.264$ at 20 °C. The measured solubilities in various solvents are given in Table 13.3.

The ideal law of solubility is frequently in error if the temperature of interest is far below the melting point of the solid, since the assumption that ΔH_{fus} is independent of temperature is not a very good one in this circumstance. The law is never accurate for

Table 13.3

Solvent	Solubility x_2	Solvent	Solubility x_2
Chlorobenzene	0.256	Aniline	0.130
Benzene	0.241	Nitrobenzene	0.243
Toluene	0.224	Acetone	0.183
CCl_4	0.205	Methyl alcohol	0.0180
Hexane	0.090	Acetic acid	0.0456

By permission from J. H. Hildebrand and R. L. Scott, *The Solubility of Nonelectrolytes*, 3d ed. New York: Reinhold, 1950, p. 283.

solutions of ionic materials in water, since the saturated solutions of these materials are far from being ideal and are far below their melting points. As the table of solubilities of naphthalene shows, hydrogen-bonded solvents are poor solvents for a substance that cannot form hydrogen bonds.

13.8 ELEVATION OF THE BOILING POINT

Consider a solution that is in equilibrium with the vapor of the pure solvent. The equilibrium condition is that

$$\mu(T, p, x) = \mu_{vap}(T, p). \tag{13.27}$$

If the solution is ideal,

$$\mu^\circ(T, p) + RT \ln x = \mu_{vap}(T, p),$$

and

$$\ln x = \frac{\mu_{vap} - \mu^\circ(T, p)}{RT}.$$

The molar Gibbs energy of vaporization is

$$\Delta G_{vap} = \mu_{vap}(T, p) - \mu^\circ(T, p),$$

so that

$$\ln x = \frac{\Delta G_{vap}}{RT}. \tag{13.28}$$

Note that Eq. (13.28) has the same functional form as Eq. (13.12) except that the sign is changed on the right-hand side. The algebra which follows is identical to that used for the derivation of the formulas for the freezing-point depression except that the sign is reversed in each term that contains either ΔG or ΔH. This difference in sign simply means that while the freezing point is depressed, the boiling point is elevated.

We can write the final equations directly. The analogues of Eqs. (13.15) and (13.16) are

$$\ln x = \frac{\Delta H_{vap}}{R}\left(\frac{1}{T} - \frac{1}{T_0}\right), \quad \text{or} \quad \frac{1}{T} = \frac{1}{T_0} + \frac{R \ln x}{\Delta H_{vap}}. \tag{13.29}$$

The boiling point T of the solution is expressed in terms of the heat of vaporization and the boiling point of the pure solvent, ΔH_{vap} and T_0, and the mole fraction x of solvent in the solution. If the solution is dilute in all solutes, then m approaches zero and T approaches T_0. The boiling-point elevation constant is defined by

$$K_b = \left(\frac{\partial T}{\partial m}\right)_{p, m = 0} = \frac{MRT_0^2}{\Delta H_{vap}}. \tag{13.30}$$

The boiling-point elevation, $\theta_b = T - T_0$, so that $d\theta_b = dT$. So long as m is small, Eq. (13.30) integrates to

$$\theta_b = K_b m. \tag{13.31}$$

For water, $M = 0.0180152$ kg/mol, $T_0 = 373.15$ K, and $\Delta H_{vap} = 40\,656$ J/mol, then $K_b = 0.51299$ K kg/mol. The relation, Eq. (13.31), between boiling-point elevation and the

Table 13.4
Boiling-point elevation constants

Compound	$M/(kg/mol)$	$t_b/°C$	$K_b/(K\ kg/mol)$
Water	0.0180	100	0.51
Methyl alcohol	0.0320	64.7	0.86
Ethyl alcohol	0.0461	78.5	1.23
Acetone	0.0581	56.1	1.71
Acetic acid	0.0600	118.3	3.07
Benzene	0.0781	80.2	2.53
Cyclohexane	0.0842	81.4	2.79
Ethyl bromide	0.1090	38.3	2.93

molality of a dilute ideal solution corresponds to that between freezing-point depression and molality; for any liquid, the constant K_b is smaller than K_f.

The elevation of the boiling point is used to determine the molecular weight of a solute in the same way as is the freezing-point depression. It is desirable to use a solvent that has a large value of K_b. In Eq. (13.30) if ΔH_{vap} is replaced by $T_0\ \Delta S_{vap}$ then

$$K_b = \frac{RMT_0}{\Delta S_{vap}}.$$

But many liquids follow Trouton's rule: $\Delta S_{vap} \approx 90$ J/K mol. Since $R = 8.3$ J/K mol, then, approximately, $K_b \approx 10^{-1}\ MT_0$. The higher the molar mass of the solvent, the larger the value of K_b. The data in Table 13.4 illustrate the relationship.

Since the boiling point T_0 is a function of pressure, K_b is a function of pressure. The effect is rather small but must be taken into account in precise measurements. The Clausius–Clapeyron equation yields the connection between T_0 and p, which is needed to calculate the magnitude of the effect.

13.9 OSMOTIC PRESSURE

The phenomenon of osmotic pressure is illustrated by the apparatus shown in Fig. 13.6. A collodion bag is tied to a rubber stopper through which a piece of glass capillary tubing is inserted. The bag is filled with a dilute solution of sugar in water and immersed in a beaker

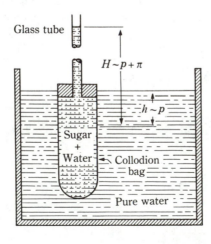

Figure 13.6 Simple osmotic pressure experiment.

of pure water. The level of the sugar solution in the tube is observed to rise until it reaches a definite height, which depends on the concentration of the solution. The hydrostatic pressure resulting from the difference in levels of the sugar solution in the tube and the surface of the pure water is the *osmotic pressure* of the solution. Observation shows that no sugar has escaped through the membrane into the pure water in the beaker. The increase in volume of the solution that caused it to rise in the tube is a result of the passage of water through the membrane into the bag. The collodion functions as a *semipermeable* membrane, which allows water to pass freely through it but does not allow sugar to pass. When the system reaches equilibrium, the sugar solution at any depth below the level of the pure water is under an excess hydrostatic pressure due to the extra height of the sugar solution in the tubing. The problem is to derive the relation between this pressure difference and the concentration of the solution.

13.9.1 The van't Hoff Equation

The equilibrium requirement is that the chemical potential of the water must have the same value on each side of the membrane at every depth in the beaker. This equality of the chemical potential is achieved by a pressure difference on the two sides of the membrane. Consider the situation at the depth h in Fig. 13.6. At this depth the solvent is under a pressure p, while the solution is under a pressure $p + \pi$. If $\mu(T, p + \pi, x)$ is the chemical potential of the solvent in the solution under the pressure $p + \pi$, and $\mu^\circ(T, p)$ that of the pure solvent under the pressure p, then the equilibrium condition is

$$\mu(T, p + \pi, x) = \mu^\circ(T, p), \tag{13.32}$$

and

$$\mu^\circ(T, p + \pi) + RT \ln x = \mu^\circ(T, p). \tag{13.33}$$

The problem is to express the μ of the solvent under a pressure $p + \pi$ in terms of the μ solvent under a pressure p. From the fundamental equation at constant T, we have $d\mu^\circ = \overline{V}^\circ \, dp$. Integrating, we have

$$\mu^\circ(T, p + \pi) - \mu^\circ(T, p) = \int_p^{p+\pi} \overline{V}^\circ \, dp. \tag{13.34}$$

This reduces Eq. (13.33) to

$$\int_p^{p+\pi} \overline{V}^\circ dp + RT \ln x = 0. \tag{13.35}$$

In Eq. (13.35), $\overline{V}^\circ$ is the molar volume of the pure solvent. If the solvent is incompressible, then $\overline{V}^\circ$ is independent of pressure and can be removed from the integral. Then

$$\overline{V}^\circ \pi + RT \ln x = 0, \tag{13.36}$$

which is the relation between the osmotic pressure π and the mole fraction of solvent in the solution. Two assumptions are involved in Eq. (13.36); the solution is ideal and the solvent is incompressible.

In terms of the solute concentration, $\ln x = \ln (1 - x_2)$. If the solution is dilute, then $x_2 \ll 1$; the logarithm may be expanded in series. Keeping only the first term, we obtain

$$\ln (1 - x_2) = -x_2 = -\frac{n_2}{n + n_2} \approx -\frac{n_2}{n},$$

since $n_2 \ll n$ in the dilute solution. Then Eq. (13.36) becomes

$$\pi = \frac{n_2 RT}{n\overline{V}^\circ}. \qquad (13.37)$$

By the addition rule the volume of the ideal solution is $V = n\overline{V}^\circ + n_2 \overline{V}_2^\circ$. If the solution is dilute, n_2 is very small, so that $V \approx n\overline{V}^\circ$. This result reduces Eq. (13.37) to

$$\pi = \frac{n_2 RT}{V} \qquad \text{or} \qquad \pi = \tilde{c}RT. \qquad (13.38)$$

In Eq. (13.38), $\tilde{c} = n_2/V$, the concentration of solute (mol/m^3) in the solution. Equation (13.38) is the van't Hoff equation for osmotic pressure.

The striking formal analogy between the van't Hoff equation and the ideal gas law should not go unnoticed. In the van't Hoff equation, n_2 is the number of moles of *solute*. The solute molecules dispersed in the solvent are analogous to the gas molecules dispersed in empty space. The solvent is analogous to the empty space between the gas molecules. In the experiment shown in Fig. 13.7, the membrane is attached to a movable piston. As the solvent diffuses through the membrane, the piston is pushed to the right; this continues until the piston is flush against the right-hand wall. The observed effect is the same *as if* the solution exerted a pressure against the membrane to push it to the right. The situation is comparable to the free expansion of a gas into vacuum. If the volume of the solution doubles in this experiment, the dilution will reduce the final osmotic pressure by half, just as the pressure of a gas is halved by doubling its volume.

In spite of the analogy, it is deceptive to consider the osmotic pressure as a sort of pressure that is somehow exerted by the solute. Osmosis, the passage of solvent through the membrane, is due to the inequality of the chemical potential on the two sides of the membrane. The kind of membrane does not matter, but it must be permeable only to the solvent. Nor does the nature of the solute matter; it is necessary only that the solvent contain dissolved foreign matter which is not passed by the membrane.

The mechanism by which the solvent permeates the membrane may be different for each different kind of membrane. A membrane could conceivably be like a sieve that allows small molecules such as water to pass through the pores while it blocks larger molecules. Another membrane might dissolve the solvent and so be permeated by it, while the solute is not soluble in the membrane. The mechanism by which a solvent passes through a membrane must be examined for every membrane–solvent pair using the methods of chemical kinetics. Thermodynamics cannot provide an answer, because the equilibrium result is the same for all membranes.

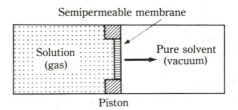

Figure 13.7   Osmotic analog of the Joule experiment.

13.9.2 Measurement of Osmotic Pressure

The measurement of osmotic pressure is useful for determining the molar masses of materials that are only slightly soluble in the solvent, or which have very high molar masses (for example, proteins, polymers of various types, colloids). These are convenient measurements because of the large magnitude of the osmotic pressure.

At 25 °C, the product $RT \approx 2480$ J/mol. Thus, for a 1 mol/L solution ($\tilde{c} = 1000$ mol/m^3), we have

$$\pi = \tilde{c}RT = 2.48 \times 10^6 \text{ Pa} = 24.5 \text{ atm}.$$

This pressure corresponds to a height of a column of water of the order of 800 ft. Simply to keep the experiment in the laboratory, the solutions must be less than 0.01 molar, and are preferably of the order of 0.001 molar. This assumes that we are using an apparatus of the type shown in Fig. 13.6. Very precise measurements of osmotic pressures up to several hundred atmospheres have been made by H. N. Morse and J. C. W. Frazer, and by Lord Berkeley and E. G. J. Hartley using special apparatus of different design.

In a molar mass determination, if w_2 is the mass of solute dissolved in the volume, V, then $\pi = w_2 RT/M_2 V$, or

$$M_2 = \frac{w_2 RT}{\pi V}.$$

Even when w_2 is small and M_2 large, the value of π is measurable and can be translated into a value of M_2.

Osmosis plays a significant role in the function of organisms. A cell that is immersed in pure water undergoes plasmolysis. The cell wall permits water to flow into it; thereupon the cell becomes distended, the wall stretches until it ultimately ruptures or becomes leaky enough to allow the solutes in the cellular material to escape from the interior. On the other hand, if the cell is immersed in a concentrated solution of salt, the water from the cell flows into the more concentrated salt solution and the cell shrinks. A salt solution which is just concentrated enough so that the cell neither shrinks nor is distended is called an *isotonic* solution.

Osmosis might be called the principle of the prune. The skin of the prune acts as a membrane permeable to water. The sugars in the prune are the solutes. Water diffuses through the skin and the fruit swells until the skin ruptures or becomes leaky. Only rarely are plant and animal membranes strictly semipermeable. Frequently, their function in the organism requires that they pass other materials, as well as water. Medicinally, the osmotic effect is utilized in, for example, the prescription of a salt-free diet in some cases of abnormally high fluid retention by the body.

QUESTIONS

13.1 Is the lowering of the chemical potential of a solvent in an ideal solution, Eq. (13.5), an enthalpy effect or an entropy effect? Explain.

13.2 Interpret (a) freezing-point depression and (b) boiling point elevation in terms of μ as a measure of "escaping tendency."

13.3 How does the temperature dependence of the solubility of a solid in a liquid illustrate LeChatelier's principle?

13.4 *Reverse* osmosis has been suggested as a means of purifying sea water (roughly an NaCl–H$_2$O solution). How could this be accomplished with an appropriate membrane, with special attention placed on the required pressure *on the solution*?

PROBLEMS

13.1 Twenty grams of a solute are added to 100 g of water at 25 °C. The vapor pressure of pure water is 23.76 mmHg; the vapor pressure of the solution is 22.41 mmHg.

a) Calculate the molar mass of the solute.
b) What mass of this solute is required in 100 g of water to reduce the vapor pressure to one-half the value for pure water?

13.2 How many grams of sucrose, $C_{12}H_{22}O_{11}$, must be dissolved in 90 g of water to produce a solution over which the relative humidity is 80%? Assume the solution is ideal.

13.3 Suppose that a series of solutions is prepared using 180 g of H_2O as a solvent and 10 g of an involatile solute. What will be the relative vapor pressure lowering if the molar mass of the solute is: 100 g/mol, 200 g/mol, 10,000 g/mol?

13.4 a) For an ideal solution plot the value of $p/p°$ as a function of x_2, the mole fraction of the solute.
b) Sketch the plot of $p/p°$ as a function of the molality of the solute, if water is the solvent.
c) Suppose the solvent (for example, toluene) has a higher molar mass. How does this affect the plot of $p/p°$ versus m? How does it affect $p/p°$ versus x_2?
d) Evaluate the derivative of $(p° - p)/p°$ with respect to m, as $m \to 0$.

13.5 A stream of air is bubbled slowly through liquid benzene in a flask at 20.0 °C against an ambient pressure of 100.56 kPa. After the passage of 4.80 L of air, measured at 20.0 °C and 100.56 kPa before it contains benzene vapor, it is found that 1.705 g of benzene have been evaporated. Assuming that the air is saturated with benzene vapor when it leaves the flask, calculate the equilibrium vapor pressure of the benzene at 20.0 °C.

13.6 Two grams of benzoic acid dissolved in 25 g of benzene, $K_f = 4.90$ K kg/mol, produce a freezing-point depression of 1.62 K. Calculate the molar mass. Compare this with the molar mass obtained from the formula for benzoic acid, C_6H_5COOH.

13.7 The heat of fusion of acetic acid is 11.72 kJ/mol at the melting point 16.61 °C. Calculate K_f for acetic acid.

13.8 The heat of fusion of water at the freezing point is 6009.5 J/mol. Calculate the freezing point of water in solutions having a mole fraction of water equal to: 1.0, 0.8, 0.6, 0.4, 0.2. Plot the values of T versus x.

13.9 Ethylene glycol, $C_2H_4(OH)_2$, is commonly used as a permanent antifreeze; assume that the mixture with water is ideal. Plot the freezing point of the mixture as a function of the volume percent of glycol in the mixture for 0%, 20%, 40%, 60%, 80%. The densities are: H_2O, 1.00 g/cm³, glycol, 1.11 g/cm³. $\Delta H_{fus}(H_2O) = 6009.5$ J/mol.

13.10 Assume that ΔH_{fus} is independent of the temperature and that the thermometer available can measure a freezing-point depression to an accuracy of ± 0.01 K. The simple law for freezing-point depression, $\theta_f = K_f m$, is based on the limiting condition that $m = 0$. At what molality will this approximation no longer predict the result within the experimental error in water?

13.11 If the heat of fusion depends on temperature through the expression

$$\Delta H_{fus} = \Delta H_0 + \Delta C_p(T - T_0),$$

where ΔC_p is constant, then the value of θ_f can be expressed in the form $\theta_f = am + bm^2 + \cdots$, where a and b are constants. Calculate the values of a and b. [*Hint:* This is a Taylor series, so evaluate $(\partial^2 \theta / \partial m^2)$ at $m = 0$.]

13.12 For CCl_4, $K_b = 5.03$ K kg/mol and $K_f = 31.8$ K kg/mol. If 3.00 g of a substance in 100 g CCl_4 raises the boiling point by 0.60 K, calculate the freezing-point depression, the relative vapor pressure lowering, the osmotic pressure at 25 °C, and the molar mass of the substance. The density of CCl_4 is 1.59 g/cm³ and the molar mass is 153.823 g/mol.

13.13 Calculate the boiling-point elevation constant for each of the following substances.

Substance	$t_b/°C$	$\Delta H_{vap}/(J/g)$
Acetone, $(CH_3)_2CO$	56.1	520.9
Benzene, C_6H_6	80.2	394.6
Chloroform, $CHCl_3$	61.5	247
Methane, CH_4	-159	577
Ethyl acetate, $CH_3CO_2C_2H_5$	77.2	426.8

Plot the values of K_b versus the product MT_b.

13.14 Since the boiling point of the liquid depends on the pressure, K_b is a function of pressure. Calculate the value of K_b for water at 750 mmHg and at 740 mmHg pressure. Use the data in the text. Assume ΔH_{vap} is constant.

13.15 a) For p-dibromobenzene, $C_6H_4Br_2$, the heat of fusion is 85.8 J/g; the melting point is 86 °C. Calculate the ideal solubility at 25 °C.
b) For p-dichlorobenzene, $C_6H_4Cl_2$, the heat of fusion is 124.3 J/g; the melting point is 52.7 °C. Calculate the ideal solubility at 25 °C.

13.16 The melting point of iodine is 113.6 °C and the heat of fusion is 15.64 kJ/mol.
a) What is the ideal solubility of iodine at 25 °C?
b) How many grams of iodine dissolve in 100 g hexane at 25 °C?

13.17 In 100.0 g benzene, 70.85 g naphthalene, $C_{10}H_8$, dissolve at 25 °C and 103.66 g dissolve at 35 °C. Assume the solution is ideal. Calculate ΔH_{fus} and T_m for naphthalene.

13.18 If 6.00 g of urea, $(NH_2)_2CO$, is dissolved in 1.00 L of solution, calculate the osmotic pressure of the solution at 27 °C.

13.19 Consider a vertical tube with a cross-sectional area of 1.00 cm^2. The bottom of the tube is closed with a semipermeable membrane and 1.00 g of glucose, $C_6H_{12}O_6$, is placed in the tube. The closed end of the tube is immersed in pure water. What will be the height of the liquid level in the tube at equilibrium? The density of the solution may be taken as 1.00 g/cm^2; the sugar concentration is assumed to be uniform in the solution. What is the osmotic pressure at equilibrium? ($t = 25°$ C; assume a negligible depth of immersion.)

13.20 At 25 °C a solution containing 2.50 g of a substance in 250.0 cm^3 of solution exerts an osmotic pressure of 400 Pa. What is the molar mass of the substance?

13.21 a) The complete expression for osmotic pressure is given by Eq. (13.36). Since $\tilde{c} = n_2/V$ and $V = nV° + n_2 V_2°$, where $V°$ and $V_2°$ are constants, the mole numbers n and n_2 can be expressed in terms of V, $V°$, $V_2°$, and $\tilde{c}$. Compute the value of $x = n/(n + n_2)$ in these terms. Then evaluate $(\partial \pi/\partial \tilde{c})_T$ at $\tilde{c} = 0$ and show that it is equal to RT.
b) By evaluating $(\partial^2 \pi/\partial \tilde{c}^2)_T$ at $\tilde{c} = 0$, show that $\pi = \tilde{c}RT(1 + V'\tilde{c})$, where $V' = V_2° - \frac{1}{2}V°$. Note that this is equivalent to writing a modified van der Waals equation, $\pi = n_2 RT/(V - n_2 V')$, and expanding it in a power series.

14

Solutions

II. More than One Volatile Component; the Ideal Dilute Solution

14.1 GENERAL CHARACTERISTICS OF THE IDEAL SOLUTION

The discussion in Chapter 13 was restricted to those ideal solutions in which the solvent was the only volatile constituent present. The concept of an ideal solution extends to solutions containing several volatile constituents. As before, the concept is based on a generalization of the experimental behavior of real solutions and represents a limiting behavior that is approached by all real solutions.

Consider a solution composed of several volatile substances in a container that is initially evacuated. Since the components are all volatile, some of the solution evaporates to fill the space above the liquid with vapor. When the solution and the vapor come to equilibrium at the temperature T, the total pressure within the container is the sum of the partial pressures of the several components of the solution:

$$p = p_1 + p_2 + \cdots + p_i + \cdots. \tag{14.1}$$

These partial pressures are measurable, as are the equilibrium mole fractions $x_1, \ldots, x_i, \ldots$, in the liquid. Let one of the components, i, be present in a relatively large amount compared with any of the others. Then it is found experimentally that

$$p_i = x_i p_i^\circ, \tag{14.2}$$

where p_i° is the vapor pressure of the pure liquid component i. Equation (14.2) is Raoult's law, and experimentally it is followed in any solution as x_i approaches unity regardless of which component is present in great excess. When any solution is dilute in all components but the solvent, the solvent always follows Raoult's law. Since all of the components are volatile, any one of them can be designated as the solvent. Therefore the ideal solution is defined by the requirement that each component obey Raoult's law, Eq. (14.2), over the entire range of composition. The significance of the symbols is worth reiterating: p_i is the *partial* pressure of i in the vapor phase; p_i° is the vapor pressure of the pure liquid i; and x_i is the mole fraction of i in the *liquid* mixture.

The ideal solution has two other important properties: The heat of mixing the pure components to form the solution is zero, and the volume of mixing is zero. These properties are observed as the limiting behavior in all real solutions. If additional solvent is added to a solution that is dilute in all of the solutes, the heat of mixing approaches zero as the solution becomes more and more dilute. In the same circumstances the volume of mixing of all real solutions approaches zero.

14.2 THE CHEMICAL POTENTIAL IN IDEAL SOLUTIONS

Consider an ideal solution in equilibrium with its vapor at a fixed temperature T. For each component, the equilibrium condition is $\mu_i = \mu_{i(\text{vap})}$, where μ_i is the chemical potential of i in the solution, $\mu_{i(\text{vap})}$ is the chemical potential of i in the vapor phase. If the vapor is ideal, then by the same argument as in Section 13.3, the value of μ_i is

$$\mu_i = \mu_i^\circ(T, p) + RT \ln x_i, \tag{14.3}$$

where $\mu_i^\circ(T, p)$ is the chemical potential of the pure liquid i at temperature T and under pressure p. The chemical potential of each and every component of the solution is given by the expression in Eq. (14.3). Figure 14.1 shows the variation of $\mu_i - \mu_i^\circ$ as a function of x_i. As x_i becomes very small, the value of μ_i decreases very rapidly. At all values of x_i, the value of μ_i is less than that of μ_i°.

Since Eq. (14.3) is formally the same as Eq. (11.14) for the μ of an ideal gas in a gas mixture, by the same reasoning as in Section 11.6 it follows that in mixing

$$\Delta G_{\text{mix}} = nRT \sum_i x_i \ln x_i, \tag{14.4}$$

$$\Delta S_{\text{mix}} = -nR \sum_i x_i \ln x_i, \tag{14.5}$$

$$\Delta H_{\text{mix}} = 0, \qquad \Delta V_{\text{mix}} = 0; \tag{14.6}$$

where n is the total number of moles in the mixture. The three properties of the ideal solution (Raoult's law, zero heat of mixing, and zero volume of mixing) are intimately

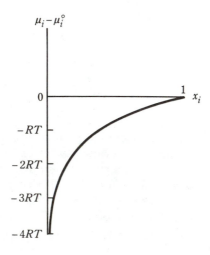

Figure 14.1 $(\mu_i - \mu_i^\circ)$ versus x_i.

related. If Raoult's law obtains for every component, then the heat and volume of mixing will be zero. (This statement cannot be reversed; if the volume of mixing and heat of mixing are both zero, it does not follow that Raoult's law will be obeyed.)

14.3 BINARY SOLUTIONS

We turn our attention now to the consequences of Raoult's law in binary solutions in which both components are volatile. In a binary solution $x_1 + x_2 = 1$. We have

$$p_1 = x_1 p_1^\circ, \tag{14.7}$$

and

$$p_2 = x_2 p_2^\circ = (1 - x_1)p_2^\circ. \tag{14.8}$$

If the total pressure over the solution is p, then

$$p = p_1 + p_2 = x_1 p_1^\circ + (1 - x_1)p_2^\circ$$
$$p = p_2^\circ + (p_1^\circ - p_2^\circ)x_1, \tag{14.9}$$

which relates the total pressure over the mixture to the mole fraction of component 1 in the liquid. It shows that p is a linear function of x_1 (Fig. 14.2a). It is clear from Fig. 14.2(a) that the addition of a solute may raise or lower the vapor pressure of the solvent depending on which is the more volatile.

The total pressure can also be expressed in terms of y_1, the mole fraction of component 1 in the vapor. From the definition of the partial pressure,

$$y_1 = \frac{p_1}{p}. \tag{14.10}$$

Using the values of p_1 and p from Eqs. (14.7) and (14.9), we obtain

$$y_1 = \frac{x_1 p_1^\circ}{p_2^\circ + (p_1^\circ - p_2^\circ)x_1}.$$

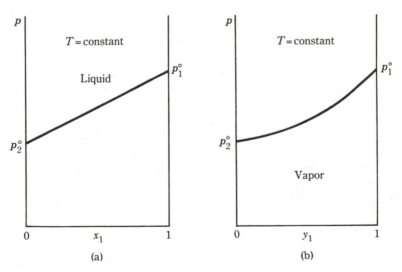

Figure 14.2 Vapor pressure as a function of composition.

Solving for x_1 yields,

$$x_1 = \frac{y_1 p_2^\circ}{p_1^\circ + (p_2^\circ - p_1^\circ)y_1}.$$

(14.11)

Using the value of x_1 from Eq. (14.11) in Eq. (14.9), we obtain, after collecting terms,

$$p = \frac{p_1^\circ p_2^\circ}{p_1^\circ + (p_2^\circ - p_1^\circ)y_1}.$$

(14.12)

Equation (14.12) expresses p as a function of y_1, the mole fraction of component 1 in the vapor. This function is plotted in Fig. 14.2(b). The relation in Eq. (14.12) can be rearranged to the more convenient, symmetrical form

$$\frac{1}{p} = \frac{y_1}{p_1^\circ} + \frac{y_2}{p_2^\circ}.$$

(14.12a)

To describe a two-component system, the phase rule shows, since $C = 2$, that $F = 4 - P$. Since P is 1 or greater, three variables at most must be specified to describe the system. Since Fig. 14.2(a) and (b) are drawn at a specified temperature, only two additional variables are required to describe completely the state of the system. These two variables may be (p, x_1) or (p, y_1). As a consequence, the points in Fig. 14.2(a) or (b) describe states of the system.

There is a difficulty here. The variable x_1, being a mole fraction in the liquid, is not capable of describing states of the system that are completely gaseous. Similarly, y_1 is incapable of describing a completely liquid state of the system. Hence, only liquid states and those states on the line in which liquid and vapor coexist are described by Fig. 14.2(a). Similarly, only gaseous states and those states, on the curve, in which vapor and liquid coexist are described by Fig. 14.2(b). The completely liquid states are those at high pressures, that is, those above the line in Fig. 14.2(a). The completely gaseous states are stable at low pressures, those below the curve in Fig. 14.2(b). These regions of stability have been so labeled on the diagrams.

Life would be much simpler if we could represent all the states on one diagram. If only liquid is present, x_1 describes the composition of the liquid and also the composition of the entire system. If only vapor is present, y_1 describes the composition of the vapor and at the same time the composition of the entire system. In view of that, it seems reasonable to plot the pressure against X_1, the mole fraction of component 1 in the entire system. In Fig. 14.3(a), p is plotted against X_1; the two curves of Fig. 14.2(a) and (b) are drawn in. The upper curve is called the liquid curve; the lower curve is the vapor curve. The system is neatly represented by one diagram: The liquid is stable above the liquid curve; the vapor is stable below the vapor curve. What significance is attributed to the points that lie between the curves? The points lying just above the liquid curve correspond to the lowest pressures at which liquid can exist by itself, since vapor appears if the point lies on the curve. Liquid cannot be present alone below the liquid curve. By the same argument vapor cannot be present alone above the vapor curve. The only possible meaning to the points between the curves is that they represent states of the system in which liquid and vapor coexist in equilibrium. The enclosed region is the *liquid–vapor* region.

Consider the point a in the liquid–vapor region (Fig. 14.3b). The value X_1 corresponding to a is the mole fraction of component 1 in the entire system, liquid + vapor. What composition of liquid can coexist with vapor at the pressure p in question? The

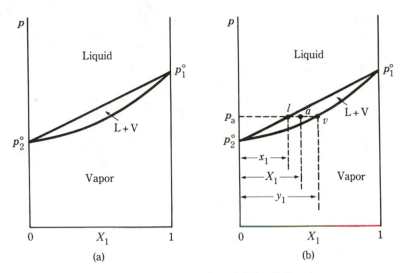

Figure 14.3   Interpretation of the p–X diagram.

intersection of a horizontal line, a *tie line*, at constant pressure, with the liquid curve at l yields the value of x_1, which describes the composition of the liquid; its intersection with the vapor curve at v yields the value of y_1, which describes the composition of the vapor.

If two phases—liquid and vapor—are present in equilibrium, the variance of the system is $F = 4 - 2 = 2$. Since the temperature is fixed, one other variable, any one of p, x_1, y_1, suffices to describe the system. So far we have used x_1 or y_1 to describe the system; since $x_1 + x_2 = 1$, and $y_1 + y_2 = 1$, we could equally well have chosen x_2 and y_2. If the pressure is chosen to describe the two-phase system, the intersections of the horizontal line at that pressure with the liquid and vapor curves yield the values of x_1 and y_1 directly. If x_1 is the describing variable, the intersection of the vertical line at x_1 with the liquid curve yields the value of p; from p the value of y_1 is obtained immediately.

14.4   THE LEVER RULE

In any two-phase region, such as L–V in Fig. 14.3(b), the composition of the entire system may vary between the limits x_1 and y_1, depending on the relative amounts of liquid and vapor present. If the state point a is very near the liquid line, the system would consist of a large amount of liquid and a relatively small amount of vapor. If a is near the vapor line, the amount of liquid present is relatively small compared with the amount of vapor present.

The relative amounts of liquid and vapor present are calculated by the lever rule. Let the length of the line segment between a and l in Fig. 14.3(b) be $(\overline{al})$ and that between a and v be $(\overline{av})$; then let $n_{1(\text{liq})}$ and $n_{1(\text{vap})}$ be the number of moles of component 1 in the liquid and in the vapor, respectively; let $n_1 = n_{1(\text{liq})} + n_{1(\text{vap})}$. If n_{liq} and n_{vap} are the total number of moles of liquid and vapor present, respectively, and if $n = n_{\text{liq}} + n_{\text{vap}}$, then from Fig. 14.3(b), we have

$$(\overline{al}) = X_1 - x_1 = \frac{n_1}{n} - \frac{n_{1(\text{liq})}}{n_{\text{liq}}}, \qquad (\overline{av}) = y_1 - X_1 = \frac{n_{1(\text{vap})}}{n_{\text{vap}}} - \frac{n_1}{n}.$$

Multiply $(\overline{al})$ by n_{liq} and $(\overline{av})$ by n_{vap} and subtract:

$$n_{\text{liq}}(\overline{al}) - n_{\text{vap}}(\overline{av}) = \frac{n_1}{n}\left(n_{\text{liq}} + n_{\text{vap}}\right) - \left(n_{1\,(\text{liq})} + n_{1\,(\text{vap})}\right) = n_1 - n_1 = 0.$$

Therefore

$$n_{\text{liq}}(\overline{al}) = n_{\text{vap}}(\overline{av}) \qquad \text{or} \qquad \frac{n_{\text{liq}}}{n_{\text{vap}}} = \frac{(\overline{av})}{(\overline{al})}. \tag{14.13}$$

This is called the lever rule, point a being the fulcrum of the lever; the number of moles of liquid times the length $(\overline{al})$ from a to the liquid line is equal to the number of moles of vapor times the length, $(\overline{av})$, from a to the vapor line. The ratio of the number of moles of liquid to the number of moles of vapor is given by the ratio of lengths of the line segments connecting a to v and l. Thus if a lies very close to v, $(\overline{av})$ is very small and $n_{\text{liq}} \ll n_{\text{vap}}$; the system consists mainly of vapor. Similarly when a lies close to l, $n_{\text{vap}} \ll n_{\text{liq}}$; the system consists mainly of liquid.

Since the derivation of the lever rule depends only on a mass balance, the rule is valid for calculating the relative amounts of the two phases present in any two-phase region of a two-component system. If the diagram is drawn in terms of mass fraction instead of mole fraction, the level rule is valid and yields the relative masses of the two phases rather than the relative mole numbers.

14.5 CHANGES IN STATE AS THE PRESSURE IS REDUCED ISOTHERMALLY

The behavior of the system is now examined as the pressure is reduced from a high to a low value, keeping the overall composition constant at a mole fraction of component 1 equal to X. At point a, Fig. 14.4, the system is entirely liquid and remains so as the pressure is reduced until the point l is reached; at point l, the first trace of vapor appears, having a composition y. Note that the first vapor to appear is considerably richer in 1 than the liquid; component 1 is the more volatile. As the pressure is reduced further, the point reaches a'; during this reduction of pressure, the composition of the liquid moves along the line ll', while the composition of the vapor moves along vv'. At a', liquid has the

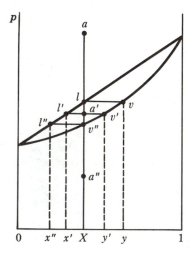

Figure 14.4 Isothermal change in pressure.

composition x' while vapor has the composition y'. The ratio of number of moles of liquid to vapor at point a' is $(\overline{a'v'})/(\overline{a'l'})$, from the lever rule. Continued reduction of pressure brings the state point to v''; at this point, only a trace of liquid of composition x'' remains; the vapor has the composition X. Note that the liquid which remains is richer in the less volatile component 2. As the pressure is reduced, the state point moves into the vapor region, and the reduction of pressure from v'' to a'' simply corresponds to an expansion of the vapor. In the final state, a'', the vapor has, of course, the same composition as the original liquid.

The vapor that forms over a liquid as the pressure is reduced is richer in a particular component than the liquid. This fact is the basis of a method of separation: isothermal distillation. The method is useful for those mixtures that would decompose if distilled by the ordinary method; it is sufficiently inconvenient so that it is used only if other methods are not suitable.

The system described above is an ideal solution. If the deviations from ideality are not very large, the figure will appear much the same except that the liquid composition curve is not a straight line. The interpretation is precisely the same as for the ideal solution.

14.6 TEMPERATURE–COMPOSITION DIAGRAMS

In the diagrams shown in Section 14.5, the temperature was constant. The equilibrium pressure of the system was then a function of either x_1 or y_1, according to Eqs. (14.9) or (14.12). In those equations, the values of p_1° and p_2° are functions of temperature. If, in Eqs. (14.9) and (14.12), we consider the total pressure p to be constant, then the equations are relations between the equilibrium temperature, the boiling point, and either x_1 or y_1. The relations $T = f(x_1)$ and $T = g(y_1)$ are not such simple ones as between pressure and composition, but they may be determined theoretically through the Clapeyron equation or, ordinarily, experimentally through determination of the boiling points and vapor compositions corresponding to liquid mixtures of various compositions.

The plot at constant pressure of boiling points versus compositions for the ideal solution corresponding to that in Fig. 14.3 is shown in Fig. 14.5. Neither the liquid nor the vapor curve is a straight line; otherwise, the figure resembles Fig. 14.3. However, the lenticular liquid–vapor region is tilted down from left to right. This corresponds to the fact

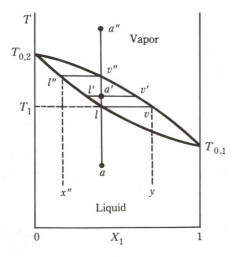

Figure 14.5   Isobaric change in temperature.

that component 1 had the higher vapor pressure; therefore it has the lower boiling point. Also in Fig. 14.5 the liquid region is at the bottom of the diagram, since under a constant pressure the liquid is stable at low temperatures. The lower curve describes the liquid composition; the upper curve describes the vapor composition. The regions in the p–X diagram are sometimes thoughtlessly confused with those in the T–X diagram. A little common sense tells us that the liquid is stable at low temperatures, the lower part of the T–X diagram, and under high pressures, the upper part of the p–X diagram. Attempting to memorize the location of the liquid or vapor regions is foolish when it is so easy to figure it out.

The principles applied to the discussion of the p–X diagram can be applied in much the same way to the T–X diagram. The pressure on the system is constant; from the phase rule, two additional variables at most are needed to describe the system. Every point in the T–X diagram describes a state of the system. The points in the uppermost portion of the diagram are gaseous states of the system; those points in the lowest part are liquid states. The points in the middle region describe states in which liquid and vapor coexist in equilibrium. The tie line in the liquid–vapor region connects the composition of vapor and the composition of liquid that coexist at that temperature. The lever rule applies to the T–X diagram, of course.

14.7 CHANGES IN STATE WITH INCREASE IN TEMPERATURE

We examine now the sequence of events as a liquid mixture under a constant pressure is heated from a low temperature, corresponding to point a, Fig. 14.5, to a high temperature corresponding to point a''. At a, the system consists entirely of liquid; as the temperature rises, the system remains entirely liquid until point l is reached; at this temperature T_1 the first trace of vapor appears, having composition y. The vapor is much richer than the liquid in component 1, the lower boiling component. This fact is the basis for the separation of volatile mixtures by distillation. As the temperature is raised, the state point moves to a', and the liquid composition changes continuously along line ll'; the vapor composition changes continuously along line vv'. At a', the relative number of moles of liquid and vapor present is given by the ratio $(a'v')/(\overline{a'l'})$. If the temperature is raised further, at v'' the last trace of liquid, of composition x'', disappears. At a'' the system exists entirely as a vapor.

14.8 FRACTIONAL DISTILLATION

The sequence of events described in Section 14.7 is observed if no material is removed from the system as the temperature is increased. If some of the vapor formed in the early stages of the process is removed from the system and condensed, the condensate, or distillate, is enriched in the more volatile constituent, while the residue is improverished in the more volatile constituent. Suppose that the temperature of a mixture M is increased until half the material is present as vapor and half remains as liquid (Fig. 14.6a). The vapor has the composition v; the residue R has the composition l. The vapor is removed and condensed, yielding a distillate D of composition v. Then the distillate is heated until half exists as vapor and half as liquid (Fig. 14.6b). The vapor is removed and condensed, yielding distillate D' with composition v' and residue R' with composition l'. The original residue R is treated in the same way to yield distillate D'' and residue R''. Since D'' and R' have about the same composition, they are combined; the process is now repeated on the

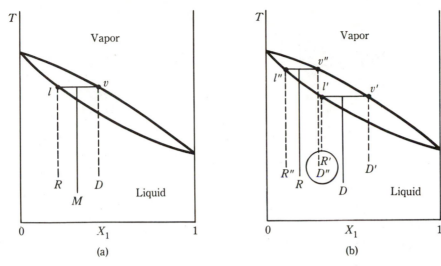

Figure 14.6   Distillation.

three fractions, R'', $(D'' + R')$, and D'; continuation of this process ultimately yields a distillate that approaches the composition of the more volatile liquid and a residue close to the composition of the less volatile liquid, together with a series of fractions of intermediate composition.

The time and labor involved in this batch type of separation is prohibitive and is eliminated through the use of a continuous method using a *fractionating column* (Fig. 14.7). The type of column illustrated is a bubble-cap column. The column is heated at the bottom; there is a temperature gradient along the length of the column, the top being cooler than the bottom. Let us suppose that the temperature at the top of the column is T_1, and the vapor issuing at this point is in equilibrium with the liquid held up on the top plate, plate 1; the compositions of liquid and vapor are shown in Fig. 14.8 as l_1, and v_1. On the next plate, plate 2, the temperature is slightly higher, T_2, and the vapor issuing from it has the composition v_2. As this vapor passes upward to plate 1, it is cooled to temperature T_1, to point a. This means that some of the vapor v_2 condenses to form l_1; since l_1 is richer in the less volatile constituent, the remaining vapor is richer in the more volatile constituent and at equilibrium attains the composition v_1. This happens at every plate in the column. As the vapor moves up the column, it cools; this cooling condenses the less volatile component preferentially, so the vapor becomes increasingly enriched in the more volatile component as it moves upward. If at each position in the column the liquid is in equilibrium with vapor, then the composition of the vapor will be given by the vapor composition curve in Fig. 14.8. It is understood that the temperature is some function of the position in the column.

As the liquid l_1 on the top plate flows down to the next plate, the temperature rises to T_2, and the state point of the liquid reaches b (Fig. 14.8). Some of the more volatile component vaporizes to yield v_2; the liquid shifts to the composition l_2. As it flows downward through the column, the liquid becomes richer in the less volatile component.

As vapor moves up the column and the liquid moves down, there is a continuous redistribution of the two components between the liquid and vapor phases to establish equilibrium at each position (that is, each temperature) in the column. This redistribution must take place quickly if the equilibrium is to be established at every position. There

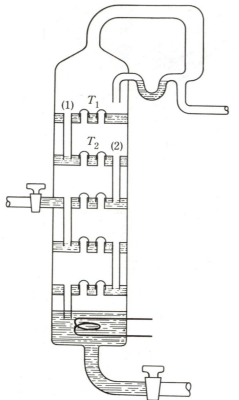

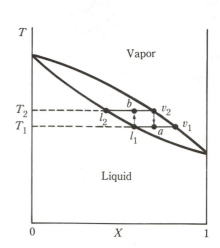

Figure 14.8 Liquid and vapor exchange in a distilling column.

Figure 14.7 Bubble-cap distilling column. (By permission from Findlay, Campbell, Smith, *The Phase Rule and Its Applications*, 9th ed. New York: Dover, 1951.)

must be efficient contact between the liquid and vapor. In the bubble-cap column, efficient contact is obtained by forcing the ascending vapor to bubble through the liquid on each plate. In the laboratory Hempel column, the liquid is spread out over glass beads and the vapor is forced upward through the spaces between the beads; intimate contact is achieved in this way. Industrial stills use a variety of packings, saddle-shaped pieces of ceramic being frequently used. Packing materials or arrangements that permit the liquid to channel, that is, to flow downward through the column along special paths, must be avoided. The aim is to spread liquid evenly in relatively thin layers so that redistribution of the components may occur quickly.

It should be noted that if a certain portion of the column is held at a particular temperature, then, at equilibrium, the composition of liquid and vapor have the values appropriate to that temperature. Under constant pressure, the variance of the system is $F = 3 - P$; since two phases are present, $F = 1$. Consequently, fixing the temperature at every position in the column fixes the liquid and vapor composition at every position in the column at equilibrium. Therefore, by imposing an arbitrary temperature distribution along the column, an equally arbitrary composition distribution of vapor and liquid along the column results "at equilibrium."

The phrase "at equilibrium" or "in equilibrium" is commonly used to describe a distilling column that is not in equilibrium at all but rather in a *steady state*. Since there are inequalities of temperature along the column, the system cannot be truly in equilibrium in the thermodynamic sense. For this reason, the phase rule does not apply rigorously to this situation. It can be used as a guide, however. Other difficulties occur as well: the pressure is higher at the bottom of the column than at the top; the countercurrent flow of liquid and vapor is an additional nonequilibrium phenomenon.

In practice, equilibrium is not established at every position of the column, but rather the vapor at any position has a composition in equilibrium with the liquid at a slightly lower position. If the distance between these two positions is h, the column is said to have one *theoretical plate* in the length h. The number of theoretical plates in a column depends on its geometry, the kind and arrangement of the packing, and the manner in which the column is operated. It must be determined experimentally for a given set of operating conditions.

If the individual components have boiling points that are far apart, a distilling column with only a few theoretical plates suffices to separate the mixture. On the other hand, if the boiling points are close together, a column with a large number of theoretical plates is required.

14.9 AZEOTROPES

Mixtures that are ideal or depart only slightly from ideality can be separated into their constituents by fractional distillation. On the other hand, if the deviations from Raoult's law are so large as to produce a maximum or a minimum in the vapor pressure curve, then a corresponding minimum or maximum appears in the boiling point curve. Such mixtures cannot be completely separated into the constituents by fractional distillation. It can be shown that if the vapor pressure curve has a maximum or minimum, then at that point the liquid and vapor curves must be tangent to one another and the liquid and vapor must have the same composition (Gibbs–Konovalov theorem). The mixture having the maximum or minimum vapor pressure is called an *azeotrope* (from the Greek: to boil unchanged).

Consider the system shown in Fig. 14.9, which exhibits a maximum boiling point. If a mixture described by point a, having the azeotropic composition, is heated, the vapor will first form at temperature t; that vapor has the same composition as the liquid;

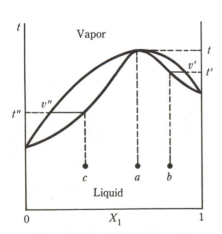

Figure 14.9 t–X diagram with maximum boiling point.

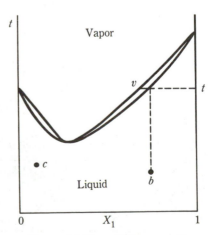

Figure 14.10 t–X diagram with minimum boiling point.

consequently, the distillate obtained has exactly the same composition as the original liquid; no separation is effected. If a mixture described by b in Fig. 14.9 is heated, the first vapor forms at t', and has a composition v'. This vapor is richer in the *higher boiling component*. Fractionation would separate the mixture into pure component 1 in the distillate and leave the azeotropic mixture in the pot. A mixture described by c would boil first at t''; the vapor would have the composition v''. Fractionation of this mixture would yield pure component 2 in the distillate and azeotrope in the pot.

The behavior of minimum boiling azeotropes shown in Fig. 14.10 is analogous. The azeotrope itself distills unchanged. A mixture described by b boils first at temperature t, the vapor having a composition v. Fractionation of this mixture produces azeotrope in the distillate; pure component 1 remains in the pot. Similarly, fractionation of a mixture described by c will produce azeotrope in the distillate and leave pure component 2 in the pot.

In Table 14.1, a number of azeotropic mixtures are listed, together with their properties. The azeotrope resembles a pure compound in the property of boiling at a constant temperature, while ordinary mixtures boil over a range of temperatures. However, changes in pressure produce changes in the *composition* of the azeotrope, as well as changes in the boiling point, and so it cannot be a pure compound. The constant boiling hydrochloric acid is a case in point. The variation in composition with pressure is illustrated by the data in Table 14.2. These compositions have been determined accurately enough that a standard HCl solution may be prepared by dilution of the constant boiling acid.

Table 14.1(a)
Minimum boiling azeotropes (1 atm)

Component A	$t_b/°C$	Component B	$t_b/°C$	Azeotrope	
				Mass % A	$t_b/°C$
H_2O	100	C_2H_5OH	78.3	4.0	78.174
H_2O	100	$CH_3COC_2H_5$	79.6	11.3	73.41
CCl_4	76.75	CH_3OH	64.7	79.44	55.7
CS_2	46.25	CH_3COCH_3	56.15	67	39.25
$CHCl_3$	61.2	CH_3OH	64.7	87.4	53.43

Table 14.1(b)
Maximum boiling azeotropes (1 atm)

Component A	$t_b/°C$	Component B	$t_b/°C$	Azeotrope	
				Mass % A	$t_b/°C$
H_2O	100	HCl	-80	79.778	108.584
H_2O	100	HNO_3	86	32	120.5
$CHCl_3$	61.2	CH_3COCH_3	56.10	78.5	64.43
C_6H_5OH	182.2	$C_6H_5NH_2$	184.35	42	186.2

By permission from *Azeotropic Data*; Advances in Chemistry Series No. 6. Washington, D.C.: American Chemical Society, 1952.

Table 14.2
Dependence of azeotropic temperature and composition on pressure

Pressure/mmHg	Mass % HCl	$t_b/°C$
500	20.916	97.578
700	20.360	106.424
760	20.222	108.584
800	20.155	110.007

W. D. Bonner, R. E. Wallace, *J. Amer. Chem. Soc.*
52:1747 (1930).

14.10 THE IDEAL DILUTE SOLUTION

The rigid requirement of the ideal solution that every component obey Raoult's law over the entire range of composition is relaxed in the definition of the *ideal dilute solution*. To arrive at the laws governing dilute solutions, we must examine the experimental behavior of these solutions. The vapor–pressure curves for three systems are described below.

14.10.1 Benzene–Toluene

Figure 14.11 shows the vapor pressure versus mole fraction of benzene for the benzene–toluene system, which behaves ideally to a good degree of accuracy over the entire range of composition. The partial pressures of benzene and toluene, also shown in the figure, are linear functions of the mole fraction of benzene, since Raoult's law is obeyed.

14.10.2 Acetone–Carbon Disulfide

Figure 14.12(a) shows the partial-pressure curves and the total vapor pressure of mixtures of carbon disulfide and acetone. In this system the individual partial-pressure curves fall well above the Raoult's law predictions indicated by the dashed lines. The system exhibits positive deviations from Raoult's law. The total vapor pressure exhibits a maximum that lies above the vapor pressure of either component.

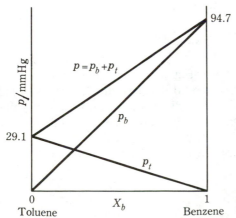

Figure 14.11   Vapor pressures in the benzene–toluene system.

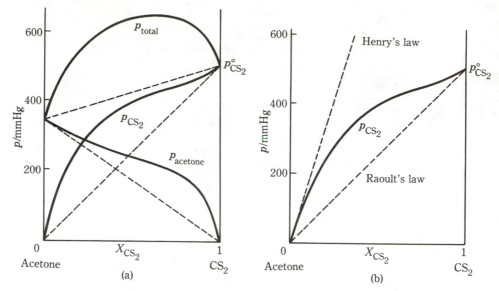

Figure 14.12 Vapor pressure in the acetone–carbon disulfide system (35.17 °C). [J. v. Zawidski, *Z. physik Chem.*, **35**:129 (1900).]

Figure 14.12(b) displays another interesting feature of this system. In this figure only the partial pressure of carbon disulfide is shown; in the region near $X_{CS_2} = 1$, when CS_2 is the solvent, the partial-pressure curve is tangent to the Raoult's law line. However, in the region near $X_{CS_2} = 0$, when CS_2 is the solute present in low concentration, the partial-pressure curve is linear.

$$p_{CS_2} = K_{CS_2} X_{CS_2}, \tag{14.14}$$

where K_{CS_2} is a constant. The slope of the line in this region is different from the Raoult's law slope. The solute obeys *Henry's law*, Eq. (14.14), where K_{CS_2} is the Henry's law constant. Inspection of the partial-pressure curve of the acetone discloses the same type of behavior:

$$p_{acetone} = X_{acetone} p_{acetone}^\circ \qquad \text{near } X_{acetone} = 1;$$

$$p_{acetone} = K_{acetone} X_{acetone} \qquad \text{near } X_{acetone} = 0.$$

Note that if the solution were ideal, then K would equal p° and both Henry's law and Raoult's law would convey the same information.

14.10.3 Acetone–Chloroform

In the acetone–chloroform system shown in Fig. 14.13, the vapor pressure curves fall below the Raoult's law predictions. This system exhibits *negative* deviations from Raoult's law. The total vapor pressure has a minimum value that lies below the vapor pressure of either of the pure components. The Henry's law lines, the fine dashed lines in the figure, also lie below the Raoult's law lines for this system.

Algebraically, we can express the properties of the ideal dilute solution by the following equations:

$$\text{Solvent (Raoult's law):} \qquad p_1 = x_1 p_1^\circ, \tag{14.15}$$

$$\text{Solutes (Henry's law):} \qquad p_j = K_j x_j, \tag{14.16}$$

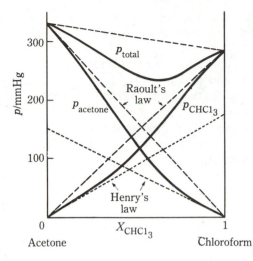

Figure 14.13 Vapor pressure in the acetone–chloroform system (35.17 °C). [J. v. Zawidski, *Z. physik Chem.*, **35**:129 (1900).]

where the subscript j denotes any of the solutes, and the subscript 1 denotes the solvent. All real solutions approach the behavior described by Eqs. (14.15) and (14.16), provided that the solution is sufficiently dilute. The same is true if several solutes are present, but the solution must be dilute in all solutes; each solute has a different value of K_j.

14.11 THE CHEMICAL POTENTIALS IN THE IDEAL DILUTE SOLUTION

Since the solvent follows Raoult's law, the chemical potential of the solvent is given by Eq. (14.3), repeated here for easy comparison:

$$\mu_1 = \mu_1^\circ(T, p) + RT \ln x_1.$$

For the solutes we require, as usual, equality of the chemical potential in the liquid, $\mu_j(l)$, with that in the gas phase, $\mu_j(g)$:

$$\mu_j(l) = \mu_j(g) = \mu_j^\circ(g) + RT \ln p_j.$$

Using Henry's law, Eq. (14.16), for p_j, this becomes

$$\mu_j(l) = \mu_j^\circ(g) + RT \ln K_j + RT \ln x_j$$

We define a standard free energy, $\mu_j^*(l)$, by

$$\mu_j^*(l) = \mu_j^\circ(g) + RT \ln K_j \tag{14.17}$$

where μ_j^* is a function of temperature and pressure but not of composition. The final expression for μ_j in the liquid is

$$\mu_j = \mu_j^* + RT \ln x_j \tag{14.18}$$

According to Eq. (14.18), μ_j^* is the chemical potential the solute j would have in the hypothetical state in which $x_j = 1$ if Henry's law were obeyed over the entire range, $0 \le x_j \le 1$.

The concept of the ideal dilute solution is extended to include nonvolatile solutes by requiring that the chemical potential of such solutes also have the form given by Eq. (14.18).

The mole fractions, x_j, often are not convenient measures for the concentration of solutes in dilute solution. Molalities, m_j, and molarities, c_j, are more commonly used. We can use Eq. (14.18) to obtain expressions for the chemical potential in terms of m_j or c_j. To do this we must write x_j in terms of m_j or c_j.

By definition, $x_j = n_j/(n + \Sigma_j n_j)$, where n is the number of moles of solvent. Also by definition, the molality of j is the number of moles of j per unit mass (1 kg) of solvent. Thus, if M is the molar mass (kg/mol) of the solvent, we have

$$m_j = \frac{n_j}{nM} \quad \text{or} \quad n_j = nMm_j. \tag{14.19}$$

Using this result for n_j in the expression for x_j, we obtain

$$x_j = \frac{Mm_j}{1 + Mm}, \tag{14.20}$$

where $m = \Sigma_j m_j$, the total molality of all the solutes. In dilute solution as m approaches zero, we have

$$\lim_{m=0} \left(\frac{x_j}{m_j}\right) = \lim_{m=0} \frac{M}{1 + mM} = M,$$

so that near $m = 0$,

$$x_j = Mm_j. \tag{14.21}$$

This can be written in the form

$$x_j = Mm^\circ\left(\frac{m_j}{m^\circ}\right) \tag{14.22}$$

where m° is the standard molal concentration, $m^\circ = 1$ mol/kg. This value for x_j may be used in Eq. (14.18), which becomes

$$\mu_j = \mu_j^* + RT \ln Mm^\circ + RT \ln \left(\frac{m_j}{m^\circ}\right)$$

Defining $\mu_j^{**} = \mu_j^* + RT \ln Mm^\circ$, this becomes

$$\mu_j = \mu_j^{**} + RT \ln m_j \tag{14.23}$$

in which we understand m_j as an abbreviation for the pure number, $m_j/(1 \text{ mol/kg})$. Equation (14.23) expresses the μ_j in a dilute solution as a convenient function of m_j. The standard value, μ_j^{**}, is the value μ_j would have in the hypothetical state of unit molality if the solution had the properties of the ideal dilute solution in the entire range, $0 \leq m_j \leq 1$.

To express μ_j in terms of c_j, we first establish the relation between m_j and $\tilde{c}_j$, the concentration in SI units, mol/m^3. By definition

$$\tilde{c}_j = \frac{n_j}{V} = \frac{nMm_j}{V}$$

If ρ_s is the density of the solution, then $V = w/\rho_s$, where the mass of the solution, $w = nM + \Sigma_j n_j M_j = nM + \Sigma_j nMm_j M_j$. Thus

$$V = \frac{nM}{\rho_s}\left(1 + \sum_j m_j M_j\right)$$

and

$$\tilde{c}_j = \frac{\rho_s m_j}{1 + \sum_j m_j M_j}. \tag{14.24}$$

As all the m_j approach zero we have

$$\lim_{m_j=0} \left(\frac{\tilde{c}_j}{m_j}\right) = \lim_{m_j=0} \frac{\rho_s}{1 + \sum_j m_j M_j} = \rho,$$

where ρ is the density of the pure solvent. Thus, in dilute solution,

$$\tilde{c}_j = \rho m_j \qquad \text{or} \qquad m_j = \frac{\tilde{c}_j}{\rho} \tag{14.25}$$

Rewriting to introduce the dimensionless ratios, Eq. (14.25) becomes

$$\frac{m_j}{m^\circ} = \frac{\tilde{c}^\circ}{\rho m^\circ}\left(\frac{\tilde{c}_j}{\tilde{c}^\circ}\right) \qquad \text{or} \qquad \frac{m_j}{m^\circ} = \frac{\tilde{c}^\circ}{\rho m^\circ}\left(\frac{c_j}{c^\circ}\right),$$

since $\tilde{c}_j/\tilde{c}_0 = c_j/c^\circ$. Putting this value of m_j/m° in Eq. (14.23) yields

$$\mu_j = \mu_j^{**} + RT \ln\left(\frac{\tilde{c}^\circ}{\rho m^\circ}\right) + RT \ln\frac{c_j}{c^\circ}.$$

This can be written

$$\mu_j = \mu_j^\square + RT \ln c_j, \tag{14.26}$$

in which we understand c_j as an abbreviation for the pure number, $c_j/(1\ \text{mol/L})$. In Eq. (14.26) we have set

$$\mu_j^\square = \mu_j^{**} + RT \ln\left(\frac{\tilde{c}^\circ}{\rho m^\circ}\right). \tag{14.27}$$

Equation (14.26) relates μ_j in dilute solution to c_j, the concentration in mol/L. It is not as commonly used as Eq. (14.23); $\mu_j^\square$ is the chemical potential the solute would have at a concentration of 1 mol/L if the solution behaved ideally up to that concentration.

The difference between $\mu_j^\square$ and μ_j^{**} is not very large. Since $c^\circ = 1$ mol/L, the corresponding value of $\tilde{c}^\circ = 10^3$ mol/m³. Also, $m^\circ = 1$ mol/kg, and for water at 25 °C, $\rho = 997.044$ kg/m³. Then

$$\frac{\tilde{c}^\circ}{\rho m^\circ} = \frac{10^3\ \text{mol/m}^3}{(997.044\ \text{kg/m}^3)(1\ \text{mol/kg})} = 1.002965.$$

The second term in Eq. (14.27) becomes (8.314 J/K mol)(298.15 K) ln (1.002965) = 7.339 J/mol. In most cases, this is less than the uncertainties in the experimental values so that the difference between the m_j and c_j standard states can be ignored.

14.12 HENRY'S LAW AND THE SOLUBILITY OF GASES

Henry's law, Eq. (14.16), relates the partial pressure of the solute in the vapor phase to the mole fraction of the solute in the solution. Viewing the relation in another way, Henry's law relates the equilibrium mole fraction, the solubility of j in the solution, to the partial pressure of j in the vapor:

$$x_j = \frac{1}{K_j} p_j. \tag{14.28}$$

Equation (14.28) states that the solubility x_j of a volatile constituent is proportional to the partial pressure of that constituent in the gaseous phase in equilibrium with the liquid. Equation (14.28) is used to correlate the data on solubility of gases in liquids. If the solvent and gas do not react chemically, the solubility of gases in liquids is usually small and the condition of diluteness is fulfilled. Here we have another example of the physical significance of the partial pressure.

The solubility of gases is often expressed as the Bunsen absorption coefficient, α, which is the volume of gas, measured at 0 °C and 1 atm, dissolved by unit volume of solvent if the partial pressure of the gas is 1 atm.

$$\alpha_j = \frac{V_j^\circ(g)}{V(l)}, \tag{14.29}$$

but $V_j^\circ(g) = n_j^\circ R T_0/p_0$, while the volume of the solvent is $V(l) = nM/\rho$, where n is the number of moles of solvent, M its molar mass, and ρ, the density. Thus

$$\alpha_j = \frac{n_j^\circ R T_0/p_0}{nM/\rho}. \tag{14.30}$$

When the partial pressure of the gas, $p_j = p^\circ = 1$ atm, the solubility by Henry's law is x_j°,

$$x_j^\circ = \frac{n_j^\circ}{n + n_j^\circ} = \frac{1}{K_j}$$

If the solution is dilute, $n_j^\circ \ll n$ and we have

$$\frac{n_j^\circ}{n} = \frac{1}{K_j}. \tag{14.31}$$

Using this value of n_j°/n in Eq. (14.30) brings it to

$$\alpha_j K_j = \left(\frac{R T_0}{p_0}\right)\left(\frac{\rho}{M}\right) = (0.022414 \text{ m}^3/\text{mol}) \frac{\rho}{M}, \tag{14.32}$$

which is the relation between the Henry's law constant K_j and the Bunsen absorption coefficient α_j; knowing one, we can calculate the other. The solubility of the gas in moles per unit volume of solvent, $n_j^\circ/(nM/\rho)$, is directly proportional to α_j, Eq. (14.30); this makes α_j more convenient than K_j for the discussion of solubility.

Some values of α for various gases in water are given in Table 14.3. Note the increase in α with increase in boiling point of the gas.

Table 14.3
Bunsen absorption coefficients
in water at 25°C

Gas	$t_b/°C$	α
Helium	−268.9	0.0087
Hydrogen	−252.8	0.0175
Nitrogen	−195.8	0.0143
Oxygen	−182.96	0.0283
Methane	−161.5	0.0300
Ethane	−88.3	0.0410

14.13 DISTRIBUTION OF A SOLUTE BETWEEN TWO SOLVENTS

If a dilute solution of iodine in water is shaken with carbon tetrachloride, the iodine is distributed between the two immiscible solvents. If μ and μ' are the chemical potentials of iodine in water and carbon tetrachloride, respectively, then at equilibrium $\mu = \mu'$. If both solutions are ideal dilute solutions, then, choosing Eq. (14.18) to express μ and μ', the equilibrium condition becomes $\mu^* + RT \ln x = \mu'^* + RT \ln x'$, which can be rearranged to

$$RT \ln \frac{x'}{x} = -(\mu'^* - \mu^*). \tag{14.33}$$

Since both μ'^* and μ^* are independent of composition, it follows that

$$\frac{x'}{x} = K, \tag{14.34}$$

where K, the distribution coefficient or partition coefficient, is independent of the concentration of iodine in the two layers. The quantity $\mu'^* - \mu^*$ is the standard Gibbs energy change ΔG^* for the transformation

$$I_2 \text{ (in } H_2O) \longrightarrow I_2 \text{ (in } CCl_4).$$

Equation (14.33) becomes

$$RT \ln K = -\Delta G^*, \tag{14.35}$$

which is the usual relation between the standard Gibbs energy change and the equilibrium constant of a chemical reaction.

If the solutions are quite dilute, then the mole fractions are proportional to the molalities or the molarities; so we have

$$K' = \frac{m'}{m} \quad \text{and} \quad K'' = \frac{c'}{c}, \tag{14.36}$$

where K' and K'' are independent of the concentrations in the two layers. Equation (14.36) was originally proposed by W. Nernst and is called the Nernst distribution law.

14.14 CHEMICAL EQUILIBRIUM IN THE IDEAL SOLUTION

In Section 11.7 it was shown that the condition of chemical equilibrium is

$$\left(\sum_i v_i \mu_i \right)_{eq} = 0, \tag{14.37}$$

the v_i being the stoichiometric coefficients. To apply this condition to chemical equilibrium in the ideal solution, we simply insert the proper form of the μ_i from Eq. (14.3). This yields directly

$$\sum_i v_i \mu_i^\circ + RT \sum_i \ln (x_i)_e^{v_i} = 0,$$

which can be written in the usual way

$$\Delta G^\circ = -RT \ln K, \tag{14.38}$$

where ΔG° is the standard Gibbs energy change for the reaction, and K is the equilibrium quotient of mole fractions. Thus, in an ideal solution, the proper form of the equilibrium constant is a quotient of mole fractions.

If the solution is an ideal dilute solution, then for a reaction between solutes only, each μ_j is given by Eq. (14.18),

$$\mu_j = \mu_j^* + RT \ln x_j,$$

so that the equilibrium condition is

$$\Delta G^* = -RT \ln K, \tag{14.39}$$

K being again a quotient of equilibrium mole fractions. Obviously, we could equally well have chosen either Eq. (14.23) or (14.26) to express μ_j. In that event we would obtain

$$\Delta G^{**} = -RT \ln K' \quad \text{or} \quad \Delta G^\square = -RT \ln K''; \tag{14.40}$$

K' is a quotient of the equilibrium molalities; K'' is a quotient of equilibrium molarities; ΔG^{**} and $\Delta G^\square$ are the appropriate standard Gibbs energy changes.

Values of standard Gibbs energy changes are obtained from the measurement of equilibrium constants in the same way as were those for reactions in the gas phase. Values of individual standard Gibbs energies of solutes in solution are obtained, as they are for gaseous reactions, by combining the Gibbs energy changes for several reactions.

The temperature dependence is the same for these equilibrium constants as for any others; for example, for K' and K'',

$$\left(\frac{\partial \ln K'}{\partial T} \right)_p = \frac{\Delta H^{**}}{RT^2} \quad \text{and} \quad \left(\frac{\partial \ln K''}{\partial T} \right)_p = \frac{\Delta H^\square}{RT^2} \tag{14.41}$$

where ΔH^{**} and $\Delta H^\square$ are the appropriate standard enthalpy changes.

If the chemical reaction involves the solvent, the equilibrium constant has a slightly modified form. For example, suppose the equilibrium

$$CH_3COOH + C_2H_5OH \rightleftharpoons CH_3COOC_2H_5 + H_2O$$

is studied in water solution; then if the solution is dilute enough to use molarities to describe the Gibbs energy of the solutes, the equilibrium constant has the form

$$K'' = \frac{c_{EtAc} \, x_{H_2O}}{c_{HAc} \, c_{EtOH}}, \tag{14.42}$$

since in dilute solution Raoult's law holds for the solvent. In dilute solution $x_{H_2O} \approx 1$, so K'' becomes

$$K'' = \frac{c_{EtAc}}{c_{HAc} \, c_{EtOH}}. \tag{14.43}$$

The standard Gibbs energy change for K'' is $\Delta G^\square$, by Eq. (14.40), and must include $\mu_{H_2O}^\circ$; therefore

$$\Delta G^\square = \mu_{EtAc}^\square + \mu_{H_2O}^\circ - \mu_{HAc}^\square - \mu_{EtHO}^\square. \tag{14.44}$$

The $\mu_{H_2O}^\circ$ is the molar Gibbs energy of pure water; the $\mu_j^\square$ are the chemical potentials of the solutes in the hypothetical ideal solution of unit molarity.

QUESTIONS

14.1 The heat of vaporization increases in the normal alkane series $C_6H_{14}, C_8H_{18}, C_{10}H_{22}$. If octane is the solvent, should hexane or decane be added to decrease the vapor pressure?

14.2 If you want to prepare pure methanol distillate by fractional distillation of a CCl_4–CH_3OH solution, should the initial solution consist of more than, exactly, or less than 79.44 wt % CCl_4? [Consult Table 14.1(a).]

14.3 Consider a solution of molecular liquids A and B. If the intermolecular interactions between molecules A, between molecules B, and between molecules A and B are all comparable, the ideal solution conditions, Eqs. (14.4)–(14.6), are usually satisfied. Why? On this basis suggest why the solution benzene–toluene exhibits nearly ideal behavior (Fig. 14.11).

14.4 Fairly strong hydrogen bond interactions exist between acetone and chloroform molecules, but are absent in the pure liquids. What is a molecular explanation for the negative deviations displayed in Fig. 14.13?

14.5 Dissolving a gas in a liquid is an exothermic process. Assuming an ideal gas, account for this in terms of molecular forces. Suggest a molecular explanation of the Bunsen coefficient α increase with increasing gas boiling point.

14.6 Many organic reactions are effected between dilute solutions of reactants in inert organic solvents. Which of the relations (14.38) or (14.39) is appropriate to describe equilibria in such reactions?

PROBLEMS

14.1 Benzene and toluene form nearly ideal solutions. At 300 K, $p^\circ_{toluene} = 32.06$ mmHg and $p^\circ_{benzene} = 103.01$ mmHg.

a) A liquid mixture is composed of 3 mol of toluene and 2 mol of benzene. If the pressure over the mixture at 300 K is reduced, at what pressure does the first vapor form?
b) What is the composition of the first trace of vapor formed?
c) If the pressure is reduced further, at what pressure does the last trace of liquid disappear?
d) What is the composition of the last trace of liquid?
e) What will be the pressure, the composition of the liquid, and the composition of the vapor when 1 mol of the mixture has been vaporized? (*Hint:* Lever rule.)

14.2 Two liquids, A and B, form an ideal solution. At the specified temperature, the vapor pressure of pure A is 200 mmHg while that of pure B is 75 mmHg. If the vapor over the mixture consists of 50 mol percent A, what is the mole percent A in the liquid?

14.3 At $-31.2\,°C$, we have the data

Compound	Propane	n-butane
Vapor pressure, p°/mmHg	1200	200

a) Calculate the mole fraction of propane in the liquid mixture that boils at $-31.2\,°C$ under 760 mmHg pressure.
b) Calculate the mole fraction of propane in the vapor in equilibrium with the liquid in (a).

14.4 At $-47\,°C$ the vapor pressure of ethyl bromide is 10 mmHg, while that of ethyl chloride is 40 mmHg. Assume that the mixture is ideal. If there is only a trace of liquid present and the mole fraction of ethyl chloride in the vapor is 0.80,

a) What is the total pressure and the mole fraction of ethyl chloride in the liquid?
b) If there are 5 mol of liquid and 3 mol of vapor present at the same pressure as in (a), what is the overall composition of the system?

14.5 A gaseous mixture of two substances under a total pressure of 0.8 atm is in equilibrium with an ideal liquid solution. The mole fraction of substance A is 0.5 in the vapor phase and 0.2 in the liquid phase. What are the vapor pressures of the two pure liquids?

14.6 The composition of the vapor over a binary ideal solution is determined by the composition of the liquid. If x_1 and y_1 are the mole fractions of 1 in the liquid and vapor, respectively, find the value of x_1 for which $y_1 - x_1$ has a maximum. What is the value of the pressure at this composition?

14.7 Suppose that the vapor over an ideal solution contains n_1 mol of 1 and n_2 mol of 2 and occupies a volume V under the pressure $p = p_1 + p_2$. If we define $\overline{V}_2^\circ = RT/p_2^\circ$ and $\overline{V}_1^\circ = RT/p_1^\circ$, show that Raoult's law implies $V = n_1 \overline{V}_1^\circ + n_2 \overline{V}_2^\circ$.

14.8 Show that, while the vapor pressure in a binary ideal solution is a linear function of the mole fraction of either component in the liquid, the reciprocal of the pressure is a linear function of the mole fraction of either component in the vapor.

14.9 Given the vapor pressures of the pure liquids, and the overall composition of the system, what are the upper and lower limits of pressure between which liquid and vapor coexist in equilibrium?

14.10 a) The boiling points of pure benzene and pure toluene are 80.1 °C and 110.6 °C under 1 atm. Assuming the entropies of vaporization at the boiling points are the same, 90 J/K mol, by applying the Clausius–Clapeyron equation to each, derive an implicit expression for the boiling point of a mixture of the two liquids as a function of the mole fraction of benzene, x_b.
b) What is the composition of the liquid that boils at 95 °C?

14.11 Some nonideal systems can be represented by the equations $p_1 = x_1^a p_1^\circ$ and $p_2 = x_2^a p_2^\circ$. Show that if the constant a is greater than unity, the total pressure exhibits a minimum, while if a is less than unity, the total pressure exhibits a maximum.

14.12 a) In an ideal dilute solution, if p_1° is the vapor pressure of the solvent and K_h is the Henry's law constant for the solute, write the expression for the total pressure over the solution as a function of x_2, the mole fraction of the solute.
b) Find the relation between y_1 and the total pressure of the vapor.

14.13 The Bunsen absorption coefficients of oxygen and nitrogen in water are 0.0283 and 0.0143, respectively, at 25 °C. Suppose that air is 20% oxygen and 80% nitrogen. How many cubic centimetres of gas, measured at STP, will be dissolved by 100 cm^3 of water in equilibrium with air at 1 atm pressure? How many will be dissolved if the pressure is 10 atm? What is the mole ratio, N_2/O_2, of the dissolved gas?

14.14 The Henry's law constant for argon in water is 2.17×10^4 at 0 °C and 3.97×10^4 at 30 °C. Calculate the standard heat of solution of argon in water.

14.15 Suppose that a 250-cm^3 bottle of carbonated water at 25 °C contains CO_2 under 2 atm pressure. If the Bunsen absorption coefficient of CO_2 is 0.76, what is the total volume of CO_2, measured at STP, that is dissolved in the water?

14.16 At 25 °C, for $CO_2(g)$, $\mu^\circ(g) = -394.36$ kJ/mol and $\mu^{**}(aq) = -386.02$ kJ/mol, while $\overline{H}^\circ(g) = -393.51$ kJ/mol and $\overline{H}^{**}(aq) = -413.80$ kJ/mol. For the equilibrium, $CO_2(g) \rightleftharpoons CO_2(aq)$, calculate

a) the molality of CO_2 in water under 1 atm pressure at 25 °C and at 35 °C;
b) the Bunsen absorption coefficient for CO_2 in water at 25 °C and 35 °C; $\rho_{H_2O} = 1.00$ g/cm^3.

14.17 At 25 °C, the standard Gibbs energies of formation of the inert gases in aqueous solution at unit molality are

Gas	He	Ne	Ar	Kr	Xe
μ_j^{**}/(kJ/mol)	19.2	19.2	16.3	15.1	13.4

Calculate the Bunsen adsorption coefficient for each of these gases; $\rho_{H_2O} = 1.00$ g/cm^3.

14.18 The Bunsen adsorption coefficient for hydrogen in nickel at 725 °C is 62. The equilibrium is

$$H_2(g) \rightleftharpoons 2H(Ni)$$

a) Show that the solubility of hydrogen in nickel follows Sieverts's law, $x_H = K_s p_{H_2}^{1/2}$; calculate the Sieverts's law constant, K_s.
b) Calculate the solubility of hydrogen in nickel (atoms H per atom Ni) at $p_{H_2} = 1$ atm and 4 atm; $\rho_{Ni} = 8.7$ g/cm^3.

14.19 At 800 °C, 1.6×10^{-4} mole O_2 dissolves in 1 mole of silver. Calculate the Bunsen adsorption coefficient for oxygen in silver; $\rho(Ag) = 10.0 \text{ g/cm}^3$.

14.20 The distribution coefficient of iodine between CCl_4 and H_2O is $c_{CCl_4}/c_{H_2O} = K = 85$, where c_S is the concentration (mol/L) of iodine in the solvent S.

   a) If 90 % of the iodine in 100 cm^3 of aqueous solution is to be extracted in one step, what volume of CCl_4 is required?
   b) What volume of CCl_4 is required if two extractions, using equal volumes, are permitted?
   c) If β is the fraction of the original amount of I_2 that is to remain in the water layer after n extractions using equal volumes of CCl_4, show that the limiting total volume of CCl_4 needed as $n \to \infty$ is $K^{-1} \ln(1/\beta)$ per unit volume of the aqueous layer.

14.21 The equilibrium constant for the reaction at 25 °C

$$CO_2(aq) + H_2O(l) \;\rightleftharpoons\; H_2CO_3(aq)$$

is $K = 2.58 \times 10^{-3}$. If $\Delta G_f^\circ(CO_2, aq) = -386.0 \text{ kJ/mol}$ and $\Delta G_f^\circ(H_2O, l) = -237.18 \text{ kJ/mol}$, calculate $\Delta G_f^\circ(H_2CO_3, aq)$.

14.22 Evaluate the difference, $\mu_j^{**} - \mu_j^*$, in aqueous solution at 25 °C.

14.23 Suppose we were to use for a solute in the ideal dilute solution, $\mu_j = \mu_j^{\circ\circ} + RT \ln \tilde{c}_j$, where $\tilde{c}_j$ is the abbreviation for $\tilde{c}_j/(1 \text{ mol/m}^3)$. Find the difference between $\mu_j^{\circ\circ}$ and μ_j° and evaluate it at 25 °C for aqueous solutions.

15

Equilibria Between Condensed Phases

15.1 LIQUID–LIQUID EQUILIBRIA

If small amounts of toluene are added to a beaker containing pure benzene we observe that, regardless of the amount of toluene added, the mixture obtained remains as one liquid phase. The two liquids are *completely miscible*. In contrast to this behavior, if water is added to nitrobenzene, two separate liquid layers are formed; the water layer contains only a trace of dissolved nitrobenzene, while the nitrobenzene layer contains only a trace of dissolved water. Such liquids are *immiscible*. If small amounts of phenol are added to water, at first the phenol dissolves to yield one phase; however, at some point in the addition the water becomes saturated and further addition of phenol yields two liquid layers, one rich in water, the other rich in phenol. Such liquids are *partially miscible*. It is these systems that presently engage our attention.

Consider a system in equilibrium that contains two liquid layers, two liquid phases. Let one of these liquid layers consist of pure liquid A, the other layer is a saturated solution of A in liquid B. The thermodynamic requirement for equilibrium is that the chemical potential of A in the solution, μ_A, be equal to that in the pure liquid, μ_A°. So $\mu_A = \mu_A^\circ$, or

$$\mu_A - \mu_A^\circ = 0. \tag{15.1}$$

First, we ask whether Eq. (15.1) can be satisfied for an ideal solution. In an ideal solution, by Eq. (14.3),

$$\mu_A - \mu_A^\circ = RT \ln x_A. \tag{15.2}$$

It is clear from Eq. (15.2) that $RT \ln x_A$ is never zero unless the mixture of A and B has $x_A = 1$, that is, unless the mixture contains no B. In Fig. 15.1, $\mu_A - \mu_A^\circ$ is plotted against x_A for the ideal solution (full line). The value of $\mu_A - \mu_A^\circ$ is negative for all compositions of the ideal solution. This implies that pure A can always be transferred into an ideal solution

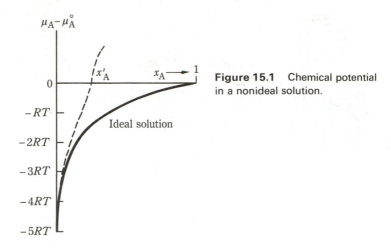

Figure 15.1   Chemical potential in a nonideal solution.

with a decrease in Gibbs energy. Consequently, substances that form ideal solutions are completely miscible in each other.

For partial miscibility the value for $\mu_A - \mu_A^\circ$ must be zero at some intermediate composition of the solution; thus $\mu_A - \mu_A^\circ$ must follow some such curve as the dashed line in Fig. 15.1. At the point x_A', the value of $\mu_A - \mu_A^\circ$ is zero, and the system can exist as a solution having mole fraction of A $= x_A'$ and a separate layer of pure liquid A. The value x_A' is the solubility of A in B expressed as a mole fraction. If the mole fraction of A in B were to exceed this value, then Fig. 15.1 shows that $\mu_A - \mu_A^\circ$ would be positive, that is $\mu_A > \mu_A^\circ$. In this circumstance, A would flow spontaneously from the solution into the pure liquid A, thus reducing x_A until it reached the equilibrium value x_A'.

Liquids that are only partially miscible form solutions which are far from ideal, as the curves in Fig. 15.1 show. Rather than explore the mathematical side of this situation in great detail, we restrict ourselves to a description of the experimental results interpreted in the light of the phase rule.

Suppose that at a given temperature T_1, small amounts of liquid A are added successively to liquid B. The first amount of A dissolves completely, as do the second and the third; the state points can be represented on a $T-X$ diagram such as Fig. 15.2(a), which is drawn at constant pressure. The points a, b, c represent the composition after the addition of three amounts of A to pure B. Since all the A dissolves, these points lie in a one-phase region. After a certain amount of A has been added, the solubility limit is reached, point l_1. If more A is added, a second layer forms, since no more A will dissolve. The region to the right of point l_1 is therefore a two-phase region.

The same could be done on the right side by adding B to A. At first B dissolves to yield a homogeneous (one-phase) system, points d, e, f. The solubility limit of B in A is reached at l_2. Points to the left of l_2 represent a two-phase system. In the region between l_1 and l_2 two liquid layers, called *conjugate solutions*, coexist. Layer l_1 is a saturated solution of A in B in equilibrium with layer l_2, which is a saturated solution of B in A. If the experiment were done at a higher temperature, different values of the solubility limits, l_1' and l_2', would be obtained.

The T versus X diagram for the system phenol–water is shown in Fig. 15.2(b). As the temperature increases, the solubility of each component in the other increases. The solubility curves join smoothly at the *upper consolute temperature*, also called the *critical*

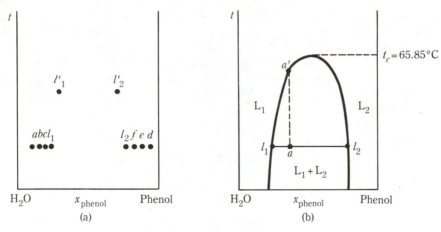

Figure 15.2 Water–phenol system.

solution temperature, t_c. Above t_c, water and phenol are completely miscible. Any point a under the loop is the state point of a system consisting of two liquid layers: L_1 of composition l_1 and L_2 of composition l_2. The relative mass of the two layers is given by the lever rule, by the ratio of the segments of the tie line $(\overline{l_1 l_2})$.

$$\frac{\text{moles of } l_1}{\text{moles of } l_2} = \frac{(\overline{a l_2})}{(\overline{a l_1})}.$$

If the temperature of this system is raised, the state point follows the dashed line aa'; L_1 becomes richer in phenol, while L_2 becomes richer in water. As the temperature increases, the ratio $(\overline{a l_2})/(\overline{a l_1})$ becomes larger; the amount of L_2 decreases. At point a' the last trace of L_2 disappears and the system becomes homogeneous.

Systems are known in which the solubility *decreases* with increase in temperature. In some of these systems, a *lower consolute temperature* is observed; Fig. 15.3(a) shows schematically the triethylamine–water system. The lower consolute temperature is at 18.5 °C. The curve is so flat that it is difficult to determine the composition of the solution

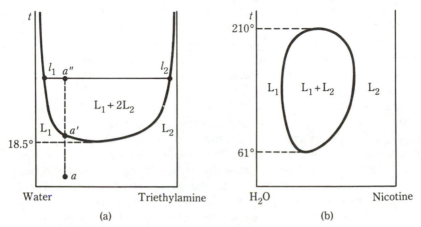

Figure 15.3 (a) Lower consolute temperature. (b) Upper and lower consolute temperature.

corresponding to the consolute temperature; it seems to be about 30% by weight of triethylamine. If a solution having a state point a is heated, it remains homogeneous until the temperature is slightly above 18.5 °C; at this point, a', it splits into two layers. At a higher temperature a'', the solutions have the compositions given by l_1 and l_2. In view of the lever rule, l_1 will be present in somewhat greater amount than l_2. As a rule, the liquid pairs that have solubility diagrams of this type tend to form loosely bound compounds with each other; this enhances solubility at low temperatures. As the temperature is increased, the compound is dissociated and the mutual solubility is diminished.

Some substances exhibit both upper and lower consolute temperatures. The diagram for the system nicotine–water is shown schematically in Fig. 15.3(b). The lower consolute temperature is about 61 °C, the upper one about 210 °C. At all points in the closed loop two phases are present, while the points outside the loop represent homogeneous states of the system.

The phase rule for a system at constant pressure is $F' = C - P + 1$, in which F' is the number of variables in addition to the pressure needed to describe the system. For two-component systems, $F' = 3 - P$. If two phases are present, only one variable is required to describe the system. In the two-phase region, if the temperature is described, then the intersections of the tie line with the curve yield the compositions of *both* conjugate solutions. Similarly, the composition of one of the conjugate solutions is sufficient to determine the temperature and the composition of the other conjugate solution. If only one phase is present, $F' = 2$ and both the temperature and the composition of the solution must be specified.

15.2 DISTILLATION OF PARTIALLY MISCIBLE AND IMMISCIBLE LIQUIDS

The discussion in Section 15.1 assumed that the pressure is high enough so that vapor does not form in the temperature range of interest. For this reason the liquid–vapor curves were omitted from the diagrams. A typical situation at lower pressures is shown in Fig. 15.4(a) in which the liquid–vapor curves are also shown, still with the assumption that the pressure is fairly high. Figure 15.4(a) presents no new problem in interpretation. The upper and lower portions of the diagram can be discussed independently using the principles described before. Partial miscibility at low temperatures usually, though not always, implies a minimum boiling azeotrope, as is shown in Fig. 15.4(a). The partial miscibility implies that when mixed the two components have a greater escaping tendency than in an ideal solution. This greater escaping tendency may lead to a maximum in the vapor pressure–composition curve, and correspondingly to a minimum in the boiling point–composition curve.

If the pressure on the system shown in Fig. 15.4(a) is lowered, the boiling points will all be shifted downward. At a low enough pressure, the boiling point curves will intersect the liquid–liquid solubility curves. The result is shown in Fig. 15.4(b), which represents schematically the system water–n-butanol under 1 atm pressure.

Figure 15.4(b) presents several new features. If the temperature of a homogeneous liquid, point a, is increased, vapor having the composition b forms at t_A. This behavior is ordinary enough; however, if this vapor is chilled and brought to point c, the condensate will consist of two liquid layers, since c lies in the two-liquid region. So the first distillate produced by the distillation of the homogeneous liquid a will separate into two liquid layers having compositions d and e. Similar behavior is exhibited by mixtures having compositions in the region L_1.

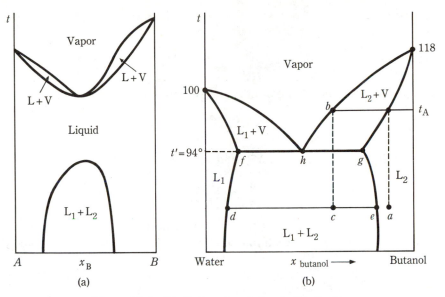

Figure 15.4 Distillation of partially miscible liquids.

As the temperature of the two-liquid system of overall composition c is increased, the compositions of the conjugate solutions shift slightly. The system is univariant, $F' = 3 - P = 1$ in this region. At the temperature t', the conjugate solutions have the compositions f and g, and vapor, composition h, appears. Three phases are present, liquids f and g, and vapor h. Then $F' = 0$; the system is invariant. As long as these three phases remain, their compositions and the temperature are fixed. For example, the flow of heat into the system does not change the temperature, but simply produces more vapor at the expense of the two solutions. The vapor, h, that forms is richer in water than the original composition c; therefore the water-rich layer evaporates preferentially. After the water-rich layer disappears, the temperature rises and the vapor composition changes along the curve hb. The last liquid, which has the composition a, disappears at t_A.

If a two-phase system in the composition range between f and h is heated, then at t' liquids f and g are present, and vapor h appears. The system at t' is invariant. Since the vapor is richer in butanol than the original overall composition, the butanol-rich layer evaporates preferentially, leaving liquid f and vapor h. As the temperature rises, the liquid is depleted in butanol; finally only vapor remains.

The point h has the azeotropic property; a system of this composition distills unchanged. It cannot be separated into its components by distillation.

The distillation of *immiscible* substances is most easily discussed from a different standpoint. Consider two immiscible liquids in equilibrium with vapor at a specified temperature (Fig. 15.5). The barrier only keeps the liquids apart; since they are immiscible, removing the barrier would not change anything. The total vapor pressure is the sum of the

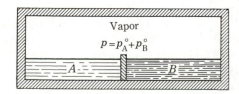

Figure 15.5 Immiscible liquids in equilibrium with vapor.

vapor pressures of the pure liquids: $p = p_A^\circ + p_B^\circ$. The mole fractions y_A and y_B in the vapor are

$$y_A = \frac{p_A^\circ}{p} \qquad y_B = \frac{p_B^\circ}{p}.$$

If n_A and n_B are the number of moles of A and B in the vapor, then

$$\frac{n_A}{n_B} = \frac{y_A}{y_B} = \frac{p_A^\circ/p}{p_B^\circ/p} = \frac{p_A^\circ}{p_B^\circ}.$$

The masses of A and B are $w_A = n_A M_A$, and $w_B = n_B M_B$, so that

$$\frac{w_A}{w_B} = \frac{M_A p_A^\circ}{M_B p_B^\circ}, \tag{15.3}$$

which relates the relative masses of the two substances present in the vapor to their molar masses and vapor pressures. If this vapor were condensed, Eq. (15.3) would express the relative masses of A and B in the condensate. Suppose we choose the system aniline (A)–water (B) at 98.4 °C. The vapor pressure of aniline at this temperature is about 42 mmHg, while that of water is about 718 mmHg. The total vapor pressure is $718 + 42 = 760$ mmHg, so this mixture boils at 98.4 °C under 1 atm pressure. The mass of aniline that distills for each 100 g of water which comes over is

$$w_A = 100 \text{ g} \frac{(94 \text{ g/mol})(42 \text{ mmHg})}{(18 \text{ g/mol})(718 \text{ mmHg})} \approx 31 \text{ g}.$$

Equation (15.3) can be applied to the steam distillation of liquids. Some liquids that decompose if distilled in the ordinary way can be steam distilled if they have fair volatility near the boiling point of water. In the laboratory, steam is passed through the liquid to be steam distilled. Since the vapor pressure is greater than that of either component, it follows that the boiling point is below the boiling points of both liquids. Furthermore, the boiling point is an invariant temperature so long as the two liquid phases and the vapor are present.

If the vapor pressure of the substance is known over a range of temperatures near 100 °C, measurement of the temperature at which the steam distillation occurs and the mass ratio in the distillate yield, through Eq. (15.3), a value of the molar mass of the substance.

15.3 SOLID–LIQUID EQUILIBRIA; THE SIMPLE EUTECTIC DIAGRAM

If a liquid solution of two substances A and B is cooled to a sufficiently low temperature, a solid will appear. This temperature is the freezing point of the solution, which depends on the composition. In the discussion of freezing-point depression, Section 13.6, we obtained the equation

$$\ln x_A = -\frac{\Delta H_{\text{fus, A}}}{R} \left(\frac{1}{T} - \frac{1}{T_{0A}} \right), \tag{15.4}$$

assuming that pure solid A is in equilibrium with an ideal liquid solution. Equation (15.4) relates the freezing point of the solution to x_A, the mole fraction of A in the solution. A plot of this function is shown in Fig. 15.6(a). The points above the curve represent liquid states of the system; those below the curve represent states in which pure solid A coexists in equilibrium with solution. The curve is called the *liquidus* curve.

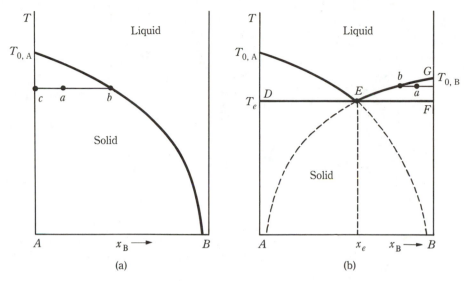

Figure 15.6 Solid–liquid equilibria in a two-component system.

A point such as a represents solution of composition b in equilibrium with solid of composition c, that is, pure A. By the lever rule, the ratio of the number of moles of solution to the number of moles of solid A present is equal to the ratio of segments of the tie line $\overline{ac}/\overline{ab}$. The lower the temperature, the greater the relative amount of solid for a specified overall composition.

This curve cannot represent the situation over the entire range of composition. As $x_B \to 1$, we would expect solid B to freeze out far above the temperatures indicated by the curve in this region. If the solution is ideal, the same law holds for substance B:

$$\ln x_B = -\frac{\Delta H_{\text{fus,B}}}{R}\left(\frac{1}{T} - \frac{1}{T_{0B}}\right), \tag{15.5}$$

where T is the freezing point of B in the solution. This curve is drawn in Fig. 15.6(b) together with the curve for A from Fig. 15.6(a). The curves intersect at a temperature T_e, the *eutectic* temperature. The composition x_e is the eutectic composition. The line GE is the freezing point versus composition curve for B. Points such as a below this curve represent states in which pure solid B is in equilibrium with solution of composition b. A point on EF represents pure solid B in equilibrium with solution of composition x_e. However, a point on DE represents pure solid A in equilibrium with solution of composition x_e. *Therefore the solution having the eutectic composition x_e is in equilibrium with both pure solid A and pure solid B.* If three phases are present, then $F' = 3 - P = 3 - 3 = 0$; the system is invariant at this temperature. If heat flows out of such a system, the temperature remains the same until one phase disappears; thus the relative amounts of the three phases change as heat is withdrawn. The amount of liquid diminishes while the amounts of the two solids present increase. Below the line DEF are the states of the system in which only the two solids, two phases, pure A and pure B, are present.

15.3.1 The Lead–Antimony System

The lead–antimony system has the simple eutectic type of phase diagram (Fig. 15.7). The regions are labeled; L signifies liquid, Sb or Pb signifies pure solid antimony or pure solid

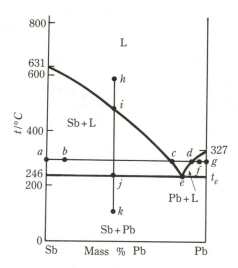

Figure 15.7   The antimony–lead system.

lead. The eutectic temperature is 246 °C; the eutectic composition is 87 mass percent lead. In the lead–antimony system, the values of t_e and x_e calculated from Eqs. (15.4) and (15.5) agree satisfactorily with the experimental values. This implies that the liquid is nearly an ideal solution.

Consider the isothermal behavior of the system at 300 °C, the horizontal line, *abcdfg*. The point *a* represents pure solid antimony at 300 °C. Suppose sufficient solid lead is added to bring the composition to point *b*. This point *b* lies in the region Sb + L, therefore solid antimony coexists with liquid of composition *c*. All the added lead melts and the molten lead dissolves enough of the solid antimony to bring the liquid to the composition *c*. The lever rule shows that the relative amount of liquid present at point *b* is quite small, so the liquid may not be visible; nonetheless it is present at equilibrium. On further addition of lead, the lead continues to melt and dissolve more of the solid antimony to form solution *c*; meanwhile the state point moves from *b* to *c*. When the state point reaches *c*, sufficient lead has been added to dissolve all of the original antimony present to form the saturated solution of antimony in lead. Further addition of lead simply dilutes this solution as the state point moves through the liquid region *c* to *d*. At *d* the solution is saturated with lead; further addition of lead produces no change. The state point meanwhile has moved to *f*. If we had reached *f* by starting with pure lead at *g* and adding antimony, all of the antimony would have melted, 330 °C below its melting point, and dissolved sufficient lead to form the solution *d*.

An *isopleth* is a line of constant composition such as *hijk* in Fig. 15.7. At *h*, the system is entirely liquid. As the system cools, solid antimony appears at *i*; as the antimony crystallizes out, the saturated liquid becomes richer in lead, and the liquid composition moves along the curve *ice*. At *j* the solution has the eutectic composition *e* and is saturated with respect to lead also, so lead begins to precipitate. The temperature remains constant even though heat flows out since, in this condition, the system is invariant. The amount of liquid diminishes and the amounts of solid lead and antimony increase. Finally the liquid solidifies, and the temperature of the mixed solids decreases along the line *jk*. If the process is done in reverse, heating a mixture of solid lead and solid antimony from *k*, the state point moves from *k* to *j*. At *j*, liquid forms having the composition *e*. Note that the liquid formed has a different composition than the solid mixture. The system is invariant, so the temperature remains at 246 °C until all of the lead melts; since the liquid was richer in lead than the original mixture, the lead melts completely leaving a residue of solid anti-mony. After the lead has melted the temperature rises, and the antimony that melts moves the liquid composition from e to i. At i the last bit of antimony melts and the system is homogeneous above i.

The eutectic (Greek: easily melted) point gets its name from the fact that the eutectic composition has the lowest melting point. The eutectic mixture melts sharply at t_e to form a liquid of the same composition, while other mixtures melt over a range of temperature. Because of the sharp melting point, the eutectic mixture was originally thought to be a compound. In aqueous systems, this "compound" was called a cryohydrate; the eutectic point was called the cryohydric point. Microscopic examination of the eutectic under high magnification discloses its heterogeneous character; it is a mixture, not a compound. In alloy systems, such as the lead–antimony system, the eutectic is often particularly fine-grained; however, under the microscope the separate crystals of lead and antimony can be discerned.

15.3.2 Thermal Analysis

The shape of the freezing point curves can be determined experimentally by *thermal analysis*. In this method, a mixture of known composition is heated to a high enough temperature so that it is homogeneous. Then it is allowed to cool at a regulated rate. The temperature is plotted as a function of time. The curves obtained for various com-positions are shown schematically for a system A–B in Fig. 15.8. In the first curve the homogeneous liquid cools along the curve ab; at b the primary crystals of component A form. This releases the latent heat of fusion: the rate of cooling slows, and a kink in the curve appears at b. The temperature t_1 is a point on the liquidus curve for this composition. The cooling continues along bc; at c the liquid has the eutectic composition, and solid B appears. Since the system is invariant, the temperature remains constant at the eutectic temperature until all the liquid solidifies at d. The horizontal plateau cd is called the *eutectic halt*. After the liquid solidifies the two solids cool quickly along the curve df. The

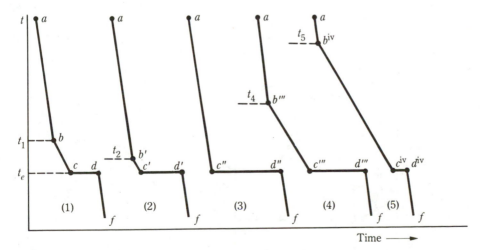

Figure 15.8 Cooling curves.

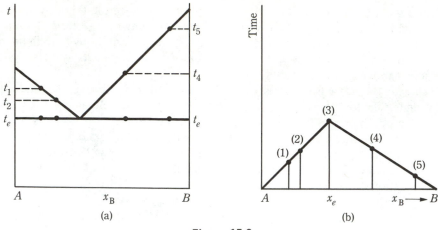

Figure 15.9

second curve is for a liquid somewhat richer in **B**; the interpretation is the same; however, the eutectic halt is longer; t_2 is the point on the liquidus curve. The third curve illustrates the cooling of the eutectic mixture; the eutectic halt has its maximum length. The fourth and fifth curves are for compositions on the **B**-rich side of the eutectic point; t_4 and t_5 are the corresponding points on the liquidus curve. The length of the eutectic halt diminishes as the composition departs from the eutectic composition. The temperatures t_1, t_2, t_4, t_5, and t_e are plotted against composition in Fig. 15.9(a). The eutectic composition can be determined as the intersection of the two solubility curves if sufficient points are taken; otherwise the length (in time) of the eutectic halt is plotted as a function of composition, Fig. 15.9(b). The intersection of the two curves yields the maximum value of the eutectic halt, and thus the eutectic composition.

★15.3.3   Other Simple Eutectic Systems

Many binary systems, both ideal and nonideal, have phase diagrams of the simple eutectic type. The phase diagram, water–salt, is the simple eutectic type if the salt does not form a stable hydrate. The diagram for H_2O–NaCl is shown in Fig. 15.10. The curve ae is the freezing-point curve for water, while ef is the solubility curve, or the freezing-point curve, for sodium chloride.

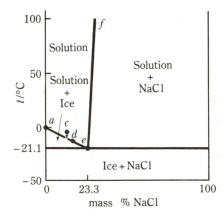

Figure 15.10   Freezing points in the H_2O–NaCl system.

Table 15.1

Salt	Eutectic temperature °C	Mass percent anhydrous salt in eutectic
Sodium chloride	-21.1	23.3
Sodium bromide	-28.0	40.3
Sodium sulfate	-1.1	3.84
Potassium chloride	-10.7	19.7
Ammonium chloride	-15.4	19.7

By permission from A. Findlay, A. N. Campbell, and N. O. Smith, *The Phase Rule and Its Applications*, 9th ed. New York: Dover, 1951, p. 141.

The invariance of the system at the eutectic point allows eutectic mixtures to be used as constant temperature baths. Suppose solid sodium chloride is mixed with ice at 0 °C in a vacuum flask. The composition point moves from 0% NaCl to some positive value. However, at this composition the freezing point of ice is below 0 °C; hence, some ice melts. Since the system is in an insulated flask, the melting of the ice reduces the temperature of the mixture. If sufficient NaCl has been added, the temperature will drop to the eutectic temperature, -21.1 °C. At the eutectic temperature, ice, solid salt, and saturated solution can coexist in equilibrium. The temperature remains at the eutectic temperature until the remainder of the ice is melted by the heat that leaks slowly into the flask.

The action of rock salt or calcium chloride in melting ice on sidewalks and streets can be interpreted by the phase diagram. Suppose sufficient solid salt is added to ice at -5 °C to move the state point of the system to c (Fig. 15.10). At c the solution is stable; the ice will melt completely if the system is isothermal. If the system were adiabatic, the temperature would fall until the state point reached d. The eutectic temperatures of a few ice–salt systems are given in Table 15.1.

15.4 FREEZING-POINT DIAGRAM WITH COMPOUND FORMATION

If two substances form one or more compounds, the freezing-point diagram often has the appearance of two or more simple eutectic diagrams in juxtaposition. Figure 15.11 is the freezing-point–composition diagram for the system in which a compound, AB_2, is formed. We can consider this diagram as two simple eutectic diagrams joined at the position of the arrows in Fig. 15.11. If the state point lies to the right of the arrows, the interpretation is based on the simple eutectic diagram for the system AB_2–B; if it lies to the left of the arrows, we discuss the system A–AB_2. In the composite diagram there are two eutectics: one of the A–AB_2–liquid; the other of AB_2–B–liquid. The melting point of the compound is a maximum in the curve; a maximum in the melting-point–composition curve is almost always indicative of compound formation. Only a few systems are known in which the maximum occurs for other reasons. The first solid deposited on cooling a melt of any composition between the two eutectic compositions is the solid compound.

It is conceivable that more than one compound is formed between the two substances; this is often the case with salts and water. The salt forms several hydrates. An extreme example of this behavior is shown by the system ferric chloride–water; Fig. 15.12. This diagram could be split into five simple eutectic diagrams.

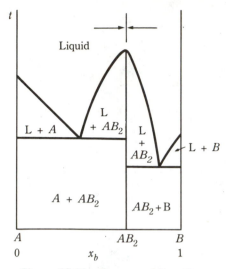

Figure 15.11 Compound formation.

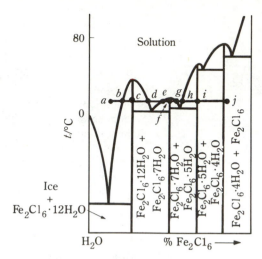

Figure 15.12 Freezing points in the system $H_2O-Fe_2Cl_6$ (schematic).

15.5 COMPOUNDS HAVING INCONGRUENT MELTING POINTS

In the system in Fig. 15.11, the compound has a higher melting point than either component. In this situation the diagram always has the shape shown in Fig. 15.11; two eutectics appear on the diagram. However, if the melting point of the compound lies below the melting point of one of the constituents, two possibilities arise. The first of these is illustrated in Fig. 15.12; each part of the diagram is a simple eutectic diagram just as in the simpler case in Fig. 15.11. The second possibility is illustrated by the alloy system potassium–sodium shown schematically in Fig. 15.13. In this system, the solubility curve of sodium does not drop rapidly enough to intersect the other curve between the composition of Na_2K and pure Na. Instead it swings to the left of the composition Na_2K and intersects the other solubility curve at point c, the *peritectic* point. For the system Na–K it is at 7 °C.

First we examine the behavior of the pure solid compound. If the temperature is raised, the state point moves along the line ab. At b liquid having the composition c forms. Since this liquid is richer in potassium than the original compound, some solid sodium d is left unmelted. Thus, on melting, the compound undergoes the reaction

$$Na_2K(s) \longrightarrow Na(s) + c(l).$$

This is a *peritectic reaction* or a *phase reaction*. The compound is said to melt *incongruently*, since the melt differs from the compound in composition. (The compounds illustrated in Figs. 15.11 and 15.12 melt *congruently*, without change in composition.) Since three phases, solid Na_2K, solid sodium and liquid are present, the system is invariant; as heat flows into the system, the temperature remains the same until the solid compound melts completely. Then the temperature rises; the state point moves along the line bef and the system consists of solid sodium plus liquid. At f the last trace of solid sodium melts, and above f the system consists of one liquid phase. Cooling the composition g reverses these changes. At f solid sodium appears; the liquid composition moves along fc. At b liquid of composition c coexists with solid sodium and solid Na_2K. The reverse of the phase reaction occurs until liquid and solid sodium are both consumed simultaneously; only Na_2K remains and the state point moves along ba.

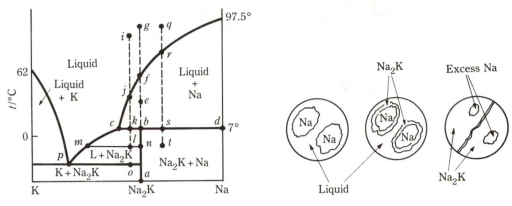

Figure 15.13 Compound with incongruent melting point.

Figure 15.14 Peritectic crystallization with excess Na.

If a system of composition i is cooled, primary crystals of sodium form at j; the liquid composition moves along jc as more sodium crystallizes. At k solid Na_2K forms because of the peritectic reaction.

$$c(l) + Na(s) \longrightarrow Na_2K(s).$$

The amount of sodium in the composition i is insufficient to convert the liquid c completely into compound. Hence the primary crystals of sodium are consumed completely. After the sodium is consumed, the temperature drops, Na_2K crystallizes, and the liquid composition moves along cm; at l, the tie line shows that Na_2K, n, coexists with liquid m. When the temperature reaches o, pure potassium begins to crystallize; the liquid has the eutectic composition p; the system is invariant until the liquid disappears, leaving a mixture of solid potassium and solid Na_2K.

If liquid of composition q is cooled, primary crystals of sodium form at r; continued cooling crystallizes more sodium, the liquid composition moves along rc. At s, solid Na_2K forms by the peritectic reaction. The liquid is consumed entirely, and the state point drops to t, the system consisting of a mixture of solids, Na_2K and sodium. Because the compound is formed by the reaction of liquid with the primary crystals of sodium, the structure of the solid mixture is unusual. The steps in the reaction are illustrated in Fig. 15.14. The final mixture has kernels of the primary sodium crystals within a shell of the compound. Since the phase reaction occurs between the primary crystal, which is shielded from the liquid by a layer of compound, it is difficult to establish equilibrium in a system such as this unless the experiments are prolonged to allow time for one reactant or the other to diffuse through the layer of compound. An interesting sidelight on this particular system is the wide range of composition in which the alloys of sodium and potassium are liquid at room temperature.

★**15.5.1   The Sodium Sulfate–Water System**

The sodium sulfate–water system forms an incongruently melting compound, $Na_2SO_4 \cdot 10H_2O$ (Fig. 15.15a). The line eb is the solubility curve for the decahydrate, while the line ba is the solubility curve for the anhydrous salt. The figure shows that the solubility of the decahydrate increases, while that of the anhydrous salt decreases with

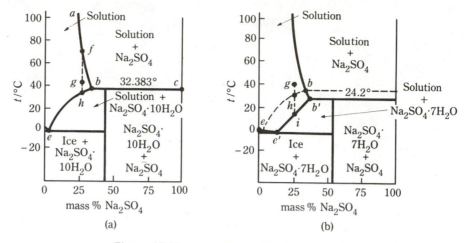

Figure 15.15   The sodium sulfate–water system.

temperature. The peritectic point is at b. On the line bc, three phases coexist: Na_2SO_4, $Na_2SO_4 \cdot 10H_2O$, saturated solution; the system is invariant and the peritectic temperature, 32.383 °C, is fixed. This temperature is frequently used as a calibration point for thermometers. If a small amount of water is added to anhydrous Na_2SO_4 in a vacuum bottle at room temperature, the salt and water react to form the decahydrate; this reaction is exothermic so that the temperature of the system rises to 32.383 °C and remains at that temperature as long as the three phases are present.

If an unsaturated solution of composition g is heated, anhydrous salt will crystallize at f; if it is cooled, the decahydrate will crystallize at h. It is possible to supercool the solution to a temperature below h; then the heptahydrate will crystallize at i; Fig. 15–15(b). The curve $e'b'$ is the solubility curve for the heptahydrate, $Na_2SO_4 \cdot 7H_2O$. The peritectic temperature for anhydrous salt–heptahydrate-saturated solution is at 24.2 °C. In Fig. 15.15(b), the dashed lines are the curves for the decahydrate. The solubility curve for the heptahydrate lies for the most part in the region of stability of solid decahydrate-saturated solution. Therefore the equilibrium between solid heptahydrate and its saturated solution is a metastable one; the system in such a state can precipitate the less soluble decahydrate spontaneously.

★15.6 MISCIBILITY IN THE SOLID STATE

In the systems described so far, only pure solids have been involved. Many solids are capable of dissolving other materials to form *solid solutions*. Copper and nickel, for example, are soluble in each other in all proportions in the solid state. The phase diagram for the copper–nickel system is shown in Fig. 15.16.

The upper curve in Fig. 15.16 is the *liquidus* curve; the lower curve, the *solidus* curve. If a system represented by point a is cooled to b, a solid solution of composition c appears. At point d the system consists of liquid of composition b' in equilibrium with solid solution of composition c'. The interpretation of the diagram is similar to the interpretation of the liquid–vapor diagrams in Section 14.6. An experimental difficulty arises in working with this type of system. Suppose the system were chilled quickly from a to e. If the system managed to stay in equilibrium, then the last vestige of liquid b'' would be in contact with a

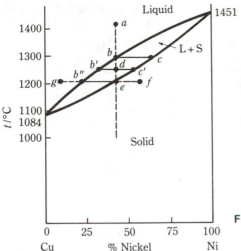

Figure 15.16   The copper–nickel system.

solid having a uniform composition e throughout. However, in a sudden chilling there is not time for the composition of the solid to become uniform throughout. The first crystal had the composition c and layers having compositions from c to e are built up on the outside of the first crystal. The average composition of the solid that has crystallized lies perhaps at the point f; the solid is richer in nickel than it should be; it lies to the right of e. Hence the liquid is richer in copper than it should be; its composition point lies perhaps at g. Thus some liquid is left at this temperature and further cooling is required before the system solidifies completely. This difficulty poses a severe experimental problem. The system must be cooled extremely slowly to allow time for the solid to adjust its composition at each temperature to a uniform value. In the discussion of these diagrams we assume that equilibrium has been attained and disregard the experimental difficulty which this implies.

Binary systems are known that form solid solutions over the entire range of composition and which exhibit either a maximum or a minimum in the melting point. The liquidus–solidus curves have an appearance similar to that of the liquid–vapor curves in systems which form azeotropes. The mixture having the composition at the maximum or minimum of the curve melts sharply and simulates a pure substance in this respect just as an azeotrope boils at a definite temperature and distills unchanged. Mixtures having a maximum in the melting-point curve are comparatively rare.

★15.7 FREEZING-POINT ELEVATION

In Section 13.6 we showed that the addition of a foreign substance always lowered the melting point of a pure solid. Figure 15.16 illustrates a system in which the melting point of one component, copper, is *increased* by the addition of a foreign substance. This increase in the melting point can only occur if the solid in equilibrium with the liquid is not pure but is a solid solution.

Suppose that the solid solution is an *ideal solid solution*, defined, in analogy to ideal gaseous and ideal liquid solutions, by requiring that for every component, $\mu_i = \mu_i^\circ + RT \ln x_i$, where μ_i° is the chemical potential of the pure solid, x_i its mole fraction in the solid solution. The equilibrium condition for solid solution in equilibrium with liquid solution for one of the components is $\mu_1(s) = \mu_1(l)$. Assuming both solutions are ideal, we obtain

$$\mu_1^\circ(s) + RT \ln x_1(s) = \mu_1^\circ(l) + RT \ln x_1(l). \tag{15.6}$$

Let $\Delta G_1^\circ = \mu_1^\circ(l) - \mu_1^\circ(s)$, the Gibbs energy of fusion of the pure component at temperature T. Then, Eq. (15.6) becomes

$$\ln \left(\frac{x_1(l)}{x_1(s)} \right) = - \frac{\Delta G_1^\circ}{RT}. \tag{15.7}$$

Since $\Delta G_1^\circ = \Delta H_1^\circ - T \, \Delta S_1^\circ$; and at the melting point, T_{01}, of the pure substance, $\Delta S_1^\circ = \Delta H_1^\circ/T_{01}$, this equation becomes

$$\ln \left(\frac{x_1(l)}{x_1(s)} \right) = - \frac{\Delta H^0}{R} \left(\frac{1}{T} - \frac{1}{T_{01}} \right).$$

Solving this equation for T, we obtain

$$T = T_{01} \left\{ \frac{\Delta H^\circ}{\Delta H^\circ + RT_{01} \ln [x_1(s)/x_1(l)]} \right\}. \tag{15.8}$$

If the *pure* solid were present, then $x_1(s) = 1$; in this case the second term of the denominator in Eq. (15.8) would be positive so that the fraction in the braces would be less than unity. The freezing point T is therefore less than T_{01}. If a solid solution is present in equilibrium then if $x_1(s) < x_1(l)$, the second term in the denominator will be negative, the fraction in the braces will be greater than unity and the melting point will be greater than T_{01}.

Figure 15.16 shows that the mole fraction of copper in the solid solution $x_{Cu}(s)$ is always less than the mole fraction of copper in the liquid solution $x_{Cu}(l)$. Consequently, the melting point of copper is elevated. An analogous set of equations can be written for the second component, from which we would conclude that the melting point of nickel is depressed. In the argument we have assumed that the ΔH° and ΔS° do not vary with temperature; this is incorrect but does not affect the general conclusion.

★15.8 PARTIAL MISCIBILITY IN THE SOLID STATE

It is usual to find that two substances are neither completely miscible nor immiscible in the solid state, but rather each substance has a limited solubility in the other. For this case, the most common type of phase diagram is shown in Fig. 15.17. The points in region α describe solid solutions of B in A, while those in β describe solid solutions of A in B. The points in region $\alpha + \beta$ describe states in which the two saturated solid solutions, *two phases*, α and β, coexist in equilibrium. If we cool a system described by point a, then at point b crystals of solid solution α having the composition c appear. As the temperature drops, the compositions of solid and liquid shift; at d compositions f and g are in equilibrium. At h the liquid has the eutectic composition e; solid β appears, α, β, and liquid coexist, and the system is invariant. On cooling to i, two solid solutions coexist: α of composition j, β of composition k.

A different type of system in which solid solutions appear is shown in Fig. 15.18. This system has a transition point rather than a eutectic point. Any point on the line abc describes an invariant system in which α, β, and melt of composition c coexist. The temperature of abc is the transition temperature. If the point lies between a and b, cooling will cause melt to disappear, $\alpha + \beta$ remaining. If the point lies between b and c, cooling first causes α to disappear, $\beta + L$ remaining; further cooling causes liquid to disappear and only β remains. If the temperature increases, any point on abc goes into $\alpha + L$; β disappears.

An interesting example of a system in which many solid solutions occur is the Cu–Zn diagram (brass diagram) in Fig. 15.19. The symbols α, β, γ, δ, ϵ, η refer to homogeneous solid solutions, while regions labeled $\alpha + \beta$, $\beta + \gamma$ indicate regions in which two solid solutions coexist. Note that there is a whole series of transition temperatures and no eutectic temperatures in this diagram.

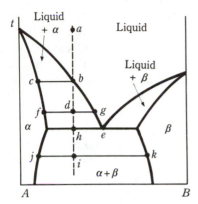

Figure 15.17   Partial miscibility in the solid state.

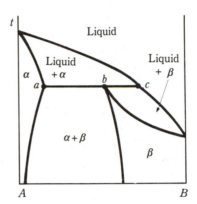

Figure 15.18   System with transition point.

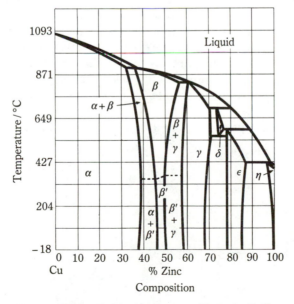

Figure 15.19   The brass diagram. (From A. G. Guy, *Physical Metallurgy for Engineers*. Reading, Mass.: Addison-Wesley, 1962.)

It is usual for phase diagrams to contain several features: solid solutions, compound formation, eutectic points, transition points, and the like. Once the interpretation of the individual features is understood, the interpretation of complex diagrams poses no difficulty.

★15.9 GAS–SOLID EQUILIBRIA; VAPOR PRESSURE OF SALT HYDRATES

In describing the equilibria between solids and liquids, we assumed implicitly that the pressure on the system was high enough to prevent the appearance of vapor in the system. At lower pressures, if one or more of the components of the system is volatile, vapor may be present at equilibrium. A common and important example of the equilibrium between solid and vapor is the equilibrium between salt hydrates and water vapor.

We examine the vapor pressure of the system, water–$CuSO_4$, at a fixed temperature. Figure 15.20 shows schematically the vapor pressure as a function of the concentration of copper sulfate. As anhydrous $CuSO_4$ is added to liquid water, the vapor pressure of the system drops (Raoult's law) along the curve ab. At b the solution is saturated with respect to the pentahydrate, $CuSO_4 \cdot 5H_2O$. The system is invariant along bc, since three phases (saturated solution, solid $CuSO_4 \cdot 5H_2O$, and vapor) are present at constant temperature. Addition of anhydrous $CuSO_4$ does not change the pressure but converts some of the solution to pentahydrate. At c all of the water has been combined with $CuSO_4$ to form pentahydrate. Further addition of $CuSO_4$ drops the pressure to the value at de, with the formation of some trihydrate:

$$2\,CuSO_4 + 3\,CuSO_4 \cdot 5H_2O \longrightarrow 5\,CuSO_4 \cdot 3H_2O.$$

The system is invariant along de; the three phases present are: vapor, $CuSO_4 \cdot 5H_2O$, $CuSO_4 \cdot 3H_2O$. At e the system consists entirely of $CuSO_4 \cdot 3H_2O$; addition of $CuSO_4$ converts some of the trihydrate to monohydrate; the pressure drops to the value at fg. Finally along hi the invariant system is vapor, $CuSO_4 \cdot H_2O$, $CuSO_4$.

The establishment of a constant pressure in a salt hydrate system requires the presence of three phases; a single hydrate does not have a definite vapor pressure. For example, the trihydrate can coexist in equilibrium with any water vapor pressure in the range from e to f. If the pentahydrate *and* the trihydrate are present, then the pressure is fixed at the value de.

As we have seen in Chapter 11, the equilibrium constant for the reaction

$$CuSO_4 \cdot 5H_2O(s) \longrightarrow CuSO_4 \cdot 3H_2O(s) + 2\,H_2O(g)$$

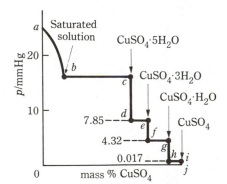

Figure 15.20 Vapor pressure of $CuSO_4$–H_2O (25 °C).

can be written $K = p_e^2$, where p_e is the equilibrium vapor pressure of water over the mixture of tri- and pentahydrates. The dependence of the vapor pressure on temperature is readily obtained from this equation combined with the Gibbs–Helmholtz equation.

★15.10 SYSTEMS OF THREE COMPONENTS

In a system of three components the variance is $F = 3 - P + 2 = 5 - P$. If the system consists only of one phase, four variables are required to describe the system; these may conveniently be taken as T, p, x_1, x_2. It is not possible to give a complete graphic representation of these systems in three dimensions, much less in two dimensions. Consequently, it is customary to represent the system at constant pressure and at constant temperature. The variance then becomes $F' = 3 - P$, so that the system has, at most, a variance of two, and can be represented in the plane. After fixing the temperature and pressure, the remaining variables are composition variables x_1, x_2, x_3, related by $x_1 + x_2 + x_3 = 1$. Specifying any two of them fixes the value of the third. The method of Gibbs and Roozeboom uses an equilateral triangle for graphic representation. Figure 15.21 illustrates the principle of the method. The points A, B, C at the apices of the triangle represent 100% of A, 100% B, 100% C. The lines parallel to AB represent the various percentages of C. Any point on the line AB represents a system containing 0% C; any point on xy represents a system containing 10% C, etc. Point P represents a system containing 30% C. The length perpendicular to a given side of the triangle represents the percent of the component at the vertex opposite to that side. Thus the length PM represents the percent of C, the length PN represents the percent of A, the length PL represents the percent of B. (The lines parallel to AC and CB have been omitted for clarity.) The sum of the lengths of these perpendiculars is always equal to the length of the height of the triangle which is taken as 100%. By this method any composition of a three-component system can be represented by a point within the triangle.

Two other properties of this diagram are important. The first is illustrated in Fig. 15.22(a). If two systems with compositions represented by P and Q are mixed together, the composition of the mixture obtained will be represented by a point x somewhere on the

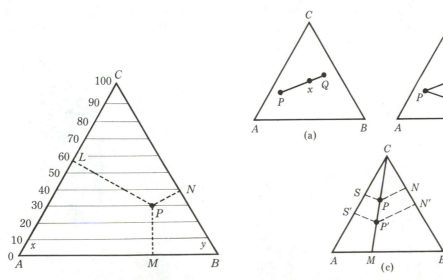

Figure 15.21 The triangular diagram.

Figure 15.22 Properties of the triangular diagram.

line connecting points P and Q. It follows immediately that if three systems, represented by points P, Q, R, are mixed, the composition of the mixture will lie within the triangle PQR. The second important property is that all systems represented by points on a line through a vertex contain the other two components in the same ratio. For example, all systems represented by points on CM contain A and B in the same ratio. In Fig. 15.22(c), by erecting the perpendiculars from two points P and P' and using the properties of similar triangles, we obtain:

$$\frac{PS}{P'S'} = \frac{CP}{CP'} \quad \text{and} \quad \frac{PN}{P'N'} = \frac{CP}{CP'}.$$

Therefore

$$\frac{PS}{P'S'} = \frac{PN}{P'N'} \quad \text{or} \quad \frac{PS}{PN} = \frac{P'S'}{P'N'},$$

which was to be proved. This property is important in discussing the addition or removal of a component to the system without change in the amount of the other two components present.

★15.11 LIQUID–LIQUID EQUILIBRIA

Among the simplest examples of the behavior of three-component systems is the chloroform–water–acetic acid system. The pairs chloroform–acetic acid and water–acetic acid are completely miscible. The pair chloroform–water is not. Figure 15.23 shows schematically the liquid–liquid equilibrium for this system. Points a and b represent the conjugate liquid layers in the absence of acetic acid. Suppose that the overall composition of the system is c so that by the lever rule there is more of layer b than layer a. If a little acetic acid is added to the system, the composition moves along the line connecting c with the acetic acid apex to the point c'. The addition of acetic acid changes the composition of the two layers to a' and b'. Note that the acetic acid goes preferentially into the water-rich layer b', so that tie line connecting the conjugate solutions a' and b' is not parallel to ab. The relative amounts of a' and b' are given by the lever rule; that is, by the ratio of the segments of the tie line $a'b'$. Continued addition of acetic acid moves the composition farther along the dashed line cC; the water-rich layer grows in size while the chloroform-rich layer

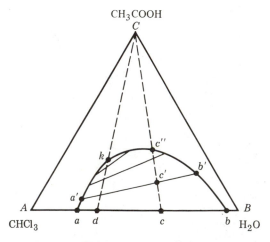

Figure 15.23 Two partially miscible liquids.

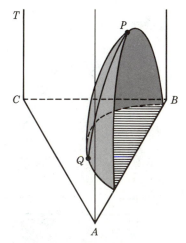

Figure 15.24 Effect of temperature on a partially miscible pair.

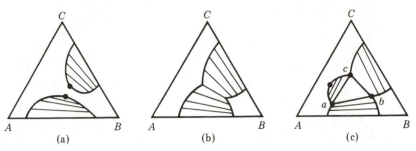

Figure 15.25 Two partially miscible pairs.

diminishes. At c'' only a trace of the chloroform-rich layer remains, while above c'' the system is homogeneous.

Since the tie lines are not parallel, the point at which the two conjugate solutions have the same composition does not lie at the top of the binodal curve but off to one side at the point k, the *plait* point. If the system has the composition d and acetic acid is added, the composition will move along dk; just below k the two layers will *both* be present in *comparable* amounts; at k the boundary between the two solutions vanishes as the system becomes homogeneous. Compare this behavior with that at c'' where only a trace of one of the conjugate layers remained.

If the temperature is increased, the shape and extent of the two-phase region alters. A typical example for a system in which increase in temperature increases the mutual solubility is shown in Fig. 15.24. If temperature were plotted as a third coordinate, the two-phase region would be a loaf-shaped region. In the figure, P is the consolute temperature for the two-component system $A-B$. The line PQ connects the plait points at the various temperatures.

If two of the pairs $A-B$ and $B-C$ are partially miscible, the situation becomes more complex. Two binodal curves can appear as in Fig. 15.25(a). At lower temperatures, the two binodal curves in Fig. 15.25(a) may overlap. If they do so in such a way that the plait points join one another, then the two-phase region becomes a band, as in Fig. 15.25(b). If the binodal curves do not join at the plait points, the resultant diagram has the form shown in Fig. 15.25(c). Points within the triangle abc represent states of the system in which *three* liquid layers having compositions a, b, and c coexist. Such a system is isothermally invariant.

★ 15.12 SOLUBILITY OF SALTS; COMMON-ION EFFECT

Systems that contain two salts with a common ion and water have great interest from a practical standpoint. Each salt influences the solubility of the other. The schematic diagram for NH_4Cl, $(NH_4)_2SO_4$, H_2O at 30 °C is shown in Fig. 15.26. Point a represents the saturated solution of NH_4Cl in water in the absence of $(NH_4)_2SO_4$. Points between A and a represent various amounts of solid NH_4Cl in equilibrium with saturated solution a. Points between a and C represent the unsaturated solution of NH_4Cl. Similarly, b represents the solubility of $(NH_4)_2SO_4$ in the absence of NH_4Cl. Points on Cb represent the unsaturated solution, while those on bB represent solid $(NH_4)_2SO_4$ in equilibrium with saturated solution. The presence of $(NH_4)_2SO_4$ changes the solubility of NH_4Cl along the line ac, while the presence of NH_4Cl changes the solubility of $(NH_4)_2SO_4$ along the line bc. Point c represents a solution that is saturated with respect to both NH_4Cl and $(NH_4)_2SO_4$. The tie lines connect the composition of the saturated solution and the solid in equilibrium with it. The regions of stability are shown in Table 15.2.

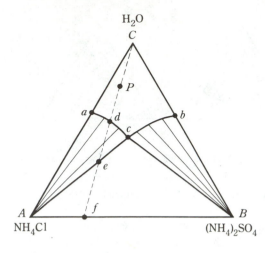

Figure 15.26 Common-ion effect.

Table 15.2

Region	System	Variance
$Cacb$	Unsaturated solution	2
Aac	NH_4Cl + saturated solution	1
Bbc	$(NH_4)_2SO_4$ + saturated solution	1
AcB	NH_4Cl + $(NH_4)_2SO_4$ + saturated solution c	0

Suppose an unsaturated solution represented by P is evaporated isothermally; the state point must move along the line $Pdef$, which has been drawn through the apex C and the point P. At d, NH_4Cl crystallizes; the composition of the solution moves along the line dc. At point e, the solution composition is c, and $(NH_4)_2SO_4$ begins to crystallize. Continued evaporation deposits both NH_4Cl and $(NH_4)_2SO_4$ until the point f is reached, where the solution disappears completely.

★ 15.13 DOUBLE-SALT FORMATION

If it happens that the two salts can form a compound, a double salt, then the solubility of the compound may also appear as an equilibrium line in the diagram. Figure 15.27 shows two typical cases of compound formation. In both figures, ab is the solubility of A; bc that of the compound AB, cd that of B. The regions and what they represent are tabulated in Table 15.3.

The difference in behavior of the two systems can be demonstrated in two ways. First begin with the dry solid compound and add water; the state point moves along the line DC. In Fig. 15.27(a), this moves the point into the region of the compound plus saturated solution of the compound. Hence, this compound is said to be *congruently saturating*. Addition of water to the compound AB in Fig. 15.27(b) moves the state point along DC into the region of stability of $A + AB$ + saturated solution b. The addition of water, therefore, decomposes the compound into solid A and a solution. This compound is said to be *incongruently saturating*. Similarly the compound in Fig. 15.27(b) cannot be prepared by evaporating a solution containing A and B in the equimolar ratio. Evaporation crystallizes solid A at point e; at point f the solid A reacts with the solution b to precipitate AB. When D is reached, all of A has disappeared and only the compound remains. If the

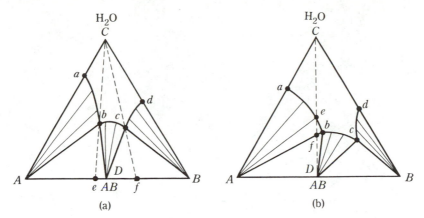

Figure 15.27 (a) Congruently saturating compound. (b) Incongruently saturating compound.

Table 15.3

Region	System	Variance
Cabcd	Unsaturated solution	2
abA	A + saturated solution	1
AbD	A + AB + saturated solution b	0
Dbc	AB + saturated solution	1
DcB	AB + B + saturated solution c	0
cdB	B + saturated solution	1

solids are filtered off when the state point is between f and D, the crystals of compound will be mixed with crystals of A. It is understandable how annoying this is in the laboratory. A knowledge of the phase diagram or the double-salt system is very helpful in preparative problems.

If one of the salts forms a hydrate of composition D, then the diagram will have the appearance shown in Fig. 15.28(a). The interesting feature of this diagram is that if the state point lies in the triangle ADB, the system consists exclusively of the three solids, D, B, A. Under appropriate conditions, usually at a higher temperature, the anhydrous salt may make its appearance on the diagram as in Fig. 15.28(b).

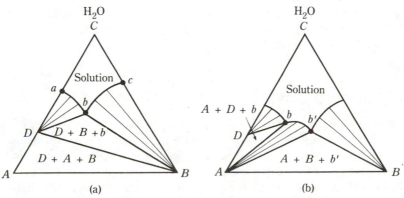

Figure 15.28 Hydrate formation.

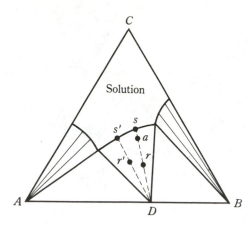

Figure 15.29 The method of wet residues.

★ 15.14 THE METHOD OF "WET RESIDUES"

The determination of equilibrium curves in three-component systems in some respects is simpler than in two-component systems. Consider the diagram in Fig. 15.29. Suppose that the system consists of a solution in equilibrium with solid and that the state point is at a. We do not know the location of a, but we do know that it lies on a tie line connecting the solid composition with the liquid composition. We proceed as follows: some of the saturated liquid is removed and analyzed for A and B. This fixes the point s on the equilibrium line. After the removal of some of the saturated solution, the state point of the remainder of the system must lie at point r. So the remainder, that is, the solids together with the supernatant liquid, called the "wet residue," is analyzed for two of the components. This analysis determines the point r. A tie line is drawn through s and r. The procedure is repeated on a system that contains a slightly different ratio of two of the components. The solution analysis yields the point s', while the analysis of the wet residue yields a point r'. The tie line is drawn through s' and r'. These two tie lines must intersect at the composition of the solid that is present. In this system, they would intersect at point D. This intersection point yields the composition of the solid phase D, which is in equilibrium with the liquid.

The method of wet residues is superior to the procedure necessary in two-component systems, where the liquid and solid phase must be separated and analyzed individually. It is a practical impossibility to separate the solid phase from the liquid without some of the liquid adhering to the solid and thus contaminating it. For this reason, it is frequently easier to add a third substance to a two-component system, determine the equilibrium lines, as well as the composition of the solid phases by the method of wet residues, and infer the composition of the solid in the two-component system from the features of the triangular diagram.

★ 15.15 "SALTING OUT"

In the practice of organic chemistry, it is common procedure to separate a mixture of an organic liquid in water by adding salt. For example, if the organic liquid and water are completely miscible, addition of salt to the system may produce a separation into two liquid layers—one rich in the organic liquid, the other rich in water. The phase relations may be illustrated as in Table 15.4 and by the diagram for K_2CO_3–H_2O–CH_3OH, Fig. 15.30, which is typical of the system salt–water–alcohol.

Table 15.4

Region	System
Aab	K_2CO_3 in equilibrium with water-rich saturated solution
Aed	K_2CO_3 in equilibrium with alcohol-rich saturated solution
bcd	two conjugate liquids joined by tie lines
Abd	K_2CO_3 in equilibrium with conjugate liquids b and d

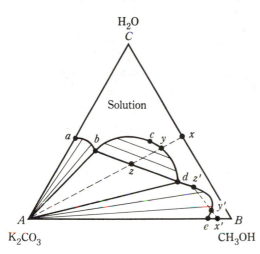

Figure 15.30   Salt–alcohol–water diagram.

   The system is distinguished by the appearance of the two-liquid region bcd. Suppose that solid K_2CO_3 is added to a mixture of water and alcohol of composition x. The state point will move along the line $xyzA$. At y two layers form; at z K_2CO_3 ceases to dissolve so that solid K_2CO_3 and liquids b and d coexist. The liquid d is the alcohol-rich layer and may be separated from b, the water-rich layer. Note that addition of salt after the solid ceases to dissolve does not produce any change in the composition of the layers b and d; this must be so since the system is isothermally invariant in the triangle Abd.

   This diagram can also be used to show how additional salt may be precipitated by the addition of alcohol to a saturated solution; the state point moves from a, let us say, along a line connecting a and B. Since, in this particular case, only a little more salt is precipitated before two liquid layers form, the trick is not particularly useful. This system is curious in the effect of the addition of water to an unsaturated solution of K_2CO_3 in alcohol of composition x'. The line $x'y'z'$ connecting x' and C shows that K_2CO_3 will precipitate at y' if water is added to the alcoholic solution. Further addition of water will redissolve the K_2CO_3 at z'.

QUESTIONS

15.1 Describe the similarities of the solution upper consolute point and the liquid–gas critical point.

15.2 Will there be a lower or an upper consolute point if the solution process for two liquids is exothermic? For an endothermic process?

15.3 The hardness of an alloy is greater the more fine grained the alloy is. Why should eutectic alloys be especially hard?

15.4 Cu and Ni have nearly the same atomic radius and crystallize in the same solid lattice structure. With this information and the solid solution analogue of Eq. (14.6), suggest why Cu and Ni form a nearly ideal solid solution.

15.5 Interpret the freezing-point elevation in solid solutions in terms of the "escaping tendency" of the solid in the solid solution.

PROBLEMS

15.1 The vapor pressures of chlorobenzene and water at different temperatures are

$t/°C$	90	100	110
$p°(\phi Cl)/mmHg$	204	289	402
$p°(H_2O)/mmHg$	526	760	1075

a) At what pressure will ϕCl steam-distill at 90 °C?
b) At what temperature will ϕCl steam-distill under a total pressure of 800 mmHg?
c) How many grams of steam are required to distill 10.0 g of ϕCl (a) at 90 °C and (b) under 800 Torr total pressure?

15.2 A mixture of 100 g water and 80 g of phenol separates into two layers at 60 °C. One layer, L_1, consists of 44.9 % water by mass; the other, L_2, consists of 83.2 % water by mass.

a) What are the masses of L_1 and L_2?
b) What are the total number of moles in L_1 and L_2?

15.3 The melting point and heats of fusion of lead and antimony are

	Pb	Sb
$t_m/°C$	327.4	630.5
$\Delta H_{fus}/(kJ/mol)$	5.10	20.1

Calculate the solid–liquid equilibrium lines; estimate the eutectic composition graphically; then calculate the eutectic temperature. Compare the result with the values given by Fig. 15.7.

15.4 From the melting points of the mixtures of Al and Cu, sketch the melting-point curve.

a)

mass % Cu	0	20	40	60	80	100
$t/°C$	660	600	540	610	930	1083

b) For copper, $T_m/K = 1356$ and $\Delta H°_{fus}(Cu) = 13.05$ kJ/mol; for aluminum, $T_m/K = 933$, and $\Delta H°_{fus}(Al) = 10.75$ kJ/mol. Sketch the ideal solubility curves and compare with the experimental curve in (a).

15.5 The solubility of KBr in water is

$t/°C$	0	20	40	60	80	100
g KBr/g H_2O	0.54	0.64	0.76	0.86	0.95	1.04

In a one molal solution, KBr depresses the freezing point of water by 3.29 °C. Estimate the eutectic temperature for the system KBr–H_2O graphically.

15.6 KBr is recrystallized from water by saturating the solution at 100 °C, then cooling to 20 °C; the crystals obtained are redissolved in water and the solution evaporated until it is saturated at 100 °C. Cooling to 20 °C produces a second crop of crystals. What is the percent yield of pure KBr after these two crystallizations? Use data in Problem 15.5.

15.7 Two crops of KBr crystals are obtained as follows. A solution saturated at 100 °C is cooled to 20 °C; after filtering off the first crop, the mother liquor is evaporated until the solution is saturated at 100 °C; cooling to 20 °C produces the second crop. What fraction of the KBr is recovered in the two crops by this method? (Use data in Problem 15.5.)

15.8 Figure 15.16 shows the equilibrium between liquid and solid solutions in the copper–nickel system. If we suppose that both the solid and liquid solutions are ideal, then the equilibrium conditions lead to two equations of the form of Eq. (15.8); one of these applies to copper, the other to nickel. If we invert the equations they become

$$\frac{1}{T} = \left(\frac{1}{T_{Cu}}\right)\left[1 + \left(\frac{R}{\Delta S_{Cu}}\right)\ln\left(\frac{x'_{Cu}}{x_{Cu}}\right)\right]$$

and

$$\frac{1}{T} = \left(\frac{1}{T_{Ni}}\right)\left[1 + \left(\frac{R}{\Delta S_{Ni}}\right)\ln\left(\frac{x'_{Ni}}{x_{Ni}}\right)\right],$$

where x' is the mole fraction in the solid solution, x that in the liquid. In addition we have the relations, $x'_{Cu} + x'_{Ni} = 1$, and $x_{Cu} + x_{Ni} = 1$. There are four equations with five variables, $T, x'_{Cu}, x'_{Ni}, x_{Cu}, x_{Ni}$. Suppose that $x_{Cu} = 0.1$; calculate values for all the other variables. $T_{Cu} = 1356.2$ K, $T_{Ni} = 1728$ K; assume that $\Delta S_{Cu} = \Delta S_{Ni} = 9.83$ J/K mol. (*Hint:* Use value of x_{Cu} in the first two equations, then eliminate T between them. By trial and error solve the resulting equation for either x'_{Cu} or x'_{Ni}. Then T is easily calculated. Repetition of this procedure for other values of x_{Cu} would yield the entire diagram.)

15.9 In Fig. 15.18, what is the variance in each region of the diagram? Keep in mind that the pressure is constant. What is the variance on the line *abc*?

15.10 What is the variance in each region of Fig. 15.30?

15.11 a) Using Fig. 15.30, what changes will be observed if water is added to a system containing 50% K_2CO_3 and 50% CH_3OH?
   b) What is observed if methanol is added to a system containing 90% water and 10% K_2CO_3? (or 30% water and 70% K_2CO_3?)

15.12 a) What is the variance in each of the regions of Fig. 15.15(a)?
   b) Describe the changes that occur if an unsaturated solution of Na_2SO_4 is evaporated at 25 °C; at 35 °C.

15.13 Describe the changes that occur if water is evaporated isothermally along the line *aj* from the system in Fig. 15.12.

16

Equilibria in Nonideal Systems

16.1 THE CONCEPT OF ACTIVITY

The mathematical discussions in the preceding chapters have been limited to systems that behave ideally; the systems were either pure ideal gases, or ideal mixtures (gaseous, liquid, solid). Many of the systems described in Chapter 15 are not ideal; the question that arises is how are we to deal mathematically with nonideal systems. These systems are handled conveniently using the concepts of fugacity and activity first introduced by G. N. Lewis.

The chemical potential of a component in a mixture is in general a function of temperature, pressure, and the composition of the mixture. In gaseous mixtures we write the chemical potential of each component as a sum of two terms:

$$\mu_i = \mu_i^{\circ}(T) + RT \ln f_i. \tag{16.1}$$

The first term, μ_i°, is a function of temperature only, while the fugacity f_i in the second term may depend on the temperature, pressure, and composition of the mixture. The fugacity is a measure of the chemical potential of the gas i in the mixture. In Section 10.9 a method of evaluating the fugacity for a pure gas was described.

Now we will confine our attention to liquid solutions, although most of what is said can be applied to solid solutions as well. For any component i in any liquid mixture, we write

$$\mu_i = g_i(T, p) + RT \ln a_i, \tag{16.2}$$

where $g_i(T, p)$ is a function only of temperature and pressure while a_i, the *activity of i*, may be a function of temperature, pressure, and composition. As it stands, Eq. (16.2) is not particularly informative; however, it does indicate that at a specified temperature and pressure an increase in the activity of a substance means an increase in the chemical potential of that substance. *The equivalence of the activity to the chemical potential, through an equation having the form of Eq. (16.2), is the fundamental property of the activity.* The theory of equilibrium could be developed entirely in terms of the activities of substances instead of in terms of chemical potentials.

To use Eq. (16.2), the significance of the function $g_i(T, p)$ must be accurately described; then a_i has a precise meaning. Two ways of describing $g_i(T, p)$ are in common use; each leads to a different system of activities. In either system the activity of a component is still a measure of its chemical potential.

16.2 THE RATIONAL SYSTEM OF ACTIVITIES

In the rational system of activities, $g_i(T, p)$ is identified with the chemical potential of the pure liquid, $\mu_i^\circ(T, p)$:

$$g_i(T, p) = \mu_i^\circ(T, p). \tag{16.3}$$

Then Eq. (16.2) becomes

$$\mu_i = \mu_i^\circ + RT \ln a_i. \tag{16.4}$$

As $x_i \to 1$, the system comes nearer to being pure i, and μ_i must approach μ_i°, so that

$$\mu_i - \mu_i^\circ = 0 \quad \text{as} \quad x_i \to 1.$$

Using this fact in Eq. (16.4), we have $\ln a_i = 0$, as $x_i \to 1$, or

$$a_i = 1 \quad \text{as} \quad x_i \to 1. \tag{16.5}$$

Therefore the activity of the pure liquid is equal to unity.

If we compare Eq. (16.4) with the μ_i in an ideal liquid solution,

$$\mu_i^{id} = \mu_i^\circ + RT \ln x_i; \tag{16.6}$$

by subtracting Eq. (16.6) from Eq. (16.4), we obtain

$$\mu_i - \mu_i^{id} = RT \ln \frac{a_i}{x_i}. \tag{16.7}$$

The *rational activity coefficient of i*, γ_i, is defined by

$$\gamma_i = \frac{a_i}{x_i}. \tag{16.8}$$

With this definition, Eq. (16.7) becomes

$$\mu_i = \mu_i^{id} + RT \ln \gamma_i, \tag{16.9}$$

which shows that $\ln \gamma_i$ measures the extent of the deviation from ideality. From the relation in Eq. (16.5), and the definition of γ_i, we have

$$\gamma_i = 1 \quad \text{as} \quad x_i \to 1. \tag{16.10}$$

The rational activity coefficients are convenient for those systems in which the mole fraction of any component may vary from zero to unity; mixtures of liquids such as acetone and chloroform, for example.

16.2.1 Rational Activities; Volatile Substances

The rational activity of volatile constituents in a liquid mixture can be readily measured by measuring the partial pressure of the constituent in the vapor phase in equilibrium with the liquid. Since at equilibrium the chemical potentials of each constituent must be equal in the liquid and the vapor phase, we have $\mu_i(l) = \mu_i(g)$. Using Eq. (16.4) for $\mu_i(l)$ and assuming the gas is ideal, component i having a partial pressure p_i, we can write

$$\mu_i^\circ(l) + RT \ln a_i = \mu_i^\circ(g) + RT \ln p_i.$$

For the pure liquid,

$$\mu_i^\circ(l) = \mu_i^\circ(g) + RT \ln p_i^\circ,$$

where p_i° is the vapor pressure of the pure liquid. Subtracting the last two equations and dividing by RT, we obtain $\ln a_i = \ln (p_i/p_i^\circ)$, or

$$a_i = \frac{p_i}{p_i^\circ}, \tag{16.11}$$

which is the analogue of Raoult's law for a nonideal solution. Thus a measurement of p_i over the solution together with a knowledge of p_i° yields the value of a_i. From measurements at various values of x_i, the value of a_i can either be plotted or tabulated as a function of x_i. Similarly, the activity coefficient can be calculated using Eq. (16.8) and plotted as function of x_i. In Figs. 16.1 and 16.2, plots of a_i and γ_i versus x_i are shown for binary systems that exhibit positive and negative deviations from Raoult's law. If the solutions were ideal, then $a_i = x_i$, and $\gamma_i = 1$, for all values of x_i.

Depending on the system, the activity coefficient of a component may be greater or less than unity. In a system showing positive deviations from ideality, the activity coefficient, and therefore the escaping tendency, is greater than in an ideal solution of the same concentration. In a solution exhibiting negative deviations from Raoult's law, the substance has a lower escaping tendency than in an ideal solution of the same concentration, γ is less than unity.

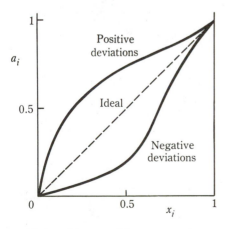

Figure 16.1 Activity versus mole fraction.

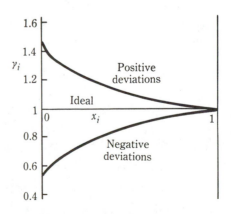

Figure 16.2 Activity coefficient versus mole fraction.

16.3 COLLIGATIVE PROPERTIES

The colligative properties of a solution of involatile solutes are simply expressed in terms of the rational activity of the solvent.

16.3.1 Vapor Pressure

If the vapor pressure of the solvent over the solution is p, and the activity of the solvent is a, then from Eq. (16.11),

$$a = \frac{p}{p^\circ} \tag{16.11a}$$

If a is evaluated from measurements of vapor pressure at various concentrations, these values can be used to calculate the freezing-point depression, boiling-point elevation, and osmotic pressure for any concentration.

16.3.2 Freezing-Point Depression

If pure solid solvent is in equilibrium with solution, the equilibrium condition $\mu(l) = \mu^\circ(s)$ becomes, using Eq. (16.4), $\mu^\circ(l) + RT \ln a = \mu^\circ(s)$; or,

$$\ln a = -\frac{\Delta G^\circ_{\text{fus}}}{RT}.$$

Repetition of the argument in Section 13.6 yields, finally,

$$\ln a = -\frac{\Delta H^\circ_{\text{fus}}}{R}\left(\frac{1}{T} - \frac{1}{T_0}\right), \tag{16.12}$$

which is the analogue of Eq. (13.15) for the ideal solution. Knowing a from vapor pressure measurements, the freezing point can be calculated from Eq. (16.12); conversely, if the freezing point T is measured, a can be evaluated from Eq. (16.12).

16.3.3 Boiling-Point Elevation

The analogous argument shows that the boiling point is related to $\Delta H^\circ_{\text{vap}}$ and T_0, the heat of vaporization and the boiling point of the pure solvent, by

$$\ln a = \frac{\Delta H^\circ_{\text{vap}}}{R}\left(\frac{1}{T} - \frac{1}{T_0}\right), \tag{16.13}$$

which is the analogue of Eq. (13.29) for the ideal solution.

16.3.4 Osmotic Pressure

The osmotic pressure is given by

$$\bar{V}^\circ \pi = -RT \ln a, \tag{16.14}$$

which is the analogue of Eq. (13.36).

In Eqs. (16.11a), (16.12), (16.13), and (16.14), a is the rational activity of the solvent. Measurements of any colligative property yield values of a through these equations.

16.4 THE PRACTICAL SYSTEM

The practical system of activities and activity coefficients is useful for solutions in which only the solvent has a mole fraction near unity; all of the solutes are present in relatively small amounts. For such a system we use the rational system for the *solvent* and the practical system for the *solutes*. As the concentration of solutes becomes very small, the behavior of any real solution approaches that of the *ideal dilute* solution. Using a subscript j to identify the solutes, then in the ideal dilute solution (Section 14.11)

$$\mu_j^{id} = \mu_j^{**} + RT \ln m_j, \tag{16.15}$$

For a solute, Eq. (16.2) becomes

$$\mu_j = g_j(T, p) + RT \ln a_j. \tag{16.16}$$

If we subtract Eq. (16.15) from Eq. (16.16) and set $g_j(T, p) = \mu_j^{**}$, then

$$\mu_j - \mu_j^{id} = RT \ln \frac{a_j}{m_j}. \tag{16.17}$$

The identification of $g_j(T, p)$ with μ_j^{**} defines the practical system of activities; the practical activity coefficient γ_j is defined by

$$\gamma_j = \frac{a_j}{m_j}. \tag{16.18}$$

Equations (16.17) and (16.18) show that $\ln \gamma_j$ is a measure of the departure of a solute from its behavior in an ideal dilute solution. Finally, as $m_j \to 0$, the solute must behave in the ideal dilute way so that

$$\gamma_j = 1 \quad \text{as} \quad m_j \to 0. \tag{16.19}$$

It follows that $a_j = m_j$ as $m_j = 0$. Thus, for the chemical potential of a solute in the practical system, we have

$$\mu_j = \mu_j^{**} + RT \ln a_j. \tag{16.20}$$

The μ_j^{**} is the chemical potential the solute would have in a 1 molal solution if that solution behaved according to the ideal dilute rule. This standard state is called the ideal solution of unit molality. It is a hypothetical state of a system. According to Eq. (16.20) the practical activity measures the chemical potential of the substance relative to the chemical potential in this hypothetical ideal solution of unit molality. Equation (16.20) is applicable to either volatile or involatile solutes.

16.4.1 Volatile Solute

The equilibrium condition for the distribution of a volatile solute j between solution and vapor is $\mu_j(g) = \mu_j(l)$. Using Eq. (16.20) and assuming that the vapor is ideal, we have

$$\mu_j^{\circ} + RT \ln p_j = \mu_j^{**} + RT \ln a_j.$$

Since μ_j° and μ_j^{**} depend only on T and p and not on composition, we can define a constant K_j', which is independent of composition, by

$$RT \ln K_j' = -(\mu_j^{\circ} - \mu_j^{**}).$$

The relation between p_j and a_j becomes

$$p_j = K'_j a_j. \tag{16.21}$$

The constant K'_j is a modified Henry's law constant. If K'_j is known, values of a_j can be computed immediately from the measured values of p_j. Dividing Eq. (16.21) by m_j, we obtain

$$\frac{p_j}{m_j} = K'_j \left(\frac{a_j}{m_j}\right). \tag{16.22}$$

Measured values of the ratio p_j/m_j are plotted as a function of m_j. The curve is extrapolated to $m_j = 0$. The extrapolated value of p_j/m_j is equal to K'_j, since $a_j/m_j = 1$ as $m_j \to 0$. Thus

$$\left(\frac{p_j}{m_j}\right)_{m_j = 0} = K'_j.$$

Having obtained the value of K'_j, the values of a_j are computed from the measured p_j by Eq. (16.21).

16.4.2 Involatile Solute; Colligative Properties and the Activity of the Solute

In Section 16.3 we related the colligative properties to the rational activity of the *solvent*. These properties can also be related to the activity of the solute. Symbols without subscripts refer to the solvent; symbols with a subscript 2 refer to the solute, except that the molality m of the *solute* will not bear a subscript. For simplicity we assume that only one solute is present. The chemical potentials are

Solvent: $\mu = \mu^\circ + RT \ln a,$

Solute: $\mu_2 = \mu_2^{**} + RT \ln a_2.$

These are related by the Gibbs–Duhem equation, Eq. (11.97),

$$d\mu = -\frac{n_2}{n} d\mu_2 \qquad (T, p \text{ constant}).$$

Differentiating μ and μ_2, keeping T and p constant, we obtain

$$d\mu = RT \, d \ln a \qquad \text{and} \qquad d\mu_2 = RT \, d \ln a_2.$$

Using these values in the Gibbs–Duhem equation, we have

$$d \ln a = -\frac{n_2}{n} d \ln a_2.$$

But $n_2/n = Mm$, where M is the molar mass of the solvent, and m is the molality of the solute. Therefore

$$d \ln a = -Mm \, d \ln a_2, \tag{16.23}$$

which is the required relation between the activities of solvent and solute.

16.4.3 Freezing-Point Depression

Differentiating Eq. (16.12) and using the value for $d \ln a$ given by Eq. (16.23), we obtain

$$d \ln a_2 = - \frac{\Delta H^\circ_{\text{fus}}}{MRT^2 m} \, dT = \frac{d\theta}{K_f m(1 - \theta/T_0)^2},$$

where $K_f = MRT_0^2/\Delta H^\circ_{\text{fus}}$, and the freezing-point depression, $\theta = T_0 - T$, $d\theta = -dT$, have been introduced. If $\theta/T_0 \ll 1$, then

$$d \ln a_2 = \frac{d\theta}{K_f m}. \tag{16.24}$$

A similar equation could be derived for the boiling-point elevation.

As is, Eq. (16.24) is not very sensitive to deviations from ideality. To arrange it in terms of more responsive functions, we introduce the *osmotic coefficient*, $1 - j$, defined by

$$\theta = K_f m(1 - j). \tag{16.25}$$

In an ideal dilute solution, $\theta = K_f m$, so that $j = 0$. In a nonideal solution, j is not zero. Differentiating Eq. (16.25), we have

$$d\theta = K_f[(1 - j) \, dm - m \, dj].$$

Using Eq. (16.18), we set $a_2 = \gamma_2 m$; and differentiate $\ln a_2$:

$$d \ln a_2 = d \ln \gamma_2 + d \ln m = d \ln \gamma_2 + \frac{dm}{m}.$$

Using these two relations in Eq. (16.24), it becomes

$$d \ln \gamma_2 = -dj - \left(\frac{j}{m}\right) dm.$$

This equation is integrated from $m = 0$ to m; at $m = 0$, $\gamma_2 = 1$, and $j = 0$; we obtain

$$\int_0^{\ln \gamma_2} d \ln \gamma_2 = - \int_0^j dj - \int_0^m \left(\frac{j}{m}\right) dm,$$

$$\ln \gamma_2 = -j - \int_0^m \left(\frac{j}{m}\right) dm. \tag{16.26}$$

The integral in Eq. (16.26) is evaluated graphically. From experimental values of θ and m, j is calculated from Eq. (16.25); j/m is plotted versus m; the area under the curve is the value of the integral. After obtaining the value of γ_2, the activity a_2 is obtained from the relation $a_2 = \gamma_2 m$.

We have assumed that $\Delta H^\circ_{\text{fus}}$ is independent of temperature and that θ is much less than T_0. In precision work, more elaborate equations not restricted by these assumptions, are used. Any of the colligative properties can be interpreted in terms of the activity of the solute.

16.5 ACTIVITIES AND REACTION EQUILIBRIUM

If a chemical reaction takes place in a nonideal solution, the chemical potentials in the form given by Eq. (16.4) or (16.20) must be used in the equation of reaction equilibrium.

The practical system, Eq. (16.20), is more commonly used. The condition of equilibrium becomes

$$\Delta G^{**} = -RT \ln K_a, \tag{16.27}$$

where ΔG^{**} is the standard Gibbs energy change, and K_a is the proper quotient of equilibrium activities. Since ΔG^{**} is a function only of T and p, K_a is a function only of T and p, and is independent of the composition. Since each activity has the form $a_i = \gamma_i m_i$, we can write

$$K_a = K_\gamma K_m, \tag{16.28}$$

where K_γ and K_m are proper quotients of activity coefficients and of molalities, respectively. Since the γ's depend on composition, Eq. (16.28) shows that K_m depends on composition. In dilute real solutions all the γ's approach unity, K_γ approaches unity, and K_m approaches K_a. Except when we are particularly interested in the evaluation of activity coefficients, we shall treat K_m as if it were independent of composition; doing so greatly simplifies the discussion of equilibria.

In most elementary treatments of equilibria in solution, the equilibrium constant is usually written as a quotient of equilibrium concentrations expressed as molarities, K_c. It is possible to develop an entire system of activities and activity coefficients using molar rather than molal concentrations. We could write $a = \gamma_c c$, where c is the molar concentration and γ_c the corresponding activity coefficient; as c approaches zero, γ_c must approach unity. We will not dwell on the details of this system except to show that in dilute aqueous solution the systems based on molarity and on molality are nearly the same. We have seen, Eq. (14.25), that in dilute solution, $\tilde{c}_j = \rho m_j$, or $c_j = \rho m_j/(1000 \text{ L/m}^3)$, where ρ is the density of the pure solvent. At 25 °C the density of water is 997.044 kg/m^3. The error made by replacing molalities by molarities is therefore insignificant in ordinary circumstances. The concomitant error in the standard Gibbs energy is well below the experimental error. In more concentrated solutions the relation between c_j and m_j is not so simple, Eq. (14.24), and the two systems of activities are different.

Ordinarily for purposes of illustration we shall use molar concentrations in the equilibrium constant, realizing that to be precise we should use the activities. One misunderstanding that arises because of this replacement of activity by concentration should be avoided. The activity is sometimes regarded as if it were an "effective concentration." This is a legitimate formal point of view; however, it is deceptive in that it conveys the incorrect notion that activity is designed to measure the concentration of a substance in a mixture. The activity is designed for one purpose only, namely to provide a convenient *measure of the chemical potential* of a substance in a mixture. The connection between activity and concentration in dilute solutions is not that one is a measure of the other, but that *either* one is a measure of the chemical potential of the substance. It would be better to think of the concentration in an ideal solution as being the effective activity.

16.6 ACTIVITIES IN ELECTROLYTIC SOLUTIONS

The problem of defining activities is somewhat more complicated in electrolytic solutions than in solutions of nonelectrolytes. Solutions of strong electrolytes exhibit marked deviations from ideal behavior even at concentrations well below those at which a solution of a nonelectrolyte would behave in the ideal dilute way. The determination of activities and activity coefficients has a correspondingly greater importance for solutions of strong electrolytes. To simplify the notation as much as possible a subscript s will be used for the properties of the solvent; symbols without subscript refer to the solute; subscripts $+$ and $-$ refer to the properties of the positive and negative ions.

Consider a solution of an electrolyte that is completely dissociated into ions. By the additivity rule the Gibbs energy of the solution should be the sum of the Gibbs energies of the solvent, the positive and the negative ions:

$$G = n_s \mu_s + n_+ \mu_+ + n_- \mu_-. \tag{16.29}$$

If each mole of the electrolyte dissociates into ν_+ positive ions and ν_- negative ions, then $n_+ = \nu_+ n$, and $n_- = \nu_- n$, where n is the number of moles of electrolyte in the solution. Equation (16.29) becomes

$$G = n_s \mu_s + n(\nu_+ \mu_+ + \nu_- \mu_-). \tag{16.30}$$

If μ is the chemical potential of the electrolyte in the solution, then we should also have

$$G = n_s \mu_s + n\mu. \tag{16.31}$$

Comparing Eqs. (16.30) and (16.31), we see that

$$\mu = \nu_+ \mu_+ + \nu_- \mu_-. \tag{16.32}$$

Let the total number of moles of ions produced by one mole of electrolyte be $\nu = \nu_+ + \nu_-$. Then the mean ionic chemical potential $\mu_\pm$ is defined by

$$\nu\mu_\pm = \nu_+ \mu_+ + \nu_- \mu_-. \tag{16.33}$$

Now we can proceed in a purely formal way to define the various activities. We write*

$$\mu = \mu^\circ + RT \ln a; \tag{16.34}$$

$$\mu_\pm = \mu_\pm^\circ + RT \ln a_\pm; \tag{16.35}$$

$$\mu_+ = \mu_+^\circ + RT \ln a_+; \tag{16.36}$$

$$\mu_- = \mu_-^\circ + RT \ln a_-. \tag{16.37}$$

In these relations, a is the activity of the electrolyte, $a_\pm$ is the mean ionic activity, and a_+ and a_- are the individual ion activities. To define the various activities completely we require the additional relations

$$\mu^\circ = \nu_+ \mu_+^\circ + \nu_- \mu_-^\circ; \tag{16.38}$$

$$\nu\mu_\pm^\circ = \nu_+ \mu_+^\circ + \nu_- \mu_-^\circ. \tag{16.39}$$

First we work out the relation between a and $a_\pm$. From Eqs. (16.32) and (16.33) we have $\mu = \nu\mu_\pm$. Using the values for μ and $\mu_\pm$ from Eqs. (16.34) and (16.35), we get

$$\mu^\circ + RT \ln a = \nu\mu_\pm^\circ + \nu RT \ln a_\pm.$$

Using Eqs. (16.38) and (16.39) this reduces to

$$a = a_\pm^\nu. \tag{16.40}$$

Next we want the relation between $a_\pm$, a_+, and a_-. Using the values of $\mu_\pm$, μ_+, and μ_- given by Eqs. (16.35), (16.36), and (16.37) in Eq. (16.33), we obtain

$$\nu\mu_\pm^\circ + \nu RT \ln a_\pm = \nu_+ \mu_+^\circ + \nu_- \mu_-^\circ + RT(\nu_+ \ln a_+ + \nu_- \ln a_-).$$

---

*   Since we are using molalities, for consistency we should write μ^{**} for the standard value of μ, but this would make the symbolism too forbidding.

From this equation we subtract Eq. (16.39); then it reduces to

$$a_\pm^v = a_+^{v_+} a_-^{v_-}.$$ (16.41)

The mean ionic activity is the geometric mean of the individual ion activities.

The various activity coefficients are defined by the relations

$$a_\pm = \gamma_\pm m_\pm;$$ (16.42)

$$a_+ = \gamma_+ m_+;$$ (16.43)

$$a_- = \gamma_- m_-;$$ (16.44)

where $\gamma_\pm$ is the mean ionic activity coefficient, $m_\pm$ is the mean ionic molality, and so on. Using the values of $a_\pm$, a_+, and a_- from Eqs. (16.42), (16.43), and (16.44) in Eq. (16.41), we obtain

$$\gamma_\pm^v m_\pm^v = \gamma_+^{v_+} \gamma_-^{v_-} m_+^{v_+} m_-^{v_-}.$$

We then require that

$$\gamma_\pm^v = \gamma_+^{v_+} \gamma_-^{v_-};$$ (16.45)

$$m_\pm^v = m_+^{v_+} m_-^{v_-}.$$ (16.46)

These equations show that $\gamma_\pm$ and $m_\pm$ are also geometric means of the individual ionic quantities. In terms of the molality of the electrolyte we have

$$m_+ = v_+ m \quad \text{and} \quad m_- = v_- m,$$

so that the mean ionic molality is

$$m_\pm = (v_+^{v_+} v_-^{v_-})^{1/v} m.$$ (16.47)

Knowing the formula of the electrolyte, we obtain $m_\pm$ immediately in terms of m.

■ **EXAMPLE 16.1**   In a $1:1$ electrolyte such as NaCl, or in a $2:2$ electrolyte such as $MgSO_4$

$$v_+ = v_- = 1, \quad v = 2, \quad m_\pm = m.$$

In a $1:2$ electrolyte such as Na_2SO_4

$$v_+ = 2, \quad v_- = 1, \quad v = 3, \quad m_\pm = (2^2 \cdot 1^1)^{1/3} m = \sqrt[3]{4} m = 1.587 m.$$

The expression for the chemical potential in terms of the mean ionic activity, from Eqs. (16.34) and (16.40), is

$$\mu = \mu^\circ + RT \ln a_\pm^v.$$ (16.48)

Using Eqs. (16.42) and (16.47) this becomes

$$\mu = \mu^\circ + RT \ln [\gamma_\pm^v (v_+^{v_+} v_-^{v_-}) m^v],$$

which can be written in the form

$$\mu = \mu^\circ + RT \ln (v_+^{v_+} v_-^{v_-}) + vRT \ln m + vRT \ln \gamma_\pm.$$ (16.49)

In Eq. (16.49), the second term on the right is a constant, evaluated from the formula of the electrolyte; the third term depends on the molality; the fourth can be determined from measurements of the freezing point, or any other colligative property of the solution.

★ 16.6.1 Freezing-Point Depression and the Mean Ionic Activity Coefficient

The relation between the freezing-point depression θ and the mean ionic activity coefficient is obtained easily. Writing Eq. (16.24) using a for the activity of the *solute*, we have

$$d \ln a = \frac{d\theta}{K_f m}. \tag{16.50}$$

But from Section 16.6, we have

$$a = a_\pm^\nu = \gamma_\pm^\nu m_\pm^\nu = \gamma_\pm^\nu (\nu_+^{\nu_+} \nu_-^{\nu_-}) m^\nu.$$

Then

$$d \ln a = \nu \, d \ln m + \nu \, d \ln \gamma_\pm. \tag{16.51}$$

So that Eq. (16.50) becomes

$$\frac{\nu \, dm}{m} + \nu \, d \ln \gamma_\pm = \frac{d\theta}{K_f m}. \tag{16.52}$$

For an ideal solution, $\gamma_\pm = 1$, and Eq. (16.52) becomes

$$d\theta = \nu K_f \, dm,$$

$$\theta = \nu K_f m, \tag{16.53}$$

which shows that the freezing-point depression in a very dilute solution of an electrolyte is the value for a nonelectrolyte multiplied by ν, the number of ions produced by the dissociation of one mole of the electrolyte.

The osmotic coefficient for an electrolytic solution is defined by

$$\theta = \nu K_j m(1 - j). \tag{16.54}$$

With this definition of j, Eq. (16.52) becomes, after repetition of the algebra in Section 16.4.3,

$$\ln \gamma_\pm = -j - \int_0^m \left(\frac{j}{m}\right) dm, \tag{16.55}$$

which has the same form as Eq. (16.26).

Values of the mean ionic activity coefficients for several electrolytes in water at 25 °C are given in Table 16.1. Figure 16.3 shows a plot of $\gamma_\pm$ versus $m^{1/2}$ for different electrolytes in water at 25 °C.

The values of $\gamma_\pm$ are nearly independent of the kind of ions in the compound so long as the compounds are of the same valence type. For example, KCl and NaBr have nearly the same activity coefficients at the same concentration, as do K_2SO_4 and $Ca(NO_3)_2$. In Section 16.7 we shall see that the theory of Debye and Hückel predicts that in a sufficiently dilute solution the mean ionic activity coefficient should depend only on the charges on the ions and their concentration, but not on any other individual characteristics of the ions.

Any of the colligative properties could be used to determine the activity coefficients of a dissolved substance whether it is an electrolyte or nonelectrolyte. The freezing-point depression is much used, because this experiment requires somewhat less elaborate equipment than any of the others. It has the disadvantage that the values of γ can be

Table 16.1
Mean ionic activity coefficients of strong electrolytes

m	0.001	0.005	0.01	0.05	0.1	0.5	1.0
HCl	0.966	0.928	0.904	0.830	0.796	0.758	0.809
NaOH	—	—	—	0.82	—	0.69	0.68
KOH	—	0.92	0.90	0.82	0.80	0.73	0.76
KCl	0.965	0.927	0.901	0.815	0.769	0.651	0.606
NaBr	0.966	0.934	0.914	0.844	0.800	0.695	0.686
H_2SO_4	0.830	0.639	0.544	0.340	0.265	0.154	0.130
K_2SO_4	0.89	0.78	0.71	0.52	0.43	—	—
$Ca(NO_3)_2$	0.88	0.77	0.71	0.54	0.48	0.38	0.35
$CuSO_4$	0.74	0.53	0.41	0.21	0.16	0.068	0.047
$MgSO_4$	—	—	0.40	0.22	0.18	0.088	0.064
$La(NO_3)_3$	—	—	0.57	0.39	0.33	—	—
$In_2(SO_4)_3$	—	—	0.142	0.054	0.035	—	—

By permission from Wendell M. Latimer, *The Oxidation States of the Elements and Their Potentials in Aqueous Solutions*, 2d ed. Englewood Cliffs, N.J.: Prentice-Hall, 1952, pp. 354–356.

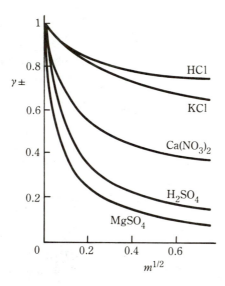

Figure 16.3 Mean ionic activity coefficients as functions of $m^{1/2}$.

obtained only near the freezing point of the solvent. The measurement of vapor pressure does not have this drawback, but is more difficult to handle experimentally. In Chapter 17 the method of obtaining mean ionic activity coefficients from measurements of the potentials of electrochemical cells is described. The electrochemical method is easily handled experimentally, and it can be used at any convenient temperature.

16.7 THE DEBYE–HÜCKEL THEORY OF THE STRUCTURE OF DILUTE IONIC SOLUTIONS

At this stage it is worthwhile to describe the constitution of ionic solutions in some detail. The solute in dilute solutions of nonelectrolytes is adequately described thermodynamically by the equation,

$$\mu = \mu^\circ + RT \ln m. \tag{16.56}$$

The chemical potential is a sum of two terms: the first, μ°, is independent of composition, and the second depends on the composition. Equation (16.56) is fairly good for most nonelectrolytes up to concentrations as high as 0.1 m, and for many others it does well at even higher concentrations. The simple expression in Eq. (16.56) is not adequate for electrolytic solutions; deviations are pronounced even at concentrations of 0.001 m. This is true even if Eq. (16.56) is modified to take into account the several ions produced.

To describe the behavior of an electrolyte in a dilute solution, the chemical potential must be written in the form, see Eq. (16.49),

$$\mu = \mu^\circ + vRT \ln m + vRT \ln \gamma_\pm. \tag{16.57}$$

In Eq. (16.57) the second term on the right of Eq. (16.49) has been absorbed into the μ°. The μ° is independent of the composition; the second and third terms depend on the composition.

The extra Gibbs energy represented by the term $vRT \ln \gamma_\pm$ in Eq. (16.57) is mainly the result of the energy of interaction of the electrical charges on the ions; since in one mole of the electrolyte there are vN_A ions, this interaction energy is, on the average, $kT \ln \gamma_\pm$ per ion, where the Boltzmann constant $k = R/N_A$. The van der Waals forces acting between neutral particles of solvent and nonelectrolyte are weak and are effective only over very short distances, while the coulombic forces that act between ions and those between ions and neutral molecules of solvent are much stronger and act over greater distances. This difference in range of action accounts for the large deviations from ideality in ionic solutions even at high dilutions where the ions are far apart. Our object is to calculate this electrical contribution to the Gibbs energy.

For a model of the electrolyte solution we imagine that the ions are electrically charged, conducting spheres having a radius a, immersed in a solvent of permittivity ϵ. Let the charge on the ion be q. If the ion were not charged, $q = 0$, its μ could be represented by Eq. (16.56); since it is charged, its μ must incorporate an extra term, $kT \ln \gamma_\pm$. The extra term, which we are trying to calculate, must be the work expended in charging the ion, bringing q from zero to q. Let the electrical potential at the surface of the sphere be ϕ_a, a function of q. By definition, the potential of the sphere is the work that must be expended to bring a unit positive charge from infinity to the surface of the sphere; if we bring a charge dq from infinity to the surface, the work will be $dW = \phi_a\,dq$. Integrating from zero to q, we obtain the work expended in charging the ion:

$$W = \int_0^q \phi_a\,dq, \tag{16.58}$$

where W is the extra energy possessed by the ion in virtue of its charge; the Gibbs energy of an ion is greater than that of a neutral particle by W. This additional energy is made up of two contributions:

$$W = W_s + W_i. \tag{16.59}$$

The energy required to charge an *isolated* sphere immersed in a dielectric medium is the *self-energy* of the charged sphere, W_s. Since W_s does not depend on the concentration of the ions, it will be absorbed in the value of μ°. The additional energy beyond W_s needed to charge the ion in the presence of all the other ions is the interaction energy W_i, whose value depends very much on the concentration of the ions. It is W_i which we identify with the term, $kT \ln \gamma_\pm$:

$$kT \ln \gamma_\pm = W_i = W - W_s. \tag{16.60}$$

The potential of an isolated conducting sphere immersed in a medium having a permittivity ϵ is given by the formula from classical electrostatics: $\phi_a = q/4\pi\epsilon a$. Using this value in the integral of Eq. (16.58), we obtain for W_s

$$W_s = \int_0^q \frac{q}{4\pi\epsilon a}\, dq = \frac{q^2}{8\pi\epsilon a}. \tag{16.61}$$

Having this value of W_s, we can obtain a value for W_i if we succeed in calculating W. To calculate W we must first calculate ϕ_a; see Eq. (16.58). Before doing the calculation we can guess reasonably that W_i will be negative. Consider a positive ion: It attracts negative ions and repels other positive ions. As a result negative ions will be, on the average, a little closer to the positive ion than will be the other positive ions. This in turn gives the ion a lower Gibbs energy than it would have if it were not charged; since we are interested in the energy relative to that of the uncharged species, W_i is negative. In 1923 P. Debye and E. Hückel succeeded in obtaining a value of ϕ_a. The following is an abbreviated version of the method they used.

We locate the origin of a spherical coordinate system at the center of a positive ion (Fig. 16.4). Consider a point P at a distance r from the center of the ion. The potential ϕ at the point P is related to the charge density ρ, the charge per unit volume, by the Poisson equation (for the derivation, see Appendix II):

$$\frac{1}{r^2}\frac{d}{dr}\left(r^2\frac{d\phi}{dr}\right) = -\frac{\rho}{\epsilon}. \tag{16.62}$$

If ρ can be expressed as a function of either ϕ or r, then Eq. (16.62) can be integrated to yield ϕ as a function of r, from which we can get ϕ_a directly.

To calculate ρ we proceed as follows. Let $\tilde{c}_+$ and $\tilde{c}_-$ be the concentrations of positive and negative ions, respectively. If z_+ and z_- are the valences (complete with sign) of the ions and e is the magnitude of the charge on the electron, then the charge on one mole of positive ions is $z_+ F$, and the positive charge in unit volume is $\tilde{c}_+ z_+ F$, in which F is the Faraday constant; $F = 96\,484.56$ C/mol. The charge density, ρ, is the total charge, positive plus negative, in unit volume; therefore

$$\rho = \tilde{c}_+ z_+ F + \tilde{c}_- z_- F = F(\tilde{c}_+ z_+ + \tilde{c}_- z_-) \tag{16.63}$$

If the electrical potential at P is ϕ, then the potential energies of the positive and negative

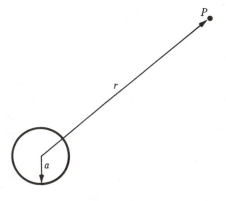

Figure 16.4

ions at P are $ez_+\phi$ and $ez_-\phi$, respectively. Debye and Hückel assumed that the distribution of the ions is a Boltzmann distribution (Section 4.13). Then

$$\tilde{c}_+ = \tilde{c}^\circ_+ e^{-z_+ e\phi/kT} \qquad \text{and} \qquad \tilde{c}_- = \tilde{c}^\circ_- e^{-z_- e\phi/kT},$$

where c°_+ and c°_- are the concentrations in the region where $\phi = 0$; but where $\phi = 0$, the distribution is uniform and the solution must be electrically neutral; ρ must be zero. This requires that

$$\tilde{c}^\circ_+ z_+ + \tilde{c}^\circ_- z_- = 0.$$

Putting the values of $\tilde{c}_+$ and $\tilde{c}_-$ in the expression for ρ yields

$$\rho = F(z_+ \tilde{c}^\circ_+ e^{-z_+ e\phi/kT} + z_- \tilde{c}^\circ_- e^{-z_- e\phi/kT}).$$

Assuming that $ze\phi/kT \ll 1$, the exponentials are expanded in series; $e^{-x} = 1 - x + \cdots$. This reduces ρ to

$$\rho = F\left[\tilde{c}^\circ_+ z_+ + \tilde{c}^\circ_- z_- - \frac{e\phi}{kT}(\tilde{c}^\circ_+ z^2_+ + \tilde{c}^\circ_- z^2_-) \right].$$

The condition of electrical neutrality drops out the first two terms; then, since $e/k = F/R$, we have

$$\rho = -\frac{F^2\phi}{RT} \sum_i \tilde{c}^\circ_i z^2_i, \tag{16.64}$$

where the sum is over all the kinds of ions in the solution, in this case, two kinds of ions. Using this relation, we have

$$-\frac{\rho}{\epsilon} = \left(\frac{F^2}{\epsilon RT} \sum_i \tilde{c}^\circ_i z^2_i \right)\phi = \varkappa^2\phi, \tag{16.65}$$

where we have defined $\varkappa^2$ as

$$\varkappa^2 \equiv \frac{F^2}{\epsilon RT} \sum_i \tilde{c}^\circ_i z^2_i. \tag{16.66}$$

Using this value of $-\rho/\epsilon$, the Poisson equation, Eq. (16.62), becomes

$$\frac{1}{r^2}\frac{d}{dr}\left(r^2 \frac{d\phi}{dr} \right) - \varkappa^2\phi = 0. \tag{16.67}$$

If we substitute $\phi = v/r$ in Eq. (16.67), it reduces to

$$\frac{d^2v}{dr^2} - \varkappa^2 v = 0, \tag{16.68}$$

which has the solution*

$$v = Ae^{-\varkappa r} + Be^{\varkappa r},$$

where A and B are arbitrary constants. The value of ϕ is

$$\phi = A\frac{e^{-\varkappa r}}{r} + B\frac{e^{\varkappa r}}{r}. \tag{16.69}$$

---

* You should verify this by substitution and work out the transformation of Eq. (16.67) into (16.68) in detail.

As $r \to \infty$, the second term on the right approaches infinity.* The potential must remain finite as $r \to \infty$, so this second term cannot be part of the physical solution; therefore we set $B = 0$ and obtain

$$\phi = A \frac{e^{-\varkappa r}}{r}. \tag{16.70}$$

Expanding the exponential in series and retaining only the first two terms, we have

$$\phi = A\left(\frac{1 - \varkappa r}{r}\right) = \frac{A}{r} - A\varkappa. \tag{16.71}$$

If the concentration is zero, then $\varkappa = 0$, and the potential at point P should be that due to the central positive ion only; $\phi = z_+ e/4\pi\epsilon r$. But when $\varkappa = 0$, Eq. (16.71) reduces to $\phi = A/r$; hence, $A = z_+ e/4\pi\epsilon$; Eq. (16.71) becomes:

$$\phi = \frac{z_+ e}{4\pi\epsilon r} - \frac{z_+ e\varkappa}{4\pi\epsilon}. \tag{16.72}$$

At $r = a$, we have

$$\phi_a = \frac{z_+ e}{4\pi\epsilon a} - \frac{z_+ e\varkappa}{4\pi\epsilon}. \tag{16.73}$$

If, with the exception of our central positive ion, all other ions in the solution are fully charged, then the work to charge this positive ion in the presence of all the others is, Eq. (16.58),

$$W_+ = \int_0^q \phi_a \, dq;$$

but $q = z_+ e$, so that $dq = e \, dz_+$. Using Eq. (16.73) for ϕ_a, we obtain

$$W_+ = \int_0^{z_+} \left(\frac{z_+ e^2}{4\pi\epsilon a} - \frac{z_+ e^2 \varkappa}{4\pi\epsilon}\right) dz_+ = \left(\frac{e^2}{4\pi\epsilon a} - \frac{e^2 \varkappa}{4\pi\epsilon}\right) \int_0^{z_+} z_+ \, dz_+,$$

$$W_+ = \frac{(z_+ e)^2}{8\pi\epsilon a} - \frac{(z_+ e)^2 \varkappa}{8\pi\epsilon}, \tag{16.74}$$

where the first term is the self-energy $W_{s,+}$, and the second is the interaction energy $W_{i,+}$, the extra Gibbs energy of a single positive ion that is due to the presence of the others. Using Eq. (16.60), we have

$$kT \ln \gamma_+ = -\frac{(z_+ e)^2 \varkappa}{8\pi\epsilon}. \tag{16.75}$$

For a negative ion we would get

$$kT \ln \gamma_- = -\frac{(z_- e)^2 \varkappa}{8\pi\epsilon}. \tag{16.76}$$

The mean ionic activity coefficient can be calculated using Eq. (16.45):

$$\gamma_\pm^\nu = \gamma_+^{\nu_+} \gamma_-^{\nu_-}.$$

Taking logarithms, we obtain

$$\nu \ln \gamma_\pm = \nu_+ \ln \gamma_+ + \nu_- \ln \gamma_-.$$

---

*   Verify using L'Hopital's rule.

Using Eqs. (16.75) and (16.76) this becomes

$$v \ln \gamma_\pm = -\frac{e^2 \varkappa}{8\pi\epsilon kT}(v_+ z_+^2 + v_- z_-^2).$$

Since the electrolyte itself is electrically neutral, we must have

$$v_+ z_+ + v_- z_- = 0:$$

Multiplying by z_+: $\qquad v_+ z_+^2 = -v_- z_+ z_-$

Multiplying by z_-: $\qquad \underline{v_- z_-^2 = -v_+ z_+ z_-}$

Adding: $\qquad v_+ z_+^2 + v_- z_-^2 = -(v_+ + v_-)z_+ z_- = -v z_+ z_-.$

Using this result we obtain finally:

$$\ln \gamma_\pm = \frac{e^2 \varkappa}{8\pi\epsilon kT} z_+ z_- = \frac{F^2 \varkappa}{8\pi\epsilon N_A RT} z_+ z_-. \tag{16.77}$$

Converting to common logarithms and introducing the value of $\varkappa$ from Eq. (16.66), we obtain

$$\log_{10} \gamma_\pm = \frac{1}{(2.303)8\pi N_A}\left(\frac{F^2}{\epsilon RT}\right)^{3/2}\left(\sum_i \tilde{c}_i^\circ z_i^2\right)^{1/2} z_+ z_-. \tag{16.78}$$

The ionic strength, I_c, is defined by

$$I_c = \frac{1}{2}\sum_i c_i z_i^2 \tag{16.79}$$

where c_i is the concentration of the ith ion in mol/L. Since $\tilde{c}_i^\circ = (1000 \text{ L/m}^3)c_i$, we have

$$\sum_i \tilde{c}_i^\circ z_i^2 = (1000 \text{ L/m}^3)\sum_i c_i z_i^2 = 2(1000 \text{ L/m}^3)I_c.$$

Finally, we obtain

$$\log_{10} \gamma_\pm = \left[\frac{(2000 \text{ L/m}^3)^{1/2}}{(2.303)8\pi N_A}\left(\frac{F^2}{\epsilon RT}\right)^{3/2}\right]z_+ z_- I_c^{1/2} \tag{16.80}$$

The factor enclosed in the brackets is made up of universal constants and the values of ϵ and T. For a continuous medium, $\epsilon = \epsilon_r\epsilon_0$, where ϵ_r is the dielectric constant of the medium. Introducing the values of the constants, we obtain

$$\log_{10} \gamma_\pm = \frac{(1.8248 \times 10^6 \text{ K}^{3/2} \text{ L}^{1/2}/\text{mol}^{1/2})}{(\epsilon_r T)^{3/2}} z_+ z_- I_c^{1/2}. \tag{16.81}$$

In water at 25 °C, $\epsilon_r = 78.54$; then we have

$$\log_{10} \gamma_\pm = (0.5092 \text{ L}^{1/2}/\text{mol}^{1/2})z_+ z_- I_c^{1/2} \tag{16.82}$$

Either of Eqs. (16.81) or (16.82) is the *Debye–Hückel limiting law*. The limiting law predicts that the logarithm of the mean ionic activity coefficient should be a linear function of the square root of the ionic strength and that the slope of the line should be proportional to the product of the valences of the positive and negative ions. (The slope is negative, since z_- is negative.) These predictions are confirmed by experiment in dilute solutions of strong electrolytes. Figure 16.5 shows the variation of $\log_{10} \gamma_\pm$ with I_c; the solid curves are the experimental data; the dashed lines are the values predicted by the limiting law, Eq. (16.82).

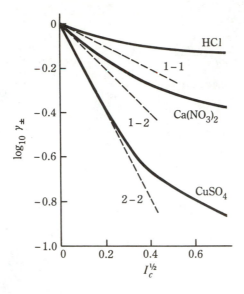

Figure 16.5 $\log_{10} \gamma_{\pm}$ versus $I_c^{1/2}$.

The approximations required in the theory restrict its validity to solutions that are very dilute. In practice, deviations from the limiting law become appreciable in the concentration range from 0.005 to 0.01 mol/L. More accurate equations have been derived that extend the theory to slightly higher concentrations. However, as yet there is no satisfactory theoretical equation that can predict the behavior in solutions of concentration higher than 0.01 mol/L.

The Debye–Hückel theory provides an accurate representation of the limiting behavior of the activity coefficient in dilute ionic solutions. In addition, it yields a picture of the structure of the ionic solution. We have alluded to the fact that the negative ions cluster a little closer to a positive ion than do positive ions, which are pushed away. In this sense every ion is surrounded by an atmosphere of oppositely charged ions; the total charge on this atmosphere is equal, but opposite in sign, to that on the ion. The mean radius of the ionic atmosphere is given by $1/\varkappa$, which is called *the Debye length*. Since $\varkappa$ is proportional to the square root of the ionic strength, at high ionic strengths the atmosphere is closer to the ion than at low ionic strengths. This concept of the ionic atmosphere and the mathematics associated with it have been extraordinarily fruitful in clarifying many aspects of the behavior of electrolytic solutions.

The concept of the ionic atmosphere can be made clearer by calculating the charge density as a function of the distance from the ion. By combining the final expression for the charge density in terms of ϕ with Eq. (16.70) and the value of A, we obtain

$$\rho = -\frac{z_+ e\varkappa^2}{4\pi} \frac{e^{-\varkappa r}}{r}.$$ (16.83)

The total charge contained in a spherical shell bounded by spheres of radii r and $r + dr$ is the charge density multiplied by the volume of the shell, $4\pi r^2 dr$:

$$-z_+ e\varkappa^2 r e^{-\varkappa r}\, dr.$$

By integrating this quantity from zero to infinity we obtain the total charge on the atmosphere which is $-z_+ e$. The fraction of this total charge that is in the spherical shell, per

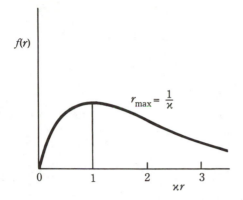

Figure 16.6 Charge distribution in the ionic atmosphere.

unit width dr of the shell, we will call $f(r)$. Then

$$f(r) = \varkappa^2 r e^{-\varkappa r}. \tag{16.84}$$

The function $f(r)$ is the distribution function for the charge in the atmosphere. A plot of $f(r)$ versus r is shown in Fig. 16.6. The maximum in the curve appears at $r_{max} = 1/\varkappa$, the Debye length. In an electrolyte of a symmetrical valence type, $1:1$, $2:2$, and so on, we may say that $f(r)$ represents the probability per unit width dr of finding the balancing ion in the spherical shell at the distance r from the central ion. In solutions of high ionic strength the mate to the central ion is very close, $1/\varkappa$ is small; at lower ionic strengths $1/\varkappa$ is large and the mate is far away.

16.8 EQUILIBRIA IN IONIC SOLUTIONS

From the Debye–Hückel limiting law, Eq. (16.78), we find a negative value of $\ln \gamma_\pm$, which confirms the physical argument that the interaction with other ions lowers the Gibbs energy of an ion in an electrolytic solution. This lower Gibbs energy means that the ion is more stable in solution than it would be if it were not charged. The extra stability is measured by the term, $kT \ln \gamma_\pm$, in the expression for the chemical potential. Now we examine the consequences of this extra stability in two simple cases: the ionization of a weak acid, and the solubility of a sparingly soluble salt.

Consider the dissociation equilibrium of a weak acid, HA:

$$HA \; \rightleftharpoons \; H^+ + A^-.$$

The equilibrium constant is the quotient of the activities,

$$K = \frac{a_{H^+} a_{A^-}}{a_{HA}}. \tag{16.85}$$

By definition, we have

$$a_{H^+} = \gamma_+ n_{H^+}, \qquad a_{A^-} = \gamma_- m_{A^-}, \qquad a_{HA} = \gamma_{HA} m_{HA},$$

so that

$$K = \left(\frac{\gamma_+ \gamma_-}{\gamma_{HA}}\right) \frac{m_{H^+} m_{A^-}}{m_{HA}} = \frac{\gamma_\pm^2}{\gamma_{HA}} \frac{m_{H^+} m_{A^-}}{m_{HA}},$$

where we have used the relation, $\gamma_+ \gamma_- = \gamma_\pm^2$. If the total molality of the acid is m and the degree of dissociation is α,

$$m_{H^+} = \alpha m, \qquad m_{A^-} = \alpha m, \qquad m_{HA} = (1 - \alpha)m.$$

Then,

$$K = \frac{\gamma_\pm^2 \alpha^2 m}{\gamma_{HA}(1 - \alpha)}. \tag{16.86}$$

If the solution is dilute, we may set $\gamma_{HA} = 1$, since HA is an uncharged species. Also if K is small, $1 - \alpha \approx 1$. Then, Eq. (16.86) yields

$$\alpha = \left(\frac{K}{m}\right)^{1/2} \frac{1}{\gamma_\pm}. \tag{16.87}$$

If we ignored the ionic interactions, we would set $\gamma_\pm = 1$, and calculate $\alpha_0 = (K/m)^{1/2}$. Then Eq. (16.87) becomes

$$\alpha = \frac{\alpha_0}{\gamma_\pm}. \tag{16.88}$$

From the limiting law, $\gamma_\pm < 1$; hence the correct value of α given by Eq. (16.88) is greater than the rough value α_0, which ignored the ionic interactions. The stabilization of the ions by the presence of the other ions shifts the equilibrium to produce more ions; hence the degree of dissociation is increased.

If the solution is dilute enough in ions, $\gamma_\pm$ can be obtained from the limiting law, Eq. (16.82), which for a 1:1 electrolyte can be written as

$$\gamma_\pm = 10^{-0.51(\alpha_0 m)^{1/2}} = e^{-1.17(\alpha_0 m)^{1/2}},$$

where the ionic strength $I_c = \alpha_0 m$. (We have ignored the difference between c and m.) The value of α_0 can be used to compute I_c, since α and α_0 are not greatly different. Using this expression, Eq. (16.88) becomes

$$\alpha = \alpha_0 e^{1.17(\alpha_0 m)^{1/2}} = \alpha_0[1 + 1.17(\alpha_0 m)^{1/2}].$$

In the last equality, the exponential has been expanded in series. The computation for 0.1 molal acetic acid, $K = 1.75 \times 10^{-5}$, shows that the degree of dissociation is increased by about 4%. The effect is small because the dissociation does not produce many ions.

If a large amount of an inert electrolyte, one that does not contain either H^+ or A^- ions, is added to the solution of the weak acid, then a comparatively large effect on the dissociation is produced. Consider a solution of a weak acid in 0.1 m KCl, for example. The ionic strength of this solution is too large to use the limiting law, but the value of $\gamma_\pm$ can be estimated from Table 16.1. The table shows that for 1:1 electrolytes the value of $\gamma_\pm$ in 0.1 molal solution is about 0.8. We may assume that this is a reasonable value for H^+ and A^- ions in the 0.1 molal KCl solution. Then by Eq. (16.88),

$$\alpha = \frac{\alpha_0}{0.8} = 1.25\alpha_0.$$

Thus the presence of a large amount of inert electrolyte exerts an appreciable influence, the *salt effect*, on the degree of dissociation. The salt effect is larger the higher the concentration of the electrolyte.

Consider the equilibrium of a slightly soluble salt, such as silver chloride, with its ions:

$$AgCl(s) \; \rightleftharpoons \; Ag^+ + Cl^-.$$

The solubility product constant is

$$K_{sp} = a_{Ag^+} a_{Cl^-} = (\gamma_+ m_+)(\gamma_- m_-).$$

If s is the solubility of the salt in moles per kilogram of water, then $m_+ = m_- = s$, and

$$K_{sp} = \gamma_\pm^2 s^2.$$

If s_0 is the solubility calculated ignoring ionic interaction, then $s_0^2 = K_{sp}$, and we have

$$s = \frac{s_0}{\gamma_\pm}, \tag{16.89}$$

which shows that the solubility is increased by the ionic interaction. By the same reasoning as we used in discussing the dissociation of a weak acid, we can show that in 0.1 molal solution of an inert electrolyte such as KNO_3 the solubility would be increased by 25%. This increase in solubility produced by an inert electrolyte is sometimes called "salting in." The effect of an inert electrolyte on the solubility of a salt such as $BaSO_4$ would be much larger because of the larger charges on the Ba^{2+} and SO_4^{2-} ions. The salt effect on solubility produced by an inert electrolyte should not be confused with the *decrease* in solubility effected by an electrolyte that contains an ion in common with the sparingly soluble salt. In addition to acting in the opposite sense, the "common ion" effect is enormous compared to the effect of an inert electrolyte.

QUESTIONS

16.1 What is the activity? How is it related to, but distinguished from, the concentration?

16.2 What is the direction of the influence of nonideality (for example, positive deviations from Raoult's law) on (a) freezing-point depression, (b) boiling-point elevation, and (c) osmotic pressure compared to the ideal solution case?

16.3 Why do deviations from ideality begin to occur at much lower concentrations for electrolyte solutions than for nonelectrolyte solutions?

16.4 Discuss and interpret the trends of the Debye length with increasing (a) temperature, (b) dielectric constant, and (c) ionic strength.

16.5 What is the correct order of the following inert electrolytes in terms of increasing enhancement of acetic acid dissociation: 0.01 molal NaCl, 0.001 molal KBr, 0.01 molal $CuCl_2$?

PROBLEMS

16.1 The *apparent* value of K_f in sucrose ($C_{12}H_{22}O_{11}$) solutions of various concentrations

$m/(\text{mol/kg})$	0.10	0.20	0.50	1.00	1.50	2.00
$K_f/(\text{K kg/mol})$	1.88	1.90	1.96	2.06	2.17	2.30

a) Calculate the activity of water in each solution.
b) Calculate the activity coefficient of water in each solution.
c) Plot the values of a and γ against the mole fraction of water in the solution.
d) Calculate the activity and the activity coefficient of sucrose in a 1 molal solution.

16.2 The Henry's law constant for chloroform in acetone at 35.17 °C is 0.199 if the vapor pressure is in atm, and concentration of chloroform is in mole fraction. The partial pressure of chloroform at several values of mole fraction is:

x_{CHCl_3}	0.059	0.123	0.185
p_{CHCl_3}/mmHg	9.2	20.4	31.9

If $a = \gamma x$, and $\gamma \to 1$ as $x \to 0$, calculate the values of a and γ for chloroform in the three solutions.

16.3 At the same concentrations as in Problem 16.2, the partial pressures of acetone are 323.2, 299.3, and 275.4 mmHg, respectively. The vapor pressure of pure acetone is 344.5 mmHg. Calculate the activities of acetone and the activity coefficients in these three solutions ($a = \gamma x$; $\gamma \to 1$ as $x \to 1$).

16.4 The liquid–vapor equilibrium in the system, isopropyl alcohol-benzene, was studied over a range of compositions at 25 °C. The vapor may be assumed to be an ideal gas. Let x_1 be the mole fraction of the isopropyl alcohol in the liquid, and p_1 be the partial pressure of the alcohol in the vapor. The data are:

x_1	1.000	0.924	0.836
p_1/mmHg	44.0	42.2	39.5

a) Calculate the rational activity of the isopropyl alcohol at $x_1 = 1.000$, $x_1 = 0.924$, and $x_1 = 0.836$.
b) Calculate the rational activity coefficient of the isopropyl alcohol at the three compositions in (a).
c) At $x_1 = 0.836$ calculate the amount by which the chemical potential of the alcohol differs from that in an ideal solution.

16.5 A regular binary liquid solution is defined by the equation

$$\mu_i = \mu_i^\circ + RT \ln x_i + w(1 - x_i)^2,$$

where w is a constant.

a) What is the significance of the function μ_i°?
b) Express $\ln \gamma_i$ in terms of w; γ_i is the *rational* activity coefficient.
c) At 25 °C, $w = 324$ J/mol for mixtures of benzene and carbon tetrachloride. Calculate γ for CCl_4 in solutions with $x_{CCl_4} = 0, 0.25, 0.50, 0.75$, and 1.0.

16.6 The freezing point depression of solutions of ethanol in water is given by

$m/(mol/kg\ H_2O)$	θ/K	$m/(mol/kg\ H_2O)$	θ/K
0.074 23	0.137 08	0.134 77	0.248 21
0.095 17	0.175 52	0.166 68	0.306 54
0.109 44	0.201 72	0.230 7	0.423 53

Calculate the activity and the activity coefficient of ethanol in 0.10 and 0.20 molal solution.

16.7 The freezing-point depression of aqueous solutions of NaCl is:

$m/(mol/kg)$	0.001	0.002	0.005	0.01	0.02	0.05	0.1
θ/K	0.003676	0.007322	0.01817	0.03606	0.07144	0.1758	0.3470

  a) Calculate the value of j for each of these solutions.
  b) Plot j/m versus m, and evaluate $-\log_{10}\gamma_{\pm}$ for each solution. $K_f = 1.8597$ K kg/mol. From the Debye–Hückel limiting law it can be shown that $\int_0^{0.001} (j/m)\,dm = 0.0226$. [G. Scatchard and S. S. Prentice, *J.A.C.S.*, **55**: 4355 (1933).]

16.8 From the data in Table 16.1, calculate the activity of the electrolyte and the mean activity of the ions in 0.1 molal solutions of

  a) KCl, b) H_2SO_4, c) $CuSO_4$, d) $La(NO_3)_3$, e) $In_2(SO_4)_3$.

16.9 a) Calculate the mean ionic molality, $m_{\pm}$, in 0.05 molal solutions of $Ca(NO_3)_2$, NaOH, $MgSO_4$, $AlCl_3$.
  b) What is the ionic strength of each of the solutions in (a)?

16.10 Using the limiting law, calculate the value of $\gamma_{\pm}$ in 10^{-4} and 10^{-3} molal solutions of HCl, $CaCl_2$, and $ZnSO_4$ at 25 °C.

16.11 Calculate the values of $1/\varkappa$ at 25 °C, in 0.01 and 1 molal solutions of KBr. For water, $\epsilon_r = 78.54$.

16.12 a) What is the total probability of finding the balancing ion at a distance greater than $1/\varkappa$ from the central ion?
  b) What is the radius of the sphere around the central ion such that the probability of finding the balancing ion within the sphere is 0.5?

16.13 At 25 °C the dissociation constant for acetic acid is 1.75×10^{-5}. Using the limiting law, calculate the degree of dissociation in 0.010, 0.10, and 1.0 molal solutions. Compare these values with the value obtained by ignoring ionic interaction.

16.14 Estimate the degree of dissociation of 0.10 molal acetic acid, $K = 1.75 \times 10^{-5}$, in 0.5 molal KCl, in 0.5 molal $Ca(NO_3)_2$, and in 0.5 molal $MgSO_4$ solution.

16.15 For silver chloride at 25 °C, $K_{sp} = 1.56 \times 10^{-10}$. Using the data in Table 16.1, estimate the solubility of AgCl in 0.001, 0.01, and 1.0 molal KNO_3 solution. Plot $\log_{10} s$ against $m^{1/2}$.

16.16 Estimate the solubility of $BaSO_4$, $K_{sp} = 1.08 \times 10^{-10}$, in (a) 0.1 molal NaBr and (b) 0.1 molal $Ca(NO_3)_2$ solution.

17

Equilibria in Electrochemical Cells

17.1 INTRODUCTION

An electrochemical cell is a device that can produce electrical work in the surroundings. For example, the commercial dry cell is a sealed cylinder with two brass connecting terminals protruding from it. One terminal is stamped with a plus sign and the other with a minus sign. If the two terminals are connected to a small motor, electrons flow through the motor from the negative to the positive terminal of the cell. Work is produced in the surroundings and a chemical reaction, the *cell reaction*, occurs within the cell. By Eq. (10.14), the electrical work produced, W_{el}, is less than or equal to the decrease in the Gibbs energy of the cell reaction, $-\Delta G$.

$$W_{el} \leq -\Delta G \tag{17.1}$$

Before continuing the thermodynamic development we pause to look at some fundamentals of electrostatics.

17.2 DEFINITIONS

The electric potential of a point in space is defined as the work expended in bringing a unit positive charge from infinity, where the electric potential is zero, to the point in question. Thus if ϕ is the electric potential at the point and W the work required to bring a charge Q from infinity to that point, then

$$\phi = \frac{W}{Q}. \tag{17.2}$$

Similarly, if ϕ_1 and ϕ_2 are the electric potentials of two points in space, and W_1 and W_2 are the corresponding quantities of work required to bring the charge Q to these points, then

$$W_1 + W_{12} = W_2, \tag{17.3}$$

where W_{12} is the work to bring Q from point 1 to point 2. This relation exists since the electric field is conservative. Thus the same quantity of work must be expended to bring Q to point 2, whether we bring it directly, W_2, or bring it first to point 1 and then to point 2, $W_1 + W_{12}$. Therefore $W_{12} = W_2 - W_1$, and, from Eq. (17.2),

$$\phi_2 - \phi_1 = \frac{W_{12}}{Q}. \tag{17.4}$$

The difference in electric potential between two points is the work expended in taking a unit positive charge from point 1 to point 2.

Applying Eq. (17.) to the transfer of an infinitesimal quantity of charge, we obtain the element of work expended on the system.

$$W_{12} = -dW_{el} = \mathscr{E}\, dQ, \tag{17.5}$$

where $\mathscr{E}$ has been written for the potential difference $\phi_2 - \phi_1$, and dW_{el} is the work *produced*.

17.3 THE CHEMICAL POTENTIAL OF CHARGED SPECIES

The escaping tendency of a charged particle, an ion or an electron, in a phase depends on the electric potential of that phase. Clearly, if we impress a large negative electric potential on a piece of metal, the escaping tendency of negative particles will be increased. To find the relation between the electric potential and the escaping tendency, the chemical potential, we consider a system of two balls M and M' of the same metal. Let their electric potentials be ϕ and ϕ'. If we transfer a number of electrons carrying a charge, dQ, from M to M', the work expended on the system is given by Eq. (17.4): $-dW_{el} = (\phi' - \phi)\, dQ$. The work *produced* is dW_{el}. If the transfer is done reversibly, then by Eq. (10.13), the work produced is equal to the decrease in Gibbs energy of the system; $dW_{el} = -dG$, so that

$$dG = (\phi' - \phi)\, dQ.$$

But, in terms of the chemical potential of the electrons, $\tilde{\mu}_{e^-}$, if dn moles of electrons were transferred, we have

$$dG = \tilde{\mu}'_{e^-}\, dn - \tilde{\mu}_{e^-}\, dn.$$

The dn moles of electrons carry a negative charge $dQ = -F\, dn$, where F is the charge per mole of electrons, $F = 96\,484.56$ C/mol. Combining these two equations yields, after division by dn,

$$\tilde{\mu}'_{e^-} - \tilde{\mu}_{e^-} = -F(\phi' - \phi),$$

which rearranges to

$$\tilde{\mu}_{e^-} = \tilde{\mu}'_{e^-} + F\phi' - F\phi.$$

Let μ_{e^-} be the chemical potential of the electrons in M when ϕ is zero; then, $\mu_{e^-} = \tilde{\mu}'_{e^-} + F\phi'$. Subtracting this equation from the preceding one, we obtain

$$\tilde{\mu}_{e^-} = \mu_{e^-} - F\phi. \tag{17.6}$$

Equation (17.6) is the relation between the escaping tendency of the electrons, $\tilde{\mu}_{e^-}$, in a phase and the electrical potential of the phase, ϕ. The escaping tendency is a linear function of ϕ. Note that Eq. (17.6) shows that if ϕ is negative, $\tilde{\mu}_{e^-}$ is larger than when ϕ is positive.

By a similar argument, it may be shown that for any charged species in a phase

$$\tilde{\mu}_i = \mu_i + z_i F\phi, \tag{17.7}$$

where z_i is the charge on the species. For electrons, $z_{e^-} = -1$, so Eq. (17.7) would reduce to Eq. (17.6). Equation (17.7) divides the chemical potential $\tilde{\mu}_i$ of a charged species into two terms. The first term, μ_i, is the "chemical" contribution to the escaping tendency. The chemical contribution is produced by the chemical environment in which the charged species exists, and is the same in two phases of the same chemical composition since it is a function only of T, p, and composition. The second term, $z_i F\phi$, is the "electrical" contribution to the escaping tendency; it depends on the electrical condition of the phase that is manifested by the value of ϕ. Because it is convenient to divide the chemical potential into these two contributions, $\tilde{\mu}_i$, the *electrochemical potential*, has been introduced to preserve μ_i for the ordinary chemical potential.

17.3.1 Conventions for the Chemical Potential of Charged Species

IONS IN AQUEOUS SOLUTION

For ions in aqueous solution we assign $\phi = 0$ in the solution; then $\tilde{\mu}_i = \mu_i$, and we can use the ordinary μ_i for these ions. This assignment is justified by the fact that the value of ϕ in the electrolytic solution will drop out of our calculations; we have no way of determining its value, and so it might as well be zero and save us a little algebraic labor.

ELECTRONS IN METALS

In the metallic parts of our system we cannot throw away the electrical potential, since we often compare the electrical potentials of two different wires of the same composition (the two terminals of the cell). However, within a *single* piece of a metal it is evident that the division of the chemical potential into a "chemical" part and an "electrical" part is a purely arbitrary one, justified only by convenience. Since the "chemical" part of the escaping tendency arises from the interactions of the electrically charged particles that compose any piece of matter, there is no way to determine in a single piece of matter where the "chemical" part leaves off and the "electrical" part begins.

To make the arbitrary division of $\tilde{\mu}_i$ as convenient as possible, we assign the "chemical" part of the $\tilde{\mu}_{e^-}$ the most convenient value, zero, in every metal. Thus in every metal, by convention,

$$\mu_{e^-} = 0. \tag{17.8}$$

Then, for electrons in every metal, Eq. (17.6) becomes

$$\tilde{\mu}_{e^-} = -F\phi. \tag{17.9}$$

IONS IN PURE METALS

The arbitrary definition in Eq. (17.9) simplifies the form of the chemical potential of the metal ion in a metal. Within any metal there exists an equilibrium between metal atoms M, metal ions M^{z+}, and electrons:

$$M \rightleftharpoons M^{z+} + ze^-.$$

The equilibrium condition is

$$\mu_M = \tilde{\mu}_{M^{z+}} + z\tilde{\mu}_{e^-}.$$

Using Eq. (17.7) for $\tilde{\mu}_{M^{z+}}$ and Eq. (17.9) for $\tilde{\mu}_{e^-}$, we obtain $\mu_M = \mu_{M^{z+}} + zF\phi - zF\phi$, or $\mu_M = \mu_{M^{z+}}$. For a *pure* metal at 1 atm and 25 °C, we have $\mu_M^\circ = \mu_{M^{z+}}^\circ$; by our earlier convention that $\mu^\circ = 0$ for elements under these conditions, we obtain

$$\mu_{M^{z+}}^\circ = 0. \tag{17.10}$$

The "chemical" part of the escaping tendency of the metal ion is zero in a pure metal under standard conditions; then using Eq. (17.7),

$$\tilde{\mu}_{M^{z+}} = zF\phi. \tag{17.11}$$

Equations (17.9) and (17.11) are the conventional values of the chemical potential of the electrons and the metal ions *within any pure metal*.

THE STANDARD HYDROGEN ELECTRODE

A piece of platinum in contact with hydrogen gas at unit fugacity and an acid solution in which the hydrogen ion has unit activity is called a standard hydrogen electrode (SHE). The electric potential of the SHE is assigned the conventional value, zero.

$$\phi_{H^+, H_2}^\circ = \phi_{SHE} = 0. \tag{17.12}$$

As we will show later, this choice implies that the standard Gibbs energy of hydrogen ion in aqueous solution is zero.

$$\mu_{H^+}^\circ = 0. \tag{17.13}$$

This gives us a reference value against which we can measure the Gibbs energy of other ions in aqueous solution.

SUMMARY OF CONVENTIONS AND STANDARD STATES ($T = 298.15$ K; $p = 1$ atm.)

Elements in their stable state of aggregation:

Standard state $\qquad\qquad \mu_{\text{elements}}^\circ = 0$

Charged particles:

General form $\qquad\qquad \tilde{\mu}_i = \mu_i + z_i F\phi \qquad\qquad (17.7)$

a) Ions in aqueous solution $\qquad \phi_{\text{aq}} = 0$

$\quad$ Standard state $\qquad\qquad \mu_{H^+}^\circ = 0 \qquad\qquad\qquad (17.13)$

$\quad$ General form $\qquad\qquad \tilde{\mu}_i = \mu_i = \mu_i^\circ + RT \ln a_i$

b) Electrons in any metal

$\quad$ Standard state $\qquad\qquad \tilde{\mu}_{e^- (\text{SHE})} = 0 \quad$ or $\quad \phi_{\text{SHE}} = 0 \qquad (17.12)$

$\quad$ General form $\qquad\qquad \tilde{\mu}_{e^-} = -F\phi \qquad\qquad\qquad (17.9)$

c) Ions in a pure metal

$\quad$ Standard state $\qquad\qquad \mu_{M^{z+}}^\circ = 0 \qquad\qquad\qquad (17.10)$

$\quad$ General form $\qquad\qquad \tilde{\mu}_{M^{z+}} = zF\phi \qquad\qquad\qquad (17.11)$

17.4 CELL DIAGRAMS

The electrochemical cell is represented by a diagram that shows the oxidized and reduced forms of the electroactive substance, as well as any other species that may be involved in the electrode reaction. The metal electrodes (or inert metal collectors) are placed at the ends of the diagram; insoluble substances and/or gases are placed in interior positions adjacent to the metals, and soluble species are placed near the middle of the diagram. In a complete diagram the states of aggregation of all the substances are described and the concentrations or activities of the soluble materials are given. In an abbreviated diagram some or all of this information may be omitted if it is not needed and if no misunderstanding is likely. A phase boundary is indicated by a solid vertical bar; a single dashed vertical bar indicates a junction between two miscible liquid phases; a double dashed vertical bar indicates a junction between two miscible liquid phases at which the junction potential has been eliminated. (A salt bridge, such as an agar jelly saturated with KCl, is often used between the two solutions to eliminate the junction potential.) Commas separate different soluble species in the same phase. The following examples illustrate these conventions.

Complete	$Pt_I(s)\|Zn(s)\|Zn^{2+}(a_{Zn^{2+}} = 0.35)\|\!\|Cu^{2+}(a_{Cu^{2+}} = 0.49)\|Cu(s)\|Pt_{II}(s)$
Abbreviated	$Zn\|Zn^{2+}\|\!\|Cu^{2+}\|Cu$
Complete	$Pt\|H_2(g, p = 0.80)\|H_2SO_4(aq, a = 0.42)\|Hg_2SO_4(s)\|Hg(l)$
Abbreviated	$Pt\|H_2\|H_2SO_4(aq)\|Hg_2SO_4(s)\|Hg$
Complete	$Ag(s)\|AgCl(s)\|FeCl_2(m = 0.540), FeCl_3(m = 0.221)\|Pt$
Abbreviated	$Ag\|AgCl(s)\|FeCl_2(aq), FeCl_3(aq)\|Pt$

17.5 THE DANIELL CELL

Consider the electrochemical cell, the Daniell cell, shown in Fig. 17.1. It consists of two electrode systems—two *half-cells*—separated by a salt bridge, which prevents the two solutions from mixing but allows the current to flow between the two compartments. Each half-cell consists of a metal, zinc or copper, immersed in a solution of a highly soluble salt of the metal such as $ZnSO_4$ or $CuSO_4$. The electrodes are connected to the exterior by

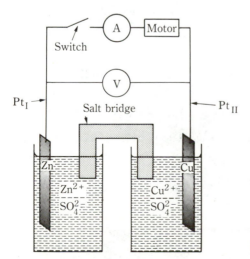

Figure 17.1   The Daniell cell.

two platinum leads. The cell diagram is

$$Pt_I(s)|Zn(s)|Zn^{2+}(aq) \vdots\vdots Cu^{2+}(aq)|Cu(s)|Pt_{II}(s).$$

Assume that the switch in the external circuit is open and that the local electrochemical equilibria are established at the phase boundaries and within the bulk phases. At the $Pt_I|Zn$ and the $Cu|Pt_{II}$ interfaces the equilibrium is established by the free passage of electrons across the interface. The equilibrium conditions at these interfaces are

$$\tilde{\mu}_{e^-}(Pt_I) = \tilde{\mu}_{e^-}(Zn) \quad\text{and}\quad \tilde{\mu}_{e^-}(Cu) = \tilde{\mu}_{e^-}(Pt_{II}). \tag{17.14}$$

Using Eq. (17.9), we obtain

$$\phi_I = \phi_{Zn}, \qquad \phi_{Cu} = \phi_{II}, \tag{17.15}$$

where ϕ_I and ϕ_{II} are the potentials of the two pieces of platinum and ϕ_{Zn} is the potential of the zinc electrode in contact with a solution containing zinc ion. The electric potential difference of any cell (the cell potential) is defined by

$$\mathscr{E} = \phi_{right} - \phi_{left}. \tag{17.16}$$

For this case, the potential of the cell is

$$\mathscr{E} = \phi_{II} - \phi_I = \phi_{Cu} - \phi_{Zn}. \tag{17.17}$$

The difference $\phi_{II} - \phi_I$ is measurable since it is a difference of potential between two phases having the same chemical composition (both are platinum).

Suppose we connect the two platinum wires through an ammeter to a small motor: we observe that (1) some zinc dissolves, (2) some copper is deposited on the copper electrode, (3) electrons flow in the external circuit from the zinc to the copper electrode, and (4) the motor runs. The changes in the cell can be summarized as:

At the left electrode $\qquad\qquad\qquad$ $Zn(s) \longrightarrow Zn^{2+}(aq) + 2e^-(Zn);$

In the external circuit $\qquad\qquad$ $2e^-(Zn) \longrightarrow 2e^-(Cu);$

At the right electrode $\qquad$ $Cu^{2+}(aq) + 2e^-(Cu) \longrightarrow Cu(s).$

The overall transformation is the sum of these changes:

$$Zn(s) + Cu^{2+}(aq) \longrightarrow Zn^{2+}(aq) + Cu(s).$$

This chemical reaction is the *cell reaction*; ΔG for this reaction is

$$\Delta G = \Delta G^\circ + RT \ln \frac{a_{Zn^{2+}}}{a_{Cu^{2+}}}. \tag{17.18}$$

The work expended on the system to move the electrons from the zinc electrode to the copper electrode is $-W_{el}$, where

$$-W_{el} = Q(\phi_{II} - \phi_I) = -2F\mathscr{E},$$

in which Eq. (17.17) has been used for $\phi_{II} - \phi_I$. The work *produced* is

$$W_{el} = 2F\mathscr{E}. \tag{17.19}$$

Using this value in Eq. (17.1) for W_{el}, it becomes

$$2F\mathscr{E} \leq -\Delta G, \tag{17.20}$$

where ΔG is the change in Gibbs energy for the cell reaction.

If the transformation is done reversibly, the work produced is equal to the decrease in the Gibbs energy: $W_{el} = -\Delta G$. We have then,

$$2F\mathscr{E} = -\Delta G, \tag{17.21}$$

which in view of Eq. (17.18) becomes

$$2F\mathscr{E} = -\Delta G^\circ - RT \ln \frac{a_{Zn^{2+}}}{a_{Cu^{2+}}}.$$

If both electrodes are in their standard states, $a_{Zn^{2+}} = 1$ and $a_{Cu^{2+}} = 1$, the cell potential is the standard cell potential, $\mathscr{E}^\circ$. Thus, after we divide by $2F$, the equation becomes

$$\mathscr{E} = \mathscr{E}^\circ - \frac{RT}{2F} \ln \frac{a_{Zn^{2+}}}{a_{Cu^{2+}}}, \tag{17.22}$$

which is the Nernst equation for the cell. This equation relates the cell potential to a standard value and the proper quotient of activities of the substances in the cell reaction.

17.6 GIBBS ENERGY AND THE CELL POTENTIAL

The result obtained for the Daniell cell in Eq. (17.20) is quite general. If the cell reaction as written involves n electrons rather than two electrons, the relation is

$$nF\mathscr{E} \leq -\Delta G. \tag{17.23}$$

Equation (17.23) is the fundamental relation between the cell potential and the Gibbs energy change accompanying the cell reaction.

Observation shows that the value of $\mathscr{E}$ depends on the current drawn in the external circuit. The limiting value of $\mathscr{E}$ measured as the current goes to zero is called the *electromotive force* of the cell (the cell emf) or the reversible cell potential, $\mathscr{E}_{rev}$:

$$\lim_{I \to 0} \mathscr{E} = \mathscr{E}_{rev}.$$

Then Eq. (17.23) becomes

$$nF\mathscr{E}_{rev} = -\Delta G. \tag{17.24}$$

We see that the cell emf is proportional to $(-\Delta G/n)$, the decrease in Gibbs energy of the cell reaction *per electron transferred*. The cell emf is therefore an *intensive* property of the system; it does not depend on the size of the cell or on the coefficients chosen to balance the chemical equation for the cell reaction.

To avoid a cumbersome notation we will suppress the subscript, rev, on the cell potential; we do so with the understanding that the thermodynamic equality (as distinct from the inequality) holds only for the reversible cell potentials (cell emfs).

The spontaneity of a reaction can be judged by the corresponding cell potential. Through Eq. (17.24) it follows that if ΔG is negative, $\mathscr{E}$ is positive. Thus we have the criteria:

ΔG	$\mathscr{E}$	Cell reaction is
−	+	Spontaneous
+	−	Nonspontaneous
0	0	At equilibrium

17.7 THE NERNST EQUATION

For any chemical reaction the reaction Gibbs energy is written

$$\Delta G = \Delta G^\circ + RT \ln Q, \tag{17.25}$$

where Q is the proper quotient of activities. Combining this with Eq. (17.24), we obtain

$$-nF\mathscr{E} = \Delta G^\circ + RT \ln Q.$$

The standard potential of the cell is defined by

$$-nF\mathscr{E}^\circ = \Delta G^\circ. \tag{17.26}$$

Introducing this value of ΔG° and dividing by $-nF$, we obtain

$$\mathscr{E} = \mathscr{E}^\circ - \frac{RT}{nF} \ln Q; \tag{17.27a}$$

$$\mathscr{E} = \mathscr{E}^\circ - \frac{2.303\,RT}{nF} \log_{10} Q; \tag{17.27b}$$

$$\mathscr{E} = \mathscr{E}^\circ - \frac{0.05916\text{ V}}{n} \log_{10} Q \qquad \text{(at 25 °C)}. \tag{17.27c}$$

Equations (17.27) are various forms of the Nernst equation for the cell. The Nernst equation relates the cell potential to a standard value, $\mathscr{E}^\circ$, and the activities of the species taking part in the cell reaction. Knowing the values of $\mathscr{E}^\circ$ and the activities, we can calculate the cell potential.

17.8 THE HYDROGEN ELECTRODE

The definition of the cell potential requires that we label one electrode as the right-hand and the other as the left-hand electrode. The cell potential is defined as in Eq. (17.16) by

$$\mathscr{E} = \phi_{\text{right}} - \phi_{\text{left}}.$$

It is customary, but not necessary, to place the more positive electrode in the right-hand position. As we pointed out, this cell potential is always measurable as a difference in potential between two wires (for example, Pt) having the same composition. The measurement also establishes which electrode is positive relative to the other; in our example, the copper is positive relative to the zinc. It does not, however, yield any hint as to the absolute value of the potential of either electrode. It is useful to establish an arbitrary zero of potential; we do this by assigning the value zero to the potential of the hydrogen electrode in its standard state.

The hydrogen electrode is illustrated in Fig. 17.2. Purified hydrogen gas is passed over a platinum electrode which is in contact with an acid solution. At the electrode surface the equilibrium

$$\text{H}^+(\text{aq}) + \text{e}^-(\text{Pt}) \;\Longleftrightarrow\; \tfrac{1}{2}\text{H}_2(\text{g})$$

is established. The equilibrium condition is the usual one:

$$\mu_{\text{H}^+(\text{aq})} + \tilde{\mu}_{\text{e}^-(\text{Pt})} = \tfrac{1}{2}\mu_{\text{H}_2(\text{g})}.$$

Using Eq. (17.9) for $\tilde{\mu}_{\text{e}^-(\text{Pt})}$ and the usual forms of $\mu_{\text{H}^+(\text{aq})}$ and $\mu_{\text{H}_2(\text{g})}$ we obtain

$$\mu_{\text{H}^+}^\circ + RT \ln a_{\text{H}^+} - F\phi_{\text{H}^+/\text{H}_2} = \tfrac{1}{2}\mu_{\text{H}_2}^\circ + \tfrac{1}{2}RT \ln f,$$

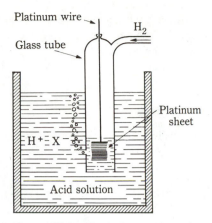

Figure 17.2   The hydrogen electrode.

where f is the fugacity of H_2 and a_{H^+} is the activity of the hydrogen ion in the aqueous solution. Then

$$\phi_{H^+/H_2} = \frac{\mu^\circ_{H^+} - \frac{1}{2}\mu^\circ_{H_2}}{F} - \frac{RT}{F}\ln\frac{f^{1/2}}{a_{H^+}}. \tag{17.28}$$

If the fugacity of the gas is unity, and the activity of H^+ in solution is unity, the electrode is in its standard state, and the potential is the standard potential, $\phi^\circ_{H^+/H_2}$. Letting $f = 1$ and $a_{H^+} = 1$ in Eq. (17.28), we obtain

$$\phi^\circ_{H^+/H_2} = \frac{\mu^\circ_{H^+} - \frac{1}{2}\mu^\circ_{H_2}}{F} = \frac{\mu^\circ_{H^+}}{F}, \tag{17.29}$$

since $\mu^\circ_{H_2} = 0$. Subtracting Eq. (17.29) from Eq. (17.28) yields

$$\phi_{H^+/H_2} = \phi^\circ_{H^+/H_2} - \frac{RT}{F}\ln\frac{f^{1/2}}{a_{H^+}}, \tag{17.30}$$

which is the Nernst equation for the hydrogen electrode; it relates the electrode potential to a_{H^+} and f. Now the electrons in platinum of the standard hydrogen electrode are in a definite standard state. We choose the standard state of zero Gibbs energy for electrons as this state in the SHE. Since, by Eq. (17.9), $\tilde{\mu}_{e^-} = -F\phi$, we have

$$\tilde{\mu}_{e^-\text{(SHE)}} = 0 \qquad \text{and} \qquad \phi^\circ_{H^+/H_2} = 0. \tag{17.31}$$

The Gibbs energy of the electrons in any metal is measured relative to that in the standard hydrogen electrode. The assignment in Eq. (17.31) yields the conventional zero of Gibbs energy for ions in aqueous solution. Using Eq. (17.31) in Eq. (17.29), we obtain

$$\mu^\circ_{H^+} = 0. \tag{17.32}$$

Standard Gibbs energies of other ions in aqueous solution are measured relative to that of the H^+ ion, which has a standard Gibbs energy equal to zero.

The Nernst equation, Eq. (17.30), for the hydrogen electrode becomes

$$\phi_{H^+/H_2} = -\frac{RT}{F}\ln\frac{f^{1/2}}{a_{H^+}}. \tag{17.33}$$

Note that the argument of the logarithm is a proper quotient of fugacity and activity for the electrode reaction if the presence of the electrons is ignored in constructing the quotient. From Eq. (17.33) we can calculate the potential, relative to SHE, of a hydrogen electrode at which f_{H_2} and a_{H^+} have any values.

17.9 ELECTRODE POTENTIALS

Having assigned the hydrogen electrode a zero potential, we next compare the potentials of all other electrode systems to that of the standard hydrogen electrode. For example, the potential of the cell

$$Pt_I | H_2(g, f = 1) | H^+(a_{H^+} = 1) \vdots\vdots Cu^{2+}(a_{Cu^{2+}}) | Cu | Pt_{II}$$

is designated by $\mathscr{E}_{Cu^{2+}/Cu}$:

$$\mathscr{E}_{Cu^{2+}/Cu} = \phi_{II} - \phi_I = \phi_{Cu} - \phi_{SHE} = \phi_{Cu}. \tag{17.34}$$

Note that $\mathscr{E}_{Cu^{2+}/Cu}$ is equal to the conventional potential of the copper electrode, ϕ_{Cu}. The cell reaction is

$$H_2(f = 1) + Cu^{2+}(a_{Cu^{2+}}) \rightleftharpoons 2H^+(a_{H^+} = 1) + Cu. \tag{17.35}$$

The equilibrium at the SHE is:

$$H_2(f = 1) \rightleftharpoons 2H^+(a_{H^+} = 1) + 2e^-_{SHE} \tag{17.36}$$

All of the species in this reaction have zero Gibbs energy by our conventional assignments. If we subtract the equilibrium in Eq. (17.36) from that in Eq. (17.35), we obtain

$$Cu^{2+}(a_{Cu^{2+}}) + 2e^-_{SHE} \rightleftharpoons Cu, \tag{17.37}$$

which is simply a shorthand way of writing Eq. (17.35). Equation (17.37) is called a half-cell reaction. Since the potential of this cell depends only on the conventional Gibbs energies of the copper and the copper ion, it is called the *half-cell potential*, or the *electrode potential* of the $Cu^{2+} | Cu$ electrode.

This half-cell potential is related to the Gibbs energy change in reaction (17.37) by

$$2F\mathscr{E} = -\Delta G;$$

keep in mind that for electrons in SHE the Gibbs energy is zero. Using Eq. (17.37), the Nernst equation for the electrode becomes

$$\mathscr{E}_{Cu^{2+}/Cu} = \mathscr{E}^{\circ}_{Cu^{2+}/Cu} - \frac{RT}{2F} \ln \frac{1}{a_{Cu^{2+}}}. \tag{17.38}$$

By measuring the cell potential at various concentrations of Cu^{2+}, we can determine $\mathscr{E}^{\circ}_{Cu^{2+}/Cu} = \phi^{\circ}_{Cu^{2+}/Cu}$. This standard potential is tabulated along with the standard potentials of other half-cells in Table 17.1. Such a table of half-cell potentials, or electrode potentials, is equivalent to a table of standard Gibbs energies from which we can calculate values of equilibrium constants for chemical reactions in solution. Note that the standard potential is the potential of the electrode when all of the reactive species are present with unit activity, $a = 1$.

The situation may be summarized as follows: if the half-cell reaction is written with the electrons in SHE *on the reactant side*, any electrode system can be represented by

$$\text{oxidized species} + ne^-_{SHE} \rightleftharpoons \text{reduced species}.$$

Table 17.1
Standard electrode potentials at 25 °C

Electrode reaction	$\phi°/V$
$K^+ + e^- = K$	-2.925
$Na^+ + e^- = Na$	-2.714
$H_2 + 2e^- = 2H^-$	-2.25
$Al^{3+} + 3e^- = Al$	-1.66
$Zn(CN)_4^{2-} + 2e^- = Zn + 4CN^-$	-1.26
$ZnO_2^{2-} + 2H_2O + 2e^- = Zn + 4OH^-$	-1.216
$Zn(NH_3)_4^{2+} + 2e^- = Zn + 4NH_3$	-1.03
$Sn(OH)_6^{2-} + 2e^- = HSnO_2^- + H_2O + 3OH^-$	-0.90
$Fe(OH)_2 + 2e^- = Fe + 2OH^-$	-0.877
$2H_2O + 2e^- = H_2 + 2OH^-$	-0.828
$Fe(OH)_3 + 3e^- = Fe + 3OH^-$	-0.77
$Zn^{2+} + 2e^- = Zn$	-0.763
$Ag_2S + 2e^- = 2Ag + S^{2-}$	-0.69
$Fe^{2+} + 2e^- = Fe$	-0.440
$Bi_2O_3 + 3H_2O + 6e^- = 2Bi + 6OH^-$	-0.44
$PbSO_4 + 2e^- = Pb + SO_4^{2-}$	-0.356
$Ag(CN)_2^- + e^- = Ag + 2CN^-$	-0.31
$Ni^{2+} + 2e^- = Ni$	-0.250
$AgI + e^- = Ag + I^-$	-0.151
$Sn^{2+} + 2e^- = Sn$	-0.136
$Pb^{2+} + 2e^- = Pb$	-0.126
$Cu(NH_3)_4^{2+} + 2e^- = Cu + 4NH_3$	-0.12
$Fe^{3+} + 3e^- = Fe$	-0.036
$2H^+ + 2e^- = H_2$	0.000
$AgBr + e^- = Ag + Br^-$	$+0.095$
$HgO(r) + H_2O + 2e^- = Hg + 2OH^-$	$+0.098$
$Sn^{4+} + 2e^- = Sn^{2+}$	$+0.15$
$AgCl + e^- = Ag + Cl^-$	$+0.222$
$Hg_2Cl_2 + 2e^- = 2Hg + 2Cl^-$	$+0.2676$
$Cu^{2+} + 2e^- = Cu$	$+0.337$
$Ag(NH_3)_2^+ + e^- = Ag + 2NH_3$	$+0.373$
$Hg_2SO_4 + 2e^- = 2Hg + SO_4^{2-}$	$+0.6151$
$Fe^{3+} + e^- = Fe^{2+}$	$+0.771$
$Ag^+ + e^- = Ag$	$+0.7991$
$O_2 + 4H^+ + 4e^- = 2H_2O$	$+1.229$
$PbO_2 + SO_4^{2-} + 4H^+ + 2e^- = PbSO_4 + 2H_2O$	$+1.685$
$O_3 + 2H^+ + 2e^- = O_2 + H_2O$	$+2.07$

Values in this table are printed by permission from W. M. Latimer, *The Oxidation States of the Elements and Their Potentials in Aqueous Solution*, 2d ed. Englewood Cliffs, N.J.: Prentice-Hall, 1952.

Then the following relations obtain:

$$\mathscr{E} = \phi; \tag{17.39}$$

$$\Delta G = -nF\mathscr{E}; \tag{17.40}$$

$$\phi = \phi° - \frac{RT}{nF} \ln Q. \tag{17.41}$$

■ **EXAMPLE 17.1**   For the copper ion/copper electrode we have explicitly

$$2F\phi^{\circ}_{Cu^{2+}/Cu} = -\Delta G^{\circ} = -(\mu^{\circ}_{Cu} - \mu^{\circ}_{Cu^{2+}})$$

Since $\mu^{\circ}_{Cu} = 0$, this becomes

$$\mu^{\circ}_{Cu^{2+}} = 2F\phi^{\circ}_{Cu^{2+}/Cu}.$$

Since $\phi^{\circ}_{Cu^{2+}/Cu} = +0.337$ V, we find

$$\mu^{\circ}_{Cu^{2+}} = 2(96\,484 \text{ C/mol})(+0.337 \text{ V}) = 65.0 \times 10^3 \text{ J/mol} = 65.0 \text{ kJ/mol}.$$

■ **EXAMPLE 17.2**   For the Sn^{4+}/Sn^{2+} electrode $\phi^{\circ}_{Sn^{4+}/Sn^{2+}} = 0.15$ V; for the Sn^{2+}/Sn electrode $\phi^{\circ}_{Sn^{2+}/Sn} = -0.136$ V. Calculate $\mu^{\circ}_{Sn^{4+}}$, $\mu^{\circ}_{Sn^{2+}}$, and $\phi^{\circ}_{Sn^{4+}/Sn}$.

The reactions are:

$$Sn^{4+} + 2e^- \rightleftharpoons Sn^{2+} \qquad 2F(0.15 \text{ V}) = -(\mu_{Sn^{2+}} - \mu_{Sn^{4+}})$$

$$Sn^{2+} + 2e^- \rightleftharpoons Sn \qquad 2F(-0.136 \text{ V}) = -(\mu_{Sn} - \mu_{Sn^{2+}})$$

The second equation yields:

$$\mu_{Sn^{2+}} = 2(96\,484 \text{ J/mol})(-0.136 \text{ V})(10^{-3} \text{ kJ/J}) = -26.2 \text{ kJ/mol}.$$

The first equation yields:

$$\mu_{Sn^{4+}} - \mu_{Sn^{2+}} = 2(96\,484 \text{ C/mol})(0.15 \text{ V})(10^{-3} \text{ kJ/J}) = 29 \text{ kJ/mol}.$$

Then

$$\mu_{Sn^{4+}} = 29 \text{ kJ/mol} + \mu_{Sn^{2+}} = 29 - 26.2 = 3 \text{ kJ/mol}.$$

To find $\phi^{\circ}_{Sn^{4+}/Sn}$, write the half-cell reaction:

$$Sn^{4+} + 4e^- \rightleftharpoons Sn.$$

Then

$$4F\phi^{\circ}_{Sn^{4+}/Sn} = -(\mu_{Sn} - \mu_{Sn^{4+}}) = \mu_{Sn^{4+}},$$

and

$$\phi^{\circ}_{Sn^{4+}/Sn} = \frac{3000 \text{ J/mol}}{4(96\,484 \text{ C/mol})} = 0.008 \text{ V}.$$

17.10 TEMPERATURE DEPENDENCE OF THE CELL POTENTIAL

By differentiating the equation, $nF\mathscr{E} = -\Delta G$ with respect to temperature, we obtain

$$nF\left(\frac{\partial \mathscr{E}}{\partial T}\right)_p = -\left(\frac{\partial \Delta G}{\partial T}\right)_p = \Delta S,$$

$$\left(\frac{\partial \mathscr{E}}{\partial T}\right)_p = \frac{\Delta S}{nF} \qquad (17.42)$$

If the cell does not contain a gas electrode, then since the entropy changes of reactions in solution are frequently rather small, less than 50 J/K, the temperature coefficient of the cell potential is usually of the order of 10^{-4} or 10^{-5} V/K. As a consequence, if only routine equipment is being used to measure the cell potential and the temperature coefficient is sought, the measurements should cover as wide a range of temperature as is feasible.

The value of ΔS is independent of temperature to a good approximation; by integrating Eq. (17.42) between a reference temperature T_0 and any temperature T, we obtain

$$\mathscr{E} = \mathscr{E}_{T_0} + \frac{\Delta S}{nF}(T - T_0) \quad \text{or} \quad \mathscr{E} = \mathscr{E}_{25\,°C} + \frac{\Delta S}{nF}(t - 25) \tag{17.43}$$

where t is in °C. The cell potential is a linear function of temperature.

Through Eq. (17.42), the temperature coefficient of the cell potential yields the value of ΔS. From this and the value of $\mathscr{E}$ at any temperature we can calculate ΔH for the cell reaction. Since $\Delta H = \Delta G + T\,\Delta S$, then

$$\Delta H = -nF\left[\mathscr{E} - T\left(\frac{\partial \mathscr{E}}{\partial T}\right)_p\right]. \tag{17.44}$$

Thus, by measuring $\mathscr{E}$ and $(\partial \mathscr{E}/\partial T)_p$ we can obtain the thermodynamic properties of the cell reaction, ΔG, ΔH, ΔS.

■ **EXAMPLE 17.3**   For the cell reaction

$$Hg_2Cl_2(s) + H_2(1\text{ atm}) \longrightarrow 2\,Hg(l) + 2\,H^+(a = 1) + 2\,Cl^-(a = 1),$$
$$\mathscr{E}^{\circ}_{298} = +0.2676\text{ V} \quad \text{and} \quad (\partial \mathscr{E}^{\circ}/\partial T)_p = -3.19 \times 10^{-4}\text{ V/K}.$$

Since $n = 2$,

$\Delta G^{\circ} = -2(96\,484\text{ C/mol})(0.2676\text{ V})(10^{-3}\text{ kJ/J}) = -51.64\text{ kJ/mol};$

$\Delta H^{\circ} = -2(96\,484\text{ C/mol})[0.2676\text{ V} - 298.15\text{ K}(-3.19 \times 10^{-4}\text{ V/K})](10^{-3}\text{ kJ/J})$
$\quad\quad = -69.99\text{ kJ/mol};$

$\Delta S^{\circ} = 2(96\,484\text{ C/mol})(-3.19 \times 10^{-4}\text{ V/K}) = -61.6\text{ J/K mol}.$

★ 17.10.1   Heat Effects in the Operation of a Reversible Cell

In Example 17.3, we computed the ΔH° for the cell reaction from the cell potential and its temperature coefficient. If the reaction were carried out irreversibly by simply mixing the reactants together, ΔH° is the heat that flows into the system in the transformation by the usual relation, $\Delta H = Q_p$. However, if the reaction is brought about reversibly in the cell, electrical work in the amount $W_{el,\,rev}$ is produced. Then, by Eq. (9.4), the definition of ΔS,

$$Q_{p(rev)} = T\,\Delta S. \tag{17.45}$$

Using Example 17.3, we have $Q_{p(rev)} = 298.15\text{ K}(-61.6\text{ J/K mol}) = -18\,350\text{ J/mol}$. Consequently, in the operation of the cell only 18.35 kJ/mol of heat flow to the surroundings, while if the reactants are mixed directly, 69.99 kJ/mol of heat pass to the surroundings. The ΔH° for the transformation is -69.99 kJ/mol and is independent of the way the reaction is carried out.

17.11   KINDS OF ELECTRODES

At this point we describe briefly some important kinds of electrodes, and present the half-cell reactions and the Nernst equation for each.

17.11.1   Gas–Ion Electrodes

The gas–ion electrode consists of an inert collector of electrons, such as platinum or graphite, in contact with a gas and a soluble ion. The $H_2\,|\,H^+$ electrode, discussed in detail in Section

17.8, is one example. Another example is the chlorine electrode, $Cl_2 | Cl^- |$ graphite:

$$Cl_2(g) + 2e^- \ \rightleftharpoons \ 2Cl^-(aq) \qquad \phi = \phi^\circ - \frac{RT}{2F} \ln \frac{a_{Cl^-}^2}{p_{Cl_2}} \qquad (17.46)$$

17.11.2 Metal Ion–Metal Electrodes

The electrode consists of a piece of the metal immersed in a solution containing the metal ion. The $Zn^{2+} | Zn$ and $Cu^{2+} | Cu$ electrodes described earlier are examples.

$$M^{n+} + ne^- \ \rightleftharpoons \ M \qquad \phi = \phi^\circ - \frac{RT}{nF} \ln \frac{1}{a_{M^{n+}}} \qquad (17.47)$$

17.11.3 Metal–Insoluble Salt-Anion Electrodes

This electrode is sometimes called an "electrode of the second kind." It consists of a bar of metal immersed in a solution containing a solid insoluble salt of the metal and anions of the salt. There are a dozen common electrodes of this kind; we cite only a few examples.

The Silver–Silver Chloride Electrode. $Cl^- | AgCl(s) | Ag(s)$: (Fig. 17.3).

$$AgCl(s) + e^- \ \rightleftharpoons \ Ag(s) + Cl^-(aq) \qquad \phi = \phi^\circ - \frac{RT}{F} \ln a_{Cl^-} \qquad (17.48)$$

The activity of AgCl does not appear in the quotient, since AgCl is a pure solid. Since the potential is sensitive to the concentration of chloride ion, it can be used to measure that concentration. The silver–silver chloride electrode is a very commonly used reference electrode.

A number of commonly used reference electrodes based on mercury belong to this class of electrodes.

The Calomel Electrode. A pool of mercury covered with a paste of calomel (mercurous chloride) and a solution of KCl.

$$Hg_2Cl_2(s) + 2e^- \ \rightleftharpoons \ 2Hg(l) + 2Cl^-(aq) \qquad \phi = \phi^\circ - \frac{RT}{2F} \ln a_{Cl^-}^2$$

The Mercury–Mercuric Oxide Electrode. A pool of mercury covered with a paste of mercuric oxide and a solution of a base.

$$HgO(s) + H_2O(l) + 2e^- \ \rightleftharpoons \ Hg(l) + 2OH^-(aq) \qquad \phi = \phi^\circ - \frac{RT}{2F} \ln a_{OH^-}^2$$

The Mercury–Mercurous Sulfate Electrode. A pool of mercury covered with a paste of mercurous sulfate and a solution containing sulfate.

$$Hg_2SO_4(s) + 2e^- \ \rightleftharpoons \ 2Hg(l) + SO_4^{2-}(aq) \qquad \phi = \phi^\circ - \frac{RT}{2F} \ln a_{SO_4^{2-}}$$

17.11.4 "Oxidation–Reduction" Electrodes

Any electrode involves oxidation and reduction in its operation, but these electrodes have had that superfluous phrase attached to them. An oxidation–reduction electrode has an inert metal collector, usually platinum, immersed in a solution that contains two soluble

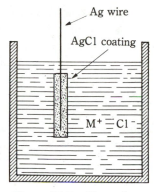

Figure 17.3   Silver–silver chloride electrode.

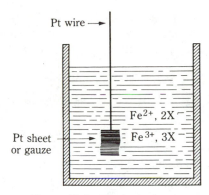

Figure 17.4   The ferric–ferrous electrode.

species in different states of oxidation. An example is the ferric–ferrous ion electrode (Fig. 17.4):

$$Fe^{3+} + e^- \rightleftharpoons Fe^{2+} \qquad \phi = \phi° - \frac{RT}{F} \ln \frac{a_{Fe^{2+}}}{a_{Fe^{3+}}} \qquad (17.49)$$

17.12 EQUILIBRIUM CONSTANTS FROM STANDARD HALF–CELL POTENTIALS

Any chemical reaction can be written as a combination of two half-cell reactions so that a cell potential can be associated with it. The value of $\mathscr{E}$ is determined by the relation, $nF\mathscr{E} = -\Delta G$. The equilibrium condition for any chemical reaction is $\Delta G° = -RT \ln K$. Since $\Delta G° = -nF\mathscr{E}°$, we can write

$$RT \ln K = nF\mathscr{E}°, \qquad \text{or at 25 °C} \qquad \log_{10} K = \frac{n\mathscr{E}°}{0.05916 \text{ V}} \qquad (17.50)$$

Using Eq. (17.50), we can calculate the equilibrium constant for any reaction from the standard cell potential which, in turn, can be obtained from the tabulated values of the standard half-cell potentials. The following method and examples illustrate a procedure that will ensure obtaining the $\mathscr{E}°$ with both a correct sign and magnitude.

Step 1. Break the cell reaction into two half-cell reactions.

a. For the first half-cell reaction (the right-hand electrode) choose the oxidized species that appears on the reactant side of the cell reaction and write the equilibrium with the appropriate reduced species.

b. For the second half-cell reaction (the left-hand electrode) choose the oxidized species that appears on the product side of the cell reaction and write the equilibrium with the appropriate reduced species.

Write *both* half-cell reactions *with the electrons on the reactant side.*

Step 2. Balance the half-cell reactions with the same number of electrons, *n*, in each.

Step 3. If the second half-cell reaction is subtracted from the first one, the overall cell reaction is regenerated; check to be sure that it is. Subtract the electrode potentials in the same sense (first minus second) to obtain the standard potential of the cell, $\mathscr{E}°$.

Step 4. Use Eq. (17.50) to calculate K.

■ **EXAMPLE 17.4** $\quad 2\,Fe^{3+} + Sn^{2+} \rightleftharpoons 2\,Fe^{2+} + Sn^{4+}$

Step 1. Choose the oxidized species, Fe^{3+}, on the reactant side for the first half-cell reaction; choose the oxidized species, Sn^{4+}, on the product side for the second half-reaction. The half-cell reactions are:

$$Fe^{3+} + e^- \rightleftharpoons Fe^{2+} \qquad \phi° = 0.771 \text{ V}$$

$$Sn^{4+} + 2e^- \rightleftharpoons Sn^{2+} \qquad \phi° = 0.15 \text{ V}$$

Step 2. Multiply the first half-cell reaction by 2 so that each will involve the same number of electrons.

Step 3. Subtract the second reaction from the first; this regenerates the original reaction. Subtracting the second potential from the first yields $\mathscr{E}°$. $\mathscr{E}° = 0.771 - 0.15 = 0.62$ V.

Step 4. Since $n = 2$, we find

$$\log_{10} K = \frac{n\mathscr{E}°}{0.05916 \text{ V}} = \frac{2(0.62 \text{ V})}{0.05916 \text{ V}} = 21 \qquad \text{so that} \qquad K = 10^{21}.$$

■ **EXAMPLE 17.5** $\quad 2\,MnO_4^- + 6\,H^+ + 5\,H_2C_2O_4 \rightleftharpoons 2\,Mn^{2+} + 8\,H_2O + 10\,CO_2$.

The half-reactions are (choose the oxidized species, MnO_4^-, on reactant side for the first half-reaction)

$$MnO_4^- + 8\,H^+ + 5e^- \rightleftharpoons Mn^{2+} + 4\,H_2O, \qquad \mathscr{E}° = 1.51 \text{ V};$$

$$2\,CO_2 + 2\,H^+ + 2e^- \rightleftharpoons H_2C_2O_4, \qquad \mathscr{E}° = -0.49 \text{ V}.$$

Multiplying the coefficients of the first reaction by 2 and those in the second reaction by 5, we obtain

$$2\,MnO_4^- + 16\,H^+ + 10\,e^- \rightleftharpoons 2\,Mn^{2+} + 8\,H_2O, \qquad \mathscr{E}° = 1.51 \text{ V};$$

$$10\,CO_2 + 10\,H^+ + 10\,e^- \rightleftharpoons 5\,H_2C_2O_4, \qquad \mathscr{E}° = -0.49 \text{ V}.$$

Subtracting, we have

$$2\,MnO_4^- + 6\,H^+ + 5\,H_2C_2O_4 \rightleftharpoons 2\,Mn^{2+} + 8\,H_2O + 10\,CO_2,$$

$$\mathscr{E}° = 1.51 \text{ V} - (-0.49 \text{ V}) = 1.51 \text{ V} + 0.49 \text{ V} = 2.00 \text{ V}.$$

Since $n = 10$,

$$\log_{10} K = \frac{10(2.00 \text{ V})}{0.05916 \text{ V}} = 338 \qquad \text{or} \qquad K = 10^{338}.$$

■ **EXAMPLE 17.6** $\quad Cd^{2+} + 4\,NH_3 \rightleftharpoons Cd(NH_3)_4^{2+}$.

This reaction is not an oxidation–reduction reaction; nonetheless, it may be decomposed into two half-cell reactions. Choosing Cd^{2+} as the oxidized species for the first half-cell reaction, we suddenly realize that there is no corresponding reduced species. The same situation prevails when we select $Cd(NH_3)_4^{2+}$ as the oxidized species for the second half-cell reaction. Arbitrarily, we introduce the same reduced species for both reactions; cadmium metal seems a logical choice. Then the half-cell reactions are:

$$Cd^{2+} + 2e^- \rightleftharpoons Cd, \qquad \mathscr{E}° = -0.40 \text{ V};$$

$$Cd(NH_3)_4^{2+} + 2e^- \rightleftharpoons Cd + 4NH_3, \qquad \mathscr{E}° = -0.61 \text{ V}.$$

Subtracting, we obtain

$$Cd^{2+} + 4NH_3 \rightleftharpoons Cd(NH_3)_4^{2+}, \qquad \mathscr{E}° \doteq -0.40 \text{ V} - (-0.61 \text{ V}) = +0.21 \text{ V},$$

$$\log_{10} K = \frac{2(0.21 \text{ V})}{0.05916 \text{ V}} = 7.1, \qquad \text{or} \qquad K = 1.3 \times 10^7.$$

This is the *stability* constant of the complex ion.

■ **EXAMPLE 17.7** $Cu(OH)_2 \rightleftharpoons Cu^{2+} + 2OH^-,$

$$Cu(OH)_2 + 2e^- \rightleftharpoons Cu + 2OH^-, \qquad \mathscr{E}° = -0.224 \text{ V};$$

$$Cu^{2+} + 2e^- \rightleftharpoons Cu, \qquad \mathscr{E}° = +0.337 \text{ V}.$$

Subtracting, we obtain

$$Cu(OH)_2 \rightleftharpoons Cu^{2+} + 2OH^-, \qquad \mathscr{E}° = -0.224 \text{ V} - (+0.337 \text{ V}) = -0.561 \text{ V},$$

$$\log_{10} K = \frac{2(-0.561 \text{ V})}{0.05916 \text{ V}} = -18.97, \qquad \text{or} \qquad K = 1.1 \times 10^{-19}.$$

This is the solubility product constant of copper hydroxide.

17.13 SIGNIFICANCE OF THE HALF-CELL POTENTIAL

In the case of the metal ion/metal electrode, the half-cell potential is a measure of the tendency of the reaction $M^{n+} + ne^- \rightleftharpoons M$ to occur. It is thus a measure of the tendency of M^{n+} to be reduced by H_2 at unit fugacity to form the metal and H^+ ion at unit activity. In Example 17.1 we showed that for the $M^{n+}|M$ electrode

$$\mu°_{M^{n+}} = nF\phi°_{M^{n+}/M}. \tag{17.51}$$

Thus the standard electrode potential is a measure of the standard molar Gibbs energy of the metal ion relative to the hydrogen ion.

Active metals such as Zn, Na, or Mg have highly negative standard potentials. Their compounds are not reduced by hydrogen, but rather the metal itself can be oxidized by H^+ to yield H_2. Noble metals such as Cu and Ag have positive $\phi°$'s. Compounds of these metals are readily reduced by H_2; the metals themselves are not oxidized by hydrogen ion.

Since the potential of a metal depends on the activity of the metal ion in solution, factors that influence the activity of the ion will *ipso facto* influence the electrode potential. For silver, the Nernst equation is

$$\phi_{Ag^+/Ag} = 0.7991 \text{ V} - (0.05916 \text{ V})\log_{10} \frac{1}{a_{Ag^+}}. \tag{17.52}$$

As the value of a_{Ag^+} decreases, the value of $\phi_{Ag^+/Ag}$ also decreases. Using different values of a_{Ag^+} in Eq. (17.52) we obtain:

a_{Ag^+}	1.0	10^{-2}	10^{-4}	10^{-6}	10^{-8}	10^{-10}
$\phi°_{Ag^+/Ag}/V$	0.7991	0.6808	0.5625	0.4441	0.3258	0.2075

For each power of ten by which the activity of the silver ion decreases, the potential drops by 59.16 mV.

Rather than simply diluting the solution to reduce the activity of silver ion, if we add a precipitating agent—or a complexing agent that combines strongly with silver ion—then both the activity of the silver ion and the electrode potential will be drastically reduced.

For example, if we add sufficient HCl to the $AgNO_3$ solution in the $Ag^+|Ag$ electrode, not only to completely precipitate the silver ion as AgCl but also to bring the activity of the chloride ion to unity, the electrode will be converted to the standard $Ag|AgCl|Cl^-$ electrode. For this electrode the equilibrium is

$$AgCl(s) + e^- \rightleftharpoons Ag(s) + Cl^-; \qquad \phi° = 0.222 \text{ V}.$$

This potential, if we use the Nernst equation for the $Ag^+|Ag$ electrode, corresponds to a silver ion activity given by

$$0.222 \text{ V} = 0.799 \text{ V} - (0.05916 \text{ V})\log_{10}\frac{1}{a_{Ag^+}} \qquad \text{or} \qquad a_{Ag^+} = 1.8 \times 10^{-10}.$$

At the same time, the solubility equilibrium must be satisfied. Thus

$$AgCl(s) \rightleftharpoons Ag^+ + Cl^-; \qquad K_{SP} = a_{Ag^+}a_{Cl^-}.$$

Since $a_{Ag^+} = 1.8 \times 10^{-10}$ and $a_{Cl^-} = 1$, we conclude that

$$K_{SP} = a_{Ag^+}a_{Cl^-} = 1.8(10^{-10})(1) = 1.8 \times 10^{-10}.$$

It follows that we can determine the solubility product constants for slightly soluble materials by measuring the standard potential of the appropriate electrochemical cell. (Compare to Examples 17.6 and 17.7, Section 17.12.)

From the argument above it can be seen that the more stable the species in which the silver ion is bound, the lower will be the electrode potential of the silver. A group of $\phi°$'s for various silver couples is given in Table 17.2. From the values in Table 17.2, it is clear that iodide ion ties up Ag^+ more effectively than bromide or chloride; AgI is less soluble than AgCl or AgBr. The fact that the silver iodide–silver couple has a negative potential means that silver should dissolve in HI with the liberation of hydrogen. This occurs in fact, but the action ceases promptly due to the layer of insoluble AgI that forms and protects the Ag surface from further attack.

Table 17.2

Couples	$\phi°/V$
$Ag^+ + e^- \rightleftharpoons Ag$	0.7991
$AgCl(s) + e^- \rightleftharpoons Ag + Cl^-$	0.2222
$AgBr(s) + e^- \rightleftharpoons Ag + Br^-$	0.03
$AgI(s) + e^- \rightleftharpoons Ag + I^-$	−0.151
$Ag_2S(s) + 2e^- \rightleftharpoons 2Ag + S^=$	−0.69

Substances that form soluble complexes with the metal ion also lower the electrode potential. Two examples are

$$Ag(NH_3)_2^+ + e^- \rightleftharpoons Ag + 2NH_3, \qquad \phi^\circ = +0.373 \text{ V};$$

$$Ag(CN)_2^- + e^- \rightleftharpoons Ag + 2CN^-, \qquad \phi^\circ = -0.31 \text{ V}.$$

Whether a metal is a noble metal or an active metal depends on its environment. Ordinarily silver is a noble metal; in the presence of iodide, sulfide, or cyanide ion, it is an active metal (if we consider the zero of potential as the dividing line between active and noble metals).

17.14 THE MEASUREMENT OF CELL POTENTIALS

The simplest method of measuring the potential of an electrochemical cell is to balance it against an equal and opposite potential difference in the slidewire of a potentiometer. Figure 17.5 shows a potentiometer circuit with the cell connected in it. The battery B sends a current i through the slidewire R. The contact S is adjusted until no deflection is observed on the galvanometer G. At the null point, the potential of the cell is balanced by the potential difference between the points S and P of the slidewire. The slidewire is calibrated so that the potential drop ir between the points S and P can be read directly. If the resistance of the cell is very large, the potentiometer setting may be moved over a wide range without producing a sensible deflection on the galvanometer. In this case a high impedance electronic voltmeter must be used.

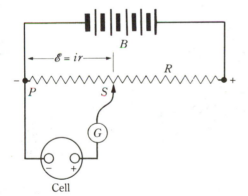

Figure 17.5 Potentiometer circuit.

17.15 REVERSIBILITY

In the foregoing treatment of electrodes and cells we assumed implicitly that the electrode or cell was in equilibrium with respect to certain chemical and electrical transformations. By definition such an electrode or cell is *reversible*. To correlate measured values of cell potentials with the ones calculated by the Nernst equation, the measured values must be equilibrium or reversible values; the potentiometric measurement in which no current is drawn from the cell is ideally suited for the measurement of reversible potentials.

Consider the cell, $Pt|H_2|H^+ \vdots Cu^{2+}|Cu$, which we discussed in Sect. 17.9. The cell reaction is

$$Cu^{2+} + H_2 \rightleftharpoons 2H^+ + Cu.$$

The copper is the positive electrode and the platinum is negative. Suppose that the cell is in balance with a potentiometer, as shown in Fig. 17.5. Now if we move the sliding contact, S, to the right of the balance point, that will make the copper more positive; Cu will then leave the electrode as Cu^{2+} and the electrons will move from right to left in the external circuit. On the platinum the electrons will combine with H^+ to form H_2. The entire reaction goes in the reverse direction from right to left. Conversely, if the slider is moved to the left. the electrons will move from left to right in the external circuit; H_2 will ionize to H^+ and Cu^{2+} will be reduced to copper. In this situation the cell produces work, while in the earlier circumstances work was destroyed.

The cell behaves reversibly if moving the potentiometer contact slightly to one side of the balance point and then to the other reverses the current and the direction of the chemical reaction. In practice it is not necessary to analyze for the amounts of the reactants and products after each of the adjustments to decide whether the reaction is behaving in the required way. If the cell is irreversible, throwing the potentiometer slightly out of balance ordinarily results in the flow of a comparatively large current, while reversibility demands that only a small current flow when the imbalance between the potentials is slight. Furthermore, in the irreversible cell, after disturbing the balance in the circuit slightly, the new balance point is usually significantly different from the original one. For these reasons, the irreversible cell exhibits what is apparently an erratic behavior and often it is impossible to bring a potentiometer into balance with such a cell.

17.16   THE DETERMINATION OF THE $\mathscr{E}°$ OF A HALF-CELL

Since the values of equilibrium constants are obtained from the standard half-cell potentials, the method of obtaining the $\mathscr{E}°$ of a half-cell has great importance. Suppose we wish to determine the $\mathscr{E}°$ of the silver–silver ion electrode. Then we set up a cell that includes this electrode and another electrode the potential of which is known; for simplicity we choose the SHE as the other electrode. Then the cell is

$$SHE \vdots\vdots Ag^+ | Ag.$$

The cell reaction is $Ag^+ + e^-_{SHE} \rightleftharpoons Ag$, and the cell potential is

$$\mathscr{E} = \mathscr{E}_{Ag^+/Ag} = \mathscr{E}°_{Ag^+/Ag} - \frac{RT}{F} \ln \frac{1}{a_{Ag^+}}.$$

At 25 °C

$$\mathscr{E} = \mathscr{E}°_{Ag^+/Ag} + (0.05916 \text{ V}) \log_{10} a_{Ag^+}. \tag{17.53}$$

If the solution were an ideal dilute solution, we could replace a_{Ag^+} by $m_+ = m$, the molality of the silver salt. Equation (17.53) would become

$$\mathscr{E} = \mathscr{E}°_{Ag^+/Ag} + (0.05916 \text{ V}) \log_{10} m.$$

By measuring $\mathscr{E}$ at several values of m and plotting $\mathscr{E}$ versus $\log_{10} m$, a straight line of slope 0.05916 V would be obtained, as in Fig. 17.6(a). The intercept on the vertical axis, $m = 1$, would be the value of $\mathscr{E}°$. However, life is not so simple. We cannot replace a_{Ag^+} by m and hope for any real accuracy in our equation. In an ionic solution, the activity of an ion can be replaced by the mean ionic activity $a_{\pm} = \gamma_{\pm} m_{\pm}$. If the solution contains only silver nitrate, then $m_{\pm} = m$; and Eq. (17.53) becomes

$$\mathscr{E} = \mathscr{E}°_{Ag^+/Ag} + (0.05916 \text{ V}) \log_{10} m + (0.05916 \text{ V}) \log_{10} \gamma_{\pm}.$$

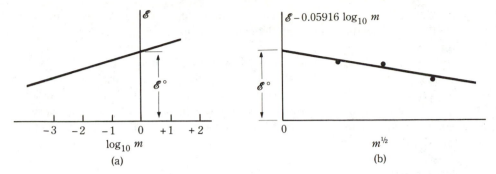

Figure 17.6 (a) "Ideal" dependence of $\mathscr{E}$ on m. (b) Plot to obtain $\mathscr{E}^\circ$ by extrapolation.

If the measurements are carried to solutions dilute enough so that the Debye–Hückel limiting law, Eq. (16.82), is valid, then $\log_{10} \gamma_\pm = -(0.5092 \text{ V kg}^{1/2}/\text{mol}^{1/2})m^{1/2}$, and we can reduce the equation to

$$\mathscr{E} - (0.05916 \text{ V})\log_{10} m = \mathscr{E}^\circ_{Ag^+/Ag} - (0.03012 \text{ V kg}^{1/2}/\text{mol}^{1/2})m^{1/2}. \qquad (17.54)$$

From the measured values of $\mathscr{E}$ and m, the left-hand side of this equation can be calculated. The left-hand side is plotted against $m^{1/2}$; extrapolation of the curve to $m^{1/2} = 0$ yields an intercept equal to $\mathscr{E}^\circ_{Ag^+/Ag}$. The plot is shown schematically in Fig. 17.6(b). It is by this method that accurate values of $\mathscr{E}^\circ$ are obtained from the measured values of the $\mathscr{E}$ of any half-cell.

17.17 THE DETERMINATION OF ACTIVITIES AND ACTIVITY COEFFICIENTS FROM CELL POTENTIALS

Once an accurate value of $\mathscr{E}^\circ$ has been obtained for a cell, then the potential measurements yield values of the activity coefficients directly. Consider the cell

$$Pt\,|\,H_2(f = 1)\,|\,H^+, Cl^-\,|\,AgCl\,|\,Ag.$$

The cell reaction is

$$AgCl(s) + \tfrac{1}{2}H_2(f = 1) \;\rightleftharpoons\; Ag + H^+ + Cl^-.$$

The cell potential is

$$\mathscr{E} = \mathscr{E}^\circ - \frac{RT}{F} \ln (a_{H^+}\, a_{Cl^-}). \qquad (17.55)$$

According to Eq. (17.55), the potential of the cell does not depend on the individual ion activities but on the product $a_{H^+}\, a_{Cl^-}$. As it turns out there is no measurable quantity that depends on an individual ion activity. Consequently, we replace the product $a_{H^+}\, a_{Cl^-}$ by $a_\pm^2$. Since in HCl, $m_\pm = m$, we have $a_\pm^2 = (\gamma_\pm m)^2$; this reduces Eq. (17.55) to

$$\mathscr{E} = \mathscr{E}^\circ - \frac{2RT}{F} \ln m - \frac{2RT}{F} \ln \gamma_\pm. \qquad (17.56)$$

At 25 °C

$$\mathscr{E} = \mathscr{E}^\circ - (0.1183 \text{ V})\log_{10} m - (0.1183 \text{ V})\log_{10} \gamma_\pm. \qquad (17.57)$$

Having determined $\mathscr{E}^\circ$ by the extrapolation described in Section 17.16, we see that the values of $\mathscr{E}$ determine the values of $\gamma_\pm$ at every value of m. Conversely, if the value of $\gamma_\pm$ is known at all values of m, the cell potential $\mathscr{E}$ can be calculated from Eq. (17.56) or (17.57) as a function of m.

The measurement of cell potentials is the most powerful method of obtaining values of activities of electrolytes. Experimentally it is, in many cases at least, much easier to handle than measurements of colligative properties. It has the additional advantage that it can be used over a wide range of temperatures. Although cell potentials can be measured in nonaqueous solvents, the electrode equilibria often are not as easily established so that the experimental difficulties are much greater.

★ 17.18 CONCENTRATION CELLS

If the two electrode systems that compose a cell involve electrolytic solutions of different composition, there will be a potential difference across the boundary between the two solutions. This potential difference is called the liquid junction potential, or the diffusion potential. To illustrate how such a potential difference arises, consider two silver–silver chloride electrodes, one in contact with a concentrated HCl solution, activity $= a_1$, the other in contact with a dilute HCl solution, activity $= a_2$; Fig. 17.7(a). If the boundary between the two solutions is open, the H^+ and Cl^- ions in the more concentrated solution diffuse into the more dilute solution. The H^+ ion diffuses much more rapidly than does the Cl^- ion (Fig. 17.7b). As the H^+ ion begins to outdistance the Cl^- ion, an electrical double layer develops at the interface between the two solutions (Fig. 17.7c). The potential difference across the double layer produces an electrical field that slows the faster moving ion and speeds the slower moving ion. A steady state is established in which the two ions migrate at the same speed; the ion that moved faster initially leads the march.

The diffusion from the concentrated to the dilute solution is an irreversible change; however, if it is very slow—slow enough that the interface does not move appreciably in the time we require to make the measurements—then we may consider the system at "equilibrium" and ignore the motion of the boundary. However, the additional potential difference in the liquid junction will show up in the measurements of the cell potential.

Choosing the lower electrode as the left-hand electrode, the symbol for this cell is

$$Ag\,|\,AgCl\,|\,Cl^-(a_1)\,\vdots\,Cl^-(a_2)\,|\,AgCl\,|\,Ag,$$

where the dashed vertical bar represents the junction between the two aqueous phases.

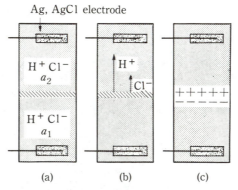

Figure 17.7 Development of the junction potential.

We can calculate the potential of the cell if we assume that on the passage of one mole of electrical charge through the cell all of the changes take place reversibly. Then the potential of the cell is given by

$$-F\mathscr{E} = \sum_i \Delta G_i, \qquad (17.58)$$

where $\sum \Delta G_i$ is the sum of all the Gibbs energy changes in the cell that accompany the passage of one mole of positive charge upward through the cell. These Gibbs energy changes are:

Lower electrode	$Ag(s) + Cl^-(a_1) \longrightarrow$	$AgCl(s) + e^-$
Upper electrode	$AgCl(s) + e^- \longrightarrow$	$Cl^-(a_2) + Ag(s)$
Net change at two electrodes	$Cl^-(a_1) \longrightarrow$	$Cl^-(a_2)$

In addition, at the boundary of the two solutions a fraction t_+ of the charge is carried by H^+ and a fraction t_- is carried by Cl^-. The fractions t_+ and t_- are the transference numbers, or transport numbers, of the ions. One mole of positive charge passing through the boundary requires that t_+ moles of H^+ ion are moved upward from the solution a_1 to the solution a_2, and t_- moles of Cl^- are moved downward from a_2 to a_1. Thus at the boundary:

$$t_+ H^+(a_1) \longrightarrow t_+ H^+(a_2), \quad \text{and} \quad t_- Cl^-(a_2) \longrightarrow t_- Cl^-(a_1).$$

The total change within the cell is the sum of the changes at the electrodes and at the boundary:

$$t_+ H^+(a_1) + Cl^-(a_1) + t_- Cl^-(a_2) \longrightarrow t_+ H^+(a_2) + Cl^-(a_2) + t_- Cl^-(a_1).$$

The sum of the fractions must be unity, so that $t_- = 1 - t_+$. Using this value of t_- in the equation, after some rearrangement, reduces it to

$$t_+ H^+(a_1) + t_+ Cl^-(a_1) \longrightarrow t_+ H^+(a_2) + t_+ Cl^-(a_2). \qquad (17.59)$$

The cell reaction (17.59) is the transfer of t_+ moles of HCl from the solution a_1 to the solution a_2. The total Gibbs energy change is

$$\Delta G = t_+ [\mu^\circ_{H^+} + RT \ln (a_{H^+})_2 + \mu^\circ_{Cl^-} + RT \ln (a_{Cl^-})_2$$
$$- \mu^\circ_{H^+} - RT \ln (a_{H^+})_1 - \mu^\circ_{Cl^-} - RT \ln (a_{Cl^-})_1],$$

$$\Delta G = t_+ RT \ln \frac{(a_{H^+} a_{Cl^-})_2}{(a_{H^+} a_{Cl^-})_1} = 2t_+ RT \ln \frac{(a_\pm)_2}{(a_\pm)_1},$$

since $a_{H^+} a_{Cl^-} = a_\pm^2$. Using Eq. (17.58), we have for the potential of the cell *with transference*,

$$\mathscr{E}_{wt} = -\frac{2t_+ RT}{F} \ln \frac{(a_\pm)_2}{(a_\pm)_1}. \qquad (17.60)$$

If the boundary between the two solutions did not contribute to the cell potential, then the only change would be that contributed by the electrodes, which is

$$Cl^-(a_1) \longrightarrow Cl^-(a_2).$$

The corresponding value of ΔG is

$$\Delta G = \mu^\circ_{Cl^-} + RT \ln (a_{Cl^-})_2 - \mu^\circ_{Cl^-} - RT \ln (a_{Cl^-})_1 = RT \ln \frac{(a_\pm)_2}{(a_\pm)_1},$$

where a_{Cl^-} has been replaced by the mean ionic activity $a_\pm$. This cell is without transference and has the potential,

$$\mathscr{E}_{wot} = -\frac{\Delta G}{F} = -\frac{RT}{F}\ln\frac{(a_\pm)_2}{(a_\pm)_1}. \tag{17.61}$$

The total potential of the cell with transference is that of the cell without transference plus the junction potential, $\mathscr{E}_j$. Thus, $\mathscr{E}_{wt} = \mathscr{E}_{wot} + \mathscr{E}_j$, so that

$$\mathscr{E}_j = \mathscr{E}_{wt} - \mathscr{E}_{wot}, \tag{17.62}$$

Using Eqs. (17.60) and (17.61), this becomes

$$\mathscr{E}_j = (1 - 2t_+)\frac{RT}{F}\ln\frac{(a_\pm)_2}{(a_\pm)_1}. \tag{17.63}$$

From Eq. (17.63) it is apparent that if t_+ is near 0.5, the liquid junction potential will be small; this relation is correct only if the two electrolytes in the cell produce two ions in solution. By measuring the potential of the cells with and without transference it is possible to evaluate $\mathscr{E}_j$ and t_+. Note, by comparing Eqs. (17.60) and (17.61), that

$$\mathscr{E}_{wt} = 2t_+\mathscr{E}_{wot}. \tag{17.64}$$

The trick in all of this is to be able to establish a sharp boundary so as to obtain reproducible measurements of $\mathscr{E}_{wt}$ and to be able to construct a cell that eliminates $\mathscr{E}_j$ so that $\mathscr{E}_{wot}$ can be measured. There are several clever ways of establishing a sharp boundary between the two solutions; however, they will not be described here. The second problem of constructing a cell without a liquid boundary is more pertinent to our discussion.

A concentration cell without transference (that is, without a liquid junction) is shown in Fig. 17.8. The cell consists of two cells in series, which can be symbolized by

$$\text{Pt}|\text{H}_2(p)|\text{H}^+, \text{Cl}^-(a_\pm)_1|\text{AgCl}|\overline{\text{Ag} \qquad \text{Ag}}|\text{AgCl}|\text{Cl}^-, \text{H}^+(a_\pm)_2|\text{H}_2(p)|\text{Pt}.$$

The potential is the sum of the potentials of the two cells separately:

$$\mathscr{E} = [\phi(\text{AgCl/Ag}) - \phi(\text{H}^+/\text{H}_2)]_1 + [\phi(\text{H}^+/\text{H}_2) - (\text{AgCl/Ag})]_2.$$

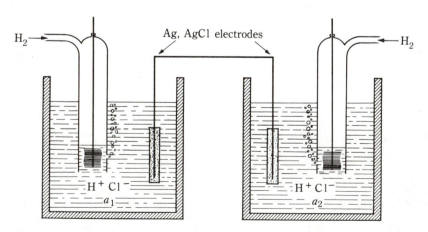

Figure 17.8 Concentration cell without transference.

Writing the Nernst equation for each potential, we obtain

$$\mathscr{E} = \left[\phi^{\circ}_{AgCl/Ag/Cl^-} - \frac{RT}{F} \ln (a_{Cl^-})_1 + \frac{RT}{F} \ln \frac{p^{1/2}}{(a_{H^+})_1} \right]$$
$$+ \left[-\frac{RT}{F} \ln \frac{p^{1/2}}{(a_{H^+})_2} - \phi^{\circ}_{AgCl/Ag/Cl^-} + \frac{RT}{F} \ln (a_{Cl^-})_2 \right],$$
$$\mathscr{E} = \frac{RT}{F} \ln \frac{(a_{H^+} a_{Cl^-})_2}{(a_{H^+} a_{Cl^-})_1} = \frac{2RT}{F} \ln \frac{(a_{\pm})_2}{(a_{\pm})_1}.$$

By comparison with Eq. (17.61), we see that

$$\mathscr{E} = -2\mathscr{E}_{wot}. \qquad (17.65)$$

Measurement of the potential of this double cell yields the value of $\mathscr{E}_{wot}$ through Eq. (17.65).

Every measurement of the potential of a cell whose two electrodes require different electrolytes raises the problem of the liquid junction potential between the electrolytes. The problem can be solved in two ways: Either measure the junction potential or eliminate it. The junction potential can be eliminated by designing the experiment, as above, so that no liquid junction appears. Or, rather than using two cells, choose a reference electrode that uses the same electrolyte as the electrode being investigated. This is often the best way to eliminate the liquid junction; however, it is not always feasible.

The salt bridge, an agar jelly saturated with either KCl or NH_4NO_3, is often used to connect the two electrode compartments. This device introduces two liquid junctions, whose potentials are often opposed to one another, and the net junction potential is very small. The physical reason for the cancellation of the two potentials is complex. The use of a jelly has some advantages in itself: It prevents siphoning if the electrolyte levels differ in the two electrode compartments, and it slows the ionic diffusion very much so that the junction potentials, whatever they may be, settle down to reproducible values very quickly.

17.19 TECHNICAL ELECTROCHEMICAL PROCESSES

Practical electrochemical processes divide naturally into power-consuming processes and power-producing processes. The industrial electrolytic preparative processes consume electrical power and produce high-energy substances. Typical of substances produced at the cathode are: hydrogen and sodium hydroxide in the electrolysis of brine; aluminum, magnesium, and the alkali and alkaline earth metals in the electrolysis of molten salts. Electroplating and electrorefining of metals are important technical cathodic processes. Substances produced at the anode are: oxygen in water electrolysis, and chlorine in the electrolysis of brine and molten chlorides; hydrogen peroxide; potassium perchlorate; and oxide coatings for decorative finishes on anodized aluminum. Anodic dissolution of a metal is important in the electrorefining and electromachining of metals.

The power-producing processes occur in the electrochemical cell; these processes consume high-energy substances and produce electrical power. Two important devices are described in Section 17.21.

It is interesting to note that the invention of the electrochemical cell by Alessandro Volta in 1800 is, in fact, a re-invention. Recently, archaeological excavations in the Near East unearthed what is apparently an electrochemical cell based on iron and copper electrodes; the device is dated somewhere between 300 B.C. and 300 A.D. There is also some evidence that, as early as 2500 B.C., the Egyptians knew how to electroplate objects.

17.20 ELECTROCHEMICAL CELLS AS POWER SOURCES

It is remarkable that, in principle, any chemical reaction can be harnessed to perform work in an electrochemical cell. If the cell operates reversibly, the electrical work obtained is $W_{el} = -\Delta G$, or

$$W_{el} = -\Delta H + T\Delta S = -\Delta H + Q_{rev}$$

$$= -\Delta H\left(1 - \frac{Q_{rev}}{\Delta H}\right).$$

In many practical cases the increase in entropy is not very large, so that $T\Delta S/\Delta H$ is relatively small and

$$W_{el} \approx -\Delta H.$$

This means that the electrical work that is produced is only slightly less than the decrease in enthalpy in the reaction. Note that if we simply let the reaction occur without producing work, the quantity of heat, $-\Delta H$, would be released. This could be used to heat a boiler which in turn could run a turbine. But this heat engine is subject to the Carnot restriction; the electrical work that could be produced by a generator operated by a turbine would be

$$W_{el} = -\Delta H\left(\frac{T_1 - T_2}{T_1}\right).$$

This amount of work is substantially less (often three to five times less) than could be obtained electrochemically from the same reaction. Thus the electrochemical cell offers possibilities for efficient production of electrical energy from chemical sources that are unequalled by any other device.

17.20.1 Classification of Electrochemical Cells

We can classify electrochemical cells that provide electrical energy into three general types.

1. Primary cells. These are constructed of high-energy materials which react chemically and produce electrical power. The cell reaction is not reversible, and when the materials are consumed the device must be discarded. Typical examples of the primary cell are the ordinary flashlight battery (the LeClanché cell), and the zinc–mercury cells used in cameras, clocks, hearing aids, watches, and other familiar articles.

2. Secondary cells. These devices are reversible. After providing power, the high-energy materials can be reconstituted by imposing a current from an exterior power source in the reverse direction. The cell reaction is thus reversed and the device is "recharged".

The most important example of a secondary cell is the lead storage battery used in automobiles. Other examples of secondary cells are the Edison cell and the nickel–cadmium rechargeable cells used in calculators and flash lamps.

3. Fuel cells. The fuel cell, like the primary cell, is designed to use high-energy materials to produce power. It differs from the primary cell in that it is designed to accept a continuing supply of the "fuel," and the "fuels" are materials that we would commonly regard as fuels, such as hydrogen, carbon, and hydrocarbons. Ultimately, we might even hope to use raw coal and petroleum.

17.20.2   Requirements for a Power Source

If we are to draw power from an electrochemical cell, since

$$P = \mathscr{E}I, \tag{17.66}$$

it follows that the product of the cell potential and the current must remain at a reasonable value over the useful life of the cell. The current, I, is distributed over the entire area of the electrode, A. The current into or out of a unit area of the electrode surface is the current density, i. Thus

$$i = \frac{I}{A}. \tag{17.67}$$

This current density implies a definite rate of reaction on each unit of electrode area. Suppose we draw a current, I, from the cell. For purposes of argument, suppose that the negative electrode is a hydrogen electrode. Then charge is drained away from each unit of electrode area at the rate, $i = (1/A) \, dQ/dt = I/A$. As the electrons leave the platinum of the H^+/H_2 electrode, more H_2 must ionize, $H_2 \rightarrow 2H^+ + 2e^-$, or the potential of the electrode will move to a less negative value. If the rate at which electrons are produced by the ionization of hydrogen is comparable to the rate at which electrons leave the platinum to enter the external circuit, then the potential of the electrode will be near the open-circuit potential. On the other hand, if the electrode reaction is so slow that the electrons are not quickly replenished when they are drained away into the external circuit, then the potential of the electrode will depart substantially from the open-circuit potential. Similarly, if the electrode reaction on the positive electrode is slow, the electrons that arrive from the external circuit are not quickly consumed by the electrode reaction and the potential of the positive electrode becomes much less positive. We conclude that when a cell provides power, the cell potential decreases since the positive electrode becomes less positive and the negative electrode becomes less negative.

The curves in Fig. 17.9 show the cell potential versus time for various cells after connection to a load that draws a current density i_1. The electrode reactions in cells A and B are too slow and cannot keep up with the current drain. The cell potential falls quickly to zero and the power, $\mathscr{E}I$, also goes to zero. Both cells provide a small amount of power initially, but neither cell is capable of being a practical power source. On the other hand,

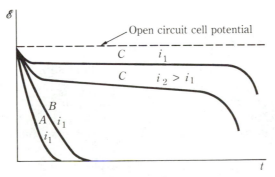

Figure 17.9   Cell potential under load as a function of time.

the electrode reactions in cell C are fast enough to restore the charge on the electrodes. The cell potential drops slightly but then stays steady at a relatively high value for a long period of time, so that the power, $\mathscr{E}I$, provided is substantial. If a larger current is drawn from cell C ($i_2 > i_1$), the potential drops a bit more but is still relatively high. Even in this circumstance cell C is a practical source of power. The rapid drop of the cell potential as at the end of curves C, signals the exhaustion of the active materials, the "fuel." If more "fuel" is supplied, the curve will remain flat, and the cell will continue to provide power.

We conclude that if a cell is to be practical as a power source the electrode reactions must be fast. The reactions must occur quickly enough so that the potential of the cell drops only slightly below its open-circuit potential. The problem in devising a fuel cell to burn coal lies in finding electrode surfaces on which the appropriate reactions will occur rapidly at reasonable temperatures. Can we invent the appropriate catalysts? Time will tell.

17.21 TWO PRACTICAL POWER SOURCES

17.21.1 The Lead Storage Cell

Consider first the lead–acid storage cell. As we draw current from the cell, at the positive plate, the cathode, PbO_2 is reduced to $PbSO_4$:

$$PbO_2(s) + 4H^+ + SO_4^{2-} + 2e^- \longrightarrow PbSO_4(s) + 2H_2O,$$

while at the negative plate, the anode, lead is oxidized to $PbSO_4$,

$$Pb(s) + SO_4^{2-} \longrightarrow PbSO_4(s) + 2e^-.$$

The potential of the cell is 2.0 volts. As current is drawn from the cell, the cell potential does not drop very much so the power, $\mathscr{E}I$, is near the reversible value, $\mathscr{E}_{rev}I$. Rather large currents—hundreds of amperes—can be drawn from the fully charged device without dropping the potential excessively.

When the cell needs to be recharged, we use an external power source to force current through the cell in the reverse direction; the positive plate is now the anode on which $PbSO_4$ is oxidized to PbO_2; the negative plate is the cathode on which $PbSO_4$ is reduced to Pb. The potential difference that must be impressed to recharge the cell has to be greater than the potential difference during discharge, but not excessively larger. The voltage efficiency of the cell is defined as:

$$\text{Voltage efficiency} = \frac{\text{average voltage during discharge}}{\text{average voltage during charge}}.$$

The voltage efficiency of the lead–acid cell is about 80%. This near reversibility is a consequence of the rapidity of the chemical reactions in the cell. As we have seen, the ability to supply large currents at potentials near the open-circuit potential means that the chemical reactions at the electrodes are fast; as the charge is drained away by the current, the potential should drop, but the chemical reaction occurs rapidly enough to rebuild the potential.

If we compare the quantity of charge obtained from the lead–acid cell to the quantity that must be passed in to charge the cell, we obtain values of 90 to 95%, or even higher in special circumstances. This means that very little of the charging current is dissipated in side reactions (such as electrolysis of water). Overall, the lead storage cell is an extra-

---

ELECTROCHEMICAL TERMINOLOGY

Once a cell is described, we can measure its potential and decide once and for all which electrode is the positive electrode (positive plate) and which is the negative electrode (negative plate). Nothing that happens later will change that.

Furthermore, oxidation *always* occurs at the anode and reduction *always* occurs at the cathode. Whether an electrode is a cathode or an anode depends on the direction of the current flow. In any secondary cell, the relations are

Plate	Discharge	Charge
Positive	Cathode	Anode
Negative	Anode	Cathode

In a primary cell, only discharge occurs so only the entries under "Discharge" are pertinent.

---

ordinary device: It is highly efficient; its larger versions can last 20 to 30 years (if carefully attended); and it can be cycled thousands of times. Its chief disadvantages are its great weight (low energy storage to weight ratio), and that if left unused in partially charged condition it can be ruined in a short time by the growth of relatively large $PbSO_4$ crystals, which are not easily reduced or oxidized by the charging current; this disaster is known as "sulfation."

For the standard Gibbs energy change in the lead–acid cell we have (for a two-electron change):

$$\Delta G^\circ = -376.97 \text{ kJ/mol};$$

$$\Delta H^\circ = -227.58 \text{ kJ/mol};$$

$$Q_{rev} = T \Delta S^\circ = +149.39 \text{ kJ/mol}.$$

Note that the reaction is endothermic if the cell performs reversibly. These figures mean that not only is the energy change, the ΔH, available to provide electrical work but also the quantity of heat, $Q_{rev} = T \Delta S$, that flows from the surroundings to keep the cell isothermal can be converted to electrical work. The ratio

$$\frac{-\Delta G^\circ}{-\Delta H^\circ} = \frac{376.97}{277.58} = 1.36$$

compares the electrical work that can be produced to the decrease in enthalpy of the materials. The extra 36% is the energy that flows in from the surroundings.

17.21.2 The Fuel Cell

The question is whether the kinds of reactions and the kinds of substances we commonly regard as "fuels," (coal, petroleum, natural gas) can be combined in the usual fuel burning reactions in an electrochemical way.

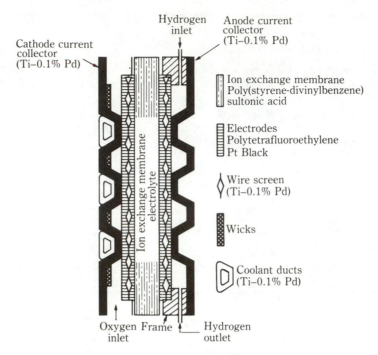

Figure 17.10   Schematic representation of a single Gemini hydrogen—oxygen fuel cell. (From H. A. Leibhafsky and E. J. Cairns, *Fuel Cells and Fuel Batteries*. New York, Wiley, 1968.)

Probably the most successful fuel cell thus far is the hydrogen–oxygen cell, which has been used in spacecraft. The electrodes consist of porous screens of titanium coated with a layer of a platinum catalyst. The electrolyte is a cation exchange resin that is mixed with a plastic material and formed into a thin sheet. The entire combination of two electrodes with the plastic membrane between them is only about 0.5 mm thick. The device is shown schematically in Fig. 17.10. The resin is kept saturated with water by means of a wick; the water formed by the operation of the cell drains out through the wick and is collected for drinking water. Connecting several of these cells raises the voltage to a practical value, while increasing the active area increases the current that can be drawn from the cell. This cell has been built to supply power of about 1 kilowatt.

The power available is limited by the relatively slow reduction of oxygen at the cathode surface, $O_2 + 4H^+ + 4e^- \rightarrow 2H_2O$; this problem exists with any fuel cell that uses an oxygen electrode. At present, platinum seems to be the best catalyst, but even platinum is not nearly as good as we would like. The rate of the anodic reaction, $H_2 \rightarrow 2H^+ + 2e^-$, the oxidation of hydrogen at the platinum surface, is relatively rapid. However, it would be nice if we could use something less expensive than platinum as a catalyst. At higher temperatures, the reaction rates are faster and the cell performance is better.

In Table 17.3 we have listed the thermodynamic properties (at 25 °C) of several reactions that would be desirable as fuel cell reactions. Each of the oxidizable substances

Table 17.3
Thermodynamic properties of possible fuel cell reactions at 25 °C

Reaction	$-\Delta G$ kJ/mol	$-\Delta H$ kJ/mol	$\dfrac{-\Delta G°}{-\Delta H°}$	$T\Delta S°$ kJ/mol	$\mathscr{E}°$ V
$H_2 + \frac{1}{2}O_2 \rightarrow H_2O$	237.178	285.830	0.83	-48.651	1.23
$C + O_2 \rightarrow CO_2$	394.359	393.509	1.002	$+0.857$	1.02
$C + \frac{1}{2}O_2 \rightarrow CO$	137.152	110.524	1.24	26.628	1.42
$CO + \frac{1}{2}O_2 \rightarrow CO_2$	257.207	282.985	0.91	-25.77	1.33
$CH_4 + 2O_2 \rightarrow CO_2 + 2H_2O$	817.96	890.36	0.92	-72.38	1.06
$CH_3OH + \frac{3}{2}O_2 \rightarrow CO_2 + 2H_2O$	702.36	726.51	0.97	-24.11	1.21
$C_8H_{18} + \frac{25}{2}O_2 \rightarrow 8CO_2 + 9H_2O$	5306.80	5512.10	0.96	-205.19	1.10
$C_2H_5OH + 3O_2 \rightarrow 2CO_2 + 3H_2O$	1325.36	1366.82	0.97	-41.36	1.15

can, in principle, be brought to equilibrium on an electrode. For example, the methanol oxidation can be written

$$CH_3OH + H_2O \longrightarrow CO_2 + 6H^+ + 6e^-.$$

This electrode, when combined with an oxygen electrode would yield a cell with an open-circuit potential of 1.21 V. A fuel cell based on methanol and air in KOH solution has been used to power television relay stations. All the reactions in Table 17.3 would yield cells with potentials of about one volt.

Cells have been built based on the oxidation of carbon to carbon dioxide. Relatively high temperatures are required (500 to 700 °C). One version uses a molten sodium carbonate electrolyte. The reactions are:

$$\text{Anode} \qquad C + 2CO_3^{2-} \longrightarrow 3CO_2 + 4e^-$$

$$\text{Cathode} \qquad O_2 + 2CO_2 + 4e^- \longrightarrow 2CO_3^{2-}$$

The overall reaction is simply

$$C + O_2 \longrightarrow CO_2.$$

One of the difficulties with high-temperature cells is that the construction materials may corrode very rapidly. This disadvantage has to be weighed against the increase in available power at the higher temperature.

Hydrocarbons such as methane, propane, and decane have been successfully oxidized in fuel cells, even at temperatures below 100 °C. We can reasonably expect that these devices will be much improved in the future.

As an alternative to the direct oxidation of the hydrocarbon at an electrode, the substance can be reformed at high temperatures by the reaction

$$CH_4 + 2H_2O \longrightarrow CO_2 + 4H_2.$$

The hydrogen is then oxidized at the anode. This method may ultimately be the most successful one for using hydrocarbons and carbon itself as electrochemical fuels.

QUESTIONS

17.1 Explain the meaning of Eq. (17.11), in terms of the reversible work required to bring a metal ion M^{+z} from infinity into the metal M maintained at potential ϕ.

17.2 Sketch the potential ϕ_{H^+/H_2} versus a_{H^+} for the hydrogen electrode; assume that $f = p = 1$ for H_2. Explain why the potential increases for increasing a_{H^+}, in terms of the "escaping tendency" of the Pt electrons and the aqueous H^+ ions.

17.3 Outline the logic leading to the conclusion that K is the most "active" alkali metal in Table 17.1.

17.4 Consider a cell composed of the two half-cells of Example 17.4. At what ionic activities will the measured cell potential be $\mathscr{E}° = \mathscr{E}°_{Fe^{3+}/Fe^{2+}} - \mathscr{E}°_{Sn^{4+}/Sn^{2+}}$? How would the overall reaction equilibrium constant be calculated? Contrast this procedure to the difficulty of direct measurement of K.

17.5 Use Table 17.1 to decide if it is likely that metallic zinc reduces the copper ion, $Zn(s) + Cu^{2+}(aq) \rightarrow Zn^{2+}(aq) + Cu(s)$.

17.6 Electrochemical cells can perform work. Imagine two hydrogen electrodes A and B connected by an external wire, with appropriate electrical contact between the two acid solutions. Assume that $a_{H^+}(A) = a_{H^+}(B)$, and that $f = p$ for both A and B. If $p_{H_2}(B) = 2p_{H_2}(A)$, show that the net cell reaction corresponds to a gas expansion, which *outside* of the cell could produce work. Discuss the work performed *by* the cell in terms of the current produced (how?) in the external wire.

17.7 What is the fate of the energy that does *not* flow to the surroundings in the cell-reaction example of Section 17.10.1?

PROBLEMS

Unless otherwise noted, the temperature is to be taken as 25 °C in the problems below.

17.1 Calculate the cell potential and find the cell reaction for each of the cells (data in Table 17.1):

    a) $Ag(s)|Ag^+(aq, a_\pm = 0.01)\vdots Zn^{2+}(a_\pm = 0.1)|Zn(s)$;
    b) $Pt(s)|Fe^{2+}(aq, a_\pm = 1.0), Fe^{3+}(aq, a_\pm = 0.1)\vdots Cl^-(aq, a_\pm = 0.001)|AgCl(s)|Ag(s)$;
    c) $Zn(s)|ZnO_2^{2-}(aq, a_\pm = 0.1), OH^-(aq, a_\pm = 1)|HgO(s)|Hg(1)$.

In each case is the cell reaction as written spontaneous or not?

17.2 Calculate the equilibrium constant for each of the cell reactions in Problem 17.1.

17.3 From the data in Table 17.1 calculate the equilibrium constant for each of the reactions:

    a) $Cu^{2+} + Zn \rightleftharpoons Cu + Zn^{2+}$;
    b) $Zn^{2+} + 4CN^- \rightleftharpoons Zn(CN)_4^{2-}$;
    c) $3H_2O + Fe = Fe(OH)_3(s) + \frac{3}{2}H_2$;
    d) $Fe + 2Fe^{3+} \rightleftharpoons 3Fe^{2+}$;
    e) $3HSnO_2^- + Bi_2O_3 + 6H_2O + 3OH^- \rightleftharpoons 2Bi + 3Sn(OH)_6^{2-}$;
    f) $PbSO_4(s) \rightleftharpoons Pb^{2+} + SO_4^{2-}$.

17.4 The Edison storage cell is symbolized

$$Fe(s)|FeO(s)|KOH(aq, a)|Ni_2O_3(s)|NiO(s)|Ni(s)$$

The half-cell reactions are

$$Ni_2O_3(s) + H_2O(l) + 2e^- \;\rlap{\longrightarrow}{\longleftarrow}\; 2NiO(s) + 2OH^-, \qquad \phi° = 0.4 \text{ V};$$
$$FeO(s) + H_2O(l) + 2e^- \;\rlap{\longrightarrow}{\longleftarrow}\; Fe(s) + 2OH^-, \qquad \phi° = -0.87 \text{ V}.$$

a) What is the cell reaction?

b) How does the cell potential depend on the activity of the KOH?

c) How much electrical energy can be obtained per kilogram of the active materials in the cell?

17.5 Consider the lead storage cell

$$Pb(s)|PbSO_4(s)|H_2SO_4(aq, a)|PbSO_4(s)|PbO_2(s)|Pb(s),$$

in which $\phi^\circ_{SO_4^{2-}/PbSO_4/Pb} = -0.356$ V, and $\phi^\circ_{SO_4^{2-}/PbO_2/PbSO_4/Pb} = +1.685$ V.

a) If the cell potential is 2.016 volts, compute the activity of the sulfuric acid.

b) Write the cell reaction. Is this reaction spontaneous?

c) If the cell produces work (discharge) the reaction goes in one direction, while if work is destroyed (charge) the reaction goes in the opposite direction. How much work must be destroyed per mole of PbO_2 produced if the average potential during charge is 2.15 volts?

d) Sketch the dependence of the cell potential on the activity of the sulfuric acid.

e) How much electrical energy can be obtained per kilogram of the active materials in the cell?

17.6 Consider the cell

$$Hg(l)|Hg_2SO_4(s)|FeSO_4(aq, a = 0.01)|Fe(s)$$

a) Write the cell reaction.

b) Calculate the cell potential, the equilibrium constant for the cell reaction, and the standard Gibbs energy change, ΔG°, at 25 °C. (Data in Table 17.1.)

17.7 For the electrode $SO_4^{2-}(aq, a_{SO_4^{2-}})|PbSO_4(s)|Pb(s)$, $\phi^\circ = -0.356$ V.

a) If this electrode is the right-hand electrode and the SHE is the left-hand electrode, the cell potential is -0.245 volt. What is the activity of the sulfate ion in this cell?

b) Calculate the mean ionic activity of the sulfuric acid in the cell

$$Pt(s)|H_2(g, 1\text{ atm})|H_2SO_4(aq, a)|PbSO_4(s)|Pb(s)$$

if the cell potential is -0.220 V. (*Note:* the left-hand electrode is not the SHE.)

17.8 Consider the cell

$$Pt(s)|H_2(g, 1\text{ atm})|H^+(aq, a = 1), Fe^{3+}(aq), Fe^{2+}(aq)|Pt(s),$$

given $Fe^{3+} + e^- \rightleftharpoons Fe^{2+}$, $\phi^\circ = 0.771$ V.

a) If the potential of the cell is 0.712 V, what is the ratio of concentration of Fe^{2+} to Fe^{3+}?

b) What is the ratio of these concentrations if the potential of the cell is 0.830 V?

c) Calculate the fraction of the total iron present as Fe^{3+} at $\phi = 0.650$ V, 0.700 V, 0.750 V, 0.771 V, 0.800 V, 0.850 V, and 0.900 V. Sketch this fraction as a function of ϕ.

17.9 The standard potentials at 25 °C are:

$$Pd^{2+}(aq) + 2e^- \rightleftharpoons Pd(s), \qquad\qquad \phi^\circ = 0.83 \text{ V};$$
$$PdCl_4^{2-}(aq) + 2e^- \rightleftharpoons Pd(s) + 4Cl^-(aq), \qquad \phi^\circ = 0.64 \text{ V}.$$

a) Calculate the equilibrium constant for the reaction $Pd^{2+} + 4Cl^- \rightleftharpoons PdCl_4^{2-}$.

b) Calculate the ΔG° for this reaction.

17.10 a) Calculate the potential of the $Ag^+|Ag$ electrode; $\phi^\circ = 0.7991$ V, for activities of $Ag^+ = 1$, 0.1, 0.01, and 0.001.

b) For AgI, $K_{sp} = 8.7 \times 10^{-17}$; what will be the potential of the $Ag^+|Ag$ electrode in a saturated solution of AgI?

c) Calculate the standard potential of the $I^-|AgI|Ag$ electrode.

17.11 A 0.1 mol/L solution of NaCl is titrated with $AgNO_3$. The titration is followed potentiometrically, using a silver wire as the indicating electrode and a suitable reference electrode. Calculate the potential of the silver wire when the amount of $AgNO_3$ added is 50%, 90%, 99%, 99.9%, 100%, 100.1%, 101%, 110%, and 150% of the stoichiometric requirement (ignore the change in volume of the solution).

$$\phi^{\circ}_{Cl^-/AgCl/Ag} = 0.222 \text{ V}, \qquad \phi^{\circ}_{Ag^+/Ag} = 0.799 \text{ V}.$$

For silver chloride, $K_{sp} = 1.7 \times 10^{-10}$.

17.12 Consider the couple $O + e^- \rightleftharpoons R$, with all of the oxidized and reduced species at unit activity. What must be the value of ϕ° of the couple if the reductant R is to liberate hydrogen at 1 atm from

a) an acid solution, $a_{H^+} = 1$?
b) water at pH $= 7$?
c) Is hydrogen a better reducing agent in acid or in basic solution?

17.13 Consider the same couple under the same conditions as in Problem 17.12. What must be the value of ϕ° of the couple if the oxidant is to liberate oxygen at 1 atm by the half-cell reaction,

$$O_2(g) + 2H_2O(l) + 4e^- \quad \rightleftharpoons \quad 4OH^-, \qquad \phi^{\circ} = 0.401 \text{ V},$$

a) from a basic solution, $a_{OH^-} = 1$?
b) from an acid solution, $a_{H^+} = 1$?
c) from water at pH $= 7$?
d) Is oxygen a better oxidizing agent in acid or in basic solution?

17.14 From the values of the standard potentials in Table 17.1, calculate the standard molar Gibbs energy μ° of the ions Na^+, Pb^{2+}, Ag^+.

17.15 Calculate $\mu^{\circ}_{Fe^{3+}}$ from the data: $\phi^{\circ}_{Fe^{3+}/Fe^{2+}} = +0.771 \text{ V}$, $\phi^{\circ}_{Fe^{2+}/Fe} = -0.440 \text{ V}$.

17.16 Consider the half-cell reaction

$$AgCl(s) + e^- \quad \rightleftharpoons \quad Ag(s) + Cl^-(aq).$$

If $\mu^{\circ}(AgCl) = -109.721 \text{ kJ/mol}$, and if $\phi^{\circ} = +0.222 \text{ V}$ for this half-cell, calculate the standard Gibbs energy of $Cl^-(aq)$.

17.17 At 25 °C for the potential of the cell,

$$Pt|H_2(g, f = 1)|HCl(aq, m)|AgCl(s)|Ag(s),$$

as a function of m, the molality of HCl, we have

m/(mol/kg)	$\mathscr{E}$/V	m/(mol/kg)	$\mathscr{E}$/V	m/(mol/kg)	$\mathscr{E}$/V
0.001	0.579 15	0.02	0.430 24	0.5	0.272 31
0.002	0.544 25	0.05	0.385 88	1	0.233 28
0.005	0.498 46	0.1	0.352 41	1.5	0.207 19
0.01	0.464 17	0.2	0.318 74	2	0.186 31
				3	0.151 83

Calculate $\mathscr{E}^{\circ}$ and $\gamma_{\pm}$ for HCl at $m = 0.001, 0.01, 0.1, 1, 3$.

17.18 The standard potential of the quinhydrone electrode is $\phi^{\circ} = 0.6994 \text{ V}$. The half-cell reaction is

$$Q(s) + 2H^+ + 2e^- \quad \rightleftharpoons \quad QH_2(s).$$

Using a calomel electrode as a reference electrode, $\phi^{\circ}_{Cl^-/Hg_2Cl_2/Hg} = 0.2676 \text{ V}$, we have the cell

$$Hg(l)|Hg_2Cl_2(s)|HCl(aq, a)|Q \cdot QH_2(s)|Au(s).$$

The compound $Q \cdot QH_2$, quinhydrone, is sparingly soluble in water, producing equal concentrations of Q, quinone, and QH_2, hydroquinone. Using the values of the mean ionic activity coefficients for HCl given in Table 16.1, calculate the potential of this cell at $m_{HCl} = 0.001$, 0.005, 0.01.

17.19 H. S. Harned and W. J. Hamer [*J. Amer. Chem. Soc.* **57**; 33 (1935)] present values for the potential of the cell,

$$Pb(s)\,|\,PbSO_4(s)\,|\,H_2SO_4(aq, a)\,|\,PbSO_4(s)\,|\,PbO_2(s)\,|\,Pt(s),$$

over a wide range of temperature and concentration of H_2SO_4. In $1\,m\,H_2SO_4$ they found, between 0 °C and 60 °C,

$$\mathscr{E}/V = 1.91737 + 56.1(10^{-6})t + 108(10^{-8})t^2,$$

where t is the Celsius temperature.

a) Calculate ΔG, ΔH, and ΔS for the cell reaction at 0 °C and 25 °C.

b) For the half-cells at 25 °C

$$PbO_2(s) + SO_4^{2-} + 4H^+ + 2e^- \;\rightleftharpoons\; PbSO_4(s) + 2H_2O, \qquad \phi^\circ = \quad 1.6849 \text{ V};$$

$$PbSO_4(s) + 2e^- \;\rightleftharpoons\; Pb(s) + SO_4^{2-}, \qquad \phi^\circ = -0.3553 \text{ V}.$$

Calculate the mean ionic activity coefficient in $1\,m\,H_2SO_4$ at 25 °C. Assume that the activity of water is unity.

17.20 At 25 °C the potential of the cell,

$$Pt(s)\,|\,H_2(g, f = 1)\,|\,H_2SO_4(aq, a)\,|\,Hg_2SO_4(s)\,|\,Hg(l),$$

is 0.61201 V in $4\,m\,H_2SO_4$; $\mathscr{E}^\circ = 0.61515$ V. Calculate the mean ionic activity coefficient in $4\,m\,H_2SO_4$. [H. S. Harned and W. J. Hamer, *J. Amer. Chem. Soc.* **57**; 27 (1933)].

17.21 In $4\,m\,H_2SO_4$, the potential of the cell in Problem 17.19 is 2.0529 V at 25 °C. Calculate the value of the activity of water in $4\,m\,H_2SO_4$ using the result in Problem 17.20.

17.22 Between 0 °C and 90 °C, the potential of the cell,

$$Pt(s)\,|\,H_2(g, f = 1)\,|\,HCl(aq, m = 0.1)\,|\,AgCl(s)\,|\,Ag(s),$$

is given by

$$\mathscr{E}/V = 0.35510 - 0.3422(10^{-4})t - 3.2347(10^{-6})t^2 + 6.314(10^{-9})t^3,$$

where t is the Celsius temperature. Write the cell reaction and calculate ΔG, ΔH, and ΔS for the cell at 50 °C.

17.23 Write the cell reaction and calculate the potential of the following cells without transference.

a) $Pt(s)\,|\,H_2(g, p = 1 \text{ atm})\,|\,HCl(aq, a)\,|\,H_2(g, p = 0.5 \text{ atm})\,|\,Pt(s)$

b) $Zn(s)\,|\,Zn^{2+}(aq, a = 0.01)\,\vdots\vdots\,Zn^{2+}(aq, a = 0.1)\,|\,Zn(s)$.

17.24 At 25 °C the potential of the cell with transference,

$$Pt(s)\,|\,H_2(g, f = 1)\,|\,HCl(aq, a_\pm = 0.009048)\,\vdots\,HCl(aq, a_\pm = 0.01751)\,|\,H_2(g, f = 1)\,|\,Pt(s),$$

is 0.02802 V. The corresponding cell without transference has a potential of 0.01696 V. Calculate the transference number of H^+ ion and the value of the junction potential.

17.25 Consider the reaction

$$Sn + Sn^{4+} \;\rightleftharpoons\; 2\,Sn^{2+}.$$

If metallic tin is in equilibrium with a solution of Sn^{2+} in which $a_{Sn^{2+}} = 0.100$, what is the equilibrium activity of Sn^{4+} ion? Use data in Table 17.1.

17.26 Consider a Daniell cell that has 100 cm^3 of 1.00 mol/L $CuSO_4$ solution in the positive electrode compartment and 100 cm^3 of 1.00 mol/L $ZnSO_4$ in the negative electrode compartment. The zinc electrode is sufficiently large that it does not limit the reaction.

a) Calculate the cell potential after 0%, 50%, 90%, 99%, 99.9%, and 99.99% of the available copper sulfate has been consumed.

b) What is the total electrical energy that can be drawn from the cell? *Note:* $\Delta G_{total} = \int_0^{\xi_e} (\partial G / \partial \xi)_{T, p} \, d\xi$.

c) Plot the cell potential as a function of the fraction of the total energy that has been delivered.

17.27 A platinum electrode is immersed in 100 mL of a solution in which the sum of the concentrations of the Fe^{2+} and Fe^{3+} ions is 0.100 mol/L.

a) Sketch the fraction of the ions that are present as Fe^{3+} as a function of the potential of the electrode.

b) If Sn^{2+} is added to the solution, the reaction $2\,Fe^{3+} + Sn^{2+} \rightleftharpoons 2\,Fe^{2+} + Sn^{4+}$ occurs. Assume that initially all the iron is present as Fe^{3+}. Plot the potential of the platinum after the addition of 40 mL, 49.0 mL, 49.9 mL, 49.99 mL, 50.0 mL, 50.01 mL, 50.10 mL, 51.0 mL, and 60 mL of 0.100 mol/L Sn^{2+} solution.

18

Surface Phenomena

18.1 SURFACE ENERGY AND SURFACE TENSION

Consider a solid composed of spherical molecules in a close-packed arrangement. The molecules are bound by a cohesive energy E per mole and $\epsilon = E/N$ per molecule. Each molecule is bonded to twelve others; the bond strength is $\epsilon/12$. If the surface layer is close packed, a molecule on the surface is bonded to a total of only nine neighbors. Then the total binding energy of the surface molecule is $9\epsilon/12 = \frac{3}{4}\epsilon$. From this rather crude picture we conclude that the surface molecule is bound with only 75% of the binding energy of a molecule in the bulk. The energy of a surface molecule is therefore higher than that of a molecule in the interior of the solid and energy must be expended to move a molecule from the interior to the surface of a solid; this is also true of liquids.

Suppose that a film of liquid is stretched on a wire frame having a movable member (Fig. 18.1). To increase the area of the film by dA, a proportionate amount of work must be done. The Gibbs energy of the film increases by $\gamma \, dA$, where γ is the surface Gibbs energy per unit area. The Gibbs energy increase implies that the motion of the wire is opposed by a force f; if the wire moves a distance dx, the work expended is $f \, dx$. These two energy increments are equal, so that

$$f \, dx = \gamma \, dA$$

Liquid film

Figure 18.1   Stretched film.

Table 18.1
Surface tension of liquids at 20 °C

Liquid	$\gamma/(10^{-3}\,\text{N/m})$	Liquid	$\gamma/(10^{-3}\,\text{N/m})$
Acetone	23.70	Ethyl ether	17.01
Benzene	28.85	n-Hexane	18.43
Carbon tetrachloride	26.95	Methyl alcohol	22.61
Ethyl acetate	23.9	Toluene	28.5
Ethyl alcohol	22.75	Water	72.75

If l is the length of the movable member, the increase in area is $2(l\,dx)$; the factor two appears because the film has two sides. Thus

$$f\,dx = \gamma(2l)\,dx \qquad \text{or} \qquad f = 2l\gamma.$$

The length of the film in contact with the wire is l on each side, or a total length is $2l$; the force acting per unit length of the wire in contact with the film is the *surface tension* of the liquid, $f/2l = \gamma$. The surface tension acts as a force that opposes the increase in area of the liquid. The SI unit for surface tension is the newton per metre, which is numerically equal to the rate of increase of the surface Gibbs energy with area, in joules per square metre. The magnitude of the surface tension of common liquids is of the order of tens of millinewtons per metre. Some values are given in Table 18.1.

18.2 MAGNITUDE OF SURFACE TENSION

By the argument used in Section 18.1 we estimated that the surface atoms have an energy roughly 25% above that of those in the bulk. This excess energy does not show up in systems of ordinary size since the number of molecules on the surface is an insignificant fraction of the total number of molecules present. Consider a cube having an edge of length a. If the molecules are 10^{-10} m in diameter, then $10^{10}a$ molecules can be placed on an edge; the number of molecules in the cube is $(10^{10}a)^3 = 10^{30}a^3$. On each face there will be $(10^{10}a)^2 = 10^{20}a^2$ molecules; there are six faces, making a total of $6(10^{20}a^2)$ molecules on the surface of the cube. The fraction of molecules on the surface is $6(10^{20}a^2)/10^{30}a^3 = 6 \times 10^{-10}/a$. If $a = 1$ metre, then only six molecules in every ten billion are on the surface; or if $a = 1$ centimetre, then only six molecules in every 100 million are on the surface. Consequently, unless we make special efforts to observe the surface energy, we may ignore its presence as we have in all the earlier thermodynamic discussions.

If the ratio of surface to volume of the system is very large, the surface energy shows up willy nilly. We can calculate the size of particle for which the surface energy will contribute, let us say 1% of the total energy. We write the energy in the form,

$$E = E_v V + E_s A,$$

where V and A are the volume and area, E_v and E_s are the energy per unit volume and the energy per unit area. But, $E_v = \epsilon_v N_v$, and $E_s = \epsilon_s N_s$, where ϵ_v and ϵ_s are energies per molecule in the bulk and per molecule on the surface, respectively; N_s and N_v are the number of molecules per unit area and per unit volume, respectively. Then

$$E = E_v V\left(1 + \frac{E_s A}{E_v V}\right) = E_v V\left(1 + \frac{N_s \epsilon_s A}{N_v \epsilon_v V}\right).$$

But $N_s = 10^{20}\,m^{-2}$ and $N_v = 10^{30}\,m^{-3}$, so that $N_s/N_v = 10^{-10}\,m$; also the ratio $(\epsilon_s/\epsilon_v) = 1.25 \approx 1$. So we have

$$E = E_v V\left(1 + 10^{-10}\frac{A}{V}\right).$$

If the second term is to have 1% of the value of the first, then $0.01 = 10^{-10}A/V$. This requires that $A/V = 10^8$. If a cube has a side a, the area is $6a^2$, and the volume is a^3, so that $A/V = 6/a$. Therefore $6/a = 10^8$, and $a = 6 \times 10^{-8}\,m = 0.06\,\mu m$. This gives us a very rough, but reasonable, estimate of the maximum size of particle for which the effect of the surface energy becomes noticeable. In practice, surface effects are significant for particles having diameters less than about $0.5\,\mu m$.

18.3 MEASUREMENT OF SURFACE TENSION

In principle, by measuring the force needed to extend the film, the wire frame shown in Fig. 18.1 could be used to measure the surface tension. In practice, other devices are more convenient. The ring-pull device (called the duNouy tensiometer) shown in Fig. 18.2 is one of the simplest of these. We can calibrate the torsion wire by adding tiny masses to the end of the beam and determining the setting of the torsion scale required to keep the beam level. To make the measurement, we place the ring on the beam and raise the liquid to be

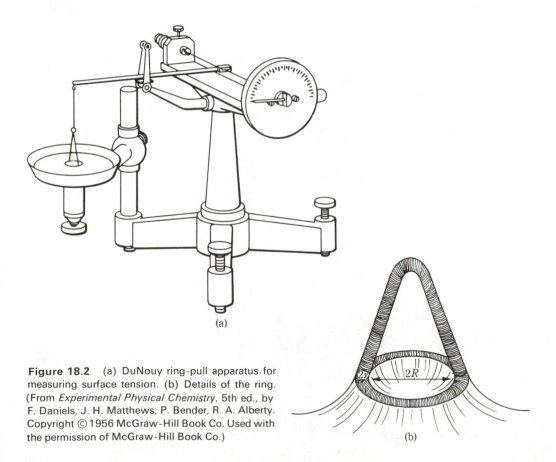

(a)

Figure 18.2 (a) DuNouy ring-pull apparatus for measuring surface tension. (b) Details of the ring. (From *Experimental Physical Chemistry*, 5th ed., by F. Daniels, J. H. Matthews, P. Bender, R. A. Alberty. Copyright © 1956 McGraw-Hill Book Co. Used with the permission of McGraw-Hill Book Co.)

$2r$ $2R$

(b)

studied on the platform until the ring is immersed and the beam is level (for a zero setting of the torsion wire). We pull the ring out slowly by turning the torsion wire and at the same time lower the height of the platform so that the beam remains level. When the ring pulls free, we take the reading on the torsion scale; using the calibration, we convert the reading into an equivalent force, F. This force is equal to the length of the wire in contact with the ring $2(2\pi R)$ times γ, the force per unit length. Thus

$$F = 2(2\pi R)\gamma. \tag{18.1}$$

The length is twice the circumference since the liquid is in contact with both the inside and the outside of the ring (Fig. 18.2b). This method requires an empirical correction factor, f, which accounts for the shape of the liquid pulled up and for the fact that the diameter of the wire itself, $2r$, is not zero. Then Eq. (18.1) can be written as

$$F = 4\pi R\gamma f. \tag{18.1a}$$

Extensive tables of f as a function of R and r are available in the literature. The method is highly accurate if we use Eq. (18.1a); Eq. (18.1) is much too crude for accurate work.

The Wilhelmy slide method is somewhat similar to the ring-pull method. A very thin plate, such as a microscope cover glass or a sheet of mica, is hung from one arm of a balance and allowed to dip in the solution (Fig. 18.3). If p is the perimeter of the slide, the downward pull on the slide due to surface tension is γp. If F and F_a are the forces acting downward when the slide is touching the surface and when it is suspended freely in air respectively, then

$$F = F_a + \gamma p \tag{18.2}$$

assuming that the depth of immersion is negligible. If the depth of immersion is not negligible, the buoyant force must be subtracted from the right-hand side of Eq. (18.2). This method is particularly convenient for measuring differences in γ (for example, in measurements on the Langmuir tray since the depth of immersion is constant).

The drop-weight method depends, as do all of the detachment methods, on the assumption that the circumference times the surface tension is the force holding two parts of a liquid column together. When this force is balanced by the mass of the lower portion, a drop breaks off (Fig. 18.4a) and

$$2\pi R\gamma = mg, \tag{18.3}$$

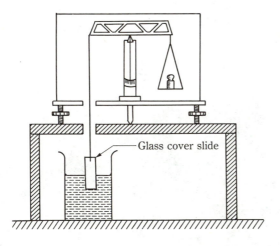

Figure 18.3   Wilhelmy method for measuring surface tension.

Glass cover slide

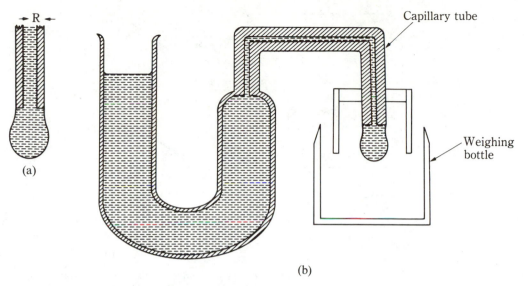

Figure 18.4 The drop-weight method for measuring surface tension. (Adapted from *Experimental Physical Chemistry*, 5th ed., by F. Daniels, J. H. Matthews, P. Bender, R. A. Alberty. Copyright © 1956 McGraw-Hill Book Co. Used with the permission of McGraw-Hill Book Co.)

in which m is the mass of the drop. By adjusting the amount of liquid in the apparatus (Fig. 18.4b) the time for formation of the drop can be controlled. The drop must form slowly if the method is to yield accurate results, but even then an empirical correction factor must be used. Tables of these correction factors are available in the literature.

Before considering other methods of measurement we need to understand the thermodynamic relations for the system.

18.4 THERMODYNAMIC FORMULATION

Consider two phases and the interface between them. We choose as the system the portions of the two phases M_1 and M_2, and the portion of the interface I enclosed by a cylindrical bounding surface B (Fig. 18.5a). Suppose that the interface is displaced slightly to a new position I'. The changes in energy are:

$$\text{For } M_1 \qquad dU_1 = TdS_1 - p_1 dV_1; \qquad (18.4)$$

$$\text{For } M_2 \qquad dU_2 = TdS_2 - p_2 dV_2; \qquad (18.5)$$

$$\text{For the surface} \qquad dU^\sigma = TdS^\sigma + \gamma dA. \qquad (18.6)$$

The last equation is written in analogy to the others, since $dW = -\gamma dA$. There is no pdV term for the surface, since the surface obviously has no volume. The total change in energy is

$$dU = dU_1 + dU_2 + dU^\sigma = Td(S_1 + S_2 + S^\sigma) - p_1 dV_1 - p_2 dV_2 + \gamma dA$$

$$= TdS - p_1 dV_1 - p_2 dV_2 + \gamma dA.$$

Since the total volume $V = V_1 + V_2$, then $dV_1 = dV - dV_2$, and

$$dU = TdS - p_1 dV + (p_1 - p_2)dV_2 + \gamma dA. \qquad (18.7)$$

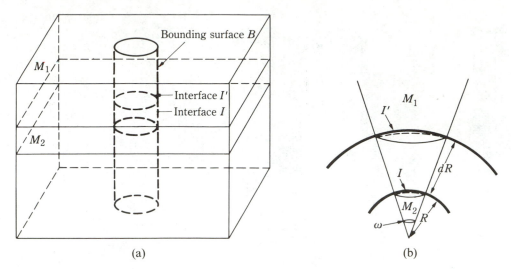

Figure 18.5   Displacement of the interface. (a) Planar interface. (b) Spherical interface.

If the entropy and volume are constant, $dS = 0$ and $dV = 0$. Then at equilibrium the energy is a minimum, $dU = 0$. This reduces the equation to

$$(p_1 - p_2)\,dV_2 + \gamma\,dA = 0. \tag{18.8}$$

If, as is shown in Fig. 18.5(a), the interface is plane and the bounding surface B is a cylinder having sides perpendicular to the interface, the area of the interface does not change, $dA = 0$. Since $dV_2 \neq 0$, Eq. (18.8) requires that $p_1 = p_2$. Consequently, the pressure is the same in two phases that are separated by a *plane* dividing surface.

If the interface is not planar, a displacement of the interface will involve a change in area. This implies an inequality of the pressures in the two phases. Suppose that the bounding surface is conical and that the interface is a spherical cap having a radius of curvature R (Fig. 18.5b). Then the area of the cap is $A = \omega R^2$, and the volume of M_2 enclosed by the cone and the cap is $V_2 = \omega R^3 / 3$, where ω is the solid angle subtended by the cap. But $dV_2 = \omega R^2\,dR$ and $dA = 2\omega R\,dR$; therefore, Eq. (18.8) becomes

$$(p_2 - p_1)\omega R^2\,dR = \gamma 2\omega R\,dR,$$

which reduces immediately to

$$p_2 = p_1 + \frac{2\gamma}{R}. \tag{18.9}$$

Equation (18.9) expresses the fundamental result that the pressure inside a phase which has a convex surface is greater than that outside. The difference in pressure in passing across a curved surface is the physical reason for capillary rise and capillary depression, which we consider in the next section. Note that in the case of a bubble the increment in pressure in moving from the outside to the inside is $4\gamma/R$, or twice the value given by Eq. (18.9), because two convex interfaces are traversed.

If the interface is not spherical but has principal radii of curvature R and R', then Eq. (18.9) would have the form

$$p_2 = p_1 + \gamma\left(\frac{1}{R} + \frac{1}{R'}\right). \tag{18.10}$$

18.5 CAPILLARY RISE AND CAPILLARY DEPRESSION

If a capillary tube is partially immersed in a liquid, the liquid stands at different levels inside and outside the tube, because the liquid–vapor interface is curved inside the tube and flat outside. By considering Eq. (18.9) and the effect of gravity on the system, we can determine the relation between the difference in liquid levels, the surface tension, and the relative densities of the two phases.

Figure 18.6 shows two phases, 1 and 2, separated by an interface that is plane for the most part but has a portion in which phase 2 is convex; the levels of the interface are different under the plane and curved portions. The densities of the two phases are ρ_1 and ρ_2. Let p_1 be the pressure in phase 1 at the plane surface separating the two phases; this position is taken as the origin ($z = 0$) of the z-axis, which is directed downward. The pressures at the other positions are as indicated in the figure: p_1' and p_2' are the pressures just inside phases 1 and 2 at the curved interface; p_1' and p_2' are related by Eq. (18.9). The condition of equilibrium is that the pressure at the depth z, which lies below both the plane and curved parts of the interface, must have the same value everywhere. Otherwise, at depth z, a flow of material would occur from one region to another. Equality of the pressures at the depth z requires that

$$p_1 + \rho_2 gz = p_2' + \rho_2 g(z - h). \tag{18.11}$$

Since $p_2' = p_1' + 2\gamma/R$, and $p_1' = p_1 + \rho_1 gh$, Eq. (18.11) reduces to

$$(\rho_2 - \rho_1)gh = \frac{2\gamma}{R}, \tag{18.12}$$

which relates the capillary depression h to the surface tension, the densities of the two phases, and the radius of curvature of the curved surface. We have assumed that the surface of phase 2, the liquid phase, is convex. In this case there is a capillary depression. If the surface of the liquid is concave, this is equivalent to R being negative, which makes the capillary depression h negative. Therefore a liquid that has a concave surface will exhibit a capillary elevation. Water rises in a glass capillary, while mercury in a glass tube is depressed.

The use of Eq. (18.12) to calculate the surface tension from the capillary depression requires knowing how the radius of curvature is related to the radius of the tube. Figure 18.7 shows the relation between the radius of curvature R, the radius of the tube r, and the contact angle θ, which is the angle within the liquid between the wall of the tube and the tangent to the liquid surface at the wall of the tube. From Fig. 18.7, we have

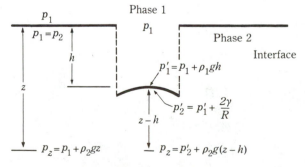

Figure 18.6   Pressures under plane and curved portions of a surface.

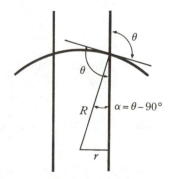

Figure 18.7   Contact angle.

$r/R = \sin \phi = \sin (\theta - 90°) = -\cos \theta$, so that $R = -r/\cos \theta$. In terms of the radius of the tube, Eq. (18.12) becomes

$$-\gamma \cos \theta = \tfrac{1}{2}(\rho_2 - \rho_1)grh.$$

Since h is the capillary depression, it is convenient to replace it by the capillary rise $-h$. This removes the negative sign and we have

$$\gamma \cos \theta = \tfrac{1}{2}(\rho_2 - \rho_1)grH. \tag{18.13}$$

In Eq. (18.13), H is the capillary rise. If $\theta < 90°$, the liquid meniscus is concave and H is positive. When $\theta > 90°$, the meniscus is convex and $\cos \theta$ and H are negative.

Liquids that wet the tube have values of θ less than 90°, while those that do not wet the tube have values greater than 90°. For making measurements we choose a tube narrow enough that $\theta = 0°$ (or 180°). This is necessary because it is difficult to establish other values of θ reproducibly.

18.6 PROPERTIES OF VERY SMALL PARTICLES

If a particle is small enough, the surface energy produces measurable effects on the observable properties of a substance. Two examples are the enhanced vapor pressure of small droplets and the increased solubility of fine particles.

18.6.1 Enhanced Vapor Pressure

Consider a liquid in equilibrium with vapor, with a plane interface between the two phases. Let the vapor pressure in this circumstance be p_o. The pressure just inside the liquid phase is also p_o, since the interface is plane, by Eq. (18.9). If, on the other hand, we suspend a small droplet of radius, r, then the pressure inside the droplet is higher than in the gas phase because of the curvature of the surface, also by Eq. (18.9). This increase in pressure increases the chemical potential by an amount $d\mu^1 = \overline{V}^1 \, dp^1$, where $\overline{V}^1$ is the molar volume of the liquid. If the vapor is to remain in equilibrium, the chemical potential of the vapor must increase by an equal amount, or

$$d\mu^g = d\mu^1.$$

Using the fundamental equation, Eq. (10.22), at constant T,

$$\overline{V}^g \, dp = \overline{V}^1 \, dp^1,$$

where p is the pressure of the vapor. Let's assume that the vapor is ideal and integrate:

$$\int_{p_1}^{p} \frac{RT}{p} \, dp = \int_{p_1}^{p_2} \overline{V}^1 \, dp^1.$$

If $\overline{V}^1$ is constant, we have

$$RT \ln \frac{p}{p_1} = V_m^1(p_2 - p_1).$$

Using Eq. (18.9) for the pressure jump across the interface, we have

$$RT \ln \frac{p}{p_1} = \overline{V}^1 \left(\frac{2\gamma}{r}\right).$$

When $r \to \infty$, the interface is planar, and $p = p_1 = p_o$. Thus we can write

$$\ln \frac{p}{p_o} = \frac{\bar{V}^1}{RT} \frac{2\gamma}{r}. \qquad (18.14)$$

If M is the molar mass and ρ the density, then $\bar{V}^1 = M/\rho$. For water at 25 °C we have $M = 0.018$ kg/mol, $\rho = 1.0 \times 10^3$ kg/m^3, $\gamma = 72 \times 10^{-3}$ N/m. Then

$$\ln \frac{p}{p_o} = \left(\frac{0.018 \text{ kg/mol}}{1.0 \times 10^3 \text{ kg/m}^3} \right)\left(\frac{2(72 \times 10^{-3} \text{ N/m})}{8.314 \text{ J K}^{-1} \text{ mol}^{-1}(298 \text{ K})r} \right) = \frac{1.0 \times 10^{-9} \text{ m}}{r}.$$

Values of p/p_o as a function of r are:

r/m	10^{-6}	10^{-7}	10^{-8}	10^{-9}
p/p_o	1.0010	1.010	1.11	2.7

A drop of radius 10^{-9} m has about ten molecules across its diameter and perhaps 100 molecules in it. This calculation indicates that if we compress water vapor in the absence of a liquid phase, we can bring it to 2.7 times its saturation pressure before it comes into equilibrium with a drop having 100 molecules in it. Thus, in the absence of foreign nuclei on which the vapor can condense, considerable supersaturation of the vapor can occur before droplets form. This effect is used in the Wilson cloud chamber in which super-saturation is induced by cooling the saturated vapor by an adiabatic expansion. Condensation does not occur until the passage of a charged particle (an α-ray or β-ray) produces gaseous ions that provide the nuclei on which droplets of water condense, leaving a visible trail to mark the path of the particle. Similarly, the fine particles of AgI, which are used in cloud seeding, provide the nuclei on which the water in a supersaturated atmosphere can condense and thus produce rain or show.

Another consequence of Eq. (18.14) is that a vapor condenses in a fine capillary at pressures below the saturation pressure if the liquid wets the capillary. In this situation, r is negative; the liquid surface is concave. Similarly, if the liquid is to evaporate from the capillary, the pressure must be below the saturation pressure.

18.6.2 Increased Solubility

The solubility of solids depends on particle size in a similar way. The solubility equilibrium condition is

$$\mu^{\text{sln}} = \mu^{\text{s}},$$

where sln = solution. If the solution is ideal, then

$$\mu^{\text{sln}} = \mu^{\text{ol}} + RT \ln x.$$

where x is the mole fraction solubility. For the solid,

$$\mu^{\text{s}} = \mu^{\text{os}} + \gamma \bar{A},$$

in which $\bar{A}$ is the area per mole of the solid. If one mole of the solid consists of n small cubes of edge a, then the molar volume of the solid, $\overline{V}^s$ is

$$\overline{V}^s = na^3 \quad \text{or} \quad n = \frac{\overline{V}^s}{a^3},$$

but the molar area, $\overline{A}$, is

$$\overline{A} = n(6a^2) = \frac{\overline{V}^s}{a^3} 6a^2 = \frac{6\overline{V}^s}{a}.$$

Using this value for $\overline{A}$, the equilibrium condition becomes

$$\mu^{ol} + RT \ln x = \mu^{os} + \overline{V}^s\left(\frac{6\gamma}{a}\right).$$

As $a \to \infty$, $x \to x_o$, the solubility of large crystals. Thus

$$\mu^{ol} + RT \ln x_o = \mu^{os}.$$

Subtracting this equation from the preceding one and dividing by RT yields

$$\ln \frac{x}{x_o} = \frac{\overline{V}^s}{RT}\left(\frac{6\gamma}{a}\right). \tag{18.15}$$

This equation differs from Eq. (18.14) only in that the factor, $(6/a)$, replaces $(2/r)$. Since the crystal may not indeed be cubical, in general the factor $(6/a)$ could be replaced by a factor (α/a) where α is a numerical factor of the order of unity, which depends on the shape of the crystal and a is the average diameter of the crystals. Just as Eq. (18.14) predicts an increased vapor pressure for fine droplets of a liquid, so Eq. (18.15) predicts an enhanced solubility for finely divided solids. Since the surface tension of some solids may be five to twenty times larger than that of common liquids, the enhanced solubility is noticeable for somewhat larger particles than those for which the enhanced vapor pressure is observable.

If a freshly precipitated sample of AgCl or $BaSO_4$ is allowed to stand for a period of time, or better yet, if it is held at a high temperature for some hours in contact with the saturated solution, we observe that the average particle size increases. The more highly soluble fine particles produce a solution that is supersaturated with respect to the solubility of the larger particles. Thus the large particles grow larger and the fine particles ultimately disappear.

VON WEIMARN'S LAW

A related effect, the von Weimarn effect, is important in crystal growth. If a high degree of supersaturation occurs before nuclei appear in the solution, then large numbers of nuclei appear at once. This produces a heavy crop of very small crystals. However, if little supersaturation occurs before nucleation, a few large crystals form. In the limiting case, we can immerse a single seed crystal in a saturated solution; then, on extremely slow cooling, no supersaturation occurs and one large crystal grows.

Von Weimarn's law states that the average size of the crystals is inversely proportional to the supersaturation ratio; that is, the ratio of the concentration at which crystallization begins to the saturation concentration at the same temperature. For example, if hot, dilute solutions of $CaCl_2$ and Na_2CO_3 are mixed, there is relatively little supersaturation before the precipitate of $CaCO_3$ forms and the precipitate consists of relatively large crystals. On the other hand, if cold, concentrated solutions of the same reagents are mixed, there is a high degree of supersaturation and a very large number of nuclei are formed. The system sets to a gel; the particles of $CaCO_3$ are colloidal in size. After standing for a period of time, these crystals grow, the gel collapses, and the particles drop to the bottom of the container. This behavior is a classic example of von Weimarn's law.

18.7 BUBBLES; SESSILE DROPS

It is possible to determine the surface tension from the maximum pressure required to blow a bubble at the end of a capillary tube immersed in a liquid. In Fig. 18.8, three stages of a bubble are shown. In the first stage the radius of curvature is very large, so that the difference in pressure across the interface is small. As the bubble grows, R decreases and the pressure in the bubble increases until the bubble is hemispherical with $R = r$, the radius of the capillary. Beyond this point, as the bubble enlarges, R becomes greater than r; the pressure drops and air rushes in. The bubble is unstable. Thus the situation in Fig. 18.8(b) represents a minimum radius and therefore a maximum bubble pressure, by Eq. (18.9). From a measurement of the maximum bubble pressure the value of γ can be obtained. If p_{max} is the maximum pressure required to blow the bubble and p_h is the pressure at the depth of the tip, h, then

$$p_{max} = p_h + \frac{2\gamma}{r}.$$

Again, for large values of r, corrections must be applied.

Since the shape of a drop sitting (sessile) on a surface that it does not wet depends on the surface tension, we can measure the surface tension by making an accurate measurement of the parameters that characterize the shape of the drop. The profile of a drop is shown in Fig. 18.9. For large drops it can be shown that

$$\gamma = \tfrac{1}{2}(\rho_2 - \rho_1)gh^2, \tag{18.16}$$

where h is the distance between the top of the drop and the "equator," the point where $dy/dx \rightarrow \infty$. The function $y = y(x)$ is the equation of the profile of the drop. Measurements

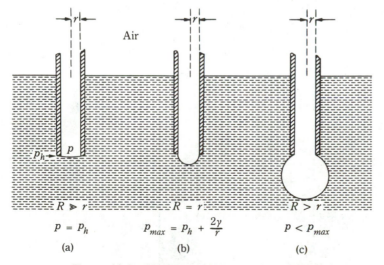

$$p = p_h \qquad\qquad p_{max} = p_h + \frac{2\gamma}{r} \qquad\qquad p < p_{max}$$

(a)   (b)   (c)

Figure 18.8 Maximum bubble-pressure method for measuring surface tension.

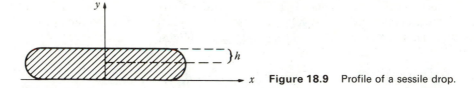

Figure 18.9   Profile of a sessile drop.

on a photograph of the drop profile yield the surface tension. The differential equation that describes $y(x)$ apparently does not have a solution in closed form. Numerical integrations and approximations of various types abound in the literature.

★ 18.8   LIQUID–LIQUID AND SOLID–LIQUID INTERFACES

The interfacial tension between two liquid phases, α and β, is designated by $\gamma^{\alpha\beta}$. Suppose that the interface has unit area; then if we pull the two phases apart we will form 1 m² of a surface of pure phase α with surface Gibbs energy, $\gamma^{\alpha v}$, and 1 m² of a surface of pure phase β with surface Gibbs energy, $\gamma^{\beta v}$ (Fig. 18.10). The increase in Gibbs energy in this transformation is

$$\Delta G = w_A^{\alpha\beta} = \gamma^{\alpha v} + \gamma^{\beta v} - \gamma^{\alpha\beta}. \tag{18.17}$$

This increase in Gibbs energy is called the work of adhesion, $w_A^{\alpha\beta}$, between the phases α and β. Note that since the pure phases α and β are in contact with the vapor phase, we have written $\gamma^{\alpha v}$ for the interfacial tension between α and the vapor phase. Similarly, $\gamma^{\beta v}$ is the interfacial tension between phase β and the equilibrium vapor phase.

If we pull apart a column of pure phase α, 2 m² of surface are formed, and

$$\Delta G = w_C^{\alpha} = 2\gamma^{\alpha v}.$$

This increase in Gibbs energy, w_C^{α}, is called the work of cohesion of α. Similarly, $w_C^{\beta} = 2\gamma^{\beta v}$. Then

$$w_A^{\alpha\beta} = \tfrac{1}{2}w_C^{\alpha} + \tfrac{1}{2}w_C^{\beta} - \gamma^{\alpha\beta}$$

or

$$\gamma^{\alpha\beta} = \tfrac{1}{2}(w_C^{\alpha} + w_C^{\beta}) - w_C^{\alpha\beta}. \tag{18.18}$$

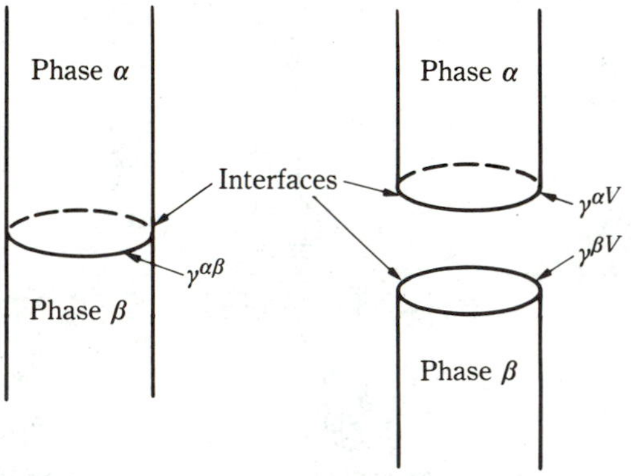

Phase α

Phase α

Interfaces

$\gamma^{\alpha V}$

$\gamma^{\beta V}$

$\gamma^{\alpha\beta}$

Phase β

Phase β

Figure 18.10   Interfacial tension.

Table 18.2
Interfacial tension between water (α) and various liquids (β) at 20 °C

Liquid	$\gamma^{\alpha\beta}/(10^{-3}\ \text{N/m})$	Liquid	$\gamma^{\alpha\beta}/(10^{-3}\ \text{N/m})$
Hg	375		
$n\text{-}C_6H_{14}$	51.1	$C_2H_5OC_2H_5$	10.7
$n\text{-}C_7H_{16}$	50.2	$n\text{-}C_8H_{17}OH$	8.5
$n\text{-}C_8H_{18}$	50.8	$C_6H_{13}COOH$	7.0
C_6H_6	35.0	$CH_3COOC_2H_5$	6.8
C_6H_5CHO	15.5	$n\text{-}C_4H_9OH$	1.8

As the Gibbs energy of adhesion between the phases α and β increases, $\gamma^{\alpha\beta}$ decreases. When $\gamma^{\alpha\beta} = 0$, there is no resistance to the extension of the interface between phases α and β; the two liquids mix spontaneously. In this case, the work of adhesion is the average of the work of cohesion of the two liquids.

$$w_A^{\alpha\beta} = \tfrac{1}{2}(w_C^\alpha + w_C^\beta) \tag{18.19}$$

Table 18.2 shows values of the interfacial tensions between water and various liquids. Note that the interfacial tensions between water and those liquids that are close to being completely miscible in water (for example, n-butyl alcohol) have very low values.

The same argument holds for the interfacial tension between a solid and a liquid. Thus, in analogy to Eq. (18.17), we have

$$w_A^{sl} = \gamma^{sv} + \gamma^{lv} - \gamma^{sl}. \tag{18.20}$$

Although γ^{sv} and γ^{sl} are not measurable, it is possible to obtain a relation between $\gamma^{sv} - \gamma^{sl}$, the contact angle, θ, and γ^{lv}. To do this, we consider the liquid drop resting on a solid surface as in Fig. 18.11.

If we deform the liquid surface slightly so that the area of the solid–liquid interface increases by dA_{sl}, then the Gibbs energy change is

$$dG = \gamma^{sl}\,dA_{sl} + \gamma^{sv}\,dA_{sv} + \gamma^{lv}\,dA_{lv}.$$

From Fig. 18.11 we have

$$dA_{sv} = -dA_{sl} \quad \text{and} \quad dA_{lv} = dA_{sl}\cos\theta;$$

then

$$dG = (\gamma^{sl} - \gamma^{sv} + \gamma^{lv}\cos\theta)\,dA_{sl}. \tag{18.21}$$

It can be shown that it is not necessary to allow for a change in θ since this would contribute

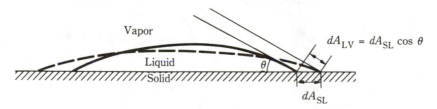

Figure 18.11   Spreading a liquid on a solid.

only a second-order term. Then we can define σ^{ls}, the spreading coefficient for the liquid on the solid, as

$$\sigma^{ls} = -\frac{\partial G}{\partial A_{sl}}. \tag{18.22}$$

Thus, if σ^{ls} is positive, $(\partial G/\partial A_{sl})$ is negative, and the Gibbs energy will decrease as the solid–liquid interface enlarges; the liquid will spread spontaneously. If $\sigma^{ls} = 0$, the configuration is stable (in equilibrium) with respect to variations in the area of the solid–liquid interface. If σ^{ls} is negative, the liquid will contract and decrease A_{sl} spontaneously. Combining Eqs. (18.21) and (18.22) we get

$$\sigma^{ls} = \gamma^{sv} - \gamma^{sl} - \gamma^{lv} \cos\theta. \tag{18.23}$$

If the liquid is to be stable against variations in its area, $\sigma^{ls} = 0$, and we have

$$\gamma^{sv} - \gamma^{sl} = \gamma^{lv} \cos\theta \tag{18.24}$$

This is combined with Eq. (18.20) to eliminate $\gamma^{sv} - \gamma^{sl}$ and obtain

$$w_A^{sl} = \gamma^{lv}(1 + \cos\theta) \tag{18.25}$$

If $\theta = 0$, then $w_A^{sl} = 2\gamma^{lv}$; that is, the work of adhesion between solid and liquid is equal to the work of cohesion of the liquid. Thus the liquid can spread indefinitely over the surface, since energetically the system is indifferent to whether the liquid is in contact with itself or with the solid. On the other hand, if $\theta = 180°$, $\cos\theta = -1$, and $w_A^{sl} = 0$. No Gibbs energy expenditure is required to separate the solid and the liquid. The liquid does not wet the solid and does not spread on it. The spreading coefficient for one liquid on another is defined in the same way as for a liquid on a solid, Eq. (18.23), except that $\cos\theta = 1$. Thus

$$\sigma^{\alpha\beta} = \gamma^{\beta v} - \gamma^{\alpha\beta} - \gamma^{\alpha v}.$$

Note that as a liquid spreads on a surface the interfacial tensions change, with the result that the spreading coefficient changes. For example, benzene spreads on a pure water surface, $\sigma^{BW} \approx 9 \times 10^{-3}$ N/m initially. When the water is saturated with benzene and the benzene saturated with water $(\sigma^{BW})_{sat} \approx -2 \times 10^{-3}$ N/m and any additional benzene collects as a lens on the surface.

18.9 SURFACE TENSION AND ADSORPTION

Consider the system of the type shown in Fig. 18.5(a): two phases with a plane interface between them. Since the interface is plane, we have $p_1 = p_2 = p$ and the Gibbs energy becomes a convenient function. If we have a multicomponent system the chemical potential of each component must have the same value in each phase and at the interface. The variation in total Gibbs energy of the system is given by

$$dG = -SdT + Vdp + \gamma dA + \sum_i \mu_i dn_i, \tag{18.26}$$

in which $\gamma\,dA$ is the increase in Gibbs energy of the system associated with a variation in area. The Gibbs energy increments for the two phases are given by

$$dG_1 = -S_1 dT + V_1 dp + \sum_i \mu_i dn_i^{(1)}$$

and

$$dG_2 = -S_2 dT + V_2 dp + \sum_i \mu_i dn_i^{(2)},$$

in which $n_i^{(1)}$ and $n_i^{(2)}$ are the number of moles of i in phases 1 and 2, respectively. Subtracting these two equations from the equation for the change in total Gibbs energy yields

$$d(G - G_1 - G_2) = -(S - S_1 - S_2)dT + (V - V_1 - V_2)dp + \gamma dA$$
$$+ \sum_i \mu_i d(n_i - n_i^{(1)} - n_i^{(2)}).$$

If the presence of the interface produced no physical effect, then the difference between the total Gibbs energy, G, and the sum of the Gibbs energies of the bulk phases, $G_1 + G_2$, would be zero. Since the presence of the interface does produce physical effects, we ascribe the difference $G - (G_1 + G_2)$ to the presence of the surface and define it as the surface Gibbs energy, G^σ. Then,

$$G^\sigma = G - G_1 - G_2, \qquad S^\sigma = S - S_1 - S_2, \qquad n_i^\sigma = n_i - n_i^{(1)} - n_i^{(2)}.$$

Note that the presence of the interface cannot affect the geometric requirement that $V = V_1 + V_2$. The differential equation becomes

$$dG^\sigma = -S^\sigma dT + \gamma dA + \sum_i \mu_i dn_i^\sigma. \tag{18.27}$$

At constant temperature, pressure, and composition, let the bounding surface, the cylinder B in Fig. 18.5(a), increase in radius from zero to some finite value. Then the interfacial area increases from zero to A and the n_i^σ increase from zero to n_i^σ, while γ and all the μ_i are constants. Then Eq. (18.27) integrates to

$$\int_0^{G^\sigma} dG^\sigma = \gamma \int_0^A dA + \sum_i \mu_i \int_0^{n_i^\sigma} dn_i^\sigma$$

$$G^\sigma = \gamma A + \sum_i \mu_i n_i^\sigma. \tag{18.28}$$

This equation is similar to the usual additivity rule for Gibbs energy, but contains the additional term, γA. Dividing by A and introducing the Gibbs energy per unit area, $g^\sigma = G^\sigma/A$, and the surface excesses, Γ_i, defined by

$$\Gamma_i = \frac{n_i^\sigma}{A}, \tag{18.29}$$

yields

$$g^\sigma = \gamma + \sum_i \mu_i \Gamma_i, \tag{18.30}$$

which is similar to the additivity rule for bulk phases but contains the additional term, γ.

Differentiating Eq. (18.28) yields

$$dG^\sigma = \gamma dA + A d\gamma + \sum_i \mu_i dn_i^\sigma + \sum_i n_i^\sigma d\mu_i. \tag{18.31}$$

By subtracting Eq. (18.27) from Eq. (18.31) we obtain an analogue of the Gibbs–Duhem equation,

$$0 = S^\sigma dT + A d\gamma + \sum_i n_i^\sigma d\mu_i.$$

Division by A, and introduction of the entropy per unit area $s^\sigma = S^\sigma/A$, and the surface excess, Γ_i, reduces this relation to

$$dy = -s^\sigma dT - \sum_i \Gamma_i d\mu_i. \tag{18.32}$$

At constant temperature this becomes

$$dy = -(\Gamma_1 d\mu_1 + \Gamma_2 d\mu_2 + \cdots). \tag{18.33}$$

This equation relates the change in surface tension, γ, to change in the μ_i which, at constant T and p, are determined by the variation in composition.

As we will show below, in a single-component system it is always possible to choose the position of the interfacial surface so that the surface excess, $\Gamma_1 = 0$. Then, Eqs. (18.30) and (18.32) become

$$g^\sigma = \gamma \quad \text{and} \quad s^\sigma = -\left(\frac{\partial \gamma}{\partial T}\right)_A. \tag{18.34a, b}$$

Since $g^\sigma = u^\sigma - Ts^\sigma$, we obtain for u^σ, the surface energy per unit area,

$$u^\sigma = \gamma - T\left(\frac{\partial \gamma}{\partial T}\right)_A. \tag{18.35}$$

To obtain a clearer meaning for the surface excesses, consider a column having a constant cross-sectional area, A. Phase 1 fills the space between height $z = 0$ and z_0, and has a volume, $V_1 = Az_0$. Phase 2 fills from z_0 to Z, and has a volume $V_2 = A(Z - z_0)$. The molar concentration, c_i, of species i is shown (by the solid curve) as a function of height, z, in Fig. 18.12. The interface between the two phases is located approximately at z_0. In the region near z_0 the concentration changes smoothly from $c_i^{(1)}$, the value in the bulk of phase 1, to $c_i^{(2)}$, the value in the bulk of phase 2; the width of this region has been enormously exaggerated in Fig. 18.12. To calculate the actual number of moles of species i in the system, we multiply c_i by the volume element, $dV = A\, dz$ and integrate over the entire length of the system from zero to Z:

$$n_i = \int_0^Z c_i A\, dz = A \int_0^Z c_i dz. \tag{18.36}$$

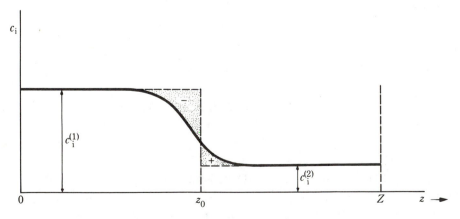

Figure 18.12   Concentration as a function of position.

The concentration c_i is the function of z shown in Fig. 18.12. It is clear that the value of n_i calculated in this way is the correct value and does not depend in the least on the position chosen for the reference surface, z_0.

Now if we define the total number of moles of i in phase 1, $n_i^{(1)}$ and the total number in phase 2, $n_i^{(2)}$, in terms of the *bulk* concentrations, $c_i^{(1)}$ and $c_i^{(2)}$, we obtain

$$n_i^{(1)} = c_i^{(1)}V_1 = c_i^{(1)}Az_0 = A \int_0^{z_0} c_i^{(1)}dz;$$

$$n_i^{(2)} = c_i^{(2)}V_2 = c_i^{(2)}A(Z - z_0) = A \int_{z_0}^{Z} c_i^{(2)}dz.$$

Using these equations, Eq. (18.36), and the definition of n_i^σ, we find that

$$n_i^\sigma = n_i - n_i^{(1)} - n_i^{(2)} = A \left[\int_0^Z c_i\,dz - \int_0^{z_0} c_i^{(1)}dz - \int_{z_0}^{Z} c_i^{(2)}dz \right].$$

Since $\Gamma_i = n_i^\sigma/A$ and

$$\int_0^Z c_i\,dz = \int_0^{z_0} c_i\,dz + \int_{z_0}^{Z} c_i\,dz,$$

we have

$$\Gamma_i = \int_0^{z_0} (c_i - c_i^{(1)})dz + \int_{z_0}^{Z} (c_i - c_i^{(2)})dz. \tag{18.37}$$

The first of these integrals is the negative of the shaded area to the left of the line z_0 in Fig. 18.12, while the second integral is the shaded area to the right of z_0. It is clear from the manner in which this figure is drawn that Γ_i, the sum of the two integrals, is negative. However, it is also clear that this value of Γ_i depends critically on the position chosen for the reference plane, z_0. By moving z_0 slightly to the left, Γ_i would have a positive value; moving z_0 to the right would decrease the value to zero; moving z_0 farther to the right would make Γ_i negative. We may vary the numerical values of the surface excesses arbitrarily by adjusting the position of the reference surface z_0. Suppose we adjust the position of the reference surface in such a way that the surface excess of one of the components is made equal to zero. This component is usually chosen as the solvent and labeled component 1. Then, by this adjustment,

$$\Gamma_1 = 0.$$

However, in general this location for the reference surface will not yield zero values for the surface excesses for the other components. Hence, Eq. (18.33) for a two-component system takes the form

$$-d\gamma = \Gamma_2 d\mu_2. \tag{18.38}$$

In an ideal dilute solution, $\mu_2 = \mu_2^\circ + RT \ln c_2$, and $d\mu_2 = RT (dc_2/c_2)$, so that

$$-\left(\frac{\partial \gamma}{\partial c_2}\right)_{T,p} = \Gamma_2 \frac{RT}{c_2}$$

or

$$\Gamma_2 = -\frac{1}{RT}\left(\frac{\partial \gamma}{\partial \ln c_2}\right)_{T,p}. \tag{18.39}$$

This is the Gibbs adsorption isotherm. If the surface tension of the solution decreases with increase in concentration of solute, then $(\partial\gamma/\partial c_2)$ is negative and Γ_2 is positive; there is an excess of solute at the interface. This is the usual situation with surface active materials; if they accumulate at the interface, they lower the surface tension. The Langmuir surface films described in the following section are a classic example of this.

18.10 SURFACE FILMS

Certain insoluble substances will spread on the surface of a liquid such as water until they form a monomolecular layer. Long-chain fatty acids, stearic acid and oleic acid, are classical examples. The —COOH group at one end of the molecule is strongly attracted to the water, while the long hydrocarbon chain is hydrophobic.

A shallow tray, the Langmuir tray, is filled to the brim with water (Fig. 18.13). The film is spread in the area between the float and the barrier by adding a drop of a dilute solution of stearic acid in benzene. The benzene evaporates leaving the stearic acid on the surface. The float is attached rigidly to a superstructure that allows any lateral force, indicated by the arrow, to be measured by means of a torsion wire.

By moving the barrier, we can vary the area confining the film. If the area is reduced, the force on the barrier is practically zero until a critical area is reached, whereupon the force rises rapidly (Fig. 18.14a). The extrapolated value of the critical area is 0.205 nm² per molecule. This is the area at which the film becomes close packed. In this state the molecules in the film have the polar heads attached to the surface and the hydrocarbon tails extended upward. The cross-sectional area of the molecule is therefore 0.205 nm².

The force F is a consequence of the lower surface tension on the film-covered surface as compared with that of the clean surface. If the length of the barrier is l, and it moves a

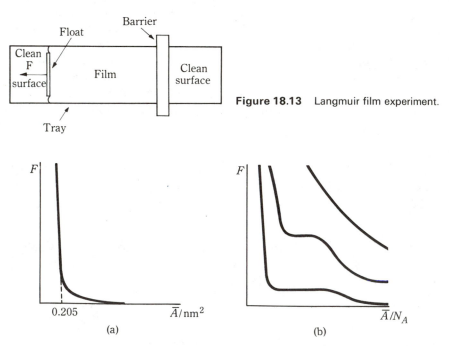

Figure 18.13 Langmuir film experiment.

Figure 18.14 Force–area curves. (a) High surface pressure. (b) Low surface pressure.

distance dx, then the area of the film decreases by $l\,dx$ and that of the clean surface behind the barrier increases by $l\,dx$. The energy increase is $\gamma_0 l\,dx - \gamma l\,dx$, where γ_0 and γ are the surface tensions of the water and the film-covered surface. This energy is supplied by the barrier moving a distance dx against a force Fl, so that $Fl\,dx = (\gamma_0 - \gamma)l\,dx$, or

$$F = \gamma_0 - \gamma. \tag{18.40}$$

Note that F is a force per unit length of the barrier, which is equal to that on the float. From curve 1 in Fig. 18.14(a) and Eq. (18.40), we see that the surface tension of the film-covered surface is not very different from that of the clean surface until the film becomes close packed.

Figure 18.14(b) shows the behavior of the surface pressure at very high areas and very low surface pressures F. The curves look very much like the isotherms of a real gas. In fact, the uppermost curve follows a law that is much like the ideal gas law,

$$FA = n_2^{\sigma} RT, \tag{18.41}$$

where A is the area and n_2^{σ} is the number of moles of the substance in the surface film. Equation (18.41) is easily derived from kinetic theory by supposing that the "gas" is two dimensional. The plateaus in Fig. 18.14(b) represent a phenomenon that is analogous to liquefaction.

We can obtain Eq. (18.41) by writing the Gibbs adsorption isotherm in the form

$$d\gamma = -RT\Gamma_2 \frac{dc_2}{c_2}$$

and considering the difference in surface tension in comparing the film-covered surface, γ, with the clean surface, γ_0. At low concentrations, the surface excess is proportional to the bulk concentration, so that $\Gamma_2 = Kc_2$. Using this in the Gibbs adsorption isotherm, we obtain $d\gamma = -RTK\,dc_2$; integrating, we have $\gamma - \gamma_0 = -RTKc_2$, or

$$\gamma - \gamma_0 = -RT\Gamma_2.$$

Since $F = \gamma_0 - \gamma$, we have

$$F = RT\Gamma_2.$$

But $\Gamma_2 = n_2^{\sigma}/A$; inserting this value, we get

$$FA = n_2^{\sigma} RT,$$

which is the result in Eq. (18.41). If the area per mole is $\bar{A}$, then

$$F\bar{A} = RT \tag{18.42}$$

If a glass slide is dipped through the close-packed film, as it is withdrawn the polar heads of the stearic acid molecules attach themselves to the glass. Pushing the slide back in allows the hydrocarbon tails on the water surface to join with the tails on the glass slide. Figure 18.15 shows the arrangement of molecules on the surface and on the slide. By repeated dipping, a layer of stearic acid containing a known number of molecular layers can be built up on the slide. After about twenty dippings the layer is thick enough to show interference colors, from which the thickness of the layer is calculated. Knowing the number of molecular layers on the slide from the number of dippings, we can calculate the length of the molecule. This method of Langmuir and Blodgett is an incredibly simple method— and was one of the first methods—for the direct measurement of the size of molecules. The results agree well with those obtained from x-ray diffraction.

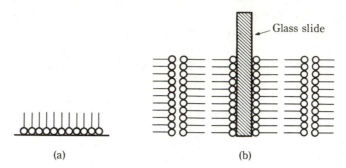

Figure 18.15   Surface films. (a) Monolayer of stearic acid on a surface. (b) Multilayer obtained by dipping a glass slide through a monolayer.

The study of surface films of the Langmuir type covers an extremely diverse group of phenomena. Measurements of film viscosity, diffusion on the surface, diffusion through the surface film, surface potentials, the spreading of monolayers, and chemical reactions in monolayers are just a few of the topics that have been studied. One interesting application is the use of long-chain alcohols to retard evaporation from reservoirs and thus conserve water. The phrase "to pour oil on the troubled waters" reflects the ability of a mono-molecular film to damp out ripples, apparently by distributing the force of the wind more evenly. There are also several different types of surface films; only the simplest was discussed in this section.

18.11 ADSORPTION ON SOLIDS

If a finely divided solid is stirred into a dilute solution of a dye, we observe that the depth of color in the solution is much decreased. If a finely divided solid is exposed to a gas at low pressure, the pressure decreases noticeably. In these situations the dye or the gas is *adsorbed* on the surface. The magnitude of the effect depends on the temperature, the nature of the adsorbed substance (the adsorbate), the nature and state of subdivision of the adsorbent (the finely divided solid), and the concentration of the dye or pressure of the gas.

The Freundlich isotherm is one of the first equations proposed to relate the amount of material adsorbed to the concentration of the material in the solution:

$$m = kc^{1/n}, \tag{18.43}$$

where m is the mass adsorbed per unit mass of adsorbent, c is the concentration, and k and n are constants. By measuring m as a function of c and plotting $\log_{10} m$ versus $\log_{10} c$, the values of n and k can be determined from the slope and intercept of the line. The Freundlich isotherm fails if the concentration (or pressure) of the adsorbate is too high.

We can represent the process of adsorption by a chemical equation. If the adsorbate is a gas, then we write the equilibrium

$$A(g) + S \rightleftharpoons AS,$$

where A is the gaseous adsorbate, S is a vacant site on the surface, and AS represents an adsorbed molecule of A or an occupied site on the surface. The equilibrium constant can be written

$$K = \frac{x_{AS}}{x_S p}, \tag{18.44}$$

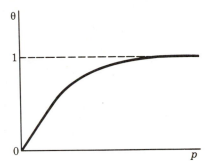

Figure 18.16   Langmuir isotherm.

where x_{AS} is the mole fraction of occupied sites on the surface, x_S is the mole fraction of vacant sites on the surface, and p is the pressure of the gas. It is more common to use θ for x_{AS}. Then $x_S = (1 - \theta)$ and the equation can be written

$$\frac{\theta}{1 - \theta} = Kp, \tag{18.45}$$

which is the Langmuir isotherm; K is the equilibrium constant for the adsorption. Solving for θ, we obtain

$$\theta = \frac{Kp}{1 + Kp}. \tag{18.46}$$

If we are speaking of adsorption of a substance from solution, Eq. (18.46) is correct if p is replaced by the molar concentration c.

The amount of the substance adsorbed, m, will be proportional to θ for a specified adsorbent, so $m = b\theta$, where b is a constant. Then

$$m = \frac{bKp}{1 + Kp}, \tag{18.47}$$

which, if inverted, yields

$$\frac{1}{m} = \frac{1}{b} + \frac{1}{bKp}. \tag{18.48}$$

By plotting $1/m$ against $1/p$, the constants K and b can be determined from the slope and intercept of the line. Knowing K, we can calculate the fraction of the surface covered from Eq. (18.46).

The Langmuir isotherm, in the form of Eq. (18.46), is generally more successful in interpreting the data than is the Freundlich isotherm if only a monolayer is formed. A plot of θ versus p is shown in Fig. 18.16. At low pressures, $Kp \ll 1$ and $\theta = Kp$, so that θ increases linearly with pressure. At high pressures, $Kp \gg 1$, so that $\theta \approx 1$. The surface is nearly covered with a monomolecular layer at high pressures, so that change in pressure produces little change in the amount adsorbed.

18.12 PHYSICAL AND CHEMISORPTION

If the adsorbate and the surface of the adsorbent interact only by van der Waals forces, then we speak of physical adsorption, or van der Waals adsorption. The adsorbed molecules are weakly bound to the surface and heats of adsorption are low (a few kilojoules at most)

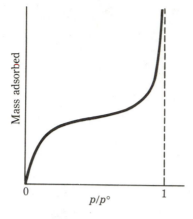

Figure 18.17   Multilayer adsorption.

and are comparable to the heat of vaporization of the adsorbate. Increase in temperature markedly decreases the amount of adsorption.

Since the van der Waals forces are the same as those that produce liquefaction, adsorption does not occur at temperatures that are much above the critical temperature of the gaseous adsorbate. Also, if the pressure of the gas has values near the equilibrium vapor pressure of the liquid adsorbate, then a more extensive adsorption multilayer adsorption— occurs. A plot of the amount of material adsorbed versus $p/p°$, where $p°$ is the vapor pressure of the liquid, is shown in Fig. 18.17. Near $p/p° = 1$ more and more of the gas is adsorbed; this large increase in adsorption is a preliminary to outright liquefaction of the gas, which occurs at $p°$ in the absence of the solid.

If the adsorbed molecules react chemically with the surface, the phenomenon is called *chemisorption*. Since chemical bonds are broken and formed in the process of chemisorption, the heat of adsorption has the same range of values as for chemical reactions: from a few kilojoules to as high as 400 kJ. Chemisorption does not go beyond the formation of a monolayer on the surface. For this reason an isotherm of the Langmuir type, which predicts a monolayer and nothing more, is well suited for interpreting the data. The Langmuir adsorption isotherm predicts a heat of adsorption that is independent of θ, the fraction of the surface covered at equilibrium. For many systems the heat of adsorption decreases with increasing coverage of the surface. If the heat adsorption depends on the coverage, then we must use an isotherm more elaborate than the Langmuir isotherm.

The difference between physical and chemisorption is typified by the behavior of nitrogen on iron. At the temperature of liquid nitrogen, $-190\ °C$, nitrogen is adsorbed physically on iron as nitrogen molecules, N_2. The amount of N_2 adsorbed decreases rapidly as the temperature rises. At room temperature iron does not adsorb nitrogen at all. At high temperatures, $\sim 500\ °C$, nitrogen is chemisorbed on the iron surface as nitrogen atoms.

18.13   THE BRUNAUER, EMMET, AND TELLER (BET) ISOTHERM

Brunauer, Emmet, and Teller have worked out a model for multilayer adsorption. They assumed that the first step in the adsorption is

$$A(g) + S \; \Longrightarrow \; AS \qquad K_1 = \frac{\theta_1}{\theta_v p} \qquad (18.49)$$

where K_1 is the equilibrium constant, θ_1 is the fraction of the surface sites covered by a single molecule, and θ_v is the fraction of vacant sites. If nothing else occurred, this would simply be the Langmuir isotherm (Section 18.11).

Next they assumed that additional molecules sit on top of one another to form a variety of multilayers. They interpreted the process as a sequence of chemical reactions, each with an appropriate equilibrium constant:

$$A(g) + AS \rightleftharpoons A_2S, \qquad K_2 = \frac{\theta_2}{\theta_1 p};$$

$$A(g) + A_2S \rightleftharpoons A_3S, \qquad K_3 = \frac{\theta_3}{\theta_2 p};$$

$$A(g) + A_{n-1}S \rightleftharpoons A_nS, \qquad K_n = \frac{\theta_n}{\theta_{n-1} p};$$

where the symbol A_3S indicates a surface site that has a stack of three A molecules piled up on it. The θ_i is the fraction of sites on which the stack of A molecules is i layers deep. The interaction between the first A molecule and the surface site is unique, depending on the particular nature of the A molecule and the surface. However, when the second A molecule sits on the first A molecule, the interaction cannot be very different from the interaction of two A molecules in the liquid; the same is true when the third sits on the second. All of these processes except the first can be regarded as being essentially equivalent to liquefaction, and so they should have the same equilibrium constant, K. Thus the BET treatment assumes that

$$K_2 = K_3 = K_4 = \cdots = K_n = K \tag{18.50}$$

where K is the equilibrium constant for the reaction $A(g) \rightleftharpoons A(\text{liquid})$. Then

$$K = \frac{1}{p^\circ}, \tag{18.51}$$

where p° is the equilibrium vapor pressure of the liquid.

We can use the equilibrium conditions to calculate the values of the various θ_i. We have

$$\theta_2 = \theta_1 Kp, \qquad \theta_3 = \theta_2 Kp, \qquad \theta_4 = \theta_3 Kp \cdots. \tag{18.52}$$

Combining the first two we have, $\theta_3 = \theta_1(Kp)^2$. Repeating the operation, we find

$$\theta_i = \theta_1(Kp)^{i-1}. \tag{18.53}$$

The sum of all these fractions must be equal to unity:

$$1 = \theta_v + \sum_{i=1} \theta_i = \theta_v + \sum_i \theta_1(Kp)^{i-1}.$$

In the second writing we replaced θ_i by its equal from Eq. (18.53). If we temporarily set $Kp = x$, this becomes

$$1 = \theta_v + \theta_1(1 + x + x^2 + x^3 + \cdots).$$

If we now assume that the process can go on indefinitely, then $n \to \infty$, and the series is simply the expansion of $1/(1 - x) = 1 + x + x^2 + \cdots$. Thus

$$1 = \theta_v + \frac{\theta_1}{1 - x}. \tag{18.54}$$

Using the equilibrium condition for the first adsorption, we find $\theta_v = \theta_1/K_1p$. We define a new constant, $c = K_1/K$; then

$$\theta_v = \frac{\theta_1}{cx}$$

and Eq. (18.54) becomes

$$1 = \theta_1\left(\frac{1}{cx} + \frac{1}{1-x}\right),$$

$$\theta_1 = \frac{cx(1-x)}{1+(c-1)x}. \tag{18.55}$$

Let N be the total number of molecules adsorbed per unit mass of adsorbent and c_s be the total number of surface sites per unit mass. Then $c_s\theta_1$ is the number of sites carrying one molecule, $c_s\theta_2$ is the number carrying two molecules, and so on. Then

$$N = c_s(1\theta_1 + 2\theta_2 + 3\theta_3 + \cdots) = c_s\sum_i i\theta_i.$$

From Eq. (18.53) we have $\theta_i = \theta_1 x^{i-1}$; this brings N to the form

$$N = c_s\theta_1\sum_{i=1} ix^{i-1} = c_s\theta_1(1 + 2x + 3x^2 + \cdots).$$

We recognize this series as the derivative of the earlier one:

$$1 + 2x + 3x^2 + \cdots = \frac{d}{dx}(1 + x + x^2 + x^3 + \cdots)$$

$$= \frac{d}{dx}\left(\frac{1}{1-x}\right) = \frac{1}{(1-x)^2}.$$

Using this result in the expression for N, we obtain

$$N = \frac{c_s\theta_1}{(1-x)^2}.$$

If the entire surface were covered with a monolayer, then N_m molecules would be adsorbed; $N_m = c_s$ and

$$N = \frac{N_m\theta_1}{(1-x)^2}.$$

Using the value for θ_1 from Eq. (18.55), this becomes

$$N = \frac{N_m cx}{(1-x)[1+(c-1)x]}. \tag{18.56}$$

The amount adsorbed is usually reported as the volume of the gas adsorbed, measured at STP. The volume is, of course, proportional to N so we have $N/N_m = v/v_m$, or

$$v = \frac{v_m cx}{(1-x)[1+(c-1)x]}. \tag{18.57}$$

Recalling that $x = Kp$ and that $K = 1/p°$, we have finally the BET isotherm:

$$v = \frac{v_m cp}{(p° - p)[1 + (c-1)(p/p°)]}. \tag{18.58}$$

The volume, v, is measured as a function of p. From the data we can obtain the value of v_m and c. Note that when $p = p°$, the equation has a singularity and $v \to \infty$. This accounts for the steep rise of the isotherm (Fig. 18.17) as the pressure approaches $p°$.

To obtain the constants c and v_m we multiply both sides of Eq. (18.58) by $(p° - p)/p$:

$$\frac{v(p° - p)}{p} = \frac{v_m c}{1 + (c - 1)(p/p°)}.$$

Next we take the reciprocal of both sides:

$$\frac{p}{v(p° - p)} = \frac{1}{v_m c} + \left(\frac{c - 1}{v_m c}\right)\left(\frac{p}{p°}\right). \tag{18.59}$$

The combination of measured quantities on the left is plotted against p. The result in many instances is a straight line. From the intercept, $(1/v_m c)$, and the slope, $(c - 1)/v_m c p°$, we can calculate values of v_m and of c. The reasonable values obtained confirm the validity of the approach.

From the value of v_m at STP, we can calculate N_m.

$$N_m = N_A \frac{v_m}{0.022414 \text{ m}^3/\text{mol}}. \tag{18.60}$$

Since N_m is the number of molecules required to cover a unit mass with a monolayer, then if we know the area covered by one molecule, a, we can calculate the area of unit mass of material:

$$\text{Area/unit mass} = N_m a. \tag{18.61}$$

This method is a useful way to determine the surface area of a finely divided solid.

If we write the equilibrium constants, K_1 and K, in terms of the standard differences in Gibbs energy for the transformations, then

$$K_1 = e^{-\Delta G_1°/RT} \quad \text{and} \quad K = e^{-\Delta G_{\text{liq}}°/RT}, \tag{18.62}$$

where $\Delta G_1°$ is the standard Gibbs energy of adsorption of the first layer and $\Delta G_{\text{liq}}°$ is the standard Gibbs energy of liquefaction. Dividing the first of Eqs. (18.62) by the second, we obtain c.

$$c = \frac{K_1}{K} = e^{-(\Delta G_1° - \Delta G_{\text{liq}}°)/RT}. \tag{18.63}$$

Using the relations,

$$\Delta G_1° = \Delta H_1° - T\,\Delta S_1° \quad \text{and} \quad \Delta G_{\text{liq}}° = \Delta H_{\text{liq}}° - T\,\Delta S_{\text{liq}}°,$$

and assuming that $\Delta S_1° \approx \Delta S_{\text{liq}}°$ (that is, that the loss in entropy is the same regardless of which layer the molecule sits in), Eq. (18.63) becomes

$$c = e^{-(\Delta H_1° - \Delta H_{\text{liq}}°)/RT}. \tag{18.64}$$

Note that the heat of liquefaction, $\Delta H_{\text{liq}}°$, is the negative of the heat of vaporization, $\Delta H_{\text{vap}}°$, so that we have $\Delta H_{\text{liq}}° = -\Delta H_{\text{vap}}°$ and

$$c = e^{-(\Delta H_1° + \Delta H_{\text{vap}}°)RT}.$$

Taking logarithms and rearranging,

$$\Delta H_1° = -\Delta H_{\text{vap}}° - RT \ln c.$$

Since we know the value of $\Delta H_{\text{vap}}^\circ$ of the adsorbate, the value of ΔH_1° can be calculated from the measured value of c. In all cases, it is found that $c > 1$, which implies that $\Delta H_1^\circ < \Delta H_{\text{liq}}^\circ$. The adsorption in the first layer is more exothermic than liquefaction.

The measurement of surface areas and ΔH_1° has increased our knowledge of surface structure enormously and is particularly valuable in the study of catalysts. One important point to note is that the actual area of any solid surface is substantially greater than its apparent geometric area. Even a mirror-smooth surface has hills and valleys on the atomic scale; the actual area is perhaps 2 to 3 times the apparent area. For finely divided powders or porous spongy material the ratio is often much higher: 10 to 1000 times in some instances.

18.14 ELECTRICAL PHENOMENA AT INTERFACES; THE DOUBLE LAYER

If two phases of different chemical composition are in contact, an electric potential difference develops between them. This potential difference is accompanied by a charge separation, one side of the interface being positively charged and the other being negatively charged.

For simplicity we will assume that one phase is a metal and the other is an electrolytic solution (Fig. 18.18a). Suppose that the metal is positively charged and the electrolytic solution has a matching negative charge. Then several charge distributions corresponding to different potential fields are possible, as shown in Fig. 18.18. The metal is in the region $x \leq 0$, and the electrolytic solution is in the region $x \geq 0$. The electric potential on the vertical axis is the value relative to that in the solution. The first possibility was proposed by Helmholtz: that the matching negative charge is located in a plane a short distance, δ, from the metal surface. Fig. 18.18(b) shows the variation of the potential in the solution as a function of x. This double layer, composed of charges at a fixed distance, is called the Helmholtz double layer. The second possibility, proposed by Gouy and Chapman, is that the matching negative charge is distributed in a diffuse way throughout the solution (much like the diffuse atmosphere around an ion in solution). The potential variation for this situation is shown in Fig. 18.18(c). This diffuse layer is called the Gouy layer, or Gouy–Chapman layer.

In concentrated solutions, $c \geq 1 \text{ mol/dm}^3$, the Helmholtz model is reasonably successful; in more dilute solutions, neither model is adequate. Stern proposed a combination of the fixed and diffuse layers. At the distance δ there is a fixed layer of negative charge insufficient to balance the positive charge on the metal. Beyond the distance δ, a diffuse layer contains the remainder of the negative charge (Fig. 18.18d). The fixed layer can also carry more than enough negative charge to balance the positive charge on the metal. When that happens, the diffuse layer will be positively charged; the potential variation is shown in Fig. 18.18(e). Either of these composite layers is called a Stern double layer. Stern's theory also includes the possibility of specific adsorption of anions or cations on the surface. If the metal were negatively charged, four additional possibilities analogous to these could be realized (Fig. 18.18f, g, h, and i).

In an elegant and successful model, Grahame distinguished between two planes of ions. Nearest the surface is the plane at the distance of closest approach of the centers of chemisorbed anions to the metal surface; this is called the inner-Helmholtz plane. Somewhat beyond this plane is the outer-Helmholtz plane, which is at the distance of closest approach of the centers of hydrated cations. The diffuse layer begins at the outer Helmholtz plane. This model, shown in Fig. 18.19, has been used very successfully in interpreting the phenomena associated with the double layer.

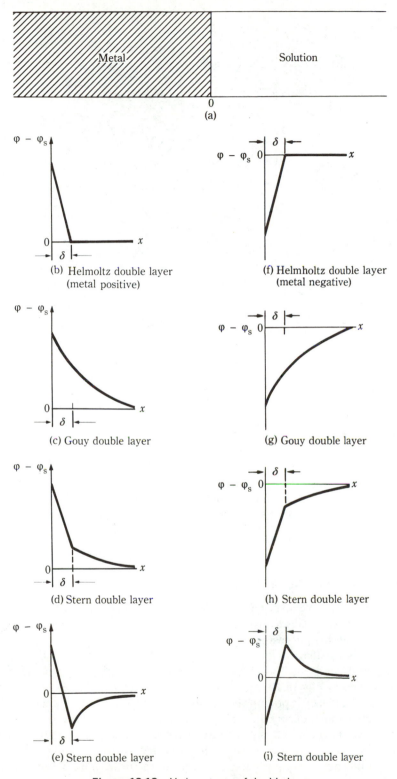

Figure 18.18 Various types of double layer.

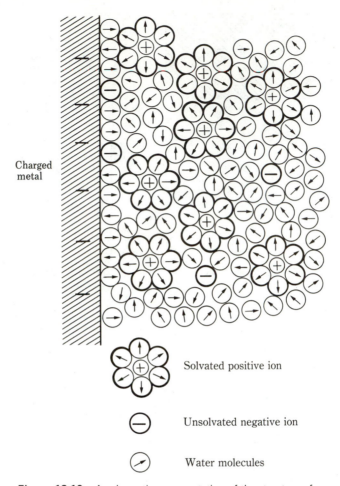

Charged
metal

Solvated positive ion

Unsolvated negative ion

Water molecules

Figure 18.19   A schematic representation of the structure of an
electrified interface. The small positive ions tend to be solvated
while the larger negative ions are usually unsolvated. (From
J. O'M. Bockris and A. K. V. Reddy, *Modern Electrochemistry*, vol.
1. New York: Plenum, 1970.)

18.15 ELECTROKINETIC EFFECTS

The existence of the double layer has four electrokinetic effects as consequences: electro-
osmosis, streaming potential, electro-osmotic counterpressure, and the streaming current.
Two other effects, electrophoresis and the sedimentation potential (Dorn effect) are also
consequences of the existence of the double layer. All of these effects depend on the fact that
part of the double layer is only loosely attached to the solid surface and therefore is mobile.
Consider the device in Fig. 18.20, which has a porous quartz disc fixed in position and is
filled with water. If an electric potential is applied between the electrodes, a flow of water to
the cathode compartment occurs. In the case of quartz and water, the diffuse (mobile) part
of the double layer in the liquid is positively charged. This positive charge moves to the
negative electrode and the water flows with it (*electro-osmosis*). Conversely, if water is
forced through fine pores of a plug, it carries the charge from one side of the plug to the

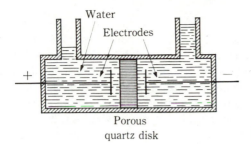

Water
Electrodes
+ —
Porous
quartz disk

Figure 18.20 Electro-osmosis.

other, and a potential difference, the *streaming potential*, develops between the electrodes.

Very finely divided particles suspended in a liquid carry an electrical charge which is equivalent to the charge on the particle itself plus the charge on the fixed portion of the double layer. If an electrical field is applied to such a suspension, the particles move in the field in the direction determined by the charge on the particle (*electrophoresis*). The diffuse part of the double layer, since it is mobile, has the opposite sign and is attracted to the other electrode. Conversely, if a suspension of particles is allowed to settle, they carry their charge toward the bottom of the vessel and leave the charge on the diffuse layer in the upper portion of the vessel. A potential difference, the *sedimentation potential*, develops between the top and bottom of the container.

The magnitude of all of the electrokinetic effects depends on how much of the electrical charge resides in the mobile part of the double layer. The potential at the surface of shear, the dividing line between the fixed and mobile portions of the double layer, is called the *zeta potential* (ζ potential). The charge in the mobile portion of the double layer depends on the ζ potential and therefore the magnitude of all of the electrokinetic effects depends on ζ. It is commonly assumed that the entire diffuse portion of the double layer is mobile; if this is so, then the ζ potential is the value of ϕ at the position $x = \delta$ in Fig. 18.19. It is more likely that part of the diffuse layer is fixed so that the value of ζ corresponds to the value of ϕ at a distance of perhaps two or three times δ. In any case, ζ has the same sign and same general magnitude as the value of ϕ at $x = \delta$.

18.16 COLLOIDS

A colloidal dispersion has traditionally been defined as a suspension of small particles in a continuous medium. Because of their ability to scatter light and the apparent lack of osmotic pressure, these particles were recognized to be much larger than simple small molecules such as water, alcohol, or benzene and simple salts like NaCl. It was assumed that they were aggregates of many small molecules, held together in a kind of amorphous state quite different from the usual crystalline state of these substances. Today we recognize that many of these "aggregates" are in fact single molecules that have a very high molar mass. The size limits are difficult to specify but if the dispersed particles are between 1 μm and 1 nm, we might say that the system is a colloidal dispersion. The anthracene molecule, which is 1.091 nm across the wide dimension, is one example of the specification problems. It is not clear that we would describe all anthracene solutions as colloids. However, a sphere with this same diameter could contain an aggregate of about 27 water molecules. It might be useful to call that aggregate a colloidal particle.

There are two classical subdivisions of colloidal systems: (1) *lyophilic*, or *solvent-loving* colloids (also called *gels*) and (2) *lyophobic*, or *solvent-fearing* colloids (also called *sols*).

18.16.1 Lyophilic Colloids

The lyophilic colloids are invariably polymeric molecules of one sort or another, so that the solution consists of a dispersion of single molecules. The stability of the lyophilic colloid is a consequence of the strong, favorable solvent–solute interactions. Typical lyophilic systems would be proteins (especially gelatin) or starch in water, rubber in benzene, and cellulose nitrate or cellulose acetate in acetone. The process of solution may be rather slow. The first additions of solvent are slowly absorbed by the solid, which swells as a result (this stage is called *imbibition*). Further addition of solvent together with mechanical kneading (as in the case of rubber) slowly distributes the solvent and solute uniformly. In the case of ordinary gelatin, the solution process is aided considerably by raising the temperature. As the solution cools, the long and twisted protein molecules become entangled in a network with much open space between the molecules. The presence of the protein induces some structure in the water, which is physically trapped in the interstices of the network. The result is a gel. The addition of gross amounts of salts to a hydrophilic gel will ultimately precipitate the protein. However, this is a consequence of competition between the protein and the salt for the solvent, water. Lithium salts are particularly effective because of the large amount of water than can be bound by the lithium ion. The charge of the ion is not a primary determinant of its effectiveness as a precipitant. We will deal in detail with properties such as light scattering, sedimentation, precipitation, and the osmotic properties of lyophilic colloids in Chapter 35 where we discuss polymeric molecules.

18.16.2 Lyophobic Colloids

The lyophobic colloids are invariably substances that are highly insoluble in the dispersing medium. The lyophobic colloids are usually aggregates of small molecules (or in cases where a molecule is not defined, such as AgI, they consist of a rather large number of units of formula). The lyophobic dispersion can be prepared by grinding the solid with the dispersing medium in a "colloid mill," a ball mill, which over a prolonged period of time reduces the substance to a size in the colloidal range, $< 1 \ \mu m$. More often the lyophobic dispersion, the sol, is produced by precipitation under special conditions in which a large number of nuclei are produced while limiting their growth. Typical chemical reactions for the production of sols are:

Hydrolysis $FeCl_3 + 3H_2O \longrightarrow Fe(OH)_3(colloid) + 3H^+ + 3Cl^-.$

Pouring a solution of $FeCl_3$ into a beaker of boiling water produces a deep red sol of $Fe(OH)_3$.

Metathesis $AgNO_3 + KI \longrightarrow AgI(colloid) + K^+ + NO_3^-$

Reduction $SO_2 + 2H_2S \longrightarrow 2S(colloid) + 2H_2O$

$2AuCl_3 + 3H_2O + 3CH_2O \longrightarrow 2Au(colloid) + 3HCOOH + 6H^+ + 6Cl^-$

One classic method for producing metal sols is to pass an arc between electrodes of the desired metal immersed in water (Bredig arc). The vaporized metal forms aggregates of colloidal size.

Since the sols are extremely sensitive to the presence of electrolytes, preparative reactions that do not produce electrolytes are better than those that do. To avoid precipitation of the sol by the electrolyte, the sol can be purified by dialysis. The sol is placed in a collodion bag and the bag is immersed in a stream of flowing water. The small ions can diffuse through the collodion and be washed away, while the larger colloid particles are retained in the bag. The porosity of the collodion bag can be adjusted over a fairly wide range by varying the preparation method. A bare trace of electrolyte is needed to stabilize the colloid since sols derive their stability from the presence of the electrical double layer on the particle. If AgI is washed too clean, the sol precipitates. Addition of a trace of either $AgNO_3$ to provide a layer of adsorbed Ag^+ ion or KI to provide a layer of adsorbed I^- ions will often resuspend the colloid; this process is called *peptization*.

18.16.3   Electrical Double Layer and Stability of Lyophobic Colloids

The stability of a lyophobic colloid is a consequence of the electrical double layer at the surface of the colloidal particles. For example, if two particles of an insoluble material do not have a double layer, they can come close enough that the attractice van der Waals force can pull them together. In contrast to this behavior suppose that the particles do have a double layer, as shown in Fig. 18.21. The overall effect is that the particles repel one another at large distances of separation since, as two particles approach, the distance between like charges (on the average) is less than that between unlike charges. This repulsion prevents close approach of the particles and stabilizes the colloid. Curve (a) in Fig. 18.22 shows the potential energy due to the van der Waals attractive force as a function of the distance of separation between the two particles; curve (b) shows the repulsion energy. The combined

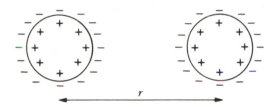

Figure 18.21   Double layer on two particles.

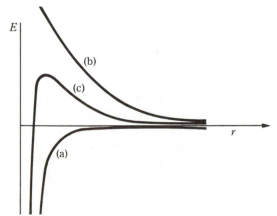

Figure 18.22   Energy of interaction of colloidal particles as a function of distance of separation.

curve for double-layer repulsion and van der Waals attraction is shown by curve (c). So long as curve (c) has a maximum, the colloid will have some stability.

The addition of electrolytes to the sol suppresses the diffuse double layer and reduces the zeta potential. This drastically decreases the electrostatic repulsion between the particles and precipitates the colloid. The colloid is particularly sensitive to ions of the opposite sign. A positively charged sol such as ferric oxide is precipitated by negative ions such as Cl^- and SO_4^{2-}. These ions are incorporated into the fixed portion of the double layer, reducing the net charge on the particle. This lowers the ζ potential, which reduces the repulsion between the particles. Similarly, a negative sol will be destabilized by positive ions. The higher the charge on the ion the more effective it is in coagulating the colloid (the Schulz–Hardy rule). The minimum concentration of electrolyte needed to produce rapid coagulation is roughly in the ratio of 1:10:500 for triply, doubly, and singly charged ions. The ion having the same charge as the colloidal particle does not have much effect on the coagulation, except for its assistance in suppressing the diffuse part of the double layer. Since the double layer contains very few ions, only a small concentration of electrolyte is needed to suppress the double layer and precipitate the colloid.

18.17 COLLOIDAL ELECTROLYTES; SOAPS AND DETERGENTS

The metal salt of a long-chain fatty acid is a soap, the most common example being sodium stearate, $C_{17}H_{35}COO^-Na^+$. At low concentrations the solution of sodium stearate consists of individual sodium and stearate ions dispersed throughout the solution in the same way as in any ordinary salt solution. At a rather definite concentration, the critical micelle concentration, the stearate ions aggregate into clumps, called *micelles* (Fig. 18.23). The micelle contains perhaps 50 to 100 individual stearate ions. The micelle is roughly spherical and the hydrocarbon chains are in the interior, leaving the polar $—COO^-$ groups on the outer surface. It is the outer surface that is in contact with the water, and the polar groups on the outer surface stabilize the micelle in the water solution. The micelle is the size of a colloidal particle; since it is charged, it is a colloidal ion. The micelle binds a fairly large number of positive ions to its surface as counter ions which reduces its charge considerably.

The formation of micelles results in a sharp drop in the electrical conductivity per mole of the electrolyte. Suppose 100 sodium and 100 stearate ions were present individually. If the stearate ions aggregate into a micelle and the micelle binds $70\,Na^+$ as counter ions, then there will be $30\,Na^+$ ions and 1 micellar ion having a charge of -30 units; a total of 31 ions. The same quantity of sodium stearate would produce 200 ions as individuals but only 31 ions if the micelle is formed. This reduction in the number of ions sharply reduces the conductivity. The formation of micelles also reduces the osmotic pressure of the solution. The average molar mass, and thus an estimate of the average number of stearate ions in the micelle, can be obtained from the osmotic pressure.

By incorporating molecules of hydrocarbon into the hydrocarbon interior of the micelle, the soap solution can act as a solvent for hydrocarbons. The action of soap as a cleanser depends in part on this ability to hold grease in suspension.

The detergents are similar in structure to the soaps. The typical anionic detergent is an alkyl sulfonate, $ROSO_3^-Na^+$. For good detergent action, R should have at least 16 carbon atoms. Cationic detergents are often quaternary ammonium salts, in which one alkyl group is a long chain; $(CH_3)_3RN^+Cl^-$ is a typical example if R has between 12 and 18 carbon atoms.

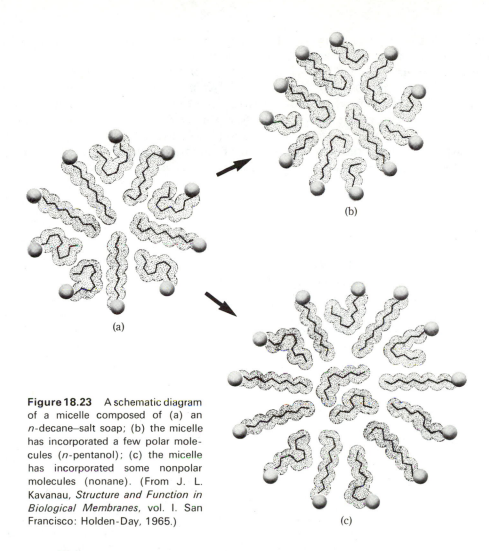

Figure 18.23 A schematic diagram of a micelle composed of (a) an *n*-decane–salt soap; (b) the micelle has incorporated a few polar molecules (*n*-pentanol); (c) the micelle has incorporated some nonpolar molecules (nonane). (From J. L. Kavanau, *Structure and Function in Biological Membranes*, vol. I. San Francisco: Holden-Day, 1965.)

18.18 EMULSIONS AND FOAMS

Water and oil can be whipped or beaten mechanically to produce a suspension of finely divided oil droplets in water, an *emulsion*. Mayonnaise is a common household example. It is also possible to produce an emulsion consisting of water droplets in a continuous oil phase (for example, butter). In either type of emulsion, the large interfacial tension between water and oil coupled with the very large interfacial area implies that the emulsion has a high Gibbs energy compared with the separated phases. To supply this Gibbs energy an equal amount of mechanical work must be expended in the whipping or beating.

The addition of a surface active agent, such as a soap or detergent, or any molecule with a polar end and a large hydrocarbon end, to the separated system of oil and water lowers the interfacial tension markedly. In this way the Gibbs-energy requirement for formation of the emulsion can be lowered. Such additives are called emulsifying agents. The interfacial tension is lowered because of the adsorption of the surface active agent at the interface with the polar end in the water and the hydrocarbon end in the oil. The interfacial tension decreases just as it does when a monomolecular film of stearic acid is spread on a water surface in the Langmuir experiment.

Foams consist of a large number of tiny gas bubbles in a continuous liquid phase. A thin film of liquid separates any two gas bubbles. As in the case of emulsions, the surface energy is high and foaming agents are added to lower the interfacial tension between liquid and gas. The foaming agents are the same type of surface active agents as the emulsifying agents. Since the bubbles in the foam are fragile, other additives are needed to give the foam an elasticity to stabilize the foam against mechanical shock. Long-chain alcohols (or if a soap is the foaming agent, the undissociated acid) can serve as foam stabilizers.

QUESTIONS

18.1 Suggest a Gibbs-energy argument for why a liquid drop is spherical.

18.2 What happens to the surface tension at the gas–liquid critical point?

18.3 Why should the Langmuir adsorption isotherm be more reliable, at high gas pressures, for chemisorption than for physical adsorption?

18.4 Colloidal particles of the same charge immersed in an electrolyte solution attract each other by van der Waals forces and repel each other by Debye screened interactions (see Eq. 16.70). Why does the ease of coagulation increase rapidly with increasing solution ionic strength?

18.5 Describe the roles of both the inner and outer portions of the micelle in the action of soap.

PROBLEMS

18.1 One cm^3 of water is broken into droplets having a radius of 10^{-5} cm. If the surface tension of water is $72.75(10^{-3})$ N/m at 20 °C, calculate the Gibbs energy of the fine droplets relative to that of the water.

18.2 An emulsion of toluene in water can be prepared by pouring a toluene–alcohol solution into water. The alcohol diffuses into the water and leaves the toluene behind in small droplets. If 10 g of a solution that is 15% ethanol and 85% toluene by mass is poured into 10 g of water, an emulsion forms spontaneously. The interfacial tension between the suspended toluene droplets and the water–alcohol mixture is 0.036 N/m, the average diameter of the droplets is 10^{-4} cm, and the density of toluene is 0.87 g/cm^3. Calculate the increase in Gibbs energy associated with the formation of the droplets. Compare this increase with the Gibbs energy of mixing of the alcohol and water at 25 °C.

18.3 As a vapor condenses to liquid and a droplet grows in size, the Gibbs energy of the droplet varies with its size. For a bulk liquid, $G_{vap} - G_{liq} = \Delta H_{vap} - T\Delta S_{vap}$; if ΔH_{vap} and ΔS_{vap} are independent of temperature, then $\Delta S_{vap} = \Delta H_{vap}/T_b$, where T_b is the boiling point. If we take $G_{vap} = 0$, then $G_{liq} = -\Delta H_{vap}(1 - T/T_b)$. If G_{liq} and ΔH_{vap} refer to the values per unit volume of liquid, then the total Gibbs energy of the volume V of bulk liquid is $G' = VG_{liq} = -V\Delta H_{vap}(1 - T/T_b)$. If we speak of a fine droplet rather than the bulk liquid then a term γA, where A is the area of the droplet, must be added to this expression $G' = -V\Delta H_{vap}(1 - T/T_b) + \gamma A$.

a) Show that for a spherical droplet, the Gibbs energy of the droplet is positive when the drop is small, then passes through a maximum, and then decreases rapidly as the radius increases. If $T < T_b$, at what value of the radius r does $G' = 0$? Show that at larger values of r, G' is negative. Keeping in mind that we chose $G_{vap} = 0$, what radius must the droplet have before it can grow spontaneously by condensation from the vapor?

b) At 25 °C for water $\gamma = 71.97 \times 10^{-3}$ J/m^2, $\Delta H_{vap} = 2443.3$ J/g, and the density is 0.9970 g/cm^3. What radius must a water droplet have before it grows spontaneously?

18.4 In the duNouy tensiometer, the force required to pull up a ring of fine wire lying in the surface of the liquid is measured. If the diameter of the ring is 1.0 cm and the force needed to pull the ring up (with the surface of the liquid attached to the inner *and* outer periphery of the ring) is 6.77 mN, what is the surface tension of the liquid?

18.5 At 25 °C, the density of mercury is 13.53 g/cm^3 and $\gamma = 0.484$ N/m. What would be the capillary depression of mercury in a glass tube of 1 mm inner diameter if we assume that $\theta = 180°$? Neglect the density of air.

18.6 In a glass tube, water exhibits a capillary rise of 2.0 cm at 20 °C. If $\rho = 0.9982$ g/cm^3 and $\gamma = 72.75 \times 10^{-3}$ N/m, calculate the diameter of the tube ($\theta = 0°$).

18.7 If a 30-metre-tall tree were supplied by sap that is drawn up solely by capillary elevation, what would the radius of the channels have to be? Assume that the density of the sap is 1.0 g/cm^3, $\theta = 0°$, and $\gamma = 73 \times 10^{-3}$ N/m. Neglect the density of air. (Note: Sap rises mainly by osmotic pressure.)

18.8 A microscope-cover glass with a perimeter of 2.100 cm is used in the Wilhelmy apparatus. A 10.00 mL sample of water is placed in the container and the beam is balanced. The water is removed and is replaced by 10.00 mL samples of 5.00%, 10.00%, and 20.00% acetone (mass %) in the same container. To restore the balance in each case, the following masses had to be removed: 35.27 mg, 49.40 mg, and 66.11 mg. Calculate the surface tension of each solution if the surface tension of water is 71.97×10^{-3} N/m. The effect of density differences can be neglected.

18.9 Consider a fine-capillary tube of radius $= 0.0500$ cm, which just dips into a liquid with a surface tension equal to 0.0720 N/m. What excess pressure is required to blow a bubble with a radius equal to that of the capillary? Assume that the depth of immersion is negligible.

18.10 An excess pressure of 364 Pa is required to produce a hemispherical bubble at the end of a capillary tube of 0.300 mm diameter immersed in acetone. Calculate γ.

18.11 Consider two soap bubbles, one with a radius $r_1 = 1.00$ cm and the other with a radius $r_2 = 2.00$ cm. What is the excess pressure inside each bubble if $\gamma = 0.030$ N/m for the soap solution? If the bubbles collide and stick together with a film between them, what is the radius of curvature of this film? On which side is the center of curvature? Keep in mind that in going from the outside to the inside of a soap bubble, *two* interfaces are passed.

18.12 Two bubbles of different radii are connected by a hollow tube. What happens?

18.13 At 20 °C the interfacial tension between water and benzene is 35 mN/m. If $\gamma = 28.85$ mN/m for benzene and 72.75 mN/m for water (assuming that $\theta = 0$), calculate

a) the work of adhesion between water and benzene;
b) the work of cohesion for benzene and for water;
c) the spreading coefficient for benzene on water.

18.14 If, at 20 °C, for pure CH_2I_2 $\gamma = 50.76$ mJ/m^2 and for pure water $\gamma = 72.75$ mJ/m^2, and the interfacial tension is 45.9 mJ/m^2, calculate

a) the spreading coefficient for CH_2I_2 on water;
b) the work of adhesion between CH_2I_2 and H_2O.

18.15 Assuming that crystals form as tiny cubes having edge length δ, calculate the freezing point of ice consisting of small crystals relative to the freezing point of infinitely large crystals; $T_0 = 273.15$ K. Assume that the interfacial tension is 25 mN/m; $\Delta H_{fus}^\circ = 6.0$ kJ/mol; $\overline{V}^s = 20$ cm^3/mol. Calculate for $\delta = 10~\mu$m, $1~\mu$m, $0.1~\mu$m, $0.01~\mu$m, and $0.001~\mu$m.

18.16 Calculate the solubility of crystals of $BaSO_4$ having edge lengths of $1~\mu$m, $0.1~\mu$m, and $0.01~\mu$m, relative to the solubility of coarse crystals at 20 °C. Assume $\gamma = 500$ mJ/m^2; $\rho = 4.50$ g/cm^3.

18.17 At 20 °C the density of CCl_4 is 1.59 g/cm^3, $\gamma = 26.95$ mN/m. The vapor pressure is 11.50 kPa. Calculate the vapor pressure of droplets with radii of $0.1~\mu$m, $0.01~\mu$m, and $0.001~\mu$m.

18.18 For water the surface tension depends on temperature according to the rule

$$\gamma = \gamma_o\left(1 - \frac{t}{368}\right)^{1.2}$$

where t is the Celsius temperature and $\gamma_o = 75.5 \times 10^{-3}$ J/m². Calculate the value of g^σ, s^σ, and u^σ at 30-degree intervals from 0 °C to 368 °C. Plot these values as a function of t. (*Note:* The critical temperature of water is 374 °C.)

18.19 Stearic acid, $C_{17}H_{35}COOH$, has a density of 0.85 g/cm³. The molecule occupies an area of 0.205 nm² in a close-packed surface film. Calculate the length of the molecule.

18.20 Hexadecanol, $C_{16}H_{33}OH$, has been used to produce monomolecular films on reservoirs to retard the evaporation of water. If the cross-sectional area of the alcohol in the close-packed layer is 0.20 nm², how many grams of the alcohol are required to cover a 10-acre ($\approx$ 40,000 m²) lake?

18.21 The number of cubic centimetres of methane, measured at STP, adsorbed on 1 g of charcoal at 0 °C and several different pressures is

p/mmHg	100	200	300	400
cm³ adsorbed	9.75	14.5	18.2	21.4

Plot the data using the Freundlich isotherm and determine the constants k and $1/n$.

18.22 a) The adsorption of ethyl chloride on a sample of charcoal at 0 °C and at several different pressures is

p/mmHg	20	50	100	200	300
grams adsorbed	3.0	3.8	4.3	4.7	4.8

Using the Langmuir isotherm, determine the fraction of the surface covered at each pressure.
   b) If the area of the ethyl chloride molecule is 0.260 nm², what is the area of the charcoal?

18.23 The adsorption of butane on an NiO powder was measured at 0 °C; the volumes of butane at STP adsorbed per gram of NiO are:

p/kPa	7.543	11.852	16.448	20.260	22.959
v/(cm³/g)	16.46	20.72	24.38	27.13	29.08

   a) Using the BET isotherm, calculate the volume at STP adsorbed per gram when the powder is covered by a monolayer; $p^\circ = 103.24$ kPa.
   b) If the cross-sectional area of a single butane molecule is 44.6×10^{-20} m², what is the area per gram of the powder?
   c) Calculate θ_1, θ_2, θ_3, and θ_v at 10 kPa and 20 kPa.
   d) Using the Langmuir isotherm, calculate θ at 10 kPa and 20 kPa and estimate the surface area. Compare with the area in (b).

18.24 By considering the derivation of the Langmuir isotherm on the basis of a chemical reaction between the gas and the surface, show that if a diatomic gas is adsorbed as atoms on the surface, then $\theta = K^{1/2}p^{1/2}/(1 + K^{1/2}p^{1/2})$.

18.25 a) At 30 °C, the surface tensions of acetic acid solutions in water are

wt % acid	2.475	5.001	10.01	30.09	49.96	69.91
$\gamma/(10^{-3} \text{ N/m})$	64.40	60.10	54.60	43.60	38.40	34.30

Plot γ versus ln m and determine the surface excess of acetic acid using the Gibbs adsorption isotherm. (*Note:* We can use the molality, m, in the isotherm instead of c_2, the molarity.)
b) At 25 °C, the surface tensions of propionic acid solutions in water are

wt % acid	1.91	5.84	9.80	21.70
$\gamma/(10^{-3} \text{ N/m})$	60.00	49.00	44.00	36.00

Calculate the surface excess of propionic acid.

18.26 Consider the two systems, 10 cm³ of liquid water and 10 cm³ of liquid mercury, each in a separate 200 mL beaker. For water on glass, $\theta = 0°$; for mercury on glass $\theta = 180°$. If we turned off the gravity field, how would each system behave?

19

The Structure of Matter

19.1 INTRODUCTION

The notion that matter consists of discrete, indivisible particles (atoms) is quite ancient. The pre-Christian writers Lucretius and Democritus constructed elaborate speculative natural philosophies based on the supposition of the atomicity of matter. In the absence of experimental evidence to support them, these early atomic theories bore no fruit.

Modern atomic theory is based on the quantitative observation of nature; its first proposal by Dalton came after a period in which quantitative measurement had risen to importance in scientific investigation. In contrast to the ancient theories, modern atomic theory has been exceedingly fruitful.

To put modern theory in some perspective, it is worthwhile to trace some of its development, at least in bare outline. We shall not attempt anything that could be dignified by the name of history, but only call attention to some major mileposts and courses of thought.

19.2 NINETEENTH CENTURY

In the period 1775–1780, Lavoisier established chemistry as a quantitative science by proving that in the course of a chemical reaction the total mass is unaltered. The conservation of mass in chemical reactions proved ultimately to be a death blow to the phlogiston theory. Shortly after Lavoisier, Proust and Dalton proposed the laws of definite and multiple proportions. In 1803 Dalton proposed his atomic theory. Matter was made up of very small particles called atoms. Every kind of atom has a definite weight. The atoms of different elements have different weights. Compounds are formed by atoms which combine in definite ratios of (usually small) whole numbers. This theory could give a satisfying interpretation of the quantitative data available at the time.

Gay-Lussac's experiments on gas volumes in 1808 led to the law of combining volumes. The volumes of the reactant gases are related to those of the product gases by simple ratios of whole numbers. Gay-Lussac suggested that equal volumes of different gases contained the same number of atoms. This suggestion was rejected. At that time attempts to construct a table of atomic weights were mired in contradictions, since it was supposed that the "atom" of the simplest compound of two elements was formed by combination of two single atoms of the elements; the formation of water and of ammonia would be written

$$H + O \longrightarrow OH \quad \text{and} \quad H + N \longrightarrow NH.$$

This would require a ratio of atomic weights $N/O = 7/12$. No compound of nitrogen and oxygen exhibiting such a ratio of combining weights was known.

By distinguishing between an atom, the smallest particle that can take part in a chemical change, and a molecule, the smallest particle that can exist permanently, Avogadro (1811) removed the contradictions in the weight ratios by supposing that the molecules of certain elementary gases were diatomic; for example, H_2, N_2, O_2, Cl_2. He also proposed what is now Avogadro's law: under the same conditions of temperature and pressure equal volumes of all gases contain the same number of molecules. These ideas were ignored and forgotten until 1858 when Cannizaro used them with the law of Dulong and Petit (c. 1816) to establish the first consistent table of atomic weights. Chaos reigned in the realm of chemical formulas in the fifty-five year interval between the announcement of the atomic theory and the construction of a table of atomic weights substantially the same as the modern one.

A parallel development began in 1832 when Faraday announced the laws of electrolysis. First law: the weight of material formed at an electrode is proportional to the quantity of electricity passed through the electrolyte. Second law: the weights of different materials formed at an electrode by the same quantity of electricity are in the same ratio as their chemical equivalent weights. It was not until 1881 when Helmholtz wrote that acceptance of the atomic hypothesis and Faraday's laws compelled the conclusion that both positive and negative electricity were divided into definite elementary portions, "atoms" of electricity; a conclusion that today seems obvious waited fifty years to be drawn. From 1880 onward, intensive study of electrical conduction in gases led to the discovery of the free electron (J. J. Thomson, 1897), positive rays, and x-rays (Roentgen, 1895). The direct measurement of the charge on the electron was made by Millikan, 1913.

Another parallel development began with Count Rumford's experiment (c. 1798) of rubbing a blunt boring tool against a solid plate. (He was supposed to be boring cannon at the time; no doubt his assistants thought him a bit odd.) The tool and plate were immersed in water and the water finally boiled. This suggested to Rumford that "heat" was not a fluid, "caloric," but a form of motion. Later experiments, particularly the careful work of Joule in the 1840s, culminated in the recognition of the first law of thermodynamics in 1847. Independently, Helmholtz in 1847 proposed the law of conservation of energy. The second law of thermodynamics, founded on the work of Carnot in 1824, was formulated by Kelvin and Clausius in the 1850s.

In the late 1850s the kinetic theory of gases was intensively developed and met with phenomenal success. Kinetic theory is based on the atomic hypothesis and depends importantly on Rumford's idea of the relation between "heat" and motion.

The chemical achievements, particularly in synthetic and analytical chemistry, in the 19th century are staggering in number; we mention only a few. The growth of organic chemistry after Wöhler's synthesis of urea, 1824. The stereochemical studies of van't Hoff, LeBel, and Pasteur. The chemical proof of the tetrahedral arrangement of the bonds about the carbon atom; Kekule's structure for benzene. Werner's work on the stereochemistry of inorganic complexes. The work of Stas on exact atomic weights. The Arrhenius theory of electrolytic solutions. Gibbs's treatise on heterogeneous equilibria and the phase rule. And in fitting conclusion, the observation of chemical periodicity: Döbereiner's triads, Newland's octaves, climaxed by the periodic law of Mendeleev and Meyer, 1869–1870.

In the preceding chronicles, those developments that supported the atomic idea were stressed. On the other hand, Maxwell's development of electromagnetic theory, an undulatory theory, is an important link in the chain. Another fact of great consequence is that in the latter part of the 19th century a great amount of experimental work was devoted to the study of spectra.

Today it is difficult to imagine the complacency of the physicist of 1890. Classical physics was a house in order: mechanics, thermodynamics, kinetic theory, optics, and electromagnetic theory were the main foundations—an imposing display. By choosing tools from the appropriate discipline any problem could be solved. Of course, there were one or two problems that were giving some trouble, but everyone was confident that these would soon yield under the usual attack. There were two parts in this house of physics: the corpuscular and the undulatory, or the domain of the particle and the domain of the wave. Matter was corpuscular, light was undulatory, and that was that. The joint between matter and light did not seem very smooth.

19.3 THE EARTHQUAKE

It is difficult to describe what happened next because everything happened so quickly. Within thirty-five years classical physics was shaken to the very foot of the cellar stairs. When the dust settled, the main foundations remained, not too much the worse for wear. But entirely new areas of physics were opened. Again only the barest mention of these events must suffice for the moment: the discovery of the photoelectric effect by Hertz in 1887. The discovery of x-rays by Roentgen in 1895. The discovery of radioactivity by Becquerel in 1896. The discovery of the electron by J. J. Thomson in 1897. The quantum hypothesis in blackbody radiation by Planck in 1900. The quantum hypothesis in the photoelectric effect by Einstein in 1905. Thomson's model of the atom in 1907. The scattering experiment with α-particles by Geiger, Marsden, and Rutherford in 1909. The nuclear model of the atom of Rutherford in 1911. Quantitative confirmation of Rutherford's calculations on scattering by Geiger and Marsden in 1913. The quantum hypothesis applied to the atom, the Bohr model of the atom, in 1913. Another development in the first decade which does not concern us directly here is the Einstein theory of relativity.

The year 1913 marks a major climax in the history of science. The application of Planck's quantum hypothesis to blackbody radiation, and later by Einstein to the photoelectric effect, had met with disbelief and in some quarters even with scorn. Bohr's application to the theory of the hydrogen atom compelled belief and worked a revolution in thought. In the following ten years this new knowledge was quickly assimilated and applied with spectacular success to the interpretation of spectra and chemical periodicity.

A new series of discoveries was made in the third decade of the 20th century. The theoretical prediction of the wave nature of matter by de Broglie in 1924. Experimental verification; measurement of the wavelength of electrons by Davisson and Germer in 1927. The quantum mechanics of Heisenberg and Schrödinger in 1925–1926. Since then quantum mechanics has been successful in all of its applications to atomic problems. In principle any chemical problem can be solved on paper using the Schrödinger equation. In practice, the computations are so laborious for most chemical problems that experimental chemistry is, and will be for many years, a very active field. This attitude must be distinguished from that of the complacent physicist of 1890. Although the theoretical basis for attacking chemical problems is well understood today—and it is unlikely that this foundation will be overturned—we recognize our limitations. We break off the chronology in 1927. Those discoveries since 1927 that concern us will be dealt with as they are needed.

Looking back on the developments before 1927 we see two main consequences. Radiation, which was a wave phenomenon in classical physics, was endowed with a particle aspect by the work of Planck, Einstein, and Bohr. Electrons and atoms, which were particles in the classical view, were given a wave aspect by the work of de Broglie, Schrödinger, and Heisenberg. The two parts of classical physics that did not join smoothly are brought together in a unified way in the quantum mechanics. The dual nature of matter and of light, the wave-particle nature, permits this unification.

19.4 DISCOVERY OF THE ELECTRON

From the time of Dalton, atoms were indivisible. The discovery of the electron by J. J. Thomson in 1897 was the first hint of the existence of particles smaller than atoms. Thomson's discovery allowed speculation about the interior structure of the atom and extended the hope that such speculation could be verified experimentally.

The studies of electrical conduction in gases had led to the discovery of cathode rays. If a glass tube fitted with two electrodes connected to a source of high potential is evacuated, a spark will jump between the electrodes. At lower pressures the spark broadens to a glow that fills the tube; at still lower pressure various dark spaces appear in the glowing gas. At very low pressures the interior of the tube is dark, but its walls emit a fluorescence, the color of which depends only on the kind of glass. It was soon decided that the cathode was emitting some kind of ray, a cathode ray, which impinged on the glass wall and produced the fluorescence. Objects placed in the path of these cathode rays cast a shadow on the walls of the tube; the rays are deflected by electric and magnetic fields. Figure 19.1 shows the device used by J. J. Thomson in his famous experiments which showed that the cathode ray was a stream of particles, later called electrons.

In the highly evacuated tube, cathode rays are emitted from the cathode C. Two slotted metal plates A and A' serve as anodes. Passage through the two slots collimates the beam, which then moves in a straight line to hit the spot P at which the fluorescence appears. An

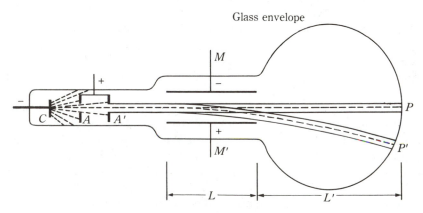

Figure 19.1   Device to measure e/m for cathode rays.

electric field can be applied between the plates M and M'; a magnetic field can be applied in the region of M and M' but perpendicular to the plane of the drawing. The forces produced on the ray by the fields act in the vertical direction only; the horizontal component of velocity is unaffected by the fields. Two experiments are done.

The electrical field E is applied, which pulls the beam downward and deflects the spot to P'; the magnetic field, with a flux density B is applied and adjusted so that the spot returns to the original position P. If the beam consists of particles of charge e and mass m, then the force on the beam due to the electrical field is eE, and that due to the magnetic field is Bev, where v is the horizontal component of velocity of the particle. Since these forces are in balance, $eE = Bev$, and we obtain the horizontal velocity component in terms of E and B:

$$v = \frac{E}{B}. \tag{19.1}$$

In the second experiment, the magnetic field is turned off, and the deflection PP' under the electrical field only is measured. Since the force is eE, the vertical acceleration is eE/m. The time to pass through the field is $t = L/v$. After this time, the vertical component of velocity $w = (eE/m)t$; in this same time the vertical displacement is $s = \frac{1}{2}(eE/m)t^2$. The value of s can be calculated from the displacement PP' and the length L'. Using the value for t, we have $e/m = 2sv^2/EL^2$, and using the value for v from Eq. (19.1)

$$\frac{e}{m} = \frac{2sE}{B^2L^2}. \tag{19.2}$$

The experiment yields the value of e/m for the particles. The present value of this ratio is

$$\frac{e}{m} = 1.758804 \times 10^{11} \text{ C/kg}.$$

From the direction of the deflection it is apparent that e is negative.

Earlier experiments on electrolysis had measured the ratio of charge to mass of hydrogen, the lightest atom. The present value is

$$\left(\frac{e}{m}\right)_{\text{H}} = 9.57354 \times 10^7 \text{ C/kg}.$$

The e/m for the cathode particles was about 1837 times larger than that of hydrogen. At the time it was not known whether this was because of a difference in charge or mass or both. In 1913, R. A. Millikan measured the charge on the electron directly, the "oil-drop" experiment. The present value is

$$e = 1.6021892 \times 10^{-19} \text{ C}.$$

Combining this with the e/m value, we obtain for the electron mass

$$m = 9.109534 \times 10^{-31} \text{ kg}.$$

From the atomic weight of hydrogen, and the value of the Avogadro number from kinetic theory, the average mass of the hydrogen atom could be estimated. The present value is

$$m_{\text{H}} = 1.6737 \times 10^{-27} \text{ kg}.$$

It was finally apparent that the charge on the hydrogen ion was equal and opposite to that on the electron, while the mass of the electron was very much less (1837.151 times less) than that of the hydrogen atom. Being less massive, the electron was a more elementary particle than the atom. Presumably atoms were composites of negative electrons and positively charged matter, which was much more massive. After Thomson's work it was possible to think of how atoms could be built with such materials.

19.5 POSITIVE RAYS AND ISOTOPES

The discovery of positive rays, canal rays, by Goldstein in 1886 is another important result of the studies of electrical condition in gases. The device is shown in Fig. 19.2.

The cathode C has a hole, a canal, drilled through it. In addition to the usual discharge between A and C, a luminous stream emerges from the canal to the left of the cathode. This ray is positively charged and, reasonably enough, is called a positive ray. The systematic study of positive rays was long delayed, but it was determined at an early date that their characteristics depended on the kind of residual gas in the tube. In contrast, the cathode ray did not depend on the residual gas.

Thomson was engaged in the measurement of the e/m of positive rays by the same general method as he used for the electron when, in 1913, he discovered that neon consisted of two different kinds of atoms: one having a mass of 20, the other having a mass of 22. These different atoms of the same element are called *isotopes*, meaning "same place" (that is, in the periodic table). Since this discovery that an element may contain atoms of different mass, the isotopic constitution of all the elements has been determined. Moreover, as is well known, in recent years many artificial isotopes have been synthesized by the high-energy techniques of physics.

Isotopes of an element are almost indistinguishable chemically, since the external electron configurations are the same. Their physical properties differ slightly because of the difference in mass. The differences are most pronounced with the lightest elements, since the relative difference in mass is greatest. The differences in properties of isotopes are most marked in the positive-ray tube itself, where the strengths of the applied electrical and magnetic fields can be adjusted to spread the rays having different values of e/m into a pattern resembling a spectrum, called a mass spectrum. The modern mass spectrometer is a descendant of Thomson's e/m apparatus.

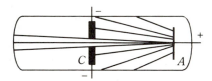

Figure 19.2   Simple positive-ray tube.

19.6 RADIOACTIVITY

In 1896, shortly after the discovery of x-rays, H. Becquerel tried to discover if fluorescent substances emitted x-rays. He found that a fluorescent salt of uranium emitted a penetrating radiation that was not connected with the fluorescence of the salt. The radiation could pass through several thicknesses of the black paper used to protect photographic plates and through thin metal foils. The radiation differed from x-radiation in that it could be resolved into three components, α-, β-, and γ-rays, by the imposition of a strong magnetic field. The β-ray has the same e/m as the electron; the γ-ray is undeflected in the field; the α-ray is positively charged, with an e/m value of one-half that of hydrogen. The β-ray is a stream of electrons; the α-ray is a stream of helium nuclei; the γ-ray is a light ray of extremely short wavelength. A great deal of effort was devoted to the study of radioactivity in the years that followed. The discovery of two new elements, polonium and radium, by Pierre and Marie Curie, was one of the notable accomplishments.

The striking fact about radioactivity is that the rate of emission of the rays is completely unaffected by even the most drastic changes in external conditions such as chemical environment, temperature, pressure, and electrical and magnetic fields. The rays are emitted from the nucleus; the lack of influence of external variables on this process shows that the situation in the nucleus is independent of these variables. Secondly, the energies of the emitted rays are of the order of one-million electron volts, (1 eV $\approx$ 96 kJ). This energy is enormously greater than that associated with any chemical transformation.

The rate law for the radioactive decay of a nucleus is described in Section 32.4.1.

19.7 ALPHA-RAY SCATTERING

In 1908 Thomson proposed a model of the atom: the positive charge was uniformly spread throughout a sphere of definite radius; to confer electrical neutrality, electrons were imbedded in the sphere. For stability according to classical theory, the electrons had to be at rest. This requirement could be met for the hydrogen atom by having the electron at the center of the sphere. This model failed the crucial test provided by the scattering of α-rays by thin metal foils.

In 1909 Geiger and Marsden discovered that if a beam of α-particles was directed at a thin gold foil, some of the α-particles were scattered back toward the source. Figure 19.3 shows the experiment schematically. The majority of the α-particles pass through the foil and can be detected at A. Some are scattered through small angles θ and are detected at A'; remarkably, quite a few are scattered through large angles such as θ' and can be detected at A''.

The scattering occurs because of the repulsion between the positive charge on the α-particle and the positive charges on the atoms of the foil. If the positive charges on the atoms were spread uniformly, as in the Thomson model, the scattering would be the result of a gradual deflection of the particle as it progressed through the foil. The scattering angle would be very small. Rutherford reasoned that the scattering at large angles was due to a very close approach of the α-particle to a positively charged center with subsequent rebound; a single scattering event. By calculation he could show that to be scattered through a large angle in a single event the α-particle would have to approach the positive

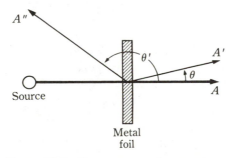

Figure 19.3 The α-ray scattering experiment.

part of the scattering atom very closely, to within 10^{-14} m. The sizes of atoms were known to be about 10^{-10} m. Since the mass of the atom is associated with the positively charged part of the atom, Rutherford's calculation implied that the positive charge and the mass of the atom are concentrated in a space which is very much smaller than that occupied by the atom as a whole.

The nuclear model of the atom proposed by Rutherford supposed that the atom was a sphere of negative charge, not having much mass but having a tiny kernel or nucleus at the center in which the mass and positive charge are concentrated. Using the nuclear model, Rutherford calculated the angular distribution of scattered α-particles. Later experiments of Geiger and Marsden confirmed the predicted distribution in all its particulars.

The Rutherford model had its difficulties. The sphere, uniformly filled with negative charge, was incompatible with the concept of the electron as a particle that should be localized in space. But it is not possible to take a positive discrete particle and a negative discrete particle, place them a certain distance apart, and ask them to stay put. Being oppositely charged, they will attract one another; the electron will fall into the nucleus. Thomson's model did not have this type of instability. Matters are not helped by whirling the electron around in an orbit to achieve the stability of a satellite in orbit around a planet. The electrical attractive force could be balanced by the centrifugal force, but a fundamental objection arises. An electron in orbit is subject to a continual acceleration toward the center; otherwise, the orbit would not be stable. Classical electromagnetic theory, confirmed by Hertz's discovery of radio waves, required an accelerated electrical charge to emit radiation. The consequent loss of energy should bring the electron down in a spiral to the nucleus. This difficulty seemed insuperable. But less than two years later Niels Bohr found a way out. To appreciate Bohr's contribution we must return to 1900 and follow the course of another series of discoveries.

19.8 RADIATION AND MATTER

By 1900 the success of Maxwell's electromagnetic theory had firmly established the wave nature of light. One puzzle that remained was the distribution of wavelengths in a cavity, or blackbody; the observed distribution had eluded explanation on accepted principles. In 1900 Max Planck calculated the distribution, within the experimental error, in a completely mysterious way. Planck's work proved ultimately to be the key to the entire problem of atomic structure; yet at first glance it seems to have little bearing on that problem.

A perfect blackbody is one which adsorbs all the radiation, light, that falls on it. Experimentally the most nearly perfect blackbody is a pinhole in a hollow object. Radiation falling on the pinhole enters the cavity and is trapped (absorbed) within the cavity. Let the radiation in the cavity be brought to thermal equilibrium with the walls at a temperature T. Since there is energy in the radiation, there is a certain energy density in the cavity, $u = U/V$, where U is the energy, V the volume, and u the energy density. From electromagnetic theory, the pressure exerted by the radiation is $p = \frac{1}{3}u$, and experiment shows that the energy density is independent of the volume; that is, $u = u(T)$. The relation between u and T is obtained from the thermodynamic equation of state, Eq. (10.31):

$$\left(\frac{\partial U}{\partial V}\right)_T = T\left(\frac{\partial p}{\partial T}\right)_V - p.$$

Since $U = u(T)V$, $(\partial U/\partial V)_T = u(T)$. Also $p = \frac{1}{3}u(T)$, so that $(\partial p/\partial T)_V = \frac{1}{3}(du/dT)$.

The equation of state becomes $du/dT = 4u/T$. Integration yields

$$u = \alpha T^4, \tag{19.3}$$

where the constant $\alpha = 7.5657 \times 10^{-16}$ J/m^3 K^4. The rate of emission of energy from a cavity per unit area of opening is proportional to the energy density within; this rate is the total emissive power, e_t; thus,

$$e_t = \tfrac{1}{4}cu = \sigma T^4, \tag{19.4}$$

where the Stefan–Boltzmann constant $\sigma = \tfrac{1}{4}c\alpha = 5.6703 \times 10^{-8}$ J/m^2 s K^4. Equation (19.4) is the Stefan–Boltzmann law; among other things it is used to establish the absolute temperature scale at very high temperatures.

So far everything is fine; we may keep our confidence in the second law of thermodynamics. The difficulty is this: the energy in the cavity is the sum of the energies of light waves of many different wavelengths. Let $u_\lambda \, d\lambda$ be the energy density contributed by light waves having wavelengths in the range λ to $\lambda + d\lambda$. Then the total energy density u is

$$u = \int_0^\infty u_\lambda \, d\lambda, \tag{19.5}$$

where we sum the contribution of all wavelengths from zero to infinity. It is rather easy to measure the distribution function u_λ, shown in Fig. 19.4. Experimentally it has been shown that the wavelength at the maximum of this spectral distribution is inversely proportional to the temperature:

$$\lambda_m T = 2.8979 \times 10^{-3} \text{ m K}.$$

This is Wien's displacement law. Classical principles had failed to explain the shape of the curve in Fig. 19.4 and failed to predict the displacement law. The application of the classical law of equipartition of energy between the various degrees of freedom by Rayleigh and Jeans was satisfactory at long wavelengths but failed at short wavelengths, in the ultraviolet ("ultraviolet catastrophe").

The Rayleigh–Jeans treatment assigned the classical value kT to the average energy of each mode of oscillation in the cavity; $\tfrac{1}{2}kT$ for kinetic and $\tfrac{1}{2}kT$ for potential energy. The number of modes of oscillation dn in the wavelength range from λ to $\lambda + d\lambda$ per unit

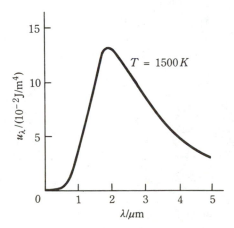

Figure 19.4 Spectral distribution in blackbody radiation.

volume of the cavity is* $dn = 8\pi\, d\lambda/\lambda^4$. The energy density in the same wavelength range is $u_\lambda\, d\lambda$ and is equal to the number of modes of oscillation multiplied by kT. Therefore, $u_\lambda\, d\lambda = 8\pi kT\, d\lambda/\lambda^4$, so that

$$u_\lambda = \frac{8\pi kT}{\lambda^4}, \tag{19.6}$$

which is the Rayleigh–Jeans formula. It predicts an infinite energy density as $\lambda \to 0$; hence an infinite value of the total energy density in the cavity, an absurdity.

If a mode of oscillation can possess any arbitrary amount of energy from zero to infinity, there is no reason for the Rayleigh–Jeans formula to be incorrect. Let us suppose, for the sake of argument, that an oscillator cannot have any arbitrary energy but may have energy only in integral multiples of a certain unit of energy ϵ. Then the distribution of a collection of these oscillators is discrete and we can represent it by

Energy	0	ϵ	2ϵ	3ϵ	4ϵ	$\cdots$
Number	n_0	n_1	n_2	n_3	n_4	$\cdots$

We further suppose that the distribution is governed by the Boltzmann law: $n_i = n_0\, e^{-\epsilon_i/kT}$. Using these ideas we calculate the total number of particles N and the total energy:

$$N = \sum_i n_i = n_0 + n_0 e^{-\epsilon/kT} + n_0 e^{-2\epsilon/kT} + n_0 e^{-3\epsilon/kT} + \cdots .$$

If we set $x = e^{-\epsilon/kT}$, this expression becomes

$$N = n_0(1 + x + x^2 + x^3 + \cdots).$$

The series is the expansion of $1/(1 - x)$, so we obtain

$$N = \frac{n_0}{1 - x}. \tag{19.7}$$

The average energy $\langle U \rangle$ is given by

$$N\langle U \rangle = n_0(0) + n_1\epsilon + n_2(2\epsilon) + n_3(3\epsilon) + \cdots$$
$$= n_0\,\epsilon(x + 2x^2 + 3x^3 + \cdots) = n_0\,\epsilon x(1 + 2x + 3x^2 + \cdots).$$

But $(1 + 2x + 3x^2 + \cdots) = d(1 + x + x^2 + x^3 + \cdots)/dx = d[1/(1 - x)]/dx = 1/(1 - x)^2$, so that

$$N\langle U \rangle = \frac{n_0\,\epsilon x}{(1 - x)^2}.$$

Putting in the values of N and x, this becomes

$$\langle U \rangle = \frac{\epsilon e^{-\epsilon/kT}}{1 - e^{-\epsilon/kT}} = \frac{\epsilon}{e^{\epsilon/kT} - 1}. \tag{19.8}$$

---

*   The derivation of this formula is beyond the scope of the treatment here.

If we use the value given by Eq. (19.8) for the average energy in a mode of vibration, then multiply it by the number of modes in the wavelength range to calculate the spectral distribution, we obtain for u_λ

$$\mu_\lambda = \frac{8\pi}{\lambda^4}\left(\frac{\epsilon}{e^{\epsilon/kT}-1}\right). \tag{19.9}$$

Now if ϵ is a constant in Eq. (19.9), we are no better off than were Rayleigh and Jeans.

Planck took the extraordinary step of setting ϵ inversely proportional to the wavelength, recognizing that the Wien displacement law would come out of the resulting equation. Since the frequency times the wavelength is equal to the velocity, we have $1/\lambda = v/c$, where v is the frequency and c the velocity. Setting ϵ proportional to $1/\lambda$ is equivalent to setting it proportional to the frequency:

$$\epsilon = hv = \frac{hc}{\lambda}, \tag{19.10}$$

where Planck's constant $h = 6.626176 \times 10^{-34}$ J s. Putting the value of ϵ from Eq. (19.10) into Eq. (19.9) yields the distribution function

$$u_\lambda = \frac{8\pi hc}{\lambda^5}\frac{1}{e^{hc/\lambda kT}-1}. \tag{19.11}$$

By properly choosing the value of the constant h, Planck found that Eq. (19.11) agreed with the measured distribution within the experimental error! To find the maximum, we set $du_\lambda/d\lambda = 0$; the Wien displacement law is obtained in the form

$$\lambda_m T = \frac{hc}{4.965k}. \tag{19.12}$$

The value of ϵ is the energy gap separating the energies of the various groups of oscillators; classically this gap should be zero to yield a continuous energy distribution. Planck's assumption that $\epsilon = hv$ required the gap to be finite, approaching zero, the classical value, only at infinitely long wavelengths.

The worst part of this is that it lacks the logic of classical physics and it has far-reaching implications. If the radiation in a cavity can possess energy only in multiples of a certain unit hv, then it can exchange energy with the oscillators in the cavity walls only in multiples of this unit. Therefore the interchange of energy had to be discontinuous also; energy had to be exchanged in small lumps or bundles called *quanta*. The quantum of energy for an oscillator is hv.

The nature of light seemed no longer to be simple. Light was a wave motion, but with Planck's work it acquired a corpuscular aspect. The light wave contains energy in elementary discrete units, quanta.

19.9 THE PHOTOELECTRIC EFFECT

As may be imagined, Planck's discovery excited very little interest and no controversy. The prevailing attitude seemed to be "if we ignore it, it will go away." Perhaps it might have gone away but for Einstein's interpretation of the photoelectric effect, another longstanding thorn in the side of classical physics.

If a beam of light falls on a clean metal plate in vacuum, the plate emits electrons. This effect, discovered by Hertz in 1887, had been thoroughly investigated. Two aspects of the phenomenon were the rocks on which classical physics foundered.

1. Whether or not electrons are emitted from the plate depends only on the frequency of the light and not at all on the intensity of the beam. The number of electrons emitted is proportional to the intensity.
2. There is no time lag between the light beam striking the plate and the emission of the electrons.

An electron in a metal is bound by a potential energy ω, which must be supplied to bring the electron outside the metal. If, in addition, the electron outside the metal has kinetic energy, then the total energy of the electron is

$$E_t = \tfrac{1}{2}mv^2 + \omega. \tag{19.13}$$

Presumably the electron acquires this energy from the beam of light. Classically, the energy of the light beam depends on its intensity, and that energy should be absorbed continuously by the metal plate. It can be shown that for weak intensities and reasonable values of ω that after the onset of illumination, a long time period (days or even years) should intervene before any electron would soak up enough energy to be kicked out of the metal. After this time interval many electrons should be energetic enough to escape and a steady current should flow from then on. Increasing the intensity should lessen the time interval. No time interval has ever been observed. The proportionality of the current to intensity is reasonable on classical grounds, but the absence of a time interval could not be explained.

In 1905 Albert Einstein took a different view of the problem. Classically, the energy of the light beam is absorbed continuously by the metal and divided among all the electrons in the plate, each electron receiving only a tiny share of the total. Suppose that the energy of the light beam is concentrated in Planck's quanta of energy $h\nu$, and further that the entire quantum of energy must be accepted by a single electron, and cannot be divided among all the electrons present. Then the energy of the electron after accepting the quantum must be $h\nu$, and this must be the total energy after emission, Eq. (19.13). Therefore

$$h\nu = \tfrac{1}{2}mv^2 + \omega. \tag{19.14}$$

Equation (19.14) is the Einstein photoelectric equation. It is apparent from the equation that below a critical frequency, ν_0, given by $h\nu_0 = \omega$, the electron does not gain sufficient energy from the light quantum to escape the metal. This explains the "cut-off" frequency ν_0 that is observed. A greater light intensity means only that more quanta are absorbed per unit time and more electrons are emitted; the energies of the emitted electrons are completely independent of the intensity.

Using the same value of h as had been obtained by Planck in the treatment of blackbody radiation, the Einstein equation provided a completely satisfactory explanation of the photoelectric effect. Satisfactory? Yes, but very unsettling! Einstein spoke of photons, corpuscles of light, each carrying energy $h\nu$. Planck's idea seemed to be gaining ground, a most distressing turn of events.

19.10 BOHR'S MODEL OF THE ATOM

Throughout the 19th century, spectroscopy was a very popular field of study. A great number of precise measurements of wavelengths of lines had been made and catalogued.

Regularities in the spacing between lines had been observed and correlated by empirical formulas. One of the most famous of these formulas is that given by Balmer in 1885. Balmer found that the wavelengths of nine lines in the visible and near-ultraviolet spectrum of hydrogen could be expressed by the formula

$$\frac{\lambda}{10^{-10} \text{ m}} = 3645.6\left(\frac{n^2}{n^2 - 2^2}\right), \tag{19.15}$$

where n had the integral values 3, 4, 5, ..., 11. Each integral value of n corresponds, through Eq. (19.15), to a wavelength. The computed wavelengths agreed excellently with the measured values. Somewhat later, Ritz proposed a more general formula which, for hydrogen, takes the form

$$\frac{1}{\lambda} = \tilde{v} = R_{\text{H}}\left(\frac{1}{k^2} - \frac{1}{n^2}\right), \tag{19.16}$$

where both k and n are integers, R_{H} is the Rydberg constant for hydrogen. The wave number $\tilde{v}$ is the reciprocal of the wavelength. If $k = 2$, the Rydberg formula reduces to the Balmer formula. The Rydberg formula is remarkably accurate, and with slight modification it represents the wavelengths in the spectra of many different atoms. Because of the accuracy of the formula and the precision with which wavelengths can be measured, the Rydberg constant was known with great accuracy. Today the value is known to within less than one part in ten million. The present value is $R_{\text{H}} = 10\,967\,758.5 \pm 0.8 \text{ m}^{-1}$.

The spectrum emitted by an atom presumably is related to the structure of the atom. Until 1913, attempts to relate the spectrum to a definite atomic model were unsuccessful. By 1913 it was known that the atom had a positively charged nucleus, but the nuclear model of Rutherford was unstable according to classical electromagnetic theory. This Gordian knot was cut by Niels Bohr in 1913.

In the Bohr model the hydrogen atom consists of a central nucleus with a charge $+e$, and an electron of charge $-e$ whirling about the nucleus with velocity v in an orbit of radius r (Fig. 19.5). For mechanical stability, the electrical force of attraction $-e^2/4\pi\epsilon_0 r^2$ must balance the centrifugal force mv^2/r:

$$-\frac{e^2}{4\pi\epsilon_0 r^2} + \frac{mv^2}{r} = 0,$$

or

$$mv^2 = \frac{e^2}{4\pi\epsilon_0 r}. \tag{19.17}$$

Figure 19.5

The total energy E is the sum of the kinetic energy $\frac{1}{2}mv^2$, and the potential energy $-e^2/4\pi\epsilon_0 r$:

$$E = \frac{mv^2}{2} - \frac{e^2}{4\pi\epsilon_0 r}.$$

Using Eq. (19.17), we obtain

$$E = -\frac{e^2}{8\pi\epsilon_0 r}. \tag{19.18}$$

Classically, since the electron is accelerated, this system should radiate. To avoid this difficulty, Bohr broke completely with tradition. Bohr assumed: (1) that the electron can move around the nucleus only in certain orbits, and not in others (classically, no particular orbit is preferable to any other); (2) that these allowed orbits correspond to definite stationary states of the atom, and in such a stationary state the atom is stable and does not radiate (Bohr avoided the classical difficulty by simply *assuming* that it was not a difficulty in these special circumstances!*); and (3) that in the transition of the electron from one stable orbit to another, radiation is emitted or absorbed, the frequency of the radiation being given by

$$h\nu = \Delta E,$$

where ΔE is the energy difference between the two stationary states and h is Planck's constant. (There was nothing quite so nonclassical as a formula with Planck's constant in it.)

The problem of choosing these special orbits out of all the possible ones remained. Bohr's condition is that the angular momentum mvr be an integral multiple of $\hbar = h/2\pi$:

$$mvr = n\hbar, \qquad n = 1, 2, \ldots \tag{19.19}$$

This condition is equivalent in a certain sense to Planck's condition on an oscillator.

Solving Eqs. (19.17) and (19.19) for v and r, we obtain

$$v = \frac{e^2}{4\pi\epsilon_0 n\hbar} \qquad \text{and} \qquad r = \frac{4\pi\epsilon_0 n^2\hbar^2}{me^2}.$$

If $n = 1$, then $r = a_0$, the radius of the first Bohr orbit;

$$a_0 = \frac{4\pi\epsilon_0 \hbar^2}{me^2} = 0.529177 \times 10^{-10} \text{ m} = 52.9177 \text{ pm} \tag{19.20}$$

then

$$r = n^2 a_0. \tag{19.21}$$

Using this value of r in Eq. (19.18) for the total energy yields

$$E_n = -\frac{e^2}{8\pi\epsilon_0 a_0}\left(\frac{1}{n^2}\right), \tag{19.22}$$

where the subscript on E indicates that the energy depends on the integer n. Equation (19.22) expresses the energy entirely in terms of fundamental constants, e, h, m, and the integer n. Consider two stationary states, one described by the integer n and the other by

---

* Bohr's approach cannot be recommended for solving *standard* problems in physical chemistry!

the integer k. The difference in energy of these two states is

$$\Delta E_{nk} = E_n - E_k = \frac{e^2}{8\pi\epsilon_0 a_0} \left(\frac{1}{k^2} - \frac{1}{n^2}\right).$$

By Bohr's third assumption, this difference should equal $h\nu$:

$$h\nu = \frac{e^2}{8\pi\epsilon_0 a_0} \left(\frac{1}{k^2} - \frac{1}{n^2}\right).$$

If we replace ν by $\nu = c/\lambda = c\tilde{\nu}$, the equation becomes

$$\tilde{\nu} = \frac{e^2}{8\pi\epsilon_0 a_0 hc} \left(\frac{1}{k^2} - \frac{1}{n^2}\right), \tag{19.23}$$

which is the Rydberg formula. Bohr's argument yields a value of the Rydberg constant:

$$R_{\mathrm{H}} = \frac{e^2}{8\pi\epsilon_0 a_0 hc}. \tag{19.24}$$

Calculating the value of R_{H} from Eq. (19.24), Bohr obtained a value of R_{H} which agreed with the empirical value within the uncertainty of the knowledge of the constants.

Bohr had calculated the most accurately known experimental constant in physics by a method which was, to use a mild description, simply an outrage! The corpuscular nature of light had come to stay; it could no longer be ignored. No evangelist ever made so many converts in so short a time as did Bohr.

The connection between matter and radiation soon became firmly established. In the decade following Bohr's discovery, what is now called the quantum theory or the "old quantum theory" burst into full flower. The systematic interpretation of the data in the catalogues of spectra went forward by leaps and bounds. The Bohr–Sommerfeld atom model, which used elliptical as well as circular orbits, was introduced and found useful. From studies of spectra Bohr constructed a theoretical periodic chart which agreed with that of the chemists. A detail was different: According to Bohr, element 72, which chemists had sought among the rare earths, was not a rare earth, but a member of the fourth family, with titanium and zirconium. Shortly thereafter, von Hevesy looked at the spectrum of zirconium and found that many of the lines should be ascribed to element 72. Therefore, zirconium was a mixture of zirconium and element 72. The new element was named hafnium, after the ancient name of Copenhagen, in honor of Bohr who is Danish. The discovery of hafnium ended a long controversy over the atomic weight of zirconium; samples used by different investigators contained different amounts of hafnium, so the discrepancies were rather large.

The Bohr theory of the atom destroyed the last pockets of resistance to the quantum concept. Yet the wave attributes of light were there too. The nature of light took on a dual aspect. This duality in the nature of light is accepted now, though to some in the beginning it was a bitter pill.

19.11 PARTICLES AND LOUIS DE BROGLIE

In 1924, Louis de Broglie argued on theoretical grounds that particles should have a wavelength associated with them. The de Broglie formula for the wavelength is

$$\lambda = \frac{h}{p} = \frac{h}{mv}, \tag{19.25}$$

where $p = mv$ is the momentum of the particle. We cannot reproduce de Broglie's argument here, since it requires some knowledge of electromagnetic theory as well as relativity theory. However, if a particle does have a wavelength, then that fact must be capable of experimental demonstration.

A demonstration of the wave property of electrons was provided by the experimental work of Davisson and Germer in 1927. A beam of light reflected from a ruled grating produces a diffraction pattern; diffraction is a property exclusively of wave motion. Davisson and Germer directed an electron beam at a nickel crystal. The rows of nickel atoms served as the ruling. The intensity of the diffracted beam was measured as a function of the diffraction angle. They found maxima in the intensity at special values of the diffraction angle. From these values of the diffraction angle and the usual diffraction formula they computed the wavelength of the electrons. This value of the wavelength agreed with that predicted by the de Broglie formula for electrons having the experimental velocity.

This confirmation of de Broglie's prediction brought duality into the concept of the nature of fundamental particles. A particle was not simply a particle but had a wave aspect to its nature. This idea led very quickly to the development of wave mechanics, or quantum mechanics, by Heisenberg and Schrödinger. All of our modern ideas on atomic and molecular structure are based on wave mechanics.

The distinct concepts, wave or particle, of 19th century physics are now inseparably mingled. Wave mechanics, so essential to our ideas now, would have been a contradiction in terms in the 19th century. The question of whether an electron, or a photon, is a wave or particle has lost all meaning. We can say with precision in what circumstances it is useful to treat the electron as a classical particle or the photon as a classical wave. We know when we must consider the wave aspects of the classical particles and the particle aspects of the classical wave. Any final classification into particle or wave would be artificial. Both particles and waves have a more general nature than their names indicate. We use the old names, fully realizing the more general character of the entity in question.

In 1925 Werner Heisenberg and in 1926 Erwin Schrödinger independently formulated the law that governs the motion of a particle. The discussion here will be more closely related to Schrödinger's treatment.

★ 19.12   THE CLASSICAL WAVE EQUATION

The classical law governing wave motion is the wave equation

$$\frac{\partial^2 D}{\partial x^2} + \frac{\partial^2 D}{\partial y^2} + \frac{\partial^2 D}{\partial z^2} = \frac{1}{v^2}\frac{\partial^2 D}{\partial t^2}, \tag{19.26}$$

where x, y, z are the coordinates, t the time, v the velocity of propagation, and D the displacement of the wave. If v does not depend on the time, then the displacement is the product of a function of the coordinates only, $\psi(x, y, z)$, and a periodic function of time, $e^{i2\pi vt}$, where v is the frequency of the wave, and $i = \sqrt{-1}$. Then

$$D = \psi(x, y, z)e^{i2\pi vt}, \tag{19.27}$$

which means that if we sit at a fixed position x, y, z, and observe the value of D, then at an arbitrary time $t = 0$, $e^{i2\pi v0} = 1$ and $D = \psi$. At a later time,

$$t = 0 + 1/v, \qquad e^{i2\pi v/v} = e^{i2\pi} = 1*$$

---

*   Remember that $e^{ix} = \cos x + i \sin x$. Therefore e^{ix} has the period 2π.

and the displacement $D = \psi$, the same value as at $t = 0$. Thus the value of D at any point varies with a period, $t_0 = 1/v$. By the mean value theorem, we can calculate the average value of the displacement D in the time interval t_0:

$$\langle D \rangle = \frac{1}{t_0} \int_0^{t_0} D \, dt = \frac{1}{t_0} \psi(x, y, z) \int_0^{t_0} e^{i2\pi vt} \, dt.$$

Evaluation of the integral yields $\langle D \rangle = 0$. This result is physically obvious; since in one complete period D is positive for half the time and negative for half the time, the values sum to zero. We avoid this difficulty by computing the average value of the square of the absolute value of the function:

$$\langle |D|^2 \rangle = \frac{1}{t_0} \int_0^{t_0} |D|^2 \, dt.$$

Since D may be a complex function, to compute $|D|^2$ we use the formula $|D|^2 = D^*D$, where D^* is the complex conjugate of D obtained by replacing i in the function by $-i$. Then

$$|D|^2 = \psi^*\psi e^{i2\pi vt} e^{-i2\pi vt} = \psi^*\psi.$$

Putting this value in the integral, we obtain

$$\langle |D|^2 \rangle = \psi^*\psi \frac{1}{t_0} \int_0^{t_0} dt = \psi^*\psi = |\psi|^2. \tag{19.28}$$

By Eq. (19.28) the *time average* of the square of the absolute value of the displacement is equal to the square of the absolute value of the space-dependent part ψ. The function ψ is called the *amplitude* of the wave.

Using the value of D given by Eq. (19.27), we can form the second derivatives and put them into the wave equation, Eq. (19.26). The result, after division by $e^{i2\pi vt}$, is

$$\frac{\partial^2 \psi}{\partial x^2} + \frac{\partial^2 \psi}{\partial y^2} + \frac{\partial^2 \psi}{\partial z^2} + \frac{4\pi^2 v^2 \psi}{v^2} = 0. \tag{19.29}$$

In Eq. (19.29) only the space coordinates appear; this equation governs the spatial dependence of the amplitude ψ. After solving this differential equation for ψ, we can write down the value of ψ^* immediately. Multiplying ψ and ψ^* yields, through Eq. (19.28), a value of the time average of the square of the absolute value of the displacement. It is the time average that is of interest in the discussion of stationary states.

★ 19.13 THE SCHRÖDINGER EQUATION

Now we can make an argument for the Schrödinger equation.* If, as de Broglie says, the particle has some of the properties of a wave, then it likely will have some property that is analogous to the displacement of a classical wave. Since in an atomic system we cannot follow the detailed motion of a particle in time, perhaps we should concentrate on the time average value of the displacement analogue, which can be calculated if we know the amplitude. The classical amplitude ψ is governed by Eq. (19.29); we can translate this equation into a nonclassical one by using de Broglie's equation, $\lambda = h/mv$. But for any

---

*   This must be regarded as "argument for," not "proof of" or "derivation of."

wave, the frequency times the wavelength is the velocity, $\lambda v = v$. Combining this with de Broglie's relation, we have $v/v = mv/h$. Using this in Eq. (19.29), we obtain

$$\frac{\partial^2 \psi}{\partial x^2} + \frac{\partial^2 \psi}{\partial y^2} + \frac{\partial^2 \psi}{\partial z^2} + \frac{4\pi^2 m^2 v^2}{h^2} \psi = 0. \tag{19.30}$$

For the moment we will think of ψ as some analogue of the classical amplitude. As yet we have no meaning for ψ, but using Eq. (19.30), we can express a familiar mechanical variable, the velocity v, in terms of ψ and its derivatives. Suppose we solve Eq. (19.30) algebraically for v and then calculate the kinetic energy, $\frac{1}{2}mv^2$. This yields

$$E_{\text{kin}} \psi = \frac{1}{2m} \left[-\hbar^2 \frac{\partial^2 \psi}{\partial x^2} - \hbar^2 \frac{\partial^2 \psi}{\partial y^2} - \hbar^2 \frac{\partial^2 \psi}{\partial x^2} \right], \tag{19.31}$$

where we have used $\hbar = h/2\pi$. If it were not for the ψ function in Eq. (19.31), we would be tempted to see a similarity between this equation and the classical one,

$$E_{\text{kin}} = \frac{1}{2m} (p_x^2 + p_y^2 + p_z^2). \tag{19.32}$$

Since ψ bothers us, suppose we just leave a blank where it appears in Eq. (19.31); then

$$\mathbf{K} = \frac{1}{2m} \left[-\hbar^2 \frac{\partial^2}{\partial x^2} - \hbar^2 \frac{\partial^2}{\partial y^2} - \hbar^2 \frac{\partial^2}{\partial z^2} \right], \tag{19.33}$$

where $\mathbf{K}$ has replaced E_{kin}, since the right-hand side of Eq. (19.33) is an *operator*; this operator tells us to perform the operation of taking the second partial derivatives of some function, multiply each by $-\hbar^2/2m$, and add them together. We can see an analogy between the classical equation, Eq. (19.32), and Eq. (19.33). Corresponding to the classical E_{kin}, in the quantum mechanics there is an operator $\mathbf{K}$. Corresponding to the classical momentum p_x, there is an operator $\mathbf{p}_x$. We find the momentum operator easily using p_x as an example. Comparing Eqs. (19.32) and (19.33), we get

$$\mathbf{p}_x^2 = -\hbar^2 \frac{\partial^2}{\partial x^2} = i^2 \hbar^2 \frac{\partial^2}{\partial x^2} = \left(-i\hbar \frac{\partial}{\partial x} \right) \left(-i\hbar \frac{\partial}{\partial x} \right),$$

where $i = \sqrt{-1}$. Therefore the operators corresponding to the momenta in the three directions are

$$\mathbf{p}_x = -i\hbar \frac{\partial}{\partial x}, \qquad \mathbf{p}_y = -i\hbar \frac{\partial}{\partial y}, \qquad \mathbf{p}_z = -i\hbar \frac{\partial}{\partial z}. \tag{19.34}$$

All of this is very puzzling, but let's go a step further. Using Eq. (19.31), we calculate the total energy $E = E_{\text{kin}} + V$, where V is the potential energy and is usually a function of the coordinates; then we have

$$E\psi = \frac{1}{2m} \left[-\hbar^2 \frac{\partial^2 \psi}{\partial x^2} - \hbar^2 \frac{\partial^2 \psi}{\partial y^2} - \hbar^2 \frac{\partial^2 \psi}{\partial z^2} \right] + V(x, y, z)\psi, \tag{19.35}$$

which could be written in operator form as

$$E\psi = \mathbf{K}\psi + \mathbf{V}\psi. \tag{19.36}$$

Either of Eqs. (19.35) or (19.36) is the Schrödinger equation, which bears some similarity to the classical equation for the conservation of energy:

$$E = E_{\text{kin}} + E_{\text{pot}}. \tag{19.37}$$

Indeed, the Schrödinger equation is the quantum-mechanical analogue of this classical equation.

★ 19.14 THE INTERPRETATION OF ψ

A problem with Eq. (19.36) is that this partial differential equation often has complex solutions, solutions with real and imaginary parts. The physical quantities we measure are real quantities. We can convert the equation into one containing only real quantities if we multiply both sides by the complex conjugate of ψ. Then we have (E is a constant!)

$$\psi^* E\psi = \psi^* \mathbf{K}\psi + \psi^* \mathbf{V}\psi.$$

Since $\psi^*\psi = |\psi|^2$, and $\mathbf{V}\psi = V(x, y, z)\psi$, this equation can be written as

$$E|\psi|^2 = \psi^* \mathbf{K}\psi + V(x, y, z)|\psi|^2. \tag{19.38}$$

Having derived this equation, which contains only real quantities, let us divert our attention for a few moments.

Suppose we wish to compute the average potential energy of a classical particle (moving in one dimension) in the interval X. Then, if V is a function only of x, by the mean value theorem of the calculus,

$$\langle V \rangle = \frac{1}{X} \int_0^X V \, dx = \int_0^X V\left(\frac{1}{X}\right) dx,$$

where dx/X is the probability of finding the particle between x and $x + dx$, multiplied by the value of the potential energy at x.

If V is a function of the three coordinates, we would have an expression such as

$$\langle V \rangle = \int V\left(\frac{1}{\phi}\right) d\tau,$$

where $d\tau$ is the small element of volume, and $1/\phi$ is the probability per unit volume of finding the particle at the position x, y, z.

Now it seems that we could do something similar with Eq. (19.38). First multiply by the small volume element $d\tau$,

$$E|\psi|^2 \, d\tau = \psi^* \mathbf{K}\psi \, d\tau + V(x, y, z)|\psi|^2 \, d\tau.$$

If we interpret $|\psi|^2$ as a probability per unit volume, the second term on the right is the potential energy of the particle at the position x, y, z, multiplied by the probability of finding it in the volume element at that position. If we integrate over the entire coordinate space, that second term should be the average potential energy. Thus

$$E \int |\psi|^2 \, d\tau = \int \psi^* \mathbf{K}\psi \, d\tau + \int V(x, y, z)|\psi|^2 \, d\tau. \tag{19.39}$$

The integration extends over all space, and E is removed from the integral, since it is a constant. The sum of the probability density $|\psi|^2$ times $d\tau$ must be the total probability of finding the particle and this must be unity; the particle must be somewhere!

$$\int |\psi|^2 \, d\tau = 1. \tag{19.40}$$

Equation (19.40) is called the normalization condition; if ψ fulfills this condition, ψ is called a normalized wave function. Then Eq. (19.39) becomes

$$E = \int \psi^* \mathbf{K} \psi \, d\tau + \int V |\psi|^2 \, d\tau,$$

or, more symmetrically,

$$E = \int \psi^* \mathbf{K} \psi \, d\tau + \int \psi^* \mathbf{V} \psi \, d\tau. \tag{19.41}$$

This equation can be written

$$E = \langle E_{\text{kin}} \rangle + \langle V \rangle \tag{19.42}$$

where

$$\langle E_{\text{kin}} \rangle = \int \psi^* \mathbf{K} \psi \, d\tau \quad \text{and} \quad \langle V \rangle = \int \psi^* \mathbf{V} \psi \, d\tau. \tag{19.43}$$

The $\langle E_{\text{kin}} \rangle$ and $\langle V \rangle$ are average values, and these are sensible analogues of the classical mechanical properties. Now we have a name for the wavefunction. Since $|\psi|^2$ is a probability density, ψ is called a *probability amplitude*.

The Schrödinger equation, Eq. (19.36), can be written in the form

$$\mathbf{H} \psi = E \psi, \tag{19.44}$$

in which

$$\mathbf{H} = \mathbf{K} + \mathbf{V}. \tag{19.45}$$

The Hamiltonian operator, $\mathbf{H}$, is the sum of the operators for the kinetic energy and the potential energy. The Hamiltonian operator is the operator for the total energy of the system.

19.15 RETROSPECTION

The Schrödinger equation can be written in the form

$$\mathbf{H} \psi = E \psi.$$

This differential equation can be solved for a property of the particle, ψ, the wave function of the particle. The wave function itself does not have any immediate physical significance. However, the quantity $|\psi|^2 \, d\tau$ can be interpreted as the probability of finding the particle in the volume element $d\tau$ at the position x, y, z. Since the wave function is continuous, this interpretation yields the extraordinary result that everywhere in space there is a finite probability of finding the particle. This is in marked contrast to the classical picture of a strictly localized particle. For example, the Bohr model and the Schrödinger model of the hydrogen atom are quite different pictorially. If we could shrink ourselves and sit on the nucleus of the Bohr atom, we would see the electron moving around the nucleus in a circular orbit having a radius of precisely a_0. Sitting on the nucleus of the Schrödinger atom, we would see a fog of negative charge. (This supposes that our eyes take a "time exposure" of the electron motion.) Near the nucleus the density of the fog is high, but if we walk out along a radius, the fog thins out. When we are several atomic diameters away from the nucleus, we look back and see a spherical cloud of negative charge; with keen eyesight we can discern the nucleus in the center of the cloud. (Rutherford's model of a uniform sphere of negative charge with the nucleus imbedded in it was not so far wrong after all!) The electron in the Bohr atom moves in an orbit much like a satellite about a planet; in the Schrödinger atom the electron is smeared out into an electron cloud, much like a puff of cotton candy.

Another important property of the wave function is its relation to average properties of physically observable quantities, illustrated by Eq. (19.43). Knowing the wave function of a system, we can calculate the average value of any measurable quantity. Thus, in a somewhat mysterious way, the wave function has hidden in it all of the physically important properties of the system.

The Schrödinger equation opened the way to the systematic mathematical treatment of all atomic and molecular phenomena. The predictions of this equation for atoms and molecules have been confirmed without exception. It is therefore the basis for any modern discussion of atomic and molecular structure.

QUESTIONS

19.1 Describe the "ultraviolet catastrophe" and its empirical resolution by the Planck radiation law.

19.2 Describe how the Rutherford α-particle scattering experiment (a) overturned the Thomson model of the atom and (b) led to a conflict with the predictions of classical physics.

19.3 What is a photon? What are its properties?

19.4 In the Bohr model of the atom, why is the energy higher for larger values of n?

19.5 In what way does Eq. (19.41) differ from a standard average over the probability density $|\psi|^2$?

PROBLEMS

19.1 Calculate the energy density of the radiation in a cavity at 100 K, 300 K, and 1000 K.

19.2 a) In a cavity at 1000 K estimate the fraction of the energy density that is provided by light in the region between 780 nm and 800 nm.
 b) Repeat the calculation for 2500 K.

19.3 For parts (a) through (c) estimate the fraction of the radiant energy in the visible range; that is, between 400 and 800 nm. Assume that the object is a blackbody. (*Hint:* In the denominator of u_λ, set $\exp(hc/\lambda kT) - 1 \approx \exp(hc/\lambda kT)$. The integration is simpler if λ is replaced by c/v.)

 a) The glowing coals in a fireplace; $T = 1000$ K.
 b) The filament of an incandescent bulb; $T = 2800$ K.
 c) The surface of the sun; $T = 6000$ K.

19.4 At what wavelength does the maximum in the energy density distribution function for a blackbody occur if

 a) $T = 300$ K?
 b) $T = 500$ K?

19.5 If the energy density distribution function has a maximum at 600 nm, what is the temperature?

19.6 Assume that a flat sheet of iron at 300 K does not receive any radiation from its surroundings. The metal is 10 cm × 10 cm × 0.1 cm; the heat capacity is 25 J/K mol; the density is 7.86 g/cm^3.

    a) How many joules per second are lost by radiation initially? Assume that one flat side only radiates into space.

    b) After 20 minutes, what is the temperature of the metal plate?

19.7 The radius of the sun is 7×10^5 km, and the surface temperature is 6000 K. What is the total energy loss per second from the sun's surface? How many metric tons of matter are consumed per second in the nuclear reactions that supply this energy?

19.8 Derive the Wien displacement law from Eq. (19.11).

19.9 About 5 eV are required to remove an electron from the interior of platinum.

    a) What is the minimum frequency of light required to observe the photoelectric effect using platinum?

    b) If light with $\lambda = 200$ nm strikes the platinum, what will be the velocity of the emitted electron?

19.10 a) Calculate the wavelength of the line emitted as n changes from 3 to 2 in the hydrogen atom.

    b) What is the angular momentum in the lowest energy state of the Bohr atom?

    c) What is the velocity of the electron in the state with $n = 1$?

19.11 If an electron falls through an electric potential difference of one volt, it acquires an energy of one electron volt.

    a) If the electron is to have a wavelength of 0.1 nm, what potential difference must it pass through?

    b) What is the velocity of this electron?

19.12 Through what electric potential difference must a proton pass if it is to have a wavelength of 0.1 nm?

19.13 Show that the wavelength of a charged particle is inversely proportional to the square root of the potential difference through which it is accelerated.

19.14 What is the wavelength of a ball bearing having $m = 10$ g, and $v = 10$ cm/s?

19.15 What is the kinetic energy of an electron that has a wavelength of 10 nm?

19.16 Use the value of D given by Eq. (19.27) in Eq. (19.26) and prove Eq. (19.29).

20

Introduction to Quantum Mechanical Principles

20.1 INTRODUCTION

Although in Chapter 19 we patched the De Broglie relation onto the equation for classical wave motion to lend plausibility to the time-independent Schrödinger equation, this procedure has dubious merit. In a certain sense it is as if we were to attempt to justify Newtonian mechanics by appealing to the Pythagorean music of the spheres. Experience has shown that the Schrödinger equation is the correct one for atomic and molecular problems; whether or not it is "derivable" from other equations is not a matter that need concern us greatly here. For a systematic treatment of atomic and molecular problems we can most easily proceed by stating a series of postulates and using them to discuss the behavior of a system.

20.2 POSTULATES OF THE QUANTUM MECHANICS

20.2.1 Postulate I

There exists a function, $\Psi(x, y, z, t)$, of the coordinates and time that we call a wave function and describe as a probability amplitude. This wave function is in general a complex function; that is,

$$\Psi(x, y, z, t) = u(x, y, z, t) + iv(x, y, z, t), \tag{20.1}$$

where $i = \sqrt{-1}$, and u and v are real functions of coordinates and time. The complex conjugate of Ψ is designated by Ψ^* and is obtained from Ψ by replacing i by $-i$;

$$\Psi^*(x, y, z, t) = u(x, y, z, t) - iv(x, y, z, t). \tag{20.2}$$

The product, $\Psi^*\Psi$, is a real function of x, y, z and t,

$$\Psi^*\Psi = |\Psi|^2 = u^2(x, y, z, t) + v^2(x, y, z, t), \tag{20.3}$$

and is equal to the square of the absolute value of Ψ. The product, $\Psi^*\Psi \, dx \, dy \, dz = \Psi^*\Psi \, d\tau$, is the probability at time t that the system will be in the volume element $d\tau$ at the position x, y, z. Thus, $\Psi^*\Psi$ is a *probability density*. In view of this, if the probability of finding the system in the volume element $d\tau$ is summed over all possible positions of the volume element, the result is unity. We must have unit probability of finding the system somewhere. Thus we have the property

$$\int \Psi^*\Psi \, d\tau = \int_{-\infty}^{\infty} dx \int_{-\infty}^{\infty} dy \int_{-\infty}^{\infty} \Psi^*(x, y, z, t)\Psi(x, y, z, t) \, dz = 1, \tag{20.4}$$

where the limits in the integral are understood to be such as to cover the entire coordinate space.

The value of the integral in Eq. (20.4) must be independent of the time, t. This implies that the time dependence of the wave function must have the form

$$\Psi(x, y, z, t) = \psi(x, y, z)e^{if(q, t)}, \tag{20.5}$$

in which $f(q, t)$ is some real function of the coordinates, symbolized by q, and time. Using Eq. (20.5) in Eq. (20.4) yields

$$\int \psi^*(x, y, z)\psi(x, y, z) \, dz = 1. \tag{20.6}$$

The requirement expressed by Eqs. (20.4) and (20.6), namely that the wave function be quadratically integrable, imposes severe restrictions on ψ. The wave function must be single-valued, continuous, and may not have singularities anywhere of a character that result in the nonconvergence of the integral in Eq. (20.6). In particular, at the extremes of the Cartesian coordinates, $x = \pm\infty$, $y = \pm\infty$, and $z = \pm\infty$, the wave function, as well as well as its first derivative, must vanish.

20.2.2 Postulate II

The expectation value, $\langle A \rangle$, of any observable is related to the wave function of the system by

$$\langle A \rangle = \frac{\int \psi^* \mathbf{A}\psi \, d\tau}{\int \psi^*\psi \, d\tau}, \tag{20.7}$$

in which $\mathbf{A}$ is an operator corresponding to the observable A. If the wave functions have been *normalized*, then Eq. (20.6) is fulfilled; Eq. (20.7) becomes simply

$$\langle A \rangle = \int \psi^* \mathbf{A}\psi \, d\tau. \tag{20.8}$$

Since we are dealing with wave functions that are functions only of coordinates, then, as was pointed out in Section 19.14, to obtain the expectation value of any function of the coordinates we multiply that function by $\psi^*\psi$, the probability density, and integrate over the entire space. Thus, for a function, $f(x, y, z)$, we have

$$\langle f \rangle = \int f(x, y, z)\psi^*\psi \, d\tau = \int \psi^* f(x, y, z)\psi \, d\tau = \int \psi^* \mathbf{f}\psi \, d\tau. \tag{20.9}$$

So we may conclude that the operation corresponding to any function of the coordinates only is multiplication by that function; for example,

$$\mathbf{f}\psi = f(x, y, z)\psi. \tag{20.10}$$

The situation is not quite so simple if the observable is a momentum component. In this case, it turns out that we must have, for example,

$$\mathbf{p}_x \psi = -i\hbar \frac{\partial \psi}{\partial x}. \tag{20.11}$$

The proof of this statement is beyond our scope here; an argument for plausibility appeared in Section 19.13. For the component of momentum along any Cartesian coordinate q, we associate the differential operator,

$$\mathbf{p}_q = \frac{h}{2\pi i}\frac{\partial}{\partial q} = -i\hbar\frac{\partial}{\partial q}. \tag{20.12}$$

The quantity $\langle A \rangle$ in Eq. (20.8) is sometimes called the "average" value of A; for example, in Section 19.14 we used this terminology. The meaning of "average" in this context can be misinterpreted; see the discussion of this point in Section 20.4 in relation to Eqs. (20.29) through (20.31).

20.3 MATHEMATICAL INTERLUDE: OPERATOR ALGEBRA

An operator changes a function into another function according to a rule. Suppose we have a function $w(x, y, z)$ and an operator $\mathbf{x}$, defined by

$$\mathbf{x}w(x, y, z) = xw(x, y, z). \tag{20.13}$$

The function $w(x, y, z)$ is the *operand*; the operator acting on w changes w into xw. This is one example of the type of operator described in Eq. (20.10). Similarly we may have a differential operator such as the one in Eq. (20.12):

$$\mathbf{p}_x w = -i\hbar\frac{\partial w}{\partial x}. \tag{20.14}$$

This operator replaces the function w by its partial derivative with respect to x multiplied by the constant, $-i\hbar$.

Operators may be combined by addition; if $\boldsymbol{\alpha}$ and $\boldsymbol{\beta}$ are two operators on the same function, then

$$(\boldsymbol{\alpha} + \boldsymbol{\beta})w = \boldsymbol{\alpha}w + \boldsymbol{\beta}w, \tag{20.15}$$

since the new functions $\boldsymbol{\alpha}w$ and $\boldsymbol{\beta}w$ are simply functions, it is clear that operator addition is commutative; that is,

$$(\boldsymbol{\alpha} + \boldsymbol{\beta})w = (\boldsymbol{\beta} + \boldsymbol{\alpha})w. \tag{20.16}$$

Operators may also be combined by multiplication, which may be defined by

$$\boldsymbol{\alpha}\boldsymbol{\beta}w = \boldsymbol{\alpha}(\boldsymbol{\beta}w). \tag{20.17}$$

This equation states that to form the function corresponding to $\boldsymbol{\alpha}\boldsymbol{\beta}w$, we first form the function $\boldsymbol{\beta}w$, then perform the operation $\boldsymbol{\alpha}$ on the new function $\boldsymbol{\beta}w$. For example, suppose that for $w(x, y, z)$

$$\boldsymbol{\alpha}w = xw \qquad \text{and} \qquad \boldsymbol{\beta}w = \frac{\partial w}{\partial x};$$

then

$$\alpha\beta w = \alpha(\beta w) = \alpha\left(\frac{\partial w}{\partial x}\right) = x\,\frac{\partial w}{\partial x}; \qquad (20.18)$$

but, note that

$$\beta\alpha w = \beta(\alpha w) = \beta(xw) = \frac{\partial(xw)}{\partial x} = w + x\,\frac{\partial w}{\partial x}. \qquad (20.19)$$

It is apparent that in general operator multiplication is not commutative.

$$\alpha\beta w \neq \beta\alpha w.$$

The *commutator* γ is defined by $\alpha\beta - \beta\alpha = \gamma$. In this example, $(\alpha\beta - \beta\alpha)w = -w$ so that the commutator, γ, is multiplication by -1; $\gamma w = -w$. If, for all w,

$$(\alpha\beta - \beta\alpha)w = 0 \qquad (20.20)$$

then the operators commute. As will be seen in Section 20.4, the commutation properties of quantum-mechanical operators have great significance for the properties of a system.

Repeated applications of an operator are handled in the manner of Eq. (20.17):

$$\alpha^2 w = \alpha(\alpha w).$$

If, for example, $\alpha w = xw$, then

$$\alpha^2 w = \alpha(\alpha w) = \alpha(xw) = x(xw) = x^2 w.$$

If $\alpha w = (\partial w/\partial x)$, then

$$\alpha^2 w = \alpha(\alpha w) = \alpha\left(\frac{\partial w}{\partial x}\right) = \frac{\partial}{\partial x}\left(\frac{\partial w}{\partial x}\right) = \frac{\partial^2 w}{\partial x^2}.$$

If an operator α is such that in operating on two different operands, v and w, the relation

$$\alpha[c_1 v + c_2 w] = c_1 \alpha v + c_2 \alpha w \qquad (20.21)$$

is fulfilled, in which c_1 and c_2 are constants, the operator is said to be *linear*. For example, let $\alpha = \partial/\partial x$, then

$$\frac{\partial}{\partial x}(c_1 v + c_2 w) = c_1 \frac{\partial v}{\partial x} + c_2 \frac{\partial w}{\partial x},$$

so that the differential operator is linear. All the quantum-mechanical operators are linear.

As an example of a nonlinear operator, suppose that $\alpha f = e^f$ and $\alpha g = e^g$. Then $\alpha(c_1 f + c_2 g) = e^{c_1 f + c_2 g}$, which is obviously not equal to $c_1 e^f + c_2 e^g$. This operator is therefore not linear.

20.4   THE SCHRÖDINGER EQUATION

20.4.1   Postulate III

The probability amplitude, $\Psi(x, y, z, t)$, must satisfy the differential equation

$$\mathbf{H}\Psi - i\hbar\,\frac{\partial\Psi}{\partial t} = 0, \qquad (20.22)$$

in which $\mathbf{H}$ is the Hamiltonian operator.

To construct the Hamiltonian operator for one particle we write down the classical expression for the total energy,

$$\frac{1}{2m}(p_x^2 + p_y^2 + p_z^2) + V(x, y, z).$$

The first three terms are the kinetic energy; the fourth term is the potential energy. Then we replace the classical momentum components by their quantum-mechanical operators:

$$p_x \rightarrow \mathbf{p}_x = -i\hbar \frac{\partial}{\partial x}$$

$$p_x^2 \rightarrow \mathbf{p}_x^2 = -i\hbar \frac{\partial}{\partial x}\left(-i\hbar \frac{\partial}{\partial x}\right) = -\hbar^2 \frac{\partial^2}{\partial x^2},$$

so that

$$\mathbf{H} = -\frac{\hbar^2}{2m}\left[\frac{\partial^2}{\partial x^2} + \frac{\partial^2}{\partial y^2} + \frac{\partial^2}{\partial z^2}\right] + V(x, y, z). \tag{20.23}$$

We can conveniently introduce the abbreviation

$$\nabla^2 \equiv \frac{\partial^2}{\partial x^2} + \frac{\partial^2}{\partial y^2} + \frac{\partial^2}{\partial z^2}, \tag{20.24}$$

where ∇^2 is the Laplacian operator, to obtain

$$\mathbf{H} = -\frac{\hbar^2}{2m}\nabla^2 + V(x, y, z). \tag{20.25}$$

The first term on the right is the operator for the kinetic energy of the particle,

$$\mathbf{K} = -\frac{\hbar^2}{2m}\nabla^2,$$

while the second term is the operator for the potential energy,

$$\mathbf{V} = V(x, y, z).$$

If there are two particles, masses m_1 and m_2, with coordinates x_1, y_1, z_1 and x_2, y_2, z_2, then the classical energy would have the form

$$E = \frac{1}{2m_1}(p_{x_1}^2 + p_{y_1}^2 + p_{z_1}^2) + \frac{1}{2m_2}(p_{x_2}^2 + p_{y_2}^2 + p_{z_2}^2) + V(x_1, y_1, z_1, x_2, y_2, z_2).$$

The corresponding Hamiltonian operator would be

$$\mathbf{H} = -\frac{\hbar^2}{2m_1}\nabla_1^2 - \frac{\hbar^2}{2m_2}\nabla_2^2 + V(x_1, y_1, z_1, x_2, y_2, z_2),$$

in which ∇_1^2 is the Laplacian operator containing the coordinates of the first particle and ∇_2^2 is the operator containing the coordinates of the second particle. The extension to a system of many particles is obvious.

In the simple case where the potential energy is time independent, the solution to Eq. (20.22) has the form

$$\Psi(x, y, z, t) = \psi(x, y, z)f(t). \tag{20.26}$$

Then, $\mathbf{H}\Psi = f(t)\mathbf{H}\psi$, and the Schrödinger equation, Eq. (20.22), becomes

$$f(t)\mathbf{H}\psi - i\hbar\psi\frac{df}{dt} = 0.$$

Dividing by ψf and transposing we obtain

$$\frac{1}{\psi}\mathbf{H}\psi = \frac{i\hbar}{f(t)}\frac{df}{dt}. \tag{20.27}$$

The left-hand side of this equation is a function only of the coordinates, while the right-hand side is a function only of time. If we vary the coordinates keeping time constant, the left-hand side would appear to vary but in fact it does not since the right-hand side remains constant. It follows that both the members of Eq. (20.27) are equal to a constant, which we designate by E. Then

$$\frac{1}{f}\frac{df}{dt} = \frac{E}{i\hbar} = -\frac{iE}{\hbar}$$

and therefore

$$f = Ae^{-iEt/\hbar}, \tag{20.28}$$

where A is a constant; also,

$$\mathbf{H}\psi = E\psi. \tag{20.29}$$

Equation (20.29) is a differential equation, the time-independent Schrödinger equation. By solving the Schrödinger equation, we obtain the function $\psi(x, y, z)$ from which we can calculate expectation values of the observables associated with the system by use of the appropriate forms of Eq. (20.8).

The Schrödinger equation has a special form. Whenever we have an operator, $\boldsymbol{\alpha}$, such that

$$\boldsymbol{\alpha}f = af \tag{20.30}$$

where a is a constant, then a is called an *eigenvalue* of the operator and f is called an *eigenfunction* of the operator $\boldsymbol{\alpha}$. (Rather less frequently these are called characteristic values and characteristic functions.)

Thus, by Eq. (20.29), ψ is an eigenfunction of the Hamiltonian operator and E is the corresponding eigenvalue. This means that the energy of the system has the exact value E. Since the Hamiltonian operator corresponds to the total energy of the system, we calculate the expectation value of the total energy, $\langle E \rangle$, by applying Eq. (20.8):

$$\langle E \rangle = \int \psi^*\mathbf{H}\psi \, d\tau.$$

Substituting for $\mathbf{H}$ from Eq. (20.29) we obtain

$$\langle E \rangle = \int \psi^*E\psi \, d\tau = E \int \psi^*\psi \, d\tau;$$

$$\langle E \rangle = E, \tag{20.31}$$

where the second form follows since E is constant and the final form since $\int \psi^*\psi \, d\tau = 1$. Eq. (20.31) says that the expectation value for the total energy of the system is precisely equal to the constant E introduced in solving Eq. (20.22), or Eq. (20.27).

The apparently obvious result in Eq. (20.31) is important enough to make us digress for a moment. Suppose we calculate the expectation value of the square of the energy of the system:

$$\langle E^2 \rangle = \int \psi^* \mathbf{H}^2 \psi \, d\tau = \int \psi^* \mathbf{H}(\mathbf{H}\psi) \, d\tau$$

$$= \int \psi^* \mathbf{H}(E\psi) \, d\tau = E \int \psi^*(\mathbf{H}\psi) \, d\tau = E \int \psi^* E\psi \, d\tau;$$

$$\langle E^2 \rangle = E^2 = \langle E \rangle^2. \tag{20.32}$$

In Eq. (20.32) we have the result that the expectation value of the square of the energy is equal to the square of the expectation value of the energy. This could not be correct if the energy were in some way distributed. The reader will recall that in dealing with the Maxwell distribution of molecular speeds we found that

$$\langle c^2 \rangle = \frac{3RT}{M}, \qquad \langle c \rangle = \sqrt{\frac{8RT}{\pi M}}.$$

Of these, the first, $\langle c^2 \rangle$, is not equal to the square, $\langle c \rangle^2$, of the second. In averaging the square of a value over a distribution, the higher values are always accentuated. The fact that Eq. (20.32) is correct means that we are not averaging bits of energy here and there in space using $\psi^*\psi$ as some kind of distribution function; if we were, Eq. (20.32) could not be correct. The energy of the system in the state described by ψ has a precise value, determined by the Schrödinger equation as well as additional conditions that we will discuss further. Such a state of the system is called an eigenstate (or characteristic state) of the system.

This result is general in the following sense. Whenever the wave function of a system is an eigenfunction of an operator corresponding to an observable, then that observable has a precise value.

In general, solution of the Schrödinger equation for a system yields a set of functions, $\psi_j \ (j = 1, 2, \ldots)$, each of which describes a particular state of the system. Each such state would have its particular expectation values of the observables. Assume that in any state of the system two different observables, A and B, have precise values. This implies that ψ_j is an eigenfunction of *both* of the operators $\mathbf{A}$ and $\mathbf{B}$; that is

$$\mathbf{A}\psi_j = a_j\psi_j, \qquad \mathbf{B}\psi_j = b_j\psi_j,$$

where a_j and b_j are the eigenvalues. Then if ψ_j is normalized,

$$\langle A \rangle = \int \psi_j^*(\mathbf{A}\psi_j) \, d\tau = \int \psi_j^* a_j \psi_j \, d\tau = a_j.$$

and

$$\langle B \rangle = \int \psi_j^*(\mathbf{B}\psi_j) \, d\tau = \int \psi_j^* b_j \psi_j \, d\tau = b_j.$$

If we construct the commutator, $\mathbf{AB} - \mathbf{BA}$, and operate on the wave function, we have

$$(\mathbf{AB} - \mathbf{BA})\psi_j = \mathbf{A}(\mathbf{B}\psi_j) - \mathbf{B}(\mathbf{A}\psi_j) = \mathbf{A}(b_j\psi_j) - \mathbf{B}(a_j\psi_j)$$

$$= b_j(\mathbf{A}\psi_j) - a_j(\mathbf{B}\psi_j) = b_j a_j \psi_j - a_j b_j \psi_j = (b_j a_j - a_j b_j)\psi_j = 0.$$

Thus the operators $\mathbf{A}$ and $\mathbf{B}$ commute:

$$\mathbf{AB}\psi_j = \mathbf{BA}\psi_j.$$

Conversely, it can be shown that if the operators for two observables commute then the two observables can have precise values simultaneously. If the two operators do not commute, then it is not possible for the corresponding observables to have precise values simultaneously. Consider the operators for the x coordinate and the x component of momentum; then $\mathbf{x}\psi = x\psi$ and $\mathbf{p}_x\psi = -i\hbar\,\partial\psi/\partial x$, so that

$$\mathbf{x}\mathbf{p}_x\psi - \mathbf{p}_x\mathbf{x}\psi = -i\hbar\left[x\frac{\partial\psi}{\partial x} - \frac{\partial}{\partial x}(x\psi)\right] = -i\hbar[-\psi] = i\hbar\psi$$

These operators do not commute; consequently, x and p_x cannot simultaneously have precise values. This is the basis for the Heisenberg uncertainty principle, which we will discuss in more detail later.

One final remark on the Schrödinger equation. For physical sense we require that the constant E, the energy, be a real quantity. This requires that $E^* = E$; consequently, since

$$E = \int \psi^*(\mathbf{H}\psi)\,d\tau, \quad \text{and} \quad E^* = \int \psi(\mathbf{H}\psi)^*\,d\tau,$$

then we must have

$$\int \psi^*(\mathbf{H}\psi)\,d\tau = \int \psi(\mathbf{H}\psi)^*\,d\tau. \tag{20.23}$$

An operator that satisfies the condition in Eq. (20.33) is said to be *Hermitian*. Conversely, *the eigenvalues of Hermitian operators are all real quantities.* In quantum mechanics we deal only with Hermitian operators. The definition of an Hermitian operator is somewhat more general than Eq. (20.33) would imply. The operator $\mathbf{H}$ is Hermitian if

$$\int \psi_1^*(\mathbf{H}\psi_2)\,d\tau = \int \psi_2(\mathbf{H}\psi_1)^*\,d\tau \tag{20.34}$$

in which ψ_1 and ψ_2 are the same or different operands.

20.5 THE EIGENVALUE SPECTRUM

The Schrödinger equation

$$\mathbf{H}\psi = \mathbf{K}\psi + \mathbf{V}\psi = E\psi$$

of itself admits a variety of solutions, depending on the nature of the potential function V. However, we must impose further restrictions. Even with a particular potential function, among the possible solutions we frequently find solutions that are inadmissible; for example, the requirement that the wave function be single valued everywhere eliminates some functions; the requirement of quadratic integrability is a very restrictive condition; the imposition of particular boundary conditions in some problems reduces the number of acceptable solutions.

The net effect of these restrictions is that in some cases, E may have any value; then we speak of a continuous spectrum of eigenvalues of E. In other cases, E may be restricted to certain particular values; then we have a discrete spectrum of eigenvalues. In these latter cases we say that E is *quantized*. Ordinarily, for each boundary condition we impose, we introduce a quantization of some observable.

If the energy is quantized and restricted to certain values, $E_1, E_2, E_3, \ldots, E_n, \ldots$, then corresponding to each of these values there is at least one eigenfunction ψ_n, such that

$$\mathbf{H}\psi_n = E_n\psi_n,$$

so that in general we deal with a set of eigenfunctions, $\psi_1, \psi_2, \ldots, \psi_n$. If for each energy level (eigenvalue), there is only one eigenfunction, then the set of eigenfunctions and the set of eigenstates is *nondegenerate*. If for the nth eigenstate there are g_n eigenfunctions, $\psi_{n1}, \psi_{n2}, \ldots$, such that

$$\mathbf{H}\psi_{nk} = E_n\psi_{nk}, \qquad k = 1, 2, \ldots, g_n, \tag{20.35}$$

then the nth eigenstate is g_n-fold degenerate. To speak of a three-fold degenerate level simply means that corresponding to one particular energy value there are three distinct eigenfunctions.

The existence of a degeneracy poses a problem in describing the state. Suppose the eigenstate with energy E is three-fold degenerate, with eigenfunctions ψ_1, ψ_2, and ψ_3. This means that

$$\mathbf{H}\psi_1 = E\psi_1, \qquad \mathbf{H}\psi_2 = E\psi_2 = E\psi_2, \qquad \mathbf{H}\psi_3 = E\psi_3.$$

Since the Hamiltonian operator is linear, this implies that if we construct a linear combination,

$$\phi_1 = c_{11}\psi_1 + c_{12}\psi_2 + c_{13}\psi_3,$$

where c_{11}, c_{12}, and c_{13} are constants, then in view of the linear character of $\mathbf{H}$,

$$\begin{aligned}
\mathbf{H}\phi_1 &= c_{11}\mathbf{H}\psi_1 + c_{12}\mathbf{H}\psi_2 + c_{13}\mathbf{H}\psi_3 \\
&= c_{11}E\psi_1 + c_{12}E\psi_2 + c_{13}E\psi_3 \\
&= E(c_{11}\psi_1 + c_{12}\psi_2 + c_{13}\psi_3) \\
\mathbf{H}\phi_1 &= E\phi_1.
\end{aligned}$$

The linear combination ϕ_1 is also an eigenfunction of the Hamiltonian operator with E as an eigenvalue; therefore ϕ_1 is an appropriate description of the eigenstate. We may construct two additional independent linear combinations of the same type:

$$\phi_2 = c_{21}\psi_1 + c_{22}\psi_2 + c_{23}\psi_3;$$

$$\phi_3 = c_{31}\psi_1 + c_{32}\psi_2 + c_{33}\psi_3.$$

In general, we cannot prefer the description ψ_1, ψ_2, ψ_3 to the description ϕ_1, ϕ_2, ϕ_3. Out of the entire collection of possibilities, we are required for completeness to choose any three linearly independent eigenfunctions.* Beyond that, it is a matter only of convenience. As we shall see later, in systems with certain symmetry properties, some combinations are more convenient than others.

The foregoing gives an example of the *principle of superposition*. Because of the linear character of the Schrödinger equation (linear character of the Hamiltonian), if a state is equally well described by either of two functions, for example, then it is equally well described by any two independent linear combinations of those functions.

---

*   Linearly independent functions are such that no relation $a_1\phi_1 + a_2\phi_2 + a_3\phi_3 = 0$ exists (the a_i are constants) other than the trivial one, $a_1 = a_2 = a_3 = 0$.

★ 20.6 EXPANSION THEOREM

The Hermitian property of quantum-mechanical operators leads to an important result. Consider two eigenfunctions, ψ_n and ψ_k, of the Hamiltonian operator; we have

$$\mathbf{H}\psi_n = E_n\psi_n \quad \text{and} \quad \mathbf{H}\psi_k = E_k\psi_k. \tag{20.36}$$

We take the complex conjugate of the second equation, $(\mathbf{H}\psi_k)^* = E_k\psi_k^*$; then we multiply the first equation by ψ_k^* and the second by ψ_n, and integrate over all space. This yields

$$\int \psi_k^* \mathbf{H}\psi_n \, d\tau = E_n \int \psi_k^* \psi_n \, d\tau, \qquad \int \psi_n(\mathbf{H}\psi_k)^* \, d\tau = E_k \int \psi_k^* \psi_n \, d\tau.$$

Subtracting these two equations and transposing,

$$(E_n - E_k) \int \psi_k^* \psi_n \, d\tau = \int \psi_k^*(\mathbf{H}\psi_n) \, d\tau - \int \psi_n(\mathbf{H}\psi_k)^* \, d\tau.$$

By Eq. (20.34), the Hermitian property, the two integrals on the right are equal; hence

$$(E_n - E_k) \int \psi_k^* \psi_n \, d\tau = 0.$$

If $E_n \neq E_k$, then

$$\int \psi_k^* \psi_n \, d\tau = 0 \qquad k \neq n. \tag{20.37}$$

Equation (20.37) is the orthogonality relation. Two eigenfunctions of a linear Hermitian operator corresponding to distinct eigenvalues are *orthogonal*.* Note that if $k = n$, our normalization requirement is

$$\int \psi_n^* \psi_n \, d\tau = 1. \tag{20.38}$$

These two conditions are usually written

$$\int \psi_k^* \psi_n \, d\tau = \delta_{nk}, \tag{20.39}$$

where the function δ_{nk} (of n and k) is called the Kronecker delta and is defined by

$$\begin{aligned} \delta_{nk} &= 1, \qquad n = k, \\ \delta_{nk} &= 0, \qquad n \neq k. \end{aligned} \tag{20.40}$$

---

*   This concept of orthogonality can be obtained by extension from the concept of orthogonality of two ordinary vectors in three-dimensional space. If the x, y, and z components are a_x, a_y, a_z for the first vector and b_x, b_y, b_z for the second vector, then the condition for orthogonality is

$$a_x b_x + a_y b_y + a_z b_z = 0 \qquad \text{or} \qquad \sum_{i=1}^{3} a_i b_i = 0.$$

As a simple illustration, take the first vector along the x-axis and the second along the y-axis, then $a_y = a_z = 0$ and $b_x = b_z = 0$; the left-hand side becomes $a_x \cdot 0 + 0 \cdot b_y + 0 \cdot 0$ which is clearly equal to zero. The sum of products on the left-hand side is called the scalar product of the two vectors, the summation is taken over the components. The integral in Eq. (20–37) is called the Hermitian scalar product of the two functions ψ_k and ψ_n; summation over the components in the ordinary vector is replaced by integration over the variables of the functions.

(Note that $\delta_{nk} = \delta_{kn}$.) The set of functions ψ_n that satisfies Eq. (20.39) is called an ortho-normal set. This property of the eigenfunctions of Hermitian operators allows us to expand an arbitrary function in the domain of definition of the orthonormal set in terms of members of that set.

Suppose ϕ is a function defined in the domain of the orthonormal set; then assume that we can write ϕ as a series with c_n as constant coefficients

$$\phi = \sum_{n=1}^{\infty} c_n \psi_n. \tag{20.41}$$

To determine the coefficients of this series we multiply by ψ_k^* and integrate over the entire space, so that

$$\int \psi_k^* \phi \, d\tau = \sum_{n=1}^{\infty} c_n \int \psi_k^* \psi_n \, d\tau.$$

By Eq. (20.39) this becomes

$$\int \psi_k^* \phi \, d\tau = \sum_{n=1}^{\infty} c_n \delta_{nk}.$$

In view of the properties of δ_{nk}, the sum on the right-hand side reduces to one term, c_k, so we have for the coefficients:

$$c_k = \int \psi_k^* \phi \, d\tau. \tag{20.42}$$

This very simple and very elegant means of expanding a function in terms of an ortho-normal set is extremely useful not only in quantum mechanics but in many other areas of theoretical physics.

20.7 CONCLUDING REMARKS ON THE GENERAL EQUATIONS

Thus far we have developed equations that are not restricted to particular systems, although in the main we have kept to systems that are in stationary states. These are systems having energy that is precise and unchanging in time.

A great deal more could be said in general; nothing has been said of the uncertainty principle, for example. The postulates of quantum mechanics have not been exhausted by those presented in this chapter. However, at this point we take up particular examples, in the belief that the skeleton presented so far will be less repulsive if it is fleshed out a bit.

Finally, a remark or two about the treatment at the end of Chapter 19, which attempts to relate the classical wave equation and the Schrödinger equation. It should be clear that whether or not the Schrödinger equation is correct depends only on its predictions of behavior and not in the least on whether or not there is some means of transforming the classical wave equation into the Schrödinger equation. On the other hand, the Schrödinger treatment of a system is required to reduce to Newtonian mechanics in the limit as Planck's constant approaches zero, or in the limit of large masses and distances. Suffice it to say that the Schrödinger equation does reduce properly in these circumstances.

In passing, it should be mentioned that since $|\psi|^2$ is a probability per unit volume, it follows that ψ has dimension $(length)^{-3/2}$ in three-dimensional space. For a one-dimensional problem, the volume element is simply a length, so the dimension of ψ is $(length)^{-1/2}$.

QUESTIONS

20.1 What are the consequences for measurement when two operators commute? When they do not commute?

20.2 Show why the expectation value $\langle p_x \rangle$ of the x component of momentum cannot be written either as $\int (-i\hbar\, \partial/\partial x)(\psi^*\psi)\, d\tau$ or $\int \psi^*\psi(-i\hbar\, \partial/\partial x)\, d\tau$.

20.3 What is the distinction between an operator and its associated eigenvalues?

20.4 Why are quantum-mechanical operators Hermitian?

20.5 Criticize the statement: If wave functions ψ_1 and ψ_2 are orthogonal, then ψ_1 vanishes everywhere that ψ_2 is finite.

PROBLEMS

20.1 Consider the differential equation, $d^2u/dx^2 + k^2u = 0$. Show that two possible solutions are: $u_1 = \sin kx$ and $u_2 = \cos kx$. Then show that, if a_1 and a_2 are constants, $a_1u_1 + a_2u_2$ is also a solution.

20.2 Show that the function Ae^{ax} is an eigenfunction of the differential operator, (d/dx).

20.3 If $f = x^n$, show that $x(df/dx) = nf$, and thus that f is an eigenfunction of the operator $x(d/dx)$.

20.4 Find the commutator for the operators, x^2 and d^2/dx^2.

20.5 The operators for the components of angular momentum are:

$$\mathbf{M}_x = -i\hbar\left(y\frac{\partial}{\partial z} - z\frac{\partial}{\partial y}\right), \qquad \mathbf{M}_y = -i\hbar\left(z\frac{\partial}{\partial x} - x\frac{\partial}{\partial z}\right), \qquad \mathbf{M}_z = -i\hbar\left(x\frac{\partial}{\partial y} - y\frac{\partial}{\partial x}\right).$$

Show that: $\mathbf{M}_x\mathbf{M}_y - \mathbf{M}_y\mathbf{M}_x = i\hbar\mathbf{M}_z$, and that $\mathbf{M}^2\mathbf{M}_z = \mathbf{M}_z\mathbf{M}^2$, in which $\mathbf{M}^2 \equiv \mathbf{M}_x^2 + \mathbf{M}_y^2 + \mathbf{M}_z^2$. Derive the corresponding commutation rules for $\mathbf{M}_y$ and $\mathbf{M}_z$, for $\mathbf{M}_z$ and $\mathbf{M}_x$, and for $\mathbf{M}^2$ with $\mathbf{M}_x$ and with $\mathbf{M}_y$.

20.6 The function, $f(x) = 3x^2 - 1$, is an eigenfunction of the operator, $\mathbf{A} = -(1 - x^2)(d^2/dx^2) + 2x(d/dx)$. Find the eigenvalue corresponding to this eigenfunction.

20.7 Show that the function, $f(y) = (16y^4 - 48y^2 + 12)e^{-y^2/2}$ is an eigenfunction of the operator, $\mathbf{B} = -(d^2/dy^2) + y^2$, and calculate the eigenvalue.

20.8 Show that the function, $f(y) = y(6 - y)e^{-y/3}$ is an eigenfunction of the operator, $\mathbf{F} = -(d^2/dy^2) - (2/y)(d/dy) - (2/y) + (2/y^2)$, and calculate the eigenvalue.

20.9 Show that in the interval $-1 \le x \le +1$ the polynomials, $P_0(x) = a_0$, $P_1(x) = a_1 + b_1x$, $P_2(x) = a_2 + b_2x + c_2x^2$ are the first members of an orthogonal set of functions. Evaluate the constants $a_0, a_1, b_1, \ldots$, and so on.

20.10 Show that in the interval $0 \le \phi \le 2\pi$ the functions, $e^{in\phi}$, where $n = 0, \pm 1, \pm 2, \ldots$, form an orthogonal set.

21

The Quantum Mechanics of Some Simple Systems

21.1 INTRODUCTION

In the quantum mechanical discussion of a system the following general scheme should be kept in mind. To begin, in principle we obtain the wave function for the system by solving the Schrödinger equation for that system. In practice we may have to guess at the form of the wave function. After obtaining the wave function we calculate expectation values for any observable by application of the equation

$$\langle a \rangle = \int \psi^* \alpha \psi \, d\tau, \tag{21.1}$$

in which α is the operator corresponding to the observable a. It follows that, in spite of its lack of direct physical significance, *the wave function is implicitly a complete description of the system*. In fact, we will often refer to ψ as "the description of the system" rather than the "wave function of the system."

The wave function is such that the product $\psi^* \psi = |\psi|^2$ is the probability density. Therefore $|\psi|^2 \, d\tau$ is the probability of finding the particle in the volume element $d\tau$. Since the particle has unit probability of being somewhere,

$$\int |\psi|^2 \, d\tau = 1, \tag{21.2}$$

where the integration is carried over the entire coordinate space.

Finally, the Schrödinger equation is a linear differential equation. The solutions of the equation are an entire set of functions: $\psi_1, \psi_2, \psi_3, \psi_4, \ldots$, each of which describes a state of the system. The property of linearity in the differential equation means that linear combinations of these descriptions, or of a subset of them, are also descriptions of the system. The descriptions of a system can thus be superposed to obtain new descriptions of the system (the principle of superposition). For example, if ψ_1 and ψ_2 are two descriptions of the system, then the linear combinations,

$$\phi_1 = a_1\psi_1 + a_2\psi_2, \qquad \phi_2 = b_1\psi_1 + b_2\psi_2, \tag{21.3}$$

where the a's and the b's are arbitrary constants, are also descriptions of the system.

With these four fundamental properties in mind, we can understand a great deal of the consequences of the quantum mechanics. We begin by discussing a few simple systems in detail.

21.2 THE FREE PARTICLE

Consider a particle of mass m which moves in the absence of external forces along the x-axis only; the absence of forces implies that the potential energy is constant, so for convenience we may choose $V = 0$. The components of momentum along the y- and z-axes are zero; that along the x-axis is p_x. The total energy is a constant and is equal to the kinetic energy; the classical description is

$$E = \frac{p_x^2}{2m}. \tag{21.4}$$

Replacing p_x by $\mathbf{p}_x$, as in Section 20.4, we obtain the Schrödinger equation for this system,

$$E\psi = -\frac{\hbar^2}{2m}\frac{d^2\psi}{dx^2},$$

or

$$\frac{d^2\psi}{dx^2} + \left(\frac{2mE}{\hbar^2}\right)\psi = 0. \tag{21.5}$$

Since E is a constant, this differential equation has two solutions:*

$$\psi_1 = Ae^{i\sqrt{2mE}x/\hbar} \qquad \text{and} \qquad \psi_2 = Be^{-i\sqrt{2mE}x/\hbar},$$

where A and B are arbitrary constants.

If we operate on ψ_1 with the momentum operator, we obtain

$$\mathbf{p}_x\psi_1 = -i\hbar\frac{d\psi_1}{dx} = -i\hbar(i\sqrt{2mE}/\hbar)\psi_1$$

$$= \sqrt{2mE}\psi_1.$$

Similarly,

$$\mathbf{p}_x\psi_2 = -\sqrt{2mE}\psi_2.$$

These equations are typical eigenvalue relations; the constant $\sqrt{2mE}$ appearing on the right is an eigenvalue of the momentum operator.

The interpretation is that in a state described by ψ_1 the momentum of the particle has a fixed precise value, $\sqrt{2mE}$. The classical values of momentum according to Eq. (21.4) are $p_x = \pm\sqrt{2mE}$. Thus ψ_1 describes a particle moving in the $+x$ direction (p_x is positive) with

---

* The solutions are readily verified by substituting them into the equation.

the classical momentum. On the other hand, ψ_2 describes the particle moving in the $-x$ direction (p_x is negative) with the classical momentum. Since no other conditions are specified, the energy may have any value, and so may the momentum. The spectrum of energy eigenvalues is continuous, as is that of the momentum eigenvalues.

Using ψ_1, suppose that we calculate the probability density $\psi_1^*\psi_1$. Since

$$\psi_1 = Ae^{i\sqrt{2mE}x/\hbar}, \quad \text{then} \quad \psi_1^* = A^*e^{-i\sqrt{2mE}x/\hbar}.$$

Therefore

$$|\psi_1|^2 = \psi_1^*\psi_1 = A^*e^{-i\sqrt{2mE}x/\hbar}Ae^{i\sqrt{2mE}x/\hbar} = A^*A = |A|^2. \tag{21.6}$$

But A is a constant; therefore $|A|^2$ is a constant and is *independent of the value of x*. This means that the probability of finding the particle is the same everywhere along its path. Therefore, we can make no statement about its position. The momentum has a definite value, p_x, but we can know nothing about the position of the particle. If we use ψ_2, the momentum is definite, $-p_x$, but as with ψ_1 the position is completely indefinite.

21.3 PARTICLE IN A "BOX"

21.3.1 Wave Functions

In view of the indefiniteness in the position of the free particle, suppose that we enclose the particle in a "box" so that we know that its position lies within the boundaries of the "box." The "box" is made in the following way: let the potential energy of the particle be zero inside the box and infinitely large at the walls and everywhere outside the box. Since the particle cannot have an infinite potential energy, it will stay in the box where its potential energy is zero. Again we restrict the particle to move only along the x-axis. A plot of the potential energy as a function of x is shown in Fig. 21.1; L is the width of the box.

Since $V = 0$ inside the box, the Schrödinger equation has the same form as for the free particle, so the solutions are

$$\psi_1 = Ae^{i\sqrt{2mE}x/\hbar}, \quad \psi_2 = Be^{-i\sqrt{2mE}x/\hbar}.$$

However, we must place the following boundary conditions on the wave function (Fig. 21.1):

$$\psi = 0, \quad \text{when} \quad x \le 0,$$

$$\psi = 0, \quad \text{when} \quad x \ge L.$$

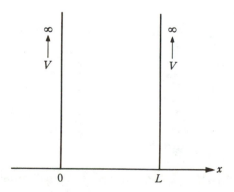

Figure 21.1 The potential-energy "box."

If these conditions were not fulfilled, the probability density $\psi^*\psi$ would be finite at the edges and outside of the box, where the potential energy is infinite. This is not possible.

These conditions cannot be satisfied by ψ_1 or ψ_2 individually. For example, at $x = 0$, $\psi_1(0) = A$; to satisfy the condition $\psi_1 = 0$, A would have to be zero. But if A is zero, then ψ_1 is zero everywhere. This would mean that $|\psi_1|^2 = 0$ everywhere; the particle isn't anywhere! The same difficulty appears with ψ_2. To avoid this we use the principle of superposition to construct a more general description,

$$\psi = A e^{i\sqrt{2mE}x/\hbar} + B e^{-i\sqrt{2mE}x/\hbar}, \tag{21.7}$$

and apply our conditions to ψ.

At $x = 0$, $\psi = 0$, so Eq. (21.7) becomes $0 = A + B$, or $B = -A$. Using this result, we obtain

$$\psi = A(e^{i\sqrt{2mE}x/\hbar} - e^{-i\sqrt{2mE}x/\hbar}).$$

By the Euler equation, $e^{iy} - e^{-iy} = 2i \sin y$, this becomes

$$\psi = 2iA \sin\left(\frac{\sqrt{2mE}\,x}{\hbar}\right) = C \sin\left(\frac{\sqrt{2mE}\,x}{\hbar}\right). \tag{21.8}$$

The first condition being satisfied, we look to the second: that $\psi = 0$ when $x = L$. At $x = L$, ψ becomes

$$\psi(L) = C \sin\left(\frac{\sqrt{2mE}\,L}{\hbar}\right) = 0.$$

This condition cannot be met by setting $C = 0$, because again the particle would be nowhere. The condition must be met by requiring that $\sin(\sqrt{2mE}\,L/\hbar) = 0$. If the argument of the sine is an integral multiple of π, then the sine is zero: that is,

$$\sin(\pm n\pi) = 0, \qquad n = 1, 2, 3, \ldots$$

The boundary condition is therefore fulfilled if, and only if,

$$\frac{\sqrt{2mE_n}\,L}{\hbar} = \pm n\pi, \qquad n = 1, 2, 3, \ldots \tag{21.9}$$

Since Eq. (21.9) indicates that the energy depends on n, E has been replaced by E_n. Using this result in Eq. (21.8), we obtain

$$\psi_n = \pm C \sin\frac{n\pi x}{L}. \tag{21.10}$$

The final condition is that the total probability of finding the particle in the box is unity:

$$\int_0^L \psi_n^*\psi_n \, dx = 1.$$

Since $\psi_n^* = \pm C^* \sin(n\pi x/L)$, we have

$$C^*C \int_0^L \sin^2\frac{n\pi x}{L} \, dx = 1.$$

The integral is equal to $L/2$ so $C^*CL/2 = 1$ or $|C|^2 = 2/L$. Choosing C as a real number, we have $C = \sqrt{2/L}$. The final description is

$$\psi_n = \sqrt{\frac{2}{L}} \sin \frac{n\pi x}{L}. \tag{21.11}$$

There are several curious things about this problem.

First of all, we write Eq. (21.9) in the form

$$E_n = \frac{n^2\pi^2\hbar^2}{2mL^2}, \qquad n = 1, 2, 3, \ldots \tag{21.12}$$

Since n may have only integral values, E may have only the special value given by Eq. (21.12) and may not have any other value. The energy in this system is *quantized*, and the integer n is called a quantum number. The spectrum of energy eigenvalues is discrete. In contrast, the energy of the free particle could have any value. In retrospect, we see that the quantization entered when we restricted the particle to the interior of the box.

The classical momentum corresponding to the energy value E_n is given formally by

$$p_x = \pm \frac{n\pi\hbar}{L} = \pm \frac{nh}{2L}, \qquad n = 1, 2, 3, \ldots \tag{21.13}$$

Another aspect of the situation is displayed if the de Broglie wavelength is introduced in Eq. (21.13). The de Broglie relation is $\lambda = h/|p_x|$, where we have used the absolute value of p_x, since λ is not a vector quantity; then Eq. (21.13) becomes

$$L = n\left(\frac{\lambda}{2}\right), \tag{21.14}$$

which requires that an integral number of half-wavelengths fit exactly in the length L. This situation is analogous to the possible vibrations of a string that is clamped at positions 0 and L. The permissible modes of vibration of the string are given by the same formula as Eq. (21.14). The value of the wave function for several values of n is shown in Fig. 21.2(a) and the probability density $\psi^*\psi$ is shown in Fig. 21.2(b). Note that the values of ψ in Fig. 21.2(a) look exactly like the displacements of a vibrating string clamped at 0 and L in its various modes of vibration.

The probability density in Fig. 21.2(b) is curious. When $n = 1$ the most probable position is at $L/2$, but all positions in the box have fairly large probability density. When $n = 2$, $\psi^*\psi$ vanishes at $L/2$! The particle has zero probability of being at $L/2$. Yet it has high probability of being at either side of the midpoint. This situation can be legitimately viewed in two ways.

1. The particle can get from one side of the midpoint to the other without ever passing through the midpoint. Surprising as this may seem, it is correct.

2. The particle is "smeared" over the entire space, the density of the smear being large on both sides and exactly zero at the center. This is the point of view we will usually adopt. After getting used to the notion of a "smeared" particle, this idea is quite comfortable. It is also a bit easier to become accustomed to than the idea of the particle being now here, then there, but never in between. This latter notion also tends to keep us entangled with classical ideas of mechanical motion. The smearing of the particle may be regarded as what we would see if we attached a light to the particle and took a time exposure of the motion. The probability densities in Fig. 21.2(b) represent this time average.

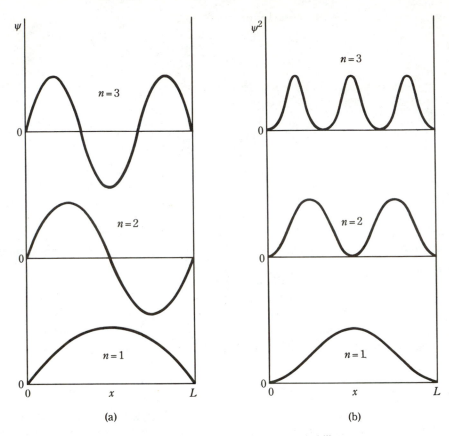

(a)   (b)

Figure 21.2   Wave functions and probability
densities for the particle in the box.

21.3.2 Energy Levels

The allowed values of energy given by Eq. (21.12) are called the energy levels of the system. The least energy the particle may possess is called the zero-point energy, or the ground-state energy E_0. Note that if $n = 0$, the wave function vanishes everywhere and we lose our particle. For the particle in the box the least energy value is that for $n = 1$, so that $E_0 = \pi^2\hbar^2/2mL^2 = h^2/8mL^2$. Then

$$E_n = n^2 E_0, \tag{21.15}$$

and the spacing, or energy gap, between the levels $n + 1$ and n is

$$E_{n+1} - E_n = [(n + 1)^2 - n^2]E_0 = (2n + 1)E_0. \tag{21.16}$$

Since $h = 6.626 \times 10^{-34}$ J s,

$$E_0 = \frac{5.49 \times 10^{-68} \text{ J}^2 \text{ s}^2}{mL^2}. \tag{21.17}$$

We apply Eqs. (21.15) through (21.17) to three examples.

■ **EXAMPLE 21.1**   Consider a ball bearing, $m = 1$ g (0.001 kg), in a box 10 cm (0.1 m) in length; then

$$E_0 = 5.49 \times 10^{-63} \text{ J}.$$

If the ball bearing has a velocity of 1 cm/s, then its kinetic energy is $\frac{1}{2}(0.001$ kg$)$ $(0.01 \text{ m/s})^2 = 5 \times 10^{-8}$ J. The quantum number n would be

$$n^2 = \frac{E_n}{E_0} = \frac{5 \times 10^{-8} \text{ J}}{5.49 \times 10^{-63} \text{ J}} \approx 10^{55},$$

so that $n = 3 \times 10^{27}$. The spacing between levels for this value of n is

$$[2(3 \times 10^{27}) + 1]5.49(10^{-63} \text{ J}) \approx 3 \times 10^{-35} \text{ J}.$$

Thus to observe the quantization in this system would require us to distinguish between an energy of 5×10^{-8} J and $5 \times 10^{-8} \pm 3 \times 10^{-35}$ J. This type of precision is impossible, of course, so we do not observe the quantization in the kinetic energy of the ball bearing; it behaves as though it could have any value of kinetic energy, and the most convenient way to treat this system is to use the classical mechanical laws of Newton. This example illustrates the Bohr "correspondence principle," a rough statement of which is: in the limit of high quantum numbers, the results of quantum mechanics approach those of classical mechanics. Whenever we deal with massive particles such as ball bearings and golf balls, quantum mechanics reduces to classical mechanics.

■ **EXAMPLE 21.2**   Consider an electron, $m \approx 10^{-30}$ kg, in a box the size of an atom, 10^{-10} m; then

$$E_0 = 5.5 \times 10^{-18} \text{ J} \approx 34 \text{ eV}.$$

This 34 eV is in the energy region of very short x-rays. The spacing between levels with $n = 1$ and $n = 2$ is $E_2 - E_1 = 102$ eV. If the electron dropped from level 2 to level 1, a quantum of x-radiation of energy $= 102$ eV would be emitted. The energy quantization is readily observed here.

■ **EXAMPLE 21.3**   Consider an electron, $m \approx 10^{-30}$ kg, in a box the size of the nucleus, 10^{-14} m; then

$$E_0 = 5.5 \times 10^{-10} \text{ J} = 3.5 \times 10^9 \text{ eV} = 3500 \text{ MeV}.$$

This is a fantastic energy. The coulombic energy that might be expected to hold an electron in the nucleus would be

$$V = \frac{-e^2}{4\pi\epsilon_0 r} = \frac{-(1.6 \times 10^{-19} \text{ C})^2}{4\pi(8.85 \times 10^{-12} \text{ C/V m})(10^{-14} \text{ m})} = -2.3 \times 10^{-14} \text{ J} = -0.14 \text{ MeV}.$$

It is clear that if the electron were confined in the nucleus, it would have an inordinately large kinetic energy, which would not be compensated by the electrostatic attraction of the positive charge. It could not be held there. For this reason, among others, only heavy particles such as protons, neutrons, and so on, are supposed to be present within the nucleus. This argument also explains why the electron in the hydrogen atom does not fall into the nucleus; for it to exist in the nucleus this fantastically high kinetic energy is required.

Finally, we observe that for the allowed energy levels in the box, the half-wavelength of the particle must fit in the box exactly an integral number of times.

If we are to fit the electron in the nucleus, its half-wavelength must be the size of the nucleus or smaller. This very small wavelength implies, by the de Broglie equation $p = h/\lambda$, a very high momentum and consequently the inordinately high kinetic energy.

21.3.3 Expectation Values of Position and Momentum

We return now to the composite description of the system given by Eq. (21.7). The first term in Eq. (21.7) represents motion of the particle along the $+x$-axis with momentum $+p_x$, while the second term represents motion of the particle in the $-x$ direction with momentum $-p_x$. Confining the particle in the box forces us to use a composite description embodying motion in both directions. In a certain sense the particle is moving in both directions! Or rather, we cannot decide from the description whether the particle is coming or going!

To calculate the expectation value of the momentum of the particle in the box, we rewrite Eq. (21.10) in the form

$$\psi_n = A(e^{in\pi x/L} - e^{-in\pi x/L}).$$

Normalization requires $A = (1\sqrt{2L})$. Hence

$$\psi_n = \frac{1}{\sqrt{2L}}(e^{in\pi x/L} - e^{-in\pi x/L}). \tag{21.18}$$

At this point we observe that the functions

$$\phi_n = \frac{1}{\sqrt{L}}e^{in\pi x/L}, \qquad \phi_{-n} = \frac{1}{\sqrt{L}}e^{-in\pi x/L} \tag{21.19}$$

are eigenfunctions of the momentum operator corresponding to different discrete eigenvalues; that is,

$$\mathbf{p}_x\phi_n = -i\hbar\frac{\partial\phi_n}{\partial x} = \frac{n\pi\hbar}{L}\phi_n = p_n\phi_n,$$

$$\mathbf{p}_x\phi_{-n} = -i\hbar\frac{\partial\phi_{-n}}{\partial x} = -\frac{n\pi\hbar}{L}\phi_{-n} = -p_n\phi_{-n}.$$

Consequently, ϕ_n and ϕ_{-n} are members of an orthonormal set in the interval $0 \le x \le L$. (This was not true in the case of the free particle where the spectrum of eigenvalues of $\mathbf{p}_x$ was continuous.) Thus we can write Eq. (21.18) in the form

$$\psi_n = \frac{1}{\sqrt{2}}(\phi_n - \phi_{-n}),$$

$$\psi_n = c_n\phi_n + c_{-n}\phi_{-n},$$

which is the series expansion of ψ_n in terms of the appropriate orthonormal set of functions; the coefficients, $c_n = -c_{-n} = 1/\sqrt{2}$, are the appropriate series coefficients required to normalize ψ_n.

■ **EXAMPLE 21.4** Find the expectation value of p_x. The expectation value of p_x is given by

$$\langle p_x \rangle = \int_0^L \psi_n^* \mathbf{p}_x \psi_n \, dx = \int_0^L (c_n^* \phi_n^* + c_{-n}^* \phi_{-n}^*)(c_n \mathbf{p}_x \phi_n + c_{-n} \mathbf{p}_x \phi_{-n}) \, dx$$

$$= \int_0^L (c_n^* \phi_n^* + c_{-n}^* \phi_{-n})(c_n p_n \phi_n + c_{-n}(-p_n)\phi_{-n}) \, dx$$

$$= c_n^* c_n p_n \int_0^L \phi_n^* \phi_n \, dx + c_n^* c_{-n}(-p_n) \int_0^L \phi_n^* \phi_{-n} \, dx$$

$$+ c_{-n}^* c_n p_n \int_0^L \phi_{-n}^* \phi_n \, dx + c_{-n}^* c_{-n}(-p_n) \int_0^L \phi_{-n}^* \phi_{-n} \, dx.$$

In view of the orthonormality condition, we have

$$\int_0^L \phi_n^* \phi_n \, dx = \int_0^L \phi_{-n}^* \phi_{-n} \, dx = 1 \quad \text{and} \quad \int_0^L \phi_{-n}^* \phi_n \, dx = \int_0^L \phi_n^* \phi_{-n} \, dx = 0.$$

This reduces the expression for $\langle p_x \rangle$ to

$$\langle p_x \rangle = c_n^* c_n p_n + c_{-n}^* c_{-n}(-p_n) = |c_n|^2 p_n + |c_{-n}|^2(-p_n).$$

Putting in the values of $|c_n|^2 = |c_{-n}|^2 = \frac{1}{2}$, we have

$$\langle p_x \rangle = \tfrac{1}{2} p_n + \tfrac{1}{2}(-p_n) = 0. \tag{21.20}$$

The expectation value of p_x is zero; we could have shown this more easily by using the value of ψ_n from Eq. (21.11):

$$\langle p_x \rangle = \int_0^L \sqrt{\frac{2}{L}} \sin\left(\frac{n\pi x}{L}\right)\left(-i\hbar \frac{d[\sqrt{2/L}\,\sin{(n\pi x/L)}]}{dx}\right)dx = -\frac{i\hbar}{L}\sin^2\left(\frac{n\pi x}{L}\right)\Big|_0^L = 0.$$

However, the lengthier procedure given above is more instructive. The short calculation might erroneously be taken to mean that the momentum of the particle is precisely zero, while the longer calculation shows quite clearly that the zero value is composed equally of a 50% probability of the particle moving with $p_x = +n\pi\hbar/L$ and a 50% probability of the particle moving with $p_x = -n\pi\hbar/L$. (The square of the absolute value of the series coefficient, $|c_n|^2$, is the probability of finding the situation described by the function ϕ_n.)

Compare this situation with the classical description in which at some time, t_0, we would specify precisely a value of momentum, p_x. At any subsequent time the momentum can be calculated from the classical laws. In quantum mechanics this precision is replaced by a probability. At any randomly chosen time, if by some magic we could ask the particle in the box for its momentum, it could only reply "Plus $n\pi\hbar/L$, with 50% probability," or "Minus $n\pi\hbar/L$, with 50% probability." Thus a statistical element is present in quantum mechanics. Certainty in classical mechanics is replaced by probability, or uncertainty, in quantum mechanics.

We can give precision to the meaning of uncertainty in an observable by defining it as the root-mean-square deviation from the expectation value. Thus, if Δp_x is the uncertainty in p_x, then

$$(\Delta p_x)^2 = \langle (p_x - \langle p_x \rangle)^2 \rangle \tag{21.21}$$
$$= \langle (p_x^2 - 2p_x\langle p_x \rangle + \langle p_x \rangle^2) \rangle = \langle p_x^2 \rangle - \langle p_x \rangle^2;$$
$$\Delta p_x = \sqrt{\langle p_x^2 \rangle - \langle p_x \rangle^2}.$$

■ **EXAMPLE 21.5**   Find the uncertainty in p_x. In view of the fact that $\langle p_x \rangle = 0$, and that $\mathbf{p}_x^2 = 2m\mathbf{H}$, it follows that

$$\langle p_x^2 \rangle = \int_0^L \psi_n^* \mathbf{p}_x^2 \psi_n \, dx = 2m \int_0^L \psi_n^* \mathbf{H} \psi_n \, dx = 2mE_n \int_0^L \psi_n^* \psi_n \, dx$$

$$= 2mE_n = \frac{n^2\pi^2\hbar^2}{L^2},$$

and we have

$$\Delta p_x = \sqrt{\frac{n^2\pi^2\hbar^2}{L^2}} = \frac{n\pi\hbar}{L} = \frac{nh}{2L}. \tag{21.22}$$

■ **EXAMPLE 21.6**   Find $\langle x \rangle$ and $\langle x^2 \rangle$. In a similar way we calculate the expectation value of the particle position,

$$\langle x \rangle = \int_0^L \psi_n^*(\mathbf{x}\psi_n) \, dx = \int_0^L \psi_n^* x \psi_n \, dx.$$

Introducing ψ_n and ψ_n^* from Eq. (21.11), we find

$$\langle x \rangle = \frac{2}{L} \int_0^L x \sin^2 \frac{n\pi x}{L} \, dx,$$

which integrates to

$$\langle x \rangle = \tfrac{1}{2}L. \tag{21.23}$$

Not surprisingly, the expectation value of the position of the particle is at the middle of the box. If we calculate $\langle x^2 \rangle$, using Eq. (21.11) for ψ_n, we obtain

$$\langle x^2 \rangle = \int_0^L \psi_n^* x^2 \psi_n \, dx = \frac{2}{L} \int_0^L x^2 \sin^2 \left(\frac{n\pi x}{L} \right) dx.$$

Direct evaluation of the integral yields the expression

$$\langle x^2 \rangle = \left(\frac{L}{2n\pi} \right)^2 \left[\frac{(2n\pi)^2}{3} - 2 \right].$$

Then, defining the uncertainty in the position in the analogous way to the definition of Δp_x,

$$\Delta x = \sqrt{\langle x^2 \rangle - \langle x \rangle^2}$$

$$\Delta x = \sqrt{\left(\frac{L}{2n\pi} \right)^2 \left[\frac{(2n\pi)^2}{3} - 2 \right] - \frac{L^2(n\pi)^2}{4(n\pi)^2}} = \frac{L}{2n\pi} \sqrt{\frac{(n\pi)^2}{3} - 2} = \frac{L}{2n\pi} \sqrt{1 + \frac{n^2\pi^2 - 3^2}{3}}$$

Multiplying Δp_x by Δx, we obtain

$$\Delta p_x \cdot \Delta x = \frac{h}{4\pi} \sqrt{1 + \frac{n^2\pi^2 - 3^2}{3}}.$$

Since the radical is greater than unity for all values of n, we have the result,

$$\Delta p_x \cdot \Delta x > \frac{h}{4\pi}. \tag{21.24}$$

The inequality (21.24) is the statement of the Heisenberg uncertainty principle for the particle in the box.

21.4 THE UNCERTAINTY PRINCIPLE

The situation for the free particle compared with the particle in the box may be summarized as follows.

1. The free particle has an exactly defined momentum, but the position is completely indefinite.
2. When we try to gain information about the position of the particle by confining it within the length L, an indefiniteness or uncertainty is introduced in the momentum. The product of these uncertainties is given by the inequality (21.24) $\Delta p_x \Delta x > h/4\pi$.
3. If we attempt to give the particle a precise position by letting $L \to 0$, then to satisfy (21.24), $\Delta p_x \to \infty$; the momentum becomes completely indefinite.

These facts are given general expression by the Heisenberg uncertainty principle, which we may state in the form: the product of the uncertainty in a coordinate and the uncertainty in the conjugate momentum is at least as large as $h/4\pi$. (By the conjugate momentum of a coordinate we mean the component of momentum along that coordinate.) In Cartesian coordinates we can state the uncertainty principle by the relations

$$\Delta p_x \cdot \Delta x \geq \frac{h}{4\pi}, \qquad \Delta p_y \cdot \Delta y \geq \frac{h}{4\pi}, \qquad \Delta p_z \cdot \Delta z \geq \frac{h}{4\pi}. \tag{21.25}$$

In passing, we reiterate that the operators for p_x and x do not commute. Variables having operators that do not commute are subject to uncertainties that are related as in (21.25). It follows from this principle that it is not possible to measure exactly and simultaneously both the x position and the x component of the momentum of a particle. *Either* the position *or* the conjugate momentum may be measured as precisely as we please, but increase in precision in the knowledge of one results in a loss of precision in the knowledge of the other.

Suppose that we attempt a precise measurement of the position of a particle using a microscope. The resolving power of a microscope is limited by the wavelength of the light used to illuminate the object; the shorter the wavelength (the higher the frequency) of the light used, the more accurately the position of the particle can be defined. If we wish to measure the position very accurately, then light of very high frequency would be required; a γ-ray, for example. To be seen, the γ-ray must be scattered from the particle into the objective of the microscope. However, a γ-ray of such high frequency has a large momentum; if it hits the particle, some of this momentum will be imparted to the particle, which will be kicked off in an arbitrary direction. The very process of measurement of position introduces an uncertainty in the momentum of the particle.

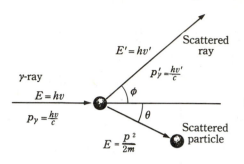

Figure 21.3 The Compton effect.

The scattering of a γ-ray by a small particle and the accompanying recoil of the particle is called the Compton effect (Fig. 21.3). Let p be the momentum of the particle after the collision; the momentum of the γ-ray is obtained from the energy $h\nu$, which according to the Einstein equation, must be equal to $m_\gamma c^2$. Therefore the momentum of the γ-ray is $m_\gamma c = h\nu/c$ before the collision and $h\nu'/c$ after the collision. Energy conservation requires that

$$h\nu = h\nu' + \frac{p^2}{2m},$$

while momentum conservation requires that for the x component

$$\frac{h\nu}{c} = \frac{h\nu'}{c}\cos\phi + p\cos\theta,$$

and for the y component

$$0 = \frac{h\nu'}{c}\sin\phi - p\sin\theta.$$

In addition to the original frequency ν, these three equations involve four variables: ν', p, ϕ, and θ. Using two of these equations, we can eliminate θ and ν', and reduce the third to a relation between p and ϕ. If the particle is to be seen, the γ-ray must be scattered into the objective of the microscope, that is, within a range $\Delta\phi$. Since $\Delta\phi$ is finite, there is a corresponding finite range of values, an uncertainty, in the particle momentum p. The process of measurement itself perturbs the system so that the momentum becomes indefinite even if it were not indefinite before the measurement. Since no method of measurement has been devised that is free from this difficulty, the uncertainty principle is an accepted physical principle. Examination of the equations for the Compton effect shows that this difficulty is a practical one only for particles with a mass of the order of that of the electron or of that of individual atoms. It does not give trouble with golf balls.

Another important uncertainty relation occurs in time-dependent systems. We can take the result for the particle in a box, which classically is written $E = p_x^2/2m$; if the momentum has an uncertainty Δp_x, then there is a corresponding uncertainty in the energy, $\Delta E = (\partial E/\partial p_x)\Delta p_x$, and thus

$$\Delta E = (1/m)p_x\,\Delta p_x = (1/m)(mv_x)\,\Delta p_x = v_x\,\Delta p_x.$$

However, the velocity, v_x, can be written $v_x = \Delta x/\Delta t$; using this result in the equation for ΔE yields $\Delta E \cdot \Delta t = \Delta p_x \cdot \Delta x$; extending this argument to the general case we have

$$\Delta E \cdot \Delta t \geq \frac{h}{4\pi}. \tag{21.26}$$

This relation says that there is an uncertainty in the energy of a particle and an uncertainty in the time at which the particle passes a given point in space; the product of these uncertainties must equal or exceed $h/4\pi$.

Heisenberg's development of quantum mechanics began with the uncertainty relations and led to the quantum mechanical equation. Schrödinger's treatment began with the wave equation and, as we have seen, we can argue from that to the uncertainty relations.

21.5 THE HARMONIC OSCILLATOR

The particle in the "box" was strictly confined to a particular region of space by the infinitely high potential energy "walls" erected at the boundaries. We now ask how the system behaves if the walls are not infinitely high at any particular point in space but rise gradually to infinity. The simplest potential energy function that has this property is $V(x) = \frac{1}{2}kx^2$, in which k is a constant. The potential energy is parabolic (Fig. 21.4). Choice of this potential-energy function has a double advantage. It displays the behavior if the walls are not infinitely high and, since this is the potential function for the harmonic oscillator, the results will be applicable to real physical oscillators insofar as they are harmonic oscillators. For example, the vibration of a diatomic molecule such as N_2, or O_2, in the lower energy states is nearly harmonic. We begin with a brief outline of the classical mechanical problem and then discuss the quantum mechanical behavior.

21.5.1 Classical Mechanics

Consider a particle moving in one dimension, along the x-axis, and bound to the origin ($x = 0$) by a Hooke's law restoring force, $-kx$. Newton's law, $ma = F$, then reads $m(d^2x/dt^2) = -kx$, or

$$\frac{d^2x}{dt^2} + \frac{k}{m}x = 0. \tag{21.27}$$

Define a circular frequency, ω, such that $\omega^2 = k/m$; this is related to the frequency v, by $\omega = 2\pi v$ or $v = (1/2\pi)\sqrt{k/m}$. Equation (21.27) becomes:

$$\frac{d^2x}{dt^2} + \omega^2x = 0.$$

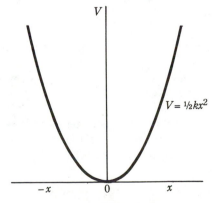

Figure 21.4   Potential energy for the harmonic oscillator.

This equation has the solution,

$$x = Ae^{i\omega t} + Be^{-i\omega t}. \tag{21.28}$$

The velocity, $v = dx/dt$, is obtained by differentiating Eq. (21.28):

$$v = i\omega[Ae^{i\omega t} - Be^{-i\omega t}]. \tag{21.29}$$

The constants A and B are determined by specifying the position, x, and the velocity, v, at some time t_0. For a simple solution suppose that at $t = t_0$, $x = x_0$, and $v = 0$; then Eqs. (21.28) and (21.29) become

$$x_0 = Ae^{i\omega t_0} + Be^{-i\omega t_0},$$

$$0 = Ae^{i\omega t_0} - Be^{-i\omega t_0}.$$

Solving for A and B, we obtain

$$A = \tfrac{1}{2}x_0 e^{-i\omega t_0} \quad \text{and} \quad B = \tfrac{1}{2}x_0 e^{i\omega t_0}.$$

Then, using these values in Eq. (21.28) gives

$$x = \tfrac{1}{2}x_0(e^{i\omega(t - t_0)} + e^{-i\omega(t - t_0)}).$$

Since $\cos y = (e^{iy} + e^{-iy})/2$ and $\sin y = (e^{iy} - e^{-iy})/2i$, we obtain

$$x = x_0 \cos \omega(t - t_0) \tag{21.30}$$

and

$$v = \frac{dx}{dt} = -\omega x_0 \sin \omega(t - t_0), \tag{21.31}$$

Equation (21.30) shows that the particle moves between $-x_0$ and $+x_0$ with a sinusoidal motion. Equation (21.31) shows that the velocity varies between $-\omega x_0$ and $+\omega x_0$ and is 90° out of phase with respect to the position. The total energy is

$$E = \tfrac{1}{2}mv^2 + \tfrac{1}{2}kx^2, \tag{21.32}$$

or, using Eqs. (21.30) and (21.31), we have

$$E = \tfrac{1}{2}m[\omega^2 x_0^2 \sin^2 \omega(t - t_0)] + \tfrac{1}{2}kx_0^2 \cos^2 \omega(t - t_0),$$

which, since $\omega^2 = k/m$, reduces to

$$E = \tfrac{1}{2}kx_0^2. \tag{21.33}$$

Note that the energy of the classical oscillator depends only on the force constant, k, and on the maximum displacement, x_0, which is an arbitrary quantity. The oscillator may have any total energy, depending on how large we make x_0.

In terms of the momentum, $p = mv$, we can write the total energy at any time in the form

$$E = \frac{p^2}{2m} + \frac{kx^2}{2}. \tag{21.34}$$

The total energy, E, is a constant throughout the motion. When the particle reaches the extreme value, $\pm x_0$, then the momentum, $p = 0$. When $x = 0$, the momentum has its largest value. Thus the particle moves very slowly at the extremes of the displacement and moves very quickly near $x = 0$.

21.5.2 Quantum Mechanics

The classical energy of the harmonic oscillator is given by Eq. (21.34). We obtain the Hamiltonian operator by replacing p by $\mathbf{p} = -i\hbar(d/dx)$. Then $\mathbf{H} = (-\hbar^2/2m)(d^2/dx^2) + \frac{1}{2}kx^2$. The Schrödinger equation becomes

$$-\frac{\hbar^2}{2m}\frac{d^2\psi}{dx^2} + \frac{1}{2}kx^2\psi = E\psi. \tag{21.35}$$

Before we can deal with a differential equation such as Eq. (21.35) we must remove the garbage. To do this we introduce dimensionless variables. Let $x = \beta\xi$, where the constant, β, is some unit of length and ξ is dimensionless. Then Eq. (21.35) becomes

$$-\left(\frac{\hbar^2}{2m\beta^2}\right)\frac{d^2\psi}{d\xi^2} + (\tfrac{1}{2}k\beta^2)\xi^2\psi = E\psi.$$

Now we observe that each term in the equation contains ψ so the dimensions of ψ, whatever they are, do not matter. However, since ψ is multiplied by E on the right side, it follows that every term in the equation is multiplied by a quantity of energy. Thus both $\hbar^2/2m\beta^2$ and $\frac{1}{2}k\beta^2$ must have the dimensions of energy. We observe that by adjusting the value of β properly we can make these two quantities of energy equal; therefore we determine β by the condition

$$\frac{\hbar^2}{2m\beta^2} = \tfrac{1}{2}k\beta^2 \qquad \text{or} \qquad \beta^4 = \frac{\hbar^2}{mk} \qquad \text{or} \qquad \beta^2 = \frac{\hbar}{\sqrt{mk}}. \tag{21.36}$$

Then we find that our unit of energy is

$$\tfrac{1}{2}k\beta^2 = \tfrac{1}{2}k\left(\frac{h}{2\pi}\right)\frac{1}{\sqrt{mk}} = \left(\frac{h}{2}\right)\frac{1}{2\pi}\sqrt{\frac{k}{m}} = \tfrac{1}{2}h\nu. \tag{21.37}$$

Next we write E as a multiple of this unit of energy;

$$E = \tfrac{1}{2}h\nu(2n + 1) \tag{21.38}$$

in which n is a dimensionless parameter. (We could have written $E = \frac{1}{2}h\nu\alpha$, with α dimensionless, of course. The choice of $2n + 1$ rather than α is convenient because it reduces the equation to a well-known standard form.) Introducing Eqs. (21.36) through (21.38) into the Schrödinger equation, dividing by $\frac{1}{2}h\nu$, and rearranging yields

$$\frac{d^2\psi}{d\xi^2} + (2n + 1 - \xi^2)\psi = 0, \tag{21.39}$$

which is considerably less cumbersome than Eq. (21.35).

To solve this equation we observe that at very large values of ξ, such that $\xi^2 \gg 2n + 1$, it becomes approximately

$$\frac{d^2\psi}{d\xi^2} = \xi^2\psi.$$

This equation has approximate solutions

$$\psi_1 = e^{\xi^2/2} \qquad \text{and} \qquad \psi_2 = e^{-\xi^2/2}.$$

The first solution, ψ_1, is unacceptable as a wave function since ψ_1^2 is not quadratically integrable. The second solution, ψ_2, becomes zero at $\xi = \pm\infty$ and is quadratically integrable.

Having an approximate solution, $e^{-\xi^2/2}$, we attempt an exact solution of Eq. (21.39) by choosing a solution in the form

$$\psi = u(\xi)e^{-\xi^2/2}. \tag{21.40}$$

Then, using prime and double prime for first and second derivatives, we have

$$\psi'' = u''e^{-\xi^2/2} - 2\xi u'e^{-\xi^2/2} + u(-1 + \xi^2)e^{-\xi^2/2}.$$

If we use this value for ψ'' and the value of ψ from Eq. (21.40), Eq. (21.39) becomes, after dividing out $e^{-\xi^2/2}$,

$$u'' - 2\xi u' + 2nu = 0. \tag{21.41}$$

This is Hermite's differential equation. Two cases occur.

Case I. The parameter n is either nonintegral or is a negative integer. If n is nonintegral, the solution of Eq. (21-41) behaves as e^{ξ^2} for large values of ξ and we would have $\psi = e^{\xi^2}e^{-\xi^2/2} = e^{\xi^2/2}$ for large values of ξ. This function is not quadratically integrable and is therefore unacceptable. The case for negative integral values of n also yields functions that are not quadratically integrable.

Case II. The parameter n is either zero or a positive integer. The solutions of Eq. (21-41) are polynomials of nth degree, the Hermite polynomials, usually written $H_n(\xi)$. Then the wave function has the form

$$\psi_n = \beta^{-1/2}A_n H_n(\xi)e^{-\xi^2/2}, \tag{21.42}$$

in which A_n is a constant. The integral

$$\int_{-\infty}^{\infty} \psi_n^2 \, dx = A_n^2 \int_{-\infty}^{\infty} H_n^2(\xi)e^{-\xi^2} \, d\xi = 1$$

converges since $H_n^2(\xi)$ is a polynomial of $2n$th degree and any polynomial multiplied by $e^{-\xi^2}$ will yield a convergent integral. The constant A_n is determined by requiring that the integral equal unity. Evaluation of the integral yields

$$A_n = \left(\frac{1}{\sqrt{\pi}\,2^n n!}\right)^{1/2}. \tag{21.43}$$

The grand conclusion from all of this is that the condition of quadratic integrability requires n to be a positive integer or zero. Looking back, we realize that n governs the energy through Eq. (21.38), which can be written

$$E_n = (n + \tfrac{1}{2})h\nu, \qquad n = 0, 1, 2, \dots \tag{21.44}$$

As in the case of the particle in a box, the energy is quantized. Only the special values given by Eq. (21.44) are permitted; these are the energy levels of the harmonic oscillator. Note that the energy levels are evenly spaced, $E_{n+1} - E_n = h\nu$, while in the case of the particle in a box the spacing increased with the value of n.

The lowest permissible value of the energy is the zero-point energy, obtained by setting $n = 0$ in Eq. (21.44):

$$E_0 = \tfrac{1}{2}h\nu_0.$$

In this lowest energy state, there is still some motion of the oscillator. If the motion ceased altogether, this would require $p_x = 0$ and $x = 0$ precisely. This is not permitted. A compromise is reached, which leaves a small residual motion and uncertainties in both position and momentum in conformity with the uncertainty principle.

The general expression for the Hermite polynomials is

$$H_n(\xi) = (-1)^n e^{\xi^2} \frac{d^n(e^{-\xi^2})}{d\xi^n}. \tag{21.45}$$

Explicitly, we may write

$$H_n(\xi) = \sum_{k=0} \frac{(-1)^k n!(2\xi)^{n-2k}}{(n-2k)!k!}. \tag{21.46}$$

Since the factorial of a negative integer is infinite, all terms for which $2k > n$ have vanishing coefficients so the series reduces to a polynomial. When n is even, the upper limit of k is $n/2$; when n is odd, the upper limit is $(n-1)/2$. The first few Hermite polynomials are given in Table 21.1.

Since the functions $\psi_n = \beta^{-1/2} A_n H_n(\xi) e^{-\xi^2/2}$ are eigenfunctions of an Hermitian operator, they form an orthonormal set in the interval $-\infty < \xi < +\infty$. Thus

$$\int_{-\infty}^{\infty} \psi_m \psi_n \, dx = \delta_{mn}$$

or

$$\int_{-\infty}^{\infty} A_m A_n H_m(\xi) H_n(\xi) e^{-\xi^2} \, d\xi = \delta_{mn}. \tag{21.47}$$

Note that the product of two Hermite polynomials by themselves would not yield a convergent integral, but when each is weighted by the function $e^{-\xi^2/2}$, the integral does converge. The polynomials are said to be orthonormal with respect to the weighting function $e^{-\xi^2/2}$.

We can easily show, using either Eq. (21.45) or (21.46), that $H_n(-\xi) = (-1)^n H_n(\xi)$; the polynomial is even or odd as n is even or odd.

For the evaluation of integrals involving Hermite polynomials two formulas are very useful: the differential relation

$$\frac{dH_n(\xi)}{d\xi} = 2nH_{n-1}(\xi), \tag{21.48}$$

and the recurrence formula

$$\xi H_n(\xi) = nH_{n-1}(\xi) + \tfrac{1}{2}H_{n+1}(\xi). \tag{21.49}$$

Table 21.1
The Hermite polynomials

n even	n odd
$H_0(\xi) = 1$	$H_1(\xi) = 2\xi$
$H_2(\xi) = 4\xi^2 - 2$	$H_3(\xi) = 8\xi^3 - 12\xi$
$H_4(\xi) = 16\xi^4 - 48\xi^2 + 12$	$H_5(\xi) = 32\xi^5 - 160\xi^3 + 120\xi$
$H_6(\xi) = 64\xi^6 - 480\xi^4 + 720\xi^2 - 120$	$H_7(\xi) = 128\xi^7 - 1344\xi^5 + 3360\xi^3 - 1680\xi$

Equation (21.48) is easily derived from either Eq. (21.45) or Eq. (21.46); Eq. (21.49) is then obtained by evaluating derivatives, using Eq. (21.48) and combining with the Hermite equation, Eq. (21.41).

■ **EXAMPLE 21.7**   Find the value of $\langle \xi \rangle$. To evaluate $\langle \xi \rangle$ we need an integral of the form

$$\langle \xi \rangle = \int_{-\infty}^{\infty} \psi_n \xi \psi_n \, dx = \int_{-\infty}^{\infty} A_n^2 H_n(\xi) \xi H_n(\xi) e^{-\xi^2} \, d\xi.$$

Using the value for $\xi H_n(\xi)$ from Eq. (21.49), we obtain

$$\langle \xi \rangle = A_n^2 \int_{-\infty}^{\infty} H_n(\xi) [n H_{n-1}(\xi) + \tfrac{1}{2} H_{n+1}(\xi)] e^{-\xi^2} \, d\xi,$$

$$\langle \xi \rangle = n A_n^2 \int_{-\infty}^{\infty} H_n(\xi) H_{n-1}(\xi) e^{-\xi^2} \, d\xi + \tfrac{1}{2} A_n^2 \int_{-\infty}^{\infty} H_n(\xi) H_{n+1}(\xi) e^{-\xi^2} \, d\xi.$$

By the orthogonality relation, Eq. (21.47), both of these integrals vanish, so $\langle \xi \rangle = 0$. We could have predicted this from the first form of the integral since $H_n^2(\xi) e^{-\xi^2}$ is an even function of ξ, so that when multiplied by ξ it becomes an odd function, which vanishes on integration over the symmetrical interval.

The wave functions shown in Fig. 21.5 indicate that there is a finite probability of finding the particle at very large distances from $x = 0$. Classically the particle may not pass beyond the point at which the kinetic energy is zero; that is, where the potential energy is equal to the total energy. If x_0 is the maximum displacement allowed classically, then

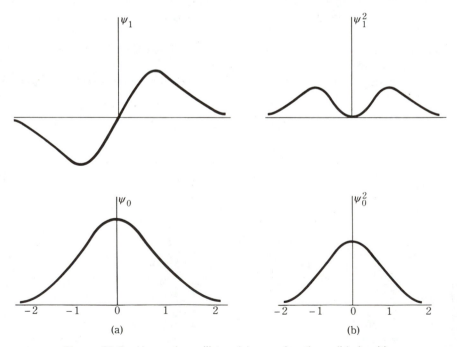

Figure 21.5   Harmonic oscillator: (a) wave functions; (b) densities.

$\frac{1}{2}kx_0^2 = E$, or $\frac{1}{2}k\beta^2\xi_0^2 = (n + \frac{1}{2})h\nu$, which yields $\xi_0^2 = 2n + 1$. Thus, for $n = 0$, the maximum displacement allowed classically would correspond to $\xi_0 = \pm 1$. The density function shown in Fig. 21.5(b) is substantial at $\xi_0 = \pm 1$.

■ **EXAMPLE 21.8**   What is the total probability, $P(n)$, of finding the particle in the classically forbidden region? We obtain $P(n)$ by integrating the probability density over the forbidden region.

$$P(n) = \int_{-\infty}^{-\sqrt{2n+1}} A_n^2 H_n^2(\xi)e^{-\xi^2}\,d\xi + \int_{+\sqrt{2n+1}}^{\infty} A_n^2 H_n^2(\xi)e^{-\xi^2}\,d\xi.$$

From symmetry these two integrals are equal, so we have

$$P(n) = 2A_n^2 \int_{\sqrt{2n+1}}^{\infty} H_n^2(\xi)e^{-\xi^2}\,d\xi.$$

If $n = 0$, we have $A_0^2 = 1/\sqrt{\pi}$, and $H_0^2(\xi) = 1$,

$$P(0) = \frac{2}{\sqrt{\pi}} \int_1^{\infty} e^{-\xi^2}\,d\xi = 1 - \text{erf}(1),$$

in which $\text{erf}(x)$ is the error function of x. From Table 4.2 we find $\text{erf}(1) = 0.8427$, so that $P(0) = 0.1573$. This indicates that in the ground state the particle spends 15.73% of its time in the classically forbidden region. This is by no means a negligible figure. The values for some other values of n are:

n	0	1	2	3	4
$P(n)$	0.1573	0.1116	0.0951	0.0855	0.0785

As n increases, $P(n)$ decreases and approaches the classical value, zero, as $n \to \infty$.

The ability of a particle to penetrate into a classically forbidden region is the basis of the quantum mechanical "tunnel effect." Consider a particle for which the potential function looks like that in Fig. 21.6. Two regions of low potential energy are separated by a

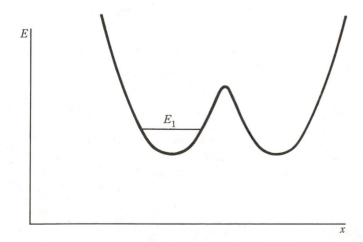

Figure 21.6   Barrier for tunnel effect.

barrier. Assume that the particle has a total energy corresponding to E_1 and is in the left-hand region. The energy E_1 is less than the height of the potential barrier so that classically the particle would be confined to the left side. However, quantum mechanically there is a finite probability of finding the particle in the forbidden region and therefore there is a probability of the particle leaking or "tunneling" through the barrier. The probability of this event decreases as the mass of the particle increases and as the barrier gets higher and wider.

21.6 MULTIDIMENSIONAL PROBLEMS

21.6.1 Particle in a Three-Dimensional Box

The majority of interesting problems involve more than one coordinate and momentum. Immediately the Schrödinger equation becomes a partial differential equation and the solutions become more complicated. One of the simplest cases that illustrates a general method of solving the partial differential equation is the example of the particle in a three-dimensional box.

We assume that the potential energy is defined by

$$V = 0 \qquad\qquad 0 < x < L_1 \qquad\qquad 0 < y < L_2 \qquad\qquad 0 < z < L_3$$

$$V \to +\infty \qquad x \le 0, x \ge L_1 \qquad y \le 0, y \ge L_2 \qquad z \le 0, z \ge L_3$$

Since the particles cannot exist in a region of infinite potential energy, we know that $\psi = 0$ outside of and at the walls of the box. Since $V = 0$ in the interior of the box, we have the Schrödinger equation in the form

$$-\frac{\hbar^2}{2m}\left(\frac{\partial^2\psi}{\partial x^2} + \frac{\partial^2\psi}{\partial y^2} + \frac{\partial^2\psi}{\partial z^2}\right) = E\psi. \tag{21.50}$$

We now assume that ψ is a product of functions of the individual coordinates; that is,

$$\psi(x, y, z) = X(x) \cdot Y(y) \cdot Z(z), \tag{21.51}$$

then

$$\frac{\partial^2\psi}{\partial x^2} = Y(y) \cdot Z(z) \cdot \frac{d^2 X}{dx^2}, \qquad \frac{\partial^2\psi}{\partial y^2} = X(x) \cdot Z(z) \cdot \frac{d^2 Y}{dy^2}, \qquad \text{and so on.}$$

We insert these expressions for the partial derivatives in the Schrödinger equation and divide through by ψ; this reduces the equation to the form

$$-\frac{\hbar^2}{2m}\left(\frac{1}{X(x)}\frac{d^2 X}{dx^2}\right) - \frac{\hbar^2}{2m}\left(\frac{1}{Y(y)}\frac{d^2 Y}{dy^2}\right) - \frac{\hbar^2}{2m}\left(\frac{1}{Z(z)}\frac{d^2 Z}{dz^2}\right) = E. \tag{21.52}$$

Now suppose we keep x and y constant; then the first terms in the equation, since they depend on x and on y respectively, remain constant. If we vary z, the third term would appear to vary since it depends on z. But in fact it cannot vary, since the addition of a varying third term to the two constant ones would make E vary and E is a constant. Thus, we may write

$$-\frac{\hbar^2}{2m}\left(\frac{1}{Z(z)}\frac{d^2 Z}{dz^2}\right) = E_z,$$

where E_z is a constant. The analogous argument may be applied to show that the first and second terms in Eq. (21.52) must also be constants, E_x and E_y. The partial differential Schrödinger equation has thus been reduced to three ordinary differential equations, which can be written

$$\frac{d^2X}{dx^2} + \frac{2mE_x}{\hbar^2}X = 0, \qquad \frac{d^2Y}{dy^2} + \frac{2mE_y}{\hbar^2}Y = 0, \qquad \frac{d^2Z}{dz^2} + \frac{2mE_z}{\hbar^2}Z = 0. \quad (21.53)$$

Comparison of these equations with that for the one-dimensional particle in the box shows that both the equation and the boundary conditions are the same. The solutions are therefore

$$X(x) = \sqrt{\frac{2}{L_1}} \sin\left(\frac{n_x \pi x}{L_1}\right) \qquad n_x = 1, 2, 3, \ldots ;$$

$$Y(y) = \sqrt{\frac{2}{L_2}} \sin\left(\frac{n_y \pi y}{L_2}\right) \qquad n_y = 1, 2, 3, \ldots ;$$

$$Z(z) = \sqrt{\frac{2}{L_3}} \sin\left(\frac{n_z \pi z}{L_3}\right) \qquad n_z = 1, 2, 3, \ldots$$

Thus, by Eq. (21.51),

$$\psi(x, y, z) = \sqrt{\frac{8}{L_1 L_2 L_3}} \sin\left(\frac{n_x \pi x}{L_1}\right)\sin\left(\frac{n_y \pi y}{L_2}\right)\sin\left(\frac{n_z \pi z}{L_3}\right). \qquad (21.54)$$

The energy is given by $E = E_x + E_y + E_z$, and since each term has the form for a particle in a box, we have

$$E_{n_x, n_y, n_z} = \frac{n_x^2 h^2}{8mL_1^2} + \frac{n_y^2 h^2}{8mL_2^2} + \frac{n_z^2 h^2}{8mL_3^2}, \qquad \left.\begin{array}{c} n_x \\ n_y \\ n_z \end{array}\right\} = 1, 2, 3, \ldots \qquad (21.55)$$

If the dimensions of the box are all equal, that is, if $L_1 = L_2 = L_3 = L$, then

$$E_{n_x, n_y, n_z} = \frac{(n_x^2 + n_y^2 + n_z^2)h^2}{8mL^2}.$$

In this case, we have an interesting case of degeneracy; for example, the quantum number combinations $(n_x, n_y, n_z) = (112), (121), (211)$ represent different states of the system having the same energy. This energy state is three-fold degenerate.

21.6.2 Separation of Variables

We may generalize the result for the particle in the three-dimensional box in the following way. If the Hamiltonian operator can be written as a *sum* of groups of terms, each of which depends only on one coordinate or one set of coordinates, then the wave function can be written as a *product* of functions each of which depends only on the one coordinate or the one set of coordinates; correspondingly, the total energy is the sum of the energies associated with each coordinate or each set of coordinates.

For example, suppose there are two sets of coordinates, q_1 and q_2; further, suppose that the Hamiltonian operator can be arranged in the form

$$\mathbf{H} = \mathbf{H}_1 + \mathbf{H}_2,$$

where $\mathbf{H}_1$ depends only on the first set of coordinates q_1 and $\mathbf{H}_2$ depends only on a second set q_2. Then the wave function will have the form $\psi = f_1(q_1) \cdot f_2(q_2)$ and the energy will have the form $E = E_1 + E_2$. The proof is simple:

$$\mathbf{H}\psi = E\psi$$

$$(\mathbf{H}_1 + \mathbf{H}_2)f_1 f_2 = Ef_1 f_2$$

$$f_2(\mathbf{H}_1 f_1) + f_1(\mathbf{H}_2 f_2) = Ef_1 f_2.$$

Dividing by $f_1 f_2$ yields:

$$\frac{1}{f_1}(\mathbf{H}_1 f_1) + \frac{1}{f_2}(\mathbf{H}_2 f_2) = E.$$

The first term depends only on the set q_1; the second term depends only on the set q_2. Keeping the members of the set q_1 constant and varying the members of q_2 shows that the second term is a constant, E_2. Similarly, the first term must be a constant, E_1. Thus we can write

$$\mathbf{H}_1 f_1 = E_1 f_1, \qquad \mathbf{H}_2 f_2 = E_2 f_2; \qquad (21.56)$$

$$E = E_1 + E_2. \qquad (21.57)$$

We see that if the energy is made up of contributions from independent modes of motion, the wave function will be a product of functions each of which is a wave function describing an independent mode of motion. If, for example, as a first approximation, the internal energy of a diatomic molecule is made up of a sum of contributions,

$$E = E_{\text{electronic}} + E_{\text{rotational}} + E_{\text{vibrational}},$$

then, in the first approximation, the wave function will be a product,

$$\psi = \psi_{\text{electronic}} \cdot \psi_{\text{rotational}} \cdot \psi_{\text{vibrational}},$$

where $\psi_{\text{electronic}}$ depends only on the electronic coordinates; $\psi_{\text{rotational}}$ depends only on the angular coordinates; and $\psi_{\text{vibrational}}$ depends only on the vibrational coordinates, in this case the internuclear distance. We shall have numerous examples of the use of this theorem.

21.7 THE TWO-BODY PROBLEM

The classical energy of a system of two point masses, m_1 and m_2, has the form

$$E = \frac{1}{2m_1}(p_{x_1}^2 + p_{y_1}^2 + p_{z_1}^2) + \frac{1}{2m_2}(p_{x_2}^2 + p_{y_2}^2 + p_{z_2}^2) + V(x_1, y_1, z_1, x_2, y_2, z_2),$$

in which x_1, y_1, z_1 and x_2, y_2, z_2 are the coordinates of the masses m_1 and m_2 respectively. Replacing the momenta by the corresponding quantum mechanical operators yields the Hamiltonian operator

$$\mathbf{H} = -\frac{\hbar^2}{2m_1}\left(\frac{\partial^2}{\partial x_1^2} + \frac{\partial^2}{\partial y_1^2} + \frac{\partial^2}{\partial z_1^2}\right) - \frac{\hbar^2}{2m_2}\left(\frac{\partial^2}{\partial x_2^2} + \frac{\partial^2}{\partial y_2^2} + \frac{\partial^2}{\partial z_2^2}\right) + V(x_1, y_1, z_1, x_2, y_2, z_2).$$

In general, the potential energy is not separable into terms involving only certain sets of the six Cartesian coordinates but depends on the internal coordinates of the system. So to simplify the problem we transform these six Cartesian coordinates into three coordinates for the center of mass X, Y, and Z and three internal coordinates, x, y, z.

★ 21.7.1   Mathematical Details; Change in Variables

The center-of-mass coordinates (X, Y, Z) are determined by the condition that the sum of the first moments of mass about the center of mass vanish for each axis; that is,

$$m_1(x_1 - X) + m_2(x_2 - X) = 0 \qquad X = \frac{m_1 x_1 + m_2 x_2}{m_1 + m_2}$$

$$m_1(y_1 - Y) + m_2(y_2 - Y) = 0 \quad \text{or} \quad Y = \frac{m_1 y_1 + m_2 y_2}{m_1 + m_2}$$

$$m_1(z_1 - Z) + m_2(z_2 - Z) = 0 \qquad Z = \frac{m_1 z_1 + m_2 z_2}{m_1 + m_2}.$$

In addition, we define the three internal coordinates, x, y, z, by

$$x = x_2 - x_1, \qquad y = y_2 - y_1, \qquad z = z_2 - z_1.$$

Since x_1 and x_2 depend only on X and x, we have

$$\frac{\partial}{\partial x_1} = \left(\frac{\partial X}{\partial x_1}\right)\frac{\partial}{\partial X} + \left(\frac{\partial x}{\partial x_1}\right)\frac{\partial}{\partial x} = \left(\frac{m_1}{m_1 + m_2}\right)\frac{\partial}{\partial X} - \frac{\partial}{\partial x}.$$

In the second expression, the derivatives have been evaluated from the definitions of X and x. Then,

$$\frac{\partial^2}{\partial x_1^2} = \left[\left(\frac{m_1}{m_1 + m_2}\right)\frac{\partial}{\partial X} - \frac{\partial}{\partial x}\right]\left[\left(\frac{m_1}{m_1 + m_2}\right)\frac{\partial}{\partial X} - \frac{\partial}{\partial x}\right]$$

$$= \left(\frac{m_1}{m_1 + m_2}\right)^2 \frac{\partial^2}{\partial X^2} - \frac{2m_1}{m_1 + m_2}\frac{\partial^2}{\partial X \partial x} + \frac{\partial^2}{\partial x^2}.$$

Similarly, we find

$$\frac{\partial}{\partial x_2} = \left(\frac{m_2}{m_1 + m_2}\right)\frac{\partial}{\partial X} + \frac{\partial}{\partial x},$$

and

$$\frac{\partial^2}{\partial x_2^2} = \left(\frac{m_2}{m_1 + m_2}\right)^2 \frac{\partial^2}{\partial X^2} + \left(\frac{2m_2}{m_1 + m_2}\right)\frac{\partial^2}{\partial X \partial x} + \frac{\partial^2}{\partial x^2}.$$

These two terms are combined as they appear in the Hamiltonian,

$$\frac{1}{m_1}\frac{\partial^2}{\partial x_1^2} + \frac{1}{m_2}\frac{\partial^2}{\partial x_2^2} = \frac{m_1}{(m_1 + m_2)^2}\frac{\partial^2}{\partial X^2} - \frac{2}{m_1 + m_2}\frac{\partial^2}{\partial X \partial x} + \frac{1}{m_1}\frac{\partial^2}{\partial x^2}$$

$$+ \frac{m_2}{(m_1 + m_2)^2}\frac{\partial^2}{\partial X^2} + \frac{2}{m_1 + m_2}\frac{\partial^2}{\partial X \partial x} + \frac{1}{m_2}\frac{\partial^2}{\partial x^2}$$

$$= \frac{1}{m_1 + m_2}\frac{\partial^2}{\partial X^2} + \left(\frac{1}{m_1} + \frac{1}{m_2}\right)\frac{\partial^2}{\partial x^2}.$$

The algebra for y_1, y_2 and z_1, z_2 proceeds in exactly the same fashion, so the Hamiltonian becomes

$$\mathbf{H} = -\frac{\hbar^2}{2(m_1 + m_2)}\left(\frac{\partial^2}{\partial X^2} + \frac{\partial^2}{\partial Y^2} + \frac{\partial^2}{\partial Z^2}\right) - \frac{\hbar^2}{2\mu}\left(\frac{\partial^2}{\partial x^2} + \frac{\partial^2}{\partial y^2} + \frac{\partial^2}{\partial z^2}\right) + V(x, y, z),$$

in which μ is the "reduced mass" of the system, defined by $1/\mu = 1/m_1 + 1/m_2$. In the problems of interest here, the potential energy will be independent of the position of the center of mass, hence we have written $V(x, y, z)$.

21.7.2   The Wave Equation for the Internal Motion

The transformation has separated the Hamiltonian into two groups of terms, the first group depending only on X, Y, Z, the second group depending on x, y, z. Thus we may write by the theorem, Eq. (21.56),

$$\psi_{\text{total}} = \psi_{\text{trans}}(X, Y, Z) \cdot \psi(x, y, z),$$

$$E_{\text{total}} = E_{\text{trans}} + E,$$

$$-\frac{\hbar^2}{2(m_1 + m_2)}\left(\frac{\partial^2}{\partial X^2} + \frac{\partial^2}{\partial Y^2} + \frac{\partial^2}{\partial Z^2}\right)\psi_{\text{trans}}(X, Y, Z) = E_{\text{trans}}\psi_{\text{trans}}(X, Y, Z),$$

and

$$-\frac{\hbar^2}{2\mu}\left(\frac{\partial^2\psi}{\partial x^2} + \frac{\partial^2\psi}{\partial y^2} + \frac{\partial^2\psi}{\partial z^2}\right) + V(x, y, z)\psi = E\psi. \tag{21.58}$$

The wave function, ψ_{trans}, is the wave function of a free particle of mass $m_1 + m_2$, moving with the center of mass of the system. Correspondingly, E_{trans} is the translational energy of the total system. This motion has no particular interest to us, so we may ignore it. We are interested in $\psi = \psi(x, y, z)$, which will provide a description of the internal motions of the system; E is the energy of these internal motions. For any system we can always separate out the center-of-mass coordinates in this way, discard the energy associated with them and consider only the internal coordinates. In the future we will assume that this has been done.

It is convenient now to transform Eq. (21.58) into spherical coordinates with m_1 at the center and m_2 at the position r, θ, ϕ. (Fig. 21.7.) The transformation equations are

$$x = r \sin \theta \cos \phi \qquad\qquad r = \sqrt{x^2 + y^2 + z^2}$$

$$y = r \sin \theta \sin \phi \qquad \text{or} \qquad \cos \theta = \frac{z}{\sqrt{x^2 + y^2 + z^2}} \tag{21.59}$$

$$z = r \cos \theta \qquad\qquad \tan \phi = \frac{y}{x}.$$

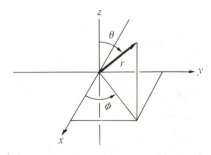

Figure 21.7   Spherical coordinates.

The calculation of the differential operators goes as above; for example,

$$\frac{\partial}{\partial x} = \frac{\partial r}{\partial x}\frac{\partial}{\partial r} + \frac{\partial \theta}{\partial x}\frac{\partial}{\partial \theta} + \frac{\partial \phi}{\partial x}\frac{\partial}{\partial \phi},$$

but is tedious. The final result takes the form

$$-\frac{\hbar^2}{2\mu}\left[\frac{1}{r^2}\frac{\partial}{\partial r}\left(r^2\frac{\partial\psi}{\partial r}\right) + \frac{1}{r^2\sin\theta}\frac{\partial}{\partial\theta}\left(\sin\theta\frac{\partial\psi}{\partial\theta}\right) + \frac{1}{r^2\sin^2\theta}\frac{\partial^2\psi}{\partial\phi^2}\right] + V(r,\theta,\phi)\psi = E\psi.$$

$$(21.60)$$

Depending on the form of the potential, this equation is applicable to a number of problems. We consider them in turn.

The volume element in spherical coordinates is

$$d\tau = r^2\sin\theta\,d\theta\,d\phi\,dr \tag{21.61}$$

If the entire coordinate space is to be covered, the limits of integration are $0 \le \theta \le \pi$; $0 \le \phi \le 2\pi$; $0 \le r \le \infty$.

21.8 THE RIGID ROTOR

Suppose the two masses are held rigidly apart at some fixed distance r_0. Since there is no momentum in the r direction, the derivative with respect to r cannot appear in the equation. The potential energy is equal to zero since the system rotates freely. Then Eq. (21.60) becomes

$$-\frac{\hbar^2}{2\mu r_0^2}\left[\frac{1}{\sin\theta}\frac{\partial}{\partial\theta}\left(\sin\theta\frac{\partial Y(\theta,\phi)}{\partial\theta}\right) + \frac{1}{\sin^2\theta}\frac{\partial^2 Y(\theta,\phi)}{\partial\phi^2}\right] = EY(\theta,\phi). \tag{21.62}$$

In which we have written $Y(\theta,\phi)$ for the wave function.

Suppose the center of mass is at the position R; then the sum of the second moments of mass about the center of mass is the moment of inertia, I, about any axis perpendicular to the axis of the rotor (Fig. 21.8). From the figure, we have $I = m_1(0 - R)^2 + m_2(r_0 - R)^2$.

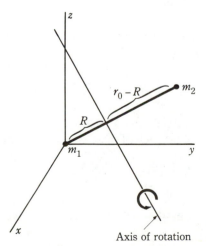

Figure 21.8 Coordinates for calculation of moment of inertia.

As usual, R is determined by the vanishing of the sum of the first moments.

$$0 = m_1(-R) + m_2(r_0 - R) \quad \text{or} \quad R = \frac{m_2}{m_1 + m_2} r_0;$$

$$I = m_1\left(\frac{m_2}{m_1 + m_2}\right)^2 r_0^2 + m_2\left(1 - \frac{m_2}{m_1 + m_2}\right)^2 r_0^2 = \frac{m_1 m_2}{m_1 + m_2} r_0^2 = \mu r_0^2.$$

Introducing I for μr_0^2, Eq. (21.62) becomes

$$-\frac{\hbar^2}{2I}\left[\frac{1}{\sin\theta}\frac{\partial}{\partial\theta}\left(\sin\theta\frac{\partial Y}{\partial\theta}\right) + \frac{1}{\sin^2\theta}\frac{\partial^2 Y}{\partial\phi^2}\right] = EY. \tag{21.63}$$

The classical energy of a rigid rotor is given by

$$\frac{M^2}{2I} = E,$$

where M^2 is the square of the total angular momentum of the system. Comparing these equations, we conclude that the operator for the square of the total angular momentum is

$$\mathbf{M}^2 = -\hbar^2\left[\frac{1}{\sin\theta}\frac{\partial}{\partial\theta}\left(\sin\theta\frac{\partial}{\partial\theta}\right) + \frac{1}{\sin^2\theta}\frac{\partial^2}{\partial\phi^2}\right]. \tag{21.64}$$

We further observe from dimensionality that the energy E must be some multiple of $\hbar^2/2I$, so we write

$$E_J = J(J + 1)\frac{\hbar^2}{2I}, \tag{21.65}$$

in which J is a dimensionless parameter. Equation (21.63) becomes

$$\frac{1}{\sin\theta}\frac{\partial}{\partial\theta}\left(\sin\theta\frac{\partial Y}{\partial\theta}\right) + \frac{1}{\sin^2\theta}\frac{\partial^2 Y}{\partial\phi^2} + J(J + 1)Y = 0.$$

We observe that multiplying the equation through by $\sin^2\theta$ brings it to the form

$$\sin\theta\frac{\partial}{\partial\theta}\left(\sin\theta\frac{\partial Y}{\partial\theta}\right) + J(J + 1)\sin^2\theta Y + \frac{\partial^2 Y}{\partial\phi^2} = 0.$$

By our earlier arguments, since only the last term depends on ϕ, we can substitute $Y(\theta, \phi) = \Theta(\theta)\Phi(\phi)$, and divide through by $\Theta(\theta)\Phi(\phi)$ to obtain

$$\frac{1}{\Theta(\theta)}\sin\theta\frac{d}{d\theta}\left(\sin\theta\frac{d\Theta}{d\theta}\right) + J(J + 1)\sin^2\theta + \frac{1}{\Phi(\phi)}\frac{d^2\Phi}{d\phi^2} = 0.$$

The terms in this partial differential equation have been separated into a group depending on θ and a single term depending on ϕ. It follows that each of these sets of terms must be a constant so we write

$$\frac{1}{\Phi}\frac{d^2\Phi}{d\phi^2} = -m^2 \quad \text{or} \quad \frac{d^2\Phi}{d\phi^2} = -m^2\Phi.$$

This equation has the solution $\Phi = Ae^{im\phi}$. We note that when $\phi \to \phi + 2\pi$ we return to the same set of points in space so that our boundary condition must be

$$\Phi(\phi + 2\pi) = \Phi(\phi);$$

$$Ae^{im(\phi + 2\pi)} = Ae^{im\phi},$$

$$e^{im2\pi} = 1.$$

This relation is satisfied only if m is zero or a positive or negative integer. We may write

$$\Phi_m(\phi) = Ae^{im\phi}, \qquad m = 0, \ \pm 1, \ \pm 2, \dots$$

For normalization we require, since ϕ varies from 0 to 2π,·

$$\int_0^{2\pi} \Phi_m^* \Phi_m \, d\phi = 1 \qquad A^*A \int_0^{2\pi} e^{-im\phi} e^{im\phi} \, d\phi = 1$$

$$|A|^2 \int_0^{2\pi} d\phi = 1 \qquad\qquad A = \frac{1}{\sqrt{2\pi}}$$

where we have chosen A as a real number; then

$$\Phi_m(\phi) = \frac{1}{\sqrt{2\pi}} e^{im\phi}, \qquad m = 0, \pm 1, \pm 2, \pm 3, \dots \tag{21.66}$$

The remaining part of the equation can be written

$$\sin\theta \frac{d}{d\theta}\left(\sin\theta \frac{d\Theta}{d\theta}\right) + [J(J+1)\sin^2\theta - m^2]\Theta = 0. \tag{21.67}$$

In this equation we change variable to $\xi = \cos\theta$; then $d/d\theta = (d\xi/d\theta)(d/d\xi)$, but $d\xi/d\theta = -\sin\theta$, and $\sin^2\theta = 1 - \cos^2\theta = 1 - \xi^2$; then Eq. (21.67) becomes, with $\Theta(\theta) = P(\xi)$,

$$(1 - \xi^2)\frac{d}{d\xi}\left[(1 - \xi^2)\frac{dP(\xi)}{d\xi}\right] + [J(J+1)(1 - \xi^2) - m^2]P(\xi) = 0.$$

Dividing by $(1 - \xi^2)$ we get

$$(1 - \xi^2)\frac{d^2P}{d\xi^2} - 2\xi\frac{dP}{d\xi} + \left[J(J+1) - \frac{m^2}{(1 - \xi^2)}\right]P = 0. \tag{21.68}$$

Since $\xi = \cos\theta$, and the limits of θ are 0 and π, the corresponding limits of ξ are $+1$ and -1. Equation (21.68) is the associated Legendre equation and it may be shown that the only case in which this equation possesses continuous, single-valued, and quadratically integrable solutions in the interval $-1 \leq \xi \leq +1$ is that in which J is a positive integer or zero. The solutions depend on the integers J and m and are written $P_J^{|m|}(\xi)$. It is clear from Eq. (21.68), which contains only m^2, that the function $P_J^{|m|}(\xi)$ can therefore depend only on the absolute value of m. The function of $P_J^{|m|}(\xi)$ is the associated Legendre function of degree J and order $|m|$. The first few of these functions are shown in Table 21.2.

Using this result in Eq. (21.65) we find that the energy of the rigid rotor is quantized:

$$E_J = J(J+1)\frac{\hbar^2}{2I}, \qquad J = 0, 1, 2, 3, \dots \tag{21.69}$$

Table 21.2
The associated Legendre functions

m \ n	0	1	2	3	4
0	1	ξ	$\frac{1}{2}(3\xi^2 - 1)$	$\frac{1}{2}(5\xi^3 - 3\xi)$	$\frac{1}{8}(35\xi^4 - 30\xi^2 + 3)$
1	—	$(1 - \xi^2)^{1/2}$	$3\xi(1 - \xi^2)^{1/2}$	$\frac{3}{2}(5\xi^2 - 1)(1 - \xi^2)^{1/2}$	$\frac{5}{2}(7\xi^3 - 3\xi)(1 - \xi^2)^{1/2}$
2	—	—	$3(1 - \xi^2)$	$15\xi(1 - \xi^2)$	$\frac{15}{2}(7\xi^2 - 1)(1 - \xi^2)$
3	—	—	—	$15(1 - \xi^2)^{3/2}$	$105\xi(1 - \xi^2)^{3/2}$
4	—	—	—	—	$105(1 - \xi^2)^2$

Similarly since $E = M^2/2I$, it follows that the square of the total angular momentum is quantized

$$M^2 = J(J + 1)\hbar^2, \qquad J = 0, 1, 2, 3, \ldots \tag{21.70}$$

The interesting result here is that, in contrast to the case of the particle in the box and the harmonic oscillator, when $J = 0$ the lowest energy is zero, corresponding to a precise value, zero, for the angular momentum. This is possible since the angles of orientation, θ and ϕ, are completely unspecified, in conformity with the uncertainty principle.

The meaning of the integer m remains to be investigated. To do this we consider the classical z component of angular momentum in Cartesian coordinates. This has the form

$$M_z = xp_y - yp_x.$$

The quantum mechanical operator for the z-component is then

$$\mathbf{M}_z = -i\hbar\left(x\frac{\partial}{\partial y} - y\frac{\partial}{\partial x}\right). \tag{21.71}$$

If we transform $\mathbf{M}_z$ into spherical coordinates using the same method as above, we obtain the very simple result

$$\mathbf{M}_z = -i\hbar\frac{\partial}{\partial \phi}. \tag{21.72}$$

The wave function for the rigid rotor has the form

$$Y_{J,m}(\xi, \phi) = A_{J,m}P_J^{|m|}(\xi)\frac{3}{\sqrt{2\pi}}e^{im\phi}, \tag{21.73}$$

in which the normalization constant, $A_{J,m}$, is

$$A_{J,m} = \left[\left(\frac{2J + 1}{2}\right)\frac{(J - |m|)!}{(J + |m|)!}\right]^{1/2}. \tag{21.74}$$

Then

$$\mathbf{M}_z Y_{J,m} = A_{J,m}P_J^{|m|}(\xi)\frac{1}{\sqrt{2\pi}}\mathbf{M}_z e^{im\phi}$$

$$= -i\hbar(im)Y_{J,m};$$

$$\mathbf{M}_z Y_{J,m} = m\hbar Y_{J,m}. \tag{21.75}$$

Therefore we find that $Y_{J,m}$ is an eigenfunction of $\mathbf{M}_z$ with the eigenvalue $m\hbar$. The z component of the angular momentum is therefore quantized; the quantum number is $m = 0$, $\pm 1, \pm 2, \cdots$. Again, precise values of the z component of angular momentum are permitted since the angle ϕ is totally unspecifiable. Repeating the application of $\mathbf{M}_z$ on Eq. (21.75), we obtain

$$\mathbf{M}_z^2 Y_{J,m} = m\hbar(\mathbf{M}_z Y_{J,m}) = m^2\hbar^2 Y_{J,m}, \tag{21.76}$$

so that the square of the z component of angular momentum is $m^2\hbar^2$. The total angular momentum is the sum of squares of the components:

$$M^2 = M_x^2 + M_y^2 + M_z^2. \tag{21.77}$$

Replacing M^2 and M_z^2 by their values from Eqs. (21.70) and (21.76), we obtain after rearranging

$$[J(J + 1) - m^2]\hbar^2 = M_x^2 + M_y^2.$$

Since the right-hand side is a sum of squares, it cannot be negative; hence we have the condition

$$J(J + 1) - m^2 \geq 0.$$

It is apparent that this condition is fulfilled so long as $|m| \leq J$. The values of m are therefore restricted to $m = 0, \pm 1, \pm 2, \ldots, \pm J$. For a given value of J, there are $2J + 1$ values of m. The energy is determined by J only, hence each energy level has a degeneracy of $2J + 1$.

The possible orientations of the angular-momentum vector for $J = 1$ are shown in Fig. 21.9. When $J = 1$,

$$\sqrt{M^2} = \sqrt{J(J + 1)}\hbar$$
$$= \sqrt{2}\hbar.$$

Any position of the vector in the conical surface above the xy-plane as shown in Fig. 21.9(c) will yield a projection of the vector equal to $+\hbar$ on the z-axis, while any vector in the conical surface below the plane will have a projection $-\hbar$ on the z-axis. The permissible projections for $J = 2$ are 2, 1 and 0.

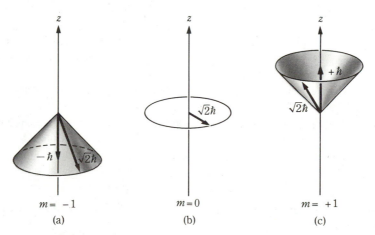

$$m = -1 \qquad\qquad m = 0 \qquad\qquad m = +1$$
$$(a) \qquad\qquad\qquad (b) \qquad\qquad\qquad (c)$$

Figure 21.9 The z components of angular momentum for $J = 1$:
(a) $m = -1$; (b) $m = 0$; (c) $m = +1$.

QUESTIONS

21.1 What is the physical origin of quantization for the particle in a box? For a harmonic oscillator? For a rigid rotor?

21.2 The eigenfunctions for a particle in a box oscillate more rapidly in space as n increases. Connect this behavior with the increasing momentum magnitude as n increases.

21.3 For the ground state of a particle in a one dimensional box, $(\Delta p_x)^2 = \langle p_x^2 \rangle = 2mE_0$. Use the uncertainty principle to estimate the zero point energy E_0.

21.4 Why is there no zero point energy for a free particle?

21.5 Why is there no uncertainty principle for the product $\Delta p_x \cdot \Delta y$?

21.6 Sketch ψ_n^2 for the harmonic oscillator versus displacement for $n = 0$ to $n = 5$. What is the connection between increasing oscillation of ψ_n^2 as n increases and the increasing momentum magnitude?

2.17 Give a sketch illustrating the orthogonality of ψ_0 and ψ_1 for a harmonic oscillator.

21.8 In Eq. (21.61), $\sin \theta \, d\theta \, d\phi$ gives the area element dA on a sphere of unit radius. Sketch this element on a sphere for θ near $0°$ and for θ near $90°$ to see why the $\sin \theta$ term is present in dA.

21.9 Why are there two quantum numbers for a rigid rotor? What is their meaning?

21.10 Classically, a rigid rotor that has its angular momentum directed *exclusively* along the z-axis is rotating solely in the xy-plane. Argue that this would violate the uncertainty principle for a quantum rotor.

21.11 The wavefunction ψ of a certain system is the linear combination

$$\psi = \sqrt{\frac{1}{4}}\psi_1 + \sqrt{\frac{3}{4}}\psi_2,$$

where ψ_1 and ψ_2 are energy eigenfunctions with (nondegenerate) energy eigenvalues E_1 and E_2 respectively. What is the probability that the system energy will be observed to be E_1? To be E_2?

PROBLEMS

21.1 Calculate the probability of finding the particle in the "box" in the region between $\frac{1}{4}L$ and $\frac{3}{4}L$.

21.2 The electrons in a vacuum tube are confined in a "box" between filament and plate that is perhaps 0.1 cm in width. Compute the spacing between the energy levels in this situation. Do the electrons behave more like waves or like golf balls? In a simple tube the energy of the electron is about 100 eV. What is the quantum number of the electrons?

21.3 If the energy of the electron is 5 eV, what size box must confine it so that the wave property will be exhibited? Assume we can observe 0.1 % of the total energy.

21.4 The muzzle velocity of a rifle bullet is about 900 m/s. If the rifle bullet weighs 30 g, with what accuracy can the position be measured without perturbing the momentum by more than one part in a million?

21.5 Evaluate $\langle x^2 \rangle$ for the harmonic oscillator and from this value obtain

$$\Delta x = [\langle x^2 \rangle - \langle x \rangle^2]^{1/2}.$$

21.6 Evaluate $\langle p_x^2 \rangle$ and $\langle p_x \rangle^2$ for the harmonic oscillator and calculate the uncertainty in the momentum, Δp_x.

21.7 By combining the results of Problems 21.6 and 21.7 find the uncertainty relation for the harmonic oscillator.

21.8 Calculate the expectation value for the kinetic energy and for the potential energy of the harmonic oscillator, in the states $n = 0$ and $n = 1$.

21.9 Calculate the uncertainty in the value of the kinetic energy in the states $n = 0$ and $n = 1$.

21.10 Show that for $n = 1$, the probability of finding the harmonic oscillator in the classically forbidden region is 0.1116.

21.11 Derive Eq. (21.49) using the hint suggested in the text.

21.12 For a particle in a cubical box, $L_1 = L_2 = L_3 = L$, tabulate the energy values in the lowest eight energy levels, (as multiples of $h^2/8mL^2$), and the degeneracy of each level.

21.13 Calculate the moment of inertia, the angular momentum, and the energy in the first rotational state above the ground level, $J = 1$, for

a) H_2 in which $r_0 = 74.6$ pm; $M_H = 1.007825$ g/mol.
b) O_2 in which $r_0 = 120.8$ pm; $M_O = 15.99491$ g/mol.

21.14 From the definitions in Eqs. (21.71) and (21.59), prove Eq. (21.72).

22

The Hydrogen Atom

22.1 THE CENTRAL–FIELD PROBLEM

Returning to Eq. (21.60), we consider the case in which $V(r, \theta, \phi)$ is, in fact, a function only of r, the distance between the two bodies. Then any forces act only along the line of centers of the two bodies; this defines a "central-field" problem. To discuss the central-field problem, we multiply Eq. (21.60) by $2\mu r^2$, and rearrange to

$$-\hbar^2 \frac{\partial}{\partial r}\left(r^2 \frac{\partial \psi}{\partial r}\right) - \hbar^2\left[\frac{1}{\sin\theta}\frac{\partial}{\partial\theta}\left(\sin\theta\frac{\partial\psi}{\partial\theta}\right) + \frac{1}{\sin^2\theta}\frac{\partial^2\psi}{\partial\phi^2}\right] + 2\mu r^2[V(r) - E]\psi = 0. \quad (22.1)$$

We recognize immediately from Eq. (21.64) the operator for the square of the total angular momentum; then Eq. (22.1) becomes

$$-\hbar^2 \frac{\partial}{\partial r}\left(r^2 \frac{\partial \psi}{\partial r}\right) + \mathbf{M}^2\psi + 2\mu r^2[V(r) - E]\psi = 0.$$

Since the set of terms, $\mathbf{M}^2\psi$, depends only on θ and ϕ and not on r, we may write

$$\psi(r, \theta, \phi) = R(r)Y_{J, m}(\theta, \phi).$$

Then, since $\mathbf{M}^2 Y_{J, m} = J(J + 1)\hbar^2 Y_{J, m}$, we find that $\mathbf{M}^2\psi = R(r)\mathbf{M}^2 Y_{J, m}(\theta, \phi) = J(J + 1)\hbar^2 R(r)Y_{J, m}(\theta, \phi)$. Using this result in the equation and dividing through by $Y_{J, m}(\theta, \phi)$ and $2\mu r^2$ yields

$$-\frac{\hbar^2}{2\mu r^2}\frac{d}{dr}\left(r^2 \frac{dR}{dr}\right) + \left[\frac{J(J + 1)\hbar^2}{2\mu r^2} + V(r) - E\right]R(r) = 0. \quad (22.2)$$

This result simply tells us that the angular momentum in the presence of a central field is quantized in exactly the same way as for the rigid rotor. The rotational energy, on the other hand, is affected by the fact that r is not constant; consequently, the moment of inertia is not constant.

Equation (22.2) also shows that the total energy is made up of three contributions: the first term in the equation is the contribution of the kinetic energy of the motion along the line of centers; the second term is the kinetic energy associated with the rotation; the third term is the potential energy, $V(r)$.

22.2 THE HYDROGEN ATOM

The hydrogen atom is a typical case of the central-field problem. As was shown in Fig. 19.5, the proton is at the center with a charge $+e$ while the electron is at a distance r with a charge $-e$. The coulombic force acts along the line of centers and corresponds to a potential energy, $V(r) = -e^2/4\pi\epsilon_0 r$.

We can rewrite Eq. (22.2) in the form

$$-\frac{\hbar^2}{2\mu r^2}\frac{d}{dr}\left(r^2\frac{dR}{dr}\right) + \left[\frac{l(l+1)\hbar^2}{2\mu r^2} - \frac{e^2}{4\pi\epsilon_0 r}\right]R = ER. \qquad (22.3)$$

We have replaced the rotational quantum number J by l, since this is the usual notation in atomic systems. The quantum number l is called the *azimuthal* quantum number and characterizes the total angular momentum of the atom,

$$M^2 = l(l+1)\hbar^2, \qquad l = 0, 1, 2, 3, \ldots$$

The quantum number m has the same interpretation as before; it characterizes the z component of the angular momentum.

$$M_z = m\hbar, \qquad m = -l, -(l-1), \ldots, -1, 0, 1, \ldots, l-1, l.$$

In this situation, m is called the *magnetic* quantum number, for reasons that will be apparent later.

★ 22.2.1 Mathematical Details; Solution of the Radial Equation

Again we introduce dimensionless variables; $r = \beta\rho$ and $E = -(1/n^2)(\hbar^2/2\mu\beta^2)$, where n is a parameter characterizing the energy. Note that the energy has been chosen as a negative quantity; this implies that the discussion will deal only with the bound states of the hydrogen atom. The zero of potential energy is at $r \to \infty$, in which state the two particles move independently. The equation becomes

$$-\frac{\hbar^2}{2\mu\beta^2}\left[\frac{1}{\rho^2}\frac{d}{d\rho}\left(\rho^2\frac{dR}{d\rho}\right)\right] + \left\{\frac{l(l+1)}{\rho^2}\left(\frac{\hbar^2}{2\mu\beta^2}\right) - \frac{2}{\rho}\left(\frac{e^2}{8\pi\epsilon_0\beta}\right)\right\}R = -\frac{1}{n^2}\left(\frac{\hbar^2}{2\mu\beta^2}\right)R.$$

It is convenient to determine β by requiring that

$$\frac{\hbar^2}{2\mu\beta^2} = \frac{e^2}{8\pi\epsilon_0\beta} \qquad \text{or} \qquad \beta = \frac{4\pi\epsilon_0\hbar^2}{\mu e^2} = a_0,$$

in which a_0 is the first Bohr radius; compare to Eq. (19.20). Then we have

$$\frac{1}{\rho^2}\frac{d}{d\rho}\left(\rho^2\frac{dR}{d\rho}\right) + \left(\frac{2}{\rho} - \frac{l(l+1)}{\rho^2} - \frac{1}{n^2}\right)R = 0,$$

or

$$\frac{d^2R}{d\rho^2} + \frac{2}{\rho}\frac{dR}{d\rho} + \left(\frac{2}{\rho} - \frac{l(l+1)}{\rho^2} - \frac{1}{n^2}\right)R = 0. \tag{22.4}$$

As $\rho \to \infty$, this equation becomes $d^2R/d\rho^2 = (1/n^2)R$, which yields the values

$$R_\infty = e^{\rho/n} \quad \text{and} \quad R_\infty = e^{-\rho/n};$$

only the second value is finite at $\rho = \infty$, so we choose our solution in the form

$$R = u(\rho)e^{-\rho/n}. \tag{22.5}$$

Calculation of $d^2R/d\rho^2$ and $dR/d\rho$ and substitution in Eq. (22.4) yields, after dividing out $e^{-\rho/n}$,

$$\frac{d^2u}{d\rho^2} + \frac{1}{\rho}\left(2 - \frac{2\rho}{n}\right)\frac{du}{d\rho} + \frac{1}{\rho^2}\left[\frac{2\rho}{n}(n-1) - l(l+1)\right]u = 0.$$

If we set $(2\rho/n) = x$ in this equation it simplifies immediately to

$$\frac{d^2u}{dx^2} + \left(\frac{2-x}{x}\right)\frac{du}{dx} + \frac{1}{x^2}[(n-1)x - l(l+1)]u = 0. \tag{22.6}$$

Solutions to this equation have the form

$$u(x) = x^l L(x); \tag{22.7}$$

then

$$\frac{du}{dx} = x^l\frac{dL}{dx} + lx^{l-1}L;$$

$$\frac{d^2u}{dx^2} = x^l\frac{d^2L}{dx^2} + 2lx^{l-1}\frac{dL}{dx} + l(l-1)x^{l-2}L.$$

These values reduce the equation, after division by x^{l-1}, to

$$x\frac{d^2L}{dx^2} + [2(l+1) - x]\frac{dL}{dx} + [n - (l+1)]L = 0. \tag{22.8}$$

The only solutions, $L(x)$, of this equation that are quadratically integrable are those for which the coefficient of L is zero or a positive integer; this condition requires that the parameter n be an integer such that $n - (l+1) \geq 0$ or $n \geq l + 1$. Since the least value of l is zero we have the quantization conditions

$$n = 1, 2, 3, \ldots, \quad 0 \leq l \leq n - 1. \tag{22.9}$$

We recognize these conditions as the familiar requirements on the value of n, *the principal quantum number*, of the hydrogen atom.

The functions $L(x)$ are the associated Laguerre polynomials, which depend on n and l; if we write $s = n + l$ and $t = 2l + 1$, then the equation becomes

$$\frac{xd^2L_s^t(x)}{dx^2} + (t + 1 - x)\frac{dL_s^t(x)}{dx} + (s - t)L_s^t(x) = 0. \tag{22.10}$$

The general form for the polynomial $L_s^t(x)$ is

$$L_s^t(x) = -\sum_{k=0}^{s-t}(-1)^k\frac{(s!)^2x^k}{(s-t-k)!(t+k)!k!}. \tag{22.11}$$

Table 22.1
The associated Laguerre polynomials

$n = 1$;	$l = 0$	$L_1^1(x) = -1$	$x = 2\rho$
$n = 2$;	$l = 0$	$L_2^1(x) = -2!(2 - x)$	$x = \rho$
	$l = 1$	$L_3^3(x) = -3!$	
$n = 3$;	$l = 0$	$L_3^1(x) = -3!(3 - 3x + \frac{1}{2}x^2)$	$x = \frac{2}{3}\rho$
	$l = 1$	$L_4^3(x) = -4!(4 - x)$	
	$l = 2$	$L_5^5(x) = -5!$	
$n = 4$;	$l = 0$	$L_4^1(x) = -4!(4 - 6x + 2x^2 - \frac{1}{6}x^3)$	$x = \frac{1}{2}\rho$
	$l = 1$	$L_5^3(x) = -5!(10 - 5x + \frac{1}{2}x^2)$	
	$l = 2$	$L_6^5(x) = -6!(6 - x)$	
	$l = 3$	$L_7^7(x) = -7!$	
$n = 5$;	$l = 0$	$L_5^1(x) = -5!(5 - 10x + 5x^2 - \frac{5}{6}x^3 + \frac{1}{24}x^4)$	$x = \frac{2}{5}\rho$
	$l = 1$	$L_6^3(x) = -6!(20 - 15x + 3x^2 - \frac{1}{6}x^3)$	
	$l = 2$	$L_7^5(x) = -7!(21 - 7x + \frac{1}{2}x^2)$	
	$l = 3$	$L_8^7(x) = -8!(8 - x)$	
	$l = 4$	$L_9^9(x) = -9!$	

It should be observed that in contrast to the Hermite and Legendre polynomials, the Laguerre polynomials contain both odd and even powers of x. The first few are given in Table 22.1.

22.2.2 Wave Functions for the Hydrogen Atom

The normalized radial wave functions have the form

$$R_{nl}(r) = -\frac{2}{n^2 a_0^{3/2}} \left\{ \frac{(n - l - 1)!}{[(n + l)!]^3} \right\}^{1/2} e^{-x/2} x^l L_{n+l}^{2l+1}(x) \tag{22.12}$$

in which $x = 2r/na_0$.

The complete wave function for the hydrogen atom has the form

$$\psi_{n, l, m}(r, \theta, \phi)$$

$$= -\frac{2}{n^3 a_0^{3/2}} \left\{ \frac{(n - l - 1)!(2l + 1)(l - |m|)!}{[(n + l)!]^3 4\pi(l + |m|)!} \right\}^{1/2} e^{-x/2} x^l L_{n+l}^{2l+1}(x) P_l^{|m|}(\cos\theta) e^{im\phi}. \tag{22.13}$$

A list of the complete hydrogen atom wave functions is given in Table 22.2.

22.2.3 Recapitulation on the Hydrogen Atom

The hydrogen atom consists of two particles, a proton and an electron. Six coordinates and six momenta, three for each particle, are needed to describe the mechanical state of this system. The six coordinates of such a system can always be transformed to three coordinates of the center of mass of the system and three internal coordinates. After this is done the Schrödinger equation separates into two independent equations. The first involves only the coordinates of the center of mass and the translational energy of the atom as a whole; since the translational part has no interest to us, we discard it. The remaining equation involves the internal coordinates of the atom and the internal energy. It is this energy and this part of the description that is of interest. We will refer to this internal energy simply as the energy of the atom.

The internal coordinates are the usual spherical coordinates r, θ, and ϕ, displayed in their relation to the Cartesian coordinates in Fig. 21.7. The nucleus is at the origin, r is the distance between the nucleus and the electron, θ is the angle between the z-axis and the radius vector connecting nucleus and electron, and ϕ is the angle between the $+$ x-axis and the projection of the radius vector on the xy-plane. The potential energy is $-e^2/4\pi\epsilon_0 r$, resulting from the electrical attraction of the charges $+e$ on the nucleus and $-e$ on the electron. The Schrödinger equation can be solved for ψ as a function of the coordinates

Table 22.2
Complete hydrogen-atom wave functions, $\psi_{n,l,m}(r, \theta, \phi)$ ($\rho = r/a_0$)

$n = 1$,	$l = 0$,	$m = 0$.	$\psi_{100} = \left(\dfrac{1}{\pi a_0^3}\right)^{1/2} e^{-\rho}$
$n = 2$,	$l = 0$,	$m = 0$.	$\psi_{200} = \dfrac{1}{8}\left(\dfrac{2}{\pi a_0^3}\right)^{1/2}(2 - \rho)e^{-\rho/2}$
$n = 2$,	$l = 1$,	$m = 0$.	$\psi_{210} = \dfrac{1}{8}\left(\dfrac{2}{\pi a_0^3}\right)^{1/2}\rho e^{-\rho/2}\cos\theta$
$n = 2$,	$l = 1$,	$m = \pm 1$.	$\psi_{211} = \dfrac{1}{8}\left(\dfrac{1}{\pi a_0^3}\right)^{1/2}\rho e^{-\rho/2}\sin\theta\, e^{i\phi}$
			$\psi_{21-1} = \dfrac{1}{8}\left(\dfrac{1}{\pi a_0^3}\right)^{1/2}\rho e^{-\rho/2}\sin\theta\, e^{-i\phi}$
$n = 3$,	$l = 0$,	$m = 0$.	$\psi_{300} = \dfrac{1}{243}\left(\dfrac{3}{\pi a_0^3}\right)^{1/2}(27 - 18\rho + 2\rho^2)e^{-\rho/3}$
$n = 3$,	$l = 1$,	$m = 0$.	$\psi_{310} = \dfrac{1}{81}\left(\dfrac{2}{\pi a_0^3}\right)^{1/2}\rho(6 - \rho)e^{-\rho/3}\cos\theta$
$n = 3$,	$l = 1$,	$m = \pm 1$.	$\psi_{311} = \dfrac{1}{81}\left(\dfrac{1}{\pi a_0^3}\right)^{1/2}\rho(6 - \rho)e^{-\rho/3}\sin\theta\, e^{i\phi}$
			$\psi_{31-1} = \dfrac{1}{81}\left(\dfrac{1}{\pi a_0^3}\right)^{1/2}\rho(6 - \rho)e^{-\rho/3}\sin\theta\, e^{-i\phi}$
$n = 3$,	$l = 2$,	$m = 0$.	$\psi_{320} = \dfrac{1}{486}\left(\dfrac{6}{\pi a_0^3}\right)^{1/2}\rho^2 e^{-\rho/3}(3\cos^2\theta - 1)$
$n = 3$,	$l = 2$,	$m = \pm 1$.	$\psi_{321} = \dfrac{1}{81}\left(\dfrac{1}{\pi a_0^3}\right)^{1/2}\rho^2 e^{-\rho/3}\sin\theta\cos\theta\, e^{i\phi}$
			$\psi_{32-1} = \dfrac{1}{81}\left(\dfrac{1}{\pi a_0^3}\right)^{1/2}\rho^2 e^{-\rho/3}\sin\theta\cos\theta\, e^{-i\phi}$
$n = 3$,	$l = 2$,	$m = \pm 2$.	$\psi_{322} = \dfrac{1}{162}\left(\dfrac{1}{\pi a_0^3}\right)^{1/2}\rho^2 e^{-\rho/3}\sin^2\theta\, e^{i2\phi}$
			$\psi_{32-2} = \dfrac{1}{162}\left(\dfrac{1}{\pi a_0^3}\right)^{1/2}\rho^2 e^{-\rho/3}\sin^2\theta\, e^{-i2\phi}$

r, θ, and ϕ. The wave functions obtained by solving this equation are descriptions of the states of the hydrogen atom.

If the solutions of the Schrödinger equation are to make physical sense, certain integers, quantum numbers, must be introduced. Just as in the case of the particle in the box, these integers enter because of the constraints that are placed on the system. For example, if the probability density $|\psi|^2$ is to have a unique value at every point in space, the description ψ must have the same value at $\phi = 2\pi$ and at $\phi = 0$, since these values of ϕ correspond to the same set of points in space. This restriction together with the form of the equation requires ψ to depend on ϕ through either $e^{im\phi}$ or $e^{-im\phi}$, where m is an integer. Two other integers, n and l, are introduced by the requirement that the probability density be finite everywhere. It is clear that we may not have an infinite probability of finding the electron at any point in space.

The final description, $\psi_{nlm}(r, \theta, \phi)$ or, more concisely, ψ_{nlm}, is a function of the co-ordinates r, θ, and ϕ, and of the quantum numbers n, l, and m. Since ψ_{nlm} depends on the integers in a unique way, the integers by themselves constitute a convenient, abbreviated description of the system. Knowing the integers, we can look up the corresponding ψ_{nlm} in a table such as Table 22.2 if we need it. For the most part we shall use only the quantum numbers to describe the system.

22.3 SIGNIFICANCE OF THE QUANTUM NUMBERS IN THE HYDROGEN ATOM

22.3.1 The Principal Quantum Number

The integer n, the principal quantum number, describes the energy of the hydrogen atom through the equation

$$E_n = -\frac{1}{2n^2}\left(\frac{e^2}{4\pi\epsilon_0 a_0}\right) = -\left(\frac{1}{2n^2}\right)E_{\text{h}} \tag{22.14}$$

with allowed values $n = 1, 2, 3, \ldots$ In Eq. (22.14), a_0 is the first Bohr radius and E_{h} is the hartree energy defined by $E_{\text{h}} = e^2/4\pi\epsilon_0 a_0$. (In atomic and molecular problems it is convenient to express energies as multiples of the hartree energy.) The energies given by Eq. (22.14) are the same as those given by Bohr's initial calculation, Eq. (19.22). However, in the Schrödinger model, n has nothing directly to do with angular momentum, while in the Bohr model, n was a measure first of the angular momentum of the system. This difference in interpretation should be kept in mind.

The energy of the hydrogen atom is quantized, so it has a system of energy levels. The lowest permissible energy is that corresponding to $n = 1$ in Eq. (22.14).

$$-\tfrac{1}{2}E_{\text{h}} = -\frac{e^2}{8\pi\epsilon_0 a_0} = -2.17872 \times 10^{-18} \text{ J}.$$

Then the permitted energies are $-\tfrac{1}{2}E_{\text{h}}$, $-\tfrac{1}{8}E_{\text{h}}$, $-\tfrac{1}{18}E_{\text{h}}$, $-\tfrac{1}{32}E_{\text{h}}$, $\ldots$ The energy levels and the possible transitions between them are shown in Fig. 22.1.

In making a transition between a high energy state and one of lower energy, the atom emits a quantum of light having a frequency determined by $h\nu = \Delta E$, where ΔE is the difference in energy of the two states. The spectrum of the atom therefore consists of series of lines having frequencies corresponding to the possible values of the energy differences, represented by the lengths of the arrows in Fig. 22.1. There are several series of lines in the

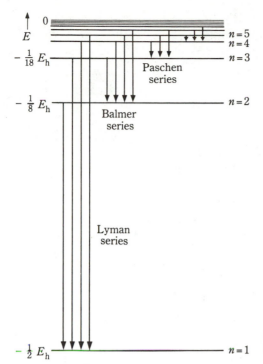

Figure 22.1 Energy levels in the hydrogen atom.

spectrum. The general form for the energy difference between two states, from Eq. (22.14), is

$$\Delta E = h\nu_{nk} = \tfrac{1}{2}E_h\left(\frac{1}{n^2} - \frac{1}{k^2}\right). \tag{22.15}$$

This is equivalent to the Rydberg formula, Eq. (19.16).

Transitions from the upper states to the ground state, $n = 1$, involve large differences in energy; the Lyman series of lines is in the ultraviolet region of the spectrum. Transitions from higher states to the level $n = 2$ involve smaller differences in energy; the Balmer series of lines lies in the visible and near ultraviolet. Transitions to the level $n = 3$ yield the Paschen series of lines in the infrared. Transitions to $n = 4$ and to $n = 5$ yield the Brackett and Pfund series in the far infrared. Note that as the transitions to any particular level occur from higher and higher levels, the energy difference, and therefore the frequency, does not change much. The lines in the spectrum come closer together and approach a *series limit*.

If a hydrogen atom absorbs light, only those frequencies that match the allowed energy differences will be absorbed. The absorption spectrum and the emission spectrum therefore have the same lines. The absorbed quantum lifts the hydrogen atom from one permitted energy level to a higher one.

22.3.2 The Azimuthal Quantum Number

The integer l, the azimuthal quantum number, describes the total angular momentum of the hydrogen atom through the equation

$$M^2 = l(l + 1)\hbar^2, \tag{22.16}$$

with allowed values $l = 0, 1, 2, \ldots, n - 1$; M^2 is the square of the total angular momentum. The principal quantum number n could have any positive, nonzero, integral value; in contrast, l may be zero, and may not exceed $n - 1$. It is customary to designate the values of l by letters; the correspondence is:

Value of l	0	1	2	3	4	5
Letter designation	s	p	d	f	g	h

(The letters s, p, d, f originated in the initial letters of *sharp*, *principal*, *diffuse*, and *fundamental*; words originally used to describe lines and series in spectra.) After $l = 3$, the letter designation proceeds alphabetically. The possible combinations of values of n and l for $n = 1$ to $n = 4$ are:

Value of n	1	2		3			4			
Value of l	0	0	1	0	1	2	0	1	2	3
Notation	$1s$	$2s$	$2p$	$3s$	$3p$	$3d$	$4s$	$4p$	$4d$	$4f$

The notation on the third line is that usually employed for the particular combination of values of n and l; the number is the value of n, the letter is the letter designation of the value of l.

In making a transition from one state to another in the absorption or emission of radiation, there is a restriction on l, called a selection rule. The value of l must change by ± 1. Thus, if a hydrogen atom in the ground state, $1s$ state, absorbs radiation and goes to level $n = 2$, then it must be finally in the $2p$ state. Any other transition between levels 1 and 2 is forbidden by the selection rule. The existence of selection rules helps enormously in the interpretation of spectra.

Although we know the magnitude of the total angular momentum from Eq. (22.16), we do not know the sign. Therefore the orientation of the angular momentum vector is indefinite, and so the orientation of the orbit is indefinite.

Suppose we compare the values of angular momentum in the Bohr and the Schrödinger models:

Bohr $\qquad M^2 = n^2\hbar^2, \qquad n = 1, 2, \ldots$

Schrödinger $\qquad M^2 = l(l + 1)\hbar^2, \qquad l = 0, 1, 2, \ldots$

(Incidentally, the Bohr–Sommerfeld model required the Schrödinger value of the angular momentum.) In the Bohr atom, the angular momentum always had a nonzero value, while in the modern theory the angular momentum is zero in the s states for which $l = 0$. The absence of angular momentum in the s states makes it impossible to imagine the motion of the electron in these states in terms of a classical orbital motion. It is better not to try. Where it may help to visualize it, the electronic motion will sometimes be described *as if* it were moving in a classical orbit; this description must not be accepted literally, but analogically.

22.3.3 The Magnetic Quantum Number

The integer m, the magnetic quantum number, describes the z component of the angular momentum M_z through the equation

$$M_z = m\hbar, \tag{22.17}$$

with allowed values $m = -l, -l + 1, -l + 2, \ldots, -1, 0, +1, +2, \ldots, +l$. Any integral value from $-l$ to $+l$ including zero is a permitted value for m. There are $2l + 1$ values of m for a given value of l. If $l = 0$, then $m = 0$. But if $l = 1$, then m may be $-1, 0, +1$. If $l = 2$, then m may be $-2, -1, 0, +1, +2$. In the absorption or emission of a light quantum, the selection rules require either $\Delta m = 0$, or $\Delta m = \pm 1$.

22.4 PROBABILITY DISTRIBUTION OF THE ELECTRON CLOUD IN THE HYDROGEN ATOM

22.4.1 Probability Distribution in *s* States

The requirements of the uncertainty principle make it necessary to visualize the hydrogen atom as a nucleus imbedded in a "fog" of negative charge. This electron cloud has a different shape in the different states of the atom. To discover the shape of the cloud, we construct the probability density $|\psi_{nlm}|^2$, for the state in question.

For the ground state, 1s state, of the hydrogen atom, the wave function is

$$\psi_{1s} = \frac{1}{(\pi a_0^3)^{1/2}} e^{-r/a_0}. \tag{22.18}$$

This equation shows that in the 1s state, the wave function and the probability density are independent of the angles θ and ϕ. Consequently, the electron cloud is spherically symmetric. Let $P(r)$ be the probability density; then

$$P_{1s}(r) = \psi_{1s}^2 = \frac{e^{-2r/a_0}}{\pi a_0^3}. \tag{22.19}$$

This function is shown in Fig. 22.2(a). The probability density is high near the nucleus and decreases rapidly as r increases. Since the volume near the nucleus is very small, the total amount of the cloud near the nucleus is very small. So we ask a different question. How much of the cloud is contained in the spherical shell bounded by the spheres of radius r and $r + dr$?

The volume of this spherical shell is $dV_{\text{shell}} = 4\pi r^2 \, dr$, so that the amount of the cloud in the shell is $P_{1s}(r)4\pi r^2 \, dr$. The function $4\pi r^2 P(r) = f(r)$ is the radial distribution

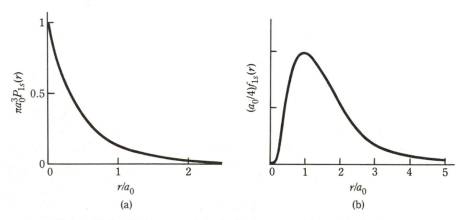

Figure 22.2 The 1s state of the hydrogen atom. (a) Probability density. (b) Radial distribution function.

function:

$$f_{1s}(r) = 4\pi r^2 P_{1s}(r) = \frac{4r^2}{a_0^3} e^{-2r/a_0}. \tag{22.20}$$

The radial distribution function is the total probability of finding the electron in the spherical shell (Fig. 22.2b). In the 1s state the probability of finding the electron is a maximum in the spherical shell which has a radius a_0, the radius of the first Bohr orbit. The probability of finding the electron in a spherical shell near the nucleus is very small, as is that of finding it very far away from the nucleus.

The distance of the electron from the nucleus in the 1s state can be calculated using the theorem on expectation values, Eq. (20.7). For an s state the volume element may be taken as the volume of the spherical shell; the operator for r is simply multiplication of the wave function by r; so we get

$$\langle r_{1s} \rangle = \int \psi_{1s} r \psi_{1s} \, d\tau = \int \psi_{1s} r \psi_{1s} \, d\tau.$$

Putting in the values of ψ_{1s} and the volume element, this becomes

$$\langle r_{1s} \rangle = \frac{a_0}{4} \int_0^\infty \left(\frac{2r}{a_0}\right)^3 e^{-2r/a_0} \, d\left(\frac{2r}{a_0}\right) = \frac{a_0}{4} 3! = \tfrac{3}{2}a_0. \tag{22.21}$$

By the same method, using the appropriate wave function, we can show that for any s state of principal quantum number n,

$$\langle r_{ns} \rangle = \tfrac{3}{2}a_0 n^2. \tag{22.22}$$

In states with larger values of n (higher energies) the average distance of the electron from the nucleus is larger. This is apparent in Fig. 22.3, which shows the radial distribution function for hydrogen is the 1s, 2s, and 3s states. Also, note that as n increases, the distribution function becomes "lumpier"; this "lumpiness" is characteristic of the larger values of the kinetic energy in these states.

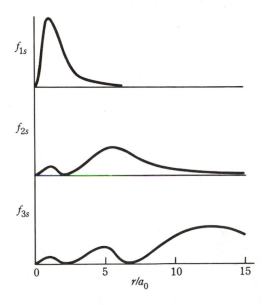

Figure 22.3   Radial distribution function for 1s, 2s, and 3s states.

22.4.2 Probability Distribution in States with Angular Momentum

In states having angular momentum, the z component has a precise value. This fact has the consequence, through the uncertainty principle, that the angle of orientation of the electron around the z-axis is completely indefinite. The electron has equal probability of having any orientation about the z-axis; therefore the charge cloud is symmetric about the z-axis. In contrast to s states, which have spherical symmetry, states with angular momentum have axial symmetry, conventionally associated with the z-axis.

To be concrete, consider the p states. In these states $l = 1$, so that by Eq. (22.16) the total angular momentum is $M = \sqrt{l(l + 1)}\hbar = \sqrt{2}\hbar$. Since m may be -1, 0, or $+1$, the possible values of the z component are, by Eq. (22.17),

$$M_z = -\hbar, \qquad 0, \qquad +\hbar.$$

Figure 22.4(a) shows the possible orientations of the angular momentum vector, of magnitude $\sqrt{2}\hbar$, which have $M_z = -\hbar$. Any vector lying in the conical surface fulfills this requirement. In Fig. 22.4(b) it is apparent that any vector lying in the xy-plane has $M_z = 0$. Any vector lying in the conical surface of Fig. 22.4(c) has $M_z = +\hbar$.

The corresponding charge density distributions are shown in Fig. 22.5. Note that a large z component, either positive or negative, squeezes the charge cloud nearer the

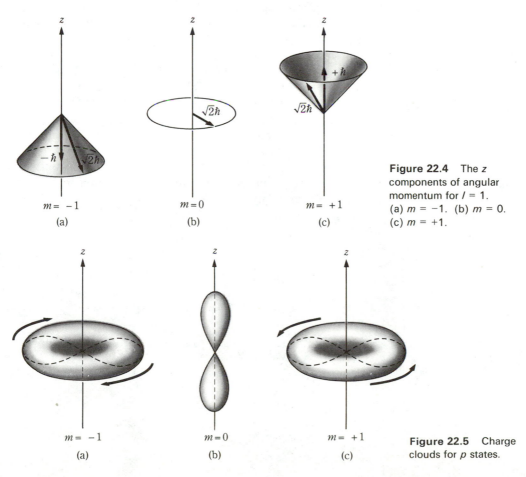

$m = -1$	$m = 0$	$m = +1$
(a)	(b)	(c)

Figure 22.4 The z components of angular momentum for $l = 1$. (a) $m = -1$. (b) $m = 0$. (c) $m = +1$.

$m = -1$	$m = 0$	$m = +1$
(a)	(b)	(c)

Figure 22.5 Charge clouds for p states.

Table 22.3

Old description	New description	Relation between old and new
$p_{+1}(m = +1)$	p_x	$p_x = \frac{1}{2}\sqrt{2}(p_{+1} + p_{-1})$
$p_{-1}(m = -1)$	p_y	$p_y = -i\frac{1}{2}\sqrt{2}(p_{+1} - p_{-1})$
$p_0(m = 0)$	p_z	$p_z = p_0$

xy-plane. If $M_z = 0$, the charge cloud density vanishes in the xy-plane. The only distinction we can make between the charge clouds for $M_z = +\hbar$ and $-\hbar$ is by supposing that for $M_z = -\hbar$ the rotation of the electron is in a clockwise sense, and for $M_z = +\hbar$ that the rotation of the electron is counterclockwise.

Since for $m = \pm 1$ we cannot distinguish the shapes of the clouds, we use the principle of superposition to construct new descriptions. Let the old p functions be designated by the proper values of m; we write p_{+1} and p_{-1}. By taking linear combinations of these descriptions, we obtain the new descriptions which we designate by p_x and p_y. This procedure is shown in Table 22.3.

The advantage of the new description is that the three charge clouds shown in Fig. 22.6 for p_x, p_y, p_z look equivalent; each consists of two lobes that lie along the x-, y-, and z-axis, respectively. The function p_x corresponds to $M_x = 0$, p_y to $M_y = 0$, and p_z to $M_z = 0$. For each of the p functions in the new description, the maximum in the probability density is along the particular axis. The probability density is zero in the coordinate plane perpendicular to that axis; this is evident in Fig. 22.6.

There are five d states corresponding to values of $m = -2, -1, 0, +1, +2$. The charge clouds for these states are shown in Fig. 22.7. The charge cloud has the same appearance for d_{+2} as for d_{-2}. The direction of motion is counterclockwise for d_{+2} and clockwise for d_{-2}; the same is true for d_{+1} and d_{-1}. We will not need alternative descriptions of the d functions. Note the axial symmetry for all values of m in Fig. 22.7 and that the higher the value of m, the closer the charge cloud is to the xy-plane.

We can compare the extension of these charge clouds in space by calculating the average distance of the electron from the nucleus for the state in question. Using the appropriate wave function, we do the calculation by the same method we used to obtain $\langle r_{1s} \rangle$ in Eq. (22.21). We omit the tedious details and write only the result, which is quite

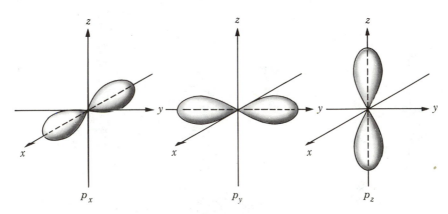

Figure 22.6   Charge clouds for p_x, p_y, and p_z.

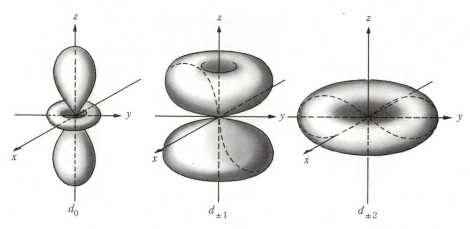

Figure 22.7   Charge clouds in the *d* states.

simple. The average distance of the electron from the nucleus depends only on *n* and *l*:

$$\langle r_{nl} \rangle = \tfrac{1}{2}a_0[3n^2 - l(l + 1)].$$ (22.23)

Equation (22.23) shows that for a specified value of *n* in states having high angular momentum, high values of *l*, the average distance of the electron from the nucleus is less than in states of low angular momentum. As we shall see later, this is the underlying reason for the great similarity in the chemistries of the rare earth elements.

22.5 ELECTRON SPIN AND THE MAGNETIC PROPERTIES OF ATOMS

Before development of the Schrödinger equation, it was shown by Uhlenbeck and Goudsmit that certain troublesome features of atomic spectra could be explained if the electron itself possessed an intrinsic angular momentum. If we do not take the picture too seriously, we may imagine the electron as a tiny ball of negative charge that is spinning on its axis. If the square of the total spin angular momentum is $M_{\text{spin}}^2 = s(s + 1)\hbar^2$, and if $s = \tfrac{1}{2}$, then the experimental data are explained. The *z* component of the spin angular momentum has the value

$$M_{z(\text{spin})} = m_s\hbar.$$ (22.24)

The quantity m_s is the *spin quantum number*; it may have only the values $+\tfrac{1}{2}$ or $-\tfrac{1}{2}$. The Schrödinger equation in its usual form gives no indication of the existence of the electron spin. However, Dirac has shown that if the Schrödinger equation is cast into a form that satisfies certain requirements of relativity theory, then four quantum numbers, the fourth being the electron spin quantum number, appear in the solution for the hydrogen atom. Thus the spin is a coherent part of the fundamental theory and is not tacked on just to patch things up.

   If the electron spins on its axis, the fact that it is electrically charged implies that there is a current flow around the axis. This flow of current gives the electron a magnetic moment, just as the flow of current in a coil of wire gives the coil a magnetic moment. The magnetic moment is perpendicular to the plane of the current flow and so is parallel to the angular momentum vector, but directed oppositely because of the negative charge on the electron. According to Eq. (22.24), the *z* component of the spin angular momentum may be either $+\tfrac{1}{2}\hbar$ or $-\tfrac{1}{2}\hbar$. Consequently, the *z* component of the magnetic moment may have either of two corresponding values, $-\mu_B$ or $+\mu_B$, where μ_B, the Bohr magneton, is a natural unit of magnetic moment. The magnetic moment of the electron made the first observations of the spin property possible.

If in an atom the orbital angular momentum is not zero, $l \neq 0$, then the atom has an orbital magnetic moment as well as a magnetic moment caused by the spin of the electron. In states having angular momentum, the electronic motion constitutes a current flowing around the atom, which produces the magnetic moment. The z component of the orbital magnetic moment is proportional to the z component of the angular momentum and is equal to $-m\mu_B$; for this reason m is called the magnetic quantum number.

Two electrons in an atom that differ only in the value of the spin quantum number are called *paired* electrons. The z component of spin angular momentum of the second electron is equal and opposite in sign to that of the first electron. The net z component of spin angular momentum of the electron pair is the sum, $\frac{1}{2}\hbar + (-\frac{1}{2})\hbar = 0$; thus the pair of electrons has no net spin angular momentum and no magnetic moment along any axis. The pairing of the spins is frequently indicated by arrows, ↑↓, representing the spin quantum numbers, m_s, of the electrons. A head-up arrow represents an electron with $m_s = +\frac{1}{2}$; a head-down arrow represents an electron with $m_s = -\frac{1}{2}$.

Similarly, the net orbital angular momentum of any filled subshell is zero. First of all, the orbital angular momentum in an s level is zero by definition. In a group of p levels, the possible values of m are $-1, 0, +1$. If we place a pair of electrons in each of these p levels, then two electrons have $m = -1$, two have $m = +1$, two have $m = 0$. The net z component for all of these is zero since $2(-1) + 2(0) + 2(+1) = 0$. A filled subshell has no net component of orbital angular momentum around any specified axis, and so it contributes no magnetic moment due to orbital motion.

The magnetic moment of an atom is due entirely to *partially* filled subshells and unpaired spins. The completed subshells do not contribute to the permanent magnetic moment of an atom. In this connection we might remark that most molecules have an even number of electrons whose spins in the ground state are all paired. Also the subshells are usually complete so that there is no contribution to the magnetic moment from the orbital motion. It is unusual for a molecule to have a permanent magnetic moment; the possession of a permanent magnetic moment is an important key to the electronic structure of the molecule.

22.6 THE STRUCTURE OF COMPLEX ATOMS

Using the hydrogen atom as a guide, we will assume that every electron in a complex atom (a complex atom is an atom with more than one electron) can be described by a set of four quantum numbers n, l, m, m_s. We also assume that the system of energy levels in a complex atom is generally similar to that of the hydrogen atom. Usually our attention will be confined to the state of lowest energy, the ground state, of the atom. It is worth noting that we are talking about the structure of the isolated atom; for example, the structure of an isolated sodium atom, not that of a sodium atom in metallic sodium, where many atoms are very close together.

If the electron is in the $1s$ state, the hydrogen atom is in its lowest state of energy. In a polyelectronic atom such as carbon (six electrons) or sodium (eleven electrons) it would not seem unreasonable if all the electrons were in the $1s$ level, thereby giving the atom the lowest possible energy. We might denote such a structure for carbon by the symbol $1s^6$ and for sodium, $1s^{11}$. This result is wrong, but from what has been said so far there is no apparent reason why it should be wrong. The reason lies in an independent and fundamental postulate of the quantum mechanics, the Pauli exclusion principle: *no two electrons may have the same set of four quantum numbers.* Only two sets of four quantum numbers exist for the 1s level; in the order $(n \; l \; m | m_s)$; these sets are $(100|\frac{1}{2})$ and $(100| -\frac{1}{2})$. If more than two electrons are placed in the 1s level, at least one of these sets would be duplicated, a situation forbidden by the Pauli principle. In this light it is clear why the structures, $1s^6$ for carbon and $1s^{11}$ for sodium, are incorrect.

The construction principle (*Aufbau Prinzip*) for the electronic structure of complex atoms is as follows.

1. Each electron in a complex atom is described by a set of four quantum numbers, the quantum numbers being the same as those used to describe the states of the hydrogen atom.
2. The relative arrangement of energy levels in the complex atom is roughly the same as that in the hydrogen atom. To make up the structure of the complex atom, the electrons are arranged in the lowest possible energy levels consistent with the restriction imposed by the Pauli principle.

We divide the levels into *shells*, those levels with the same value of the principal quantum number, and *subshells*, those within a shell that have the same value of the azimuthal quantum number. For a specified value of l, there are $2l + 1$ values of m; for a specified value of m, the electron may have two values of m_s. Hence there are $2(2l + 1)$ distinct combinations of m and m_s. This is the maximum number of electrons permitted in any subshell. For an s subshell, $l = 0$, so only two electrons may occupy the subshell. For a p subshell, $l = 1$, and six electrons are required to fill the p subshell. Ten electrons fill a d subshell, $l = 2$, and so on. The shell with $n = 1$, is the K shell; that with $n = 2$, the L shell; $n = 3$, the M shell; and so on. The number of electrons required to fill the shells is shown in Table 22.4. The numbers 2, 8, 18, 32, ... in the last column are given by $2n^2$, where n is the principal quantum number. The numbers in this famous sequence are the numbers of elements in the periods of the periodic table.

The number of electrons in a subshell is indicated by the superscript on the symbol of the subshell. Using the principles outlined above, we write the electronic configurations for hydrogen and helium as

$$\text{H:} \quad 1s, \qquad \text{He:} \quad 1s^2.$$

The K shell is complete with helium. The next electrons added must go into the shell with $n = 2$. The question is which subshell, the 2s or the 2p, fills in first? In the hydrogen atom the energy of these subshells is the same, but in complex atoms the energy depends on l as well as on n. For a specified value of n, the order of the sublevels is $s, p, d, \ldots$, where s has the lowest energy. So in lithium the 2s level lies lower than the 2p, and the structure is Li: $1s^2 2s$. Following Li we have Be: $1s^2 2s^2$. Then in the six succeeding elements the 2p shell fills; B: $1s^2 2s^2 2p$; C: $1s^2 2s^2 2p^2$, and so on until neon is reached; Ne: $1s^2 2s^2 2p^6$. With

Table 22.4

Value of n	Subshells present	Number of electrons in the filled shell
1 (K shell)	s	2
2 (L shell)	s, p	$2 + 6 = 8$
3 (M shell)	s, p, d	$2 + 6 + 10 = 18$
4 (N shell)	s, p, d, f	$2 + 6 + 10 + 14 = 32$

Table 22.5
Electronic configurations of the gaseous atoms.

1	H	$1s$	51	Sb	Kr $4d^{10}\,4s^2\,5p^3$	
2	He	$1s^2$	52	Te	Kr $4d^{10}\,5s^2\,5p^4$	
			53	I	Kr $4d^{10}\,5s^2\,5p^5$	
3	Li	He $2s$	54	Xe	Kr $4d^{10}\,5s^2\,5p^6$	
4	Be	He $2s^2$				
5	B	He $2s^2\,2p$	55	Cs	Xe	$6s$
6	C	He $2s^2\,2p^2$	56	Ba	Xe	$6s^2$
7	N	He $2s^2\,2p^3$	57	La	Xe $\quad 5d\;6s^2$	
8	O	He $2s^2\,2p^4$	58	Ce	Xe $4f^2 \quad 6s^2$	
9	F	He $2s^2\,2p^5$	59	Pr	Xe $4f^3 \quad 6s^2$	
10	Ne	He $2s^2\,2p^6$	60	Nd	Xe $4f^4 \quad 6s^2$	
			61	Pm	Xe $4f^5 \quad 6s^2$	
11	Na	Ne $3s$	62	Sm	Xe $4f^6 \quad 6s^2$	
12	Mg	Ne $3s^2$	63	Eu	Xe $4f^7 \quad 6s^2$	
13	Al	Ne $3s^2\,3p$	64	Gd	Xe $4f^7\;5d\;6s^2$	
14	Si	Ne $3s^2\,3p^2$	65	Tb	Xe $4f^9 \quad 6s^2$	
15	P	Ne $3s^2\,3p^3$	66	Dy	Xe $4f^{10} \quad 6s^2$	
16	S	Ne $3s^2\,3p^4$	67	Ho	Xe $4f^{11} \quad 6s^2$	
17	Cl	Ne $3s^2\,3p^5$	68	Er	Xe $4f^{12} \quad 6s^2$	
18	Ar	Ne $3s^2\,3p^6$	69	Tm	Xe $4f^{13} \quad 6s^2$	
			70	Yb	Xe $4f^{14} \quad 6s^2$	
19	K	Ar $\quad 4s$	71	Lu	Xe $4f^{14}\;5d\;6s^2$	
20	Ca	Ar $\quad 4s^2$	72	Hf	Xe $4f^{14}\;5d^2\,6s^2$	
21	Sc	Ar $3d\;4s^2$	73	Ta	Xe $4f^{14}\;5d^3\,6s^2$	
22	Ti	Ar $3d^2\;4s^2$	74	W	Xe $4f^{14}\;5d^4\,6s^2$	
23	V	Ar $3d^3\;4s^2$	75	Re	Xe $4f^{14}\;5d^5\,6s^2$	
24	Cr	Ar $3d^5\;4s$	76	Os	Xe $4f^{14}\;5d^6\,6s^2$	
25	Mn	Ar $3d^5\;4s^2$	77	Ir	Xe $4f^{14}\;5d^9$	
26	Fe	Ar $3d^6\;4s^2$	78	Pt	Xe $4f^{14}\;5d^9\,6s$	
27	Co	Ar $3d^7\;4s^2$	79	Au	Xe $4f^{14}\;5d^{10}\,6s$	
28	Ni	Ar $3d^8\;4s^2$	80	Hg	Xe $4f^{14}\;5d^{10}\,6s^2$	
29	Cu	Ar $3d^{10}\;4s$	81	Tl	Xe $4f^{14}\;5d^{10}\,6s^2\,6p$	
30	Zn	Ar $3d^{10}\;4s^2$	82	Pb	Xe $4f^{14}\;5d^{10}\,6s^2\,6p^2$	
31	Ga	Ar $3d^{10}\;4s^2\,4p$	83	Bi	Xe $4f^{14}\;5d^{10}\,6s^2\,6p^3$	
32	Ge	Ar $3d^{10}\;4s^2\,4p^2$	84	Po	Xe $4f^{14}\;5d^{10}\,6s^2\,6p^4$	
33	As	Ar $3d^{10}\;4s^2\,4p^3$	85	At	Xe $4f^{14}\;5d^{10}\,6s^2\,6p^5$	
34	Se	Ar $3d^{10}\;4s^2\,4p^4$	86	Rn	Xe $4f^{14}\;5d^{10}\,6s^2\,6p^6$	
35	Br	Ar $3d^{10}\;4s^2\,4p^5$				
36	Kr	Ar $3d^{10}\;4s^2\,4p^6$	87	Fr	Rn	$7s$
			88	Ra	Rn	$7s^2$
37	Rb	Kr $\quad 5s$	89	Ac	Rn $\quad 6d\;7s^2$	
38	Sr	Kr $\quad 5s^2$	90	Th	Rn $\quad 6d^2\,7s^2$	
39	Y	Kr $4d\;5s^2$	91	Pa	Rn $5f^2\;6d\;7s^2$	
40	Zr	Kr $4d^2\;5s^2$	92	U	Rn $5f^3\;6d\;7s^2$	
41	Nb	Kr $4d^4\;5s$	93	Np	Rn $5f^4\;6d\;7s^2$	
42	Mo	Kr $4d^5\;5s$	94	Pu	Rn $5f^6 \quad 7s^2$	
43	Tc	Kr $4d^6\;5s$	95	Am	Rn $5f^7 \quad 7s^2$	
44	Ru	Kr $4d^7\;5s$	96	Cm	Rn $5f^7\;6d\;7s^2$	
45	Rh	Kr $4d^8\;5s$	97	Bk	Rn $5f^8\;6d\;7s^2$	
46	Pd	Kr $4d^{10}$	98	Cf	Rn $5f^{10} \quad 7s^2$	
47	Ag	Kr $4d^{10}\;5s$	99	Es	Rn $5f^{11} \quad 7s^2$	
48	Cd	Kr $4d^{10}\;5s^2$	100	Fm	Rn $5f^{12} \quad 7s^2$	
49	In	Kr $4d^{10}\;5s^2\,5p$	101	Md	Rn $5f^{13} \quad 7s^2$	
50	Sn	Kr $4d^{10}\;5s^2\,5p^2$				

neon the L shell is filled. Atoms such as helium and neon which have filled electronic shells are chemically inert. The completed group of eight electrons ns^2np^6, the octet, is the configuration of the inert gases (except He) and is always a very stable configuration chemically. The stability of this configuration is one of the bases for the Lewis rules of chemical valency.

The electronic configurations of the elements are shown in Table 22.5, in order of the number of electrons in the atom. Examination of the table shows immediately that chemically similar atoms have similar configurations of the outer electrons. For example, all the alkali metals have the configuration ns over a shell of eight. The coinage metals have the configuration $(n-1)d^{10}ns$; the lone s electron lies over a completed d subshell rather than a shell of s^2p^6. This gives the coinage metals characteristically different properties from those of the alkalies. Other similarities can be picked out readily; for example, the halogens, the inert gases, and so on.

A number of points may be made about Table 22.5. Argon has the configuration $1s^22s^22p^63s^23p^6$. Logically one might expect that the $3d$ subshell would commence to fill with the element following argon. However, the interactions of the electrons give the $4s$ level a lower energy than the $3d$, so the $4s$ level fills first with the elements potassium and calcium; then the $3d$ levels start to fill with the *transition elements* from scandium through nickel. Transition elements have a partially completed d subshell. A similar thing happens after krypton; the $5s$ level fills first, then the $4d$. The $4f$ shell is not filled until the $5s$, $5p$, and $6s$ levels are complete. Following barium, the $5d$ level acquires one electron in lanthanum. Then the $4f$ level fills with 14 electrons in the *inner transition elements* from cerium to lutecium, the rare earths. The inner transition elements have *two* partially filled inner subshells, the $5d$ and the $4f$.

In a level of high angular momentum such as the f level, $l = 3$, the electron is much closer to the nucleus than the other electrons of the same principal quantum number, as shown by Eq. (22.23). In lanthanum and the succeeding fourteen elements, the rare earths, the high value of l and the low value of the principal quantum number, $n = 4$, compared to $n = 6$ in the valence shell both contribute to burying the $4f$ electrons in the interior of the atom. As a result, the exterior electrons that give the atom its chemical properties are not much affected by the number of f electrons present. The chemistries of the rare earths are remarkably similar, differing only by the number of $4f$ electrons present.

22.7 SOME GENERAL TRENDS IN THE PERIODIC SYSTEM

22.7.1 Atomic Radii

The diameter of the hydrogen atom in its ground state is about 100 picometres. Using Eq. (22.23), we can show that in the $3s$ state the diameter of the atom would be of the order of 2500 pm. However, this is much larger than the diameter of the largest atoms. As we progress from hydrogen to the atoms of the heavier elements, the increased nuclear charge pulls the electrons much closer to the center than is possible with the simple hydrogen atom. Equation (22.23) therefore does not give an accurate indication of the radius of atoms other than hydrogen.

If we compare the sizes of the atoms in a vertical family of the periodic table such as the alkali metals, we note that the size of the atom increases from the top to the bottom of the column. The valence electron is in the $2s$, $3s$, $4s$, $5s$, $6s$, $7s$ level as we pass from lithium to francium. Thus we may retain the general statement that the radius of an atom increases with the principal quantum number of the electrons in the valence shell. The increase is not very rapid.

In passing through a horizontal row of the periodic table, the atom size is a maximum with the alkali metal, drops quickly to a minimum in about the third group, then rises irregularly to reach another maximum in the next alkali metal. Two tendencies are operative in this behavior. In passing, let us say from lithium to beryllium, the nuclear charge increases, pulling the electrons in and making the atom smaller. But there is an additional electron, whose presence increases the mutual repulsions in the electron cloud and tends to make the atom larger. In the early part of the period, the electrons easily keep out of each other's way, and the effect of the nuclear charge is dominant; the size decreases. Toward the middle and end of a period the mutual repulsion of the electrons in the shell increases and overbalances the effect of the increased nuclear charge; the size increases.

In speaking of sizes of atoms, we should keep in mind that the electron cloud does not end at any definite distance from the nucleus. As an experimental quantity the radius of the atom is found to depend a good deal on the environment of the atom during the measurement. The following values for the molecular diameter of argon are obtained by the method indicated:

Method	Gas viscosity	Liquid density	Solid density
Diameter	297 pm	404 pm	384 pm

This sort of variation is to be expected because of the lack of definition of the radius of the atom. In making comparisons of the sizes of atoms, situations should be chosen in which the atoms have about the same environment. It is possible to construct consistent tables of ionic radii, for example, or of covalent radii or metallic radii. If possible, it is best to compare the radii for situations in which the atoms have the same number of neighbors.

22.7.2 Ionization Energies

The ionization energy of an atom is a measure of how strongly an electron is bound to the atom. The first ionization energy of an atom is the energy required to remove an electron from the atom to an infinite distance.

$$A \longrightarrow A^+ + e^-.$$

The second ionization energy of an atom is the energy required to remove an electron from the singly charged ion:

$$A^+ \longrightarrow A^{++} + e^-.$$

An atom has as many ionization energies as it has electrons.

For comparison, the first (I) and second (II) ionization energies of a number of atoms are recorded in Table 22.6. These elements have comparable electronic configurations:

Table 22.6
Ionization energies

Element	H	Li	Na	K	Rb	Cs
Configuration	$1s$	$2s$	$3s$	$4s$	$5s$	$6s$
I/(kJ/mol)	1311.7	520.1	495.7	418.6	402.9	375.6
II/(kJ/mol)		7296	4563	3069	2640	2260

a single s electron over an inert gas shell. It is clear that as the quantum number goes up, the electron is more easily removed. This is mainly because of the increase in the distance between nucleus and the outer electron as the quantum number goes up. The greatest difference is between hydrogen and lithium. Note that the second ionization energies of these atoms are 6 to 15 times larger than the first ionization energies. There are two reasons for this. First, there is the unbalanced positive charge, which always increases a second ionization energy compared with the first ionization energy. Secondly, to remove the second electron in these atoms requires that a very stable closed shell, the shell of eight, be opened. This increases the required energy enormously.

The ionization energies for the inert gases are very high:

	He	Ne	Ar	Kr	Xe
Ionization energy (kJ/mol)	2371.6	2080.1	1520.1	1350.4	1170.1

The high values account for the inability of these elements to form compounds involving ions such as He^+, Ne^+, and so on.

If in removing the second electron it is not necessary to break into a closed shell, then the second ionization energy is not enormously greater than the first; generally it is two to three times the first ionization energy. The energy required to remove the second electron is always greater than that required to remove the first; the removal of the third requires more energy than the removal of the second. Roughly, provided that a closed shell of eight is not broken into, the second electron requires two to three times as much energy as the first; the third requires 1.5 to 2 times as much as the second. If a closed shell of eight must be opened, the energy required is *very much* larger. These facts are illustrated by the data in Table 22.7.

Table 22.7

Element	H	He	Li	Be	Na
Configuration	$1s$	$1s^2$	$1s^2 2s$	$1s^2 2s^2$	$1s^2 2s^2 2p^6 3s$
I/(kJ/mol)	1311.7	2371.6	520.1	899.2	495.7
II/(kJ/mol)		5249	7296	1756.6	4563
III/(kJ/mol)			11810	14840	21000

QUESTIONS

22.1 What is the connection between the fact that the Coulomb potential binds the electron in the hydrogen atom and the origin of the quantum number n?

22.2 What is the physical origin of the quantum numbers l and m for the hydrogen atom?

22.3 Give a qualitative uncertainty principle argument that accounts for the behavior of the $1s$ state radial distribution near $r = 0$ in Fig. 22.2(b) (that is, that there is negligible probability of finding the electron near the nucleus).

22.4 Order the $n = 2$ states of H in terms of increasing kinetic energy associated with rotation.

22.5 Discuss the probability of observing M_z values of $+1$, -1 and 0 for an electron in a p_x state.

22.6 What are the similarities and differences between the (a) orbital and (b) spin angular momentum of an electron?

22.7 How many elements would be in the third row of the periodic table if m_s could have three different values rather than two?

22.8 Correlate the trends in atomic radius and first ionization energy for the rare gases.

22.9 Which ionization energy (first, second, and so on) should exhibit a large jump for each of the following species: Li, Be, B, C?

PROBLEMS

22.1 Calculate (a) the wavelengths of the first three lines of the Lyman, Balmer, and Paschen series, and (b) the series limit, the shortest wavelength line, for each series.

22.2 The radial distribution function for the 1s state of hydrogen is given by Eq. (22.20). Show that the maximum of this function occurs at $r = a_0$.

22.3 Calculate the radius of the sphere that will contain 90 % of the hydrogen atom's electron cloud, if the atom is in the (a) 1s state, (b) 2s state, (c) 2p state, and (d) 3s state.

22.4 The p_z wave function for hydrogen has the form $f(r) \cos \theta$, where θ is the angle between the radius vector and the z-axis (Fig. 21.7) and may vary from 0 to π. For a fixed value of r, sketch the probability density as a function of θ.

22.5 a) Calculate $\langle r \rangle$ in the 1s, 2s, 2p, and 3s states of the hydrogen atom; compare with the results from Eq. (22.23).
b) Calculate the expectation value of $(r - \langle r \rangle)^2$ for the same states as in (a).

22.6 The operator for the radial component of the momentum in the hydrogen atom is

$$\mathbf{p}_r = -i\hbar\left(\frac{\partial}{\partial r} + \frac{1}{r}\right)$$

a) Calculate the expectation values of p_r and p_r^2, and the uncertainty in p_r, in the 1s, 2s, 2p, and 3s states of the hydrogen atom.
b) Using the results in (a) with the result of Problem 22.5, write the expression of the Heisenberg uncertainty principle for the radial coordinate and momentum in these states of the hydrogen atom.

22.7 Calculate the expectation value of the potential energy, $V(r) = -e^2/4\pi\epsilon_0 r$, for the hydrogen atom in the 1s state, the 2s state, the 2p state, and the 3s state. Also calculate the expectation value of the kinetic energy in each of these states.

22.8 The force acting between the proton and the electron in the hydrogen atom is given by $F = -e^2/4\pi\epsilon_0 r^2$. Compute the expectation value of this force for the 1s, 2s, and 2p states of the atom.

22.9 a) Calculate the expectation value of the rotational energy, the energy associated with the angular motions of the electron, in the 2p, 3p, and 3d states of the hydrogen atom. Compare each with the expectation value of the total kinetic energy in these states.
b) Calculate the fraction of each rotational energy that is associated with the z component of the angular motion for the same three states.

22.10 Calculate the expectation value of the moment of inertia of the hydrogen atom in the 1s, 2s, 2p, and 3s, 3p, and 3d states.

22.11 The possible values of the z component of the magnetic moment of an electron in an atom are given by $\mu_z = -(m + 2m_s)\mu_B$. What are the possible values of μ_z for an s, p, d, and f electron?

23

The Covalent Bond

23.1 GENERAL REMARKS

Until the advent of quantum mechanics the reasons for the stability of molecules were unknown. The cohesive energy of ionic crystals could be adequately interpreted on the purely classical basis of the electrical attraction of the oppositely charged ions. Some attempts were made to interpret the interaction of all atoms on the basis of the electrical interaction of positive and negative charges, electrical dipoles, induced dipoles, and so on. These classical calculations indicated that the bonding between two like atoms, such as two hydrogen atoms, should be very much weaker than it is. This is another problem that classical physics failed to solve.

The quantum-mechanical problem is to calculate the energies of the individual atoms that make up the molecule, then calculate the energy of the molecule itself. The molecule is stable if the energy of the molecule is less than the sum of the energies of the individual atoms. The difference in these energies is a measure of the strength of binding in the molecule. To state the problem is easy; to do the calculation in complete detail is apparently impossible. Fortunately, there are several simplifying circumstances.

First of all, consider the hydrogen molecule H_2. It consists of four bodies: two protons and two electrons. The classical problem of the motion of three bodies, and the quantum-mechanical one as well, has escaped exact solution; a larger number of bodies only aggravates this difficulty. However, since the nuclei are so much heavier than the electrons, their motion is sluggish in comparison; the electronic motion is fast enough to adjust to any change in the position of the nuclei. Thus to a good approximation the nuclear motions, the vibrations and rotations of the molecule, can be treated as a completely separate problem; this is called the Born–Oppenheimer approximation. The internuclear distances and the relative orientation of the nuclei thus enter the problem of the electronic motion as parameters; if we wish, we can explore how changing those parameters affects the energy of the molecule.

The energy of the molecule is given by the expression

$$E = \int \psi^* \mathbf{H} \psi \, d\tau, \tag{23.1}$$

where ψ is the wave function or description of all the electrons in the molecule, and $\mathbf{H}$ is the operator for the total energy, kinetic and potential, of the electrons in the nuclear skeleton of the molecule. And now a second fortunate thing happens. If instead of using the exact wave function in Eq. (23.1), we use an approximate one (which we might even obtain by guessing!) the Schrödinger equation has the property that the value of the integral is always *greater* than the energy of the ground state of the molecule—the *variation theorem*.

Let E_0 be the lowest permitted energy of the system, the ground state energy. Then if ϕ is the approximate wave function, we have

$$\int \phi^* \mathbf{H} \phi \, d\tau > E_0 \tag{23.2}$$

This is the variation theorem. It allows us to take a guess at the description ψ, put adjustable constants in the mathematical form of the guess, and evaluate the integral. Then we vary the constants to minimize the value of the integral; this minimum value is still greater (by the theorem) than the ground state energy. With experience, our guesses become more refined and we come closer to the correct value of the energy. The theorem is helpful, since it tells us that a "guessed" description will never give us an energy below the correct value.

Having agreed to be content with approximate descriptions of the system, we can gain some insight into the nature of the chemical bond. Two main approaches to this problem can be distinguished. The *valence bond* method, developed principally by Heitler, London, Slater, and Pauling, recognizes that two electrons are usually needed to form a chemical bond and then looks at the behavior of an electron pair. Each bonding pair in the molecule is described in a simple way, and a description of the molecule is built up by a description of its parts. The *molecular orbital** method, developed by Hund and Mulliken, looks at the nuclear framework of the individual molecule and says that this framework must have a system of energy levels just as the hydrogen atom has such a system of levels. If we fit the molecule's electrons into this system of levels, observing the Pauli principle, we obtain a description of the molecular electronic structure. This approach is the method, modified appropriately for molecules, that we used to describe the electronic structure of complex atoms.

The molecular orbital theory is more satisfying esthetically, perhaps, but its lack of emphasis on a localized chemical bond has led many chemists to prefer the valence bond method, which gives them a better pictorial grasp of the situation. The above distinction between the two methods is a primitive one. If all the refinements in the present day valence bond and molecular orbital theories are included, any distinction between them is probably more imagined than real.

23.2 THE ELECTRON PAIR

To describe the electron pair in a molecule, we investigate the behavior of two identical particles in the potential field supplied by the nuclei of the molecule. We begin by over-

---

*   "Orbital" is *not* a fancy word for "orbit." Orbital and wave function are synonymous.

simplifying the problem. If we ignore the electrical repulsion between the two electrons, then each moves independently. The state of the first electron is described by a wave function $\psi_n(x_1, y_1, z_1)$; similarly, the second electron is described by a wave function $\psi_k(x_2, y_2, z_2)$. We will abbreviate these descriptions to $\psi_n(1)$ and $\psi_k(2)$, where (1) stands for (x_1, y_1, z_1), the coordinates of electron 1, and (2) stands for (x_2, y_2, z_2), the coordinates of electron 2. The subscripts n and k indicate that the states of the two electrons may be different. Since the particles move independently, the energy of the pair is the sum of the energies of the individuals: $E = E_n + E_k$. If the energy of the system is given by this sum, then the Schrödinger equation requires that the wave function for the pair be the *product* of the individual descriptions; the electron pair is described by the function

$$\psi_\mathrm{I} = \psi_n(1)\psi_k(2). \tag{23.3}$$

Since the electrons are indistinguishable, we have no way of discovering which is in state k and which in state n. An equally correct description is therefore

$$\psi_\mathrm{II} = \psi_k(1)\psi_n(2), \tag{23.4}$$

where the coordinates of the particles have been exchanged. The description in Eq. (23.4) has the same energy as that in Eq. (23.3). (States with the same energy are *degenerate* states; these two states exhibit exchange degeneracy, since they differ only in the exchange of the coordinates.) If the particles do not interact, either description, or a superposition of them, is perfectly correct.

The curious feature of the problem is that if the repulsion between the electrons is introduced, we are forced to use a superposition of these descriptions. The permissible combinations are

$$\psi_S = \frac{1}{\sqrt{2}} [\psi_n(1)\psi_k(2) + \psi_n(2)\psi_k(1)], \tag{23.5}$$

and

$$\psi_A = \frac{1}{\sqrt{2}} [\psi_n(1)\psi_k(2) - \psi_n(2)\psi_k(1)]. \tag{23.6}$$

The two functions ψ_S and ψ_A have an important symmetry property. If we exchange the coordinates of electrons 1 and 2 (interchange the 1's and 2's in the parentheses), the function ψ_S is unaffected; ψ_S is symmetric under this operation. The function ψ_A changes sign under this operation and so is antisymmetric.

Now we ask which of these descriptions is likely to describe a bond between two atoms. Consider the hydrogen molecule H_2, with two protons rather close together and two electrons. By themselves the two protons would repel one another. To form a stable molecule this repulsion must be reduced. To reduce it, the electrons must be for the most part in the small space between the two nuclei, which implies that the electrons must be rather close to one another. As the coordinates of electrons 1 and 2 approach in value,

$$\psi_n(1) \approx \psi_n(2) \qquad \text{and} \qquad \psi_k(1) \approx \psi_k(2).$$

Using these relations in Eqs. (23.5) and (23.6), we find

$$\psi_S \approx \frac{2}{\sqrt{2}} \psi_n(1)\psi_k(2) \qquad \text{and} \qquad \psi_A \approx 0.$$

Therefore, if the two electrons are described by ψ_A, the probability, $|\psi_A|^2$, of finding the two electrons close together is very small, while if they are described by ψ_S, there is a sizable probability, $|\psi_S|^2$, of finding them close together. We conclude that it is ψ_S which describes the state of the electrons in the electron-pair bond between two nuclei; this conclusion is confirmed by detailed calculation of the energy of the molecule.

However, the Pauli exclusion principle requires that the wave function of a system be *antisymmetric* under this operation of interchanging the coordinates of the particles. We save the situation by noting that the total wave function of an electron pair is the product of a space part, ψ_S or ψ_A, and a spin part. The spin part may also be symmetric, Σ_S, or antisymmetric, Σ_A, under interchange of the particles. The possible combinations of space and spin functions that yield an antisymmetric total wave function are

$$\psi_1 = \psi_S \Sigma_A \quad \text{and} \quad \psi_2 = \psi_A \Sigma_S.$$

The first, ψ_1, incorporates the function we need for the chemical bond. The antisymmetric spin function implies that the spins of the two electrons in the bonding pair have opposite orientations; hence, their magnetic moments cancel one another. For this reason, the majority of molecules have no net magnetic moment. The possession of a magnetic moment by a molecule indicates that one or more of the electrons in the molecule are unpaired.

The conclusions about the bonding electron pair can be summarized briefly. The requirement of the Pauli principle (antisymmetry of the wave function under exchange of identical particles) along with the requirement that the electrons concentrate in a small region of space between the nuclei, forces us to describe the electron pair in a chemical bond by the function

$$\psi_1 = \psi_S \Sigma_A. \tag{23.7}$$

The symmetric space function ψ_S has a large electron cloud density between the nuclei and thus prevents electrical repulsion from driving the nuclei apart. The antisymmetric spin function requires the magnetic moments of the two electrons to be oppositely oriented (paired). Thus the proposal of G. N. Lewis in 1916 that atoms are held together by electron pairs is confirmed and given deeper meaning by the quantum mechanics. Detailed calculation shows that the energy of the state described by ψ_S is very much lower than that of the state described by ψ_A. These conclusions are general and can be applied to the electron pair holding any two atoms together. First we examine the hydrogen molecule in more detail.

23.3 THE HYDROGEN MOLECULE; VALENCE BOND METHOD

We label the protons a and b, and the electrons 1 and 2. If the two hydrogen atoms are infinitely far apart, there is no interaction between the electrons or between the two protons. If electron 1 is with proton a, it is described by $\psi_a(1)$, which is any wave function of hydrogen atom a. Similarly, $\psi_b(2)$ describes electron 2 with proton b; $\psi_b(2)$ is any wave function of hydrogen atom b. Since we are concerned only with the state of lowest energy, we choose ψ_a and ψ_b as $1s$ functions on the respective atoms. As we have seen in Section 23.2, the description of the two-electron system is given by either of the products, $\psi_a(1)\psi_b(2)$ or $\psi_a(2)\psi_b(1)$. Regardless of which description is used, the energy of the system at infinite separation is $E = 2E_{1s}$, the sum of the energies of the individual atoms in the $1s$ state.

It is customary to write a "chemical" structure to correspond to each of these quantum mechanical descriptions.

Designation	"Chemical" structure	Description	Energy
I	$H_a^{\cdot 1}$ $\quad ^2{}^{\cdot}H_b$	$\psi_I = \psi_a(1)\psi_b(2)$	$E_I = 2E_{1s}$
II	$H_a^{\cdot 2}$ $\quad ^1{}^{\cdot}H_b$	$\psi_{II} = \psi_a(2)\psi_b(1)$	$E_{II} = 2E_{1s}$

As the two atoms approach one another, the electrons no longer move independently; they influence each other and are influenced by both nuclei. The descriptions ψ_I and ψ_{II} are no longer exact; furthermore, neither by itself is satisfactory as an approximate wave function. We are forced to choose between the linear combinations,

$$\psi_S = N(\psi_I + \psi_{II}) \tag{23.8}$$

and

$$\psi_A = N'(\psi_I - \psi_{II}), \tag{23.9}$$

where N and N' are normalization constants. From what has been said, ψ_S is the description of the molecule with a stable bond between the two atoms. So far no one has devised a simple chemical representation of the description ψ_S. We write the structures I and II, which are called *resonance* structures, and then describe the correct structure as a *resonance hybrid* of the two.

Using ψ_S, we can calculate the energy as a function of R, the internuclear distance; this energy, relative to that of the two atoms at infinite separation, is shown by the curve labeled ψ_S in Fig. 23.1. The wave function ψ_S predicts a minimum in the energy of the system at R_0, the equilibrium value of the internuclear distance. The existence of this minimum indicates that a stable molecule is formed; the depth of the minimum, E_D, is the binding energy or dissociation energy of the molecule.

In Fig. 23.1 the energy curve for ψ_A shows that at all values of R the energy of the system is greater than that of the separated atoms. The lowest energy is obtained if the atoms remain apart. This state is an antibonding or a repulsive state of the system.

If we ignore our principles and calculate the energy using either ψ_I or ψ_{II} by itself, we obtain the dashed curve in Fig. 23.1. The difference in energy between this curve and that

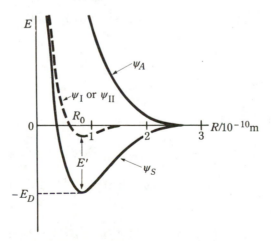

Figure 23.1 Energy of H_2 as a function of R.

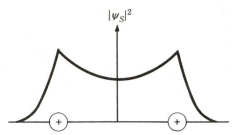
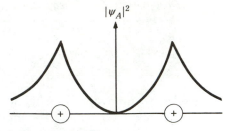

Figure 23.2   Electron densities in the two states of H_2.

for ψ_S is the *resonance stabilization energy* or the *resonance energy*. It is apparent that the resonance energy accounts for the greater part of the stability of the molecule. Physically we can understand why the simple descriptions, ψ_I and ψ_{II}, are not adequate in a molecule. Both positive nuclei attract an electron on an atom which has been brought close to another atom. Therefore the electron spreads itself over both nuclei. The remarkable thing is that spreading the electrons over both nuclei lowers the energy of the system so greatly.

The probability density of the electron cloud is obtained by squaring the wave functions. The density along the internuclear axis is shown for the two states ψ_S and ψ_A in Fig. 23.2. It is apparent that in the bound state described by ψ_S, the electron cloud is very dense in the region between the nuclei, while in the state described by ψ_A, the cloud is comparatively thin between the nuclei. The electron density that builds up between the nuclei in the bound state of the molecule can be thought of as the result of the overlapping and interpenetration of the electron clouds on the individual atoms. Qualitatively, the greater the overlapping of the two electron clouds, the stronger is the bond between the two atoms; this is Pauling's principle of maximum overlap.

Normalization of the wave functions ψ_S and ψ_A requires that

$$\int \psi_S^2 \, d\tau = 1 \quad \text{and} \quad \int \psi_A^2 \, d\tau = 1.$$

Using Eq. (23.8), we obtain for $\int \psi_S^2 \, d\tau$:

$$N^2 \int (\psi_I + \psi_{II})^2 \, d\tau = N^2 \left(\int \psi_I^2 \, d\tau + 2 \int \psi_I \psi_{II} \, d\tau + \int \psi_{II}^2 \, d\tau \right) = 1.$$

Since ψ_I and ψ_{II} are normalized, this becomes

$$N^2 \left(2 + 2 \int \psi_I \psi_{II} \, d\tau \right) = 1. \tag{23.10}$$

Using the definitions of ψ_I and ψ_{II} in the integral, we find that

$$\int \psi_I \psi_{II} \, d\tau = \int \psi_a(1) \psi_b(2) \psi_a(2) \psi_b(1) \, d\tau_1 \, d\tau_2,$$

in which the general volume element $d\tau$ has been replaced by the volume elements $d\tau_1$ and $d\tau_2$ for the two electrons. Then

$$\int \psi_I \psi_{II} \, d\tau = \int \psi_a(1) \psi_b(1) \, d\tau_1 \int \psi_a(2) \psi_b(2) \, d\tau_2.$$

Since the two integrals on the right differ only in the labeling of the coordinates, they are equal. We define S, the overlap integral, by

$$S = \int \psi_a(1)\psi_b(1)\, d\tau_1 = \int \psi_a(2)\psi_b(2)\, d\tau_2. \tag{23.11}$$

Then

$$\int \psi_I \psi_{II}\, d\tau = S^2. \tag{23.12}$$

Finally, using Eq. (23.12) in Eq. (23.10), we obtain for N^2:

$$N^2 = \frac{1}{2(1 + S^2)} \quad \text{or} \quad N = \frac{1}{\sqrt{2(1 + S^2)}}. \tag{23.13}$$

By a similar argument we can show that to normalize the function ψ_A, we must have

$$N' = \frac{1}{\sqrt{2(1 - S^2)}}. \tag{23.14}$$

★ 23.3.1   Construction of the Proper Wave Functions

Equation (23.12) shows that when S is not zero (R is not infinite), ψ_I and ψ_{II} are not orthogonal. Consequently, they are no longer proper wave functions for the system when R is not infinite. Two orthogonal wave functions can be constructed from ψ_I and ψ_{II} by taking them in linear combination. These orthogonal wave functions serve as the first approximate wave functions of the system.

We develop the proper wave functions by requiring that they be symmetric or anti-symmetric under interchange of the two electrons. If the operator, $\mathbf{I}$, interchanges the coordinates of the two electrons, then

$$\mathbf{I}\psi_I = \mathbf{I}[\psi_a(1)\psi_b(2)] = \psi_a(2)\psi_b(1) = \psi_{II}, \tag{23.15}$$

and similarly,

$$\mathbf{I}\psi_{II} = \psi_I. \tag{23.16}$$

If we construct the linear combination,

$$\psi_S = N\psi_I + \lambda\psi_{II},$$

in which N and λ are constants, and require that

$$\mathbf{I}\psi_S = \psi_S, \tag{23.17}$$

then

$$\mathbf{I}\psi_S = N(\mathbf{I}\psi_I) + \lambda(\mathbf{I}\psi_{II}) = N\psi_{II} + \lambda\psi_I.$$

If Eq. (23.17) is to be satisfied, it must be that

$$N\psi_{II} + \lambda\psi_I = N\psi_I + \lambda\psi_{II}$$

or

$$(N - \lambda)(\psi_{II} - \psi_I) = 0.$$

This condition can be satisfied for $\psi_{II} - \psi_I \neq 0$ only if $\lambda = N$. Thus we obtain

$$\psi_S = N(\psi_I + \psi_{II}). \tag{23.18}$$

where N is given by Eq. (23.13).

If we require the wave function to be antisymmetric (that is, $I\psi_A = -\psi_A$), then by a similar argument we obtain

$$\psi_A = N'(\psi_I - \psi_{II}),$$

where N' is given by Eq. (23.14). It is easy to show that ψ_A and ψ_S are orthogonal.

The energy corresponding to these wave functions is obtained by evaluating the integral

$$E = \int \psi H \psi \, d\tau \qquad (23.19)$$

for each of the wave functions. Since none of these wave functions is exact when R is finite, the variation theorem assures us that the energy obtained in this way is greater than the actual energy of the ground state of the system.

23.4 THE COVALENT BOND

The covalent bond between any two atoms A and B can be described by a wave function similar to the ψ_S used for the hydrogen molecule. Consider the structures I and II:

I   A$^{\cdot 1}$ $^{2 \cdot}$B $\psi_I = \psi_a(1)\psi_b(2),$

II   A$^{\cdot 2}$ $^{1 \cdot}$B $\psi_{II} = \psi_a(2)\psi_b(1),$

where ψ_a and ψ_b are wave functions appropriate to atoms A and B, respectively. The structure is described by the symmetric combination of ψ_I and ψ_{II}:

$$\psi_S = \frac{1}{\sqrt{2(1 + S^2)}} (\psi_I + \psi_{II}).$$

This description predicts a minimum energy corresponding to formation of a bond. The resonance energy is obtained by taking the difference between the energy computed for ψ_S and that computed for either ψ_I or ψ_{II}.

Since the stability of the bond depends principally on the resonance energy, it is important to know what factors influence the magnitude of this energy. The resonance energy has its largest possible value if the energies of the contributing structures are the same, or nearly so. The greater the energy difference between the contributing structures the less the stabilization due to resonance between them. In the case of any molecule AB, the structures I and II differ only by the exchange of the electron coordinates, so they have exactly the same energy. Consequently, the stabilization conferred by resonance between I and II is large.

The two ionic structures of the molecule AB,

III   A$^+$ $\frac{1}{2}$:B$^-$,

IV   A:$\frac{1}{2}^-$ B$^+$,

also contribute to the overall structure of the molecule; however, one of these structures is usually much lower in energy than the other; and, of course, the energies are different from the energy of I or II. In many molecules one or several ionic structures contribute to the overall structure of the molecule. In the molecule of hydrogen chloride three structures are important:

I   H$^{\cdot 1}$ $^{2 \cdot}$Cl   II   H$^{\cdot 2}$ $^{1 \cdot}$Cl   III   H$^+$ $\frac{1}{2}$:Cl$^-$.

The overall structure of the molecule is a resonance hybrid of the structures I, II, and III. The quantum-mechanical description is a linear combination,

$$\psi = c[\psi_{\text{I}} + \psi_{\text{II}} + \lambda\psi_{\text{III}}].$$

The coefficients of ψ_{I} and ψ_{II} in the composite description are equal, indicating that these two contribute equally to the structure. The coefficient of ψ_{III} differs from the other two, indicating that ψ_{III} contributes differently. The contributions of the three structures in HCl are estimated to have the values: I, 26%; II, 26%; III, 48%. The structures I and II are covalent structures, so we may say that the bond in HCl is 52% covalent and 48% ionic. A bond in which the ionic contribution is significant is called a covalent bond with partial ionic character.

Every covalent bond has more or less ionic character. Even if the two atoms are the same, there is a small contribution of ionic structures, $\sim 3\%$ in H_2. The bond between two like atoms is usually called a pure covalent bond, nonetheless.

There are restrictions on the structures that can contribute to the composite structure of a molecule. The structures that can "resonate" to produce a composite structure must: (1) have the same number of unpaired electrons; and (2) have the same arrangement of nuclei. For resonance to be effective, the structures should not differ greatly in energy.

23.5 OVERLAP AND DIRECTIONAL CHARACTER OF THE COVALENT BOND

To form a covalent bond two things are needed: a pair of electrons with spins opposed, and a stable orbital, an orbital in the valence shell, on each atom. The strength of a bond is qualitatively proportional to the extent of overlap of the charge clouds on the two atoms. The *overlap integral S* is a measure of the overlap of two charge clouds:

$$S = \int \psi_a(1)\psi_b(1)\, d\tau_1, \tag{23.20}$$

If we choose any point, the wave functions extending from nucleus a and nucleus b each have a particular value at that point. The product of these values summed over the entire coordinate space is the overlap integral. If the two nuclei are far apart, then near a, where $\psi_a(1)$ is large, $\psi_b(1)$ is extremely small and the product is extremely small; similarly, near b, $\psi_b(1)$ is large but $\psi_a(1)$ is extremely small and the product is extremely small. Thus, when the nuclei are far apart, S is very small and, indeed, is zero when R is infinite. As the nuclei approach, S gets larger. We may think of S as a measure of the interpenetration or overlapping of the electron clouds on the two nuclei; thus the name, *overlap integral*.

Consider an electron in an s orbital on an atom. The s function does not depend on the angles, θ, ϕ. As a consequence we can represent the s function as a sphere, or in two dimensions as a circle. We can make the sphere large enough so that it include any desired fraction of the charge cloud. This sphere is called a boundary surface.

Consider the $2p_z$ function; we have, from Table 22.2,

$$\psi_{2p_z} = \psi_{210} = C_{210}\rho e^{-\rho/2} \cos\theta.$$

Since there is no dependence on ϕ, the function is symmetric around the z-axis. More importantly, the sign of the function changes when $\theta = \pi/2$. Thus above the xy-plane the function is positive; below the plane it is negative. When we square the function to obtain the charge density the negative sign disappears and the charge density is positive on both

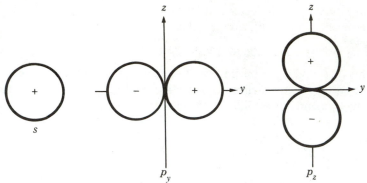

Figure 23.3 Boundary surfaces for s and p functions.

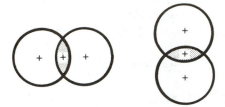

Figure 23.4 Overlap of s functions.

sides of the xy-plane. Clearly, if we set $|\psi_{2p_z}|^2$ = constant, we would obtain a relation between ρ and θ that would define a dumbbell-shaped surface on which the charge density is constant. Choosing the constant appropriately would allow us to include any desired fraction of the charge cloud within this boundary surface. Rather than attempt to draw this boundary surface accurately, we will represent the $2p_z$ wave function more conveniently by drawing two circles. Figure 23.3 shows a two-dimensional representation of the s, p_y, and p_z orbitals. For an s orbital, the boundary surface is a sphere, so the circle is the two-dimensional representation. For the p orbital, the two-lobed surface shown in Fig. 23.3 is represented in two dimensions by two circles in contact. The signs $+$ and $-$ in Fig. 23.3 are the algebraic signs of the wave function in the respective regions.

23.5.1   The Overlap of s Orbitals

If the electrons on the two atoms both occupy s orbitals, then the extent of overlapping of the two clouds is independent of the direction of approach. Figure 23.4 shows the overlapping in H_2 for two different directions of approach. Since both functions are positive, the overlap integral, Eq. (23.20), is positive.

23.5.2   Overlap between s and p Orbitals

If the electrons that will form the bond are in an s orbital on one atom and in a p orbital on the other, then the overlap depends on the relative direction of approach of the two atoms. Figure 23.5(a) shows the approach of an s electron to a p electron. The p function changes sign on passing through the plane of the nucleus. The s function is positive everywhere in space. The integral $\int \psi_s \psi_p \, d\tau$ is the sum of the values of the product $\psi_s \psi_p$

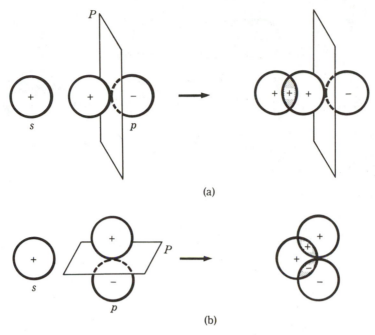

(a)

(b)

Figure 23.5   Overlap of s and p functions.

everywhere in space. In the region to the left of the plane P, the product $\psi_s\psi_p$ is always positive, since both ψ_s and ψ_p are positive in this region. The numerical value of $\psi_s\psi_p$ is moderately large, since part of this region is close to both nuclei where both wave functions have sizable values. The integration over this region yields a sum of positive contributions. To the right of plane P, ψ_p is negative and ψ_s is positive; their product is therefore negative and the integral is a sum of these negative contributions. The total integral has a positive contribution to which a small negative contribution is added. The positive contribution predominates, because the value of ψ_s is smaller the greater the distance from the nucleus on which it is centered, and so is very small to the right of P. Thus, for this direction of approach of an s electron cloud to a p electron cloud, the overlap integral is positive and bond formation is possible.

Consider the approach of an s cloud to a p cloud along the direction in Fig. 23.5(b). The p function is positive above the plane and negative below it. Therefore the product $\psi_s\psi_p$ is positive above the plane and negative below it. Because of the symmetry of the s and p clouds, the positive contributions are exactly balanced by the negative contributions. The overlap integral is equal to zero. There is no overlap and therefore no possibility of forming a bond if the clouds approach in this orientation. It is readily shown that the maximum overlapping of the two charge clouds occurs if the approach is along the axis of the p cloud (Fig. 23.5a). Therefore the strongest bond is formed in this manner.

23.5.3   Overlap of a p Orbital with a p Orbital

If we consider the approach of two atoms each having a p electron, there are several possibilities, as shown in Fig. 23.6. By the same argument as above we can show that: the overlap is zero for the approach illustrated in Fig. 23.6(b); maximum overlap is achieved

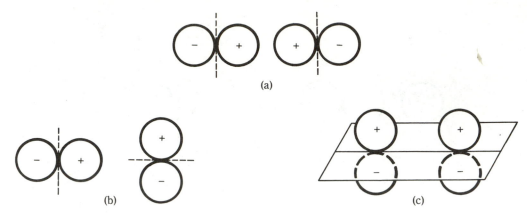

Figure 23.6   Overlap of p functions. (a) Maximum overlap.
(b) Zero overlap. (c) Moderate overlap.

in the configuration of Fig. 23.6(a); and a moderate value is obtained in the configuration of Fig. 23.6(c). It should be noted that in Fig. 23.6(c) the wave functions are both zero in the horizontal plane, and that the internuclear axis lies in this plane. This implies that the charge density is zero along the internuclear axis in this type of bond (π bond).

The way in which electron clouds overlap gives the first indication of the reason an atom forms covalent bonds in a particular relative orientation. Directional character is a distinguishing attribute of the covalent bond; other types of bonds do not prefer special directions. The ability to explain and predict the number of bonds formed and their geometric arrangement around the atom is one of the great triumphs of the quantum mechanics. In what follows note that very approximate methods suffice to provide the qualitative picture.

23.6 MOLECULAR GEOMETRY

Knowing that the amount of overlap between the orbitals on the two atoms forming the bond depends on the direction of approach, we can construct a crude theory of molecular geometry.

The elements in the first row of the periodic table have only four valence orbitals, the $2s$, $2p_x$, $2p_y$, and $2p_z$. Since for every bond formed the atom must have an orbital in the valence shell, the number of bonds formed by these elements is limited to four.

Consider the oxygen atom, which has the electron configuration,

$$\text{O:} \quad 1s^2 \quad 2s^2 \quad 2p_x^2 \quad 2p_y^1 \quad 2p_z^1.$$

To form a bond we need a valence orbital on each atom and an electron pair. The two unpaired electrons in oxygen occupy two different p orbitals that lie along perpendicular axes. If we bring up two hydrogen atoms at 90° to one another we should get a maximum overlapping and thus maximum bond strength (Fig. 23.7). This would predict a 90° bond angle in H_2O. Similarly, the three unpaired electrons in the nitrogen atom are in three different p orbitals—each at 90° to the other two—so for NH_3 we would predict 90° bond angles. The bond angles in water (104.5°) and in NH_3 (107.3°) are much larger than the predicted 90° values. Clearly, some refinement of this idea is needed.

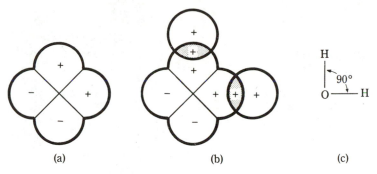

Figure 23.7   (a) Oxygen atom. (b) Water molecule.
(c) Predicted bond angle in water.

Recall that the principle of superposition allows us to construct linear combinations of the wave functions and thus find new descriptions of the system that are equally correct. Pauling formed linear combinations of the s, and p_x, p_y, and p_z orbitals, which we label t_1, t_2, t_3, and t_4. Thus

$$t_1 = a_{11}\psi_{2s} + a_{12}\psi_{2p_x} + a_{13}\psi_{2p_y} + a_{14}\psi_{2p_z};$$
$$t_2 = a_{21}\psi_{2s} + a_{22}\psi_{2p_x} + a_{23}\psi_{2p_y} + a_{24}\psi_{2p_z};$$
$$t_3 = a_{31}\psi_{2s} + \cdots\cdots\cdots\cdots\cdots + \vdots$$
$$t_4 = a_{41}\psi_{2s} + \cdots\cdots\cdots\cdots\cdots + a_{44}\psi_{2p_z}.$$

The constants a_{11}, a_{12}, ... are determined by the conditions: (1) The four new descriptions are to be equivalent in their extension in space; (2) the extension of the orbitals shall be as large as possible so that the overlap will be a maximum. It is possible to determine the coefficients a_{11}, a_{12}, ... so that four equivalent orbitals with maximum extension are formed. These four new orbitals are directed to the apices of a tetrahedron! The shape of one of these orbitals is shown in Fig. 23.8(a), and the set of four is shown in Fig. 23.8(b). The orbitals t_1, t_2, t_3, and t_4 are called hybrid (or mixed) orbitals. The process of making linear combinations is called hybridization (or mixing). These particular ones are called tetrahedral hybrids or sp^3 hybrids. The extension of the hybrid orbitals is much greater than that of either an s or p orbital by itself. The overlap and consequently the bond strength are correspondingly greater.

We may regard the oxygen atom as having the electronic configuration,

$$O: \quad 1s^2 \quad t_1^2 t_2^2 t_3 t_4.$$

Two of the tetrahedral hybrids, t_1 and t_2, are occupied by electron pairs; the remaining two, t_3 and t_4, are occupied by one electron and can form bonds with atoms such as hydrogen (Fig. 23.8c and d). The angle between these bonds should be the tetrahedral angle 109.47°. This value is much closer to the observed value in water, 104.5°, than is the 90° value predicted on basis of the angle between the simple p orbitals.

Similarly, the electron configuration in the nitrogen atom may be written

$$N: \quad 1s^2 \quad t_1^2 t_2 t_3 t_4.$$

Thus in NH_3 the $1s$ wave functions on the three hydrogen atoms overlap with three of the

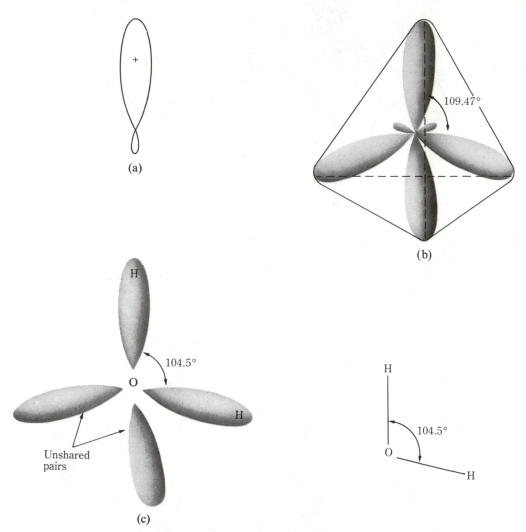

Figure 23.8   Tetrahedral orbital geometry. (a) Tetrahedral hybrid. (b) Set of four tetrahedral hybrids. (c) Water molecule. (d) Angular molecular geometry.

tetrahedral hybrid wave functions on the nitrogen atom. The fourth tetrahedral hybrid is occupied by the unshared pair. The observed bond angle, 107.3°, is quite close to the predicted tetrahedral value, 109.5°. The ammonia molecule is therefore a trigonal pyramid, with the nitrogen atom at the apex and the three hydrogen atoms defining the base. The orbital geometry and the molecular geometry are shown in Fig. 23.9.

The carbon atom has the electronic structure,

$$\text{C:}\qquad 1s^2\quad 2s^2 2p^3.$$

If we unpair the two $2s$ electrons (some energy is required for this), we can write for the excited carbon atom, C*,

$$\text{C*:}\qquad 1s^2\quad t_1 t_2 t_3 t_4.$$

That is, there is one unpaired electron in each of the tetrahedral hybrids. Carbon can form

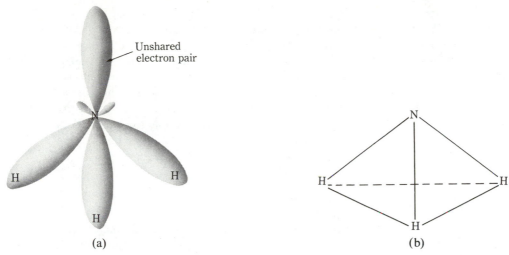

Unshared
electron pair

Figure 23.9   Geometry of ammonia. (a) Tetrahedral orbital
geometry. (b) Pyramidal molecular geometry.

a bond with each of four hydrogen atoms. The methane molecule, CH_4, is a regular tetrahedron with the carbon atom at the center and the four hydrogen atoms at the apices. The
observed value for the H—C—H angle is 109.5° within the experimental error.

Thus, for the atoms whose valence shell consists of the s, p_x, p_y, and p_z orbitals, the
geometry of compounds that involve only single bonds is based on a tetrahedral orbital
geometry. The arrangement of the nuclei in the molecule, the molecular geometry,
depends on how many of the tetrahedral orbitals are occupied by unshared pairs. The
following group of isoelectronic species illustrates the point.

Tetrahedral orbital occupancy	*Molecule*				*Molecular geometry*
4 single bonds	BH_4^-	CH_4	NH_4^+		tetrahedral
3 single bonds + 1 unshared pair			$:NH_3$	$:OH_3^+$	pyramidal
2 single bonds + 2 unshared pairs			$\overset{..}{:}NH_2^-$	$\overset{..}{:}OH_2$	angular
1 single bond + 3 unshared pairs			$\overset{..}{:}\overset{..}{N}H^{2-}$	$\overset{..}{:}\overset{..}{O}H^-$	linear

The following group of isoelectronic sulfur and chlorine oxyanions provides another
illustration of molecular geometries that are based on tetrahedral orbital geometry.

Tetrahedral	*Pyramidal*	*Angular*	*Linear*
ClO_4^-	$:ClO_3^-$	$\overset{..}{:}ClO_2^-$	$\overset{..}{:}\overset{..}{C}lO^-$
SO_4^{2-}	$:SO_3^{2-}$		

The shapes of molecules can be interpreted in terms of *valence shell electron pair repulsion* (VSEPR). This theory is based on the fact that electron pairs repel one another and states that the molecular geometry will be such that the repulsion between all pairs of electrons is minimized. Thus if a molecule has four equivalent electron pairs, they must be in orbitals that are directed to the apexes of a tetrahedron for the repulsion to be a minimum. Similarly, the deviations of the bond angle from the tetrahedral value can be interpreted in terms of the repulsion of the electron pairs. For example, in water the unshared electron pairs repel each other more than do those in the bonds; thus the bond angle closes a bit. In ammonia, the repulsion of the bonding pairs is less than that between the unshared pair and the bonding pairs, so again the bond angle closes slightly.

23.7 STRUCTURES WITH MULTIPLE BONDS

If the classical structure of a molecule involves one double bond to the central atom, the others being single bonds, then the hybrids involve only the s, p_x, and p_y orbitals; the p_z orbital is left as is. Thus we can form the trigonal hybrids, tr:

$$tr_1 = a_{11}\psi_{2s} + a_{12}\psi_{2p_x} + a_{13}\psi_{2p_y};$$
$$tr_2 = a_{21}\psi_{2s} + \quad \cdots \qquad\qquad ;$$
$$tr_3 = a_{31}\psi_{2s} + \quad \cdots \quad + a_{33}\psi_{2p_y}.$$

The requirements that the three hybrid orbitals have maximum extension and be equivalent to each other yield three orbitals that are directed to the apices of an equilateral triangle in the xy-plane. The set of sp^2 hybrids is illustrated in Fig. 23.10. The unhybridized p_z orbital has its charge density above and below the plane of the hybrid orbitals.

The electron configurations of the various atoms in the first period are

$$
\begin{array}{lll}
\text{B:} & 1s^2 & tr_1 tr_2 tr_3; \\
\text{C:} & 1s^2 & tr_1 tr_2 tr_3\, 2p_z; \\
\text{N:} & 1s^2 & tr_1^2 tr_2 tr_3\, 2p_z; \\
\text{O:} & 1s^2 & tr_1^2 tr_2^2 tr_3\, 2p_z.
\end{array}
$$

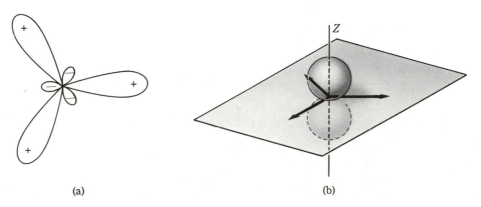

(a)   (b)

Figure 23.10   (a) sp^2 hybrids. (b) sp^2 hybrids with p_z orbital.

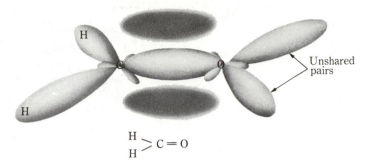

Figure 23.11   Charge clouds in formaldehyde.

The simplest example is formaldehyde, H_2CO, which has the classical structure

$$\begin{array}{c} H \\ \diagdown \\ \diagup C=O. \\ H \end{array}$$

If both the carbon atom and the oxygen are hybridized sp^2, the bond structure would appear as in Fig. 23.11. The carbon atom forms three sigma (σ) bonds with the two hydrogen atoms and with the oxygen atom. (A sigma bond has its charge density concentrated along the internuclear axis between the two bonded atoms.) The remaining bond in this molecule is formed by the overlapping of the remaining $2p_z$ electrons on carbon and on oxygen. These charge clouds do not have any density in the xy-plane; the overlap occurs above and below the plane. A bond formed in this way is called a π bond.

In ethylene, both carbon atoms are hybridized in this way; a strong σ bond is formed by the overlap of a hybrid orbital from each carbon atom. The remaining two hybrids on each carbon form σ bonds with the s orbitals of the four hydrogen atoms. All of the atoms lie in one plane. The overlap of the p_z orbitals on each carbon atom forms the second bond between the two carbon atoms, the π bond. The charge cloud of the electrons in the π bond lies above and below the plane of the atoms. Figure 23.12(a) and (b) shows the relative locations of the bonds.

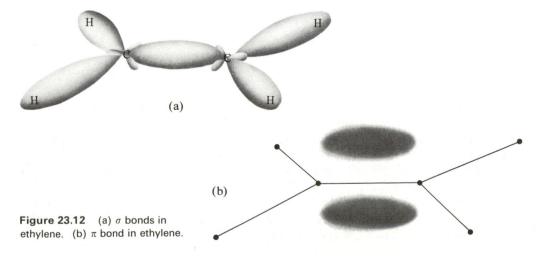

(a)

(b)

Figure 23.12   (a) σ bonds in ethylene.   (b) π bond in ethylene.

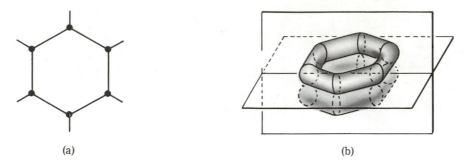

Figure 23.13   (a) Carbon skeleton in benzene. (b) π bond in benzene.

The stability gained by the molecule through the overlapping of the p_z orbitals in the π bond locks the molecule in a planar configuration. If the plane of one CH_2 group were at 90° to the plane of the other, the p_z orbitals would not overlap; the molecule would be much less stable in such a configuration. This accounts for the absence of rotation about the double bond and makes possible the existence of geometric isomers, the *cis* and *trans* forms of disubstituted ethylene.

Any carbon atom bound to three atoms in a stable compound is hybridized in this fashion. The unsaturated aliphatic hydrocarbons are only one class of compound that includes this kind of bonding. Benzene is an important example of a compound in which each carbon atom is attached to only three other atoms. Each carbon atom in benzene is hybridized sp^2 so that the bonds are at 120° in a plane. The carbon skeleton is shown in Fig. 23.13(a). The p_z orbitals of the six carbon atoms project above and below the plane of the ring. The overlap of the p_z orbitals produces a doughnut-shaped cloud above and below the plane of the ring (Fig. 23.13b). There are six electrons spread out in these "doughnuts," enough for only three bonds in the classical sense. These three bonds are spread over six positions so that each carbon–carbon bond in benzene has one-half double-bond character.

Note that in the sp^2 hybrid the bond angles need not be equal (120°), except in cases in which all three atoms bonded to the central atom are the same. In formaldehyde, for example, the H—C—H angle is 126° and the other two are 117°.

In the carbonate ion, the classical resonance structures are

$$
\begin{array}{ccc}
\overset{-}{\text{O}}\diagdown & \text{O}\diagdown & \overset{-}{\text{O}}\diagdown \\
\quad \diagup\text{C=O} & \quad \diagup\text{C—O}^- & \quad \diagup\text{C—O}^-. \\
\overset{-}{\text{O}}\diagup & \overset{-}{\text{O}}\diagup & \text{O}\diagup \\
\text{(a)} & \text{(b)} & \text{(c)}
\end{array}
$$

The carbon atom and the oxygen atom are hybridized sp^2. The electronic distribution corresponding to structure (a) is shown in Fig. 23.14(a); the overall distribution is shown in Fig. 23.14(b). This π bond distributes itself equally over the three oxygen atoms so that we have a partial double bond character of $\frac{1}{3}$ for each CO bond. Similarly, each oxygen atom has $\frac{2}{3}$ of a formal negative charge. The bond angles are 120°.

The isoelectronic species, BO_3^{3-}, CO_3^{2-}, and NO_3^- all have the same electronic structure. In these molecules the oxygen atoms are at the corners of an equilateral triangle; the remaining atom, B, C, or N, is at the center.

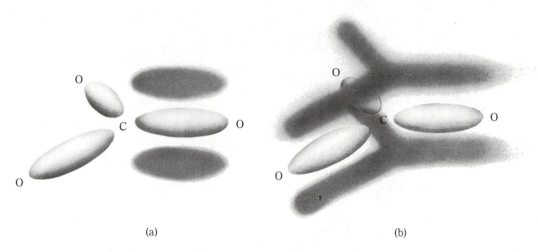

Figure 23.14   Charge clouds in carbonate ion.

The BF_3 molecule is also planar with F—B—F angle of 120°. If we assume that only six electrons are involved, the structure is understandable because the repulsion between the bonded pairs is minimized in this configuration. However, it is isoelectronic with CO_3^{2-} ion so we might regard it as a compound with CO_3^{2-} structures of the type

$$
\begin{array}{c}
F \\
\diagdown \\
\diagup \\
F
\end{array}
B^- {=} F^+.
$$

23.8 STRUCTURES INVOLVING TWO DOUBLE BONDS OR A TRIPLE BOND

If we hybridize only the s and $2p_z$ orbitals (sp hybridization) we obtain two sp hybrids, l_1 and l_2, which are oppositely directed along a straight line (Fig. 23.15a). The electron configurations for the species of interest are

$$
\begin{array}{ll}
C: & 1s^2 \quad l_1 l_2 2p_x 2p_y; \\
N^+: & 1s^2 \quad l_1 l_2 2p_x 2p_y; \\
N: & 1s^2 \quad l_1^2 l_2 2p_x 2p_y.
\end{array}
$$

The simplest example of the triple bond is in the nitrogen molecule illustrated in Fig. 23.15. Two π bonds are formed in addition to the σ bond. The unshared pairs are 180° from the sigma bond on each end of the molecule. The formation of two π bonds results in the charge cloud having the shape of a cylindrical sheath around the axis of the molecule. (Compare to Fig. 22.5 for $m = \pm 1$.) Species that are isoelectronic with nitrogen and therefore have the same electronic structure are $C{\equiv}O$, $^-C{\equiv}N$, and $^-C{\equiv}C^-$.

In the acetylene molecule, both carbon atoms are hybridized in this way. The σ bond and π bonds are shown in Fig. 23.16. The formation of two π bonds yields a cylindrical

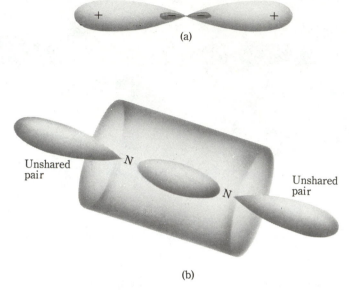

Figure 23.15   (a) Linear *sp* hybrids.
(b) The two π bonds in nitrogen.

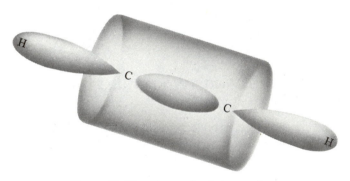

Figure 23.16   Charge clouds in acetylene.

sheath like that in the nitrogen molecule in Fig. 23.15. Note that nitrogen and acetylene are isoelectronic.

In carbon dioxide, the carbon is hybridized *sp* and the oxygens are hybridized sp^2; thus the molecule is linear. The charge clouds are shown in Fig. 23.17(a). Since the π bonds might have formed with the left-hand bond above and below and the right-hand one in front and in back of the paper (Fig. 23.17b), the result is a cylindrical sheath around the axis of the CO_2 molecule, as in N_2; the charge density in the sheath is only half as great, however (Fig. 23.17c). Species isoelectronic with CO_2 include $^-N{=}C{=}O$, $^-N{=}C{=}N^-$, $^-N{=}N^+{=}O$, $O{=}N^+{=}O$, $^-N{=}N^+{=}N^-$.

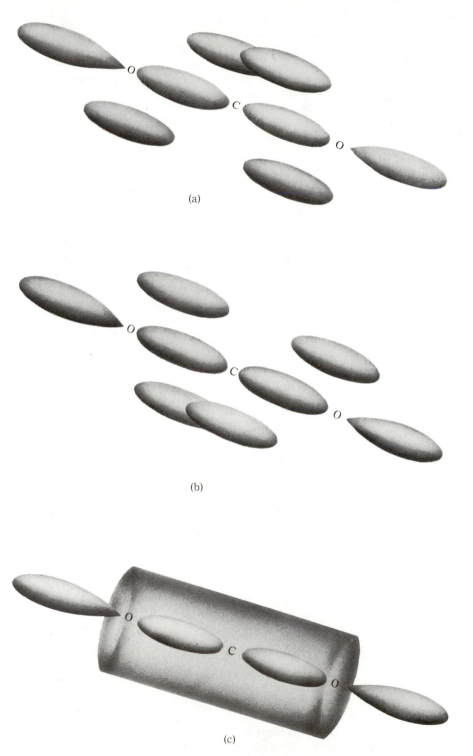

Figure 23.17   (a) and (b) Possible modes of overlap for π bonds in CO_2. (c) Actual charge cloud in CO_2.

Table 23.1

	Substance				
	Diamond	Graphite	Benzene	Ethylene	Acetylene
Bond order	1.0	1.33	1.50	2.0	3.0
C—C bond length/pm	154	142	139	135	120

23.9 BOND ORDER AND BOND LENGTH

It is a general rule that a double bond between two atoms is stronger than a single bond, and a triple bond is stronger than a double bond: the higher the bond order, the stronger the bond. It is also a general rule that increasing the strength of the bond shortens the bond length, the distance between the two atoms. This is illustrated by the carbon–carbon bond lengths and bond orders in some simple carbon compounds (see Table 23.1). A correlation such as that shown in the table is useful if we do not know what contributions the various resonance structures make to the overall structure of the molecule. If we determine the bond length, the bond order can be estimated from a plot of bond order versus bond length. The estimate of the bond order may provide a clue to the contributions of the various resonance structures. The correlation must, of course, be worked out for the particular kind of bond in question.

23.10 THE COVALENT BOND IN ELEMENTS OF THE SECOND AND HIGHER PERIODS

The elements in the second and higher periods have d orbitals in the valence shell in addition to the s and p orbitals. There are a total of nine orbitals (one s, three p, and five d orbitals) that could be used for bond formation. It is conceivable that an atom could be bonded to as many as nine other atoms. This coordination number is unknown. Ordinarily the number of atoms attached to a central atom does not exceed six, although there are a few compounds in which seven and eight atoms or groups are attached.

The most common higher coordination number in these elements is six; some of the fluorides of the first and second periods provide examples.

$$\begin{array}{lllll} \text{First period} & CF_4 & NF_3 & OF_2 & F_2 \\ \text{Second period} & SiF_4 & PF_3 & S_2F_2 & ClF_3 \\ & SiF_6^{2-} & PF_5 & SF_4 & \\ & & PF_6^- & SF_6 & \end{array}$$

The fluorides are chosen, since fluorine tends to bring out high coordination numbers.

In phosphorus, the electronic configuration is

$$P: \quad 3s^2 \quad 3p_x \quad 3p_y \quad 3p_z.$$

Since there are three unpaired electrons, the valence is three. In this state phosphorus forms the same type of trivalent compounds as nitrogen: NH_3, PH_3, NF_3, PF_3. Due to the presence of vacant d orbitals *in the valence shell* a relatively small expenditure of energy is required to form pentavalent P* with five unpaired electrons:

$$P^*: \quad 3s \quad 3p_x \quad 3p_y \quad 3p_z \quad 3d.$$

In this state phosphorus can form bonds to five neighboring atoms as in PF_5, PCl_5.

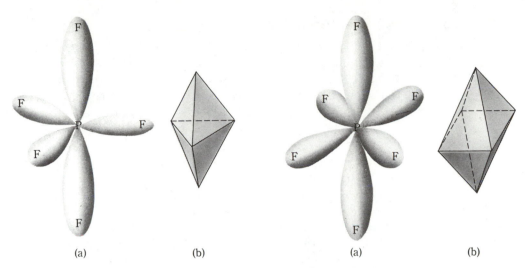

Figure 23.18 (a) PF_5. (b) Trigonal bipyramid.   **Figure 23.19** (a) PF_6^-. (b) Regular octahedron.

The orbitals used are hybridized, sp^3d hybrids, and are directed to the apices of a trigonal bipyramid (Fig. 23.18). The phosphorus and three of the fluorine atoms lie in a plane; the remaining two fluorine atoms are placed symmetrically above and below this plane. To promote an electron in nitrogen, the electron would have to be moved out of the valence shell to a shell of higher principal quantum number. The energy required would be too large to be compensated by the formation of two additional bonds.

If we add an electron to pentavalent phosphorus, the hexavalent species P^- is obtained:

$$P^-: \qquad 3s \quad 3p_x \quad 3p_y \quad 3p_z \quad 3d \quad 3d.$$

The ion PF_6^- may be regarded as a compound of P^- with six neutral fluorine atoms. The hybridization, sp^3d^2, yields six equivalent bonds directed to the apices of a regular octahedron (Fig. 23.19a). (An *octa*hedron has *eight faces*, but only *six* apices.) Similarly, the sulfur atom forms bonds to six atoms in SF_6; the hybridization is sp^3d^2 and the geometric configuration is octahedral (Fig. 23.19b).

The species PF_5, PF_6^-, SF_6 are exceptions to the octet rule. In PF_5 there are ten electrons in the valency group around the phosphorus atom; in PF_6^- and SF_6 there are twelve electrons in the valency groups. The elements in the first row, on the other hand, are bound rigidly to the "rule of eight."

Because of the availability of vacant d orbitals in the valence shell, the transition elements can form a variety of complex compounds. The electron pair for the bond is provided by a *donor* molecule or group such as NH_3 or CN^-. The most common hybridizations and their geometry are summarized in Table 23.2.

Table 23.2

Hybridization	Geometry	Hybridization	Geometry
sp	Linear	sp^2d	Square planar
sp^2	Trigonal planar	sp^3d	Trigonal bipyramidal
sp^3	Tetrahedral	sp^3d^2	Octahedral

23.11 MOLECULAR ENERGY LEVELS

In the molecular orbital method we consider the motion of one electron in the potential field due to all the nuclei of the molecule. We first arrange the nuclei in specified fixed positions; for example, in Fig. 23.20, we have four nuclei, a, b, c, d, with positive charges Z_a, Z_b, Z_c, Z_d. The Hamiltonian for one electron is written, in units of a_0 and E_h,

$$\mathbf{H} = -\tfrac{1}{2}\nabla_1^2 - \frac{Z_a}{r_{a1}} - \frac{Z_b}{r_{b1}} - \frac{Z_c}{r_{c1}} - \frac{Z_d}{r_{d1}} + C. \tag{23.21}$$

The constant C is the sum of the internuclear repulsion and is independent of the electronic coordinates; hence, if $\psi(1)$ is the appropriate wave function, the energy contribution from internuclear repulsion is $\int \psi C \psi \, d\tau = C \int \psi\psi \, d\tau = C$. As a result, at the beginning we can ignore the internuclear repulsion term in the Hamiltonian and simply add it in at the end of the calculation. In principle, we could solve the Schrödinger equation and obtain a set of wave functions and energy levels appropriate to the motion of one electron in the molecular framework. These wave functions are called *molecular orbitals* (MOs). Having such a set of wave functions for a given molecular geometry, the electronic structure of the molecule could be built up in much the same way that we build the structure of an atom on the basis of the hydrogenic wave functions. For example, we represent the structure of the carbon atom by putting two electrons in each atomic level until we have placed all six in the lowest energy orbitals.

In the same way, we represent the electronic structure of a molecule such as N_2 by putting two electrons in each molecular energy level until all fourteen electrons are placed in the lowest levels. We can proceed in a qualitative way to construct one-electron wave functions for diatomic molecules from the wave functions of the hydrogen atom. We begin by establishing such a scheme for the hydrogen molecule.

If we choose the hydrogen molecular framework of two protons, then, in Fig. 23.20, nuclei c and d are absent and $Z_a = Z_b = 1$; the Hamiltonian is (momentarily ignoring $1/R$, the internuclear repulsion)

$$\mathbf{H} = -\tfrac{1}{2}\nabla_1^2 - \frac{1}{r_{a1}} - \frac{1}{r_{b1}}. \tag{23.22}$$

Suppose that protons a and b are infinitely far apart. If we put one electron in this system, in the lowest energy state it must either be on nucleus a, with a description $\psi_a(1) = 1s_a$, or on nucleus b, with a description $\psi_b(1) = 1s_b$. Now either $\psi_a(1)$ or $\psi_b(1)$ is acceptable so long as the nuclei are infinitely far apart. As the nuclei come closer, we must have a wave function that is appropriate for the entire molecule. In particular, since protons a and b are identical, the wave function must be either symmetric or antisymmetric

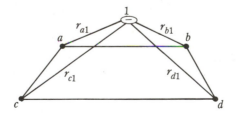

Figure 23.20   Coordinates for one electron in the field of four nuclei.

under the interchange of the nuclei. If $\mathbf{I}_N$ is the operator which interchanges the nuclear coordinates, then

$$\mathbf{I}_N \psi_a(1) = \psi_b(1) \qquad \text{and} \qquad \mathbf{I}_N \psi_b(1) = \psi_a(1).$$

Combining these two equations, we obtain the linear combinations

$$\psi_g(1) = \frac{1}{\sqrt{2(1 + S)}} [\psi_a(1) + \psi_b(1)] \tag{23.23}$$

and

$$\psi_u(1) = \frac{1}{\sqrt{2(1 - S)}} [\psi_a(1) - \psi_b(1)] \tag{23.24}$$

as the symmetric, $\psi_g(1)$ and antisymmetric, $\psi_u(1)$, functions. Let the energies of these two levels be E_g and E_u, respectively. When R, the internuclear distance, is infinite, $E_g = E_u$. As the nuclei approach one another, E_u rises and E_g falls. We have for the molecule two distinct energy levels corresponding to two different wave functions.

The wave function for a pair of electrons in the lowest energy level, E_g, is

$$\Psi = \frac{1}{2(1 + S)} [\psi_a(1) + \psi_b(1)][\psi_a(2) + \psi_b(2)], \tag{23.25}$$

if we assume that the electrons move independently.

If we expand the wave function in Eq. (23.25), we obtain

$$\Psi = \frac{1}{2(1 + S)} [\psi_a(1)\psi_a(2) + \psi_b(1)\psi_b(2) + \psi_a(1)\psi_b(2) + \psi_b(1)\psi_a(2)]. \tag{23.26}$$

The last two terms are the function we used in our earlier discussion of the electron-pair bond. The MO description contains two additional terms: the first, $\psi_a(1)\psi_a(2)$, corresponds to both electrons on proton a. The second, $\psi_b(1)\psi_b(2)$, has both electrons on proton b. The chemical stuctures would be written

$$\text{III} \qquad \text{H}_{a}^{\cdot 1 \bar{}}_{\cdot 2} \qquad \text{H}_b^+,$$

$$\text{IV} \qquad \text{H}_a^+ \qquad {\scriptstyle \frac{1}{2}}\text{:H}_b^- .$$

Structures III and IV are ionic structures. These structures do contribute slightly ($\sim 3\%$) to the overall structure of the hydrogen molecule. In this simple molecular orbital approach the ionic structures III and IV are weighted equally with the covalent structures I and II. In more refined versions their contribution is reduced to a more realistic level.

We can make linear combinations comparable to those in Eqs. (23.23) and (23.24) with any two equivalent orbitals on nucleus a and nucleus b. If we use the $1s$ orbitals on the two atoms, the notation would be

$$1s\sigma_g = 1s_a + 1s_b \tag{23.27}$$

and

$$1s\sigma_u = 1s_a - 1s_b. \tag{23.28}$$

The symbol σ is used to denote an MO that has its charge density concentrated along the internuclear axis. The symbol π is used for MOs that have zero charge density on the internuclear axis.

In Fig. 23.21 we show the symmetry of the various combinations of s and p orbitals. On the far right we have the wave function; slightly to the left, the boundary surfaces for

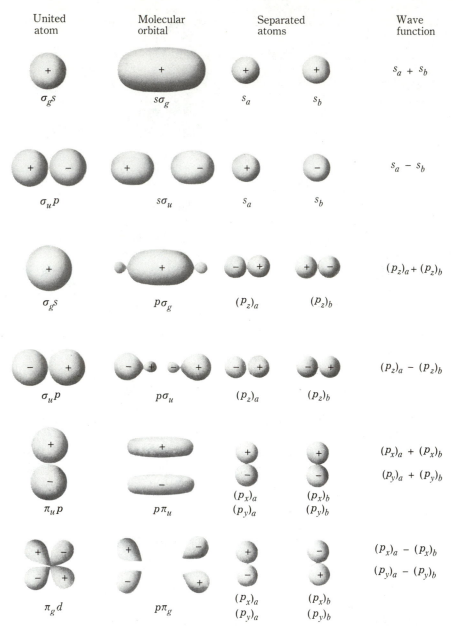

United atom	Molecular orbital	Separated atoms		Wave function
$\sigma_g s$	$s\sigma_g$	s_a	s_b	$s_a + s_b$
$\sigma_u p$	$s\sigma_u$	s_a	s_b	$s_a - s_b$
$\sigma_g s$	$p\sigma_g$	$(p_z)_a$	$(p_z)_b$	$(p_z)_a + (p_z)_b$
$\sigma_u p$	$p\sigma_u$	$(p_z)_a$	$(p_z)_b$	$(p_z)_a - (p_z)_b$
$\pi_u p$	$p\pi_u$	$(p_x)_a$ $(p_y)_a$	$(p_x)_b$ $(p_y)_b$	$(p_x)_a + (p_x)_b$ $(p_y)_a + (p_y)_b$
$\pi_g d$	$p\pi_g$	$(p_x)_a$ $(p_y)_a$	$(p_x)_b$ $(p_y)_b$	$(p_x)_a - (p_x)_b$ $(p_y)_a - (p_y)_b$

Figure 23.21

the orbitals as they appear on the two separated nuclei; in the middle, the boundary surface as it appears in the MO; at the far left, the boundary surface as it appears in the "united atom"—that is, if we push the two nuclei to the point where they coalesce. The point of Fig. 23.21 is that the states of the separated atoms on the far right must correlate with states of the united atom having the same symmetry on the far left.

We have already described the g and u aspect of the symmetry of the wave function. The correlation between the states of the separated atoms and the united atom depends on a different symmetry property: the position of the nodes in the wave function.

A node in the wave function occurs at any point or on any surface where the wave function is zero. Whenever two equivalent wave functions are combined by subtraction, a node is introduced exactly at the plane midway between the two nuclei. Furthermore, that node remains exactly midway between the nuclei regardless of whether $R = \infty$ or $R = 0$. For this reason, the function, $s_a - s_b$, looks like a p function on the united atom and consequently must correlate with a p function. If nodes are introduced by the addition of two p functions with their positive heads toward each other, as in $(p_z)_a + (p_z)_b$, the addition of the two functions moves the nodes (which were originally situated on the nuclei) slightly outward and away from the nuclear positions. As the nuclei move closer together the nodal positions move farther and farther away from the nuclei. When the two nuclei coalesce into the united atom, the nodes have moved to $+\infty$ and $-\infty$; the wave function looks like an s function. Thus $(p_z)_a + (p_z)_b$ must correlate with an s function on the united atom. Since the united atom will have a nuclear charge, $Z_a + Z_b$, the energy levels on the united atom are generally much lower than the energy of the corresponding orbitals on the separated atoms. In Fig. 23.22 we show the energy levels and the correlations between them.

An orbital that correlates with a united atom orbital having a lower energy than the separated atoms is a *bonding orbital*. An orbital that correlates with a united atom orbital having a higher energy is an *antibonding orbital*. The number of bonds is measured by the excess of the number of bonding pairs over the number of antibonding pairs. If we use the order of the energy levels at the position of the appropriate vertical line in Fig. 23.22, we can describe the bonds in a diatomic molecule by placing a pair of electrons in each level, beginning with the lowest level. The results are shown in Table 23.3. It is notable that the molecular orbital model predicts the paramagnetism of O_2. The outermost orbital, $(2p\pi_g)$, contains only two electrons. Since there are two $(2p\pi_g)$ orbitals of equal energy, differing only in their orientation, they can accommodate a total of four electrons. Consequently, the spins of the two electrons in the $(2p\pi_g)$ orbital of the oxygen molecule are not paired and the molecule has a magnetic moment.

We can extend the method to heteronuclear diatomics with relative ease. Since the heteronuclear molecule does not have the nuclear exchange symmetry the g and u character is lost, but the rest of the notation remains the same. The correlation diagram becomes more complicated because of the difference in energies of the electron on the two different separated atoms. It is worth noting that isoelectronic species such as $N{\equiv}N$, $C{\equiv}O$,

Table 23.3
Electronic configurations in homonuclear diatomic molecules

Molecule	Number of electrons	Electronic configuration	Bonding pairs	Antibonding pairs	Bonds
H_2	2	$(1s\sigma_g)^2$	1	0	1
He_2	4	$(1s\sigma_g)^2(1s\sigma_u)^2$	1	1	0
Li_2	6	$(He)_2(2s\sigma_g)^2$	2	1	1
Be_2	8	$(He)_2(2s\sigma_g)^2(2s\sigma_u)^2$	2	2	0
B_2	10	$(Be)_2(2p\sigma_g)^2$	3	2	1
C_2	12	$(Be)_2(2p\sigma_g)^2(2p\pi_u)^2$	4	2	2
N_2	14	$(Be)_2(2p\sigma_g)^2(2p\pi_u)^4$	5	2	3
O_2	16	$(Be)_2(2p\sigma_g)^2(2p\pi_u)^4(2p\pi_g)^2$	5	3	2
F_2	18	$(Be)_2(2p\sigma_g)^2(2p\pi_u)^4(2p\pi_g)^4$	5	4	1
Ne_2	20	$(Be)_2(2p\sigma_g)^2(2p\pi_u)^4(2p\pi_g)^4(2p\sigma_u)^2$	5	5	0

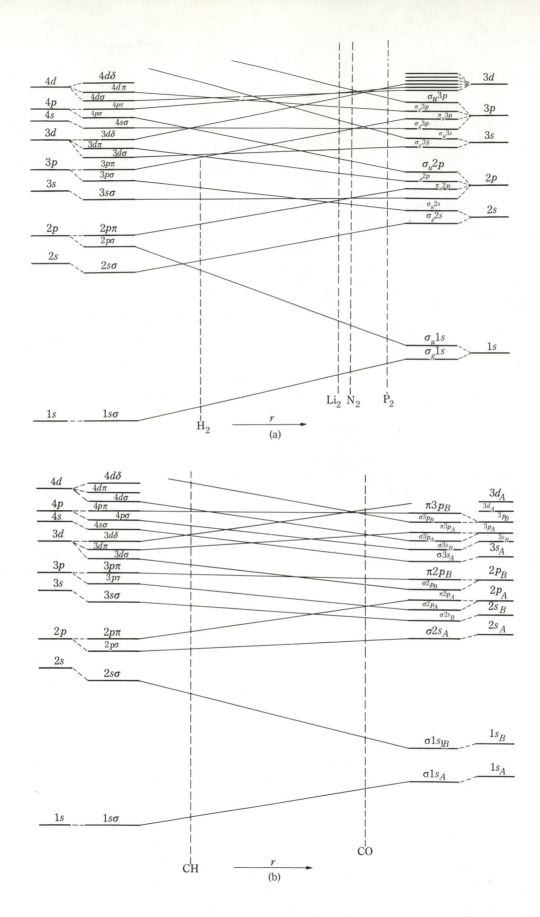

Figure 23.22 Correlation diagrams for diatomic molecules. To the extreme left and the extreme right are given the orbitals in the united and separated atoms, respectively; beside them are those in the molecule for very small and very large internuclear distances, respectively. The region between corresponds to intermediate internuclear distances. The vertical broken lines give the approximate positions in the diagram that correspond to the molecules indicated. It should be noticed that the scale of r in this figure is by no means linear but becomes rapidly smaller on the right-hand side. (From *Molecular Spectra and Molecular Structure*, Vol. 1, by G. Herzberg. © 1950 by Litton Educational Publishing, Inc. Reprinted by permission of Wadsworth Publishing Company, Belmont, California 94002.)

$^-C\equiv N$, $^-C\equiv C^-$, $N\equiv O^+$, all have essentially the same electronic configuration as $N\equiv N$ shown in Table 23.3, but without the g and u character in the case of the heteronuclear species. The extension to nonlinear molecules requires the introduction of a sophisticated way to describe the symmetry of the orbitals; the group theoretical notation for the symmetry types is used.

The system of energy levels for the molecular orbitals of diatomic molecules is often represented schematically as in Fig. 23.23, in which the energy levels of the two separated

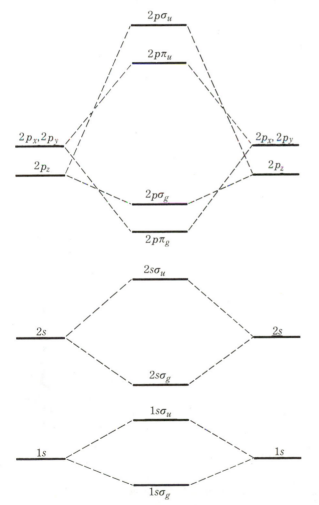

Figure 23.23 Splitting of atomic levels in a molecule.

atoms are shown at the two sides of the figure and the energy levels of the molecule are shown in the middle. This figure emphasizes the splitting of the atomic levels as the atoms approach each other. This splitting as a function of internuclear distance is shown for the $1s\sigma_g$ and $1s\sigma_u$ levels in Fig. 23.1.

23.12 WAVE FUNCTIONS AND SYMMETRY

It should be clear that a scalar physical property, such as the energy of a molecule, is independent of the coordinate system we use to describe the molecule. For example, consider the water molecule shown in Fig. 23.24, in which we have chosen two different coordinate systems to describe the molecule. Assuming that it is possible to carry out the calculation of the energy in either coordinate system, the result finally obtained must be the same. The calculation using the coordinate system in Fig. 23.24(b) would be much more complicated to do but would, nonetheless, yield the same result.

Consider first a symmetrical molecule such as H_2O, which has only nondegenerate wave functions. Suppose we have positioned this molecule advantageously in a coordinate system as illustrated in Fig. 23.23(a). If we subject the molecule to a *symmetry operation*—an operation that brings every part of the molecule into the same position or interchanges identical parts of the molecule—the Hamiltonian operator is invariant under the symmetry operation, and the wave function that describes the molecule must either be unchanged or changed only in algebraic sign by the symmetry operation. Since the energy depends on the integral of the product function, $\psi^*(H\psi)$, a change in sign will leave this integrand unchanged; the sign change will occur twice, once in ψ^* and once in $H\psi$, so that the two negative signs will yield a positive product. Thus the function $\psi^*(H\psi)$ is invariant even if the sign of ψ is changed. Consequently, the symmetry of the situation requires that either

$$R\psi = \psi \quad \text{or} \quad R\psi = -\psi, \tag{23.29}$$

where R is the operator corresponding to any symmetry operation for the molecule. If ψ

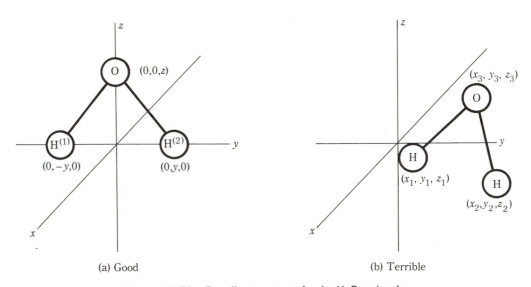

(a) Good   (b) Terrible

Figure 23.24   Coordinate systems for the H_2O molecule.

is invariant under this operation, then ψ is *symmetric* under the operation **R**. If ψ changes sign, then ψ is *antisymmetric* under **R**.

In molecules of higher symmetry the wave function may be a member of a degenerate set; if so, some symmetry operations transform the wave function into a linear combination of the wave functions in the degenerate set. The energy is still invariant under the symmetry operation. We will deal with this complication in the discussion of the NH_3 molecule.

Before considering the consequences of this behavior of the wave function under the symmetry operations, we pause to introduce some of the vocabulary of group theory.

23.13 MATHEMATICAL INTERLUDE

We say that a figure or a molecule has certain elements of symmetry. One or more symmetry operations are associated with each element of symmetry. A symmetry operation may leave some or all parts of the figure in the same position or it may interchange some or all identical parts of the figure.

There are four types of symmetry elements; these elements, with the corresponding symmetry operations and the symbols for the operations, are:

Symmetry element	*Symmetry operation*	*Symbol for the symmetry operation*
Plane of symmetry	Reflection in the plane	σ
Center of symmetry	Inversion through the center of symmetry. This changes the sign of all the coordinates of all the particles.	i
n-fold axis of symmetry	Rotation about the axis through the angles $(2\pi/n)$, $2(2\pi/n)$, $3(2\pi/n)$, ... $(n-1)(2\pi/n)$	$C_n, C_n^2, C_n^3 \cdots C_n^{n-1}$,
Improper axis of symmetry	Rotation through the angle $2\pi/n$ followed by reflection in the plane perpendicular to the rotation axis	S_n

An important consequence of symmetry is that the symmetry operations for any figure form a mathematical group. We can then use the properties of the group as a powerful aid in the calculation of properties of molecular systems and the understanding of why systems behave the way they do.

23.13.1 Group Properties

We do not need to develop all aspects of group theory to gain some understanding of its implications for the structure of molecules. However, it is useful to know some group-theoretical terminology and some of the properties of groups.

A group is defined as a set of elements having the following four properties.

1. There is a law of combination, defined by the group multiplication rule, which requires that the product of any element with itself or with any other element of the group is in turn an element of the group. The group multiplication rule defines what is meant by the "product" of one element with another. Thus, if $A, B, C, D, \ldots$ are elements of the group, we require that the product of two elements, written AB, be an element of the group.

$$AB = C, \tag{23.30}$$

in which C is some element of the group. Note that the order of multiplication is important; in general, $AB \neq BA$. We describe AB as "B premultiplied by A" and BA as "B postmultiplied by A."

2. The group contains a unit element or identity element. The unit element is always symbolized by E. By definition, when E multiplies any other element, A, of the group, we have

$$EA = AE = A. \tag{23.31}$$

3. Every element, A, has an inverse element, A^{-1}, such that either pre- or postmultiplication of A by A^{-1} produces the unit element

$$A^{-1}A = AA^{-1} = E. \tag{23.32}$$

4. The associative law holds for group multiplication; if A, B, C are elements of the group, then

$$A(BC) = (AB)C. \tag{23.33}$$

The elements in a group can be divided into classes. If we construct the product, $X^{-1}AX$, and replace X by each element of the group in turn, we obtain all the elements in the class of element A. Repeating the procedure using $X^{-1}BX$, we obtain all the elements in the class of element B, and so on. No element in the group can belong to more than one class.

23.14 THE WATER MOLECULE (GROUP C_{2v}): EXAMPLE

The symmetry operations appropriate to a figure or molecule are the elements of a group. As a first example, we consider the water molecule. The elements of symmetry are: one two-fold axis of symmetry (this is the z-axis in Fig. 23.24a); and two planes of symmetry (these are the xz-plane and the yz-plane). The four symmetry operations are:

E: The identity operation. This operation transforms every point into itself: $x_i \to x_i$; $y_i \to y_i$; $z_i \to z_i$. This may seem trivial, but the inclusion of the identity operation is crucial; without it the other symmetry operations would not form a mathematical group.

C$_2$:   Counterclockwise rotation through 180° around the z-axis. This transformation leaves the coordinates of the oxygen atom unchanged. The coordinates of the two hydrogen atoms are interchanged.

$\sigma_v(xz)$:   Reflection in the xz plane. This interchanges the coordinates of the two hydrogen atoms; it leaves the oxygen atom in place.

$\sigma_v'(yz)$:   Reflection in the yz plane. This leaves all the atoms in place.

These symmetry operations make up the symmetry group C_{2v}.

The effect of each of these symmetry operations on a point is shown in Fig. 23.25. In each case, the point P is transformed into the point P', as indicated. These transformations can be summarized by

$$\mathbf{E}(x, y, z) = (x, y, z);$$
$$\mathbf{C}_2(x, y, z) = (-x, -y, z);$$
$$\boldsymbol{\sigma}_v(xz)(x, y, z) = (x, -y, z);$$
$$\boldsymbol{\sigma}_v'(yz)(x, y, z) = (-x, y, z).$$

(23.34)

Using these relations we can work out the multiplication table for the group. For example, the product $\mathbf{C}_2\boldsymbol{\sigma}_v$ signifies the two operations (first $\boldsymbol{\sigma}_v$, then $\mathbf{C}_2$) performed on the figure. If the point P is subjected to $\boldsymbol{\sigma}_v$ then P is moved to the point P' in Fig. 23.25(b). If this point is subjected to $\mathbf{C}_2$ it is carried to the position of the point P'' in Fig. 23.25(c). Thus we conclude that $\mathbf{C}_2\boldsymbol{\sigma}_v = \boldsymbol{\sigma}_v'$ is the expression of the group multiplication rule for the operation $\boldsymbol{\sigma}_v$ and $\mathbf{C}_2$. Algebraically we can use Eq. (23.34) to obtain the same result. Namely, to

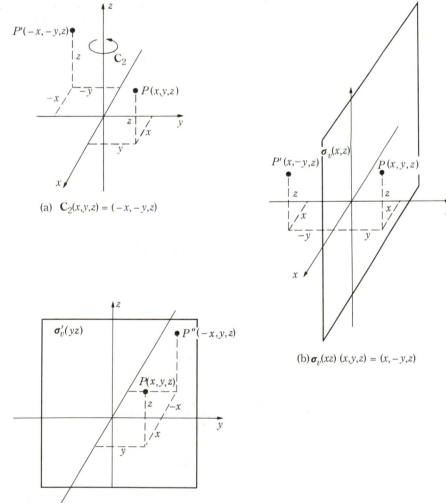

(a) $\mathbf{C}_2(x,y,z) = (-x,-y,z)$

(b) $\boldsymbol{\sigma}_v(xz)\ (x,y,z) = (x,-y,z)$

(c) $\boldsymbol{\sigma}_u'(yz)\ (x,y,z) = (-x,y,z)$

Figure 23.25 The product of two operations $\mathbf{C}_2\boldsymbol{\sigma}_v = \boldsymbol{\sigma}_v'$.

Table 23.4
Multiplication table for C_{2v}

	E	C_2	$\sigma_v(xz)$	$\sigma'_v(yz)$
E	E	C_2	σ_v	σ'_v
C_2	C_2	E	σ'_v	σ_v
$\sigma_v(xz)$	σ_v	σ'_v	E	C_2
$\sigma'_v(yz)$	σ'_v	σ_v	C_2	E

find the product, $C_2 \sigma_v$, we use Eq. (23.34) and find that

$$C_2 \sigma_v(x, y, z) = C_2(x, -y, z) = (-x, y, z).$$

Comparing with Eqs. (23.34) we see that $\sigma'_v(x, y, z) = (-x, y, z)$, and conclude that

$$C_2 \sigma_v = \sigma'_v.$$

In this way we can work out all of the products between the group elements, thus obtaining the group multiplication table, Table 23.4. We agree, quite arbitrarily, that the product $C_2 \sigma_v$ will be found at the intersection of the *row* labeled σ_v and the *column* labeled C_2.

23.15 REPRESENTATIONS OF A GROUP

Any set of numbers or any set of square matrices that has the same multiplication table as the group is called a *representation* of the group. It is possible to construct any arbitrary number of representations of a group. The representation may consist of matrices of any order, but within any one representation all must be of the same order.

The set of matrices in any given representation can be transformed to an equivalent set, usually of lower order. If it is not possible to reduce all the matrices in a representation to lower order matrices by a specified transformation, this set of matrices is called an *irreducible* representation of the group. The number of distinct irreducible representations of any group is equal to the number of classes in the group.

Once we know the matrices in an irreducible representation, we obtain the sum of the main diagonal elements in each matrix. This sum is the *character* of the matrix. The set of characters of the matrices in an irreducible representation are the characters of that representation. It is these numbers, collected in the *character table* of the group, that have prime importance in the application of group theory to molecules.

23.15.1 Characters of the Representations of the Group C_{2v}: Example

Table 23.5 is the character table for the group C_{2v}.

Table 23.5
Character table for C_{2v}

	E	C_2	$\sigma_v(xz)$	$\sigma'_v(xy)$		
a_1	1	1	1	1	z	x^2, y^2, z^2
a_2	1	1	-1	-1	R_z	xy
b_1	1	-1	1	-1	x, R_y	xz
b_2	1	-1	-1	1	y, R_x	yz

For this group, C_{2v}, there are four distinct irreducible representations. Each representation is a set of one-dimensional matrices. The first irreducible representation, a_1, consists of the set of 1×1 matrices, each of which has a $+1$ in the lone position in the array.

	E	C_2	σ_v	σ_v'
a_1	[1]	[1]	[1]	[1]

The sum of the diagonal elements is simply the element itself, $+1$. So the characters are

	E	C_2	σ_v	σ_v'
a_1	1	1	1	1

The other three irreducible representations are three other sets of one-dimensional matrices. Again the characters (Table 23.5) are simply equal to the single element in the matrices. The one-dimensional irreducible representations are given the conventional labels a_1, a_2, b_1, b_2. (Capital letters are used as often as not.)

The behavior of any coordinate under the symmetry operations of the group is described by the characters of one of the irreducible representations. For example, the coordinate x transforms as

$$\mathbf{E}x = +x \qquad \mathbf{C}_2 x = -x \qquad \sigma_v(xz)x = +x \qquad \sigma_v'(yz)x = -x.$$

The coefficients on the right-hand side are:

E	C_2	σ_v	σ_v'
1	-1	1	-1

Comparing this set of numbers with the sets in the character table we see that x belongs to the irreducible representation b_1. Similarly, y belongs to b_2. Since $\mathbf{R}z = z$, where $\mathbf{R}$ is any operation of the group, z is totally symmetric and belongs to the totally symmetric representation a_1. (All the characters of a_1 are $+1$.)

If we know the irreducible representation to which the coordinates belong (for example, x belongs to b_1 and y belongs to b_2), we can find the representation to which product functions belong by forming the *direct product* of the representations $b_1 \times b_2$. For any symmetry element we obtain the character of the direct product by multiplying the characters of the two representations. For example, the characters of the direct product $b_1 \times b_2$ are

	E	C_2	σ_v	σ_v'
$b_1 \times b_2$	$(+1)(+1)$	$(-1)(-1)$	$(+1)(-1)$	$(-1)(+1)$

or

	E	C_2	σ_v	σ_v'
$b_1 \times b_2$	1	1	-1	-1

This is the representation a_2; therefore we write

$$b_1 \times b_2 = a_2,$$

and we conclude that the product xy belongs to the representation a_2. Since the totally symmetric representation a_1 has characters all equal to $+1$, it follows that the direct product of a_1 with any other irreducible representation belongs to the latter representation:

$$a_1 \times a_1 = a_1; \quad a_1 \times a_2 = a_2; \quad a_1 \times b_1 = b_1; \quad a_1 \times b_2 = b_2.$$

Note that the direct product of any irreducible representation with itself, such as $b_1 \times b_1$, belongs to the totally symmetric representation, a_1. Hence the functions x^2, y^2, z^2 all belong to a_1. The character table includes these simple functions as shown in Table 23.5. The symbols R_x, R_y, and R_z identify the representations to which rotations around the x-, y-, and z-axes belong. Character tables for several symmetry groups are given in Appendix VI.

23.15.2   Representations of the Group C_{3v}: Example

Before applying these symmetry principles, we will discuss a slightly more general example to illustrate what occurs when one (or more) of the irreducible representations is two-dimensional (or three-dimensional). The simplest symmetry group that has a two-dimensional irreducible representation is C_{3v}. This group contains the symmetry elements appropriate to the ammonia molecule (Fig. 23.26). The projection of the atoms on the xy plane is shown in Fig. 23.26(b). The symmetry operations are:

E: The identity.

C$_3$: Rotation counterclockwise through 120° around the z-axis.

C̄$_3$: Rotation clockwise through 120° around the z-axis.

$\sigma_v^{(1)}$: Reflection in the vertical plane containing the z-axis and hydrogen atom 1.

$\sigma_v^{(2)}$: Reflection in the vertical plane containing the z-axis and hydrogen atom 2.

$\sigma_v^{(3)}$: Reflection in the vertical plane containing the z-axis and hydrogen atom 3.

In the symmetry group C_{2v}, each coordinate was either invariant or changed only in sign by the symmetry operations; in the group C_{3v} the situation is more involved. For the z coordinate we still have $\mathbf{R}z = z$, where $\mathbf{R}$ is any operation in the group, so that in C_{3v}, z belongs to the totally symmetric representation just as it did in C_{2v}. But x and y do not transform so simply. For example, under $\mathbf{C}_3$ we find that

$$\mathbf{C}_3 x = x \cos \tfrac{2}{3}\pi - y \sin \tfrac{2}{3}\pi;$$

$$\mathbf{C}_3 y = x \sin \tfrac{2}{3}\pi + y \cos \tfrac{2}{3}\pi.$$

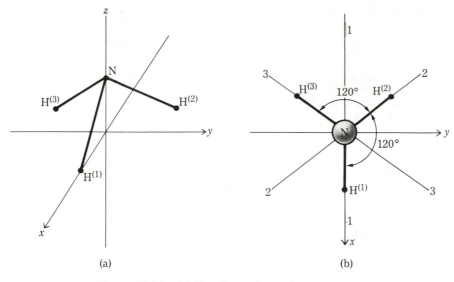

Figure 23.26   (a) Coordinates for the NH_3 molecule.
(b) Projection of atoms on xy-plane.

This can be written in matrix notation (see Appendix I, Section 8) as

$$\mathbf{C}_3 \begin{bmatrix} x \\ y \end{bmatrix} = \begin{bmatrix} \cos \frac{2}{3}\pi & -\sin \frac{2}{3}\pi \\ \sin \frac{2}{3}\pi & \cos \frac{2}{3}\pi \end{bmatrix} \begin{bmatrix} x \\ y \end{bmatrix}.$$

Similarly, we find that

$$\bar{\mathbf{C}}_3 \begin{bmatrix} x \\ y \end{bmatrix} = \begin{bmatrix} \cos \frac{2}{3}\pi & \sin \frac{2}{3}\pi \\ -\sin \frac{2}{3}\pi & \cos \frac{2}{3}\pi \end{bmatrix} \begin{bmatrix} x \\ y \end{bmatrix}.$$

Thus, the operators $\mathbf{C}_3$ or $\bar{\mathbf{C}}_3$ can be written as matrices. Since $\cos \frac{2}{3}\pi = -\frac{1}{2}$ and $\sin \frac{2}{3}\pi = \frac{1}{2}\sqrt{3}$, these matrices become:

$$\mathbf{C}_3 = \begin{bmatrix} -\frac{1}{2} & -\frac{1}{2}\sqrt{3} \\ \frac{1}{2}\sqrt{3} & -\frac{1}{2} \end{bmatrix} \quad \text{and} \quad \bar{\mathbf{C}}_3 = \begin{bmatrix} -\frac{1}{2} & \frac{1}{2}\sqrt{3} \\ -\frac{1}{2}\sqrt{3} & -\frac{1}{2} \end{bmatrix}.$$

Note that the sum of the main diagonal elements, $-\frac{1}{2} + (-\frac{1}{2}) = -1$, is the same for both $\mathbf{C}_3$ and $\bar{\mathbf{C}}_3$. These operations have equal characters.

The reflection operator $\boldsymbol{\sigma}_v^{(1)}$ has the effect

$$\boldsymbol{\sigma}_v^{(1)} x = x \quad \text{and} \quad \boldsymbol{\sigma}_v^{(1)} y = -y,$$

which can be written as a matrix transformation:

$$\boldsymbol{\sigma}_v^{(1)} \begin{bmatrix} x \\ y \end{bmatrix} = \begin{bmatrix} 1 & 0 \\ 0 & -1 \end{bmatrix} \begin{bmatrix} x \\ y \end{bmatrix} = \begin{bmatrix} x \\ -y \end{bmatrix}.$$

The matrix corresponding to $\boldsymbol{\sigma}_v^{(1)}$ has the character $1 + (-1) = 0$. The reflections in the other two planes again give more involved expressions. It can be shown that

$$\sigma_v^{(2)}x = -\tfrac{1}{2}x - \tfrac{1}{2}\sqrt{3}\,y, \qquad \sigma_v^{(3)}x = -\tfrac{1}{2}x + \tfrac{1}{2}\sqrt{3}\,y;$$

and

$$\sigma_v^{(2)}y = -\tfrac{1}{2}\sqrt{3}\,x + \tfrac{1}{2}y, \qquad \sigma_v^{(3)}y = \tfrac{1}{2}\sqrt{3}\,x + \tfrac{1}{2}y.$$

Thus, under the reflections in planes 2 and 3, both x and y are transformed into linear combinations of x and y. In matrix notation

$$\sigma_v^{(2)}\begin{bmatrix} x \\ y \end{bmatrix} = \begin{bmatrix} -\tfrac{1}{2} & -\tfrac{1}{2}\sqrt{3} \\ -\tfrac{1}{2}\sqrt{3} & \tfrac{1}{2} \end{bmatrix}\begin{bmatrix} x \\ y \end{bmatrix} \quad \text{and} \quad \sigma_v^{(3)}\begin{bmatrix} x \\ y \end{bmatrix} = \begin{bmatrix} -\tfrac{1}{2} & \tfrac{1}{2}\sqrt{3} \\ \tfrac{1}{2}\sqrt{3} & \tfrac{1}{2} \end{bmatrix}\begin{bmatrix} x \\ y \end{bmatrix}.$$

For the matrices corresponding to the reflections we have

$$\sigma_v^{(1)} = \begin{bmatrix} 1 & 0 \\ 0 & -1 \end{bmatrix}, \quad \sigma_v^{(2)} = \begin{bmatrix} -\tfrac{1}{2} & -\tfrac{1}{2}\sqrt{3} \\ -\tfrac{1}{2}\sqrt{3} & \tfrac{1}{2} \end{bmatrix}, \quad \sigma_v^{(3)} = \begin{bmatrix} -\tfrac{1}{2} & \tfrac{1}{2}\sqrt{3} \\ \tfrac{1}{2}\sqrt{3} & \tfrac{1}{2} \end{bmatrix}.$$

Note that the sum of the main diagonal elements for each of these matrices is zero. All the reflections have the same character, namely, zero.

The identity operator always has the unit matrix as a representation:

$$E = \begin{bmatrix} 1 & 0 \\ 0 & 1 \end{bmatrix}.$$

This matrix has a character equal to 2.

This set of matrices, one for each member of the group, is a two-dimensional irreducible representation of the group. The characters of the representation are summarized by

E	C_3	$\bar{C}_3$	$\sigma_v^{(1)}$	$\sigma_v^{(2)}$	$\sigma_v^{(3)}$
2	-1	-1	0	0	0

Note that the two rotations have the same character and that the three reflections have the same character. The two rotations are in one class of elements; the three reflections are in another class; the unit element is in a class by itself. Representations of elements in the same class always have the same character. For this reason it is customary to write the character table in the condensed form shown in Table 23.6. (Note that this group also has two one-dimensional representations, a_1 and a_2, that have the characters shown in Table 23.6.)

Table 23.6
Character table for C_{3v}

	E	$2C_3$	$3\sigma_v$		
a_1	1	1	1	z	$x^2 + y^2,\ z^2$
a_2	1	1	-1	R_z	
e	2	-1	0	$(x, y)(R_x, R_y)$	$(x^2 - y^2,\ xy)(xz,\ yz)$

In this table the notation (x, y) indicates that the pair (x, y) transforms into linear combinations of x and y under the operations of the group. As usual, a_1 is the totally symmetric representation.

23.16 REDUCIBLE REPRESENTATIONS; THE ORTHOGONALITY THEOREM

In forming the direct products we find

$$a_1 \times a_1 = a_1, \qquad a_1 \times a_2 = a_2, \qquad a_1 \times e = e;$$

$$a_2 \times a_2 = a_1, \qquad a_2 \times e = e.$$

All of these direct products of the irreducible representations are themselves irreducible representations. On the other hand when we form the direct product $e \times e$, we obtain for the characters:

	E	$2C_3$	$3\sigma_v$
$e \times e$	4	1	0

These numbers do not correspond to the characters of any of the irreducible representations. It follows that the direct product $e \times e$ belongs to a *reducible* representation. The character of any reducible representation is the sum of the characters of the irreducible representations that compose it. In this particularly simple case we can see by inspection of the character table that

$$e \times e = a_1 + a_2 + e.$$

We say that the direct product $e \times e$ contains the irreducible representations a_1, a_2 and e.

In more complicated cases, inspection is not a practical way to proceed. For these we use the orthogonality theorem, which we must simply state without proof. This theorem can be written in the special form

$$\sum_R \chi_i(R)\chi_j(R) = h\delta_{ij}, \tag{23.35}$$

where $\chi_i(R)$ and $\chi_j(R)$ are the characters of the two irreducible representations i and j, h is the number of elements in the group, R is any operation of the group, and δ_{ij} is the Kronecker delta. Equation (23.35) states that the sum of the products of the characters of two different $(i \neq j)$ irreducible representations over the group operations is zero, while the sum of the squares of the characters of any irreducible representation over the operations of the group is equal to h. For example, let $i = a_1$ and $j = a_2$, then

$$\begin{array}{ccc} E & 2C_3 & 3\sigma_v \end{array}$$
$$\sum_R \chi_{a_1}(R)\chi_{a_2}(R) = (1)(1) + 2(1)(1) + 3(1)(-1) = 0.$$

Note that on the right side, the second term $(1)(1)$ occurs twice, once for C_3 and once for $\bar{C}_3$ (hence the 2 multiplier); similarly the third term $(1)(-1)$ occurs three times, once for each reflection (hence the 3 multiplier). If we choose $i = j = e$, then

$$\sum_R \chi_e(R)\chi_e(R) = (2)(2) + 2(-1)(-1) + 3(0)(0) = 6.$$

Suppose that we have a reducible representation with characters $\chi(R)$. These characters are the sum of the characters of the irreducible representations contained in the reducible one. Thus for any operation

$$\chi(R) = \sum_i \alpha_i \chi_i(R), \tag{23.36}$$

which says that the ith irreducible representation is contained α_i times in the reducible representation. The sum is over all the irreducible representations. If we multiply both sides of this equation by $\chi_j(R)$, the character of the jth irreducible representation, and then sum both sides over all the operations of the group, we obtain

$$\sum_R \chi(R)\chi_j(R) = \sum_R \sum_i \alpha_i \chi_i(R)\chi_j(R)$$

$$= \sum_i \alpha_i \sum_R \chi_i(R)\chi_j(R).$$

Using the orthogonality theorem, Eq. (23.35), for the second sum on the right side, we obtain

$$\sum_R \chi(R)\chi_j(R) = \sum_i \alpha_i h \delta_{ij}.$$

Summing over i, the right side reduces to one term, $h\alpha_j$; then

$$\alpha_j = \frac{1}{h} \sum_R \chi(R)\chi_j(R) \tag{23.37}$$

Using this equation we can determine how many times the jth irreducible representation is contained in any reducible representation.

23.16.1 Resolving a Representation into its Irreducible Components: An Example

If we choose to construct molecular orbitals for NH_3 using linear combinations of the $1s$ orbitals on the three hydrogen atoms, and the $2s$, $2p_x$, $2p_y$, and $2p_z$ orbitals on the nitrogen atom, we can show that the resulting molecular orbitals belong to a reducible representation of the group C_{3v} with the characters:

	E	$2C_3$	$3\sigma_v$
$\chi(R)$	7	1	3

To which irreducible representations do these molecular orbitals belong? To answer this question we first construct the right side of Eq. (23.37) for $j = a_1$; thus

$$\alpha_{a_1} = \tfrac{1}{6}[(7)(1) + 2(1)(1) + 3(3)(1)] = \tfrac{18}{6} = 3.$$

We conclude that a_1 is contained three times in the reducible representation. Next we repeat the procedure for a_2:

$$\alpha_{a_2} = \tfrac{1}{6}[(7)(1) + 2(1)(1) + 3(3)(-1)] = 0.$$

Thus a_2 does not appear in this reducible representation. Then repeat the procedure for e.

$$\alpha_e = \tfrac{1}{6}[(7)(2) + 2(1)(-1) + 3(3)(0)] = \tfrac{12}{6} = 2.$$

The representation e is contained twice in the reducible representation. Thus the reducible representation is made up of $3a_1 + 2e$. We can easily verify that the sums of the characters for this combination of irreducible representations are equal to those of the reducible representation.

An important point is that the molecular orbitals belonging to one-dimensional representations such as a_1 or a_2 are nondegenerate. The molecular orbitals belonging to two-dimensional representations are doubly degenerate; both wave functions have the same energy. In symmetry groups in which a three-dimensional representation, t, occurs, the corresponding wave functions are triply degenerate.

Note that one-dimensional irreducible representations are always labeled a or b (A or B) with as many subscripts as are necessary; two-dimensional ones are labeled e (or E) and three-dimensional ones are labeled t (or T). A one-dimensional representation is designated a if it is symmetric under a rotation about the principal axis of symmetry, and b if it is antisymmetric under this operation.

23.16.2 Construction of Molecular Orbitals: Example

Any proper wave function is required to transform under the symmetry operations in the manner determined by the matrices in the irreducible representations of the group. This requirement severely restricts the form of the wave functions for any molecule. In the case of the water molecule, only four types of functions fulfill the condition: the types belonging to the four irreducible representations of the group C_{2v}. The phrase "symmetry species" is used as a synonym for "irreducible representation." Thus the wave functions of the water molecule may belong to one of the four symmetry species, a_1, a_2, b_1, or b_2.

For molecules with other symmetries there are other character tables and a similar notation for the symmetry species. The wave functions of asymmetric molecules cannot be classified into symmetry species.

Note that the above discussion does not depend on knowing anything about the detailed functional form of wave functions. We do not have to solve the Schrödinger equation. The statements are based solely on the symmetry properties and the consequent mathematical properties of the symmetry group.

However, if we wish to construct approximate wave functions for the system, the symmetry gives us a powerful method for doing so. For example, if we wish to construct *molecular orbitals* (MOs) for the water molecule by using *linear* combinations of *atomic orbitals* (LCAOs) we can do so with great ease. Let the atomic orbitals* (AOs) be denoted by χ_i:

$$
\begin{aligned}
\chi_1 &= (1s)_1 && \text{(the } 1s \text{ orbital on H atom 1)}; \\
\chi_2 &= (1s)_2 && \text{(the } 1s \text{ orbital on H atom 2)}; \\
\chi_3 &= (2s)_O && \text{(the } 2s \text{ orbital on the O atom)}; \\
\chi_4 &= (2p_z)_O && \text{(the } 2p_z \text{ orbital on the O atom)}; \\
\chi_5 &= (2p_x)_O && \text{(the } 2p_x \text{ orbital on the O atom)}; \\
\chi_6 &= (2p_y)_O && \text{(the } 2p_y \text{ orbital on the O atom)}.
\end{aligned}
\tag{23.38}
$$

---

* It is unfortunate that the usual notation both for atomic orbitals and for the characters uses the same Greek letter: *chi*, χ. Ordinarily this causes no difficulty, but the beginner should be on guard against confusing them.

We will assume that the $1s$ electrons on the oxygen are sunk deep in the core and are not involved in the bonding. The most general set of LCAO MOs would have the form:

$$\phi_1 = c_{11}\chi_1 + c_{12}\chi_2 + c_{13}\chi_3 + c_{14}\chi_4 + c_{15}\chi_5 + c_{16}\chi_6;$$
$$\phi_2 = c_{21}\chi_1 + c_{22}\chi_2 + \cdots$$
$$\phi_3 = \quad\vdots$$
$$\phi_4 = \quad\vdots \qquad\qquad\qquad\qquad\qquad\qquad\qquad\vdots \qquad (23.39)$$
$$\phi_5 = \quad\vdots$$
$$\phi_6 = c_{61}\chi_1 + \cdots \qquad\qquad\qquad\qquad \cdots + c_{66}\chi_6;$$

in which the c_{ij} are constants. However, if the $\phi_1, \phi_2, \ldots \phi_6$ are to have the symmetry properties required, not all of the atomic orbitals can contribute to every ϕ. Thus many of the c_{ij} are zero.

We begin by examining the symmetry behavior of each χ_i. Suppose we look at χ_3, the $2s$ function on the oxygen atom. Under each symmetry operation in the group, χ_3 is unchanged. Therefore, from the character table, we conclude that χ_3 is an a_1 type function. The behavior of χ_4, χ_5, and χ_6 under the various symmetry operations can be established by examining what happens to the figures in Fig. 23.27 under these operations. Consider χ_4, the $(2p_z)_O$ function. Then

$$\mathbf{C}_2\chi_4 = \chi_4; \qquad \sigma_v\chi_4 = \chi_4 \qquad \sigma'_v\chi_4 = \chi_4.$$

So χ_4 is also totally symmetric and is an a_1 function. The functions $\chi_5 = (2p_x)_O$ and $\chi_6 = (2p_y)_O$ transform as:

$$\mathbf{C}_2\chi_5 = -\chi_5, \qquad \sigma_v\chi_5 = \chi_5, \qquad \sigma'_v\chi_5 = -\chi_5;$$
$$\mathbf{C}_2\chi_6 = -\chi_6, \qquad \sigma_v\chi_6 = -\chi_6, \qquad \sigma'_v\chi_6 = \chi_6.$$

Comparing these results with the character table we find that χ_5 belongs to b_1, whereas χ_6 belongs to b_2. Looking at χ_1 and χ_2 we find that

$$\mathbf{C}_2\chi_1 = \chi_2, \qquad \sigma_v\chi_1 = \chi_2, \qquad \sigma'_v\chi_1 = \chi_1;$$
$$\mathbf{C}_2\chi_2 = \chi_1, \qquad \sigma_v\chi_2 = \chi_1, \qquad \sigma'_v\chi_2 = \chi_2.$$

Two of the symmetry operations interchange χ_1 and χ_2. When this happens we can construct linear combinations of the two functions that have the required symmetry properties. We form two new functions:

$$\chi'_1 = \chi_1 + \chi_2; \qquad \chi'_2 = \chi_1 - \chi_2.$$

Then

$$\mathbf{C}_2\chi'_1 = \mathbf{C}_2\chi_1 + \mathbf{C}_2\chi_2 = \chi_2 + \chi_1 = \chi'_1;$$
$$\sigma_v\chi'_1 = \sigma_v\chi_1 + \sigma_v\chi_2 = \chi_2 + \chi_1 = \chi'_1;$$
$$\sigma'_v\chi'_1 = \sigma'_v\chi_1 + \sigma'_v\chi_2 = \chi_1 + \chi_2 = \chi'_1.$$

From this we conclude that χ'_1 is an a_1 function.

Next we examine χ'_2

$$\mathbf{C}_2\chi'_2 = \mathbf{C}_2\chi_1 - \mathbf{C}_2\chi_2 = \chi_2 - \chi_1 = -(\chi_1 - \chi_2) = -\chi'_2;$$
$$\sigma_v\chi'_2 = \sigma_v\chi_1 - \sigma_v\chi_2 = \chi_2 - \chi_1 = -(\chi_1 - \chi_2) = -\chi'_2;$$
$$\sigma'_v\chi'_2 = \sigma'_v\chi_1 - \sigma'_v\chi_2 = \chi_1 - \chi_2 = \chi_1 - \chi_2 = \chi'_2$$

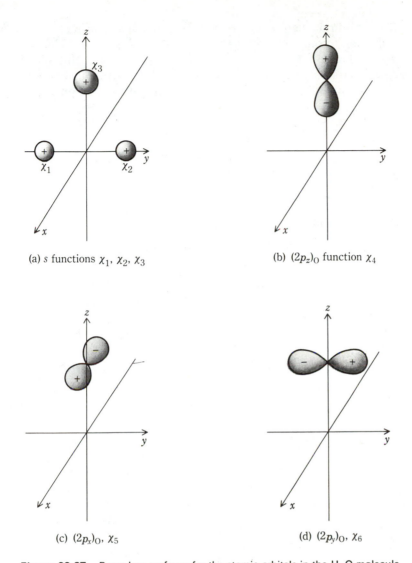

(a) s functions χ_1, χ_2, χ_3

(b) $(2p_z)_O$ function χ_4

(c) $(2p_x)_O$, χ_5

(d) $(2p_y)_O$, χ_6

Figure 23.27   Boundary surfaces for the atomic orbitals in the H_2O molecule.

These transformations show that χ_2' is a b_2 function. Summarizing, we have:

Symmetry species	Functions
a_1	$(\chi_1 + \chi_2)$, χ_3, χ_4
a_2	none
b_1	χ_5
b_2	$(\chi_1 - \chi_2)$, χ_6

Thus, from this limited group of atomic wave functions, we can form molecular wave functions belonging to only three symmetry species: a_1, b_1, and b_2. The most general a_1 functions will be linear combinations of $(\chi_1 + \chi_2)$, χ_3, and χ_4. No other function can enter these combinations without destroying the symmetry type. Since we begin with three independent functions, we end up with three independent a_1 functions:

$$a_1 \qquad \begin{aligned} \phi_1 &= c_{11}(\chi_1 + \chi_2) + c_{13}\chi_3 + c_{14}\chi_4; \\ \phi_2 &= c_{21}(\chi_1 + \chi_2) + c_{23}\chi_3 + c_{24}\chi_4; \\ \phi_3 &= c_{31}(\chi_1 + \chi_2) + c_{33}\chi_3 + c_{34}\chi_4. \end{aligned} \qquad (23.40)$$

Only one function has the b_1 symmetry; thus

$$b_1 \qquad \phi_4 = \chi_5.$$

There are two b_2 functions, $\chi_1 - \chi_2$ and χ_6.

$$b_2 \qquad \begin{aligned} \phi_5 &= c_{51}(\chi_1 - \chi_2) + c_{56}\chi_6; \\ \phi_6 &= c_{61}(\chi_1 - \chi_2) + c_{66}\chi_6. \end{aligned} \qquad (23.42)$$

A most important property of wave functions that belong to different symmetry species is that they are, *ipso facto*, orthogonal. Thus, by constructing these symmetry orbitals, ϕ_i, as they are called we automatically fulfill the condition of orthogonality, which is required of the wave functions of nondegenerate states. Thus ϕ_1, ϕ_2, and ϕ_3 are orthogonal to ϕ_4, ϕ_5, and ϕ_6, while ϕ_4 is orthogonal also to ϕ_5 and ϕ_6. By suitable adjustment of the constants c_{ij}, we can assure that ϕ_1, ϕ_2, and ϕ_3 are orthogonal to each other, and that ϕ_5 and ϕ_6 are orthogonal to each other.

The order of energies of the MOs in the water molecule is

$$1a_1 < 2a_1 < 1b_2 < 3a_1 < 1b_1 < 4a_1 < 2b_2 < 5a_1$$

$$\phi_1 \qquad \phi_5 \qquad \phi_2 \qquad \phi_4 \qquad \phi_3 \qquad \phi_6$$

The states of a given symmetry are labeled in serial order, $1a_1, 2a_1, 3a_1, \ldots$, beginning with the state of lowest energy. The $1a_1$ state is the a_1 orbital containing the $1s$ electrons on the oxygen atom. We omitted this state in our discussion above. Our three a_1 states, ϕ_1, ϕ_2, ϕ_3, are associated with the states $2a_1, 3a_1, 4a_1$; our ϕ_4 with $1b_1$; and our ϕ_5 and ϕ_6 with $1b_2$ and $2b_2$. The energy level diagram for the water molecule is shown in Chapter 25, Fig. 25.12. The electronic configuration of the water molecule is

$$\text{H}_2\text{O:} \qquad (1a_1)^2(2a_1)^2(1b_1)^2(3a_1)^2(1b_1)^2$$

or

$$\text{H}_2\text{O:} \qquad K(2a_1)^2(1b_2)^2(3a_1)^2(1b_1)^2,$$

in which the filled $(1a_1)^2$ (the $1s$ shell in oxygen) is replaced by K to indicate a filled atomic K shell. Since all the occupied orbitals are filled, the overall symmetry of the state is totally symmetric. The term symbol for the ground state of the water molecule is then $^1A_1$; the left-hand superscript indicates a singlet state. (Capital letters are used to describe the symmetry of the entire electron configuration.)

The actual determination of the order of energy levels, shown in Fig. 25.12, was done by a theoretical calculation.* Considering the complexity of the calculation and the

---

*   F. D. Ellison and H. Shull, *J. Chem. Phys.* **23**: 2348 (1955).

approximations that are necessary, the results are quite good. Comparison with experiment involves comparing the experimental and theoretical values of (a) the total binding energy of the system, (b) the ionization energies of electrons from various levels in the molecule, and (c) the frequencies of absorption and emission bands. Arriving at a complete energy level scheme for a molecule is an intricate task of fitting together many pieces of the puzzle. We will return to the subject of symmetry in the discussion of selection rules in spectroscopy.

QUESTIONS

23.1 How does the Born–Oppenheimer approximation allow the discussion of a diatomic molecule's energy as a function of only the internuclear separation (and not the coordinates of electrons)?

23.2 What is the variation theorem? What is its importance?

23.3 Chemical bonding forces saturate; that is, there are only two electrons to a bond. Discuss this for the H_2 example by examining the possibility of placing another electron in the orbital in Eq. (23.8). Pay attention to spin!

23.4 Sketch and explain the behavior of the overlap integral (Eq. 23.20) versus internuclear separation for identical nuclei when (a) $\psi_a = \psi_b = \psi_{1s}$ and (b) $\psi_a = \psi_{1s}$ and $\psi_b = \psi_{2p_z}$ (for $\theta = \pi/2$).

23.5 Discuss hybridization, formation of σ and π bonds, and the geometry of ethylene and acetylene.

23.6 Sketch and explain a plot of the energy of ethylene versus the angle between the planes of the CH_2 groups.

23.7 Predict the geometry of SF_4 and SF_6.

23.8 What is a molecular orbital wave function? How does it differ from a valence bond wave function?

23.9 Sketch the behavior of the bond order and the bond length for the homonuclear diatomic series Li_2–Ne_2, based on a molecular orbital description.

23.10 Describe the bonding in NO and CO in terms of molecular orbitals.

23.11 List the symmetry operations and the group for the molecules H_2O, NH_3, C_2H_4, and PCl_5.

PROBLEMS

23.1 Show that if ψ_n and ψ_k are orthogonal, then ψ_S and ψ_A, in Eqs. (23.5) and (23.6), are orthogonal.

23.2 Let α and β be the two spin wave functions corresponding to the two possible values of the electron spin quantum number; then $\alpha(1)$ indicates that electron 1 has spin α. The possible spin functions for two electrons are: $\sigma_1 = \alpha(1)\alpha(2); \sigma_2 = \alpha(1)\beta(2); \sigma_3 = \beta(1)\alpha(2); \sigma_4 = \beta(1)\beta(2)$. By making linear combinations where necessary, show that three functions are symmetric (triplet state) and one is antisymmetric (singlet state) under the interchange of the two electrons.

23.3 Which of the following overlap integrals are zero?

    a) an s function approaches a p_z function along the z-axis.
    b) an s function approaches a p_y function along the x-axis.
    c) a p_y function approaches a p_z function along the y-axis.
    d) two p_x functions approach each other along the y-axis.

23.4 a) Sketch the double-bond system in 1,3-butadiene.
    b) Compare the double bonds in 1,3-butadiene with those in 1,4-pentadiene.
    c) Show that any hydrocarbon containing a conjugated system of double bonds is planar over the region of conjugation.

23.5 Nitrogen forms two distinct types of compounds in which it is attached to three neighbors. In ammonia and the amines, the configuration is pyramidal, while the NO_3^- ion is planar. Sketch the hybridization possibilities for the two situations. (*Hint:* N^+ is isoelectronic with carbon.)

23.6 The "one-electron" bond is stabilized in the species H_2^+ by resonance between the structures

$$H^+ \quad \cdot H \qquad \text{and} \qquad H\cdot \quad H^+.$$

Suggest a reason why the "one-electron" bond is not observed between two *unlike* atoms A and B to yield $(A \cdot B)^+$.

23.7 Nickel ion, Ni^{2+}, forms two types of four-coordinate complex compounds. One type is tetrahedral; the other is square. Which of these types will have a magnetic moment due to unpaired electron spins?

23.8 Construct the multiplication table for the group C_{3v}.

23.9 Consider a hypothetical molecule, H_3, consisting of three hydrogens at the apices of an isosceles triangle; the symmetry group is C_{2v}, with $H^{(2)}$ and $H^{(3)}$ equivalent. Construct molecular orbitals of proper symmetry using $1s$ functions on the three hydrogen atoms.

23.10 Construct molecular orbitals of proper symmetry for the formaldehyde molecule, $H_2C{=}O$, symmetry group C_{2v}. Use $1s$ orbitals on the hydrogen atoms, and $2s$, $2p_x$, $2p_y$, $2p_z$ orbitals on the carbon and oxygen atoms. The orbitals, χ_1 to χ_{10}, are labeled in the order: $(1s)_{H(1)}$; $(1s)_{H(2)}$; $(2s)_C$; $(2p_x)_C$; $(2p_y)_C$; $(2p_z)_C$; $(2s)_O$; $(2p_x)_O$; $(2p_y)_O$; $(2p_z)_O$.

23.11 The ozone molecule is angular and therefore has the symmetry C_{2v}. Using the valence shell orbitals on each of the three oxygen atoms, construct molecular orbitals of the proper symmetry; let $O(2)$ and $O(3)$ be equivalent. The orbitals, χ_1 to χ_{12}, are labeled in the order: $(2s)_1$; $(2p_x)_1$; $(2p_y)_1$; $(2p_z)_1$; $(2s)_2$; $(2s)_3$; $(2p_x)_2$; $(2p_x)_3$; $(2p_y)_2$; $(2p_y)_3$; $(2p_z)_2$; $(2p_z)_3$.

23.12 Construct molecular orbitals of proper symmetry for the ethylene molecule, $H_2C{=}CH_2$, symmetry group D_{2h}. Use $1s$ orbitals on the hydrogen atoms; $2s$, $2p_x$, $2p_y$, $2p_z$ orbitals on the two carbon atoms. Label H atoms 1 and 2 with carbon atom 1; H atoms 3 and 4 with carbon atom 2. The character table is in Appendix VI. The operation $\mathbf{i}$ is inversion through the center. The orbitals, χ_1 to χ_{12} are labeled in the order: $(1s)_{H(1)}$; $(1s)_{H(2)}$; $(1s)_{H(3)}$; $(1s)_{H(4)}$; $(2s)_{C(1)}$; $(2s)_{C(2)}$; $(2p_x)_{C(1)}$; $(2p_x)_{C(2)}$; $(2p_y)_{C(1)}$; $(2p_y)_{C(2)}$; $(2p_z)_{C(1)}$; $(2p_z)_{C(2)}$.

23.13 Construct molecular orbitals of proper symmetry for the diborane molecule, $H_2BH_2BH_2$, which belongs to the symmetry group D_{2h}. Diborane has the same structure as ethylene except that the two additional hydrogens (bridge hydrogens) lie on an axis that is perpendicular to the plane of the rest of the molecule and bisects the B—B axis. The orbitals are labeled in the order: χ_1 to χ_4 are $1s$ orbitals on the four equivalent hydrogen atoms; χ_5 and χ_6 are $1s$ orbitals on the bridge hydrogens; χ_7 to χ_{14} are $(2s)_{B(1)}$; $(2s)_{B(2)}$; $(2p_x)_{B(1)}$; $(2p_x)_{B(2)}$; $(2p_y)_{B(1)}$; $(2p_y)_{B(2)}$; $(2p_z)_{B(1)}$; $(2p_z)_{B(2)}$.

23.14 Hydrogen peroxide, H_2O_2, in the *trans*-form with all the atoms in a plane has the symmetry C_{2h}. (Note: σ_h is reflection in the horizontal plane, the plane of the molecule.) Construct molecular orbitals of the proper symmetry. The orbitals, χ_1 to χ_{10}, are labeled in the order: $(1s)_{H(1)}$; $(1s)_{H(2)}$; $(2s)_{O(1)}$; $(2s)_{O(2)}$; and so on as in Problem 23.13.

23.15 *Trans*-difluoroethylene has the symmetry C_{2h}. Construct molecular orbitals of proper symmetry. The orbitals, χ_1 to χ_{18}, are labeled in the order: $(1s)_{H(1)}$; $(1s)_{H(2)}$; $(2s)_{C(1)}$; $(2s)_{C(2)}$; $(2p_x)_{C(1)}$; $\dots$; $(2p_x)_{C(2)}$; $(2s)_{F(1)}$; $(2s)_{F(2)}$; $(2p_x)_{F(1)}$; $\dots$; $(2p_z)_{F(2)}$.

23.16 If we use the $2s$, $2p_x$, $2p_y$, and $2p_z$ orbitals on the nitrogen and the three fluorine atoms of the NF_3 molecule as a basis for a representation of the group C_{3v}, the character of the representation is

E	$2C_3$	$3\sigma_v$
16	1	4

How is this representation composed of the irreducible representations?

23.17 If we use the valence shell orbitals on the three oxygen atoms in ozone as a basis for a representation of the group C_{2v}, the character of the representation is

E	C_2	σ_v	σ_v'
12	0	2	6

How is this representation composed of the irreducible representations?

24

Atomic Spectroscopy

24.1 SPECTRAL REGIONS

A light beam is an electromagnetic wave that has oscillating electrical and magnetic fields associated with it. Figure 24.1 shows the variation of the field vectors for a plane wave propagating in the z direction.

The properties used to characterize the light beam are: frequency, wavelength, wave number, and energy.

The *frequency* is the fundamental property of the light beam; it is the number of oscillations of the field vectors per second. The frequency is independent of the medium through which the beam is passing. The usual symbol for frequency is the Greek letter

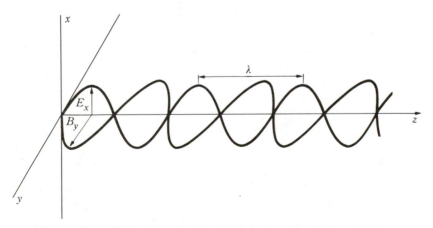

Figure 24.1 Field vectors for plane wave propagating in the z direction.

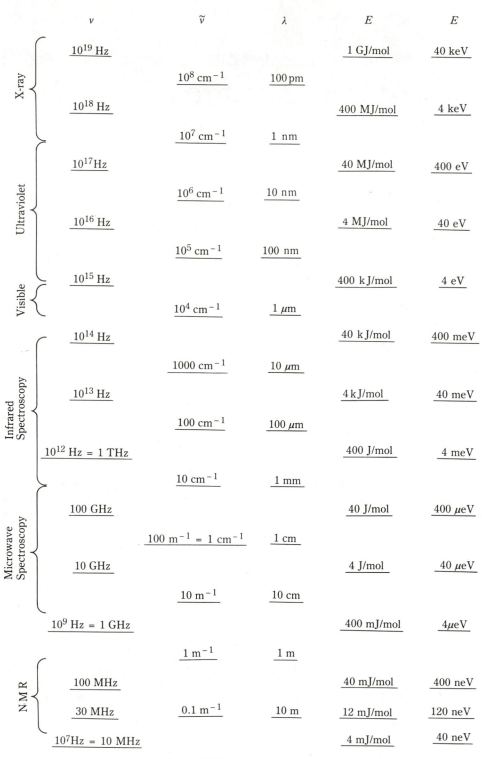

Figure 24.2 Spectral regions.

nu, v. The unit is the hertz; $1 \text{ Hz} = 1 \text{ s}^{-1}$. The reciprocal of the frequency is the *period* of oscillation, $T = 1/v$.

The *wavelength* is the distance traveled by the light wave in the time required for the electric or magnetic field vectors to complete one oscillation. The wavelength depends on the medium through which the beam is passing, and is related to the frequency by the equation

$$\lambda v = c \tag{24.1}$$

where λ (Greek: *lambda*) is the wavelength and c is the velocity of the light in the medium. In vacuum, $c = 2.99792458 \times 10^8 \text{ m/s} \approx 3 \times 10^8 \text{ m/s}$.

The *wave number* is the reciprocal of the wavelength; the symbol used is $\tilde{v}$, (*nu* tilde). Thus

$$\tilde{v} \equiv \frac{1}{\lambda} \tag{24.2}$$

The wave number is the number of oscillations of either field vector in unit distance. The SI base unit is m^{-1}, but the literature values are almost all in cm^{-1}. In view of Eq. (24.1), we have also

$$\tilde{v} = \frac{v}{c}. \tag{24.3}$$

The *energy* in a single quantum of light is given by the Planck relation

$$E = hv \qquad \text{or} \qquad E = hc\tilde{v}. \tag{24.4}$$

Thus the energy is proportional to either the frequency or the wave number. The energy of one mole of quanta is

$$E_m = N_A hv. \tag{24.5}$$

This quantity was formerly called an *einstein*. For visible light where v ranges from 4.3×10^{14} to $7.5 \times 10^{14} \text{ s}^{-1}$, the energy ranges from 170 to 300 kJ/mol, which is in the range of the energies of ordinary chemical reactions. This is the reason that these frequencies are visible.

The frequencies of interest in atomic and molecular problems range from the short x-ray region where $v \approx 3 \times 10^{19} \text{ Hz}$ and $E = 1.2 \times 10^7 \text{ kJ/mol}$ to the long radio frequency (NMR) region where $v \approx 3 \times 10^7 \text{ Hz} = 30 \text{ MHz}$ and $E \approx 0.012 \text{ J/mol}$. These regions are shown on a logarithmic scale in Fig. 24.2.

24.2 BASIC SPECTROSCOPIC EXPERIMENTS

In general, the light emitted from an excited substance consists of radiation of many frequencies. A spectroscope is designed to resolve this light into its component frequencies. The simplest kind of optical spectroscopic experiment is illustrated in Fig. 24.3.

Light from a slit is collimated by a lens and passed into a prism. Since the prism refracts the light of different frequencies by different amounts, a beam of white light, for example, is spread out into its component frequencies (colors). These different frequencies appear at different positions on the receptor. Typically a detector such as a photoelectric tube can be moved across the receptor area to measure the intensity of light of each frequency. The receptor is marked with a scale of wavelengths or frequency.

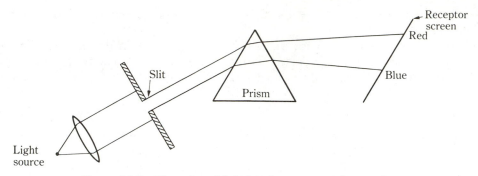

Figure 24.3   Dispersion of light into its component frequencies.

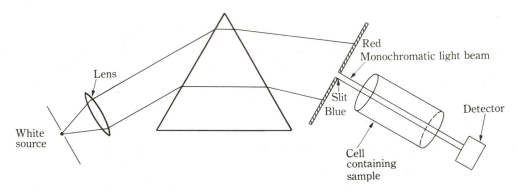

Figure 24.4   Schematic diagram for measurement of light absorption.

The basic elements of any emission spectroscope are (1) a source that contains the substance to be studied and is capable of energizing that substance so that it can emit its characteristic radiation, (2) a dispersing device to resolve the emitted radiation into its component frequencies, and (3) a detector that can measure the intensity of the radiation at the various frequencies. The choice of devices for each of these elements depends on the region of the spectrum under investigation.

For absorption spectroscopy, we could choose a source of "white" radiation, insert a sample of the substance to be studied in the light beam, pass the transmitted beam through a dispersing device, and measure the intensity of radiation as a function of frequency. Alternatively, we might disperse the white radiation into its component frequencies and, by means of a slit, select light having a narrow range of frequencies ("monochromatic" light), then pass this beam through a sample of the substance under study. The detector measures the intensity of light transmitted at the selected frequency. By suitably changing the geometry, we can focus light of a different frequency on the slit, and measure the transmitted intensity as a function of frequency. The device is illustrated in Fig. 24.4.

24.3 ORIGINS OF SPECTRA

The frequency emitted or absorbed when an atomic or molecular system undergoes a change in state is related to the absolute value of the difference in energy between the two states by

$$hv = |\Delta E|. \tag{24.6}$$

Table 24.1

Radiation	Process	Information gained
X-rays	Transition of inner electrons of an atom	Details of electronic structure
Ultraviolet and visible	Transitions of the outer (or valence) electrons in the atom or molecule	Details of electronic structure and bond energies in molecules
Infrared	Changes in vibrational–rotational state of the molecule	Internuclear distances, force constants
Far infrared and microwave	Changes in rotational state only	Internuclear distances
Radio frequency	Change of spin orientation of nucleus in a magnetic field	Magnetic environment of the spinning nucleus from which structure is inferred

The energy of the light quantum reveals the energy of the molecular transition, thus providing evidence for the type of transformation occurring. The various regions of the spectrum correspond to different kinds of transitions in the atom or molecule. The processes producing the radiation in the various regions of the spectrum are summarized in Table 24.1.

To interpret a spectrum we have to keep three fundamental ideas in mind.

1. Any atomic or molecular system possesses energy only in certain special amounts, which are called the energy levels of the system. When a system makes a transition between these energy levels, light is either emitted or absorbed. The frequency is given by Eq. (24.6), which is the fundamental equation of spectroscopy. If the energy of the system decreases in the transition, a light quantum of that energy is emitted. If a light quantum is absorbed, the energy of the atomic system increases by an equal amount.

2. There are restrictions, called *selection rules*, on the transitions that can occur between the energy levels. The selection rules are a consequence of the symmetry of the wave functions in the two states. As an example, consider the set of energy levels for the hydrogen atom shown in Fig. 24.5. The energies of these levels, using Eq. (22.14), are given by $E_m = -E_h/2n^2$, in which n is the principal quantum number and the hartree energy, $E_h = e^2/4\pi\epsilon_0 a_0$. In the cases presented farther on, we assume that we have a system composed of a very large number of hydrogen atoms only. In Fig. 24.5, the energy levels are separated into groups corresponding to the value of l, the azimuthal quantum number. These groups are labeled $S, P, D, F, \ldots$, corresponding to $l = 0, 1, 2, 3, \ldots$, and are arranged along the horizontal axis. The left-hand superscript records that the electron spin quantum number may have two values, either $+\frac{1}{2}$ or $-\frac{1}{2}$. The advantage of separating these levels into groups is that the selection rule requires that in a transition the value of l must change by unity; $\Delta l = \pm 1$. Thus transitions are allowed either way between S and P, between P and D, and between D and F states (and so on). Transitions are *not allowed* between S and D, or between P and F states, or between two different S states or two different P states.

3. If a spectral line is to be observed, there must be a substantial population of the systems in the initial energy level.

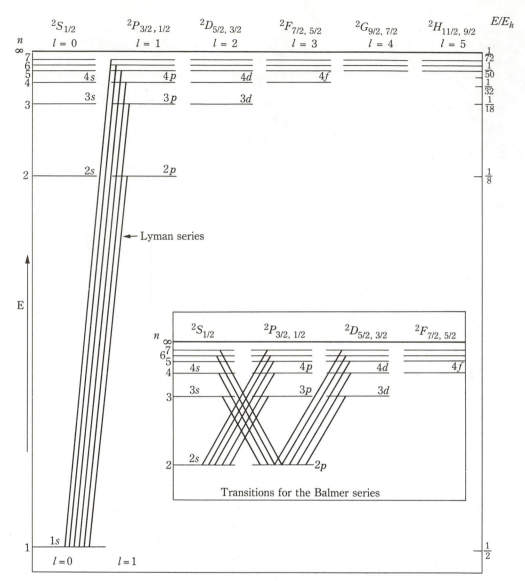

Figure 24.5 Energy-level diagram for the hydrogen atom showing transitions for the Lyman series. Inset: Balmer series transitions.

Case 1. The absorption spectrum. Assume that all the hydrogen atoms are in the ground state, the 1*S* state. Clearly, the system cannot emit light since that would require some atoms to drop into a lower energy state, which is not possible. However, the system can absorb light corresponding to any transition between the 1*S* state and any of the *P* states. These transitions are indicated by the lines between the 1*S* and the various *P* levels in Fig. 24.5. Since the energy differences are rather large, these lines lie in the ultraviolet. No other frequencies will be absorbed since none of the higher states has a significant population.

Case 2. The emission spectrum. To obtain an emission spectrum, the hydrogen atoms are energized by placing them in an electric arc, thus raising them to a very high temperature. Now there will be a significant number of hydrogen atoms in all of the higher energy levels. These atoms in the higher levels can emit light and drop to lower energy states. Those in the P states can drop into the $1S$ state and emit the Lyman series of lines in the ultraviolet. These frequencies are the same as those that were absorbed in the circumstances of Case 1. However, other series can now appear. The Balmer series of lines can arise in three ways: (1) by transitions from the higher s levels ($3s$, $4s$, . . .) to the $2p$ level; (2) by transitions from the higher p levels to the $2s$ level; and (3) by transition from the d levels to the $2p$ level. These transitions are shown in the inset in Fig. 24.5. Since the energy differences here are much smaller than for the Lyman series, the Balmer series lies in the visible and near-ultraviolet. The next series, the Paschen series, lies in the near-infrared. The Paschen series has any of the levels of principal quantum number $n = 3$ as the final level; hence there are five ways to obtain the Paschen series: higher s to $3p$, higher p to $3s$, higher d to $3p$, higher p to $3d$, and higher f to $3d$.

24.4 LIGHT ABSORPTION; BEER'S LAW

Consider a beam of monochromatic light passing through a slice of an absorber of thickness, dx. Let I be the intensity of the incident beam and $I + dI$ be the intensity of the emergent beam (Fig. 24.6). The intensity of the beam is the number of quanta of light passing through unit area of a plane perpendicular to the direction of the beam in unit time. Let this number be I. Then $-dI$ is the number of quanta absorbed in the distance, dx. The probability of absorption in the distance, dx, is $-dI/I$; if the slice is thin, the probability of absorption is proportional to the thickness of the slice and the number of absorbing molecules in the slice, that is, to the concentration of the absorbing species. We have

$$-\frac{dI}{I} = k\tilde{c} \, dx, \tag{24.7}$$

where k is the proportionality constant, $\tilde{c}$ is the concentration, (mol/m^3), and dx is the thickness of the slice.

Equation (24.7) states that the *relative* decrease in intensity of the beam is proportional to the number of absorbing molecules in the slab of material. If there are several kinds of

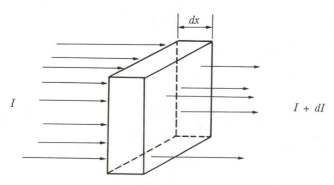

Figure 24.6 Attenuation of a light beam passing through an absorber.

molecules present, each with a different ability to absorb light of the frequency in question, then

$$-\frac{dI}{I} = (k_1 \tilde{c}_1 + k_2 \tilde{c}_2 + \cdots)\, dx. \tag{24.8}$$

The constants, $k_1, k_2, \ldots$, are characteristic of the substances in question. For any substance the value of k depends on the wavelength. If a substance is transparent at a particular wavelength, all the light goes through and $k = 0$. If at a particular wavelength all the substances are transparent except one, then Eq. (24.8) reduces to Eq. (24.7). Integration of Eq. (24.7) yields

$$\int_{I_0}^{I} \frac{dI}{I} = -k\tilde{c}\int_0^l dx,$$

where I_0 is the intensity of the incident beam and I is the intensity of the emergent beam after passing through the total cell length, l. Integrating we obtain,

$$\ln\left(\frac{I}{I_0}\right) = -k\tilde{c}l \qquad \text{or} \qquad I = I_0 e^{-k\tilde{c}l}. \tag{24.9}$$

It is customary in spectrophotometry to use common logarithms rather than natural logarithms; thus in Eq. (24.9) we replace the natural base, e, by $10^{0.43429\cdots}$ and obtain $I = I_0 10^{-0.4343k\tilde{c}l}$. We define $\tilde{\epsilon} \equiv 0.4343k$; then

$$I = I_0 10^{-\tilde{\epsilon}\tilde{c}l}. \tag{24.10}$$

The constant, $\tilde{\epsilon}$, is the *molar absorption coefficient* of the substance; $\tilde{\epsilon}$ is also called the extinction coefficient. The *transmittance*, T, is defined by

$$T \equiv \frac{I}{I_0} \tag{24.11}$$

and the absorbance, A, is defined by

$$A \equiv -\log_{10} T \qquad \text{or} \qquad T = 10^{-A} \tag{24.12}$$

If the absorbance increases by unity, the transmittance drops by a factor of ten. Equation (24.10) is an expression of the Beer–Lambert law, often called simply Beer's law. Beer's law is the basic equation for the various colorimetric and spectrophotometric methods of analysis. If Beer's law holds, then the absorbance is given by

$$A = \tilde{\epsilon}\tilde{c}l. \tag{24.13}$$

Since $\tilde{c}$ is in mol/m³, l is in metres, and A must be a pure number, we have m²/mol as the SI unit for $\tilde{\epsilon}$. The molar absorption coefficient, ϵ, has traditionally been defined by $A = \epsilon c b$, where c is in mol/L and b is the cell length in cm. This gives ϵ the pathological (but handy) unit, L mol⁻¹ cm⁻¹. Consequently, $\epsilon = 10\tilde{\epsilon}$, where ϵ and $\tilde{\epsilon}$ are the molar absorption coefficients expressed in classical and SI units, respectively.

If the composition of the system is not variable, we have $I/I_0 = e^{-\alpha x}$, where α is the absorption coefficient and x is the path length. The absorbance of a solution is given by

$$A = \tilde{\epsilon}_1 \tilde{c}_1 l + \tilde{\epsilon}_2 \tilde{c}_2 l + \cdots, \tag{24.14}$$

in which $\tilde{\epsilon}_1, \tilde{\epsilon}_2, \ldots$, are the molar absorption coefficients and $\tilde{c}_1, \tilde{c}_2, \ldots$, are the concentrations of the species $1, 2, \ldots$. A typical example of the use of this equation is in the determination of the concentration of several species in a solution. The molar absorption coefficients of each of the substances must be known as a function of wavelength. If the concentrations of two species are to be determined, the absorbance of the solution is measured at two different wavelengths.

If a chemical equilibrium is established between two different chemical species, then Beer's law ordinarily will not be obeyed, since the concentration of the absorbing species will not usually be directly proportional to the apparent total concentration. For example, consider an acid–base indicator, HX, and assume that HX is the absorbing species. We have the equilibrium,

$$\text{HX} \;\rightleftharpoons\; \text{H}^+ + \text{X}^- \quad \text{and} \quad K = \frac{c_{\text{H}^+} c_{\text{X}^-}}{c_{\text{HX}}}.$$

The total concentration of HX and X^- is $c = c_{\text{HX}} + c_{\text{X}^-}$. In the simplest case $c_{\text{H}^+} = c_{\text{X}^-} = c - c_{\text{HX}}$; then $K = (c - c_{\text{HX}})^2/c_{\text{HX}}$, which shows that c_{HX} is not simply proportional to c. Therefore Beer's law will not be obeyed. The equilibrium constant can be determined by measuring the absorbance as a function of the concentration. This assumes that only one species absorbs significantly at the wavelength in question.

24.5 THEORY OF ATOMIC SPECTRA

The simplest spectra are those obtained from excited atoms. Since all atoms except the hydrogen atom have more than one electron, we need a quantum-mechanical description for multielectron atoms.

The Schrödinger equation for the hydrogen atom can be solved exactly as we did in Chapter 22. If we attempt to solve the Schrödinger equation for the helium atom, we must deal with the mechanics of three bodies (the nucleus and two electrons), which is not solvable in closed form either in classical mechanics or in quantum mechanics. Therefore we are forced to use approximate methods.

Suppose that the nucleus has a charge $+Ze$ and is separated from electron 1 by a distance r_1 and from electron 2 by r_2; the distance between electrons 1 and 2 is r_{12} (Fig. 24.7). Since the nucleus is very massive compared to the electrons, we will regard the nucleus as being fixed at the center of mass of the system. The Hamiltonian for the system can then be written as though the system consisted of only the two electrons that are moving in the field of the nucleus and each other. If we write the energy as multiples of E_{h}, the hartree, and distances as multiples of a_0, the Hamiltonian becomes

$$\mathbf{H} = -\tfrac{1}{2}\nabla_1^2 - \tfrac{1}{2}\nabla_2^2 - \frac{Z}{r_1} - \frac{Z}{r_2} + \frac{1}{r_{12}}, \tag{24.15}$$

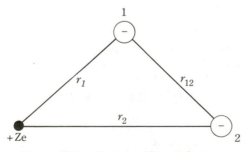

Figure 24.7   Nucleus with two electrons.

in which $-\frac{1}{2}\nabla_1^2$ is the kinetic energy operator for electron 1; it involves only the co-ordinates of electron 1, (r_1, θ_1, ϕ_1); similarly, $-\frac{1}{2}\nabla_2^2$ involves only the coordinates of electron 2, (r_2, θ_2, ϕ_2). If we define

$$\mathbf{H}_1 = -\tfrac{1}{2}\nabla_1^2 - \frac{Z}{r_1} \qquad \text{and} \qquad \mathbf{H}_2 = -\tfrac{1}{2}\nabla_2^2 - \frac{Z}{r_2}, \qquad (24.16)$$

then the Hamiltonian operator becomes

$$\mathbf{H} = \mathbf{H}_1 + \mathbf{H}_2 + \frac{1}{r_{12}}. \qquad (24.17)$$

The term, $1/r_{12}$, is the potential energy of repulsion between the two electrons. Since the electrons repel one another, this effect will tend to keep them apart; thus r_{12} will be as large as possible under the constraints that r_1 and r_2 must be small, since the electrons are both attracted to the nucleus.

We can simplify our problem by dropping out the repulsion term entirely. Similarly, in atoms with many electrons, we will ignore the electron repulsion terms in our first treatment. Then for any multielectron atom we have

$$\mathbf{H} = \mathbf{H}_1 + \mathbf{H}_2 + \mathbf{H}_3 + \cdots. \qquad (24.18)$$

Since the $\mathbf{H}_1$ is a set of terms that depends only on the coordinates of electron 1, $\mathbf{H}_2$ depends only on the coordinates of electron 2, and so on, then by the theorem in Section 21.8 we can write the wave function as a product of "one-electron" wave functions.

$$\psi = \phi_a(1)\phi_b(2)\phi_c(3)\ldots, \qquad (24.19)$$

where the (1) is an abbreviation for the coordinates of electron 1; that is,

$$\phi_a(1) = \phi_a(x_1, y_1, z_1) \qquad \text{and} \qquad \phi_a(2) = \phi_a(x_2, y_2, z_2),$$

and so on. The $\phi_a, \phi_b, \ldots,$ are the one-electron wave functions.

To fulfill the requirement that the wave function be antisymmetric under the interchange of any two electrons, a suitable linear combination of product functions must be used. The linear combination that assures antisymmetry under the exchange of any two electrons is the determinant:

$$\Psi = \frac{1}{\sqrt{N!}}
\begin{vmatrix}
\phi_a(1) & \phi_a(2) & \phi_a(3) & \cdots & \phi_a(N) \\
\phi_b(1) & \phi_b(2) & \phi_b(3) & \cdots & \phi_b(N) \\
\phi_c(1) & \phi_c(2) & \phi_c(3) & \cdots & \phi_c(N) \\
\vdots & \vdots & \vdots & & \vdots \\
\phi_N(1) & \phi_N(2) & \phi_N(3) & \cdots & \phi_N(N)
\end{vmatrix} \qquad (24.20)$$

Interchanging any two electrons interchanges the corresponding two columns of the determinant, which changes the sign of the determinant. The Pauli principle is therefore satisfied. The determinant is usually abbreviated by simply enclosing the product of the functions on the main diagonal between vertical lines; thus

$$\Psi = \frac{1}{\sqrt{N!}} |\phi_a(1)\phi_b(2)\phi_c(3)\cdots\phi_N(N)| \qquad (24.21)$$

The total energy corresponding to this wave function is the sum of the individual one-electron energies.

$$E = E_a + E_b + E_c + \cdots. \tag{24.22}$$

Each of the one-electron functions is a solution of a hydrogen-like Schrödinger equation:

$$\mathbf{H}_1 \phi_a(1) = E_a \phi_a(1). \tag{24.23}$$

This equation differs from that of the hydrogen atom only in that the factor Z, instead of unity, appears for the nuclear charge. Consequently, we have a set of quantum numbers n, l, m, m_s for each electron in the atom. The presence of Z modifies the wave function but not the quantum numbers. For example, the $1s$ wave function becomes

$$\phi_{1s} = \left(\frac{Z^3}{\pi a_0^3}\right)^{1/2} e^{-Zr/a_0} \tag{23.24}$$

We will not have use for the detailed functions, so no other example will be given. If desired, we can obtain the functions from those in Table 22.2 by replacing a_0 by a_0/Z.

In this approximation, each electron has its own set of four quantum numbers, $[n_i, l_i, m_i, (m_s)_i]$. This description is the basis for the model of the electronic structures of the atoms that we used to interpret the periodic table in Chapter 22.

In the hydrogen atom the wave functions are eigenfunctions of the angular momentum operators, M^2, M_z, and $(M_s)_z$. In the approximation we are using here the wave functions of the multielectron atom are also eigenfunctions of the angular momentum operators. We now examine the eigenvalues of the angular momentum operators in these multielectron systems.

24.6 QUANTUM NUMBERS IN MULTIELECTRON ATOMS

Corresponding to the quantum numbers, l, m, s, and m_s, for the single electron in the hydrogen atom, in multielectron atoms we have a set of quantum numbers, L, M_L, S, M_S.

The quantum number, L, describes the square of the total orbital angular momentum of all the electrons in the usual way,

$$M^2 = L(L + 1)\hbar^2 \tag{24.25}$$

The energy states of the atom, the *terms*, are described using the letter code for the different values of L.

L	0	1	2	3	4	$\cdots$
Letter code	S	P	D	F	G	$\cdots$

To obtain the allowed values of L we consider only the values of l_i for electrons outside any closed subshells. From this set of l_i we select the largest single value and from it we subtract the sum of all the remaining l_i. This difference is the *least* allowed value of L. If this difference is negative, the lowest value of L is zero. The largest value of L is the sum of all the l_i in this set. The integral values between the largest and least values are all allowed. Thus

$$L = L_{max}, L_{max} - 1, L_{max} - 2, \ldots, L_{least}$$

■ **EXAMPLE 24.1**   Suppose that we have an electron configuration, p^2f. Then $l_1 = 1$, $l_2 = 1$, $l_3 = 3$. The least value of L is $L_{least} = 3 - (1 + 1) = 1$. The largest value of L is $L_{max} = 3 + 1 + 1 = 5$. Thus the allowed values of L are $L = 5, 4, 3, 2, 1$.

The possible values of the z component of the orbital angular momentum are given by M_L,

$$M_z = M_L \hbar, \qquad M_L = 0, \pm 1, \pm 2, \ldots, \pm L. \tag{24.26}$$

The value of M_L is obtained by summing the values of m, the magnetic quantum number, for all the electrons in the atom:

$$M_L = \sum_i m_i. \tag{24.27}$$

The quantum number, S, describes the square of the total spin angular momentum through the relation

$$M_{spin}^2 = S(S + 1)\hbar^2. \tag{24.28}$$

The possible values of the z component of the spin angular momentum are described by the quantum number M_S in the equation

$$M_{spin, z} = M_S \hbar. \tag{24.29}$$

The value of M_S is obtained by summing the values of m_s, the spin quantum number, for all the electrons in the atom:

$$M_S = \sum_i (m_s)_i. \tag{24.30}$$

The *multiplicity* of the state is $2S + 1$, the number of possible values of the z component of the spin angular momentum.

The orbital angular momentum and the spin angular momentum couple to yield a total angular momentum characterized by a quantum number J.

$$M_{total}^2 = J(J + 1)\hbar^2 \tag{24.31}$$

The allowed values of J are:

$$J = L + S, L + S - 1, L + S - 2, \ldots, L - S. \tag{24.32}$$

This particular mode of combining the quantum numbers L and S is called Russell–Saunders coupling, the most common type of coupling. The orbital angular momenta of the electrons couple strongly as do the spin angular momenta of the several electrons. The total orbital angular momentum then couples more weakly with the total spin angular momentum to yield a resultant vector characterized by J. The values of J are either integral or half-integral depending on whether the number of electrons is even or odd.

The quantum number M_J characterizes the z component of the total angular momentum (orbital plus spin) through the relation

$$M_{total, z} = M_J \hbar. \tag{24.33}$$

There are $2J + 1$ allowed values of M_J:

$$M_J = 0, \pm 1, \pm 2, \ldots, \pm J, \qquad \text{(for an even number of electrons)};$$

$$M_J = \pm \tfrac{1}{2}, \pm \tfrac{3}{2}, \pm \tfrac{5}{2}, \ldots, \pm J, \qquad \text{(for an odd number of electrons)}.$$

24.7 ATOMIC SPECTROSCOPY; TERM SYMBOLS

In the hydrogen atom, the energy depends only on n, the principal quantum number, and not on the values of l, or s. Any electron with orbital or spin angular momentum possesses a magnetic moment. In atoms having more than one electron, the magnetic moments of the electrons interact (or "couple") with the result that the energy levels in the atom depend on the total orbital angular momentum and the total spin angular momentum (that is, the energy levels depend on L and S). Hund's rule is that among the states given by equivalent electrons (electrons with the same values of n and l) the state of maximum multiplicity has the lowest energy. Among states having the same multiplicity, the state with the greatest L has the lowest energy.

In atomic spectroscopy, the energy levels in the atom are called *terms* or *spectral terms*. A term is described by a *term symbol* such as:

$$^1S_0, \qquad \text{read "singlet ess zero"};$$

$$^2P_{3/2}, \qquad \text{read "doublet pee three-halves"};$$

$$^3D_2, \qquad \text{read "triplet dee two"}.$$

The letter in the term symbol is the letter code describing the value of L. The left superscript is the multiplicity of the state, $2S + 1$, and is called singlet, doublet, triplet, quartet, and so on. The right subscript is the value of J. Sometimes the principal quantum number of the state is written before the term symbol.

■ **EXAMPLE 24.2**   Consider the hydrogen atom. Since there is only one electron, $L = l$, always. Thus all s configurations yield an S term, all p configurations yield P terms, and so on. For the spin angular momentum, $s = \tfrac{1}{2}$; hence $S = \tfrac{1}{2}$ and the multiplicity, $2S + 1 = 2(\tfrac{1}{2}) + 1 = 2$. The terms are all doublets. Then $J = L + S = l + \tfrac{1}{2}$. When $l = 0$, the only allowed value of J is $J = \tfrac{1}{2}$; there is only one state, which is therefore not a true doublet. In all the other cases we have two states with different values of J:

l	0	1		2		3	
J	$\tfrac{1}{2}$	$\tfrac{3}{2}$,	$\tfrac{1}{2}$	$\tfrac{5}{2}$,	$\tfrac{3}{2}$	$\tfrac{7}{2}$,	$\tfrac{5}{2}$
Term symbol	$^2S_{1/2}$	$^2P_{3/2}$,	$^2P_{1/2}$	$^2D_{5/2}$,	$^2D_{3/2}$	$^2F_{7/2}$,	$^2F_{5/2}$

These term symbols were used to label the energy levels in Figure 24.5.

24.8 ATOMS WITH CLOSED SHELLS

The simplest method for determining the possible terms corresponding to a given electron configuration is to calculate M_L and M_S. From these values we can infer L and S.

The values of M_L and M_S are determined by the relations in Eqs. (24.27) and (24.30). The first consequence of these rules is that a closed shell of electrons (a filled subshell) contributes neither orbital angular momentum nor spin angular momentum to an atom. In a closed shell the electrons all have paired spins. Thus, in the configurations s^2, p^6, d^{10}, f^{14}, ..., the values, $m_s = +\frac{1}{2}$ and $m_s = -\frac{1}{2}$ occur equally often. For any closed shell, the resultant $M_S = \sum_i (m_s)_i = 0$. Since zero is the only possible value of M_S, it follows that the quantum number, S, must also be zero. (If S were not zero, the z component of the total spin angular momentum would have to have some nonzero value.) Since $S = 0$, the multiplicity, $2S + 1 = 1$; so the closed shell configurations are all singlets.

Similarly, in a filled subshell such as p^6 the electrons have the quantum numbers:

Electron	1	2	3	4	5	6
m	-1	-1	0	0	$+1$	$+1$
m_s	$+\frac{1}{2}$	$-\frac{1}{2}$	$+\frac{1}{2}$	$-\frac{1}{2}$	$+\frac{1}{2}$	$-\frac{1}{2}$

Clearly, the values $m = +1$ and $m = -1$ occur equally often, so that $M_L = 0$ for any filled subshell.

$$M_L = \sum_i m_i = (-1) + (-1) + 0 + 0 + 1 + 1 = 0.$$

Again, it follows that $L = 0$ in the filled subshell since only if $L = 0$ will zero be the *only possible value* of the z component of the orbital angular momentum. It follows that atoms having electrons only in filled subshells are all in ^1S states, $(L = 0, S = 0)$.

We conclude that to describe the angular momentum in an atom, we need to consider only those electrons outside of the filled subshells. For the same reasons, complementary configurations of equivalent electrons have the same terms. (Equivalent electrons are electrons having the same values of n and l.) Examples of complementary electron configurations are p^x and p^{6-x}; d^x and d^{10-x}; f^x and f^{14-x}; ... Consider the complementary configurations p^2 and p^4. Since p^6 is a *filled subshell*, we have

$$M_L = \sum_{i=1}^{6} m_i = 0 \quad \text{or} \quad \sum_{i=1}^{4} m_i + \sum_{i=5}^{6} m_i = 0.$$

Then

$$\sum_{i=1}^{4} m_i = - \sum_{i=5}^{6} m_i.$$

Thus the values for M_L for the p^4 configuration are the same as for p^2, except that they seem to differ in sign. This means only that the set of values is arranged in different order. Thus M_L for p^4 is equal to M_L for p^2. It follows that L is the same for both, as are M_S and S.

24.9 OBTAINING TERM SYMBOLS FROM THE ELECTRON CONFIGURATION

Consider the configuration, $2p^2$, for the two outer electrons in the ground state of the carbon atom. Since both electrons have $n = 2$ and $l = 1$ they are equivalent electrons. The possible one-electron eigenfunctions, described by the notation, $[m_i \ (m_s)_i]$, are:

Function	f_1	f_2	f_3	f_4	f_5	f_6
Electron 1	$(1\frac{1}{2})$	$(1\overline{\frac{1}{2}})$	$(0\frac{1}{2})$	$(0\overline{\frac{1}{2}})$	$(\overline{1}\frac{1}{2})$	$(\overline{1}\overline{\frac{1}{2}})$
Electron 2	$(1\frac{1}{2})$	$(1\overline{\frac{1}{2}})$	$(0\frac{1}{2})$	$(0\overline{\frac{1}{2}})$	$(\overline{1}\frac{1}{2})$	$(\overline{1}\overline{\frac{1}{2}})$

(The overbars indicate negative values.)

The possible product functions are $f_i f_j$, so long as $i \neq j$. If $i = j$, the two electrons would have the same set of quantum numbers, in violation of the Pauli principle. Each of the six functions can be combined with any of the other five, making a total of $6 \times 5 = 30$ possible functions. Only 15 of these product functions are independent; the other 15 are derived from the first set by interchanging the coordinates of the two electrons.

If $m_1 = +1, 0, -1$ and $m_2 = +1, 0, -1$, then the possible values of $M_L = m_1 + m_2$ are $2, 1, 0, 1, 0, -1, 0, -1, -2$. These can be arranged into the sets:

$$M_L = 2, 1, 0, -1, -2 \qquad L = 2 \qquad D;$$
$$M_L = \quad 1, 0, -1 \qquad L = 1 \qquad P;$$
$$M_L = \quad 0 \qquad L = 0 \qquad S.$$

From the values of M_L we infer the values of L shown in the middle column. The corresponding letter designation of the term symbol is in the third column.

Since, when $M_L = 2$ or -2, $m_1 = m_2 = 1$ or $m_1 = m_2 = -1$. Then to avoid violating the Pauli principle, the spins must be different. Thus, for this term, $M_S = \frac{1}{2} + (-\frac{1}{2}) = 0$. Since $M_S = 0$, it must be that $S = 0$. The D term is consequently a singlet, ^1D. Then $L + S = 2$ and $L - S = 2$, so the only possible value of $J = 2$. The complete term symbol is $^1D_2$. Since $2J + 1 = 5$, there are five product functions associated with this term. From these five product functions we could construct five determinantal wave functions. The details concerning which of the product functions belong to the term and the construction of the determinantal functions are not needed for our purposes here.

The next set of M_L values, $M_L = 1, 0, -1$, belongs to $L = 1$ and therefore to a P term. The combinations, $(1\frac{1}{2})(0\frac{1}{2})$ and $(0\frac{1}{2})(1\frac{1}{2})$, have $M_L = 1$ or -1 and $M_S = 1$. But this value of M_S occurs only in the set, $M_S = 1, 0, -1$. These values of M_S require that $S = 1$; then $2S + 1 = 3$, and the term is a triplet, ^3P. Next we find that $L + S = 1 + 1 = 2$ and $L - S = 1 - 1 = 0$. The possible values of J are therefore $J = 2, 1, 0$. The corresponding values of $2J + 1$ are $5, 3, 1$. Thus the terms are $^3P_2$(five functions); $^3P_1$(three functions); $^3P_0$(one function). These terms together with $^1D_2$ account for 14 of the 15 product functions. The remaining function has $M_L = 0$; hence $L = 0$ and the term is S. Since there is only one function, $M_S = 0$ and $S = 0$. Then, also, $J = L + S = 0$. The term symbol is $^1S_0$. Since $2J + 1 = 1$, there is one function corresponding to this term.

In a configuration of two nonequivalent p electrons, such as $2p3p$, there are six possible functions for each electron. Since there is no restriction imposed by the Pauli principle (the principal quantum numbers are different), all 36 product functions are allowed and all are independent. Again the sets of values of M_L are:

$$M_L = 2, 1, 0, -1, -2 \qquad L = 2 \qquad D$$
$$M_L = \quad 1, 0, -1, \qquad L = 1 \qquad P$$
$$M_L = \quad 0 \qquad L = 0 \qquad S$$

But in this case for each value of L both the singlet and the triplet can occur. The possible combinations of L, S, and J are:

L	S	J	$2J + 1$	$Terms$
2	1	3, 2, 1	7, 5, 3	$^3D_3, {}^3D_2, {}^3D_1$
2	0	2	5	$^1D_2$
1	1	2, 1, 0	5, 3, 1	$^3P_2, {}^3P_1, {}^3P_0$
1	0	1	3	$^1P_1$
0	1	1	3	$^3S_1$
0	0	0	1	$^1S_0$

The sum of the values of $2J + 1$ is equal to 36, the total number of product functions.

Consider the configuration, $2p^3$, which is the ground state configuration of the nitrogen atom. Since $m_1 = 1, 0, -1$, $m_2 = 1, 0, -1$, and $m_3 = 1, 0, -1$, we see that the possible values of L are 3, 2, 1, 0, corresponding to F, D, P, and S terms. However, to have $M_L = 3$ or -3 requires $m_1 = m_2 = m_3 = 1$ or $m_1 = m_2 = m_3 = -1$. Since m_s can only equal $+\frac{1}{2}$ or $-\frac{1}{2}$ we find that there is no way to satisfy the Pauli principle with $M_L = 3$, that is, with $L = 3$. Thus the F term cannot occur.

The D term, $L = 2$, requires the set of values of $M_L = 2, 1, 0, -1, -2$. The value $M_L = 2$ can occur only if two of the spins are paired, that is, in a combination such as $(1\frac{1}{2})(1\bar{\frac{1}{2}})(0\frac{1}{2})$. Thus, in the D terms, $M_S = \Sigma_i (m_s)_i = \frac{1}{2} - \frac{1}{2} + \frac{1}{2} = \frac{1}{2}$ and, of course, $-\frac{1}{2}$ can appear as well. Therefore $S = \frac{1}{2}$ and $2S + 1 = 2$; the term is a doublet, ^2D. Since $L + S = 2 + \frac{1}{2} = \frac{5}{2}$, and $L - S = 2 - \frac{1}{2} = \frac{3}{2}$, the only allowed values of J are $J = \frac{5}{2}, \frac{3}{2}$. The corresponding values of $2J + 1$ are 6 and 4. Thus there is a total of ten functions associated with the ^2D term: $^2D_{5/2}$ (six functions) and $^2D_{3/2}$ (four functions).

The P term, with $L = 1$, requires $M_L = 1, 0, -1$. The value, $M_L = 1$, can occur only in combinations such as $(1\frac{1}{2})(0\frac{1}{2})(0\bar{\frac{1}{2}})$ or $(1\frac{1}{2})(1\bar{\frac{1}{2}})(\bar{1}\frac{1}{2})$. Therefore $M_S = \frac{1}{2}$ or $-\frac{1}{2}$ and $S = \frac{1}{2}$. Again we have a doublet, ^2P. Since $L + S = 1 + \frac{1}{2} = \frac{3}{2}$, and $L - S = 1 - \frac{1}{2} = \frac{1}{2}$, the two allowed values for J are $J = \frac{3}{2}, \frac{1}{2}$. Then $2J + 1 = 4, 2$. The complete terms symbols are: $^2P_{3/2}$ (four functions) and $^2P_{1/2}$ (two functions).

The remaining values, $M_L = 0$ and $L = 0$, cannot come from $m_1 = m_2 = m_3 = 0$, because this combination violates the Pauli principle. Therefore $L = 0$ comes from combinations such as $m_1 = 1, m_2 = -1, m_3 = 0$. Since the three values of m are different, they can have the same value of the spin quantum number. Hence the possible values of M_S are $M_S = \frac{3}{2}, \frac{1}{2}, -\frac{1}{2}, -\frac{3}{2}$. The value of $S = \frac{3}{2}$. The only value of J is $\frac{3}{2}$, so that the complete term symbol is $^4S_{3/2}$. This exhausts the functions available.

The states of the nitrogen atom belonging to the configuration $2p^3$ are ^2D, ^2P and ^4S. According to Hund's rule, Section 24.7, the energies of these states are in the order $^4S < {}^2D < {}^2P$.

24.10 EXAMPLES OF ATOMIC SPECTRA

24.10.1 One-Electron Systems

We have discussed the energy of the hydrogen atom earlier. The energy level system for the lithium atom is shown in Fig. 24.8. The lowest filled level, $1s^2$, is not shown. In the ground state, the term is 2^2S. Since the selection rule requires $\Delta l = \Delta L = \pm 1$, the only

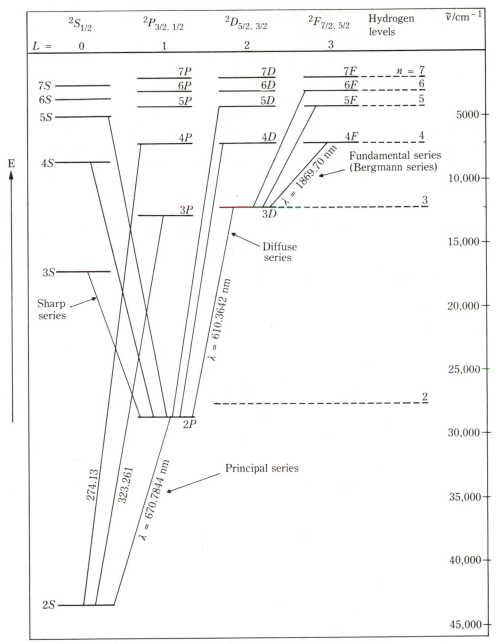

Figure 24.8 Energy-level diagram for the lithium atom. Dashed lines are the corresponding levels for the hydrogen atom. Numbers on the transition lines are the wavelengths in nanometres.

possibilities for absorbing energy are in the transitions

$$2^2S \longrightarrow 2^2P \quad \lambda = 670.7844 \text{ nm};$$
$$2^2S \longrightarrow 3^2P \quad \lambda = 323.261 \text{ nm};$$
$$2^2S \longrightarrow 4^2P \quad \lambda = 274.13 \text{ nm};$$
$$\cdots \quad \lambda_{\text{lim}} = 230 \text{ nm}.$$

This is the so-called *principal* series of lines. If the lithium atoms are excited in an arc or flame, the upper states become populated and light is emitted as the atoms undergo transitions to the lower states. Four series appear in the ultraviolet, visible, and near infrared:

1. the principal series, consisting of transitions from higher ^2P states to the 2^2S state;

2. the sharp series, consisting of transitions from the higher ^2S states to the 2^2P state;

3. the diffuse series, consisting of transitions from the higher ^2D states to the 2^2P state;

4. the fundamental series, consisting of transitions from the higher ^2F states to the 3^2D state.

Figure 24.8 includes the energy levels of the hydrogen atom for comparison. It is apparent that in high quantum states, the energy levels for the lithium atom are almost coincident with those of the hydrogen atom, at least in those states that have angular momentum. This means that, as the valence electron moves farther from the $1s^2$ core of lithium, the electronic interaction effects die away and the electron sees effectively a single excess positive charge, effectively a proton. This effect also occurs with the higher alkali metals but is less marked as a result of the larger volume occupied by the core electrons.

The spectra of all the alkali metals are similar to that of lithium; they are based on the same kinds of energy levels (terms). Close inspection of the spectral lines shows that each

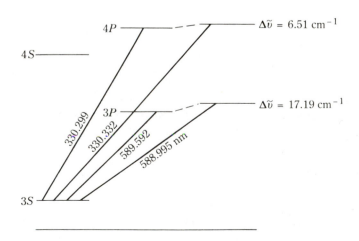

Figure 24.9 Portion of the energy-level diagram of sodium, showing the doublet splitting. The ^2P level splitting is exaggerated (not to scale). Wavelengths are in nm.

of them is double; the separation between the two lines of the doublet increases markedly in the series Li, Na, K, Rb, Cs. This separation is a consequence of the fact that, in the multielectron system, the energy depends not only on the values of n and L but also on the value of J. Thus the $^2S_{1/2}$ terms are not split, since there is only one possible value of J, namely, $J = \frac{1}{2}$. On the other hand, the ^2P terms have $J = \frac{3}{2}$ and $J = \frac{1}{2}$ so that the energy of $^2P_{3/2}$ is slightly higher than that of $^2P_{1/2}$. The upper terms, $^2D, {}^2F, \ldots$, are also split, but the magnitude of the separation is too small to be noticeable. Effectively only the ^2P level is split and any transition to or from this level results in two closely spaced lines. The classical example is the yellow sodium doublet lines (sodium D lines) at 589.592 nm and 588.995 nm, which are emitted in the transition from $3^2P \to 3^2S$ (Fig. 24.9).

24.10.2 Two-Electron Systems; the Alkaline Earths

The energy levels for the calcium atom are shown in Fig. 24.10. These levels comprise two groups, singlets and triplets. Since the selection rule, $\Delta S = 0$, permits no change in multiplicity, transitions between a singlet term and a triplet term are forbidden. Consequently, the emission spectrum exhibits two independent series of lines. One series results from transitions between singlet terms and the other from transitions between triplet terms.

The absorption spectrum of calcium consists only of the series resulting from the transitions from the 4^1S state to the n^1P state. Because of the large energy differences, these lines appear in the short wavelength end of the visible spectrum and in the ultraviolet. Thus $\lambda = 422.673$ nm, 272.165 nm, 239.858 nm, ... The series limit, $\lambda_{\text{lim}} = 152.995$ nm, corresponds to the energy that must be supplied to ionize the calcium atom, 781.898 kJ/mol.

The emission spectrum exhibits not only the principal series but also other singlet series such as:

$$\text{higher } {}^1P \text{ to } 5^1S;$$

$$\text{higher } {}^1S \text{ to } 4^1P, \qquad \text{higher } {}^1D \text{ to } 4^1P;$$

$$\text{higher } {}^1P \text{ to } 3^1D, \qquad \text{higher } {}^1F \text{ to } 3^1D.$$

Each line in these series is a single line.

In the triplet spectrum, the series lie in the longer wavelength region, ranging from the visible to the infrared. Each line in the triplet spectrum consists of a number of closely spaced lines; the separation between these lines increases rapidly as the atomic number increases in the two-electron species: He, Be, Mg, Ca, Sr, Ba, Zn, Cd, Hg. Consider the first line in the sharp series in calcium, corresponding to the transition from 5^3S to 4^3P. Since the 5^3S level is a single level while the 4^3P level is split into three levels, the line consists of three closely spaced lines; $\lambda = 610.272$ nm, 612.222 nm, and 616.218 nm. The transition from 4^3D to 4^3P yields a group of six closely spaced lines. In contrast to the alkali metal case in which the ^2D levels were not split, in calcium the ^3D levels are split into three levels. If each level of ^3D could combine with each level of ^3P, nine lines would be expected. In fact, the selection rule, $\Delta J = 0, \pm 1$, rules out three of these possibilities so that only six lines appear. This situation is illustrated in Fig. 24.11. The transitions $J = 2 \to 0, 3 \to 1$, and $3 \to 0$ are forbidden.

In systems having three, four, or more electrons there are several systems of terms. For example, in the three-electron system there are doublets and quartets; no transitions between the doublet and quartet levels are allowed. In the four-electron system there are three systems: singlets, triplets, and quintets; transitions between them are forbidden.

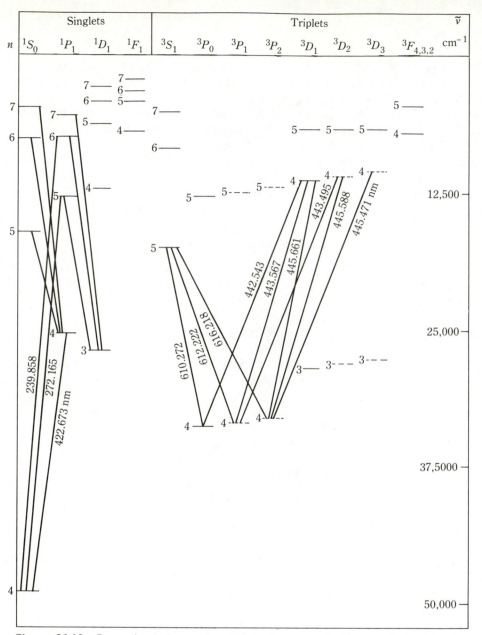

Figure 24.10 Energy level diagram for calcium. Dashed levels indicate that splitting is shown out of scale. Wavelengths are in nm. Not all the allowed transitions are indicated.

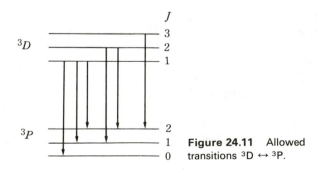

Figure 24.11 Allowed transitions $^3D \leftrightarrow {}^3P$.

24.11 THE MAGNETIC PROPERTIES OF ATOMS

If the electron spins on its axis, the fact that it is electrically charged implies that there is a current flow around the axis. This flow of current gives the electron a magnetic moment, just as the flow of current in a coil of wire gives the coil a magnetic moment. The magnetic moment is perpendicular to the plane of the current flow and therefore parallel to the angular momentum vector, but is directed oppositely because of the negative charge on the electron.

Similarly, if the angular momentum of an electron in an atom has a nonzero value, there is a flow of current about an axis and the system possesses a magnetic moment. The magnetic moment of the atom is the resultant of the moments due to the spin angular momentum and the orbital angular momentum of all the electrons in the atom.

If a current I flows in a loop of wire, the loop has a magnetic moment given by

$$\mu = IA \tag{24.34}$$

in which A is the area of the loop. If r is the radius of a circular loop, we have

$$\mu = I\pi r^2. \tag{24.35}$$

Now if we imagine an electron moving with a velocity v in an orbit of radius r, the number of times the electron passes any point in the orbit in one second is the velocity divided by the circumference of the orbit, $v/2\pi r$. If we multiply this by the charge on the electron, $-e$, we obtain the charge passing that point in one second, which is the current, I.

$$I = -e\left(\frac{v}{2\pi r}\right). \tag{24.36}$$

The angular momentum, $M_z = mvr$, if the orbit lies in the xy-plane. Since $v = M_z/mr$,

$$I = -\frac{eM_z}{2\pi mr^2}. \tag{24.37}$$

Using this value of I in Eq. (24.35), we obtain

$$\mu_z = -\left(\frac{eM_z}{2\pi mr^2}\right)\pi r^2 = -\frac{eM_z}{2m}. \tag{24.38}$$

This yields for the magnetogyric ratio, γ:

$$\gamma = \frac{\mu_z}{M_z} = -\frac{e}{2m} = -8.794024 \times 10^{10} \, \text{C kg}^{-1}. \tag{24.39}$$

This formula from classical physics is correct for the electron only for the orbital angular momentum. The negative sign simply means that the magnetic moment is opposite in direction to the angular momentum vector. For spin, it turns out that the value of γ is about twice as large (2.00232 times to be exact) as that in Eq. 24.39. Consequently, the magnetic moment of the electron is usually expressed as

$$\mu_z = -g\left(\frac{e}{2m}\right)M_z, \tag{24.40}$$

in which g is a pure number, the Lande g factor, which has a rational value that depends on the values of J, L, and S.

Introducing the value of $M_z = M_J \hbar$, the value of μ_z becomes

$$\mu_z = -g\left(\frac{e\hbar}{2m}\right)M_J, \tag{24.41}$$

in which $M_J = 0, \pm 1, \ldots, \pm J$, where J is the total angular momentum quantum number. A natural unit of magnetic moment, the Bohr magneton, μ_B, is defined by

$$\mu_B = \frac{e\hbar}{2m} = 9.274078 \times 10^{-24}\ \mathrm{m^2\ A.} \tag{24.42}$$

Then

$$\mu_z = -g\mu_B M_J. \tag{24.43}$$

If M_{total} is the total angular momentum, the magnetic moment, μ, of an atom is given by

$$\mu = -g\left(\frac{e}{2m}\right)M_{total}. \tag{24.44}$$

Since $M_{total} = \sqrt{J(J+1)}\,\hbar$, we obtain

$$\mu = -g\left(\frac{e\hbar}{2m}\right)\sqrt{J(J+1)} = -g\sqrt{J(J+1)}\mu_B. \tag{24.45}$$

If we consider a single electron that has no orbital angular momentum, then $g = 2$ and $J = S = \frac{1}{2}$. It follows from Eq. (24.45) that $\mu_{spin} = -\sqrt{3}\mu_B$. Since $J = \frac{1}{2}$, then $M_J = \pm\frac{1}{2}$. Using Eq. (24.43) we obtain for the z component of the magnetic moment, $(\mu_{spin})_z = \mp\mu_B$. Thus the allowed values of the component of the spin magnetic moment along any specified axis are ± 1 Bohr magneton.

★ 24.11.1 The Zeeman Effect

If we examine the spectrum of an atomic system, we find that the imposition of a strong magnetic field splits the line into a number of components; this is the Zeeman effect. Suppose we apply a uniform magnetic field with magnitude B to an atomic system. The field direction is chosen along the z-axis. The energy of the atom in the field will depend on the component of the magnetic moment in the field direction. The application of the field lowers the energy by an amount $\mu_z B$. If the original energy of the atom is E_0, and the energy in the presence of the field is E, we have

$$E = E_0 - \mu_z B. \tag{24.46}$$

The negative sign indicates that when the magnetic moment and the magnetic field are in the same direction, the energy is lowered. Note that the *tesla* (T) is the SI unit for B, the magnetic field (the magnetic flux density); $1\mathrm{T} = 1\ \mathrm{kg/s^2\ A}$ (1 tesla $= 10^4$ gauss). Then μ has the unit: $\mathrm{m^2\ A = J\ s^2\ A/kg = J/T}$.

Inserting the value of μ_z from Eq. (24.43) into Eq. (24.46), we obtain

$$E = E_0 + g\mu_B M_J B \tag{24.47}$$

Since there are $2J + 1$ allowed values of M_J, the original energy level is split into $2J + 1$ new levels having different energies. Since every energy level except a $^1S_0$ level is split by the field, it follows that the spectral lines are split into several components. The magnitude of the separation of the lines is proportional to the magnetic field.

Ultimately, we will be interested in the differences in energy between two states and the frequencies of the lines emitted. Therefore it is convenient to introduce the Larmor frequency, v_L, defined by

$$v_L \equiv \left(\frac{\mu_B B}{h}\right) = \frac{1}{2\pi}\left(\frac{eB}{2m}\right). \tag{24.48}$$

By replacing $\mu_B B$ by hv_L, Eq. (24.47) becomes

$$E = E_0 + gM_J hv_L. \tag{24.49}$$

We find that the displacements of the lines in the Zeeman effect are small, rational multiples of the Larmor frequency.

To use Eq. (24.49) we need the value of g. It can be shown that

$$g = 1 + \frac{J(J+1) + S(S+1) - L(L+1)}{2J(J+1)}. \tag{24.50}$$

The simplest application of Eq. (24.50) is to a singlet system. Since $S = 0$, then $J = L$ and $g = 1$ for all possible transitions within the singlet system. Equation (24.49) becomes

$$E = E_0 + M_J hv_L. \tag{24.51}$$

Then for the $^1S_0$ term, the only possible value of M_J is $M_J = 0$; the term is not split. Equation (24.51) yields

$$E(^1S_0) = E_0(^1S_0).$$

But for the $^1P_1$ term, the possible M_J values are $M_J = 0, \pm 1$, so that Eq. (24.51) yields three energy values:

$$E_3(^1P_1) = E_0(^1P_1) + hv_L;$$

$$E_2(^1P_1) = E_0(^1P_1);$$

$$E_1(^1P_1) = E_0(^1P_1) - hv_L.$$

For the $^1D_2$ term, the possible M_J values are $M_J = 0, \pm 1, \pm 2$, so that $^1D_2$ is split into five levels:

$$E_5(^1D_2) = E_0(^1D_2) + 2hv_L;$$

$$E_4(^1D_2) = E_0(^1D_2) + 1hv_L;$$

$$E_3(^1D_2) = E_0(^1D_2);$$

$$E_2(^1D_2) = E_0(^1D_2) - 1hv_L;$$

$$E_1(^1D_2) = E_0(^1D_2) - 2hv_L.$$

The energy level scheme is shown in Fig. 24.12.

Since the separation between the levels is the same in the $^1P_1$ term as in the $^1D_2$ term, and since the selection rule requires $\Delta M_J = 0, \pm 1$, it follows that each line in the spectrum is split into three lines. The middle line is at the original frequency, while the other two lines are spaced equally on either side of the original frequency. This is the *normal* Zeeman effect.

If v_0 is the frequency emitted in the transition from one state to another in the absence of the magnetic field, then the three lines in the presence of the field have the frequencies

$$v_1 = v_0 - v_L, \qquad v_2 = v_0, \qquad v_3 = v_0 + v_L.$$

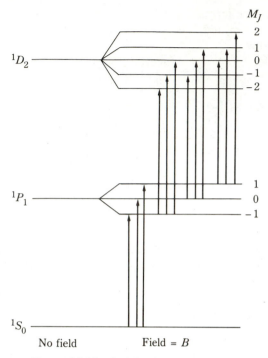

Figure 24.12   Splitting in a magnetic field.

For systems other than singlets, g is not equal to 1. In this case, a more complicated splitting occurs: the *anomalous Zeeman effect*. For example, consider the doublet system in hydrogen and the alkali metals. The lowest term is $^2S_{1/2}$. For this term, $L = 0$, $J = S = \frac{1}{2}$; therefore $M_J = \frac{1}{2}$, $-\frac{1}{2}$ and $g = 2$. The product $gM_J = 2(\pm\frac{1}{2}) = \pm 1$. Using this value in Eq. (24.49) we obtain

$$E(^2S_{1/2}) = E_0(^2S_{1/2}) \pm h\nu_L.$$

This term is split into two levels.

The ^2P term has $L = 1$, $S = \frac{1}{2}$, $J = \frac{3}{2}, \frac{1}{2}$. For the term $^2P_{1/2}$, using Eq. (24.50), we obtain $g = \frac{2}{3}$; since $M_J = \pm\frac{1}{2}$, the product $gM_J = \frac{2}{3}(\pm\frac{1}{2}) = \pm\frac{1}{3}$. Then

$$E(^2P_{1/2}) = E_0(^2P_{1/2}) \pm \tfrac{1}{3}h\nu_L.$$

Since the splitting is different in the two terms, this member of the doublet ($^2P_{1/2} \rightarrow {}^2S_{1/2}$) is split into four lines. If ν_1 is the frequency of the line in the absence of the field, we obtain

$$\nu = \nu_1 \pm \tfrac{2}{3}\nu_L \qquad \text{and} \qquad \nu = \nu_1 \pm \tfrac{4}{3}\nu_L$$

for the frequencies of the four lines. Note that the original frequency does not appear.

For the term $^2P_{3/2}$, we have $L = 1$, $S = \frac{1}{2}$, $J = \frac{3}{2}$; from this, we find that $g = \frac{4}{3}$. The possible values of M_J are $M_J = \frac{3}{2}, \frac{1}{2}, -\frac{1}{2}, -\frac{3}{2}$. Thus, the term $^2P_{3/2}$ is split into four levels. The corresponding values of gM_J are $2, \frac{2}{3}, -\frac{2}{3}, -2$. Because of the selection rule, $\Delta M_J = 0, \pm 1$, transitions may occur only between the lower $^2S_{1/2}$ level and the lowest three $^2P_{3/2}$ levels, yielding three lines; and between the upper $^2S_{1/2}$ level and the highest

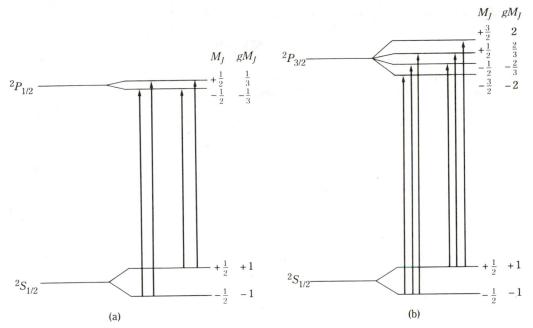

Figure 24.13 Anomalous Zeeman splitting of the sodium doublet. (a) D_1 is split into four lines; (b) D_2 into six. (From *Atomic Spectra and Atomic Structure* by G. Herzberg, 1944, Dover Publications, Inc. N.Y.)

$^2P_{3/2}$ levels, yielding another three lines. This member of the doublet ($^2P_{3/2} \rightarrow \,^2S_{1/2}$) is split into six lines. If v_2 is the frequency in the absence of the field, then in the presence of the magnetic field the frequencies are

$$v = v_2 \pm \tfrac{1}{3}v_L, \qquad v = v_2 \pm v_L, \qquad v = v_2 \pm \tfrac{5}{3}v_L.$$

This situation is illustrated in Fig. 24.13.

24.11.2 Magnetic Resonance Spectroscopy

To conclude the discussion of magnetic properties of atoms, we consider the magnetic resonance spectroscopies, electron spin resonance (ESR, EPR), and nuclear magnetic resonance (NMR), which depend on the detection of the energy differences between the quantum levels of a magnetic dipole in a magnetic field. As we pointed out in the discussion of the Zeeman effect in Section 24.11.1, the energy of a magnetic dipole in a magnetic field along the z-axis is given by Eq. (24.46), while μ_z for an electron is given by Eq. (24.43).

If we deal with nuclear magnetic moments, then, corresponding to Eq. (24.43), we have

$$\mu = g_N \mu_N I \qquad \text{and} \qquad \mu_z = g_N \mu_N M_I, \qquad (24.52)$$

where g_N is the nuclear g factor, μ_N is the nuclear magneton, I is the quantum number for the nuclear spin angular momentum, and $M_I = I, I - 1, I - 2, \dots, -I$ is the set of possible quantum numbers for the z component of the nuclear spin angular momentum.

For the proton with mass m_p, corresponding to Eq. (24.42), we have

$$\mu_N = \frac{e\hbar}{2m_p} = 5.050824 \times 10^{-27} \text{ J/T}, \tag{24.53}$$

while the empirical factor $g_N = 5.5856912$.

For a proton in a magnetic field, the energy is

$$E = E_0 + g_N \mu_N M_I B \tag{24.54}$$

Since $M_I = +\frac{1}{2}$ or $-\frac{1}{2}$, we find for the difference in energy between these two spin states,

$$\Delta E = g_N \mu_N B(\tfrac{1}{2}) - g_N \mu_N B(-\tfrac{1}{2}) = g_N \mu_N B \tag{24.55}$$

The frequency emitted or absorbed in this transition is the Larmor frequency, ν_L,

$$\nu_L = \frac{\Delta E}{h} = \frac{g_N \mu_N B}{h} = \frac{g_N eB}{4\pi m_p}. \tag{24.56}$$

The energy difference between the states, and consequently ν_L, increases linearly with the magnetic field. Suppose that the field, $B = 1.00$ tesla, then for the proton

$$\nu_L = \frac{5.586(1.602 \times 10^{-19} \text{ C})(1.00 \text{ T})}{4(3.1416)(1.673 \times 10^{-27} \text{ kg})} = 4.257 \times 10^7 \text{ Hz} = 42.6 \text{ MHz}.$$

Thus protons in a field of one tesla should absorb energy at 42.6 MHz. A popular design frequency for a nuclear magnetic resonance spectrometer is 60 MHz; for a proton to absorb at this frequency requires that $B = 1.41$ tesla.

The corresponding value of ν_L for an electron, since $g = 2$, is

$$\nu_L = \frac{2(1.602 \times 10^{-19} \text{ C})(1.00 \text{ T})}{4(3.1416)(9.1095 \times 10^{-31} \text{ kg})} = 28.0 \text{ GHz}.$$

The ratio,

$$\frac{(\nu_L)_{\text{electron}}}{(\nu_L)_{\text{proton}}} \approx 658,$$

shows that the energy change for the electron is 658 times larger than for the proton.

24.11.3 Nuclear Magnetic Resonance

We begin by describing a somewhat idealized nuclear magnetic resonance experiment. The device is shown in Fig. 24.14. The sample is placed between the poles of a very strong electromagnet. A coil is wrapped around the sample tube and connected to a radio-frequency oscillator that is capable of varying the frequency over a short range. This oscillator sends a signal to the sample; when the frequency of the oscillator reaches the Larmor frequency, the system can absorb the radiation sent by the oscillator. This absorption changes the impedance of the radio-frequency circuit; the change in impedance is measured by a bridge circuit, amplified and displayed on a chart recorder. The time axis of the recorder reflects the frequency range scanned. After passing through the resonant frequency, the impedance returns to its normal value; the trace on the chart looks somewhat like that in Fig. 24.15.

In practice it is difficult to design a variable frequency oscillator that has the required accuracy. Since the resonant frequency depends on the magnetic field, we vary the resonant

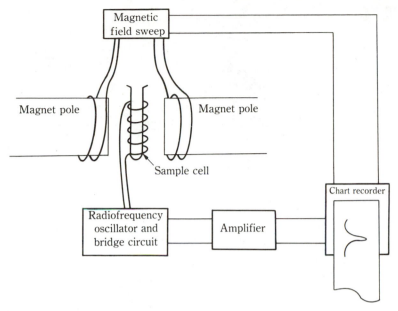

Figure 24.14   Nuclear magnetic resonance apparatus.

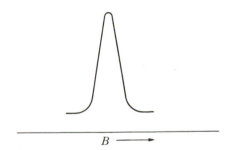

$B \longrightarrow$

Figure 24.15   Typical proton resonance peak.

frequency ν_L by changing B (instead of varying the frequency of the incident radiation). The radio-frequency oscillator is now locked to a design frequency; 60 MHz is one of the most common, but for greater sensitivity and higher resolution, 100 MHz and higher frequencies are also used. The magnet is designed to provide the high field required. Then small coils that can vary the field over a small range are added. When the magnetic field is such that the Larmor frequency matches the design frequency, energy is absorbed and the change in circuit characteristics is displayed on a chart record.

We will not discuss here the details of the instrument. Both the electronic circuitry and the magnet require a very high-level technology to produce. For example, the magnetic field must be uniform over the sample to within 1 part in 10^8; that was not easily achieved in the past, but is now routine.

The frequency at which the proton absorbs energy is not a fixed property of the proton, but depends on the magnetic environment of the proton. If the magnetic environment is

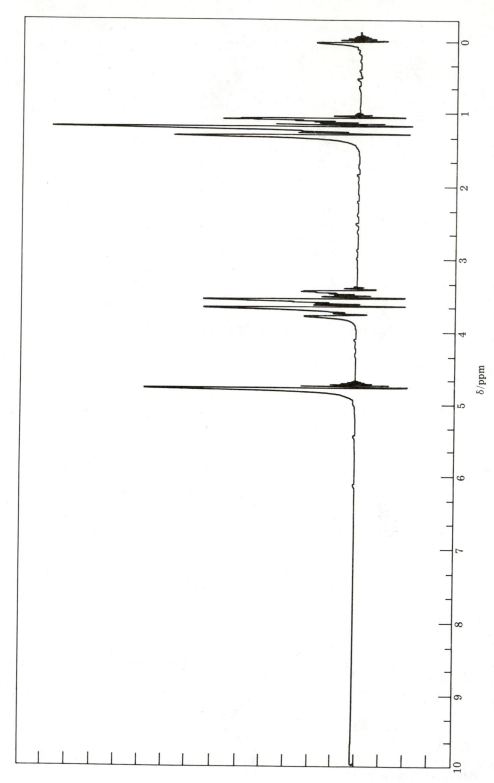

Figure 24.16   High-resolution NMR spectrum of ethanol. (Courtesy of Carl Esche, Dept. of Chemistry, University of Maryland.)

altered, the value of the field at which the proton absorbs energy will also be altered. A typical example of this is the adsorption in ethyl alcohol, CH_3CH_2OH, which shows three groups of closely spaced lines (Fig. 24.16). The magnetic environment of the CH_3 protons is different from that of the CH_2 protons, which is different from that of the OH proton. Thus the resonances appear at slightly different values of the field. The total areas under the peaks are in the ratio, 3:2:1.

The chemical shift is a measure of this difference in magnetic environment. We define the chemical shift, δ, as

$$\delta(ppm) = \frac{B_r - B}{B_r} \times 10^6. \tag{24.57}$$

This definition is a relative one; δ is measured in parts per million (ppm) displacement from the resonant field, B_r, of some reference substance. A common reference substance is tetramethyl silane (TMS), $Si(CH_3)_4$. Because the protons are all equivalent and there are twelve of them, TMS produces a single, sharp, intense resonance. Further, because the protons in TMS resonate at a higher field than the protons in almost any other compound, most chemical shifts are positive (that is, the resonance is at lower fields than for TMS). Since the absolute value of the field is very difficult to measure accurately, whereas the difference is readily measurable, it is convenient to express δ as a ratio of $\Delta B/B_r$ (multiplied by 10^6 to avoid handling very small numbers). Whether one uses B_r or B in the denominator is of no consequence since ΔB is so small; for protons the range of B is about 25 μT, which is equivalent to a frequency range of about 1000 Hz.

In the case of CH_3CH_2OH, the values of δ shown in Fig. 24.16 are:

methyl protons $\qquad \delta = 1.22$;

methylene protons $\qquad \delta = 3.70$;

hydroxyl proton $\qquad \delta = 4.80$.

In another example, the spectrum of CH_3CHO is shown in Fig. 24.17. In this case, for the methyl protons, $\delta = 2.20$, while for the —CHO proton, $\delta = 9.80$. The areas under the peaks are in the ratio 3:1.

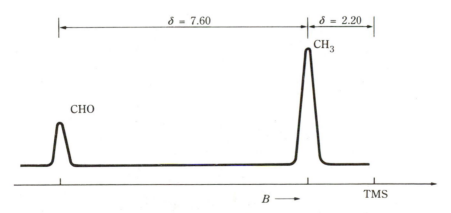

Figure 24.17   Low-resolution NMR spectrum of acetaldehyde.

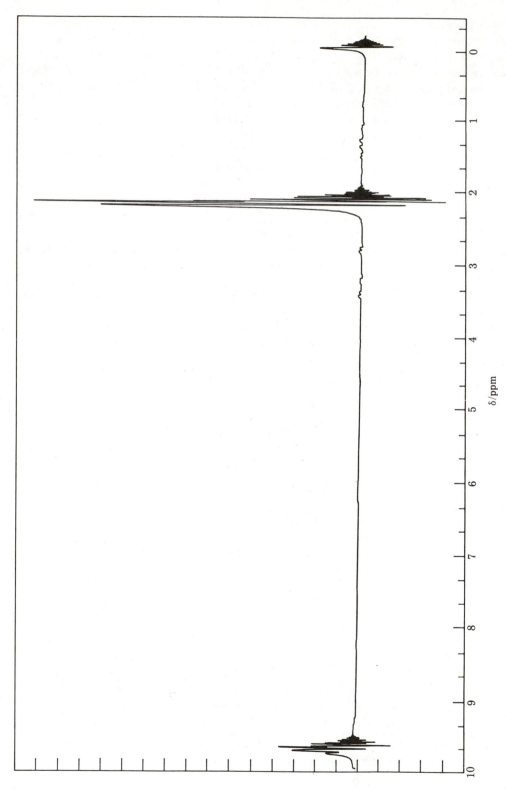

Figure 24.18   High-resolution NMR spectrum of acetaldehyde. (Courtesy of Carl Esche, Dept. of Chemistry, University of Maryland.)

δ/ppm

In high resolution we find that the peaks are often split multiplets (Fig. 24.18). For example, in the acetaldehyde case, the possible spin quantum numbers of the three methyl protons are $\frac{3}{2}$, $\frac{1}{2}$, $-\frac{1}{2}$, $-\frac{3}{2}$; these spin states have statistical weights of 1, 3, 3, 1. These four spin states exert four different effects on the remaining proton in the —CHO group. Consequently, in high resolution, four lines appear at the low-field end of the spectrum. The possible spin orientation of the aldehydic proton is $+\frac{1}{2}$ and $-\frac{1}{2}$. This proton exerts two different effects on the methyl protons, resulting in the split of the methyl proton resonance into two lines. Even after splitting into a number of lines, the total area ratio remains at 1:3 between the aldehydic proton peaks and the methyl proton peaks.

The NMR technique is routinely used to aid in establishing the structure and properties of various compounds. Because of relative ease in interpretation, it yields a wealth of information for a brief time investment.

24.12 X-RAY SPECTROSCOPY

In the ordinary x-ray tube, Fig. 24.19, a beam of electrons strikes a metal anode, such as copper or tungsten. As the electron is stopped by collision with the target, some or all of the energy of the electron may be emitted as radiation. If the electron loses all its energy in a single collision, the frequency of light emitted is given by

$$h v_m = E_{kin} \tag{24.58}$$

where E_{kin} is the kinetic energy of the electron. The kinetic energy of the electron is determined by the electric potential difference it passed through in traveling from the cathode to the anode:

$$E_{kin} = eV, \tag{24.59}$$

in which e is the electronic charge and V is the potential difference applied across the tube.

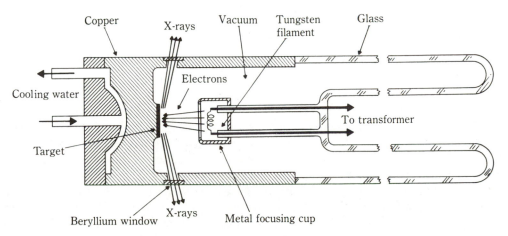

Copper   X-rays   Vacuum   Tungsten filament   Glass

Electrons

Cooling water

To transformer

Target

Beryllium window   X-rays   Metal focusing cup

Figure 24.19 Cross section of sealed-off filament x-ray tube (schematic). (From B. D. Cullity, *Elements of X-ray Diffraction.* Reading, Mass.: Addison-Wesley, 1956.)

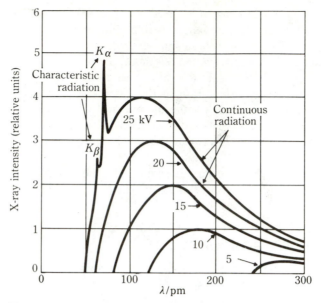

Figure 24.20   X-ray spectrum of molybdenum as a function of applied voltage (schematic). Line widths not to scale. (From B. D. Cullity, *Elements of X-ray Diffraction*. Reading, Mass.: Addison-Wesley, 1956.)

Thus we have, $h\nu_m = eV$, or

$$\nu_m = \frac{e}{h}\,V, \tag{24.60}$$

for the maximum frequency that will be emitted. The entire spectrum of lower frequencies would also appear, being emitted by electrons that lose only part of their energy in collision. This continuous x-ray spectrum is shown in Fig. 24.20, as a function of wavelength. Note that the minimum wavelength, λ_{min}, corresponds to the maximum frequency ν_m.

$$\lambda_{min} = \frac{c}{\nu_m} = \frac{hc}{eV} = \frac{1239.8 \times 10^3 \text{ pm V}}{V}. \tag{24.61}$$

Superimposed on the continuous spectrum is a line spectrum characteristic of the target element, if the applied potential is high enough (Fig. 24.21). Assume that we bombard a target with electrons that have sufficient energy to excite this line spectrum of the target. In collision with a target atom, the incident electron has sufficient energy to knock out an electron from a shell deep in the atom; for example, the 1s shell (K shell). We now have an ion in a very highly excited state, with the lowest level empty and many higher levels occupied. As the electrons rearrange within the atom, various x-ray lines are emitted. Suppose that an electron from the L-shell ($n = 2$) drops into the K shell; the frequency emitted is called the K_α line. Similarly, if an electron from the M shell ($n = 3$) drops into the K shell the K_β line is emitted. Since there is more than one level in both the L and M shells, there will be more than one K_α line and more than one K_β.

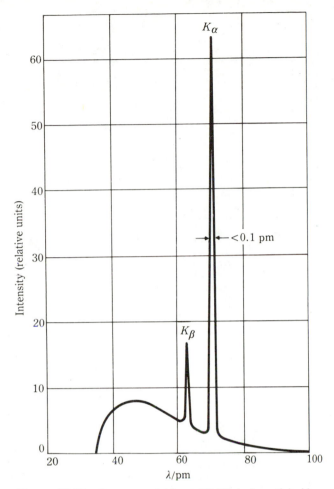

Figure 24.21 Spectrum of Mo at 35 kV (schematic). Line widths not to scale. (From B. D. Cullity, *Elements of X-ray Diffraction*. Reading, Mass.: Addison-Wesley, 1956.)

The L lines originate as electrons from the $M, N, \ldots$ shells drop into the L-shell. M lines appear as electrons from the $N, O, \ldots$ levels drop into the M level. In the sense that our prime focus is on the highest energy state, the K-state, the system of levels is inverted compared to the system in optical spectra where our prime focus was on the ground state.

Figure 24.22 shows the energy levels that are appropriate to x-ray emission. The highest energy level, the K-state, consists of the ion with one K electron missing. If an electron from the L level drops to the K level, an x-ray photon is emitted and the ion now has a lower energy and has a vacancy in the L level. This state is called an L state of the system. The K_{α_1} and K_{α_2} lines are emitted, depending on which L level the electron was in. The notation is the same as the term notation in optical spectra, and the selection rules are the same. Note that the $K_{\alpha_1}, K_{\alpha_2}$ doublet is fundamentally the same as the doublet sodium D line in the optical spectrum.

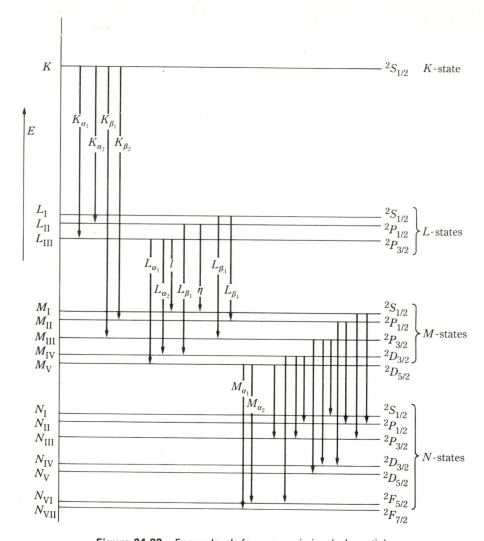

Figure 24.22   Energy levels for x-ray emission (schematic).

The other notable fact is that the x-ray spectrum of one element is very much like that of another. The K_α lines, for example, originate in the same transition for all elements; the frequencies are simply shifted; this is Moseley's law. Moseley made accurate wavelength measurements of the K_α lines of an entire series of elements. He found that the square root of the frequency of a given type of line, K_α for example, is a linear function of the atomic number of the element. Moseley's law can be written as

$$\sqrt{\nu} = 4.98 \times 10^7 (Z - 1) \text{ Hz}^{1/2}. \tag{24.62}$$

Moseley's law demonstrated that the atomic number rather than the atomic mass is the significant factor in determining this fundamental property, the x-ray line frequencies of the elements.

Only the outer electrons are involved in the optical spectrum of an atom. It is characteristic of the optical spectrum that a system that can emit a particular frequency in the transition from state m to state n can also absorb that exact frequency by the reverse transition from state n to state m. Such is not the case with x-rays. If the K_α line is emitted from an atom, a neighboring atom cannot absorb the K_α line by the reverse transition of lifting a K electron to a position in the L shell simply because L level is filled (if $Z > 10$). There is no vacant L level to which the K electron can move. On the other hand, because the energy of the K_α radiation is so large, it can be absorbed by other processes such as by knocking out a less strongly bound electron (for example, one in the L or M shell). Thus the mass absorption coefficient as a function of wavelength varies as shown in Fig. 24.23.

At the shortest wavelengths the photon is sufficiently energetic to eject the K electron. As the wavelength approaches the K absorption edge, the absorption coefficient increases since the probability of ejection of an electron is greater the more closely the incident photon's energy matches the energy required to eject the electron. At wavelengths longer than 14.088 pm, the photon's energy is insufficient to eject the K electron and the absorption coefficient drops sharply. In the sense of the photoelectric effect, the incident photon has dropped below the threshold frequency. Between 14.088 pm and 78.196 pm the absorption is due principally to ejection of electrons from the L levels, leaving the ions in either the L_I, L_{II}, or L_{III} state. As the wavelength approaches the L_I absorption edge, the absorption coefficient increases because of the closer match between the photon energy and the energy required for ejection of the L electrons. As soon as the wavelength exceeds 78.196 pm, the absorption coefficient drops sharply; the photon no longer has sufficient energy to eject the electron to form an L_I state. Even though it has dropped, the absorption stays fairly large since the photon still has energy sufficient to form both L_{II} and L_{III} states.

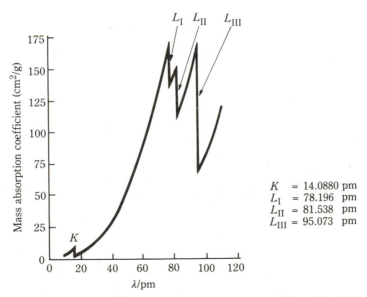

Figure 24.23 Absorption coefficients of lead, showing K and L absorption edges. (From B. D. Cullity, *Elements of X-ray Diffraction*. Reading, Mass.: Addison-Wesley, 1956.)

After passing 81.538 pm, a larger drop occurs because now only the L_{III} state and the five M states contribute to the absorption. At 95.073 pm another large drop occurs; beyond this wavelength only the M states can be formed. The absorption edges for the five M states lie between 321.7 and 495.5 pm, off the scale of Fig. 24.23.

The relative simplicity of x-ray spectra (compared to the complexity of optical spectra) makes x-ray spectroscopy an extremely important analytical and investigative tool. Three methods will be described: x-ray fluorescence spectroscopy, electron probe microanalysis, and x-ray photoelectron spectroscopy.

24.13 X-RAY FLUORESCENCE SPECTROSCOPY

In x-ray fluorescence spectroscopy, we bathe the sample in an intense, continuous and characteristic spectrum from an x-ray tube with a heavy metal target, such as tungsten. The tube is operated at the highest practicable potential, as high as 100 kV, to produce as wide a range of wavelengths as possible. If this radiation falls on an atom, the x-ray wavelengths equal to and below the K absorption edge can remove a K electron from the atom. After the K electron is removed, a cascade occurs within the atom. Suppose at first a K line is emitted as fluorescent radiation. This leaves a vacancy in an L level; an electron from a third level drops to fill this second vacancy, emitting another characteristic line; another electron drops to fill the third vacancy, and so on. Similarly, the L- and M-shell electrons can be removed, and the L lines and M lines can be emitted. The emitted radiation is passed into an analyzer capable of separating the fluorescent radiation into its component wavelengths. The analyzer is usually a crystal such as LiF, which serves as a diffraction grating. The detector then scans the various wavelengths and measures the intensity of each. The x-ray spectrometer is shown schematically in Fig. 24.24.

The radiation detected at the angle 2θ depends on the wavelength through the Bragg condition, $n\lambda = 2d \sin \theta$, where n is an integer. (See Section 27.11.) The distance d is an interplanar distance in the crystal; it can be changed if we use a different crystal or a different set of planes in the same crystal. As the crystal is moved through the angle θ, the

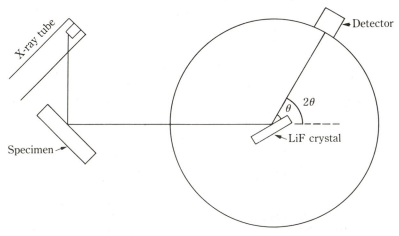

Figure 24.25   Spectrometer for x-ray fluorescence.

detector is synchronized to move through 2θ. Thus, as the crystal moves through 90°, the wavelength region from 0 to $2d$ is scanned.

The elements present can be identified from the wavelengths, and the amounts of each element can be determined from the intensities. The method is limited, as are all fluorescence methods, by the low intensity of the fluorescent radiation; nonetheless, detection of a few parts per million of an element is easily achieved. (For quantitative analysis, comparison with standard substances of known composition is necessary.) The method has the advantage of being very rapid and nondestructive.

24.14 X-RAY MICROANALYSIS WITH THE ELECTRON PROBE

In this technique, a finely focused electron beam impinges on the sample (Fig. 24.25). The width of the beam is of the order of 1 μm. The electron beam is energetic enough to excite the x-ray spectrum of the various elements in the sample. We analyze the spectrum of emitted x-rays and identify the elements, which can also be determined quantitatively. The technique is enormously useful, but it has the disadvantage that the sample must be inside the vacuum chamber of the electron gun. We mount the sample on an adjustable stage so that the position of the spot on the sample can be accurately fixed—and reproduced if necessary. By moving the sample and repeating the analysis, we can study any variation in composition with location.

In the scanning type of microprobe, the electron beam can sweep an area perhaps 100 μm by 100 μm. The variation in intensity of the image on the screen will depend on the variation in concentration of a particular element if (a) the detector is locked to a characteristic x-ray line of the element during the sweep, and (b) the intensity of an oscilloscope trace synchronized to the sweep of the electron beam is modulated by the input to the detector. In effect, we can obtain a picture of the distribution of that element in the area being scanned. By repeating the experiment while the detector is locked to the K_α line of a second element, we can obtain the distribution of the second element, and so on. Figure 24.26 on page 616 shows a series of pictures obtained from scanning a mineral sample for different elements. The inhomogeneity of the surface is brought strongly to our attention by this technique.

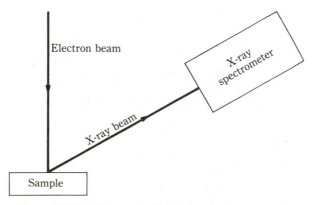

Figure 24.25 Schematic diagram for the electron probe.

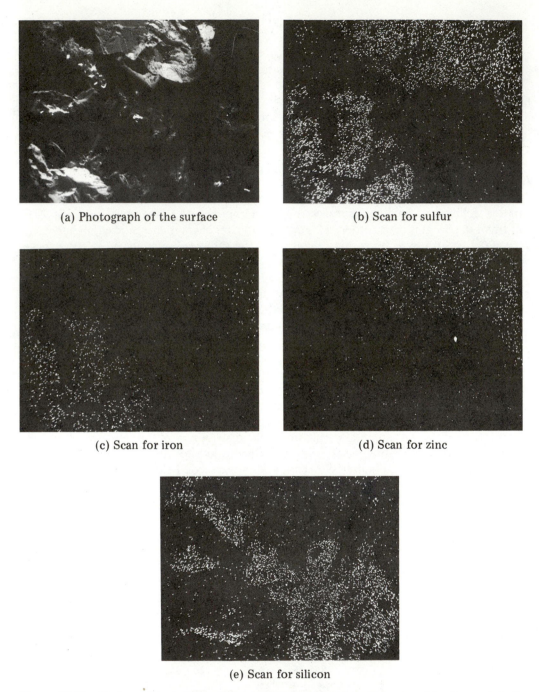

(a) Photograph of the surface   (b) Scan for sulfur

(c) Scan for iron   (d) Scan for zinc

(e) Scan for silicon

Figure 24.26 Photograph and electron beam scanning pictures of a sulfide mineral, having a sphalerite (ZnS) crystal in the upper right corner, a pyrite (FeS$_2$) crystal in the lower left corner, and a wedge of quartz (SiO$_2$) running diagonally downward from left to right. Scan for sulfur (b) shows a higher concentration (higher dot density) in the lower left and lower concentration in upper right. Scans for iron, zinc, and silicon (c), (d), and (e) show ZnS in the upper right, FeS$_2$ in the lower left, and SiO$_2$ mainly in the lower right. (Courtesy M. E. Taylor and M. C. Carney, Electron Microscope Central Facility, Institute for Physical Science and Technology, University of Maryland, College Park.)

24.15 X-RAY PHOTOELECTRON SPECTROSCOPY

In the technique of x-ray photoelectron spectroscopy we use an incident beam of monochromatic x-rays to irradiate the sample. The incident photon ejects an electron from the atom and imparts a kinetic energy to that electron. The energy required to remove the electron, E_b, plus the kinetic energy of the electron, E_k, must be equal to $h\nu$, the energy supplied by the photon. Thus

$$h\nu = E_b + E_k. \tag{24.63}$$

This is the equation for the photoelectric effect (see Section 19.9). The frequency of the incident photon is known very accurately. (The Mg $K_{\alpha_1\alpha_2}$ radiation, with $E = 1253.6$ eV, or Al $K_{\alpha_1\alpha_2}$ with $E = 1486.6$ eV, is commonly used.) We measure the kinetic energy and, from this information, obtain directly the value of E_b, the binding energy of the electron. Since the electron ejected may have originated from any energy level in the atom or molecule, the measurement yields the binding energies of all the occupied levels in the system (if the energy of the incident photon is high enough). This information is characteristic of the elements present and permits their identification; when used in this way, the technique is called ESCA, an acronym for *electron spectroscopy for chemical analysis.*

Figure 24.27 shows a schematic diagram of the device. Electrons emitted from the sample at right angles to the incident x-ray beam are collected and focused on the entrance to the kinetic energy analyzer. We can adjust the electrical field between the two concentric hemispherical electrodes so that the kinetic energy analyzer will select electrons of a particular energy, focus them on the exit slit, and send them into the detector. The output of the detector shows the number of electrons emitted as a function of their kinetic energy on a convenient recording device. The spectrum obtained displays the energy levels in the atom or molecule being studied.

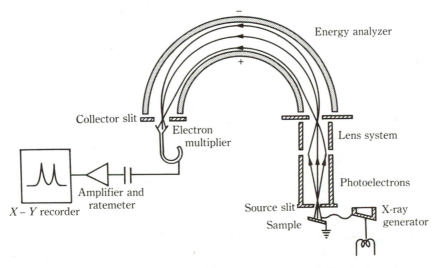

Figure 24.27   Schematic diagram illustrating the basic design of an x-ray photoelectron spectrometer using a retarding lens system and a hemispherical electrostatic analyser. (From P. M. A. Sherwood in *Spectroscopy*, vol. 3. B. P. Straughan and S. Walker, eds. London: Chapman and Hall, Ltd., 1976.)

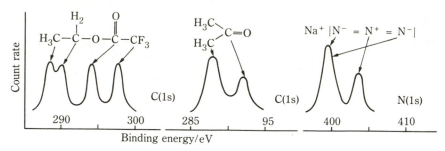

Figure 24.28 Spectra of C (1s) core electrons in ethyl trifluoroacetate, acetone, and the N (1s) electrons in sodium azide. (From W. C. Price in *Advances in Atomic and Molecular Physics*, vol. 10. New York: Academic Press, 1974.)

The binding energies of the $1s$ electron in the elements of the first row are:

Element	Li	Be	B	C	N	O	F	Ne	Na
E_b/eV	55	111	188	284	399	532	686	867	1072

The difference in the values between two neighboring elements ranges from 56 to 205 eV. It follows that we can easily distinguish one element from another by this technique. The binding energies depend slightly on the chemical environment. However, this variation is usually less than 10 eV and consequently does not hinder identification of the elements.

Figure 24.28 shows the chemical shift for the carbon atom $1s$ electrons in ethyl trifluoracetate. The area under each peak is the same; the shift indicates that each carbon atom has a different chemical environment, not surprising in view of the structure of the molecule.

In Fig. 24.28 we also see a portion of the spectrum of NaN_3. The binding energy of about 400 eV identifies the peaks as belonging to the nitrogen atoms. The appearance of two peaks indicates that there are two kinds of nitrogen atoms in the azide ion, $^-N{=}N^+{=}N^-$, with slightly different binding energies. The areas under the peaks are in the ratio 2:1 indicating that there are two atoms with the $1s$ electron having a lower binding energy and one atom with a more tightly bound $1s$ electron. In looking at the Lewis structure we would agree that the electrons on the terminal nitrogen atoms should be less strongly bound because of the negative charge. Generally speaking, the more positive the formal charge on the atom, the more strongly bound is the electron. This is obvious also in the spectrum of $CF_3COOCH_2CH_3$. Because the peak areas are proportional to the number of atoms in a given environment, the technique is very useful in analysis and in elucidating structure.

24.16 ULTRAVIOLET PHOTOELECTRON SPECTROSCOPY

The technique of ultraviolet photoelectron spectroscopy involves the same principle and the same type of measurement as x-ray photoelectron spectroscopy, but the incident radiation is in the ultraviolet region rather than the x-ray region. The exciting lines are often the helium $^1P_1 \rightarrow {}^1S_0$ line at 58.433 nm; or the helium ion, He^+, transition $^2P \rightarrow {}^2S_{1/2}$

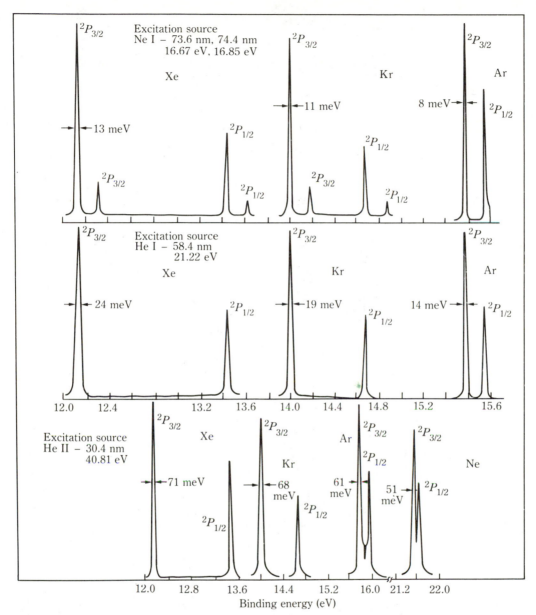

Figure 24.29   The $^2P_{3/2}$ and $^2P_{1/2}$ lines of the rare gases produced by Ne I, He I and He II resonance radiation. (From J. W. Rabelais, T. P. Debies, J. L. Berkosky, J. J. Huang, and F. O. Ellison. *J. Chem. Phys.* **61**:516, 1974.)

at 30.378 nm. The spectrum we obtain is similar to the spectrum from x-ray photoelectron spectroscopy but shows great detail in the lower energy ranges, that is, in the energy states of the valence electrons.

Figure 24.29 shows the spectrum of simple atoms such as argon, krypton, and xenon. We use these relatively sharp lines to calibrate the instrument. They correspond to the energy differences between the $^1S_0$ state of the atom and the $^2P_{3/2}$ and $^2P_{1/2}$ states of the ion.

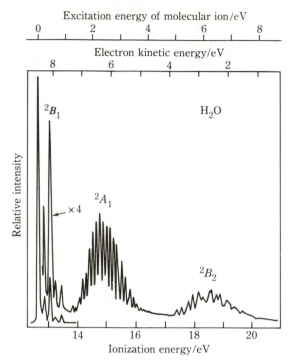

Figure 24.30   The He I photoelectron spectrum of water. (From J. W. Rabelais, *Principles of Ultraviolet Photoelectron Spectroscopy*. New York: Wiley, 1977).

Figure 24.30 shows the photoelectron spectrum of water. Note that there are three prominent bands, the energies of which are the energies needed to excite water in its ground state to H_2O^+ in its three lowest states; these molecular states are labeled $^2B_1$, $^2A_1$, and $^2B_2$, to correspond to the symmetry species of the molecular wave function. (See Section 23.16.1.) The fine structure is due to the large number of vibrational levels in each state; consequently, there are a large number of states with only slightly different energies.

QUESTIONS

24.1 Estimate the temperature required for noticeable (say 10%) probability of light absorption by the $2s$ state in a collection of hydrogen atoms.

24.2 How does the Pauli principle "couple" electron spin angular momenta?

24.3 Repelling electrons tend to occupy different orbitals (if possible) to increase their average separation. Illustrate this for the carbon $2p$ electrons. Suggest how Coulomb repulsion between electrons "couples" their orbital angular momenta.

24.4 How do Hund's rules apply to the $2p$ electrons in the carbon atom?

24.5 What are the allowed transitions from the ground states of (a) a beryllium atom and (b) a boron atom?

24.6 Sketch the allowed orientations of the electron orbital angular momentum vector and magnetic moment vector for hydrogen atom $2p$ states in a magnetic field (neglect spin). Identify the high and low energy orientations.

24.7 Draw an energy level diagram and indicate energy spacings for the $1s$ and $2p$ levels in the hydrogen atom in a magnetic field, including spin.

24.8 Why can different protons in a molecule have different magnetic environments? (Think of interacting magnets.)

24.9 Why should the x-ray line frequencies of the elements correlate with the atomic number rather than the atomic mass?

PROBLEMS

24.1 At 440 nm a glass filter, 2 mm thick, has a transmittance of 0.810.

  a) What percent of the incident light of this wavelength will be transmitted through an 8 mm thickness of the filter?
  b) What is the absorption coefficient?
  c) What is the absorbance of the 2 mm filter?

24.2 At 540 nm, the molar absorption coefficient of MnO_4^- ion is 202.5 m^2/mol. If 20.000 g of an alloy are dissolved, the manganese in the solution is oxidized to MnO_4^-, and the resulting solution is diluted in a volumetric flask to 500.0 mL, the transmittance measured at 540 nm is 0.325 in a 1.00 cm cell. What is the percent manganese in the sample?

24.3 Suppose 0.100 g of a dye having a molar mass of 425 g/mol is dissolved in 100 mL of alcohol. If 1.00 mL of the resulting solution is diluted to 250.0 mL, the transmittance measured in a 1.00 cm cell is 0.550, calculate the molar absorption coefficient of the dye at the wavelength used.

24.4 A dye has a molar absorption coefficient of 2.44×10^4 L mol^{-1} cm^{-1}. The dye is a weak monoprotic acid, and it can be shown that the absorption is due entirely to the anion. The transmittance at various values of pH is:

pH	3.01	3.64	4.12	5.23	5.79	6.17
T	0.976	0.912	0.812	0.613	0.586	0.580

Calculate the pK of the dye.

24.5 If the molar absorption coefficient of a solution is 475 m^2/mol, how thick a layer must be used to reduce the intensity of the transmitted beam to 20% of its initial value? The concentration is 0.126 mol/dm^3.

24.6 A bottle of $KMnO_4$ solution is 15 cm in diameter. What must the concentration of the salt be if 75% of the light at 540 nm is to pass through the bottle? The molar absorption coefficient of MnO_4^- is 202.5 m^2/mol at 540 nm.

24.7 The transmittance of the palladium complex of thio-Michler's ketone, $[(CH_3)_2NC_6H_4]_2CS$, at 520 nm is 0.380 at $c = 2.0 \times 10^{-6}$ mol/L and 0.617 at 1.0×10^{-6} mol/L in a 1.00 cm cell. Calculate the molar absorption coefficient.

24.8 For each of the electron configurations, s^2, sp, sd, pd, d^2, find the possible values of L, S, and J, and write the appropriate term symbols a) if the electrons have different principal quantum numbers; b) if the electrons have the same principal quantum number.

24.9 Which of the transitions are allowed in an atom?

  a) $^3P \leftrightarrow {}^1D$ e) $^2F \leftrightarrow {}^2G$ h) $^4F \leftrightarrow {}^2D$
  b) $^3P \leftrightarrow {}^3S$ f) $^3S \leftrightarrow {}^1S$ i) $^2F \leftrightarrow {}^2P$
  c) $^1P \leftrightarrow {}^1D$ g) $^3S \leftrightarrow {}^1P$ j) $^3P \leftrightarrow {}^3D$
  d) $^2F \leftrightarrow {}^1S$

24.10 Consider an atom with two electrons in the configuration $2p^2$. What are the possible values of the z component of the magnetic moment of the atom corresponding to each of the terms developed in Section 24.9?

24.11 Repeat Problem 24.10, but with two nonequivalent p electrons.

24.12 How many lines appear in the Zeeman effect for the transitions below, and what are the frequencies in terms of v_0 and the Larmor frequency? (The selection rules are: $\Delta J = 0, \pm 1$; $\Delta M_J = 0, \pm 1$; except that when $\Delta J = 0$, the combination of $M_J = 0$ with $M'_J = 0$ is not permitted.)

a) $^3S_1 \leftrightarrow {}^3P_2$; $^3S_1 \leftrightarrow {}^3P_1$; $^3S_1 \leftrightarrow {}^3P_0$.

b) $^3P_2 \leftrightarrow {}^3D_3$; $^3P_2 \leftrightarrow {}^3D_2$; $^3P_2 \leftrightarrow {}^3D_1$; $^3P_1 \leftrightarrow {}^3D_2$; $^3P_1 \leftrightarrow {}^3D_1$; $^3P_0 \leftrightarrow {}^3D_1$.

c) $^2P_{3/2} \leftrightarrow {}^2D_{5/2}$; $^2P_{3/2} \leftrightarrow {}^2D_{3/2}$.

24.13 a) In a magnetic field of 1 tesla, what are the resonant frequencies for the nuclei:

Nucleus	^{11}B	^{13}C	^{19}F	^{31}P
μ/μ_N	2.6880	0.70216	2.6273	1.1305
I	$\frac{3}{2}$	$\frac{1}{2}$	$\frac{1}{2}$	$\frac{1}{2}$

The selection rule is $\Delta M_I = \pm 1$.

b) At what value of the magnetic field will the nuclei in (a) resonate in a 60 MHz instrument? In a 100 MHz instrument?

24.14 One instrument uses a permanent magnet with $B = 0.1750$ tesla. What is the proton resonance frequency in this instrument?

24.15 Construct the nuclear spin functions for the system of 3 protons (for example, a CH_3 group) and show that the statistical weights of the four states, $\frac{3}{2}, \frac{1}{2}, -\frac{1}{2}, -\frac{3}{2}$ are 1, 3, 3, 1, respectively. Note that if α and β symbolize $M_I = +\frac{1}{2}$ and $-\frac{1}{2}$, respectively, the spin functions for the three protons are products such as $\alpha(1)\alpha(2)\alpha(3)$, $\alpha(1)\beta(2)\alpha(3)$, and so on.

24.16 A 30 kV electron strikes a target. What is the shortest wavelength x-ray that can be emitted?

24.17 The wavelengths of the K_{α_1} line for several elements are given below. Plot $\sqrt{v}$ versus Z. Determine the slope and the intercept on the horizontal axis of the straight line (Moseley's law).

Element	S	Cl	K	Ca	Sc
λ/pm	537.21	472.77	374.12	335.85	303.11

Element	Ti	V	Cr	Mn	Fe
λ/pm	274.84	250.34	228.96	210.17	193.60

24.18 The absorption of x-rays is governed by the expression, $I/I_0 = e^{-\mu x}$ where I_0 is the intensity incident on a slab of thickness x, I is the transmitted intensity, and μ is the absorption coefficient. The mass absorption coefficient is defined by μ/ρ where ρ is the density. Compare the transmittances of 1.0 cm slabs of each of the following elements. The mass absorption coefficient is for radiation with $\lambda = 20$ pm.

Element	C	Mg	Fe	Cu	Pt	Pb
$(\mu/\rho)/(\text{cm}^2/\text{g})$	0.175	0.250	1.10	1.55	4.25	4.90
$\rho/(\text{g/cm}^3)$	2.25	1.74	7.86	8.92	21.45	11.3

24.19 Using data from Problem 24.18, what thicknesses of Pb, Fe, and C would be required to reduce the intensity of 20 pm radiation to below 1 % of the incident intensity?

24.20 The K_{α_1} line of tungsten is at 208.99 pm. What is the minimum voltage that must be applied to the x-ray tube if the tungsten target is to emit the K_{α_1} line?

24.21 Oxygen gas was irradiated with Mg $K_{\alpha_1 \alpha_2}$ x-rays having an energy of 1253.6 eV. A peak appears for photoelectrons with kinetic energies of 710.5 eV. What is the binding energy of this electron in the O_2 molecule?

24.22 If water is irradiated with He I radiation having an energy of 21.22 eV, electrons with kinetic energies of about 3.0, 6.4, and 8.6 eV are emitted. (*Note:* these are approximate values of the center of the bands; compare with Fig. 24.30.) Calculate the binding energies of these electrons.

25

Molecular Spectroscopy

25.1 NUCLEAR MOTIONS; ROTATION AND VIBRATION

From the spectrum of a molecule we can obtain experimental information about the geometry of the molecule (bond lengths), and the energy states from which bond strengths are ultimately obtained. The molecular spectrum depends on the characteristics of the nuclear motions as well as on the electronic motions. In Section 23.1, by invoking the Born–Oppenheimer approximation, we discussed the electronic motion that produces the bonding between the atoms as a problem separate from that of the nuclear motions. We begin the discussion of molecular spectroscopy with a brief recapitulation of the description of the nuclear motions.

The motions of the nuclei are of three kinds: the translational motion of the molecule as a whole, which we discard as uninteresting; the rotation of the molecule; and the vibrations of the nuclei within the molecule. To a good approximation these motions are independent and can be discussed separately.

A molecule containing N atoms has $3N$ nuclear coordinates and $3N$ nuclear momenta; therefore there are $3N$ independent modes of motion or $3N$ *degrees of freedom*. Discarding three coordinates and three momenta that pertain to the translation of the whole molecule, there remain $3N - 3$ degrees of freedom. If the molecule is linear and the axis of the molecule is the z-axis, then two independent modes of rotation, about the x- and y-axis, are possible. For linear molecules the number of coordinates and momenta remaining to describe the vibrations is $3N - 3 - 2 = 3N - 5$. Nonlinear molecules have three independent modes of rotation about three mutually perpendicular axes, so the number of coordinates and momenta remaining to describe the vibrations is $3N - 3 - 3 = 3N - 6$. The number of modes of each type of motion is shown in Table 25.1.

Table 25.1

Molecule	Linear	Nonlinear
Total number of degrees of freedom	$3N$	$3N$
Number of translational degrees of freedom	3	3
Number of rotational degrees of freedom	2	3
Number of vibrational degrees of freedom	$3N - 5$	$3N - 6$

In addition to the selection rules restricting the changes in the quantum numbers, the presence or absence of a dipole moment in the molecule imposes a restriction on the appearance of lines and bands in the spectrum. If the transition between one vibrational or rotational state to another is to produce emission or absorption of radiation the vibration or rotation must be accompanied by an oscillation in the magnitude of the *dipole moment* of the molecule.

An electrical dipole consists of a positive and a negative charge, $+q$ and $-q$, separated by a distance r:

$$\overset{\textstyle r}{\underset{+q\quad -q}{\longleftarrow}}.$$

The dipole moment μ is defined by

$$\mu = qr,$$

and is a vector quantity; the direction is indicated by an arrow drawn from the negative to the positive charge. If the centers of positive and negative charge in a molecule do not coincide, the molecule has a permanent dipole moment.

Symmetrical (homonuclear) diatomic molecules such as H_2, O_2, N_2 do not have a permanent dipole moment, since an asymmetry in the electrical charge distribution is not possible. The symmetrical vibration does not alter the dipole moment, so these molecules do not emit or absorb in the infrared; the vibration is said to be *forbidden* in the infrared.

In a heteronuclear molecule such as HCl, the centers of positive and negative charge do not coincide, and the molecule has a permanent dipole moment. As this molecule vibrates, the displacement of the centers of charge varies and the magnitude of the dipole moment changes. The corresponding vibration–rotation band appears in the infrared. Rotation of the HCl molecule will produce an oscillation of the component of the dipole moment along a specified axis; hence, HCl has a pure rotational spectrum in the far infrared.

25.2 ROTATIONS

For simplicity, at first we restrict the discussion to diatomic molecules. The origin of coordinates is fixed at the center of mass of the molecule; if the nuclei lie on the z-axis, then the two independent modes of motion are rotation about the x-axis and about the y-axis. For either mode the moment of inertia is

$$I = \sum_i m_i r_i^2, \tag{25.1}$$

where m_i is the mass of the ith atom, and r_i is its perpendicular distance from the axis of rotation. Solution of the Schrödinger equation for this motion shows that the angular momentum M is quantized through the relation

$$M^2 = J(J + 1)\hbar^2, \qquad J = 0, 1, 2, \ldots, \tag{25.2}$$

where J, the rotational quantum number, may have any positive integral value or may be zero. The rotational energy is

$$E_J = \frac{M^2}{2I} = J(J+1)\frac{\hbar^2}{2I}. \qquad (25.3)$$

25.3 THE ROTATIONAL SPECTRUM

If a molecule changes from one rational state to another, the energy difference between the two states is made up by the emission or absorption of a quantum of radiation. For transitions between rotational states of linear molecules, the selection rule requires that $\Delta J = \pm 1$. The energy difference between these neighboring states is

$$E_{J+1} - E_J = \frac{\hbar^2}{2I}[(J+1)(J+2) - J(J+1)] = \frac{\hbar^2}{I}(J+1).$$

The frequency v_J associated with this transition is determined by $hv_J = E_{J+1} - E_J$; since $\hbar = h/2\pi$, we obtain

$$v_J = \frac{h}{4\pi^2 I}(J+1). \qquad (25.4)$$

It is customary to replace the frequency, v, by the equivalent wavenumber* of the light wave, $\tilde{v} = 1/\lambda = v/c$. Using this relation in Eq. (25.4) we obtain for the wavenumber

$$\tilde{v}_J = \frac{h}{4\pi^2 cI}(J+1) = 2B(J+1), \qquad (25.5)$$

where $B = h/8\pi^2 cI$ is the rotational constant for the particular molecule. The fundamental rotational frequency emitted in the transition from $J = 1$ to $J = 0$ is $2B$. For each value of J there is a line of frequency $\tilde{v}_J$ in the rotational spectrum. These lines are in the far infrared and microwave regions of the spectrum.

The spacing between the lines is $\tilde{v}_{J+1} - \tilde{v}_J = 2B$. Therefore, from the measured spacing between the rotational lines, the moment of inertia of the molecule can be determined. For a diatomic molecule the interatomic distance can be calculated immediately from the value of the moment of inertia.

Since the actual molecule is not a rigid rotor, it is necessary to provide for the effect of the rotation and vibration on the moment of inertia. The rotational energy levels are approximated by the expression

$$E_J = hcBJ(J+1) - hcD[J(J+1)]^2. \qquad (25.6)$$

The second term accounts for the increase in the moment of inertia at high rotational energies due to centrifugal stretching. The constant D is related to the vibrational frequency, $\tilde{v}_0$, by $D = 4B^3/\tilde{v}_0^2$. If the vibrational frequency is high, the atoms are tightly bound, D is small, and rotation does not change the moment of inertia very much.

---

*   The wavenumber $\tilde{v}$ is commonly referred to as a "frequency." For example, "Carbon dioxide has a characteristic frequency of 667 cm^{-1}." When the word "frequency" is used, it may refer to v in s^{-1} or to $\tilde{v}$ in cm^{-1}. Since the symbols are always different, this custom causes no difficulty. The SI unit for $\tilde{v}$ is m^{-1}; calculations using the fundamental constants therefore yield a value in m^{-1}; the literature values are almost all in cm^{-1}; 1 cm^{-1} = 100 m^{-1}.

The frequency of the line is $\tilde{v}_J = (E_{J+1} - E_J)/hc$; therefore

$$\tilde{v}_J = 2B(J + 1) - 4D(J + 1)^3. \tag{25.7}$$

The spacing between the lines is no longer constant, but decreases slightly with J.

$$\tilde{v}_{J+1} - \tilde{v}_J = 2B - 4D[3(J + 1)(J + 2) + 1]. \tag{25.8}$$

The moment of inertia also depends on the vibrational state of the molecule. To take this into account we can write

$$B_v = B_e - \alpha_e(v + \tfrac{1}{2}) \quad \text{and} \quad D_v = D_e + \beta_e(v + \tfrac{1}{2}), \tag{25.9}$$

where v is the vibrational quantum number. The higher the vibrational quantum state, the larger is the moment of inertia and the smaller is the value of B_v. Ordinarily it is sufficiently accurate to neglect the dependence of the centrifugal term on the vibrational state, so we set $\beta_e = 0$. The expression for the rotational energy becomes

$$E_J = hcB_e J(J + 1) - hc\alpha_e(v + \tfrac{1}{2})J(J + 1) - hcD[J(J + 1)]^2. \tag{25.10}$$

The frequency now depends slightly on the vibrational quantum number, $\tilde{v}_J = (E_{J+1} - E_J)/hc$.

$$\tilde{v}_J = 2B_e(J + 1) - 2\alpha_e(v + \tfrac{1}{2})(J + 1) - 4D(J + 1)^3. \tag{25.11}$$

To determine α_e and B_e we have to observe changes in the rotational state in molecules in two different vibrational states. The populations in the higher vibrational states are often very small and, consequently, the absorption lines are very weak. A high temperature measurement is required if the vibrational frequency is high.

We defer the discussion of the rotational spectra of polyatomic molecules until after our consideration of the vibrational–rotational spectra.

25.4 VIBRATIONS

For simplicity we assume that each of the molecular vibrations is a simple harmonic vibration characterized by an appropriate reduced mass μ and Hooke's law constant k. The wave functions are determined by a single quantum number v, the vibrational quantum number. The energy of the oscillator is

$$E_v = (v + \tfrac{1}{2})hv_0, \qquad v = 0, 1, 2, \ldots, \tag{25.12}$$

where $v_0 = (1/2\pi)\sqrt{k/\mu}$ is the classical vibration frequency. Each vibrational degree of freedom has a characteristic value of the fundamental frequency v_0.

Diatomic molecules provide the simplest example of molecular vibration. There is only one mode of vibration, the oscillation of the two atoms along the line of centers.

25.5 THE VIBRATION–ROTATION SPECTRUM

Molecules do not have a pure vibrational spectrum because the selection rules require a change in the vibrational state of the molecule to be accompanied by a change in the rotational state as well. As a result, in the infrared region of the spectrum there are vibration–rotation *bands*; each band consists of several closely spaced lines. The appearance of a band can be simply interpreted by supposing that the vibrational and rotational energies of the molecule are additive. For simplicity we consider a diatomic molecule; the energy is

$$E_{\text{vib-rot}} = hc\tilde{v}_0(v + \tfrac{1}{2}) + hcBJ(J + 1). \tag{25.13}$$

In the transition from the state with energy E' to that with energy E,

$$\Delta E = (E' - E)_{\text{vib-rot}} = hc\tilde{v}_0(v' - v) + hcB[J'(J' + 1) - J(J + 1)].$$

The selection rule for vibration is $\Delta v = \pm 1$; since the frequency emitted is $\tilde{v} = \Delta E/hc$, we have

$$\tilde{v} = \tilde{v}_0 + B[J'(J' + 1) - J(J + 1)]. \tag{25.14}$$

The selection rule for the rotational quantum number requires that either $J' = J + 1$ or $J' = J - 1$. Thus we obtain two sets of values for the frequency, designated by $\tilde{v}_R$ and $\tilde{v}_P$.

$$\text{If } J' = J + 1: \qquad \tilde{v}_R = \tilde{v}_0 + 2B(J + 1), \qquad J = 0, 1, 2, \ldots$$

$$\text{If } J' = J - 1: \qquad \tilde{v}_P = \tilde{v}_0 - 2BJ, \qquad J = 1, 2, 3, \ldots$$

These formulas can be simplified by writing both in the form

$$\left.\begin{array}{l} \tilde{v}_R = \tilde{v}_0 + 2BJ \\ \tilde{v}_P = \tilde{v}_0 - 2BJ \end{array}\right\}, \qquad J = 1, 2, 3, \ldots, \tag{25.15}$$

and excluding the value $J = 0$. The vibration–rotation band is made up of two sets of lines, the P branch and R branch. Since J may not be zero, the fundamental vibration frequency $\tilde{v}_0$ does not appear in the spectrum. The lines in the band appear on each side of $\tilde{v}_0$.

The vibration–rotation band for a molecule such as HCl is shown in Fig. 25.1. There is no absorption at the fundamental frequency $\tilde{v}_0$. The spacing between the lines $\Delta\tilde{v} = \tilde{v}_{J+1} - \tilde{v}_J = 2B$. Since B contains the moment of inertia, measurement of the spacing yields a value of I immediately. The spacing between the lines is the same, $2B$, in both the vibration–rotation band and in the pure rotational spectrum. The first line in the rotational spectrum is at the position $2B$. The *location* of the vibration–rotation band is determined by the vibrational frequency.

Just as we corrected the expressions for the rigid rotor to allow for the centrifugal effect and an interaction with the vibration, we also must adjust the expression for the harmonic oscillator to account for the anharmonicity in the oscillation. The potential energy surface for the molecule is not symmetrical (Fig. 25.2). The parabola (dotted figure) represents the potential energy of the harmonic oscillator. The correct potential energy is shown by the full lines; the vibration is anharmonic. The vibrational energy levels for such a system can be approximated by a series:

$$E_v = hv_0(v + \tfrac{1}{2}) - x_e hv_0(v + \tfrac{1}{2})^2 + y_e hv_0(v + \tfrac{1}{2})^3 + \cdots, \tag{25.16}$$

in which x_e and y_e are anharmonicity constants. Ordinarily the third term is negligible and will be omitted hereafter.

The anharmonicity correction reduces the energy of every level, but the reduction is greater for the higher levels. Thus the spacing between levels, $E_{v+1} - E_v$, gets smaller as v gets larger:

$$E_{v+1} - E_v = hv_0 - 2x_e hv_0(v + 1) = hv_0[1 - 2x_e(v + 1)]. \tag{25.17}$$

Combining Eq. (25.16) for vibration with Eq. (25.10) for rotation, we obtain for the vibrational–rotational energy of the molecule,

$$E_{vr} = hv_0(v + \tfrac{1}{2})[1 - x_e(v + \tfrac{1}{2})] + hc[B_e - \alpha_e(v + \tfrac{1}{2}) - DJ(J + 1)]J(J + 1). \tag{25.18}$$

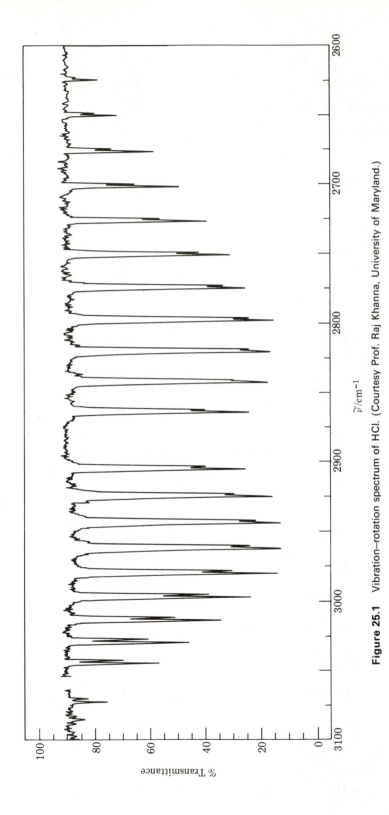

Figure 25.1   Vibration–rotation spectrum of HCl. (Courtesy Prof. Raj Khanna, University of Maryland.)

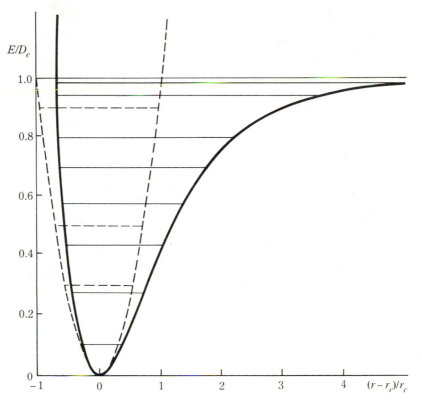

Figure 25.2 Potential energy curve for a diatomic molecule (full line) compared to that for the harmonic oscillator (dashed curves). Note differences in vibrational levels.

If the transition occurs between the state with v and J and the state with v' and J', then $\tilde{v} = [E(v', J') - E(v, J)]/hc$, or

$$\tilde{v} = \tilde{v}_0(v' - v)[1 - x_e(v' + v + 1)] + B_e(J' - J)(J' + J + 1)$$
$$-\alpha_e[J'(J' + 1)(v' + \tfrac{1}{2}) - J(J + 1)(v + \tfrac{1}{2})]$$
$$-D(J' - J)(J' + J + 1)[J'(J' + 1) + J(J + 1)]. \tag{25.19}$$

The application of this equation is conveniently divided into several cases.

Case 1. $\Delta v = 1$. The strict selection rule requires that $\Delta v = 1$, so that ordinarily the absorption is from $v = 0$ to $v' = 1$. Then

$$\tilde{v} = \tilde{v}_0(1 - 2x_e) + B_e(J' - J)(J' + J + 1) - \tfrac{1}{2}\alpha_e[3J'(J' + 1) - J(J + 1)]$$
$$-D(J' - J)(J' + J + 1)[J'(J' + 1) + J(J + 1)]. \tag{25.20}$$

The selection rule requires that $\Delta J = \pm 1$, so that if $J' = J + 1$, we have

$$\tilde{v}_R = \tilde{v}_0(1 - 2x_e) + 2B_e(J + 1) - \alpha_e(J + 1)(J + 3) - 4D(J + 1)^3, \qquad J = 0, 1, 2, \ldots \tag{25.21}$$

This corresponds to the R branch of the band, a group of lines all having frequencies above the vibrational frequency, $\tilde{v}_0(1 - 2x_e)$.

The other possibility is that $J' = J - 1$, then Eq. (25.20) becomes:

$$\tilde{v}_P = \tilde{v}_0(1 - 2x_e) - 2B_eJ - \alpha_eJ(J - 2) + 4DJ^3, \qquad J = 1, 2, 3, \ldots \qquad (25.22)$$

This is the expression for the low frequency branch of the band, the P branch, in which the group of lines all have frequencies below the vibrational frequency.

Case 2. $\Delta v \neq \pm 1$. For the anharmonic oscillator, the selection rule requiring that $\Delta v = \pm 1$ is no longer a rigid requirement. There is a small probability of transitions with $\Delta v = \pm 2$ and an even smaller probability of transitions with $\Delta v = \pm 3$. If we insert these conditions in Eq. (25.19), that is, $v' = 2, v = 0$, we can develop expressions analogous to Eqs. (25.21) and (25.22) for the R and P branches of a band centered on the first *overtone* frequency, $2\tilde{v}_0(1 - 3x_e)$, or approximately *twice* the fundamental vibrational frequency. This overtone band is much weaker than the fundamental band. If $v' = 3, v = 0$, there is a second overtone band that is much weaker than the first overtone band. The requirement, $\Delta J = \pm 1$, still applies.

25.6 ROTATIONAL AND VIBRATION–ROTATION SPECTRA OF POLYATOMIC MOLECULES

From the rotational or vibration–rotation spectra of a diatomic molecule such as HCl, we can calculate the internuclear distance directly from the measurement of the spacing of the lines. This simplicity is absent in polyatomic molecules except in certain special cases. We consider only a few simple examples to illustrate the difficulties.

■ **EXAMPLE 25.1** CO_2; triatomic, linear, symmetrical.

$$O{=}C{=}O$$

$$\vdash\!\!-\!\!-\!\!\vert\!\!-\!\!-\!\!-\!\!\dashv$$
$$r_1 = r_2$$

Since the molecule is symmetrical, the two bond distances are equal; thus only one distance appears in the moment of inertia. We can calculate this internuclear distance directly from the line spacings in the vibration–rotation band just as we did for HCl. No additional difficulty is involved.

The four vibrational modes for carbon dioxide are shown in Fig. 25.3. The first vibration is the totally symmetric stretching vibration, $\tilde{v}_1 = 1388.3$ cm^{-1}. This vibration does not produce a band in the infrared since it does not produce an oscillation in the dipole moment of the molecule. The second and third vibrations in Fig. 25.3 differ only in that the bending occurs in mutually perpendicular planes. Thus the vibration is doubly degenerate; the frequency is the same for both modes: $\tilde{v}_2 = 667.3$ cm^{-1}. The fourth mode is the asymmetric stretching vibration and has a third distinct frequency: $\tilde{v}_3 = 2349.3$ cm^{-1}.

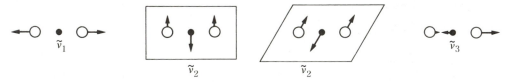

Figure 25.3   Vibrations of CO_2.

If we ignore anharmonicity, the vibrational energy of the CO_2 molecule can be written as a sum of terms:

$$E_v = hc\tilde{v}_1(v_1 + \tfrac{1}{2}) + hc\tilde{v}_2(v_2 + \tfrac{1}{2}) + hc\tilde{v}_2(v_2' + \tfrac{1}{2}) + hc\tilde{v}_3(v_3 + \tfrac{1}{2});$$

$$E_v = hc\tilde{v}_1(v_1 + \tfrac{1}{2}) + hc\tilde{v}_2(v_2 + v_2' + 1) + hc\tilde{v}_3(v_3 + \tfrac{1}{2}). \tag{25.23}$$

The quantum numbers, v_1, v_2, v_2', v_3, may have any integral value from zero upward. The expression in Eq. (25.23) is correct for any linear triatomic molecule.

The CO_2 molecule has two strong absorption bands centered on $\tilde{v}_2$ and $\tilde{v}_3$, since these modes produce an oscillating dipole moment. Since the bending vibration produces an oscillating dipole perpendicular to the molecular axis, $\Delta J = 0$ is permitted (in addition to the rotational selection rule, $\Delta J = \pm 1$). If we put $J' = J$ in Eq. (25.20), we obtain for the Q branch,

$$\tilde{v}_Q = \tilde{v}_0(1 - 2x_e) - \alpha_e J(J + 1), \qquad J = 0, 1, 2, \ldots \tag{25.24}$$

Thus the Q branch is centered on the frequency, $\tilde{v}_0(1 - 2x_e)$, which is very close to the fundamental frequency, $\tilde{v}_0$. Since α_e is very small, all the lines in the Q branch fall very close to the central one. The presence of the Q branch identifies the band as belonging to the bending vibration (Fig. 25.4).

■ **EXAMPLE 25.2** N_2O; triatomic, linear, asymmetrical.

In N_2O, there are two distances involved: r_{NN} and r_{NO}. The moment of inertia is given by $I = \Sigma\, m_i r_i^2$, in which r_i is the perpendicular distance of the mass m_i from the center of mass. The coordinate of the center of mass, X, is determined by the condition $\Sigma_i m_i(x_i - X) = 0$.

To calculate the center of mass of the molecule, we assign the origin on the x-axis to atom number 1, mass $= m_1$. Then

$$X = \frac{m_2 x_2 + m_3 x_3}{M},$$

and the moment of inertia is

$$I = m_1\left[\frac{m_2 x_2 + m_3 x_3}{M}\right]^2 + m_2\left[x_2 - \frac{m_2 x_2 + m_3 x_3}{M}\right]^2 + m_3\left[x_3 - \frac{m_2 x_2 + m_3 x_3}{M}\right]^2.$$

This can be arranged into the symmetrical form

$$I = \frac{1}{M}\left[m_1 m_2 x_{21}^2 + m_2 m_3 x_{32}^2 + m_1 m_3 x_{31}^2\right], \tag{25.25}$$

in which $M = m_1 + m_2 + m_3$, and $x_{21} = x_2 - x_1; x_{32} = x_3 - x_2; x_{31} = x_3 - x_1$. The two bond lengths are x_{21} and x_{32}, while $x_{31} = x_{32} + x_{21}$. In any event, the measurement of line spacings in the rotational spectrum yields a value of I, but this is not sufficient to determine the two bond distances. It is necessary to study an isotopic molecule to obtain another experimental value of the moment of inertia. For example, if

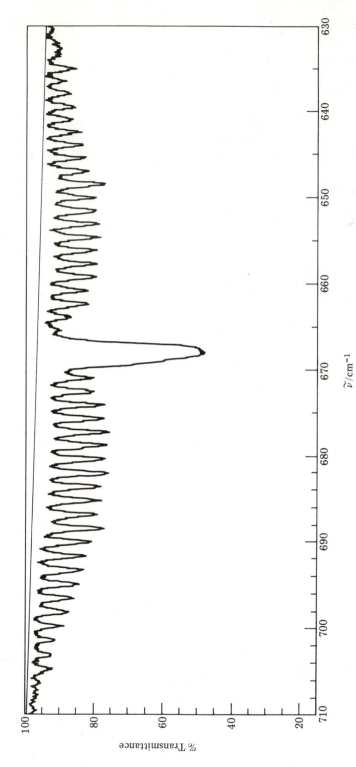

Figure 25.4   Vibration–notation spectrum for CO_2. (Courtesy Prof. Raj Khanna, University of Maryland.)

N=N=^{16}O and N=N=^{18}O are studied, it is possible to extract values for the two internuclear distances from the two measured values of I. This has been done using the microwave spectrum of N_2O.

If we look at the N_2O vibration–rotation spectrum, instead of two bands in the infrared, we find three. This is immediate evidence that the molecule does not have a center of symmetry. The frequencies are $\tilde{v}_1 = 1285.0$ cm^{-1} (symmetrical stretch), $\tilde{v}_2 = 588.8$ cm^{-1} (degenerate bending vibration), and $\tilde{v}_3 = 2223.5$ cm^{-1} (asymmetrical stretch).

■ **EXAMPLE 25.3**   Nonlinear molecules. Nonlinear molecules have three moments of inertia compared to the two equal moments possessed by linear molecules. This complicates matters considerably. We can sort these molecules into several types, depending on the relations between the three moments of inertia. For example, if I_A is the moment of inertia about the principal molecular axis while I_B and I_C are the moments about the axes perpendicular to the molecular axis, then in the linear molecule, $I_A = 0$ and $I_B = I_C = I$. For nonlinear molecules, if $I_A = I_B = I_C$, the molecule is called a spherical top. Examples are CH_4 and SF_6. Since these molecules have such high symmetry (for example, two or more threefold or higher axes), they do not have a pure rotational spectrum, since they have no dipole moment and therefore the rotation cannot produce an oscillating dipole moment.

If two of the moments of inertia are equal and the third is nonzero, the molecule is called a symmetric top. The principal axis must be threefold or higher. Symmetric-top molecules are of two types. If the moment of inertia around the axis of highest symmetry is less than the other two equal moments, the molecule is a prolate symmetric top. Examples of prolate (cigar-shaped) symmetric-top molecules are NH_3, CH_3Cl, or CH_3CN. If the moment of inertia about the axis of highest symmetry is greater than the other two equal moments, the molecule is an oblate symmetric top. Examples of oblate (pancake-shaped) symmetric top molecules are BCl_3, a trigonal planar molecule, and benzene, C_6H_6.

If no two of the moments of inertia are equal, the molecule is an asymmetric top. Most molecules, including such simple molecules as H_2O and SO_2, belong in this category.

The simplest nonlinear molecule is a symmetrical triatomic such as H_2O or SO_2. In these molecules, two parameters, the O—H or S—O bond length and the interbond angle, must be determined from the data. The relation between the three moments of inertia and the total rotational energy cannot be expressed simply. The microwave spectrum of water is complex but has been analyzed to yield the bond distance and interbond angle.

A nonlinear triatomic molecule such as H_2O has three vibrational modes. Two of these involve symmetrical motions of the three atoms; the third is an asymmetrical stretching vibration. These motions are illustrated in Fig. 25.5. Since all of these vibrations produce oscillations in the dipole moment, three fundamental bands centered on $\tilde{v}_1 = 3651.7$ cm^{-1}, $\tilde{v}_2 = 1595.0$ cm^{-1}, and $\tilde{v}_3 = 3755.8$ cm^{-1} appear in the infrared region.

Symmetrical
stretch
$\tilde{v}_1 = 3651.7$ cm^{-1}

Symmetrical
bend
$\tilde{v}_2 = 1595.0$ cm^{-1}

Asymmetrical
stretch
$\tilde{v}_3 = 3755.8$ cm^{-1} **Figure 25.5**

Table 25.2
Infrared absorption bands for water*

Final State			Frequency (cm^{-1})	Remarks
v_1	v_2	v_3		
0	1	0	1595	Fundamental, $\tilde{v}_2$
0	2	0	3152	First overtone, $2\tilde{v}_2$
1	0	0	3652	Fundamental, $\tilde{v}_1$
0	0	1	3756	Fundamental, $\tilde{v}_3$
0	1	1	5331	Combination, $\tilde{v}_2 + \tilde{v}_3$
1	1	1	8807	Combination, $\tilde{v}_1 + \tilde{v}_2 + \tilde{v}_3$
2	0	1	10613	Combination, $2\tilde{v}_1 + \tilde{v}_3$
0	0	3	11032	Second overtone, $3\tilde{v}_3$
2	1	1	12151	Combination, $2\tilde{v}_1 + \tilde{v}_2 + \tilde{v}_3$
0	1	3	12565	Combination, $\tilde{v}_2 + 3\tilde{v}_3$
3	1	1	15348	Combination, $3\tilde{v}_1 + \tilde{v}_2 + \tilde{v}_3$

*   A ground state with $v_1 = v_2 = v_3 = 0$ is assumed. Listed in order of increasing frequency. (From I. N. Levine, *Molecular Spectroscopy*, Wiley-Interscience, N.Y. 1975, p. 263.)

In addition to the bands centered on the fundamental frequencies, other bands appear in the spectra of polyatomic molecules. We have mentioned overtone bands in the spectrum of diatomic molecules due to violation of the selection rule, $\Delta v = \pm 1$, that is permitted because of anharmonicity. But in polyatomic molecules, *combination bands* also appear. For example, in the case of water if the absorbed quantum splits to raise v_1 from 0 to 1 and v_2 from $0 \rightarrow 1$, there will be a vibration–rotation band centered on the combination frequency, $\tilde{v}_1 + \tilde{v}_2$. This process is relatively less probable than the absorbtion of a single quantum at either fundamental frequency, so the intensity of the band is relatively weak. Nonetheless, combination bands appear with sufficient intensity to be an important feature of the infrared spectra of polyatomic molecules. Even in the case of a simple molecule like water, there are a large number of prominent bands, several of which are listed in Table 25.2.

25.7 APPLICATIONS OF INFRARED SPECTROSCOPY

Infrared spectroscopy is commonly used to identify and to determine quantitatively the amount of various substances present in mixtures. For the explanation of fundamental ideas we restricted our attention to the infrared spectra of very simple compounds. Even then, we observed that the spectrum can become very complicated. When a multitude of atoms is present, as in most organic compounds, the spectrum takes on a different appearance; much broader bands are in evidence.

The identification of a compound is based on the existence of characteristic group frequencies that have roughly the same value regardless of the compound in which the group appears. Typical of such groups are those with multiple bonds such as —C=O, —C≡N, —N=N—, —C=C—. The C=O group will ordinarily appear as a strong band in the region between 1650 cm^{-1} and 1850 cm^{-1}. The exact position depends on the type of compound; for example, in aliphatic ketones, the frequency is in the range from 1710 cm^{-1} to 1720 cm^{-1}. Another useful band is the —O—H stretching frequency that

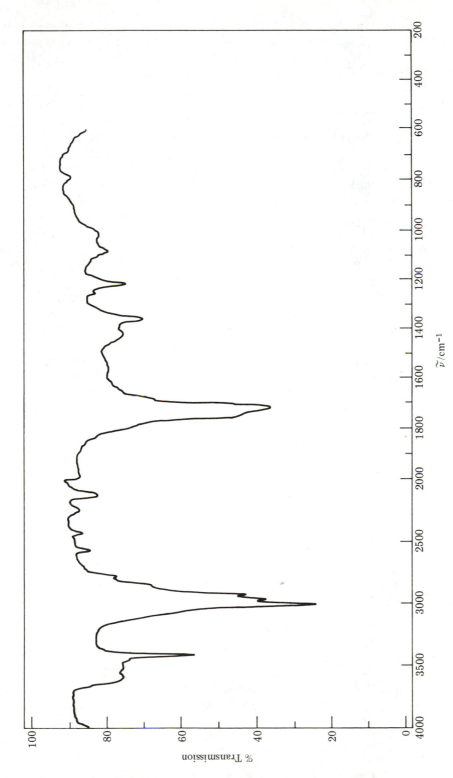

Figure 25.6 Infrared spectrum of liquid acetone. (Courtesy Carl Esche, Dept. of Chemistry, University of Maryland.)

appears between 2500 cm^{-1} and 3650 cm^{-1}. Use of the method requires some experience, since the frequencies shift about in different molecules. The spectrum of acetone is shown in Fig. 25.6. The band at 1715 cm^{-1} is the carbonyl frequency, while those at 3008, 2960, 2925, 1440, 1425 and 1361 cm^{-1} are due to the methyl group.

25.8 RAMAN EFFECT

If an intense beam of monochromatic light is passed through a substance, a small fraction of the light is scattered by the molecules in the system. The electron cloud in a molecule can be polarized (deformed) by an electric field. If we apply an oscillating electric field (the electric field vector of the light wave) to the molecule, the deformation of the electron cloud will oscillate with the frequency, v_0, of the incident light beam. This oscillation of the cloud produces an oscillating dipole that radiates at the same frequency as the incident light. This process is called Rayleigh scattering. The Rayleigh-scattered radiation is emitted in all directions; we can observe it by placing a detector on a line perpendicular to the direction of the incident light (Fig. 25.7). Since only about 0.1 % of the light is scattered, we must use as intense a source as possible. A laser fulfills this requirement admirably. The mirrors shown in the figure serve to collect as much of the scattered radiation as possible.

In addition to the Rayleigh scattering, another effect, Raman scattering, occurs. There is a small but finite probability that the incident radiation will transfer part of its energy to one of the vibrational or rotational modes of the molecule. As a result, the scattered radiation will have a frequency $v_0 - v_m$, where v_m is the molecular frequency. Similarly, there is a slight chance that molecules in excited vibrational or rotational states will give up energy to the light beam; in this case the scattered radiation will have a higher frequency, $v_0 + v_m$. Thus we observe three lines in the scattered radiation: one line at v_0 corresponding to the

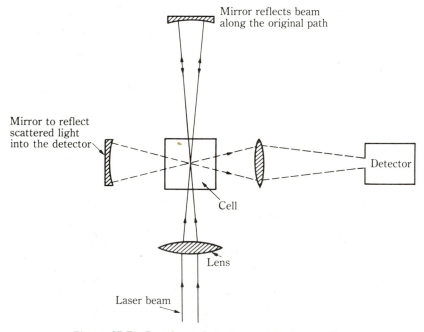

Figure 25.7   Experimental arrangement for Raman effect.

Rayleigh scattering; and two Raman lines, one at $v_0 + v_m$, the anti-Stokes line, and the other at $v_0 - v_m$, the Stokes line. The two Raman lines are extremely weak compared to the intensity of the Rayleigh scattered light. The total intensity of the scattered light is only about 1/1000 of the incident intensity. Only about 10^{-4} of the scattered intensity is in the Raman lines; thus the Raman intensity is less than 10^{-7} of the intensity of the incident light. The first observation of the Raman effect required very long exposures of a photographic plate to detect it.

The classical argument for the Raman effect can be expressed mathematically as follows. Let μ be the dipole moment induced by an electric field E; then the polarizability, α, is defined by

$$\mu = \alpha E; \tag{25.26}$$

that is, the polarizability is the dipole moment induced by unit electrical field. In a light beam, the electric field vector is given by

$$E = E^0 \sin 2\pi v_0 t, \tag{25.27}$$

where E^0 is the amplitude of the vibrating electric vector, v_0 is the frequency of the light beam, and t is the time. Then

$$\mu = \alpha E^0 \sin 2\pi v_0 t, \tag{25.28}$$

and the dipole moment oscillates with a frequency v_0 to yield the Rayleigh scattered beam. However, if the polarizability varies slightly as the molecular vibration occurs, we can write

$$\alpha = \alpha_0 + \left(\frac{\partial \alpha}{\partial q}\right) q, \tag{25.29}$$

in which q is the coordinate describing the molecular vibration; then

$$q = q_0 \sin 2\pi v_m t, \tag{25.30}$$

in which v_m is the frequency of the molecular vibration and q_0 is its amplitude. Combining this with the last equation, we have

$$\alpha = \alpha_0 + \left(\frac{\partial \alpha}{\partial q}\right) q_0 \sin 2\pi v_m t. \tag{25.31}$$

Placing this value of α in Eq. (25.28) yields

$$\mu = \alpha_0 E^0 \sin 2\pi v_0 t + \left(\frac{\partial \alpha}{\partial q}\right) q_0 E^0 \sin 2\pi v_0 t \sin 2\pi v_m t. \tag{25.32}$$

Using the identity, $\sin x \sin y = \frac{1}{2}[\cos (x - y) - \cos (x + y)]$, this becomes

$$\mu = \alpha_0 E^0 \sin 2\pi v_0 t + \frac{1}{2} \left(\frac{\partial \alpha}{\partial q}\right) q_0 E^0 \cos 2\pi(v_0 - v_m)t$$

$$- \frac{1}{2} \left(\frac{\partial \alpha}{\partial q}\right) q_0 E^0 \cos 2\pi(v_0 + v_m)t. \tag{25.33}$$

Thus the dipole moment oscillates with three distinct frequencies: v_0 with amplitude $\alpha_0 E^0$; $v_0 - v_m$ and $v_0 + v_m$ with much smaller amplitudes, $\frac{1}{2}(\partial \alpha/\partial q)q_0 E^0$. Therefore we observe a relatively intense beam at one frequency and two very weak beams at frequencies slightly above and below that of the intense one.

Quantum mechanically, as the incident photon is scattered, the molecule undergoes a transition from the lowest vibrational state to the state with $v = 1$. This requires the absorption of energy, so the scattered photon has a reduced energy, $h(v_0 - v_m)$; the scattered light is shifted toward the red end of the spectrum; this is the Stokes line. On the other hand, it may happen that as the photon is scattered the molecule make a transition from an upper state, $v = 1$, to the ground state. Then the energy of the scattered photon is greater, $h(v_0 + v_m)$; the light is shifted toward the blue end of the spectrum. Since there are fewer molecules in the upper state than in the lower state, the intensity of the blue-shifted line, the anti-Stokes line, is much less than that of the red-shifted line, the Stokes line.

Since the occurrence of the Raman effect depends on the change in polarizability as vibration occurs, the selection rules are different for the Raman effect than they are for the infrared spectrum. In particular, in molecules with a center of symmetry the totally symmetric vibration is Raman-active, but is forbidden in the infrared since it produces no change in dipole moment. Thus the homonuclear diatomic molecules, H_2, O_2, N_2, show the Raman effect but do not absorb in the infrared. There is also a purely rotational Raman effect in these molecules. However, in this case the selection rule is $\Delta J = \pm 2$. Thus we have for the rotational Stokes lines

$$\tilde{v}_J = \tilde{v}_0 - 2B(2J + 3), \qquad J = 0, 1, 2, \ldots$$

and the spacing between them,

$$\Delta\tilde{v} = 4B, \tag{25.34}$$

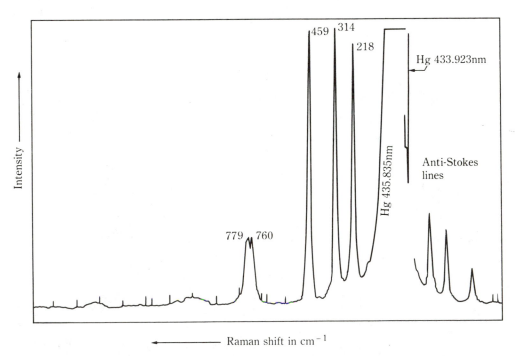

Figure 25.8 Raman spectrum of CCl_4 showing relative intensities of Stokes and anti-Stokes lines. Exciting line is the mercury 433.923 nm line. (Courtesy American Petroleum Institute Research Project; Raman Spectrum No. 351.)

which is twice the spacing in the far infrared or microwave regions for the pure rotational spectrum.

The Raman spectrum of CCl_4 is shown in Fig. 25.8.

25.9 ELECTRONIC SPECTRA

When we discussed the molecular orbital treatment for the hydrogen molecule, we obtained two energy levels of the molecule corresponding to the two wave functions:

Wave Function		Energy
$\psi_g = \dfrac{1}{\sqrt{2(1 + S)}} [\psi_a(1) + \psi_b(1)]$		E_g
$\psi_u = \dfrac{1}{\sqrt{2(1 - S)}} [\psi_a(1) - \psi_b(1)]$		E_u

This result was then generalized to the statement that we could make similar combinations of other wave functions on the atoms to yield symmetric and antisymmetric states of higher energy. We now undertake the systematic description of these electronic states.

In the case of diatomic molecules, we find that the Schrödinger equation requires that the component of the angular momentum along the molecular axis be quantized. The quantum number, Λ, describing this component is the basis for the term symbols for diatomic molecules. The quantum number Λ may have the values, $\Lambda = 0, 1, 2, \ldots$. For diatomic molecules, we use a Greek letter code for Λ.

Λ	0	1	2	3	$\cdots$
Letter code	Σ	Π	Δ	Φ	$\cdots$

The value of Λ is determined by the resultant value of L for all the electrons in the molecule; the possible values of Λ range from L to 0.

$$\Lambda = L, L - 1, L - 2, \ldots, 2, 1, 0. \tag{25.35}$$

Similarly, if we calculate the resultant value of the spin quantum number for the electrons outside the closed shells we obtain the total spin quantum number, S. The components of the total spin quantum numbers, $S, S - 1, \ldots, -(S - 1), -S$, are the possible values of the quantum number Σ, which corresponds to M_S in the atom. There are $2S + 1$ values of Σ; the multiplicity of the state is $2S + 1$. There is a total angular momentum quantum number, Ω, which has the values

$$\Omega = \Lambda + \Sigma. \tag{25.36}$$

Typical term symbols would be $^1\Sigma$, $^3\Sigma$, $^1\Pi$, $^3\Delta$, $^1\Delta$, $\ldots$, and so on. For example, for $^3\Pi$ the value of Λ is 1 and the value of Σ is 1, 0, -1. Then $\Omega = 2, 1, 0$ and the states would have the term symbols $^3\Pi_2$, $^3\Pi_1$, $^3\Pi_0$.

The Σ states, $\Lambda = 0$, are described as either Σ^+ or Σ^- depending on the behavior of the wave function upon reflection in the plane containing the internuclear axis. The Σ^+ wave function is invariant under this operation, while the Σ^- wave function changes sign on reflection in this plane. To the term symbols for homonuclear diatomic molecules, we attach a right-hand subscript, either g or u; the subscript is g if the wave function is invariant and u if it changes sign when all the electronic coordinates are subjected to inversion through the center of symmetry.

Thus the permissible kinds of functions for diatomic molecules are:

	Homonuclear		*Heteronuclear*
$\Lambda = 0$	$\Sigma_g^+, \Sigma_g^-, \Sigma_u^+, \Sigma_u^-$		Σ^+, Σ^-
$\Lambda = 1$	Π_g	Π_u	Π
$\Lambda = 2$	Δ_g	Δ_u	Δ

The selection rules are:

$$\Delta\Lambda = 0, \pm 1, \qquad \Delta S = 0.$$

Therefore transitions must occur only within the singlet system, or within the triplet system, and so on.

Between the Σ^+ and Σ^- states, the allowed transitions are

$$\Sigma^+ \leftrightarrow \Sigma^+ \qquad \text{and} \qquad \Sigma^- \leftrightarrow \Sigma^-.$$

The transition $\Sigma^+ \leftrightarrow \Sigma^-$ is forbidden. For homonuclear molecules the transition $g \leftrightarrow u$ is permitted; $g \leftrightarrow g$ and $u \leftrightarrow u$ are forbidden. These requirements also restrict the changes in the rotational quantum number to $\Delta J = \pm 1$. There is no restriction on the change in the vibrational quantum number.

If we ignore the interaction between the various modes, the energy of the molecule can be written as the sum of the electronic energy, the vibrational energy, and the rotational energy.

$$E = E_{\text{elec}} + E_{\text{vib}} + E_{\text{rot}}. \tag{25.37}$$

For a change in state, we obtain for the wave number,

$$\tilde{\nu} = \frac{\Delta E}{hc} = \frac{\Delta E_{\text{elec}}}{hc} + \frac{\Delta E_{\text{vib}}}{hc} + \frac{\Delta E_{\text{rot}}}{hc}. \tag{25.38}$$

The wave number associated with the transition can be written as a sum of wave numbers for each type of transition:

$$\tilde{\nu} = \tilde{\nu}_{\text{elec}} + \tilde{\nu}_{\text{vib}} + \tilde{\nu}_{\text{rot}} \tag{25.39}$$

Consider two electronic states between which a transition is permitted; the two curves in Fig. 25.9 show the variation in the electronic energy with internuclear separation in the two states. The vibrational energy levels are shown as the horizontal lines and are labeled with the vibrational quantum numbers. In addition, very closely spaced rotational levels are associated with every vibrational level; these are only shown next to the lowest level in each state.

If we examine the absorption of a quantum by a molecule in electronic state 1 and with vibrational quantum number zero, then the transition must occur along the vertical line *ab*. This is required by the Franck–Condon principle, which states that electronic transitions occur in times that are very short compared to the time required for the nuclei to move appreciably. This requires that the transition be "vertical" (r is constant).

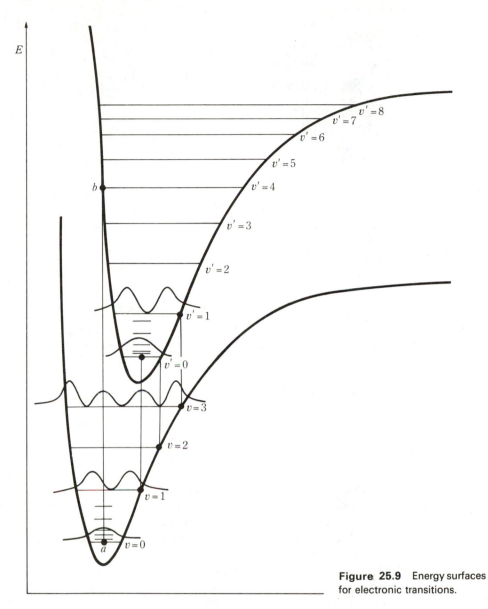

Figure 25.9 Energy surfaces for electronic transitions.

Since there is no selection rule for Δv, the jump will most likely occur between the positions at which the vibrational wave function has its largest values. For the ground state, this position is at the midpoint of the vibration, point a. For the excited states, the wave function has its largest values near the extremes of the vibration. The square of the vibrational wave function is sketched for the lower states in the figure. Thus the likely transition from the ground state (lower curve, $v = 0$) is to the vibrational level, $v' = 4$, which is nearest to where the vertical line from a intersects the energy surface of the upper state at point b. Other likely transitions are from $v = 1$ to $v' = 0$ and from $v = 3$ to $v' = 1$. Other transitions occur, but with somewhat lower probability. For example, the transitions from $v = 0$ to $v' = 0, 1, 2, 3, 5$, and higher all occur, in addition to the one from 0 to 4 shown in the figure.

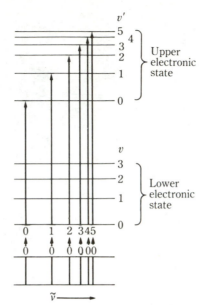

Figure 25.10

The frequency of the absorbed quantum corresponds to the energy difference represented by the vertical line. A system originating in the ground vibrational state, $v = 0$, can originate in any of the multitude of rotational states associated with $v = 0$ in the ground state and could end up in either of two matching rotational states determined by the selection rule, $\Delta J = \pm 1$, in the upper electronic and vibrational state. Thus a number of frequencies are absorbed. This group of closely spaced lines is called an *electronic band*. In any collection of molecules, there are some in various vibrational states; hence, transitions may occur between any vibrational state in the lower electronic state to whatever level matches properly in the upper state. Each one of these possible transitions produces a band; the collection of the bands is called a *band system*. Add to this the fact that there are several electronic states, not just two, between which transitions may occur. The result is that the electronic spectrum of a molecule consists of *several* band systems. By detailed analysis of the wavelengths of the lines in the spectrum, we can calculate the energy curves for the various electronic states as functions of r, the internuclear distances, the force constants of the vibrations, and bond energies.

Figure 25.10 shows schematically the absorption lines originating from the level, $v = 0$, in the ground electronic state; the rotational band around each of these lines is not shown. If this corresponds to Fig. 25.9, then by our argument the transition $0 \to 4$ should be most intense, with the intensity of the others decreasing on either side of that frequency. Note that the frequency differences become less as v' gets larger and that there is a high frequency-convergence limit.

The important point about the electronic spectrum is that we gain information about any vibration in the molecule, even though the fundamental frequency of that vibration might not appear in the infrared spectrum. For example, the vibrational frequency of homonuclear diatomic molecules does not appear in the infrared, since the oscillating dipole moment is absent; the bands resulting from that vibration do show up in the electronic spectrum.

The energies involved in electronic transitions are comparatively large, so the bands and band systems appear in the visible and ultraviolet regions of the spectrum; for example, the violet color of iodine is due to an electronic transition (Fig. 25.11).

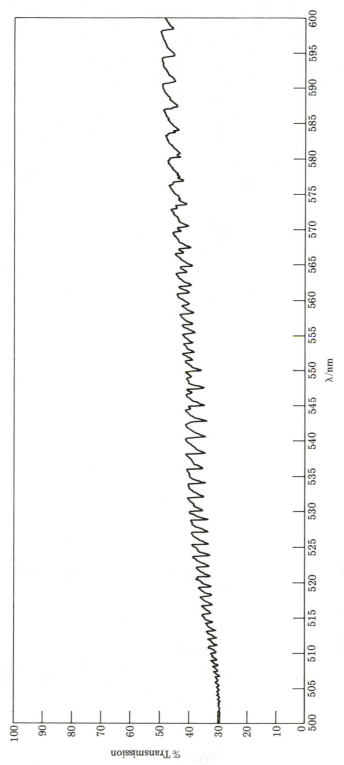

Figure 25.11   Electronic spectrum of I_2. (Courtesy Prof. Raj Khanna and Dr. David Stranz, University of Maryland.)

25.10 ELECTRONIC SPECTRA OF POLYATOMIC MOLECULES

The electronic spectrum of a nonlinear polyatomic molecule is very complicated. In addition to three modes of rotation with distinct moments of inertia, there are $3N - 6$ modes of vibration. While some of these may be forbidden in the infrared or Raman spectrum on the basis of symmetry, there is no rule to forbid their appearance in the electronic spectrum, which is extraordinarily complex as a consequence. For our purposes here, we mention only a few fundamental points and present one example.

The wave functions and levels in polyatomic molecules are described in terms of their symmetry. (See Sections 23.14 and 23.16.2). For example, if we consider the water molecule, its symmetry requires that any molecular wave function either be invariant or change only in algebraic sign under any symmetry operation. This requirement severely restricts the form of the wave functions. These wave functions can be of only four types—denoted by the letters a_1, a_2, b_1, and b_2—each of which belongs to a particular symmetry species. The symmetry properties of each type are summarized in the character table of the group C_{2v}, Table 23.5.

As we pointed out in Section 23.16.2, the ground state electronic configuration for the water molecule is

$$K(2a_1)^2(1b_2)^2(3a_1)^2(1b_1)^2,$$

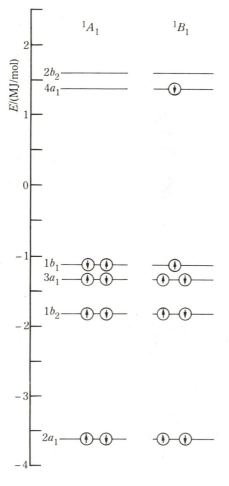

Figure 25.12 Configurations of the ground state $^1A_1$ and the first state $^1B_1$ of the H_2O molecule.

where a_1, b_1, and b_2 denote the symmetry species of the molecular orbitals. Since all the levels contain two electrons, the overall wave function for the molecule is totally symmetric and described by the symbol $^1A_1$ (read "singlet ay one"). This is the term symbol for the ground state; the left-hand superscript is the multiplicity of the state.

The lowest excited configuration for water is

$$H_2O \qquad K(2a_1)^2(1b_2)^2(3a_1)^2(1b_1)^1(4a_1)^1,$$

in which one electron has been moved from the $(1b_1)$ level to the $(4a_1)$ level (Fig. 25.12). The symmetry of this state is B_1. Since the electrons may now be either paired or unpaired, this state may be $^1B_1$ or $^3B_1$. The $^3B_1$ state has a lower energy than the $^1B_1$ state but, of course, the singlet–triplet transition from $^1A_1$ to $^3B_1$ is spin-forbidden.

The symmetry properties also enable us to establish the selection rules for transitions between states. For the group C_{2v}, transitions between a_1 and a_2 states and between b_1 and b_2 states are forbidden. All others are allowed. Thus the transition between $^1A_1 \leftrightarrow {}^1B_1$ is allowed by both spin and symmetry, and is a possible absorption mode for water. This transition involves a very large energy difference; in this connection, it is notable that the longest wavelength electronic absorption in water is in the region between 150 and 180 nm of the vacuum ultraviolet region.

★ 25.11   QUANTUM MECHANICAL DESCRIPTION OF TIME-DEPENDENT SYSTEMS

Until now in our quantum mechanical discussions we have described the stationary or time-independent states of a system. Furthermore, our language was such as to imply that we had sufficient information about the system to know that it was in a particular state described by a particular set of quantum numbers. For example, in the case of the harmonic oscillator we spoke as though we knew that the oscillator was in the vth state with wave function ψ_v, and energy $E_v = (v + \frac{1}{2})h\nu$; or, in the case of the hydrogen atom, that it was in a state described by the set of numbers n, l, m. This approach is very useful in a first discussion of quantum mechanical properties of various kinds of systems. However, we do not have reason to presuppose that a system is in a particular quantum state.

Having obtained the set of particular solutions of the Schrödinger equation, the set of ψ_n, the general solution is a linear combination of the ψ_n, namely,

$$\psi = \sum_n a_n \psi_n. \tag{25.40}$$

The normalization requirement is

$$\int \psi^* \psi \, d\tau = 1. \tag{25.41}$$

This is the total probability of finding the system in some state. If we use Eq. (25.40) in Eq. (25.41), we have

$$1 = \int \psi^* \psi \, d\tau$$

$$= \int \left(\sum_m a_m^* \psi_m^* \right) \left(\sum_n a_n \psi_n \right) d\tau$$

$$= \sum_m \sum_n a_m^* a_n \int \psi_m^* \psi_n \, d\tau.$$

Since the particular solutions are orthonormal we have

$$\int \psi_m^* \psi_n \, d\tau = \delta_{mn},$$ (25.42)

so that

$$1 = \sum_m \sum_n a_m^* a_n \delta_{mn}.$$

Summation over m yields

$$1 = \sum_n a_n^* a_n = \sum_n |a_n|^2.$$ (25.43)

Equation (25.43) says that the sum of the squares of the absolute value of the coefficients a_n in the series in Eq. (25.40) is unity. The manner in which we obtained Eq. (25.43) requires that we interpret the right-hand side as a sum of probabilities. Therefore, we interpret $|a_n|^2$ as the probability of finding the system in the state described by ψ_n. According to Eq. (25.43), the total probability of finding the system in one or another of the eigenstates is unity.

For example, suppose $|a_k|^2 = 1$, and $|a_n|^2 = 0$ for all $n \neq k$. Then the system is in state k and is described as completely as possible by ψ_k. This is the terminology, referred to above, that we have used so far. On the other hand, if we envision transitions occurring that take the system from one state to another, this implies that the a_n are functions of time, $a_n(t)$. Suppose that at $t = 0$ the system is in state k, while at some later time, t, there is a probability that the system is in state m, as a result of transitions occurring in the time interval t. This situation could be described by writing

$$|a_k(0)|^2 = 1 \qquad\qquad |a_k(t)|^2 < 1$$

$$|a_m(0)|^2 = 0 \qquad (m \neq k) \qquad |a_m(t)|^2 > 0.$$

Our problem is to discover how these coefficients, $a_n(t)$, depend on the time. Since one way for transitions to occur is by absorption or emission of radiation, this problem is central to any discussion of spectroscopy.

★ 25.12 VARIATION IN THE STATE OF A SYSTEM WITH TIME

If at time zero, the probability of the system being in the nth state is unity, and at time t the probability of the system being in the mth state is $|a_m(t)|^2$, then the task is to discover how the coefficient a_m varies with time if the system is irradiated by a light beam. In this situation it can be shown that

$$\frac{da_m}{dt} = -\frac{i}{\hbar} \int \Psi_m^{0*} \mathbf{H}' \psi_n^0 \, d\tau$$ (25.44)

in which Ψ_m^0 and Ψ_n^0 are the time-dependent wave functions,

$$\Psi_n^0(x, y, z, t) = \psi_n^0(x, y, z)e^{-iE_n t/\hbar}.$$ (25.45)

and $\mathbf{H}'$ is the operator corresponding to the energy of interaction of the charges in the molecule with the electrical field associated with the incident light wave.

The oscillating electrical field associated with the incident wave displaces the charges in the molecule slightly. This displacement in the x direction corresponds to an energy given by $E_x(\Sigma \, e_j x_j) = \mu_x E_x$, where E_x is the component of the electrical field in the x direction and $\mu_x = \Sigma \, e_j x_j$, is the induced dipole moment in the x direction. The sum is over all the charges, e_j, multiplied by their displacements in the x direction, x_j. Since the field is oscillating, we have $E_x = E_x^0 \cos \omega t$, where $\omega = 2\pi\nu$ is the angular frequency of the light wave and E_x^0 is the amplitude of the x component of the field. The total energy effect of the light wave includes the effects in the y and z directions as well and is given by

$$H' = (\mu_x E_x^0 + \mu_y E_y^0 + \mu_z E_z^0) \cos \omega t \tag{25.46}$$

Equation (25.44) tells us how the coefficient a_m depends on the time by giving an expression for the derivative (da_m/dt). From this equation we can obtain an expression for $a_m(t)$.

According to Eq. (25.46), the first term in H' is proportional to μ_x. Since the integration in Eq. (25.44) is over coordinates only, the time-dependent factors in H' and in the wave functions can be removed from the integral. Thus we see that the integrals in Eq. (25.44) are proportional to integrals of the type

$$(\mu_x)_{mn} = \int \psi_m^{0*} \mu_x \psi_n^0 \, d\tau, \quad (\mu_y)_{mn} = \int \psi_m^{0*} \mu_y \psi_n^0 \, d\tau, \quad (\mu_z)_{mn} = \int \psi_m^{0*} \mu_z \psi_n^0 \, d\tau. \tag{25.47}$$

These are called the transition-moment integrals. We define the quantity μ_{mn}^2 by

$$\mu_{mn}^2 = |(\mu_x)_{mn}|^2 + |(\mu_y)_{mn}|^2 + |(\mu_z)_{mn}|^2. \tag{25.48}$$

If we solve Eq. (25.44) under the condition that at $t = 0$ the system was in state n, we finally obtain the result that

$$|a_m(t)|^2 = \frac{\rho(\nu_{mn})\mu_{mn}^2 t}{6\epsilon_0 \hbar^2}, \tag{25.49}$$

in which $\rho(\nu_{mn})$ is the radiation density at the frequency ν_{mn} defined by $h\nu_{mn} = |E_m - E_n|$ and ϵ_0 is the permittivity of vacuum.

The probability of transition from state n to m in unit time is given by

$$\frac{\rho(\nu_{mn})\mu_{mn}^2}{6\epsilon_0 \hbar^2} = B_{nm} \rho(\nu_{mn}), \tag{25.50}$$

where

$$B_{mn} = \frac{\mu_{mn}^2}{6\epsilon_0 \hbar^2} \tag{25.51}$$

is the Einstein coefficient of absorption for radiation of frequency, ν_{mn}. Equation (25.49) is a fundamental result, crucial to the understanding of spectroscopy.

For a system which at time $t = 0$ was in state n, in the presence of a light beam a short time t later, the probability of finding the system in state m is given by the value of $|a_m(t)|^2$ in Eq. (25.49). This probability is proportional to t. It is also proportional to the energy density, $\rho(\nu_{mn})$, of the light beam at the frequency that "fits" the transition from n to m; $\nu_{mn} = |E_m - E_n|/h$. If the light beam contains only a minor component at ν_{mn}, then $\rho(\nu_{mn})$ will be small and the probability of transition from n to m will be small. Most importantly, the probability of finding the system in state m is proportional to the square of the absolute value of the *transition moment*, μ_{mn}. This last dependence is the basis for *selection rules* that govern the appearance or nonappearance of lines in a spectrum. If $\mu_{mn} = 0$, then the transition from state n to state m is a *forbidden transition*.

★ 25.13 SELECTION RULES FOR THE HARMONIC OSCILLATOR

Consider a diatomic molecule that has a dipole moment due to effective charges $-q$ and $-q$ separated by a distance r; then $\mu = qr$; we rewrite this as

$$\mu = q(r_e + r - r_e) = qr_e + q(r - r_e);$$
$$\mu = \mu_e + q(r - r_e), \tag{25.52}$$

where μ_e is the dipole moment the molecule would have if the charges were at rest at the equilibrium separation r_e, and the term $q(r - r_e)$ is the variation in the dipole moment due to the variation in r. We may regard this expression as the first two terms of a Taylor series expansion of μ in terms of $r - r_e$; that is,

$$\mu = \mu_e + \left(\frac{d\mu}{dr}\right)_{r_e} (r - r_e).$$

Then, taking r in the direction of the field, the dipole transition-moment integrals have the form,

$$\mu_{mn} = \int_{-\infty}^{\infty} \psi_m^{0*} \left[\mu_e + \left(\frac{d\mu}{dr}\right)_{r_e} (r - r_e)\right] \psi_n^0 \, dr,$$

$$\mu_{mn} = \mu_e \int_{-\infty}^{\infty} \psi_m^{0*} \psi_n^0 \, dr + \left(\frac{d\mu}{dr}\right)_{r_e} \int_{-\infty}^{\infty} \psi_m^{0*}(r - r_e)\psi_n^0 \, dr \tag{25.53}$$

Since $m \neq n$, the first integral vanishes because ψ_m^0 and ψ_n^0 are orthogonal. The normalized harmonic oscillator functions are given by Eq. (21.42),

$$\psi_n^0 = \left(\frac{1}{\beta\sqrt{\pi}\,2^n n!}\right)^{1/2} H_n(\xi)e^{-\xi^2/2} = \frac{A_n}{\beta^{1/2}} H_n(\xi)e^{-\xi^2/2}, \tag{25.54}$$

in which $\xi = (r - r_e)/\beta$. Using this value in Eq. (25.53) reduces it to

$$\mu_{mn} = \beta\left(\frac{d\mu}{dr}\right)_{r_e} A_m A_n \int_{-\infty}^{\infty} H_m(\xi)\xi H_n(\xi)e^{-\xi^2} \, d\xi.$$

To evaluate this integral we use the recurrence relation Eq. (21.49); that is,

$$\xi H_n(\xi) = nH_{n-1}(\xi) + \tfrac{1}{2}H_{n+1}(\xi).$$

This brings μ_{mn} to the form

$$\mu_{mn} = \beta\left(\frac{d\mu}{dr}\right)_{r_e} A_m A_n \left[n \int_{-\infty}^{\infty} H_m(\xi)H_{n-1}(\xi)e^{-\xi^2} \, d\xi + \frac{1}{2} \int_{-\infty}^{\infty} H_m(\xi)H_{n+1}(\xi)e^{-\xi^2} \, d\xi\right].$$

But, by Eq. (21.47),

$$\int_{-\infty}^{\infty} H_m(\xi)H_{n-1}(\xi)e^{-\xi^2} \, d\xi = \frac{\delta_{m,n-1}}{A_m A_{n-1}},$$

and

$$\int_{-\infty}^{\infty} H_m(\xi)H_{n+1}(\xi)e^{-\xi^2} \, d\xi = \frac{\delta_{m,n+1}}{A_m A_{n+1}}.$$

This yields for μ_{mn}

$$\mu_{mn} = \beta \left(\frac{d\mu}{dr}\right)_{r_e} \left[n \frac{A_n}{A_{n-1}} \delta_{m,\,n-1} + \frac{1}{2} \frac{A_n}{A_{n+1}} \delta_{m,\,n+1} \right].$$

Using the value of A_n from Eq. (25.54), we find that

$$\mu_{mn} = \beta \left(\frac{d\mu}{dr}\right)_{r_e} \left[\sqrt{\frac{n}{2}} \, \delta_{m,\,n-1} + \sqrt{\frac{n+1}{2}} \, \delta_{m,\,n+1} \right]. \tag{25.55}$$

When $m = n + 1$, the system is making the transition $n \to n + 1$, so the radiation is absorbed;

$$\mu_{n+1,\,n}^2 = \frac{\beta^2}{2} \left(\frac{d\mu}{dr}\right)_{r_e}^2 (n + 1), \qquad n = 0, 1, 2, \ldots \tag{25.56}$$

When $m = n - 1$, the final state m is lower in energy than the initial state n, so radiation is emitted (*stimulated emission*). The impinging light wave stimulates an excited molecule to emit radiation. We have

$$\mu_{n-1,\,n}^2 = \frac{\beta^2}{2} \left(\frac{d\mu}{dr}\right)_{r_e}^2 n, \qquad n = 1, 2, \ldots \tag{25.57}$$

Note that the lowest possible value of n in this last formula is $n = 1$.

The selection rule for the harmonic oscillator, Eq. (25.55), requires that $\Delta n = \pm 1$. Under the influence of a light beam the harmonic oscillator makes transitions only to states immediately above and below its original state. The existence of selection rules simplifies the interpretation of spectra enormously.

The other requirement on the transition is that the derivative $(d\mu/dr)_{r_e}$ be non-vanishing. Whether or not a molecule has a dipole moment is not significant; the dipole moment *must change as the vibration occurs*. For example, the molecule HCl has a permanent dipole moment and as the molecule vibrates this dipole moment varies. The molecule therefore has $(d\mu/dr) \neq 0$ and exhibits a vibrational spectrum. In contrast consider the CO_2 molecule, which has no permanent dipole moment because of the symmetrical distribution of positive and negative charges; in the symmetric vibration, Fig. 25.3, the charge symmetry is undisturbed and the dipole moment remains zero throughout this vibration. Hence, $d\mu/dr = 0$ and this vibration does not appear in the spectrum. In the asymmetric stretching vibration, the symmetry is destroyed. The dipole moment varies during this vibration and therefore this vibration appears in the spectrum. In the remaining two bending vibrations (degenerate since they differ only in the plane in which the vibration occurs) the symmetry is destroyed and the dipole moment varies during the oscillation. These vibrations appear in the spectrum.

★ 25.14 SELECTION RULES AND SYMMETRY

For a system in state n, Eq. (25.49) shows that the probability of finding it in state m at a later time is proportional to μ_{mn}^2. The transition moment, μ_{mn}, has components that are given by integrals such as

$$(\mu_x)_{mn} = q \int \psi_m^{0*} x \psi_n^0 \, d\tau, \tag{25.58}$$

where x could represent any one of the coordinates x, y, or z. We can frequently identify from consideration of the symmetry of the system the combinations of m and n for which the integral will be nonvanishing. These combinations of m and n are the "allowed" transitions for the system. If the integral is zero for a particular combination of n and m, the transition has zero probability of occurring; it is a "forbidden" transition. As we have seen in the preceding sections via detailed computations, Eq. (25.58) provides the basis for establishing *selection rules*. The selection rules govern the type of states between which transitions may or may not occur.

From symmetry we can establish a general selection rule with a minimum of computation. To begin, we consider a general wave function, $\psi(x, y, z)$, for any system together with the integral for the total probability of finding the particle,

$$\int \psi^*(x, y, z)\psi(x, y, z)\, d\tau = 1. \tag{25.59}$$

Since the integral in Eq. (25.59) is a real physical quantity, its value cannot depend on the orientation of the coordinate system. Consequently, the integrand, $\psi^*\psi$, must be invariant under transformations of the coordinate system. This invariance can obtain only if ψ is invariant or merely changes sign under transformations of the coordinate system.

Suppose that we subject the system to the operation of *inversion*, symbolized by the operator $\mathbf{i}$, which reverses the direction of all three axes. This operation simply changes the sign of all the coordinates; thus

$$\mathbf{i}\psi(x, y, z) = \psi(-x, -y, -z).$$

If we operate on the integrand in Eq. (25.59), we obtain

$$\mathbf{i}[\psi^*(x, y, z)\psi(x, y, z)] = [\mathbf{i}\psi^*(x, y, z)][\mathbf{i}\psi(x, y, z)]$$
$$= \psi^*(-x, -y, -z)\psi(-x, -y, -z).$$

If the integrand is to be unchanged by this operation, it is clear that the worst that may happen to the wave function is that it changes sign. Thus we have the two possibilities alluded to above:

$$\mathbf{i}\psi_g(x, y, z) = \psi_g(-x, -y, -z) = \psi_g(x, y, z), \qquad \text{(symmetric)}$$
$$\mathbf{i}\psi_u(x, y, z) = \psi_u(-x, -y, -z) = -\psi_u(x, y, z), \qquad \text{(antisymmetric)}$$

and similarly for the complex conjugate. In the first case, the wave function is said to be symmetric under inversion or is an "even" function; in the second case, the wave function is antisymmetric under inversion or is an "odd" function. The subscripts g and u (from the initial letters of the German: *gerade* = even; *ungerade* = odd) are used to describe these two kinds of wave function.

If the inversion operation is applied to the integrand of the transition-moment integral in Eq. (25.58), we note that the coordinate x (or y or z) is an odd function under inversion; $\mathbf{i}x = -x$. Thus, if $\psi_m^* x \psi_n$ is to be invariant under inversion, the product, $\psi_m^* \psi_n$ must be an odd function. This can only be so if ψ_m is even and ψ_n is odd or vice versa. Thus we have the important result that transitions are allowed only between odd and even states, $g \leftrightarrow u$. Transitions between two odd states, $u \leftrightarrow u$, or between two even states, $g \leftrightarrow g$, are forbidden. This is a fundamental selection rule for dipole radiation. If the system contains several particles, the argument is unchanged and leads to the same result.

If we seek to apply this rule in the case of the simple harmonic oscillator, we write the transition moment integral

$$\int_{-\infty}^{\infty} H_m(\xi)\xi H_n(\xi)e^{-\xi^2} \, d\xi.$$

Applying the inversion operator to the Hermite polynomial yields $iH_n(\xi) = H_n(-\xi) = (-1)^n H_n(\xi)$. The last equality is one of the properties of the Hermite polynomials that is easily obtained from the definition in Eq. (21.46). Thus, $H_n(\xi)$ is even or odd depending on whether n is even or odd. Therefore the quantum number must change from even to odd or from odd to even in an allowed transition. Detailed evaluation of the integral (Section 25.13) using the recurrence formula for Hermite polynomials shows that the integral vanishes unless $m = n \pm 1$. The selection rule is generally written $\Delta n = \pm 1$.

To obtain the selection rules for the rigid rotor we must look at the symmetry of the problem in slightly greater detail. The rotor is described by the two angles θ and ϕ and by a wave function having the form (omitting normalization constants)

$$\psi_{J,m} = P_J^m(\cos \theta)e^{im\phi}. \tag{25.60}$$

Suppose we fix the direction of the z-axis and then choose the position of the x- and y-axes so that the angle ϕ is established. Then it is clear that if we rotate the x- and y-axes about the z-axis to some new position which changes ϕ to $\phi + \alpha$ in the new coordinate system, nothing is changed physically. We symbolize this rotation operation by $\mathbf{C}_\alpha$. This operation does not affect θ in the slightest. Examining the effect on the function $e^{im\phi}$ we find

$$\mathbf{C}_\alpha e^{im\phi} = e^{im(\phi + \alpha)} = e^{im\alpha}e^{im\phi}. \tag{25.61}$$

Similarly, for the complex conjugate, $e^{-im\phi}$:

$$\mathbf{C}_\alpha e^{-im\phi} = e^{-im\alpha}e^{-im\phi}. \tag{25.62}$$

To find the effect of $\mathbf{C}_\alpha$ on $x = r \sin \theta \cos \phi$ and $y = r \sin \theta \sin \phi$, we construct the sum and difference of x and iy; since $\cos \phi + i \sin \phi = e^{i\phi}$, we have

$$x + iy = r \sin \theta e^{i\phi},$$
$$x - iy = r \sin \theta e^{-i\phi}.$$

Then

$$\mathbf{C}_\alpha(x + iy) = r \sin \theta \mathbf{C}_\alpha e^{i\phi} = r \sin \theta e^{i(\phi + \alpha)} = (x + iy)e^{i\alpha},$$
$$\mathbf{C}_\alpha(x - iy) = (x - iy)e^{-i\alpha}.$$

Next we consider the combination of transition-moment integrals, defined by

$$\int \psi_{J',m'}^*(x + iy)\psi_{J,m} \, d\tau = \mu_x + i\mu_y.$$

Then

$$\mu_x + i\mu_y = r \int_0^\pi P_{J'}^{|m'|}(\cos \theta) \sin \theta P_J^{|m|}(\cos \theta) \sin \theta \, d\theta \int_0^{2\pi} e^{-im'\phi}e^{i\phi}e^{im\phi} \, d\phi. \tag{25.63}$$

Consider the integral over ϕ in Eq. (25.63); let

$$I_{+1}(m', m) = \int_0^{2\pi} e^{i(m - m' + 1)\phi} \, d\phi.$$

Then

$$\mathbf{C}_\alpha I_{+1}(m', m) = \int_0^{2\pi} \mathbf{C}_\alpha e^{i(m - m' + 1)\phi} \, d\phi = e^{i(m - m' + 1)\alpha} I_{+1}(m', m).$$

Since the physical situation requires that this integral be independent of α, we must have

$$\mathbf{C}_\alpha I_{+1}(m', m) = I_{+1}(m', m).$$

It follows that only if $e^{i(m - m' + 1)\alpha} = 1$, that is, if $m - m' + 1 = 0$, will the integral be independent of α. Let $\Delta m = m' - m$; then $\Delta m = +1$ for the integral to be invariant under the rotation. Similarly, if we consider the integral

$$I_{-1}(m', m) = \int_0^{2\pi} e^{i(m - m' - 1)\phi} \, d\phi,$$

which will appear in the combination, $\mu_x - i\mu_y$, we find that the integral is invariant under rotation if $\Delta m = -1$. In the case of the z component of the transition moment integral, $z = r \cos \theta$; consequently, z is independent of ϕ and the integral

$$I_0(m', m) = \int_0^{2\pi} e^{i(m - m')\phi} \, d\phi$$

must be invariant under rotation. This invariance requires $\Delta m = 0$.

The selection rules on m can be restated briefly: for the x or y component of the transition moment, $\Delta m = \pm 1$; for the z component, $\Delta m = 0$.

The second consideration of symmetry which arises is that the orientation of the z-axis cannot matter. The z-axis may point up or down; this is equivalent to saying that the integrals must be independent of a reflection in the horizontal plane, an operation we symbolize by σ_h. This operation changes θ into $\pi - \theta$ or $\xi = \cos \theta$ into $\cos(\pi - \theta) = -\cos \theta = -\xi$. The integrals for the x and y components of the transition moment both have the form

$$I_{x,y} = \int_0^\pi P_{J'}^{|m'|}(\cos \theta) \sin \theta P_J^{|m|}(\cos \theta) \sin \theta \, d\theta = \int_{-1}^1 P_{J'}^{|m'|}(\xi)(1 - \xi^2)^{1/2} P_J^{|m|}(\xi) \, d\xi.$$

If we apply the horizontal reflection operator to the integral we obtain

$$\sigma_h I_{x,y} = \int_{-1}^1 P_{J'}^{|m'|}(-\xi)[1 - (-\xi)^2]^{1/2} P_J^{|m|}(-\xi) \, d\xi.$$

But $P_J^{|m|}(-\xi) = (-1)^{J - |m|} P_J^{|m|}(\xi)$. Using this relation we obtain

$$\sigma_h I_{x,y} = (-1)^{J' - |m'| + J - |m|} I_{x,y}.$$

If we replace $J' = J + \Delta J$ and $|m'| = |m| + \Delta|m|$, we obtain

$$\sigma_h I_{x,y} = (-1)^{2J + \Delta J - 2|m| - \Delta|m|} I_{x,y} = (-1)^{\Delta J - \Delta|m|} I_{x,y}.$$

If $I_{x,y}$ is to be invariant, $\Delta J - \Delta|m|$ must be an even number; since the requirement on $\Delta|m|$ for the x or y component is $\Delta|m| = \pm 1$, it follows that ΔJ must be odd. Detailed calculation using the recurrence formulas shows that $\Delta J = \pm 1$ only.

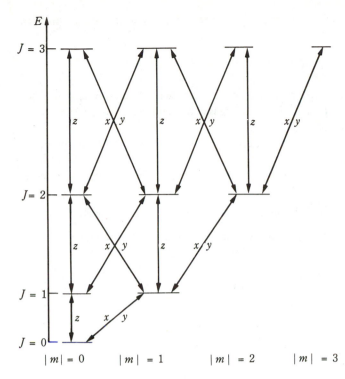

Figure 25.13 Allowed transitions for the rigid rotor for various polarizations.

For the z component, $z = r \cos \theta$, and since we must have $|m'| = |m|$, the integral has the form

$$I_z = \int_0^\pi P_{J'}^{|m|}(\cos \theta) \cos \theta P_J^{|m|}(\cos \theta) \sin \theta \, d\theta = \int_{-1}^1 P_{J'}^{|m|}(\xi) \xi P_J^{|m|}(\xi) \, d\xi;$$

then

$$\sigma_h I_z = \int_{-1}^1 P_{J'}^{|m|}(-\xi)(-\xi)P_J^{|m|}(-\xi) \, d\xi = (-1)^{J'|m| + 1 + J - |m|} I_z$$

$$= (-1)^{\Delta J + 1} I_z.$$

Again this relation requires ΔJ to be odd if the integral is to be invariant under reflection. Detailed calculation using the recurrence formula shows that $\Delta J = \pm 1$ only.

The types of transitions and the polarizations that produce them are shown in Fig. 25.13; the vertical axis represents the energy; the horizontal axis is simply used to space out the values of $|m|$ corresponding to a particular value of J.

★ 25.15 SELECTION RULES FOR THE HYDROGEN ATOM

The wave functions for the hydrogen atom have the form of a product of a radial function (a function of r only) and the rigid rotor functions. The selection rules for the rotor functions must be the same as those obtained above, namely,

$$\Delta l = \pm 1$$

$$\Delta |m| = 0 \quad \text{(z component)} \qquad \Delta |m| = \pm 1 \quad \text{(x and y component)}.$$

Thus the only remaining question concerns the radial functions. The transition-moment integral has the form

$$\int_0^\infty R_{nl}(r)rR_{n'l'}(r)r^2 \, dr.$$

However, the functions in the integral depend only on r, which is unaffected by any rotation of the axes or by any reflection in a plane. We conclude that any of the symmetry operations will leave r and functions of r unchanged.

Since symmetry imposes no restrictions, there is no selection rule for n, the principal quantum number. This result obtained so simply by symmetry considerations would be extremely cumbersome to prove by direct evaluation of the transition-moment integrals.

★ 25.16 SELECTION RULES FOR POLYATOMIC MOLECULES

In Chapter 23 we discussed the wave functions of the water molecule and concluded that they can be classified by the irreducible representations, or symmetry species, to which they belong. We also showed how to classify functions such as x, y, z, xy, yz, xz, and so on, according to symmetry species. The symmetry properties can be used to establish the selection rules for the spectral transitions.

If the transition-moment integrals

$$\int \psi_m^{0*} x \psi_n^0 \, d\tau, \qquad \int \psi_m^{0*} y \psi_n^0 \, d\tau, \qquad \int \psi_m^{0*} z \psi_n^0 \, d\tau$$

are not to be zero, then the integrands must be invariant under all the operations of the symmetry group. This requirement will be met if and only if the integrands belong to the totally symmetric representation A_1. We can conclude that if the direct product of the representations of ψ_m^{0*} and ψ_n^0 is equal to or contains the representation of one of the co-ordinates, then the direct product of $\psi_m^{0*}\psi_n^0$ with that coordinate will belong to the totally symmetric representation A_1. Using the character table (Table 23.5) for the group C_{2v}, we obtain the possible direct products between the representations of ψ_m^{0*} and ψ_n^0. These products are collected in the form of a multiplication table:

$\psi_m^* \diagdown \psi_n$	A_1	A_2	B_1	B_2
A_1	A_1	A_2	B_1	B_2
A_2	A_2	A_1	B_2	B_1
B_1	B_1	B_2	A_1	A_2
B_2	B_2	B_1	A_2	A_1

The coordinates belong to A_1, B_1, and B_2 wherever an A_1, B_1, or B_2 appears in the table; that transition is allowed, since one of the transition-moment integrals is nonvanishing. Wherever A_2 appears, that transition is forbidden since the direct product of A_2 with any of the representations belonging to the coordinates cannot be A_1. Thus we conclude that, for the water molecule, the transitions

$$A_1 \leftrightarrow A_2 \qquad \text{and} \qquad B_1 \leftrightarrow B_2$$

are forbidden. All other transitions are allowed.

QUESTIONS

25.1 The Born–Oppenheimer approximation states that a diatomic's electronic energy depends only on the internuclear separation. Use this information to sketch and explain the relative location of the first few vibrational levels for H_2 and D_2.

25.2 How do the rotational selection rules exclude absorption at the fundamental frequency $\tilde{\nu}_0$ in a diatomic vibration–rotation spectrum?

25.3 Classify the following molecules as spherical, prolate symmetric, oblate symmetric, or asymmetric tops: O_3, $AlCl_3$ NH_4^+, CH_2Cl_2, CH_3Br, $CHCl_3$, CCl_4.

25.4 Which of the following molecules exhibit infrared absorption? I_2, HBr, CH_3Cl, CCl_4, SO_2, CS_2, NH_3.

25.5 Which of the following molecules exhibit a rotational microwave spectrum? H_2O, CS_2, CH_4, OCS, HC≡CH, Br_2, HBr.

25.6 Which of the following linear molecules exhibit vibrational Raman, but not infrared, spectra? O_2, CO_2, OCS, HCN, N_2, NO.

25.7 What do overtone and combination bands reveal about anharmonicity?

25.8 State and rationalize the Franck–Condon principle and indicate its importance.

25.9 What types of spectra yield information on the vibrational states of homonuclear diatomics?

25.10 Discuss an allowed and forbidden oscillator transition by sketching the integrands of the transition-moment integrals μ_{mn} for two appropriate oscillator state pairs.

25.11 Repeat Question 25.10 for the I_z transition-moment integral for a rotor.

PROBLEMS

25.1 The wavenumbers of the rotational lines of HCl can be fitted to the expression
$$\tilde{\nu}/\text{cm}^{-1} = 20.794(J + 1) - 1.64 \times 10^{-3}(J + 1)^3.$$
Calculate the moment of inertia of the $H^{35}Cl$ molecule and the internuclear distance. H = 1.007825 g/mol; ^{35}Cl = 34.96885 g/mol. Estimate the vibrational frequency of $H^{35}Cl$.

25.2 The iodine atom has an atomic mass of 126.9045 g/mol and hydrogen has an atomic mass of 1.007825 g/mol. If the internuclear separation in HI is 160.4 pm, calculate the moment of inertia and the separation of the rotational frequencies.

25.3 Calculate the position of the center of mass, the moment of inertia, and the spacing between the rotational lines for each of the following linear molecules. The masses are ^{12}C = 12.0000; ^{14}N = 14.00307; ^{16}O = 15.99491; ^1H = 1.007825 g/mol.

    a) The asymmetrical N^-=N^+=O, which has bond lengths N^-=N^+ = 112.6 pm; and N=O = 119.1 pm.
    b) The symmetrical O=C=O, which has C=O = 116.2 pm.
    c) H—C≡C—C≡N, in which C—H = 105.7 pm; C≡C = 120.3 pm; C—C = 138.2 pm and C≡N = 115.7 pm.

25.4 The fundamental vibration frequency of the $^{35}Cl_2$ molecule is 1.6780×10^{13} s^{-1}. Calculate the energies of the first three vibrational levels and the force constant if M (^{35}Cl) = 34.96885 g/mol.

25.5 The force constant in $^{79}Br_2$ is 246.053 N/m; for ^{79}Br and ^{81}Br the masses (g/mol) are 78.9183 and 80.9163. Calculate the fundamental vibrational frequencies and the zero-point energies in $^{79}Br_2$, $^{79}Br^{81}Br$, and $^{81}Br_2$.

25.6 The first five vibrational bands ($v = 0 \to 1$, $0 \to 2, \ldots 0 \to 5$) of $H^{35}Cl$ are centered on the wavenumbers, $\tilde{\nu}/\text{cm}^{-1}$ = 2886; 5668; 8347; 10 923; 13 397. Calculate the fundamental frequency of the HCl molecule, the anharmonicity constant, the reduced mass, and the force constant of the molecule. H = 1.007825 g/mol; ^{35}Cl = 34.96885 g/mol.

25.7 An argon–krypton laser with $\lambda = 488.0$ nm is used to observe the Raman spectrum of benzene. Two of the thirty normal modes of vibration for benzene have the frequencies, $\tilde{\nu}_1/\text{cm}^{-1} = 3062$ and $\tilde{\nu}_2/\text{cm}^{-1} = 992$. At what wavelengths will the Raman lines for these vibrations be observed?

25.8 Derive the selection rules for the electronic spectrum of

a) C_2H_4 (D_{2h});   b) NH_3 (C_{3v});   c) BF_3 (D_{3h}).

25.9 The atomic masses (g/mol) are: $H = 1.007825$; $D = 2.01410$; $^{35}Cl = 34.96885$; $^{37}Cl = 36.96590$. The fundamental vibrational frequency in $H^{35}Cl$ is 2990.946 cm^{-1}.

a) Assuming that the force constant does not change, calculate the reduced masses and the fundamental vibrational frequencies in $H^{37}Cl$, $D^{35}Cl$, and $D^{37}Cl$.

b) The internuclear distance is the same (128.65 pm) for all of the compounds below. Calculate the moments of inertia and the separation of the rotational lines in $H^{35}Cl$, $H^{37}Cl$, $D^{35}Cl$, and $D^{37}Cl$.

25.10 a) Evaluate the transition moment integral for the particle in a one-dimensional box.

b) What is the selection rule for this system?

25.11 Show that for any linear molecule, $I = \Sigma_i m_i(x_i^2 - X^2)$, where X is the position of the center of the mass.

26

Intermolecular Forces

26.1 INTRODUCTION

Two chemically saturated particles, such as two molecules of methane or two atoms of argon, are subject to attractive forces as they approach one another. The *intermolecular forces* are electrical in origin; since they are responsible for the phenomena of gas imperfection and liquefaction, they are often called van der Waals forces. The energy of vaporization of a liquid provides a convenient measure of the strength of these forces, since it is the energy required to pull the molecules from the liquid, where they are in proximity, and bring them into the gas where they are widely separated. The energy of vaporization is simply related to the heat of vaporization of the liquid at constant pressure:

$$Q_{vap} = \Delta H_{vap} = \Delta U_{vap} + p(\overline{V}_{gas} - \overline{V}_{liq}).$$

Approximately, $\overline{V}_{gas} - \overline{V}_{liq} = \overline{V}_{gas} = RT/p$, so that $\Delta H_{vap} = \Delta U_{vap} + RT$. At the normal boiling point, $\Delta H_{vap} = T_b \Delta S_{vap}$; hence for ΔU_{vap} we obtain $\Delta U_{vap} = T_b(\Delta S_{vap} - R)$. For normal liquids we have the Trouton rule (see Section 9.3.1), $\Delta S_{vap} = 90$ J/K mol; hence,

$$\Delta U_{vap} \approx 82T_b \text{ J/mol} \tag{26.1}$$

Even for substances that do not obey the Trouton rule, the proportionality between ΔU_{vap} and the boiling point is roughly correct. In view of Eq. (26.1) we may take the boiling point of a liquid as a convenient measure of the cohesive energy, which in turn depends on the strength of the intermolecular forces.

Since the intermolecular forces depend on the effect on one molecule of the electrical field produced by another molecule, we begin by looking at the effect of an electrical field on matter in bulk and then at the effect of a field on individual molecules. The basic equations of electrostatics that are required for this are developed in Appendix II.

26.2 POLARIZATION IN A DIELECTRIC

If an electrical field E is applied between two parallel metal plates separated by a fixed distance (a parallel-plate capacitor), one plate acquires a positive charge and the other a negative charge (Fig. 26.1a). The charge per unit area of the plate is the charge density σ.

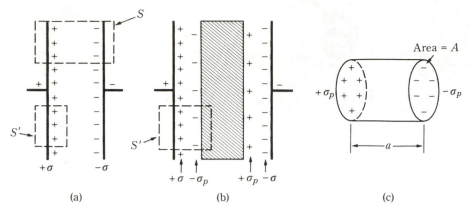

Figure 26.1   (a) Charged capacitor. (b) Capacitor with dielectric.
(c) Cylindrical section of dielectric.

Since the charge is uniformly distributed on each plate, the field is perpendicular to the plates, $E_n = E$, and the tangential components of the field are zero.

If we draw a gaussian surface, S, which encloses both plates, then the total charge enclosed by this surface is $q = \sigma A + (-\sigma)A = 0$, where A is the area of the gaussian surface parallel to the plates. Then, by Gauss's law, Eq. (A.II.14),

$$\int_S E \, da = 0,$$

which requires that $E = 0$. The electric field vanishes in the space behind the plates.

To find the field between the plates we apply Gauss's law to the surface S' (Fig. 26.1a). Then $q = \sigma A$, and since $E_n = E$,

$$\int_{S'} E \, da = \frac{\sigma A}{\epsilon_0}.$$

Since the tangential components are zero, there is no contribution to the flux from the surfaces perpendicular to the metal plates. This leaves

$$\int_{\text{outside surface}} E \, da + \int_{\text{inside surface}} E \, da = \frac{\sigma A}{\epsilon_0}.$$

But we showed that E on the outside surface is zero, so the first integral is zero. Since the charge density is uniform, the E on the inside surface is constant everywhere on the surface and can be removed from the integral. The equation becomes $EA = \sigma A/\epsilon_0$, or

$$E = \frac{\sigma}{\epsilon_0} \qquad \text{(in vacuum).} \tag{26.2}$$

If an insulating material (a dielectric) is placed between the plates, the application of the electric field produces in the dielectric a minute shift of negative charge toward the positive plate and a shift of positive charge toward the negative plate; the dielectric is *polarized*. Bound to the surface of the dielectric at the negative plate is a positive charge density, $+\sigma_p$; at the positive plate a negative charge density, $-\sigma_p$, is bound to the surface of the dielectric (Fig. 26.1b). The quantity σ_p is called the *polarization* of the dielectric.

Figure 26.1(c) shows a cylindrical element of the dielectric with its axis in the direction of the polarizing field. If the area of each face of the cylinder is A, then the charges on the faces are $+\sigma_p A$ and $-\sigma_p A$. These charges are separated by the length of the cylinder a, so that the cylinder has a dipole moment equal to $(\sigma_p A)a$. Since the volume of the cylinder is aA, the dipole moment per unit volume is

$$\frac{\text{dipole moment}}{\text{volume}} = \sigma_p. \tag{26.3}$$

Therefore the polarization σ_p in addition to being the charge density on the surface is also equal to the dipole moment per unit volume of the dielectric.

The induced surface charge on the dielectric reduces the net charge per unit area enclosed by the surface S' to $\sigma - \sigma_p$; thus for the field in the dielectric we can write

$$E = \frac{\sigma - \sigma_p}{\epsilon_0}. \tag{26.4}$$

But the polarization is itself proportional to the field within the dielectric. This proportionality is written

$$\sigma_p = (\epsilon_r - 1)\epsilon_0 E, \tag{26.5}$$

where ϵ_r is the *relative permittivity* of the medium; more commonly, ϵ_r is called the *dielectric constant* of the medium. The *permittivity* of the medium is $\epsilon = \epsilon_r \epsilon_0$.

When we put this value of σ_p into Eq. (26.4) and solve for E,

$$E = \frac{\sigma}{\epsilon_r \epsilon_0} = \frac{\sigma}{\epsilon}. \tag{26.6}$$

Comparing Eq. (26.6) with Eq. (26.2) we see that in the presence of the dielectric the effective charge density has been reduced from σ to σ/ϵ_r. This result does not depend on the geometry used in Fig. 26.1 to demonstrate it. Quite generally, the electrostatic equations for vacuum can be modified to apply in a dielectric by replacing the field producing charge, q, by q/ϵ_r. This is equivalent to replacing ϵ_0 by ϵ. (Strictly speaking this result applies only to electrically isotropic media such as liquids and gases. Since most solids are anisotropic, the situation is only crudely described by this procedure.)

Our next task is to calculate the field acting on a molecule immersed in a dielectric. Consider a parallel-plate capacitor filled with a dielectric; there is a molecule at the center of a spherical cavity in the dielectric. The cavity is small compared to the macroscopic dimensions of the capacitor but large compared to the size of the molecule (Fig. 26.2a).

The field in the cavity is the resultant of the field in the dielectric, given by Eq. (26.4), and the field due to the induced charges on the wall of the cavity. To calculate the field due to the charges on the wall, we consider the polarization vector, σ_p, in Fig. 26.2(b). This vector is parallel to the z-axis and is directed from left to right. The component of this vector on the ray OP is $\sigma_p \cos\theta$. Just as σ_p is the charge density on the vertical plane, $\sigma_p \cos\theta$ is the charge density on the spherical surface at the position, r, θ. The total charge on the element of area, da, is $\sigma_p \cos\theta \, da$. The field at the center of the cavity due to this charge is, by Coulomb's law,

$$\frac{\sigma_p \cos\theta \, da}{4\pi\epsilon_0 r^2}.$$

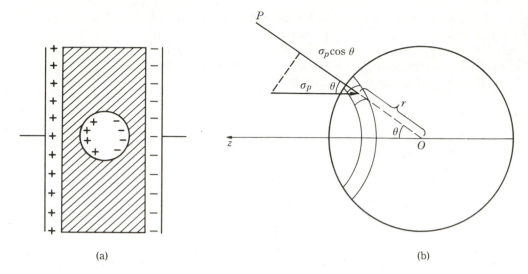

Figure 26.2   Calculation of the field inside a cavity in a dielectric.

This field vector is directed along the ray PO; to obtain the horizontal component, we multiply by $\cos\theta$; this yields

$$\frac{\sigma_p \cos^2\theta \, da}{4\pi\epsilon_0 r^2}.$$

To obtain the total effect of the positive charge, we must integrate over the left hemisphere. Since $da = (r\sin\theta \, d\phi)r \, d\theta$, we need to integrate from $\phi = 0$ to 2π, and from $\theta = 0$ to $\frac{1}{2}\pi$. Thus

$$\int_0^{\pi/2} \frac{\sigma_p \cos^2\theta r^2 \sin\theta \, d\theta}{4\pi\epsilon_0 r^2} \int_0^{2\pi} d\phi = \frac{\sigma_p}{2\epsilon_0}\int_0^{\pi/2} \cos^2\theta \, d(-\cos\theta) = \frac{\sigma_p}{6\epsilon_0}.$$

The effect of the negative charges in the right hemisphere is equal to that of the positive charges in the left hemisphere and acts in the same direction. The value is thus doubled, $2\sigma_p/6\epsilon_0 = \sigma_p/3\epsilon_0$. For the field at the center of the cavity, this yields

$$E_{\text{cav}} = E + \frac{\sigma_p}{3\epsilon_0}.$$

Using $E = \sigma_p/(\epsilon_r - 1)\epsilon_0$ from Eq. (26.5), this becomes

$$E_{\text{cav}} = \frac{\sigma_p}{(\epsilon_r - 1)\epsilon_0} + \frac{\sigma_p}{3\epsilon_0} = \frac{\sigma_p(\epsilon_r + 2)}{3\epsilon_0(\epsilon_r - 1)}.$$

Solving for σ_p, we obtain

$$\sigma_p = \frac{3\epsilon_0(\epsilon_r - 1)E_{\text{cav}}}{\epsilon_r + 2}. \tag{26.7}$$

Note that Eq. (26.7) is based on purely classical electrostatics and is in no way dependent on the atomic or molecular structure of the dielectric. The macroscopic property,

ϵ_r, is easily measurable; it is the ratio of the capacitance of a capacitor filled with the dielectric material to the capacitance in vacuum.

$$\epsilon_r = \frac{C}{C_{vac}}. \qquad (26.8)$$

26.3 MOLAR POLARIZATION

The dipole moment per unit volume of the dielectric is made up of contributions from all the molecules in the unit volume. If $\tilde{N}$ is the number of molecules per unit volume, and m is the average dipole moment per molecule induced by the field, then the dipole moment per unit volume is $m\tilde{N}$:

$$\sigma_p = m\tilde{N}. \qquad (26.9)$$

Using this result in Eq. (26.7), we obtain

$$m = \frac{3\epsilon_0(\epsilon_r - 1)E_{cav}}{\tilde{N}(\epsilon_r + 2)}, \qquad (26.10)$$

which describes the value of m in terms of the macroscopic properties E_{cav}, and ϵ_r. Having obtained such a relation, we inquire as to how the dipole moment m is produced in the direction of the field.

If a molecule that has no permanent dipole moment is placed in an electrical field, the electronic cloud will be displaced slightly toward the positive plate. This distorted molecule possesses a dipole moment m, which is proportional to the applied field,

$$m = \alpha_0 E. \qquad (26.11)$$

The constant of proportionality α_0 is the *distortion polarizability* of the molecule. The polarizability is the dipole moment produced by an applied field of unit strength.

For any substance it can be shown that

$$m = \alpha E, \qquad (26.12)$$

where α is the polarizability of the substance. If the substance has a permanent dipole moment, then the polarizability is the sum of two terms.

$$\alpha = \alpha_0 + \alpha_\mu, \qquad (26.13)$$

where α_0 is the distortion polarizability, and α_μ is the *orientation polarizability*. The orientation polarizability arises from the tendency of the permanent dipole moment μ to be oriented in the direction of the applied field.

The field that acts on a molecule in the dielectric is E_{cav}, so that $m = \alpha E_{cav}$. Using this value for m in Eq. (26.10) and rearranging, we obtain

$$\frac{\epsilon_r - 1}{\epsilon_r + 2} = \frac{\tilde{N}\alpha}{3\epsilon_0}. \qquad (26.14)$$

Since $\tilde{N} = N_A \rho/M$, where ρ is the density and M the molar mass of the substance, we can write this equation in the form

$$\left(\frac{\epsilon_r - 1}{\epsilon_r + 2}\right)\left(\frac{M}{\rho}\right) = \frac{N_A \alpha}{3\epsilon_0}, \qquad (26.15)$$

This is the Clausius–Mosotti equation. The molar polarization P is defined by

$$P = \left(\frac{\epsilon_r - 1}{\epsilon_r + 2}\right)\left(\frac{M}{\rho}\right). \tag{26.16}$$

The macroscopic quantities in P are easily measured. Then we have

$$P = \frac{N_A \alpha}{3\epsilon_0}. \tag{26.17}$$

If α is a constant characteristic of the molecule, then P is a constant, and Eq. (26.16) is a relation between the dielectric constant and the density of the substance. Furthermore, if α is characteristic of the molecule, then α and the molar polarization P should be independent of the temperature. This is substantiated experimentally for *nonpolar* molecules, those that have no permanent dipole moment.

26.3.1   Orientation Polarizability

Suppose that a large number of polar molecules, each having a permanent dipole moment μ, are placed between the plates of a capacitor. In the absence of a field and at reasonably high temperatures, the thermal motions of the molecules will produce a random orientation of the molecules so that there is no net dipole moment in any direction. However, if a field is applied across the plates of the capacitor, the dipole molecules will be oriented in the field, producing a net dipole moment in the direction of the field. The net induced dipole moment divided by the number of molecules is the average dipole moment per molecule in the direction of the field, $\bar{m}$. We can show that

$$\bar{m} = \frac{\mu^2 E}{3kT}. \tag{26.18}$$

This equation shows that $\bar{m}$ is proportional* to the field E. The orientation polarizability α_μ is defined by $\bar{m} = \alpha_\mu E$; from Eq. (26.18), we obtain

$$\alpha_\mu = \frac{\mu^2}{3kT}. \tag{26.19}$$

At high temperatures $\bar{m}$ and α_μ are much smaller than at low temperatures. At high temperature, the thermal motion is more successful in reducing the orientation in the field.

The total polarizability of any molecule is the sum of the distortion polarizability and the orientation polarizability, Eq. (26.13). Thus we have

$$\alpha = \alpha_0 + \alpha_\mu = \alpha_0 + \frac{\mu^2}{3kT}. \tag{26.20}$$

Using this result in Eq. (25.14), we obtain the Debye equation,

$$\left(\frac{\epsilon_r - 1}{\epsilon_r + 2}\right)\left(\frac{M}{\rho}\right) = \frac{N_A \alpha_0}{3\epsilon_0} + \frac{N_A \mu^2}{9\epsilon_0 kT}, \tag{26.21}$$

which is used to obtain the value of the dipole moment of a molecule from the measured value of the dielectric constant at several temperatures. From the values of ϵ_r and ρ at

---

*   It is clear that $\bar{m}$ cannot continue to be proportional to E if the field is strong. A saturation effect occurs. If all the molecules are completely oriented, increasing the field does not increase $\bar{m}$ any further.

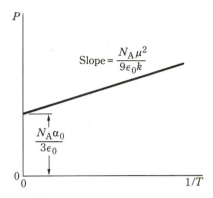

Figure 26.3 Typical Debye plot of P versus $1/T$.

Table 26.1
Dipole moments of molecules

Molecule	$\mu/10^{-30}$ C m	Molecule	$\mu/10^{-30}$ C m	Molecule	$\mu/10^{-30}$ C m
HF	6.37	H_2O	6.17	NH_3	4.90
HCl	3.57	H_2S	3.07	PH_3	1.83
HBr	2.67			AsH_3	0.53
HI	1.40				
CH_3F	6.04	CH_3OCH_3	4.34	C_6H_5Cl	5.67
CH_3OH	5.67	$CH_2\!-\!CH_2$ $\diagdown O \diagup$	6.34	C_6H_5Br	5.67
CH_3NH_2	4.20			$C_6H_5NO_2$	14.2
$o\text{-}C_6H_4Cl_2$	8.34	$o\text{-}C_6H_4(NO_2)_2$	20		
$m\text{-}C_6H_4Cl_2$	5.74	$m\text{-}C_6H_4(NO_2)_2$	13.0		
$p\text{-}C_6H_4Cl_2$	0	$p\text{-}C_6H_4(NO_2)_2$	0		

Note: The cgs unit for dipole moment is the debye: 1 debye $= 1$ D $= 10^{-18}$ esu cm $= 3.3356\ldots \times 10^{-30}$ C m.

several temperatures, the value of the molar polarization (the left-hand side of the equation) is calculated. A plot of molar polarization against the reciprocal of the temperature should be linear. By Eq. (26.21), the slope is $N_A \mu^2/9\epsilon_0 k$ and the intercept is $N_A \alpha_0/3\epsilon_0$. A typical plot is shown in Fig. 26.3. From the slope and intercept, the dipole moment μ and the distortion polarizability α_0 of the molecule are obtained. A few values of μ are shown in Table 26.1.

26.3.2 Molar Refraction

The dielectric constant is ordinarily measured in an alternating current circuit. The direction of the field across the capacitor changes back and forth with the frequency of the applied potential. If we imagine a single polar molecule between the plates of a capacitor, then if the frequency is not too high, this single molecule will flip back and forth as the field oscillates, always adjusting its orientation to match the direction of the field.

However, the molecule requires a finite time to adjust its orientation. If this time, the *relaxation time*, is very short compared with the time of one cycle of the applied field, then the molecule adjusts itself readily to the different orientations of the field. On the other hand, if the frequency of the applied field is increased, then finally a situation prevails in which the

Table 26.2
Molar refractions of ions

Ion			He	Li$^+$	Be^{2+}	B^{3+}	C^{4+}	
R_D/(cm^3/mol)			0.50	0.20	0.09	0.05	0.03	
Ion	O^{2-}	F$^-$	Ne	Na$^+$	Mg^{2+}	Al^{3+}	Si^{4+}	
R_D/(cm^3/mol)	7	2.5	1.00	0.50	0.29	0.17	0.1	
Ion	S^{2-}	Cl$^-$	Ar	K$^+$	Ca^{2+}	Sc^{3+}	Ti^{4+}	Zn^{2+}
R_D/(cm^3/mol)	15	8.7	4.20	2.2	1.35	1.0	0.7	0.3
Ion	Se^{2-}	Br$^-$	Kr	Rb$^+$	Sr^{2+}	Y^{3+}	Zr^{4+}	Cd^{2+}
R_D/(cm^3/mol)	16.3	12.2	6.37	3.6	2.3	2.6	2.0	2.4
Ion	Te^{2-}	I$^-$	Xe	Cs$^+$	Ba^{2+}	La^{3+}	Ce^{4+}	Hg^{2+}
R_D/(cm^3/mol)	24.4	18.5	10.42	6.3	4.3	4.0	3.1	5.0

By permission from C. P. Smyth, *Dielectric Behavior and Structure*. New York: McGraw-Hill, 1955.

molecule does not have time to change its orientation before the field switches back again. As a result, at very high frequencies the molecule is not oriented by the field at all, and the permanent dipole moment ceases to contribute to the molar polarization; only the distortion polarization remains.

At high frequencies, Eq. (26.15), even for molecules with a permanent dipole moment, becomes

$$\left(\frac{\epsilon_r - 1}{\epsilon_r + 2}\right)\left(\frac{M}{\rho}\right) = \frac{N_A \alpha_0}{3\epsilon_0}. \tag{26.22}$$

The distortion polarization remains because even at high frequencies the electron cloud is mobile enough to adjust to the changing field.

An electromagnetic frequency of the order of 10^{14} Hz corresponds to a light wave; then it may be shown that $\epsilon_r = r^2$, where r is the index of refraction of the substance for light of the frequency in question, and Eq. (26.22) becomes

$$\left(\frac{r^2 - 1}{r^2 + 2}\right)\left(\frac{M}{\rho}\right) = \frac{N_A \alpha_0}{3\epsilon_0}, \tag{26.23}$$

where the quantity on the left-hand side is called the molar refraction R; thus

$$R = \left(\frac{r^2 - 1}{r^2 + 2}\right)\left(\frac{M}{\rho}\right) = \frac{N_A \alpha_0}{3\epsilon_0}. \tag{26.24}$$

The value of R can be calculated from the measured value of the refractive index of the liquid or solid. Equation (26.24) can be combined with Eq. (26.21) to express the dielectric constant at low frequencies:

$$\left(\frac{\epsilon_r - 1}{\epsilon_r + 2}\right)\left(\frac{M}{\rho}\right) = R + \frac{N_A \mu^2}{9\epsilon_0 kT}. \tag{26.25}$$

The molar refraction of a substance is approximately the sum of the refractions of the electron groups within it. The molar refraction of NaCl, for example, is the sum of the refractions of the Na$^+$ ion and the Cl$^-$ ion. A few values of the molar refraction appropriate to the D line of sodium are given in Table 26.2. The refraction of the inert gases can be measured directly. To obtain the refraction of the individual ions from that of their salts, the value of the refraction of at least one ion must be known. The refraction of the fluoride ion has been accurately calculated from the quantum mechanics, and using this value, we can calculate the refractions of the ions Li^+, Na^+, and so on, from the refraction of the corresponding fluorides. Table 26.2 is built up in this way.

From the values in Table 26.2 it is apparent that the refraction of a particular group of electrons decreases greatly as the nuclear charge increases. The group of two electrons has a refraction of 0.50 cm^3/mol in helium, but only has 0.03 cm^3/mol in C^{4+}. The group of ten in the second row has a high value of 7 cm^3/mol for O^{2-} which drops to 0.1 cm^3/mol in Si^{4+}. Clearly, the contribution of the inner core of two electrons drops to a negligible value in this group of ten; the refraction is essentially the refraction of the outer group of eight electrons. The more tightly the electrons are bound to the central core, the less they are deformed in an applied field and the less contribution they will make to the refraction of the compound. If the electron cloud is large, loose, and floppy it is easily deformed in the field; correspondingly, the refraction is large. For this same reason the molar refraction (cm^3/mol) roughly parallels the molar volume (cm^3/mol) of the substance.

The same argument is applied to electron groups in covalent molecules. The refraction of methane is attributed to the refraction of four equivalent electron groups, the pair bonds between the carbon and hydrogen atoms. Thus for the carbon–hydrogen bond refraction we can assign $R_{C-H} = \frac{1}{4}R_{CH_4}$. Then for the C—C bond, we derive a value from the refraction of ethane; $R_{C_2H_6} = 6R_{C-H} + R_{C-C}$. This procedure yields $R_{C-H} = 1.70$ cm^3/mol, $R_{C-C} = 1.21$ cm^3/mol. Using these values, we can calculate the refraction of any saturated hydrocarbon.

The contribution of double and triple bonds to the refraction is found from the refractions of $H_2C{=}CH_2$ and $HC{\equiv}CH$. Table 26.3 lists a few values of bond refractions. Note in comparing the single, double, and triple carbon–carbon bond that the refraction increases with the multiplicity of the bond. The electron pairs in the π bonds are looser than those in the single bond. The values in the table for groups including an oxygen show that the refraction depends on the mode of attachment of the oxygen. The refraction, which includes the two electron pairs on oxygen as well as the bonding pairs, is different in ketones, ethers, and alcohols.

For simple compounds the sum of the group refractions yields the molar refraction of the compound with reasonable accuracy. Difficulties show up in compounds with

Table 26.3
Molar refractions of electron groups

Group	R_D/(cm^3/mol)	Group	R_D/(cm^3/mol)	Group	R_D/(cm^3/mol)
H—H	2.08	H $\diagdown$ $\ddot{O}$ $\diagup$ H	3.76	C $\diagdown$ $\ddot{O}$ $\diagup$ C	2.85
C—H	1.70				
C—C	1.21				
C=C	4.15	C $\diagdown$ $\ddot{O}$ $\diagup$ H	3.23	C=$\ddot{O}$	3.42
C≡C	6.03				

By permission from C. P. Smyth, *Dielectric Behavior and Structure*. New York: McGraw-Hill, 1955.

conjugated double bonds, which have a higher refraction than would be expected. In a conjugated system the electrons in the π bonds are free to move over the entire molecule; consequently they are "looser" and more easily deformed. The additional contribution to the refraction is called "exaltation."

26.4 INTERMOLECULAR FORCES

Having described the methods used to obtain such properties as polarizability and dipole moment, we return to the problem of intermolecular forces. If two polar molecules have the proper orientation, their positive and negative ends produce a mutual attraction between the molecules. Furthermore, since the field of one polar molecule should induce a dipole moment by distortion of the electron cloud of the other molecule, this effect leads to a mutual attraction of the molecules. It is possible to construct a purely electrostatic theory of intermolecular forces, at least for polar molecules, based on this mutual interaction.

To calculate the energy of interaction between two dipoles, consider the approach of two dipoles end to end as shown in Fig. 26.4. Let the charge on the ends of the dipoles be q and the charge separation be a. The dipole at the left, which produces a field E, is fixed at the origin of the coordinate system. The field is, by definition, the force acting on a unit positive charge at the point in question. We may form the second dipole at the distance r from the first by bringing up the two charges $+q$ and $-q$, one at a time. The work required to bring $-q$ from infinity to the position r is the integral of the force acting on the charge $E(-q)$, multiplied by $-dr$, the distance moved: $\int_{\infty}^{r} E(-q)(-dr)$. Similarly, the work required to bring $+q$ from infinity to $r + a$ is $\int_{\infty}^{r+a} E(+q)(-dr)$. The total potential energy W of the dipole at r is the sum of these two integrals:

$$W = \int_{\infty}^{r} Eq\, dr - \int_{\infty}^{r+a} Eq\, dr = -\int_{r}^{\infty} Eq\, dr + \int_{r+a}^{\infty} Eq\, dr,$$

where the change in sign in the second expression is effected by interchanging the limits of integration. The first integral may be written as the sum of two terms so that

$$W = -\left(\int_{r}^{r+a} Eq\, dr + \int_{r+a}^{\infty} Eq\, dr\right) + \int_{r+a}^{\infty} Eq\, dr = -\int_{r}^{r+a} Eq\, dr.$$

In the limit as $a \to 0$, the quantity Eq is constant in the range of integration, and we have $W = -Eqa$. But the dipole moment is $m = qa$, so that

$$W = -Em, \tag{26.26}$$

where W is the potential energy of a dipole of moment m in the field E produced by the other dipole.

So far as electrostatics is concerned, Eq. (26.26) is fine for classical dipoles, that is, disembodied electrical charges at certain fixed distances of separation. The situation with molecules is not so simple. Consider molecule 1, having a permanent dipole moment μ, and a fixed orientation in space. Molecule 2 approaches. The orientation of the dipole axis of molecule 2 relative to molecule 1 is completely random. The field of molecule 1 acts on mole-

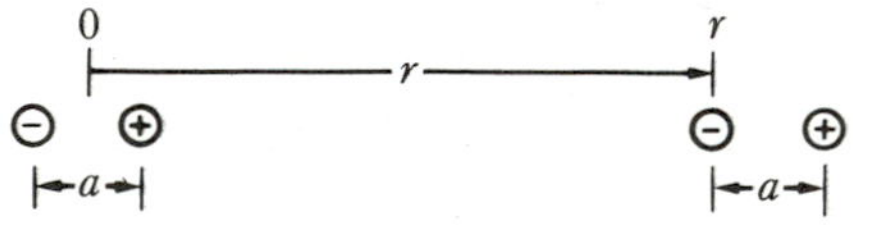

Figure 26.4

cule 2 to induce a dipole moment in molecule 2. The induced moment may result from distortion or orientation of a permanent moment. In either case, the induced moment is in the direction of the field. We write

$$m_2 = \alpha_2 E_1, \tag{26.27}$$

where m_2 is the moment induced in molecule 2 by E_1, the field of molecule 1; the quantity α_2 is the polarizability of molecule 2. [Compare with Eq. (26.12).]

To induce the dipole moment, a charge q must be moved through a distance a; the element of work dW' done in this process is the force $E_1 q$ multiplied by the distance moved da. Hence, $dW' = E_1 q\, da$. Since $m_2 = qa$, then $dm_2 = q\, da$ so that

$$W' = \int dW' = \int_0^{m_2} E_1\, dm_2.$$

From Eq. (26.27), $E_1 = m_2/\alpha_2$, so that

$$W' = \int_0^{m_2} \frac{m_2}{\alpha_2}\, dm_2 = \frac{m_2^2}{2\alpha_2}.$$

This can be expressed in terms of E_1 by using Eq. (26.27) again:

$$W' = \tfrac{1}{2}\alpha_2 E_1^2. \tag{26.28}$$

The total energy W_2 of molecule 2 in the field of molecule 1 is the sum of the potential energy due to its position, which is given by Eq. (26.26), and the energy of distortion or orientation given by Eq. (26.28):

$$W_2 = -E_1 m_2 + \tfrac{1}{2}\alpha_2 E_1^2,$$

which in view of Eq. (26.27) becomes

$$W_2 = -\alpha_2 E_1^2 + \tfrac{1}{2}\alpha_2 E_1^2 = -\tfrac{1}{2}\alpha_2 E_1^2.$$

A similar expression, $W_1 = -\tfrac{1}{2}\alpha_1 E_2^2$, can be written for the energy of molecule 1 in the field of molecule 2. The interaction energy per molecule W_i is the sum

$$W_i = \tfrac{1}{2}(W_1 + W_2) = -\tfrac{1}{4}\alpha_1 E_2^2 - \tfrac{1}{4}\alpha_2 E_1^2.$$

If the molecules are the same kind, then $\alpha_1 = \alpha_2 = \alpha$ and $E_1 = E_2 = E$. Therefore

$$W_i = -\tfrac{1}{2}\alpha E^2. \tag{26.29}$$

Since α is positive and E^2 is positive, the interaction energy is negative. The molecules are lower in energy at the distance r than at $r = \infty$ ($E = 0$ at $r = \infty$). This lower energy means that the molecules attract each other because of their mutual influence.

The remaining difficulty is that the field depends on the angle of approach of the two molecules. If E^2 is averaged over all the possible angles of approach, then $\langle E^2 \rangle = 2\mu^2/(4\pi\epsilon_0)^2 r^6$. Using this value of $\langle E^2 \rangle$ in Eq. (26.29) yields

$$W_i = -\frac{\alpha\mu^2}{(4\pi\epsilon_0)^2 r^6}. \tag{26.30}$$

But, by Eq. (26.20), $\alpha = \alpha_0 + \mu^2/3kT$, so that

$$W_i = -\frac{\mu^2}{(4\pi\epsilon_0)^2 r^6}\left(\alpha_0 + \frac{\mu^2}{3kT}\right), \tag{26.31}$$

where the first term represents the attraction resulting from the distortion of the electron cloud of one molecule by the permanent moment of the other molecule, and the second term represents the attraction resulting from the favorable induced orientation of the permanent moment of one molecule by the field of the other. The order of magnitude of W_i is easily estimated: $\mu \approx 3 \times 10^{-30}$ C m, $\alpha \approx 10^{-40}$ C m^2/V, $4\pi\epsilon_0 \approx 10^{-10}$ C V/m. If we calculate the interaction when the molecules are very close to each other, $r \approx 10^{-10}$ m. From Eq. (26.30) we obtain

$$W_i \approx -\frac{(10^{-40} \text{ C m}^2/\text{V})(3 \times 10^{-30} \text{ C m})^2}{(10^{-10} \text{ C/V m})^2(10^{-10} \text{ m})^6} = -9 \times 10^{-20} \text{ J} \approx -50 \text{ kJ/mol}.$$

Since this is the correct order of magnitude of energies of vaporization of liquids, it seems that this may be a reasonable way to explain the cohesive energies of liquids.

26.4.1 The Dispersion Energy

The treatment of intermolecular forces in Section 26.4 presupposed that the molecules possess a permanent dipole moment. We now ask how it is possible for two molecules such as H_2 or CH_4 or argon, which have no permanent dipole moment, to attract one another. This problem was first considered by F. London; the forces producing attraction are sometimes called London forces, sometimes dispersion forces.

To visualize the physical situation consider an atom of an inert gas such as helium or argon. The electron distribution around the positive nucleus is spherical so that there is no net dipole moment. However, the electron distribution is an average over time (see Section 19.12). Suppose that the electrons are moving relative to the nucleus in such a way that the time average of the electron positions yields the spherical electron cloud, yet at any instant the atom has a separation of positive and negative charge, a dipole moment. The orientation of the dipole moment vector changes constantly as the motion continues so that the average dipole moment is zero.

If two such atoms are brought near one another, each has a momentary dipole and the electronic motions in the two atoms are coupled by the electrical interaction of the momentary dipoles. The electronic motions in the two atoms synchronize so that the momentary dipoles remain in an attractive orientation, and thus lower the energy of the system. The interaction energy is

$$U_d = -\tfrac{3}{4}h v_0 \left(\frac{\alpha_0}{4\pi\epsilon_0}\right)^2 \left(\frac{1}{r^6}\right) \tag{26.32}$$

For many simple molecules the quantity $h v_0$ is equal to the ionization energy of the molecule. The polarizability α_0 can be calculated from the molar refraction of the liquid.

The values of α_0 parallel those of the volume of the molecules. Therefore the dispersion energy is greater for large than for small molecules. Comparing the large iodine molecule with the small fluorine molecule, we note that iodine is a solid at room temperature, while fluorine is a gas. This implies that the intermolecular forces are larger in iodine than in fluorine. The values of $h v_0$ are slightly different, also, but this effect is minor compared with the effect of the larger molecular volume. The dispersion interaction is usually the most important part of the interaction even if the molecules have a dipole moment. For any molecule the interaction energy per pair, U_i, is a sum of terms; Eqs. (26.31) and (26.32):

$$U_i = -\frac{2\mu^2\alpha_0}{(4\pi\epsilon_0)^2 r^6} - \frac{2\mu^4}{3(4\pi\epsilon_0)^2 kT r^6} - \frac{3\alpha_0^2 h v_0}{4(4\pi\epsilon_0)^2 r^6}. \tag{26.33}$$

★ 26.5 INTERACTION ENERGY AND THE VAN DER WAALS "*a*"

The attractive forces discussed so far are inversely proportional to the sixth power of the distance of separation r of the molecules, so we write Eq. (26.33) in the form

$$U_i = -\frac{A}{r^6},\qquad(26.34)$$

where A is a constant of proportionality having a different value for each kind of molecule. The energy U_i is the interaction energy of a pair of molecules separated by a distance r. In a gas, the distances of separation may have many different values. What is the average interaction energy between all the molecules of a gas?

To solve this problem we fix our attention on a molecule at the center of a spherical container of radius R, having a volume $v = 4\pi R^3/3$. If there are N molecules in the container, then the number per unit volume is $\tilde{N}$. How many molecules are at a distance between r and $r + dr$ from the central molecule? The volume of the spherical shell bounded by spheres of radii r and $r + dr$ is $dV_{\text{shell}} = 4\pi r^2\, dr$. The number, dN, of molecules in this shell is $dN = \tilde{N}4\pi r^2\, dr$. The energy of interaction of these molecules with the one at the center is $U_i\, dN$; the average interaction energy of all the molecules with the one at the center is

$$\langle U_i\rangle = \int \frac{U_i\, dN}{N}.$$

Using Eq. (26.34) for U_i and the value of dN, we obtain

$$\langle U_i\rangle = -\frac{4\pi A\tilde{N}}{N}\int_\sigma^R \frac{dr}{r^4} = \frac{4\pi A\tilde{N}}{3N}\left(\frac{1}{R^3} - \frac{1}{\sigma^3}\right),\qquad(26.35)$$

where the lower limit σ is the distance of closest approach of the centers of the molecules, the molecular diameter (Fig. 26.5).

In any reasonable situation, the radius R of the container is very much larger than σ, so that Eq. (26.35) reduces to

$$\langle U_i\rangle = -\frac{4\pi A\tilde{N}}{3\sigma^3 N},\qquad(26.36)$$

which is the average interaction energy *per pair* of molecules in the gas. The total interaction energy is obtained by multiplying this average by the number of pairs of molecules.

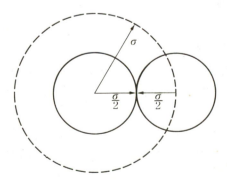

Figure 26.5   The excluded volume.

If there are N molecules, the first one of a pair may be chosen in N ways, the second member may be chosen in $N - 1$ ways, then the total number of pairs is the product of $N(N - 1)$; since N is very large, this is effectively equal to N^2. But this enumeration of the number of pairs counts both the pair between molecules a and b and the pair between molecules b and a as being different; so we divide by 2 and get $\frac{1}{2}N^2$. The total interaction energy is

$$U = \tfrac{1}{2}N^2\langle U_i\rangle = -\frac{2\pi N\tilde{N}A}{3\sigma^3}.$$

The energy per mole $\bar{U} = N_A U/N = -2\pi N_A\tilde{N}A/3\sigma^3$; the volume per mole $\bar{V} = N_A/\tilde{N}$, so that

$$\bar{U} = -\frac{2\pi N_A^2 A}{3\sigma^3\bar{V}}. \tag{26.37}$$

By differentiation we obtain

$$\left(\frac{\partial\bar{U}}{\partial\bar{V}}\right)_T = \frac{2\pi N_A^2 A}{3\sigma^3\bar{V}^2}. \tag{26.38}$$

Problem 10.1 required proof that $(\partial\bar{U}/\partial\bar{V})_T = a/\bar{V}^2$ for a van der Waals gas. Comparing this result with Eq. (26.38) shows that the van der Waals constant a is given by

$$a = \frac{2\pi N_A^2 A}{3\sigma^3}. \tag{26.39}$$

The form of Eq. (26.38) is in fact a justification of the form of the term $a/\bar{V}^2$ in the van der Waals equation. The van der Waals a is proportional to the coefficient A in the interaction energy. Comparing Eqs. (26.33) and (26.34), we see that A depends on temperature if the molecule has a permanent dipole moment. Thus we expect, correctly, that the van der Waals equation would be improved considerably if a were allowed to depend on temperature.

For the sake of completeness we note that the van der Waals b is related to the volume of the molecules and therefore to σ. Figure 26.5 shows that the center of a molecule cannot come closer than a distance σ to the center of any other. A molecule thus excludes the volume $v_x = \frac{4}{3}\pi\sigma^3$. If molecules are added to a container one by one, the volume available to the first molecule is $\bar{V}$, and that available to the second is $\bar{V} - v_x$; if v_i is the volume available to the ith molecule, then $v_i = \bar{V} - (i - 1)v_x$. The average available volume is $\bar{V} - b = (\Sigma\, v_i)/N_A$. It is easy to show that $(\Sigma\, v_i)/N_A = \bar{V} - \frac{1}{2}N_A v_x$, so that

$$b = \tfrac{1}{2}N_A v_x = \tfrac{2}{3}\pi N_A\sigma^3. \tag{26.40}$$

But the volume of the molecule $v_m = \frac{4}{3}\pi(\sigma/2)^3 = v_x/8$, so that

$$b = \tfrac{1}{2}N_A(8v_m) = 4N_A v_m. \tag{26.41}$$

Since $N_A v_m$ is the volume of the molecules, b is four times the volume of the molecules. Using Eq. (26.40) to express σ^3 in terms of b, Eq. (26.39) becomes

$$a = \frac{4\pi^2 N_A^3 A}{9b}. \tag{26.42}$$

From the molecular diameter σ, b can be calculated by Eq. (26.40). Then, knowing A, we can calculate the constant a using Eq. (26.42).

26.6 LAWS OF INTERACTION

Considering that the van der Waals equation is not a particularly good one for the interpretation of the p-V-T data, it seems a hollow victory to be able to calculate the constants a and b. Illustration of the technique involved is the more important achievement. In seeing how to proceed from the energy of interaction of two molecules to the macroscopic constants a and b, we gain insight into the refinements and modifications that could be made to yield a more accurate equation of state than the van der Waals equation. Even without that insight it is consoling to be able to calculate van der Waals constants from such seemingly unrelated properties as the refractive index, the dielectric constant, and the ionization energy of the molecule.

Returning to the energy of interaction between two molecules as a function of distance, we see that the energy at large distances decreases as $1/r^6$ until r reaches the value σ. At $r \leq \sigma$, the energy becomes infinitely positive; this is shown by the vertical line in Fig. 26.6(a). This form of the interaction energy results from the supposition that the molecules are "hard spheres" of diameter σ.

Considering the diffuse electron cloud around the molecule, we would be surprised if the molecule behaved as a hard sphere of definite diameter. The repulsion of the molecules should begin smoothly as the electron clouds of the molecules begin to encroach upon each other's domain. The repulsion energy has been given many different mathematical forms. One of the commonest forms is a term proportional to $1/r^n$, where n is a large integer. The total interaction energy would then be written

$$U_i = -\frac{A}{r^6} + \frac{B}{r^n}. \tag{26.43}$$

This form of the interaction law is called the Lennard–Jones potential. In practice the law is most easily handled if $n = 12$; it is then called a 6–12 potential. The shape of the Lennard–Jones potential is shown in Fig. 26.6(b).

Other modifications of the interaction energy can be made by adding additional attractive terms in $1/r^8$, for the dipole–quadrupole interaction, and even higher terms for the higher multipole interactions. These higher terms arise because the distance between positive and negative charges in a molecular dipole is not infinitesimal but finite. These higher terms make comparatively small contributions to the interaction energy.

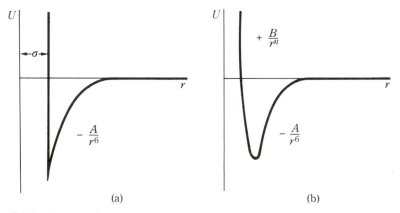

(a)   (b)

Figure 26.6 Interaction energy as a function of the distance of separation. (a) van der Waals potential. (b) Lennard–Jones potential.

26.7 COMPARISON OF THE CONTRIBUTIONS TO THE INTERACTION ENERGY

In Section 26.1 it was shown that the boiling point is a qualitative measure of the interaction energy between the molecules of the substance. Three contributions make up the interaction energy:

1. The oriéntation effect, produced by the mutual action of the permanent dipole moments of the molecules;
2. The distortion effect, produced by the interaction of an induced dipole moment of one molecule with the permanent dipole moment of another molecule;
3. The dispersion effect, produced by the synchronization of the electronic motion in two molecules, which results in momentary dipole moments oriented so as to produce an attraction between the molecules.

First we compare the interaction energy, the boiling points, of molecules that have no permanent dipole moment. The interaction energy is due solely to the dispersion effect, and this, by Eq. (26.32), depends on the polarizability α_0, and $h\nu_0$, which may be thought of as the binding energy of the least tightly bound electron in the molecule. In most molecules, $h\nu_0 \approx 10 \text{ eV} \approx 1 \text{ MJ/mol}$ and does not change very much for different molecules; we will assume that it has the same value for all the molecules under discussion. Listed in Table 26.4 are the number of electrons N, the polarizabilities α_0, and the boiling points T_b, for a number of simple molecules that have no permanent dipole moment. The boiling point increases with the value of α_0, as we expect. The atoms with more electrons have larger and floppier electron clouds that are more easily deformed in a field; the polarizability is therefore larger, and this is reflected in a larger value of the interaction energy and a higher boiling point.

As a general rule the more electrons a molecule has, the larger and less tightly held will be the electron cloud. The large loose cloud is easily deformed, so that the polarizability, the dispersion energy, and the boiling point are all large. A few more examples of this for molecules that have $\mu = 0$ are listed in Table 26.5. The increase in boiling point of hydrocarbons with increase in molecular weight is, of course, a result of the larger number of electrons, and is not immediately related to the larger mass.

Table 26.4

Molecule	He	Ne	Ar	Kr	Xe
N	2	10	18	36	54
$\alpha_0/(10^{-40} \text{ C m}^2/\text{V})$	0.226	0.436	1.81	2.74	4.46
T_b/K	4.216	27.3	87.3	119.9	165.1

Table 26.5

Molecule	H_2	N_2	O_2	CH_4	C_2H_6	C_3H_8
N	2	14	16	10	18	26
$\alpha_0/(10^{-40} \text{ C m}^2/\text{V})$	0.90	1.91	1.72	2.9	5.0	7.1
T_b/K	20.4	77.3	90.2	111.7	184.5	231

Table 26.6

Molecule	Isobutane	Isobutylene	Trimethyl amine
Formula	$(CH_3)_3CH$	$(CH_3)_2C{=}CH_2$	$(CH_3)_3N$
$\alpha_0/(10^{-40}\,C\,m^2/V)$	9.30	9.30	8.99
$\mu/10^{-30}\,C\,m$	0.440	1.63	2.23
T_b/K	263	267	278

Table 26.7

Molecule	Propane	Dimethyl ether	Ethylene oxide
Formula	$(CH_3)_2CH_2$	$(CH_3)_2O$	C_2H_4O
$\alpha_0/(10^{-40}\,C\,m^2/V)$	7.1	6.7	5.8
$\mu/10^{-30}\,C\,m$	0.28	4.34	6.34
T_b/K	231	248	284

Table 26.8

	Dichlorobenzene			Dinitrobenzene		
	para	meta	ortho	para	meta	ortho
$\mu/10^{-30}\,C\,m$	0	5.74	8.34	0	13.0	20
T_b/K	446	445	453	572(subl)	576	592

Next we examine the effect of a permanent dipole moment, choosing first a group of small molecules with about the same number of electrons so that α_0 is about the same, as in Table 26.6. The presence of the dipole moment in isobutylene and trimethyl amine results in a slight increase in boiling point. The effect is not very dramatic, but then the dipole moments are not very large. The effect of large dipole moments on moderately sized molecules is illustrated by the compounds listed in Table 26.7. In these compounds, the polarizabilities are roughly the same; the dipole moments are large, and we see a marked effect on the boiling point.

If the molecules are large (have many electrons), the presence or absence of a dipole moment makes little difference in the boiling points; see Table 26.8. The three dichlorobenzenes have about the same value of $\alpha_0 = 17 \times 10^{-40}\,C\,m^2/V$; for the three dinitrobenzenes, $\alpha_0 = 21 \times 10^{-40}\,C\,m^2/V$. In these large molecules, the dispersion interaction is the most important part of the cohesive energy. The presence or absence of a dipole moment in the dichlorobenzenes makes little difference in the boiling point. In the case of the dinitrobenzenes, the increase in dipole moment from 0 to $20 \times 10^{-30}\,C\,m$ produces only a 20 K change in boiling point. Compare this with propane and ethylene oxide, where an increase in μ from 0 to $6.34 \times 10^{-30}\,C\,m$ increases the boiling point by 53 K.

We can conclude that the larger and more complicated a molecule is, the less the presence or absence of a dipole moment matters to the interaction energy. The dipole moment must be quite large in a large molecule if it is going to affect the interaction energy.

26.7.1   Hydrogen Fluoride, Water, Alcohols, Amines

We return to the discussion of small molecules and consider ethyl alcohol, which has

$$\alpha_0 = 5.8 \times 10^{-40} \text{ C m}^2/\text{V}$$

and

$$\mu = 5.67 \times 10^{-30} \text{ C m}.$$

From our experience with dimethyl ether and ethylene oxide, we should expect a boiling point somewhere between the values for those compounds, a value somewhat less than 284 K. The actual boiling point is about 70 K higher, 352 K. We can compare HF; $\alpha_0 = 0.9 \times 10^{-40} \text{ C m}^2/\text{V}$, $\mu = 6.37 \times 10^{-30} \text{ C m}$. Looking at ethylene oxide, which has the same μ but a considerably higher α_0, we expect that HF should boil at a temperature considerably below 284 K. HF boils at 291 K. Water and ammonia behave in the same way. The properties of the several compounds are shown in Table 26.9. By comparison of any of the compounds on the left of Table 26.9 with ethylene oxide, all should have boiling points considerably lower than 284 K. Methane and neon fulfill this expectation; ammonia has a somewhat lower boiling point, but not as low as one would expect. Water boils 90 K *higher* than ethylene oxide. Since NH_3, HF, and H_2O are very small molecules, we might argue that this increase in interaction energy results from dipole–dipole interaction at very close distances. This is not so, as we can demonstrate by comparing methyl fluoride and methyl alcohol, which are essentially the same size, as in Table 26.10. Methyl alcohol has a lower value of α_0 and μ and so should have a *lower* boiling point. In fact the boiling point is higher by 143 K. A final example is provided by dimethyl ether and ethyl alcohol. These molecules are roughly the same size, as shown in Table 26.11. It is apparent that the boiling point of ethyl alcohol is about 100 K too high.

Most compounds containing OH, NH, or NH_2 groups have higher boiling points than would be predicted on the basis of their dipole moments and polarizabilities. Among the fluorides, only HF has this anomaly; other fluorides behave normally.

Table 26.9

Molecule	Neon	Methane	Ammonia	Water	Hydrogen fluoride	Ethylene oxide
Formula	Ne	CH_4	NH_3	H_2O	HF	C_2H_4O
$\alpha_0/(10^{-40} \text{ C m}^2/\text{V})$	0.436	2.88	2.60	1.77	0.89	5.8
$\mu/10^{-30} \text{ C m}$	0	0	4.87	6.17	6.37	6.34
T_b/K	27.3	111.7	240	373	293	284

Table 26.10

Formula	CH_3F	CH_3OH
$\alpha_0/(10^{-40} \text{ C m}^2/\text{V})$	4.27	3.3
$\mu/10^{-30} \text{ C m}$	6.04	5.67
T_b/K	195	338

Table 26.11

Formula	CH_3OCH_3	C_2H_5OH
$\alpha_0/(10^{-40} \text{ C m}^2/\text{V})$	6.7	5.18
$\mu/10^{-30} \text{ C m}$	4.34	5.67
T_b/K	248	352

26.8 THE HYDROGEN BOND

The anomalies observed in the boiling points of compounds containing hydroxyl, amino, and imino groups indicate that there exists in these compounds an additional interaction energy over and above the van der Waals interaction energy. The magnitude of this additional energy is comparable to the van der Waals interaction energy, being of the order of 20 to 40 kJ/mol. This additional energy is a result of the formation of a weak bond between the oxygen atoms in two molecules of ethanol, for example. A hydrogen atom lies between the two bonded oxygen atoms. This bond is called a *hydrogen bond* and is given the conventional structural representation, $O\cdots H$—O. A hydrogen bond can be formed between any two highly electronegative atoms; this requirement restricts the hydrogen bond to fluorine, oxygen, and nitrogen, so that the following types occur: F—$H\cdots F$, O—$H\cdots O$, N—$H\cdots N$; and, of course, mixed types such as N—$H\cdots O$. The effects of hydrogen bonding with S and Cl are very slight.

Although much has been written in the attempt to explain the stability of the hydrogen bond, its fundamental nature remains somewhat obscure. Because of reluctance to assign two covalent bonds to a hydrogen atom, great emphasis has been placed on interpreting the bond on a purely electrostatic basis, which has been successful in explaining some of the properties of the hydrogen bond. More recently, it has been realized that hydrogen can be connected to more than one atom by bonds that are distinct from the ordinary covalent bond, since they involve three nuclei rather than two, yet seem to have some characteristics of the covalent bond.

One of the most striking illustrations of the effect of hydrogen bonding on physical properties is shown by the plot of boiling points of the hydrides of the elements in periodic groups IV, V, VI, and VII, shown in Fig. 26.7; the horizontal axis serves only to separate one compound from the next. The melting points of these compounds exhibit a similar anomaly. The high boiling points of water, ammonia, alcohols, amines, and hydrogen fluoride are a consequence of the fact that these substances are hydrogen bonded in the liquid into polymers; for example, a liquid alcohol can be viewed as a mixture of polymers,

$$
\begin{array}{ccccc}
R & & R & & R \\
| & & | & & | \\
O\text{—}H & \cdots & O\text{—}H & \cdots & O\text{—}H
\end{array}
$$

Once we adopt the view that these kinds of compounds are capable of association, then a number of diverse observations come into focus.

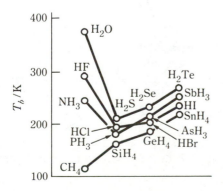

Figure 26.7 Boiling points of hydrides.

Hydrogen bonding accounts for unusually high melting and boiling points of such compounds as alcohols, sugars, organic acids, and simple inorganic acids such as H_2SO_4, HNO_3, H_3BO_3. A compound such as urea, $NH_2—CO—NH_2$, is solid at room temperature, but acetone, $CH_3—CO—CH_3$, having the same number of electrons, is a volatile liquid. Urea is hydrogen bonded; acetone can form hydrogen bonds only in the enol form, $CH_2=C(OH)CH_3$. Boric acid should be a gas or at worst a volatile liquid at room temperature; BF_3, with the same number of electrons, is a gas. In fact, boric acid is an involatile solid that melts with decomposition at 185 °C. The formula, when written correctly as $B(OH)_3$, betrays the possibility of hydrogen bonding.

The unusual values of entropies of vaporization of these associated liquids are understandable. While so-called normal liquids have entropies of vaporization of about 90 J/K mol, the values for these hydrogen-bonded compounds are usually (but not always) higher. This is illustrated by the following data.

Compound	H_2O	NH_3	CH_3OH	CH_3NH_2	HF	CH_3COOH
$\Delta S_{vap}/$(J/K mol)	109	97	105	97	26	62

In those cases for which ΔS_{vap} is higher than the normal value, the liquid, as a result of being associated, has a higher degree of order than a normal liquid. Therefore the transition from liquid to gas is attended by an unusually large entropy increase. Substances such as HF and acetic acid, which have unusually low values of ΔS_{vap}, are polymerized even in the vapor state. This polymerization reduces the disorder in the vapor and so lessens the value of ΔS_{vap}. It is known independently of this that HF is highly polymerized in the vapor state. The polymers are zigzag chains (or rings). The chain in HF has the structure

The vapor of acetic acid contains an appreciable amount of the dimer

This accounts for the low value of ΔS_{vap} for this substance.

Hydrogen bonding also affects the appearance of the infrared spectrum. The OH group absorbs at a frequency of 3500 cm^{-1}. If the spectrum of a hydroxyl compound is measured in the vapor, where it is not hydrogen bonded, this absorption shows up as a sharp peak centered at this frequency. If the spectrum of the hydrogen-bonded compound is examined, the peak is greatly broadened and the center is shifted to a lower frequency. The broadening of the absorption peak is characteristic of hydrogen bonding.

Additional evidence of hydrogen bonding in solids is provided by x-ray diffraction studies of crystals. In the crystal of a substance containing hydroxyl groups, certain of the oxygen–oxygen distances are abnormally short, indicating bond formation between these oxygens. There are just enough of these short distances to account for the hydrogen atoms in the hydroxyl groups.

QUESTIONS

26.1 Sketch Debye plots for gaseous HF, HCl, HBr and HI. Indicate the relative sizes of the intercepts and the slopes.

26.2 Explain why the net attractive forces between polar molecules decrease as T increases. Compare this with the T dependence of the molar polarization of a polar substance in an external field.

26.3 Sketch and explain the behavior of the polarization of a polar substance as a function of the frequency of an external electric field.

26.4 The larger a molecule, the larger is the dispersion energy, and the higher is the boiling point. Use this idea and the T dependence of the polar attractive energy to rationalize the relative influence of μ on the boiling points displayed in Tables 26.6, 26.7, and 26.8.

PROBLEMS

26.1 By combining the thermodynamic equation of state with the van der Waals equation, we can show that $(\partial \bar{U}/\partial \bar{V})_T = a/\bar{V}^2$. Below the critical temperature, the van der Waals equation predicts, approximately, a liquid volume equal to b, and a gas volume equal to RT/p. Assuming that the substance follows the van der Waals equation, what increase in energy attends the isothermal expansion of one mole of a substance from the liquid volume to the gaseous volume?

26.2 A.D. Buckingham and R.E. Raab [*J. Chem. Soc.* 5511 (1961)] measured the limiting value at zero pressure of the molar polarization of gaseous acetonitrile as a function of temperature.

$t/°C$	80.64	100.48	131.69	160.00
$P/(cm^3/mol)$	273.68	258.50	239.26	224.01

From a least-squares fit of the data, find α_0 and μ for CH_3CN.

26.3 From the values in Table 26.3, calculate the molar refraction of butane, propene, and acetone.

26.4 From the value of R_D in Table 26.3, calculate the polarizability of water.

26.5 Compare the magnitude of the average interaction energy between two molecules in the two situations: a liquid having a molar volume of 20 cm^3; a gas having a molar volume of 20,000 cm^3.

26.6 The Lennard–Jones potential $\epsilon = -A/r^6 + B/r^n$ can be expressed in terms of ϵ_m, the energy at the minimum, and r_0, the distance of separation at the minimum. Find A and B in terms of r_0, ϵ_m, and n. Write the potential in terms of the new parameters. If σ is the distance of separation when $\epsilon = 0$, find the relation between r_0 and σ.

26.7 Using the results in Problem 26.6, note the simplification in the form of ϵ if $n = 12$. Write ϵ in terms of ϵ_m and r_0 and in terms of ϵ_m and σ if $n = 12$.

26.8 Calculate the dispersion interaction energy at 500 pm separation between two molecules of

a) neon, $\alpha_0 = 0.436 \times 10^{-40}$ C m^2/V; $N_A h\nu_0 = 2.080$ MJ/mol;
b) argon, $\alpha_0 = 1.81 \times 10^{-40}$ C m^2/V; $N_A h\nu_0 = 1.520$ MJ/mol;
c) krypton, $\alpha_0 = 2.74 \times 10^{-40}$ C m^2/V; $N_A h\nu_0 = 1.350$ MJ/mol;
d) xenon, $\alpha_0 = 4.46 \times 10^{-40}$ C m^2/V; $N_A h\nu_0 = 1.170$ MJ/mol.
e) Plot the boiling points of each (Table 26.4) as a function of the dispersion energy.

26.9 For water, $\alpha_0 = 1.77 \times 10^{-40}$ C m^2/V, $\mu = 6.17 \times 10^{-30}$ C m, $N_A h\nu_0 = 1.216$ MJ/mol, and $\sigma = 276$ pm. At 20 °C evaluate and compare the three contributions to the interaction of a pair of water molecules at the distance of closest approach.

26.10 According to classical electrostatics, the polarizability of a perfectly conducting sphere in a vacuum is equal to $4\pi\epsilon_0 r^3$ where r is the radius of the sphere. Using this relation, and the data in Table 26.2, compare the calculated radii of the species in the second row of Table 26.2; O^{2-}, F^-, Ne, and so on.

27

Structure of Solids

27.1 THE STRUCTURAL DISTINCTION BETWEEN SOLIDS AND LIQUIDS

The word "solid" will be applied only to *crystalline* solids, since it is always possible to distinguish between a crystalline solid and noncrystalline phases such as liquid and amorphous "solids." Structurally, the constituent particles—atoms, molecules, or ions—of a crystalline solid are arranged in an orderly, repetitive pattern in three dimensions. If we observe this pattern in some small region of the crystal, we can predict accurately the positions of particles in any region of the crystal however far they may be removed from the region of observation. The crystal has *long-range order*.

The particles composing a liquid are not arranged in such a precise fashion. In a small region of a liquid, there may appear to be a pattern of arrangement. However, if we observe a neighboring region, the pattern will be somewhat different, or if it is nearly the same, it may not be accurately joined to the first region. In terms of the arrangement of the particles, a liquid has *short-range order* but lacks the long-range order that characterizes the solid. Figure 27.1 illustrates the distinction in two dimensions. The difference between solid and liquid is the difference between two ways of putting ball bearings in a box: they may be carefully packed row upon row or may simply be dumped into the box.

The irregular arrangement of particles in the liquid results in the appearance of voids or holes here and there throughout the structure. Note that a *nearly* regular arrangement exists around many of the particles in Fig. 27.1(b). The presence of the holes requires a larger volume, and in most cases the volume of the liquid is larger than that of the solid. The ease of flow is also related to this larger volume of the liquid. Amorphous "solids" have the same structural features as liquids, and are conveniently regarded as extremely viscous liquids.

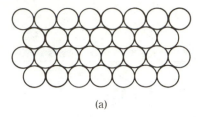

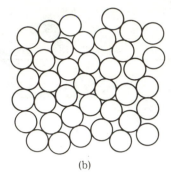

Figure 27.1   Schematic view of structure in a crystal (a) and in a liquid (b).

27.2 AN EMPIRICAL CLASSIFICATION OF SOLID TYPES

There is no unique way of classifying solids. The method chosen depends in great measure on the purpose at hand. For the present we will classify solids according to the type of bond that holds the constituent particles of the solid together.

Four types of bonds are operative in binding individual species into a crystal. On this basis we distinguish four types of solids: (1) metals, (2) ionic crystals, (3) van der Waals crystals, and (4) covalent crystals. The crystals bound by hydrogen bonds are usually classed with van der Waals crystals, since the strength of bonding is of the same order of magnitude in both types. However, since the arrangement of the hydrogen atoms in a molecule permits hydrogen bonds only in special directions, the hydrogen-bonded crystals have some of the features of covalent crystals.

In the first three solid types (metals, ionic crystals, van der Waals crystals), the forces of interaction that hold the particles together do not act in any preferred direction; in covalent crystals, the bonds are formed only in special directions because of the directional character of the covalent bond. The principles governing the direction of bond formation in covalent crystals are the same as those governing the covalent bond in molecules.

27.3 GEOMETRIC REQUIREMENTS IN THE CLOSE-PACKED STRUCTURES

If we ask how atoms or molecules can be arranged in a regular way to build a crystal, it seems that the possibilities might be unlimited. Although there are a great number of possible arrangements, a relatively small number of these recur again and again. One factor that limits the possibilities is the requirement that the arrangement be the most stable one energetically. Some departure from this principle is allowed, but generally different crystalline forms of a single substance do not differ greatly in energy. Energies of transitions between different crystalline forms of a substance are usually of the order of 1 kJ/mol.

First consider those crystals in which the energy of interaction between the particles does not depend on the direction of approach. As two particles approach, the energy of the system decreases and passes through a minimum at some distance. The two-particle system has its greatest stability at this point. If a third particle is introduced, the energy of the system decreases further. The maximum stability is attained when each particle in the aggregate is surrounded by the greatest possible number of neighbors. In brief, the particles must be as closely packed as possible. If the particles are spheres of the same size, the problem reduces to how to pack as many balls as possible in a given space.

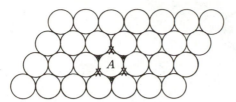

Figure 27.2

Clearly, the balls will be packed in layers, and each layer must be closely packed. We begin by arranging the layer as shown in Fig. 27.2. In the layer, each sphere has six nearest neighbors. To build the crystal in three dimensions, we stack the layers one on top of the other in a regular way. Two possibilities exist after the second layer is put on. Consider sphere A in the figure and suppose that the spheres in the next layer nestle in the notches at the positions marked with dots. Three of these will make contact with A. In this second layer there is a notch over A, but there are also notches over the positions marked with crosses. The third layer may then repeat the arrangement in the first layer or not. If the third layer repeats the first, then there are only two kinds of layers, denoted x and y, and the patterns of layers is $xyxyxy$. ... If the third layer does not repeat the first, it is denoted by z and the pattern of layers is $xyzxyzxyz$. ...* These two arrangements are common ones in metals and in van der Waals crystals composed of effectively spherical molecules such as CH_4, HCl, Ar.

The arrangement of close-packed layers in the pattern $xyxy$. .. is the hexagonal close-packed structure (hcp); the pattern $xyzxyz$... is the cubic close-packed (ccp) or face-centered cubic (fcc) structure. In each of these structures, every sphere is in contact with twelve others: six in its own layer, three in the layer above, and three in the layer below. The twelve coordination in these structures is shown by an exploded view in Fig. 27.3(a) and (b), and in a different view in Fig. 27.3(c) and (d). The hcp and fcc structures are the typical structures encountered in metals. The high coordination number (twelve) in these structures results in a crystal of comparatively high density.

In some structures the repetition pattern of the close-packed layers is more complicated, and we find the CH notation is useful to describe the structure. If we choose any close-packed layer, and if the two layers on each side of it are identical, the layer is designated by H, since repetition of these layers will build the hexagonal close-packed structure. Thus the alternation $xyxyxy$... is denoted by $HHHH$. .... If the two layers on each side of the chosen layer are not identical, the layer is designated by C, since repetition of layers of this kind will build the cubic close-packed (the fcc) structure. Thus the repetition pattern, $xyzxyz$... is denoted by $CCCC$. .... More sophisticated repetition patterns are possible; for example, in lanthanum the repetition is $xyxzxyxz$. .. , which becomes $HCHCHC$. .. , and in samarium, we have $xyxyzyzx$. .. , which becomes $HHCHHC$. .... The relative simplicity of the CH notation for these complicated repetitions is apparent.

Another common arrangement of spheres that occurs in a few metals is the body-centered cubic (bcc), which is built up of layers having the arrangement shown in Fig. 27.4. The second layer fits in the notches of the first and the pattern of layers repeats, $xyxy$. .... In these layers the number of nearest neighbors around any sphere is four, as compared with six in the close-packed layers. In the body-centered structure the overall coordination number is eight; there are four nearest neighbors within the most closely packed layer, two

---

*   Twenty or thirty marbles or coins can be helpful in working out these arrangements.

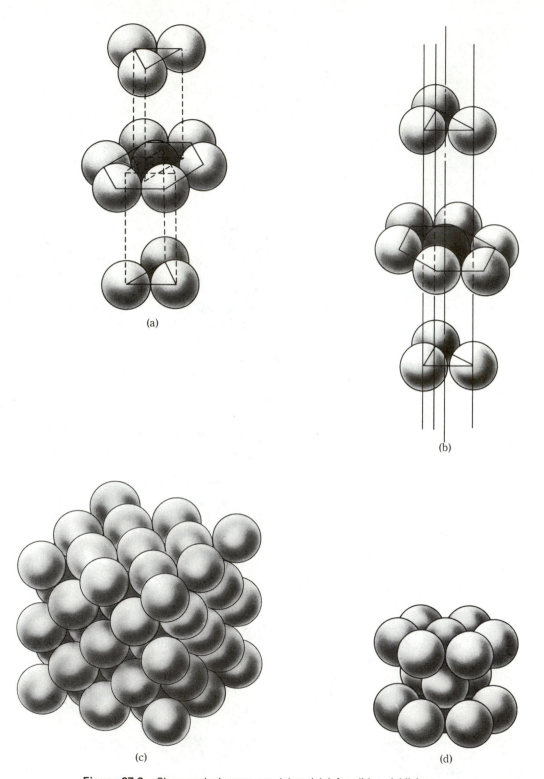

Figure 27.3   Close-packed structures. (a) and (c) fcc. (b) and (d) hcp.

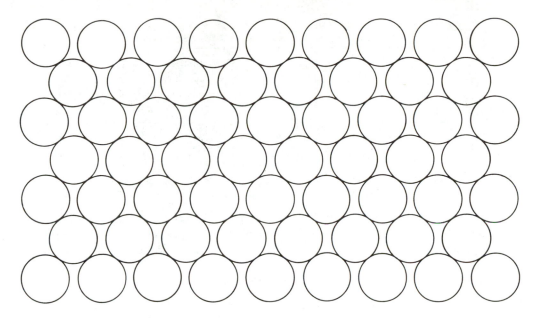

Figure 27.4   Most closely packed layer in bcc structure.

in the layer above, and two in that below. As a result of the less efficient packing, the bcc structure has an inherently lower density than the hcp or the fcc structures.

The positions of the atom centers in the three structures are shown in a different view in Fig. 27.5. To describe these structures completely requires the specification of the edge length a of the fundamental cube in the face-centered and body-centered cubic arrangements. The hexagonal close-packed structure requires the specification of two lengths, the nearest-neighbor distance a within the close-packed layer, and the distance c between the two repeating layers. If the spheres were truly rigid spheres, geometry would require $c = 1.633a$. Since the particles in a crystal are not truly rigid spheres, this relation is not exactly fulfilled; a and c must be specified separately. In metals having the hcp structure, the relation is nearly fulfilled. Table 27.1 lists a few of the metals that crystallize in the three structures, along with values of the lattice parameters.

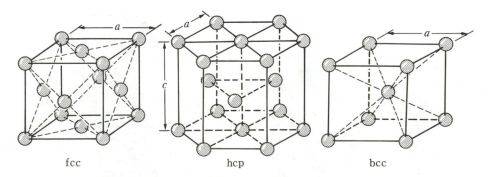

fcc   hcp   bcc

Figure 27.5   Location of atom centers in fcc, hcp, and bcc structures.

Table 27.1
Crystal structure and lattice constants for common metals

Face-centered cubic				Hexagonal close-packed			Body-centered cubic	
Metal	a/pm	Metal	a/pm	Metal	a/pm	c/pm	Metal	a/pm
Co(β)	355	Ni	352	Co(α)	251	407	Cr	288
Cu	362	Pd	389	Mg	321	521	Fe	287
Au	408	Pt	392	Ti	295	468	W	316
Pb	495	Ag	409	Zn	266	495	Na	429

Note that cobalt may be fcc or hcp depending on the temperature of crystallization. Some metals have structures that are slight distortions of these; mercury, for example, has a rhombohedral structure related to the fcc which has been compressed along one of the body diagonals.

27.3.1 Packing in Ionic Crystals

The packing of spheres in ionic crystals is complicated by the fact that the ions are positively and negatively charged. Suppose that the electrical charges on the positive and negative ions are equal, though opposite in sign. To build an electrically neutral structure requires that the number of negative ions around each positive ion be the same as the number of positive ions around each negative ion. If the positive and negative ions are the same size, we find that it is not possible to build a layer of alternating positive and negative ions with six positive ions around each negative ion, and vice versa, to yield a total coordination of twelve positive ions around each negative ion. The highest coordination possible, if the structure is to be electrically neutral, is that having the most closely packed layers built in the manner shown in Fig. 27.6a. In the layer each positive ion (shaded circles) is surrounded by four negative ions (open circles), and each negative ion by four positive ions. Piling up layers of this kind in the order $xyxy\ldots$ yields a cubic type of structure in which the central particle is an ion of one charge, with eight oppositely charged ions at the cube corners. The structure consists of two interpenetrating simple cubic lattices; (a simple cubic lattice is shown in Fig. 27.19). The positions of one lattice are occupied by positive ions, while those of the other are occupied by negative ions. The result is the cesium chloride, CsCl, structure (Fig. 27.6b).

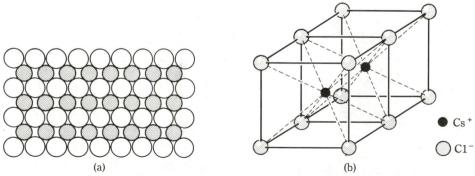

$\bullet$ Cs$^+$

$\bigcirc$ Cl$^-$

(a)   (b)

Figure 27.6   (a) Layer in CsCl structure; (b) CsCl structure.

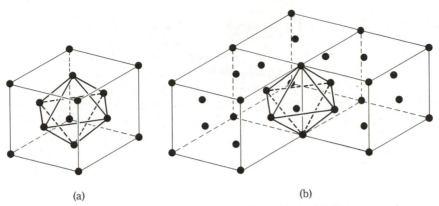

(a)   (b)

Figure 27.7   Octahedral holes in the fcc structure. (a) Central hole. (b) Hole centered on an edge.

A curious compromise is reached in many ionic crystals. The crystal NaCl, for example, is based on two interpenetrating close-packed (fcc) lattices. The positions of one lattice are occupied by positive ions, while those of the other are occupied by negative ions. Consider the unit cube of the fcc structure in Fig. 27.7(a). There is a void, or hole, outlined by the octahedron, at the center of the cube. An identical octahedral hole is centered on each edge of the unit cube (Fig. 27.7b). Each hole is at the center of an octahedron, which has atoms at each of the six apices. The centers of the octahedral holes occupy the positions of an fcc lattice, which interpenetrates the lattice on which the atoms are located. Small foreign atoms, such as H, B, C, N, can occupy these holes. Many carbides, hydrides, borides, and nitrides of the metals are *interstitial* compounds formed in this way.

If we wish to locate comparatively large particles in these holes, then all the particles of the original lattice must move apart; the structure expands to enlarge the holes to sufficient size. We can view the sodium chloride structure as an fcc arrangement of chloride ions that has expanded sufficiently to permit the sodium ions in the octahedral holes. As a result, neither of these interpenetrating lattices is closely packed in the sense of having all the particles in contact as they are in metals, but both have the symmetry of the close-packed fcc lattices. Each sodium ion is in contact with six chloride ions, and each chloride ion is in contact with six sodium ions; 6-6 coordination. Figure 27.8 shows the NaCl

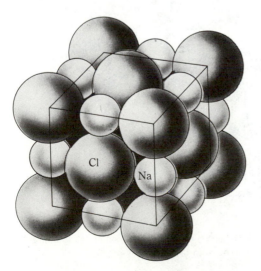

Figure 27.8   The NaCl structure.

structure; the fcc arrangement of the negative chloride ions is apparent; there is a sodium ion at the center of the cube.

In addition to the CsCl structure, 8-8 coordination, and the NaCl structure, 6-6 coordination, there are two structures having 4-4 coordination, the cubic ZnS (zinc blende) structure and the hexagonal ZnS (wurzite) structure (Fig. 27.9a and b). Note that the zinc blende structure is an fcc array of sulfide ions. There is a tetrahedrally coordinated hole at each corner of the cube; the zinc ions occupy four of the eight tetrahedral holes. In wurzite, the sulfide ions form an hcp array, and the zinc ions occupy half of the tetrahedral holes (Fig. 27.9c).

Unsymmetrical valence types of compounds such as cubic CaF_2 and Na_2O have more complicated structures, because the coordination number of the ion with the larger charge, Ca^{2+} in the case of CaF_2, must be twice that of the ion of lower charge if the crystal is to

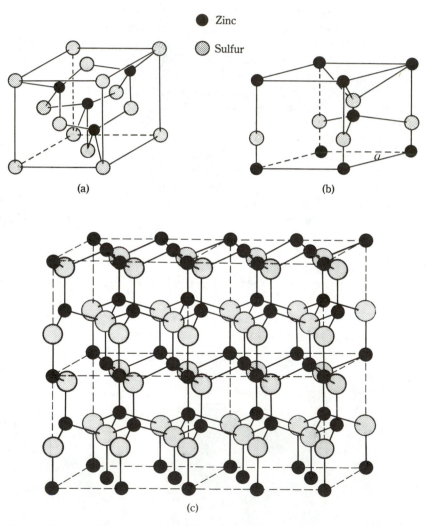

● Zinc

◉ Sulfur

(a)   (b)

(c)

Figure 27.9   (a) Unit cell in cubic ZnS, zinc blende. (b) Unit cell in hexagonal ZnS, wurzite. (c) Extended wurzite structure showing hexagonal symmetry.

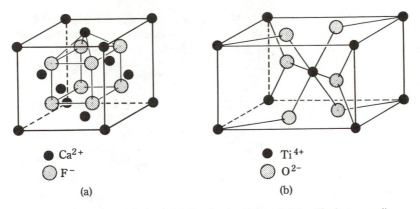

Figure 27.10 Unit cells in (a) fluorite (cubic) and (b) rutile (tetragonal).

be electrically neutral. For compounds of the 1-2 valence type, the typical structures are the cubic fluorite (CaF_2) structure and the tetragonal rutile (TiO_2) structure; Fig. 27.10.

The Ca^{2+} ions in fluorite are in a face-centered cubic arrangement. This lattice has, in addition to the octahedral holes mentioned earlier, holes that are tetrahedrally coordinated. The tetrahedral holes of the fcc structure are occupied by F^- ions in fluorite. Each F^- ion is tetrahedrally coordinated to Ca^{2+} ions. Figure 27.10(a) also shows that the Ca^{2+} ion on the top face is connected to four F^- ions below it; it is similarly connected to four F^- ions (not shown) lying above it. The coordination of the Ca^{2+} ion is eight, and the fluorite structure is described as having 8-4 coordination. Fluorite may be considered as a face-centered cubic array of Ca^{2+} ions interpenetrated by a simple cubic array of F^- ions.

In rutile, having 6-3 coordination, the octahedral coordination of Ti^{4+} to O^{2-} and the triangular coordination of O^{2-} to Ti^{4+} is evident in Fig. 27.10(b).

We have described structures of metals and ionic crystals in terms of close-packed arrangements of spheres. Clearly, if the particles are not spherical, the close packing must be done in a manner appropriate to the shape of the particle. We could scarcely expect long rod-shaped molecules to pack as spheres would; such particles pack into the crystals as matches in a box.

27.3.2 The Radius Ratio Rules

A factor of great importance in ionic crystals is the difference in size of the positive and negative ions. Pauling has shown how the geometric requirements for close packing of spheres of different sizes can be simply expressed in terms of the radius ratio $\rho = r_s/r_l$, defined as the ratio of the radius of the smaller ion, r_s, to that of the larger ion, r_l.

For simple ionic compounds of symmetrical valence type, the radius ratio rules are:

ρ	$\rho < 0.414$	$0.414 < \rho < 0.732$	$\rho > 0.732$
Coordination	4-4	6-6	8-8
Structure	zinc blende or wurzite	NaCl	CsCl

These rules enable us to predict the structure of the compound from the relative sizes of the two ions. Applied to many different ionic crystals of different valence types, the rules are quite good. There are at least two reasons for the exceptions to the radius ratio rules: (1) the ions are not rigid spheres; (2) the ions of opposite charge are not in contact.

27.4 GEOMETRIC REQUIREMENTS IN COVALENT CRYSTALS

The notable exception to the rule of close packing appears in covalent crystals in which the maximum stability is obtained, not with the greatest possible number of neighbors, but by forming the allowed number of covalent bonds in the proper directions. This requirement is peculiar to the individual substance so that a generalization of the kind embodied in the radius ratio rules is out of the question for covalent crystals. We cannot build up typical structures with the ease and confidence with which we stacked spheres into layers and layers one upon another. Rather than struggle with this host of individual problems, we will make only a few elementary remarks about the subject.

First of all, comparatively few solids are held together exclusively by covalent bonds. The majority of solids incorporating covalent bonds are bound also by either ionic or van der Waals bonds. The common occurrence is to find distinct molecules held together by covalent bonds and the molecules bound in the crystal by van der Waals bonds. The covalent bonds may hold a complex anion or cation together; the cations and anions are bound in the crystal by ionic bonds.

Only those atoms that form four covalent bonds produce a repetitive three-dimensional structure using only covalent bonds. The diamond structure, Fig. 27.11, is one of several related structures in which only covalent bonds are used to build the solid. The diamond structure is based on a face-centered cubic lattice wherein four out of the eight tetrahedral holes are occupied by carbon atoms. Every atom in this structure is surrounded tetrahedrally by four others. No discrete molecule can be discerned in diamond. The entire crystal is a giant molecule.

Generally the covalent solids have comparatively low densities as a result of the low coordination numbers. This effect is intensified in those crystals in which covalently bound structural units are bound in the crystal by van der Waals forces. The distance between two units held by van der Waals forces is significantly greater than that between units held by covalent, ionic, or metallic bonds; these large distances result in solids having comparatively low densities.

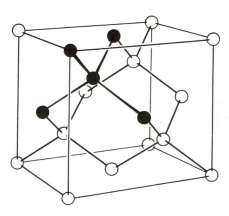

Figure 27.11 Diamond structure.

27.5 THE SYMMETRY OF CRYSTALS

The symmetry exhibited by a macroscopic crystal is a consequence of the symmetrical arrangement of the units of structure that compose the crystal. To understand the choice of a unit of structure and how a repetitive structure is built from that unit, we examine the problem in two dimensions. Any area-filling repetitive pattern on a plane surface is based on a unit of pattern that may be outlined by a parallelogram (Fig. 27.12). The entire pattern can be generated by translating this parallelogram, the unit of pattern, by definite distances parallel to its edges. We describe the unit of pattern in terms of two vectors of lengths a and b, which form two sides of the parallelogram. Starting at a point and moving any integral multiple of the distance a in the direction of the first vector, we reach an equivalent point in the pattern; similarly, moving an integral multiple of the distance b in the direction of the second vector, we reach an equivalent point in the pattern. The two vectors are the *primitive translations* of the pattern.

In two dimensions there are five possible units of pattern, unit cells, which can build a repetitive pattern by translation in directions parallel to the edges (Fig. 27.13). The unit cell with the 120° angle is interesting because it permits a threefold or sixfold axis of symmetry at a point in the pattern, which is a permissible type of symmetry in two-dimensional patterns.

To generate a repetitive pattern in three dimensions, an additional repetition vector out of the plane of the first two must be added. The three vectors define a parallelepiped. Any repetitive pattern in three dimensions has a parallelepiped as a unit cell. There are seven distinct parallelepipeds, those labeled (P) or (R) in Fig. 27.19 (on p. 696), which can generate by translation any repetitive pattern in three dimensions. Crystals are classified

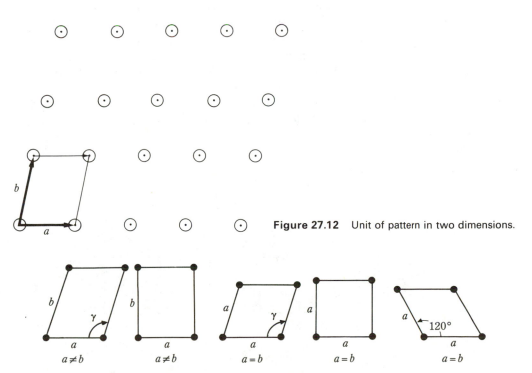

Figure 27.12   Unit of pattern in two dimensions.

Figure 27.13   The five units of pattern in two dimensions.

Table 27.2
The seven crystal systems

Axes	Angles	System
$a \neq b \neq c$	$\alpha \neq \beta \neq \gamma \neq 90°$	Triclinic
$a \neq b \neq c$	$\beta \neq \alpha = \gamma = 90°$	Monoclinic
$a \neq b \neq c$	$\alpha = \beta = \gamma = 90°$	Orthorhombic
$a = b \neq c$	$\alpha = \beta = 90°; \gamma = 120°$	Hexagonal
$a = b \neq c$	$\alpha = \beta = \gamma = 90°$	Tetragonal
$a = b = c$	$\alpha = \beta = \gamma$	Rhombohedral (trigonal)
$a = b = c$	$\alpha = \beta = \gamma = 90°$	Cubic

into seven *crystal systems* according to the shape (the lengths and inclinations of the vectors) of the unit cell. The lengths of the primitive translation vectors, the axes of the unit cell, are denoted by a, b, and c. The angle between a and b is γ, that between b and c is α, and that between c and a is β. Table 27.2 lists the relations between the lengths and between the angles for the crystal systems. In all cases the edges of the unit cell are parallel to edges or possible edges of the crystal.

27.6 THE CRYSTAL CLASSES

Having divided crystals into systems according to the possible shapes of the unit cell, we can make a further division according to the combinations of symmetry elements that are compatible with each system. An element of symmetry is an operation that brings the crystal into coincidence with itself. The elements of symmetry are: rotation about an axis, reflection in a plane, inversion through a center of symmetry, and rotation inversion. For example, consider the simple cube shown in Fig. 27.14(a). Rotation through 90° around the vertical axis brings the cube into coincidence with itself. This axis is a fourfold axis of symmetry. (If a rotation of 360°/p around an axis brings the figure into coincidence, then the axis is a $p$fold axis of symmetry.) The cube has three fourfold axes of symmetry, which are symbolized by the small squares at the ends of the axes. A twofold axis, rotation through 180°, is shown in Fig. 27.14(b), symbolized by ellipses at the ends of the axis. The cube has

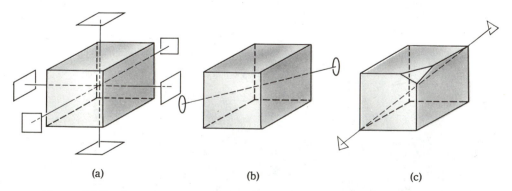

Figure 27.14 Axes of symmetry of the cube. (a) Fourfold. (b) Twofold. (c) Threefold.

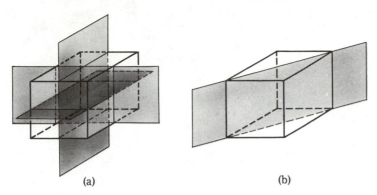

Figure 27.15 Planes of symmetry of the cube. (a) Principal planes (3).
(b) Diagonal planes (6).

six twofold axes. The four main diagonals of the cube are threefold axes of symmetry, symbolized by the triangles in Fig. 27.14(c). The threefold axis has been emphasized in Fig. 27.14(c) by truncating one corner of the cube to display the triangle.

A plane of symmetry divides any figure into mirror images. The cube has nine planes of symmetry, as shown in Fig. 27.15.

Finally, the cube has a center of symmetry. Possession of a center of symmetry, a center of inversion, means that if any point on the cube is connected to the center by a line, that line produced an equal distance beyond the center will intersect the cube at an equivalent point. More succinctly, a center of symmetry requires that diametrically opposite points in a figure be equivalent. These elements together with rotation–inversion* are the symmetry elements for crystals. The elements of symmetry found in crystals are: (a) center of symmetry; (b) planes of symmetry; (c) 2-, 3-, 4-, and 6-fold axes of symmetry; and (d) 2- and 4-fold axes of rotation–inversion. Of course, every crystal does not have all these elements of symmetry. In fact, there are only 32 possible combinations of these elements of symmetry. These possible combinations divide crystals into 32 *crystal classes*. The class to which a crystal belongs can be determined by the external symmetry of the crystal. The number of crystal classes corresponding to each crystal system are: triclinic, 2; monoclinic, 3; orthorhombic, 3; rhombohedral, 5; cubic, 5; hexagonal, 7; tetragonal, 7.

To illustrate how crystals with the same crystallographic axes (belonging to the same crystal system) can have different combinations of symmetry elements, we choose the cubic system with three equal axes at 90°. Rock salt belongs to the cubic system; the crystals have the full symmetry of the cube described above: three fourfold axes, four threefold axes, six twofold axes, nine planes of symmetry and a center of symmetry. The crystal shown in Fig. 27.16(a) also belongs to the cubic system. However, the crystallographic axes have only twofold symmetry; the principal diagonals are threefold axes of symmetry still; there are only six mirror planes. This figure has the symmetry of the tetrahedron (Fig. 27.16b). The crystal shown in Fig. 27.16(a) belongs to the tetrahedral class of the cubic system rather than the normal class, which has the full symmetry of the cube.

The most common method of determining the symmetry class is by examining the symmetry of the x-ray diffraction pattern of a small single crystal specimen. However,

---

*   The *p*fold axis of rotation–inversion: rotation through 360°/*p*, followed by inversion through the center.

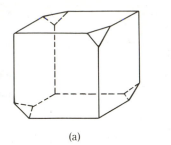 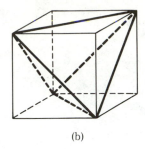

(a)   (b)

Figure 27.16 Tetrahedral class of the cubic system. (a) Truncated cube. (b) Tetrahedron developed from the cube.

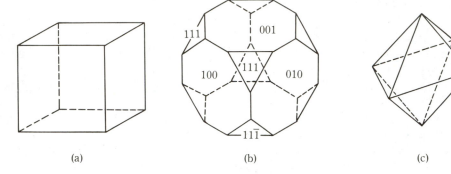

(a)   (b)   (c)

Figure 27.17 Crystal habit. (a) Cubic NaCl crystal. (b) NaCl crystal showing some development of octahedral faces. (c) NaCl crystal grown in 10 % urea solution.

it is possible to determine the symmetry class to which a given crystal belongs by examining a number of specimens of the crystal: measuring interfacial angles, etc. Measurement of optical properties, such as refractive index, which may have different values along different axes, can aid in the determination of symmetry class. It is essential to examine a number of crystals preferably grown under different conditions. The *habit* of the crystal (but not the symmetry class) depends on how the crystal is grown. Various possible *habits* of NaCl are shown in Fig. 27.17. Both the cube and the octahedron have the same combination of symmetry elements; therefore a substance such as NaCl may show the faces appropriate to either. Another variation in habit is shown by sodium chlorate, $NaClO_3$. It may grow as a cube, or a tetrahedron, or in a crystal showing a more complicated pattern of faces. The symmetry class is correctly assigned by judging the least symmetrical crystal. The cubic and tetrahedral habits of $NaClO_3$ both exhibit higher symmetry than is proper to the crystal class. The exhibition of a higher symmetry than the true symmetry is a common occurrence. If the crystals are grown quickly, they do not develop all the faces that are proper to the crystal class. The absence of these faces gives the appearance of higher symmetry.

27.7 SYMMETRY IN THE ATOMIC PATTERN

The division of crystals into crystal systems and crystal classes is based on the symmetry of the crystal as a finite object, or the symmetry of a single unit cell. In a unit cell all of the corners are equivalent points, since by translation along the axes the entire pattern can be

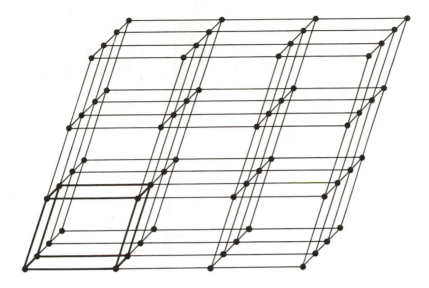

Figure 27.18 A point lattice.

generated; this is shown in Fig. 27.18, where the unit cell is heavily outlined. By translation of the unit cell, the entire pattern of equivalent points, called a *point lattice*, is generated. However, all the possible point lattices cannot be obtained if points are placed only at the corners of the unit cells in the seven systems. It was shown by Bravais that there are seven more unit cells which are required to produce every possible arrangement of equivalent points in space, that is, every point lattice. These additional lattices are conveniently described in terms of centered lattices. These fourteen unit cells are the Bravais lattices (Fig. 27.19).

In addition to the symmetry within any particular cell, the points in the neighboring cells are related by symmetry to those in that particular cell. Thus we can add symmetry operations that contain an element of translation as well as the other elements appropriate to the finite figure. The addition of translation to the possible symmetry operations greatly increases the number of possible combinations of the symmetry elements. There are 230 possible combinations (space groups); any atomic arrangement in a crystal must have the symmetry corresponding to one of these 230 combinations of symmetry operations. To determine the space group requires a detailed examination of the crystal by x-rays.

The new symmetry elements that are introduced are screw axes and glide planes. A screw axis in a pattern is exemplified in the structure of selenium, which has a threefold screw axis. The chain of selenium atoms winds around the edge of the unit cell. If we imagine a cylinder centered on the edge of a unit cell, Fig. 27.20(a), then a rotation of 120° with a translation of $\frac{1}{3}$ the height of the cell moves atom a to position b, atom b to position c, atom c to a', atom a' to a position in the next unit cell, and so on. Repetition of this operation three times moves a to a'. The unit cell has been transformed into itself, but moved upward to the position of the next unit cell.

Figure 27.20(b) shows the operation of a glide plane. If the upper layer of atoms is moved a distance $\frac{1}{2}a$ and then reflected in the plane MM', the lower layer of atoms is generated. Repetition of this operation regenerates the upper layer, but translated by the distance a.

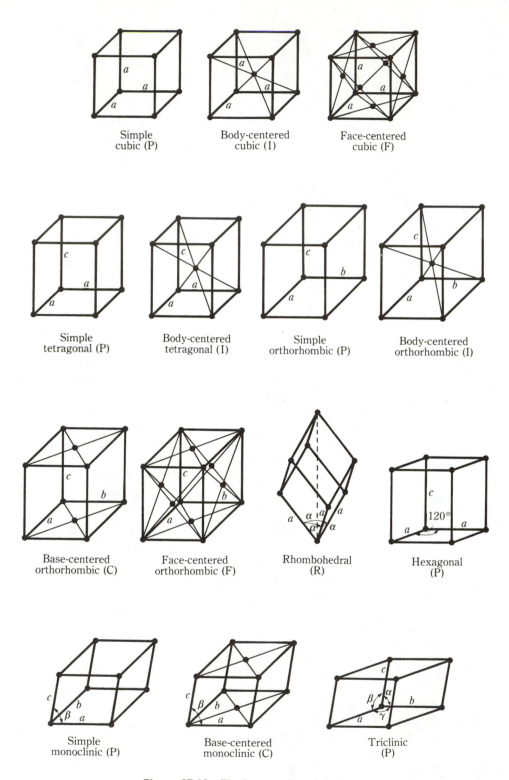

Figure 27.19   The fourteen Bravais lattices.

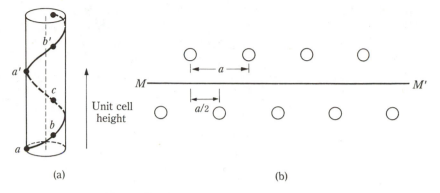

(a)   (b)

Figure 27.20   (a) Screw axis. (b) Glide plane.

★ 27.8   THE DESIGNATION OF CRYSTAL PLANES AND FACES

Knowing that a crystal is built up by the repetition of a unit cell, we can explain the development of faces of various kinds; Fig. 27.21(a) illustrates this in two dimensions. The faces F_1 and F_2 are formed by the bottoms and sides of the unit cells. Other faces F_3 and F_4 are possible, formed by the corners of the unit cells. Since the unit cell is of atomic size, we do not see the little steps but see only another face of the crystal. Because the crystal is built in this special way an important relation exists between the axial intercepts of any face and those of any other face. We compare the intercepts on the axes of the face F_3 with those of a possible face F_4. The geometry of these faces is shown in Fig. 27.21(b). The line PL is produced until it intersects the x-axis at a and the y-axis at b. In intercept form the equation of the line is

$$\frac{x}{a} + \frac{y}{b} = 1. \tag{27.1}$$

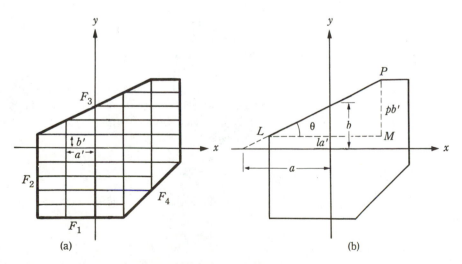

(a)   (b)

Figure 27.21   The law of rational intercepts.

Let the width of the unit cell be a', and the height of the unit cell be b'. For the face F_3, suppose MP is pb' and ML is la', where p and l are integers. Then $\tan \theta = pb'/la'$, and also $\tan \theta = b/a$; therefore, $b/a = pb'/la'$, so that $b = (p/l)(a/a')b'$ and Eq. (27.1) becomes, after multiplying by a/a',

$$\frac{x}{a'} + \frac{y}{(p/l)b'} = \frac{a}{a'}. \qquad (27.2)$$

But there are other points on F_3: $x = ma'$, $y = nb'$, where m and n are integers. Equation (27.2) must be satisfied at these points. Hence

$$m + \frac{nl}{p} = \frac{a}{a'}; \qquad (27.3)$$

the left-hand side of this equation involves only integers; hence a/a' is *rational*, expressible as a ratio of integers. From the earlier equality, $b/b' = (p/l)(a/a')$, it follows that b/b' is rational. Since the face F_3 was not a special one, it follows that the axial intercepts of any face, measured in units of the length of the unit cell, are rational numbers. The argument in three dimensions goes in the same way, except that we deal with the intercepts of planes on the three axes. Since the intercepts of any plane are rational multiples of the length of the unit cell, it follows that the intercepts of two planes are rational multiples of each other. Let the intercepts on the x-axis of two faces be a_1 and a_2; then $a_1 = r_1 a'$ and $a_2 = r_2 a'$, where r_1 and r_2 are rational. It follows that $a_2 = (r_2/r_1)a_1$. Since r_1 and r_2 are rational a_2 is a rational multiple of a_1. The same argument can be made for the y and z intercepts.

Therefore, if to a given face, F_3, of a crystal we assign intercepts, a, b, c on the coordinate axes, then the intercepts a_1, b_1, c_1 of any possible face, such as F_4, of the crystal are rational multiples of a, b, c. This is a fundamental law of crystallography, the law of rational intercepts.

Instead of describing a given face of the crystal by multiples of standard intercepts, we use the reciprocals of these multiples. That is, in terms of the intercepts a, b, c of the reference face, the intercepts of any face are given by

$$a_1 = a/h, \qquad b_1 = b/k, \qquad c_1 = c/l.$$

The numbers h, k, l are rational numbers or zero. If any of h, k, or l are fractions, the whole set is multiplied by the least common denominator to yield a set of integers h, k, l. The resulting integers h, k, l are called the Miller indices of the face. Through this process of taking reciprocals and clearing fractions, the law of rational intercepts becomes the law of rational indices. It should be clear that they are one and the same law. The indices of a face describe its orientation relative to the reference face, but do not describe the actual position. The usefulness of the indices in describing the face can be seen from writing the equation of the plane in intercept form:

$$\frac{x}{a_1} + \frac{y}{b_1} + \frac{z}{c_1} = 1.$$

But in terms of the intercepts of the reference plane, $a_1 = a/h$, and so on; hence

$$\frac{hx}{a} + \frac{ky}{b} + \frac{lz}{c} = 1.$$

If we measure distances in terms of the reference intercepts, x in units of a, y in units of b,

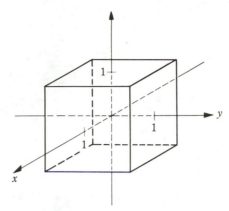

Figure 27.22

and so on, the equation becomes

$$hx' + ky' + lz' = 1, \tag{27.4}$$

where $x' = x/a$, $y' = y/b$, $z' = z/c$.

Consider the cube in Fig. 27.22. The intercepts of the right-hand side of the cube are $\infty, 1, \infty$. The reciprocals (indices) are 010. The front face of the cube has intercepts $1, \infty, \infty$, and indices 100; the left face has intercepts $\infty, -1, \infty$, and indices $0\bar{1}0$. (The minus sign is written over the number.) The rear face has indices $\bar{1}00$. The indices of the top and bottom faces of the cube are 001, and $00\bar{1}$. Any plane parallel to an axis has an intercept of ∞, and an index of zero for that axis.

The crystal consists of atoms, ions, or molecules; any face of a crystal consists of a layer of atoms, ions, or molecules. The method of describing faces of a crystal can be used to describe the planes of atoms in the crystal. Consider the body-centered cubic cell in Fig. 27.23(a); the intercepts of the shaded planes of atoms are $\infty, 1, 1$, so the indices are 011. In Fig. 27.23(b), the shaded plane has intercepts $\infty, 2, 1$, so the reciprocals are $0, \frac{1}{2}, 1$; clearing the fraction, we obtain the indices: 012. Figure 27.23(c), (d), and (e) show the 101, the 110, and the 111 planes.

Figure 27.17(b) showed the development of octahedral faces on a cube. These faces are obtained by truncating the eight corners of the cube. The crystal faces are designated by Miller indices obtained in the usual way. The angle between the 100 and 111 face is always $54° \, 44' \, 8''$.

The constancy of interfacial angles is a law of crystallography that has been recognized since the 17th century. In 1669, Nicolaus Steno observed that regardless of how the real crystal might be distorted from the ideal shape, the angles between two types of face were always constant. For example, a crystal of quartz has hexagonal symmetry, and a section cut perpendicular to the hexagonal axis should have a regular hexagonal shape, Fig. 27.24(a), each interior angle being 120°; in actual fact, because the growth of certain faces in the crystal is inhibited by the crowding of neighboring crystals and other influences, the faces of a real crystal develop unequally, and the section of real quartz crystals may have shapes such as those shown in Fig. 27.24(b) and (c). In all the possible distortions, the interfacial angle remains at 120°. This constancy of interfacial angles introduces an enormous simplification into crystallography, for it permits the recognition of the fundamental crystal symmetry from the interfacial angles of imperfect crystals.

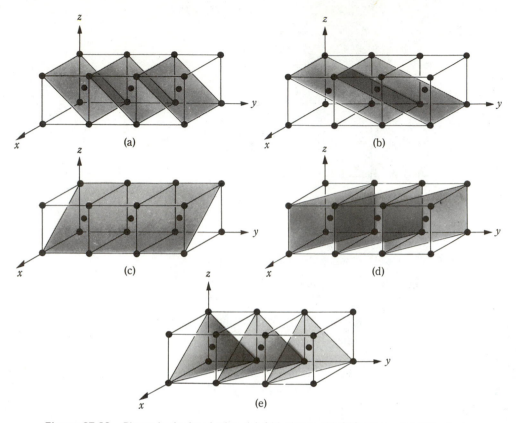

Figure 27.23 Planes in the bcc lattice. (a) 011 planes. (b) 012 planes. (c) 101 plane. (d) 110 planes. (e) 111 planes.

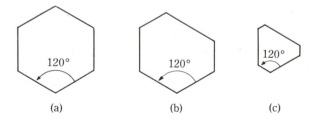

Figure 27.24 The constancy of interfacial angles. (a) Section of an ideal quartz crystal. (b) and (c) Possible shapes of the crystal section.

★ 27.9 THE X-RAY EXAMINATION OF CRYSTALS

The conclusion that x-rays were light rays of very short wavelength and the realization that a crystal consisted of a regular array of planes of atoms prompted Max von Laue in 1912 to suggest that the crystal should behave as a diffraction grating for x-rays if the wavelength were comparable to the spacing in the crystal. This suggestion was confirmed experimental-

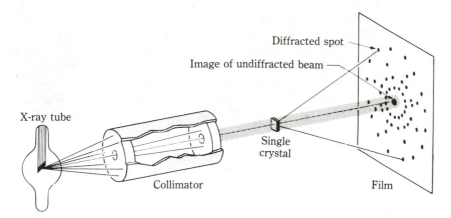

Figure 27.25   The Laue method.

ly almost immediately by Friedrich and Knipping. Figure 27.25 shows a typical setup for the Laue method. The diffraction pattern of spots produced on the photographic plate is called a Laue pattern.

An x-ray beam progressing through a crystal is reflected from every possible plane of atoms in the crystal by the usual law of specular reflection, the incidence angle equal to the reflection angle. Since there are many different planes of atoms all oriented at different angles relative to the incident beam, we might expect that the emergent beam would be completely diffused over all angles. The fact is that the emergent beam appears only at certain particular angles, and thus produces the Laue pattern. This happens because a plane of atoms is not present in the crystal by itself but with an enormous number of similar planes parallel to it. As a general rule the reflected beams from these parallel planes interfere destructively, and there is no emergent beam in most directions.

The condition for the reflected beams from a given set of parallel planes to reinforce each other and produce a spot is easily derived. Figure 27.26 shows a set of planes within a

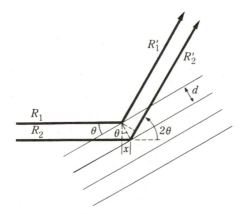

Figure 27.26   X-ray reflection from a set of planes.

crystal having an interplanar spacing d. The angle between the planes and the direction of the beam is the glancing angle θ. Ray R_1 is reflected specularly by the first plane to yield R'_1. Similarly, ray R_2 is reflected specularly from the second plane to yield R'_2. If the rays R'_1 and R'_2 are to reinforce one another, they must have the same phase; this condition is met if the extra distance traversed by $R_2 R'_2$ is equal to an integral number of wavelengths of the x-ray. The extra distance is $2x$, so that $2x = n\lambda$, where n is an integer. But from the geometry of the situation, $x = d \sin \theta$. Consequently, in terms of the interplanar spacing d, the condition for constructive interference becomes

$$2d \sin \theta = n\lambda, \qquad n = 1, 2, 3, \ldots, \qquad (27.5)$$

which is the fundamental law of x-ray crystallography, the Bragg condition, or Bragg's law: for a given wavelength of x-rays, the reflected beam will emerge only at those angles for which the condition is satisfied. This accounts for the Laue pattern of spots. Each spot is produced by a certain set of planes that fulfills the condition. Since similar sets of planes are disposed within the crystal in accordance with the crystal symmetry, the arrangement of the spots in the Laue pattern has the symmetry of the crystal to a certain degree.

The reflected beam makes an angle 2θ with the direction of the incident beam. This is shown by the simple geometry of Fig. 27.26. For a set of planes with specified values of h, k, l and a θ fixed by the geometry of the experiment, there are several wavelengths (determined by the Bragg relation) that combine constructively to produce the spots in the Laue pattern. In principle every set of planes can produce several spots. In practice, the number of spots is restricted by the limited range of wavelengths in the incident radiation and the limited range of observable values of θ. If monochromatic radiation were used in the Laue method no pattern would be produced, unless the Bragg relation were accidentally fulfilled for some set of planes.

The fact that some planes have a higher density of atoms than others produces the variation in intensity of the diffracted beam for different sets of planes. The planes of high atomic density scatter x-rays better and produce the more intense beam. If more than one kind of atom is present in the crystal, the species with the greater number of electrons has the greater scattering power. For light elements, the scattering power is proportional to the number of electrons around the atom.

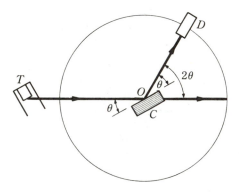

Figure 27.27   X-ray diffractometer.

In the Bragg x-ray diffractometer, Fig. 27.27, x-radiation from the tube T bathes a crystal C, which is mounted so that it may be rotated; the angle of rotation is measured on the scale of the instrument. By rotating the crystal it is possible to bring the coherent scattered beam from each set of planes into the detecting chamber D. The response of the detector, an ionization chamber or a Geiger counter, at various angles of rotation can be recorded to produce a pattern of peaks at various values of θ. From this measurement of θ, the interplanar spacing can be calculated from the Bragg equation.

★ 27.10 DEBYE–SCHERRER (POWDER) METHOD

The Bragg method of obtaining interplanar spacings has the disadvantage that mounting the crystal on a precise axis is time consuming. The *rotating crystal* method of obtaining a diffraction pattern, one of the most valuable methods for structure determination, also requires precise mounting of the crystal.

The simplest method of obtaining interplanar spacings is the Debye–Scherrer method. A sample of the crystal is ground to a powder and placed in a thin-walled glass tube mounted in the x-ray beam. Since many crystals are present, all having different orientations, some will be so oriented as to satisfy the Bragg relation for a given set of planes. Another group will be oriented so that the Bragg relation is satisfied for another set of planes, and so on. If a given set of planes satisfies the Bragg condition, the reflected ray produces a spot. If this set of planes is rotated about the axis of the incident beam, the corresponding rotation of the reflected ray generates a cone. In the powder method, a cone of reflected radiation is produced (Fig. 27.28) because all the orientations of a given set of planes about the axis of the beam are present in different particles of the powder. Figure 27.28 shows the conical reflected radiation and the position of the film that results in the line pattern. The film makes a nearly complete circle so that even the rays reflected at large angles are recorded on the film.

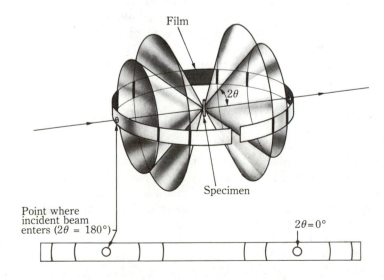

Figure 27.28 Debye–Scherrer powder method.

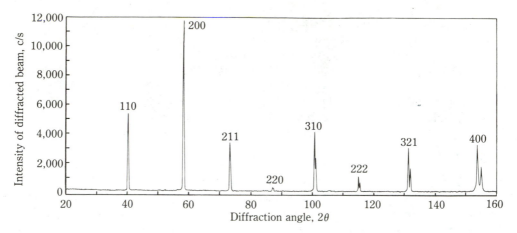

Figure 27.29 Typical powder diffraction pattern. Intensity at various angles for tungsten.

We measure accurately the distances between the lines on the film. From these and the dimensions of the camera, the diffraction angle 2θ can be calculated for each set of planes. From the Bragg angle θ the interplanar spacings d are calculated from the Bragg equation.

Through the use of a densitometer, a device for measuring the light transmission of the developed film, we can measure the intensity of the lines on a powder photograph. If, for example, the material whose structure is under study is a metal, then all the atoms are of the same kind. The most intense line will be reflected from the planes that are most closely packed. These would be the 111 planes in an fcc structure and the 001 planes in the hcp structure. Having identified the most closely packed planes and calculated their spacing from the Bragg equation, we can sometimes use other simple features of the pattern to identify the structure and establish all of the interplanar spacings by simple geometry.

Figure 27.29 shows a powder diffraction pattern obtained by mounting a capillary tube filled with fine tungsten powder in the center of the diffractometer illustrated in Fig. 27.28. The numbers on the peaks are the indices of the planes which produce that peak. By measuring the values of θ, the interplanar spacing can be calculated from the Bragg equation.

★ 27.11 INTENSITIES AND STRUCTURE DETERMINATION

Consider the face-centered cubic lattice in Fig. 27.30. The 100 planes are interleaved at just half the spacing by the 200 planes, which contain only face-centered atoms. The reflected rays from this second set of planes are 180° out of phase with those from the 100 planes. The two reflections interfere destructively so that a first-order reflection does not appear from the 100 planes in this lattice. (Higher-order reflections appear from both sets, but the intensities are much weaker.) For the same reason the first-order reflection from the 110 plane does not appear, being destroyed by the first-order reflection from 220 planes. The 111 planes are not interleaved in this way, so the first-order reflection comes through loud and clear, especially because the 111 planes are close packed. The absence of certain lines helps enormously in the assignment of indices, the indexing, of the lines that do appear.

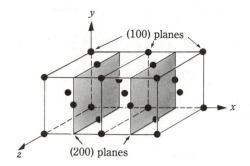

Figure 27.30 Interleaving of the 200 and 100 planes in the fcc structure.

From a complete study of line spacings and intensities in the diffraction pattern we can determine the size and shape of the unit cell and the arrangement of the kinds of atoms within the cell. In crystals of high symmetry, which have no more than two kinds of particles, it is possible to do this directly with relative ease; in metals, for example, or in crystals such as NaCl or ZnS. As we mentioned above, symmetrically interleaved layers of atoms of the same kind can completely extinguish certain reflections. Suppose, however, that the interleaved layers are not just halfway between, so that the reflected radiation is neither completely in nor out of phase with that from the first set of layers; add to this the fact that the interleaved layer may contain atoms of different scattering ability. The problem becomes quite complicated; nonetheless, for small molecules the solution can usually be accomplished by direct methods. From the positions of the lines we can establish the principal spacings in the structure, the shape and size of the unit cell, and the crystal class. That much is straightforward. Knowledge of the chemical constitution and density of the crystal establishes the number of atoms of each kind in the unit cell. Then the problem is to fix the positions of the various atoms in the unit cell. We do this by establishing the phase associated with each reflection. We then calculate the intensities of the lines in the diffraction pattern from an assumed arrangement of the atoms, a trial structure. We compare the calculated intensities with the observed intensities, and refine the trial structure by the least squares method. The procedure is repeated until reasonable agreement is obtained.

These calculations are extremely tedious; fortunately, they can be done by computer. But even before the advent of high-speed computers, the structures of hundreds of crystals had been worked out by hand calculation. The fruits of these x-ray studies are seen in the structures described earlier in this chapter.

★ 27.12 X-RAY DIFFRACTION IN LIQUIDS

The diffraction pattern of a liquid resembles a powder photograph except that the very sharp lines of the powder photograph are replaced by a few broad bands of reflected radiation. From an analysis of the intensity distribution in these broad bands, we can construct the radial distribution function for particles around a central particle in the liquid. This distribution function is interpreted in terms of the *average* number of atoms surrounding a central atom at the distance corresponding to the peak.

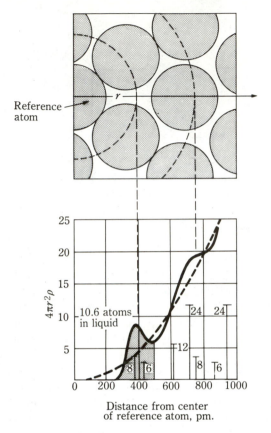

Figure 27.31   Radial distribution
curve in liquid sodium.

Figure 27.31 shows the radial distribution function, $4\pi r^2 \rho$, in liquid sodium. The upper drawing interprets the peaks in terms of "shells" of atoms around the central atom. At 400 pm from the central atom, the average number of atoms in the liquid is 10.6. This number is determined by the shaded area under the curve. The vertical lines show the number of atoms in successive "shells" in solid sodium.

QUESTIONS

27.1 Contrast the bonding and structure in (a) metals and (b) ionic crystals.

27.2 Ionic crystals are quite brittle and easily cleave when struck. Explain this by considering the electrostatic forces generated when two layers in the crystal are displaced.

27.3 Diamond is one of the hardest substances known. Account for this in terms of its structure.

27.4 Why are radio waves unsuitable for determining crystal structure?

27.5 How does Fig. 27.31 reflect the differences in order for a solid and a liquid?

PROBLEMS

27.1 Using the data in Table 27.1, compute the axial ratio c/a for the metals crystallizing in the hcp system and compare with the ideal value, 1.633.

27.2 Figure 27.5(a) and (c) show the unit cell for the fcc and bcc structures. How many atoms does the unit cell contain in each of these cases? [*Note:* an atom on a face is shared between two cells; an atom at a corner is shared between a number (how many?) of cells.]

27.3 The hexagonal cell shown in Fig. 27.5 consists of three unit cells. How many atoms are in the hexagonal cell shown and how many in the unit cell?

27.4 Referring to Fig. 27.5, which shows the location of the atom centers for the fcc, hcp, and bcc structures, if the edge length of the cube is a, compute, for the fcc and bcc structures, the volume of the cube and the volume of the cube that is actually occupied by the spheres. The spheres are in close contact; keep in mind that a sphere on a face or at a corner is only partially inside the cube. What percent of the space within the cube is empty?

27.5 Figure 27.6(b) shows two unit cells of CsCl. How many Cs^+ and Cl^- ions in the unit cell?

27.6 Figure 27.8 shows the unit cell of NaCl. How many Na^+ and Cl^- ions in the unit cell? (There is a sodium ion at the center of the cube!)

27.7 Using Fig. 27.7, how many octahedral holes per atom are present in the fcc structure? By sketching the bcc structure decide how many octahedral holes per atom are present.

27.8 Calculate what size of sphere can be accommodated in the octahedral hole of the fcc structure; cube edge $= a$, atom radius $= r_a$.

27.9 How many atoms (or ion pairs) are in the unit cell of

    a) diamond, Fig. 27.11;
    b) zinc blende, Fig. 27.9(a);
    c) wurzite, Fig. 27.9(b);
    d) fluorite, Fig. 27.10(a);
    e) rutile, Fig. 27.10(b).

27.10 Consider eight small spheres, radius $= r_s$, at the corners of a cube; they are small enough so that they are not in contact. Now place a larger sphere, radius $= r_l$, at the center of the cube. This sphere is just large enough so that it is in contact with the eight small spheres. Now let the radius of the large sphere shrink, but keep the small spheres in contact with it. What is the radius ratio, r_s/r_l, when the small spheres at the cube corners come into contact with each other? Compare your answer with the numbers in Section 27.3.2.

27.11 Consider a large sphere, radius $= r_l$, which has four small spheres, radius $= r_s$, arranged symmetrically (at the corners of a square) around its equator. There are also two small spheres in contact with the large sphere at its poles. The large spheres is now allowed to shrink, keeping the small spheres in contact with it. What is the radius ratio, r_s/r_l, when the small spheres come in contact with each other? Compare your answer with the numbers in Section 27.3.2.

27.12 What are the elements of symmetry of a tetragon?

$$a = b \neq c, \qquad \alpha = \beta = \gamma = 90°.$$

27.13 a) Sketch a cube and label each face with the proper Miller indices.
    b) Suppose that every edge of the cube is truncated by a plane perpendicular to the plane containing the edge and center of the cube. Sketch at least two of the faces exposed and find the Miller indices of these faces.

27.14 Using Fig. 27.5, sketch the fcc arrangements of atoms in 111 plane, in the 100 plane, in the 011 plane. Which plane is close packed?

27.15 Using x-rays of wavelength $\lambda = 179.0$ pm, a metal produces a reflection at $2\theta = 47.2°$. If this is a first-order reflection from the 110 planes of a body-centered cubic lattice what is the edge length of the cube?

27.16 The lattice parameter of silver, an fcc structure, is 408.6 pm. An x-ray beam produces a strong reflection from the 111 planes at $2\theta = 38.2°$. What is the wavelength of the x-ray?

27.17 Using x-radiation, $\lambda = 154.2$ pm, a face-centered cubic lattice produces reflections from the 111 and 200 planes. If the density of copper, which is face-centered cubic, is 8.935 g/cm^3, at what angles will the reflections from copper appear?

28

Electronic Structure
and Macroscopic Properties

28.1 PRELIMINARY REMARKS

In the discussions of the kinetic theory of gases and of intermolecular forces, we obtained expressions for properties of matter in bulk in terms of the properties of the individual molecules. In this chapter we will describe the cohesive energy of ionic crystals in terms of the interactions of the ions in the crystals, and some of the properties of metals and covalent crystals in terms of the quantum mechanical picture obtained from the Schrödinger equation. In Chapter 29 we will describe the method for calculating the thermodynamic properties of bulk systems from a knowledge of structure.

28.2 COHESIVE ENERGY IN IONIC CRYSTALS

A satisfactory theory of the cohesive energy of ionic crystals can be based almost exclusively on Coulomb's law. If two particles i and j, having charges z_i and z_j, are placed a distance r_{ij} apart in vacuum, the energy of interaction between them is

$$E_{ij} = \frac{z_i z_j}{4\pi\epsilon_0 r_{ij}}. \tag{28.1}$$

If z_i and z_j have the same sign, E_{ij} is positive, the particles repel one another, and the system lacks stability. If z_i and z_j are opposite in sign, the energy is negative, and the two are bound together by the energy E_{ij}. The more ions of opposite charge that surround a particular ion, the greater is the stability of the structure. We used this result implicitly in arguing for close-packed structures in ionic crystals in Section 27.3.1.

Consider a crystal, such as NaCl, in which the charges on the ions are equal and opposite in sign, $z_+ = -z_-$. To calculate the cohesive energy of a crystal that contains N ions of each sign, N ion pairs, we add the interaction of every ion with all the others. Since the interaction energies depend on the distances between the ions, we must know the geometrical arrangement of the ions in the structure. Fortunately, for symmetrical salts (rock salt, CsCl, zinc blende, and wurzite), the possible structures are all cubic, so that if we know the distance r between a cation and the nearest anion, then the distance between any two ions in the crystal can be calculated from the geometry of the structure. We write for the distance between ions i and j, $r_{ij} = \alpha_{ij}r$ in which α_{ij} is a numerical factor obtained from the geometry of the structure. The expression in Eq. (28.1) becomes

$$E_{ij} = \frac{z_i z_j}{4\pi\epsilon_0 r \alpha_{ij}}.$$

The interaction energy of ion i with all of the others, j, is obtained by summing this expression.

$$E_i = \frac{z_i}{4\pi\epsilon_0 r} \sum_{j \neq i} \frac{z_j}{\alpha_{ij}}.$$

In every term of this sum, $z_j = \pm z_i$; furthermore, each term in the sum is simply a number, determined by geometry, so the sum is a number that we write as $z_i S_i$. Then

$$E_i = \frac{z_i^2 S_i}{4\pi\epsilon_0 r}.$$

Summing the energies of all the ions in the lattice yields the total energy of interaction U_M:

$$U_M = \tfrac{1}{2} \sum_i E_i = \frac{z^2}{4\pi\epsilon_0 r} \sum_i \tfrac{1}{2} S_i,$$

where the factor $\tfrac{1}{2}$ appears because we must count the interaction between any pair of ions only once. The sum is a sum of numbers; it is negative and proportional to N, so we write $\tfrac{1}{2}\Sigma S_i = -NA$, where A is a numerical factor, the Madelung constant (named after E. Madelung, who first evaluated sums of this type). The total electrostatic energy, the Madelung energy, is

$$U_M = -\frac{NAz^2}{4\pi\epsilon_0 r}. \tag{28.2}$$

Values of the Madelung constant calculated from the geometry of the symmetrical structures are given in Table 28.1.

The cohesive energy of ionic crystals given by Eq. (28.2) is about 10% too large. This is a consequence of neglecting the repulsion that arises at close distances. Figure 28.1 shows how this comes about. As a function of r, the energy given by Eq. (28.2) follows the dotted curve in the figure. At the equilibrium separation r_0, the depth of the curve is somewhat below that of the minimum in the solid curve, which represents the actual cohesive energy.

Table 28.1

Structure	Coordination	A
CsCl	8-8	1.7627
NaCl	6-6	1.7476
Zinc blende	4-4	1.6381
Wurzite	4-4	1.641

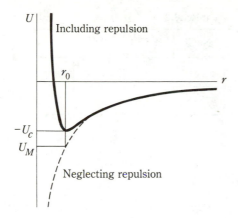

The repulsion that develops as the ions come in contact is represented, just as with neutral molecules (Section 26.6), by a term b/r^n, where b and n are constants and n is a large power, usually $n = 6$ to 12. This form of the repulsion energy was first introduced by M. Born, and is called the Born repulsion. The cohesive energy is written

$$U = -\frac{NAz^2}{4\pi\epsilon_0 r} + \frac{b}{r^n}. \tag{28.3}$$

The two empirical constants b and n are determined by two conditions. First we require that the energy have a minimum value at r_0, the equilibrium separation in the crystal; $(dU/dr)_{r=r_0} = 0$. Differentiating Eq. (28.3), we have

$$\frac{dU}{dr} = \frac{NAz^2}{4\pi\epsilon_0 r^2} - \frac{nb}{r^{n+1}}.$$

Setting this equal to zero at r_0 and solving for b, we obtain

$$b = \frac{NAz^2 r_0^{n-1}}{4\pi\epsilon_0 n}, \tag{28.4}$$

which reduces Eq. (28.3) to

$$U = -\frac{NAz^2}{4\pi\epsilon_0 r_0}\left[\frac{r_0}{r} - \frac{1}{n}\left(\frac{r_0}{r}\right)^n\right]. \tag{28.5}$$

At $r = r_0$ this becomes the negative of the cohesive energy $-U_c$:

$$-U_c = -\frac{NAz^2}{4\pi\epsilon_0 r_0}\left(1 - \frac{1}{n}\right). \tag{28.6}$$

The cohesive energy U_c is the Madelung energy at $r = r_0$ multiplied by $(1 - 1/n)$. If n is about 10, then the 10% error noted above is explained. Using Eq. (28.6) in Eq. (28.5), we obtain

$$U = -\frac{nU_c}{n-1}\left[\frac{r_0}{r} - \frac{1}{n}\left(\frac{r_0}{r}\right)^n\right]. \tag{28.7}$$

The constant n is determined from the compressibility of the crystal at 0 K; $\kappa = -(1/V_0)(\partial V/\partial p)_T$. At $T = 0$ K the thermodynamic equation of state, Eq. (10.28), becomes $p = -(\partial U/\partial V)_T$; by differentiating, $(\partial p/\partial V)_T = -(\partial^2 U/\partial V^2)_T$. Combining this result with

Table 28.2

Salt	LiCl	NaCl	KCl	RbCl	CsCl
n	7.0	8.0	9.0	9.5	10.5
U_c/kJ mol^{-1}	808.8	754.8	687.8	664.8	623.0

the definition of κ, we obtain

$$\frac{1}{V_0 \kappa} = \left(\frac{\partial^2 U}{\partial V^2}\right)_T. \tag{28.8}$$

To express the energy in terms of the volume of the crystal, we observe that the volume is proportional to the cube of the interionic distance r, so that r is proportional to the cube root of the volume; therefore, $(r_0/r) = (V_0/V)^{1/3}$, and Eq. (28.7) becomes

$$U = -\frac{nU_c}{n-1}\left[\left(\frac{V_0}{V}\right)^{1/3} - \frac{1}{n}\left(\frac{V_0}{V}\right)^{n/3}\right].$$

Differentiating, we have

$$\frac{dU}{dV} = \frac{nU_c}{3(n-1)}\left[\frac{V_0^{1/3}}{V^{4/3}} - \frac{V_0^{n/3}}{V^{1+n/3}}\right],$$

and

$$\frac{d^2U}{dV^2} = -\frac{nU_c}{9(n-1)}\left[\frac{4V_0^{1/3}}{V^{7/3}} - \frac{(n+3)V_0^{n/3}}{V^{2+n/3}}\right].$$

At $V = V_0$, this becomes $(d^2U/dV^2)_r = nU_c/9V_0^2$. Using this value and the value of U_c from Eq. (28.6) in Eq. (28.8), we obtain after solving for n,

$$n = 1 + \frac{9(4\pi\epsilon_0)V_0 r_0}{NAz^2\kappa}. \tag{28.9}$$

Having determined the value of n from the compressibility and the volume V_0, by Eq. (28.9), the cohesive energy of the crystal can be obtained from Eq. (28.6). A few values for U_c are given in Table 28.2.

The cohesive energy U_c of an ionic crystal is the energy of the crystal relative to the infinitely separated ions, the energy required for the reaction

$$MX(s) \longrightarrow M^+(g) + X^-(g).$$

The energy of this reaction is not directly measurable, and therefore is determined indirectly. The Born–Haber cycle of reactions is used.

Reaction	Energy
$M(s) \to M(g)$	S = sublimation energy
$M(g) \to M^+(g) + e^-(g)$	I = ionization energy
$\frac{1}{2}X_2(g) \to X(g)$	$\frac{1}{2}D = \frac{1}{2}$ the dissociation energy
$e^-(g) + X(g) \to X^-(g)$	$-E_A$ = minus electron affinity
$X^-(g) + M^+(g) \to MX(s)$	$-U_c$ = minus cohesive energy.

Summing these yields the formation reaction of MX(s);

$$M(s) + \tfrac{1}{2}X_2(g) \rightarrow MX(s), \qquad \Delta U_f = \text{energy of formation.}$$

Therefore

$$\Delta U_f = S + I + \tfrac{1}{2}D - E_A - U_c,$$
$$U_c = S + I + \tfrac{1}{2}D - E_A - \Delta U_f. \qquad (28.10)$$

The values computed for U_c from the experimental values of the quantities on the right of Eq. (28.10) agree with the values predicted by Eq. (28.6) to about 4 % for the alkali halides.

The theory can be refined somewhat by including the van der Waals attraction of the electron clouds of the ions; this is more important for a substance such as CsI, in which the electron clouds are large and floppy, than for LiF, in which the electron clouds are small and tightly bound. The presence of van der Waals interaction increases the cohesive energy slightly. The only important contribution of quantum mechanics to this problem is the requirement that the zero-point energy of the crystal be included in the calculation. This decreases the calculated value of the cohesive energy by about 0.5 to 1.0 %. These additional contributions do not change the values in Table 28.2 by more than 2 or 3 %.

Attention should be directed to the magnitude of the cohesive energy in uni-univalent ionic crystals, which ranges from 600 to 800 kJ/mol. This is 10 to 20 times larger than that found in van der Waals crystals. Furthermore, as the ions get larger, the cohesive energy decreases, Eq. (28.6). The extreme values are: LiF, $U_c = 1004$ kJ/mol; CsI, $U_c = 569.4$ kJ/mol. The larger ions are simply farther apart in the crystal. Finally, if we consider crystals made up of divalent ions, such as CaO and BaO, Eq. (28.6) predicts that the cohesive energy should be proportional to the square of the charge so the energies should be roughly four times greater than the energies of 1-1 salts. This is approximately correct; the values for CaO and BaO are 3523 kJ and 3125 kJ, respectively.

The increase in cohesive energy in the 2-2 salts explains the generally lower solubility of these salts (for example, the sulfides, as compared with that of the alkali halides). The greater the cohesive energy, the more difficult it is for a solvent to break up the crystal.

28.3 THE ELECTRONIC STRUCTURE OF SOLIDS

For an isolated atom quantum mechanics predicts a set of energy levels of which some, but not all, are occupied by electrons. What happens to this scheme of energy levels if many atoms are packed closely together in a solid? Consider two helium atoms, infinitely far apart; each has two electrons in the 1s level. As these two atoms approach, they attract each other slightly; the interaction energy has a shallow minimum at some distance. Since each atom is influenced by the presence of the other, the energy levels on each atom are slightly perturbed. The 1s level splits into a set of two levels, which may be thought of as the energy levels for the systems $(He)_2$. Each of these levels can accommodate two electrons; the four electrons of the system fill the two levels. The average energy of the two levels is slightly less than the energy of the 1s level of the isolated atom. This slight lowering of the average energy is the cohesive energy, the van der Waals interaction energy, of the system $(He)_2$. If three helium atoms were brought together, the system would have a set of three closely spaced 1s levels. In a system of N atoms, the 1s levels split into a group of N closely spaced levels called an *energy band*, the 1s band. For a collection of N helium atoms, since the 1s level is fully occupied in the individual atoms, the 1s band is completely occupied. For helium and for any saturated molecule that forms a van der Waals solid, the width of the band (the energy difference between the topmost and lowermost levels in the band) is very

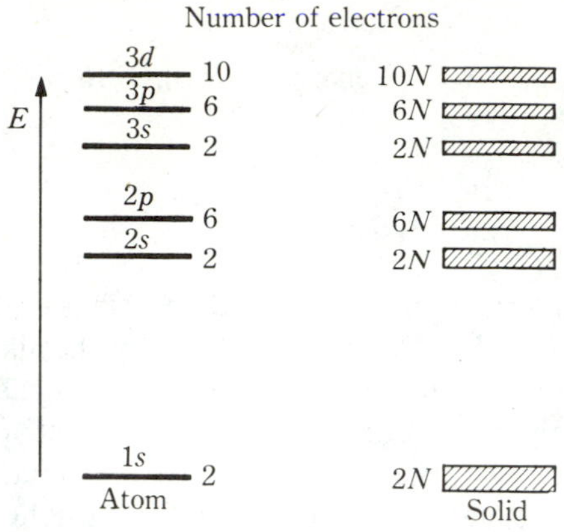

Number of electrons

$$
\begin{array}{c|cc}
 & 3d \quad 10 & 10N \\
 & 3p \quad 6 & 6N \\
E & 3s \quad 2 & 2N \\[1em]
 & 2p \quad 6 & 6N \\
 & 2s \quad 2 & 2N \\[2em]
 & 1s \quad 2 & 2N \\
 & \text{Atom} & \text{Solid}
\end{array}
$$

Figure 28.2 Correspondence between energy levels in the atom and energy bands in a solid (schematic).

small, because of the weak interaction between saturated molecules. To a good approximation, the energy level scheme in a van der Waals solid is much like that in the individual molecules which compose the solid, the filled levels being displaced downward very slightly to account for the cohesive energy of the solid.

Consider a solid that contains N atoms of one kind only. For each energy level in the isolated atom that accommodates two electrons there is in the solid an energy band containing N levels each of which can accommodate two electrons. This energy band has a definite width, a fact which implies that the N levels within the band are very closely spaced. They are so closely spaced that the band may be considered as a continuum of allowed energies; it is often called a quasi-continuous band of levels. Figure 28.2 illustrates schematically the contrast between the energy level systems in an isolated atom and in a solid. The shaded regions in the figure cover the ranges of energy permitted to an electron in the solid, the energy bands; the spaces between the bands are the values of energy that are not permitted. Figure 28.2 has been drawn so that none of the bands overlap; ordinarily the higher energy bands do overlap.

Consider metallic sodium. The sodium atom has eleven electrons in the configuration $1s^2 2s^2 2p^6 3s$. Bringing many sodium atoms together in the crystal scarcely affects the energies of the electrons in the $1s$, $2s$, and $2p$ levels, since the electrons in these levels are screened from the influence of the other atoms by the valence electron; the corresponding bands are filled. The levels in the valence shell are very much influenced by the presence of other atoms and split into bands as shown in Fig. 28.3(a). The $3s$ and $3p$ bands have been

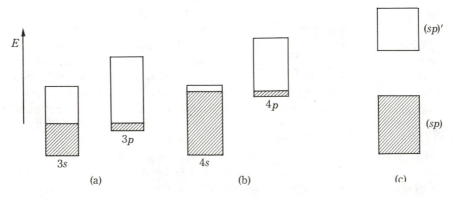

Figure 28.3 Band structure in solids. (a) Sodium. (b) Calcium. (c) Diamond.

displaced horizontally to illustrate the effect of overlapping bands. The N valence electrons fill the lowest levels available, which results in a partial filling of both the 3s and 3p bands; the filled portion of the bands is indicated by the shading. The overlapping of s and p bands is a characteristic feature of the electronic structure of metals; the merged bands are often designated as an *sp* band.

The s band can hold $2N$ electrons or 2 electrons/atom. Were it not for the fact that the p band overlaps the s band, the s band would be completely filled in divalent metals such as calcium. As we shall see shortly, if the s band were filled and a gap of forbidden energies separated the top of the s band from the bottom of the p band, then the divalent metals would be insulators. As is shown in Fig. 28.3(b) the p and s bands in calcium overlap slightly; the shaded area indicates the way in which the two electrons of calcium fill the bands.

Diamond is a crystal with filled bands. The s band, which holds 2 electrons/atom, and the p band, which holds 6 electrons/atom, interact in diamond to form two distinct bands each of which holds four electrons per atom; these bands are designated *sp* and *(sp)'* in Fig. 28.3(c). The four electrons per atom in diamond exactly fill the lower band. Diamond with this filled band is an insulator.

28.4 CONDUCTORS AND INSULATORS

A crystal with completely filled energy bands is an insulator, and one with partially filled bands is a conductor. The band in a real crystal contains as many levels as atoms in the crystal, but for argument's sake suppose we imagine that the band has only eight levels in it (Fig. 28.4a). We may suppose that half of these levels are associated with motion of the electrons in the $+x$ direction and half with motion in the $-x$ direction. This is indicated by the arrowheads on the levels. No matter how the band is filled, half of the electrons are in levels corresponding to motion in the $+x$ direction and half in levels corresponding to motion in the $-x$ direction; consquently there is no net motion in one direction and no current flow. If we apply an electric field in the $+x$ direction, the energy of one set of levels is lowered and the energy of the other set is rasied (Fig. 28.4b). If the band is full, then all levels are occupied before and after the application of the field, and there is still no net electronic motion in either direction; the crystal is an insulator. However, if the band is only partly filled, then only the lowest levels are occupied; application of the field rearranges the positions of the levels, and the electrons drop into the lowest set of levels in the presence of the field. In this lowest set of levels, the ones corresponding to motion in the $-x$ direction predominate, so there is a net flow of electrons to the left; a net current flows and the crystal is a conductor.

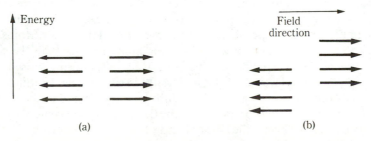

Figure 28.4 Displacement of energy levels in a band by an electric field. (a) Field off. (b) Field on.

Metals that conduct by electron flow have incompletely filled bands, while insulators such as diamond have completely filled bands. If it is possible to raise electrons from a filled band in an insulator to an empty band of higher energy, then these excited electrons can carry a current. Since the energy gap between the bands is fairly large, this ordinarily cannot be done by an increase in temperature to supply sufficient thermal energy. By using light of high enough frequency it is possible to excite the electrons. The phenomenon is called photoconductivity. Visible light will do this for selenium.

28.5 IONIC CRYSTALS

In the first approximation, the band system of a crystal containing two different kinds of atom may be regarded as a superposition of the band systems of the two individual particles. The band system for sodium chloride is shown in Fig. 28.5. The eight electrons occupy the $3s$ and $3p$ bands of the chloride ion, while the $3s$ band of the sodium ion, which has a higher energy, is vacant. This is a quantum-mechanical way of saying that the crystal is made up of sodium ions and chloride ions rather than of atoms of sodium and chlorine. The filled bands are separated from the empty bands by an energy gap so that sodium chloride is an insulator.

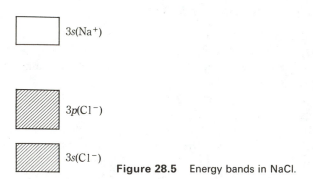

$3s(Na^+)$

$3p(Cl^-)$

$3s(Cl^-)$

Figure 28.5 Energy bands in NaCl.

28.6 SEMICONDUCTORS

Semiconductors are solids which exhibit a feeble electrical conductivity that increases with increase in temperature. (The conductivity of metals decreases with increase in temperature.) Semiconductivity appears in insulators that are slightly contaminated with foreign substances, and in compounds, such as Cu_2O and ZnO, which do not contain exactly stoichiometric amounts of metal and nonmetal.

Pure silicon is an insulator, similar to diamond in both crystal structure and electronic structure. The electronic structure in pure silicon can be represented by the filled and empty bands shown in Fig. 28.6(a). Suppose that we remove a few of the silicon atoms and replace them by phosphorus atoms, each of which has one more electron than the silicon atom. The energy levels of the phosphorus atoms, impurity levels, are superposed on the band system of the silicon; these levels do not match those in silicon exactly. (Since there are so few phosphorus atoms, the levels are not split into bands.) It is found that the extra electrons introduced by the phosphorus atoms occupy the impurity levels shown in Fig. 28.6(b), which are located slightly below the empty band of the silicon lattice. In these levels the electrons are bound to the phosphorus atoms and cannot conduct a current; since the

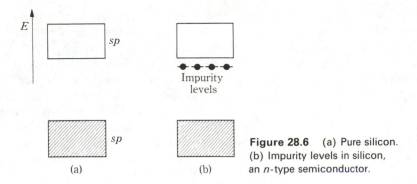

Figure 28.6 (a) Pure silicon. (b) Impurity levels in silicon, an *n*-type semiconductor.

energy gap between these levels and the empty band of silicon, the conduction band, is comparable to kT, the thermal energy, a certain fraction of these electrons are thermally excited to the conduction band in which they can move under the influence of an applied field. At higher temperatures more electrons are excited to the conduction band and the conductivity is larger. If very many phosphorus atoms are introduced in the lattice, the impurity level itself widens into a band that overlaps the conduction band of the silicon; the conductivity then becomes metallic in character. This is an example of *n*-type semi-conductivity, so-called because the carriers of the current, the electrons, are *negatively* charged.

If atoms of aluminum or boron are introduced in the silicon lattice, they also introduce their own system of levels. Since the aluminum atom has one less electron than silicon, the impurity levels are vacant. Figure 28.7(a) shows the position of the impurity levels, which in this case are only slightly above the filled band of the silicon lattice. Electrons from the filled band can be excited thermally to the impurity levels (Fig. 28.7b), where they are bound to the aluminum atoms to produce the species Al^- in the lattice. The holes left in the band effectively carry a positive charge, can move under the influence of an applied field, and thus carry a current. This is an example of *p*-type semiconductivity, since the carrier is *positively* charged.

The semiconductivity of nonstoichiometric compounds such as ZnO and Cu_2O can be explained in a similar way. If ZnO loses a little oxygen, it can be considered as ZnO with a few zinc atoms as impurities. The zinc atoms have two more electrons than the zinc ions; therefore the semiconductivity is *n*-type. Since the crystal Cu_2O may contain extra oxygen, it may be considered as Cu_2O with some Cu^{2+} ions as impurities. The Cu^{2+} ion has one less electron than the Cu^+ ion, so the conductivity is *p*-type. Sodium chloride

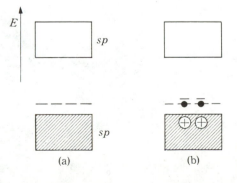

Figure 28.7 Impurity levels in a *p*-type semiconductor.

exposed at high temperatures to sodium vapor incorporates sodium atoms as impurities; the impure crystal has *n*-type semiconductivity. Excess halogen can be introduced into sodium chloride to yield a *p*-type semiconductor.

Until the late 1940s the study of semiconductivity was a frustrating occupation. Reproducible measurements were very difficult to obtain. To study the phenomenon in silicon, for example, it is necessary to begin with silicon of a fantastic degree of purity, less than one part per billion of impurity. Accurately controlled amounts of a definite type of impurity are then added. If ordinary silicon is used, the accidental impurities and their concentrations vary from sample to sample, making the experimental measurements nearly valueless. In the years since 1948 the technique of producing materials of the required degree of purity, the technique of zone refining, has been developed to such an extent that semiconductors with reproducible characteristics are produced with ease on a commercial scale. Devices made of semiconducting materials are commonplace items.

28.7 COHESIVE ENERGY IN METALS

Any detailed calculation of the cohesive energy of metals is quite complicated; however, it is possible from a qualitative examination of the band systems to gain a little insight into the problem. Consider the transition metals that as isolated atoms have partially filled *d* shells, and as solids have partially filled *d* bands. The *d* band can hold 10 electrons/atom. Since the *d* shell in the atoms is shielded somewhat by the outer electrons, the *d* band is very narrow compared with the *sp* band. Figure 28.8(b) shows the relative widths and the filling of the *d* and *sp* bands in copper. The *d* band is completely filled. In nickel, which has one less electron per atom, the *d* band is only partially filled (Fig. 28.8a). The lower cohesive energy of copper compared with that of nickel is a result of the higher average energy of the electrons in the *sp* band of copper. Zinc has one more electron than copper; adding this electron to the *sp* band fills it to a much higher level (Fig. 28.8c), resulting in a marked decrease in the cohesive energy. The cohesive energies are nickel, 425.1 kJ/mol; copper, 341.1 kJ/mol; zinc, 130.5 kJ/mol. The lower the energy of the electrons in the metal, the more stable is the system and the greater is the cohesive energy. It is evident from Fig. 28.8 that a partially empty *d* band in a metal is an indication of a large cohesive energy, since the average energy of the electrons is low. Addition of an electron to the *sp* band, as in going from copper to zinc, increases the average electronic energy rapidly, because the very wide *sp* band accommodates only 4 electrons/atom, while the very narrow *d* band accommodates 10.

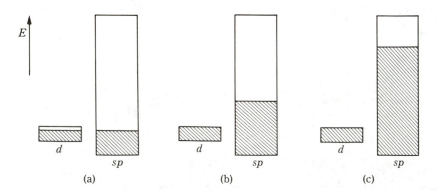

Figure 28.8   Effect of the *d* band on the electronic energy.
(a) Nickel. (b) Copper. (c) Zinc.

QUESTIONS

28.1 Why is the Madelung energy U_M (a) negative? (b) proportional to the number of ion pairs?

28.2 What is the approximate dependence of the cohesive energy on the magnitude of the ionic charges?

28.3 How is the *ionic* character of an ionic solid reflected in the band picture?

28.4 Identify the current carriers in *n*- and *p*-semiconductors.

28.5 Explain the temperature dependence of the conductivity of semiconductors.

PROBLEMS

28.1 Consider the following arrangements of ions:

The charge on the positive and negative ions is $+e$ and $-e$ respectively; the spacing in the linear arrays is r between any two neighbors. Calculate the Madelung constant for these arrangements of ions.

28.2 The ion radii for Na^+ and Cl^- are 95 pm and 181 pm. Calculate the cohesive energy neglecting repulsion; $A = 1.7476$. Calculate the cohesive energy if $n = 8.0$.

28.3 Using the Madelung constants in Table 28.1, compare the cohesive energy of RbCl in the NaCl structure and in the CsCl structure. The radii are $r_+ = 148$ pm, $r_- = 181$ pm, and are assumed to be the same in both structures.

28.4 a) Arrange the alkali metal fluorides in order of increasing cohesive energy.
b) Arrange the potassium halides in order of increasing cohesive energy.

28.5 What is the approximate ratio of cohesive energies of NaF and MgO?

28.6 The density of NaCl is 2.165 g/cm^3. Calculate the interionic, Na^+—Cl^-, distance. If $n = 8.0$, calculate the compressibility of solid NaCl.

29

Structure and Thermodynamic Properties

29.1 THE ENERGY OF A SYSTEM

The energy of an individual atom or molecule can be calculated from quantum mechanics. In a collection of a large number of molecules there is an energy distribution; some molecules have more energy and some less. The *average* energy of the collection of molecules is identified with the thermodynamic energy of the system. It is our aim to discover the relation between the properties of the individual molecule, obtained from the Schrödinger equation, and the thermodynamic properties of the bulk system, which contains many individual molecules.

Consider a system of fixed volume V, which contains a very large number N of molecules. Since the energies of the individual molecules have discrete values, the possible energies of the system have discrete values $E_1, E_2, E_3, \ldots, E_i$. We find these energy values by solving the Schrödinger equation. We specify that the temperature is constant, since the system is immersed in a heat reservoir at constant temperature. The system exchanges energy with the reservoir; thus if we make a number of observations of the system, we will find that it is in a different quantum state, that it has a different E_i, in each observation. The thermodynamic energy is the average of the energies exhibited in a large number of observations. If we wish, instead of observing one system a large number of times, we can construct a large number of identical systems, immerse them in the same temperature reservoir, and make one observation on each system. Each is found to be in a different quantum state; the energy is obtained by averaging over all the systems.

Consider a collection of a large number, $\mathbf{N}$, of identical systems, an *ensemble*. Every system in the ensemble has one of the energy values, so we may write the energy distribution as follows:

Energy	E_1	E_2	E_3	$\cdots$	E_i	$\cdots$
Number of systems	$\mathbf{n}_1$	$\mathbf{n}_2$	$\mathbf{n}_3$	$\cdots$	$\mathbf{n}_i$	$\cdots$

The probability of finding a system with the energy E_i is $P_i = \mathbf{n}_i/\mathbf{N}$. This probability depends on the energy E_i, so we write

$$P_i = f(E_i). \tag{29.1}$$

Similarly, the probability of finding a system with energy E_j is

$$P_j = f(E_j). \tag{29.2}$$

Suppose that we choose two systems from the ensemble; the probability P_{ij} that one has E_i and the other has E_j is the product of the individual probabilities,

$$P_{ij} = P_i P_j = f(E_i)f(E_j). \tag{29.3}$$

There is another way to choose two systems from the ensemble. Suppose that we pair off the systems randomly, to form $\frac{1}{2}\mathbf{N}$ paired systems. The probability that a pair has energy $E_i + E_j$ is also P_{ij} and must be the same function of the energy of the paired system as the P_i is of the energy of the single system; P_{ij} differs at most by a multiplicative constant B, since the total number of systems involved is different. Therefore

$$P_{ij} = Bf(E_i + E_j). \tag{29.4}$$

Combining this with the result in Eq. (29.3), we obtain the functional equation

$$f(E_i)f(E_j) = Bf(E_i + E_j). \tag{29.5}$$

We have met a similar equation, Eq. (4.27), in the kinetic theory of gases. Equation (29.5) is satisfied if $f(E_i)$ has the form

$$f(E_i) = Be^{-\beta E_i}, \tag{29.6}$$

where β is a positive constant, and the negative sign in the exponential was chosen to avoid predicting an infinite probability of finding systems with infinite energy. The constant β must be the same for all systems; otherwise the functional relation, Eq. (29.5), would not be fulfilled. The property common to all the systems is the temperature, so without further argument, we set

$$\beta = \frac{1}{kT}, \tag{29.7}$$

where k is the Boltzmann constant. The relation in Eq. (29.7) can be proved rigorously, of course, but to avoid rewriting many equations, we will not undertake the proof here.

Finally, the probability becomes

$$P_i = Be^{-E_i/kT}. \tag{29.8}$$

The constant B is determined by the condition that the sum of the probabilities over all possible energy states is unity:

$$\sum_i P_i = 1, \tag{29.9}$$

so that

$$B\sum_i e^{-E_i/kT} = 1. \tag{29.10}$$

The summation in Eq. (29.10) is called the *partition function*, or the *state sum*, and is given the symbol Q:

$$Q = \sum_i e^{-E_i/kT}.$$

(29.11)

Thus, $B = 1/Q$, and

$$P_i = \frac{e^{-E_i/kT}}{Q}.$$

(29.12)

Knowing the probability of finding the system with energy E_i, we can calculate the thermodynamic energy U of the system, which is the average energy of the ensemble:

$$U = \langle E \rangle = \frac{\sum_i \mathbf{n}_i E_i}{\mathbf{N}}.$$

Since $\mathbf{n}_i/\mathbf{N} = P_i$, this becomes

$$U = \sum_i P_i E_i.$$

(29.13)

By the same reasoning, any function of the energy $Y(E_i)$ has the average value $\langle Y \rangle$ given by

$$\langle Y \rangle = \sum_i P_i Y(E_i).$$

(29.14)

The argument assumes that the probabilities of choosing one system with energy E_i and another with E_j are independent; this leads to a distribution function P_i, which is of the Maxwell–Boltzmann type. The independence of the probabilities implies that the distribution is a random one.

29.2 DEFINITION OF THE ENTROPY

In the ensemble the systems are distributed over the various quantum states; every possible way of arranging the systems in the quantum states is called a *complexion* of the ensemble. The number of complexions is denoted by $\boldsymbol{\Omega}$; then the entropy of the ensemble is defined, as in Section 9.12, by

$$\mathbf{S} = k \ln \boldsymbol{\Omega}.$$

(29.15)

The entropy of the system is the entropy of the ensemble divided by the number of systems $\mathbf{N}$, so that

$$S = \frac{\mathbf{S}}{\mathbf{N}} = \frac{k \ln \boldsymbol{\Omega}}{\mathbf{N}}.$$

(29.16)

We regard the quantum states with energies E_i as boxes and the systems as balls to be distributed among the boxes. The total number of distinguishable ways of arranging the balls in the boxes (the systems in the quantum states) is the number of complexions $\boldsymbol{\Omega}$ of the ensemble. This number is given by Eq. (9.70):

$$\boldsymbol{\Omega} = \frac{\mathbf{N}!}{\mathbf{n}_1! \mathbf{n}_2! \mathbf{n}_3! \dots}.$$

(29.17)

To find S we first calculate $\ln \boldsymbol{\Omega}$.

$$\ln \boldsymbol{\Omega} = \ln \mathbf{N}! - \sum_i \ln \mathbf{n}_i!$$

If $\mathbf{N}$ is large, the Stirling formula yields $\ln \mathbf{N}! = \mathbf{N} \ln \mathbf{N} - \mathbf{N}$. Then

$$\ln \mathbf{\Omega} = \mathbf{N} \ln \mathbf{N} - \mathbf{N} - \sum_i \mathbf{n}_i \ln \mathbf{n}_i + \sum_i \mathbf{n}_i.$$

Since $\Sigma_i \, \mathbf{n}_i = \mathbf{N}$, and $\mathbf{n}_i = P_i \mathbf{N}$, this reduces to

$$\ln \mathbf{\Omega} = \mathbf{N} \ln \mathbf{N} - \mathbf{N} \sum_i P_i \ln (\mathbf{N} P_i).$$

Expanding $\ln (\mathbf{N} P_i)$ and using the fact that $\Sigma_i \, P_i = 1$, this becomes

$$\ln \mathbf{\Omega} = -\mathbf{N} \sum_i P_i \ln P_i. \tag{29.18}$$

Using this result in Eq. (29.16), we obtain for the entropy of the system

$$S = -k \sum_i P_i \ln P_i. \tag{29.19}$$

Equation (29.19) expresses the dependence of the entropy on the P_i. It is important to observe that P_i is the fraction of the systems in the state with energy E_i, so that the form of the sum in Eq. (29.19) has the appearance of an entropy of mixing. The systems in the ensemble are "mixed," or spread over the possible energy states of the system. It is this "mixing" or spreading that gives rise to the property of a system we call the entropy.

29.3 THE THERMODYNAMIC FUNCTIONS IN TERMS OF THE PARTITION FUNCTION

Equations (29.13) and (29.19) relate the energy and entropy to the P_i. From these equations, the relation between P_i and Q, Eq. (29.12), and the definition of Q, Eq. (29.11), all of the thermodynamic functions can be expressed in terms of the partition function Q and its derivatives. We begin by differentiating Eq. (29.11) with respect to temperature:

$$\left(\frac{\partial Q}{\partial T}\right)_V = \frac{1}{kT^2} \sum_i E_i e^{-E_i/kT}. \tag{29.20}$$

Since the E_i are obtained ultimately from the Schrödinger equation, they do not depend on temperature; they may, however, depend on the volume, so the derivative is a partial derivative. Using Eq. (29.12), the exponential in the sum in Eq. (29.20) may be replaced by QP_i, which brings the equation to the form

$$kT^2 \left(\frac{\partial Q}{\partial T}\right)_V = Q \sum_i P_i E_i.$$

By comparison with Eq. (29.13) this summation is equal to the energy, so we have

$$U = \frac{kT^2}{Q} \left(\frac{\partial Q}{\partial T}\right)_V = kT^2 \left(\frac{\partial \ln Q}{\partial T}\right)_V, \tag{29.21}$$

which relates the energy to the partition function.

To obtain the entropy, we calculate $\ln P_i$ using Eq. (29.12):

$$\ln P_i = -\frac{E_i}{kT} - \ln Q.$$

Putting this expression in Eq. (29.19) for the entropy yields

$$S = -k\left[-\frac{1}{kT}\sum_i P_i E_i - \ln Q \sum_i P_i\right].$$

Using Eqs. (29.9) and (29.13) we obtain

$$S = \frac{U}{T} + k \ln Q.$$

Insertion of the value of U from Eq. (29.21) yields

$$S = k \ln Q + kT\left(\frac{\partial \ln Q}{\partial T}\right)_V, \tag{29.22}$$

which expresses the entropy in terms of the partition function. Since all of the other thermodynamic functions are simply related to S, U, T, and V, it is an easy matter to calculate them. For example, the Helmholtz function $A = U - TS$. Using the values for U and S in terms of Q, we obtain for A,

$$A = -kT \ln Q. \tag{29.23}$$

From the fundamental equation, Eq. (10.21), $p = -(\partial A/\partial V)_T$. Differentiating Eq. (29.23), we obtain for the pressure,

$$p = kT\left(\frac{\partial \ln Q}{\partial V}\right)_T. \tag{29.24}$$

Then the values of H and G follow immediately from the definitions:

$$H = kT\left[T\left(\frac{\partial \ln Q}{\partial T}\right)_V + V\left(\frac{\partial \ln Q}{\partial V}\right)_T\right]; \tag{29.25}$$

$$G = -kT\left[\ln Q - V\left(\frac{\partial \ln Q}{\partial V}\right)_T\right]. \tag{29.26}$$

Finally by differentiating Eq. (29.21), we obtain the heat capacity C_v:

$$C_v = kT\left[2\left(\frac{\partial \ln Q}{\partial T}\right)_V + T\left(\frac{\partial^2 \ln Q}{\partial T^2}\right)_V\right]. \tag{29.27}$$

In a certain sense we have solved the problem of obtaining thermodynamic functions from the properties of molecules. These functions have been related to Q, which, by its definition, is related to the energy levels of the system, which are in turn related to the energy levels of the molecules in the system. To make these expressions useful, we must express the partition function in terms of the energies of the molecules.

29.4 THE MOLECULAR PARTITION FUNCTION

Consider the quantum state of the system that has the energy E_i. This energy is composed of the sum of the energies of the molecules ϵ_1, ϵ_2, ..., plus any interaction energy W between the molecules:

$$E_i = \epsilon_1 + \epsilon_2 + \cdots + W. \tag{29.28}$$

For the present we assume that the particles do not interact (ideal gas) and set $W = 0$. Each energy ϵ_i corresponds to one of the allowed quantum states of the molecule. Because the energy E_i has the form given by Eq. (29.28), it is possible to write the partition function as a product of partition functions of the individual molecules q. The final form is, for indistinguishable molecules,

$$Q = \frac{1}{N!} q^N, \tag{29.29}$$

where N is the number of molecules in the system and

$$q = \sum_i e^{-\epsilon_i/kT}. \tag{29.30}$$

The sum in Eq. (29.30) is over all the *quantum states* of the *molecule*, so q is the *molecular partition function*. If g_i quantum states have the same energy, they are said to be degenerate; degeneracy $= g_i$. The terms in the partition function can be grouped according to the energy level. The g_i states yield g_i equal terms in the partition function. The expression in Eq. (29.30) can be written as

$$q = \sum_i g_i e^{-\epsilon_i/kT}, \tag{29.30a}$$

in which the sum is taken over the different energy levels of the system.

If two kinds of molecules are present, N_a of A and N_b of B, then

$$Q = \frac{q_a^{N_a} q_b^{N_b}}{N_a! N_b!}. \tag{29.31}$$

We will not justify these equations except to say that if the Q were written simply as q^N, too many terms would be included; division by $N!$ is required to yield the correct result.

Since only $\ln Q$ appears in the formulas for the thermodynamic functions, we find, using the Stirling formula, from Eq. (29.29)

$$\ln Q = N \ln q - N \ln N + N. \tag{29.32}$$

Using the expression in Eq. (29.32), we can express all the thermodynamic functions in terms of $\ln q$ instead of $\ln Q$.

29.5 THE CHEMICAL POTENTIAL

We calculate the value of the chemical potential in a mixture by using the relation $\mu_a = (\partial A/\partial N_a)_{T,V,N_b}$. By differentiating Eq. (29.23), we obtain

$$\left(\frac{\partial A}{\partial N_a}\right)_{T,V,N_b} = -kT\left(\frac{\partial \ln Q}{\partial N_a}\right)_{T,V,N_b} = \mu_a.$$

From Eq. (29.31) we have

$$\ln Q = N_a \ln q_a + N_b \ln q_b - N_a \ln N_a + N_a - N_b \ln N_b + N_b.$$

Differentiating with respect to N_a, this becomes

$$\left(\frac{\partial \ln Q}{\partial N_a}\right)_{T,V,N_b} = \ln q_a - 1 - \ln N_a + 1 = \ln\left(\frac{q_a}{N_a}\right).$$

Thus

$$\mu_a = -kT \ln \left(\frac{q_a}{N_a}\right), \tag{29.33}$$

which expresses the chemical potential of a gas (indistinguishable molecules) in terms of the molecular partition function per molecule q_a/N_a, a result that is useful for the discussion of chemical equilibria. If we were dealing with a solid in which the molecules are locked in place and therefore are distinguishable, the factors $N_a!$ and $N_b!$ do not appear in Eq. (29.31), so we have the simpler result

$$\mu = -kT \ln (qe^{-W/NkT}), \tag{29.34}$$

where the interaction energy W appears, since W is not zero in a solid.

29.6 APPLICATION TO TRANSLATIONAL DEGREES OF FREEDOM

The application of the formulas is simplest if the molecules possess energy in only one form. Therefore we consider a system such as a monatomic gas, which has only translational energy. For the moment we ignore any contribution of the internal electronic energy of the atom to the properties of the system. The energy of translation ϵ_t is made up of the energies in each component of the motion, so we write

$$\epsilon_t = \epsilon_x + \epsilon_y + \epsilon_z.$$

Again, because these energies are additive, the translational partition function q_t factors into a product:

$$q_t = q_x q_y q_z. \tag{29.35}$$

The energy levels for translation are the energy levels for a particle in a box (Section 21.3). If the width of the box in the x direction is a, then the permitted values of the kinetic energy from the Schrödinger equation are

$$\epsilon_x = \frac{h^2 n^2}{8ma^2}, \qquad n = 1, 2, 3, \ldots \tag{29.36}$$

It was shown in Section 21.3.2 that the spacing between levels in a box of macroscopic dimensions is extremely small—too small to distinguish the levels observationally. Therefore we choose a new set of distinct levels, ϵ_i, but separated by an energy $d\epsilon$. Let there be g_i levels in the energy range $d\epsilon$ between ϵ_i and ϵ_{i+1}. All of these levels in this energy range will be assigned the single energy value ϵ_i. Then the terms in the partition function group into sets, and we can write the partition function as

$$q_x = \sum_i g_i e^{-\epsilon_i/kT}, \tag{29.37}$$

since g_i terms containing the single exponential $e^{-\epsilon_i/kT}$ appear when the sum is made over all levels.

To obtain g_i we calculate the spacing between levels:

$$\epsilon_{n+1} - \epsilon_n = (2n + 1)\frac{h^2}{8ma^2}.$$

If n is very large, then $2n + 1 \approx 2n = (4a/h)(2m\epsilon)^{1/2}$, where the last form is obtained by solving Eq. (29.36) for $2n$ (the subscript x on ϵ has been dropped). The value of the spacing becomes

$$\epsilon_{n+1} - \epsilon_n = \frac{h}{a}\left(\frac{\epsilon}{2m}\right)^{1/2}.$$

The number of levels in the range $d\epsilon$ is g and is obtained by dividing the range by the spacing between levels:

$$g = \frac{a}{h}\left(\frac{2m}{\epsilon}\right)^{1/2} d\epsilon.$$

When we put this value of g in the partition function and change the summation to integration from $\epsilon = 0$ to $\epsilon = \infty$, Eq. (29.37) becomes

$$q_x = \int_0^\infty \frac{a}{h}\left(\frac{2m}{\epsilon}\right)^{1/2} e^{-\epsilon/kT}\, d\epsilon. \tag{29.38}$$

We change variables to $y^2 = \epsilon/kT$; then $d\epsilon = 2kTy\, dy$, and the integral in Eq. (29.38) becomes

$$q_x = \frac{2a}{h}(2mkT)^{1/2}\int_0^\infty e^{-y^2}\, dy.$$

The integral has the value $\frac{1}{2}\pi^{1/2}$; hence

$$q_x = \frac{a}{h}(2\pi mkT)^{1/2}.$$

Similarly, if b and c are the widths of the box in the y and z directions, we get

$$q_y = \frac{b}{h}(2\pi mkT)^{1/2}, \qquad q_z = \frac{c}{h}(2\pi mkT)^{1/2}.$$

The translational partition function q_t is the product of these, by Eq. (29.35); we obtain

$$q_t = \left(\frac{2\pi mkT}{h^2}\right)^{3/2} V, \tag{29.39}$$

where the product of the dimensions of the box abc has been replaced by the volume V. Using Eq. (29.32), we obtain for Q

$$\ln Q = N \ln q_t - N \ln N + N. \tag{29.40}$$

The derivatives of $\ln Q$ are

$$\left(\frac{\partial \ln Q}{\partial T}\right)_V = \frac{3N}{2T} \quad \text{and} \quad \left(\frac{\partial \ln Q}{\partial V}\right)_T = \frac{N}{V}.$$

We can easily verify that the energy of the system is $\frac{3}{2}NkT$ and that $p = NkT/V$, which are the values we expect for a monatomic gas in which the interaction energy is zero. From the values of the derivatives and $\ln Q$ itself, any of the thermodynamic quantities can be calculated using the formulas in Section 29.3.

29.7 PARTITION FUNCTION OF THE HARMONIC OSCILLATOR

If the particles composing the system have only vibrational motion, the energies permitted are given by

$$\epsilon_s = (s + \tfrac{1}{2})h\nu, \qquad s = 0, 1, 2, \ldots,$$

where v is the frequency of the oscillator. Using this value for the energy, we see that the vibrational molecular partition function becomes

$$q_v = \sum e^{-\epsilon_s/kT} = \sum_{s=0}^{\infty} e^{-(s+1/2)hv/kT},$$

where the sum is over all the integral values of s from zero to infinity. To simplify, let $y = e^{-hv/kT}$, then

$$q_v = y^{1/2} \sum_{s=0}^{\infty} y^s = y^{1/2}(1 + y + y^2 + \cdots).$$

Since $1/(1 - y) = 1 + y + y^2 + y^3 + \cdots$, we see that q_v becomes

$$q_v = \frac{y^{1/2}}{1 - y} = \frac{e^{-hv/2kT}}{1 - e^{-hv/kT}}. \tag{29.41}$$

It is customary to define a characteristic temperature for the oscillator, $\theta = hv/k$. Then

$$q_v = \frac{e^{-\theta/2T}}{1 - e^{-\theta/T}}. \tag{29.42}$$

If the temperature is either very high or very low, this equation takes on a simpler form.

Case I. T is very low; $\theta/T \gg 1$. Then, $e^{-\theta/T}$ is negligible compared with unity in the denominator of Eq. (29.42), and we have

$$q_v = e^{-\theta/2T} \qquad \text{(low temperature)}. \tag{29.43}$$

Case II. T is very high; $\theta/T \ll 1$. Then the exponential in the denominator can be expanded $e^{-\theta/T} = 1 - \theta/T$, and we have

$$q_v = \frac{T}{\theta} e^{-\theta/2T} \qquad \text{(high temperature)}. \tag{29.44}$$

29.8 THE MONATOMIC SOLID

The monatomic solid has only vibrational motion. The partition function can be written as a product of the partition functions of the atoms composing the solid and an exponential factor that includes the interaction energy W of the atoms in the solid:

$$Q = e^{-W/NkT} q_1 q_2 q_3 \cdots q_N. \tag{29.45}$$

Since each particle in the solid has three vibrational degrees of freedom, the partition function for each atom is a product of three vibrational partition functions; thus Q contains a product of $3N$ vibrational partition functions:

$$Q = e^{-W/NkT} q_{v_1} q_{v_2} q_{v_3} \cdots q_{v_{3N}}. \tag{29.46}$$

Each partition function contains a frequency, so that there is a total of $3N$ frequencies and, correspondingly, $3N$ values of θ which are involved. We cannot do anything further at low temperatures without knowing something more about the frequencies.

At very high temperatures we can do a little bit. Taking the logarithm of Q from Eq. (29.46), we obtain

$$\ln Q = -\frac{W}{NkT} + \ln q_{v_1} + \ln q_{v_2} + \cdots + \ln q_{v_{3N}}. \tag{29.47}$$

But from Eq. (29.44) at high temperatures we have

$$\ln q_v = -\frac{\theta}{2T} - \ln \theta + \ln T,$$

and the temperature derivative

$$\left(\frac{\partial \ln q_v}{\partial T}\right)_V = \frac{\theta}{2T^2} + \frac{1}{T}, \tag{29.48}$$

so that the energy in the individual vibration is

$$\langle E_v \rangle = \tfrac{1}{2}k\theta + kT = \tfrac{1}{2}h\nu + kT. \tag{29.49}$$

It is apparent from the form of Eq. (29.47) that the total energy of the solid is made up of a sum of the contributions from each vibration, $\langle E_v \rangle$, and a contribution from the term $-W/NkT$. This last term contributes W, so we have for the energy of the solid,

$$U = W + \sum_{i=1}^{3N} \langle E_{v_i} \rangle$$

From Eq. (29.49) this becomes

$$U = W + \sum_{i=1}^{3N} \tfrac{1}{2}h\nu_i + \sum_{i=1}^{3N} kT.$$

The summation in the second member on the right is the sum of the zero-point energies of all the oscillators. The third member is a sum of $3N$ terms each of value kT, so that it is equal to $3NkT$. Since the first two members are constant, we combine them in a single term U_0. The final result at high temperatures is

$$U = U_0 + 3NkT. \tag{29.50}$$

The heat capacity of the solid is, by differentiation,

$$C_v = 3Nk. \tag{29.51}$$

The value of the heat capacity in Eq. (29.51) does not depend on any assumption about the frequencies in the solid. This should be the value of the heat capacity of any monatomic solid if the temperature is sufficiently high. If we deal with one mole of the solid, then $N = N_A$, and $N_A k = R$. For one mole, $C_v = 3R \approx 25$ J/K mol. This result is the law of Dulong and Petit, recognized for a century and a half.

The result in Eq. (29.51) is confirmed at high temperatures for many solids. At ordinary temperatures the heat capacity is often less than the ideal value $3R$. In diamond at room temperature the heat capacity is only 6.07 J/K mol, indicating that the vibrations are not fully excited; temperatures of the order of 2000 to 3000 K are required before diamond has the high-temperature value. A crystal such as NaCl has $2N_A$ atoms per mole and should therefore have $C_v = 6R = 49.89$ J/K mol. For NaCl at 25 °C, the value of $C_p = 49.71$ J/K mol. Ferric oxide, Fe_2O_3, with $5N_A$ atoms per mole should have $C_v = 15R = 125$ J/K mol. The 25 °C value of $C_p = 105$ J/K mol.

A salt such as $NaNO_3$ has a vibrational heat capacity of $6R$ contributed by vibrations of Na^+ and NO_3^- in the solid and an additional contribution from the vibrations within the nitrate ion, which are partly but not fully excited. Some metals, notably the transition metals, exhibit values of C_v greater than $3R$ at high temperatures; this extra contribution comes from the heat capacity of the electron gas in the metal.

To discuss the heat capacity at intermediate and low temperatures requires some additional assumption about the $3N$ frequencies. The simplest approach is that of Einstein who assumed that all the frequencies have the same value, ν_E. The partition function then takes the form

$$Q = e^{-W/NkT} q_v^{3N},\tag{29.52}$$

where q_v has the form given by Eq. (29.41) with $\nu = \nu_E$. The Einstein model agrees well with experimental values of C_v at intermediate and high temperatures, but predicts values that are too low at low temperatures.

The Debye theory assumes that there is a continuous distribution of frequencies from $\nu = 0$ to a certain maximum value $\nu = \nu_D$. The final expression obtained for the heat capacity is complicated, but succeeds in interpreting the heat capacity of many solids over the entire temperature range rather more accurately than the Einstein expression. At low temperatures, the Debye theory yields the simple result

$$C_v = \frac{12\pi^4}{5} Nk \left(\frac{T}{\theta_D}\right)^3,\tag{29.53}$$

where $\theta_D = h\nu_D/k$. This is the Debye "T-cubed" law for the heat capacity of a solid. At temperatures near the absolute zero the great majority of solids follow this law quite accurately.

29.9 THE ROTATIONAL PARTITION FUNCTION

The rotational energy of a rigid linear molecule is

$$E_J = \frac{J(J + 1)\hbar^2}{2I}.\tag{29.54}$$

Since the z component of the angular momentum may have any of the values $0, \pm 1, \pm 2, \ldots, \pm J$, there are $2J + 1$ orientations of the angular momentum vector; the degeneracy, $g_J = 2J + 1$. The rotational partition function is therefore given by

$$q_r = \sum_{J=0} (2J + 1)e^{-J(J+1)\hbar^2/2IkT}.\tag{29.55}$$

If we define a characteristic rotational temperature as

$$\theta_r = \frac{\hbar^2}{2Ik},\tag{29.56}$$

then

$$q_r = \sum_{J=0} (2J + 1)e^{-J(J+1)\theta_r/T}.\tag{29.57}$$

Unfortunately, the sum on the right side cannot be evaluated in closed form. Nonetheless, using Eqs. (29.27) and (29.32) we find that the heat capacity, $C_{v(\text{rot})}$, due to the rotation of the molecules becomes

$$C_{v(\text{rot})} = NkT\left(2\frac{d \ln q_r}{dT} + T\frac{d^2 \ln q_r}{dT^2}\right).\tag{29.58}$$

We can evaluate this expression by performing the required differentiations on the series in Eq. (29.57), inserting the numerical values, and summing term by term until the desired

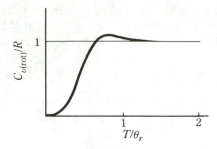

Figure 29.1   Rotational heat capacity of an unsymmetrical diatomic molecule. (From T. L. Hill, *Introduction to Statistical Mechanics*. Reading, Mass.: Addison-Wesley, 1960.)

accuracy is attained. The behavior of $\overline{C}_{v(\text{rot})}/R$ as a function of temperature is shown in Fig. 29.1. It is clear from the figure that at all values of $T/\theta_r > \sim 1.2$, the heat capacity has reached the classical value, R. Suppose that we estimate the value of θ_r for a molecule such as CO. Since $m = (0.028 \text{ kg/mol})/(6 \times 10^{23}/\text{mol}) \approx 5 \times 10^{-26}$ kg, then $I = mr^2 \approx (5 \times 10^{-26} \text{ kg})(10^{-10} \text{ m})^2 \approx 5 \times 10^{-46}$ kg m², and

$$\theta_r = \frac{(1 \times 10^{-34} \text{ J s})^2}{2(5 \times 10^{-46} \text{ kg m}^2)(1 \times 10^{-23} \text{ J/K})} \approx 1 \text{ K}.$$

Some actual values of θ_r, as well as θ_v, are given in Table 29.1. Except for those diatomic molecules containing hydrogen, the values of θ_r are indeed about 1 to 2 K. Thus these molecules have the classical value of the heat capacity at any temperature above about 2.5 K.

When $\theta_r/T \ll 1$, we can replace the sum in Eq. (29.55) by an integral. Let $y = J(J + 1)$; then $dy = (2J + 1) dJ$. But since J is an integer, the difference between successive values is 1; thus $dJ = 1$ and therefore $dy = 2J + 1$. Using this result in Eq. (29.55) and replacing summation by integration, we obtain

$$q_r = \int_0^\infty e^{-y\theta_r/T} \, dy = \frac{T}{\theta_r} = \frac{2IkT}{\hbar^2} \tag{29.59}$$

Table 29.1
Parameters for diatomic molecules

	θ_v/K	θ_r/K	r_e/pm	$D_0/10^{-19} \text{ J}$
H_2	6210	85.4	74.0	7.174
N_2	3340	2.86	109.5	15.637
O_2	2230	2.07	120.4	8.196
CO	3070	2.77	112.8	17.798
NO	2690	2.42	115.0	10.426
HCl	4140	15.2	127.5	7.109
HBr	3700	12.1	141.4	5.071
HI	3200	9.0	160.4	4.892
Cl_2	810	0.346	198.9	3.985
Br_2	470	0.116	228.4	3.158
I_2	310	0.054	266.7	2.474

From T. L. Hill, *Introduction to Statistical Thermodynamics*. Reading, Mass.: Addison-Wesley, 1960.

This is the value of the rotational partition function for unsymmetrical linear molecules (for example, heteronuclear diatomic molecules). Using this value of q_r we can calculate the values of the thermodynamic functions attributable to rotation.

For nonlinear molecules with principal moments of inertia, I_α, I_β, and I_γ, the value of the rotational partition function at high temperature is

$$q_r = \frac{\pi^{1/2}}{\sigma} \left(\frac{2I_\alpha kT}{\hbar^2}\right)^{1/2} \left(\frac{2I_\beta kT}{\hbar^2}\right)^{1/2} \left(\frac{2I_\gamma kT}{\hbar^2}\right)^{1/2}, \tag{29.60}$$

in which σ is the symmetry number.

If we rotate a homonuclear diatomic molecule through 180° around an axis perpendicular to the molecular axis, the final configuration is indistinguishable from the initial one. This means that the expression for the partition function given in Eq. (29.59) must be divided by 2 for homonuclear molecules to avoid counting indistinguishable configurations as distinct. Thus, for diatomic molecules when $\theta_r/T \ll 1$, we have the general expression

$$q_r = \frac{2IkT}{\sigma \hbar^2} \tag{29.61}$$

in which $\sigma = 1$ for a heteronuclear molecule and $\sigma = 2$ for a homonuclear molecule.

For the same reasons, the symmetry factor, σ, was introduced in Eq. (29.60) for polyatomic molecules. Consider CH_4, methane, a molecule with 12 indistinguishable configurations. Suppose that the carbon atom is at the origin and hydrogen atom number 1 is on the vertical axis; then the three hydrogen atoms in the horizontal plane can be in any of three indistinguishable configurations. But any one of the four hydrogen atoms may be on the vertical axis; thus there are $4 \times 3 = 12$ indistinguishable configurations and therefore, for methane, $\sigma = 12$ in Eq. (29.60).

29.10 THE ELECTRONIC PARTITION FUNCTION

Using Eq. (29.30a) we can write the electronic partition function in the form,

$$q_e = \sum_i g_{ei} e^{-\epsilon_{ei}/kT} = g_{e1} e^{-\epsilon_{e1}/kT} + g_{e2} e^{-\epsilon_{e2}/kT} + \cdots, \tag{29.62}$$

or, factoring the first term out of the sum,

$$q_e = g_{e1} e^{-\epsilon_{e1}/kT} \left[1 + \left(\frac{g_{e2}}{g_{e1}}\right) e^{-(\epsilon_{e2} - \epsilon_{e1})/kT} + \cdots \right].$$

For most atoms and molecules the energy of the next higher electronic state is very much greater than ϵ_{e1}, so that $(\epsilon_{e1} - \epsilon_{e2})/kT$ is very large unless the temperature is exceedingly high. It follows that the second and following terms in the sum are negligibly small. This reduces q_e to the one term

$$q_e = g_{e1} e^{-\epsilon_{e1}/kT}. \tag{29.63}$$

If these conditions are not met, we simply add as many higher terms as are required by the particular case. In many practical cases Eq. (29.63) is all that is needed.

The degeneracy of the electronic state is given by

$$\text{Atoms:} \quad g_e = 2J + 1; \quad \text{Molecules:} \quad g_e = 2\Omega + 1. \tag{29.64}$$

For atoms, J is the quantum number for the total angular momentum, orbital plus spin; $J = L + S$. The number J is the right-hand subscript in the term symbol derived in Section 24.7. For example, for the halogen atoms the term symbol for the ground state is $^2P_{3/2}$. Then $J = \frac{3}{2}$, and $g_{e1} = 2(\frac{3}{2}) + 1 = 4$. Similarly, for molecules, Ω is the quantum number for the total angular momentum of the molecule; $\Omega = \Lambda + \Sigma$ (Section 25.9). For the vast majority of chemically saturated molecules, $\Omega = 0$ in the ground state so that $g_{e1} = 1$. Oxygen is an exception; for oxygen, $\Omega = 1$ and $g_{e1} = 3$. Molecules with an odd number of electrons, such as NO and NO_2, are exceptions. For NO, $g_{e1} = 2$. Also for NO the second term in the partition function must be included even at ordinary temperatures.

As we will see below, if we are to use the expression in Eq. (29.63) in the calculation of an equilibrium constant for a chemical reaction, we must choose a common energy zero for all the species involved in the reaction. The condition of separated atoms at rest in the gas phase (that is, at $T = 0$ K) is usually the most convenient choice of energy zero. The situation is illustrated for a diatomic molecule in Fig. 29.2. The depth of the minimum is the value of ϵ_{e1}; thus $\epsilon_{e1} = -D_e$. This position is the origin for the vibrational energy of the molecule. Thus, if the dissociation energy is D_0 (a positive number), then

$$-D_0 = -D_e + \tfrac{1}{2}h\nu. \tag{29.65}$$

If the molecule has more than one vibrational degree of freedom, we have

$$-D_0 = -D_e + \sum_i \tfrac{1}{2}h\nu_i, \tag{29.66}$$

in which $\epsilon_{e1} = -D_e$, the summation is over all the vibrational modes, and D_0 is the energy required for the reaction

$$\text{molecule} \rightarrow \text{gaseous atoms} \qquad \text{(at 0 K)}.$$

Some values of D_0 for diatomic molecules are given in Table 29.1. Note that D_0 and ν_i are measurable while D_e is calculated from Eq. (29.66). For a polyatomic molecule in which the various vibrations are independent, it is convenient to combine the product of the electronic and vibrational partition functions. We have for $q_e q_v$,

$$q_v q_e = \left(\prod_i \frac{e^{-\theta_i/2T}}{1 - e^{-\theta_i/T}} \right) g_{e1} e^{-\epsilon_{e1}/kT}$$

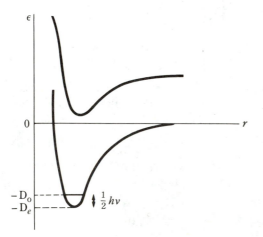

Figure 29.2   Ground state and first excited state (dashed curve) electronic energies as a function of internuclear separation. (Adapted from T. L. Hill, *Introduction to Statistical Mechanics*. Reading, Mass.: Addison-Wesley, 1960.)

where $\theta_i = h\nu_i/k$. Since

$$\prod_i e^{-\theta_i/2T} = e^{-\Sigma_i \theta_i/2T},$$

$q_v q_e$ reduces to

$$q_v q_e = \frac{g_e e^{-(\epsilon_{e1} + \Sigma_i 1/2 h\nu_i)/kT}}{\prod_i (1 - e^{-\theta_i/T})} = \frac{g_e e^{D_0/kT}}{\prod_i (1 - e^{-\theta_i/T})}, \tag{29.67}$$

where we have used Eq. (29.66) to obtain the second equality on the right. Note that at $T = 0$ K, the energy change in a chemical reaction is given by ΔD_0.

29.11 *ORTHO-* AND *PARA*-HYDROGEN

The quantum-mechanical interpretation of the appearance of the symmetry number in the rotational partition function has its basis in the symmetry of the total wave function of the molecule. We will consider only the case of the hydrogen molecule.

The total wave function, Ψ, of the hydrogen molecule is a product of the wave functions for the various modes of motion of the molecule. We write

$$\Psi = \psi_{el}\psi_{rot}\psi_{vib}\psi_{ns}, \tag{29.68}$$

in which ψ_{el} is the electronic wave function, ψ_{rot} is the rotational wave function, ψ_{vib} is the vibrational wave function, and ψ_{ns} is the nuclear spin wave function. The Pauli exclusion principle requires that the wave function be antisymmetric under the interchange of any two identical elementary particles. Consequently, if we interchange the two protons (the nuclei), the total wave function must change sign. Since ψ_{el} and ψ_{vib} are symmetric under the interchange of the nuclei, the product $\psi_{rot}\psi_{ns}$ must be antisymmetric. This requirement can be satisfied in two ways, each way corresponding to a different kind of hydrogen. If ψ_{ns} is symmetric, then ψ_{rot} must be antisymmetric and vice versa. If J is odd, ψ_{rot} is antisymmetric, while if J is even, ψ_{rot} is symmetric. Thus we have the two kinds of hydrogen molecules:

	ψ_{ns}	ψ_{rot}
ortho-hydrogen (*o*-H_2)	Symmetric	Antisymmetric $J = 1, 3, 5, \ldots$
para-hydrogen (*p*-H_2)	Antisymmetric	Symmetric $J = 0, 2, 4, \ldots$

Since the nuclear spin quantum number for the proton is $\frac{1}{2}$, and there are two protons, it follows that the spins can add to give a total spin quantum number of 1, or they can subtract to give a net spin quantum number of 0. The first case corresponds to a triplet (three symmetric spin functions) and the second corresponds to a singlet (one antisymmetric spin function). The functions are the same as those for the electron pair:

Symmetric functions	$\alpha(1)\alpha(2)$ $\qquad \beta(1)\beta(2) \qquad \alpha(1)\beta(2) + \beta(1)\alpha(2)$
Antisymmetric function	$\alpha(1)\beta(2) - \beta(1)\alpha(2)$

It follows that, in the absence of any other influence, hydrogen atoms will combine to yield o-H_2 and p-H_2 in a three to one ratio, simply because there are three symmetric nuclear spin wave functions and one antisymmetric nuclear spin wave function. This $3:1$ ratio is observed at high temperatures, where the molecules are spread out over enough rotational states that the energy difference between the populations in the odd and even rotational states is insignificant.

The rotational partition function for *para*-hydrogen is

$$q_r(para) = \sum_{J=\text{even}} (2J + 1)e^{-J(J+1)\theta_r/T}, \tag{29.69}$$

for *ortho*-hydrogen, we have

$$q_r(ortho) = 3 \sum_{J=\text{odd}} (2J + 1)e^{-J(J+1)\theta_r/T} \tag{29.70}$$

Note that for any particular hydrogen molecule, the allowed values of J are either even or odd, but not both. This limitation, which allows the occupation of only one-half of the possible states, effectively divides the partition function by a factor of two. This division by two was accomplished in our elementary argument by introducing the symmetry factor.

The weighting factor, 3, appears in front of the sum in Eq. (29.70) for o-H_2 because there are three nuclear spin states permitted for each value of J.

The rotational heat capacities, $C_{v,\text{rot}}$, for the various forms of hydrogen are shown as functions of temperature in Fig. 29.3. So-called *normal*-hydrogen, n-H_2, consists of three parts o-H_2 to one part p-H_2. This is the equilibrium ratio at high temperature. In the absence of a catalyst for the conversion between the two forms, this ratio is maintained as the temperature is lowered. Therefore the experimental heat capacity curve shown in the figure corresponds to what we would calculate for a mixture of $\frac{3}{4}$ o-H_2 and $\frac{1}{4}$ p-H_2.

On the other hand, if the $3:1$ ortho to para mixture at high temperature is cooled in the presence of a catalyst (activated charcoal), which brings the reaction o-$H_2 \rightleftharpoons p$-H_2 to

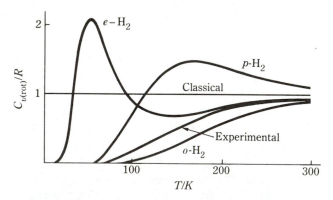

Figure 29.3 Rotational-nuclear contribution to the heat capacity for o-H_2, p-H_2, e-H_2 (equilibrium mixture at each temperature), and n-H_2 (labeled "experimental"). (From T. L. Hill, *Introduction to Statistical Mechanics*. Reading, Mass.: Addison-Wesley, 1960.)

equilibrium, then the relative amounts of the two species change with temperature. At absolute zero, all the hydrogen will be converted to *para*-hydrogen. Since p-H_2 has $J = 0$ at zero kelvin, it has a lower rotational energy than o-H_2, which has $J = 1$ in its lowest possible rotational state. Thus in the presence of a catalyst the o-H_2 is converted to p-H_2 because of the lower rotational energy of the p-H_2.

At any temperature in the presence of a catalyst, the equilibrium ratio of *ortho*- to *para*-hydrogen is given by

$$\frac{N_o}{N_p} = \frac{q_r(ortho)}{q_r(para)} = \frac{3 \sum_{J=\text{odd}} (2J + 1)e^{-J(J+1)\theta_r/T}}{\sum_{J=\text{even}} (2J + 1)e^{-J(J+1)\theta_r/T}}. \tag{29.71}$$

The slight difference in the average rotational energy of the two forms enhances the heat capacity due to the LeChatelier shift in the equilibrium position as the temperature is changed. This effect is exhibited in the curve for $C_{v,\text{rot}}$ labeled e-H_2. *Equilibrium*-hydrogen, e-H_2, is hydrogen that is kept in the presence of a catalyst to ensure that the equilibrium between o-H_2 and p-H_2 is established at all temperatures. The curve for e-H_2 is typical of the heat capacity of a reactive mixture maintained in equilibrium as the temperature is changed.

At 0 K, $J = 1$ for *ortho*-hydrogen, and $2J + 1 = 3$. These three rotational states combine with the three nuclear spin states to yield nine quantum states available to the o-H_2 molecule at 0 K. In normal hydrogen, the $\frac{3}{4}$ mole fraction of o-H_2 is evenly distributed over all nine of these states. On the other hand, for p-H_2 at 0 K, $J = 0$ and $2J + 1 = 1$. This one rotational state combines with the one nuclear spin state to yield a single quantum state that is available to the p-H_2 molecule. The entire $\frac{1}{4}$ mole fraction of p-H_2 in n-H_2 occupies this single quantum state at 0 K.

At 0 K normal hydrogen has a residual entropy due to the entropy of mixing the $\frac{1}{4}$ mole fraction of p-H_2 with the nine different states occupied by the $\frac{3}{4}$ mole fraction of o-H_2. Each of the nine states has $\frac{1}{9}(\frac{3}{4}) = \frac{1}{12}$ mole fraction in it. The ΔS of mixing is

$$\Delta S_{\text{mix}} = -R(\tfrac{1}{4} \ln \tfrac{1}{4} + \tfrac{3}{4} \ln \tfrac{1}{12}) = 2.21R = 18.38 \text{ J/K mol.}$$

The nuclear spin can be oriented in two ways; this leads to an entropy of $R \ln 2$ per nucleus or a total of $2R \ln 2$. Since this contribution to the entropy persists through all changes, it is not taken as part of the residual entropy and so must be subtracted from the entropy of mixing above. This yields for the residual entropy of hydrogen at 0 K, $18.38 - 2R \ln 2 = 6.8$ J/K mol, which is in good agreement with the observed value of 6.2 J/K mol.

29.12 GENERAL EXPRESSIONS FOR THE PARTITION FUNCTION

In general the molecules composing a system will possess energy in several ways: in translation, rotation, and vibration. In a diatomic gas, for example, there are three translational degrees of freedom, two rotational degrees of freedom, and one vibrational degree of freedom. If the energies in these various degrees of freedom are additive, the molecular partition function factors into a product of partition functions for the various degrees of freedom. For a diatomic molecule, for example,

$$q = q_t q_r q_v q_e. \tag{29.72}$$

29.13 THE EQUILIBRIUM CONSTANT IN TERMS OF THE PARTITION FUNCTIONS

Consider the chemical equilibrium in a system of ideal gases:

$$0 = \sum_i \nu_i A_i(g).$$

The equilibrium condition, Eq. (11.33), is

$$\sum_i \nu_i \mu_i = 0.$$

Using Eq. (29.33) for μ_i and dividing each term in the sum by $-kT$, we obtain

$$\sum_i \nu_i \ln\left(\frac{q_i}{N_i}\right) = 0 \quad \text{or} \quad \sum_i \ln\left(\frac{q_i}{N_i}\right)^{\nu_i} = 0.$$

But the sum of logarithms is the logarithm of a product, so that

$$\sum_i \ln\left(\frac{q_i}{N_i}\right)^{\nu_i} = \ln \prod_i \left(\frac{q_i}{N_i}\right)^{\nu_i} = 0 \quad \text{or} \quad \prod_i \left(\frac{q_i}{N_i}\right)^{\nu_i} = 1.$$

(Since the logarithm of the product is zero, the product must be equal to one.) This result can be restated in terms of the concentrations, $\tilde{N}_i = N_i/V$. Thus we replace N_i by $N_i = \tilde{N}_i V = (\tilde{N}_i/\tilde{N}^\circ)\tilde{N}^\circ V$, in which $\tilde{N}^\circ$ is a standard concentration. This yields

$$\prod_i \left(\frac{q_i}{(\tilde{N}_i/\tilde{N}^\circ)\tilde{N}^\circ V}\right)^{\nu_i} = \frac{\prod_i (q_i/\tilde{N}^\circ V)^{\nu_i}}{\prod_i (\tilde{N}_i/\tilde{N}^\circ)^{\nu_i}} = 1.$$

Then

$$\prod_i \left(\frac{\tilde{N}_i}{\tilde{N}^\circ}\right)^{\nu_i} = \prod_i \left(\frac{q_i}{\tilde{N}^\circ V}\right)^{\nu_i}. \tag{29.73}$$

The left side of this equation is the concentration equilibrium constant,

$$K_c = \prod_i \left(\frac{\tilde{N}_i}{\tilde{N}^\circ}\right)^{\nu_i} = \prod_i \left(\frac{c_i}{c^\circ}\right)^{\nu_i}. \tag{29.74}$$

The second equality in this equation obtains since $\tilde{N}_i/\tilde{N}^\circ = c_i/c^\circ$ if $\tilde{N}^\circ$ and c° refer to the same standard concentration but simply express it in different units. This K_c is the same as that defined in Section 11.11 if we choose $c^\circ = 1$ mol/L. Combining Eqs. (29.73) and (29.74) we have

$$K_c = \frac{1}{(\tilde{N}^\circ)^{\Delta\nu}} \prod_i \left(\frac{q_i}{V}\right)^{\nu_i} \tag{29.75}$$

This equation says that K_c is proportional to the proper quotient of partition functions per unit volume. As usual, we have expressed the equilibrium constant as a product of dimensionless ratios.

If the standard state of concentration is $c^\circ = 1$ mol/L, then the corresponding value of $\tilde{N}^\circ = c^\circ(1000 \text{ L/m}^3)N_A = 1000\, N_A$ mol/m^3. Equation (29.75) then becomes

$$K_c = \frac{1}{(1000\, N_A \text{ mol/m}^3)^{\Delta\nu}} \prod_i \left(\frac{q_i}{V}\right)^{\nu_i} \tag{29.75a}$$

Equation (29.75) is an important link between quantum mechanics and chemistry. Knowing the energy levels of the molecules, we can calculate the molecular partition functions. Then we use Eq. (29.75) to obtain the equilibrium constant for the chemical reaction.

For systems involving substances other than ideal gases, the equilibrium constant can be obtained in a similar way from the partition functions.

■ **EXAMPLE 29.1**   From the hydrogen spectrum we find that the ionization energy of the hydrogen atom is 2.1782×10^{-18} J. Calculate the equilibrium constant for the reaction

$$H(g) \quad \rightleftharpoons \quad H^+(g) + e^-(g)$$

at 2000 K. The term symbol for the ground state of the hydrogen atom is $^2S_{1/2}$. The electron has a spin quantum number of $\frac{1}{2}$ and thus a spin degeneracy, $g_{spin} = 2(\frac{1}{2}) + 1 = 2$.

The expression for the equilibrium constant, since $\Delta v = 1$, is

$$K_c = \frac{1}{\tilde{N}^\circ} \frac{(q_{H^+}/V)(q_{e^-}/V)}{(q_H/V)}$$

We assume that all the species are ideal gases. Then the partition functions are $q_H = (q_t)_H(q_e)_H$; $q_{H^+} = (q_t)_{H^+}$; $q_{e^-} = (q_t)_{e^-}(q_e)_{e^-}$. It is convenient to group the translational functions together and the electronic functions together in the equilibrium constant:

$$K_c = \frac{1}{\tilde{N}^\circ} \left[\frac{(q_t/V)_{H^+}(q_t/V)_{e^-}}{(q_t/V)_H} \right] \left[\frac{(q_e)_{e^-}}{(q_e)_H} \right] = \frac{1}{\tilde{N}^\circ} F_{trans} \times F_{el}$$

Since $m_{H^+} \approx m_H$, it follows from Eq. (29.39) that $(q_t/V)_{H^+} = (q_t/V)_H$; then the first part of K_c reduces to

$$\frac{1}{\tilde{N}^\circ} F_{trans} = \frac{1}{\tilde{N}^\circ} \left(\frac{q_t}{V} \right)_{e^-} = \frac{1}{\tilde{N}^\circ} \left(\frac{2\pi m_e kT}{h^2} \right)^{3/2}$$

$$= \frac{1}{6.0220 \times 10^{26} \text{ m}^{-3}} \left[\frac{2\pi(9.1095 \times 10^{-31} \text{ kg})(1.3807 \times 10^{-23} \text{ J/K})2000K}{(6.6262 \times 10^{-34} \text{ J s})^2} \right]^{3/2}$$

$$= 0.35863.$$

We arbitrarily choose the separated ions as the state of zero energy; then for the H atom we can write, $\epsilon_{e1} = -2.1782 \times 10^{-18}$ J. Since for the H atom, $g_{e1} = 2(\frac{1}{2}) + 1 = 2$, and for the electron, $g_{spin} = 2$, the partition functions become

$$(q_e)_H = g_{e1}e^{-\epsilon_{e1}/kT} = 2e^{-\epsilon_{e1}/kT} \qquad \text{and} \qquad (q_e)_{e^-} = g_{spin} = 2.$$

But

$$\frac{\epsilon_{e1}}{kT} = \frac{-2.1782 \times 10^{-18} \text{ J}}{(1.3807 \times 10^{-23} \text{ J/K})(2000 \text{ K})} = -78.882.$$

Then we find for F_{e1},

$$F_{el} = \frac{2}{2e^{-(-78.882)}} = e^{-78.882} = 5.5206 \times 10^{-35}.$$

Finally, we obtain $K_c = 0.35863(5.5206 \times 10^{-35}) = 1.9798 \times 10^{-35}$.

■ **EXAMPLE 29.2**   Using the data in Table 29.1 calculate the equilibrium constant at 500 K for the reaction

$$H(g) + I_2(g) \rightleftharpoons HI(g) + I(g).$$

The ground electronic state for the H atom is $^2S_{1/2}$; for the I atom it is $^2P_{3/2}$. For both I_2 and HI in the ground state, $g_e = 1$. For the energies relative to the separated atoms, we have $D_0(I_2) = 2.4736 \times 10^{-19}$ J and $D_0(HI) = 4.8917 \times 10^{-19}$ J.

The partition functions are

$$q_I = (q_t)_I(q_e)_I; \qquad\qquad q_H = (q_t)_H(q_e)_H;$$

$$q_{HI} = (q_t)_{HI}(q_r)_{HI}(q_v)_{HI}(q_e)_{HI}; \qquad q_{I_2} = (q_t)_{I_2}(q_r)_{I_2}(q_v)_{I_2}(q_e)_{I_2}.$$

Then

$$K_c = \frac{(q_I/V)(q_{HI}/V)}{(q_H/V)(q_{I_2}/V)}.$$

Again, we group the translational functions together, the rotational functions together, and the vibrational and electronic functions together in K_c

$$K_c = \left[\frac{(q_t/V)_I(q_t/V)_{HI}}{(q_t/V)_H(q_t/V)_{I_2}}\right]\left[\frac{(q_r)_{HI}}{(q_r)_{I_2}}\right]\left[\frac{(q_e)_I}{(q_e)_H}\right]\left[\frac{(q_v q_e)_{HI}}{(q_v q_e)_{I_2}}\right];$$

$$K_c = F_{trans} \times F_{rot} \times F_{el} \times F_{vib\text{-}el}$$

Each q_t differs from the others only in the value of the mass, Eq. (29.39). Therefore the constants and T drop out of the quotient and we have

$$F_{trans} = \left(\frac{m_I m_{HI}}{m_H m_{I_2}}\right)^{3/2} = \left(\frac{m_{HI}}{2m_H}\right)^{3/2} = \left(\frac{127.9124}{2(1.0079)}\right)^{3/2} = 505.47.$$

Since $q_r = T/\sigma\theta_r$, from Table 29.1 we obtain $\theta_r = 9.0$ K for HI and $\theta_r = 0.054$ K for I_2. Also $\sigma = 1$ for HI and $\sigma = 2$ for I_2. Then

$$F_{rot} = \frac{T/9.0\ \text{K}}{T/2(0.054\ \text{K})} = 0.012.$$

For I, $q_e = g_e = 2(\tfrac{3}{2}) + 1 = 4$; for H, $q_e = g_e = 2(\tfrac{1}{2}) + 1 = 2$; then $F_{el} = \tfrac{4}{2} = 2$.

Also from Table 29.1, for HI we have $\theta_v = 3200$ K and for I_2, $\theta_v = 310$ K. For both HI and I_2, $g_e = 1$. Then, in view of Eq. (29.67), we have

$$F_{vib\text{-}el} = \left[\frac{g_e e^{D_0/kT}}{1 - e^{-\theta/T}}\right]_{HI}\left[\frac{1 - e^{-\theta/T}}{g_e e^{D_0/kT}}\right]_{I_2}$$

For HI

$$\frac{D_0}{kT} = \frac{4.8917 \times 10^{-19}\ \text{J}}{(1.3807 \times 10^{-23}\ \text{J/K})(500\ \text{K})} = 70.860.$$

For I_2

$$\frac{D_0}{kT} = \frac{2.4736 \times 10^{-19}\ \text{J}}{(1.3807 \times 10^{-23}\ \text{J/K})(500\ \text{K})} = 35.832.$$

Thus

$$F_{\text{vib-el}} = \frac{1 \times e^{70.860}}{1 \times e^{35.832}} \frac{(1 - e^{-310/500})}{(1 - e^{-3200/500})} = 0.4621 e^{35.028} = 7.5373 \times 10^{14}.$$

The final expression for K_c is

$$K_c = 505.47(0.012)(2)(7.5373 \times 10^{14}) = 9.1 \times 10^{15}.$$

29.14 CONCLUSION

We have explored some of the simpler aspects of statistical thermodynamics, a very powerful theoretical tool. If the energy levels of the molecules composing the system can be obtained by solution of the Schrödinger equation, the partition function can be calculated; then any thermodynamic property can be evaluated. One of the great virtues of statistical thermodynamics is its ability to reveal general laws. For example, we reached the conclusion that all monatomic solids should have the same heat capacity at high temperatures. Restrictions on the laws are made apparent; for example, the heat capacity of a monatomic solid at low temperatures depends on what is assumed about the frequencies in the solid.

A number of objections may be raised at this point: difficulties in solving the Schrödinger equation, approximations that must be made in many of the mathematical steps, and so on. These are legitimate objections, but we have concentrated here in presenting the more theoretical side of the statistical thermodynamics. The actual values of the energies may be known from experiment! Analysis of the spectrum of a molecule will give us all the information about energy levels that we need. We simply insert the experimental values of the energies into the exponentials of the partition function, add all the exponentials together, and by brute force evaluate any thermodynamic quantity that happens to be of interest. This is somewhat laborious but very practical. The most difficult part is obtaining and analyzing the spectral data in terms of energy levels. To obtain the heat capacity of hydrogen at 2000 K by a calorimetric method would be a nasty job; to study the spectrum and calculate the heat capacity from the spectral data using partition functions is very much easier and yields a much more accurate result. In this connection, it should be said that a large proportion of the tabulated thermodynamic data is obtained from spectral data.

A final word should be said about the entropy. Although we can consider that the energy of a system is the sum of the energies of the individual molecules, the entropy of a system is not the sum of the entropies of individual molecules. The entropy is defined in terms of the complexions of a very large number of systems in an ensemble. The entropy of a molecule has no real meaning. We can divide the entropy of a system by the number of molecules and talk about an entropy per molecule, but in the final analysis this is a fiction. Entropy and temperature have meaning only for matter in bulk, which is composed of a very large number of individual particles. Systems containing only a few molecules need not obey the second law of thermodynamics. The number of molecules must be very large before probabilities become actualities with a negligible chance of observing a deviation.

QUESTIONS

29.1 Why can a system in contact with a thermal reservoir have various possible values of its energy?

29.2 What is an ensemble? What is identical about members of the ensemble? What might be examples of things that are *different* about members of the ensemble?

29.3 Sketch P_i versus E_i for low, intermediate, and high temperature T.

29.4 Use the answer to Question 29.3 and Eq. (29.19) to discuss qualitatively the variation of the entropy with temperature.

29.5 What is the importance of the partition function of a system? What simplifications occur if the molecules of the system interact only weakly?

29.6 Qualitatively sketch the heat capacity of an ideal gas of diatomic molecules as a function of temperature. Indicate the characteristic temperatures (in terms of vibrational frequency, moment of inertia, and so on) where various degrees of freedom begin to contribute.

29.7 Identify the features that tend to favor the products in the chemical equilibrium of Example 29.2.

PROBLEMS

29.1 Using the partition function, show that for a monatomic gas, $U = \frac{3}{2}NkT$ and that $p = NkT/V$.

29.2 Using the partition function, derive expressions for S, A, and G for a monatomic gas in terms of M, T, and V. Evaluate these functions for one mole of argon at 1 atm and 298.15 K and at 1 atm and 1000 K.

29.3 For N_2 calculate the contributions to the thermodynamic functions U, H, S, A, and G from translation, rotation, and vibration at 1 atm and 298.15 K. Use the values of θ_r and θ_v from Table 29.1. Compare with the values at 1000 K.

29.4
a) Using the complete expression for the vibrational partition function, Eq. (29.42), derive the expression for C_v as a function of θ/T.
b) Using the expression in Eq. (29.44), compute $C_v(\infty)$, the heat capacity at infinite temperature.
c) Calculate the values of $C_v/C_v(\infty)$ for $\theta/T = 0$, 0.5, 1.0, 1.5, 2.0, 3.0, 4.0, 5.0, 6.0. Plot these values against θ/T.

29.5 The bending vibration in CO_2 has a frequency of 2.001×10^{13} Hz.
a) Calculate the characteristic temperature θ for this vibration.
b) What contribution does this vibration make to the heat capacity of CO_2 at 300 K?
c) This vibration is doubly degenerate; that is, the CO_2 molecule has two vibrations of this frequency. How does this affect the heat capacity?
d) The stretching vibrations have much higher frequencies. What contribution does the frequency, the asymmetrical stretch, at 7.0430×10^{13} Hz make to the heat capacity at 300 K?

29.6 Calculate the contributions for translation, rotation, and vibration of CO_2 to the thermodynamic functions U, H, S, A, and G at 1 atm and 298.15 K. The values of θ_v are $\theta_1 = 3380.1$ K, $\theta_2 = 1997.5$ K, $\theta_3 = \theta_4 = 960.1$ K. The CO distance in CO_2 is 116.2 pm.

29.7 Calculate the populations of the lowest three rotational levels of p-hydrogen at 10 K, 50 K, and 100 K. For H_2, $\theta_r = 85.4$ K.

29.8 For hydrogen, $\theta_r = 85.4$ K. (a) Calculate the entropy and the heat capacity due to rotation for o-H_2 at 100 K, 150 K, and 200 K. (b) Repeat the calculation in (a) for p-H_2. (c) Calculate the equilibrium constant for p-$H_2 \rightleftharpoons o$-H_2 at 100 K, 150 K, and 200 K.

29.9 Consider the reaction

$$H_2(g) + Cl_2(g) \rightleftharpoons 2\,HCl(g)$$

at 25 °C. Using the data from Table 29.1,
a) Calculate the rotational partition function for Cl_2 and HCl; assume $q_r = 1.9206$ for H_2.
b) Calculate the equilibrium constant for the reaction.

29.10 Calculate the equilibrium constant for the reaction

$$H_2(g) + CO_2(g) \rightleftharpoons H_2O(g) + CO(g)$$

at 800 K, 1000 K, and 1200 K. For CO_2 the vibrational temperatures are $\theta_1 = 3380.1$ K, $\theta_2 = 1997.5$ K, and $\theta_3 = \theta_4 = 960.1$ K; the CO distance in CO_2 is 116.1 pm; $D_0(CO_2) = 2.6534 \times 10^{-18}$ J. For H_2O the vibrational frequencies are 3650 cm^{-1}, 1590 cm^{-1}, and 3760 cm^{-1}; the moments of inertia are $I/10^{-47}$ kg m$^2 = 1.024$, 1.921, and 2.947; $D_0(H_2O) = 1.5239 \times 10^{-18}$ J. For CO $\theta_v = 3070$ K and $\theta_r = 2.77$ K; $D_0(CO) = 1.7798 \times 10^{-18}$ J. For H_2, see Table 29.1.

29.11 Given the data from Table 29.1, from Problem 29.10, and keeping in mind that $g_{e1} = 3$ for O_2, calculate equilibrium constants for the reactions:

a) $H_2(g) + I_2(g) \rightleftharpoons 2HI(g)$ at 600 K;
b) $2H_2(g) + O_2(g) \rightleftharpoons 2H_2O(g)$ at 2000 K;
c) $4HCl(g) + O_2(g) \rightleftharpoons 2H_2O(g) + 2Cl_2(g)$ at 800 K;
d) $2CO(g) + O_2(g) \rightleftharpoons 2CO_2(g)$ at 900 K.

29.12 Calculate the equilibrium constant at 1000 K for the reaction

$$N_2(g) + O_2(g) \;\rightleftharpoons\; 2NO(g).$$

The data are in Table 29.1. Also, for O_2, $g_{e1} = 3$; for NO, $g_{e1} = g_{e2} = 2$; and $\epsilon_{e2} = \epsilon_{e1} + 2.46 \times 10^{-21}$ J.

29.13 For any ideal gas, $\mu = -kT\ln(q/N)$. How does the possession of rotational and vibrational degrees of freedom in addition to translational degrees of freedom affect the value of the chemical potential?

29.14 The chemical potentials of an ideal monatomic gas and a monatomic solid are given by

$$\mu_{gas} = -kT\ln(q_t/N)$$

$$\mu_{solid} = -kT\ln(q_v^3 e^{-W/NkT}),$$

if we assume that the frequencies in the solid are all the same.

a) Derive an expression for the equilibrium vapor pressure of a monatomic solid. Use the high-temperature value for q_v; $q_v = (T/\theta)\exp(-\theta/2T)$.

b) By differentiating and comparing with the Gibbs–Helmholtz equation, compute the value of the enthalpy of vaporization.

c) Other things being equal (!), which crystal will have the higher vapor pressure at a specified temperature; a crystal of a monatomic substance or a crystal of a diatomic substance? Assume that the diatomic molecules do not rotate in the solid.

30

Transport Properties

30.1 INTRODUCTORY REMARKS

We turn our attention at this point to the changes in the properties of a system with time. Thermodynamics describes the change in properties of a system in a change in state, but provides no information about the time required to effect the change. On a practical level, the time interval required for any particular change in state is of utmost importance. There is little consolation in knowing that a certain reaction can occur naturally if a million years is required for the transformation. Hydrogen and oxygen if left to themselves do not form water within any practical length of time. If a trace of platinum is added to the vessel, the conversion to water is complete within a few microseconds.

Much of the importance of thermodynamics lies in the fact that thermodynamic predictions do not depend on the detailed way in which a system is transformed from an initial to a final state. The time interval required for the transformation depends very much on these details. A certain Gibbs-energy difference exists between the initial state, $H_2 + \frac{1}{2}O_2$, and the final state, H_2O. This Gibbs energy difference has nothing to do with the presence or absence of a bit of platinum, the presence of which changes the time interval required by an enormous factor. The discussion of the rates of transformations requires us in every case to postulate some model of the structure of the system, while thermodynamics needs no model. By testing the rate predicted by a certain model against the experimental data, we can judge the adequacy of the model.

First we look at the purely empirical laws that are observed to govern the rates of various processes. Later, by application of knowledge from thermodynamics and (mainly) structure, we attempt to interpret these laws in terms of the constitution of the system and the fundamental properties of the atoms and molecules that compose it.

30.2 TRANSPORT PROPERTIES

There is a particularly simple group of processes, transport processes, in which some physical quantity such as mass or energy or momentum or electrical charge is transported from one region of a system to another. Consider a metal bar connecting two heat reservoirs at different temperatures. Heat flows through the bar from the high-temperature reservoir to the low-temperature reservoir; the heat flow is the manifestation of the transport of energy through the bar. The energy flow is easily measurable. Another example is the transport of electrical charge through a conductor by the application of an electrical potential difference between the ends of the conductor. Mass is transported in the flow of a fluid through a pipe as the result of a pressure difference between the ends of the pipe. Diffusion is the mass transport that occurs in a mixture if a concentration gradient is present. The viscosity of a fluid, its resistance to flow, is determined by the transport of momentum in a direction perpendicular to the direction of flow.

In all cases the *flow*, the amount of the physical quantity transported in unit time through a unit of area perpendicular to the direction of flow, is proportional to the negative gradient of some other physical property such as temperature, pressure, or electrical potential. Choosing the z-axis as the direction of flow, the general law for transport is

$$J_z = L\left(-\frac{\partial Y}{\partial z}\right),\tag{30.1}$$

where J_z is the flow, the amount of the quantity transported per square metre per second, L is the proportionality constant, and $(-\partial Y/\partial z)$ is the negative gradient of Y in the direction of flow; Y may be any of the quantities temperature, electrical potential, pressure, and so forth; L is called the phenomenological coefficient. Since flow occurs in a particular direction, it is a vector quantity; Eq. (30.1) describes the z component of the vector. We will not need the more general three-dimensional equations. For the examples mentioned above we have the individual equations for flow in the z direction:

$$\text{Heat flow} \qquad J_z = -\kappa_T \frac{\partial T}{\partial z} \quad \text{(Fourier's law)};\tag{30.2}$$

$$\text{Electrical current} \qquad J_z = -\kappa \frac{\partial \phi}{\partial z} \quad \text{(Ohm's law)};\tag{30.3}$$

$$\text{Fluid flow} \qquad J_z = -C \frac{\partial p}{\partial z} \quad \text{(Poiseuille's law)};\tag{30.4}$$

$$\text{Diffusion} \qquad J_z = -D \frac{\partial \tilde{N}}{\partial z} \quad \text{(Fick's law)}.\tag{30.5}$$

In these equations, κ_T is the thermal conductivity coefficient, κ is the electrical conductivity, C is a frictional coefficient related to the viscosity, and D is the diffusion coefficient.

Consider the flow of heat from one end of a metal bar to the other. If the hot end is at $z = 0$ and the cold end is at Z, then in the steady state the temperature as a function of z appears as shown in Fig. 30.1. The value of $\partial T/\partial z$ is negative, so that $(-\partial T/\partial z)$ is positive. The heat flow is in the positive direction (from the hot to the cold end). If we define

$$(X_T)_z = -\frac{\partial T}{\partial z}\tag{30.6}$$

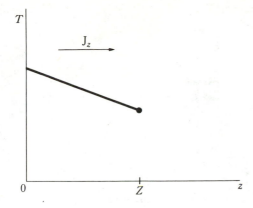

Figure 30.1   Temperature as a function of z.

as the "force" that is driving the heat flow, we can write Eq. (30.2) as

$$J_z = \kappa_T(X_T)_z \quad \text{(Fourier's Law)}, \tag{30.7}$$

and can say that the flow has the same sign as the force. Thus any of the transport laws, Eqs. (30.2), (30.3), (30.4), and (30.5), can be written in the general form

$$J = LX, \tag{30.8}$$

which says that the flow of any quantity is proportional to the force that drives the flow. This proportionality between the flux (the flow) and the force is called a linear law. The force is always the negative gradient of an intensive quantity.

These laws are well established experimentally. They were proposed initially as empirical laws, generalizations from experiment. It is our aim now to give these laws an interpretation in terms of the structure of the substance.

30.3 THE GENERAL EQUATION FOR TRANSPORT

If any physical quantity is transported, the amount transported through unit area in unit time is the number of molecules passing through the unit area in unit time multiplied by the amount of the physical quantity carried by each molecule. For any transport

$$j = N'q, \tag{30.9}$$

where j is the flow per m^2 sec, N' is the number of carriers passing through one square metre in one second, and q is the amount of the physical quantity possessed by each carrier. By calculating N' and q, we obtain the value of j. We begin with N'.

How many molecules pass the base, 1 m^2, of the parallelepiped in Fig. 30.2 in unit time? If all the molecules were moving downward with an average velocity $\langle c \rangle$, then each travels a distance $\langle c \rangle \, dt$ in the time interval dt. Therefore all the molecules in the parallelepiped of height $\langle c \rangle \, dt$ will pass the bottom face in the interval dt. The volume of the parallelepiped is $\langle c \rangle \, dt$ m^3; if $\tilde{N}$ is the number of molecules per cubic metre, then the number crossing the base in dt is $\tilde{N}\langle c \rangle \, dt$. In unit time the number crossing 1 m^2 area is

$$n' = \tilde{N}\langle c \rangle. \tag{30.10}$$

The expression for the flow, Eq. (30.9), becomes

$$j = \tilde{N}\langle c \rangle q, \tag{30.11}$$

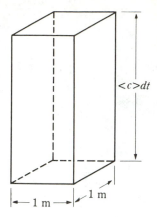

$<c>dt$

|← 1 m →|← 1 m

Figure 30.2

which is applicable to any transport process; the flow is equal to the product of the number of carriers per unit volume, the average velocity in the direction of the flow, and the amount of the physical quantity carried by each.

If not all, but only a fraction, α, of the molecules are moving downward, then the expression on the right side of Eq. (30.11) must be multiplied by that fraction!

$$j = \alpha \tilde{N} \langle c \rangle q. \tag{30.11a}$$

30.4 THERMAL CONDUCTIVITY IN A GAS

Suppose that two large metal plates parallel to the xy-plane and separated by a distance Z are at temperatures T_1 and T_2, the hotter plate (T_2) being the upper one. After some time a steady state will be established in which there is a downward flow of heat at a constant rate. This flow of heat results from the fact that the molecules at the upper levels have a greater thermal energy than those at the lower levels; the molecules moving downward carry more energy than do those moving upward.

To calculate the net energy flow in unit time through 1 m^2 parallel to the xy-plane, we imagine a large number of horizontal layers in the gas, each successive layer being at a slightly higher temperature than the one below it. The change in temperature with height is

$$\frac{\partial T}{\partial z} = \frac{\Delta T}{\Delta z} = \frac{T_2 - T_1}{Z - 0}, \tag{30.12}$$

if the lower plate lies at the position $z = 0$, the upper one at $z = Z$. The gradient, $\partial T/\partial z$, is constant, so at any height z the temperature is

$$T = T_1 + \left(\frac{\partial T}{\partial z}\right)z. \tag{30.13}$$

If the gas is monatomic with an average thermal energy $\langle \epsilon \rangle = \frac{3}{2}kT$, then the average energy of the molecules at the height z is

$$\langle \epsilon \rangle = \tfrac{3}{2}kT = \tfrac{3}{2}k\left[T_1 + \left(\frac{\partial T}{\partial z}\right)z\right]. \tag{30.14}$$

To calculate the heat flow, we consider an area 1 m^2 in a horizontal plane at the height z (Fig. 30.3). The energy carried by a molecule as it passes through the plane depends on the temperature of the layer of gas at which the molecule had its last opportunity to adjust its

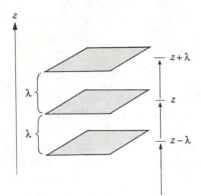

Figure 30.3

energy. This last adjustment occurred during the last collision with another molecule. Suppose that, on the average, the molecules have traveled a distance λ since their last collision. If the surface of interest lies at a height z, the molecules going down made their last collision at a height $z + \lambda$, while those going up made their last collision at a height $z - \lambda$ (Fig. 30.3). The molecules carry an amount of energy appropriate to the height where the last collision occurred. The downward flow of energy is, by Eqs. (30.11) and (30.14)

$$\epsilon\downarrow = \tfrac{1}{6}(\tilde{N}\langle c\rangle)_{z+\lambda}\tfrac{3}{2}k\left[T_1 + \left(\frac{\partial T}{\partial z}\right)(z + \lambda)\right],$$

while the upward flow is given by

$$\epsilon\uparrow = \tfrac{1}{6}(\tilde{N}\langle c\rangle)_{z-\lambda}\tfrac{3}{2}k\left[T_1 + \left(\frac{\partial T}{\partial z}\right)(z - \lambda)\right].$$

The factor $\tfrac{1}{6}$ appears since, on the average, only $\tfrac{1}{6}$ of the molecules are going down and only $\tfrac{1}{6}$ are going up. The net flow upward is denoted by J_ϵ and is

$$J_\epsilon = \epsilon\uparrow - \epsilon\downarrow. \tag{30.15}$$

Before writing out the equation in detail, we should note that if the gas is not to have net motion through the surface we require that the number of molecules going up in unit time must equal the number going down, so that

$$\tfrac{1}{6}(\tilde{N}\langle c\rangle)_{z+\lambda} = \tfrac{1}{6}(\tilde{N}\langle c\rangle)_{z-\lambda}, \tag{30.16}$$

which means that $\tilde{N}\langle c\rangle$ has the same value at every height.* Introducing the expressions for $\epsilon\uparrow$ and $\epsilon\downarrow$ into Eq. (30.15) and using Eq. (30.16), we obtain

$$J_\epsilon = \tfrac{1}{6}\tilde{N}\langle c\rangle\tfrac{3}{2}k\left(\frac{\partial T}{\partial z}\right)[z - \lambda - (z + \lambda)] = -\tfrac{1}{2}\tilde{N}\langle c\rangle k\lambda\left(\frac{\partial T}{\partial z}\right).$$

Comparing this result with the empirical law, Eq. (30.2), for thermal conductivity, $J_\epsilon = -\kappa_T(\partial T/\partial z)$, we obtain

$$\kappa_T = \tfrac{1}{2}\tilde{N}\langle c\rangle k\lambda = \frac{1}{3}\left(\frac{\tilde{N}}{N_A}\right)\bar{C}_v\langle c\rangle\lambda, \tag{30.17}$$

---

*   This is not quite correct, but to do the derivation without this assumption complicates matters considerably.

where $\bar{C}_v = \frac{3}{2}kN_A$ has been used. The factor $(\tilde{N}/N_A)$ is the concentration in moles per cubic metre. From Eq. (4.58) the average velocity is

$$\langle c \rangle = \sqrt{\frac{8kT}{\pi m}}. \tag{30.18}$$

It is therefore possible to calculate κ_T, the coefficient of thermal conductivity, if we can calculate λ, the average distance traveled by a molecule since its last collision. Or, looking at matters from the bright side, we can evaluate λ if the value of κ_T has been measured.

Unfortunately, there are several things about the above derivation that can be criticized. Both $\langle c \rangle$ and $\tilde{N}$ contain the temperature T, yet the temperature is different at different positions. Since the simple law of heat conduction is correct only if $T_2 - T_1$ is small compared with *either of the two temperatures*, it is sufficient to use the average temperature in computing $\tilde{N}$ and $\langle c \rangle$. A more serious objection is that we use quantities such as $\tilde{N}$ and $\langle c \rangle$ derived from the equilibrium distribution function and apply them to a nonequilibrium situation. The fact of the matter is that if a nonequilibrium distribution is used, the mathematical complication introduced is enormous. Happily, the result of the more accurate treatment is not substantially different but only changes the numerical constant $\frac{1}{3}$ in Eq. (30.17), assuming the absence of attractive forces. Finally, the distance λ has been introduced in a somewhat arbitrary way. To understand Eq. (30.17) we must have a more definite idea about λ.

30.5 COLLISIONS IN A GAS; THE MEAN FREE PATH

The *mean free path* λ of the molecule by definition is the average distance traveled between collisions. In one second, a molecule travels ($\langle c \rangle \times 1$ s) metres and makes Z collisions. Dividing the distance traveled by the number of collisions, we obtain the distance traveled between collisions:

$$\lambda = \frac{\langle c \rangle}{Z_1}. \tag{30.19}$$

To calculate λ we calculate Z_1.

Let σ be the diameter of the molecule and consider a cylinder (Fig. 30.4) of radius σ and height $\langle c \rangle$. In one second, the molecule travels a distance $\langle c \rangle$ and sweeps out the cylinder; it collides with all the molecules within the cylinder. (Because of collisions, the molecule follows a zig-zag path; this does not matter since the volume swept out is the same.) The number of molecules in the cylinder is $\pi\sigma^2\langle c \rangle \tilde{N}$; this is the number of collisions made by one molecule in one second. The formula $Z_1 = \pi\sigma^2\langle c \rangle \tilde{N}$ must be multiplied by the factor $\sqrt{2}$ to account for the fact that it is the average velocity along the line of centers of two molecules that matters and not the average velocity of a molecule. Consider two molecules that have their velocity vectors in the orientations shown in Fig. 30.5. For molecules moving in the same direction with the same velocity, the relative velocity of approach is zero. In the second case, where they approach head-on, the relative velocity of approach is $2\langle c \rangle$. If they approach at $90°$, the relative velocity of approach is the sum of the velocity components along the line joining the centers; this is $\frac{1}{2}\sqrt{2}\langle c \rangle + \frac{1}{2}\sqrt{2}\langle c \rangle = \sqrt{2}\langle c \rangle$. The third situation represents the average situation, so we write more exactly

$$Z_1 = \sqrt{2}\pi\sigma^2\langle c \rangle \tilde{N}. \tag{30.20}$$

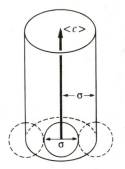

Figure 30.4 Volume swept out by a molecule in 1 sec.

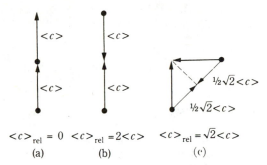

$<c>_{rel} = 0$ $<c>_{rel} = 2<c>$ $<c>_{rel} = \sqrt{2}<c>$
(a)   (b)   (c)

Figure 30.5 Relative velocity c_{rel} along the line of centers.

By combining Eqs. (30.19) and (30.20), the mean free path is

$$\lambda = \frac{1}{\sqrt{2}\pi\sigma^2 \tilde{N}}. \tag{30.21}$$

The mean free path depends on $1/\tilde{N}$ and is proportional to $1/p$ by the gas law $1/\tilde{N} = RT/N_A p$. The lower the pressure, the fewer collisions in unit time and the longer will be the mean free path.

Since there are $\tilde{N}$ molecules/m³ and each makes Z_1 collisions per second, the total number of collisions per cubic metre in one second is

$$Z_{11} = \tfrac{1}{2}Z_1\tilde{N} = \tfrac{1}{2}\sqrt{2}\pi\sigma^2\langle c\rangle\tilde{N}^2. \tag{30.22}$$

The factor $\tfrac{1}{2}$ is introduced because a simple multiplication of Z_1 by $\tilde{N}$ would count every collision twice. Without giving a detailed proof, the number of collisions in one cubic metre per second between unlike molecules in a mixture is

$$Z_{12} = \pi\sigma_{12}^2\sqrt{\frac{8kT}{\pi\mu}}\tilde{N}_1\tilde{N}_2, \tag{30.23}$$

where $\tilde{N}_1$ and $\tilde{N}_2$ are the numbers of molecules per cubic metre of kind 1 and kind 2, σ_{12} is the average of the diameters of the two kinds of molecules, and μ is the reduced mass, $1/\mu = 1/m_1 + 1/m_2$. These values for collision numbers will be useful later in the calculation of the rates of chemical reactions. A chemical reaction between two molecules can occur only when the molecules collide.

■ **EXAMPLE 30.1**   Estimate the number of collisions one molecule will make in a gas at 1 atm pressure and 25 °C. At 1 atm $= 10^5$ Pa and 25 °C,

$$\tilde{N} = \frac{pN_A}{RT} = \frac{10^5 \text{ Pa } (6 \times 10^{23} \text{ mol}^{-1})}{(8.3 \text{ J/K mol})(298 \text{ K})} \approx 2.5 \times 10^{25}/\text{m}^3,$$

Also we have $\sigma \approx 3 \times 10^{-10}$ m; $\langle c \rangle \approx 400$ m/s. Then

$$Z_1 = \sqrt{2}\pi\sigma^2\langle c\rangle\tilde{N} = 1.41(3.14)(3 \times 10^{-10} \text{ m})^2(400 \text{ m/s})(2.5 \times 10^{25}/\text{m}^3)$$
$$\approx 4 \times 10^9/\text{s}.$$

The molecule makes about 4 billion collisions per second.

■ **EXAMPLE 30.2**   Estimate the value of λ in a gas at 1 atm and 25 °C. From Example 30.1, $\tilde{N} = 2.5 \times 10^{25}/m^3$; $\sigma \approx 3 \times 10^{-10}$ m. Then

$$\lambda = \frac{1}{\sqrt{2}\,\pi\sigma^2\tilde{N}} = \frac{1}{(1.41)(3.14)(3 \times 10^{-10} \text{ m})^2(2.5 \times 10^{25}/\text{m}^3)} \approx 10^{-7} \text{ m} = 0.1 \ \mu\text{m}.$$

30.6   FINAL EXPRESSION FOR THE THERMAL CONDUCTIVITY

Having obtained an expression for the mean free path λ, we write the formula for the coefficient of thermal conductivity by combining Eqs. (30.21) and (30.17);

$$\kappa_T = \frac{\langle c \rangle \bar{C}_v}{3\sqrt{2}\pi N_A \sigma^2}. \tag{30.24}$$

This equation leads to the interesting conclusion that the thermal conductivity is independent of the pressure. This lack of dependence on pressure is a result of two compensating effects. By Eq. (30.17), κ_T is proportional to $\tilde{N}$ and to λ; but λ is inversely proportional to $\tilde{N}$ so that the product $\tilde{N}\lambda$ is independent of pressure. At lower pressures fewer molecules cross the surface in one second, but they come from a larger distance (λ is larger at lower p) and so carry a proportionately greater excess energy. Experiment confirms that κ_T is independent of pressure.

If $\bar{C}_v$ is independent of temperature, then everything on the right of Eq. (30.24) is constant except $\langle c \rangle$, which is proportional to $T^{1/2}$. Therefore, κ_T should increase as $T^{1/2}$. This is also confirmed experimentally.

In this derivation of the expression for κ_T, we have assumed that the pressure is high enough so that λ is much smaller than the distance separating the two plates. At very low pressures where λ is much larger than the distance between the plates, the molecule bounces back and forth between the plates and only rarely collides with another gas molecule. In this case the mean free path does not enter the calculation, and the value of κ_T depends on the separation of the plates. At these low pressures the thermal conductivity is proportional to the pressure, since it must be proportional to $\tilde{N}$, and λ does not appear in the formula to compensate for the pressure dependence of $\tilde{N}$.

30.7   VISCOSITY

The formula for the viscosity coefficient of a gas can be derived in a way similar to that used for heat conduction. We imagine two very large parallel flat plates, one lying in the xy-plane, the other at a distance Z above the xy-plane. We keep the lower plate stationary and pull the upper plate in the $+x$ direction with a velocity U. The viscosity of the gas exerts a drag on the moving plate. To keep the plate in uniform motion, a force must be applied to balance the viscous drag. Looking at the situation in another way, if the upper plate moves with a velocity U, the viscous force will tend to set the lower plate in motion. A force must be applied to the lower plate to keep it in place.

Again we suppose that the gas between the plates is made up of a series of horizontal layers. The layer next to the lower plate is immobile; as we move upward, each successive layer has a slightly larger component of velocity in the x direction, the topmost layer at the height Z having the velocity U. This type of flow, in which there is a regular gradation of velocity in passing from one layer to the next, is called *laminar* flow. The layer at the height z has a velocity in the x direction given by u_z:

$$u_z = \frac{\partial u}{\partial z} z. \tag{30.25}$$

At $z = Z$, $u = U$, so that

$$\frac{\partial u}{\partial z} = \frac{U}{Z}. \tag{30.26}$$

If we observe a layer at the height z, we see that molecules enter this layer from the neighboring layers. The molecules from the upper layers will bring extra x momentum to this layer, while those which come from below are deficient in x momentum. There is therefore a net downward flow of x momentum through the layer. Now we compute the rate of this flow through one square metre of the layer at the height z (Fig. 30.6).

The number of molecules passing downwards through one square metre per second is, by Eq. (30.10), $\frac{1}{6}\tilde{N}\langle c \rangle$, and as many come upward as come downward. The molecules that pass downward through the layer at z carry x momentum appropriate to the layer in which they made their last collision, the layer at height $z + \lambda$. This x momentum is

$$mu_{z+\lambda} = m\left(\frac{\partial u}{\partial z}\right)(z + \lambda).$$

So the momentum coming down through one square metre in one second is

$$(mu)\downarrow = \frac{1}{6}\tilde{N}\langle c \rangle m\left(\frac{\partial u}{\partial z}\right)(z + \lambda).$$

Similarly, the momentum coming up is

$$(mu)\uparrow = \frac{1}{6}\tilde{N}\langle c \rangle m\left(\frac{\partial u}{\partial z}\right)(z - \lambda),$$

since the molecules coming up adjusted their momentum in the layer at $z - \lambda$. The net downward flow of x momentum is

$$(mu)\downarrow - (mu)\uparrow = \frac{1}{3}\tilde{N}\langle c \rangle m\lambda \frac{\partial u}{\partial z}.$$

Since this quantity is independent of z, it must also be equal to the net x momentum transferred in one second to one square metre of the lower plate. Since the momentum transfer in

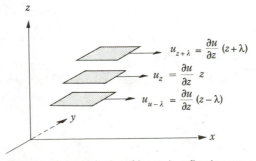

Figure 30.6 Velocity of layers in a flowing gas.

unit time is the force, the force acting in the x direction on one square metre of the lower plate is

$$f_x = \tfrac{1}{3}\tilde{N}\langle c\rangle m\lambda \frac{\partial u}{\partial z}. \tag{30.27}$$

To hold this plate stationary, we must apply an equal and opposite force f_{-x}, such that $f_x + f_{-x} = 0$. The viscosity coefficient, η, is defined by

$$f_{-x} = -\eta\frac{\partial u}{\partial z}. \tag{30.28}$$

The viscosity coefficient is the force that must be applied to hold the lower plate stationary if the velocity gradient $\partial u/\partial z$ is unity and the plate has unit area. Comparing Eqs. (30.27) and (30.28), we see that

$$\eta = \tfrac{1}{3}\tilde{N}\langle c\rangle m\lambda. \tag{30.29}$$

If the density of the gas is ρ, then $\rho = \tilde{N}m$, and

$$\eta = \tfrac{1}{3}\rho\langle c\rangle\lambda. \tag{30.30}$$

Again, the numerical factor $\tfrac{1}{3}$ is not quite correct, since the flow of gas produces a non-equilibrium situation. For elastic spheres, the factor should be $\tfrac{1}{2}$. The unit of the viscosity coefficient is 1 newton second per square metre $(\text{N s m}^{-2}) = 1$ pascal second $(\text{Pa s}) = 1\ \text{kg m}^{-1}\,\text{s}^{-1}$. (The cgs unit is 1 poise $= 1\ \text{g cm}^{-1}\,\text{s}^{-1} = 10^{-1}\ \text{kg m}^{-1}\,\text{s}^{-1}$.)

The coefficient of viscosity depends on the product $\tilde{N}\lambda$ and so is independent of pressure, Eq. (30.29). This rather surprising result, along with the similar result for thermal conductivity, was one of the great initial triumphs of the kinetic theory of gases. It seems as though the viscosity of a gas, which is a measure of its resistance to flow, ought to be greater at high pressures than at low. This contrary prediction of kinetic theory and the subsequent experimental verification gave great impetus to the further development of the theory.

Comparison of Eqs. (30.29) and (30.17) shows that since $M = N_A m$,

$$\frac{\kappa_T}{\eta} = \frac{\bar{C}_v}{M}. \tag{30.31}$$

This ratio is the heat capacity per unit mass. More accurate theory, as well as experiment, shows that for monatomic gases

$$\frac{\kappa_T}{\eta} = 2.5\,\frac{\bar{C}_v}{M}, \tag{30.32}$$

is more nearly correct.

30.8 MOLECULAR DIAMETERS

Using $\langle c\rangle = \sqrt{8kT/\pi m}$ and λ from Eq. (30.21) in Eq. (30.29), we obtain

$$\eta = \frac{2\sqrt{mkT}}{3\pi^{3/2}\sigma^2} = \frac{2\sqrt{MRT}}{3\pi^{3/2}N_A\sigma^2}, \tag{30.33}$$

which expresses η in terms of M, T, and the quantity $N_A\sigma^2$. If we know N_A, the value of the molecular diameter can be calculated from the measured values of η.

Table 30.1

Gas	$t/°C$	$\eta/\mu Pa\ s$	σ/pm
NH_3	20	9.82	361
CO_2	20	14.80	373
Ar	20	22.17	297
C_2H_4	20	10.08	404
CH_4	20	10.87	338

Alternatively, if another expression involving N_A and σ is available, values of both N_A and σ can be determined. In Section 26.8, we related N_A and σ to the van der Waals b. From Eq. (26.40), we have

$$b = \tfrac{2}{3}\pi N_A \sigma^3. \tag{30.34}$$

Eliminating N_A between Eqs. (30.33) and (30.34) and solving for σ, we have

$$\sigma = \tfrac{9}{4}\eta b \sqrt{\frac{\pi}{MRT}}. \tag{30.35}$$

Using this result in Eq. (30.33), we obtain

$$N_A = \left(\frac{32}{243\pi b^2}\right)\left(\frac{MRT}{\pi\eta^2}\right)^{3/2}. \tag{30.36}$$

It was from Eqs. (30.35) and (30.36) that the first concrete estimates of N_A and σ were obtained. In Table 30.1 we list values of σ, calculated from η and the currently accepted value of N_A using Eq. (30.33).

■ **EXAMPLE 30.3** If the viscosity coefficient of CO_2 is 14.80 $\mu Pa\ s$ at 20 °C, what is the molecular diameter? By Eq. (30.33), we have

$$\sigma^2 = \frac{2\sqrt{MRT}}{3\pi^{3/2}N_A\eta} = \frac{2[(0.04401\ kg/mol)(8.314\ J/K\ mol)(293.15\ K)]^{1/2}}{3(3.1416)^{3/2}(6.022 \times 10^{23}/mol)(14.80 \times 10^{-6}\ Pa\ s)}$$

$$= 13.91 \times 10^{-20}\ m^2;$$

$$\sigma = 3.730 \times 10^{-10}\ m = 373.0\ pm.$$

30.9 DIFFUSION

If the concentration is not uniform in a mixture of two gases, the gases diffuse into one another until the composition is uniform. The derivation of the diffusion coefficient in such a situation is lengthy and somewhat complicated, since each gas has a different value for $\langle c \rangle$ and for λ. To simplify matters we treat the case of a single gas so that there is only one value of $\langle c \rangle$ and of λ. The result obtained is very nearly correct for the diffusion of one isotope into another. To define the problem suppose that some of the molecules of the gas are painted red; the gas is confined in a vertical tube and the number $\tilde{N}_r$ of red molecules per cubic metre is greater at one end than at the other; then the number $\tilde{N}$ of unpainted molecules per cubic metre must also vary from one end to the other if the total pressure is to be uniform throughout the tube. For each species we write for the number per m^3 at the height z:

$$\tilde{N} = \tilde{N}_{\mathrm{o}} + \frac{\partial \tilde{N}}{\partial z} z, \qquad \tilde{N}_{\mathrm{r}} = \tilde{N}_{\mathrm{ro}} + \frac{\partial \tilde{N}_{\mathrm{r}}}{\partial z} z,$$

where $\tilde{N}_{\mathrm{o}}$ and $\tilde{N}_{\mathrm{ro}}$ are the numbers per cubic metre at $z = 0$.

Consider a horizontal area of 1 m^2 at the height z. The number of red molecules passing downward through this area per second is

$$\tilde{N}_{\mathrm{r}}\!\downarrow = \tfrac{1}{6}\langle c \rangle (\tilde{N}_{\mathrm{r}})_{z+\lambda} = \tfrac{1}{6}\langle c \rangle \left[\tilde{N}_{\mathrm{ro}} + \frac{\partial \tilde{N}_{\mathrm{r}}}{\partial z}(z + \lambda) \right],$$

since the molecules originate in the layer at $z + \lambda$. Similarly the number coming up from below is

$$\tilde{N}_{\mathrm{r}}\!\uparrow = \tfrac{1}{6}\langle c \rangle (\tilde{N}_{\mathrm{r}})_{z-\lambda} = \tfrac{1}{6}\langle c \rangle \left[\tilde{N}_{\mathrm{ro}} + \frac{\partial \tilde{N}_{\mathrm{r}}}{\partial z}(z - \lambda) \right].$$

The net flow upward is

$$\tilde{N}_{\mathrm{r}}\!\uparrow - \tilde{N}_{\mathrm{r}}\!\downarrow = -\tfrac{1}{3}\langle c \rangle \lambda \frac{\partial \tilde{N}_{\mathrm{r}}}{\partial z}.$$

By the law for diffusion, Eq. (30.5), the upward flow is $-D_{\mathrm{r}}(\partial \tilde{N}_{\mathrm{r}}/\partial z)$, where D_{r} is the diffusion coefficient of the red molecules. Thus we have $D_{\mathrm{r}} = \tfrac{1}{3}\langle c \rangle \lambda$; but the red molecules differ from the others only by a coat of paint, so that

$$D = \tfrac{1}{3}\langle c \rangle \lambda. \qquad (30.37)$$

(The numerical factor $\tfrac{1}{3}$ is wrong as usual!) Since $\langle c \rangle$ is inversely proportional to $M^{1/2}$, we can understand *Graham's law* which states that the rate of diffusion of a gas is inversely proportional to the square root of the molecular weight.

Since the mean free path is inversely proportional to the pressure, the diffusion coefficient decreases with increase in pressure. The molecules have to fight their way through the swarm of other molecules by making many collisions. At high pressures they make many more collisions, and their progress in any given direction is slowed. This refutes an early objection to the kinetic theory, which was that the high molecular velocities predicted by kinetic theory were obviously ridiculous since, if the molecules moved that quickly, the smell of a gas such as NH_3 or H_2S released in one corner of a room should be noticed instantly everywhere in the room, while in fact it takes some time before the odor is detected in another part of the room. In answer it was pointed out that at ordinary pressures the gas molecule makes many collisions and the path of a molecule is a fantastic zig-zag with little net motion in any particular direction in spite of the high velocity.

An important application of the concept of the mean free path and its relation to diffusion was made by Irving Langmuir. In the ordinary incandescent lamp, the passage of the current heats a tungsten filament white hot. To prevent oxidation of the filament and immediate burnout, the bulb must be evacuated. However, if the pressure is reduced too far, the mean free path of the tungsten atom becomes large compared with the size of the bulb. Tungsten atoms that are boiled off the filament can go directly to the glass wall without an intervening collision with a gas molecule. The atoms condense on the glass wall, blacken the bulb, and weaken the filament, which soon breaks. Langmuir introduced argon under a few centimetres pressure. This reduces the mean free path to something less than the diameter of the filament. In this situation, a tungsten atom that has been boiled off travels only a short distance before it hits a gas molecule. A likely result of this collision is that the tungsten atom is reflected back onto the filament. In any event the tungsten atoms must leave the region of the filament by diffusion through the argon, which is slow. The presence of argon in the bulb lengthens bulb life enormously.

30.10 SUMMARY OF TRANSPORT PROPERTIES IN A GAS

Kinetic theory interprets the phenomenological laws of transport in gases on the basis of a single mechanism, and expresses the values of κ_T, η, and D in terms of the mean free path, the density, and the average velocity of the molecules. The equations are

$$\kappa_T = \tfrac{1}{3}\tilde{N}\langle c \rangle (\bar{C}_v/N_A)\lambda, \qquad \eta = \tfrac{1}{3}\tilde{N}\langle c \rangle m\lambda, \qquad D = \tfrac{1}{3}\langle c \rangle \lambda.$$

Since all of these depend on λ, they are sometimes called *free path* phenomena.

★ 30.11 THE NONSTEADY STATE

In the preceding sections we assumed that the flow was in a steady state, where the amount of a quantity flowing into any volume element is balanced by an equal amount of the quantity flowing out in the same time interval. For diffusion this means that the concentration in any volume element is independent of time, $\partial \tilde{N}/\partial t = 0$. For thermal conductivity it means that energy does not accumulate in any volume element, or that $\partial T/\partial t = 0$.

To treat diffusion in the nonsteady state, we consider the situation shown in Fig. 30.7. Molecules diffuse in the $+x$ direction through two elements of area, each being 1 m², perpendicular to the x-axis and located at x and $x + \Delta x$. The flow through the element at x is J_x, that through the element at $x + \Delta x$ is $J_{x+\Delta x}$. The element enclosed by the parallelepiped has a volume equal to $1\text{ m}^2 \cdot \Delta x\text{ m} = \Delta x\text{ m}^3$.

In the time dt, the number of molecules entering the volume element from the left is $J_x\,dt$, while the number leaving at $x + \Delta x$ is $J_{x+\Delta x}\,dt$. If the increase in the number of molecules in the volume element in the interval dt is ΔN, the excess of what flows in at x over what flows out at $x + \Delta x$, then $\Delta N = J_x\,dt - J_{x+\Delta x}\,dt$. But $J_{x+\Delta x} = J_x + (\partial J_x/\partial x)\,\Delta x$, so that

$$\Delta N = -\frac{\partial J_x}{\partial x}\,\Delta x\,dt.$$

The increase in concentration in the volume element is $d\tilde{N} = \Delta N/\Delta x$, so that

$$d\tilde{N} = -\frac{\partial J_x}{\partial x}\,dt, \qquad \text{or} \qquad \frac{\partial \tilde{N}}{\partial t} = -\frac{\partial J_x}{\partial x}. \tag{30.38}$$

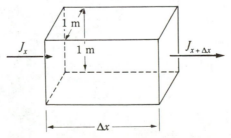

Figure 30.7 The nonsteady state of flow.

By Eq. (30.5), $J_x = -D(\partial \tilde{N}/\partial x)$. Using this in Eq. (30.38), we obtain

$$\frac{\partial \tilde{N}}{\partial t} = \frac{\partial}{\partial x}\left(D\,\frac{\partial \tilde{N}}{\partial x}\right),$$

or, if D is independent of x,

$$\frac{\partial c}{\partial t} = D\,\frac{\partial^2 c}{\partial x^2}, \tag{30.39}$$

in which we have replaced $\tilde{N}$ by c to emphasize that this equation does not depend on the concentration unit used.

The solution of Eq. (30.39), Fick's second law of diffusion, under a specified set of conditions, yields the concentration as a function of x and t. From Eq. (30.39) we discover that the condition for the steady state, $\partial c/\partial t = 0$, implies that $\partial^2 c/\partial x^2 = 0$ or $\partial c/\partial x = $ constant. Therefore in the steady state the concentration varies linearly with the coordinate.

By a similar argument using the equation for heat conduction, we obtain

$$\frac{\partial T}{\partial t} = \frac{\kappa_T}{\rho c_v}\,\frac{\partial^2 T}{\partial x^2}, \tag{30.40}$$

where the factor ρc_v appears when the energy increment in the volume element is converted to a temperature increment; c_v is the heat capacity per unit mass and ρ is the density. Equations (30.39) and (30.40) are formally the same. Solving a problem in diffusion solves an analogous problem in heat conduction.

★ 30.12 THE POISEUILLE FORMULA

The rate at which a fluid flows through a tube depends on the dimensions (radius and length) of the tube, the viscosity of the fluid, and the pressure drop between the ends of the tube. To discover the relation between these quantities, we first calculate the volume passing any point in a circular tube in unit time.

In narrow circular tubes the flow is laminar so that the cylindrical sheath at the boundary of the tube is stationary; as we move to the center each successive cylindrical sheath moves with a slightly larger velocity. Suppose the tube lies with its length along the x-axis. Then consider the sheath shown in Fig. 30.8, having an inner radius r and an outer radius $r + dr$. If the velocity of this sheath in the x direction is v m/s, then in one second the sheath moves v m and carries all the fluid in it past a given point. The volume passing any point in unit time is $2\pi r v\, dr$. The total volume passing any point in unit time is $\dot{V}$, and is the

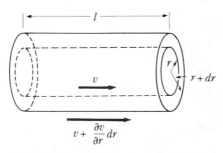

Figure 30.8   Flow in a cylindrical layer.

sum of the contribution of every sheath in the tube. Therefore

$$\dot{V} = \int_0^a 2\pi r v \, dr, \qquad (30.41)$$

where a is the radius of the tube. To obtain $\dot{V}$, the volume delivered by the tube in unit time, v must be known as a function of r.

The relation between v and r is obtained by balancing the forces due to the pressure difference and the viscosity. Let the pressure on the left end of the tube be p_1 and that on the right end be p_2. The force acting on the left end of the sheath is $p_1 2\pi r \, dr$, that on the right end is $p_2 2\pi r \, dr$. The net force in the $+x$ direction, f_x, due to the pressure difference is

$$f_x = (p_1 - p_2) 2\pi r \, dr. \qquad (30.42)$$

Each square meter of the inner surface of the sheath is subject to a viscous force in the $+x$ direction equal to $-\eta(\partial v/\partial r)$. If the area of the inner surface is $S = 2\pi r l$, then the total force acting on the inner surface is $-\eta S(\partial v/\partial r)$. This inner surface is being pulled along by the faster moving interior cylinder. The outer surface of the sheath is *retarded* by the slower moving fluid outside the sheath, the force in the x direction on the outer surface being

$$\eta S \frac{\partial v}{\partial r} + d\left(\eta S \frac{\partial v}{\partial r}\right).$$

The net viscous force is the sum of the forces on the inner and outer surfaces, f'_x:

$$f'_x = d\left(\eta S \frac{\partial v}{\partial r}\right). \qquad (30.43)$$

For balance, the sum of the forces in the $+x$ direction due to pressure difference and viscous forces must be zero: $f_x + f'_x = 0$. Using Eqs. (30.42) and (30.43) and rearranging, we obtain

$$d\left(\eta S \frac{\partial v}{\partial r}\right) = -2\pi r(p_1 - p_2) \, dr,$$

which integrates immediately to

$$\eta S \frac{\partial v}{\partial r} = -\pi(p_1 - p_2)r^2 + A,$$

where A is the integration constant. When we use the value of $S = 2\pi r l$, this becomes

$$\frac{\partial v}{\partial r} = -\frac{(p_1 - p_2)r}{2\eta l} + \frac{A}{2\pi\eta l r}.$$

Integrating again, we obtain

$$v = -\frac{(p_1 - p_2)r^2}{4\eta l} + \frac{A}{2\pi\eta l} \ln r + B, \qquad (30.44)$$

where B is another integration constant. Now the velocity must be finite at $r = 0$, and this is not possible if the logarithmic term appears in Eq. (30.44); therefore it must be that $A = 0$. Then

$$v = -\frac{(p_1 - p_2)r^2}{4\eta l} + B.$$

At the radius of the tube, $r = a$, the velocity of the fluid is zero, so we have

$$0 = -\frac{(p_1 - p_2)a^2}{4\eta l} + B.$$

Using this value of B, we can write the velocity as

$$v = \frac{(p_1 - p_2)(a^2 - r^2)}{4\eta l}, \tag{30.45}$$

which expresses the velocity as a function of r, a relation which is required to evaluate the volume delivered in unit time. Using this value of v in the integral of Eq. (30.41), we obtain

$$\dot{V} = \frac{\pi(p_1 - p_2)}{2\eta l} \int_0^a (a^2 - r^2)r \, dr = \frac{\pi a^4 (p_1 - p_2)}{8\eta l}, \tag{30.46}$$

which is Poiseuille's formula; it has been verified quite accurately for fluid flow through tubes for which $a \ll l$. Knowing the radius and length of the tube, and the pressure difference, we can calculate the value of η from the measured volume of liquid discharged in unit time. Conversely, if η is known the radius of the tube can be calculated from the volume discharged; this is useful for measuring the average cross section of a fine capillary tube.

Since the pressure gradient $\partial p/\partial x = (p_2 - p_1)/l$, Eq. (30.46) can be written in the form

$$\dot{V} = -\frac{\pi a^4}{8\eta} \frac{\partial p}{\partial x}, \tag{30.47}$$

which is again Poiseuille's law; compare this with Eq. (30.4).

30.13 THE VISCOSIMETER

The viscosimeter is an instrument for determining viscosity by measuring the time required for a fixed volume of a liquid to flow through a capillary tube, the efflux time. A simple viscosimeter is shown in Fig. 30.9. Two bulbs are connected by a length of capillary tubing.

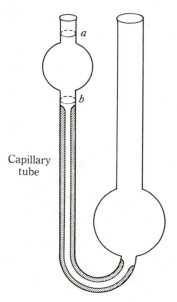

Figure 30.9   Simple viscosimeter.

The liquid is forced into the left-hand limb until it rises above the mark at a. It is then allowed to flow into the lower bulb. The time required for the liquid level to drop from a to b is measured. This is the time required for a fixed volume of liquid to flow through the capillary. The pressure difference varies with time during efflux but is proportional to the density ρ of the liquid. So if two different liquids are compared in the same viscosimeter, we have

$$\frac{\eta_1}{\eta_2} = \frac{\rho_1 t_1}{\rho_2 t_2}. \tag{30.48}$$

This is a convenient method for measuring the viscosity of one liquid relative to that of another.

The temperature dependence of the viscosity is quite different for liquids and gases. The simple derivation for gases predicts a proportionality of η to $\sqrt{T}$, and this is observed experimentally. In contrast, the viscosity of liquids *decreases* with increase in temperature. An equation, first proposed empirically and later given a theoretical foundation, represents the data reasonably well:

$$\ln \eta = a + \frac{E}{RT},$$

where a and E are constants. We may write this in the alternative form

$$\eta = Ae^{E/RT}; \tag{30.49}$$

E is called the *activation* energy for flow. We will appreciate more fully the significance of Eq. (30.49) after studying the rates of chemical reactions.

QUESTIONS

30.1 Describe the common features of the transport laws, Eqs. (30.2) through (30.5).

30.2 Why is the mean free path inversely proportional to σ^2 and to $\tilde{N}$?

30.3 What inequality between λ and the size of the container is required for the validity of the transport laws? Why?

30.4 The damping rate of a pendulum in a gas is predicted to be proportional to the viscosity η when λ is much smaller than the radius of the pendulum bob. How will the damping rate depend on the density?

30.5 The mechanism of momentum transfer in liquids involves the continual action of intermolecular forces rather than the free flight mechanism in gases. Qualitatively explain the different behavior of η with T on this basis.

30.6 Which of the gas transport coefficients will show an isotope effect, that is, depend on the molecular mass?

PROBLEMS

30.1 If the molecular diameter of H_2 is 0.292 nm, calculate the number of collisions made by a hydrogen molecule in 1 second

  a) if $T = 300$ K and $p = 1$ atm,
  b) if $T = 500$ K and $p = 1$ atm,
  c) if $T = 300$ K and $p = 10^{-4}$ atm.
  d) Calculate the total number of collisions per second occurring in 1 cm^3 for each of the cases in (a), (b), (c).

30.2 The molecular diameter of N_2 is 0.368 nm.

    a) Compute the mean free path of N_2 at 300 K and 1 atm, 0.1 atm, 0.01 atm.

    b) A reasonably good vacuum system achieves a pressure of about 10^{-9} atm. What is the mean free path at this pressure?

    c) If the diameter of the evacuated tube ($p = 10^{-9}$ atm) is 5 cm, how many times does the molecule strike the walls between two successive collisions with other gas molecules?

30.3 The collision diameters of hydrogen and nitrogen are 292 pm and 368 pm respectively. At 20 °C and 1 atm pressure,

    a) calculate the mean free path of the hydrogen molecule.

    b) How much larger is the mean free path of the hydrogen molecule compared to that of the nitrogen molecule?

30.4 Consider a gas at constant temperature. If the pressure is doubled, what effect does this have on:

    a) the number of collisions per second made by any one molecule;

    b) the total number of collisions per second occurring in 1 m^3 of gas;

    c) the mean free path of a gas molecule;

    d) the viscosity of the gas?

30.5 Suppose that there are 20 couples on a dance floor which is 50 ft × 50 ft. If the diameter of each couple is 2 ft and their velocity is 2 ft/s, derive formulas for, and then calculate the mean free path, the number of collisions per minute made by each couple and the total number of collisions per minute. (Assume the motion is chaotic.)

30.6 Compare the thermal conductivities of O_2 and H_2; ignore the difference in molecular diameter. Both have $\bar{C}_v = \frac{5}{2}R$.

30.7 Two parallel plates 0.50 cm apart are maintained at 298 K and 301 K. The space between the two plates is filled with H_2, $\sigma = 0.292$ nm, $\bar{C}_v = \frac{5}{2}R$. Calculate the heat flow between the two plates in W/cm^2.

30.8 Ethane has a molar mass of 30 g/mol, compared with 28 g/mol for N_2 and 32 g/mol for O_2. The molecular diameter is not greatly different from that of oxygen or nitrogen. The thermal conductivity of ethane is significantly larger than that of O_2 or N_2. Explain.

30.9 Since argon and neon are both monoatomic gases, they have the same heat capacity, $\bar{C}_v = \frac{3}{2}R$. The thermal conductivities at 20 °C are: neon, 11.07 mW/m K; argon, 5.236 mW/m K. (a) Calculate the ratio of the molecular diameter of argon to that of neon. (b) Calculate the molecular diameter of neon. (Use the $\frac{1}{3}$ factor rather than the exact one.)

30.10 The viscosity coefficient of methane at 280 K is 10.53×10^{-6} Pa s. Calculate the molecular diameter.

30.11 The thermal conductivity of silver is 431 W/m K. Calculate the heat flow per second through a silver disc 0.1 cm in thickness and having 2 cm^2 area if the temperature difference between the two sides of the disc is 10 K.

30.12 Fiberglass batts have a thermal conductivity of 4.6×10^{-2} W/m K. Calculate the heat flow per square meter through a batt 15 cm thick, if the temperature difference between the two sides is 10 °C.

30.13 One wall of a house has an area of 25 m^2. Moving from the inside to the outside, the wall consists of a $\frac{5}{8}$ in layer of plaster, a $3\frac{1}{2}$ in fiberglass batt, and a 4 in brick facing. The thermal conductivities are: fiberglass, 4.6×10^{-2} W/m K, brick $\approx$ plaster $\approx$ 0.60 W/m K. If the interior temperature is 20 °C and the exterior temperature is 0 °C, calculate

    a) the total rate of heat loss through the wall in watts.

    b) the temperatures at the plaster–fiberglass and the fiberglass–brick interfaces.

    (*Note*: The contributions of the stagnant air layers at the various interfaces, the wooden studs, and the exterior sheathing to the thermal resistance have been neglected in this idealized calculation.)

30.14 One end of a capillary tube 10 cm long is connected horizontally through the side of a bottle. The bottle is filled with water, (density $= 1.00 \, \text{g/cm}^3$; $\eta = 1.00 \times 10^{-3} \, \text{Pa s}$) to a depth of 25 cm above the capillary tube. The water drains by gravity through the tube. To collect 200 cm^3 of water requires 40.8 s. What is the diameter of the capillary tube? Assume the depth of the water in the bottle does not change appreciably.

30.15 A barrel is filled with olive oil ($\eta = 0.0840 \, \text{Pa s}$; density $= 0.918 \, \text{g/cm}^3$) to a depth of 1 meter. The oil flows by gravity into a bottle through a pipe attached through the side of the barrel at the bottom. The inner diameter of the pipe is 13 mm; its length is 20 cm. What time is required to collect one litre of oil in the bottle?

30.16 Oil flows from a storage tank through a 1.3 cm (inner diameter) pipe, 15 m long. The pressure difference between the two ends of the pipe is 4 atm. For the oil, $\eta = 0.40 \, \text{Pa s}$.

a) What is the flow rate through the pipe?
b) What would the flow rate be if the pipe had a 1.6 cm inner diameter?
c) If the pipe had a 1.3 cm inner diameter and was 30 m long, what would the flow rate be?

30.17 Two copper pipes, each 3 m long, the first having 2.6 cm and the second having 1.3 cm inner diameter, are connected in series. A pressure of 5 atm is supplied at the opening of the wider pipe and oil exits from the narrow end at a pressure of 1 atm. For the oil, $\eta = 0.114 \, \text{Pa s}$ at 15 °C.

a) Calculate the pressure at the point where the two pipes are joined.
b) How many litres per minute of oil can be delivered by this combination?

30.18 For the flow of a compressible fluid such as a gas, instead of Poiseuille's law we have $\dot{n} = \pi a^4 (p_1^2 - p_2^2)/16\eta RTl$, where $\dot{n}$ is the number of moles per second of gas passing through the capillary tube. What time is required to pass 200 mL of hydrogen (at 20 °C and 1.05 atm) through a capillary tube 10 cm long and 0.30 mm in diameter against an outlet pressure of 1.00 atm? ($\eta = 8.8 \times 10^{-6} \, \text{Pa s}$).

30.19 Consider the flow through a cylindrical sheath of inner radius a and outer radius b. The flow per second is given by Eq. (30.41), where the limits of integration are a and b. The velocity is given by Eq. (30.44), but in this case $A \neq 0$. The constants A and B are determined by the conditions that $v = 0$ at $r = a$ and $v = 0$ at $r = b$. Derive the formula corresponding to Poiseuille's equation for this case.

30.20 The densities of acetone and water at 20 °C are 0.792 g/cm^3 and 0.9982 g/cm^3, respectively. The viscosity of water is $1.002 \times 10^{-3} \, \text{Pa s}$ at 20 °C. If water requires 120.5 s to run between the marks on a viscosimeter and acetone requires 49.5 s, what is the viscosity of acetone?

30.21 The viscosities of acetone are

$t/°C$	-60	-30	0	30
$\eta/10^{-3} \, \text{Pa s}$	0.932	0.575	0.399	0.295

By plotting $\ln \eta$ versus $1/T$, determine the value of E in Eq. (30.49).

30.22 The diffusion coefficient for urea through a membrane is 0.97 cm^2/day. If the thickness of the membrane is 0.025 cm and the concentration of the urea solution is 0.040 mol/L on one side of the membrane while the concentration on the other side is kept at zero, what is the rate in mol/cm^2 day at which urea passes through the membrane?

30.23 Hydrogen gas diffuses through a palladium foil, 0.0050 cm thick. On the left side of the foil, the hydrogen is maintained at 25.0 °C and a pressure of 750 mm, while on the right side a good vacuum is maintained. After 24 hours the volume of hydrogen in the left compartment decreased by 14.1 cm^3. If the area of the foil through which the diffusion occurs is 0.743 cm^2, what is the diffusion coefficient of hydrogen in palladium?

31

Electrical Conduction

31.1 ELECTRICAL TRANSPORT

The quantity of electrical charge that passes any point in a conductor in unit time is the *current*. The current passing through unit area perpendicular to the direction of flow is the current density j. By the general law of transport, the current density in the x direction is proportional to the potential gradient, Eq. (30.3),

$$j = -\kappa \frac{\partial \phi}{\partial x}. \tag{31.1}$$

The constant of proportionality κ is the *conductivity* of the substance. The electric field E is defined by $E = -\partial\phi/\partial x$, so Eq. (31.1) can be written in the form

$$j = \kappa E. \tag{31.2}$$

Equations (31.1) and (31.2) are expressions of Ohm's law.

To transform Ohm's law into a more familiar form, we consider a conductor of length l and cross-sectional area A. If the electric potential difference across the ends is $\Delta\phi = \phi_2 - \phi_1$, then $E = (\phi_2 - \phi_1)/l = \Delta\phi/l$. The current I carried by the conductor is related to the current density by $I = jA$. Using these expressions for E and j in Eq. (31.2), we obtain

$$I = \frac{\kappa A \Delta\phi}{l}. \tag{31.3}$$

We define the *conductance* $L = \kappa A/l$. Then

$$I = L\,\Delta\phi. \tag{31.4}$$

The *resistance* R of the conductor is defined by $R = 1/L = l/\kappa A = \rho l/A$, where the *resistivity* $\rho = 1/\kappa$. This definition brings Ohm's law, Eq. (31.4), into its familiar form

$$\Delta\phi = IR. \tag{31.5}$$

Table 31.1

Names, symbols, and units for electrical quantities

Name	Symbol	SI unit	Abbreviation for SI unit
Current	I	ampere	A
Current density	j	ampere per square metre	A/m^2
Electric potential	ϕ	volt	V
Electric potential difference	$\Delta\phi$	volt	V
Electric field	E	volt per metre	V/m
Resistance	$R = \Delta\phi/I$	ohm = volt per ampere	Ω = V/A
Conductance	$L = R^{-1}$	siemens = ohm^{-1}	S = Ω^{-1}
Resistivity	$\rho = RA/l$	ohm metre	Ω m
Conductivity	$\kappa = \rho^{-1}$	ohm^{-1} metre^{-1} = siemens per metre	S/m
Molar conductivity	$\Lambda = \kappa/\tilde{c}$	siemens square metre per mole	S m^2/mol
Faraday constant	$F = 96\,484.56$ C/mol	coulomb per mole	C/mol
Velocity	v	metre per second	m/s
Mobility	$u = v/E$	(metre per second) per (volt per metre)	m^2/s V
Mobility (generalized)	$\tilde{u}$	(metre per second) per newton	m/s N
Magnetic flux density	B	volt second per square metre = tesla	T = V s/m^2
Viscosity coefficient	η	pascal second = kilogram per metre second = newton second per square metre	Pa s = kg/m s = N s/m^2

By putting the definition of the resistivity into Eq. (31.2), we obtain an analogue of Eq. (31.5):

$$E = j\rho. \tag{31.6}$$

Ordinarily we will use Ohm's law in the form of Eq. (31.2) or Eq. (31.6). This is convenient since κ and ρ are properties of the material composing the conductor and do not depend on its geometry. The resistance depends on the geometry of the conductor through the relation

$$R = \frac{\rho l}{A}. \tag{31.7}$$

Lengthening the conductor increases its resistance, while thickening it decreases its resistance. The symbols and units for these electrical quantities are summarized in Table 31.1.

31.2 CONDUCTION IN METALS

The current in metals is carried entirely by the electrons, each of which carries a negative charge e. Using Eq. (30.11) for the flow, the product of the number of electrons per cubic metre, their average velocity in the direction of the flow, and their charge, we obtain

$$j = \tilde{N}ve. \tag{31.8}$$

Combining this result with Eq. (31.2), the expression for the conductivity becomes

$$\kappa = \frac{\tilde{N}ve}{E}. \tag{31.9}$$

Ohm's law requires that κ be a constant; it must be independent of the field E. Therefore one of the quantities in the numerator of Eq. (31.9) must be proportional to E to compensate for the presence of E in the denominator. Obviously the charge e on the electron does not depend on the field. The number of electrons per cubic metre could conceivably depend on the field, but it can be shown that such a dependence would not be a simple proportionality. It must be that the velocity of the carrier is proportional to the field and that the number of carriers is independent of the field. This is the condition that must be satisfied if any conductor is to obey Ohm's law. Therefore we write

$$v = uE. \tag{31.10}$$

The constant of proportionality u is called the *mobility*, which is the velocity acquired by a carrier in a field of unit strength; $u = v/E$.

From the requirement that the velocity must be proportional to the field, we conclude that the main force of retardation of the carrier is due to friction. If the charge on the carrier is q, then the force due to the electrical field is qE, which must be balanced by the inertial force $ma = m(dv/dt)$, and the frictional force fv, which is proportional to the velocity. Thus

$$qE = m\frac{dv}{dt} + fv,$$

where f is a constant, called the frictional coefficient. From this equation it is clear that if the velocity is to be proportional to E, the first term, the inertial force, must be negligibly small in comparison with the second, the frictional retardation, so that we have

$$qE = fv. \tag{31.11}$$

In a metal the frictional force arises from the scattering of the electrons by collisions with the metal ions in the lattice.

In terms of the mobility, the expression in Eq. (31.9) for the conductivity becomes

$$\kappa = \tilde{N}ue. \tag{31.12}$$

From a measurement of the resistance of a metal, the resistivity and the conductivity can be determined. Since we know the value of e, the measurement yields a value of the product $\tilde{N}u$. To determine $\tilde{N}$ and u individually requires an independent measurement of some other quantity that depends on one or both of these quantities.

★ 31.3 THE HALL EFFECT

Consider the following experiment: A current having a current density j is passed through a metal strip in the x direction; simultaneously a magnetic field B is applied in the z direction. Two probes A and A' are placed on opposite sides of the strip (Fig. 31.1). The magnetic field, indicated by the dashed circle in Fig. 31.1, deflects the electron stream in the metal with the result that an electrical field E_y develops across the width of the strip and produces a potential difference ϕ_{H}, the Hall potential, between the two probes A and A'.

If v is the velocity of the electrons in the x direction, the force acting in the y direction due to the magnetic field is Bev; this force is balanced by the force from the electrical field

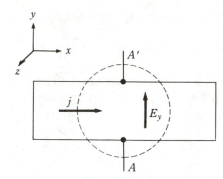

Figure 31.1   The Hall effect.

in the y direction, which is eE_y. Thus we have

$$eE_y = Bev \quad \text{or} \quad E_y = Bv,$$

When we insert the value of v from Eq. (31.8), this becomes

$$E_y = \frac{Bj}{\tilde{N}e}.$$

The Hall potential is $\phi_H = E_y w$, where w is the width of the strip; thus

$$\phi_H = \frac{wBj}{\tilde{N}e} = R_H wBj, \tag{31.13}$$

where $R_H = 1/\tilde{N}e$ is the Hall coefficient. Measurement of w, j, B, and ϕ_H suffices to determine the value of R_H. This determines $\tilde{N}$, since from the definition,

$$\tilde{N} = \frac{1}{eR_H}. \tag{31.14}$$

Combining Eq. (31.14) with Eq. (31.12), we obtain the mobility of the electrons:

$$u = \kappa R_H. \tag{31.15}$$

By measuring the conductivity and the Hall coefficient, it is possible to obtain values of the mobility and the number of carriers per cubic metre. Table 31.2 lists values of u, $\tilde{N}$, and the number of carriers per atom that contribute to the conductivity for several metals.

The values of the mobility are interesting because they are so small. This emphasizes that it is the frictional resistance that retards the motion. An electron moving in free space

Table 31.2

	Cu	Ag	Au	Li	Na	Zn*	Cd*
$\kappa/(10^7 \text{ S/m})$	6.33	6.70	4.13	1.12	1.92	1.76	1.32
$R_H/(10^{-11} \text{ m}^3/\text{C})$	-5.5	-8.4	-7.2	-17.0	-25.0	3.3	6.0
$u/(10^{-4} \text{ m}^2 \text{ V}^{-1}\text{ s}^{-1})$	34.8	56.3	29.7	19.1	48.0	5.8	7.9
$\tilde{N}/(10^{28} \text{ m}^{-3})$	11.3	7.43	8.67	3.67	2.50	18.9	10.4
Electrons/atom	1.3	1.3	1.5	0.79	0.98	2.9	2.2

*   The sign of R_H indicates the carrier is positively charged in these metals.

subject only to inertial retardation would have a mobility about one million times larger than the mobilities in metals.

Another point of interest is that not all but only about one electron per atom is free to carry the current. Only the electrons in levels near the top of the filled part of the partially filled band are free to move under the application of a field. As we saw in Section 28.4, to carry a current the electrons must be able to shift from one set of levels to another set; vacant levels that are not very much different in energy must be available. Vacant levels are available only near the top of the filled part of a partially filled band and so only these electrons contribute to the conductivity.

Finally, there is the curious result that if the electrons that carry the current are in levels near the top of a band, then the field pushes the electrons in the *wrong* direction; wrong in the sense that they are accelerated in the direction opposite to the usual one. These electrons behave as if they were positively charged. This happens with zinc and cadmium as well as a number of other metals. The effect is detected in the Hall experiment; the Hall potential for these metals has the opposite sign when compared with a metal such as copper.

Measurement of the magnitude and sign of the Hall potential in semiconductors enables us to distinguish experimentally between *p*- and *n*-type semiconductors, and to determine, knowing κ, the number and mobility of the carriers.

31.4 THE ELECTRICAL CURRENT IN IONIC SOLUTIONS

The passage of an electrical current in an ionic solution is a more complex event than the passage of a current through a metal. In the metal, the nearly weightless electrons carry all the current. In the ionic solution, the current is carried by the motion of massive positive and negative ions. Consequently, *the passage of a current is accompanied by a transport of matter.* The positive and negative ions do not carry equal portions of the current, so that a concentration gradient develops in the solution. Furthermore, transfer of the electrical charge through the solution–electrode interface is accompanied by a chemical reaction (electrolysis) at each electrode. For clarity we must keep the phenomena in the body of the solution separate from the phenomena at the electrodes. We begin by dealing briefly with the phenomena at the electrodes (electrolysis) and then describe the occurrences in the body of the solution, which are our main concern in this chapter.

If a direct current is passed between two electrodes in an electrolytic solution, a chemical reaction, electrolysis, occurs at the electrodes. After a study of various types of electrolytic reactions, Faraday (1834) discovered two simple and fundamental rules of behavior, now called Faraday's laws of electrolysis. Faraday's first law states that the amount of chemical reaction that occurs at any electrode is proportional to the quantity Q of electricity passed; Q is the product of the current and the time, $Q = It$. The second law states that the passage of a fixed quantity of electricity produces amounts of two different substances in proportion to their chemical equivalent weights. Faraday's experiments showed that these rules were followed with great accuracy. So far as we know these laws are *exact*.

Any electrolytic reaction can be written in the form,

$$0 = \sum_i \nu_i A_i + (\pm 1)e^-,$$

in which the A_i are the formulas of the substances taking part in the reaction and the ν_i are the stoichiometric coefficients; the ν_i are positive for products and negative for reactants.

The equation has been balanced so that one mole of electrons is either consumed at the cathode ($v_e = -1$) or produced at the anode ($v_e = +1$). This equation states that for each mole of electrons that passes, $|v_i|$ moles of A_i are produced or consumed. If a quantity of electricity, $Q = It$, is passed, then the number of moles of A_i produced or consumed is

$$n_i = \frac{|v_i|Q}{F} = \frac{|v_i|It}{F}$$

where $F = 96\ 484.56$ C/mol. If m_i is the mass of A_i produced or consumed and M_i is the molar mass, then

$$m_i = \frac{|v_i|M_iIt}{F} \tag{31.16}$$

The quantity, $|v_i|M_i$, defines the "equivalent weight" of A_i. Thus, if one "equivalent" (96 485 coulombs) of electricity is passed, one "equivalent" of each substance in the reaction is either produced or consumed. Equation (31.16) expresses both of Faraday's laws. The use of "equivalent weights" is becoming obsolete (the SI does not recognize them); a formulation such as that in Eq. (31.16) is preferable.

31.5 THE MEASUREMENT OF CONDUCTIVITY IN ELECTROLYTIC SOLUTIONS

A simple conductivity cell is shown in Fig. 31.2. Two platinum electrodes are sealed in the ends of the cell. These are usually coated with a deposit of finely divided platinum, platinum black, to eliminate some of the effects of electrolysis. The cell is filled with the solution, and the resistance is measured by placing the cell in one arm of the alternating current version of a Wheatstone bridge. The frequency ordinarily used is about 1000 Hz.

From Eq. (31.7) the resistance is

$$R = \frac{\rho l}{A} = \frac{l}{\kappa A},$$

since $\rho = 1/\kappa$. For κ we obtain

$$\kappa = \frac{l}{RA}. \tag{31.17}$$

The cell constant $K = l/A$ depends on the geometry of the cell; it can be determined for cells of special design by measuring the distance l between the electrodes and the area A of the electrodes. In routine measurements the cell constant is determined indirectly by measuring the resistance of the cell containing a standard solution of known conductivity. Solutions of potassium chloride are commonly used for this purpose. Some values of κ for KCl solutions are given in Table 31.3.

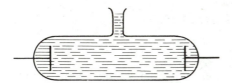

Figure 31.2 A simple conductivity cell.

Table 31.3
Conductivity of KCl solutions

g KCl/kg sln	Concentration $(\text{mol/dm}^3)_{0\,°C}$	$\kappa/(S/m)$ 0 °C	$\kappa/(S/m)$ 18 °C	$\kappa/(S/m)$ 25 °C
71.1352	1	6.517_6	9.783_8	11.134_2
7.41913	0.1	0.7137_9	1.1166_7	1.2856_0
0.745263	0.01	0.07736_4	0.12205_2	0.14087_7

G. Jones and B. C. Bradshaw, *J. Amer. Chem. Soc.* **55**, 1780 (1933).

If R_s is the resistance of the cell containing a solution of known conductivity κ_s, then

$$K = \frac{l}{A} = \kappa_s R_s, \tag{31.18}$$

so that by Eq. (31.17)

$$\kappa = \kappa_s \left(\frac{R_s}{R}\right). \tag{31.19}$$

In precision work great care must be taken to eliminate effects due to electrolysis and those due to variation in temperature. Controlling the temperature is a particularly difficult problem because of the heating effect of the current. Water of extreme purity (conductivity water) must be used, since stray impurities in the water can produce sensible variations in the value of the conductivity of the solution. The contribution of water itself to the conductivity must be subtracted from the measured value for the solution.

31.6 THE MIGRATION OF IONS

Kohlrausch established that electrolytic solutions obeyed Ohm's law accurately once the effect of the electrolysis products was eliminated by using high-frequency alternating current. Kohlrausch also showed from the experimental data that the conductivity of a solution could be composed of separate contributions from each ion; this is known as Kohlrausch's *law of the independent migration of ions.*

Consider an electrolyte with the formula, $A_{v_+} B_{v_-}$, which is completely dissociated into v_+ positive ions and v_- negative ions.

$$A_{v_+} B_{v_-} \longrightarrow v_+ A^{z+} + v_- B^{z-}.$$

Let $\tilde{N}_+$ and $\tilde{N}_-$ be the number of positive and negative ions per cubic metre, respectively. Let their velocities be v_+ and v_-, and their charges be $z_+ e$ and $z_- e$. Then by the fundamental law of transport, Eq. (30.11), the current density is

$$j = \tilde{N}_+ v_+ z_+ e + \tilde{N}_- v_- z_- e. \tag{31.20}$$

(Note that both the velocity and the charge of the negative ion are opposite in sign to those of the positive ion. However, the product $v_- z_- e$ has the same sign as that product for the positive ion, so the terms add together in Eq. (31.20). For convenience we will take all the quantities positively, since this will not affect the final result.) Physically, Eq. (31.20) states that the effects of positive ions moving in one direction and negative ions moving in the other add up to produce the total flow of charge.

If $\tilde{c}$ moles of the compound are present per cubic metre, then the composition of the compound requires

$$\tilde{N}_+ = v_+ \tilde{c} N_A \quad \text{and} \quad \tilde{N}_- = v_- \tilde{c} N_A.$$

Since $N_A e = F$, the expression for j becomes

$$j = \tilde{c} F (v_+ z_+ v_+ + v_- z_- v_-). \tag{31.21}$$

Introducing the mobilities defined in Eq. (31.10), we can write

$$j = \tilde{c} F (v_+ z_+ u_+ + v_- z_- u_-) E \tag{31.22}$$

Comparing this with Ohm's law, Eq. (31.2), we have for the conductivity

$$\kappa = \tilde{c} F (v_+ z_+ u_+ + v_- z_- u_-) \tag{31.23}$$

We observe that in the first approximation, κ is proportional to the concentration of the solution, $\tilde{c}$. The other quantities are all constants except the mobilities, u_+ and u_-, which have a slight dependence on concentration, reaching limiting values as the concentration goes to zero.

We define the molar conductivity of the electrolyte by

$$\Lambda \equiv \frac{\kappa}{\tilde{c}}. \tag{31.24}$$

The molar conductivity is the conductivity the solution would have if there were one mole of the substance in one cubic metre of the solution. Combining this definition with Eq. (31.23), we obtain

$$\Lambda = v_+ (z_+ F u_+) + v_- (z_- F u_-). \tag{31.25}$$

The molar conductivities of the ions are defined by

$$\lambda_+ \equiv z_+ F u_+ \quad \text{and} \quad \lambda_- \equiv z_- F u_-. \tag{31.26}$$

Then we can write

$$\Lambda = v_+ \lambda_+ + v_- \lambda_-. \tag{31.27}$$

Equation (31.27) expresses the molar conductivity as the sum of independent contributions from each kind of ion present; this is Kohlrausch's law; it is strictly correct only if the electrolytic solution is infinitely dilute, $\tilde{c} = 0$. This is not surprising, since the electrically charged ions should exert a mutual influence on each other, especially if they are present in appreciable concentration. Thus, if Λ^∞ is the molar conductivity at infinite dilution, then the expression for Kohlrausch's law is

$$\Lambda^\infty = v_+ \lambda_+^\infty + v_- \lambda_-^\infty. \tag{31.28}$$

In a mixture of several electrolytes, we can generalize Eq. (31.23) to

$$\kappa = \sum_i \tilde{c}_i \lambda_i, \tag{31.29}$$

in which $\tilde{c}_i$ is the concentration (mol/m³) of the ith ion, and $\lambda_i = z_i F u_i$, is its molar conductivity. The summation is over all the ions present. An important application of Eq. (31.29) is in taking account of the ionization of the solvent as a contribution to the conductivity of the solution. For example, in any aqueous salt solution, the conductivity is given by

$$\kappa = \tilde{c}_{H^+} \lambda_{H^+} + \tilde{c}_{OH^-} \lambda_{OH^-} + \tilde{c}_{salt}(v_+ \lambda_+ + v_- \lambda_-).$$

If the concentration of the salt is not high enough to affect the dissociation of water, then the first two terms are simply the conductivity of pure water, κ_w.

$$\kappa_w = \tilde{c}_{H^+}\lambda_{H^+} + \tilde{c}_{OH^-}\lambda_{OH^-}.$$ (31.30)

Then

$$\kappa = \kappa_w + \tilde{c}_{salt}(\nu_+\lambda_+ + \nu_-\lambda_-),$$ (31.31)

or

$$\Lambda_{salt} = \frac{\kappa - \kappa_w}{\tilde{c}_{salt}} = \nu_+\lambda_+ + \nu_-\lambda_-.$$ (31.32)

31.7 THE DETERMINATION OF Λ^∞

Kohlrausch found that the molar conductivity depends on the concentration of the electrolyte, and that in dilute solutions of strong electrolytes this dependence could be expressed by the equation

$$\Lambda = \Lambda^\infty - b\sqrt{c},$$ (31.33)

where Λ^∞ and b are constants. Plotting the value of Λ against the square root of the concentration yields a straight line at low concentrations. The line can be extrapolated to $c = 0$ to yield Λ^∞, the value of Λ at infinite dilution.

The molar conductivity of weak electrolytes falls off much more rapidly with increasing concentration than Eq. (31.33) predicts. The comparative behavior of KCl and acetic acid is shown schematically in Fig. 31.3. Arrhenius suggested that the degree of dissociation of an electrolyte was related to the molar conductivity by

$$\alpha = \frac{\Lambda}{\Lambda^\infty}.$$ (31.34)

Ostwald used this relation in conjunction with the law of mass action to explain the variation of the molar conductivity of weak electrolytes with concentration. Consider the dissociation of acetic acid:

$$HAc \;\rightleftharpoons\; H^+ + Ac^-;$$

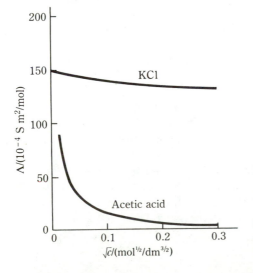

Figure 31.3 Molar conductivity of strong and weak electrolytes (c in mol/dm³).

if α is the degree of dissociation, then $c_{H^+} = c_{Ac^-} = \alpha c$, and $c_{HAc} = (1 - \alpha)c$. The equilibrium constant is

$$K = \frac{\alpha^2 c}{1 - \alpha}.$$

Using $\alpha = \Lambda/\Lambda^\infty$, we obtain

$$K = \frac{c\Lambda^2}{\Lambda^\infty(\Lambda^\infty - \Lambda)} \tag{31.35}$$

as the relation between Λ and c; the *Ostwald dilution law*. Using the values of Λ at various concentrations and the value of Λ^∞, it is found that the right-hand side of Eq. (31.35) is very nearly constant, and in fact, this is a reasonable way to determine the value of the dissociation constant of a weak electrolyte.

To use Eq. (31.35) in the way described above, we must know the value of Λ^∞. The extrapolation used for strong electrolytes is useless for a weak electrolyte. Because of the steepness of the curve near $c = 0$, any extrapolated value would be subject to gross errors. To obtain Λ^∞ for a weak electrolyte we use Kohlrausch's law. Using acetic acid as an example, we have at infinite dilution,

$$\Lambda^\infty_{HAc} = \lambda^\infty_{H^+} + \lambda^\infty_{Ac^-}.$$

To each side of this equation we add the Λ^∞ of the salt of a strong acid and strong base, such as NaCl,

$$\Lambda^\infty_{HAc} + \Lambda^\infty_{NaCl} = \lambda^\infty_{H^+} + \lambda^\infty_{Cl^-} + \lambda^\infty_{Na^+} + \lambda^\infty_{Ac^-},$$

which can be written in the form

$$\Lambda^\infty_{HAc} + \Lambda^\infty_{NaCl} = \Lambda^\infty_{HCl} + \Lambda^\infty_{NaAc};$$

hence

$$\Lambda^\infty_{HAc} = \Lambda^\infty_{HCl} + \Lambda^\infty_{NaAc} - \Lambda^\infty_{NaCl}. \tag{31.36}$$

The molar conductivities on the right-hand side can all be obtained by the extrapolation of a Λ versus $c^{1/2}$ plot, since the substances involved are all strong electrolytes.

An alternative method of obtaining K and Λ^∞ for a weak electrolyte utilizes a rearrangement of Eq. (31.35). Clearing Eq. (31.35) of fractions and removing the parenthesis,

$$K\Lambda^{\infty 2} - K\Lambda\Lambda^\infty = c\Lambda^2$$

Dividing every term by $K\Lambda^{\infty 2}\Lambda$ and transposing the second term to the right-hand side of the equation, we get

$$\frac{1}{\Lambda} = \frac{1}{\Lambda^\infty} + \frac{c\Lambda}{K\Lambda^{\infty 2}} \tag{31.37}$$

If $1/\Lambda$ is plotted against $c\Lambda$, a straight line is obtained, which has an intercept equal to $1/\Lambda^\infty$ and a slope equal to $1/K\Lambda^{\infty 2}$. From the values of the slope and intercept, the individual values of K and Λ^∞ can be obtained. This method requires only data on the conductivity of the weak electrolyte itself.

After the Arrhenius theory was first proposed, an attempt was made to fit all conductance data to the Ostwald dilution law. It soon became apparent that many substances did not conform to this law. These substances are the *strong* electrolytes, which are completely dissociated into ions. The discussion of the dependence of the molar conductivity of strong electrolytes on concentration is based on the ideas contained in the Debye–Hückel theory.

31.8 TRANSFERENCE NUMBERS

The measurement of the conductivity yields the sum of the positive and negative ion conductivities. To obtain the individual ion conductivities, an additional independent measurement is necessary. Even before Kohlrausch demonstrated the law of independent migration of ions, it was commonly supposed that each ion contributed to the flow of current. In 1853 Hittorf devised a method to measure the contribution of the individual ions.

The transference number of an ion is defined as the fraction of the current carried by that ion. By Eq. (31.29) the conductivity of a solution containing any number of electrolytes is $\kappa = \Sigma_i \tilde{c}_i \lambda_i$; then by definition the transference number of the kth ion is

$$t_k = \frac{\tilde{c}_k \lambda_k}{\kappa} = \frac{\tilde{c}_k \lambda_k}{\sum_i \tilde{c}_i \lambda_i}. \tag{31.38}$$

The transference number of an ion is not a simple property of the ion itself; it depends on which other ions are present and on their relative concentrations. It is apparent that the sum of the transference numbers of all the ions in the solution must equal unity.

In a solution containing only one electrolyte, it follows from Eq. (31.38) that the transference numbers, t_+ and t_-, are defined by

$$t_+ = \frac{v_+ \lambda_+}{\Lambda} = \frac{v_+ \lambda_+}{v_+ \lambda_+ + v_- \lambda_-}; \qquad t_- = \frac{v_- \lambda}{\Lambda} = \frac{v_- \lambda_-}{v_+ \lambda_+ + v_- \lambda_-}. \tag{31.39}$$

Obviously, $t_+ + t_- = 1$. If we replace λ_+ and λ_- by the values given in Eq. (31.26), we obtain

$$t_+ = \frac{v_+ z_+ u_+}{v_+ z_+ u_+ + v_- z_- u_-} \quad \text{and} \quad t_- = \frac{v_- z_- u_-}{v_+ z_+ u_+ + v_- z_- u_-}.$$

But electrical neutrality in the compound requires that $v_+ z_+ = v_- z_-$; thus we see that

$$t_+ = \frac{u_+}{u_+ + u_-} \quad \text{and} \quad t_- = \frac{u_-}{u_+ + u_-} \tag{31.40}$$

Since the mobilities are proportional to the velocities, $u = v/E$, we can also write

$$t_+ = \frac{v_+}{v_+ + v_-} \quad \text{and} \quad t_- = \frac{v_-}{v_+ + v_-} \tag{31.41}$$

If one mole of electrical charge is passed in the solution, then through every plane perpendicular to the current path, t_+ moles of charge are carried by the positive ions and t_- moles of charge are carried by the negative ions. Since each mole of positive ions carries z_+ moles of charge, to pass t_+ moles of charge requires the passage of t_+/z_+ moles of positive ions. Similarly, to pass t_- moles of charge requires the passage of t_-/z_- moles of negative ions.

31.8.1 Hittorf Method

To illustrate the Hittorf method for measuring the contribution of the individual ions to the current, we consider the electrolysis cell shown in Fig. 31.4. Suppose that the solution contains copper sulfate and that the anode is copper. We examine the changes that occur in each compartment if *one mole* of electricity passes. These changes are summarized in Table 31.4. If a quantity of electricity Q passes, this is Q/F moles, so all of the changes are

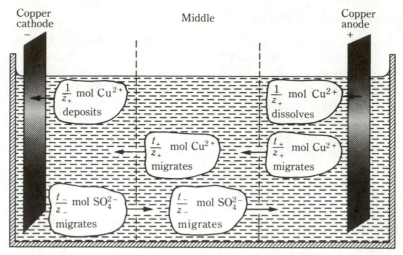

Figure 31.4 Transference in $CuSO_4$ solution using copper electrodes.

Table 31.4

Cathode compartment	Middle compartment	Anode compartment
$(1/z_+)$ mol Cu^{2+} plate out on cathode	(t_+/z_+) mol of Cu^{2+} migrate out at A	$(1/z_+)$ mol Cu^{2+} dissolve from anode
(t_+/z_+) mol Cu^{2+} migrate in	(t_+/z_+) mol of Cu^{2+} migrate in at B	(t_+/z_+) mol Cu^{2+} migrate out
(t_-/z_-) mol SO_4^{2-} migrate out	(t_-/z_-) mol of SO_4^{2-} migrate in at A	(t_-/z_-) mol SO_4^{2-} migrate in
	(t_-/z_-) mol of SO_4^{2-} migrate out at B	
Net change	**Net change**	**Net change**
$(\Delta n_{Cu^{2+}})_c = (t_+/z_+) - (1/z_+)$ mol $= -(t_-/z_+)$ mol	$\Delta n_{Cu^{2+}} = 0$	$(\Delta n_{Cu^{2+}})_a = (1/z_+) - (t_+/z_+)$ mol $= (t_-/z_+)$ mol
$(\Delta n_{SO_4^{2-}})_c = -(t_-/z_-)$ mol	$\Delta n_{SO_4^{2-}} = 0$	$(\Delta n_{SO_4^{2-}})_a = (t_-/z_-)$ mol

multiplied by Q/F. In this experiment, the amount of $CuSO_4$ in the cathode compartment decreases by $(t_-/z_+)(Q/F)$ moles, while in the anode compartment the number of moles increases by $(t_-/z_+)(Q/F)$. The concentration in the middle part of the cell is unchanged by the passage of the current. By arranging the apparatus properly, the boundaries indicated at A and B in Fig. 31.4 can be replaced by stopcocks (Fig. 31.5), so that the three portions of the solution can be drawn off separately after the experiment. The weight and concentration of electrolyte in each portion is measured after the experiment. Knowing the original concentration, we can calculate the changes in number of moles of electrolyte in each compartment. Analysis of the middle compartment is used as a check to determine if any interfering effects have occurred. The changes in numbers of moles of electrolyte in the compartments can be related to the transference numbers of the ions by a procedure such as the one given above. It is not possible to write a general formula relating the changes to the transference numbers, since what happens in every case depends on the chemical effect

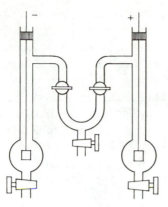

Figure 31.5 The Hittorf cell.

produced by the electrode reactions. The changes must be figured out using the above method for each combination of electrodes.

The Hittorf experiment is subject to many difficulties in practice. The development of a concentration gradient by the flow of current results in diffusion of the electrolyte from the more concentrated to the less concentrated regions. This tends to undo the effect to be measured; to minimize diffusion the experiment must not extend over too long a time. On the other hand, if the time is short, the concentration changes are small because a small current must be used. If large currents are used, heating effects occur unevenly and produce convection in the solution; this mixes the solution up again. In addition to all this, density differences that develop with the concentration differences between the parts of the solution may also produce convection. In spite of all these difficulties, reasonably good measurements of the transference numbers can be made using the Hittorf method. A difficulty in interpretation arises because the ions are solvated, and in their motion they carry solvent from one compartment to another. We will return to this problem in Section 31.11.

31.8.2 The Moving-Boundary Method

The moving-boundary method for the measurement of transference numbers has been brought to a high state of perfection. A schematic diagram of the apparatus is shown in Fig. 31.6. A tube has two electrodes fixed at the ends and contains two solutions having an

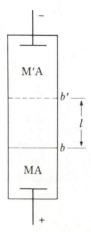

Figure 31.6 The moving-boundary method.

ion in common, one of a compound M'A and another of a compound MA. The system is arranged so that the boundary between the solutions is reasonably sharp; the position of the boundary is visible because of a difference in refractive index of the solutions, or in some cases a difference in color. To avoid mixing and destruction of the boundary, the denser solution is placed beneath the less dense. Suppose that the boundary between the two solutions is initially at b, and that Q/F moles of charge are passed. The M^{z+} ion carries $(t_+/z_+)(Q/F)$ moles of charge past the plane at b. The boundary must move up far enough (to b') so that $(t_+/z_+)(Q/F)$ moles of electrolyte may be accommodated in the volume between b and b'. If l is the length between b and b', and a is the cross-sectional area of the tube, then the volume displaced is la. If $\tilde{c}$ is the concentration of MA in mol/m³, the number of moles that can be contained in la is $\tilde{c}la$; but this is simply the number of moles passing the plane at b. Thus $\tilde{c}la = (t_+/z_+)(Q/F)$, so that

$$\frac{t_+}{z_+} = \frac{\tilde{c}laF}{Q}, \tag{31.42}$$

which assumes that the volume displaced, la, is small compared with the total volume of the solution of MA; in precise work a correction must be applied.

The moving-boundary method yields more accurate data on transference numbers than does the Hittorf method. Experimentally it is easier to handle. The difficulties lie in the establishment of a sharp boundary, the necessity of avoiding convection currents, and excessive heating by the current. However, once the boundary is established, the flow of current sharpens the boundary, making this a minor difficulty. The relative concentrations of the two solutes are important in maintaining a sharp boundary. The faster moving ion, M' in this example, does not lead by more than a few atomic diameters, since a potential difference develops in such a sense as to slow it down; in the steady state the two ions move with the same velocity, but M' is always a little bit ahead of M.

The measurements of the transference number are made over a range of concentration of electrolyte; the plot of t versus $c^{1/2}$ is linear in dilute solution and can be extrapolated to $c = 0$ to obtain the value of the transference number at infinite dilution, t^∞.

31.9 MOLAR ION CONDUCTIVITIES

Once we have measured transference numbers, we can calculate the values of the molar ionic conductivities using Eq. (31.39)

$$v_+\lambda_+^\infty = t_+^\infty \Lambda^\infty, \quad \text{and} \quad v_-\lambda_-^\infty = (1 - t_+^\infty)\Lambda^\infty.$$

Values of λ_+^∞ and λ_-^∞ for a number of ions are given in Table 31.5.

31.10 APPLICATIONS OF CONDUCTANCE MEASUREMENTS

31.10.1 Determination of the Ion Product of Water

The ion product of water is $K_w = a_{H^+}a_{OH^-}$. Since in pure water the concentrations of the ions are exceedingly small, we may set the activities equal to the concentrations of the species present; so $K_w = (\tilde{c}_{H^+}/\tilde{c}°)(\tilde{c}_{OH^-}/\tilde{c}°)$. In pure water, $\tilde{c}_{H^+} = \tilde{c}_{OH^-} = \tilde{c}°K_w^{1/2}$.

The conductivity of pure water κ_w is related to the concentrations by the equation

$$\kappa_w = \tilde{c}_{H^+}\lambda_{H^+} + \tilde{c}_{OH^-}\lambda_{OH^-},$$

Table 31.5
Limiting molar conductivities of ions at 25 °C: $\lambda^\infty/10^{-4}$ S m²/mol

Ion	λ^∞	$\dfrac{\lambda^\infty}{z^2}$	Ion	λ^∞	$\dfrac{\lambda^\infty}{z^2}$
H^+	349.81	349.81	OH^-	198.3	198.3
Li^+	38.68	38.68	F^-	55.4	55.4
Na^+	50.10	50.10	Cl^-	76.35	76.35
K^+	73.50	73.50	Br^-	78.14	78.14
Rb^+	77.81	77.81	I^-	76.84	76.84
Cs^+	77.26	77.26	NO_3^-	71.46	71.46
Ag^+	61.90	61.90	ClO_3^-	64.6	64.6
NH_4^+	73.55	73.55	BrO_3^-	55.74	55.74
$(CH_3)_4N^+$	44.92	44.92	IO_3^-	40.54	40.54
$(C_2H_5)_4N^+$	32.66	32.66	ClO_4^-	67.36	67.36
$(C_3H_7)_4N^+$	23.42	23.42	IO_4^-	54.55	54.55
Be^{2+}	90	22.5	HCO_3^-	44.50	44.50
Mg^{2+}	106.10	26.52	$HCOO^-$	54.59	54.59
Ca^{2+}	119.00	29.75	CH_3COO^-	40.90	40.90
Sr^{2+}	118.90	29.72	CH_2BrCOO^-	39.22	39.22
Ba^{2+}	127.26	31.82	$(NO_2)_3C_6H_2O^-$	30.39	30.39
Cu^{2+}	107.2	26.80	SO_4^{2-}	160.04	40.01
Zn^{2+}	105.6	26.40	$C_2O_4^{2-}$	148.30	37.08
Co^{2+}	110	27.5	CO_3^{2-}	138.6	34.65
Pb^{2+}	139.0	34.75	$Fe(CN)_6^{3-}$	302.7	33.63
La^{3+}	209.1	23.23	$P_3O_9^{3-}$	250.8	27.87
Ce^{3+}	209.4	23.26	$Fe(CN)_6^{4-}$	442.0	27.63
$[Co(NH_3)_6]^{3+}$	305.7	33.97	$P_4O_{12}^{4-}$	374.8	23.43
$[Ni_2tri\text{-}en_3]^{4+}$	210.0	13.13	$P_2O_7^{4-}$	383.6	23.98
$[Co_2tri\text{-}en_3]^{6+}$	412.2	11.45	$P_3O_{10}^{5-}$	545	21.8

By permission from R. A. Robinson and R. H. Stokes, *Electrolyte Solutions*. 2d ed. (rev.) London: Butterworths, 1959.

which becomes

$$\kappa_w = \tilde{c}^\circ K_w^{1/2}(\lambda_{H^+} + \lambda_{OH^-}).$$

The concentrations are so low that the values of the ion conductivities at infinite dilution may be used (Table 31.5). Then $\lambda_{H^+}^\infty + \lambda_{OH^-}^\infty = 548.1 \times 10^{-4}$ S m²/mol. The value of κ_w at 25 °C obtained by Kohlrausch and Heydweiller (1894) is 5.5×10^{-6} S/m. Using this value, we obtain for K_w

$$K_w = \left[\frac{\kappa_w}{\tilde{c}^\circ(\lambda_{H^+} + \lambda_{OH^-})}\right]^2 = \left[\frac{5.5 \times 10^{-6} \text{ S/m}}{(1000 \text{ mol/m}^3)(548.1 \times 10^{-4} \text{ S m}^2/\text{mol})}\right]^2 = 1.01 \times 10^{-14}.$$

The best values of K_w are obtained from measurements of electrochemical cell potentials, and these agree well with the best values from conductivity measurements. At 25 °C the most reliable value of K_w is 1.008×10^{-14}. Values of K_w at several temperatures are given in Table 31.6. The variation with temperature should be noted.

Table 31.6
The ion product of water

$t/°C$	0	10	20	25	30	40	50	60
$K_w/10^{-14}$	0.1139	0.2920	0.6809	1.008	1.469	2.919	5.474	9.614

By permission from H. S. Harned and B. B. Owen, *The Physical Chemistry of Electrolyte Solutions.* 3d ed. New York: Reinhold, 1958.

31.10.2   Determination of Solubility Products

Another application of conductance measurements is in the determination of the solubility of a slightly soluble salt. For example, a saturated solution of silver chloride in water has a conductivity which is given by

$$\kappa = \tilde{c}_{Ag^+}\lambda_{Ag^+} + \tilde{c}_{Cl^-}\lambda_{Cl^-} + \tilde{c}_{H^+}\lambda_{H^+} + \tilde{c}_{OH^-}\lambda_{OH^-}.$$

If the salt dissolves to only a small extent, then the ionization of water will be unaffected by the presence of the salt, and the last two terms in the equation are simply the conductivity of the water. Therefore

$$\kappa - \kappa_w = \tilde{c}_{Ag^+}\lambda_{Ag^+} + \tilde{c}_{Cl^-}\lambda_{Cl^-}.$$

If $\tilde{s}$ is the solubility in moles per cubic metre, then $\tilde{s} = \tilde{c}_{Ag^+} = \tilde{c}_{Cl^-}$. Then

$$\kappa - \kappa_w = \tilde{s}(\lambda_{Ag^+} + \lambda_{Cl^-}).$$

If the solution is very dilute, the values of λ^∞ may be used from Table 31.5; then

$$\tilde{s} = \frac{\kappa - \kappa_w}{\Lambda^\infty_{AgCl}}.$$

The solubility product constant is given by $K_{sp} = a_{Ag^+}a_{Cl^-}$. If the solution is dilute enough to regard the activity coefficients as unity, then $K_{sp} = (\tilde{s}/\tilde{c}^\circ)^2$. In the case of silver chloride $\kappa - \kappa_w = 1.802 \times 10^{-4}$ S/m, so that

$$K_{sp} = \left[\frac{1.802 \times 10^{-4}\ \text{S/m}}{(1000\ \text{mol/m}^3)(138.27 \times 10^{-4}\ \text{S m}^2/\text{mol})}\right]^2 = 1.698 \times 10^{-10}.$$

This value is in excellent agreement with that obtained from cell potential measurements.

31.10.3   Conductometric Titrations

The variation of the conductance of a solution during a titration can serve as a useful method of following the course of the reaction. Consider a solution of a strong acid, HA, to which a solution of a strong base, MOH, is added. The reaction

$$H^+ + OH^- \longrightarrow H_2O$$

occurs. For each equivalent of MOH added, one equivalent of hydrogen ion is removed. Effectively, the faster-moving H^+ ion is replaced by the slower-moving M^+ ion, and the conductance of the solution falls. This continues until the equivalence point is reached, at which we have a solution of the salt MA. If more base is added, the conductance of the

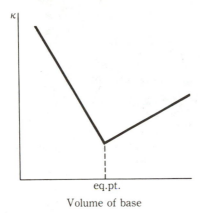

eq.pt.

Volume of base

Figure 31.7 Conductometric titration of a strong acid with a strong base.

solution increases, since more ions are being added and the reaction no longer removes an appreciable number of them. Consequently, in the titration of a strong acid with a strong base, the conductance has a minimum at the equivalence point. This minimum can be used instead of an indicator dye to determine the endpoint of the titration. A schematic plot of the conductance of the solution against the number of milliliters of base added is shown in Fig. 31.7. This technique is applicable to any titration that involves a sharp change in conductivity at the equivalence point.

Consider the titration of a silver nitrate solution with sodium chloride. In the precipitation reaction $Ag^+ + Cl^- \rightarrow AgCl$, the sodium ion replaces the silver ion in solution. This in itself produces little change in conductance, so that the plot of conductance versus number of milliliters of titrant is nearly horizontal. However, after the equivalence point is passed, the conductance increases sharply because of the additional ions. The endpoint can be determined easily.

The equation for the conductance is simple and is given in the case of the acid–base titration by

$$\kappa = \tilde{c}_{H^+} \lambda_{H^+} + \tilde{c}_{A^-} \lambda_{A^-} + \tilde{c}_{M^+} \lambda_{M^+} + \tilde{c}_{OH^-} \lambda_{OH^-},$$

where the concentrations are in moles per cubic metre. Using this equation and knowing the concentrations of the acid (HA) and base (MOH) solutions, we can easily calculate the conductance of the solution as a function of the volume of base added.

31.11 STOKES'S LAW

The simplest interpretation of the values of the ion conductivities is obtained if we imagine a single ion immersed in a fluid and subjected to an electrical field. The only retardation the ion experiences is that due to the viscosity of the fluid. If the ion is a sphere of radius r_i, the frictional force opposing its motion is given by Stokes's law:

$$f = 6\pi\eta r_i v_i, \tag{31.43}$$

where η is the viscosity coefficient of the medium, and v_i is the velocity of the ion. We balance this by the electrical force acting on the ion, $z_i eE$:

$$z_i eE = 6\pi\eta r_i v_i.$$

From this equation we obtain the mobility,

$$u_i = \frac{v_i}{E} = \frac{z_i e}{6\pi \eta r_i}. \tag{31.44}$$

Combining this result with the definition of the ion conductivity, $\lambda_i = z_i F u_i$, we obtain

$$\lambda_i = \frac{F e z_i^2}{6\pi \eta r_i}, \tag{31.45}$$

which is the value predicted by Stokes's law for the molar ion conductivity. The molar conductivity of an electrolyte is given by $\Lambda = v_+ \lambda_+ + v_- \lambda_-$, so that

$$\Lambda = \frac{Fe}{6\pi \eta} \left(\frac{v_+ z_+^2}{r_+} + \frac{v_- z_-^2}{r_-} \right) \tag{31.46}$$

It is informative to compare the predictions of Eq. (31.45) with the values of λ_i in Table 31.5. First we note the proportionality of λ_i to z_i^2. If we divide each of the values of λ_i for the polyvalent ions by z_i^2, we obtain numbers that are comparable to the values for the large, monovalent ions. For example, $\lambda^\infty(Mg^{++})/(2)^2 = 26.52 \times 10^{-4}$ S m^2/mol, a value that lies between $\lambda^\infty/(1)^2 = 32.66 \times 10^{-4}$ S m^2/mol for $(C_2H_5)_4N^+$ and $\lambda^\infty/(1)^2 = 23.42 \times 10^{-4}$ S m^2/mol for $(C_3H_7)_4N^+$. Both of these quaternary ammonium ions are quite large. Since the viscosity of water is the same in all cases, the only remaining factor is the radius of the ion. We are left to conclude that the radius of the magnesium ion is comparable to that of large quaternary ammonium ions.

Similarly, if we compare the conductivity of the alkali metal ions (Table 31.5), in the light of Eq. (31.45) we would be forced to conclude that the radius of the lithium ion is larger than that of the cesium ion. Since the crystallographic radius of lithium ion is much smaller than that of cesium ion, this indicates a difficulty with the Stokes's law interpretation of λ_i.

The only quantity on the right-hand side of Eq. (31.46) that depends on the medium is η, so that for a given ion in different solvents we should have the relation, $\lambda_i \eta = $ constant, which is Walden's rule. In particular, at infinite dilution,

$$\lambda_i^\infty \eta_0 = \text{constant}, \tag{31.47}$$

where η_0 is the viscosity coefficient for the pure solvent. If we compare the $\lambda_i^\infty \eta_0$ product for a specified ion in several different solvents, we find that the product is constant only for rather large ions such as the tetramethylammonium ion, $(CH_3)_4N^+$, and the picrate ion, $C_6H_2(NO_2)_3O^-$. For these the constancy of the $\lambda_i^\infty \eta_0$ product is very good. If we exclude water, the constancy of $\lambda_i^\infty \eta_0$ for smaller ions is only fair, being perhaps within 20 percent of an average value.

The difficulty with the small ions arises from the fact that the ions are solvated. An ion is attached to molecules of solvent that are carried along with the ion as it moves. The effective radius of the ion is therefore larger than its crystallographic radius and is different in each solvent. The amount of solvent held to the ion is less with larger ions (since the electrical field due to the ion itself is smaller), so that the effective radius is more nearly the same in various solvents; consequently, Walden's rule is more accurate for large ions. If water is included in the solvents under comparison, the $\lambda \eta$ product in water is usually quite different than for the others, indicating more marked solvation in water. If conductivities in H_2O and D_2O are compared, the $\lambda \eta$ products are very nearly equal.

The transport of water by the ions was first measured by Washburn. Using the Hittorf method, a reference substance such as sugar or urea is added to the solution. Presumably the reference substance does not move in the field, and the transport of the solvent can be calculated from the analysis of the solution in the three compartments. If a value is assumed for the number of water molecules attached to one ion, a value for the number attached to the other ion can be calculated. Presently other methods for evaluation of hydration numbers are preferred—from measurements of the partial molar volume of the salt in the solution, for example. The different methods are internally consistent but often do not agree well with each other. It is generally assumed that the negative ions are not hydrated. Then the hydration numbers are, approximately: Li^+, 6; Na^+, 4; K^+, 2; Rb^+, 1.

31.12 CONDUCTIVITIES OF THE HYDROGEN AND HYDROXYL IONS

The data in Table 31.5 also show that the molar conductivities of the hydrogen ion and the hydroxyl ion are much larger than those of other ions. While the other ions move like a sphere pushing through a viscous medium, the very large values of the molar ionic conductivity observed for H^+ and OH^- have been explained on the basis of a proton jump from one species to another. For conduction by H^+ ion, we have the scheme shown in Fig. 31.8. A proton is transferred from the H_3O^+ ion to an adjacent water molecule, thereby converting the water molecule to an H_3O^+ ion. The process is repeated, the newly formed H_3O^+ ion handing on a proton to the next water molecule, and so on. The occurrence of this process leaves the water molecules in an unfavorable orientation; for the process to happen again, they must rotate through 90°. The initial stage is shown in Fig. 31.8(a), an intermediate stage in Fig. 31.8(b), and the final stage in Fig. 31.8(c). The analogous process for the hydroxyl ion is shown in Fig. 31.9.

The process of proton transfer results in a more rapid transfer of positive charge from one region of the solution to another than would be possible if the ion H_3O^+ has to push its way through the solution as other ions must. For this reason also the conductivities of H^+ and OH^- ions are not related to the viscosity of the solution.

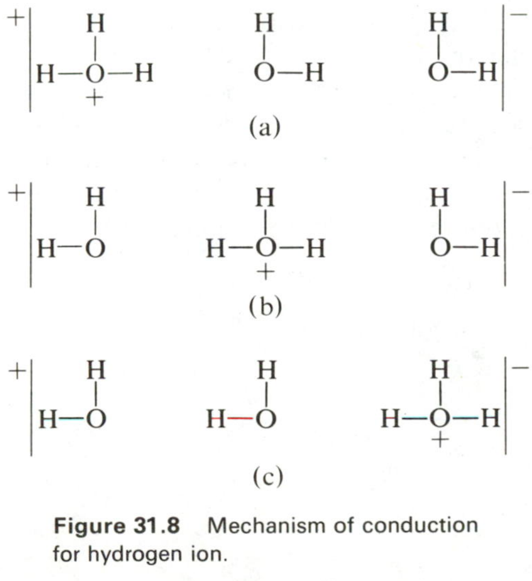

Figure 31.8 Mechanism of conduction for hydrogen ion.

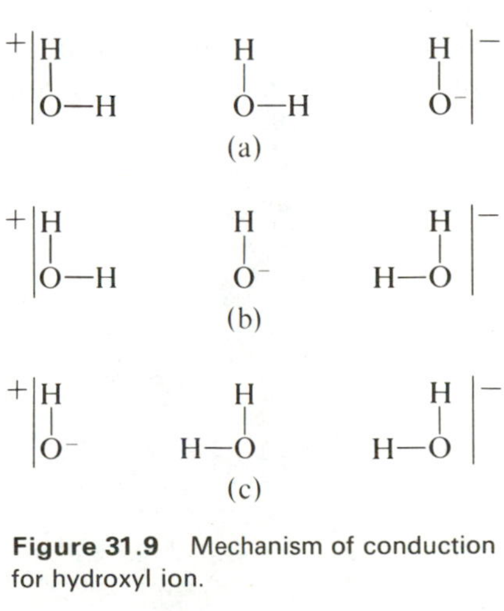

Figure 31.9 Mechanism of conduction for hydroxyl ion.

★ 31.13 TEMPERATURE DEPENDENCE OF THE ION CONDUCTIVITIES

The ion conductivities increase markedly with increase in the temperature. For ions other than H^+ and OH^- this increase is principally the consequence of the decrease in the viscosity of the medium. In water solutions in the range from 0 to 100 °C the change in molar conductivities for ions other than H^+ and OH^- averages about 2% per degree. The conductivities of H^+ and OH^- have larger temperature coefficients (about 14% and 16%, respectively) because of the difference in the conduction mechanism. The decrease in ion conductivities with increase in pressure is also mainly a result of the increase in viscosity with pressure.

★ 31.14 THE ONSAGER EQUATION

If the solution of electrolyte is not infinitely dilute, the ion is retarded in its motion because of the electrical attraction between ions of opposite sign (*asymmetry effect*), and because the positive and negative ions are moving in opposite directions each carrying some solvent (*electrophoretic effect*). Both of these effects are intensified as the concentration of the electrolyte increases so that the retarding forces increase and the conductivity decreases.

The Debye–Hückel theory of ionic solutions provides the concept of an ionic atmosphere surrounding each ion. In the absence of an applied field, this atmosphere can be imagined as a sphere of opposite charge with radius $r_a = 1/\varkappa$, the Debye length. In the absence of a field (Fig. 31.10a), the atmosphere is symmetrically disposed about the ion, so that it exerts no net force on the ion. In the presence of a field (Fig. 31.10b), as the ion moves in one direction the atmosphere does not have time to adjust itself to remain spherically disposed about the ion, and it lags behind. As a result, the ion is retarded in its motion by the atmosphere, which cannot keep up. The effect of the ionic atmosphere is less when $1/\varkappa$ is large, that is, when the atmosphere is far away; less in solvents of high dielectric constant, because the force between ions is reduced by high dielectric constant; and less when kT is large, since increase in temperature yields a less coherent atmosphere. The asymmetry effect reduces Λ by a term of the form (for uni-univalent electrolytes), $B\Lambda^\infty c^{1/2}$. The constant $B = 8.20 \times 10^5/(\epsilon_r T)^{3/2}$, where ϵ_r is the dielectric constant of the solvent.

The electrophoretic effect arises from the motion of the atmosphere in the direction opposite to that of the ion. Both the atmosphere and the ion pull solvent with them and each is, in effect, swimming upstream against the solvent pulled along by the motion of the other. This retardation is less in very viscous solvents because the motion of both the atmosphere and the ion is slowed down. The expression for the electrophoretic retardation has the form, for uni-univalent electrolytes, $Ac^{1/2}$ where $A = 8.249 \times 10^{-4}/(\epsilon_r T)^{1/2}\eta$, where η is the viscosity coefficient of the solvent. When written out in detail the electrophoretic retarda-

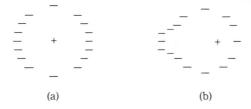

(a)   (b)

Figure 31.10 Asymmetry effect. (a) Field off. (b) Field on.

tion resembles Eq. (31.46); it has the form

$$\frac{Fe}{6\pi\eta}\left(\frac{v_+ z_+^2}{r_a} + \frac{v_- z_-^2}{r_a}\right) = \frac{Fe(v_+ z_+^2 + v_- z_-^2)\varkappa}{6\pi\eta},$$

since the radius of the atmosphere is $r_a = 1/\varkappa$. If we subtract this term from Λ^∞ to obtain Λ, and use Eq. (31.46) for Λ^∞, we get

$$\Lambda = \frac{Fe}{6\pi\eta}\left(\frac{v_+ z_+^2}{r_+} - \frac{v_+ z_+^2}{r_a} + \frac{v_- z_-^2}{r_-} - \frac{v_- z_-^2}{r_a}\right).$$

This equation implies that the effect of the motion of the atmosphere is to increase the effective radius of each ion to

$$(r_+)_{\text{eff}} = r_+\left(1 + \frac{r_+}{r_a - r_+}\right) \quad \text{and} \quad (r_-)_{\text{eff}} = r_-\left(1 + \frac{r_-}{r_a - r_-}\right).$$

As the radius of the atmosphere decreases, the effective radius of the ion increases, and the motion of the ion is slowed.

The final expression for Λ, which includes both the asymmetry effect and the electrophoretic effect, is (for uni-univalent electrolytes)

$$\Lambda = \Lambda^\infty - \left[\frac{8.249 \times 10^{-4}}{(\epsilon_r T)^{1/2}\eta} + \frac{8.20 \times 10^5 \Lambda^\infty}{(\epsilon_r T)^{3/2}}\right]\sqrt{c}, \tag{31.48}$$

which is the Onsager equation; it is usually abbreviated to

$$\Lambda = \Lambda^\infty - (A + B\Lambda^\infty)\sqrt{c}, \tag{31.49}$$

where c is the concentration in mol/L. The test of Eq. (31.48) is whether the limiting slope of a plot of experimental values of Λ versus $c^{1/2}$ has the value predicted by the equation. A comparison of the data with the values predicted by the Onsager equation is shown for several salts in water in Fig. 31.11. The agreement is usually excellent in very dilute solutions up to about 0.02 molar. In more concentrated solutions the conductivity is usually higher than we would predict from the Onsager equation.

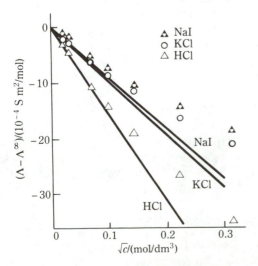

Figure 31.11   Test of the Onsager equation. The lines are the limiting slopes.

★ 31.15 CONDUCTANCE AT HIGH FIELDS AND HIGH FREQUENCIES

The concept of the ion atmosphere is further substantiated by the Wien effect and the Debye–Falkenhagen effect. In very high fields, $E > 10^7$ V/m, an increase in conductivity is observed (Wien effect), resulting from the fact that a finite time (the relaxation time) is required for the atmosphere to form about an ion. In very high fields the ion moves so quickly that it effectively loses its atmosphere; the atmosphere does not have time to form and so cannot slow the ion. The asymmetry effect disappears and the conductance increases.

For the same reason the conductivity increases at high frequencies, 3×10^6 Hz (Debye–Falkenhagen effect). The ion changes its direction of motion so quickly that the more sluggish atmosphere cannot adjust and follow the motion of the ion. The ion moves as if it had no atmosphere, and the conductivity increases. At high frequencies both the asymmetry and electrophoretic effects are absent.

★ 31.16 CONDUCTANCE IN NONAQUEOUS SOLVENTS

The principles governing conductivity in nonaqueous solvents are the same as those for aqueous solutions, of course. The dependence of the conductivity on the viscosity of the solvent was discussed in Section 31.11. However, in solvents having low dielectric constants, there is a lessening of the degree of ionization of many substances. Electrolytes that are completely dissociated in water may be only partially dissociated in a low dielectric constant solvent. Hydrochloric acid is completely dissociated in water; HCl is a "strong" acid. In ethyl alcohol, however, HCl is a "half-strong" acid, with a dissociation constant of about 1.5×10^{-2}.

Suppose we compare the energy of interaction of two ions having charges $+ze$ and $-ze$ at a distance r in a medium of dielectric constant ϵ_r. This energy is

$$V = -\frac{z^2 e^2}{4\pi\epsilon_0 \epsilon_r r}. \tag{31.50}$$

If ϵ_r is large (in H_2O, $\epsilon_r \approx 80$), the ions must come rather close together before the energy of interaction becomes appreciable. If we choose ethyl alcohol ($\epsilon_r \approx 24$), then at the same distance of approach, the interaction energy will be $\frac{80}{24} = 3.3$ times greater; or, put in another way, the energy of interaction becomes appreciable at a distance 3.3 times greater than in water. As a result, since most solvents have much lower dielectric constants than water, the effects due to ionic interaction are much larger than in water.

The large ionic interaction often renders the Onsager equation useless (it is still presumably correct) for the extrapolation to obtain Λ^∞. The solutions for which the Onsager relation is valid are so dilute that it is not possible to obtain reliable measurements of their conductivity. In these cases, special methods of obtaining Λ^∞ are used. If the electrolyte is weakly dissociated, then the Λ^∞ can be obtained by application of the Ostwald dilution law, modifying it in precise work to correct for the interionic forces.

In solvents of low dielectric constant, ion association occurs. The appearance of ion pairs A^+B^- and ion triplets $A^+B^-A^+$ and $B^-A^+B^-$ results in a very rapid variation of conductivity with concentration.

★ 31.17 DIFFUSION AND CHARGE TRANSPORT

There is an intimate connection between the mobility of an ion in an electrical field and the rate at which the ion diffuses under the influence of a concentration gradient. On general thermodynamic grounds, we expect a particle to move spontaneously from a region of high chemical potential to one of low chemical potential. The driving force for this motion is the negative gradient of the chemical potential; the velocity of the particle is proportional to this driving force. Considering the one-dimensional case for which the gradient is along the x-axis, we write for the velocity of the particle, v_i,

$$v_i = \tilde{u}_i \left(-\frac{\partial(\tilde{\mu}_i/N_A)}{\partial x} \right)_{T, p} \tag{31.51}$$

Since the ion is electrically charged, we use the *electrochemical* potential, $\tilde{\mu}_i$, in this equation. We divide by N_A to obtain the force acting on a single particle. The proportionality factor, $\tilde{u}_i$, is a generalized mobility; it is the velocity attained by the ion under a unit value of the generalized driving force, $-[\partial(\tilde{\mu}_i/N_A)/\partial x]_{T, p}$.

Using the electrochemical potential given by Eq. (17.7) and dividing by N_A, we obtain

$$\frac{\tilde{\mu}_i}{N_A} = \frac{\mu_i}{N_A} + z_i \left(\frac{F}{N_A} \right) \phi = \frac{\mu_i^0(T, p)}{N_A} + kT \ln a_i + z_i e\phi.$$

The second equality is obtained because $F/N_A = e$ and $\mu_i/N_A = \mu_i^0(T, p)/N_A + kT \ln a_i$. Differentiating with respect to x at constant T and p, then changing signs throughout, we have

$$-\frac{\partial(\tilde{\mu}_i/N_A)}{\partial x} = -kT \frac{\partial \ln a_i}{\partial x} + z_i e \left(-\frac{\partial \phi}{\partial x} \right).$$

First, we note that $-\partial \phi/\partial x = E_x$, the electric field in the x direction; then we use this expression for the driving force in Eq. (31.51). The result is

$$v_i = -\tilde{u}_i kT \frac{\partial \ln a_i}{\partial x} + \tilde{u}_i z_i eE_x. \tag{31.52}$$

We can convert the activity gradient into a concentration gradient by writing

$$\frac{\partial \ln a_i}{\partial x} = \frac{\partial \ln a_i}{\partial \ln c_i} \frac{\partial \ln c_i}{\partial x}, \tag{31.53}$$

where $a_i = \gamma_i c_i$; the c_i is the concentration in moles of ions per litre, and γ_i is the corresponding activity coefficient. Then

$$\frac{\partial \ln a_i}{\partial \ln c_i} = 1 + \frac{\partial \ln \gamma_i}{\partial \ln c_i}.$$

Since $\tilde{N}_i = (1000 \text{ L/m}^3) N_A c_i$, where $\tilde{N}_i$ is the concentration in ions/m³, we have

$$d \ln c_i = d \ln \tilde{N}_i = \frac{d\tilde{N}_i}{\tilde{N}_i}.$$

When we use these two expressions in Eq. (31.53), it becomes

$$\frac{\partial \ln a_i}{\partial x} = \left(1 + \frac{\partial \ln \gamma_i}{\partial \ln c_i} \right) \frac{1}{\tilde{N}_i} \frac{d\tilde{N}_i}{dx}. \tag{31.54}$$

For the moment we will assume that the solution is dilute enough that $\gamma_i = 1$. (The results we obtain will be valid only at infinite dilution.) Then the second term in parentheses in Eq. (31.54) vanishes. Putting the resulting value for $\partial \ln a_i/\partial x$ into Eq. (31.52), we have for the velocity,

$$v_i = -\frac{\tilde{u}_i kT}{\tilde{N}_i}\frac{\partial \tilde{N}_i}{\partial x} + \tilde{u}_i z_i eE_x. \tag{31.55}$$

Thus the velocity is a sum of two contributions: a chemical one, which is proportional to the concentration gradient, $\partial \tilde{N}_i/\partial x$: and an electrical one, which is proportional to the electric field strength, E_x. We now examine the meaning of Eq. (31.55) in various circumstances.

If the concentration is uniform, $\partial \tilde{N}_i/\partial x = 0$, and we have only the effect of the electric field on the velocity. This relation is

$$v_i = \tilde{u}_i z_i eE_x. \tag{31.56}$$

Comparing this with Eq. (31.10) yields the relation between the velocity under unit force, $\tilde{u}_i$, and the velocity under unit electric field, u_i:

$$u_i = \tilde{u}_i z_i e \quad \text{or} \quad \tilde{u}_i = \frac{u_i}{z_i e}. \tag{31.57}$$

The ion flow is given by the general law of transport, $j_i = \tilde{N}_i v_i$, Eq. (30.11). Inserting v_i from Eq. (31.55) we obtain

$$j_i = -\tilde{u}_i kT\frac{\partial \tilde{N}_i}{\partial x} + \tilde{N}_i \tilde{u}_i z_i eE. \tag{31.58}$$

The first term on the right is the diffusion flow; comparing this term with Fick's law, Eq. (30.5), we find that the diffusion coefficient is given by

$$D_i^\infty = \tilde{u}_i^\infty kT \tag{31.59}$$

or, if we use Eq. (31.57), by

$$D_i^\infty = \frac{u_i^\infty kT}{z_i e} \quad \text{or} \quad D_i^\infty = \frac{u_i^\infty RT}{z_i F}. \tag{31.60}$$

Equation (31.60) is the Einstein relation between the mobility in unit electrical field and the diffusion coefficient. If we replace u_i in the Einstein equation by the Stokes's law value, Eq. (31.44), we obtain a relation between the diffusion coefficient, the ion radius, and the viscosity of the medium.

$$D_i^\infty = \frac{kT}{6\pi\eta_0 r_i} \tag{31.61}$$

This is the Stokes–Einstein equation. Since this equation does not involve the charge of the particle, it applies to neutral particles as well. Einstein's original derivation, done in a different way, was for neutral particles, not ions.

■ **EXAMPLE 31.1**   It is informative to calculate the magnitude of D_i using Eq. (31.61). If we let $T = 300$ K, then for water $\eta_0 \approx 9 \times 10^{-4}$ kg/m s. If we choose $r_i \approx 10^{-10}$ m, we have

$$D_i^\infty = \frac{(1.38 \times 10^{-23}\text{ J/K})(300\text{ K})}{6(3.14)(9 \times 10^{-4}\text{ kg/m s})10^{-10}\text{ m}} \approx 2 \times 10^{-9}\text{ m}^2/\text{s}.$$

This is the correct order of magnitude for the diffusion coefficient of an atom-sized particle in water.

By replacing u_i in the Einstein equation by its value from Eq. (31.26), $u_i = \lambda_i/z_i F$, we obtain the relation between the molar conductivity and the diffusion coefficient.

$$D_i^\infty = \frac{RT\lambda_i^\infty}{F^2 z_i^2} \quad \text{or} \quad \lambda_i^\infty = \frac{F^2 z_i^2 D_i^\infty}{RT} \tag{31.62}$$

Combining this expression with that in Eq. (31.27) for the molar conductivity, we obtain

$$\Lambda^\infty = \frac{F^2}{RT}(v_+ z_+^2 D_+^\infty + v_- z_-^2 D_-^\infty). \tag{31.63}$$

This is the Nernst–Einstein equation, which relates the conductivity to the diffusion coefficients of the ions. The form of the expression in Eq. (31.63) raises the question of how the diffusion coefficients of the ions combine to yield a diffusion coefficient for the electrolyte. It is to this question that we now turn our attention.

★ 31.17.1 Diffusion of an Electrolyte

Consider the simple diffusion of an electrolyte in the absence of an external electric field. The diffusion occurs because of a concentration gradient. The situation shown in Fig. 31.12(a) illustrates the initial condition of an electrolytic solution over which there is a layer of pure water. We assume that initially the boundary between the two layers is sharp. Suppose that the A^{z+} ion moves more rapidly than the B^{z-} ion. Then we soon have the situation illustrated in Fig. 31.12(b). In the first few moments of the process the positive ions outdistance the negative ions. An electrical double layer forms, with an associated electric field. The effect of this electric field is to speed up the slower ion and to slow down the faster ion. The system quickly adjusts so that both ions move *in the same direction with the same velocity.* If this adjustment did not occur, large departures from electrical neutrality would occur because of the difference in velocity between the positive and negative ions. Correspondingly enormous electric potential differences would develop in the direction of diffusion. In fact, the potential difference that develops and that equalizes the velocities of the ions is rather small ($< \sim 100$ mV); it is the *diffusion potential* and is responsible for the liquid junction potential that was described in Section 17.18.

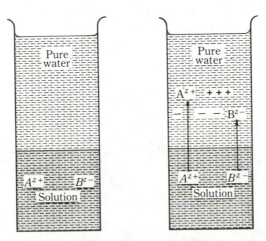

Figure 31.12 Diffusion of an electrolyte.

To deal with this situation algebraically, we write Eq. (31.52) for each ion, requiring that $v_+ = v_- = v$:

$$v_+ = v = -\tilde{u}_+ kT \frac{\partial \ln a_+}{\partial x} + \tilde{u}_+ z_+ eE_x; \tag{31.64}$$

and

$$v_- = v = -\tilde{u}_- kT \frac{\partial \ln a_-}{\partial x} + \tilde{u}_- z_- eE_x. \tag{31.65}$$

We have suppressed the subscripts T, p on this partial derivative; T and p are constant, nonetheless. These two equations determine v and E_x. We first eliminate E_x and solve for v. This is most easily done through the use of the electroneutrality condition. Since the formula of the electrolyte is $A_{v_+}^{z_+} B_{v_-}^{z_-}$, the electroneutrality condition is

$$v_+ z_+ + v_- z_- = 0. \tag{31.66}$$

If we multiply Eq. (31.64) by $v_+/\tilde{u}_+$ and Eq. (31.65) by $v_-/\tilde{u}_-$, we have

$$\frac{v_+ v}{\tilde{u}_+} = -kT \frac{v_+ \partial \ln a_+}{\partial x} + v_+ z_+ eE_x$$

and

$$\frac{v_- v}{\tilde{u}_-} = -kT \frac{v_- \partial \ln a_-}{\partial x} + v_- z_- eE_x.$$

Addition of these two equations yields

$$\left(\frac{v_+}{\tilde{u}_+} + \frac{v_-}{\tilde{u}_-}\right) v = -kT \frac{\partial \ln a_+^{v_+} a_-^{v_-}}{\partial x} = -vkT \frac{\partial \ln a_\pm}{\partial x}$$

The term in eE_x vanishes because of Eq. (31.66). To obtain the second equality we have used the definition of the mean ionic activity and $v = v_+ + v_-$. Following reasoning similar to that used in developing Eq. (31.54), and setting $\gamma_\pm = 1$, we find that we can replace $\partial \ln a_\pm$ by $\partial \ln \tilde{N} = \partial \tilde{N}/\tilde{N}$. Then, solving for v, we have

$$v = -\frac{vkT}{\dfrac{v_+}{\tilde{u}_+^\infty} + \dfrac{v_-}{\tilde{u}_-^\infty}} \frac{1}{\tilde{N}} \frac{\partial \tilde{N}}{\partial x}.$$

For the flow, $j = v\tilde{N}$, we obtain for the infinitely dilute solution,

$$j = -\frac{vkT}{\dfrac{v_+}{\tilde{u}_+^\infty} + \dfrac{v_-}{\tilde{u}_-^\infty}} \frac{\partial \tilde{N}}{\partial x}, \tag{31.67}$$

which is the equation for the diffusion of a simple electrolyte first obtained by Nernst. Comparing this equation with Fick's law, we see that

$$D^\infty = \frac{vkT}{\dfrac{v_+}{\tilde{u}_+^\infty} + \dfrac{v_-}{\tilde{u}_-^\infty}} \quad \text{or} \quad \frac{v}{D^\infty} = \frac{v_+}{kT\tilde{u}_+^\infty} + \frac{v_-}{kT\tilde{u}_-^\infty}, \tag{31.68}$$

where D^∞ is the diffusion coefficient of the electrolyte. Since by Eq. (31.59), $D_i^\infty = u_i^\infty kT,$

Eq. (31.68) reduces to

$$\frac{v}{D^\infty} = \frac{v_+}{D_+^\infty} + \frac{v_-}{D_-^\infty}. \tag{31.69}$$

This is the rule by which the individual ionic diffusion coefficients combine to yield the diffusion coefficient of the electrolyte. The ∞ superscript emphasizes that the relation is correct only at infinite dilution.

The form of Eq. (31.69) shows that the diffusion coefficient of the electrolyte is closer to the diffusion coefficient of the slower moving ion.

■ **EXAMPLE 31.2**   Using Eq. (31.62) to calculate the H^+ and Cl^- ion diffusion coefficients from their molar conductivities, we find

$$D^\infty(H^+) = \frac{(8.314 \text{ J/K mol})(298.15 \text{ K})}{(96\ 485 \text{ C/mol})^2(+1)^2} (349.8 \times 10^{-4} \text{ S m}^2/\text{mol}) = 9.315 \times 10^{-9} \text{ m}^2/\text{s}$$

and

$$D^\infty(Cl^-) = \frac{(8.314 \text{ J/K mol})(298.15 \text{ K})(76.35 \times 10^{-4} \text{ S m}^2/\text{mol})}{(96\ 485 \text{ C/mol})^2(-1)^2} = 2.033 \times 10^{-9} \text{ m}^2/\text{s}.$$

Then

$$\frac{2}{D^\infty(\text{HCl})} = \frac{1}{D^\infty(H^+)} + \frac{1}{D^\infty(Cl^-)} = \frac{1}{9.315 \times 10^{-9} \text{ m}^2/\text{s}} + \frac{1}{2.033 \times 10^{-9} \text{ m}^2/\text{s}};$$

$$D^\infty(\text{HCl}) = 3.338 \times 10^{-9} \text{ m}^2/\text{s}.$$

The diffusion coefficient for HCl is slightly larger than that of Cl^- because the faster moving H^+ ion pulls the slower Cl^- ion along.

It is interesting to compare the combinations, D_+^∞ and D_-^∞, in Eqs. (31.69) and (31.63). The difference between them is a consequence of the positive and negative ions moving in opposite directions in conduction, while in diffusion they both move in the same direction.

If we use Eq. (31.62) for the values of D_+ and D_- and substitute in Equation (31.69), we obtain

$$\frac{v}{D^\infty} = \frac{F^2}{RT} \left(\frac{v_+ z_+^2}{\lambda_+^\infty} + \frac{v_- z_-^2}{\lambda_-^\infty} \right). \tag{31.70}$$

This relation, first obtained by Nernst, expresses the diffusion coefficient in terms of the ion conductivities. Some values of the diffusion coefficients for several electrolytes are given in Table 31.7.

Table 31.7
Limiting diffusion coefficients, D^∞, for electrolytes at 25 °C

Compound	$D^\infty/(10^{-9} \text{ m}^2/\text{s})$	Compound	$D^\infty/(10^{-9} \text{ m}^2/\text{s})$
HCl	3.336	$CaCl_2$	1.335
LiCl	1.367	$SrCl_2$	1.335
NaCl	1.611	$BaCl_2$	1.386
KCl	1.994	Na_2SO_4	1.229
KNO_3	1.929	$LaCl_3$	1.293

★ 31.17.2 The Diffusion Potential

Consider two solutions of the same electrolyte with different concentrations that are in contact through a liquid junction. We can calculate the potential difference across this simple diffusion boundary by eliminating v between Eqs. (31.64) and (31.65). If we multiply Eq. (31.64) by $v_+ z_+$ and Eq. (31.65) by $v_- z_-$, then add them, we obtain

$$0 = -kT\left(v_+ z_+ \tilde{u}_+ \frac{\partial \ln a_+}{\partial x} + v_- z_- \tilde{u}_- \frac{\partial \ln a_-}{\partial x}\right) + (v_+ z_+^2 \tilde{u}_+ + v_- z_-^2 \tilde{u}_-)eE_x.$$

The left-hand side vanishes because of the electroneutrality condition, Eq. (31.66). Eliminating D_i between Eqs. (31.59) and (31.62) yields $\tilde{u}_i = \lambda_i/z_i^2 eF$. Using this value of $\tilde{u}_i$ everywhere in the equation brings it, after rearrangement, to the form

$$(v_+ \lambda_+ + v_- \lambda_-)E_x = \frac{RT}{F}\left(\frac{v_+ \lambda_+}{z_+}\frac{\partial \ln a_+}{\partial x} + \frac{v_- \lambda_-}{z_-}\frac{\partial \ln a_-}{\partial x}\right).$$

Next we introduce $E_x = -\partial\phi/\partial x$, $\Lambda = v_+\lambda_+ + v_-\lambda_-$, $v_+\lambda_+ = t_+\Lambda$, and $v_-\lambda_- = t_-\Lambda$. The equation becomes

$$-\frac{\partial\phi}{\partial x} = \frac{RT}{F}\left(\frac{t_+}{z_+}\frac{\partial \ln a_+}{\partial x} + \frac{t_-}{z_+}\frac{\partial \ln a_-}{\partial x}\right).$$

If we assume that $a_+ = a_- = a_\pm$, this becomes

$$-\frac{\partial\phi}{\partial x} = \frac{RT}{F}\left(\frac{t_+}{z_+} + \frac{t_-}{z_-}\right)\frac{\partial \ln a_\pm}{\partial x}. \tag{31.71}$$

This result relates the diffusion potential gradient to the gradient of the mean ionic activity. In symmetrical electrolytes, AB, $z_- = -z_+$ and the equation becomes

$$-\frac{\partial\phi}{\partial x} = \frac{RT}{z_+ F}(t_+ - t_-)\frac{\partial \ln a_\pm}{\partial x}$$

This formula shows clearly that if we choose an electrolyte in which $t_+ \approx t_-$ (such as KCl) the diffusion potential will be very small. It is for this reason that KCl is used in salt bridges that are intended to minimize the liquid junction potentials. The equation also indicates that more highly charged ions will produce a lower diffusion potential gradient.

Calculation of the diffusion potential requires the integration of Eq. (31.71) over the diffusion region:

$$\int_{x_1}^{x_2} -\frac{\partial\phi}{\partial x}\,dx = \frac{RT}{F}\int_{x_1}^{x_2}\left(\frac{t_+}{z_+} + \frac{t_-}{z_-}\right)\frac{\partial \ln a_\pm}{\partial x}\,dx.$$

If we have an ordinary concentration cell in which the activity varies from $(a_\pm)_1$ to $(a_\pm)_2$—and if we assume for simplicity that the expression in parentheses on the right-hand side of the equation is independent of concentration—we have

$$\Delta\phi_{\text{diff}} = -\frac{RT}{F}\left(\frac{t_+}{z_+} + \frac{t_-}{z_-}\right)\ln\frac{(a_\pm)_2}{(a_\pm)_1}, \tag{31.72}$$

an equation equivalent to Eq. (17.63) for the junction potential.

★ 31.17.3 The Ion Distribution in the Steady-State

A final remark is needed concerning Eq. (31.55). We can imagine a situation in which the imposed external electric field is in such a direction and of such a magnitude that it exactly counterbalances the effect of the concentration gradient on the motion of the ion. Then $v_i = 0$ and Eq. (31.55) becomes, after dividing by $\tilde{u}_i$,

$$0 = \frac{kT}{\tilde{N}_i} \frac{\partial \tilde{N}_i}{\partial x} + z_i e \left(-\frac{\partial \phi}{\partial x} \right).$$

Rearranging, and multiplying by dx, we have

$$\frac{d\tilde{N}_i}{\tilde{N}_i} = \frac{z_i e \, d\phi}{kT}$$

When we integrate between a position where the potential is zero and one where the potential is ϕ, this becomes

$$\ln \frac{\tilde{N}_i}{\tilde{N}_{i0}} = -\frac{z_i e \phi}{kT} = -\frac{z_i F \phi}{RT},$$

or

$$\tilde{N}_i = \tilde{N}_{i0} e^{-z_i F \phi / RT}. \tag{31.73}$$

This is the Boltzmann distribution. This derivation shows that in the nonequilibrium region, in which the linear laws hold, the equilibrium distribution can still describe the system to a first approximation. It is fortunate that this is so; if it were not so, the mathematical complications would be enormous.

QUESTIONS

31.1 Identify the charge carriers in metals and in ionic solutions.

31.2 Why should Faraday's laws of electrolysis be exact?

31.3 Why is alternating current used in the measurement of conductivities of ions in solution?

31.4 Why should ions of a strong electrolyte migrate independently at low concentration?

31.5 The equivalent conductivity of a weak electrolyte varies approximately with $c^{-1/2}$ (Fig. 31.3). Explain this in terms of the equilibrium constant for small degree of dissociation.

31.6 Identify the two effects that lead to a decrease in Λ with concentration for strong electrolytes.

31.7 What aspect of Eq. (31.71) shows that the diffusion potential is a steady state—and not an equilibrium—potential?

PROBLEMS

31.1 A potential difference of 100 V is applied across a wire 2.0 m long and 0.050 cm in diameter. If the current is 25 A, calculate

   a) the resistance and conductance of the wire;
   b) the field strength;
   c) the current density;
   d) the resistivity and conductivity of the wire.

31.2 A metal wire carries a current of 1 A. How many electrons pass a point in the wire in 1 second?

31.3 The resistivity of copper is 1.72×10^{-8} Ω m. Calculate the current if 20.0 V is impressed on a wire 6.0 m long and 2.0×10^{-5} m in diameter.

31.4 A silver foil, 0.00254 cm thick and 0.50 mm wide, connects two points that are 4.2 cm apart. If the current passing in the foil is 1.5 mA, what is the potential drop between the two points? For silver, $\kappa = 6.30 \times 10^7$ S/m.

31.5 If a potential difference of 10.0 mV is imposed between the ends of a piece of iron wire 0.1024 cm in diameter and 58.4 cm long, a current of 145 mA flows. Calculate the resistivity of the iron wire.

31.6 For platinum, the Hall coefficient is -2.00×10^{-11} m^3/C, the resistivity is 10.6×10^{-8} Ω m, and the density is 21.45 g/cm^3. Calculate

a) the mobility of the electrons;

b) the number of electrons per atom.

c) If a current of 122 mA is passed in a foil that is 0.00508 cm thick and 2.12 cm wide, what is the value of the Hall potential in a magnetic field of 0.500 tesla?

31.7 In a measurement of the Hall effect, a current of 2.00 A was passed through a strip of silver 1.50 cm wide and 0.0127 cm thick. A transverse potential of 1.32 μV was produced using a magnetic field of 0.750 tesla. Calculate the Hall coefficient for silver.

31.8 A solution of sulfuric acid is electrolyzed using a current of 0.10 A for three hours. How many cm^3 (at STP) of hydrogen and oxygen are produced?

31.9 Potassium chlorate is prepared by the electrolysis of KCl in basic solution:

$$6\,OH^- + Cl^- \longrightarrow ClO_3^- + 3\,H_2O + 6e^-$$

If only 60% of the current is utilized in this reaction, what time will be required to produce 10 g of KClO$_3$ using a current of 2 A?

31.10 What mass of AgCl is produced at a silver anode electrolyzed in HCl solution by a current of 0.50 A passing for 2.5 hours?

31.11 If 0.4793 g of silver is deposited at the cathode during the electrolysis of a silver nitrate solution lasting 4 hours, 27 minutes and 35 seconds, what was the average current passing during the experiment?

31.12 A current is passed for 3 hours, 10 minutes and 18 seconds through a solution of KI; 34.62 mL of 0.1046 mol/L Na$_2$S$_2$O$_3$ solution are required to titrate the liberated iodine according to the reaction

$$I_2 + 2\,S_2O_3^{2-} \longrightarrow 2\,I^- + S_4O_6^{2-}$$

What is the average current passed during the experiment?

31.13 Nitrobenzene, $C_6H_5NO_2$, can be reduced to aniline, $C_6H_5NH_2$, at a mercury cathode. If the current efficiency is 80%, how long must a current of 3.0 A flow in the cell to produce 1.0 kg of aniline?

31.14 A solution of KCl has a conductivity of 0.14088 S/m at 25 °C. A cell filled with this solution has a resistance of 4.2156 Ω.

a) What is the cell constant?

b) The same cell filled with a solution of HCl has a resistance of 1.0326 Ω. What is the conductivity of the HCl solution?

31.15 For mercury at 0 °C, $\kappa = 1.062963 \times 10^6$ S/m.

a) If the resistance of a cell containing mercury is 0.243166 Ω, what is the cell constant of the cell?

b) If the same cell is filled with potassium chloride solution at 0 °C, the resistance of the cell is 3.966×10^4 Ω. What is the conductivity of the KCl solution?

c) If the average cross-sectional area of the cell is 0.9643 mm^2, what is the effective distance between the electrodes?

31.16 Using the values of λ^∞ from Table 31.5 for H^+, Na^+, Ca^{2+}, La^{3+}, OH^-, Br^-, SO_4^{2-}, and $P_2O_7^{4-}$, calculate

a) the mobilities of the ions;

b) the velocities of these ions in a cell that has electrodes 5.00 cm apart to which a potential difference of 2.00 V is applied.

31.17 The mobility of the NH_4^+ ion is 7.623×10^{-8} m^2/V s. Calculate

a) the molar conductivity of the NH_4^+ ion;

b) the velocity of the ion if 15.0 volts are applied across electrodes 25 cm apart;

c) the transport number of the ion in $NH_4C_2H_3O_2$ solution if the mobility of the $C_2H_3O_2^-$ ion is 4.239×10^{-8} m^2/V s.

31.18 Use the molar conductivities in Table 31.5, assuming that they do not vary with concentration, to

a) estimate the conductivity of 0.0100 mol/L solutions of $AgNO_3$; HCl; $CaCl_2$; $MgSO_4$; $La_2(SO_4)_3$;

b) estimate the resistance for each case in (a) in a cell where the distance between the electrodes is 8.0 cm and the effective area of the conducting path is 1.6 cm^2.

31.19 a) Relate the changes in concentration in the Hittorf cell to the transference number of the positive ion and the quantity of electricity passed if the cell is filled with hydrochloric acid and both electrodes are silver–silver chloride electrodes.

b) What relation is obtained if the cathode is replaced by a platinum electrode so that H_2 is evolved?

c) What relation is obtained if the anode is replaced by a platinum electrode and oxygen is evolved? (*Note:* Using an electrode that allows gas evolution would be a very bad way to do a Hittorf experiment! Why?)

31.20 A Hittorf cell fitted with silver–silver chloride electrodes is filled with HCl solution that contains 0.3856×10^{-3} g HCl/g water. A current of 2.00 mA is passed for exactly 3 hours. The solutions are withdrawn, weighed, and analyzed. The total weight of the cathode solution is 51.7436 g; it contains 0.0267 g of HCl. The anode solution weighs 52.0461 g and contains 0.0133 g of HCl. What is the transference number of the hydrogen ion?

31.21 In a Hittorf experiment to determine the transference numbers in KCl solution, the following data were obtained. (D. A. MacInnes and M. Dole. *J.A.C.S.* **53**, 357 [1931].) Mass of the anode solution, 117.79 g; mass of the cathode solution, 120.99 g. Percent KCl in anode portion, 0.10336%; percent KCl in cathode portion, 0.19398%. The percent KCl in the middle portion was 0.14948%. Calculate t_+ from the amounts of KCl transferred from the anode compartment and to the cathode compartment and the average value of t_+. (*Note:* 0.16034 g of silver was deposited in a silver coulometer in series with the cell. The concentration of KCl was 0.2 mol/L.) Silver–silver chloride electrodes were used.

31.22 A moving-boundary experiment is done to measure the transference number of Li^+ in 0.01 mol/L LiCl. In a tube having a cross-sectional area of 0.125 cm^2, the boundary moves 7.3 cm in 1490 s using a current of 1.80×10^{-3} A. Calculate t_+.

31.23 In a moving-boundary experiment to determine the transference number of chloride ion in 0.010 mol/L sodium chloride solution, the chloride ion moved a distance of 3.0 cm in 976 s. The cross section of the tube was 0.427 cm^2 and the current 2.08×10^{-3} A. Calculate t_-.

31.24 From the data in Table 31.5 calculate the transference number of the chloride ion in each of the infinitely dilute solutions, HCl, NaCl, KCl, $CaCl_2$, and $LaCl_3$.

31.25 The conductivity of any solution is given by Eq. (31.29). Calculate the transference number of each ion in a solution that contains 0.10 mol/L $CaCl_2$ and 0.010 mol/L HCl. Use values of λ^∞ in Table 31.5.

31.26 What is the ratio of concentrations of HCl and NaCl in a solution if the transference number of the hydrogen ion is 0.5? (Use the data in Table 31.5.)

31.27 The equivalent conductivity of LiCl at infinite dilution is 115.03×10^{-4} S m^2/mol. The transference number of the cation is 0.336.

a) Calculate the mobility of the cation.
b) Calculate the velocity of the cation if 6.0 volts are applied across electrodes 4.0 cm apart.

31.28 At 18 °C, the data for KNO$_3$ solutions are

$c/(10^{-3}$ mol/L)	0.20	0.50	1.0	2.0	5.0
$\Lambda/(10^{-4}$ S m^2/mol)	125.2	124.5	123.7	122.6	120.5

Using an appropriate plot, find Λ^∞.

31.29 The data for HCl solutions at 25 °C are

$c/(10^{-3}$ mol/L)	2.8408	8.1181	17.743	31.863
$\Lambda/(10^{-4}$ S m^2/mol)	425.13	424.87	423.94	423.55

Using an appropriate plot, find Λ^∞.

31.30 At 25 °C, for water, $\eta = 8.949 \times 10^{-4}$ Pa s and $\epsilon_r = 78.54$. Using the Onsager equation and the data in Table 31.5, calculate the molar conductivity of HCl, KCl, and LiCl solutions having $c/(\text{mol/L}) = 0.0001, 0.001, 0.01$.

31.31 The crystallographic radii of Na$^+$ and Cl$^-$ are 95 pm and 181 pm. Estimate the ion conductivities using Stokes's law and compare with the values in Table 31.5 ($\eta = 0.89 \times 10^{-3}$ Pa s).

31.32 At 25 °C the values of $\Lambda^\infty/10^{-4}$ S m^2/mol are: sodium benzoate, 82.48; hydrochloric acid, 426.16; and sodium chloride, 126.45. Calculate Λ^∞ for benzoic acid.

31.33 The molar conductivity of acetic acid is

$\Lambda/(10^{-4}$ S m^2/mol)	49.50	35.67	25.60
$c/(\text{mol/L})$	9.88×10^{-4}	19.76×10^{-4}	39.52×10^{-4}

Using an appropriate plot, find

a) Λ^∞;
b) the dissociation constant;
c) the degree of dissociation at each concentration.

31.34 At 25 °C, a solution of KCl having a conductivity of 0.14088 S/m exhibits a resistance of 654 Ω in a particular conductivity cell. In this same cell, a 0.10 mol/L solution of NH$_4$OH has a resistance of 2524 Ω. The limiting molar ionic conductivities are available in Table 31.5. Calculate

a) the cell constant;
b) the molar conductivity of the NH$_4$OH solution;
c) the degree of dissociation of the 0.10 mol/L NH$_4$OH;
d) the dissociation constant of NH$_4$OH.

31.35 The conductivity of a saturated solution of $BaSO_4$ is 3.48×10^{-4} S/m. The conductivity of pure water is 0.50×10^{-4} S/m. Calculate the solubility product of $BaSO_4$. (Use Table 31.5.)

31.36 A saturated solution of MgF_2 has $\kappa = 0.02538$ S/m. Calculate the solubility product of MgF_2. (Use κ_w from Problem 31.35 and data from Table 31.5.)

31.37 a) Suppose that a strong acid, HA, having a concentration $\tilde{c}_a$ (mol/m^3) is titrated with a strong base, MOH, having a concentration $\tilde{c}_b$ (mol/m^3). If v_o is the volume of the acid, and v the volume of base added at any stage of the titration, show that the conductivity before the equivalence point is reached is given as a function of v by

$$\kappa = \left(\frac{v_o}{v_o + v}\right)\left[\kappa_a - \tilde{c}_b\left(\frac{v}{v_o}\right)(\lambda_{H^+} - \lambda_{M^+})\right],$$

where κ_a is the conductivity of the acid solution before any base has been added. Assume that the values of λ do not change with the volume of the solution. (Before the equivalence point is reached, the concentration of OH^- is negligible compared with that of H^+; this situation is reversed after the equivalence point.)

b) Show that after the equivalence point is passed

$$\kappa = \left(\frac{v_o + v_e}{v_o + v}\right)\kappa_e + \left(\frac{v - v_e}{v_o + v}\right)\kappa_b,$$

where κ_b is the conductivity of the basic solution, κ_e that of the solution at the equivalence point, and v_e the volume of base added at the equivalence point.

c) To 50 cm^3 of 0.100 mol/L HCl solution, portions of a 0.100 mol/L NaOH solution are added. Calculate the conductivity for volumes of base, $v/cm^3 = 0, 10, 25, 40, 45, 50, 55, 60, 75, 90,$ and 100. Sketch κ versus v. (Use data in Table 31.5.)

31.38 Suppose that acetic acid, $K_a = 1.8 \times 10^{-5}$, is titrated using a strong base. To 50 cm^3 of the 0.10 mol/L acid, 0, 10, 40, 45, 50, 55, 60, 90, and 100 cm^3 of 0.10 mol/L sodium hydroxide are added. Calculate the conductivity after each addition of base and plot the conductivity versus v. (Use data in Table 31.5.)

31.39 a) Using values of λ_i^∞ from Table 31.5, calculate the diffusion coefficient for each of the ions $H^+, OH^-, Na^+, Ag^+, Ca^{2+}, Cu^{2+}, NO_3^-, SO_4^{2-}$, and $Co(NH_3)_6^{3+}$.

b) Calculate the mobility in unit force field, $\tilde{u}_i$, and compare it with u_i for Ag^+ and Cu^{2+}.

31.40 From the data in Table 31.5 calculate the diffusion coefficients for HI, $ZnCl_2$, $MgSO_4$, and $La_2(SO_4)_3$.

31.41 Using the data in Tables 31.5 and 16.1, calculate the junction potential between two solutions having concentrations of 0.010 molal and 0.10 molal for each of the electrolytes $CuSO_4$, HCl, K_2SO_4, and $La(NO_3)_3$.

32

Chemical Kinetics

I. Empirical Laws and Mechanism

32.1 INTRODUCTION

The rates of chemical reactions form the subject matter of chemical kinetics. Experimentally it is found that the rate of a chemical reaction is dependent on the temperature, pressure, and the concentrations of the species involved. The presence of a catalyst or inhibitor can change the rate by many powers of ten. From the study of the rate of a reaction and its dependence on all these factors, much can be learned about the detailed steps by which the reactants are transformed to products.

32.2 RATE MEASUREMENTS

In the course of a chemical reaction the concentrations of all the species present change with time, and so the properties of the system change. The rate of the reaction is measured by measuring the value of any convenient property that can be related to the composition of the system as a function of time. The property chosen should be reasonably easy to measure; it should change sufficiently in the course of the reaction to permit an accurate distinction to be made between the various compositions of the system as time passes. The property chosen depends on the individual reaction. In one of the first quantitative studies of reaction rates, Wilhelmy (1850) measured the rate of inversion of sucrose by measuring the change with time of the angle of rotation of a beam of plane-polarized light passed through the sugar solution.

There are many methods of following a reaction with time. Some of them are: changes in pressure, changes in pH, changes in refractive index, changes in absorbance at one or more wavelengths, changes in thermal conductivity, changes in volume, changes in electrical resistance. If physical methods such as these can be applied, they are usually more convenient than chemical methods.

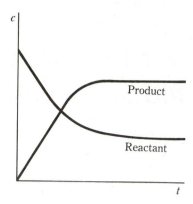

Figure 32.1   Variation of concentration with time.

Since the rate of the majority of chemical reactions is very sensitive to temperature, the reaction vessel must be kept in a thermostat so that a constant temperature can be accurately maintained. In some cases it is necessary also to control the pressure.

No matter what property we choose to measure, the data can ultimately be translated into a variation of the concentration of a reactant or a product with time. Figure 32.1 shows this variation schematically for a reactant and for a product. The concentration of any reactant decreases from its initial to the equilibrium value and the concentration of any product increases from its initial value (usually zero) to the equilibrium value.

Our task now is to describe the curves in Fig. 32.1 more accurately. We begin by describing the various rate laws that have been found experimentally. Later we will interpret these laws in terms of the molecular processes involved.

Consider a chemical equation written in the general form:

$$0 = \sum_i v_i A_i, \tag{32.1}$$

where A_i is the chemical formula of the ith species participating in the chemical reaction and v_i is the corresponding stoichiometric coefficient. For any reactant, v_i is negative, for any product, v_i is positive. The number of moles of the ith species, n_i, is given by

$$n_i = n_i^0 + v_i \xi, \tag{32.2}$$

where ξ is the advancement of the reaction, n_i^0 is the number of moles of the ith species initially, that is, when $\xi = 0$. Then, differentiating with respect to time, we obtain

$$\frac{dn_i}{dt} = v_i \frac{d\xi}{dt}. \tag{32.3}$$

We define the rate of reaction as the rate of increase of the advancement with time:

$$\text{Rate of reaction} \equiv \frac{d\xi}{dt}. \tag{32.4}$$

The rate of change of the number of moles of the ith species is given by Eq. (32.3). Inverting Eq. (32.3) we have

$$\frac{d\xi}{dt} = \frac{1}{v_i} \frac{dn_i}{dt} \tag{32.5}$$

Consider the reaction

$$2\,N_2O_5 \longrightarrow 4\,NO_2 + O_2.$$

Using Eq. (32.5) we relate the rate of reaction to the changes in the mole numbers by

$$\frac{d\xi}{dt} = -\frac{1}{2}\frac{dn_{N_2O_5}}{dt} = \frac{1}{4}\frac{dn_{NO_2}}{dt} = \frac{dn_{O_2}}{dt}. \tag{32.6}$$

Since

$$-\frac{dn_{N_2O_5}}{dt} = \text{the rate of disappearance of } N_2O_5,$$

$$\frac{dn_{NO_2}}{dt} = \text{the rate of formation of } NO_2, \text{ and}$$

$$\frac{dn_{O_2}}{dt} = \text{the rate of formation of } O_2,$$

we can say that the rate of reaction is equal to one-half the rate of disappearance of N_2O_5 or one-fourth the rate of formation of NO_2 or the rate of formation of O_2. Note that the relation of the rate of reaction to the rate of change of the mole numbers depends on the particular set of coefficients used in balancing the chemical equation. There is no unique way to choose this set of coefficients; we simply decide on a convenient set at the beginning, and then use that set consistently throughout the problem. Keep in mind that the rate of reaction, $d\xi/dt$, is not defined until we have written a balanced chemical equation.

Finally, we select some easily measurable property of the system, Z, with a known dependence on the mole numbers or concentrations of the various substances participating in the reaction; that is, we know the functionality:

$$Z = Z(n_1, n_2, n_3, \ldots).$$

Then

$$\frac{dZ}{dt} = \frac{\partial Z}{\partial n_1}\frac{\partial n_1}{\partial t} + \frac{\partial Z}{\partial n_2}\frac{\partial n_2}{\partial t} + \cdots.$$

Using the values given by Eq. (32.3) for dn_i/dt, this becomes

$$\frac{dZ}{dt} = \left(v_1\frac{\partial Z}{\partial n_1} + v_2\frac{\partial Z}{\partial n_2} + \cdots\right)\frac{d\xi}{dt}, \tag{32.7a}$$

or

$$\frac{d\xi}{dt} = \frac{\dfrac{dZ}{dt}}{v_1\dfrac{\partial Z}{\partial n_1} + v_2\dfrac{\partial Z}{\partial n_2} + \cdots}. \tag{32.7b}$$

This is the required relation between the rate of the reaction, $d\xi/dt$, and the rate of change of the measurable property with time, dZ/dt. If the volume of the system is constant, then $n_i = \tilde{c}_i V$, and we obtain

$$\frac{1}{V}\frac{d\xi}{dt} = \frac{\dfrac{dZ}{dt}}{v_1\dfrac{\partial Z}{\partial \tilde{c}_1} + v_2\dfrac{\partial Z}{\partial \tilde{c}_2} + \cdots} \tag{32.7c}$$

for the rate of the reaction per unit volume.

In Eq. (32.7c) we chose to write the volume in cubic metres and the concentration in mol/m^3. In these equations we could equally well choose to use litres for the volume unit and mol/L for the concentration unit. All that is required is that the two be in conformable units; that is, the product of concentration times volume must equal the amount of substance. For the remainder of this chapter we will use concentrations in mol/L, which is the customary unit.

■ **EXAMPLE 32.1**   The rate of decomposition of acetaldehyde can be studied by measuring the pressure in a system at constant volume and temperature. Express the rate of reaction in terms of the rate of change of the pressure. The overall reaction is

$$CH_3CHO(g) \longrightarrow CH_4(g) + CO(g)$$

with mole numbers

$$n_1 = n^0 - \xi; \quad n_2 = 0 + \xi; \quad n_3 = 0 + \xi.$$

Then

$$p = (n_1 + n_2 + n_3)\frac{RT}{V} = (n^0 - \xi + \xi + \xi)\frac{RT}{V} = (n^0 + \xi)\frac{RT}{V}.$$

The initial pressure, $p^0 = n^0 RT/V$; then $p = p^0 + (RT/V)\xi$; and we obtain

$$\frac{dp}{dt} = \frac{RT}{V}\frac{d\xi}{dt} \quad \text{or} \quad \frac{1}{V}\frac{d\xi}{dt} = \frac{1}{RT}\frac{dp}{dt}.$$

Alternatively, we can use Eq. (32.7a). We see that: $\partial p/\partial n_i = RT/V$ for all i; then Eq. (32.7a) becomes

$$\frac{dp}{dt} = \left(v_1\frac{RT}{V} + v_2\frac{RT}{V} + v_3\frac{RT}{V}\right)\frac{d\xi}{dt} = \frac{\Delta v RT}{V} = (-1 + 1 + 1)\frac{RT}{V}\frac{d\xi}{dt} = \frac{RT}{V}\frac{d\xi}{dt}.$$

32.3 RATE LAWS

The rate of reaction will be a function of temperature, pressure, and the concentrations of the various species in the reaction, c_i, and may depend on the concentrations, c_x, of species such as catalysts or inhibitors that may not appear in the overall reaction. Furthermore, if the reaction occurs homogeneously (that is, exclusively within a single phase), the rate is proportional to the volume of the phase, V. If the reaction occurs on an active surface the rate is proportional to the area of the active surface, A. Thus, in a very general way we can write the rate of reaction as the sum of the rates of the homogeneous and surface reaction:

$$\frac{d\xi}{dt} = Vf(T, p, c_i, c_x) + AF(T, p, c_i, c_x), \tag{32.8}$$

where $f(T, p, c_i, c_x)$ and $F(T, p, c_i, c_x)$ are functions to be determined from the experimental data. Equation (32.8) is the *rate law* for the reaction.

Reactions are classified kinetically as homogeneous or heterogeneous. A homogeneous reaction occurs entirely in one phase; a heterogeneous reaction occurs, at least in part, in more than one phase. A common type of heterogeneous reaction has a rate which depends on the area of a surface that is exposed to the reaction mixture. This surface may be the interior wall of the reaction vessel or it may be the surface of a solid catalyst. At some stage in any kinetic study it is necessary to find out if the reaction is influenced by the walls of the vessel. If the vessel is made of glass, it is usually packed with glass wool or beads or many fine glass tubes so as to increase the exposed area. Any effect on the rate of the reaction is noted. If the reaction is strictly homogeneous, the rate will not be affected by packing the vessel in this way. In this chapter the discussion will be restricted almost entirely to homogeneous reactions.

For homogeneous reactions, the second term on the right-hand side of Eq. (32.8) is negligible and we have

$$\frac{d\xi}{dt} = Vf(T, p, c_i, c_x).$$ (32.9a)

In this situation it is convenient to deal with the rate of the reaction per unit volume, $(1/V)(d\xi/dt)$. In view of Eq. (32.8), the rate per unit volume becomes

$$\frac{d(\xi/V)}{dt} = f(T, p, c_i, c_x),$$ (32.9b)

which is the rate law for a homogeneous reaction.

Dividing Eq. (32.5) by the volume, we obtain

$$\frac{1}{V}\frac{d\xi}{dt} = \frac{1}{v_i V}\frac{dn_i}{dt}.$$

If the volume does not change with time, this equation takes the form

$$\frac{d(\xi/V)}{dt} = \frac{1}{v_i}\frac{dc_i}{dt},$$ (32.10)

in which c_i is the concentration of the ith species; $c_i = n_i/V$.

In many cases, the rate law has the simple form

$$\frac{d(\xi/V)}{dt} = kc_A^\alpha c_B^\beta c_C^\gamma \cdots,$$ (32.11)

in which $c_A, c_B, c_C, \ldots$, denote the concentrations of the participating species, and k, α, β, and γ are constants. The constant k is the *rate constant* of the reaction, or the *specific rate* of the reaction, since k is the rate if all the concentrations are unity. In general the rate constant depends on temperature and pressure. The constant α is the *reaction order* with respect to A, β is the reaction order with respect to B, and γ is the reaction order with respect to C. The *overall reaction order* is the sum: $\alpha + \beta + \gamma$.

The order of the reaction governs the mathematical form of the rate law and therefore the variation in concentration of all the species with time. The order of the reaction with respect to the various species must be discovered *from experiment*. The experimental determination of the order of the reaction with respect to the various substances taking part is one of the first objectives of a kinetic investigation. It cannot be emphasized too strongly that the order of the reaction with respect to a given substance has *no relation whatsoever* to the stoichiometric coefficient of that substance in the chemical equation. For example, in the chemical equation above, the coefficient of N_2O_5 is 2. We cannot infer from this that the reaction is second order with respect to N_2O_5. (*Elementary reactions—*reactions that take place in a single act—are excepted from this statement.)

32.4 FIRST-ORDER REACTIONS

Consider a simple decomposition reaction of the type

$$A \longrightarrow \text{Products.}$$

Since substance A is the only reactant, we choose to balance the equation with the coefficient of A equal to unity. Suppose that the reaction is first-order with respect to A and that the rate does not depend on the concentrations of any products; then the rate law, Eq. (32.11), becomes

$$\frac{d(\xi/V)}{dt} = kc \tag{32.12}$$

where c is the concentration of A.

To integrate this equation we must either express c as a function of ξ/V or ξ/V as a function of c. In either case, we obtain the relation by dividing Eq. (32.2) by V,

$$c = c_0 - \frac{\xi}{V}, \tag{32.13}$$

and then differentiate with respect to time:

$$\frac{dc}{dt} = -\frac{d(\xi/V)}{dt}. \tag{32.14}$$

Using this value for $d(\xi/V)/dt$ in Eq. (32.12), we find that

$$-\frac{dc}{dt} = kc. \tag{32.15}$$

By rearranging we can separate the variables,

$$\frac{dc}{c} = -k\,dt,$$

and integrate from $t = 0$ when $c = c_0$ to t; then

$$\int_{c_0}^{c} \frac{dc}{c} = -k \int_0^t dt,$$

or

$$\ln \frac{c}{c_0} = -kt, \tag{32.16}$$

which can also be written

$$c = c_0 e^{-kt}. \tag{32.17}$$

Thus for a first-order decomposition, the concentration of A decreases exponentially with time. After measuring c as a function of time we can test whether the reaction is first order in A by plotting $\ln (c/c_0)$ versus t. According to Eq. (32.16) this plot should be a straight line if the reaction is first order in A. If we find that our experimental points lie on a straight line we conclude that the reaction is first order in A. The slope of this line is equal to $-k$.

The half-life, τ, of the reaction is the time required for the concentration of A to reach one-half of its initial value. Therefore, when $t = \tau$, $c = \frac{1}{2}c_0$. Putting these values into

Eq. (32.16), we obtain $\ln \frac{1}{2} = -k\tau$, so that

$$\tau = \frac{\ln 2}{k} = \frac{0.693}{k}. \tag{32.18}$$

One way to evaluate the rate constant of a reaction is to determine the half-life for various initial concentrations of the reactant A. If the half-life is independent of the initial concentration, then the reaction is first order, and the rate constant is calculated using Eq. (32.18). It is only for first-order reactions that the half-life is independent of the initial concentration.

The decomposition of N_2O_5 is an example of a first-order reaction. The stoichiometry is represented by

$$2\,N_2O_5 \longrightarrow 4\,NO_2 + O_2,$$

and the rate law is

$$-\frac{dc_{N_2O_5}}{dt} = kc_{N_2O_5}.$$

At 25 °C the rate constant is $3.38 \times 10^{-5}\ \text{s}^{-1}$. Note the absence of any relation between the order of the reaction and the stoichiometric coefficient of N_2O_5 in the chemical equation.

■ **EXAMPLE 32.2**   Calculate the half-life for N_2O_5 at 25 °C and the fraction decomposed after 8 hours; $k = 3.38 \times 10^{-5}\ \text{s}^{-1}$.

$$\tau = \frac{0.693}{k} = \frac{0.693}{3.38 \times 10^{-5}\ \text{s}^{-1}} = 20\,500\ \text{s}.$$

After 8 hours the fraction remaining is given by Eq. (32.17); since 8 hr = 8 hr (60 min/hr)(60 s/min) = 28 800 s, we have

$$\frac{c}{c_0} = e^{-kt} = e^{-3.38 \times 10^{-5}\,\text{s}^{-1}(28800\,\text{s})} = e^{-0.973} = 0.378.$$

Therefore the fraction decomposed is $1.000 - 0.378 = 0.622$

32.4.1 Radioactive Decay

The radioactive decay of an unstable nucleus is an important example of a process that follows a first-order rate law. Choosing Cu^{64} as an example, we have the transformation

$$Cu^{64} \longrightarrow Zn^{64} + \beta^-, \qquad \tau = 12.8\ \text{hr}.$$

The emission of a β-particle occurs with the formation of a stable isotope of zinc. The probability of this occurrence in the time interval dt is simply proportional to dt. Therefore

$$-\frac{dN}{N} = \lambda\,dt, \tag{32.19}$$

where $-dN$ is the number of copper nuclei that disintegrate in the interval dt. Equation (32.19) is a first-order law, and can be integrated to the form

$$N = N_0 e^{-\lambda t}, \tag{32.20}$$

N_0 being the number of Cu^{64} nuclei present at $t = 0$, N the number at any time t. The

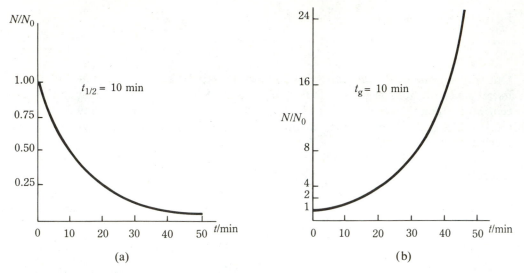

Figure 32.2   (a) Radioactive decay. (b) Bacterial growth.

constant λ is the *decay constant* and is related to the half-life by

$$\lambda = \frac{\ln 2}{\tau}. \tag{32.21}$$

In contrast to the rate constant of a chemical reaction, the decay constant λ is completely independent of any external influence such as temperature or pressure. Using the value of λ from Eq. (32.21) in Eq. (32.20), we obtain, since $\exp(\ln 2) = 2$,

$$N = N_0 e^{-(\ln 2)t/\tau} = N_0(2)^{-t/\tau} = N_0(\tfrac{1}{2})^{t/\tau}. \tag{32.22}$$

From Eq. (32.22) it is clear that after the elapse of a period equal to two half-lives, $(\tfrac{1}{2})^2 = \tfrac{1}{4}$ of the substance remains. After three half-lives have elapsed, $\tfrac{1}{8}$ remains, after 4 half-lives, $\tfrac{1}{16}$, and so on. The mathematics is the same as that of the barometric distribution (Section 2.9). The number, N, as a function of t is shown in Fig. 32.2(a).

32.4.2   Bacterial Growth

A bacterial colony grows most commonly by cell division. In an actively growing colony the probability of cell division in a time interval dt is proportional to dt; thus

$$\frac{dN}{N} = \lambda_g\, dt \tag{32.23}$$

where dN is the number of cells that divide in the time interval dt, and λ_g is a constant. This growth law is very similar to the law of radioactive decay in Eq. (32.19), except that the negative sign is missing. Upon integration we obtain

$$N = N_0 e^{\lambda_g t}. \tag{32.24}$$

Figure 32.2(b) shows N/N_0 as a function of time.

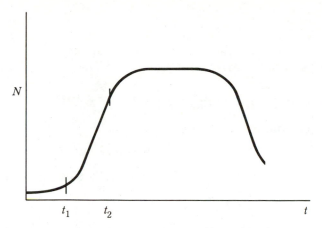

Figure 32.3   Growth and decay of bacterial colony.

The generation time, t_g, is the time required for the population to double; that is, when $t = t_g$, $N_0 = 2N_0$; thus Eq. (32.24) becomes

$$2N_0 = N_0 e^{\lambda_g t_g},$$

or

$$\lambda_g = \frac{\ln 2}{t_g}. \tag{32.25}$$

Using this value for λ_g in Eq. (32.24) we have

$$\frac{N}{N_0} = e^{(\ln 2)t/t_g} = 2^{t/t_g}, \tag{32.26}$$

since $e^{\ln 2} = 2$.

The growth law, Eq. (32.26), is not applicable during the entire history of a bacterial colony. A typical population curve, N versus t, is shown in Fig. 32.3. There is an initial induction period, followed by a period between t_1 and t_2 during which the exponential growth occurs, as described by Eq. (32.26). The population growth slows, then stops; in the final phase the population drops as the bacteria die off more rapidly than they are produced. Equation (32.26) describes the growth only during the exponential phase in the interval from t_1 to t_2. The leveling off occurs as the supply of nutrients is exhausted. Finally, if the environment becomes sufficiently hostile (due to lack of nutrients or increased concentrations of toxic substances), the colony dies.

32.4.3   Compound Interest

The law of compound interest on investment is the same as the law of bacterial growth. If P_0 is the initial value of the principal amount, t_1 the interval at which compounding occurs, and r the interest rate for the interval t_1, expressed as a fraction, the principal at time t will be

$$P = P_0(1 + rt_1)^{t/t_1}. \tag{32.27}$$

If the compounding occurs instantaneously, then we have

$$\lim_{t_1 = 0} (1 + rt_1)^{1/t_1} = e^r, \tag{32.28}$$

and thus

$$P = P_0 e^{rt}, \tag{32.29}$$

which is the same growth law as for bacteria.

32.5 SECOND-ORDER REACTIONS

We return to the decomposition reaction,

$$A \longrightarrow Products,$$

but now assume that the reaction is second order. If c is the concentration of A at any time, the rate law is

$$\frac{d(\xi/V)}{dt} = kc^2, \tag{32.30}$$

which, in view of Eq. (32.14), becomes

$$-\frac{dc}{dt} = kc^2. \tag{32.31}$$

Separating variables, we have

$$-\frac{dc}{c^2} = k\,dt.$$

Integrating from $(c_0, 0)$ to (c, t) we obtain

$$\int_{c_0}^{c} -\frac{dc}{c^2} = k \int_0^t dt;$$

$$\frac{1}{c} = \frac{1}{c_0} + kt. \tag{32.32}$$

This is the integrated rate law for a second-order reaction. To discover whether the reaction is second order, we test the data by plotting $1/c$ versus t. Equation (32.32) requires that this plot be linear. If the data fall on a straight line, this is evidence that the reaction is second order. The slope of the line is equal to the rate constant.

The half-life is defined as before. When $t = \tau, c = \frac{1}{2}c_0$. Using these values in Eq. (32.32), we obtain

$$\tau = \frac{1}{kc_0}. \tag{32.33}$$

For a second-order reaction, the half-life depends on the initial concentration of the reactant. If the initial concentration is doubled, the time required for half of A to react will be reduced by one-half. If the half-life for various initial concentrations is plotted against $1/c_0$, the rate constant is the reciprocal of the slope of the line.

■ **EXAMPLE 32.3**   Suppose that the decomposition of acetaldehyde is second-order. (See Example 32.1.) Formulate the rate law in terms of the total pressure of the system and integrate the result to express the pressure as a function of time.

In Example 32.1, we showed that the rate of reaction,

$$\frac{1}{V}\frac{d\xi}{dt} = \frac{1}{RT}\frac{dp}{dt},$$

and that $p = (n^0 + \xi)(RT/V) = p^0 + (RT/V)\xi$, where $p^0 = n^0 RT/V$. If the reaction is second order, then

$$\frac{1}{V}\frac{d\xi}{dt} = kc_1^2 \qquad \text{where} \qquad c_1 = n_1/V.$$

Since $\xi = (V/RT)(p - p^0)$, we find that $c_1 = (n^0/V) - (\xi/V) = (p^0/RT) - (p - p^0)/RT = (2p^0 - p)/RT$. Then the rate law becomes

$$\frac{1}{RT}\frac{dp}{dt} = k\left(\frac{2p^0 - p}{RT}\right)^2 \qquad \text{or} \qquad \frac{dp}{(2p^0 - p)^2} = \frac{k}{RT}dt$$

Integrating

$$\int_{p^0}^{p} \frac{dp}{(2p^0 - p)^2} = \frac{k}{RT}\int_0^t dt \qquad \text{yields} \qquad \frac{1}{2p^0 - p} = \frac{1}{p^0} + \frac{kt}{RT}.$$

This last result can also be written in terms of the final pressure, p^∞; when $t = \infty$, we have $p^\infty = 2p^0$; thus, we get

$$\frac{1}{p^\infty - p} = \frac{2}{p^\infty} + \frac{kt}{RT}$$

The left-hand side of this equation can be plotted against t to obtain the rate constant.

32.5.1 Second-Order Reactions with Two Reactants

Consider a reaction of the type

$$(-v_A)A + (-v_B)B \longrightarrow \text{Products.} \tag{32.34}$$

Keep in mind that the stoichiometric coefficients, v_A and v_B, are negative; thus $-v_A$ and $-v_B$ are positive numbers. If the instantaneous concentrations of A and B are c_A and c_B, and assuming that the reaction is first order with respect to both A and B, the overall order is second and the rate law can be written

$$\frac{d(\xi/V)}{dt} = kc_A c_B. \tag{32.35}$$

Note that

$$\frac{d(\xi/V)}{dt} = \frac{1}{v_A}\frac{dc_A}{dt} = \frac{1}{v_B}\frac{dc_B}{dt}, \tag{32.36}$$

so that the rate law in Eq. (32.35) could be written in terms of the rate of disappearance of A or of B:

$$\frac{1}{v_A}\frac{dc_A}{dt} = kc_A c_B \qquad \text{or} \qquad \frac{1}{v_B}\frac{dc_B}{dt} = kc_A c_B \tag{32.37}$$

Although these forms are physically meaningful, they are not suited to the integration of the rate law.

To bring Eq. (32.35) into an integrable form we express c_A and c_B in terms of ξ/V by dividing Eq. (32.2) by V; this yields

$$c_A = c_A^0 + v_A\frac{\xi}{V} \qquad \text{and} \qquad c_B = c_B^0 + v_B\frac{\xi}{V}, \tag{32.38}$$

in which c_A^0 and c_B^0 are the initial concentrations of A and B. Putting these values of c_A and c_B in Eq. (32.35) yields

$$\frac{d(\xi/V)}{dt} = k\left(c_A^0 + \nu_A \frac{\xi}{V}\right)\left(c_B^0 + \nu_B \frac{\xi}{V}\right).$$

We next factor out the product $(-\nu_A)(-\nu_B)$ on the right-hand side to obtain

$$\frac{d(\xi/V)}{dt} = (-\nu_A)(-\nu_B)k\left(\frac{c_A^0}{-\nu_A} - \frac{\xi}{V}\right)\left(\frac{c_B^0}{-\nu_B} - \frac{\xi}{V}\right).$$

To simplify the notation, we define

$$y_A^0 = \frac{c_A^0}{-\nu_A}; \qquad y_B^0 = \frac{c_B^0}{-\nu_B}; \qquad y = \frac{\xi}{V}.$$

Then the rate law becomes

$$\frac{dy}{dt} = \nu_A \nu_B k(y_A^0 - y)(y_B^0 - y).$$

Separating variables, we obtain

$$\frac{dy}{(y_A^0 - y)(y_B^0 - y)} = \nu_A \nu_B k \, dt. \tag{32.39}$$

We distinguish two cases.

Case 1. $y_A^0 = y_B^0$. In this case, the substances A and B are present in the required stoichiometric ratio; Eq. (32.39) becomes

$$\frac{dy}{(y_A^0 - y)^2} = \nu_A \nu_B k \, dt \tag{32.40}$$

Integrating from $(y = 0, t = 0)$ to (y, t) yields

$$\int_0^y \frac{dy}{(y_A^0 - y)^2} = \nu_A \nu_B k \int_0^t dt,$$

or

$$\frac{1}{y_A^0 - y} - \frac{1}{y_A^0} = \nu_A \nu_B kt. \tag{32.41}$$

This can be written in terms of the concentration of either A or B:

$$\frac{1}{c_A} = \frac{1}{c_A^0} + (-\nu_B)kt \qquad \text{or} \qquad \frac{1}{c_B} = \frac{1}{c_B^0} + (-\nu_A)kt. \tag{32.42}$$

In this circumstance, the rate law is very similar to the second-order law with only one reactant. The reciprocal of either concentration is plotted against t to determine the rate constant.

Case 2. $y_A^0 \neq y_B^0$. In this case, the two reactants are present in an arbitrary ratio, not the required stoichiometric ratio. Using the method of partial fractions,* we can rewrite

---

*   The method of partial fractions is described in elementary calculus texts.

Eq. (32.39):

$$\left(\frac{1}{y_B^0 - y_A^0}\right)\left(\frac{-dy}{y_A^0 - y}\right) + \left(\frac{1}{y_A^0 - y_B^0}\right)\left(\frac{-dy}{y_B^0 - y}\right) = -v_A v_B k\, dt.$$

A minus sign has been introduced into every term for mathematical convenience. Multiplying each side by $y_B^0 - y_A^0$ and integrating from ($y = 0, t = 0$) to (y, t) gives

$$\int_0^y \frac{-dy}{y_A^0 - y} - \int_0^y \frac{-dy}{y_B^0 - y} = -v_A v_B k(y_B^0 - y_A^0)\int_0^t dt,$$

which becomes

$$\ln\frac{y_A^0 - y}{y_A^0} - \ln\frac{y_B^0 - y}{y_B^0} = -v_A v_B k(y_B^0 - y_A^0)t. \tag{32.43}$$

When we replace the y's by their equivalents in concentrations, this equation becomes

$$\ln\frac{c_A}{c_A^0} - \ln\frac{c_B}{c_B^0} = -k(v_B c_A^0 - v_A c_B^0)t. \tag{32.44}$$

This equation strongly resembles the first-order law in Eq. (32.16) and reduces to it in limiting circumstances. For example, suppose B is present in very great excess—so great that $y_B^0 - y_A^0 \approx y_B^0$ and $c_B/c_B^0 \approx 1$ throughout the course of the reaction. Equation (32.44) then reduces to

$$\ln\frac{c_A}{c_A^0} = -(-v_A k c_B^0)t, \tag{32.45}$$

which is the first-order law, Eq. (32.16), with an effective first-order rate constant equal to $-v_A k c_B^0$. Similarly, if A is present in very great excess, Eq. (32.44) reduces to

$$\ln\frac{c_B}{c_B^0} = -(-v_B k c_A^0)t, \tag{32.46}$$

with an effective first-order rate constant equal to $-v_B k c_A^0$. When the concentrations of the reactants have been adjusted so that the reaction follows a rate law such as Eq. (32.45) or Eq. (32.46), the rate law is sometimes called a "pseudo-first-order" law; the quantities, $-v_A k c_B^0$ and $-v_B k c_A^0$, are called "pseudo-first-order" rate constants.

If both A and B are present in comparable concentrations, we can plot the quantity on the left-hand side of Eq. (32.44) against t to determine the rate constant. This plot should be a straight line with a slope of $-k(v_B c_A^0 - v_A c_B^0)$. All other quantities are known, so k can be obtained from the slope.

32.6 HIGHER-ORDER REACTIONS

Reactions of order higher than second are occasionally important. A third-order rate law may have any of the forms

$$\frac{d(\xi/V)}{dt} = kc_A^3, \qquad \frac{d(\xi/V)}{dt} = kc_A^2 c_B, \qquad \frac{d(\xi/V)}{dt} = kc_A c_B^2, \qquad \text{or} \qquad \frac{d(\xi/V)}{dt} = kc_A c_B c_C,$$

and so on. We can integrate these equations either directly or after expressing all the concentrations in terms of a single variable, as in the preceding example. The procedure is straightforward, but the results are not of sufficiently general interest to be included in detail here. The most common third-order reactions are several which involve nitric oxide; for example,

$$2\,NO + O_2 \longrightarrow 2\,NO_2, \qquad \frac{d(\xi/V)}{dt} = kc_{NO}^2 c_{O_2},$$

$$2\,NO + Cl_2 \longrightarrow 2\,NOCl, \qquad \frac{d(\xi/V)}{dt} = kc_{NO}^2 c_{Cl_2}.$$

32.7 DETERMINING THE ORDER OF A REACTION

Since the rate of a reaction may be proportional to different powers of the concentrations of the several reactants, we need to determine the dependence of the rate on each of these concentrations. If, for example, the rate is $kc_A^\alpha c_B^\beta c_C^\gamma$, then if B and C are present in great excess (while the concentration of A is very small), the concentrations of B and C will remain effectively constant throughout the reaction. The rate will then be proportional only to c_A^α. By altering the initial concentration of A, we can determine the order α. The procedure is repeated by having A and C present in excess to determine β, and so on. This is the isolation method for determining the order of a reaction. We used this idea in deriving Eqs. (32.45) and (32.46).

Suppose that a reaction is αth order with respect to the reactant A and that all the other reactants are present in great excess. Then the rate law is

$$\frac{d(\xi/V)}{dt} = kc_A^\alpha c_B^\beta c_C^\gamma, \ldots,$$

which can be written as

$$\frac{d(\xi/V)}{dt} = k'c_A^\alpha,$$

where $k' = (kc_B^\beta c_C^\gamma \ldots)$ is effectively a constant. We can replace $d(\xi/V)/dt$ by $(1/v_A)(dc_A/dt)$, which brings the equation to

$$-\frac{dc_A}{dt} = (-v_A)k'c_A^\alpha.$$

Taking the logarithm of both sides of this equation, we obtain

$$\log_{10}\left(-\frac{dc_A}{dt}\right) = \log_{10}(-v_A k') + \alpha \log_{10} c_A, \tag{32.47}$$

which can be used in the following way. A plot of c_A versus t is constructed from the data. The slope of the curve, dc_A/dt, is measured at several different values of t; the corresponding value of c_A is read from the plot. The logarithm of $(-dc_A/dt)$ is then plotted against $\log_{10} c_A$. The slope of the line is the order of the reaction.

Equation (32.47) can be used in another way. The initial slope of the curve of c_A versus t is measured for several different initial concentrations. Then the logarithm of the negative of the initial slope is plotted against the logarithm of the initial concentration. The slope of this plot, according to Eq. (32.47), is the order of the reaction.

It should be mentioned that considering the uncertainties in the data it is sometimes quite difficult to decide whether a reaction is first or second order.

32.8 THE DEPENDENCE OF RATE OF REACTION ON TEMPERATURE

With very few exceptions the rate of reaction increases (often very sharply) with increase in temperature. The relation between the rate constant k and temperature was first proposed by Arrhenius:

$$k = Ae^{-E^*/RT}. \tag{32.48}$$

The constant A is called the *frequency factor*, or pre-exponential factor; E^* is the *activation energy*. Converting Eq. (32.48) to logarithmic form, we have

$$\log_{10} k = \log_{10} A - \frac{E^*}{2.303RT}; \tag{32.49}$$

it is apparent that by determining the value of k at several temperatures, the plot of $\log_{10} k$ versus $1/T$ will yield the activation energy from the slope of the curve and the frequency factor from the intercept. Although the frequency factor may depend slightly on the temperature, unless the temperature range is very great, this effect can ordinarily be ignored. The determination of the activation energy is an important objective of any kinetic investigation.

The justification of the Arrhenius equation on theoretical grounds will be discussed in the next chapter. We can give a qualitative idea of the meaning of the equation for a reaction that occurs when two molecules collide. In such a case, the reaction rate should be proportional to Z, the number of collisions per second. Furthermore, if we assume that not all collisions, but only those collisions involving an energy greater than some critical value E^*, are effective, then the rate of the reaction will have the form

$$N_A \frac{d(\xi/V)}{dt} = Ze^{-E^*/RT}, \tag{32.50}$$

since the fraction of collisions having energies greater than E^* is $\exp(-E^*/RT)$ so long as $E^* \gg RT$. The form of Eq. (32.50) is that required to yield the Arrhenius equation for the rate constant in this case.

32.9 MECHANISM

It was pointed out in Section 32.3 that the exponents of the concentrations in the rate law in general do not bear any relation to the stoichiometric coefficients in the balanced chemical equation. This is so because the overall chemical equation yields no information about the *mechanism* of the reaction. By the mechanism of a reaction we mean the detailed way by which the reactants are converted into products. The rate at which equilibrium is attained in a system depends on the mechanism of the process while the equilibrium state itself is independent of the mechanism and depends only on the relative Gibbs energies. From a study of the position of equilibrium, values of changes of Gibbs energy, entropy, and enthalpy can be obtained. From a study of the rate of reaction under various conditions, information about the mechanism can be gained. The kinetic study is generally complicated and often requires a great deal of ingenuity in the interpretation of the data simply because it is as likely as not that the mechanism is complicated. Also, from kinetic data alone, it often is not possible to decide which of several reasonable mechanisms is the actual mechanism of the reaction. All too often it is not possible to distinguish on any basis which of, let us say, two mechanisms is the actual one. We may be reduced to saying that one seems more plausible than the other.

The attack on the problem of mechanism in a chemical reaction begins with the resolution of the reaction into a postulated sequence of *elementary* reactions. An elementary reaction is one that occurs in a single act. As an example, consider the reaction

$$H_2 + I_2 \longrightarrow HI + HI.$$

As a hydrogen molecule and an iodine molecule collide, we may assume that they momentarily have the configuration

$$\begin{array}{c} H \cdots I \\ | \quad | \\ H \cdots I \end{array}$$

and that this complex can then dissociate into two molecules of HI. The sequence of events is illustrated as follows:

$$\begin{array}{ccccc} \begin{array}{c} H \\ | \\ H \end{array} + \begin{array}{c} I \\ | \\ I \end{array} & \longrightarrow & \begin{array}{c} H \cdots I \\ | \quad | \\ H \cdots I \end{array} & \longrightarrow & \begin{array}{c} HI \\ + \\ HI \end{array} \\ \text{approaching} & & \text{collision} & & \text{separating} \end{array}$$

Thus, in this single act of collision, the reactants disappear and the products are formed. The reverse of this reaction is also an elementary reaction, the collision of two molecules of HI to form H_2 and I_2.

An elementary reaction that involves two molecules, such as the one above, is a *bimolecular* reaction. A *unimolecular* reaction is an elementary reaction that involves only one molecule; for example, the dissociation of a molecule such as HO_2:

$$HO_2 \longrightarrow H + O_2.$$

In a single act the HO_2 molecule simply falls apart into two fragments. The reverse reaction,

$$H + O_2 \longrightarrow HO_2,$$

is an elementary reaction and, since it involves two molecules, is a bimolecular reaction. Only elementary reactions can be characterized by their *molecularity*; the adjectives "unimolecular" and "bimolecular" do not have meaning for complex reactions that involve a sequence of many elementary steps.

The rate laws for elementary reactions can be written immediately. Under any prescribed set of conditions, the probability of a molecule A falling into fragments in unit time is a constant. So for the unimolecular elementary reaction

$$A \longrightarrow \text{fragments},$$

the rate law is

$$\frac{d(\xi/V)}{dt} = kc_A. \tag{32.51}$$

Since the probability of falling apart in unit time is constant, the greater the number of molecules present, the greater will be the rate of disappearance; hence the rate law, Eq. (32.51).

For a bimolecular reaction, the rate depends on the number of collisions in unit time; in Section 30.5 it was shown that the number of collisions between like molecules is proportional to the square of the concentration; therefore, for a bimolecular elementary reaction of the type

$$2\,A \longrightarrow \text{Products,}$$

the rate law is

$$\frac{d(\xi/V)}{dt} = kc_A^2. \tag{32.52}$$

Similarly, the number of collisions per second between unlike molecules is proportional to the product of the concentrations of the two kinds of molecules; hence for the bimolecular elementary reaction of the type

$$A + B \longrightarrow \text{Products,}$$

the rate law is

$$\frac{d(\xi/V)}{dt} = kc_A c_B. \tag{32.53}$$

A termolecular reaction is an elementary reaction that involves the simultaneous collision of three molecules:

$$A + B + C \longrightarrow \text{Products;}$$

the rate law is

$$\frac{d(\xi/V)}{dt} = kc_A c_B c_C. \tag{32.54}$$

The frequency of occurrence of three-body collisions is very much smaller than that of two-body collisions. Consequently, if a termolecular step is essential to the progress of the reaction, the reaction is very slow.

Examination of Eqs. (32.51), (32.52), and (32.53) shows that for elementary reactions the order of the reaction can be inferred from the stoichiometric coefficients. This is true *only for elementary reactions.*

32.10 OPPOSING REACTIONS; THE HYDROGEN–IODINE REACTION

The gas phase reaction of hydrogen with iodine, investigated extensively by Bodenstein in the 1890s, provides a classic example of opposing second-order reactions. Between 300 and 500 °C the reaction proceeds at conveniently measurable rates. If we assume that the mechanism is simple, consisting of one elementary reaction and its reverse,*

$$H_2 + I_2 \underset{k_{-1}}{\overset{k_1}{\rightleftharpoons}} 2\,HI,$$

then the net rate of reaction is the rate of the forward reaction minus the rate of the reverse action. Since both are elementary reactions, we have

$$\frac{d(\xi/V)}{dt} = k_1 c_{H_2} c_{I_2} - k_{-1} c_{HI}^2. \tag{32.55}$$

---

* It is customary to write the rate constant for the forward reaction over the arrow, and that for the reverse reaction under the arrow.

At equilibrium $d(\xi/V)/dt = 0$, and Eq. (32.55) can be written in the form

$$\frac{(c_{HI})_e^2}{(c_{H_2})_e(c_{I_2})_e} = \frac{k_1}{k_{-1}},$$

The left-hand side of this equation is the proper quotient of equilibrium concentrations, the equilibrium constant; therefore

$$K = \frac{k_1}{k_{-1}}. \tag{32.56}$$

The equilibrium constant of an *elementary* reaction is equal to the ratio of the rate constants of the forward and reverse reactions. This relation is correct *only* for elementary reactions.

If ξ is the advancement of the reaction at time t, we can write

$$c_{H_2} = a - y, \qquad c_{I_2} = b - y, \qquad c_{HI} = 2y,$$

where a and b are the initial concentrations of H_2 and I_2 respectively, and $y = \xi/V$. (We assume that there is no HI present initially.) When we use this notation, the rate law becomes

$$\frac{dy}{dt} = k_1(a - y)(b - y) - k_{-1}(4y^2). \tag{32.57}$$

Equation (32.56) yields $k_{-1} = k_1/K$; then the rate law, after we clear the parentheses, becomes

$$\frac{dy}{dt} = k_1\left(1 - \frac{4}{K}\right)\left[y^2 - \frac{(a + b)y}{1 - 4/K} + \frac{ab}{1 - 4/K}\right]. \tag{32.57a}$$

We can write the expression inside the brackets as a product, $(y_1 - y)(y_2 - y)$, where y_1 and y_2 are the roots of the expression. Then

$$y_1 = \frac{a + b + m}{2(1 - 4/K)}, \qquad y_2 = \frac{a + b - m}{2(1 - 4/K)}, \qquad m = \sqrt{(a - b)^2 + 16ab/K}.$$

Note that m, y_1, and y_2 are known quantities that are calculated from the initial concentrations a and b, and the equilibrium constant K. The roots, y_1 and y_2, are the possible equilibrium values of y; y_1 is an extraneous root and y_2 is the equilibrium value.

The rate equation becomes

$$\frac{dy}{(y_1 - y)(y_2 - y)} = k_1\left(1 - \frac{4}{K}\right) dt.$$

Using the partial fraction method, we can write

$$\frac{-dy}{y_1 - y} - \frac{-dy}{y_2 - y} = -k_1\left(1 - \frac{4}{K}\right)(y_2 - y_1) \, dt.$$

Integrating from $(y = 0, t = 0)$ to (y, t) and using $y_2 - y_1 = -m/(1 - 4/K)$, we have

$$\int_0^y \frac{-dy}{y_1 - y} - \int_0^y \frac{-dy}{y_2 - y} = k_1 m \int_0^t dt;$$

and finally,

$$\ln \frac{y_1 - y}{y_1} - \ln \frac{y_2 - y}{y_2} = k_1 mt. \tag{32.58}$$

Using the known parameters K, a, b, and the measured values of y as a function of t, we can plot the left-hand side of Eq. (32.58) against t to obtain the value of the rate constant k_1 from the slope. The value of K is measured independently. Using Eq. (32.58), Bodenstein obtained satisfactory values of the rate constant at several temperatures.

Equation (32.58) should be compared to the second-order rate law without the reverse reaction, Eq. (32.43), which with $v_A = v_B = -1$ can be written as

$$\ln \frac{a - y}{a} - \ln \frac{b - y}{b} = k(a - b)t. \tag{32.59}$$

The similarity between the two equations is apparent. In the limit, as $K \to \infty$, Eq. (32.58) reduces to Eq. (32.59).

For many years the hydrogen–iodine reaction had been the traditional example of opposing second-order reactions. Recent work by J. H. Sullivan indicates that the mechanism is not as simple as we have assumed here; in fact, the mechanism now seems to be unresolved. For a discussion and references see R. M. Noyes, *J. Chem. Phys.* **48**, 323 (1968).

32.11 CONSECUTIVE REACTIONS

When it is necessary for a reaction to proceed through several successive elementary steps before the product is formed, the rate of the reaction is determined by the rates of all these steps. If one of these reactions is much slower than any of the others, then the rate will depend on the rate of this single slowest step. The slow step is the *rate-determining* step. The situation is analogous to water flowing through a series of pipes of different diameters. The rate of delivery of the water will depend on the rate at which it can pass through the narrowest pipe. An apt illustration of this feature of consecutive reactions is offered by the Lindemann mechanism of activation for unimolecular decompositions.

32.12 UNIMOLECULAR DECOMPOSITIONS; LINDEMANN MECHANISM

Before 1922 the existence of unimolecular decompositions posed a severe problem in interpretation. The unimolecular elementary step consists of the breaking of a molecule into fragments:

$$A \longrightarrow \text{Fragments.}$$

If this occurs, it does so because the energy content of the molecule is too large. Too much energy somehow gets into a particular vibrational degree of freedom; this vibration then produces dissociation of the molecule into fragments.

The molecules that have this excess energy decompose. If the decomposition is to continue, other molecules must gain an excess energy. How do the molecules acquire this extra energy? In 1919, Perrin suggested that this energy was supplied by radiation, that is, by the absorption of light. This radiation hypothesis implies that in the absence of light the reaction will not occur. Immediate experimental tests of this hypothesis proved it wrong, and the puzzle remained. It appeared that the molecules could not gain the needed energy by collisions with other molecules, since the collision rate depends on the square of the concentration; this would make the reaction second order, whereas it is observed to be first order.

In 1922, Lindemann proposed a mechanism by which the molecules could be activated by collision and yet the reaction could, nonetheless, be first order. The activation of the molecule is by collision

$$A + A \xrightarrow{k_1} A^* + A,$$

where A is a normal molecule, and A* an activated molecule. The collision between two normal A molecules produces an activated molecule A*, which has an excess energy in the various vibrational degrees of freedom; the remaining molecule is deficient in energy.

Once the activated molecule is formed, it may suffer either of two fates: it may be deactivated by collision,

$$A^* + A \xrightarrow{k_{-1}} A + A,$$

or it may decompose into products,

$$A^* \xrightarrow{k_2} \text{Products.}$$

The rate of disappearance of A is the rate of the last reaction:

$$-\frac{dc_A}{dt} = k_2 c_{A^*}. \tag{32.60}$$

With this equation we are faced with the problem of expressing the concentration of an active species in terms of the concentration of normal species. We assume that, shortly after the reaction starts, a steady state is reached in which the concentration of the activated molecules does not change very much, so that $(dc_{A^*}/dt) = 0$. This is the *steady-state approximation*. Since A* is formed in the first reaction and removed in the others, we have

$$\frac{dc_{A^*}}{dt} = 0 = k_1 c_A^2 - k_{-1} c_A c_{A^*} - k_2 c_{A^*}.$$

Using this equation we can express c_{A^*} in terms of c_A, the concentration of the normal molecules,

$$c_{A^*} = \frac{k_1 c_A^2}{k_{-1} c_A + k_2}.$$

This value of c_{A^*} brings the rate law, Eq. (32.60), to the form

$$-\frac{dc_A}{dt} = \frac{k_2 k_1 c_A^2}{k_{-1} c_A + k_2}. \tag{32.61}$$

There are two important limiting forms of Eq. (32.61).

Case 1. $k_{-1} c_A \ll k_2$. Suppose that the rate of decomposition, $k_2 c_{A^*}$, is extremely fast—so fast that as soon as the activated molecule is formed it falls apart. Then there is no time for a deactivating collision to occur, and the rate of deactivation is very small compared with the rate of decomposition. Then $k_{-1} c_A c_{A^*} \ll k_2 c_{A^*}$, or $k_{-1} c_A \ll k_2$.

Hence the denominator $k_{-1}c_A + k_2 \approx k_2$, and Eq. (32.61) becomes

$$-\frac{dc_A}{dt} = k_1 c_A^2. \tag{32.62}$$

The rate of the reaction is equal to the rate at which the activated molecules are formed, since the activated molecule decomposes immediately. The kinetics are second order, since the collision is a second-order process.

Case 2. $k_{-1}c_A \gg k_2$. If after activation there is an appreciable time lag before the molecule falls apart, then there is opportunity for the activated molecule to make a number of collisions that may deactivate it. If the time lag is long, then the rate of deactivation, $k_{-1}c_A c_{A*}$, is much greater than the rate of decomposition, $k_2 c_{A*}$. This means that $k_{-1}c_A \gg k_2$, and $k_{-1}c_A + k_2 \approx k_{-1}c_A$. This brings Eq. (32.61) to the form

$$-\frac{dc_A}{dt} = k_2\left(\frac{k_1}{k_{-1}}\right)c_A, \tag{32.63}$$

and the rate law is first order. The usual fate of an activated molecule is deactivation by collision. A very small fraction of the activated molecules decompose to yield products.

In a gas-phase reaction, high pressures increase the number of collisions so that $k_{-1}c_A$ is large and the rate is first order. The supply of activated molecules is adequate, and the rate at which they fall apart limits the rate of the reaction. At lower pressures the number of collisions decreases, $k_{-1}c_A$ is small, and the rate is second order. The rate then depends on the rate at which activated molecules are produced by collisions.

The apparent first-order rate "constant" decreases at low pressures. Physically the decrease in value of the rate constant at lower pressures is a result of the decrease in number of activating collisions. If the pressure is increased by addition of an inert gas, the rate constant increases again in value, showing that the molecules can be activated by collision with a molecule of an inert gas as well as by collision with one of their own kind. Several first-order reactions have been investigated over a sufficiently wide range of pressure to confirm the general form of Eq. (32.61). The Lindemann mechanism is accepted as the mechanism of activation of the molecule.

32.13 COMPLEX REACTIONS: THE HYDROGEN–BROMINE REACTION

The kinetic law for the hydrogen–bromine reaction is considerably more complicated than that for the hydrogen–iodine reaction. The stoichiometry is the same,

$$H_2 + Br_2 \longrightarrow 2\,HBr,$$

but the rate law established by M. Bodenstein and S. C. Lind in 1906 is expressed by the equation

$$\frac{d[HBr]}{dt} = \frac{k[H_2][Br_2]^{1/2}}{1 + \dfrac{m[HBr]}{[Br_2]}}, \tag{32.64}$$

where k and m are constants, and we have used brackets to indicate the concentration of the species. The appearance of the term $[HBr]/[Br_2]$ in the denominator implies that the presence of the product decreases the rate of the reaction; the product acts as an inhibitor. However, the inhibition is less if the concentration of bromine is high.

The expression in Eq. (32.64) was not explained until 1919, when J. A. Christiansen, K. F. Herzfeld, and M. Polanyi independently proposed the correct mechanism. The mechanism consists of five elementary reactions:

1) $$Br_2 \xrightarrow{k_1} 2Br,$$

2) $$Br + H_2 \xrightarrow{k_2} HBr + H,$$

3) $$H + Br_2 \xrightarrow{k_3} HBr + Br,$$

4) $$H + HBr \xrightarrow{k_4} H_2 + Br,$$

5) $$2Br \xrightarrow{k_5} Br_2.$$

The HBr is formed in reactions (2) and (3) and removed in reaction (4), so we have for the rate of formation of HBr

$$\frac{d[HBr]}{dt} = k_2[Br][H_2] + k_3[H][Br_2] - k_4[H][HBr]. \tag{32.65}$$

The difficulty with this expression is that it involves the concentrations of H atoms and Br atoms, which are not readily measurable; thus the equation is useless unless we can express the concentrations of the atoms in terms of the concentrations of the molecules, H_2, Br_2, and HBr. Since the atom concentrations are, in any case, very small, it is assumed that a steady state is reached in which the concentration of the atoms does not change with time; the atoms are removed at the same rate as they are formed. From the elementary reactions, the rates of formation of bromine atoms and of hydrogen atoms are

$$\frac{d[Br]}{dt} = 2k_1[Br_2] - k_2[Br][H_2] + k_3[H][Br_2] + k_4[H][HBr] - 2k_5[Br]^2;$$

$$\frac{d[H]}{dt} = k_2[Br][H_2] - k_3[H][Br_2] - k_4[H][HBr].$$

The steady-state conditions are $d[Br]/dt = 0$ and $d[H]/dt = 0$, so these equations become

$$0 = 2k_1[Br_2] - k_2[Br][H_2] + k_3[H][Br_2] + k_4[H][HBr] - 2k_5[Br]^2;$$

$$0 = k_2[Br][H_2] - k_3[H][Br_2] - k_4[H][HBr].$$

By adding these two equations, we obtain $0 = 2k_1[Br_2] - 2k_5[Br]^2$, which yields

$$[Br] = \left(\frac{k_1}{k_5}\right)^{1/2}[Br_2]^{1/2}. \tag{32.66}$$

From the second equation,

$$[H] = \frac{k_2\left(\dfrac{k_1}{k_5}\right)^{1/2}[H_2][Br_2]^{1/2}}{k_3[Br_2] + k_4[HBr]}. \tag{32.67}$$

By using these values for [Br] and [H] in Eq. (32.65), we obtain, after collecting terms and dividing numerator and denominator by $k_3(Br_2)$,

$$\frac{d[HBr]}{dt} = \frac{2k_2\left(\dfrac{k_1}{k_5}\right)^{1/2}[H_2][Br_2]^{1/2}}{1 + \dfrac{k_4[HBr]}{k_3[Br_2]}}. \tag{32.68}$$

This equation has the same form as Eq. (32.64), the empirical equation of Bodenstein and Lind. (The integrated form of this equation has no particular utility.)

There are several points of interest in this mechanism. First of all, the reaction is initiated by the dissociation of a bromine molecule into atoms. Once bromine atoms are formed, a single bromine atom can produce a large number of molecules of HBr through the sequence of reactions (2) and (3). These reactions form a *chain* in which an active species such as a Br or H atom is consumed, product is formed, and the active species regenerated. These reactions are *chain-propagating* reactions. Reaction (4) propagates the chain in the sense that the active species, H, is replaced by another active species, Br, but the product, HBr, is removed by this reaction, thus decreasing the net rate of formation of HBr. Reaction (4) is an example of an *inhibiting* reaction. The final reaction, (5), removes active species and therefore is a *chain-terminating* reaction.

If we compare reactions (3) and (4), it is apparent that Br_2 and HBr are competing for the H atoms; the success of HBr in this competition, determined by the relative rates of reactions (4) and (3), will determine the extent of the inhibition:

$$\frac{(rate)_4}{(rate)_3} = \frac{k_4[H][HBr]}{k_3[H][Br_2]} = \frac{k_4[HBr]}{k_3[Br_2]}.$$

This accounts for the form of the second term in the denominator of Eq. (32.68).

Since [HBr] = 0 at $t = 0$, the initial rate of formation of HBr is given by

$$\left[\frac{d[HBr]}{dt}\right]_0 = 2k_2\left(\frac{k_1}{k_5}\right)^{1/2}[H_2]_0[Br_2]_0^{1/2}.$$

By plotting [HBr] versus t, we obtain the limiting value of the slope $(d[HBr]/dt)_0$ at $t = 0$. By doing this for several different values of the initial concentrations $[H_2]_0$ and $[Br_2]_0$, we can determine the constant $k = 2k_2(k_1/k_5)^{1/2}$.

32.14 FREE-RADICAL MECHANISMS

In 1929, F. Paneth and W. Hofeditz detected the presence of free methyl radicals from the thermal decomposition of lead tetramethyl. The apparatus they used is shown in Fig. 32.4. Lead tetramethyl is a volatile liquid. After evacuating the apparatus, a stream of H_2 under about 100 Pa pressure is passed over the liquid where it entrains the vapor of $Pb(CH_3)_4$ and carries it through the tube. The gases are removed by a high-speed vacuum pump at the other end. The furnace is at position M. After a short period, a lead mirror deposits in the tube at M, formed by the decomposition of the $Pb(CH_3)_4$. If the furnace is moved upstream to position M', a new mirror forms at M', while the original mirror at M slowly disappears.

This phenomenon can be explained by the fact that $Pb(CH_3)_4$ decomposes on heating to form lead and free methyl radicals:

$$Pb(CH_3)_4 \longrightarrow Pb + 4\,CH_3.$$

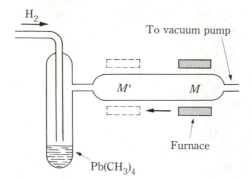

Figure 32.4   Detection of free radicals.

The lead deposits as a mirror on the wall of the tube. The methyl radicals are swept down the tube mixed in the stream of carrier gas. If the radicals find a mirror downstream, as at M, they can remove it by the reaction

$$Pb + 4CH_3 \longrightarrow Pb(CH_3)_4.$$

Following the discovery by Paneth, the technique was extensively developed, especially by F. O. Rice and his co-workers.

In 1934, F. O. Rice and K. F. Herzfeld were able to show that the kinetic laws observed for many organic reactions could be interpreted on the basis of mechanisms involving free radicals. They showed that, although the mechanism might be complex, the kinetic law could be quite simple. The mechanisms proposed were also capable of predicting the products formed in the reaction.

32.14.1   The Decomposition of Ethane

For illustration, the Rice–Herzfeld mechanism for the decomposition of ethane is

1)
$$C_2H_6 \xrightarrow{k_1} 2CH_3,$$

2)
$$CH_3 + C_2H_6 \xrightarrow{k_2} CH_4 + C_2H_5,$$

3)
$$C_2H_5 \xrightarrow{k_3} C_2H_4 + H,$$

4)
$$H + C_2H_6 \xrightarrow{k_4} H_2 + C_2H_5,$$

5)
$$H + C_2H_5 \xrightarrow{k_5} C_2H_6.$$

Reactions (1) and (2) are required for initiation, (3) and (4) constitute the chain, and (5) is the termination step. The principal products are those that are formed in the chain, so that the overall reaction can be written as

$$C_2H_6 \longrightarrow C_2H_4 + H_2.$$

A very small amount of CH_4 is produced.

The rate of disappearance of C_2H_6 is

$$-\frac{d[C_2H_6]}{dt} = k_1[C_2H_6] + k_2[CH_3][C_2H_6] + k_4[H][C_2H_6] - k_5[H][C_2H_5].$$

$$(32.69)$$

The steady-state conditions are:

for CH_3,

$$0 = \frac{d[CH_3]}{dt} = 2k_1[C_2H_6] - k_2[CH_3][C_2H_6];$$

for C_2H_5,

$$0 = \frac{d[C_2H_5]}{dt} = k_2[CH_3][C_2H_6] - k_3[C_2H_5] + k_4[H][C_2H_6] - k_5[H][C_2H_5];$$

for H,

$$0 = \frac{d[H]}{dt} = k_3[C_2H_5] - k_4[H][C_2H_6] - k_5[H][C_2H_5].$$

Solution of the first equation yields

$$[CH_3] = \left(\frac{2k_1}{k_2}\right). \tag{32.70}$$

Addition of the three equations yields $0 = 2k_1[C_2H_6] - 2k_5[H][C_2H_5]$, or

$$[H] = \left(\frac{k_1}{k_5}\right)\frac{[C_2H_6]}{[C_2H_5]}. \tag{32.71}$$

Using this result in the last steady-state equation yields

$$[C_2H_5]^2 - \left(\frac{k_1}{k_3}\right)[C_2H_6][C_2H_5] - \left(\frac{k_1k_4}{k_3k_5}\right)[C_2H_6]^2 = 0,$$

which must be solved for $[C_2H_5]$:

$$[C_2H_5] = [C_2H_6]\left[\frac{k_1}{2k_3} + \sqrt{\left(\frac{k_1}{2k_3}\right)^2 + \left(\frac{k_1k_4}{k_3k_5}\right)}\right].$$

Since k_1, the rate constant for the initiation step, is very small, the higher powers of it are negligible; thus we have

$$[C_2H_5] = \left(\frac{k_1k_4}{k_3k_5}\right)^{1/2}[C_2H_6]. \tag{32.72}$$

Then the value of [H] is

$$[H] = \left(\frac{k_1k_3}{k_4k_5}\right)^{1/2}. \tag{32.73}$$

Combining Eq. (32.71) with Eq. (32.69) yields

$$-\frac{d[C_2H_6]}{dt} = \{k_2[CH_3] + k_4[H]\}[C_2H_6]. \tag{32.74}$$

Using the values of $[CH_3]$ and [H] from Eqs. (32.70) and (32.73) in Eq. (32.74) reduces it to

$$-\frac{d[C_2H_6]}{dt} = \left[2k_1 + \left(\frac{k_1k_3k_4}{k_5}\right)^{1/2}\right][C_2H_6],$$

or, neglecting the higher power of k_1,

$$-\frac{d[C_2H_6]}{dt} = \left(\frac{k_1 k_3 k_4}{k_5}\right)^{1/2} [C_2H_6]. \tag{32.75}$$

Equation (32.75) is the rate law. In spite of the complexity of the mechanism, the reaction is a first-order reaction. The rate constant is a composite of the rate constants of the individual elementary steps.

The Rice–Herzfeld mechanisms usually yield simple rate laws; the reaction orders predicted for various reactions are $\frac{1}{2}$, 1, $\frac{3}{2}$, and 2.

The rate of decomposition of organic compounds can often be increased by the addition of compounds such as $Pb(CH_3)_4$ or $Hg(CH_3)_2$, which introduce free radicals into the system. These compounds are said to *sensitize* the decomposition of the organic compound. In contrary fashion a compound such as nitric oxide combines with free radicals to remove them from the system. This inhibits the reaction by breaking the chains.

32.15 THE TEMPERATURE DEPENDENCE OF THE RATE CONSTANT FOR A COMPLEX REACTION

The rate constant of any chemical reaction depends on temperature through the Arrhenius equation, Eq. (32.48). For a complex reaction such as the thermal decomposition of ethane, in which, by Eq. (32.75),

$$k = \left(\frac{k_1 k_3 k_4}{k_5}\right)^{1/2},$$

the rate constant for each elementary reaction can be replaced by its value from the Arrhenius equation; $k_1 = A_1 \exp(-E_1^*/RT)$, and so on. Then

$$k = \left(\frac{A_1 A_3 A_4}{A_5}\right)^{1/2} e^{-(1/2)(E_1^* + E_3^* + E_4^* - E_5^*)/RT}.$$

This is equivalent to the Arrhenius equation for the complex reaction

$$k = Ae^{-E^*/RT},$$

so that, by comparison, we have

$$A = \left(\frac{A_1 A_3 A_4}{A_5}\right)^{1/2}, \tag{32.76}$$

and

$$E^* = \tfrac{1}{2}(E_1^* + E_3^* + E_4^* - E_5^*). \tag{32.77}$$

Therefore if we know the values of A and E^* for each elementary step, the values of A and E^* can be calculated for the reaction. For the ethane decomposition, $E_1^* = 350$ kJ/mol, $E_3^* = 170$ kJ/mol, $E_4^* = 30$ kJ/mol, and $E_5^* = 0$. The activation energy for the reaction should be

$$E^* = \tfrac{1}{2}(350 + 170 + 30) = 275 \text{ kJ/mol}.$$

The experimental values found for the activation energy are about 290 kJ/mol. The agreement between the experimental value and that predicted by the mechanism is quite reasonable.

★ 32.16 BRANCHING CHAINS; EXPLOSIONS

A highly exothermic reaction which goes at a rate that intrinsically is only moderate may, nonetheless, explode. If the heat liberated is not dissipated, the temperature rises rapidly and the rate increases very rapidly. The ultimate result is a *thermal* explosion.

Another type of explosion is due to *chain branching.* In the treatment of chain reactions we employed the steady-state assumption, and balanced the rate of production of active species against their rate of destruction. In the cases described so far, this treatment yielded values for the concentration of radicals that were finite and small in all circumstances. Two things are clear about the steady-state assumption. First, it is obvious that it cannot be *exactly* correct, and second, it must be very nearly correct. If it were not very nearly correct, then the concentration of active species would change appreciably as time passed. If the concentration of active species decreased appreciably, the reaction would slow down and come to a halt before reaching the equilibrium position. On the other hand, if the concentration of active species increased appreciably with time, the rate of the reaction would increase very rapidly. This in turn would further increase the concentration of active species. The reaction would go at an explosive rate. In fact, explosions do occur if active species such as atoms or radicals are produced more quickly than they can be removed.

If in some elementary reaction an active species reacts to produce more than one active species, then the chain is said to branch. For example,

$$H + O_2 \longrightarrow OH + O.$$

In this reaction the H atom is destroyed, but two active species, OH and O, which can propagate the chain, are generated. Since one active species produces two, there are circumstances in which the destruction cannot keep up with the production. The concentration of radicals increases rapidly, thus producing an explosion.

The mechanism of the hydrogen–oxygen reaction is probably not fully understood even today. Much of the modern work has been done by C. N. Hinschelwood and his co-workers. The steps in the chain reaction are

1) $\quad\quad\quad\quad\quad H_2 \longrightarrow 2\,H$ $\quad\quad\quad$ Initiation,

2) $\quad\quad\quad\quad H + O_2 \longrightarrow OH + O$
3) $\quad\quad\quad\quad O + H_2 \longrightarrow OH + H$ $\quad\quad$ Branching,

4) $\quad\quad OH + H_2 \longrightarrow H_2O + H$ $\quad\quad$ Propagation,

The reactions that multiply radicals or atoms must be balanced by processes that destroy them. At very low pressures the radicals diffuse quickly to the walls of the vesssel and are destroyed at the surface. The destruction reactions can be written

$$H \longrightarrow \text{destruction at the surface,}$$

$$OH \longrightarrow \text{destruction at the surface,}$$

$$O \longrightarrow \text{destruction at the surface.}$$

If the pressure is low, the radicals reach the surface quickly and are destroyed. The production rate and destruction rate can balance and the reaction goes smoothly. The rate of these destruction reactions depends very much on the size and shape of the reaction vessel, of course.

As the pressure is increased, the branching rate and propagation rate increase, but the higher pressure slows the rate of diffusion of the radicals to the surface so the destruction

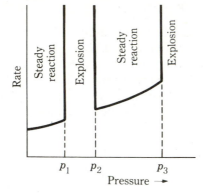

Figure 32.5 Explosion limits.

rate falls. Above a certain critical pressure, the lower explosion limit, it is not possible to maintain a steady concentration of atoms and radicals; the concentration of active species increases rapidly with time, which increases the rate of the reaction enormously. The system explodes; the lower explosion limit depends on the size and shape of the containing vessel.

At higher pressures, three-body collisions that can remove the radicals become more frequent. The reaction,

$$H + O_2 + M \longrightarrow HO_2,$$

where M is O_2 or H_2 or a foreign gas, competes with the branching reactions. Since the species HO_2 does not contribute to the reaction, radicals are effectively removed and at high enough pressures a balance between radical production and destruction can be established. Above a second critical pressure, the upper explosion limit, the reaction goes smoothly rather than explosively. There is a third explosion limit at high pressures above which the reaction again goes explosively.

The rate of the reaction as a function of pressure is shown schematically in Fig. 32.5. The rate is very slow at pressures below p_1, the lower explosion limit. Between p_1 and p_2 the reaction velocity is infinite, or explosive. Above p_2, the upper explosion limit, the reaction goes smoothly, the rate increasing with pressure. Above p_3, the third explosion limit, the reaction is explosive.

The explosion limits depend on temperature. Below about 460 °C explosion does not occur in the low-pressure region.

★ 32.17 NUCLEAR FISSION; THE NUCLEAR REACTOR AND THE "ATOMIC" BOMB

The explosion of the "atomic" bomb depends on the same general kinetic principles as the $H_2 + O_2$ explosion. The situation in the bomb is somewhat simpler.

If the nucleus of ^{235}U absorbs a thermal neutron, the nucleus splits into two fragments of unequal mass and releases several neutrons. If we add the rest masses of the products and compare this sum with the rest masses of the original ^{235}U and the neutron, there is a discrepancy. The products have less mass than do the reactants. The difference in mass, Δm, is equivalent to an amount of energy by the Einstein equation $E = (\Delta m)c^2$, where c is the velocity of light. This is the energy released in the reaction. Only a small fraction ($<1\%$) of the total mass is converted to energy, but the equivalence factor c^2 is so large that the energy released is enormous.

The fission reaction can be written

$$n + {}^{235}\text{U} \longrightarrow \text{X} + \text{Y} + \alpha n.$$

The atoms X and Y are the fission products, α is the number of neutrons released, and is, on the average, between 2 and 3. This is the same type of chain branching reaction as was encountered in the hydrogen–oxygen reaction. Here the action of one neutron can produce several. If the size and shape of the uranium is such that most of the neutrons escape before they hit another uranium nucleus, the reaction cannot sustain itself. However, in a large chunk of ^{235}U, the neutrons hit other uranium nuclei before escape is possible, and the number of neutrons multiplies rapidly, thus producing an explosive reaction. The awe-inspiring appearance of the explosion of the bomb results from the enormous amount of energy which is released, this energy being, gram for gram, some 10 to 50 million times greater than that released in any chemical reaction.

The fission reaction occurs in a controlled way in the nuclear pile. Here rods of ordinary ^{238}U which has been enriched with ^{235}U are built into a structure with a moderator such as graphite or D_2O. The neutrons that are emitted at high speeds from the fission of ^{235}U are slowed to thermal speeds by the moderator. The thermal neutrons suffer three important fates: some continue the chain to produce the fission of more ^{235}U, others are captured by ^{238}U, and some are absorbed by the control rods of the reactor. The neutron flux in the reactor is monitored constantly. Moving the absorbing control rods into or out of the pile reduces or increases the neutron flux. In this way sufficient neutrons are permitted to maintain the chain reaction at a smooth rate, but enough are absorbed to prevent an explosion.

The ^{238}U absorbs a thermal neutron and by radioactive decay yields neptunium and plutonium. The sequence is

$$^{238}_{92}\text{U} + {}^{1}_{0}n \longrightarrow {}^{239}_{92}\text{U} \xrightarrow[t_{1/2}\,=\,23\ \text{min}]{} {}^{239}_{93}\text{Np} + \beta^-,$$

$$^{239}_{93}\text{Np} \xrightarrow[t_{1/2}\,=\,2.3\ \text{day}]{} {}^{239}_{94}\text{Pu} + \beta^-,$$

$$^{239}_{93}\text{Pu} \xrightarrow[t_{1/2}\,=\,24,000\ \text{yr}]{} {}^{235}_{92}\text{U} + \alpha.$$

The plutonium produced can contribute to the chain reaction since it is fissionable by thermal neutrons.

32.18 REACTIONS IN SOLUTION

The empirical rate laws found for reactions in solution are the same as those for reactions in the gas phase. An intriguing fact about reactions that can be studied in both solution and the gas phase is that quite often the mechanism is the same, and the rate constant has the same value in both situations. This indicates that in such reactions the solvent plays no part, but serves only as a medium to separate the reactants and products. Reactions in solution may well be faster than in the gas phase because of our tendency to use comparatively concentrated solutions. For example, in a gas at 1 atm pressure, the molar concentration is about 10^{-4} mol/L. In making up solutions, our first tendency would be to make up a 0.1 or 0.01 molar solution. The reaction would go faster in solution simply because of the increased concentration, not because of a different rate constant. In those cases in which the solvent does not affect the rate constant, it is found that the frequency factors and activation energy have essentially the same values in solution as in the gas phase.

★ 32.19   RELAXATION METHODS

Since 1953, Manfred Eigen and his colleagues have invented and developed several power-ful techniques for the measurement of the rates of very fast reactions, reactions that are effectively complete within a time period of less than about 10 μsec. These techniques are called relaxation techniques. If we attempt to measure the rate of a very fast reaction by traditional methods, it is clear that the time required to mix the reactants will be a limiting factor. Many devices have been designed to produce rapid mixing of reactants. The best of these cannot mix two solutions in a time shorter than a few hundred microseconds. Any method requiring mixing of the reactants cannot succeed with reactions that take place in times shorter than the mixing time. The relaxation methods avoid the mixing problem completely.

Suppose that in the chemical reaction of interest we are able to monitor the con-centration of a colored species by passing light of the appropriate frequency through the mixture and observing the intensity of the transmitted beam. Consider a chemical reaction at equilibrium and suppose that the species we are monitoring has the concentration c (Fig. 32.6). Suppose that at time t_0 one of the parameters on which the equilibrium depends (for example, temperature) is instantaneously brought to some new value. Then the con-centration of the species we are observing must achieve some new equilibrium value $\bar{c}$. Since chemical reactions occur at a finite rate, the concentration of the species will not change instantaneously to the new value, but will follow the course indicated by the dashed curve in Fig. 32.6. The system, having been perturbed from its old equilibrium position, *relaxes* to its new equilibrium condition. As we will show, if the difference in concentration between the two states is not too large, then the curve in Fig. 32.6 is a simple exponential function, characterized by a single constant, the relaxation time τ. The relaxation time is the time required for the difference in concentration between the two states to decay to $1/e$ of its initial value.

The apparatus for the "temperature-jump" method is shown schematically in Fig. 32.7. A high-voltage power supply charges a capacitor, C. At a certain voltage the spark gap, G, breaks down and the capacitor discharges, sending a heavy current through the cell that contains the reactive system at equilibrium in a conductive aqueous solution. The passage of the current raises the temperature of the system about 10 °C in a few micro-seconds. In the following time interval the concentration of the absorbing species adjusts to

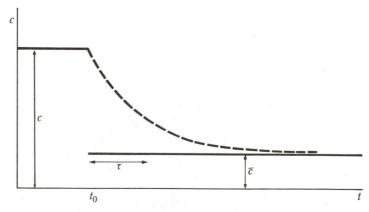

Figure 32.6   Concentration change after impulse.

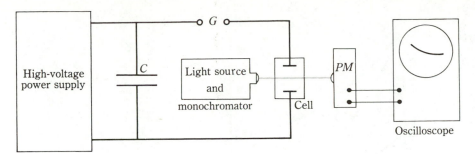

Figure 32.7   Temperature-jump apparatus.

the equilibrium value appropriate to the higher temperature. This changes the intensity of the light beam emerging from the cell into the detecting photomultiplier tube, *PM*. The output of the photomultiplier tube is displayed on the vertical axis of an oscilloscope; the horizontal sweep of the oscilloscope is triggered by the spark discharge. In this way the concentration versus time curve is displayed on the oscilloscope screen.

The relaxation methods have the advantage that the mathematical interpretation is exceptionally simple. This simplicity is a consequence of arranging matters so that the displacement from the original equilibrium position is small.

Consider the elementary reaction

$$A + B \; \rightleftharpoons \; C.$$

The rate equation for this reaction can be written

$$\frac{d(\xi/V)}{dt} = k_f c_A c_B + (-k_r c_C), \tag{32.78}$$

in which we have expressed the net rate as the *sum* of the forward rate ($k_f c_A c_B$) and the reverse rate ($-k_r c_C$). It is convenient for graphical representation to give the reverse rate a negative sign here. The mole numbers and the concentrations of each species are expressed in terms of the advancement of the reaction, ξ:

$$n_A = n_A^0 - \xi, \qquad n_B = n_B^0 - \xi, \qquad n_C = n_C^0 + \xi.$$

$$c_A = c_A^0 - \frac{\xi}{V}, \qquad c_B = c_B^0 - \frac{\xi}{V}, \qquad c_C = c_C^0 + \frac{\xi}{V}.$$

The mole numbers, n^0, and the concentrations, c^0, are the values of these quantities at $\xi = 0$. When we use these values for the concentrations, Eq. (32.78) becomes

$$\frac{1}{V}\frac{d\xi}{dt} = k_f\left(c_A^0 - \frac{\xi}{V}\right)\left(c_B^0 + \frac{\xi}{V}\right) + \left[-k_r\left(c_C^0 + \frac{\xi}{V}\right)\right].$$

From this equation it is apparent that the forward rate is a quadratic function of ξ, while the reverse rate is a linear function of ξ; these rates are shown as functions of ξ in Fig. 32.8. The sum of the two functions is the net rate, indicated by the dashed line in Fig. 32.8.

At $\bar{\xi}$, the equilibrium value of the advancement, the net rate is zero, and we have

$$r = k_f \bar{c}_A \bar{c}_B = k_r \bar{c}_C, \tag{32.79}$$

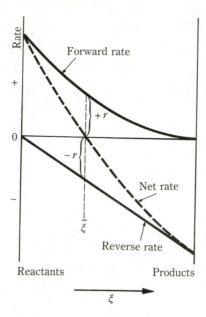

Figure 32.8   Forward, reverse, and net rates of reaction.

in which the bar over the concentration indicates the equilibrium value. The exchange rate, r, is the rate of either the forward or the reverse reaction (without the minus sign) at equilibrium. Equation (32.79) can be rearranged to the form

$$K = \frac{k_f}{k_r} = \frac{\bar{c}_C}{\bar{c}_A \bar{c}_B}, \tag{32.80}$$

which is the equilibrium relation for the elementary reaction; K is the equilibrium constant.

Although the detailed shapes of the curves will depend on the order of the reaction, the forward rate, the reverse rate, and the net rate of any elementary reaction will be related in the general way indicated in Fig. 32.8. Most importantly, it is apparent that the net rate can be approximated by a straight line over a narrow range near the equilibrium position. Let the net rate, $(1/V)(d\xi/dt) = v$. Then we expand v in a Taylor series about the equilibrium value of ξ:

$$v = v_{\bar{\xi}} + \left(\frac{dv}{d\xi}\right)_{\bar{\xi}} (\xi - \bar{\xi}).$$

However, $v_{\bar{\xi}}$ is the net rate at equilibrium, which is zero. Introducing the definition of v and multiplying by the volume, the equation becomes

$$\frac{d\xi}{dt} = V \left(\frac{dv}{d\xi}\right)_{\bar{\xi}} (\xi - \bar{\xi}). \tag{32.81}$$

We note that $V(dv/d\xi)_{\bar{\xi}}$ has the dimensions of a reciprocal time, and depends only upon $\bar{\xi}$, that is, only upon equilibrium values of concentration, not upon ξ or t. We define the constant τ, the relaxation time, by

$$\frac{1}{\tau} = -V \left(\frac{dv}{d\xi}\right)_{\bar{\xi}}. \tag{32.82}$$

The minus sign is introduced to compensate the negative sign of the derivative; see Fig. 32.8.

The introduction of τ brings the rate equation, Eq. (32.81), to the form

$$\frac{d\xi}{dt} = -\frac{1}{\tau}(\xi - \bar{\xi}), \tag{32.83}$$

in which τ is independent of ξ or t. This equation has the form of a first-order law and integrates immediately to

$$\xi - \bar{\xi} = (\xi - \bar{\xi})_0 e^{-t/\tau}, \tag{32.84}$$

in which $(\xi - \bar{\xi})_0$ is the initial displacement (at $t = 0$) from equilibrium. Since the displacement of the concentration of any species from its equilibrium value is $\Delta c_i = c_i - \bar{c}_i$, and since $c_i = c_i^0 + (v_i/V)\xi$, where v_i is the stoichiometric coefficient of the species in the reaction, we obtain directly $\Delta c_i = (v_i/V)(\xi - \bar{\xi})$. Thus the displacement of the concentration of any species from the equilibrium value is proportional to the displacement of the advancement. Consequently, the time dependence of the concentration of any species is given by the same relation as in Eq. (32.84).

$$(c_i - \bar{c}_i) = (c_i - \bar{c}_i)_0 e^{-t/\tau}. \tag{32.85}$$

The pattern that appears on the oscilloscope screen in the temperature-jump experiment is therefore a simple exponential one, provided only one reaction is involved. The value of τ can be obtained by measuring the horizontal distance (time axis) required for the value of the vertical displacement to fall to $1/e = 0.3679$ of its initial value (Fig. 32.9).

It must be emphasized that Eqs. (32.81) through (32.85) are quite general; they do not depend on the order of the reaction and most particularly they do not depend on the example we chose for illustration. Equation (32.85) is a typical example of a relaxation law. It implies that any small perturbation from equilibrium in a chemical system disappears exponentially with time. If there are several elementary steps in the mechanism of a reaction then there will be several relaxation times. In this event, the expression for $c_i - \bar{c}_i$ is a sum of exponential terms such as that in Eq. (32.85). There is one such term for each relaxation time. The coefficient of each term and the relaxation times are determined by a computer fit of the data.

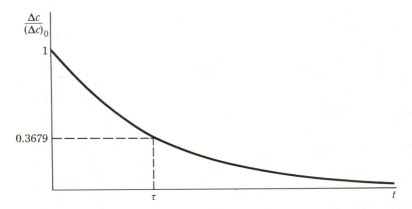

Figure 32.9   The relaxation time.

Table 32.1
Rate constants of some very rapid reactions at 25 °C

Reaction	$k_f/(L\,mol^{-1}\,s^{-1})$	k_r/s^{-1}
$H^+ + OH^- \rightleftharpoons H_2O$	1.4×10^{11}	2.5×10^{-5}
$H^+ + F^- \rightleftharpoons HF$	1.0×10^{11}	7×10^7
$H^+ + HCO_3^- \rightleftharpoons H_2CO_3$	4.7×10^{10}	$\sim 8 \times 10^6$
$OH^- + NH_4^+ \rightleftharpoons NH_3 + H_2O$ (22 °C)	3.4×10^{10}	6×10^5
$OH^- + HCO_3^- \rightleftharpoons CO_3^{2-} + H_2O$ (20 °C)	$\sim 6.0 \times 10^9$	—

From M. Eigen and L. DeMaeyer in *Techniques of Organic Chemistry*, Vol. VIII, part II. S. L. Friess, E. S. Lewis and A. Weissburger, eds. New York: Interscience, 1963.

We can evaluate the relaxation time for the example above by evaluating the derivative, $dv/d\xi$, at $\xi = \bar{\xi}$. Since $v = k_f c_A c_B - k_r c_C$, then

$$\frac{dv}{d\xi} = k_f c_A \frac{dc_B}{d\xi} + k_f c_B \frac{dc_A}{d\xi} - k_r \frac{dc_C}{d\xi}.$$

But $dc_A/d\xi = -1/V = dc_B/d\xi$, and $dc_C/d\xi = 1/V$. Thus, at $\xi = \bar{\xi}$, this becomes

$$\left(\frac{dv}{d\xi}\right)_{\bar{\xi}} = -\frac{k_f(\bar{c}_A + \bar{c}_B) + k_r}{V}.$$

Then, by the definition of τ, Eq. (32.82),

$$\frac{1}{\tau} = k_f(\bar{c}_A + \bar{c}_B) + k_r. \tag{32.86}$$

By making measurements on the system with different values of the equilibrium concentrations, we can evaluate both k_f and k_r. Knowledge of the equilibrium constant, in view of Eq. (32.80), provides additional information about k_f and k_r.

The relaxation method is not restricted to the study of very fast reactions. With appropriate choices of sensing and recording devices, we could use it to study the rate of any reaction. The value of the relaxation technique for the study of very fast reactions lies in the fact that ordinarily it is the only technique available for measuring the rate of these reactions.

A few rate constants that have been measured by relaxation techniques are given in Table 32.1. It should be noted that the rate constant k_f for the combination of two oppositely charged ions is very large. This process is always very fast since it is limited only by the rate at which the two ions can diffuse through the medium and get close enough to each other to combine. It should be mentioned that the reaction, $H^+ + OH^- \rightarrow H_2O$, has the largest known second-order rate constant.

★ **32.20 CATALYSIS**

A catalyst is a substance that increases the rate of a reaction and can itself be recovered unchanged at the end of the reaction. If a substance slows a reaction, it is called an inhibitor or a negative catalyst.

As we have seen, the rate of a reaction is determined by rates of the several reactions in the mechanism. The general function of a catalyst is simply to provide an additional

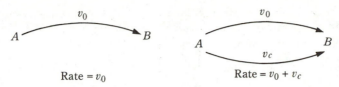

Figure 32.10   (a) Uncatalyzed reaction. (b) Catalyzed reaction.

mechanism by which reactants can be converted to products. This alternative mechanism has a lower activation energy than that for the mechanism in the absence of a catalyst, so that the catalyzed reaction is faster. Consider reactants A going to products B by an uncatalyzed mechanism at a rate v_0 (Fig. 32.10a). If an additional mechanism is provided by a catalyst, Fig. 32.10(b), so that B is formed at a rate v_c by the catalytic mechanism, then the total rate of formation of B is the sum of the rates of formation by each path.

For a catalyzed reaction, we have

$$v = v_0 + v_c. \tag{32.87}$$

In the absence of a catalyst the reaction is often immeasurably slow, $v_0 = 0$; then, $v = v_c$. The rate v_c is usually proportional to the concentration of the catalyst. The analogy to an electrical network of parallel resistances (Fig. 32.11a) or to parallel pipes carrying a fluid (Fig. 32.11b) is apparent. In each case the flow through the network is the sum of that passing through each branch.

For a catalyst to function in this way, the catalyst must enter into chemical combination either with one or more of the reactants or at least with one of the intermediate species involved. Since it must be regenerated after a sequence of reactions, the catalyst is free to act again and again. As a result, a little catalyst produces a great deal of reaction, just as a minute concentration of radicals in a chain reaction produces a lot of product.

The action of inhibitors is not so simply described, since they may act in a number of different ways. An inhibitor may slow a radical chain reaction by combining with the radicals; nitric oxide functions in this way. In other cases, the inhibitor is consumed by combination with one of the reactants and only delays the reaction until it is used up. Some inhibitors may simply "poison" a trace of catalyst whose presence is unsuspected.

The simplest mechanism by which a catalyst can act is given by the reactions

$$S + C \underset{k_{-1}}{\overset{k_1}{\rightleftharpoons}} SC$$

$$SC \xrightarrow{k_2} P + C.$$

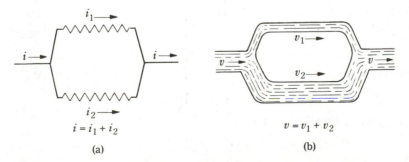

Figure 32.11   Electrical and hydraulic analogs of catalyzed reaction.

The reactant S is called the *substrate*; C is the catalyst, P is the product, and SC is an intermediate compound. The rate of the reaction per unit volume, v, is equal to the rate of formation of the product in unit volume:

$$v = \frac{1}{V}\frac{d\xi_2}{dt} = \frac{d[P]}{dt}.$$

Since the product is formed in the second reaction, the rate law is

$$v = k_2[SC] \tag{32.88}$$

The steady-state condition for the intermediate is

$$\frac{d[SC]}{dt} = 0 = k_1[S][C] - k_{-1}[SC] - k_2[SC]. \tag{32.89}$$

Dividing the equation by k_1 and solving for [SC], we obtain

$$[SC] = \frac{[S][C]}{K_m}, \tag{32.90}$$

in which the composite constant, K_m, is defined by

$$K_m = \frac{k_{-1} + k_2}{k_1}. \tag{32.91}$$

Using this steady-state value of [SC] in the rate law, we obtain

$$v = \frac{k_2[S][C]}{K_m}. \tag{32.92}$$

This expression illustrates the usual proportionality of the rate to the concentration of catalyst.

Equation (32.92) is very cumbersome if we attempt to use it over the entire course of the reaction. Therefore we consider the behavior of the rate expression only in the initial stage of the reaction. We can write the concentrations of all the species in terms of the advancements per unit volume of the two reactions: $y_1 = \xi_1/V$ and $y_2 = \xi_2/V$. Then

$$[S] = [S]_0 - y_1$$
$$[C] = [C]_0 - y_1 + y_2$$
$$[SC] = y_1 - y_2$$
$$[P] = y_2$$

in which $[S]_0$ and $[C]_0$ are the initial concentrations of substrate and catalyst, respectively. By adding the first, third, and fourth of these relations, we get

$$[S] + [SC] + [P] = [S]_0;$$

by adding the second and third,

$$[C] + [SC] = [C]_0.$$

Solving these equations for [S] and [C] yields

$$[S] = [S]_0 - [SC] - [P],$$
$$[C] = [C]_0 - [SC].$$

Using these in the steady-state expression, Eq. (32.89), recognizing that in the initial stage, $[P] = 0$, we obtain

$$0 = \{[S]_0 - [SC]\}\{[C]_0 - [SC]\} - K_m[SC]$$

or

$$0 = [S]_0[C]_0 - \{[S]_0 + [C]_0 + K_m\}[SC] + [SC]^2.$$

This expression is quadratic in $[SC]$; however, the concentration of SC is limited by which of S or C is present in the smaller amount. We always arrange conditions so that either $[S]$ or $[C]$ is present in much lower concentration than the other; thus the term in $[SC]^2$ is always negligible. We solve the resulting linear equation for $[SC]$:

$$[SC] = \frac{[S]_0[C]_0}{[S]_0 + [C]_0 + K_m}. \tag{32.93}$$

Using this value of $[SC]$ in the rate law, Eq. (32.88), we obtain for the initial rate, v_0,

$$v_0 = \frac{k_2[S]_0[C]_0}{[S]_0 + [C]_0 + K_m}. \tag{32.94}$$

Two limiting cases of Eq. (32.94) are important.

Case 1. $[C]_0 \ll [S]$. In this case, $[C]_0$ is dropped from the denominator and we have

$$v_0 = \frac{k_2[S]_0[C]_0}{[S]_0 + K_m}. \tag{32.95}$$

Note that the initial rate is proportional to the catalyst concentration.

If we invert Eq. (32.95),

$$\frac{1}{v_0} = \frac{1}{k_2[C]_0} + \left(\frac{K_m}{k_2[C]_0}\right)\frac{1}{[S]_0}. \tag{32.96}$$

A plot of $1/v_0$ against $1/[S]_0$ is linear and enables us to calculate $k_2[C]_0$ and K_m from the intercept and slope.

The dependence of the initial rate on $[S]_0$ is interesting. If $[S]_0 \ll K_m$, then

$$[S]_0 + K_m \approx K_m$$

and the rate is first order in $[S]_0$:

$$v_0 = \frac{k_2[S]_0[C]_0}{K_m} = k_{cat}[S]_0 \tag{32.97}$$

where

$$k_{cat} = \frac{k_2[C]_0}{K_m} = k_C[C]_0;$$

the constant k_C is called the catalytic coefficient for the catalyst C.

However, if $[S]_0 \gg K_m$, then

$$[S]_0 + K_m \approx [S]_0$$

and the rate is zero order in $[S]_0$,

$$v_0 = k_2[C]_0. \tag{32.98}$$

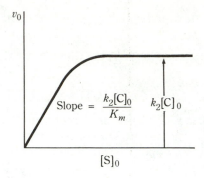

Figure 32.12   Initial rate versus initial concentration of substrate.

The initial rate as a function of $[S]_0$ is shown in Fig. 32.12. The limiting value of the rate is a result of the limited amount of catalyst present. The catalyst is needed to produce the reactive compound SC. As soon as the concentration of S reaches the point where essentially all of the catalyst is found in the complex SC, then further increase in [S] produces no change in the initial rate.

Case 2. $[S]_0 \ll [C]_0$. In this case Eq. (32.94) becomes

$$v_0 = \frac{k_2[S]_0[C]_0}{[C]_0 + K_m} \tag{32.99}$$

The reaction is always first order in $[S]_0$ in this case, but may be first order or zero order in $[C]_0$, depending on the value of $[C]_0$. This case is not usually as convenient experimentally as Case 1.

★ 32.21   ENZYME CATALYSIS

Enzymes are protein molecules that catalyze the myriads of chemical reactions required for a living organism to function. The most remarkable feature of enzyme catalysis is the specificity of the enzyme to a particular reaction. For example, urease catalyzes the hydrolysis of urea,

$$(NH_2)_2CO + H_2O \longrightarrow CO_2 + 2NH_3,$$

and no other reaction. Consequently, there are nearly as many enzymes as there are chemical reactions occurring in the organism. Not all enzymes are restricted to one reaction. Some will catalyze a class of reactions; for example, phosphatases catalyze the hydrolysis of many different phosphate esters.

The specificity of the enzyme led to the postulate of a "lock-and-key" type of mechanism. The substrate molecule, by combining in a special way with the *active site* on the enzyme, is activated for the reaction that it is to undergo. The active site on an enzyme may consist of more than one "site" on the protein molecule; one site may attach to one part of the substrate molecule, while another site binds another part of the substrate molecule. The lock-and-key model seems to be generally correct, but the details of the action are different for different enzymes.

The simplest enzyme mechanism is the same as the simple catalytic mechanism described in Section 32.23; namely,

$$E + S \underset{k_{-1}}{\overset{k_1}{\rightleftharpoons}} ES$$

$$ES \xrightarrow{k_2} P + E,$$

where E is the enzyme catalyst. Therefore we can take the result in Eq. (32.95) for the rate of reaction:

$$v_0 = \frac{k_2[E]_0[S]_0}{[S]_0 + K_m}. \tag{32.100}$$

In this context the composite constant K_m is called the Michaelis constant and the rate law, Eq. (32.100) is called the Michaelis–Menten law. Here, again, we note that as $[S]_0$ becomes very large the rate approaches a limiting value, v_{max};

$$\lim_{[S]_0 \to \infty} v_0 = k_2[E]_0 = v_{max}. \tag{32.101}$$

When we use this notation for $k_2[E]_0$, Eq. (32.100) becomes

$$v_0 = \frac{v_{max}[S]_0}{[S]_0 + K_m}. \tag{32.102}$$

Inverting both sides of this equation yields

$$\frac{1}{v_0} = \frac{1}{v_{max}} + \left(\frac{K_m}{v_{max}}\right)\frac{1}{[S]_0}. \tag{32.103}$$

This is the Lineweaver–Burk equation; a plot of $(1/v_0)$ versus $1/[S]_0$, a Lineweaver–Burk plot, yields a straight line with intercept equal to $1/v_{max}$ and slope equal to K_m/v_{max} (Fig. 32.13).

Since $v_{max} = k_2[E]_0$, if we know $[E]_0$ we can calculate k_2 from v_{max}. The constant, k_2, is called the *turnover number* of the enzyme. The turnover number is the number of mole-

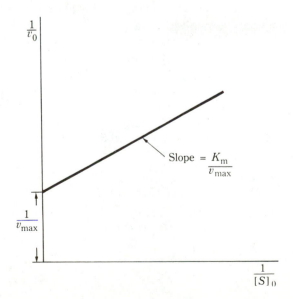

Slope $= \dfrac{K_m}{v_{max}}$

$\dfrac{1}{v_{max}}$

$\dfrac{1}{v_0}$

$\dfrac{1}{[S]_0}$

Figure 32.13   Lineweaver–Burk plot.

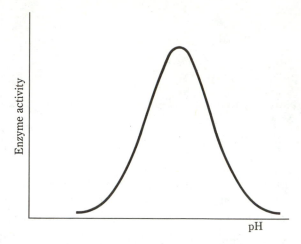

Figure 32.14   Dependence of enzyme activity on pH.

cules converted in unit time by one molecule of enzyme. Typical values of k_2 are 100 to 1000 per second, with some as large as 10^5 to 10^6 per second.

The activity of an enzyme generally passes through a maximum at a particular pH. This can be interpreted by assuming that there are three forms of the enzyme in equilibrium,

$$EH_2 \rightleftharpoons EH \rightleftharpoons E,$$

of which only EH can combine with substrate to yield an intermediate, EHS, that can react to form products. The other intermediates EH_2S and ES do not form products. Since the concentration of EH passes through a maximum at a particular pH, the activity of the enzyme has a maximum also (Fig. 32.14).

★ 32.22   ACID–BASE CATALYSIS

There are many chemical reactions that are catalyzed by acids or bases, or by both. The most common acid catalyst in water solution is the hydronium ion and the most common base is hydroxyl ion. However, some reactions are catalyzed by any acid or by any base. If any acid catalyzes the reaction, the reaction is said to be subject to *general* acid catalysis. Similarly, *general* base catalysis refers to catalysis by any base. If only hydronium or hydroxyl ions are effective, the phenomenon is called *specific* acid or base catalysis.

A classical example of specific acid–base catalysis is the hydrolysis of esters. The hydrolysis is catalyzed by H_3O^+ and OH^- but not by other acids or bases. The rate of hydrolysis in the absence of acid or base is extremely slow.

The mechanism of acid hydrolysis of an ester may be illustrated as follows:

The base-catalyzed reaction has the mechanism

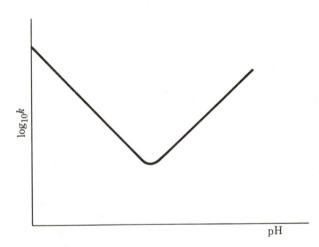

The rate of the reaction is

$$v = \{k_{H^+}[H^+] + k_{OH^-}[OH^-]\}[RCOOR'], \qquad (32.104)$$

in which k_{H^+} and k_{OH^-} are the catalytic coefficients for H^+ and OH^-, respectively.

The concentration of water does not appear in the rate law, since it is effectively constant during the course of the reaction in aqueous solution. Because of the relation $[H^+][OH^-] = K_w$, the rate constant $k = k_{H^+}[H^+] + k_{OH^-}[OH^-]$ has a minimum at a pH that depends on K_w, k_{H^+}, and k_{OH^-}. The dependence of $\log_{10} k$ on pH is shown schematically in Fig. 32.15.

Figure 32.15 Logarithm of the rate constant versus pH for a reaction catalyzed by H^+ and OH^-.

QUESTIONS

32.1 Describe the application of the isolation method to determine the rate law, Eq. (32.37).

32.2 What is a "pseudo-first-order" rate constant? How do its dimensions differ from those of a second-order rate constant?

32.3 Describe how the activation energy of the reaction in Problem 32.21 could be determined by appropriate measurements of concentration, time, and temperature.

32.4 Discuss how the idea of "rate-limiting step" applies in the (a) low-pressure and (b) high-pressure regions of unimolecular reactions.

32.5 Give several examples of the distinction between the "order" and the "molecularity" (unimolecular, bimolecular, and so on) of reactions.

32.6 What is the steady-state approximation? Use the Lindemann mechanism example to discuss its validity in terms of opposing gain and loss mechanisms for A*.

32.7 Why are chain mechanisms so common when species with *unpaired* electrons (such as H, Br, CH_3) are generated in an initiation step?

32.8 Apply the chemical relaxation Eq. (32.82) to the reaction $A \rightleftharpoons B$; Prob. 32.38a. Why does the relaxation time involve the *sum* of the forward and reverse rate constants?

32.9 Discuss the similarities and differences between the Lindemann rate law, Eq. (32.61), and the Case 1 catalysis rate law, Eq. (32.95).

32.10 Sketch and explain the variation of the logarithm of the rate constant with pH for specific acid hydrolysis of an ester.

PROBLEMS

32.1 Consider the decomposition of cyclobutane at 438 °C

$$C_4H_8 \longrightarrow 2C_2H_4.$$

The rate is to be measured by observing the pressure change in a constant volume system; assume that the gas mixture is ideal.

a) Express the rate of reaction $d(\xi/V)/dt$, in terms of dp/dt.

b) Let p^∞ be the pressure in the system after the C_4H_8 is completely decomposed (at $t = \infty$). If the reaction is first order in the concentration of C_4H_8, derive the relation between the pressure and time. What function of pressure should be plotted against time to determine the rate constant?

c) If the rate constant is $2.48 \times 10^{-4}\,s^{-1}$, calculate the half-life, and the time required for 98% of the C_4H_8 to decompose.

d) What will the value of p/p^∞ be after 2.0 hours?

32.2 The bleaching of bromophenol blue (BPB) by OH^- can be followed by measuring the absorbance at a particular wavelength. Note that $A = \epsilon l \tilde{c}$, where ϵ is the molar absorptivity; l is the length of the cell; $\tilde{c}$ is the concentration of the absorbing species. The reaction is

$$BPB + OH^- \longrightarrow BPBOH^-.$$

The product does not absorb at the wavelength used.

a) Express the rate of reaction per unit volume in terms of the change of absorbance with time, dA/dt.

b) If A_0 is the absorbance of the solution at $t = 0$, derive the relation between A and t. What quantity should be plotted against time to determine the rate constant? Assume that the reaction is first order with respect to each of the reactants and that they are mixed in the stoichiometric ratio.

32.3 a) Consider a reaction, $A \rightarrow$ Products, which is one-half order with respect to A. Integrate the rate equation and decide what function should be plotted from the data to determine the rate constant.

b) Repeat the calculation in (a) for a reaction that is three-halves order and nth order.

c) Derive the relation between the half-life, the rate constant, and the initial concentration of A for an nth-order reaction.

32.4 A certain reaction is first order; after 540 s, 32.5% of the reactant remains.

a) Calculate the rate constant for the reaction.

b) What length of time would be required for 25% of the reactant to be decomposed?

32.5 The half-life of a first-order reaction is 30 min.

a) Calculate the rate constant of the reaction.

b) What fraction of the reactant remains after 70 min?

32.6 At 25 °C the half-life for the decomposition of N_2O_5 is 2.05×10^4 s and is independent of the initial concentration of N_2O_5.

    a) What is the order of the reaction?

    b) What length of time is required for 80% of the N_2O_5 to decompose?

32.7 The gaseous reaction, $A_2 \rightarrow 2A$, is first order in A_2. After 751 seconds, 64.7% of A_2 remains undecomposed. Calculate

    a) the half-life;

    b) the length of time required to decompose 90% of A_2.

32.8 Copper-64 emits a β-particle. The half-life is 12.8 hr. At the time you received a sample of this radioactive isotope it had a certain initial activity (disintegrations/min). To do the experiment you have in mind, you have calculated that the activity must not go below 2% of the initial value. How much time do you have to complete your experiment?

32.9 Zinc-65 has a half-life of 245 days.

    a) What percentage of the original activity remains after 100 days?

    b) How much time is required for the activity to decrease to 5% of the initial activity?

32.10 The half-life of ^{238}U is 4.5×10^9 yr. How many disintegrations would occur in one minute in a 10 mg sample of ^{238}U?

32.11 Uranium-238 undergoes radioactive decay through a series of steps, ultimately producing lead-206. In a certain rock there are 0.228 g of ^{206}Pb per gram of ^{238}U. If we assume that all of the ^{206}Pb had its origin in the ^{238}U, how much time has elapsed since the rock was first formed? The decay constant for ^{238}U is 1.54×10^{-10} yr^{-1}; this isotope has the longest life in the series of radioactive elements that finally produce the ^{206}Pb.

32.12 Carbon-14 is radioactive with a half-life of 5760 years. Cosmic radiation in the upper atmosphere synthesizes ^{14}C which balances the loss through radioactive decay. Living matter maintains a level of ^{14}C that produces 15.3 disintegrations per minute for each gram of carbon. Dead organisms no longer exchange carbon with CO_2 in the atmosphere, so that the amount of ^{14}C in dead material decreases with time due to the decay. A 0.402 g sample of carbon from wheat taken from an Egyptian excavation exhibited 3.0 disintegrations per minute. How long ago did the wheat die?

32.13 A 1 mL sample of a bacterial culture at 37 °C is taken, and diluted to 10 L. A 1 mL sample of the diluted culture is spread on a culture plate. Ten minutes later, another 1 mL sample taken from the original culture is diluted and spread in the same way. The two plates are incubated for 24 hours. The first exhibits 48 colonies of bacteria, the second 72 colonies. If we assume that each colony originates with a single bacterium, what is the generation time?

32.14 In milk at 37 °C *lactobacillus acidophilus* has a generation time of about 75 minutes. Calculate the population relative to the initial value at 30, 60, 75, 90, and 150 minutes.

32.15 What must the interest rate be if an investment is to double in ten years if compounding occurs

    (a) yearly, (b) quarterly, and (c) instantaneously? (d) Derive Eq. (32.27).

32.16 A substance decomposes according to a second-order rate law. If the rate constant is 6.8×10^{-4} L/mol s, calculate the half-life of the substance

    a) if the initial concentration is 0.05 mol/L;

    b) if it is 0.01 mol/L.

32.17 A second-order reaction of the type, $A + B \rightarrow$ Products, is 40% complete in 120 minutes, when the initial concentrations of both A and B are 0.02 mol/L. Calculate

    a) the rate constant and the half-life;

    b) the time required for the reaction to be 40% complete if the initial concentrations of both A and B are 0.1 mol/L.

32.18 The rate of the reaction, $2NO + 2H_2 \rightarrow N_2 + 2H_2O$, has been studied at 826 °C. Some of the data are:

Run	Initial pressure H_2 $(p_{H_2})_0/kPa$	Initial pressure NO $(p_{NO})_0/kPa$	Initial rate $(-dp/dt)/(kPa/s)$
1	53.3	40.0	0.137
2	53.3	20.3	0.033
3	38.5	53.3	0.213
4	19.6	53.3	0.105

a) What are the orders of the reaction with respect to NO and with respect to H_2?

b) Assume that the gas mixture is ideal and find the relation between the rate of reaction per unit volume and dp/dt, where p is the total pressure. The volume is constant.

c) Combine the results of (a) and (b) to find the relation between dp/dt and the pressure. Initially, the total pressure is p_0, the mole fraction of NO is x_0, that of H_2 is $1 - x_0$.

32.19 From the following data for a reaction between A and B find the order of the reaction with respect to A and with respect to B, and calculate the rate constant.

[A]/(mol/L)	[B]/(mol/L)	Initial rate/(mol/L s)
2.3×10^{-4}	3.1×10^{-5}	5.2×10^{-4}
4.6×10^{-4}	6.2×10^{-5}	4.2×10^{-3}
9.2×10^{-4}	6.2×10^{-5}	1.7×10^{-2}

32.20 The decomposition of acetaldehyde was studied in the gas phase at 791 K. The results of two measurements are:

Initial concentration/(mol/L)	$9.72(10^{-3})$	$4.56(10^{-3})$
Half-life/s	328	572

a) What is the order of the reaction?

b) Calculate the rate constant for the reaction.

32.21 At 24.8 °C, the reaction,

$$C_6H_5N(CH_3)_2 + CH_3I \longrightarrow C_6H_5\overset{+}{N}(CH_3)_3 + I^-,$$

has a rate constant $k = 8.39 \times 10^{-5}$ L/mol s in nitrobenzene. The reaction is first order with respect to each of the reactants.

a) If equal volumes of solutions that are 0.12 mol/L in dimethylaniline and methyl iodide are mixed, how much time is required for 70% of the reactants to disappear?

b) If the concentration of each reagent is doubled, what length of time is required for 70% to disappear?

32.22 Assume that the decomposition of HI is an elementary reaction,

$$2HI \xrightarrow[k_{-1}]{k_1} H_2 + I_2.$$

The rate of the opposing reaction must be included in the rate expression. Integrate the rate equation if the initial concentrations of H_2 and I_2 are zero and that of HI is a.

32.23 Consider the opposing reactions,

$$A \underset{k_{-1}}{\overset{k_1}{\rightleftharpoons}} B,$$

both of which are first order. If the initial concentration of A is a and that of B is zero—and if y mol/L of A have reacted at time t—integrate the rate expression. Express k_{-1} in terms of the equilibrium constant K, and arrange the result in a form which resembles that for a first-order reaction in which the opposing reaction does not appear.

32.24 Consider the opposing elementary reactions,

$$A_2 \underset{k_{-1}}{\overset{k_1}{\rightleftharpoons}} 2A.$$

Integrate the rate expression if the initial concentration of A_2 is a and that of A is zero. $K = k_1/k_{-1}$. Compare this result with the result in Problem 32.22.

32.25 Near room temperature, 300 K, an old chemical rule of thumb is that the rate of a reaction doubles if the temperature is increased by 10 K. Assuming that it is the rate constant that doubles, calculate the value the activation energy must have if this rule is to hold exactly.

32.26 For the reaction of hydrogen with iodine, the rate constant is 2.45×10^{-4} L/mol s at 302 °C and 0.950 L/mol s at 508 °C.

a) Calculate the activation energy and the frequency factor for this reaction.
b) What is the value of the rate constant at 400 °C?

32.27 At 552.3 K, the rate constant for the decomposition of SO_2Cl_2 is 6.09×10^{-5} min^{-1}. If the activation energy is 210 kJ/mol, calculate the frequency factor and the rate constant at 600 K.

32.28 The activation energy for a certain reaction is 80 kJ/mol. How many times larger is the rate constant at 50 °C than the rate constant at 0 °C?

32.29 The decomposition of ethyl bromide in the gas phase is a first-order reaction. The data are:

Temperature	800 K	900 K
Rate constant	0.0361 s^{-1}	1.410 s^{-1}

What is the activation energy for the reaction?

32.30 In the Lindemann mechanism, $k_{app} = k_2 k_1 c/(k_{-1} c + k_2)$ is the "apparent" first-order rate constant. At low concentrations, the value of k_{app} decreases. If, when the concentration is 10^{-5} mol/L, the value of k_{app} reaches 90% of its limiting value at $c = \infty$, what is the ratio of k_2/k_{-1}?

32.31 Using the steady-state treatment, develop the rate expression for the following hypothetical mechanisms of formation of HBr:

a)
$$Br_2 \xrightarrow{k_1} 2\,Br,$$

$$Br + H_2 \xrightarrow{k_2} HBr + H.$$

b)
$$Br_2 \xrightarrow{k_1} 2\,Br,$$

$$Br + H_2 \xrightarrow{k_2} HBr + H,$$

$$Br + HBr \xrightarrow{k_3} Br_2 + H.$$

(Note that these are not chain mechanisms.)

32.32 The Rice–Herzfeld mechanism for the thermal decomposition of acetaldehyde is:

1) $$CH_3CHO \xrightarrow{k_1} CH_3 + CHO,$$

2) $$CH_3 + CH_3CHO \xrightarrow{k_2} CH_4 + CH_2CHO,$$

3) $$CH_2CHO \xrightarrow{k_3} CO + CH_3,$$

4) $$CH_3 + CH_3 \xrightarrow{k_4} C_2H_6.$$

Using the steady-state treatment, obtain the rate of formation of CH_4.

32.33 The activation energies for the elementary reactions in Problem 32.32 are $E_1^* = 320$ kJ/mol, $E_2^* = 40$ kJ/mol, $E_3^* = 75$ kJ/mol, and $E_4^* = 0$. Calculate the overall activation energy for the formation of methane.

32.34 The initial rate of the hydrogen–bromine reaction is given by

$$\left(\frac{d[HBr]}{dt}\right)_0 = 2k_2 \left(\frac{k_1}{k_5}\right)^{1/2} [H_2]_0[Br]_0^{1/2},$$

if we assume that no HBr is present initially. The activation energies for the reactions are:

Reaction	Rate constant	E^*/(kJ/mol)
$Br_2 \longrightarrow Br + Br$	k_1	192
$Br + Br \longrightarrow Br_2$	k_5	0
$Br + H_2 \longrightarrow HBr + H$	k_2	74

a) Calculate the overall activation energy for the initial rate.
b) Calculate the initial rate at 300 °C relative to that at 250 °C.

32.35 Consider the following hypothetical mechanism for the thermal decomposition of acetone.

Reaction	E^*/(kJ/mol)
$CH_3COCH_3 \xrightarrow{k_1} 2CH_3 + CO$	290
$CH_3 + CH_3COCH_3 \xrightarrow{k_2} CH_4 + CH_2COCH_3$	63
$CH_2COCH_3 \xrightarrow{k_3} CH_3 + CH_2CO$	200
$CH_3 + CH_2COCH_3 \xrightarrow{k_4} CH_3COC_2H_5$	33

a) What are the principal products predicted by this mechanism?
b) Show that the rate of formation of CH_4 is first order in acetone with an overall rate constant given by $k = (k_1k_2k_3/k_4)^{1/2}$. (Note: k_1 is very small.)
c) What is the overall activation energy for the reaction?

32.36 Consider the following mechanism for the decomposition of ozone into oxygen:

$$O_3 \xrightarrow{k_1} O_2 + O,$$

$$O + O_2 \xrightarrow{k_{-1}} O_3,$$

$$O_3 + O \xrightarrow{k_2} 2O_2.$$

a) Derive the rate expression for $-d[O_3]/dt$.

b) Under what condition will the reaction be first order with respect to ozone? Show how the equation reduces in this situation.

32.37 The mechanism proposed for the decomposition of N_2O_5 is:

$$N_2O_5 \xrightarrow{k_1} NO_2 + NO_3,$$

$$NO_2 + NO_3 \xrightarrow{k_{-1}} N_2O_5,$$

$$NO_2 + NO_3 \xrightarrow{k_3} NO + O_2 + NO_2,$$

$$NO + NO_3 \xrightarrow{k_4} 2NO_2.$$

Derive the expression for the rate of disappearance of N_2O_5 based on the steady-state approximation for the concentrations of NO_3 and NO.

32.38 Derive the expressions for the relaxation time for each of the reactions:

a)
$$A \underset{k_r}{\overset{k_f}{\rightleftharpoons}} B;$$

b)
$$A_2 \underset{k_r}{\overset{k_f}{\rightleftharpoons}} 2A.$$

32.39 Consider the two consecutive first-order reactions

$$A \xrightarrow{k_1} B, \quad B \xrightarrow{k_2} C.$$

Integrate the rate equations to obtain expressions for [A], [B], and [C] as functions of time. If $k_1 = 1\ \text{s}^{-1}$, sketch each of these functions for the cases $k_2/k_1 = 0.1$, 1, and 10. Assume that only A is present initially with a concentration c_0.

32.40 The reaction between iodine and acetone,

$$CH_3COCH_3 + I_2 \longrightarrow CH_3COCH_2I + HI,$$

is catalyzed by H^+ ion and by other acids. In the presence of monochloroacetic acid the rate constant is given by

$$k = k_{H^+}[H^+] + k_{ClCH_2COOH}[ClCH_2COOH],$$

where $[ClCH_2COOH]$ is the concentration of undissociated $ClCH_2COOH$. If the dissociation constant for the acid is 1.55×10^{-3}, calculate k_{H^+} and k_{ClCH_2COOH} from the following data.

$c_A/(\text{mol/L})$	0.05	0.10	0.20	0.50	1.00
$k/10^{-6}\ \text{min}^{-1}$	4.6	7.6	11.9	23.8	40.1

In this case, c_A is the total analytical concentration of $ClCH_2COOH$. (*Note:* Plot $k/[H^+]$ versus $[ClCH_2COOH]/[H^+]$ and determine slope and intercept.) (Data from K. J. Laidler, *Chemical Kinetics*, 2d ed. New York: McGraw-Hill, 1965, p. 456.)

32.41 The enzyme catalase catalyzes the decomposition of H_2O_2. The data are:

$[H_2O_2]/(\text{mol/L})$	0.001	0.002	0.005
Initial rate/(mol/L s)	1.38×10^{-3}	2.67×10^{-3}	6.00×10^{-3}

If the concentration of catalase is 4.0×10^{-9} mol/L, plot the data to determine v_{max}, the constant K_m, and the turnover number, k_2.

32.42 If an inhibitor, I, binds to an enzyme through the equilibrium, $E + I \rightleftharpoons EI$, and the dissociation constant of the species EI is K_I, then $[EI] = [E][I]/K_I$. If $[EI] \ll [I]$, then $[I] \approx [I]_0$ and $[E] = [E]_0 - [ES] - [EI]$. Show that the steady-state treatment under the condition that $[E]_0 \ll [S]_0$, yields a Lineweaver–Burk equation having the form

$$\frac{1}{v} = \frac{1}{v_{max}} + \frac{K_m(1 + [I]_0/K_I)}{v_{max}} \frac{1}{[S]_0}$$

32.43 The turnover number of the enzyme fumarase that catalyzes the reaction,

$$\text{Fumarate} + H_2O \longrightarrow \text{L-malate,}$$

is 2.5×10^3 s^{-1} and $K_m = 4.0 \times 10^{-6}$ mol/L. Calculate the rate of conversion of fumarate to L-malate if the fumarase concentration is 1.0×10^{-6} mol/L and the fumarate concentration is 2.04×10^{-4} mol/L.

32.44 If the second step in the enzyme catalysis mechanism is reversible, that is,

$$\text{ES} \underset{k_{-2}}{\overset{k_2}{\rightleftharpoons}} P + E,$$

derive the expression for the Michaelis–Menten law when $[E]_0 \ll [S]_0$.

33

Chemical Kinetics

II. Theoretical Aspects

33.1 INTRODUCTION

The ultimate goal of theoretical chemical kinetics is the calculation of the rate of any reaction from a knowledge of the fundamental properties of the reacting molecules; properties such as the masses, diameters, moments of inertia, vibrational frequencies, binding energies, and so on. At present this problem must be regarded as incompletely solved from the practical standpoint. Two approaches will be described here: the collision theory and the theory of absolute reaction rates. The collision theory is intuitively appealing and can be expressed in very simple terms. The theory of absolute reaction rates is more elegant. Neither theory is able to account for the magnitude of the activation energy except by approximations of questionable validity. The accurate calculation of activation energies from theory is a problem of extreme complexity and has been done for only a few very simple systems.

If we succeed in calculating the rate constant k for a reaction, we will have an interpretation of the Arrhenius equation,

$$k = Ae^{-E^*/RT}. \tag{33.1}$$

We begin by looking a little more closely into the meaning of the activation energy of a reaction.

33.2 THE ACTIVATION ENERGY

The expression in Eq. (33.1) is reminiscent of the form of the equation for the equilibrium constant of a reaction. Since

$$\frac{d \ln K}{dT} = \frac{\Delta H^0}{RT^2},$$

we have after integrating,

$$\ln K = -\frac{\Delta H^0}{RT} + \ln K^\infty, \qquad (33.2)$$

where $\ln K^\infty$ is the integration constant. For an elementary reaction, $K = k_f/k_r$ and $K^\infty = k_f^\infty/k_r^\infty$. Furthermore, $\Delta H^0 = H_P^0 - H_R^0$, where H_P^0 and H_R^0 are the total enthalpies of the products and the reactants, respectively. Using these values in Eq. (33.2) and rearranging, we have

$$\ln \frac{k_f}{k_f^\infty} - \frac{H_R^0}{RT} = \ln \frac{k_r}{k_r^\infty} - \frac{H_P^0}{RT}. \qquad (33.3)$$

The rate constant for the forward reaction presumably depends only on the properties of the reactants, while that of the reverse reaction depends only on the properties of the products. The left-hand side of Eq. (33.3) apparently depends only on the properties of reactants, while the right-hand side depends only on products. Each side must therefore be equal to a constant, which may be written $-H^*/RT$; then

$$\ln \frac{k_f}{k_f^\infty} = -\frac{H^* - H_R^0}{RT} \quad \text{and} \quad \ln \frac{k_r}{k_r^\infty} = -\frac{H^* - H_P^0}{RT}.$$

So that

$$k_f = k_f^\infty e^{-(H^* - H_R^0)/RT} \quad \text{and} \quad k_r = k_r^\infty e^{-(H^* - H_P^0)/RT}.$$

This argument can rationalize the form of the Arrhenius equation for the rate constants of any elementary reaction in either direction. The quantity $H^* - H_R^0$ is the energy quantity which the Arrhenius equation writes as E_f^*. Since we observe experimentally that E_f^* is positive, if follows that $H^* - H_R^0$ is positive, and that $H^* > H_R^0$. Similar argument shows that H^* is also greater than H_P^0.

The variation in enthalpy through the course of the elementary step, as reactants are converted to products, is shown in Fig. 33.1. According to this view of the situation, an energy barrier separates the reactant state from the product state. The reactants upon collision must have sufficient energy to surmount this barrier if products are to be formed.

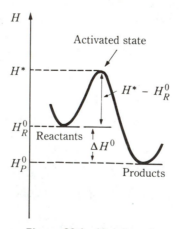

Figure 33.1   Variation of enthalpy in a reaction.

The height of this barrier is $H^* - H_R^0$; this is the activation energy* for the reaction in the forward direction E_f^*. Reactants which, upon collision, do not have sufficient energy to surmount the barrier will remain as reactants.

Viewed from the product side, the height of the barrier is $H^* - H_P^0$. This is the activation energy for the reverse reaction E_r^*. The relation between the two activation energies is obtained very simply. We write

$$H^* - H_P^0 = H^* - H_R^0 + H_R^0 - H_P^0 = H^* - H_R^0 - \Delta H^0.$$

Thus

$$E_r^* = E_f^* - \Delta H^0, \tag{33.4}$$

which is the general relation between the activation energies and the energy change in the reaction. If the activation energy for the reaction in the forward direction is known, that for the reverse reaction can be calculated directly from Eq. (33.4) if ΔH^0 is known.

33.3 THE COLLISION THEORY OF REACTION RATES

In its simplest form the collision theory is applicable only to bimolecular elementary reactions. With additional assumptions it can be applied to first-order reactions, and with some elaboration it is applicable to termolecular elementary reactions. As an example, we choose an elementary reaction of the type

$$A + B \longrightarrow C + D.$$

It is obvious that this reaction cannot occur more often than the number of times molecules A and B collide. The number of collisions between molecules A and B in one cubic metre per second is given by Eq. (30.23):

$$Z_{AB} = \pi \sigma_{AB}^2 \sqrt{\frac{8kT}{\pi\mu}} \, \tilde{N}_A \tilde{N}_B,$$

in which $\sigma_{AB} = \frac{1}{2}(\sigma_A + \sigma_B)$, and $(1/\mu) = (1/m_A) + (1/m_B)$, where σ_A and σ_B are the molecular diameters, m_A and m_B the molecular masses, $\tilde{N}_A$ and $\tilde{N}_B$ the number of molecules of A and B per cubic metre, and k the Boltzmann constant, printed in boldface in this chapter to avoid confusion with the rate constant, k. If reaction occurred with every collision, then this would be equal to the rate of disappearance of either A or B per cubic metre:

$$-\frac{d\tilde{N}_A}{dt} = -\frac{d\tilde{N}_B}{dt} = \pi \sigma_{AB}^2 \sqrt{\frac{8kT}{\pi\mu}} \, \tilde{N}_A \tilde{N}_B.$$

Every collision does not, in fact, result in the reaction of A and B, but only those collisions in which the energy of the colliding molecules exceeds E^*. The fraction of collisions in which the energy exceeds E^* is proportional to $\exp(-E^*/RT)$ so that the rate of the reaction is (after setting $\tilde{N} = \tilde{c}N_A$)

$$-\frac{d\tilde{c}_A}{dt} = N_A \pi \sigma_{AB}^2 \sqrt{\frac{8kT}{\pi\mu}} \, e^{-E^*/RT} \tilde{c}_A \tilde{c}_B. \tag{33.5}$$

---

*   There is a distinction between activation energy and activation enthalpy; however, the relation between them depends on the type of reaction in question. We will use the term "activation energy" loosely here to describe whichever one we are interested in at the moment.

The empirical law for the rate of the elementary reaction is $-d\tilde{c}_A/dt = k\tilde{c}_A\tilde{c}_B$, so for the rate constant we obtain

$$k = N_A \pi\sigma_{AB}^2 \sqrt{\frac{8\mathbf{k}T}{\pi\mu}}\, e^{-E^*/RT}, \tag{33.6}$$

$$k = N_A z_{AB} e^{-E^*/RT}, \tag{33.7}$$

where $z_{AB} = Z_{AB}/\tilde{N}_A\tilde{N}_B$.

The Arrhenius equation has the same form as Eq. (33.6), so the collision theory predicts for the frequency factor

$$A = N_A z_{AB} = N_A \pi\sigma_{AB}^2 \sqrt{\frac{8\mathbf{k}T}{\pi\mu}}. \tag{33.8}$$

Strictly speaking, A should be independent of temperature. However, the square-root dependence in Eq. (33.8) is rather slight, so a weak dependence on temperature is not really a difficulty. The order of magnitude of A can be readily estimated. The value of the radical in Eq. (33.8) is a molecular speed which, at ordinary temperatures, is about 400 m/s. The value of σ is about 3×10^{-10} m, so we have

$$A = (6 \times 10^{23}/\text{mol})\pi(3 \times 10^{-10} \text{ m})^2(400 \text{ m/s}) \approx 7 \times 10^7 \text{ m}^3/\text{mol s}.$$

If the concentration unit is mol/L, this must be multiplied by 1000 so that $A = 7 \times 10^{10}$ L/mol s. The order of magnitude of the frequency factor for bimolecular reactions is 10^9 to 10^{10} L/mol s if $T \approx 300$ K. The frequency factors for reactions involving a light molecule such as H_2 are larger: about 10^{11} L/mol s.

The collision theory predicts the value of the rate constant satisfactorily for reactions that involve relatively simple molecules if the activation energy is known. Difficulties are encountered with reactions between complicated molecules. The rates tend to be smaller than the collision theory predicts, in many cases by a factor of 10^5 or more. To account for this, an additional factor P, called the *probability factor* or the *steric factor*, is inserted in the expression for k:

$$k = N_A P z e^{-E^*/RT}. \tag{33.9}$$

The idea behind this is that even those collisions having the requisite energy may not produce reaction; originally it was supposed that the molecules had to collide in a particular configuration, hence the name *steric factor*. This idea has some validity, especially since the low rates of reaction are usually observed with complex molecules. Presumably two complex molecules will have less chance of colliding in the correct orientation for reaction than will two simple molecules. We will see shortly that the probability factor receives a more acceptable interpretation in terms of the entropy of activation of a reaction. In particular, the collision theory offers no explanation for abnormally fast reactions in which P would have to be greater than unity.

33.4 TERMOLECULAR REACTIONS

The problem of termolecular reactions can be treated by collision theory also. A number of such reactions are known; reactions of NO with H_2, O_2, Cl_2 are famous examples. If we choose the reaction with oxygen,

$$2\,NO + O_2 \longrightarrow 2\,NO_2,$$

the rate of the reaction is

$$\frac{d(\xi/V)}{dt} = k(NO)^2(O_2).$$

Apparently the reaction as written is elementary and involves the simultaneous collision of two molecules of NO with one molecule of O_2. A remarkable feature of this reaction is that the rate of the reaction *decreases* with *increase* in temperature. This behavior is exhibited by only a very few reactions.

A difficulty in the treatment by collision theory arises as soon as we attempt to define a triple collision. If the molecules are hard spheres, the time in which they are in contact is zero. The probability of being hit by a third molecule during the collision is therefore zero. A finite time interval must be specified for the collision of two molecules if a third is to collide with the two. The time interval is arbitrary; a common specification is that the molecules are in collision so long as the distance between them is less than the molecular diameter. With this specification we can show that, approximately, $Z_3/Z_2 = \sigma/\lambda$, where Z_3 and Z_2 are the numbers of triple collisions and binary collisions per cubic metre per second, σ is the molecular diameter, and λ is the mean free path. Since λ is inversely proportional to the number of molecules per cubic metre, it follows that the number of triple collisions increases as the cube of the number of molecules per cubic metre. At ordinary· pressures $\lambda \approx 10^{-6}$ m so that $Z_3 \approx Z_2(10^{-10}/10^{-6}) = 10^{-4}Z_2$. Roughly speaking, there is one triple collision for every 10,000 ordinary collisions. Therefore reactions requiring triple collisions are slower, other things being equal, than those involving binary collisions. The rate constant calculated on the basis of triple collisions is much larger than the experimental value, indicating that the probability factor is quite small.

An alternative mechanism has been proposed for these reactions. The equilibrium

$$NO + X_2 \;\rightleftharpoons\; NOX_2$$

is assumed to be established very rapidly. Then

$$(NOX_2) = K(NO)(X_2),$$

where K is the equilibrium constant. The slow reaction follows:

$$NOX_2 + NO \;\longrightarrow\; 2\,NOX.$$

The rate of this reaction is

$$\frac{d(\xi/V)}{dt} = k(NO)(NOX_2) = kK(NO)^2(X_2).$$

This mechanism accounts for the rate law. It is apparent that the equilibrium

$$2\,NO \;\rightleftharpoons\; N_2O_2$$

followed by the slow step

$$N_2O_2 + X_2 \;\longrightarrow\; 2\,NOX$$

would also account for the empirical rate law. The difference between the triple collision viewpoint and these mechanisms is not very great. The molecules NOX_2 or N_2O_2 can be thought of as two molecules that are involved in a "sticky" collision. The equilibrium assumption explains the negative temperature coefficient, since it implies that at higher temperatures more double molecules, NOX_2 or N_2O_2, are dissociated; the lower concentration of double molecules results in a lower rate. This implies that the activation energy is negative. This explanation has its problems, since in the reaction of NO with O_2, the activation energy is zero; the decrease in rate constant with temperature is due to the frequency factor, which is inversely proportional to T^3.

33.5 UNIMOLECULAR REACTIONS

Collision theory does not deal directly with unimolecular reactions but touches on the subject through the Lindemann mechanism. Once the molecule has been provided with sufficient energy by collision, the problem is to calculate the rate constant for the unimolecular decomposition,

$$A^* \longrightarrow \text{Products.}$$

The theory of this type of decomposition has been developed by O. K. Rice, H. C. Ramsberger, and L. S. Kassel; more recently, N. B. Slater has treated the problem in more exact and elegant terms. The treatment is based on the supposition that if too much energy gets into a particular mode of vibration, then vibration of the molecule in this mode leads to dissociation of the molecule.

The Rice–Ramsberger–Kassel approach assumes that the activated molecule has a certain amount of vibrational energy spread among the various vibrational degrees of freedom of the molecule. Then the probability of one particular mode of vibration acquiring so much of this energy that the vibration leads to dissociation into fragments is calculated.

We assume that there are s vibrational degrees of freedom and that the molecule has j quanta of energy distributed in the s degrees of freedom. Let N_j be the number of ways of distributing the j quanta in the s degrees of freedom. Let N_m be the number of ways of distributing the j quanta in the s degrees, so that a particular degree of freedom has m quanta. Then the probability that that particular degree of freedom has m quanta is N_m/N_j. If j and $j - m$ are large compared with s, we can show* that, approximately,

$$\frac{N_m}{N_j} = \left(\frac{j - m}{j}\right)^{s-1}. \tag{33.10}$$

Since j is the total number of quanta, it is proportional to the total vibrational energy of the molecule E; the number m is proportional to E_c, the critical minimum energy required for dissociation to occur. Therefore we can write the probability of the particular degree of freedom having the critical energy in the form

$$\frac{N_m}{N_j} = \left(\frac{E - E_c}{E}\right)^{s-1}. \tag{33.11}$$

The rate of dissociation of the molecule is proportional to this probability, so that the rate constant is given by

$$k = k'\left(\frac{E - E_c}{E}\right)^{s-1}, \tag{33.12}$$

where k' is a constant. Since E may have any value from E_c to infinity, we must average the rate constant, using a Boltzmann distribution, over all values of E from E_c to infinity. We

---

*   The proof is elementary but is too lengthy to be included here.

evaluate the integral graphically for particular values of s, and E_c. Reasonable agreement with experiment is obtained if we use values of s comparable to the number of vibrational degrees of freedom in the molecule.

Thus we see that for unimolecular reactions, as well as for others, the molecule must have at least a critical minimum energy for reaction to occur. The interpretation of the Arrhenius equation for unimolecular reactions is more complex, however. The pre-exponential factor A is a function of the number of degrees of vibrational freedom s, as well as E_c and T. Note that Eq. (33.11) implies that the higher the energy E in the vibrational modes and the greater the number s of these modes, the greater is the probability that the molecule will have the required m quanta in the critical vibration. A related fact is that the rate of activation may be larger than the collision rate predicted by the Lindemann mechanism. Some molecules may be "self-activated" in the sense that quanta of vibrational energy which are spread over the various modes of vibration may flow into the critical mode and supply it with the critical energy. This process enhances the rate of activation.

★ 33.6 IRREVERSIBLE THERMODYNAMICS

Considerable effort has been expended in the attempt to develop a general theory of reaction rates through some extension of thermodynamics or statistical mechanics. Since neither of these sciences can, by themselves, yield any information about rates of reactions, some additional assumptions or postulates must be introduced. An important method of treating systems that are not in equilibrium has acquired the title of *irreversible thermodynamics*. Irreversible thermodynamics can be applied to those systems that are "not too far" from equilibrium. The theory is based on the thermodynamic principle that in every irreversible process, that is, in every process proceeding at a finite rate, entropy is created. This principle is used together with the fact that the entropy of an isolated system is a maximum at equilibrium, and with the principle of microscopic reversibility.* The additional assumption involved is that systems that are slightly removed from equilibrium may be described statistically in much the same way as systems in equilibrium.

An outstanding success of the theory has been the general derivation of the relations between certain pairs of rate constants in transport processes, the Onsager reciprocal relations. Although these relations were known before, the derivations were individualized and in certain cases the validity of the derivation was suspect. The theory is not applicable to the data obtained from the usual type of investigation in chemical kinetics in which the system is far removed from equilibrium. Investigations specifically designed to test the theory have supported its conclusions. An interesting aspect of the theory is that it requires certain relations between the rate constants of coupled reactions in systems that are "not too far" from equilibrium.

Central to the thermodynamic discussion of irreversible processes is the concept of entropy production. Consider the Clausius inequality, $dS \geq dQ/T$, which we can rearrange to the form

$$dS - \frac{dQ}{T} \geq 0.$$

---

* Principle of microscopic reversibility: at equilibrium, any molecular process occurs at the same rate as the reverse of that process.

The quantity on the left is greater than or equal to zero, so we may write

$$dS - \frac{dQ}{T} = d\sigma, \qquad (33.13)$$

if we insist that $d\sigma$ be either zero or positive.

If we suppose that the system is in contact with a reservoir at T, and a quantity of heat dQ flows into the system, then a quantity, $-dQ$, flows into the reservoir. If the quantity, $-dQ$, is transferred reversibly to the reservoir, then the entropy change of the reservoir is $dS_{res} = -dQ/T$, and we can write Eq. (33.13) as

$$dS + dS_{res} = d\sigma.$$

The quantity $d\sigma$ is the entropy increase of the system plus that of the surroundings (the reservoir); $d\sigma$ is called the *entropy production* of the process. For any irreversible transformation, the entropy production is positive, while for a reversible transformation the entropy production is zero.

We may write Eq. (33.13) in the form

$$T \, d\sigma = T \, dS - dQ. \qquad (33.14)$$

If we apply this equation to a transformation at constant T and p, we have $dQ_p = dH$, and $T \, dS = d(TS)$, so that $T \, dS - dQ_p = d(TS) - dH = -d(H - TS)$; then

$$T \, d\sigma = -dG. \qquad (33.15)$$

For a chemical reaction at constant T and p, we have $dG = (\partial G/\partial \xi)_{T,p} \, d\xi$, and therefore

$$T \, d\sigma = -\left(\frac{\partial G}{\partial \xi}\right)_{T,p} d\xi. \qquad (33.16)$$

DeDonder has introduced A, the affinity of the reaction, for the quantity, $-(\partial G/\partial \xi)_{T,p}$.

$$A \equiv -\left(\frac{\partial G}{\partial \xi}\right)_{T,p} \qquad (33.17)$$

Combining this definition with Eq. (33.16) yields

$$T \, d\sigma = A \, d\xi. \qquad (33.18)$$

Note that for the spontaneous direction of a reaction, $(\partial G/\partial \xi)_{T,p}$ is negative, so that the affinity is positive. Dividing by dt, we obtain the *rate* of entropy production, $d\sigma/dt$.

$$T \frac{d\sigma}{dt} = A \frac{d\xi}{dt}. \qquad (33.19)$$

Since the rate of entropy production by the second law must always be positive or zero, it follows from Eq. (33.19) that the product of the affinity and the rate of reaction, $d\xi/dt$, must always be positive or zero. This result

$$A \frac{d\xi}{dt} \geq 0 \qquad (33.20)$$

is known as DeDonder's inequality.

There is an important general relation that we can obtain with relative ease by combining a rate equation with a thermodynamic equation. Consider the reaction

$$A + B \; \rightleftharpoons \; C.$$

We write the rate equation as in Section 32.19.

$$\frac{1}{V}\frac{d\xi}{dt} = k_f c_A c_B - k_r c_C;$$

$$\frac{1}{V}\frac{d\xi}{dt} = k_f c_A c_B\left(1 - \frac{k_r c_C}{k_f c_A c_B}\right).$$

We recognize that $c_C/c_A c_B = Q$, the proper quotient of concentrations for the reaction, and $k_f/k_r = K$, the equilibrium constant for the reaction. Then we have

$$\frac{1}{V}\frac{d\xi}{dt} = k_f c_A c_B\left(1 - \frac{Q}{K}\right). \tag{33.21}$$

Near equilibrium, the quantity $k_f c_A c_B$ approaches $r = k_f \bar{c}_A \bar{c}_B$, the exchange rate of the reaction. Then

$$\frac{1}{V}\frac{d\xi}{dt} = r\left(1 - \frac{Q}{K}\right).$$

However, we have the relations

$$\left(\frac{\partial G}{\partial \xi}\right)_{T,p} = \Delta G^\circ + RT \ln Q \quad\text{and}\quad \Delta G^\circ = -RT \ln K,$$

which combine to yield

$$\left(\frac{\partial G}{\partial \xi}\right)_{T,p} = RT \ln \frac{Q}{K} = -\mathsf{A},$$

or

$$\frac{Q}{K} = e^{-\mathsf{A}/RT}.$$

Since A is very small near equilibrium, we can expand the exponential in series to obtain: $Q/K = 1 - \mathsf{A}/RT + \cdots$. This brings the rate equation to the form

$$\frac{d\xi}{dt} = Vr\frac{\mathsf{A}}{RT}. \tag{33.22}$$

Equation (33.22) expresses the important result that the rate of a reaction near equilibrium is proportional to the affinity of the reaction.

Equation (33.22) is a chemical example of a linear law analogous to those mentioned in Section 30.2. In each of those cases, a flow, such as a heat flow, an electrical current, a fluid flow, or a diffusive flow, was proportional to a driving force such as a temperature gradient, an electrical potential gradient, a pressure gradient or a concentration gradient. In the chemical case, Eq. (33.22), the "flow" is the rate of the reaction, while the driving force is the affinity of the reaction divided by T.

If we combine the result of Eq. (33.22) with Eq. (33.19) for the rate of entropy production we obtain

$$\frac{d\sigma}{dt} = RVr\left(\frac{\mathsf{A}}{RT}\right)^2 = \frac{R}{Vr}\left(\frac{\partial \xi}{\partial t}\right)^2. \tag{33.23}$$

This shows the positive character of $d\sigma/dt$ since it is proportional to the square of the affinity or to the square of the reaction rate.

The two equations, Eqs. (33.22) and (33.23) are typical of the application of thermodynamics to irreversible processes. We obtain, or assume, a linear rate law such as the one in Eq. (33.22) in which the flow is proportional to the driving force; and we obtain a quadratic law for the entropy production in which, as in Eq. (33.23), the rate of entropy production is proportional to the square of the driving force.

33.7 THE THEORY OF ABSOLUTE REACTION RATES

The theory of absolute reaction rates, which is based on statistical mechanics, was developed in full generality by H. Eyring in 1935, although it was foreshadowed in kinetic theory investigations as early as 1915. A simplified development of the equations will be given here. In this theory, we have a postulate of "equilibrium" away from equilibrium, applied more broadly here than in the irreversible thermodynamics.

The fundamental postulate of the theory of absolute reaction rates is that the reactants are always in equilibrium with activated complexes. The activated complex is that configuration of the atoms which corresponds energetically to the top of the energy barrier separating the reactants from the products (Fig. 33.2). We write the equilibrium

$$A + B \rightleftharpoons M^{\ddagger},$$

and the equilibrium constant is

$$K_{\ddagger} = \frac{(\tilde{c}^{\ddagger}/\tilde{c}^{\circ})}{(\tilde{c}_A/\tilde{c}^{\circ})(\tilde{c}_B/\tilde{c}^{\circ})}. \tag{33.24}$$

in which the $\tilde{c}$'s are concentrations in moles per cubic metre. The standard concentration, $\tilde{c}^{\circ} = 1000 \text{ mol/m}^3$. The concentration of activated complexes is

$$\tilde{c}^{\ddagger} = \left(\frac{K_{\ddagger}}{\tilde{c}^{\circ}}\right)\tilde{c}_A \tilde{c}_B. \tag{33.25}$$

If we know the concentration of activated complexes, the problem resolves into the calculation of the rate at which these complexes decompose into products; that is, we must calculate the rate of the reaction

$$M^{\ddagger} \longrightarrow \text{Products}.$$

The activated complex is an aggregate of atoms, which may be thought of as being similar to an ordinary molecule except that it has one special vibration with respect to which it is unstable. This vibration leads to dissociation of the complex into products. If the frequency

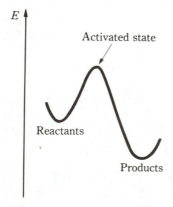

Reactants

Activated state

Products

Figure 33.2 Energy variation in the transformation from reactants to products.

of this vibration is v, then the rate, in moles per unit volume per second, at which products are formed is

$$\frac{d(\xi/V)}{dt} = v\tilde{c}^{\ddagger}. \tag{33.26}$$

Using Eq. (33.25) we can write

$$\frac{d(\xi/V)}{dt} = \frac{vK_{\ddagger}}{\tilde{c}^{\circ}} \tilde{c}_A \tilde{c}_B. \tag{33.27}$$

But the elementary reaction, $A + B \rightarrow$ Products, has the rate,

$$\frac{d(\xi/V)}{dt} = k\tilde{c}_A \tilde{c}_B \tag{33.28}$$

Comparing Eqs. (33.27) and (33.28), we find that the rate constant is given by

$$k = \frac{vK_{\ddagger}}{\tilde{c}^{\circ}}. \tag{33.29}$$

A review of the steps involved in deriving Eq. (33.29) shows that it is not restricted by the choice of two reactants, but is correct for any elementary reaction, if $\tilde{c}^{\circ}$ is replaced by $(\tilde{c}^{\circ})^{-\Delta v}$ where Δv is the net increase in the stoichometric coefficients in the elementary reaction; for the above case, $\Delta v = -1$.

The values of v and $K_{\ddagger}$ can be calculated if we write the equilibrium constant in terms of molecular partition functions per unit volume, q/V. [See Eq. (29.75).] Then

$$K_{\ddagger} = \frac{\tilde{N}^{\circ}(q_{\ddagger}/V)}{(q_A/V)(q_B/V)}. \tag{33.30}$$

Any molecular partition function can be written in the form $(q/V) = fe^{-\epsilon_0/kT}$. The function f is the partition function per unit volume evaluated using energies relative to the zero-point energy ϵ_0 of the molecule. Then Eq. (33.30) becomes

$$K_{\ddagger} = \frac{\tilde{N}^{\circ}f_{\ddagger}}{f_A f_B} e^{-(\epsilon_{0\ddagger} - \epsilon_{0A} - \epsilon_{0B})/kT} = \frac{\tilde{N}^{\circ}f_{\ddagger}}{f_A f_B} e^{-\Delta E_0^{\ddagger}/RT}. \tag{33.31}$$

The activation energy $\Delta E_0^{\ddagger}$ is defined as the difference in zero-point energies between the activated complex and the reactants: $\Delta E_0^{\ddagger} = N_A(\epsilon_{0\ddagger} - \epsilon_{0A} - \epsilon_{0B})$.

As we have seen in Section 29.12, the partition function can be written as a product of partition functions for translation, rotation, and vibration. We direct our attention to that particular vibration which dissociates the activated complex into products, and factor that vibrational partition function out of $f_{\ddagger}$. Let

$$f_{\ddagger} = f_v f^{\ddagger}, \tag{33.32}$$

where $f^{\ddagger}$ is what remains of $f_{\ddagger}$ after f_v has been factored out. If the frequency of this vibration is v, then by Eq. (29.44),

$$f_v = \frac{kT}{hv} e^{-hv/2kT},$$

if v is small and $hv/kT \ll 1$. Since the exponential is about equal to unity, $f_v = kT/hv$ and Eq. (33.32) becomes

$$f_{\ddagger} = \frac{kT}{hv} f^{\ddagger}. \tag{33.33}$$

Using this value of $f_{\ddagger}$ in Eq. (33.31), we obtain

$$K_{\ddagger} = \frac{\mathbf{k}T}{h\nu}\frac{\tilde{N}°f^{\ddagger}}{f_A f_B}e^{-\Delta E_0^{\ddagger}/RT}. \tag{33.34}$$

Define $K^{\ddagger}$ by

$$K^{\ddagger} = \frac{\tilde{N}°f^{\ddagger}}{f_A f_B}e^{-\Delta E_0^{\ddagger}/RT}; \tag{33.35}$$

then $K_{\ddagger} = (\mathbf{k}T/h\nu)K^{\ddagger}$. Using this value in Eq. (33.29) yields, for the rate constant,

$$k = \frac{\mathbf{k}T}{h}\frac{K^{\ddagger}}{\tilde{c}°}, \tag{33.36}$$

which is the Eyring equation for the rate constant of a reaction. The value of $K^{\ddagger}$ can be calculated from the partition functions of the reactants and the activated complex using an equation having the form of Eq. (33.35). If we use Eq. (33.35) in Eq. (33.36), we obtain

$$k = \frac{\mathbf{k}T}{h}\left(\frac{\tilde{N}°}{\tilde{c}°}\right)\frac{f^{\ddagger}}{f_A f_B}e^{-\Delta E_0^{\ddagger}/RT}. \tag{33.37}$$

Comparing this result with the Arrhenius equation, we see that the frequency factor A is given by

$$A = \frac{\mathbf{k}T}{h}\left(\frac{\tilde{N}°}{\tilde{c}°}\right)\frac{f^{\ddagger}}{f_A f_B}. \tag{33.38}$$

Note that $(\tilde{N}°/\tilde{c}°) = N_A$.

The expression in Eq. (33.38) is interesting because the partition functions depend on the translational degrees of freedom of the molecules and the activated complex, and on the internal degrees of freedom as well. The collision theory cannot take into account the internal degrees of freedom without becoming incredibly complicated mathematically. The Eyring theory includes the internal degrees of freedom in a very simple way.

Two things are required to calculate the rate constant by Eq. (33.37). First, the activated complex must be specified sufficiently so that $f^{\ddagger}$ can be calculated; this implies knowing its size and shape so that the moments of inertia can be calculated. The calculation of the vibrational frequencies can be done quantum mechanically, but is quite complicated. Second, $\Delta E_0^{\ddagger}$ must be known. The calculation of $\Delta E_0^{\ddagger}$ from quantum mechanics is quite complicated unless drastic approximations are made. This procedure has been carried out in full detail for a number of reactions, particularly for reactions involving hydrogen atoms and hydrogen molecules. Considering the approximations involved, the results are very good.

It is relatively easy to obtain a rough estimate of the order of magnitude of the frequency factor using Eq. (33.38). Consider the reaction

$$H_2 + I_2 \longrightarrow 2HI.$$

The frequency factor is

$$A = \frac{\mathbf{k}T}{h}\frac{N_A f^{\ddagger}}{f_{H_2} f_{I_2}}.$$

The partition function f_{H_2} can be written as a product of partition functions for three translational, two rotational, and one vibrational degree of freedom:

$$f_{I_2} = f_{H_2} = f_t^3 f_r^2 f_v.$$

We use the same value for f_{t_2} since we wish to make only a calculation of the order of magnitude. The complex $(HI)_2$ has three translational, three rotational, and five vibrational degrees (one vibrational degree was removed in the early part of the derivation); thus we write $f^{\ddagger} = f_t^3 f_r^3 f_v^5$. Using these values, we find for A,

$$A = \frac{kT}{h} \frac{N_A f_t^3 f_r^3 f_v^5}{f_t^3 f_r^2 f_v f_t^3 f_r^2 f_v} = \frac{kT}{h} \left(\frac{N_A f_v^3}{f_t^3 f_r}\right).$$

At ordinary temperatures the usual magnitudes of these quantities are $f_t \approx 10^{10}/m$, $f_r \approx 10$, $f_v \approx 1$, and $(kT/h) \approx 10^{13}/s$. Using these values, we obtain

$$A \approx \frac{(10^{13}/s)}{(10^{10}/m)^3 10} (6 \times 10^{23}/mol) \approx 10^6 \; m^3/mol \; s,$$

which is in rough agreement with the value we calculated using Eq. (30.23) for the collision frequency. This value of the frequency factor usually agrees roughly with the values A found for bimolecular reactions, which are often between 10^6 and 10^8 m^3/mol s. Considering the very approximate values used for the partition functions, the agreement is good.

33.8 COMPARISON OF THE COLLISION THEORY WITH THE ABSOLUTE REACTION RATE THEORY

For bimolecular reactions, we can easily compare collision theory with absolute reaction rate theory, using the results of the preceding section. Consider the bimolecular reaction between two polyatomic molecules A and B to yield a complex

$$A + B \longrightarrow (AB)^{\ddagger}.$$

If n_A and n_B are the number of atoms in A and B and if both molecules are nonlinear, then

$$f_A = f_t^3 f_r^3 f_v^{3n_A - 6}, \qquad f_B = f_t^3 f_r^3 f_v^{3n_B - 6}, \qquad f^{\ddagger} = f_t^3 f_r^3 f_v^{3(n_A + n_B) - 7},$$

since the complex contains $n_A + n_B$ atoms. Using these values in Eq. (33.37), we obtain

$$k = \frac{kT}{h} \left(\frac{N_A f_t^3 f_r^3 f_v^{3(n_A + n_B) - 7}}{f_t^3 f_r^3 f_v^{3n_A - 6} f_t^3 f_r^3 f_v^{3n_B - 6}}\right) e^{-\Delta E_0^{\ddagger}/RT} = \frac{kT}{h} \frac{N_A f_v^5}{f_t^3 f_r^3} e^{-\Delta E_0^{\ddagger}/RT}. \tag{33.39}$$

Now the Eyring equation yields the same result as the collision theory if we treat A and B as if they were atoms and $(AB)^{\ddagger}$ as if it were diatomic; then we would have $f_A = f_B = f_t^3$ and $f^{\ddagger} = f_t^3 f_r^2$, and

$$k_{collis} = \frac{kT}{h} \frac{N_A f_t^3 f_r^2}{f_t^3 f_t^3} e^{-\Delta E_0^{\ddagger}/RT} = \frac{kT}{h} \frac{N_A f_r^2}{f_t^3} e^{-\Delta E_0^{\ddagger}/RT}. \tag{33.40}$$

Comparing Eqs. (33.39) and (33.40), we obtain

$$k = \left(\frac{f_v}{f_r}\right)^5 k_{collis}. \tag{33.41}$$

The probability factor P, which must be introduced arbitrarily in the collision theory, is according to Eq. (33.41),

$$P = \left(\frac{f_v}{f_r}\right)^5. \tag{33.42}$$

If $f_v = 1$ and $f_r = 10$, then $P = 10^{-5}$, which is a not uncommon value of P. This equation gives us a little insight into the effect of the internal degrees of freedom that make reactions between polyatomic molecules very much slower than those between simple molecules.

The great advantage of the theory of absolute reaction rates is that the postulated equilibrium between activated complex and reactants evades entirely the question of just how the complex is formed; the theory resembles thermodynamics in this regard. For the same reason a trace of dissatisfaction with the theory can be voiced, for the object of kinetics is to look into the details of how the reaction goes. The great simplicity introduced by the equilibrium assumption enables us to suppress this feeling of dissatisfaction to some extent. The collision theory could take into account the presence of internal degrees of freedom in the molecule. However, the mathematical treatment using this approach would be intolerably complex.

33.9 GIBBS ENERGY AND ENTROPY OF ACTIVATION

Equation (33.36), the Eyring equation, we can generalize to apply to any elementary reaction,

$$A + B + C + \cdots \;\rightleftharpoons\; M^{\ddagger}$$

by writing

$$k = \frac{\mathbf{k}T}{h}(\tilde{c}^{\circ})^{\Delta \nu}K^{\ddagger}$$

where

$$K^{\ddagger} = \frac{f^{\ddagger}}{(\tilde{N}^{\circ})^{\Delta \nu}f_A\,f_B\,f_C\cdots}\,e^{-\Delta E_0^{\ddagger}/RT}.$$

The equilibrium constant $K^{\ddagger}$ can be written in terms of a standard* Gibbs energy of activation $\Delta G^{\ddagger}$:

$$K^{\ddagger} = e^{-\Delta G^{\ddagger}/RT}. \tag{33.43}$$

Then we obtain for the rate constant

$$k = \frac{\mathbf{k}T}{h}(\tilde{c}^{\circ})^{\Delta \nu}e^{-\Delta G^{\ddagger}/RT}, \tag{33.44}$$

which emphasizes the Gibbs energy of activation as the fundamental quantity, rather than the energy of activation. Then we can write $\Delta G^{\ddagger} = \Delta H^{\ddagger} - T\,\Delta S^{\ddagger}$, so that

$$k = \frac{\mathbf{k}T}{h}(\tilde{c}^{\circ})^{\Delta \nu}e^{\Delta S^{\ddagger}/R}e^{-\Delta H^{\ddagger}/RT}, \tag{33.45}$$

which resembles the Arrhenius equation, except that $\Delta H^{\ddagger}$ appears instead of E^*. The quantity $\Delta H^{\ddagger}$ is often called the heat of activation. The frequency factor A, according to Eq. (33.45), is

$$A = \frac{\mathbf{k}T}{h}(\tilde{c}^{\circ})^{\Delta \nu}e^{\Delta S^{\ddagger}/R}. \tag{33.46}$$

A negative entropy of activation will result in a quite low frequency factor, while a positive entropy of activation will raise its value. The probability factor introduced in the collision

---

*   The usual degree superscript used to designate a standard Gibbs energy is omitted on the symbol $\Delta G^{\ddagger}$ to avoid a cumbersome symbol.

theory can be interpreted in terms of the entropy of activation. We write

$$N_A P z = \frac{kT}{h} \frac{e^{\Delta S^\ddagger/R}}{\tilde{c}^\circ}. \tag{33.47}$$

The collision frequency z is readily calculated so that from the value of P the value of $\Delta S^\ddagger$, the entropy of activation, can be calculated.

If the activated complex resembles the products more than the reactant, then the entropy of activation may be nearly equal to the ΔS° of the overall reaction. The values of $\Delta S^\ddagger$ and ΔS° do seem to parallel one another in many cases, although it is rare that they are equal.

★ 33.10 REACTIONS IN SOLUTION

In Section 32.18 we mentioned the fact that the rate constant for a reaction in solution is often very nearly the same as that for the same reaction in the gas phase. A rate constant in solution which is very much different from that in the gas indicates a comparatively strong interaction between the solvent and the reactants or the activated complex.

The reason for the equality between the rate of reaction in the gas and that in the solution can be explained rather simply in terms of the collision theory. Suppose a reaction requires the collision of two molecules in a pure gas A:

$$A + A \longrightarrow \text{Products.}$$

The number of collisions per cubic metre per second can be written as

$$Z_{11} = Z\tilde{N}_A^2. \tag{33.48}$$

If foreign molecules B are introduced, there will be collisions between A and B and collisions between B and B. This fact does not change the number of collisions between A and A, which is still given by Eq. (33.48). Thus, even if all the intervening space in the gas is filled with foreign molecules B, the rate constant should not change. On this basis, the rate constant in solution should have the same value as in the gas. The argument is correct only for reasonably ideal solutions. Nonideality in the solution implies solvation effects of the type alluded to in the preceding paragraph.

Consider the reaction in solution:

$$A + B \rightleftharpoons M^\ddagger \longrightarrow \text{Products.}$$

The rate of this reaction is given by Eq. (33.26),

$$\frac{d(\xi/V)}{dt} = v\tilde{c}^\ddagger$$

But in solution,

$$K_\ddagger = \frac{a^\ddagger}{a_A a_B} = \left(\frac{\gamma^\ddagger}{\gamma_A \gamma_B}\right) \frac{(\tilde{c}^\ddagger/\tilde{c}^\circ)}{(\tilde{c}_A/\tilde{c}^\circ)(\tilde{c}_B/\tilde{c}^\circ)}, \tag{33.49}$$

where $K_\ddagger$ is the equilibrium constant, $\tilde{c}^\circ = 1000 \text{ mol/m}^3$, the a's are the activities and the γ's are activity coefficients. Then the rate becomes

$$\frac{d(\xi/V)}{dt} = \frac{vK_\ddagger}{\tilde{c}^\circ} \left(\frac{\gamma_A \gamma_B}{\gamma^\ddagger}\right) \tilde{c}_A \tilde{c}_B,$$

and the rate constant is

$$k = \frac{\nu K_{\ddagger}}{\tilde{c}^{\circ}}\left(\frac{\gamma_A \gamma_B}{\gamma^{\ddagger}}\right).$$

Comparison of Eqs. (33.29) and (33.36) shows that $\nu K_{\ddagger} = (\mathbf{k}T/h)K^{\ddagger}$, so we have

$$k = \frac{\mathbf{k}T}{h}\frac{K^{\ddagger}}{\tilde{c}^{\circ}}\left(\frac{\gamma_A \gamma_B}{\gamma^{\ddagger}}\right), \qquad (33.50)$$

which is the equation we must use to discuss reactions in solution.

The definition of the activity coefficient depends on the choice of the reference state in which $\gamma = 1$. If we wish to compare the rate of reaction with the rate in the gas phase, then we will choose the reference state of unit activity coefficient as the ideal gas. This reduces Eq. (33.50) to

$$k_g = \frac{\mathbf{k}T}{h}\frac{K^{\ddagger}}{\tilde{c}^{\circ}}, \qquad (33.51)$$

where k_g is the rate constant for the reaction in the gas phase. Then in solution

$$k = k_g\left(\frac{\gamma_A \gamma_B}{\gamma^{\ddagger}}\right)_g. \qquad (33.52)$$

In Eq. (33.52) the subscript g on the activity coefficient ratio indicates the choice of reference state. The deviation of the value of the rate constant in the solution from that in the gas depends on this ratio of activity coefficients. If the solvent lowers the Gibbs energy of the reactants more than it does that of the activated complex (if the reactants are strongly solvated), then γ_A and γ_B will be small, while $\gamma^{\ddagger}$ will not be so small. The rate constant in this case will be smaller in solution than in the gas. Conversely, if the activated complex is strongly solvated while the reactants are not, the rate constant will be larger in solution than in the gas.

If the reaction is one that does not take place in the gas phase, then it is more useful to choose the infinitely dilute solution as the reference state of unit activity coefficient. If the rate constant in infinitely dilute solution is k_0, since the γ's are unity, we have

$$k_0 = \frac{\mathbf{k}T}{h}\frac{K^{\ddagger}}{\tilde{c}^{\circ}}$$

and

$$k = k_0\left(\frac{\gamma_A \gamma_B}{\gamma^{\ddagger}}\right)_0. \qquad (33.53)$$

Here the subscript zero on the activity coefficient ratio indicates the infinitely dilute solution as the reference state.

★ 33.11 IONIC REACTIONS; SALT EFFECTS

The majority of reactions between ions in solution, particularly between simple ions of opposite charge, occur so rapidly that until recently it was impossible to measure the rates of these reactions. Relaxation techniques such as those described in Section 32.19 are now used to determine the rate of reactions such as $H_3O^+ + OH^- \rightarrow 2H_2O$. The rate constant of this particular reaction is 1.4×10^{11} L/mol s.

There are some reactions between ions and neutral molecules that progress slowly enough that ordinary methods of measurement can be used. The rate constants of these reactions depend on the ionic strength of the solution. Equation (33.53) was first derived by Brønsted and Bjerrum before the theory of absolute reaction rates was developed; applied to ionic reactions, Eq. (33.53) is called the Brønsted–Bjerrum equation. By combining Eq. (33.53) with the Debye–Hückel limiting law for ionic activity coefficients, we can deduce the dependence of the rate constant on the ionic strength. Writing Eq. (33.53) in logarithmic form, we have

$$\log_{10} k = \log_{10} k_0 + \log_{10} \gamma_A + \log_{10} \gamma_B - \log_{10} \gamma^{\ddagger}. \tag{33.54}$$

The value of $\ln \gamma$ is given by Eq. (16.75); by comparing Eqs. (16.75) and (16.77) we find that for a single ion the Debye–Hückel limiting law is

$$\log_{10} \gamma_i = -A z_i^2 I_c^{1/2}, \tag{33.55}$$

where A is a constant; in water at 25 °C, $A = 0.50$ (L/mol)$^{1/2}$. Using the limiting law in Eq. (33.54), and realizing that $z^{\ddagger} = z_A + z_B$, we can write

$$\log_{10} k = \log_{10} k_0 - A[z_A^2 + z_B^2 - (z_A + z_B)^2]I_c^{1/2} = \log_{10} k_0 + 2A z_A z_B I_c^{1/2}.$$

Using $A = 0.50$ (L/mol)$^{1/2}$, we have

$$\log_{10} k = \log_{10} k_0 + z_A z_B (I_c \, \text{L/mol})^{1/2}. \tag{33.56}$$

A plot of $\log_{10} k$ against the square root of the ionic strength should yield, in dilute solution, a straight line with a slope equal to $z_A z_B$.

If the ions have like signs, $z_A z_B$ is positive and the rate constant increases with increase in ionic strength. If the ions are oppositely charged, the rate constant decreases with increase in ionic strength. Equation (33.56) is a description of the primary kinetic salt effect or, more simply, the primary salt effect. Figure 33.3 shows a verification of this equation by

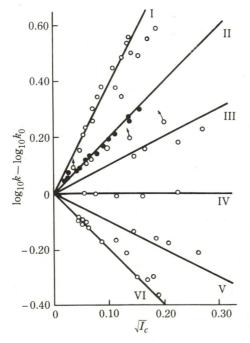

Figure 33.3   Primary salt effect. [Redrawn by permission from V. K. LeMer, *Chem. Revs.* **10**, 179 (1932).]

LaMer. The agreement is eminently satisfactory. The reactions for Fig. 33.3 are

$$
\begin{array}{lll}
\text{I.} & Co(NH_3)_5Br^{2+} + Hg^{2+}, & z_A z_B = 4, \\
\text{II.} & S_2O_8^{2-} + I^-, & z_A z_B = 2, \\
\text{III.} & NO_2NCO_2C_2H_5^- + OH^-, & z_A z_B = 1, \\
\text{IV.} & C_{12}H_{22}O_{11} + OH^-, & z_A z_B = 0, \\
\text{V.} & H_2O_2 + H^+ + Br^-, & z_A z_B = -1, \\
\text{VI.} & Co(NH_3)_5Br^{2+} + OH^-, & z_A z_B = -2.
\end{array}
$$

The physical reason for the behavior in the cases of like and unlike charges on the ions is a result of the relative net charge on the complex. The value of the activity coefficient goes down exponentially with z^2. If both ions have the same sign, the complex has a net charge that is high compared with either one. This makes $\gamma^{\ddagger}$ very small, and the ratio $(\gamma_A\gamma_B/\gamma^{\ddagger})_0$ is very large. If the ions differ in sign (the net charge on the complex is less than that on either ion), $\gamma^{\ddagger}$ is then much larger than γ_A and γ_B; the net result is that the ratio $(\gamma_A\gamma_B/\gamma^{\ddagger})_0$ and the reaction rates are small. If one species is uncharged (B for example), then A and the complex have the same charge and the ratio $\gamma_A/\gamma^{\ddagger}$ is unity and independent of $I_c^{1/2}$. The value of γ_B is not much affected by changes in I_c because B is a neutral molecule.

QUESTIONS

33.1 What is the lowest possible value of the activation energy for an endothermic reaction?

33.2 The reaction $2CH_3 \xrightarrow{k} C_2H_6$ proceeds with negligible activation energy. Estimate k via the collision theory.

33.3 Long range electrical forces operate between the reactants in ion-polarizable molecule reactions. What are they? Would Eq. (33.9) apply to such reactions?

33.4 Would a steric factor probably be required for the reaction $CH_3I + Rb \rightarrow RbI + CH_3$? Explain.

33.5 The reaction $I + I + M \rightarrow I_2 + M$, where M is a buffer gas molecule, is termolecular. Consider whether isolated collision of two I atoms can lead to energetically *stable* I_2. Suggest the role of M in the reaction.

33.6 The unimolecular rate constant Eq. (33.12) predicts that at fixed energy E, k decreases as the number of vibrational degrees of freedom increases. Rationalize this trend.

33.7 What is the activation energy E_0 in Eq. (33.31) for the reaction $H_2 + D_2 \rightarrow 2HD$?

33.8 Two ions form a very weakly polar activated complex. What is the expected effect on the rate constant as a moderately polar solvent is replaced by a highly polar solvent?

PROBLEMS

33.1 At 700 K the rate constant for the reaction $H_2 + I_2 \rightarrow 2HI$ is 6.42×10^{-2} L/mol s. The activation energy, $E^* = 167$ kJ.

    a) Calculate the rate constant predicted by the collision theory, using $\sigma_{H_2} = 225$ pm and $\sigma_{I_2} = 559$ pm, which are obtained from viscosity measurements. Compare with the experimental value.

    b) What would $\sigma_A + \sigma_B$ have to be if the collision theory prediction is to agree with the experimental value?

33.2 If the activation energy for the reaction $H_2 + I_2 \to 2HI$ is 167 kJ and the ΔE for the reaction is -8.2 kJ, what is the activation energy for the decomposition of HI?

33.3 If the diameter of HI (obtained from viscosity) is $\sigma_{HI} = 435$ pm, estimate the rate of decomposition of HI at 700 K using the kinetic theory expression for the number of collisions between like molecules and the values of the activation energy obtained in Problem 33.2.

33.4 The internuclear distances in the molecules H_2, I_2, and HI are 74.0 pm, 266.7 pm and 160.4 pm respectively. Use these values instead of the molecular diameters in the calculations in Problems 33.1 and 33.3 and observe the differences in the results.

33.5 If the molecules of a gas have a diameter of 3×10^{-10} m, calculate the number of triple collisions compared with the number of binary collisions in the gas at 300 K and 0.1, 1, 10, and 100 atm pressure. What would the values be at 600 K?

33.6 Suppose that a molecule that decomposes unimolecularly has four vibrational degrees of freedom. If 30 quanta of energy are distributed among these degrees of freedom, what is the probability that 10 quanta will be found in a particular degree of freedom? What is the probability that 20 quanta will be in a particular degree of freedom?

33.7 Combine Eq. (33.23) with Eqs. (32.83) and (32.84) and show that the rate of entropy production per unit volume is

$$\frac{d(\sigma/V)}{dt} = R\left[\frac{(\xi - \bar{\xi})_0}{V}\right]^2 \frac{e^{-2t/\tau}}{r\tau^2}$$

Integrate this result from $t = 0$ to $t = \infty$ to obtain the total entropy production per unit volume. Show that the total entropy production, σ/V, for the chemical reaction $A + B \rightleftharpoons C$, discussed in Section 32.19, is given by

$$\frac{\sigma}{V} = \tfrac{1}{2}R\left[\frac{(c_A - \bar{c}_A)_0^2}{\bar{c}_A} + \frac{(c_B - \bar{c}_B)_0^2}{\bar{c}_B} + \frac{(c_C - \bar{c}_C)_0^2}{\bar{c}_C}\right].$$

Note that this result is general; for any reaction, $\sigma/V = \tfrac{1}{2}R\,\Sigma_i\,(\Delta c_i)_0^2/\bar{c}_i$. The total entropy production near equilibrium does not depend on the rate of reaction but only on the displacements and the equilibrium concentrations.

33.8 Estimate the frequency factor at 300 K for the reaction between an atom and a diatomic molecule, $A + BC \to AB + C$, using the values of the partition functions given in Section 33.7.

33.9 For the reaction between ethyl iodide and triethylamine, the frequency factor in various solvents, at 100 °C, ranges between 2×10^3 and 1×10^5 L/mol s. Calculate the range of $\Delta S^{\ddagger}$ for the reaction.

33.10 Given the data:

Reactants	A/(L/mol s)
$Cr(H_2O)_6^{3+} + CNS^-$	10^{19}
$Co(NH_3)_5Br^{2+} + OH^-$	5×10^{17}
$ClO^- + ClO_2^-$	9×10^8

Assuming that $T \approx 300$ K, calculate $\Delta S^{\ddagger}$ for each reaction and compare. (The effect is interpreted in terms of a greater loosening of the solvent sheaths of the two ions when two ions of opposite charge form an activated complex.)

33.11 Predict the effect of increase in ionic strength on the rate constant for each of the following reactions.

a) $Pt(NH_3)_3Cl^+ + NO_2^-$;   b) $PtCl_4^{2-} + OH^-$;   c) $Pt(NH_3)_2Cl_2 + OH^-$.

33.12 Consider the reaction of two atoms to form an activated complex: $A + B \rightleftharpoons (AB)$. Write the partition functions for the atoms and the diatomic complex and show that the frequency factor predicted by the Eyring equation is identical to that predicted by the collision theory if r_{AB}, the interatomic distance in the complex, is identified with σ_{AB}.

33.13 Consider the reaction $NO + Cl_2 \rightleftharpoons NOCl + Cl$. The values for θ_v, θ_r, and r_e for NO and Cl_2 are given in Table 29.1. Estimate the frequency factor for this reaction at 300 K using the Eyring equation. Assume that the activated complex is linear: Cl—Cl—N—O and that the N—Cl distance is 200 pm while the Cl—Cl and N—O distances are the same as in the separated molecules. The degeneracy of the electronic state is the same in the initial and in the activated state. Assume that for all the vibrational degrees of freedom, $f_v = 1$.

34

Chemical Kinetics

III. Heterogeneous Reactions, Electrolysis, Photochemistry

34.1 HETEROGENEOUS REACTIONS

Very early in the development of the art of chemistry finely divided powders of various sorts were recognized as catalysts for many reactions. Only relatively recently have the details of the mechanism of reactions on surfaces been elucidated. For a long time it was thought that the function of the surface was simply to concentrate the reactants on it; the increased rate was attributed to the increase in "concentration." It can be shown that this certainly is not correct for the great majority of reactions. Calculation shows that for a concentration effect of this type to produce the increases in rate ordinarily observed would require surface areas per gram of catalyst that are impossible to attain.

In the majority of cases the increased rate of reaction on a surface is the result of the surface reaction having a lower activation energy than that of the homogeneous reaction. At ordinary temperatures, each kilojoule difference between the activation energies means a factor of 1.5 in the rate. The mode of action of the surface therefore is the same as that of other catalysts (see Section 32.20) in its provision of an alternative path of lower activation energy for the reaction.

34.2 STEPS IN THE MECHANISM OF SURFACE REACTIONS

For a reaction to occur on a surface the following sequence of steps is required.

1. Diffusion of reactants to the surface.
2. Adsorption of the reactants on the surface.
3. Reaction on the surface.
4. Desorption of products.
5. Diffusion of products from the surface.

Any one or a combination of these steps may be slow and therefore be rate determining.

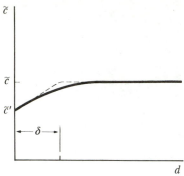

Figure 34.1   The Nernst diffusion layer.

In gaseous reactions the diffusion steps (1) and (5) are very fast and are rarely, if ever, rate determining. For very fast reactions in solution the rate may be limited by diffusion to or from the surface of the catalyst. If diffusion is the slow step, then the concentration $\tilde{c}'$ of the diffusing species at the surface will differ from the concentration $\tilde{c}$ in the bulk. In Fig. 34.1, the concentration is plotted as a function of the distance from the surface. This curve is conveniently approximated by the two dashed lines. The distance δ is the thickness of the diffusion layer. This approximation was introduced by Nernst, and the layer in which the concentration differs appreciably from that in the bulk is called the Nernst diffusion layer. The concentration gradient across the diffusion layer is given by $(\tilde{c} - \tilde{c}')/\delta$, so that the rate of transport per square metre of the surface is

$$\frac{-D(\tilde{c} - \tilde{c}')}{\delta},$$

where D is the diffusion coefficient. This approximation is a simple correction to the kinetic equations when diffusion is slow enough to matter. The rate of diffusion can be enhanced considerably by vigorous stirring, which thins the diffusion layer. The thickness δ in a well-stirred solution is of the order of 0.001 cm. In less well-stirred solutions the thickness is of the order of 0.005 to 0.010 cm.

It is more commonly observed that the rate of reaction is determined by step (2), or by a combination of steps (3) and (4). We consider these cases in order.

34.3   SIMPLE DECOMPOSITIONS ON SURFACES

In the case of the simple decomposition of a molecule on a surface, the process can be represented as a chemical reaction between the reactant A and a vacant site S on the surface. After adsorption, the molecule A may desorb unchanged or may decompose to products. The elementary steps are written

Adsorption	$A + S \xrightarrow{k_1}$	AS;
Desorption	$AS \xrightarrow{k_{-1}}$	$A + S$;
Decomposition	$AS \xrightarrow{k_2}$	Products.

If v is the rate of reaction per square metre of surface, then

$$\frac{d(\xi/A)}{dt} = v = k_2 c_{AS}, \tag{34.1}$$

where c_{AS} is the concentration (mol/m^2) of A on the surface.

Let c_s be the total concentration of surface sites per square metre and let θ be the fraction of these sites that are covered by A. Then $c_{AS} = c_s\theta$, and $c_s(1 - \theta) = c_S$, the concentration of vacant sites on the surface. Then the rate of the reaction can be written

$$v = k_2 c_s \theta. \tag{34.2}$$

The value of θ is obtained by applying the steady-state condition to the rate of formation of AS:

$$\frac{dc_{AS}}{dt} = 0 = k_1 c_a c_s (1 - \theta) - k_{-1} c_s \theta - k_2 c_s \theta, \tag{34.3}$$

where c_a is the concentration of the reactant A either in the gas or in solution. From Eq. (34.3) we obtain

$$\theta = \frac{k_1 c_a}{k_1 c_a + k_{-1} + k_2}. \tag{34.4}$$

This value of θ in the rate law, Eq. (34.2), yields

$$v = \frac{k_2 k_1 c_s c_a}{k_1 c_a + k_{-1} + k_2}. \tag{34.5}$$

If Eq. (34.5) must be considered in full, then it is convenient to invert it:

$$\frac{1}{v} = \frac{1}{k_2 c_s} + \frac{k_{-1} + k_2}{k_2 k_1 c_s c_a}. \tag{34.6}$$

A plot of $1/v$ versus $1/c_a$ yields $1/k_2 c_s$ as the intercept and $(k_{-1} + k_2)/k_2 k_1 c_s$ as the slope. Usually it is more convenient to consider the limiting cases of Eq. (34.5).

Case 1. The rate of decomposition is very large compared with the rates of absorption and desorption. In this case, $k_2 \gg k_1 c_a + k_{-1}$, and the denominator in Eq. (34.5) is equal to k_2; then the rate is given by

$$v = k_1 c_s c_a. \tag{34.7}$$

This is simply the rate of adsorption. Physically, the assumption that k_2 is large implies that an adsorbed molecule decomposes immediately, so that the rate of decomposition depends on how quickly the molecules can be adsorbed. From Eq. (34.4) and the assumption that $k_2 \gg k_1 c_a$, it follows that $\theta \ll 1$. The surface is sparsely covered with the reactant. The reaction is first order in the concentration of the reactant A. This situation is realized in the decomposition of N_2O on gold, and of HI on platinum.

Case 2. The rate of decomposition is very small compared with the rate of absorption and desorption. In this case, k_2 is very small so that the denominator of Eqs. (34.4) and (34.5) is $k_1 c_a + k_{-1}$. Introducing the absorption equilibrium constant $K = k_1/k_{-1}$, Eq. (34.4) becomes

$$\theta = \frac{K c_a}{K c_a + 1}, \tag{34.8}$$

which is the Langmuir adsorption isotherm. The occurrence of the decomposition does not affect the adsorption equilibrium at all. The rate becomes

$$v = \frac{k_2 K c_s c_a}{K c_a + 1}. \tag{34.9}$$

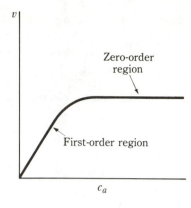

Figure 34.2 Rate of a surface reaction as a function of reactant concentration.

In this case both the surface coverage θ and the rate depend on the concentration c_a. At low concentrations, $Kc_a \ll 1$, and $\theta \approx Kc_a$; the coverage is small. Then

$$v = k_2 K c_s c_a, \tag{34.10}$$

and the reaction is first order in the concentration of A. At high concentrations, $Kc_a \gg 1$, and $\theta \approx 1$; the surface is nearly fully covered with A. Then

$$v = k_2 c_s \tag{34.11}$$

and the reaction is zero order with respect to A. Since the surface coverage ceases to vary significantly with the concentration of A at high concentrations, the reaction rate becomes independent of the concentration of A. The decomposition of HI on gold and of NH_3 on molybdenum are zero order at high pressures of HI and NH_3.

The typical variation in the rate of reaction as a function of the concentration of the reactant is shown in Fig. 34.2. This figure should be compared with Fig. 32.12, which shows the same behavior for a homogeneous catalyst. Note that Eq. (34.5) has the same form as Eq. (32.95), the equation for homogeneous catalysis, which is the same as the Michaelis–Menten law, Eq. (32.100), for enzymes. Also, Eq. (34.6) has the same form as the Lineweaver–Burk equation for enzymes.

34.4 BIMOLECULAR REACTIONS ON SURFACES

Two molecules A and B can react on a surface if they occupy neighboring sites of the surface. Let θ_a and θ_b be the fractions of the surface sites covered by A and B, respectively, and let θ_v be the fraction of sites that are vacant; $\theta_v = 1 - \theta_a - \theta_b$. We represent the reaction by

$$AS + BS \xrightarrow{\ k\ } \text{Products.}$$

The rate per unit area, v, is

$$v = kc_s^2 \theta_a \theta_b. \tag{34.12}$$

To evaluate θ_a and θ_b we consider the two adsorption reactions

$$A + S \underset{k_{-1}}{\overset{k_1}{\rightleftharpoons}} AS \quad \text{and} \quad B + S \underset{k_{-2}}{\overset{k_2}{\rightleftharpoons}} BS.$$

The steady-state equations are

$$\frac{dc_{AS}}{dt} = 0 = k_1 c_s c_a \theta_v - k_{-1} c_s \theta_a - k c_s^2 \theta_a \theta_b;$$

$$\frac{dc_{BS}}{dt} = 0 = k_2 c_s c_b \theta_v - k_{-2} c_s \theta_b - k c_s^2 \theta_a \theta_b.$$

Since $\theta_v = 1 - \theta_a - \theta_b$, these two equations can be solved for θ_a and θ_b. We will consider only the case for which k is very small; if we set $k = 0$, these equations reduce to

$$\theta_a = K_1 c_a \theta_v, \qquad \theta_b = K_2 c_b \theta_v, \tag{34.13}$$

where $K_1 = k_1/k_{-1}$ and $K_2 = k_2/k_{-2}$. Using these values of θ_a and θ_b in $\theta_v = 1 - \theta_a - \theta_b$, we obtain

$$\theta_v = 1 - K_1 c_a \theta_v - K_2 c_b \theta_v,$$

so that

$$\theta_v = \frac{1}{1 + K_1 c_a + K_2 c_b}. \tag{34.14}$$

This value of θ_v brings Eqs. (34.13) to the form

$$\theta_a = \frac{K_1 c_a}{1 + K_1 c_a + K_2 c_b}, \qquad \theta_b = \frac{K_2 c_b}{1 + K_1 c_a + K_2 c_b}. \tag{34.15}$$

These values used in Eq. (34.12) yield the rate law

$$v = \frac{k K_1 K_2 c_s^2 c_a c_b}{(1 + K_1 c_a + K_2 c_b)^2}, \tag{34.16}$$

which has some unusual characteristics. We examine each case separately.

Case 1. Both A and B are weakly adsorbed; the surface is sparsely covered. In this case, $K_1 c_a \ll 1$ and $K_2 c_b \ll 1$. The denominator of Eq. (34.16) is about equal to unity and the rate law is

$$v = k K_1 K_2 c_s^2 c_a c_b. \tag{34.17}$$

The reaction is second order overall, and is first order with respect to both A and B.

Case 2. One reactant, A, more strongly adsorbed than the other. In this case, $K_1 c_a \gg K_2 c_b$; the denominator is about equal to $1 + K_1 c_a$, and Eq. (34.16) takes the form

$$v = \frac{k K_1 K_2 c_s^2 c_a c_b}{(1 + K_1 c_a)^2}. \tag{34.18}$$

The rate is first order with respect to the less strongly adsorbed reactant; the dependence of the rate on the concentration of the more strongly adsorbed reactant is more complicated. At low values of c_a, the rate increases as c_a increases, passes through a maximum value at $c_a = 1/K_1$, and then decreases with further increase in c_a. At very high values of c_a, the rate becomes inversely proportional to c_a (see Case 3).

Case 3. One reactant very strongly absorbed. If A is very strongly adsorbed, we have the same situation as in Case 2 but with the additional condition that $K_1 c_a \gg 1$, so that the denominator of Eq. (34.18) is $(K_1 c_a)^2$. Then the rate is

$$v = \frac{k K_2 c_s^2 c_b}{K_1 c_a}. \tag{34.19}$$

The rate of the reaction is inversely proportional to the concentration of the strongly adsorbed species. This is an example of inhibition, or poisoning. In this case one of the reactants itself inhibits the reaction. The reaction between ethylene and hydrogen on copper is of this type. At low temperatures the rate is given by

$$v = k \, \frac{p_{H_2}}{p_{C_2H_4}},$$

the ethylene being strongly adsorbed. At higher temperatures the ethylene is less strongly adsorbed, the surface is sparsely covered, and the rate expression reduces to that given by Eq. (34.17):

$$v = k' p_{H_2} p_{C_2H_4}.$$

It is generally true that if one substance is strongly adsorbed on the surface (whether it be reactant, product, or a foreign material) the rate is inversely proportional to the concentration of the strongly adsorbed substance; this substance inhibits the reaction.

34.5 THE ROLE OF THE SURFACE IN CATALYSIS

In homogeneous catalysis the catalyst combines chemically with one of the reactants to form a compound that reacts readily to form products. The same is true of a surface acting as a catalyst. One or more of the reactants are chemisorbed on the surface; this is equivalent to the formation of the chemical intermediate in the homogeneous case. In both cases, the effect of the catalyst is to provide an alternative path of lower activation energy. This lower energy is the principal reason for the increased rate of reaction. Figure 34.3 shows schematically the energy variation as the reactants pass to products. It is apparent from the figure that if the activation energy for the forward reaction is lowered, then that for the reverse reaction is lowered by the same amount. The catalyst therefore increases the rate of the forward *and the reverse reaction* by the same factor.

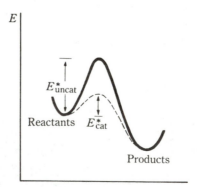

Figure 34.3 Energy surfaces for uncatalyzed and catalyzed reactions.

**Table 34.1
Activation energies for catalyzed and
uncatalyzed reactions**

Decomposition of	Surface	$E^*_{cat}/(kJ/mol)$	$E^*_{uncat}/(kJ/mol)$
HI	Au	105	184
	Pt	59	
N_2O	Au	121	245
	Pt	136	
NH_3	W	163	330
	Os	197	
	Mo	130–180	
CH_4	Pt	230–250	330

By permission from K. J. Laidler, *Chemical Kinetics*. New York: McGraw-Hill, 1950.

Table 34.1 lists a few values of the activation energies for various reactions on surfaces, and the corresponding values for the uncatalyzed reaction.

An important fact about surface reactions is that the surface sites on a catalyst differ in their ability to adsorb the reactant molecules. This is demonstrated by the action of catalytic poisons. In the preceding section, the effect of strong adsorption of one reactant was to inhibit reaction or poison the catalyst. Foreign molecules that do not take part in the reaction can also poison the surface if they are strongly adsorbed. The algebraic effect on the rate equation is to make the rate inversely proportional to some power, usually the first power, of the concentration of the poison.

It has been shown that the amount of poison required to stop the reaction is ordinarily significantly smaller than the amount needed to form a monolayer of poison on the surface. This observation led H. S. Taylor to postulate that the adsorption and subsequent reaction takes place preferentially on certain parts of the surface, which he called "active centers." The active centers may constitute only a small fraction of the total number of surface sites. If these active centers are covered by molecules of the poison, the reaction is unable to proceed except at an extremely slow rate.

Imagine the appearance of a surface on the atomic scale. There are cracks, hills and valleys, boundaries between individual grains, different crystal faces exposed, edges, points, and so on. It is not surprising that adsorption takes place more easily in some places than in others. The chemical kinetic consequences of this lack of uniformity in the surface have been explored extensively, both from the theoretical and the experimental standpoints.

The chemical nature of the surface determines its ability to act as a catalyst for a particular type of reaction. For illustration, two reactions of an alcohol can be considered. On metals of the platinum group such as Ni, Pd, and Pt, the alcohol is dehydrogenated.

$$CH_3CH_2OH \longrightarrow CH_3CHO + H_2.$$

On a surface such as alumina, dehydration occurs:

$$CH_3CH_2OH \longrightarrow CH_2CH_2 + H_2O.$$

In the two cases the mode of attachment is different.

Nickel has a strong affinity for hydrogen so that on nickel the attachment is presumably to the hydrogen atoms:

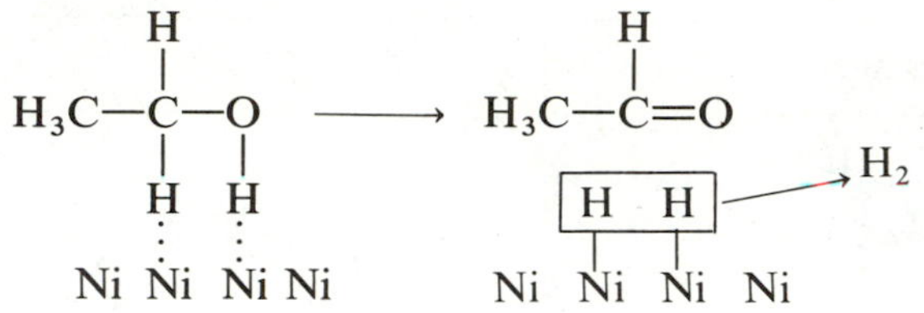

On alumina, there are hydroxyl groups at the surface as well as oxide groups. The surface could be imagined as having the configuration

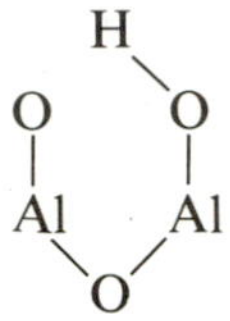

Then the attachment of the alcohol could be

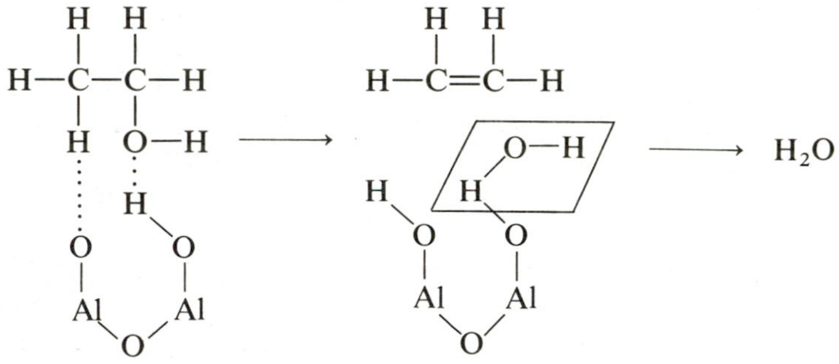

After desorption of the water molecule the surface is left unchanged. Note that these diagrams are intended to represent nothing more than plausible suppositions about the surface structure and the mode of attachment of the molecule.

34.6 ELECTROLYSIS AND POLARIZATION

Electrolysis refers to the chemical reaction or reactions that accompany the passage of a current supplied by an external source through an electrolytic solution. An electrochemical cell through which a current is passing is said to be *polarized*. Polarization is a general term that refers to any or all of the phenomena associated with the passage of a current through a cell.

We can write any electrolytic half-reaction in the general form:

$$0 = \sum_i v_i A_i + v_e e^-.$$

The quantity of charge that passes the electrode as the reaction advances by $d\xi$ is dQ, where

$$dQ = v_e F \, d\xi. \tag{34.20}$$

The current is given by $I = dQ/dt$, so that

$$I = v_e F \frac{d\xi}{dt}.$$ (34.21)

The current is proportional to the rate, $d\xi/dt$, of the reaction (or vice versa) so that the rate is usually expressed in amperes. If A is the area of the electrode, then the current density, i, is

$$i = \frac{I}{A} = v_e F\left(\frac{1}{A}\frac{d\xi}{dt}\right);$$

$$i = v_e F v_A.$$ (34.22)

where v_A is the rate of reaction per unit area.

$$v_A = \frac{1}{A}\frac{d\xi}{dt}$$ (34.23)

The significant quantity is the rate per unit area; therefore, we will use current densities to describe the rates, the usual units being A/cm^2 or mA/cm^2.

The sign of the current density follows the sign of the stoichiometric coefficient v_e. If v_e is plus, electrons appear on the product side, and the reaction is an oxidation. The current is an *anodic* current and has a positive sign. The symbol for an anodic current density is i_+ or i_a. If v_e is minus, electrons appear on the reactant side, and the reaction is a reduction. The current is a *cathodic* current and has a negative sign. The symbol for a cathodic current density is i_- or i_c.

The total current density at an electrode is the algebraic sum of the anodic and cathodic current densities for the reaction taking place on that electrode:

$$i = i_a + i_c.$$ (34.24)

If more than one electrolytic reaction is occurring on the electrode, the total current density is the algebraic sum of the current densities for all the anodic and cathodic reactions taking place on that electrode.

The study of electrode reactions is unique in the sense that within limits the rate of the reaction can be controlled by simply increasing or decreasing the current through the cell. The electrolysis reaction also differs from other chemical reactions in that "half" of it occurs at one electrode and the other "half" occurs at the second electrode, which may be spatially distant from the first. For example, the electrolysis of water,

$$H_2O \longrightarrow H_2 + \tfrac{1}{2}O_2,$$

can be broken down into two "half" reactions:

At the cathode $\qquad 2H^+ + 2e^- \longrightarrow H_2,$

At the anode $\qquad H_2O \longrightarrow \tfrac{1}{2}O_2 + 2H^+ + 2e^-$

Each of these reactions is proceeding at the same rate I, the current being passed. If the area of the cathode is A_c and that of the anode is A_a, then the rate of the cathodic reaction per unit area of cathode is $i_c = I/A_c$, and that of the anodic reaction per unit area of anode is $i_a = I/A_a$. The current density at either electrode depends on the concentrations of reactants and products near the electrode, just as any reaction rate depends on concentrations. In addition, the current density depends on the electrode material and very strongly on the potential of the electrode. The phenomena associated with electrolysis are properly linked with the kinetics of reactions on surfaces. Because of great experimental difficulties, particularly the problem of controlling impurities in liquid solutions, the study of electrode kinetics has become reasonably scientific only relatively recently. Some of the earlier work is excellent, but much of it is erroneous.

34.7 POLARIZATION AT AN ELECTRODE

Rather than describe the electrolysis of any solution with any two electrodes, we begin by considering a single reversible electrode at equilibrium and then ask what happens if we pass a current into the electrode.

Consider a hydrogen electrode in equilibrium with H^+ ion at a concentration c and hydrogen gas at a pressure p. The equilibrium potential of this electrode is denoted by ϕ_0. The equilibrium is $H_2 \rightleftharpoons 2H^+ + 2e^-$. If the potential of the electrode is increased (made more positive), this equilibrium will be disturbed. The reaction from left to right will predominate, H_2 will be oxidized, and a positive current will flow into the solution. If the potential of the electrode is lowered (made more negative), the equilibrium will be disturbed. The reaction from right to left will predominate, H_2 will be liberated, and a positive current will flow into the electrode or a negative current will flow into the solution. The current that flows to the electrode, therefore, depends on the departure of the potential from the equilibrium value, $\phi - \phi_0$. This difference between the applied potential ϕ and the equilibrium potential ϕ is the overpotential, or overvoltage, η:

$$\eta \equiv \phi - \phi_0. \tag{34.25}$$

Since the current varies continuously with the potential, and therefore with the overpotential, we can expand the current in a Taylor series. Since $i = 0$ when $\eta = 0$, the series becomes

$$i = \left(\frac{di}{d\eta}\right)_{\eta=0} \eta + \frac{1}{2}\left(\frac{d^2 i}{d\eta^2}\right)_{\eta=0} \eta^2 + \cdots. \tag{34.26}$$

We can write this in a slightly different way, using only the first term,

$$i = \left(\frac{i_0 F}{RT}\right)\eta. \tag{34.27}$$

For small values of η, i is proportional to η. Note also that the sign of i depends on the sign of η. The *exchange current density* for the reaction, i_0, defined by Eq. (34.27), is the equilibrium value of either the anodic *or* cathodic current density. The value of i_0 depends on the concentrations of the electroactive materials, H^+ and H_2 in this case, and on the composition of the electrode surface.

For the hydrogen evolution reaction on platinum, for example, $i_0 \approx 10^{-2}$ A/cm^2, but on mercury, $i_0 \approx 10^{-14}$ A/cm^2. It is, in fact, these values of i_0 that allow us to use platinum as the electron collector for a reversible hydrogen electrode and prevent our using mercury for this purpose.

In the kinetic sense there is a gradation between what are called reversible electrodes and irreversible electrodes. The reversible potential of an electrode is measured by balancing a cell in a potentiometer circuit. This involves detecting the point at which the current flow to the electrode is zero. Suppose the galvanometer registers "zero current" for any value of current between i' and $-i'$. The magnitude of i' depends on the sensitivity of the galvanometer. The characteristics of the "very reversible" electrode and the "very

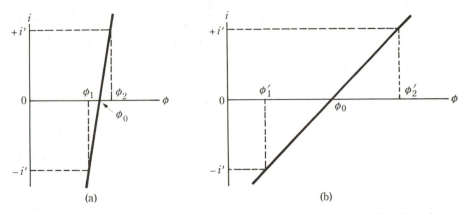

Figure 34.4 Current–potential relation at (a) reversible and (b) irreversible electrodes.

irreversible" electrode are shown in Fig. 34.4. The balance will be observed for the reversible electrode anywhere in the range between ϕ_2 and ϕ_1. This uncertainty in the measurement of ϕ_0 is very small.

For the irreversible electrode (Fig. 34.4b), the null point is registered anywhere in the wide range of potential between ϕ'_1 and ϕ'_2. The slope of the curve is very small; that is, i_0 very small. An electrode with a large i_0 is therefore "more reversible" than one with a small i_0.

34.8 MEASUREMENT OF OVERVOLTAGE

Before considering the theoretical ideas that relate the current to the overvoltage, we should understand the principle of the measurement of overvoltage. A cell is shown schematically in Fig. 34.5. A measured current is passed between the two electrodes A and B. The reference electrode R is the same kind of electrode as B. Matters are arranged so that the same electrode equilibrium is established at both B and R. When $i = 0$, B and R both have the same potential. When the current passes into B, this electrode has a potential measured on the potentiometer P that is different from that of R, which carries no current. This difference in potential is the measured overvoltage, $\eta_m = \phi_B - \phi_R$. The value of η_m is measured for various values of the current density.

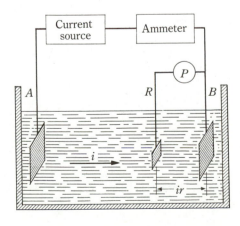

Figure 34.5 Cell for the measurement of overvoltage.

As the experiment stands, the measured value contains an ohmic component from the *ir* drop between *R* and *B*, a concentration component resulting from concentration changes in the vicinity of the electrode, and a component, denoted by η, which is related to the rate constant of the reaction. Thus

$$\eta_m = \eta_{\text{ohmic}} + \eta_{\text{conc}} + \eta. \tag{34.28}$$

There are methods for measuring η_{ohmic} separately; η_{conc} can usually be reduced to a negligible value by vigorous stirring. Thus from η_m the value of η can be found as a function of the current density. This η, which is related to the rate constant of the reaction, is often called the *activation overvoltage*.

★ 34.9 THE CURRENT–POTENTIAL RELATION

To derive the equation connecting the current with the overpotential we consider the dissolution of a metal; the reaction is

$$\text{M(metal)} \quad \longrightarrow \quad \text{M}^{z+}\text{(aq)} + ze^-\text{(metal)}$$

The reverse reaction is the metal–deposition reaction. Within the metal, the equilibrium

$$M^{z+}\text{(metal)} + ze^-\text{(metal)} \quad \rightleftharpoons \quad \text{M(metal)}$$

is established very rapidly. Adding these two equations together, we obtain

$$\text{M}^{z+}\text{(metal)} \quad \longrightarrow \quad \text{M}^{z+}\text{(aq)}$$

as the effective charge transfer reaction. Thus we may regard the metal dissolution reaction as one in which a metal ion in the metal passes over an activation potential barrier and becomes a metal ion in the aqueous solution. Similarly, the electrodeposition reaction is the transfer of the metal ion from the aqueous phase over a potential barrier to the metal. The dependence of the current on the potential is a consequence of the dependence of the barrier height on the potential. The relationships are shown in Fig. 34.6(a) and (b).

We consider first the hypothetical case in which there is no electrical potential difference between the metal and the solution. Then the Gibbs energy of activation is

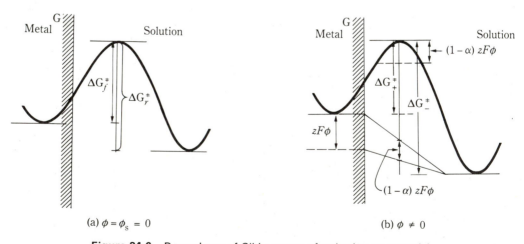

(a) $\phi = \phi_s = 0$ (b) $\phi \neq 0$

Figure 34.6   Dependence of Gibbs energy of activation on potential.

given simply by $\Delta G_f^{\ddagger}$ for the anodic reaction and by $\Delta G_r^{\ddagger}$ for the cathodic reaction (Fig. 34.6a). Next we apply a potential, ϕ, to the metal, keeping the electrical potential of the solution, ϕ_s, at its original conventional value, $\phi_s = 0$. The Gibbs energy of the metal ion in the metal is raised by an amount $zF\phi$ (Fig. 34.6b). The Gibbs energy of the metal ion in the activated state is also raised by an amount $(1 - \alpha)zF\phi$, where $0 \le \alpha \le 1$. The net result is that the Gibbs energy of activation for the anodic reaction, $\Delta G_+^{\ddagger}$, in the presence of the potential ϕ, is given by

$$\Delta G_+^{\ddagger} = \Delta G_f^{\ddagger} + (1 - \alpha)zF\phi - zF\phi = \Delta G_f^{\ddagger} - \alpha zF\phi, \tag{34.29}$$

while the Gibbs energy of activation for the cathodic reaction is

$$\Delta G_-^{\ddagger} = \Delta G_r^{\ddagger} + (1 - \alpha)zF\phi. \tag{34.30}$$

The anodic and cathodic current densities are given by

$$i_+ = zFk_f c_{red}, \qquad i_- = -zFk_r c_{ox}, \tag{34.31}$$

in which k_f and k_r are the rate constants per unit area for the forward (anodic) and reverse (cathodic) reactions; c_{red} is the concentration of the reduced species and c_{ox} the concentration of the oxidized species. In this case, c_{red} is the concentration of the metal ion in the metal, which is a constant. The c_{ox} is the concentration of the metal ion in the aqueous solution.

Since, by Eq. (33.44),

$$k_f = \left(\frac{kT}{h}\right)e^{-\Delta G_+^{\ddagger}/RT} \qquad \text{and} \qquad k_r = \left(\frac{kT}{h}\right)e^{-\Delta G_-^{\ddagger}/RT}, \tag{34.32}$$

we can use these values for k_f and k_r in the expressions for the current densities and, at the same time, use Eqs. (34.29) and (34.30) for the values of ΔG_+ and ΔG_-; this yields

$$i_+ = zFc_{red}\left(\frac{kT}{h}\right)e^{-(\Delta G_f^{\ddagger} - \alpha zF\phi)/RT} \tag{34.33}$$

$$i_- = -zFc_{ox}\left(\frac{kT}{h}\right)e^{-[\Delta G_r^{\ddagger} + (1 - \alpha)zF\phi]/RT}. \tag{34.34}$$

These can be abbreviated to

$$i_+ = k_+ c_{red}\, e^{\alpha zF\phi/RT} \qquad \text{and} \qquad i_- = -k_- c_{ox}\, e^{-(1-\alpha)zF\phi/RT}, \tag{34.35}$$

where k_+ and k_- are the parts of the rate constants that do not depend on the potential, ϕ; that is,

$$k_+ = zF\left(\frac{kT}{h}\right)e^{-\Delta G_f^{\ddagger}/RT} \qquad \text{and} \qquad k_- = zF\left(\frac{kT}{h}\right)e^{-\Delta G_r^{\ddagger}/RT} \tag{34.36}$$

The net current density is given by:

$$i = i_+ + i_-,$$

$$i = k_+ c_{red}\, e^{\alpha zF\phi/RT} - k_- c_{ox}\, e^{-(1-\alpha)zF\phi/RT}. \tag{34.37}$$

The rate of dissolution is therefore equal to the difference of the rates of the forward and reverse reactions as in ordinary kinetics. When the electrochemical reaction is at equilibrium, $i = 0$, and

$$(i_+)_0 = -(i_-)_0 = i_0, \tag{34.38}$$

where $(i_+)_0$ and $(i_-)_0$ are the (equal) rates of the anodic and cathodic reactions at equilibrium. They are both equal to i_0, the *exchange current* for the electrochemical reaction. Note that

$$i_0 = k_+(c_{\text{red}})_0\, e^{\alpha z F \phi_0/RT} = k_-(c_{\text{ox}})_0\, e^{-(1-\alpha)zF\phi_0/RT}, \tag{34.39}$$

where $(c_{\text{red}})_0$ and $(c_{\text{ox}})_0$ are the equilibrium values of the concentrations and ϕ_0 is the equilibrium value of the potential. If we now divide each term in Eq. (34.37) by i_0, we get

$$\frac{i}{i_0} = \frac{c_{\text{red}}}{(c_{\text{red}})_0}\, e^{\alpha z F(\phi-\phi_0)/RT} - \frac{c_{\text{ox}}}{(c_{\text{ox}})_0}\, e^{-(1-\alpha)zF(\phi-\phi_0)}, \tag{34.40}$$

in which the values of i_0 from Eq. (34.39) have been used to eliminate the individual rate constants, k_+ and k_-, on the right-hand side. Since $\phi - \phi_0 = \eta$ by the definition in Eq. (34.25), the current density can be written in terms of the overpotential and the exchange current:

$$i = i_0\left[\frac{c_{\text{red}}}{(c_{\text{red}})_0}\, e^{\alpha z F\eta/RT} - \frac{c_{\text{ox}}}{(c_{\text{ox}})_0}\, e^{-(1-\alpha)zF\eta/RT}\right]. \tag{34.41}$$

This equation is a typical example of a rate equation for an electrochemical reaction at an electrode.

In the particular instance of metal dissolution, the concentration of metal ion in the metal is a constant; hence in all circumstances $c_{\text{red}} = (c_{\text{red}})_0$. If, in addition, the solution is stirred well, the passage of the current does not affect the concentration of the metal ion just outside the double layer; then $c_{\text{ox}} = (c_{\text{ox}})_0$. The equation becomes

$$i = i_0[e^{\alpha z F\eta/RT} - e^{-(1-\alpha)zF\eta/RT}]. \tag{34.42}$$

We can always write Eq. (34.42) in the form:

$$i = 2i_0\, e^{(\alpha-1/2)zF\eta/RT} \sinh\left(\frac{zF\eta}{2RT}\right). \tag{34.43}$$

For many reactions $\alpha = \frac{1}{2}$, and for many others α is remarkably close to $\frac{1}{2}$. When $\alpha = \frac{1}{2}$, we have the simple symmetric form

$$i = 2i_0 \sinh\left(\frac{zF\eta}{2RT}\right). \tag{34.44}$$

These curves are shown in Fig. 34.7 for $z = 1$ and for several values of i_0. It is apparent that if i_0 is small, a large value of η is required to produce even a small current. This equation is correct for the $\text{Cd} \rightarrow \text{Cd}^{2+} + 2e^-$ reaction; $\alpha = 0.5$, $i_0 = 1.5$ ma/cm^2 at $[\text{Cd}^{2+}] = 0.01$ mol/L.

On the other hand, for $\text{Ag} \rightarrow \text{Ag}^+ + e^-$, $\alpha = 0.74$ and $i_0 = 4.5$ A/cm^2 at $[\text{Ag}^+] = 0.1$ mol/L. Consequently, Eq. (34.43) is the appropriate one for silver.

★ 34.9.1 The Tafel Equation

When the overpotential is large (either positively or negatively), that is, when $|\eta| > \sim 75$ mV, then one term in the rate equation may be neglected in comparison to the other.

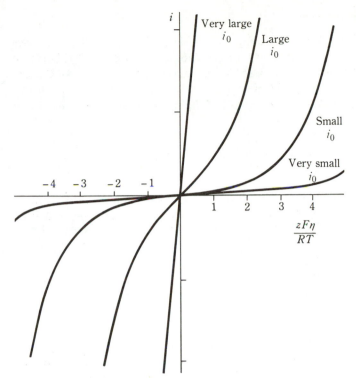

Figure 34.7 i_0 versus $zF\eta/RT$.

Thus Eq. (34.42) becomes

$$i = i_a = i_0 e^{\alpha zF\eta/RT}, \qquad (\eta > \sim 75 \text{ mV}) \tag{34.45}$$

or

$$-i = -i_c = i_0 e^{-(1-\alpha)zF\eta/RT}, \qquad (\eta < \sim -75 \text{ mV}). \tag{34.46}$$

If we take the logarithm of both sides of either of these equations we obtain an equation of the form

$$|\eta| = -b \log_{10} i_0 + b \log_{10}|i| \tag{34.47}$$

This is the Tafel equation. On the anodic side, $b = 2.303RT/\alpha zF$, while on the cathodic side, $b = 2.303RT/(1-\alpha)zF$. Thus from a Tafel plot, a plot of $|\eta|$ against $\log_{10}|i|$, we obtain the two important parameters, α and i_0. The α is obtained from the slope and the i_0 from the intercept on the horizontal axis (Fig. 34.8).

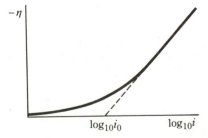

Figure 34.8 Tafel plot for cathodic overvoltage.

34.9.2 The Hydrogen Electrode

The hydrogen electrode provides a simple example of an electrode reaction that involves a two-step mechanism. We write the equations for the anodic oxidation of H_2, but keep the expressions for the forward and reverse reactions, so that the equations also describe the cathodic reduction of H^+ ion to H_2. Of the several possible mechanisms, we choose only the most obvious one for illustration. It consists of the two elementary reactions,

(1) $$H_2(aq) + 2V = 2H(ads)$$

(2) $$H(ads) = H^+ + V + e^-$$

in which V is a vacant site on the surface.

Although the first reaction does not pass electric charge across the interface directly, each time it occurs the second reaction must pass two electrons across the interface. Consequently, we can write the rate of reaction (1) in terms of an equivalent current density, i_1

$$i_1 = 2Fk_1 c_{H_2} c_V^2 - 2Fk_{-1} c_H^2. \tag{34.48}$$

At equilibrium, $i_1 = 0$, and the exchange current is given by

$$i = 2Fk_1 (c_{H_2})_0 (c_V)_0^2 = 2Fk_{-1}(c_H)_0^2.$$

Using these equations to eliminate the rate constants in Eq. (34.48) and writing $c_H = \theta c_s$ and $c_V = (1 - \theta)c_s$ (where θ is the fraction of the surface sites occupied by hydrogen atoms and c_s is the total number of surface sites per unit area), we can reduce Eq. (34.48) to the form

$$i_1 = i_{10}\left[\frac{c_{H_2}}{(c_{H_2})_0}\left(\frac{1-\theta}{1-\theta_0}\right)^2 - \left(\frac{\theta}{\theta_0}\right)^2\right], \tag{34.49}$$

where θ_0 and $(c_{H_2})_0$ are the equilibrium values ($i_1 = 0$) of θ and c_{H_2}.

Using Eq. (34.41) as a guide, we can write the rate of the second reaction as

$$i_2 = i_{20}\left[\frac{c_H}{(c_H)_0} e^{\alpha F\eta/RT} - \frac{c_{H^+}}{(c_{H^+})_0}\frac{c_V}{(c_V)_0} e^{-(1-\alpha)F\eta/RT}\right].$$

We assume that the solution is stirred well enough that the concentration of H^+ ion near the electrode is maintained at the equilibrium value, that is, $c_{H^+}/(c_{H^+})_0 = 1$; then, when we introduce the value of the surface coverage, this equation becomes

$$i_2 = i_{20}\left[\frac{\theta}{\theta_0} e^{\alpha F\eta/RT} - \frac{1-\theta}{1-\theta_0} e^{-(1-\alpha)F\eta/RT}\right]. \tag{34.50}$$

Since only the second reaction passes an electrical charge to the electrode, we have for the current, i,

$$i = i_2. \tag{34.51}$$

When the electrode reaches a steady state, the values of η and i are constant; the steady-state condition is $dc_H/dt = 0$. In terms of the currents this condition is

$$i_1 = i_2. \tag{34.52}$$

Using the values of i_1 and i_2 from Eqs. (34.49) and (34.50), and assuming that the supply of hydrogen or the stirring is such that $c_{H_2}/(c_{H_2})_0 = 1$, we obtain

$$i_{10}\left[\left(\frac{1-\theta}{1-\theta_0}\right)^2 - \left(\frac{\theta}{\theta_0}\right)^2\right] = i_{20}\left[\frac{\theta}{\theta_0}e^{\alpha F\eta/RT} - \frac{1-\theta}{1-\theta_0}e^{-(1-\alpha)F\eta/RT}\right].$$

This equation can be solved for θ as a function of η and the result placed in Eq. (34.50) to yield the relation between i and η. In practice this involves a general solution of a quadratic and the result is cumbersome. Two extreme cases can be easily distinguished.

Case 1. $i_{10} \gg i_{20}$. In this case reaction (1) is very nearly in equilibrium, thus, we can set

$$\frac{i_1}{i_{10}} \approx 0 = \left(\frac{1-\theta}{1-\theta_0}\right)^2 - \left(\frac{\theta}{\theta_0}\right)^2$$

This yields $\theta = \theta_0$. Using this value in Eq. (34.50), we obtain

$$i = i_{20}[e^{\alpha F\eta/RT} - e^{-(1-\alpha)F\eta/RT}], \tag{34.53}$$

which is the rate equation for the simple charge-transfer reaction.

Case 2. $i_{20} \gg i_{10}$. In this case the charge-transfer reaction is essentially in equilibrium and we have

$$\frac{i_2}{i_{20}} \approx 0 = \frac{\theta}{\theta_0}e^{\alpha F\eta/RT} - \frac{1-\theta}{1-\theta_0}e^{-(1-\alpha)F\eta/RT}.$$

Solving for θ yields

$$\frac{\theta}{\theta_0} = \frac{e^{-F\eta/RT}}{1-\theta_0 + \theta_0 e^{-F\eta/RT}}.$$

Using this value in the equation, $i = i_1$, yields

$$i = i_{10}\frac{1 - e^{-2F\eta/RT}}{(1-\theta_0 + \theta_0 e^{-F\eta/RT})^2}. \tag{34.54}$$

At very small coverages, $\theta_0 \ll 1$, and the equation becomes

$$i = i_{10}(1 - e^{-2F\eta/RT}). \tag{34.55}$$

On the cathodic branch, when $\eta < -75$ mV, we can neglect the first term in Eqs. (34.53) and (34.55) to obtain

$$-i = i_{20}e^{-(1-\alpha)F\eta/RT}, \qquad \text{(slow discharge of } H^+ \text{ ion)};$$

$$-i = i_{10}e^{-2F\eta/RT}, \qquad \text{(slow combination of H atoms)}.$$

If $\alpha = \frac{1}{2}$, the Tafel slope corresponding to each of these cases at $25\,°C$ is

$$b = \frac{2(2.303)RT}{F} = 0.118 \text{ V}, \qquad \text{(slow discharge)};$$

$$b = \frac{2.303RT}{F} = 0.030 \text{ V}, \qquad \text{(slow combination)}.$$

These Tafel slopes are characteristic of the two limiting cases of this mechanism. Many metals including Pb, Hg, Ag, Cu, Pt, and Pd exhibit the 0.118 V Tafel slope.

34.10   GENERAL CONSEQUENCES OF THE CURRENT–POTENTIAL RELATION

Rather than discuss the mechanisms of electrode reactions in further detail, we will describe some general implications of the current–potential relation.

We ask whether it is possible to plate zinc onto a platinum electrode. Consider a solution containing HCl and $ZnCl_2$ in which $a_{H^+} = 1$ and $a_{Zn^{2+}} = 1$. Suppose we electrolyze this solution using a platinum cathode. As soon as hydrogen starts to evolve at the cathode, it behaves as a hydrogen electrode. Since the pressure of H_2 in the vicinity is very close to 1 atm and $a_{H^+} = 1$, the potential of the electrode on the conventional scale is zero if no current flows, and slightly less than zero if current is flowing. The dependence of the potential of the electrode as a function of current density is shown in Fig. 34.9(a). Since the hydrogen overvoltage on platinum is very small, the potential decreases very slowly as the current density is increased. If zinc is to deposit on the electrode, the potential must be more negative than the value of the reversible potential of the Zn^{2+}, Zn couple, which is -0.763 V in this situation. It is clear from the figure that a very high current density (~ 400 A/cm^2) will be required to bring the potential to a value below -0.763 V and thus permit the deposition of zinc. The current density required is so large that, as a practical matter, zinc cannot be plated onto a platinum surface.

The electrolysis of this solution proceeds quite differently if a lead cathode is used. The hydrogen overpotential on lead is much larger than that on platinum at every current density. The current–potential relation is shown schematically for the system with the lead electrode in Fig. 34.9(b). Only a very small current density (~ 0.4 μA/cm^2) is required to bring the potential down far enough so that zinc will deposit. After this value is reached, the potential does not drop very much with increase in current density because the overvoltage for deposition of zinc on a lead (or zinc) surface is very small. (After the lead is coated with zinc, of course, the electrode is a zinc electrode. The hydrogen overvoltage on zinc is also quite high, so the shape of the curve is essentially the same as for lead.) Figure 34.9(b) can be interpreted as follows. In the region from A to B, all of the current goes into hydrogen evolution. At B, zinc deposition commences. At any point beyond B, both

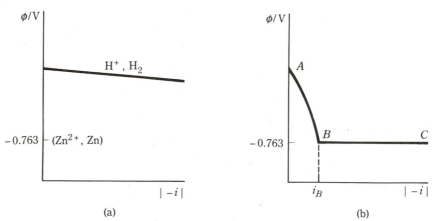

Figure 34.9   Current–potential curves for the deposition of zinc on (a) a platinum electrode and (b) a lead electrode.

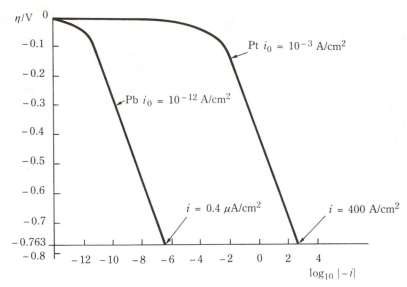

Figure 34.10 Current versus potential for deposition of zinc platinum and on lead.

hydrogen evolution and metal deposition occur. The rate of hydrogen evolution is i_B, and this rate remains nearly constant in the region from B to C, since the potential is effectively constant in this range. The rate of metal deposition, i_M, is therefore

$$i_M = i - i_B \tag{34.56}$$

and increases as i increases. The fraction of the current used in metal deposition is $i_M/i = 1 - i_B/i$. The ratio i_M/i is the current efficiency for metal deposition. Since i_B is very small, the current efficiency is nearly unity at high values of i. In Fig. 34.10 we illustrate the situation more realistically by using a logarithmic scale for the current.

A very active metal such as sodium cannot be deposited from aqueous solutions except under special circumstances. The reversible potential for the reduction of Na^+ is -2.714 V. Even with a lead cathode an enormous current density would be required to bring the cathode below this potential; the current efficiency for sodium deposition would be exceedingly small. Sodium can be deposited into mercury, which has a high hydrogen overpotential, if a highly alkaline solution is used. High current densities are required and the current efficiency is very low. Three factors influence the process.

1. The alkaline solution, which brings the potential at which hydrogen is deposited closer to the potential for sodium deposition.
2. The high hydrogen overvoltage on mercury.
3. The fact that metallic sodium will dissolve in mercury; this brings the sodium deposition potential nearer the hydrogen value and also keeps the sodium that has been deposited from reacting with water.

It is worthwhile mentioning that charging the lead storage battery would not be possible if it were not for a high hydrogen overvoltage on the negative plate, which permits the reaction

$$PbSO_4 + 2e^- \longrightarrow Pb + SO_4^{2-}$$

to occur with high efficiency. A high oxygen overvoltage on the positive plate, PbO_2, is required so that the reaction

$$PbSO_4 + 2H_2O \longrightarrow PbO_2 + SO_4^{2-} + 4H^+ + 2e^-$$

can occur with high efficiency. If these overvoltages were not large, it would not be possible to polarize the plates to the potentials required for the charging reactions to occur, and passage of the charging current would only decompose the water into hydrogen and oxygen.

Corrosion reactions are another group of reactions that depend critically on the presence or absence of a significant hydrogen overvoltage on the surface.

★ 34.11 CORROSION

34.11.1 Corrosion of Metals in Acids

To discuss the corrosion of metals in acid solution, we consider the cell shown in Fig. 34.11(a). It consists of a zinc electrode and a hydrogen electrode immersed in a $ZnSO_4$ solution that contains sufficient sulfuric acid to prevent the precipitation of $Zn(OH)_2$. If this cell is short-circuited, the reaction

$$Zn + 2H^+ \longrightarrow Zn^{2+} + H_2$$

occurs. Zinc dissolves and hydrogen is liberated on the platinum electrode. Now suppose that instead of constructing the cell, we simply attach a piece of platinum to the zinc and immerse the composite in the solution, as in Fig. 34.11(b). The result is the same: the zinc dissolves and hydrogen is liberated on the platinum. The anodic current from the zinc is balanced by the cathodic current to the platinum

$$I_a(Zn) + I_c(Pt) = 0$$

or

$$i_a(Zn)A_{Zn} + i_c(Pt)A_{Pt} = 0.$$

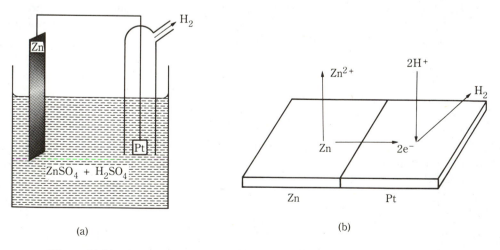

(a)

(b)

Figure 34.11   (a) Short-circuited Zn–H_2 cell. (b) Zn–Pt couple in acid solution.

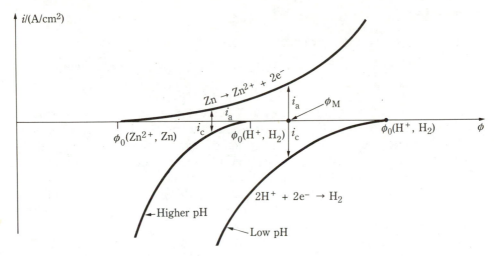

Figure 34.12   Corrosion current for zinc in acid solution. (Adapted from K. J. Vetter, *Electro-chemical Kinetics*, New York: Academic Press, 1967, pp. 737–738.)

If we make the areas of the zinc and platinum equal, $A_{Zn} = A_{Pt}$, the current densities balance.

$$i_a(\text{Zn}) + i_c(\text{Pt}) = 0.$$

This requirement is shown graphically in Fig. 34.12, which shows the i versus η curves for the zinc dissolution reaction and for the hydrogen evolution reaction. At the potential ϕ_M, the current densities sum to zero. At this point, $i_a = i_{\text{corr}}$, the corrosion current density. The potential of the Zn–Pt composite is a *mixed* potential, ϕ_M. Since ϕ_M is determined by the relative areas and the kinetics of the two electrode processes, it is very sensitive to the character of the surface and does not have any thermodynamic significance. Note that at the higher pH, since the equilibrium potentials of the Zn^{2+}, Zn and the H^+, H_2 electrodes are closer together, the corrosion current is less. This suggests that we can control corrosion by controlling the potential of the metal. This is, in fact, done in certain industrial situations. If, from an external power source, we impress a potential on the metal that is more negative than ϕ_M then by Fig. 34.12, i_a will be reduced and i_c will be increased. A sufficiently negative potential can reduce the corrosion current to a negligible value.

If we remove the platinum and simply immerse a piece of zinc in acid, what is the mechanism of corrosion? In any single piece of zinc metal there are some regions in which the crystals are under greater strain than in others. This strain is reflected in a higher average Gibbs energy of the zinc atoms. These atoms dissolve to form zinc ions in solution more readily than the atoms in the low strain areas. Thus there are in any piece of metal anodic areas in which the anodic reaction

$$\text{Zn} \longrightarrow \text{Zn}^{2+} + 2e^-$$

takes place more readily than in the other areas, which become *cathodic* areas. The H_2 evolution reaction occurs on the cathodic areas. Since a zinc surface has a much lower i_0 for hydrogen evolution than does a platinum surface, the rate of corrosion of pure zinc is much smaller than the rate of corrosion when it is in contact with a surface, such as platinum, which has a low hydrogen overvoltage.

34.11.2 Corrosion by Oxygen

Corrosion by oxygen requires the reaction,

$$O_2 + 4H^+ + 4e^- \longrightarrow 2H_2O, \qquad \phi^\circ = 1.23 \text{ V} \qquad \text{(acid solution);}$$

$$O_2 + 2H_2O + 4e^- \longrightarrow 4OH^-, \qquad \phi^\circ = 0.401 \text{ V} \qquad \text{(basic solution).}$$

at the cathodic areas of the metal. The i versus η curves for oxygen corrosion in acid solution are shown schematically in Fig. 34.13 for several different metals. The figure shows that the corrosion current for zinc is larger than that for iron, which is larger than that for copper. If Fig. 34-13 had been drawn to scale, the differences in corrosion current would be even larger than are shown in the figure. At high overpotentials the cathodic current is limited by the rate of supply of oxygen to the surface. This is indicated by the plateau in the oxygen polarization curve in Fig. 34.13. The limiting current is proportional to the oxygen concentration in the solution.

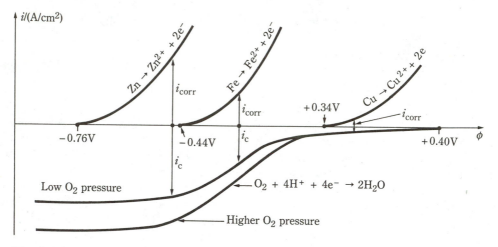

Figure 34.13   Oxygen corrosion in acid solution. (Adapted from K. J. Vetter, *Electrochemical Kinetics*, New York: Academic Press, 1967, pp. 737–738.)

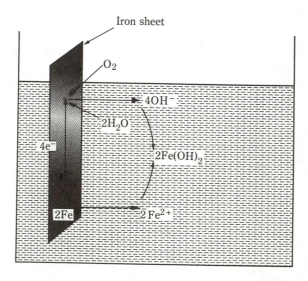

Figure 34.14   Differential aeration.

A difference in oxygen concentration is sufficient to set up a corrosion current. Consider a piece of iron partially immersed in an aqueous salt solution (Fig. 34.14). Near the air interface there is an abundance of oxygen; this makes the area more positive than the deeply immersed portions. Electrons flow in the metal from the deeply immersed region to the region near the air interface. This releases Fe^{2+} ions deep in the solution. Thus the deeply immersed portions corrode by the reaction $Fe \rightarrow Fe^{2+} + 2e^-$, while the oxygen is reduced to OH^- near the water line. The pH increases as the process goes on. As the Fe^{2+} and OH^- ions diffuse toward each other, $Fe(OH)_2$ is precipitated at an intermediate position. This is called "oxygen defect corrosion"; the corrosion occurs where there is no oxygen. This results in some unusual effects such as corrosion under a bolt head, for example. The metal dissolves underneath the bolt head where it is protected from oxygen. Pitting of metals under a particle of dust or an imperfect coat of paint or protective metal occurs because of this differential aeration effect.

34.11.3   Corrosion by Metal Contact; Corrosion Inhibition

The contact of dissimilar metals commonly results in corrosion. For example, if a piece of iron in contact with a piece of copper is immersed in a conducting solution that does not contain appreciable amounts of either Fe^{2+} or Cu^{2+}, the potential of the iron will be negative (anodic) relative to the copper; the electrons produced by the oxidation of the iron

$$Fe \longrightarrow Fe^{2+} + 2e^-$$

move to the copper, where they are removed either by hydrogen evolution or by oxygen reduction. The result is a rapid corrosion of the less noble metal, iron. If the two metals were not connected, both would corrode in an oxygenated solution; but if the two metals are electrically connected, only the more active metal corrodes. This is the basis of corrosion protection using a sacrificial anode. A bar of zinc or magnesium is electrically connected to the steel that is to be protected. In domestic hot water heaters, for example, a magnesium rod is attached to the interior of the steel tank. The magnesium rod corrodes but the steel tank is protected. Similarly, ropes to which zinc bars are attached are hung over the side of steel ships to lessen the attack by salt water. In some instances the metal to be protected is connected to an external source of power; a sufficiently cathodic potential is impressed to reduce the corrosion current to zero.

Corrosion can also be reduced through the use of coatings of various kinds. To be effective the coating must be as impermeable as possible or be capable of resealing itself if it is perforated. The protective oxide on aluminum reseals itself in air if it is broken. Some anodic inhibitors, alkalis, and oxidizing agents inhibit corrosion by assisting the formation of a relatively adherent film of oxide on the metal and assisting the repair of breaks in the coating. Some cathodic inhibitors on the other hand act by poisoning the hydrogen evolution on the cathodic surface, thus increasing the hydrogen overvoltage and reducing the corrosion current. Other cathodic inhibitors, Cu^{2+} for example, by producing a film on the surface simply block the access of oxygen to the cathode surface and thus shut off the corrosion current.

34.12   PHOTOCHEMISTRY

The study of photochemistry embraces all of the phenomena associated with the absorption and emission of radiation by chemical systems. It includes phenomena that are mainly spectroscopic, such as fluorescence and phosphorescence; luminescent chemical reactions, such as flames and the gleam of the firefly; and photostimulated reactions, such as photographic, photosynthetic, and photolytic reactions of various kinds.

The influence of light on chemical systems may be trivial or profound. If the light quanta are not energetic enough to produce a profound effect such as the dissociation of a molecule, the energy may simply be degraded into thermal energy. This latter effect may be regarded as trivial in a photochemical sense, since the same result could be achieved by raising the temperature by any means.

Any effect of light, whether trivial or profound, can be produced only by light that is absorbed by the system in question. This fact, which today seems obvious, was first recognized at the beginning of the 19th century by Grotthuss and Draper, and is called the law of Grotthuss and Draper.

34.13 THE STARK–EINSTEIN LAW OF PHOTOCHEMICAL EQUIVALENCE

The Stark–Einstein law of photochemical equivalence is in a sense simply a quantum-mechanical statement of the Grotthuss–Draper law. The Stark–Einstein law (1905) is another example of the break with classical physics. It states that each molecule which takes part in the photochemical reaction absorbs one quantum of the light which induces the reaction; that is, one molecule absorbs the entire quantum; the energy of the light beam is not spread continuously over a number of molecules.

If we define the primary act of the photochemical reaction as the absorption of the quantum, then the quantum efficiency for the primary act is, by the Stark–Einstein law, equal to unity. For each quantum absorbed, one primary act occurs. For any substance X taking part in a photochemical reaction, the quantum efficiency or quantum yield for the formation (or decomposition) of X is ϕ_X and is defined by

$$\phi_X = \frac{\text{number of molecules of X formed (or decomposed)}}{\text{number of quanta absorbed}}. \tag{34.57}$$

More conveniently, if we measure the rate of formation of X in molecules per second, dN_X/dt, then the quantum yield is

$$\phi_X = \frac{dN_X/dt}{\text{number of quanta absorbed/second}}.$$

The number of quanta absorbed per second is the absorbed intensity, so that

$$\phi_X = \frac{dN_X/dt}{I_a}. \tag{34.58}$$

To determine the quantum yield of the reaction it is necessary to measure the rate of reaction and the amount of radiation absorbed. The rate of the reaction is measured in any convenient way. Figure 34.15 shows a typical arrangement for measuring the absorbed intensity. The reacting system is confined to a cell. The intensity of the transmitted beam is measured with the reaction cell empty and with the cell filled with the reaction mixture. The detector may be a thermopile, which is a set of junctions of dissimilar metals covered with a blackened metal foil. All the radiation is absorbed on the blackened metal and the energy of the radiation is converted to a temperature increase; the temperature increase is converted to a potential difference by the thermopile. The device must be calibrated against a standard light source. It has the advantage of being

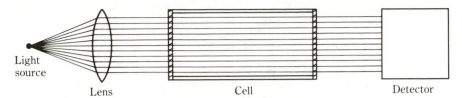

Figure 34.15   Schematic diagrams of apparatus for measurement of light intensity.

usable for light of any frequency. Photoelectric cells are convenient detectors but, since the response varies with frequency, they must be calibrated for each frequency.

A chemical actinometer can be used as the detector. The chemical actinometer utilizes a chemical reaction whose photochemical behavior has been accurately investigated. For example, the compound, $K_3Fe(C_2O_4)_3$, dissolved in a sulfuric acid solution decomposes when irradiated; the $C_2O_4^{2-}$ is oxidized to CO_2 and Fe^{3+} is reduced to Fe^{2+}. For wavelengths between 509 nm and 254 nm the quantum yield for the production of Fe^{2+} is known as a function of wavelength, varying between 0.86 and 1.25. The Fe^{2+} formed can be determined very accurately by adding 1,10-phenanthroline, which forms a complex with Fe^{2+}. The absorbance of the complex is measured colorimetrically. Two measurements are made, one with the cell empty and one with the reactive mixture in the cell. The first yields the incident intensity, I_0; the second yields the transmitted intensity, I. The difference, $I_0 - I = I_a$, is the absorbed intensity.

★ 34.14 PHOTOPHYSICAL PROCESSES; FLUORESCENCE AND PHOSPHORESCENCE

Much of what is called photochemistry is in fact concerned with the phenomena of fluorescence and phosphorescence in systems that do not undergo any chemical change. We will first describe these photophysical phenomena and later discuss photochemical processes. Since all of the electrons are paired in the ground state of most molecules, the result of the absorption of a quantum of radiation is to unpair two electrons and produce an excited electronic state that is either a singlet or a triplet. The energy levels of the molecule then divide, much like the levels of the alkaline earth atoms, into a system of singlet levels and a system of triplet levels. Bear in mind that at least some of these excited states are bound states that will have vibrational and rotational levels associated with them. A typical arrangement of the molecular levels is shown schematically in a Jablonski diagram (Fig. 34.16). The vertical axis measures the energy of the system; the horizontal axis simply spreads the figure for the sake of clarity.

The singlet and triplet electronic levels are labeled S and T, respectively. Subscripts indicate the order of increasing energy; the superscript v indicates that a molecule has an excess vibrational energy; absence of a superscript indicates that the vibrational energy of the molecule is in thermal equilibrium; a zero superscript indicates that a molecule is in the lowest vibrational level. For clarity, the vibrational and rotational levels are shown equally spaced.

If the system is initially in the ground state, S_0, the only quanta that it can absorb are those which raise it to some level in another singlet state, S_1 or S_2 in the diagram. (Radiative transitions are indicated by solid lines, nonradiative transitions by wavy lines.) Because

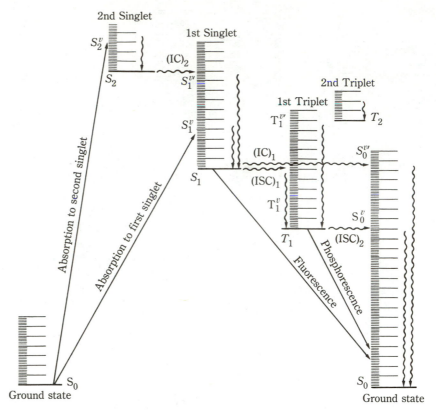

Figure 34.16   Excited states and photophysical transitions between these states in a "typical" organic molecule. Radiative transitions between states are given by solid lines, radiationless processes by wavy lines; IC = internal conversion, ISC = intersystem crossing. Vertical wavy lines are vibrational relaxation processes. Vibrational and rotational levels are shown approximately equally spaced for convenience in presentation. (By permission from J. G. Calvert and James N. Pitts, Jr. *Photochemistry*, New York: John Wiley, 1966.)

of the Franck–Condon requirement, the molecule will most likely be in an excited vibrational state such as S_1^v or S_2^v. Let us assume that the molecule is in the S_2^v state and examine the various possibilities.

The thermal equilibration of the vibrational energy within the electronic level S_2 occurs rapidly; this is represented by the wavy arrow ending at S_2. The nonradiative transition from S_2 to $S_1^{v'}$ is also rapid; this is represented by the wavy horizontal arrow and is called internal conversion (*IC*). Note that there is no change in energy in this process, whereas the equilibration of the vibrational energy entails a loss of energy and therefore requires one or more collisions to carry off the excess energy. From the level of $S_1^{v'}$ there is again a rapid equilibration of the vibrational energy via collisions. In practice, this means that after equilibration the system will most probably be in the ground vibrational level of S_1 since the vibrational level spacing is large enough to keep all but a few of the molecules from occupying the upper states.

Having reached the lowest level in S_1, the system has three paths available for return to the ground state, S_0.

Path 1: Radiative transition with emission of a quantum of fluorescent radiation, $h\nu'$. The fluorescent radiation has a lower frequency than that of the absorbed light which raised the system from S_0 to S_2^v. Since the transition, $S_1 \to S_0$, is permitted by the selection rules, it is very rapid. Since it drains the excited level very rapidly, it ceases almost immediately after the exciting radiation, which supplies population to the upper state, is extinguished.

Path 2: Nonradiative crossing to T_1^v followed by rapid vibrational equilibration to T_1. This is followed by a radiative transition $T_1 \to S_0$. The radiation emitted is called phosphorescence. The nonradiative intersystem crossing (*ISC*) is much slower than the vibrational equilibrations, but competes with the fluorescent emission in the molecules that exhibit phosphorescence. The radiative transition, $T_1 \to S_0$, is usually very slow since the triplet–singlet transition is spin-forbidden by the selection rules. Consequently, the phosphorescence persists for some time after the exciting radiation is turned off.

Path 3a: Nonradiative internal conversion to $S_0^{v'}$ and rapid thermalization of the vibrational energy to bring the system to S_0.

3b: Nonradiative quenching of S_1 by collision.

We can describe the system under steady illumination by writing the rate equation for each process.

Process	*Reaction*	*Rate*	
Excitation	$S_0 + h\nu \to S_1$	I_a	(34.59)
Fluorescence	$S_1 \to S_0 + h\nu'$	$A_{10}[S_1]$	(34.60)
Fluorescence quenching	$S_1 + M \to S_0 + M$	$k_q^F[S_1][M]$	(34.61)
Internal conversion	$S_1 \to S_0$	$k_{IC}^S[S_1]$	(34.62)
Intersystem crossing	$S_1 \to T_1$	$k_{ISC}^S[S_1]$	(34.63)
Phosphorescence	$T_1 \to S_0 + h\nu''$	$A_{TS}[T_1]$	(34.64)
Phosphorescence quenching	$T_1 + M \to S_0 + M$	$k_q^P[T_1][M]$	(34.65)
Intersystem crossing	$T_1 \to S_0$	$k_{ISC}^T[T_1]$	(34.66)

The k's are the rate constants for the various processes; the A_{10} and A_{TS} are the Einstein coefficients for spontaneous emission. In this mechanism, M is intended to represent any atom or molecule that may be present. Then [M], the concentration of M, is proportional to the total concentration of all the species in solution; in the gas phase, [M] is proportional to the total pressure.

The emitted intensity of fluorescence, I_{em}^F, is given by

$$I_{em}^F = A_{10}[S_1].$$ (34.67)

If the system is under steady illumination, the $[S_1]$ and $[T_1]$ do not vary with time; the steady-state conditions are

$$\frac{d[S_1]}{dt} = 0 = I_a - A_{10}[S_1] - k_q^F[S_1][M] - k_{IC}^S[S_1] - k_{ISC}^S[S_1]$$ (34.68)

and

$$\frac{d[T_1]}{dt} = 0 = k_{ISC}^S[S_1] - A_{TS}[T_1] - k_q^P[T_1][M] - k_{ISC}^T[T_1]. \tag{34.69}$$

We define τ_F, the fluorescence lifetime, and τ_P, the phosphorescence lifetime, by

$$\frac{1}{\tau_F} = A_{10} + k_{IC}^S + k_{ISC}^S + k_q^F[M] \tag{34.70}$$

and

$$\frac{1}{\tau_P} = A_{TS} + k_{ISC}^T + k_q^P[M]. \tag{34.71}$$

These definitions reduce Eqs. (34.68) and (34.69) to

$$\frac{d[S_1]}{dt} = 0 = I_a - \frac{[S_1]}{\tau_F} \tag{34.72}$$

and

$$\frac{d[T_1]}{dt} = 0 = k_{ISC}^S[S_1] - \frac{[T_1]}{\tau_P}. \tag{34.73}$$

Solving these two equations for $[S_1]$ and $[T_1]$ yields

$$[S_1] = \tau_F I_a; \qquad [T_1] = k_{ISC}^S \tau_F \tau_P I_a. \tag{34.74}$$

Using this value of $[S_1]$ in the expression in Eq. (34.67), we obtain for the fluorescence intensity,

$$I_{em}^F = A_{10} \tau_F I_a, \tag{34.75}$$

and for the quantum yield,

$$\phi_F = \frac{I_{em}^F}{I_a} = A_{10} \tau_F. \tag{34.76}$$

If we invert Eq. (34.75) and use the value in Eq. (34.70) for τ_F, we find that

$$\frac{1}{I_{em}^F} = \frac{1}{I_a}\left(1 + \frac{k_{IC}^S + k_{ISC}^S}{A_{10}}\right) + \frac{k_q^F[M]}{A_{10}I_a}. \tag{34.77}$$

A plot of $1/I_{em}^F$ versus $[M]$, called a Stern–Volmer plot, should yield a straight line. From the measured value of I_a and a value of A_{10} we can obtain the quenching constant k_q^F. The constant A_{10} can be calculated from the measurement of the molar absorption coefficient of the absorption band.

If we assume that every collision is effective in quenching the fluorescence, then the rate of quenching is the number of collisions between the excited species, S_1, and all other species, M. The number of collisions is given by Eq. (30.23).

$$Z_{12} = \pi\sigma_{12}^2\sqrt{\frac{8kT}{\pi\mu}}\,\tilde{N}_1\tilde{N}_2. \tag{34.78}$$

Thus we expect k_q^F will have the value

$$k_q^F = \pi\sigma_{12}^2\sqrt{\frac{8kT}{\pi\mu}}, \tag{34.79}$$

in which σ_{12} is the average·of the two molecular diameters. The factor, $\pi\sigma_{12}^2$, is the collision cross section. In fact, the values of k_q^F calculated by Eq. (34.79) are usually larger than the experimental values. If we define a quenching cross section, $\pi\sigma_q^2$ such that

$$(k_q^F)_{\text{meas}} = \pi\sigma_q^2 \sqrt{\frac{8kT}{\pi\mu}}, \tag{34.80}$$

then the ratio, $(k_q^F)_{\text{meas}}/(k_q^F)_{\text{coll}}$, is equal to the ratio of the quenching cross section to the collision cross section:

$$\frac{(k_q^F)_{\text{meas}}}{(k_q^F)_{\text{coll}}} = \frac{\pi\sigma_q^2}{\pi\sigma_{12}^2}.$$

Some values of the ratio $(\pi\sigma_q^2)/(\pi\sigma_{12}^2)$ for quenching the fluorescence of NO_2 are given in Table 34.2. These values indicate that the effectiveness of a molecule in quenching fluorescence increases as the number of internal modes of motion increases. For example, if a helium atom or argon atom collides with an activated molecule, the excess energy can only be drained off into translational energy of the colliding atom. This process is not very effective in dissipating the excitation energy and, as a result, the relative cross sections of He and Ar are quite small. The value 0.04 can be taken to mean that $1/0.04 = 25$ collisions are required to quench the excitation. Looking at the diatomics, we find that H_2 is relatively ineffective while O_2 and N_2 are only slightly more effective. Although these molecules do have a vibrational mode, the vibrational quantum is so large that it is not effective in quenching. Molecules with small spacings between the vibrational and rotational energy levels are more effective than those with larger spacings. Consequently, molecules with more modes of motion and heavier atoms are noticeably more effective. Heavy atoms lower

Table 34.2
Cross sections for quenching NO_2 fluorescence by different gases (excitation wavelength = 435.8 nm)

Quenching gas	$\pi\sigma_q^2/10^{-20}$ m^2	$\pi\sigma_{12}^2/10^{-20}$ m^2	$\dfrac{\pi\sigma_q^2}{\pi\sigma_{12}^2}$
He	1.6	34.2	0.04
Ar	3.8	44.0	0.08
H_2	2.5	38.3	0.06
N_2	5.0	48.7	0.10
O_2	6.0	46.2	0.13
NO	9.7	47.4	0.20
CH_4	7.9	50.2	0.16
N_2O	11.9	50.2	0.24
NO_2	13.5	55.3	0.24
CO_2	13.8	51.5	0.27
SF_6	25.7	72.5	0.35
CF_4	24.5	61.9	0.39
H_2O	28.3	59.3	0.48

From G. H. Myers, D. M. Silver, and F. Kaufman, *J. Chem. Phys.* **44**, 718 (1966).

the sizes of the vibrational and rotational energy quanta; similarly, complex molecules have bending vibrational modes, which have low energy quanta. Polar molecules, such as water, are particularly effective.

Note that if the system does not exhibit phosphorescence this implies that $k_{ISC}^S = 0$. Then the value of k_{IC}^S can be determined from the intercept of the Stern–Volmer plot if A_{10} is known.

If phosphorescence does occur, then

$$I_{em}^P = A_{TS}[T_1]. \tag{34.81}$$

The value of $[T_1]$ is given by Eq. (34.74), so that

$$I_{em}^P = A_{TS} k_{ISC}^S \tau_F \tau_P I_a. \tag{34.82}$$

The quantum yield for phosphorescence is

$$\phi_P = \frac{I_{em}^P}{I_a} = A_{TS} k_{ISC}^S \tau_F \tau_P. \tag{34.83}$$

Using Eq. (34.76) we can write this in the form

$$\phi_P = \phi_F \left(\frac{A_{TS}}{A_{10}}\right) k_{ISC}^S \tau_P. \tag{34.84}$$

By inverting this equation and using the value in Eq. (34.71) for τ_P, we have

$$\frac{1}{\phi_P} = \frac{1}{\phi_F} \left(\frac{A_{10}}{A_{TS}}\right) \frac{\{A_{TS} + k_{ISC}^T + k_q^P[M]\}}{k_{ISC}^S}. \tag{34.85}$$

If both ϕ_F and ϕ_P have been measured as a function of the total concentration, $[M]$, a Stern–Volmer plot of $1/\phi_P$ versus $[M]$ can yield the quenching cross section for phosphorescence and the ratio, k_{ISC}^T/k_{ISC}^S, if the Einstein coefficients are known.

★ 34.15 FLASH PHOTOLYSIS

While the steady-state measurements made under continuous illumination yield ratios of rate constants, the absolute values of the rate constants can be determined from a study of the transient phenomena (such as the decay of fluorescence or phosphorescence) that appear after the interruption of the incident radiation. Flash photolysis is an important method for studying transients in a photochemical system.

The apparatus for a flash photolysis experiment is shown in Fig. 34.17. The reactive materials are irradiated with an intense flash of "white" light, which provides a continuum of wavelengths from the infrared to the vacuum ultraviolet. The energy for the flash is the energy stored in a large capacitor ($\sim 10 \ \mu F$) charged to about 10 kV. When the capacitor is discharged through a tube containing an inert gas (such as krypton) at about 100 mm pressure, a brilliant flash of light is produced. This intense light falling on the system raises the material to a variety of excited states and produces a variety of active intermediates. The duration of the flash ranges from 1 to 100 μs. The energy is determined by the capacity and the voltage, $U = \frac{1}{2}CV^2$; the time of the flash depends on the electrical resistance, which is proportional to the length of the flash tube, the capacitance, and the inductance of the circuit. Energies of the order of 1000 J per flash are usual.

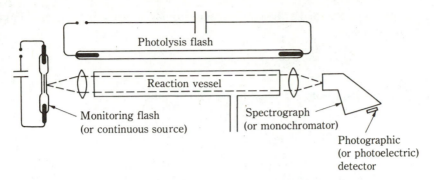

Figure 34.17 Schematic diagram illustrating the principle of the flash photolysis technique. (By permission from G. Porter, Flash photolysis, *Technique of Organic Chemistry*, vol. VIII part II. S. L. Friess, E. S. Lewis, and A. Weissburger, eds. New York: Interscience, Div. of John Wiley, 1963.)

After the system is exposed to the flash, a time-delay circuit fires a low-energy flash from the monitoring flash tube. This beam of light passes through the excited species produced by the first flash and an absorption spectrum is recorded on the spectrograph. By varying the delay time, we can observe the rate of decay of the various species. Alternatively, after making an initial spectrographic study to determine the identity of the species present, we can choose a particular wavelength and measure the absorption at this wavelength as a function of time; we do this by replacing the monitoring flash tube with a continuous source such as a tungsten filament lamp. When we place a monochromator in the optical path, the transmitted radiation is collected in a phototube whose output we can monitor on an oscilloscope. By this means we can follow the concentration of a selected species as a function of time.

We can interpret the transient behavior of S_1 by using the mechanism given above but modifying it by setting $I_a = 0$ since the source has been extinguished. Then instead of the steady-state condition, Eq. (34.72), we have

$$\frac{d[S_1]}{dt} = -\frac{[S_1]}{\tau_F}. \tag{34.86}$$

This is the first-order rate law, which on integration yields

$$[S_1] = [S_1]_0 e^{-t/\tau_F}. \tag{34.87}$$

Using this result in Eq. (34.67), we obtain the expression for the decay of the emitted fluorescence intensity with time.

$$I_{em}^F = A_{10}[S_1]_0 e^{-t/\tau_F} = (I_{em}^F)_0 e^{-t/\tau_F}, \tag{34.88}$$

where $(I_{em}^F)_0$ is the emitted fluorescence intensity at the time when the incident radiation is turned off. Measurement of τ_F as a function of concentration yields the value of k_q^F from the the plot of $1/\tau_F$ versus $[M]$, and the sum, $A_{10} + k_{IC}^S + k_{ISC}^S$, from the intercept; see Eq. (34.70).

We can treat the decay of the phosphorescence similarly. Instead of setting $d[T_1]/dt = 0$ as in Eq. (34.73), we write

$$\frac{d[T_1]}{dt} = k_{ISC}^S[S_1] - \frac{[T_1]}{\tau_P}. \tag{34.89}$$

Using the value of $[S_1]$ from Eq. (34.87) and rearranging, we have

$$\frac{d[T_1]}{dt} + \frac{[T_1]}{\tau_P} = k_{ISC}^S[S_1]_0 e^{-t/\tau_F}.$$

Since the initial concentration of S_1 is the steady-state concentration under illumination, we may use Eq. (34.73) and set $k_{ISC}^S[S_1]_0 = [T_1]_0/\tau_P$, so that this equation becomes

$$\frac{d[T_1]}{dt} + \frac{[T_1]}{\tau_P} = \frac{[T_1]_0 e^{-t/\tau_F}}{\tau_P}. \tag{34.90}$$

This equation can be easily integrated if we substitute for $[T_1]$:

$$[T_1] = F(t)e^{-t/\tau_P}. \tag{34.91}$$

It then becomes

$$\frac{dF}{dt} = \frac{[T_1]_0 e^{-t/\tau}}{\tau_P}, \qquad \text{in which} \qquad \frac{1}{\tau} = \frac{1}{\tau_F} - \frac{1}{\tau_P}.$$

This equation integrates directly to

$$F = [T_1]_0\left(1 + \frac{\tau}{\tau_P} - \frac{\tau e^{-t/\tau}}{\tau_P}\right) = [T_1]_0\left(\frac{\tau_P - \tau_F e^{-t/\tau}}{\tau_P - \tau_F}\right)$$

since $F(0) = [T_1]_0$. Then Eq. (34.91) becomes

$$[T_1] = [T_1]_0\left(\frac{\tau_P e^{-t/\tau_P} - \tau_F e^{-t/\tau_F}}{\tau_P - \tau_F}\right). \tag{34.92}$$

Inserting this expression in Eq. (34.81), we obtain

$$I_{em}^P = (I_{em}^P)_0\left(\frac{\tau_P e^{-t/\tau_P} - \tau_F e^{-t/\tau_F}}{\tau_P - \tau_F}\right), \tag{34.93}$$

in which we have set $A_{TS}[T_1]_0 = (I_{em}^P)_0$, the initial value of the phosphorescence intensity.

Ordinarily τ_F is many orders of magnitude smaller than τ_P. Typically, $\tau_F < 10^{-7}$ s, while $\tau_P \approx 1$ to 10 s. This means that $\tau_P - \tau_F \approx \tau_P$. It also means that after a few multiples of τ_F (a few microseconds) have elapsed, the term in $\exp(-t/\tau_F)$ has decayed to a negligible value and we are left with a simple exponential decay for the phosphorescence:

$$I_{em}^P = (I_{em}^P)_0 e^{-t/\tau_P}, \qquad (t \gg \tau_F). \tag{34.94}$$

The existence of the nonzero lifetime, τ_F, introduces a very slight deformation in the curve of I_{em}^P versus t at very short times but does not affect the rate of decay at longer times.

Since the flash produces relatively large populations in the excited states, it is possible to observe the absorption spectrum of some of these excited species; for example, transitions from an excited singlet state to higher singlet states, and transitions from the triplet state to the higher triplet states. This method has yielded an abundance of information about the energy levels of molecules.

34.16 ABSORPTION AND EMISSION SPECTRA OF ORGANIC MOLECULES

When the molecules in a given electronic state are thermally equilibrated, most of them are in the lowest vibrational level of that state. Therefore the absorption spectrum consists of a band that originates in the lowest vibrational level of the ground electronic state. Conversely, since the internal conversion process after excitation is very rapid, all the excited species are quickly drained down to the lowest vibrational level of S_1. The fluorescence emitted originates in the lowest vibrational level of S_1 and terminates in the various vibrational levels of S_0. The energy levels for these transitions are shown in Fig. 34.18. It is clear that absorption involves larger energy quanta than does emission; consequently, the absorption band is in a shorter wavelength region than the emission band. The ground-state to ground-state transition is the same for both. If the vibrational-level spacings are about the same in the two states, the emission spectrum will appear to be the mirror image of the absorption spectrum (at least the position of the lines on a frequency scale will appear so). Whether the intensities match depends on the overlap of the wave functions in the two states. Figure 34.19 shows the effect for anthracene in three different physical states. The numbers in the figure for the vapor phase are the vibrational quantum numbers in the upper (prime) and lower (double prime) states. The 0–0 bands for absorption and fluorescence are separated because the molecule in the excited state has a long enough lifetime to equilibrate with its neighbors and thus lower its energy slightly before the fluorescence occurs. There is not enough time for this energy adjustment to occur during the absorption process. The shift to lower frequencies (red shift) in solution and in the solid phase is typical; it is the result of the interaction of the molecule with its neighbors.

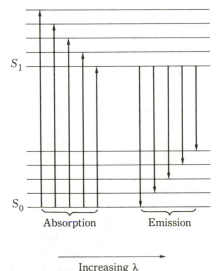

Absorption   Emission

Increasing λ

Figure 34.18 Transitions in absorption and emission.

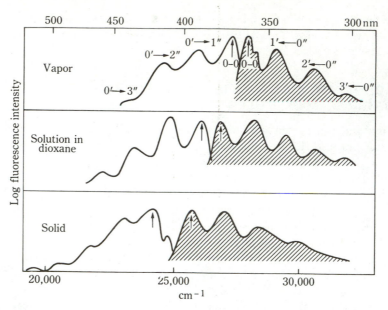

Figure 34.19 Absorption (shaded curves) and fluorescence spectra of anthracene in vapor, solution, and solid phases. Arrows indicate the 0–0 bands. Note the red shift and increased separation of the 0–0 bands in going to the solid phase. Other vibronic bands are shown for the vapor. (From J. G. Calvert and J. N. Pitts, Jr. *Photochemistry*. New York: John Wiley, 1966.)

34.17 ABSORPTION WITH DISSOCIATION

A common class of photolytic reactions consists of those in which the primary photochemical act is absorption of the quantum by a molecule followed by dissociation of the molecule:

$$M + h\nu \longrightarrow \text{Fragments.}$$

Since the fragments produced are often atoms or free radicals, this primary step frequently initiates a chain mechanism, whose occurrence may be indicated by quite large values of the quantum yield, although a chain reaction can occur with very small values of the quantum yield.

Ordinarily the energy required for an absorbed light quantum to produce dissociation of a molecule is significantly greater than the thermodynamic dissociation energy of the molecule. The reason is simply that for radiation to be absorbed there must be an upper electronic state to which the system can be transported by the absorbed quantum. The potential energy diagram for I_2 is shown in Fig. 34.20. The absorption band of I_2 is due to the transition from the lowest vibrational level of the $^1\Sigma_g^+$ state to the levels in the $^3\Pi_{0u}^+$ state. The band converges to a continuum at 499.0 nm. At this wavelength the energy of the transition is sufficient to bring the molecule to that required to dissociate the molecule into two iodine atoms, one of which is in the excited state, $^2P_{1/2}$. Strictly speaking, the transition from $^1\Sigma_g^+$ to $^3\Pi_{0u}^+$ should be spin-forbidden, but because of the large number of electrons in the I_2 molecule this prohibition is relaxed. The transitions from the ground state to $^3\Pi_{1u}$ and $^3\Sigma_u^+$ are very weak. From the diagram we see that the thermodynamic dissociation energy is only 148.71 kJ/mol while the energy of a 499 nm quantum is 239.73 kJ/mol. It

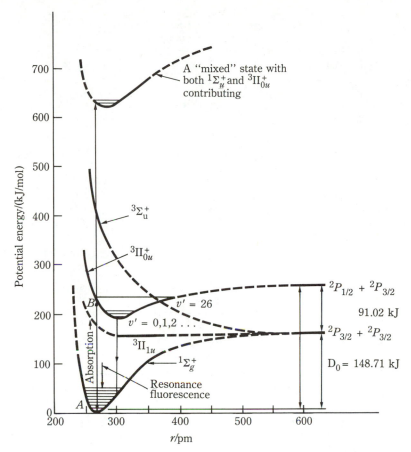

Figure 34.20 Potential energy diagram for several states of the $I_2(g)$ molecule. (From J. G. Calvert and J. N. Pitts, Jr. *Photochemistry*. New York: John Wiley, 1966.)

follows that the excitation energy for an iodine atom from $^2P_{3/2}$ to $^2P_{1/2}$ is 91.02 kJ/mol. The dissociation process can be written

$$I_2 + h\nu \longrightarrow I(^2P_{3/2}) + I(^2P_{1/2}), \qquad \lambda < 499 \text{ nm.}$$

More often we would simply write

$$I_2 + h\nu \longrightarrow I + I^*, \qquad \lambda < 499 \text{ nm,}$$

indicating by the asterisk that the iodine atom is in an excited state. Table 34.3 shows some reactions together with the wavelengths below which the dissociation occurs.

In the case of oxygen the continuum begins at 175.9 nm corresponding to a transition from the $^3\Sigma_g^-$ state to the $^3\Sigma_u^-$ state. As indicated, the dissociation produces one normal and one excited oxygen atom.

The case of HI is unusual in the sense that all the upper states of HI are repulsive. Depending on the wavelength, the iodine atom may or may not be in an excited state. But in all instances the product atoms have a large amount of excess kinetic energy. Because of the great differences in the masses of the H atom and the I atom, most of this excess energy

Table 34.3
Wavelengths required to dissociate
simple molecules

Reaction	λ/nm
$H_2 + h\nu \rightarrow H(1s) + H^*(2s \text{ or } 2p)$	< 84.5
$O_2 + h\nu \rightarrow O(^3P) + O^*(^1D)$	<175.9
$Cl_2 + h\nu \rightarrow Cl + Cl^*$	<478.5
$Br_2 + h\nu \rightarrow Br + Br^*$	<511.0
$I_2 + h\nu \rightarrow I(^2P_{3/2}) + I^*(^2P_{1/2})$	<499.0
$HI + h\nu \rightarrow H + I$	<327
$NO_2 + h\nu \rightarrow NO + O$	<366
$NH_3 + h\nu \rightarrow NH_2 + H$	$< \sim 220$
$H_2O + h\nu \rightarrow H + OH$	<242
$R\overset{\displaystyle O}{\overset{\|}{-C}}-H + h\nu \rightarrow R + CHO$	$< \sim 330$
$R\overset{\displaystyle O}{\overset{\|}{-C}}-R + h\nu \rightarrow R + RCO$	$< \sim 330$

is given to the hydrogen atom, which is then referred to as a "hot" atom. Figure 34.21 shows the transitions from a bound state to two different repulsive excited states. When the molecule is excited from the ground vibrational state to the state at point B, the excess energy labeled ΔE is an amount beyond that needed to dissociate the molecule to normal H and I atoms. This excess energy winds up as the kinetic energy of a hydrogen atom in its ground electronic state. Depending on which states are involved, the excess kinetic energy can be very large.

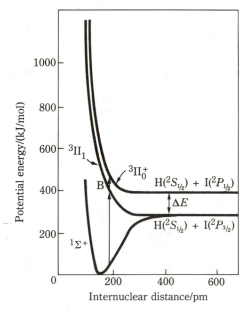

Figure 34.21 Potential energy curves for the lowest electronic states of HI. (Adapted from J. G. Calvert and J. N. Pitts, Jr. *Photochemistry*. New York: John Wiley, 1966.)

34.18 EXAMPLES OF PHOTOCHEMICAL REACTIONS

34.18.1 The Photolysis of HI

The kinetics of two photochemical reactions will be compared with the kinetics of the thermal reactions. In the absence of light, hydrogen iodide decomposes by the elementary reaction:

$$2\,HI \longrightarrow H_2 + I_2,$$

or possibly

$$2\,HI \longrightarrow H_2 + 2I.$$

In the initial stages the reverse reaction can be ignored. In either case the rate of reaction can be written

$$\frac{1}{V}\frac{d\xi}{dt} = -\frac{1}{2}\frac{d[HI]}{dt} = k[HI]^2.$$

In the photochemical reaction, at wavelengths below about 327 nm, the mechanism is

$$HI + h\nu \longrightarrow H + I, \qquad Rate = I_a,$$

$$H + HI \longrightarrow H_2 + I, \qquad Rate = k_2[H][HI],$$

$$2I \longrightarrow I_2, \qquad Rate = k_3[I]^2.$$

Other possible elementary reactions either have much higher activation energies or require three-body collisions. The rate of disappearance of HI is

$$\frac{-d[HI]}{dt} = I_a + k_2[H][HI].$$

The steady-state requirement is

$$\frac{d[H]}{dt} = 0 = I_a - k_2[H][HI].$$

Combining these two equations, we obtain

$$\frac{-d[HI]}{dt} = 2I_a \tag{34.95}$$

By definition the quantum yield is $\phi = (-d[HI]/dt)/I_a$, so that, from Eq. (34.95) we find that $\phi = 2$. In a variety of experimental situations, the observed value of ϕ is 2.

The interesting point about the photochemical reaction is that the rate, by Eq. (34.95), is simply twice the absorbed intensity and is not directly dependent on the concentration of HI. This fact implies that the reaction is very slow, since even fairly intense light sources do not produce a very large number of quanta per second. The dependence of rate on intensity can be readily verified by altering the distance between the system and the light source. The incident intensity varies inversely as the square of the distance, so for a given cell and given concentration of HI, the absorbed intensity must vary in the same way. Indirectly, the rate depends on the concentration of HI, since the absorbed intensity is dependent on concentration through Beer's law.

34.18.2 The Photochemical Reaction between H_2 and Br_2

The photochemical reaction between H_2 and Br_2 follows a kinetic law which resembles that for the thermal reaction, in contrast to the decomposition of HI, where the kinetics are quite different. Using light of wavelength less than 511 nm, the mechanism of the photochemical reaction between H_2 and Br_2 is

$$Br_2 + h\nu \longrightarrow 2\,Br,$$

$$Br + H_2 \xrightarrow{k_2} HBr + H,$$

$$H + Br_2 \xrightarrow{k_3} HBr + Br,$$

$$H + HBr \xrightarrow{k_4} H_2 + Br,$$

$$2\,Br \xrightarrow{k_5} Br_2.$$

The rate of formation of HBr is the same as that given in Eq. (32.65) for the thermal reaction

$$\frac{d[HBr]}{dt} = k_2[Br][H_2] + k_3[H][Br_2] - k_4[H][HBr]. \tag{34.96}$$

The steady-state conditions for H atoms and Br atoms are

$$\frac{d[H]}{dt} = 0 = k_2[Br][H_2] - k_3[H][Br_2] - k_4[H][HBr],$$

$$\frac{d[Br]}{dt} = 0 = 2I_a - k_2[Br][H_2] + k_3[H][Br_2] + k_4[H][HBr] - 2k_5[Br]^2.$$

Addition of these two equations yields $2k_5[Br]^2 = 2I_a$, so that

$$[Br] = \left(\frac{I_a}{k_5}\right)^{1/2}.$$

This result in either of the steady-state equations yields ultimately

$$[H] = \frac{\left(\dfrac{k_2}{k_5^{1/2}}\right)I_a^{1/2}}{k_3[Br_2] + k_4[HBr]}.$$

These values in Eq. (34.96) bring it to the form

$$\frac{d[HBr]}{dt} = \frac{\left(\dfrac{2k_2}{k_5^{1/2}}\right)I_a^{1/2}[H_2]}{1 + \dfrac{k_4[HBr]}{k_3[Br_2]}}. \tag{34.97}$$

The expression in Eq. (34.97) is very similar to that for the thermal reaction where the factor $I_a^{1/2}$ is replaced by $k_1^{1/2}(Br_2)^{1/2}$. This means that the bromine atom concentration is maintained by the photochemical dissociation of bromine rather than the thermal dissociation. The dependence on the square root of the intensity is notable, since it has the consequence that the quantum yield is inversely proportional to the square root of the intensity:

$$\phi = \frac{\dfrac{d[HBr]}{dt}}{I_a} = \frac{\left(\dfrac{2k_2}{k_5^{1/2}}\right)[H_2]}{I_a^{1/2}\left\{1 + \dfrac{k_4[HBr]}{k_2[Br_2]}\right\}}$$

As the intensity increases, a greater proportion of the bromine atoms formed are converted to Br_2 instead of entering the chain; most of the additional quanta therefore are wasted, and the process is less efficient. Because k_2 is very small, the quantum yield is less than unity at room temperature in spite of the fact that the HBr is formed in a chain reaction. As the temperature increases, the increase in k_2 increases the quantum yield (k_5 is nearly independent of temperature).

34.19 PHOTOSENSITIZED REACTIONS

Photosensitized reactions make up an important class of photochemical reactions. In these reactions the reactants are mixed with a foreign gas; mercury or cadmium vapor are often used. The primary photochemical act is the absorption of the quantum by the foreign atom or molecule.

If a mixture of hydrogen, oxygen, and mercury vapor is exposed to ultraviolet light, the mercury vapor absorbs strongly at 253.7 nm with the formation of an excited mercury atom, Hg^*:

$$Hg + h\nu \longrightarrow Hg^*.$$

The energy corresponding to this wavelength is 471.5 kJ/mol. The energy required to dissociate a molecule of hydrogen in its ground state to two hydrogen atoms in their ground state is 432.0 kJ/mol. The dissociation of oxygen requires 490.2 kJ/mol. The energy possessed by the excited mercury atom is more than enough to dissociate H_2 but not enough to dissociate O_2. The quenching reaction

$$Hg^* + H_2 \longrightarrow Hg + H + H$$

introduces H atoms into the mixture, which can initiate chains to form H_2O by the usual mechanism. A reaction that is initiated by light in this way is a photosensitized reaction; the mercury is called a sensitizer.

The importance of photosensitization derives from the fact that reaction is produced in the presence of the sensitizer in circumstances where the direct photochemical dissociation is not possible. The example just cited is a case in point. Radiation of wavelength 253.7 nm was absorbed by a mercury atom. The excited mercury atom dissociated a molecule of hydrogen by transferring the excitation energy in a collision. The mercury atom had 471.5 kJ; of this 432.0 kJ were needed for the dissociation; 39.5 kJ are left over and go into additional translational energy of the two hydrogen atoms and the mercury atom. If the attempt is made to dissociate H_2 directly by the process

$$H_2 + h\nu \longrightarrow H + H,$$

we find that light of $\lambda = 253.7$ nm will not produce any dissociation even though it still has the 471.5 kJ, which is more than enough if we consider only the thermodynamics of the process. For the direct absorption to produce dissociation, the wavelength must lie in the absorption continuum; for H_2 the continuum begins at 84.9 nm. The absorption of light in the continuum produces at least one atom in an excited state:

$$H_2 + h\nu \longrightarrow H + H^*, \qquad \lambda \leq 84.9 \text{ nm}.$$

As we have seen, the selection rules that govern the absorption of radiation forbid the direct absorption of a quantum by H_2 to yield two H atoms in their ground state; thus a quantum of light does not necessarily produce dissociation even though it may have sufficient energy. The quantum must have enough energy to produce the dissociation in the special way required by the selection rules. The transfer of energy in a collision is not limited by this requirement, and so sensitization can produce dissociation.

The decomposition of ozone sensitized by chlorine is another example of photo-sensitization. Ozone is stable under irradiation by visible light. The absorption continuum begins at about 290 nm. In the presence of a little chlorine, ozone decomposes rapidly. The chlorine absorbs continuously below 478.5 nm:

$$Cl_2 + h\nu \longrightarrow 2Cl, \qquad \lambda \leq 478.5 \text{ nm}.$$

The chlorine atoms react with ozone in a complex chain mechanism. Bromine is also effective as a sensitizer for this decomposition.

The decomposition of oxalic acid photosensitized by uranyl ion is a common actino-metric reaction. The light may have any wavelength between 250 and 450 nm. The absorption of the light quantum activates the uranyl ion, which transfers its energy to a molecule of oxalic acid, which then decomposes. The reaction may be written

$$UO_2^{2+} + h\nu \longrightarrow (UO_2^{2+})^*$$

$$(UO_2^{2+})^* + H_2C_2O_4 \longrightarrow UO_2^{2+} + H_2O + CO_2 + CO.$$

The quantum yield depends on the frequency of the light, varying between about 0.5 and 0.6. After a measured period of time, the light is turned off and the residual oxalic acid determined by titration with $KMnO_4$. From these data and knowing the original amount of oxalic acid, the intensity is calculated using the value of the quantum yield appropriate to the frequency of light used.

A practical example of photosensitization is in the use of certain dyes in photographic film to render the emulsion sensitive to wavelengths that are longer than those to which it ordinarily responds; for example, infrared photography, which permits photographs of objects in the absence of visible light.

34.20 PHOTOSYNTHESIS

The synthesis of the compounds required for the structure and function of plants is sensitized by chlorophyll, the principal constituent of green-plant pigments. Chlorophyll exhibits two strong absorption bands in the visible region, a band between about 600 and 680 nm and another between 380 and 480 nm; it is the long wavelength region that apparently is most efficient in photosynthesis. In addition, there is a molecule, apparently a particular form of chlorophyll-a, which absorbs at 700 nm. This form is crucial to the photosynthetic mechanism. The other plant pigments such as the carotenes and the phyco-bilins contribute through their ability to absorb light and transfer the excitation energy to chlorophyll-a.

The photosynthesis reaction of classical chemistry was written in the form of the over-all transformation

$$nCO_2 + nH_2O + xh\nu \longrightarrow (CH_2O)_n + nO_2.$$

The $(CH_2O)_n$ is the general formula for a carbohydrate. It was then assumed that the energy stored in the carbohydrate was used in other chemical reactions to synthesize all the other plant materials (proteins, lipids, fats, and so on). It is now clear that amino acids, for example, are immediate products of the photosynthetic reduction of carbon dioxide, and that carbohydrate need not be synthesized first. This is not to minimize the importance of the photosynthesis of carbohydrate, but only to note that many other types of compounds are produced photosynthetically. The overall mechanism and many of the details of the carbon reduction cycle, CO_2 to carbohydrate, were worked out by Melvin Calvin and his colleagues, for which he received the Nobel Prize.

The gathering of energy occurs via the plant pigments; the energy is transferred to chlorophyll-a, which in one form transfers the energy to oxidize water to oxygen and in the other form ultimately reduces the compound nicotinamide adenine dinucleotide phosphate, NADP, to NADPH. It is the reducing ability of NADPH that ultimately brings the CO_2 to $(CH_2O)_n$.

These reactions are mediated through many steps by enzymes, and labile intermediate compounds. It is apparent that between 6 and 12 moles of quanta are required to convert one mole of carbon to carbohydrate. It is an intricate mechanism requiring several stages of pumping energy uphill.

34.21 THE PHOTOSTATIONARY STATE

Absorbed light has an interesting effect on a system in chemical equilibrium. The absorption of light by a reactant can increase the rate of the forward reaction without directly influencing the rate of the reverse action; this disturbs the equilibrium. The concentration of products increases somewhat, increasing the rate of the reverse reaction. In this way the rates of the forward and reverse reaction can be brought into balance with the system having a higher concentration of products than that in the equilibrium system. This new state is not an equilibrium state but a stationary state, called a *photostationary* state.

The dimerization of anthracene offers a convenient example. The reaction

$$2A \longrightarrow A_2$$

occurs upon irradiation of a solution of anthracene by ultraviolet light. A plausible mechanism is

$$A + h\nu \longrightarrow A^*, \qquad \text{Rate} = I_a,$$
$$A^* + A \longrightarrow A_2, \qquad \text{Rate} = k_2[A^*][A],$$
$$A^* \longrightarrow A + h\nu', \qquad \text{Rate} = k_3[A^*], \qquad \text{(fluorescence)}$$
$$A_2 \longrightarrow 2A, \qquad \text{Rate} = k_4[A_2].$$

The net rate of formation of A_2 is

$$\frac{d[A_2]}{dt} = k_2[A^*][A] - k_4[A_2]. \tag{34.98}$$

In the steady state, $d[A^*]/dt = 0$, so that

$$0 = I_a - k_2[A^*][A] - k_3[A^*] \qquad \text{or} \qquad [A^*] = \frac{I_a}{k_2[A] + k_3}. \tag{34.99}$$

In the photostationary state we have the additional requirement that $d[A_2]/dt = 0$; so that Eq. (34.98) yields for the concentration of A_2,

$$[A_2] = \frac{k_2[A^*][A]}{k_4}.$$

Using the value for $[A^*]$ from Eq. (34.99), we obtain

$$[A_2] = \frac{k_2[A]I_a}{k_4(k_2[A] + k_3)} = \frac{I_a}{k_4\{1 + (k_3/k_2[A])\}}.$$

If the concentration of monomer, $[A]$, is very large, then this becomes

$$[A_2] = \frac{I_a}{k_4}. \tag{34.100}$$

The difference between Eq. (34.100) and the usual equilibrium expression $[A_2] = K[A]^2$, should be noted. In the photostationary state in the condition for which Eq. (34.100) is appropriate, the concentration of dimer is independent of the concentration of monomer.

Many other examples of the photostationary states are known. The decomposition of NO_2 occurs photochemically below ~ 400 nm.

$$NO_2 + h\nu \longrightarrow NO + \tfrac{1}{2}O_2,$$

while the reverse reaction is a dark reaction.

The maintenance of a certain amount of ozone in the upper atmosphere is the result of a complex photostationary state. The ozone filters the sun's rays so that no radiation of wavelength shorter than about 290 nm reaches the earth's surface. Ozone absorbs strongly at wavelengths shorter than this. This fortunate circumstance of the ozone layer makes life possible on earth. Radiation of wavelengths shorter than this produces severe damage and in many cases has a lethal effect on living cells. The effective thickness of the ozone layer is estimated at about 3 mm if the gas were at standard pressure and temperature. Since halogen atoms catalyze the decomposition of ozone, the accumulation of halogen atoms in the upper atmosphere from the photolytic decomposition of fluorocarbons has prompted grave concern over the massive commercial use of these compounds. A significant depletion of the ozone layer could have serious adverse effects on life forms on earth.

34.22 CHEMILUMINESCENCE

Reactions of the ordinary thermal type in which some intermediate or the product itself are formed in an electronically excited state exhibit *chemiluminescence*. The excited molecule emits a quantum of light, usually in the visible spectrum. Since the reaction may be proceeding at ordinary temperatures, the light emitted is sometimes called "cold light," presumably to contrast it with the "hot light" emitted by a flame or incandescent body. The oxidation of 3-amino phthalic cyclic hydrazide, luminol, in alkaline solution by hydrogen peroxide is a classic example. A bright green light is emitted.

The greenish glow of slowly oxidizing phosphorus is apparently due to the formation of an oxide in an excited state. The light of the firefly and the light emitted by some microorganisms in the course of metabolism, bioluminescence, are other examples of chemiluminescence. The phosphorescence observed in marshy areas, the will o' the wisp, is apparently due to a slow oxidation of rotting organic material.

QUESTIONS

34.1 Discuss the similarities and differences between the surface decomposition rate law, Eq. (34.5), and the Lindemann rate law, Eq. (32.61).

34.2 Discuss the Case 3 surface decomposition rate, Eq. (34.19), in terms of the competition of reactants A and B for the available surface sites.

34.3 Estimate the activation energy decrease by a catalyst required to triple a reaction rate at 300 K.

34.4 Argue that the Eq. (34.37), when applied at equilibrium, is consistent with the Nernst equation.

34.5 Why doesn't the exchange current i_0 vanish at equilibrium?

34.6 What is the meaning of α in Eqs. (34.29) and (34.30)? If the potential ϕ is approximately linear in the separation of a metal ion from the metal through the double layer, what does the typical value $\alpha = 1/2$ suggest about the symmetry of the chemical Gibbs energy ($\Delta G^{\ddagger}$) behavior as a function of this separation?

34.7 Contrast fluorescence and phosphorescence in terms of (a) routes available to relevant electronic states, (b) the wavelengths involved, and (c) radiative lifetimes.

34.8 Would the fluorescence emission spectrum of a molecule depend on time if internal conversion processes were not rapid? Explain.

34.9 How could absorption with dissociation of a diatomic combined with fragment energy measurements be used to determine dissociation energies?

34.10 Organic singlet S_1 and triplet T_1 states often undergo different reactions. T_1 can sometimes be produced directly from S_0 by a so-called triplet sensitizer. What advantage does this have over irradiation of S_0 if only the product from T_1 is desired?

PROBLEMS

34.1 The diffusion coefficient of many species in aqueous solutions is of the order of 10^{-9} m^2/s. In a well-stirred solution, $\delta = 0.001$ cm. If the concentration of the reactant molecule is 0.01 mole/litre, what will be the rate of the reaction if the slow step is diffusion of the reactant to the surface? The concentration of the reactant at the surface may be taken as zero, since it reacts very rapidly on arrival at the surface.

34.2 a) Using data from Table 34.1, compare the relative rates of the uncatalyzed decomposition of HI at 400 K and 500 K.

   b) Compare the relative rates of the catalyzed decomposition of HI on platinum at 400 K and 500 K.

34.3 What conclusions can be reached about absorption on the surface from each of the following facts?

   a) The rate of decomposition of HI on platinum is proportional to the concentration of HI.
   b) On gold the rate of decomposition of HI is independent of the pressure of HI.
   c) On platinum, the rate of the reaction $SO_2 + \frac{1}{2}O_2 \rightarrow SO_3$ is inversely proportional to the pressure of SO_3.
   d) On platinum the rate of the reaction $CO_2 + H_2 \rightarrow H_2O + CO$ is proportional to the pressure of CO_2 at low CO_2 pressures and is inversely proportional to the pressure of CO_2 at high CO_2 pressures.

34.4 The galvanometer in a potentiometer circuit can detect $\pm 10^{-6}$ A. The i_0 for hydrogen evolution is 10^{-14} A/cm^2 on mercury and 10^{-2} A/cm^2 on platinum. If the electrode area is 1 cm^2, over what range of potential will the potentiometer appear to be in balance (a) if platinum is used as a hydrogen electrode? (b) if mercury is used as a hydrogen electrode? (Assume that the relation between i and η is linear; $t = 20\,°$C.)

34.5 By passing a current through a ferric sulfate solution, 15 cm^3 of O_2 at STP is liberated at the anode and the equivalent quantity of ferric ion is reduced to ferrous ion at the cathode. If the anode area is 3.0 cm^2 and the cathode area is 1.2 cm^2, what are the rates of the anodic and cathodic reactions in A/cm^2? The current passes for 3.5 min.

34.6 The exchange current measures the rate at which the forward and reverse reactions occur at equilibrium. The exchange current for the reaction $\frac{1}{2}H_2 \rightleftharpoons H^+ + e^-$ on platinum is 10^{-2} A/cm^2.

a) How many hydrogen ions are formed on 1 cm^2 of a platinum surface per second?
b) If there are 10^{15} sites/cm^2 for absorption of H atoms, how many times is the surface occupied and vacated in 1 second?

34.7 Consider the oxidation reaction $Fe \rightarrow Fe^{+2} + 2e^-$.

a) By how much does the activation Gibbs energy change from its equilibrium value if an over-potential of $+0.100$ V is applied to the anode? Assume $\alpha = \frac{1}{2}$, and $t = 25\,°C$.
b) By what factor does this increase i_+ over the i_0?

34.8 Silver is deposited from a 0.10 mol/L Ag^+ solution. For the reaction $Ag^+ + e^- \rightarrow Ag$, $\alpha = 0.74$, $i_0 = 4.5$ A/cm^2 when $[Ag^+] = 0.10$ mol/L at 20 °C. Calculate the overpotential for current densities of 10^{-3}, 10^{-2}, 10^{-1}, and 1 A/cm^2.

34.9 A solution contains 0.01 mol/L Cd^{2+} and 0.10 mol/L H^+. For hydrogen deposition on cadmium, Eq. (34.42) represents the situation if $i_0 = 10^{-12}$ A/cm^2, $z = +1$, and $\alpha = \frac{1}{2}$; the same equation represents the current voltage curve for cadmium deposition if $\alpha = \frac{1}{2}$ and $i_0 = 1.5$ mA/cm^2. The equilibrium potential of the $Cd^{2+} + 2e^- \rightleftharpoons Cd$ reaction is -0.462 V, in 0.01 mol/L Cd^{2+} solution and the equilibrium potential of the $2H^+ + 2e^- \rightleftharpoons H_2$ couple is -0.060 V in 0.10 mol/L H^+ solution at 25 °C.

a) At what current density will cadmium deposition commence?
b) When the current density is -1.0×10^{-3} A/cm^2 what fraction of the current goes into hydrogen evolution?

34.10 Suppose that a piece of cadmium is touched to a piece of platinum, and the metals are immersed in a 0.1 mol/L acid solution. Calculate the corrosion potential and the rate of dissolution of the cadmium for various ratios of the areas: $A_{Pt}/A_{Cd} = 0.01$, 0.10, 1.0, 10, and 100. For hydrogen evolution on platinum, Eq. (34.42) may be used with $i_0 = 0.10$ mA/cm^2, $z = +1$, and $\alpha = \frac{1}{2}$. For cadmium dissolution $i_0 = 1.5$ mA/cm^2 and $\alpha = \frac{1}{2}$. The equilibrium potentials are: $\phi_{0(H^+, H_2)} = -0.060$ V; $\phi_{0(Cd^{2+}, Cd)} = -0.462$ V at 25 °C.

34.11 Use the data in Problems 34.9 and 34.10 and suppose that a piece of cadmium is immersed in an acid solution. What is the corrosion potential and what is the rate of dissolution of the cadmium if the area ratios are: $A_c/A_a = 1.0$, 10^3, 10^6, and 10^9?

34.12 A 0.01 molar solution of a compound transmits 20% of the sodium D line when the absorbing path is 1.50 cm. What is the molar absorption coefficient of the substance? The solvent is assumed to be completely transparent.

34.13 If 10% of the energy of a 100 W incandescent bulb goes into visible light having an average wavelength of 600 nm, how many quanta of light are emitted per second?

34.14 The temperature of the sun's surface is 6000 K. What proportion of the sun's radiant energy is contained in the spectrum in the wavelength range $0 \leq \lambda \leq 300$ nm? (See Problem 19.3.)

34.15 The ozone layer is estimated to be 3 mm thick if the gas were at 1 atm and 0 °C. Given the absorption coefficient α (defined by $I = I_0 10^{-\alpha p l}$ where p is the pressure in atm and l is the length in cm). What is the transmittance to the earth's surface at each of the following wavelengths?

λ/nm	340	320	310	300	290	280	260	240	220
α/cm^{-1} atm^{-1}	0.02	0.3	1.2	4.4	12	48	130	92	20

34.16 At 480 nm, the quantum yield for the production of Fe^{+2} in the photolysis of $K_3Fe(C_2O_4)_3$ in 0.05 mol/L sulfuric acid solution is 0.94. After 20 min irradiation in a cell containing 57.4 cm^3 of solution, the solution is mixed thoroughly and a 10.00 mL sample is pipetted into a 25.00 mL volumetric flask. A quantity of 1,10-phenanthroline is added and the flask filled to the mark with a buffer solution. A sample of this solution is placed in a 1.00 cm colorimeter cell and the trans-mittance measured relative to a blank containing no iron. The value of $I/I_0 = 0.543$. If the molar absorption coefficient of the complex solution is 1.11×10^3 m^2/mol, how many quanta were absorbed by the solution? What was the absorbed intensity?

34.17 The quantum yield of CO in the photolysis of gaseous acetone ($p < 6$ kPa) at wavelengths between 250–320 nm is unity. After 20 min irradiation with light of 313 nm wavelength, 18.4 cm^3 of CO (measured at 1008 Pa and 22 °C) is produced. Calculate the number of quanta absorbed and the absorbed intensity in joules per second.

34.18 A substance has $A_{10} = 2 \times 10^6$ s^{-1} and $k_{ISC}^S = 4.0 \times 10^6$ s^{-1}. Assume that $k_{IC}^S = 0$ and that there is no quenching. Calculate ϕ_F and τ_F.

34.19 If $\tau_F = 2.5 \times 10^{-7}$ s and $A_{10} = 1 \times 10^6$ s^{-1} calculate the k_{ISC}^S and ϕ_F, assuming that quench-ing does not occur and that $k_{IC}^S = 0$.

34.20 For napthalene, $\tau_P = 2.5$ s in a mixture of ether, isopentane, and ethanol (EPA). If $\phi_F = 0.55$ and $\phi_P = 0.05$, calculate A_{TS}, k_{ISC}^T, and k_{ISC}^S/A_{10}, assuming no quenching and $k_{IC}^S = 0$.

34.21 For phenanthrene, $\tau_P = 3.3$ s, $\phi_F = 0.12$, $\phi_P = 0.13$ in an alcohol–ether glass at 77 K. Assume no quenching and no internal conversion, $k_{IC}^S = 0$. Calculate A_{TS}, k_{ISC}^T, and k_{ISC}^S/A_{10}.

34.22 Naphthalene in an ether–alcohol glass at 77 K absorbs light below 315 nm and exhibits fluores-cence and phosphorescence. The quantum yields are $\phi_F = 0.29$ and $\phi_P = 0.026$. The lifetimes are $\tau_F = 2.9 \times 10^{-7}$ s and $\tau_P = 2.3$ s. Calculate A_{10}, A_{TS}, k_{ISC}^T, and k_{ISC}^S, assuming that no quenching occurs and that $k_{IC}^S = 0$.

34.23 a) Using the mechanism for the formation of dianthracene described in Section 34.21, write the expression for the quantum yield in the initial stage of the reaction when $[A_2] = 0$.

b) The observed value of $\phi \approx 1$. What conclusion can be reached regarding the fluorescence of A*?

34.24 A likely mechanism for the photolysis of acetaldehyde is:

$$CH_3CHO + h\nu \longrightarrow CH_3 + CHO,$$

$$CH_3 + CH_3CHO \xrightarrow{k_2} CH_4 + CH_3CO,$$

$$CH_3CO \xrightarrow{k_3} CO + CH_3,$$

$$CH_3 + CH_3 \xrightarrow{k_4} C_2H_6.$$

Derive the expressions for the rate of formation of CO and the quantum yield for CO.

34.25 A suggested mechanism for the photolysis of ozone in low-energy light (red light) is:

1) $$O_3 + h\nu \longrightarrow O_2 + O,$$

2) $$O + O_3 \xrightarrow{k_2} 2O_2,$$

3) $$O + O_2 + M \xrightarrow{k_3} O_3 + M.$$

The quantum yield for reaction (1) is ϕ_1.
a) Derive an expression for the overall rate of disappearance of ozone.
b) Write the expression for the overall quantum yield for the disappearance of ozone, ϕ_0.
c) At low total pressures $\phi_0 = 2$. What is the value of ϕ_1?

34.26 A possible mechanism for the photolysis of CH_2O vapor at 313 nm includes the following steps:

1) $$CH_2O + hv \longrightarrow CO + H_2,$$

2) $$CH_2O + hv \longrightarrow H + CHO,$$

3) $$H + CH_2O \xrightarrow{\ k_3\ } H_2 + CHO,$$

4) $$CHO + M \xrightarrow{\ k_4\ } CO + H + M,$$

5) $$CHO + wall \xrightarrow{\ k_5\ } \tfrac{1}{2}CO + \tfrac{1}{2}CH_2O.$$

The rate of the last reaction can be written $k_5[CHO]$. The quantum yields for reactions (1) and (2) are ϕ_1 and ϕ_2, respectively. Derive the expression for $d[H_2]/dt$ and for the quantum yield for H_2.

35

Polymers

35.1 INTRODUCTION

Our time has been called the "plastic age," more often than not with a derogatory sneer. Certainly the burgeoning use of polymeric materials of all kinds has brought some curses along with its multitude of blessings. The widespread use of these materials is remarkable, considering that barely a half-century has passed since the existence of macromolecules became commonly accepted. Before the pioneering work of Staudinger, beginning in 1920, polymeric materials were classified as colloids and were considered to be physical aggregates of small molecules, much as droplets in a mist or fog are physical aggregates of water molecules. Staudinger's insistence on and demonstration of the validity of the macromolecular concept ultimately led to its acceptance and to the rapid development of the science and its applications.

35.2 TYPES OF MACROMOLECULES

Among the natural macromolecules in the organic world are various gums, resins, rubber, cellulose, starch, proteins, enzymes, and nucleic acids. Inorganic polymeric substances include silicates, the polyphosphates, red phosphorus, the $PNCl_2$ polymers, and plastic sulfur, to name only a few.

Although we will refer to all macromolecules as "polymers," many of them are not simply multiples of a monomeric unit. For example, polyethylene can be described as $(-CH_2-)_n$, a simple "linear" structure with $-CH_2-$ as the repeating unit. On the other hand, a protein has the general structure

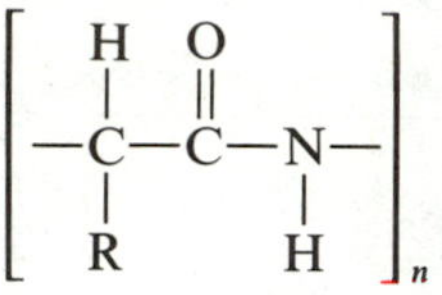

$$
\begin{bmatrix}
& H & O & \\
& | & \| & \\
-C & -C & -N- \\
& | & & | \\
& R & & H
\end{bmatrix}_n
$$

and the R group is different, as we move along the chain. Each (monomeric) segment in the chain is usually the residue of one of the 20 common amino acids. The exact sequence of amino acid residues is important to the biological function of the protein. Similarly, the DNA molecule is a polymeric ester of phosphoric acid and deoxyribose. But along this polymeric backbone, a base is attached to each unit of the polymer. The base may be any one of four: adenine, guanine, cytosine, and thymine. Here, too, the order in which the bases are attached is of overwhelming importance to the organism. A portion of a DNA molecule has the structure:

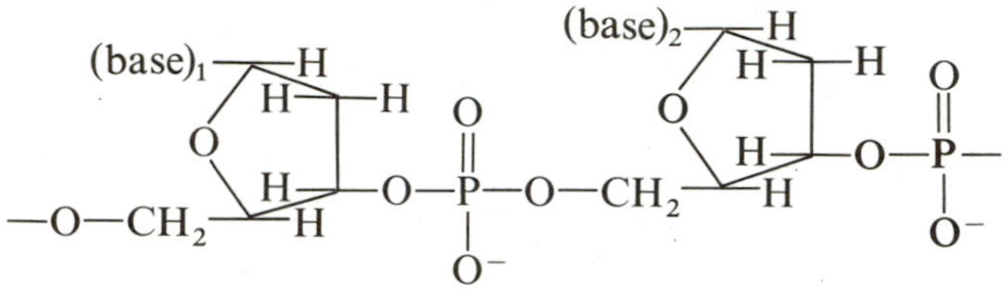

The process of polymerization can conveniently be regarded as belonging to one of two types. If the repeating unit in the polymer has the same chemical composition as the monomer from which it is formed then the process is called addition and the polymer is an *addition polymer*; for example, polyethylene:

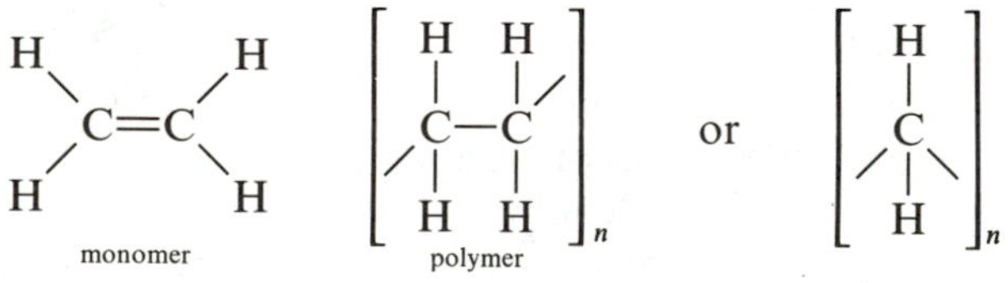

However, if the repeating unit is different in composition from the monomer, the process is called condensation and the polymer is a *condensation polymer*. Typical are the polyesters or polyamides, which eliminate water in the condensation reaction.

Type	Monomer	Polymer
Polyester	$HO-(CH_2)_x-C\!\!\diagdown^{\displaystyle O}_{\displaystyle OH}$	$\left[-O-(CH_2)_x-C\!\!\diagdown^{\displaystyle O}\right]_n$
Polyamide	$H_2N-(CH_2)_x-C\!\!\diagdown^{\displaystyle O}_{\displaystyle OH}$	$\left[\begin{matrix}H \\ | \\ -N-(CH_2)_x-C\!\!\diagdown^{\displaystyle O}\end{matrix}\right]_n$

These particular examples are linear polymers. The materials and the reaction from which the polymer is made allow no deviation from linearity, so long as no side reactions occur. For example, at low temperatures ethylene polymerizes to yield a linear polymer through the propagation of a free radical chain:

$$R \cdot + H_2C = CH_2 \longrightarrow R-CH_2-CH_2 \cdot.$$

The product radical can add another ethylene molecule:

$$R-CH_2-CH_2 \cdot + H_2C = CH_2 \longrightarrow R-CH_2-CH_2-CH_2-CH_2 \cdot.$$

Continuation in this way yields a strictly linear molecule. However, it is possible for this radical to transfer a hydrogen atom from within the chain to the end carbon atom.

This radical can now add monomer at the carbon atom next to the R group and thus produce a polymer with a short side chain. (This branched structure is typical of the ordinary polyethylene used in squeeze bottles.) Linear polyethylene produced at low temperatures has a much more rigid structure. Generally speaking, the molecules with a more regular structure produce a more rigid bulk material.

If the monomer has two double bonds as in isoprene,

the polymer has the possibility of adding a monomer at position 2 to begin a side chain and thus produce a branched molecule. Crosslinking between two polymer chains can also occur in this way.

Natural rubber is almost exclusively the head-to-tail polymer of isoprene with the H atom and the CH_3 group in the *cis* configuration while gutta-percha, the sap from another type of rubber tree, is the head-to-tail polymer having H and CH_3 in the *trans* position.

natural rubber (hevea)   gutta-percha

The synthetic polyisoprenes are not purely *cis* or *trans* but exhibit branching. If a side chain growing on a diene polymer combines with one growing from another molecule, the result is a crosslink between the two molecules. Extensive crosslinking between polymer molecules produces a network polymer that is highly insoluble and infusible. For example, the process of vulcanization introduces sulfur chains as crosslinks between two linear chains of the polyisoprene:

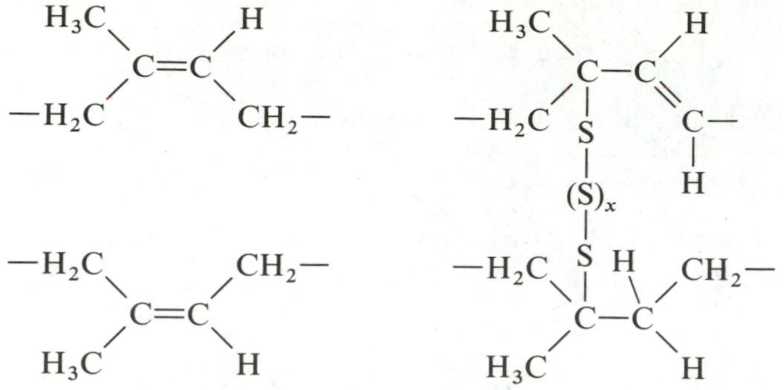

There are additional complexities in the polymerization of substituted vinyl monomers such as $H_2C=CRR'$. At alternate carbon atoms, there are two possibilities for the arrangement of the two R groups. If we draw the carbon chain in the plane perpendicular to the plane of the paper, then the atoms attached to any carbon atom are above and below that plane. If all of the R groups are above the plane and all the R′ groups below, the polymer is *isotactic* (Fig. 35.1). If every second R group is above the plane and the alternate one below, the polymer is *syndiotactic*. If the arrangement of the R groups is random, the polymer is *atactic*. Using special catalysts it is possible to synthesize isotactic and syndiotactic polymers, a feat first accomplished by G. Natta and K. Ziegler.

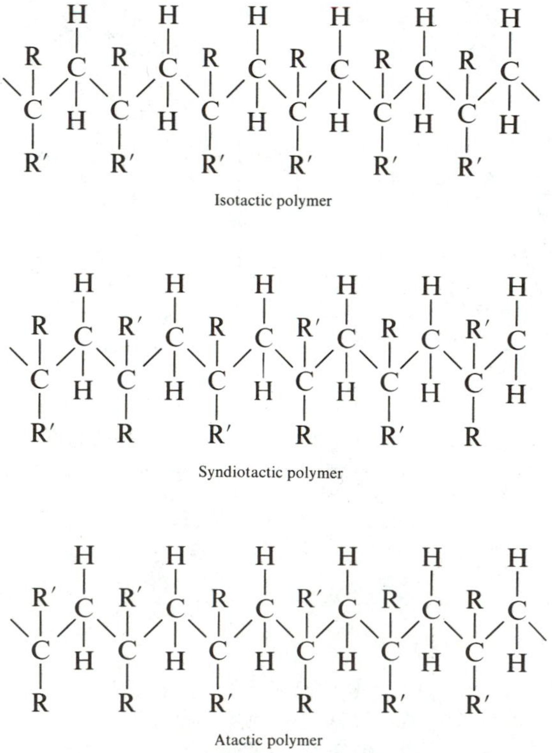

Isotactic polymer

Syndiotactic polymer

Atactic polymer

Figure 35.1

The primary structure of a polymer describes the way in which the atoms are covalently bound within the molecules. There is a secondary structure that describes the conformation of the entire molecule. For example, linear polyethylene in the crystalline solid consists of a zigzag carbon skeleton that is planar; these zigzag chains then pack into the crystal. But in polypropylene, which is polyethylene with a methyl group on every second carbon atom, the steric effect of the methyl group is to force the molecule into a helical configuration instead of a zigzag chain. There are three monomer units in one turn of the helix. With very large substituents, the helix enlarges and may incorporate 3.5 or 4 monomer units per turn. A classic example of the helical secondary structure is the α-helix exhibited by proteins, shown in Fig. 35.2. The peptide unit in the protein is

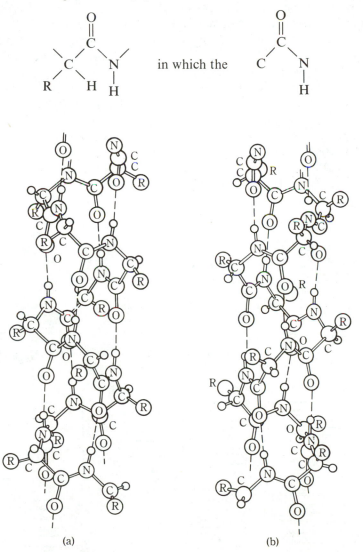

Figure 35.2 Two possible forms of the alpha helix. The one on the left is a left-handed helix, the one on the right is a right-handed helix. The amino acid residues have the L-configuration in each case. (From L. Pauling, *The Nature of the Chemical Bond*, 3d ed. Ithaca, N.Y.: Cornell University Press, 1960.)

atoms lie in one plane. If the molecule is twisted into a spiral the N—H group is in a position to form a hydrogen bond with the oxygen atom in the fourth residue preceding it in the protein chain.

In addition to this secondary structure, polymers possess a tertiary structure. In the case of proteins, the tertiary structure describes the way in which the helix is folded around itself.

35.3 POLYMER SOLUTIONS

The process of dissolving a polymer is usually a slow one. Frequently—and particularly for highly crosslinked network polymers—the addition of a solvent results only in swelling as the solvent permeates the polymer matrix. For other polymers solution takes place over a prolonged period of time after the first swelling occurs. In general, the portions with lower molar mass are more soluble; this property can be used to separate the polymer into fractions of different average molar mass.

The interactions between solvent and solute are relatively large compared to those between smaller molecules. As a result, the behavior of polymer solutions, even when very dilute, may be far from ideal.

The configuration of a polymer in solution depends markedly on the solvent. In a "good" solvent a stronger interaction occurs between solvent and polymer than between solvent and solvent, or between various segments of the polymer. The polymer stretches out in the solution (uncoils), as illustrated in Fig. 35.3(a).

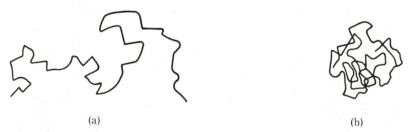

(a)   (b)

Figure 35.3 Polymer configurations in solvents. (a) Uncoiled in a good solvent. (b) Coiled in a poor solvent.

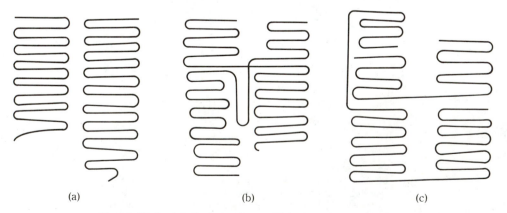

(a)   (b)   (c)

Figure 35.4 (a) Regions of crystallinity for a linear polymer. (b) and (c) show possible mistakes.

In a poor solvent, the polymer segments prefer to remain attached to other segments of the polymer molecule; thus while separating from other polymer molecules in the solid, the molecule coils upon itself (Fig. 35.3b). These different conformations have enormous influence on the viscosity, for example. The viscosity of a solution of long uncoiled chains is very much larger than that of a solution containing the coiled molecules.

The solid phase of a linear polymer, or one with branches that are not too long, may be crystalline. For example, solid linear polyethylene is mainly crystalline, consisting of regions in which the linear molecule has been neatly folded as in Fig. 35.4(a). However, such long molecules can easily make a variety of mistakes, and disordered regions appear as in Fig. 35.4(b) and (c). The mistakes do not differ very much in energy from the perfectly ordered arrangement and consequently occur frequently. However, since there are ordered regions in the solid we can describe it as at least a partly crystalline material.

35.4 THE THERMODYNAMICS OF POLYMER SOLUTIONS

The equation for the Gibbs energy of mixing of any solution is given by

$$\Delta G_{\text{mix}} = \sum_i n_i(\mu_i - \mu_i^\circ),$$

in which μ_i° is the chemical potential of pure component i. If we differentiate this equation with respect to n_k keeping T, p, and all the other n_i constant we obtain

$$\left(\frac{\partial \Delta G_{\text{mix}}}{\partial n_k}\right)_{T, p, n_i \neq k} = \mu_k - \mu_k^\circ + \sum_i n_i \frac{\partial(\mu_i - \mu_i^\circ)}{\partial n_k}$$

Since the Gibbs–Duhem equation requires $\sum_i n_i \, d\mu_i = 0$, the sum is zero and we have

$$\left(\frac{\partial \Delta G_{\text{mix}}}{\partial n_k}\right)_{T, p, n_i \neq k} = \mu_k - \mu_k^\circ = RT \ln a_k \qquad (35.1)$$

For a long time it was thought that if there was no heat of mixing a mixture would behave ideally. However, even if the heat of mixing is zero, if there are large differences between the molar volumes of the two constituents the mixture will not be ideal.

By considering the number of arrangements of polymer and solvent molecules on a lattice, we can calculate the entropy of the mixture and from that the Gibbs energy (if we assume some value for the heat of mixing). A simplified two-dimensional model of a polymer molecule arranged on a lattice is shown in Fig. 35.5. We assume that a solvent molecule

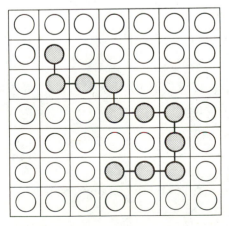

Figure 35.5 Lattice model (schematic, in two dimensions here) for polymer molecule in a solution. Sites not occupied by polymer segments are occupied by solvent molecules (one per site). (From T. L. Hill, *Introduction to Statistical Mechanics.* Reading, Mass.: Addison-Wesley, 1960.)

occupies one site while a polymer molecule occupies r sites. The calculation of the number of ways of arranging N_1 molecules of solvent and N molecules of polymer having r segments yields, after assuming that $r \gg 1$, a remarkably simple result for the Gibbs energy of mixing:

$$\Delta G_{\text{mix}} = RT(n_1 \ln \phi_1 + n \ln \phi). \tag{35.2}$$

In this equation, ϕ_1 and ϕ are the *volume fractions* of solvent and polymer, respectively; n_1 and n are the corresponding number of moles of solvent and polymer. If the solution were ideal, the expression for ΔG_{mix} would have been

$$\Delta G_{\text{mix}} = RT[n_1 \ln x_1 + n \ln x].$$

We find that replacing mole fraction by volume fraction in the logarithmic factors is sufficient to give us an equation that can begin to represent the behavior of a polymer solution.

To derive the expression for $\ln a_1$, we differentiate Eq. (35.2) with respect to n_1, using the relation in Eq. (35.1), and obtain, after dividing by RT,

$$\ln a_1 = \ln \phi_1 + n_1 \frac{\partial \ln \phi_1}{\partial n_1} + n \frac{\partial \ln \phi}{\partial n_1}. \tag{35.3}$$

The volume fractions are defined by

$$\phi_1 = \frac{n_1 \overline{V}_1^\circ}{n_1 \overline{V}_1^\circ + n \overline{V}^\circ} \quad \text{and} \quad \phi = \frac{n \overline{V}^\circ}{n_1 \overline{V}_1^\circ + n \overline{V}^\circ}, \tag{35.4}$$

in which $\overline{V}_1^\circ$ and $\overline{V}^\circ$ are the molar volumes of pure solvent and pure polymer. It is convenient to define $\rho \equiv \overline{V}^\circ / \overline{V}_1^\circ$, the ratio of the molar volumes. Then Eq. (35.4) reduces to

$$\phi_1 = \frac{n_1}{n_1 + n\rho} \quad \text{and} \quad \phi = \frac{n\rho}{n_1 + n\rho}. \tag{35.5}$$

Since $n_1 = n_t(1 - x)$ and $n = n_t x$, where x is the mole fraction of the polymer and n_t is the total number of moles, Eq. (35.5) can also be written as

$$\phi_1 = \frac{1 - x}{1 + (\rho - 1)x} \quad \text{and} \quad \phi = \frac{x\rho}{1 + (\rho - 1)x}. \tag{35.6}$$

When we use the expressions in Eq. (35.5) to evaluate the derivatives in Eq. (35.3), and keep in mind that $\phi_1 = 1 - \phi$, Eq. (35.3) becomes

$$\ln a_1 = \ln(1 - \phi) + \left(1 - \frac{1}{\rho}\right)\phi. \tag{35.7}$$

Since $\rho \gg 1$, then $1/\rho \ll 1$, and we can write

$$\ln a_1 = \ln(1 - \phi) + \phi \quad \text{or} \quad a_1 = (1 - \phi)e^\phi. \tag{35.8}$$

If we compare the solvent vapor pressure over the solution, p_1, with that over the pure solvent, p_1°, we have since $a_1 = p_1/p_1^\circ$,

$$\frac{p_1}{p_1^\circ} = (1 - \phi)e^\phi, \tag{35.9}$$

which is Flory's equation for the vapor pressure. Raoult's law for the vapor pressure is, if x

is the mole fraction of solute,

$$\frac{p_1}{p_1^\circ} = (1 - x). \tag{35.10}$$

We can rewrite this in terms of ϕ since, using Eq. (35.6), we find that $1 - x = (1 - \phi)/[1 - (1 - 1/\rho)\phi]$; thus, Eq. (35.10) becomes

$$\frac{p_1}{p_1^\circ} = \frac{1 - \phi}{1 - (1 - 1/\rho)\phi}. \tag{35.11}$$

The curves marked a, b, and c in Fig. 35.6 are plots of this function for different values of ρ. Note that for very large values of ρ (that is, as $\rho \to \infty$), Raoult's law predicts:

$$\frac{p_1}{p_1^\circ} = 1, \quad 0 \leq \phi < 1; \quad \text{and} \quad \frac{p_1}{p_1^\circ} = 0, \quad \phi = 1. \tag{35.12}$$

Figure 35.6 also shows the experimental data for the system polystyrene–toluene at three different temperatures. Note that there is not even approximate agreement with the Raoult's law predictions, neither at $\rho = 1$, which is not reasonable, nor at $\rho = 100$, which

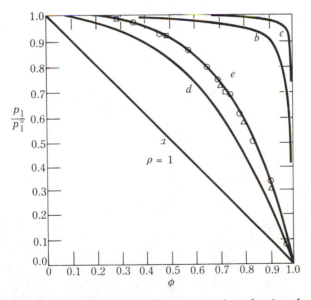

Figure 35.6   Dependence of p_1/p_1° on volume fraction of polymer. Curves a, b, and c are Raoult's law, Eq. (35.11), for $\rho = 1$, 100, and 1000, respectively. Curve d is Flory's equation, Eq. (35.9). Curve e is Eq. (35.14) with $w/kT = 0.38$. The experimental points are for the system polystyrene/toluene: $\bigcirc$, 25°C; $\triangle$, 60°C; $\square$, 80°C. (Adapted from E. A. Guggenheim, *Mixtures*. London: Oxford University Press, 1952. Data from Bawn, Freeman, and Kamaliddin, *Trans. Faraday Soc.* **46**: 677, 1950.)

still is not reasonable but at least is closer to reality. Flory's equation is substantially better, but is by no means perfect. If an adjustable parameter, w, is added to the equation, very close agreement with experiment can be obtained. This term can be added to Eq. (35.8) in the form,

$$\ln a_1 = \ln(1 - \phi) + \phi + \frac{w}{kT}\phi^2. \tag{35.13}$$

The parameter, w, represents the excess of the cohesive energy of the two pure liquids over that of the mixture. Then

$$\frac{p_1}{p_1^\circ} = (1 - \phi)e^{(\phi + w\phi^2/kT)}. \tag{35.14}$$

This is shown as curve e in Fig. 35.6 ($w/kT = 0.38$).

To obtain the expression for the osmotic pressure, we use Eq. (16.14). To conform to the notation of this chapter, we change a to a_1 and $\overline{V}^\circ$ to $\overline{V}_1^\circ$; then $\pi\overline{V}_1^\circ = -RT \ln a_1$. If we expand the logarithm on the right-hand side of Eq. (35.7) in terms of ϕ, and add the term $(w/RT)\phi^2$, we obtain

$$\ln a_1 = -\phi - \frac{1}{2}\phi^2 - \frac{1}{3}\phi^3 - \cdots + \left(1 - \frac{1}{\rho}\right)\phi + \frac{w}{RT}\phi^2,$$

$$\ln a_1 = -\frac{\phi}{\rho}\left[1 + \left(\frac{1}{2} - \frac{w}{RT}\right)\rho^2\left(\frac{\phi}{\rho}\right) + \frac{\rho^3}{3}\left(\frac{\phi}{\rho}\right)^2 + \cdots\right]. \tag{35.15}$$

Using this value for $\ln a_1$ in the expression for the osmotic pressure, we obtain

$$\pi = \frac{RT}{\overline{V}_1^\circ}\left(\frac{\phi}{\rho}\right)\left[1 + \left(\frac{1}{2} - \frac{w}{RT}\right)\rho^2\left(\frac{\phi}{\rho}\right) + \frac{\rho^3}{3}\left(\frac{\phi}{\rho}\right)^2 + \cdots\right]. \tag{35.16}$$

It is usual to express the concentrations of polymer in terms of mass per unit volume, $\tilde{c}_w$. If M is the molar mass of the polymer, we have

$$\tilde{c}_w = \frac{nM}{V}. \tag{35.17}$$

But, by Eq. (35.5), $\phi = n\overline{V}^\circ/V = n\rho\overline{V}_1^\circ/V$; thus, $n/V = \phi/\rho\overline{V}_1^\circ$, and

$$\tilde{c}_w = \frac{M\phi}{\overline{V}_1^\circ\rho} \quad \text{or} \quad \frac{\phi}{\rho} = \frac{\overline{V}_1^\circ\tilde{c}_w}{M}. \tag{35.18}$$

Using this value of ϕ/ρ in Eq. (35.16) yields

$$\pi = \frac{\tilde{c}_w RT}{M}\left[1 + \left(\frac{1}{2} - \frac{w}{kT}\right)\frac{\rho^2\overline{V}_1^\circ}{M}\tilde{c}_w + \frac{\rho^3\overline{V}_1^{\circ 2}}{3M^2}\tilde{c}_w^2 + \cdots\right]. \tag{35.19}$$

In general, we can write

$$\frac{\pi}{\tilde{c}_w} = \frac{RT}{M}(1 + \Gamma_2\tilde{c}_w + \Gamma_3\tilde{c}_w^2 + \cdots), \tag{35.20}$$

in which Γ_2 and Γ_3 are functions of temperature. Equation (35.20) is analogous to the equation for a nonideal gas. In practice the quadratic term is often negligible, and a plot of $(\pi/\tilde{c}_w)$ versus $\tilde{c}_w$ extrapolated linearly to $\tilde{c}_w = 0$ yields RT/M as the intercept (Fig. 35.7).

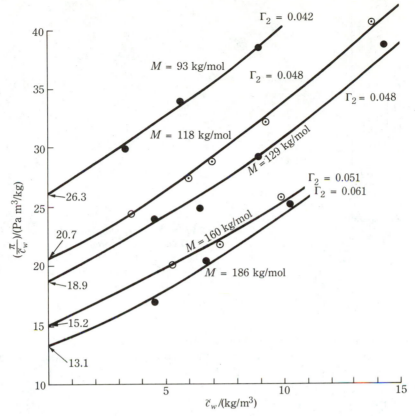

Figure 35.7 Plots of $\pi/\tilde{c}_w$ versus $\tilde{c}_w$ for polyvinyl acetates in benzene at 20°C. Data are from C. R. Masson and H. W. Melville, *J. Poly. Sci.*, **4**, 337 (1949). Curves are drawn using Eq. (35.20) with $\Gamma_3 = \frac{5}{3}\Gamma_2^2$; values for $(\pi/\tilde{c}_w)_0$ and Γ_2 were calculated from parameters given by T. G. Fox, Jr., P. J. Flory, and A. M. Bueche, *J.A.C.S.* **73**, 285 (1951). (Units for Γ_2 are m³/kg.)

From this intercept we obtain the value of M:

$$\left(\frac{\pi}{\tilde{c}_w}\right)_0 = \frac{RT}{M} \tag{35.21}$$

The measurement of colligative properties is one of the classical methods for determining the molar mass of solute. Although all of these properties have been used at one time or another to measure the molar mass of a polymer, only the osmotic effect is large enough to be generally useful.

■ **EXAMPLE 35.1**   If we choose a solution containing 5 g of polymer per litre then $\tilde{c}_w = 5$ kg/m³; $RT \approx 2500$ J/mol at 25 °C. If we assume that $M = 25$ kg/mol, then by Eq. (35.21)

$$\pi = \frac{(2500 \text{ J/mol})(5 \text{ kg/m}^3)}{25 \text{ kg/mol}} \approx 500 \text{ Pa} = 0.005 \text{ atm.}$$

This would correspond to about 4 mmHg or about 50 mm of water. If the solvent were less dense than water, the column of solvent would be higher.

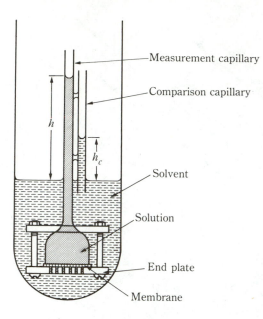

Measurement capillary

Comparison capillary

h

h_c

Solvent

Solution

End plate

Membrane

Figure 35.8 A simple osmometer; $\Delta h = h - h_c$ is the internal head corrected for capillary rise. (From D. P. Shoemaker, C. W. Garland, J. I. Steinfeld, J. W. Nibler, *Experiments in Physical Chemistry*, 4th ed. New York: McGraw-Hill, 1981.)

Figure 35.8 shows a simple osmometer that has a semipermeable membrane (a cellulose membrane is commonly used) clamped to the end of a wide cylinder from which a capillary tube extends. After it is filled, the lower part of the device is immersed in a container of solvent, which is itself immersed in a thermostat. Comparison of the liquid level in the capillary containing the solution with the level in the capillary immersed in the solvent yields the value of the osmotic pressure.

The measurements are often complicated by diffusion of the lower molar mass species through the membrane. As a consequence, the values of M obtained by osmometry may be substantially higher than those measured by other methods. We can show that the value of M obtained is the number-average molar mass, $\langle M \rangle_N$. (See Section 35.5 and Problem 35.6.)

For the boiling point elevation or freezing point depression, we have

$$\ln a_1 = \frac{\Delta H^\circ \theta}{RT_0^2}. \tag{35.22}$$

But, using Eq. (35.18) for ϕ, we can express Eq. (35.15) in the form

$$\ln a_1 = -\frac{\overline{V}_1^0}{M}\tilde{c}_w(1 + \Gamma_2\tilde{c}_w + \Gamma_2\tilde{c}_w^2 + \cdots).$$

Using the value for $\ln a_1$ from Eq. (35.22), we have

$$\frac{\theta}{\tilde{c}_w} = \frac{V_1^\circ RT_0^2}{M\Delta H^\circ}(1 + \Gamma_2\tilde{c}_w + \Gamma_3\tilde{c}_w^2 + \cdots). \tag{35.23}$$

The accuracy in the temperature measurements hardly justifies using the correction factor in the brackets. Nonetheless, a plot of $\theta/\tilde{c}_w$ versus $\tilde{c}_w$ yields an extrapolated value of $(V_1^\circ RT_0^2/M\Delta H^\circ)$, from which M can be calculated. Because the effects are very small, freezing point depression and boiling point elevation are not often used for molar mass determinations. In any event they cannot be used if M is greater than 10 kg/mol. For benzene, for example, if $M = 10$ kg/mol, then $\theta = 0.0031$ K for a concentration of 1 g/100 mL $= 10$ kg/m^3. The precision of measurement is only about ± 0.001 K.

35.5 MOLAR MASSES AND MOLAR MASS DISTRIBUTIONS

One of the important properties of any polymeric molecule is its molar mass. Furthermore, since the polymeric material does not consist of molecules of the same length, it is important to know the molar mass distribution. To illustrate the typical calculation of the distribution we choose a linear polymer that might be produced by the condensation of an hydroxy acid to produce a polyester. Suppose that we consider the monomer

$$HO-(CH_2)_n-COOH,$$

which we abbreviate to AB to symbolize the two functional end groups. Then, if we look at a polymer

$$\underset{1}{AB}-\underset{2}{AB}-AB-\cdots A-B,$$

the bond (—) indicates that the end-group B (a COOH group) is attached through an ester linkage to the end-group A (the OH) on another molecule. Then we ask what the probability is that the polymer contains k units. Let p be the probability that the end-group B is esterified, and let us assume that this probability does not depend on how many AB units are attached to the AB unit of interest. Then the probability of an ester linkage at position 1 is p, the probability of an ester linkage at position 2 is also p. The probability that both linkages are present is the product of the independent probabilities or p^2. If there are k units in the polymer, there are $k - 1$ ester linkages and the probability is p^{k-1}. However, the probability that end-group B is not linked is $1 - p$. Thus, if the molecule is to terminate after $k - 1$ links, the probability must be $p^{k-1}(1 - p)$. This probability must be equal to N_k/N, where N_k is the number of molecules that are k units long and N is the total number of molecules. Then the mole fraction, x_k, of $k$mers, is

$$x_k = \frac{N_k}{N} = p^{k-1}(1-p) = \frac{(1-p)p^k}{p}. \tag{35.24}$$

The average value of k is given by

$$\langle k \rangle = \frac{\sum\limits_{k=1}^{\infty} kN_k}{N}.$$

If we use N_k from Eq. (35.24), the expression becomes

$$\langle k \rangle = (1-p)\sum_{k=1}^{\infty} kp^{k-1}. \tag{35.25}$$

The series, $\Sigma_{k=0}^{\infty} p^k = 1 + p + p^2 + \cdots$, is the series expansion of $1/(1 - p)$. Thus

$$\sum_{k=0}^{\infty} p^k = \frac{1}{1-p}. \tag{35.26}$$

Differentiating both sides yields

$$\sum_{k=1}^{\infty} kp^{k-1} = (1-p)^{-2}. \tag{35.27}$$

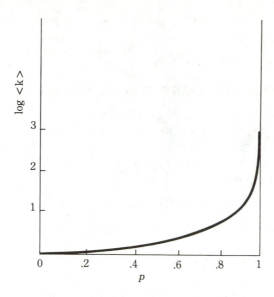

Figure 35.9 Log $\langle k \rangle$ versus p.

Using this result in Eq. (35.25), we obtain

$$\langle k \rangle = \frac{1}{1 - p}. \tag{35.28}$$

The higher the value of p, the probability of the link, the smaller is the value of $1 - p$ and thus the greater is the value of $\langle k \rangle$. If $\langle k \rangle = 50$, then $p = 1 - 1/50 = 0.98$; if $\langle k \rangle = 100$, then $p = 0.99$. It is clear that high degrees of polymerization will exist only when the probability of linkage is very near unity (Fig. 35.9). Even with $p = 0.90$, $\langle k \rangle$ is only 10.

To calculate the total number of monomer units, N_1, present in all the species we multiply N_k by k; thus

$$N_1 = \sum_{k=1}^{\infty} k N_k = N\langle k \rangle = \frac{N}{1 - p}.$$

In terms of monomer units present, since $N_k = Np^{k-1}(1 - p)$, we have

$$N_k = N_1 p^{k-1}(1 - p)^2. \tag{35.29}$$

The molar mass of a $k$mer is

$$M_k = kM_1 + M_e, \tag{35.30}$$

in which M_1 is the molar mass of the repeating unit and M_e is the excess mass due to the presence of the end groups. When k is large, M_e may be neglected. The number-average molar mass, $\langle M \rangle_N$, is defined as

$$\langle M \rangle_N = \frac{\sum_{k=1}^{\infty} N_k M_k}{N} = \frac{\sum_{k=1}^{\infty} N_k k M_1 + M_e \sum_{k=1}^{\infty} N_k}{N};$$

but $\Sigma N_k = N$ and $\Sigma k N_k = \langle k \rangle N$, so we obtain

$$\langle M \rangle_N = \langle k \rangle M_1 + M_e = \frac{M_1}{1 - p} + M_e. \tag{35.32}$$

The total mass of this system is given by

$$\sum_{k=1}^{\infty} \frac{N_k M_k}{N_A} = \frac{N\langle M\rangle_N}{N_A} = \frac{N_1(1-p)\langle M\rangle_N}{N_A},$$

so that the mass fraction of molecules having k units is

$$w_k = \frac{N_k M_k}{\displaystyle\sum_{k=1}^{\infty} N_k M_k} = \frac{N_1 p^{k-1}(1-p)^2(kM_1 + M_e)}{N_1(1-p)\langle M\rangle_N}$$

$$= \frac{p^{k-1}(1-p)^2(kM_1 + M_e)}{M_1 + (1-p)M_e}.$$

Neglecting M_e, which is negligible except when $\langle k\rangle$ is very small, we have

$$w_k = kp^{k-1}(1-p)^2. \tag{35.33}$$

The mass fraction as a function of k is shown for several values of p in Fig. 35.10. The mass-average molar mass is defined by

$$\langle M\rangle_w = \sum_{k=1}^{\infty} w_k M_k. \tag{35.34}$$

Using Eq. (35.33) for w_k, Eq. (35.30) for M_k, and neglecting M_e, we obtain

$$\langle M\rangle_w = M_1(1-p)^2 \sum_{k=1}^{\infty} k^2 p^{k-1}.$$

Equation (35.27) can then be written as $\Sigma\, kp^k = p/(1-p)^2$. Differentiating with respect

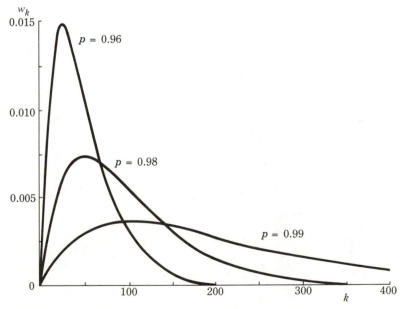

Figure 35.10 Mass fraction distributions calculated from Eq. (35.33) for several values of p.

to p, we get

$$\sum_{k=1}^{\infty} k^2 p^{k-1} = \frac{1+p}{(1-p)^3}.$$ (35.35)

Introducing this result in the expression for $\langle M \rangle_w$, we find

$$\langle M \rangle_w = \frac{(1+p)M_1}{(1-p)}.$$ (35.36)

From this we can calculate the ratio

$$\frac{\langle M \rangle_w}{\langle M \rangle_N} = 1 + p.$$ (35.37)

Since $p \approx 1$, we conclude that the mass-average molar mass is twice the number-average molar mass.

Determination of the molar mass distribution in a polymer sample was formerly a tedious task. For example, consider a solution of polymer in a solvent. If a precipitant that decreases the polymer's solubility is added, some polymer precipitates out; the material having the highest molar mass precipitates first. Removal of the precipitated polymer and addition of more precipitant causes another precipitate to be formed. In this way, the dissolved polymer is separated into fractions, whose molar masses can then be determined. A comparison of the mass fraction versus size determined in this manner agrees reasonably well with the hypothesis underlying Eq. (35.33): that the probability of forming an ester linkage does not depend on how many links have already been formed. Figure 35.11 shows a molar mass distribution so determined, compared to the calculation from Eq. (35.33).

The molar mass distribution can also be determined by gel permeation chromatography. In this method a solution of a polymer is forced through a gel that contains a

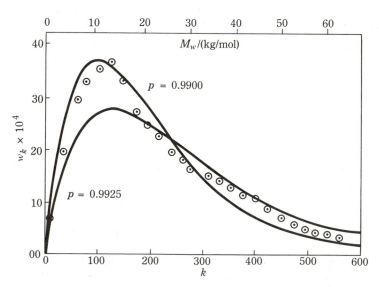

Figure 35.11   Molar mass distribution in nylon. Curves are calculated from Eq. (35.33). Data are from G. B. Taylor, *J.A.C.S.* **69**, 639 (1947).

network of pores of various sizes. Smaller molecules diffuse into the network with greater ease than larger molecules. Consequently, the larger molecules pass through the column more quickly than do the smaller ones, which become trapped in the network and require more time to become disentangled. This method is considerably more convenient than the classical precipitation method described.

The molar mass distribution can also be determined by ultracentrifugation, which we will describe later.

35.6 METHODS OF MEASURING MOLAR MASSES

35.6.1 End-Group Analysis

In addition to the measurement of colligative properties for the determination of the molar mass of a polymer discussed in Section 35.4, end-group analysis is a classical method that can be used for low-molar-mass polymers. If the molar mass is not too high ($< \sim 25\,000$), and if the molecule has reactive end groups that can be combined with another reagent, the average molar mass can be determined chemically. For example, suppose that the end group is a carboxyl group acidic enough to be titrated with NaOH. If n_k moles of NaOH is needed to titrate the $k$mer, then the total moles of NaOH required to titrate the mixture will be n_b.

$$n_b = \sum_{k=1}^{\infty} \frac{N_k}{N_A} = \frac{N}{N_A},$$

since each mole of polymer requires one mole of NaOH. The total mass of polymer in the sample is

$$w = \sum_{k=1}^{\infty} \frac{N_k M_k}{N_A} = \frac{N}{N_A} \langle M \rangle_N$$

and the number-average molar mass is

$$\langle M \rangle_N = \frac{w}{n_b}, \tag{35.38}$$

where both w and n_b are measured quantities.

★ 35.6.2 Light Scattering

If a beam of light passes through a solution of suspended particles the light is scattered; this phenomenon is called Rayleigh scattering. From the theory of scattering we can calculate the molar mass of the particle and obtain some idea about its shape. First, we consider the scattering by a particle that is small compared to the wavelength of the light. A diagram of the experiment is shown in Fig. 35.12. We pass a beam of monochromatic light into the cell and measure the intensity of the scattered light as a function of the scattering angle and as a function of the wavelength. The filter selects the wavelength to be used; the detector can be moved through an angle centered on the 90° position so that any dissymmetry can be measured. The scattering for large particles is not symmetric about the 90° angle.

Whenever an electrical charge is accelerated, it radiates energy. If the oscillating electric field of a light beam acts on a charge, the charge oscillates, is accelerated, and

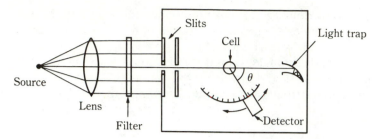

Figure 35.12   Light scattering apparatus (schematic).

radiates a light beam of the same frequency. This oscillating charge is equivalent to an oscillating dipole moment, a charge q displaced through a distance x.

Consider a charge q at the origin of the coordinate system in Fig. 35.13. The incident light beam moves from left to right along the z-axis; we assume that it is polarized and that its electric field vector oscillates in the x direction. We observe at point P, looking along the line from P to the origin, O. The component of the oscillating dipole moment in the plane perpendicular to the line of sight produces the oscillating electric field at P. This field, E_s, is given by

$$E_s = - \frac{qa \sin \beta}{4\pi\epsilon_0 c^2 r},\tag{35.39}$$

in which q is the charge that is accelerated, a is the magnitude of the acceleration vector, $\sin \beta$ is the component of a on the plane perpendicular to the line of sight, c is the velocity of light, r is the distance of P from the origin, and β is the angle between the x-axis and OP. The electric field vector of the scattered beam oscillates in the plane containing the x-axis and the line of sight, OP. If the incident electric field displaces the charge q by a distance x, the induced dipole moment, μ, is

$$\mu = qx = \alpha E_x,\tag{35.40}$$

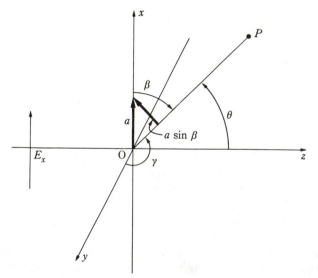

Figure 35.13   Geometry for light scattering derivation.

in which α is the polarizability of the molecule. Then

$$x = \frac{\alpha}{q} E_x. \tag{35.41}$$

If we write

$$E_x = E_x^0 \cos \omega t, \tag{35.42}$$

in which ω is the angular frequency, then using Eq. (35.41) we have

$$x = \frac{\alpha}{q} E_x^0 \cos \omega t. \tag{35.43}$$

The acceleration is given by

$$a = \frac{d^2 x}{dt^2} = \frac{d^2}{dt^2} \left(\frac{\alpha}{q} E_x^0 \cos \omega t \right) = -\frac{\alpha}{q} \omega^2 E_x^0 \cos \omega t.$$

Then

$$a = -\frac{\alpha \omega^2 E_x}{q}. \tag{35.44}$$

Using this value for a in Eq. (35.39) yields

$$E_s = \frac{\alpha \omega^2 E_x \sin \beta}{4\pi \epsilon_0 c^2 r} = \frac{\pi \alpha E_x \sin \beta}{\epsilon_0 r \lambda^2}. \tag{35.45}$$

In the second writing we set $\omega/c = 2\pi v/c = 2\pi/\lambda$.

The rate at which energy passes in unit time through unit area perpendicular to the direction of propagation is the power passing through unit area (watts/m²). This is the intensity of the beam. The intensity is proportional to the average of the square of the amplitude of the oscillating electrical field vector according to the equation

$$I_x = \epsilon_0 c \langle E_x^2 \rangle \tag{34.46}$$

and

$$I_\beta = \epsilon_0 c \langle E_s^2 \rangle = \epsilon_0 c \left(\frac{\pi \alpha \sin \beta}{\epsilon_0 r \lambda^2} \right)^2 \langle E_x^2 \rangle, \tag{35.47}$$

in which I_x is the intensity of the incoming beam, I_β is the intensity scattered along the ray OP. The quantity $\langle E_x^2 \rangle$ is the average of the square of the electric field vector. In the second writing of Eq. (35.47) we have replaced the value of E_s by its equal from Eq. (35.45). The only quantity of the right-hand side of Eq. (35.47) subject to averaging is E_x. Combining these last two equations yields

$$I_\beta = I_x \left(\frac{\pi \alpha}{\epsilon_0 r \lambda^2} \right)^2 \sin^2 \beta. \tag{35.48}$$

If the incoming light beam were not polarized but also had a component E_y in the y direction, there would be an additional contribution to the scattered intensity along OP given by

$$I_\gamma = I_y \left(\frac{\pi \alpha}{\epsilon_0 r \lambda^2} \right) \sin^2 \gamma, \tag{35.49}$$

in which γ is the angle between the y-axis and the ray OP. The total scattered intensity, I_θ, is the sum

$$I_\theta = I_\beta + I_\gamma \tag{35.50}$$

and

$$I_\theta = \left(\frac{\pi\alpha}{\epsilon_0 r\lambda^2}\right)^2 (I_x \sin^2 \beta + I_y \sin^2 \gamma). \tag{35.51}$$

The intensity of the incident beam is $I_0 = I_x + I_y$. If $I_x = I_y$, then $I_0 = 2I_x$. When we use these relations, the equation for I_θ reduces to

$$I_\theta = \tfrac{1}{2}I_0 \left(\frac{\pi\alpha}{\epsilon_0 r\lambda^2}\right)^2 (\sin^2 \beta + \sin^2 \gamma). \tag{35.52}$$

The angles β, γ, and θ are the angles between the ray OP and the x-, y-, and z-axis, respectively. The cosines of these angles are the direction cosines of the line OP and satisfy the relation

$$\cos^2 \beta + \cos^2 \gamma + \cos^2 \theta = 1.$$

Using the trigonometric identities, $\cos^2 \beta = 1 - \sin^2 \beta$ and $\cos^2 \gamma = 1 - \sin^2 \gamma$ in this equation and rearranging, we obtain

$$\sin^2 \beta + \sin^2 \gamma = 1 + \cos^2 \theta.$$

Using this result in Eq. (35.52), we have

$$I_\theta = \tfrac{1}{2}I_0 \left(\frac{\pi\alpha}{\epsilon_0 r\lambda^2}\right)^2 (1 + \cos^2 \theta). \tag{35.53}$$

This is the flux through unit area of surface at the end of the ray OP that lies at an angle θ from the z-axis. If $d\Omega$ is the differential solid angle, then the area element is $dA = r^2\,d\Omega$ (Fig. 35.14), and the total radiation through dA is

$$I_\theta\,dA = r^2 I_\theta\,d\Omega.$$

Dividing by $d\Omega$ we obtain $r^2 I_\theta$, which is the total flux through unit solid angle:

$$r^2 I_\theta = \tfrac{1}{2}I_0 \left(\frac{\pi\alpha}{\epsilon_0 \lambda^2}\right)^2 (1 + \cos^2 \theta). \tag{35.54}$$

It is important to note that this quantity does not depend on the distance of P from the

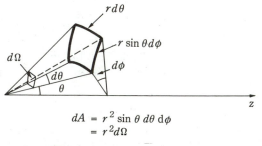

$$dA = r^2 \sin\theta\,d\theta\,d\phi$$
$$= r^2 d\Omega$$

Figure 35.14   Geometry for area of spherical cap.

scattering center. This expression is appropriate for the scattering of one molecule. We multiply this by $N/V = \tilde{N}$, the number of molecules in unit volume, to obtain the scattering from unit volume of the system. The Rayleigh ratio is defined by

$$R_\theta = \frac{r^2 I_\theta \tilde{N}}{I_0} \tag{35.55}$$

Using Eq. (35.54) we obtain

$$R_\theta = \frac{1}{2} \left(\frac{\pi \alpha}{\epsilon_0}\right)^2 \frac{\tilde{N}}{\lambda^4} (1 + \cos^2 \theta) \tag{35.56}$$

The Rayleigh ratio is the fraction of the incident energy that is scattered from unit volume of solution per unit solid angle at an angle θ to the direction of propagation.

Note that that fraction scattered is inversely proportional to the fourth power of the wavelength. This means that the intensity of scattered blue light (short wavelength) is much higher than that of scattered red light (long wavelength). This accounts for the blue sky during the day and a red sunset in the evening (particularly if there is some, but not too much, fine dust in the air). At sunset we look at the sun through a long section of the atmosphere; the reduction in intensity of the blue light due to scattering is much greater (roughly 16 times greater) than that of the red light; the red comes through, and the sun appears to be red. In looking at a portion of the sky at an angle to the sun we see the scattered light, which is more blue than red.

In the light scattering experiment, by measuring the intensity of the incident beam, I_0, and the scattered beam, I_θ, we obtain the Rayleigh ratio. Before this equation can be used to determine the molar mass we must have a relation between α and the molar mass. We obtain it by using Eq. (26.14), but we replace ϵ_r by n^2, the square of the refractive index and note that the polarizability is the sum of two contributions, $\tilde{N}_0 \alpha_0$ from the solvent and $\tilde{N}\alpha$ from the solute; that is, we rewrite Eq. (26.14) in the form

$$\frac{n^2 - 1}{n^2 + 2} = \frac{\tilde{N}_0 \alpha_0 + \tilde{N}\alpha}{3\epsilon_0}. \tag{35.57}$$

Ordinarily $n^2 \approx 1$, so we can set $n^2 + 2 \approx 3$; then

$$n^2 = 1 + \frac{\tilde{N}_0 \alpha_0 + \tilde{N}\alpha}{\epsilon_0}. \tag{35.58}$$

Differentiating with respect to $\tilde{c}_w$ where $\tilde{N} = N_A \tilde{c}_w / M$, we find

$$2n \frac{dn}{d\tilde{c}_w} = \frac{\alpha}{\epsilon_0} \frac{d\tilde{N}}{d\tilde{c}_w} = \frac{\alpha N_A}{\epsilon_0 M}.$$

In the limit as $\tilde{c}_w = 0$, $n = n_0$, the refractive index of the solvent, and

$$\frac{\alpha}{\epsilon_0} = 2n_0 \left(\frac{dn}{d\tilde{c}_w}\right)_0 \frac{M}{N_A} \tag{35.59}$$

We have taken the limiting slope of n versus $\tilde{c}_w$ since $\tilde{c}_w$ is very small. Using this result in the Rayleigh ratio, Eq. (35.56), we find

$$R_\theta = 2\pi^2 n_0^2 \left(\frac{dn}{d\tilde{c}_w}\right)_0^2 \frac{M \tilde{c}_w}{N_A \lambda^4} (1 + \cos^2 \theta). \tag{35.60}$$

We define a scattering constant, K, as

$$K \equiv \frac{2\pi^2 n_0^2}{N_A \lambda^4} \left(\frac{dn}{d\tilde{c}_w}\right)_0^2 \tag{35.61}$$

and Eq. 35.60 becomes

$$R_\theta = KM\tilde{c}_w(1 + \cos^2 \theta). \tag{35.62}$$

Since the molar mass is not uniform we should replace $M\tilde{c}_w$ by $\Sigma M_k \tilde{c}_k$. But, since $\tilde{c}_k = w_k \tilde{c}_w$, we find $\Sigma M_k \tilde{c}_k = \Sigma M_k w_k \tilde{c}_w$. Since $\langle M \rangle_w = \Sigma w_k M_k$, it follows that $\Sigma M_k \tilde{c}_k = \langle M \rangle_w \tilde{c}_w$. Consequently, it is clear that this method measures the mass-average molar mass. The constant K consists of factors that are known or measurable, such as n_0, λ, and $(dn/d\tilde{c}_w)_0$.

If the polymer particles are large enough to be comparable to the wavelength of the light, the scattering no longer is symmetric around $\theta = 90°$, and a factor $P(\theta)$ must be introduced. This factor accounts for the interference of the light scattered from different parts of the large molecule, an effect which depends on the shape of the molecule. Some conclusions about the shape of the molecule can be reached, if the measurements are made with polarized light and the change in the polarization of the scattered beam is measured.

Another use of the scattering formula is in the calculation of the turbidity of a solution. The method is similar to that of measuring the absorbance of a solution. The intensity of light is measured after it is passed through a dilute solution of a polymeric material; the wavelength used must not be absorbed by any component of the solution. Thus the decrease in intensity is due entirely to the light being scattered away (Fig. 35.15).

The turbidity, τ, is defined by

$$\tau = \lim_{\Delta x = 0} \frac{-\Delta I}{I \Delta x} = -\frac{dI}{I dx} \tag{35.63}$$

Integration of this equation yields

$$I = I_0 e^{-\tau x}, \tag{35.64}$$

an equation which is analogous to that involving the absorbance of a solution. The scattering effect is so small that it is not possible, as a practical matter, to measure the diminution in intensity of the transmitted beam. However, we can relate the turbidity to the Rayleigh ratio and measure it by measuring the intensity of the scattered light at an angle such as 90°.

Consider the cubical volume element (enclosed by the dashed lines in Fig. 35.15) having an edge length Δx and a volume, $(\Delta x)^3$. The beam entering the left-hand face and that leaving the right-hand face have intensities I and $I + \Delta I$, respectively. Since the intensity is the energy flow per second per square metre (watts/metre^2) the total energy

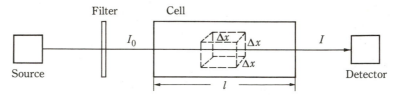

Figure 35.15   Apparatus for measuring turbidity.

fluxes into and out of the cubical element are the intensities multiplied by the area of the face.

$$\text{energy flux in} = I(\Delta x)^2 \tag{35.65}$$

$$\text{energy flux out} = (I + \Delta I)(\Delta x)^2 \tag{35.66}$$

The energy scattered by the volume element per unit solid angle is IR_θ multiplied by the volume

$$\text{energy scattered/steradian} = IR_\theta(\Delta x)^3 \tag{35.67}$$

Using R_θ from Eq. (35.62) this becomes

$$\text{energy scattered/steradian} = IKM\tilde{c}_w(1 + \cos^2 \theta)(\Delta x)^3$$

If we multiply this quantity by the element of solid angle, $d\Omega = \sin \theta \, d\theta \, d\phi$ steradians, and integrate over all values of θ and ϕ, we obtain the total scattered energy. The scattered energy is equal to the difference between the energy flux in and the energy flux out. Thus

$$I(\Delta x)^2 - (I + \Delta I)(\Delta x)^2 = IKM\tilde{c}_w(\Delta x)^3 \int_0^\pi (1 + \cos^2 \theta) \sin \theta \, d\theta \int_0^{2\pi} d\phi \tag{35.68}$$

The integral over θ is equal to $\frac{8}{3}$ and that over ϕ is equal to 2π. Then, after dividing by $I(\Delta x)^3$, we obtain

$$-\frac{\Delta I}{I \Delta x} = \frac{16}{3}\pi KM\tilde{c}_w \tag{35.69}$$

Comparing this with Eq. (35.63), we find

$$\tau = \frac{16}{3}\pi KM\tilde{c}_w, \tag{35.70}$$

which can also be written, in view of Eqs. (35.62) and (35.56),

$$\tau = \frac{8\pi^3}{3}\left(\frac{\alpha}{\epsilon_0}\right)^2 \frac{\tilde{N}}{\lambda^4} \tag{35.71}$$

or

$$\tau = \frac{16\pi R_\theta}{3(1 + \cos^2 \theta)} \tag{35.72}$$

Bear in mind that from a measurement of R_θ we can calculate τ, or from a measurement of τ we can obtain $R_\theta/(1 + \cos^2 \theta)$. Because the size of the particles may be comparable to the wavelength of light, we must use a "nonideal" equation such as

$$\frac{\tilde{c}_w}{\tau} = \left(\frac{\tilde{c}_w}{\tau}\right)_0 (1 + 2\Gamma_2\tilde{c}_w + \tfrac{3}{4}\Gamma_2^2\tilde{c}_w^2 + \cdots).$$

The basis for this equation is in the various interactions between the polymer molecules, which are similar to the interactions between the molecules of a nonideal gas. The value of τ is measured as a function of $\tilde{c}_w$ and $\tilde{c}_w/\tau$ plotted as a function of $\tilde{c}_w$. The intercept at $\tilde{c}_w = 0$ yields the value of the mass-average molar mass:

$$\left(\frac{\tilde{c}_w}{\tau}\right)_0 = \frac{3}{16\pi K\langle M\rangle_w}. \tag{35.73}$$

★ 35.6.3 Sedimentation; the Ultracentrifuge

Consider a particle of mass m falling in a gravity field through a viscous medium. The gravitational force is opposed by the inertial force, $ma = m(dv/dt)$, and the frictional force, which is proportional to the velocity of the particle. Balancing these forces we have

$$m\frac{dv}{dt} + fv = mg \qquad (35.74)$$

or

$$\frac{dv}{dt} = g - \frac{f}{m}v. \qquad (35.75)$$

If there were no friction, $dv/dt = g$, and the velocity, v, would increase linearly with time. If we now introduce the term fv, we find that this term increases as the velocity increases but that, in turn, decreases dv/dt. Ultimately v reaches a value large enough that the velocity no longer changes with time. This is the terminal velocity of the particle, v_e, and is obtained from the equation by setting $dv/dt = 0$. This behavior is shown in Fig. 35.16; we can write

$$v_e = \frac{mg}{f}. \qquad (35.76)$$

If we could measure v_e and f, we would have a method of determining m. In the ordinary gravity field v_e is far too small to be measurable, even for macromolecules. However, the sedimentation velocity of a macromolecule can be measured in the ultracentrifuge.

The ultracentrifuge has a rotor that can be driven at speeds of the order of 1000 revolutions per second. If a tube containing a solution of macromolecules is placed in this rotor, the centrifugal force, F, acting on the molecule is

$$F = \frac{mv^2}{r}, \qquad (35.77)$$

in which m is the mass and r is the distance of the molecule from the axis of rotation. If v is the number of revolutions per second, the angular velocity is $\omega = 2\pi v$ and the linear velocity is $v = \omega r$. Then the centrifugal force becomes

$$F = m\omega^2 r$$

If the specific volume (the volume per unit mass) of the molecule is $\bar{v}$, the volume displaced by the molecule is $m\bar{v}$ and the mass of the solvent displaced is $m\bar{v}\rho$, if ρ is the density of

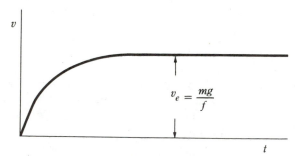

Figure 35.16 Velocity of a particle subject to frictional retardation.

the solvent. Thus the effective mass of the particle suspended in the solvent is $m - m\bar{v}\rho = m(1 - \bar{v}\rho)$. The centrifugal force acting on the particle is $F = m(1 - \bar{v}\rho)\omega^2 r$. Setting this equal to the frictional force, we obtain

$$m(1 - \bar{v}\rho)\omega^2 r = f\frac{dr}{dt}. \tag{35.78}$$

Since the particle moves in the radial direction, the velocity is dr/dt. If we assume that f is independent of r, we can write

$$\frac{1}{\omega^2 r}\frac{dr}{dt} = \frac{m(1 - \bar{v}\rho)}{f} = s. \tag{35.79}$$

In the first approximation we assume that f does not depend on the concentration. Then s is a constant called the sedimentation constant. The value of s is obtained by integrating Eq. (35.79) between (r_1, t_1) and (r_2, t_2). Then

$$\ln\frac{r_2}{r_1} = \omega^2 s(t_2 - t_1)$$

or

$$s = \frac{\ln(r_2/r_1)}{\omega^2(t_2 - t_1)}. \tag{35.80}$$

The cell in the rotor is sector shaped (Fig. 35.17). As the rotor spins, the macromolecule moves outward leaving the pure solvent behind. A moving boundary is established between the solution and the solvent. The position of the boundary can be detected by the difference in refractive index between the two parts of the cell. Using measurements of the positions r_1 and r_2 at times t_1 and t_2, we can calculate the value of s from Eq. (35.80). Alternatively, if $\ln r$ is plotted against t, we can obtain the value of s from the slope.

The molar mass can be obtained from s, using Eq. (35.79):

$$M = \frac{N_A f s}{(1 - \bar{v}\rho)}. \tag{35.81}$$

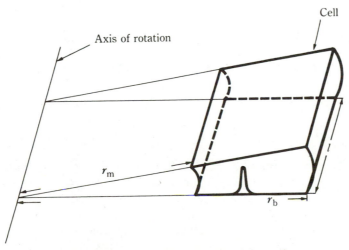

Figure 35.17 Sector-shaped cell for ultracentrifuge.

The difficulty is that the frictional coefficient, f, may not be known. If the particles are spherical, then by Stokes's law, $f = 6\pi\eta a$, in which η is the coefficient of viscosity of the solvent and a is the radius of the macromolecule. The problem then reduces to that of determining the radius of the macromolecule.

It is possible to establish a relation between the frictional coefficient, f, and the diffusion coefficient of the macromolecule. Consider a solution containing the macromolecules in equilibrium in a gravity field acting in the negative z direction. The molecules are moving downward with the terminal velocity, $v_e = mg/f$. By the general law of transport, Eq. (30.11), the number of molecules moving through unit area in unit time is the number per unit volume, $\tilde{N}$, times the velocity. Thus

$$j_{down} = \tilde{N}v_e = \frac{mg\tilde{N}}{f}.$$

When the system is in sedimentation equilibrium, this downward flow due to gravity is balanced by the upward diffusion flow given by Fick's law, Eq. (30.5),

$$j_{up} = -D\frac{\partial\tilde{N}}{\partial z},$$

in which D is the diffusion coefficient of the macromolecule. Setting the two flows equal, we obtain

$$\frac{mg\tilde{N}}{f} = -D\frac{\partial\tilde{N}}{\partial z}. \tag{35.82}$$

This equation can be integrated directly to yield

$$\tilde{N} = \tilde{N}_0 e^{-mgz/Df}.$$

But in a gravity field the Boltzmann distribution, Eq. (2.44), requires that

$$\tilde{N} = \tilde{N}_0 e^{-mgz/kT}.$$

Comparison of these two equations shows that

$$D = \frac{kT}{f}. \tag{35.83}$$

This is the Stokes–Einstein equation, which relates the diffusion coefficient and the frictional coefficient. Although we have derived this relation using a gravity field, it is correct for any conservative field. We derived this equation by a different route in Chapter 31; see Eq. (31.61).

If we assume that the particle is spherical, then by Stokes's law, $f = 6\pi\eta a$, and

$$D = \frac{kT}{6\pi\eta a}.$$

We can estimate a, the radius of the particle, from a knowledge of the diffusion coefficient. If the particle is not spherical, the frictional coefficient is larger than that given by the Stokes's law expression; the nonspherical particles exert a frictional effect larger than that exerted by an equivalent spherical particle.

Replacing f in Eq. (35.81) by the Stokes–Einstein value we obtain

$$M = \frac{RTs}{D(1 - \bar{v}\rho)}. \tag{35.84}$$

This is the Svedberg equation for the molar mass. If an independent value of D is available, the measurement of s suffices to determine M. Later we will show that the value of D/s can be obtained from an ultracentrifugation experiment.

If we consider the movement of particles in the centrifugal field through an element of the cell between positions r and $r + dr$, then we have for the flow through unit area in unit time in the radial direction at the position r:

$$j = v_r \tilde{N} - D \frac{\partial \tilde{N}}{\partial r}.$$

The velocity v_r is the radial velocity of the particle due to the centrifugal field (Fig. 35.17). Thus, by the general law of transport, Eq. (30.11), the flux is $v_r \tilde{N}$. By Eq. (35.79), $v_r = dr/dt = \omega^2 sr$, so that the expression for j becomes

$$j = \omega^2 sr\tilde{N} - D \frac{\partial \tilde{N}}{\partial r}. \tag{35.85}$$

The second term $-D(\partial N/\partial r)$ is the diffusion flow in the radial direction. The boundary conditions require that the flow be zero at the top of the solution (at the meniscus) and at the bottom. Setting $j = 0$ in Eq. (35.85), we have

$$\omega^2 sr_m \tilde{N}_m = D\left(\frac{\partial \tilde{N}}{\partial r}\right)_m \qquad \text{and} \qquad \omega^2 sr_b \tilde{N}_b = D\left(\frac{\partial \tilde{N}}{\partial r}\right)_b,$$

in which the subscripts m and b designate the values at the meniscus and at the bottom of the cell, respectively. Since we are interested in s/D, these equations should be written in the form

$$\frac{\omega^2 s}{D} = \frac{1}{r_m \tilde{N}_m}\left(\frac{\partial \tilde{N}}{\partial r}\right)_m = \frac{1}{r_b \tilde{N}_b}\left(\frac{\partial \tilde{N}}{\partial r}\right)_b. \tag{35.86}$$

From the usual optical patterns produced, both $\tilde{N}$ and $\partial \tilde{N}/\partial r$ can be obtained. Since r_m and r_b are easily measured, s/D can be calculated using Eq. (35.86). Then M can be obtained from the Svedberg equation. The conditions, Eq. (35.86), must be correct at all times; however, when the system reaches a steady state, that is when it is in sedimentation equilibrium, the flow must be zero everywhere so that

$$\omega^2 \frac{s}{D} = \frac{1}{r\tilde{N}} \frac{\partial \tilde{N}}{\partial r} \tag{35.87}$$

is correct for all values of r. This experiment may require two or three days to establish sedimentation equilibrium. At equilibrium, we can calculate s/D and, therefore, M, by integrating Eq. (35.87). Replacing $d\tilde{N}/\tilde{N}$ by dc/c, we obtain:

$$\int_{c_1}^{c_2} \frac{dc}{c} = \frac{\omega^2 s}{D} \int_{r_1}^{r_2} r\, dr,$$

which yields

$$\ln \frac{c_2}{c_1} = \frac{\omega^2 s}{2D}(r_2^2 - r_1^2) \tag{35.88}$$

or, using the Svedberg equation for s/D,

$$M = \frac{2RT}{\omega^2(1 - \bar{v}\rho)} \frac{\ln(c_2/c_1)}{(r_2^2 - r_1^2)}. \tag{35.89}$$

This equation yields the mass-average molar mass. We can derive it from thermodynamics exclusively, using the Boltzmann distribution for the centrifugal potential field. It can be shown that the ratio D/s depends on concentration, because of nonideality in the solutions, through a relation similar to Eq. (35.20).

$$\frac{D}{s} = \frac{RT}{M(1 - \bar{v}\rho)} (1 + 2\Gamma_2\tilde{c}_w + \tfrac{3}{4}\Gamma_2^2\tilde{c}_w^2). \tag{35.90}$$

If we plot D/s against $\tilde{N}$ and extrapolate to $\tilde{N} = 0$, we obtain the molar mass

$$M = \frac{RT}{\left(\dfrac{D}{s}\right)_0 (1 - \bar{v}\rho)}. \tag{35.91}$$

35.6.4 Viscosity Measurements

The presence of a polymer molecule in a solvent ordinarily results in a relatively large increase in viscosity of the solution even at low concentrations. This effect is strongly dependent on the concentration and on the solvent. In a poor solvent the long chain molecule is for the most part coiled upon itself; in a good solvent the molecule may be largely uncoiled and stretched out. Both configurations produce large changes in the viscosity of the solution. As the molar mass increases, the effect on the viscosity increases. The viscosity measurement is probably the simplest way to obtain a value for the molar mass of a macromolecule.

The viscosity of the solution and that of the pure solvent are measured in a precision capillary viscometer. The specific viscosity η_{sp} is defined as

$$\eta_{sp} = \frac{\eta_{\text{solution}} - \eta_{\text{solvent}}}{\eta_{\text{solvent}}} = \eta_r - 1, \tag{35.92}$$

in which the relative viscosity, η_r, is the ratio of the viscosity of the solution to the viscosity of the pure solvent. The specific viscosity is the relative contribution of the polymer to the viscosity of the solution.

If we divide η_{sp} by the concentration of polymer, $\tilde{c}_w$, we obtain a quantity that increases with concentration. Measurements of η are made as a function of concentration and $\eta_{sp}/\tilde{c}_w$ is plotted versus concentration. The curve is approximately linear and is extrapolated to $\tilde{c}_w = 0$ to obtain the intrinsic viscosity, $[\eta]$.

$$[\eta] = \lim_{\tilde{c}_w = 0} \left(\frac{\eta_{sp}}{\tilde{c}_w}\right). \tag{35.93}$$

Thus

$$\left(\frac{\eta_{sp}}{\tilde{c}_w}\right) = [\eta] + b\tilde{c}_w, \tag{35.94}$$

in which b is a constant. It is found empirically that the intrinsic viscosity depends on the molar mass according to the formula

$$[\eta] = KM^a \tag{35.95}$$

We determine the values of the empirical constants K and a by plotting $\log_{10}[\eta]$ versus $\log_{10} M$ for a given polymer–solvent pair, using measurements on fractions having known values of M. When we know K and a for the polymer–solvent pair, we can easily determine the molar mass of any sample of polymer from the measured value of the intrinsic viscosity. The unit commonly used for c_w in these equations is g/100 mL = g/dL.

QUESTIONS

35.1 Describe what is meant by primary, secondary, and tertiary structure of a polymer.

35.2 Why should the cohesive energy term in Eq. (35.14) depend on the *square* of ϕ?

35.3 Use Eq. (35.24) to argue why very high linkage probability is required for high degrees of polymerization.

35.4 Give a short qualitative argument for the λ^{-4} dependence of Rayleigh scattering by noting (a) the dependence of the scattered intensity on the scattered field and (b) the scattered field dependence on the charge acceleration induced by the incident field.

35.5 Diffusion is driven by a concentration gradient; sedimentation is driven by a centrifugal field. Why should the transport coefficient ratio s/D depend only on nontransport quantities?

35.6 Describe how the transition from a poor to a good solvent should affect the diffusion constant of a polymer.

PROBLEMS

35.1 If the number-average molar mass of a polystyrene sample is 45 kg/mol, what is the probability that a given monomeric unit will be an end group?

35.2 Consider a polymer mixture composed of 5 molecules of molar mass 1 kg/mol, 5 molecules of molar mass 2 kg/mol, 5 molecules of molar mass 3 kg/mol, and 5 molecules of molar mass 4 kg/mol.

a) Calculate the number-average molar mass.
b) Calculate the mass-average molar mass.

35.3 Using 0.903 g/cm^3 for the density of polystyrene, a molar mass of 50 kg/mol, and 0.867 g/cm^3 for the density of toluene, calculate the entropy of mixing based on Eq. (35.2) for values of $\phi = 0, 0.2, 0.4, 0.6, 0.8$, and 1.0. Sketch ΔS_{mix} versus ϕ. To what values of the mole fraction do these values of ϕ correspond?

35.4 Calculate the osmotic pressure of a solution of 1.5 g of bovine serum albumin in 500 mL of water at 25 °C; $M = 66.5$ kg/mol.

35.5 A solution of polystyrene in toluene at 25 °C exhibits the following osmotic pressures:

$c_w/(\text{g}/100\text{ cm}^3)$	0.155	0.256	0.293	0.380	0.538
π/Pa	16	28	32	46	76

Plot $\pi/\tilde{c}_w$ versus $\tilde{c}_w$ and determine the value of the average molar mass from the value of $\pi/\tilde{c}_w$ extrapolated to $\tilde{c}_w = 0$. Using data from Problem 35.3, calculate w.

35.6 In dilute solutions the osmotic pressure is given by $\pi = (\Sigma N_k/N_A)(RT/V)$ in which ΣN_k is the total number of molecules. Show that in view of Eq. (35.31), the definition of $\langle M \rangle_N$, that we can write $\pi = (wRT/V)/\langle M \rangle_N$ where w is the total mass and $\langle M \rangle_N$ is the number-average molar mass.

35.7 Calculate the relative vapor pressure of a solution consisting of 2 g of polystyrene ($M = 50$ kg/mol) and 100 g of toluene. The density of polystyrene is 0.903 g/cm^3 and that of toluene is 0.867 g/cm^3. Compare the predictions of Raoult's law, Flory's equation, and Flory's equation modified by the addition of the energy parameter, Eq. (35.14). Assume that $w/kT = 0.38$. Repeat the calculation for a solution having $\phi = 0.50$.

35.8 A 1.105 g sample of a polyester requires 10.80 mL of 0.01015 mol/L NaOH to titrate the end groups. What is the average molar mass of the polyester?

35.9 In a polyanhydride (a condensation polymer of a dicarboxylic acid) the end-group titration requires 36.72 mL of 0.008964 mol/L NaOH for 1.625 g of polymer. What is the average molar mass of the polymer?

35.10 The viscosities of polystyrene solutions in benzene at 25 °C are

$\tilde{c}_w/(\text{kg/m}^3)$	21.4	10.7	5.35
$\eta/\text{mPa s}$	1.35	0.932	0.757

Calculate the intrinsic viscosity and the molar mass of the polymer. For benzene, $\eta = 0.606$ mPa s. For this system the constants for $[\eta] = KM^a$ are $K = 1.71 \times 10^{-3}$ m^3/kg and $a = 0.74$.

35.11 The intrinsic viscosity of a solution of polyisobutylene in cyclohexane is 0.0248 m^3/kg. If $K = 3.3 \times 10^{-3}$ m^3/kg and $a = 0.70$, calculate the molar mass of the polymer.

35.12 A measurement of the intrinsic viscosities of polymer solutions containing fractions of polyisobutylene with known molar masses dissolved in diisobutylene yielded the data:

$M/(\text{kg/mol})$	6.2	10.4	124	856
$[\eta]/(\text{m}^3/\text{kg})$	0.00963	0.0134	0.0655	0.225

Calculate K and a for this polymer.

35.13 Sketch the fraction of the light scattered from a polymer solution as a function of θ.

35.14 Compare the scattered intensity of a light beam having $\lambda = 540$ nm and one for which $\lambda = 750$ nm in the same polymer–solvent system. Assume that $dn/d\tilde{c}_w$ does not depend on λ.

35.15 For a solution of polystyrene in methylethyl ketone, $(dn/dc_w)_0 = 0.220$ mL/g at 25 °C and a wavelength of 546 nm. The refractive index of methylethyl ketone is 1.377. Calculate the scattering constant for this solution and the molar mass of the polystyrene, if the Rayleigh ratio at 90° is 0.0942 m^{-1} for a solution containing 3.56 g of polystyrene in 100 mL of solution.

35.16 At 25 °C the refractive index of benzene is 1.498. A solution of polymethylmethacrylate in benzene has $(dn/dc_w)_0 = 0.0110$ mL/g at 25 °C, and a wavelength of 546 nm. Calculate the scattering constant and the Rayleigh ratio at 90° if the molar mass is 2050 kg/mol and the solution contains 12.47 g of polymer in 100 mL of solution.

35.17 A 1.22 g sample of polystyrene is dissolved in 100 mL of methylethyl ketone at 25 °C. The intensity of a 546 nm light beam is decreased to 0.9907 of the incident intensity in a turbidity measurement in which the cell length is 1.00 cm. Calculate the turbidity, τ, and the molar mass of the polymer. The scattering constant is 3.73×10^{-5} m^2 mol/kg^2.

35.18 At 25 °C for polystyrene in methylethyl ketone we have $(dn/dc_w)_0 = 0.231$ mL/g at $\lambda = 436$ nm. The refractive index of the methylethyl ketone is 1.377.

    a) Calculate the scattering constant for this solution and compare it with that in Problem 35.17.

    b) What would be the turbidity of the solution in Problem 35.17, if the wavelength used were $\lambda = 436$ nm instead of 546 nm?

35.19 The Rayleigh ratio at 45° is 0.141 m^{-1} for a solution of polystyrene in methylethyl ketone, at 25 °C and 546 nm wavelength.

    a) Calculate the decrease in intensity of a 546 nm light beam in a 1.00 cm cell.

    b) If the solution contains 2.14 g of polymer in 100 mL what is the molar mass of the polystyrene? (Use data from Problem 35.18.)

35.20 At 20 °C, the protein, γ-globulin, has a sedimentation constant of 7.75×10^{-13} s, a diffusion coefficient of 4.80×10^{-11} m^2 s^{-1} in water and a specific volume of 0.739 cm^3/g. The density of water is 0.998 g/cm^3.

    a) Calculate the molar mass of γ-globulin.

    b) Estimate the radius of the γ-globulin, assuming that it is spherical. For water, $\eta = 1.002 \times 10^{-3}$ Pa s at 20 °C.

35.21 If, at 20 °C, the sedimentation constant for a globular (roughly spherical) protein is 3.50×10^{-13} s,

    a) what is the radial velocity of the particle in a centrifuge turning at 50 000 rpm, at a distance of 6 cm from the axis?

    b) What length of time would it take for the particle to move from $r = 6.0$ cm to $r = 7.0$ cm?

    c) The specific volume of the protein is 0.731 cm^3/g and the density of water is 0.998 g/cm^3. What is the frictional coefficient, if $M = 42.62$ kg/mol.

    d) Calculate the diffusion coefficient.

35.22 In an ultracentrifuge rotating at 975 revolutions per second, the boundary moves from 6.187 cm to 6.297 cm in 150 minutes. Calculate the sedimentation constant.

35.23 In an ultracentrifugation experiment $D/s = 128$ m^2/s^2 is obtained for canine serum albumin at 20 °C. Calculate the molar mass, if the specific volume of this protein is 0.729 cm^3/g and the density of water is 0.998 g/cm^3.

35.24 At 20 °C, a polymer having a molar mass of 170 kg/mol and a specific volume of 0.773 cm^3/g is centrifuged in water, $\rho = 0.998$ g/cm^3. What is the relative concentration at $r = 7$ cm compared to $r = 6$ cm, if the centrifuge is turning at 60 000 revolutions per minute?

35.25 Compare the result in Problem 35.24 with the relative concentrations of the same polymer at heights of 7 cm and 6 cm in a gravity field having $g = 9.80$ m/s^2 ($t = 20$ °C).

Some Useful Mathematics

AI.1 FUNCTION AND DERIVATIVE

The symbol $f(x)$ signifies that f is a function of x. To say that f is a function of x means that if a value of x is chosen, this value determines a corresponding value of the function. The x is called the *independent variable*; f is called the *dependent variable*.

The volume of a given mass of liquid depends on the temperature. Translating this statement into mathematics, we say that the volume is a function of temperature, or simply write the symbol $V(t)$.

Since the value of f depends on the value of x, if the value of x changes, the value of f will change. Consequently, it is of interest to ask how rapidly f changes with a change in x. This information about the function is given by the derivative of the function with respect to x.

The derivative is the rate of change of the value of the function with change in the value of x. It should be noted that in general the derivative is also a function of x. To emphasize this aspect, the symbol $f'(x)$ is often used for the derivative, and we write

$$\frac{df}{dx} = f'(x).$$

If the derivative is positive, the value of the function increases with the value of x; if the derivative is negative, the value of the function decreases as x increases. If the derivative is zero, the curve of the function has a horizontal tangent; the function has a maximum or a minimum value. The fundamental definition of the derivative

$$\frac{df}{dx} = \lim_{\Delta x \to 0} \frac{\Delta f}{\Delta x}$$

leads to the geometric interpretation of the derivative as the slope of the tangent to the curve at any point.

If we ask what change in f accompanies the change in x from x_1 to x_2, the value of the derivative provides the answer. From the identity,

$$\frac{df}{dx} = \frac{df}{dx},$$

we can write

$$df = \frac{df}{dx}\,dx.$$

This equation says that the change in the value of f, df, is equal to the rate of change with respect to x, df/dx, multiplied by dx, the change in x. If a finite change is made in x from x_1 to x_2, then the total change in f is obtained by integration:

$$\int_{f_1}^{f_2} df = \int_{x_1}^{x_2} \frac{df}{dx}\,dx, \qquad f_2 - f_1 = \int_{x_1}^{x_2} f'(x)\,dx,$$

where f_2 and f_1 are the values of f corresponding to x_2 and x_1.

AI.2 THE INTEGRAL

a. The integral is the limit of a sum. In the preceding paragraph, the total change in f was found by adding together (that is, integrating) all of the small changes in the interval between x_1 and x_2.

b. The indefinite integral $\int g(x)\,dx$ always has a constant of integration associated with it. For example: Evaluate the integral $\int (1/x)\,dx$. A table of integrals gives $\ln x$ as the value of the integral; the integration constant C must be added to this, so we obtain

$$\int \frac{1}{x}\,dx = \ln x + C.$$

c. The definite integral $\int_a^b g(x)\,dx$ does not have a constant of integration associated with it. If from a table of integrals we find that

$$\int g(x)\,dx = G(x) + C, \qquad \text{then} \qquad \int_a^b g(x)\,dx = G(b) - G(a).$$

The definite integral is a function only of the limits a and b and of any parameters other than the variable of integration which may be contained in the integrand. For example, the integral $\int_a^b g(x, \alpha)\,dx$ is a function only of a, b, and α, *and is not a function of x.*

d. The integral of a function can be represented graphically as an area. The integral $\int_a^b g(x)\,dx$ is equal to the area included between the curve of the function $g(x)$ and the x-axis and between the lines $x = a$, and $x = b$.

AI.3 THE MEAN VALUE THEOREM

The mean value of any function of x over the interval, (a, b), is given by

$$\langle f \rangle = \frac{1}{b - a}\int_a^b f(x)\,dx.$$

AI.4 TAYLOR'S THEOREM

If we do not know the analytical form of a function, but know the values of its derivatives at some point, let us say at $x = 0$, then it is often possible to express the function as an infinite series. Assume that the function $f(x)$ can be expressed as a series:

$$f(x) = a_0 + a_1 x + \frac{a_2}{2!}x^2 + \frac{a_3}{3!}x^3 + \cdots.$$

By differentiating, we obtain

$$f'(x) = a_1 + a_2 x + \frac{a_3}{2!}x^2 + \cdots,$$

$$f''(x) = a_2 + a_3 x + \cdots,$$

$$f'''(x) = a_3 + \cdots.$$

At $x = 0$ these expressions reduce to

$$f(0) = a_0, \quad f'(0) = a_1, \quad f''(0) = a_2, \quad f'''(0) = a_3, \quad \cdots.$$

Thus, the values of the unknown coefficients in the series are expressed in terms of the values of the derivatives of the function at $x = 0$; the series becomes

$$f(x) = f(0) + f'(0) + \frac{f''(0)}{2!}x^2 + \frac{f'''(0)}{3!}x^3 + \cdots, \tag{AI.1}$$

which is Taylor's theorem. Not all functions can be expressed as a series in this way, but this expansion is often used for those functions that do behave properly. Usually, only the first two terms of the infinite series will be needed; the rest will be neglected. Some common and useful series are given in Appendix IV.

AI.5 FUNCTIONS OF MORE THAN ONE VARIABLE

We frequently use functions of two variables; the molar volume of a gas, for example, depends on temperature and pressure, $\overline{V} = \overline{V}(T,p)$. Written in this way, T and p are the *independent* variables, and $\overline{V}$ is the *dependent* variable. If we mentally assign a constant value to the pressure, the volume becomes a function of the temperature only. We calculate the derivative of this function in the same way we calculate the derivative of any function of one variable, but we write this derivative with a curly dee, ∂. Similarly, if we imagine that the temperature is constant, the volume becomes a function of the pressure only, and again we can form the derivative using the usual rules for a function of one variable. Thus the function, $\overline{V}(T, p)$, has two first derivatives. These are called partial derivatives and are written

$$\left(\frac{\partial \overline{V}}{\partial T}\right)_p \quad \text{and} \quad \left(\frac{\partial \overline{V}}{\partial p}\right)_T.$$

The subscript outside the parentheses in each of these symbols indicates the variable that is kept constant in the differentiation. If we ask how the molar volume changes if the temperature changes slightly at constant pressure, the answer is given by

$$d\overline{V} = \left(\frac{\partial \overline{V}}{\partial T}\right)_p dT.$$

The change in volume with change in pressure at constant temperature is given by

$$d\overline{V} = \left(\frac{\partial \overline{V}}{\partial p}\right)_T dp.$$

If *both* temperature and pressure change, then the total change in volume is the sum of the change due to temperature and the change due to pressure:

$$d\overline{V} = \left(\frac{\partial \overline{V}}{\partial T}\right)_p dT + \left(\frac{\partial \overline{V}}{\partial p}\right)_T dp.$$

This is called the *total differential* of the function. Any function of two variables $f(x, y)$ has a total differential df given by

$$df = \left(\frac{\partial f}{\partial x}\right)_y dx + \left(\frac{\partial f}{\partial y}\right)_x dy. \tag{AI.2}$$

AI.6 SOLUTIONS OF EQ. (4.27)

Equation (4.27) can be written in the form

$$Af(z) = f(x)f(y),$$

where $z = x + y$. We differentiate this equation with respect to x:

$$A\frac{df(z)}{dz}\left(\frac{\partial z}{\partial x}\right) = f'(x)f(y),$$

and then with respect to y:

$$A\frac{df(z)}{dz}\left(\frac{\partial z}{\partial y}\right) = f(x)f'(y).$$

But $\partial z/\partial x = \partial(x + y)/\partial x = 1$, and also $\partial z/\partial y = 1$, so these two equations become

$$Af'(z) = f'(x)f(y), \qquad Af'(z) = f(x)f'(y).$$

The left-hand sides are the same, so $f'(x)f(y) = f'(y)f(x)$. Dividing by $f(x)f(y)$, this becomes

$$\frac{f'(x)}{f(x)} = \frac{f'(y)}{f(y)}.$$

The left-hand side of this equation is apparently a function only of x, while the right-hand side does not depend on x but only on y. This is possible only if each side of the equation is a constant, β. Thus

$$\frac{f'(x)}{f(x)} = \beta \quad \text{and therefore} \quad \frac{df(x)}{f(x)} = \beta\,dx.$$

Integrating, we obtain $\ln f(x) = \beta x + \ln A$, and therefore, $f(x) = A \exp(\beta x)$, which is the solution of the functional equation, Eq. (4.27).

AI.7 THE METHOD OF LEAST SQUARES

It is often desirable to fit a set of data points or a set of derived quantities calculated from experimental data to a mathematical function containing parameters that can be adjusted so that the resulting curve is a "best" fit of the data. The least-squares method is a systematic way of determining the values of the parameters that results in a best fit of the data by a particular function.

In principle the method can be used for any kind of function; in practice we find that unless the function is a polynomial, the amount of numerical work required is prohibitive. For this reason we restrict our consideration here to polynomial functions.

Suppose that we have a set of N data points, (x_i, y_i), that we wish to approximate by a curve of the form

$$y = a + bx + cx^2 + \cdots. \tag{AI.3}$$

The deviation of the experimental value of y from the calculated value is d_i, where

$$d_i = y_i - (y_i)_{\text{calc}}. \tag{AI.4}$$

Since the calculated value of y is obtained from Eq. (AI.3), we have

$$d_i = y_i - (a + bx_i + cx_i^2 + \cdots). \tag{AI.5}$$

If the function is a good representation of the data and the errors are random, d_i will be negative as often as it is positive and the sum of the d_i over all the data points will be near zero.

$$\sum_i d_i \approx 0.$$

A better way to measure the closeness of fit is to square the d_i and then sum d_i^2. Now the positive and negative deviations do not cancel each other. This sum of d_i^2 is a quantity that indicates how closely the curve fits the data. We define σ^2, the variance, as

$$\sigma^2 = \frac{1}{N} \sum_{i=1}^{N} d_i^2 \tag{AI.6}$$

The smaller the value of σ^2, the better the curve fits the data. Since σ^2 depends on the constants: $a, b, c, \ldots$, we choose these constants so that σ^2 is minimized. Thus the value of the sum of squares is a least value; hence the name, least-squares method.

Using Eq. (AI.5) in Eq. (AI.6), we obtain

$$\sigma^2 = \frac{1}{N} \sum_{i=1}^{N} (y_i - a - bx_i - cx_i^2)^2. \tag{AI.7}$$

To minimize this expression, we differentiate σ^2 with respect to a, then with respect to b, and so on. (Note: It is understood in all that follows that all the sums are from $i = 1$ to $i = N$.)

$$\frac{\partial \sigma^2}{\partial a} = \frac{1}{N} \sum 2(y_i - a - bx_i - cx_i^2)(-1);$$

$$\frac{\partial \sigma^2}{\partial b} = \frac{1}{N} \sum 2(y_i - a - bx_i - cx_i^2)(-x_i);$$

$$\frac{\partial \sigma^2}{\partial c} = \frac{1}{N} \sum 2(y_i - a - bx_i - cx_i^2)(-x_i^2).$$

If we set each of these derivatives equal to zero, then divide by $(-2/N)$, the conditions for the minimum are

$$\sum (y_i - a - bx_i - cx_i^2) = 0;$$

$$\sum (y_i - a - bx_i - cx_i^2)x_i = 0; \qquad \text{(AI.8)}$$

$$\sum (y_i - a - bx_i - cx_i^2)x_i^2 = 0.$$

These equations can be solved explicitly for a, b, and c. We will do only the straight-line case here.

For the straight line, we set $c = 0$ and use only the first two equations, which can be written as

$$\sum y_i - a \sum 1 - b \sum x_i = 0$$

and

$$\sum x_i y_i - a \sum x_i - b \sum x_i^2 = 0.$$

Remembering that $\Sigma_{i=1}^{N} 1 = N$, we can solve these two equations for a and b to yield

$$a = \frac{(\sum x_i^2)(\sum y_i) - (\sum x_i)(\sum x_i y_i)}{N \sum x_i^2 - (\sum x_i)^2} \qquad \text{(AI.9)}$$

and

$$b = \frac{N \sum x_i y_i - (\sum x_i)(\sum y_i)}{N \sum x_i^2 - (\sum x_i)^2} \qquad \text{(AI.10)}$$

To use these equations we have to construct the quantities Σx_i; Σy_i; $\Sigma x_i y_i$; and Σx_i^2. Fortunately, many of the new hand calculators have a linear regression program built into them that calculates a and b from Eqs. (AI.9) and (AI.10) if we simply key in the data properly. Some calculators do not have the linear regression program; instead they have a statistical function that calculates these sums and stores them in accessible registers. In this case the sums can be recalled from the registers and used in Eqs. (AI.9) and (AI.10) to calculate a and b. (A lot of time can be saved by reading the instruction booklet for your calculator!)

These equations can also be written in terms of the average values of the various quantities $\Sigma x_i = N\langle x \rangle$; $\Sigma x_i^2 = N\langle x^2 \rangle$; and so on. Then

$$a = \frac{\langle x^2 \rangle \langle y \rangle - \langle x \rangle \langle xy \rangle}{\langle x^2 \rangle - \langle x \rangle^2} \qquad \text{(AI.11)}$$

and

$$b = \frac{\langle xy \rangle - \langle x \rangle \langle y \rangle}{\langle x^2 \rangle - \langle x \rangle^2}. \qquad \text{(AI.12)}$$

The probable errors in a and b are given by X_a and X_b.

$$X_a = r\sqrt{\frac{\langle x^2 \rangle}{N[\langle x^2 \rangle - \langle x \rangle^2]}}, \qquad X_b = r\sqrt{\frac{1}{N[\langle x^2 \rangle - \langle x \rangle^2]}},$$

where $r = 0.6745 \ldots \sqrt{N\sigma^2/(N-2)}$. The value of σ^2 can be obtained by using Eq. (AI.5) to calculate d_i, then squaring and summing, or from the relation

$$\sigma^2 = \langle y^2 \rangle - \langle y \rangle^2 - b^2(\langle x^2 \rangle - \langle x \rangle^2). \qquad \text{(AI.13)}$$

AI.8 VECTORS AND MATRICES

AI.8.1 Vectors

In three dimensions we describe a vector as a quantity having magnitude and direction. A vector can be written as a sum of three terms; each term is a scalar quantity multiplied by a unit vector that is in the direction of one of three mutually perpendicular axes. In cartesian coordinates, for example, the vector $\mathbf{a}$ is written

$$\mathbf{a} = a_x\mathbf{i} + a_y\mathbf{j} + a_z\mathbf{k}, \tag{AI.14}$$

in which $\mathbf{i}$, $\mathbf{j}$, and $\mathbf{k}$ are unit vectors in the x, y, and z directions respectively, and a_x, a_y, and a_z are the scalar quantities, the *components* of the vector.

The scalar product of two vectors, $\mathbf{a}$ and $\mathbf{b}$, is written $\mathbf{a} \cdot \mathbf{b}$ and is defined as

$$\mathbf{a} \cdot \mathbf{b} = a_xb_x + a_yb_y + a_zb_z. \tag{AI.15}$$

It is a scalar quantity, of course. From this definition, it follows that the scalar product of two orthogonal vectors (perpendicular vectors) is zero. For example, let $\mathbf{a} = a_x\mathbf{i} + (0)\mathbf{j} + (0)\mathbf{k}$, and $\mathbf{b} = (0)\mathbf{i} + b_y\mathbf{j} + (0)\mathbf{k}$. Then $\mathbf{a}$ is parallel to the x-axis and $\mathbf{b}$ is parallel to the y-axis; $\mathbf{a}$ and $\mathbf{b}$ are orthogonal to each other. Then by Eq. (AI.15) for the scalar product, we have

$$\mathbf{a} \cdot \mathbf{b} = a_x(0) + (0)b_y + (0)(0) = 0$$

The scalar product of a vector with itself is equal to the square of the length of the vector:

$$\mathbf{a} \cdot \mathbf{a} = a_x^2 + a_y^2 + a_z^2 = |a|^2. \tag{AI.16}$$

The length of the vector is

$$|a| = (a_x^2 + a_y^2 + a_z^2)^{1/2} \tag{AI.17}$$

These expressions are not limited to 3-dimensional vectors but can be written generally for n-dimensional vectors:

$$\mathbf{a} \cdot \mathbf{b} = \sum_{i=1}^{n} a_ib_i$$

$$|a| = \left(\sum_{i=1}^{n} a_i^2\right)^{1/2}. \tag{AI.18}$$

AI.8.2 Matrices

A matrix is an array of quantities, most commonly a rectangular or square array. For example, we might have

$$[a_x \quad a_y \quad a_z].$$

This matrix is a 1×3 matrix, having one row and three columns. It is also called a *row vector* because its elements can be thought of as the components of a three-dimensional vector. We could also write the components of a vector as a column,

$$\begin{bmatrix} b_x \\ b_y \\ b_z \end{bmatrix}$$

This is a 3×1 matrix, having three rows and one column. It is also called a *column vector*.

The matrix product of a row vector by a column vector has the same form as the scalar product of the two vectors.

$$[a_x \quad a_y \quad a_z] \begin{bmatrix} b_x \\ b_y \\ b_z \end{bmatrix} = [a_x b_x + a_y b_y + a_z b_z].$$

The result of this matrix multiplication is a single quantity (a scalar), which is the sole element in a 1×1 matrix.

Any matrix can be regarded as consisting of a set of row vectors and/or a set of column vectors; for example, consider the matrices $\mathbf{A}$ and $\mathbf{B}$,

$$\mathbf{A} = \begin{bmatrix} a_{11} & a_{12} & a_{13} \\ a_{21} & a_{22} & a_{23} \end{bmatrix} \qquad \mathbf{B} = \begin{bmatrix} b_{11} & b_{12} \\ b_{21} & b_{22} \\ b_{31} & b_{32} \end{bmatrix}.$$

The position of the element in the matrix is indicated by two subscripts: The first describes the row number; the second describes the column number.

The matrix $\mathbf{A}$ is made up of two row vectors or three (two-dimensional) column vectors, while $\mathbf{B}$ is made up of three row vectors or two column vectors. The matrix product of $\mathbf{A}$ and $\mathbf{B}$ is a matrix having elements that are the scalar products between the vectors which compose $\mathbf{A}$ and $\mathbf{B}$. Thus

$$\mathbf{AB} = \begin{bmatrix} a_{11} & a_{12} & a_{13} \\ a_{21} & a_{22} & a_{23} \end{bmatrix} \begin{bmatrix} b_{11} & b_{12} \\ b_{21} & b_{22} \\ b_{31} & b_{32} \end{bmatrix} = \begin{bmatrix} c_{11} & c_{12} \\ c_{21} & c_{22} \end{bmatrix}.$$

The scalar product of the first row vector of $\mathbf{A}$ with first column vector of $\mathbf{B}$ yields the element in the first row and first column of the product matrix, $\mathbf{C}$. Thus

$$c_{11} = a_{11} b_{11} + a_{12} b_{21} + a_{13} b_{31}.$$

Similarly, the product between the first row of $\mathbf{A}$ and the second column of $\mathbf{B}$ yields the element for the first row and the second column of the product matrix,

$$c_{12} = a_{11} b_{12} + a_{12} b_{22} + a_{13} b_{32}.$$

For the element in the ith row and kth column of the product matrix, we have

$$c_{ik} = \sum_{j=1}^{3} a_{ij} b_{jk}. \tag{AI.19}$$

Clearly, for this operation to make sense, the number of columns in $\mathbf{A}$ must be equal to the number of rows in $\mathbf{B}$; if this condition is not met, the product $\mathbf{AB}$ is not defined. Thus, if $\mathbf{A}$ is an $m \times n$ matrix (m rows, n columns) and $\mathbf{B}$ is a $p \times q$ matrix (p rows, q columns) the product, $\mathbf{AB}$, is defined only if $n = p$. The product matrix is an $m \times q$ matrix. In the same way, the product $\mathbf{BA}$ is defined only if $q = m$. The product matrix is a $p \times n$ matrix.

The character of a square matrix, $\chi(\mathbf{A})$, is the sum of the elements on the main diagonal,

$$\chi(\mathbf{A}) = \sum_{i=1}^{n} a_{ii} \tag{AI.20}$$

The character of a rectangular (not square) matrix is not defined.

AI.8.3 Symmetry Operations as Matrices

INTERCHANGES

In Section 23.15.2 we used matrices to represent the symmetry operations of the group. These matrices are quite easy to construct. Consider the operations in the group, C_{2v}. In Eq. (23.34) we summarized the effects of the operations on the coordinates of a point. For example, for the operator C_2, we have

$$C_2(x, y, z) = (-x, -y, z).$$

We can write this as a matrix multiplication:

$$\begin{bmatrix} a_{11} & a_{12} & a_{13} \\ a_{21} & a_{22} & a_{23} \\ a_{31} & a_{32} & a_{33} \end{bmatrix} \begin{bmatrix} x \\ y \\ z \end{bmatrix} = \begin{bmatrix} -x \\ -y \\ z \end{bmatrix}.$$

What elements must appear in the top row if x is to be replaced by $-x$? Writing the product of the top row vector with the column vector we obtain

$$a_{11}x + a_{12}y + a_{13}z = -x.$$

Thus $a_{11} = -1$; $a_{12} = 0$; $a_{13} = 0$. After doing one or two products by this method, we soon learn to do it by inspection. The complete matrix is

$$\begin{bmatrix} -1 & 0 & 0 \\ 0 & -1 & 0 \\ 0 & 0 & 1 \end{bmatrix} \begin{bmatrix} x \\ y \\ z \end{bmatrix} = \begin{bmatrix} -x \\ -y \\ z \end{bmatrix}$$

So the operator, C_2, can be represented by the matrix,

$$C_2 = \begin{bmatrix} -1 & 0 & 0 \\ 0 & -1 & 0 \\ 0 & 0 & 1 \end{bmatrix}.$$

In the same way,

$$\sigma_v \begin{bmatrix} x \\ y \\ z \end{bmatrix} = \begin{bmatrix} x \\ -y \\ z \end{bmatrix} \quad \text{so that} \quad \sigma_v = \begin{bmatrix} 1 & 0 & 0 \\ 0 & -1 & 0 \\ 0 & 0 & 1 \end{bmatrix}.$$

Since the group multiplication table requires that $C_2\sigma_v = \sigma'_v$, it must be that

$$\sigma'_v = \begin{bmatrix} -1 & 0 & 0 \\ 0 & -1 & 0 \\ 0 & 0 & 1 \end{bmatrix} \begin{bmatrix} 1 & 0 & 0 \\ 0 & -1 & 0 \\ 0 & 0 & 1 \end{bmatrix} = \begin{bmatrix} -1 & 0 & 0 \\ 0 & 1 & 0 \\ 0 & 0 & 1 \end{bmatrix}.$$

ROTATIONS

The matrix that describes the transformation of a two-dimensional vector under rotation through an angle ϕ can be developed by using the Argand diagram (Fig. AI.1). The point (x, y), when subjected to a counterclockwise rotation through the angle ϕ, is transformed into the point (x', y'). How are the coordinates (x', y') related to (x, y)? Since the length of the vector from the origin is the same in both cases, we can write

$$x + iy = re^{i\theta} \quad \text{and} \quad x' + iy' = re^{i(\theta + \phi)}.$$

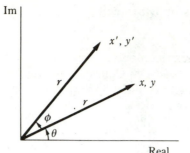

Real   **Figure AI-1**

Combining these to eliminate $re^{i\theta}$, we find that

$$x' + iy' = (x + iy)e^{i\phi} = (x + iy)(\cos \phi + i \sin \phi)$$
$$x' + iy' = x \cos \phi + y(-\sin \phi) + i(x \sin \phi + y \cos \phi).$$

Setting real and imaginary parts equal on both sides of the equation, we obtain the two relations

$$x' = x \cos \phi + y(-\sin \phi)$$

and

$$y' = x \sin \phi + y \cos \phi,$$

which in matrix notation becomes

$$\begin{bmatrix} x' \\ y' \end{bmatrix} = \begin{bmatrix} \cos \phi & -\sin \phi \\ \sin \phi & \cos \phi \end{bmatrix} \begin{bmatrix} x \\ y \end{bmatrix} = \mathbf{C}_\phi \begin{bmatrix} x \\ y \end{bmatrix}.$$

Thus for the operator, $\mathbf{C}_\phi$, counterclockwise rotation through the angle ϕ we have

$$\mathbf{C}_\phi = \begin{bmatrix} \cos \phi & -\sin \phi \\ \sin \phi & \cos \phi \end{bmatrix}.$$

This matrix was used in Section 23.15.2 for $\mathbf{C}_3$, for which $\phi = \frac{2}{3}\pi$.

APPENDIX II

Some Fundamentals of Electrostatics

AII.1 COULOMB'S LAW

Consider a charge, q, at the origin of the coordinate system and a charge, q', at a point P, which is a distance r from the origin. The force acting between the two charges in vacuum is given by Coulomb's law,

$$F = \frac{qq'}{4\pi\epsilon_0 r^2}.$$

(AII.1)

This says that the force is proportional to the product of the charges and inversely proportional to the square of the radius. The proportionality constant in the SI system is defined as $1/4\pi\epsilon_0$, where ϵ_0 is called the permittivity of vacuum. By definition,

$$\frac{1}{4\pi\epsilon_0} = c^2(10^{-7} \text{ N s}^2 \text{ C}^{-2}) \quad \text{(exactly)},$$

(AII.2)

in which $c = 2.99792458 \times 10^8$ m/s, the speed of light in a vacuum. Introducing this value for c, we find

$$\frac{1}{4\pi\epsilon_0} = 8.98755179 \times 10^9 \text{ newton meter}^2 \text{ coulomb}^{-2}$$

$$\approx 9 \times 10^9 \text{ N m}^2 \text{ C}^{-2}.$$

AII.2 THE ELECTRIC FIELD

The electric field, E, at any point is defined as the force acting on a unit positive charge at that point. In terms of the charges described in Section (AII.1), $E = F/q'$, or

$$E = \frac{q}{4\pi\epsilon_0 r^2}.$$

(AII.3)

The field due to a charge is directed radially: If q is positive, the field is positive and is directed outwardly along the radius; if q is negative, the field is directed inwardly toward the charge.

If several charges, q_j, are present, the field is the vector sum of the field produced at P by each of the charges. We can write this as

$$\mathbf{E} = \frac{1}{4\pi\epsilon_0} \sum_j \frac{q_j \mathbf{e}_j}{r_j^2}, \tag{AII.4}$$

where r_j is the distance between point P and the position of the charge, q_j, and $\mathbf{e}_j$ is a unit vector in the direction from the charge q_j to point P. This is the principle of superposition. (In this particular expression, we have emphasized the vector character of $\mathbf{E}$ by printing it in bold face.)

AII.3 THE ELECTRIC POTENTIAL

The electric potential, ϕ, at any point is the work required to move a unit positive charge from infinite distance to the point in question. Since E is the force acting on unit positive charge, we can write

$$\phi = \int_\infty^r E(-dr) = \int_\infty^r (-E)dr. \tag{AII.5}$$

For an example, we take E from Eq. (AII.3) and find

$$\phi = -\int_\infty^r \frac{q\,dr}{4\pi\epsilon_0 r^2} = \frac{q}{4\pi\epsilon_0 r} \tag{AII.6}$$

From Eq. (AII.5), it follows that $E = -\partial\phi/\partial r$. Thus, if we know the potential, we can obtain the radial component of the field by differentiation with respect to r. More generally, we can write

$$E_x = -\frac{\partial\phi}{\partial x}; \qquad E_y = -\frac{\partial\phi}{\partial y}; \qquad E_z = -\frac{\partial\phi}{\partial z} \tag{AII.7}$$

The components of the field are obtained by differentiating the potential with respect to the coordinates.

By the principle of superposition, the potential at any point is the sum of the potentials produced by all the charges. Thus

$$\phi = \sum_j \frac{q_j}{4\pi\epsilon_0 r_j}. \tag{AII.8}$$

Since this formula involves only the addition of scalar quantities, it is easier to construct than the sum in Eq. (AII.4). After the potential has been calculated by this formula, the field component in any desired direction can be calculated by differentiating with respect to the coordinate; use Eq. (AII.7).

AII.4 THE FLUX

We divert our attention temporarily to consider the flow of an incompressible fluid having a velocity v through a surface, S (Fig. AII.1a). To begin, we assume that the velocity vector is perpendicular to the surface. Then in a time interval Δt, all the fluid in the cylinder of

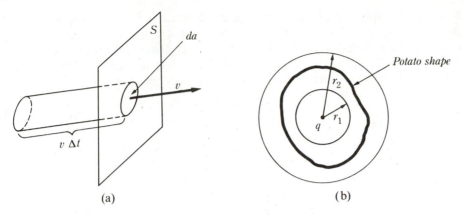

Figure AII.1

length $v\,\Delta t$, will pass through the element of surface, da. The flow through the element da is $v\,\Delta t\,da$. The flux is defined as the flow per unit time, so the flux is $v\,da$. Clearly, if the flow were tangential to the surface, no fluid would pass *through* the surface. Thus, in the general case for which the velocity vector is not perpendicular to the surface, the normal component of the velocity, v_n, determines the flux through the surface. The total flux is obtained by integrating the product $v_n\,da$ over the entire surface. (The integration over the surface is symbolized by an S written adjacent to the integral sign.) Thus we obtain for the flux,

$$\text{Flux} = \int_S v_n\,da. \tag{AII.9}$$

We can define the flux in any vector field in the same way. For the electric field vector we can write for the flux of E,

$$\text{Flux} = \int_S E_n\,da \tag{AII.10}$$

Suppose that we wish to calculate the flux through the surface of a sphere of radius r, which has a charge q at its center. The field at the surface of the sphere is given by Eq. (AII.3). Since the field is directed radially, $E_n = E$. Also the field is constant on the surface of the sphere; consequently, we can remove it from the integral and obtain

$$\text{Flux} = E_n \int_S da = E_n(\text{area}) = E_n 4\pi r^2 = \frac{q}{4\pi\epsilon_0 r^2}\,(4\pi r^2);$$

$$\text{Flux} = \frac{q}{\epsilon_0}. \tag{AII.11}$$

This is Gauss's law.

The important point about the result in Eq. (AII.11) is that the flux through the surface of a sphere is independent of the size of the sphere. Suppose we consider two concentric spheres having radii r_1 and r_2 with a charge q at their common center (Fig. AII.1b). By Eq. (AII.11), the flux out of sphere 1 and the flux out of sphere 2 are both equal to q/ϵ_0. This means that the flux is conserved; what flows out of the small sphere also flows out of the big sphere. But this implies that the flux does not depend on the *shape* of the surface. Imagine a wrinkled, potato-shaped surface that lies entirely within the annular space between the two spheres. The flux through this surface must also be q/ϵ_0. But then it follows that the charge q does not have to be at the center of the enclosure. (The potato's center is difficult to define.) Equation (AII.11) is correct if the charge is anywhere inside the surface. Beyond that it follows that several charges could be inside the surface; from the superposition law, Eq. (AII.4), we would obtain Gauss's law in the form

$$\text{Flux} = \frac{Q}{\epsilon_0}, \tag{AII.12}$$

where Q is the algebraic sum of the charges within the surface without regard to their locations. The charge may even be continuously distributed. If ρ is the charge density, the charge per unit volume, then

$$Q = \int_V \rho \, dV. \tag{AII.13}$$

The integration is over the volume enclosed by the surface. The final form of Gauss's law is

$$\int_S E_n \, da = \frac{Q}{\epsilon_0}, \tag{AII.14}$$

in which it is understood that S is *any closed* surface.

Note that the net flux from the annulus defined by the two spheres is zero. What flows in at r_1 flows out again at r_2. This is consonant with applying Gauss's law directly to the annulus. Since there is no charge within it, $q = 0$ and the flux is zero.

AII.5 THE POISSON EQUATION

Next we calculate the flux through a small cubical surface. Consider first the flux in the x direction, through the faces located at $x + \Delta x$, and at x (Fig. AII.2). The x component of the field vector has the value E_x at x, and $E_x + (\partial E_x/\partial x)\Delta x$ at $x + \Delta x$. Then the flux out of the cube in the x direction is the x component of E multiplied by the area of the face,

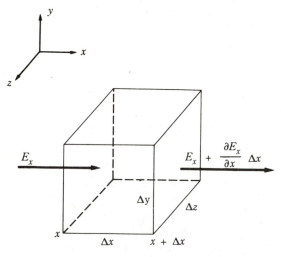

Figure AII.2

$\Delta y \, \Delta z$. Adding the fluxes through the surfaces at x and $x + \Delta x$, we have

$$(\text{Flux})_x = - E_x \, \Delta y \, \Delta z + \left(E_x + \frac{\partial E_x}{\partial x} \Delta x \right) \Delta y \, \Delta z = \frac{\partial E_x}{\partial x} \Delta x \, \Delta y \, \Delta z.$$

A similar argument shows that the fluxes in the y and z directions are given by

$$(\text{Flux})_y = \frac{\partial E_y}{\partial y} \Delta x \, \Delta y \, \Delta z; \qquad \text{and} \qquad (\text{Flux})_z = \frac{\partial E_z}{\partial z} \Delta x \, \Delta y \, \Delta z.$$

The total flux out of this small cube is the sum of the three terms

$$\left(\frac{\partial E_x}{\partial x} + \frac{\partial E_y}{\partial y} + \frac{\partial E_z}{\partial z} \right) \Delta V,$$

where we have set $\Delta x \, \Delta y \, \Delta z = \Delta V$, the volume of the cube. If we integrate this expression over the entire volume we obtain the total flux:

$$\text{Flux} = \int_V \left(\frac{\partial E_x}{\partial x} + \frac{\partial E_y}{\partial y} + \frac{\partial E_z}{\partial z} \right) dV. \tag{AII.15}$$

Using Eqs. (AII.12) and (AII.13), we have

$$\int_V \left(\frac{\partial E_x}{\partial x} + \frac{\partial E_y}{\partial y} + \frac{\partial E_z}{\partial z} \right) dV = \int_V \frac{\rho}{\epsilon_0} \, dV. \tag{AII.16}$$

Equation (AII.16) can be correct only if the integrands on the two sides are equal; then,

$$\frac{\partial E_x}{\partial x} + \frac{\partial E_y}{\partial y} + \frac{\partial E_z}{\partial z} = \frac{\rho}{\epsilon_0}. \tag{AII.17}$$

If we replace E_x, E_y, and E_z by their values from Eq. (AII.7), we obtain

$$\frac{\partial^2 \phi}{\partial x^2} + \frac{\partial^2 \phi}{\partial y^2} + \frac{\partial^2 \phi}{\partial z^2} = - \frac{\rho}{\epsilon_0}. \tag{AII.18}$$

This is Poisson's equation, which relates the electric potential to the charge density in the space. The Laplacian operator, ∇^2, is defined by

$$\nabla^2 = \frac{\partial^2}{\partial x^2} + \frac{\partial^2}{\partial y^2} + \frac{\partial^2}{\partial z^2},$$

so that we can write the Poisson equation in the form

$$\nabla^2 \phi = - \frac{\rho}{\epsilon_0}. \tag{AII.19}$$

This equation combines the two laws of electrostatics: Gauss's law and the fact that E is derivable from a scalar potential through Eq. (AII.7).

In a spherically symmetric situation, such as that in the Debye–Hückel model described in Section 16.7, the potential is a function only of r and the Poisson equation, Eq. (AII.19), becomes

$$\frac{1}{r^2} \frac{\partial}{\partial r} \left(r^2 \frac{\partial \phi}{\partial r} \right) = - \frac{\rho}{\epsilon_0}. \tag{AII.20}$$

In Section 26.2 we showed that in the presence of a dielectric the vacuum equations apply if the field-producing charge, q, is replaced by q/ϵ_r. This is equivalent to replacing ϵ_0 by $\epsilon = \epsilon_r \epsilon_0$. The ϵ is the permittivity of the medium; ϵ_r is the relative permittivity of the medium, the dielectric constant. This result is correct for electrically isotropic media such as liquids and gases, but is only crudely applicable to anisotropic materials (most solids).

You will find a very clear discussion of this entire subject in R. P. Feynman, R. B. Leighton, and M. Sands, *The Feynman Lectures on Physics.* Reading, Mass.: Addison-Wesley, 1964. Volume II, Chapters 4, 5, 10, and 11.

The International System of Units;
Le Système International d'Unités;
SI

AIII.1 THE SI BASE QUANTITIES AND UNITS

A *physical quantity* is a product of a *numerical value* (a pure number) and a *unit*.

The seven dimensionally independent base units in the SI are given in Table AIII.1. We will not have occasion to use luminous intensity.

Table AIII.1
Base physical quantities and units

Physical quantity	Symbol for quantity	Name of SI unit	Symbol for SI unit
length	l	metre	m
mass	m	kilogram	kg
time	t	second	s
electric current	I	ampere	A
thermodynamic temperature	T	kelvin	K
amount of substance	n	mole	mol
luminous intensity	I_v	candela	cd

* The quotations in Appendix III are from *Pure and Applied Chemistry*, **51**: 1–41 1979. The remaining parts follow the contents of this publication closely. Published with the permission of IUPAC. Symbols, units, and nomenclature recommendations issued periodically by IUPAC are published in the journal *Pure and Applied Chemistry*, available from Pergamon Press, Oxford.

AIII.2 DEFINITIONS OF THE SI BASE UNITS

metre: The metre is the length equal to 1 650 763.73 wavelengths in vacuum of the radiation corresponding to the transition between the levels $2p_{10}$ and $5d_5$ of the krypton-86 atom.

kilogram: The kilogram is the unit of mass; it is equal to the mass of the international prototype of the kilogram.

second: The second is the duration of 9 192 631 770 periods of the radiation corresponding to the transition between the two hyperfine levels of the ground state of the caesium-133 atom.

ampere: The ampere is that constant current which, if maintained in two straight parallel conductors of infinite length, of negligible cross-section, and placed 1 metre apart in vacuum, would produce between these conductors a force equal to 2×10^{-7} newton per metre of length.

kelvin: The kelvin, unit of thermodynamic temperature, is the fraction 1/273.16 of the thermodynamic temperature of the triple point of water.

candela: The candela is the luminous intensity, in the perpendicular direction, of a surface of 1/600 000 square metre of a black body at the temperature of freezing platinum under a pressure of 101 325 newtons per square metre.

mole: The mole is the amount of substance of a system which contains as many elementary entities as there are atoms in 0.012 kilograms of carbon-12. When the mole is used, the elementary entities must be specified and may be atoms, molecules, ions, electrons, other particles, or specified groups of such particles.

AIII.3 DERIVED PHYSICAL QUANTITIES

All other physical quantities are regarded as being derived from, and as having dimensions derived from, the seven independent base physical quantities by definitions involving only multiplication, division, differentiation, and/or integration.

For example, the velocity of a particle is defined by $v = ds/dt$; the velocity has the dimension, length/time (l/t); the SI unit is metre per second (m/s).

Table AIII.2 lists a number of common derived quantities and their units; these units do not have special names. Table AIII.3 lists a number of common derived quantities whose units have special names.

Table AIII.2
SI derived units without special names

Physical quantity	Symbol for quantity	Name of SI unit	Symbol for SI unit
area	A	square metre	m^2
volume	V	cubic metre	m^3
density	ρ	kilogram per cubic metre	$kg\ m^{-3}$
velocity	u,v,w,c	metre per second	$m\ s^{-1}$
concentration*	$\tilde{c}$	mole per cubic metre	$mol\ m^{-3}$
electric field strength	E	volt per metre	$V\ m^{-1}$

*   For concentration in mol/L = mol/dm³ we use c. Thus, $\tilde{c} = 1000c$.

Table AIII.3
Special names and symbols for certain SI derived units

Physical quantity	Name of SI unit	Symbol for SI unit	Definition of SI unit
force	newton	N	kg m s^{-2}
pressure = force/area	pascal	Pa	$\text{N m}^{-2} = \text{kg m}^{-1}\text{ s}^{-2}$
energy	joule	J	$\text{N m} = \text{kg m}^2\text{ s}^{-2}$
power = energy/time	watt	W	$\text{J s}^{-1} = \text{kg m}^2\text{ s}^{-3}$
electric charge	coulomb	C	A s
electric potential difference	volt	V	$\text{J C}^{-1} = \text{kg m}^2\text{ s}^{-3}\text{ A}^{-1}$
electric resistance	ohm	Ω	$\text{V A}^{-1} = \text{kg m}^2\text{ s}^{-3}\text{ A}^{-2}$
electric conductance	siemens	S	$\text{A V}^{-1} = \text{kg}^{-1}\text{ m}^{-2}\text{ s}^3\text{ A}^2$
electric capacitance	farad	F	$\text{C V}^{-1} = \text{kg}^{-1}\text{ m}^{-2}\text{ s}^4\text{ A}^2$
magnetic flux	weber	Wb	$\text{V s} = \text{kg m}^2\text{ s}^{-2}\text{ A}^{-1}$
magnetic flux density	tesla	T	$\text{Wb m}^{-2} = \text{kg s}^{-2}\text{ A}^{-1}$
frequency	hertz	Hz	s^{-1}

AIII.4 SI PREFIXES

To designate multiples and submultiples of the base unit, we use a standard prefix to the symbol for the unit. These prefixes are listed in Table AIII.4.

Examples: $1\text{ km} = 10^3\text{ m}$; $1\text{ ns} = 10^{-9}\text{ s}$.

Note that the base unit for mass, 1 kg, is already prefixed. We do not add a second prefix, but rather use a single prefix on the unit gram. Thus we use: ng, not pkg; mg, not μkg; Mg, not kkg. When we take the power or root of a prefixed unit, the entire unit is raised to the power.

examples: 1 dm^3 means 1 (dm)^3, not $1\text{ d(m}^3)$; 1 km^2 means 1 (km)^2, not $1\text{ k(m}^2)$.

Table AIII.4
SI prefixes

Submultiple	Prefix	Symbol	Multiple	Prefix	Symbol
10^{-1}	deci	d	10	deca	da
10^{-2}	centi	c	10^2	hecto	h
10^{-3}	milli	m	10^3	kilo	k
10^{-6}	micro	μ	10^6	mega	M
10^{-9}	nano	n	10^9	giga	G
10^{-12}	pico	p	10^{12}	tera	T
10^{-15}	femto	f	10^{15}	peta	P
10^{-18}	atto	a	10^{18}	exa	E

AIII.5   SOME GRAMMATICAL RULES

AIII.5.1   Printing of Symbols for Units and Prefixes

The symbols for the unit and any prefix are both printed in roman type with no space between them. The symbol for the unit is followed by a period only if it ends a sentence and is never plural. Thus

$$10 \text{ kg, not } 10 \text{ kg., and not } 10 \text{ kgs.}$$

Symbols for a unit derived from a proper name are capitalized, but the name of the unit is not capitalized.

examples: 10 volts is symbolized by 10 V; 100 joules is symbolized by 100 J.

AIII.5.2   SYMBOLS FOR DERIVED QUANTITIES

A product of two different units may be written in the following ways: N m   or   N × m or N · m. In this book we use the first method, simply putting a space between the symbols.

A quotient may be written as kJ/mol or kJ mol^{-1} or in any other way that is not ambiguous. More than one solidus (/) should not be used in an expression unless parentheses are included to prevent ambiguity: write (m/s)/(V/m), not m/s/V/m.

AIII.5.3   Table Headings and Graph Labels

The entries in the tables in this book are all pure numbers. The table heading contains the quotient of the physical quantity divided by the unit. For example, in Table AIV.1, the second column heading is $\Delta H_f^\circ/\text{kJ mol}^{-1}$. We obtain the value for ΔH_f° by setting the table heading equal to the pure number in the table; for $O_3(g)$ we find $\Delta H_f^\circ/\text{kJ mol}^{-1}$ = 142.7; therefore ΔH_f° = 142.7 kJ mol^{-1} = 142.7 kJ/mol.

In Table 7.1, values of the constants in the expression, $C_p/R = a + bT + cT^2 + \cdots$, are tabulated. The third column heading is $b/10^{-3}\text{ K}^{-1}$; to obtain the value of b for oxygen we set the number in the table equal to the column heading: $b/10^{-3}\text{ K}^{-1}$ = 1.6371; then $b = 1.6371 \times 10^{-3}\text{ K}^{-1}$. Note that this table heading could also be written as: $1000b/\text{K}^{-1}$, or $1000b$ K, or even (God forbid!) b kK. By an appropriate choice of the table heading, the entries are presented in a convenient form.

The same usage has been followed in labeling the coordinate axes on graphs; in a plot of pressure against temperature, the vertical axis might be labeled p/MPa, and the horizontal axis T/K.

AIII.6   EQUATIONS WITH DIMENSIONAL PROBLEMS

The relation between the Celsius temperature, t, and the thermodynamic temperature, T, provides an illustration of the quantity calculus. Heretofore we have commonly written

$$T = t + T_0,$$

where $T_0 = 273.15$ K. But, as it stands, this equation poses a dimensional problem. The unit for t is °C while the unit for T is K. Strictly speaking we should write

$$T/\text{K} = t/\text{°C} + T_0/\text{K} = t/\text{°C} + 273.15.$$

Each term in this statement is a pure number and the difficulty (such as it is) vanishes. Then we could write for t,

$$t = (T/K - 273.15)°C$$

Since $1°C = 1 K$ (exactly) this difficulty is as much imagined as real, but the point should be kept in mind.

Equation (11.50) is another example of a dimensional difficulty. Formerly, it was customary to write the relation between the pressure equilibrium constant and the concentration equilibrium constant as

$$K_p = K_c(RT)^{\Delta v},$$

which is dimensionally nonsensical. The current Eq. (11.50) gives two correct ways to write the equation; namely,

$$K_p = K_c\left(\frac{RT}{101.325 \text{ J/mol}}\right)^{\Delta v} = K_c(0.08206 \text{ } T/K)^{\Delta v}.$$

Both of these ways make sense dimensionally. The presence of the awkward numerical factor is the price we pay for using atmospheres and mol/L as units of concentration!

AIII.7 ONE SYMBOL—ONE QUANTITY

As a general rule the symbol for a physical quantity should not be different for different units of the quantity. Thus, m should be the mass, not the number of kilograms or the number of grams; l should be the length, not the number of metres or the number of feet.

I have deliberately violated this rule in using two different symbols for the volume concentration. The traditional symbol for the volume concentration in units of mol per litre is c. This is a non-SI unit in the sense that it is not derived directly from the base units, but involves a numerical factor as well. It is an SI unit in the sense that it involves legitimate SI units for amount of substance (mol) and volume (dm^3). As a laboratory unit of concentration c is going to be with us for a long time. Rather than use this symbol for the SI unit of concentration derived simply from the base units, mol/m^3, I have introduced the symbol $\tilde{c}$ for the SI base unit concentration. The use of $\tilde{c}$ in the equations avoids both cumbersome numerical factors and confusion with the moles-per-litre concentration. Similarly, I have used $\tilde{N}$ for number of molecules per cubic metre.

APPENDIX IV

Table AIV.1
Fundamental constants

Quantity	Symbol and equivalences	Value
Gas constant	R	$8.31441(26)$ J K^{-1} mol^{-1}
Zero of the Celsius scale	T_0	273.15 K exactly
Standard atmosphere	p_0	1.01325×10^5 Pa exactly
Standard molar volume of ideal gas	$V_0 = RT_0/p_0$	$22.41383(70) \times 10^{-3}$ m^3 mol^{-1}
Avogadro constant	N_A	$6.022045(31) \times 10^{23}$ mol^{-1}
Boltzmann constant	$k = R/N_A$	$1.380662(44) \times 10^{-23}$ J K^{-1}
Standard acceleration of gravity	g	9.80665 m s^{-2} exactly
Elementary charge	e	$1.6021892(46) \times 10^{-19}$ C
Faraday constant	$F = N_A e$	$96484.56(27)$ C mol^{-1}
Speed of light in vacuum	c	$2.99792458(1) \times 10^8$ m s^{-1}
Planck constant	h	$6.626176(36) \times 10^{-34}$ J s
	$\hbar = h/2\pi$	$1.0545887(57) \times 10^{-34}$ J s
Rest mass of electron	m	$9.109534(47) \times 10^{-31}$ kg
Rest mass of proton	m_p	$1.6726485(86) \times 10^{-27}$ kg
Rest mass of neutron	m_n	$1.6749543(86) \times 10^{-27}$ kg
Atomic mass unit	$1\,u = 10^{-3}$ kg mol$^{-1}/N_A$	$1.6605655(86) \times 10^{-27}$ kg

Table AIV.1 (*Continued*)
Less frequently used constants and composite quantities

Permeability of vacuum	μ_0	$4\pi \times 10^{-7}$ V s^2 C^{-1} m^{-1} exactly
Permittivity of vacuum	$\epsilon_0 = 1/\mu_0 c^2$	$8.85418782(5) \times 10^{-12}$ C V^{-1} m^{-1}
	$4\pi\epsilon_0$	$1.11265006(6) \times 10^{-10}$ C^2 N^{-1} m^{-2}
	$1/4\pi\epsilon_0$	$8.9875518(5) \times 10^9$ N m^2 C^{-2}
Gravitational constant	G	$6.6720(27) \times 10^{-11}$ m^3 kg^{-1} s^{-2}
Rydberg constant	$R_\infty = me^4/8\epsilon_0^2 h^3 c$	$1.097373177(83) \times 10^7$ m^{-1}
Bohr radius	$a_0 = 4\pi\epsilon_0 \hbar^2/me^2$	$5.2917706(44) \times 10^{-11}$ m
Hartree energy	$E_h = 2hcR_\infty = e^2/4\pi\epsilon_0 a_0$	$4.359814(24) \times 10^{-18}$ J
Bohr magneton	$\mu_B = e\hbar/2m$	$9.274078(36) \times 10^{-24}$ J T^{-1}
Nuclear magneton	$\mu_N = e\hbar/2m_p$	$5.050824(20) \times 10^{-27}$ J T^{-1}
Electron magnetic moment	μ_e	$9.284832(36) \times 10^{-24}$ J T^{-1}
Lande g factor for free electron	$g_e = 2\mu_e/\mu_B$	$2.0023193134(70)$
Proton gyromagnetic ratio	γ_p	$2.6751987(75) \times 10^8$ s^{-1} T^{-1}

From *Pure and Applied Chemistry*, **51**: 1, 1979. Printed by permission. The number in parentheses is the estimated uncertainty of the last figure(s) of the value. This estimate is three times the standard deviation.

Table AIV.2
Mathematical constants and series

$\pi = 3.14159265\ldots$
$e = 2.7182818\ldots$
$\ln x = 2.302585\ldots \log_{10} x$

$$\sin x = x - \frac{x^3}{3!} + \frac{x^5}{5!} - \frac{x^7}{7!} + \cdots \qquad \text{(all } x\text{)}$$

$$\cos x = 1 - \frac{x^2}{2!} + \frac{x^4}{4!} - \frac{x^6}{6!} + \cdots \qquad \text{(all } x\text{)}$$

$$e^x = 1 + x + \frac{x^2}{2!} + \frac{x^3}{3!} + \frac{x^4}{4!} + \cdots \qquad \text{(all } x\text{)}$$

$\ln(1 + x) = x - \frac{1}{2}x^2 + \frac{1}{3}x^3 - \frac{1}{4}x^4 + \cdots \qquad x^2 < 1$
$(1 + x)^{-1} = 1 - x + x^2 - x^3 + \cdots \qquad x^2 < 1$
$(1 - x)^{-1} = 1 + x + x^2 + x^3 + \cdots \qquad x^2 < 1$
$(1 - x)^{-2} = 1 + 2x + 3x^2 + 4x^3 + \cdots \qquad x^2 < 1$

Chemical Thermodynamic Properties at 298.15 K

Table AV.1

Substance	ΔH_f°/kJ mol^{-1}	ΔG_f°/kJ mol^{-1}	S°/J K^{-1} mol^{-1}	C_p°/J K^{-1} mol^{-1}
O(g)	249.17	231.75	160.946	21.91
O_2(g)	0	0	205.037	29.35
O_3(g)	142.7	163.2	238.82	39.20
H(g)	217.997	203.26	114.604	20.786
H_2(g)	0	0	130.570	28.82
OH(g)	38.95	34.23	183.64	29.89
H_2O(l)	−285.830	−237.178	69.950	75.291
H_2O(g)	−241.814	−228.589	188.724	33.577
H_2O_2(l)	−187.78	−120.42	109.6	89.1
F(g)	79.39	61.92	158.640	22.74
F_2(g)	0	0	202.685	31.30
HF(g)	−273.30	−275.40	173.665	29.13
Cl(g)	121.302	105.70	165.076	21.84
Cl_2(g)	0	0	222.965	33.91
HCl(g)	−92.31	−95.299	186.786	29.1
Br(g)	111.86	82.429	174.904	20.79
Br_2(l)	0	0	152.210	75.69
Br_2(g)	30.91	3.14	245.350	36.02
HBr(g)	−36.38	−53.43	198.585	29.14
I(g)	106.762	70.28	180.673	20.79
I_2(c)	0	0	116.139	54.44
I_2(g)	62.421	19.36	260.567	36.9
HI(g)	26.36	1.72	206.480	29.16
S(c, rhombic)	0	0	32.054	22.6
S(c, monoclinic)	0.33			
S(g)	276.98	238.27	167.715	23.67

Table AV.1 (*Continued*)

Substance	$\Delta H_f°$/kJ mol^{-1}	$\Delta G_f°$/kJ mol^{-1}	$S°$/J K^{-1} mol^{-1}	$C_p°$/J K^{-1} mol^{-1}
$SO_2(g)$	-296.81	-300.19	248.11	39.9
$SO_3(g)$	-395.7	-371.1	256.6	50.7
$H_2S(g)$	-20.6	-33.6	205.7	34.2
$H_2SO_4(l)$	-813.99	-690.10	156.90	138.9
$N(g)$	472.68	455.57	153.189	20.79
$N_2(g)$	0	0	191.502	29.12
$NO(g)$	90.25	86.57	210.65	29.84
$NO_2(g)$	33.18	51.30	240.0	37.2
$N_2O(g)$	82.0	104.2	219.7	38.45
$N_2O_3(g)$	83.7	139.4	312.2	65.6
$N_2O_4(g)$	9.16	97.8	304.2	77.3
$N_2O_5(g)$	11	115	356	85
$NH_3(g)$	-45.94	-16.5	192.67	35.1
$HNO_3(l)$	-174.1	-80.8	155.6	109.9
$NOCl(g)$	51.7	66.1	261.6	44.69
$NH_4Cl(c)$	-314.4	-203.0	94.6	84.1
$P(g)$	316.5	278.3	163.085	20.79
$P_2(g)$	144.0	104	218.01	32.0
$P_4(c, \alpha, white)$	0	0	164.4	95.36
$P_4(g)$	58.9	24.5	279.9	67.15
$PCl_3(g)$	-287	-268	311.7	71.8
$PCl_5(g)$	-375	-305	364.5	112.8
$C(c, graphite)$	0	0	5.74	8.53
$C(c, diamond)$	1.897	2.900	2.38	6.12
$C(g)$	716.67	671.29	157.988	20.84
$CO(g)$	-110.53	-137.15	197.556	29.12
$CO_2(g)$	-393.51	-394.36	213.677	37.11
$CH_4(g)$	-74.8	-50.8	186.15	35.31
$HCHO(g)$	-117	-113	218.7	35.4
$CH_3OH(l)$	-238.7	-166.4	127	82
$C_2H_2(g)$	226.7	209.2	200.8	43.9
$C_2H_4(g)$	52.3	68.1	219.5	43.6
$C_2H_6(g)$	-84.7	-32.9	229.5	52.6
$CH_3COOH(l)$	-485	-390	160	124
$C_2H_5OH(l)$	-277.7	-174.9	161	111.5
$C_6H_6(g)$	82.93	129.66	26.92	85.29
$Si(c)$	0	0	18.81	20
$Si(g)$	450	411	167.870	22.25
$SiO_2(c, \alpha-quartz)$	-910.7	-856.7	41.46	44.4
$SiH_4(g)$	34	57	204.5	42.8
$SiF_4(g)$	-1614.95	-1572.7	282.65	73.6
$Pb(c)$	0	0	64.80	26.4
$PbO(c, red)$	-219.0	-188.9	66.5	45.8
$PbO_2(c)$	-277	-217.4	68.6	64.6
$PbS(c)$	-100	-99	91	49.5
$PbSO_4(c)$	-919.94	-813.2	148.49	103.21
$Al(c)$	0	0	28.35	24.4
$Al_2O_3 (c, \alpha-corundum)$	-1675.7	-1582	50.92	79.0
$Zn(c)$	0	0	41.63	25.4
$Zn(g)$	130.42	95.18	160.875	20.79
$ZnO(c)$	-350.46	-318.3	43.64	40.3

Table AV.1 (Continued)

Substance	$\Delta H_f^\circ/\text{kJ mol}^{-1}$	$\Delta G_f^\circ/\text{kJ mol}^{-1}$	$S^\circ/\text{J K}^{-1}\text{ mol}^{-1}$	$C_p^\circ/\text{J K}^{-1}\text{ mol}^{-1}$
Hg(l)	0	0	75.90	27.98
Hg(g)	61.38	31.85	174.860	20.79
HgO(c, red)	−90.8	−58.56	70.3	44.1
Cu(c)	0	0	33.15	24.43
CuO(c)	−157	−130	42.6	42.3
Cu_2O(c)	−169	−146	93.1	63.6
Ag(c)	0	0	42.55	25.35
Ag_2O(c)	−31.0	−11.2	121	65.9
AgCl(c)	−127.070	−109.80	96.23	50.8
Ag_2S(c, α)	−32.6	−40.7	144.0	76.5
Fe(c, α)	0	0	27.3	25.1
$Fe_{0.947}O$(c, wustite)	−266.3	−245.1	57.5	48.1
Fe_2O_3(c, hematite)	−824.2	−742.2	87.4	103.8
Fe_3O_4(c, magnetite)	−1118	−1015	146	143.4
FeS(c, α)	−100	−100.4	60.3	50.5
FeS_2(c, pyrite)	−178	−167	52.9	62.2
Ti(c)	0	0	30.6	25.0
TiO_2(c, rutile)	−945	−890	50.3	55.0
$TiCl_4$(l)	−803	−737	252.3	145.2
Mg(c)	0	0	32.68	24.9
MgO(c)	−601.5	−569.4	26.95	37.2
$MgCO_3$(c)	−1096	−1012	66	75.5
Ca(c)	0	0	41.6	25.3
CaO(c)	−635.09	−604.0	38.1	42.8
$Ca(OH)_2$(c)	−986.1	−898.6	83.4	87.5
CaC_2(c)	−60	−65	70.0	62.7
$CaCO_3$(c, calcite)	−1206.9	−1128.8	93	81.9
SrO(c)	−592	−562	54	45.0
$SrCO_3$(c)	−1220	−1140	97	81.4
BaO(c)	−554	−525	70.4	47.8
$BaCO_3$(c)	−1216	−1138	112	85.4
Na_2O(c)	−414.2	−375.5	75.1	69.1
NaOH(c)	−425.61	−379.53	64.45	59.5
NaF(c)	−573.65	−543.51	51.5	46.9
NaCl(c)	−411.15	−384.15	72.1	50.5
NaBr(c)	−361.06	−348.98	86.8	51.4
NaI(c)	−287.8	−286.1	98.5	52.1
Na_2SO_4(c)	−1387.1	−1270.2	149.6	128.2
$Na_2SO_4 \cdot 10H_2O$	−4327.3	−3647.4	592	
$NaNO_3$(c)	−467.9	−367.1	116.5	92.9
KF(c)	−567.3	−537.8	66.6	49.0
KCl(c)	−436.75	−409.2	82.6	51.3
$KClO_3$(c)	−397.7	−296.3	143	100.2
$KClO_4$(c)	−432.8	−303.2	151	112.4
KBr(c)	−393.80	−380.7	95.9	52.3
KI(c)	−327.90	−324.89	106.3	52.9

g: gas; l: liquid; c: crystal.

The values in Table AV.1 were calculated from the data in D. D. Wagman, W. H. Evans, V. B. Parker, I. Halow, S. M. Bailey, and R. H. Schumm, *Selected Values of Chemical Thermodynamic Properties*, NBS Technical Notes 270-3, 4, 5, 6, 7, and 8.

To obtain the values in joules, the tabulated values in calories were multiplied by 4.184. The product was then rounded to avoid giving the impression of an accuracy higher than the original entries would justify. For example, the NBS entry for the entropy of HgO(red) is 16.80 cal/K mol; then $16.80 \times 4.184 = 70.291$ J/K mol. This has been rounded to 70.3 J/K mol rather than to 70.29. Consequently, the values may indeed be known to a higher degree of accuracy than the values given here indicate.

A few values are taken from CODATA Bulletin No. 28, *Recommended Key Values of Thermodynamics*, 1977.

Group Character Tables

C_{2v}	E	C_2	$\sigma_v(xz)$	$\sigma_v'(yz)$		
a_1	1	1	1	1	z	x^2, y^2, z^2
a_2	1	1	-1	-1	R_z	xy
b_1	1	-1	1	-1	x, R_y	xz
b_2	1	-1	-1	1	y, R_x	yz

C_{3v}	E	$2C_3$	$3\sigma_v$		
a_1	1	1	1	z	$x^2 + y^2, z^2$
a_2	1	1	-1	R_z	
e	2	-1	0	$(x, y)(R_x, R_y)$	$(x^2 - y^2, xy)(xz, yz)$

C_{2h}	E	C_2	i	σ_h		
a_g	1	1	1	1	R_z	x^2, y^2, z^2, xy
b_g	1	-1	1	-1	R_x, R_y	xz, yz
a_u	1	1	-1	-1	z	
b_u	1	-1	-1	1	x, y	

D_{2h}	E	$C_2(z)$	$C_2(y)$	$C_2(x)$	i	$\sigma(xy)$	$\sigma(xz)$	$\sigma(yz)$		
a_g	1	1	1	1	1	1	1	1		x^2, y^2, z^2
b_{1g}	1	1	-1	-1	1	1	-1	-1	R_z	xy
b_{2g}	1	-1	1	-1	1	-1	1	-1	R_y	xz
b_{3g}	1	-1	-1	1	1	-1	-1	1	R_x	yz
a_u	1	1	1	1	-1	-1	-1	-1		
b_{1u}	1	1	-1	-1	-1	-1	1	1	z	
b_{2u}	1	-1	1	-1	-1	1	-1	1	y	
b_{3u}	1	-1	-1	1	-1	1	1	-1	x	

D_{3h}	E	$2C_3$	$3C_2$	σ_h	$2S_3$	$3\sigma_v$		
a_1'	1	1	1	1	1	1		$x^2 + y^2, z^2$
a_2'	1	1	-1	1	1	-1	R_z	
e'	2	-1	0	2	-1	0	(x, y)	$(x^2 - y^2, xy)$
a_1''	1	1	1	-1	-1	-1		
a_2''	1	1	-1	-1	-1	1	z	
e''	2	-1	0	-2	1	0	(R_x, R_y)	(xz, yz)

Answers to Problems

Note 1. Significant figures. In those problems in which nominal values of variables are assigned to illustrate a calculation, I have decided (quite arbitrarily) to record the numerical answers to three significant figures unless there is some obvious reason in a particular problem to do otherwise. For example, "What volume does one mole of an ideal gas occupy at 2 atm pressure and 20 °C?" The answer is given as 12.0 L. The "one mole," the "2 atm," and the "20 °C" are nominal values given to illustrate the use of the ideal gas law. They may be taken as exact in the calculation. I see no need to write 1.00 mole, 2.00 atm, and 20.0 °C every time such a calculation is proposed. On the other hand, if the statement of the problem implies a measurement, then the rules for significant figures must be observed. For example, "A sample of methane is confined under a pressure of 745 mmHg at a temperature of 22.0 °C in a volume of 175 mL. What is the mass of the gas, assuming ideal behavior?" The answer is 0.113 g, but this time three figures are recorded because of the three significant figures in 175 mL. The student should be alert to instances in which the number of significant figures decreases as often happens in subtractions. I have tried to specify the data carefully in such problems so that there is no ambiguity in the calculation.

Note 2. All the problems were worked using a continuous-memory, programmable calculator; all the fundamental constants were stored in the calculator to their full accuracy. Thus, the value of R used was always 8.31441 J/K mol, that of N_A was always 6.022045×10^{23}/mol, and T was always calculated as $T = 273.15 + t$. As a consequence, the answers in the book may show trivial differences in the last recorded figure from the values calculated using $R = 8.31$ J/K mol, $T_0 = 273$ K, and $N_A = 6.02 \times 10^{23}$/mol. I trust that no one will lose any sleep over that.

Note 3. In those problems in which quantities are determined from the slope and/or intercept of a straight line plot, the slope and intercept were calculated using a least-squares fit of the data (the preprogrammed linear-regression program in the calculator).

Chapter 2

2.1 449 °C   **2.2** 300 mol; 9.6 kg   **2.3** 892.1 μg   **2.4** (a) 818.3 μg   (b) 142.2 cm^3

2.5 "R" = 10.1325 J/K mol; "N_A" = 7.339 × 10^{23}/"mol"; "M_H" = 1.228 g/"mol";
"M_O" = 19.50 g/"mol"

2.6 $\alpha = 1/T$ **2.7** $\kappa = 1/p$ **2.8** $(\partial p/\partial T)_V = \alpha/\kappa$

2.9 (p/atm, mol%) (a) H$_2$: 6.15, 94.2%; O$_2$: 0.38, 5.8%; p_t/atm = 6.53
   (b) N$_2$: 0.440, 53.3%; O$_2$: 0.385, 46.7%; p_t/atm = 0.825
   (c) CH$_4$: 0.769, 51.5%; NH$_3$: 0.724, 48.5%; p_t/atm = 1.493
   (d) H$_2$: 6.15, 97.3%; Cl$_2$: 0.17, 2.7%; p_t/atm = 6.32

2.10 (p/atm: N$_2$, O$_2$, Ar) 0.762, 0.205, 0.0098; (x_i: N$_2$, O$_2$, Ar, H$_2$O) 0.762, 0.205, 0.0098, 0.023

2.11 (a) 69% N$_2$, 18% O$_2$, 0.88% Ar, 12% H$_2$O
   (b) 12.2 L   (c) 0.993

2.12 30 L   **2.13** 20% O$_2$, 80% H$_2$ **2.14** (a) 98.0% N$_2$, 2.0% H$_2$O   (b) 10.2 L

2.15 10.3 mol% H$_2$ **2.16** 0.747 N$_2$, 0.101 O$_2$, 0.086 H$_2$O, 0.058 CO$_2$, 0.010 Ar

2.17 59.9% butane   **2.18** 154.7 g/mol

2.19 (a) 5.64 g/mol (b) 56.4 g/mol (c) 56.4 kg/mol (d) polymers

2.20 633 Torr; 462 Torr

2.21 (a) 5.8 × 10^{-20} m   (b) Yes. The potatoes try to get as close to the bottom as possible.

2.22 +0.024 atm

2.23 (p_i/atm) N$_2$: 2.44 × 10^{-3}; CO$_2$: 0.0701 × 10^{-3}; p_t/atm = 2.51 × 10^{-3}; x_{N_2} = 0.972
   (b) 1.10 × 10^6 mol

2.24 (p/atm: 50 km, 100 km; mol %: 50 km, 100 km)
   N$_2$: (3.06 × 10^{-3}, 1.20 × 10^{-5}; 89.0%, 87.7%)
   O$_2$: (3.73 × 10^{-4}, 6.66 × 10^{-7}; 10.8%, 4.86%)
   Ar: (3.44 × 10^{-6}, 1.27 × 10^{-9}; 0.100%, 0.00930%)
   CO$_2$: (5.0 × 10^{-8}, 8.2 × 10^{-12}; 0.0014%, 6.0 × 10^{-5}%)
   Ne: (3.3 × 10^{-7}, 6.1 × 10^{-9}; 0.0097%, 0.045%)
   He: (2.3 × 10^{-6}, 1.0 × 10^{-6}; 0.066%, 7.5%)
   p_{total}/atm: 50 km: 3.44 × 10^{-3}; 100 km: 1.37 × 10^{-5}

2.25 0.924 c_0 **2.26** (a) 38 cm (b) 9.71 × 10^{-4} mol/L (c) 1.94 × 10^{-4} mol

2.27 c_{top} = 0.098 mol/m^3, c_0 = 0.102 mol/m^3

2.28 (a) 65.2 kg/mol (b) 6.36 g (c) 0.244 mol/m^3 **2.29** 1.41 km   **2.30** 53

2.32 (a) $p_i = \tilde{c}_i RT$ (b) If $r_i = n_i/n_1$, then $p_1 = p/(1 + \Sigma r_i)$; $p_i = r_i p/(1 + \Sigma r_i)$

2.34 10 km, 0.81; 15 km, 0.73

2.35 (a) $N = \tilde{N}_0 ART/Mg$, where A = earth's area, $\tilde{N}_0$ = number of molecules/m^3 at ground level
   (c) 5.27 × 10^{18} kg

2.36 (a) $\langle x_i \rangle = (x_i^0/M_i)/\Sigma x_i^0/M_i$ (b) N$_2$: 0.804; O$_2$: 0.189; Ar: 0.007

2.37 $Z = (RT/Mg)\ln 2$

2.38 [$(V/n)/$(L/mol): α_e, $\alpha = 0$] 2 atm: 13.7, 12.2; 1 atm: 28.6, 24.5; $\frac{1}{2}$ atm: 60.3, 48.9

2.39 $Z = 1 + \alpha$; as $p \to 0$, $\alpha \to 1$ and $Z \to 2$; N$_2$O$_4$ becomes 2 NO$_2$.

Chapter 3

3.1 12.1 cm^3/mol   **3.2** a = 0.018 Pa m^6/mol^2, b = 2.0 × 10^{-5} m^3/mol

3.3 a = 0.212 Pa m^6/mol^2, b = 1.89 × 10^{-5} m^3/mol, R = 5.15 J/K mol;
   a = 0.553 Pa m^6/mol^2, b = 3.04 × 10^{-5} m^3/mol, $\overline{V}_c$ = 9.13 × 10^{-5} m^3/mol

3.4 $a = 3p_c \overline{V}_c^2 T_c$; $b = \frac{1}{3}\overline{V}_c$; $R = 8p_c \overline{V}_c/T_c$ **3.5** $a = e^2 p_c \overline{V}_c^2$; $b = \frac{1}{2}\overline{V}_c$; $R = \frac{1}{2}e^2 p_c \overline{V}_c/T_c$

3.6 (a) 0.520 L/mol (b) 0.195 L/mol (c) 0.146 L/mol

3.7 From 100 °C to 25 °C, p decreases thirty-fold, $1/T$ increases by only 1.2.

3.8 $(p/\text{atm}, Z)$; 200 K: (100, 0.513); (200, 0.270); (400, 0.954); (600, 3.91); (800, 10.01); (1000, 20.12); 1000 K: (100, 1.0218); (200, 1.0500); (400, 1.1288); (600, 1.244); (800, 1.401); (1000, 1.608).

3.9 0.1942 L/mol   **3.10** B-B: 0.2673 L/mol; vdW: 0.3818 L/mol

3.11 (B-B; vdW); O_2: (399.5 K, 522 K); CO_2: (867.8 K, 1026 K)

3.12 (a) 7.914×10^{-5} m^3/mol   (b) 311.3 atm

3.13 $\alpha = (1/T)[1 + (2a/RT^2)(p/RT)]/[1 + (b - a/RT^2)(p/RT)]$; $T_B = (a/Rb)^{1/2}$

3.15 $(-dp/p) = (Mg/ZRT)\,dz$; $\ln(p/p_0) + B(p - p_0) = -Mgz/RT$

3.18 $T = 2a/Rb$; $(\partial Z/\partial p)_{\text{max}} = b^2/4a$

Chapter 4

4.1 (m/s; 300 K, 500 K); c_{rms}: 484, 624; $\langle c \rangle$: 446, 575; c_{mp}: 395, 510; speeds of H_2 are four times greater.

4.2 (a) $\langle c_{O_2} \rangle = 440$ m/s; $\langle c_{O_2} \rangle / \langle c_{CCl_4} \rangle = 2.19$ (b) 6.07×10^{-21} J; same K.E.

4.3 (a) 3.74 kJ/mol, 6.24 kJ/mol   (b) 6.21×10^{-21} J   **4.4** 3.24×10^{-10} m/s; 98 yr

4.5 10 km; 12 km   **4.6** 96.6 K; 0.00925   **4.7** $(3 - 8/\pi)^{1/2}(kT/m)^{1/2}$ **4.8** $(\tfrac{3}{2})^{1/2}kT$

4.9 (a) $\langle t \rangle = (2m/\pi kT)^{1/2}$ (b) $(1 - 2/\pi)^{1/2}(m/kT)^{1/2}$ (c) 0.333   **4.10** 0.310   **4.11** $\tfrac{1}{2}kT$

4.12 kT: 0.572; $2kT$: 0.262; $5kT$: 0.018; $10kT$: 1.62×10^{-4} **4.13** 0.766   **4.14** 0.676

4.15 (a) 9.48×10^{-22} (b) 3.0×10^{-304} (c) 0.196   (d) 4.33×10^{-14}

4.16 0.0661; 0.198; 0.314   **4.17** $(\bar{C}_v/R)_{\text{total}} = 3.059$; 3.307; 3.396

4.18 $[\theta_s, (\bar{C}_v/R)_{\text{vib}}]$: (3.360 kK, 0.001618); (1.890 kK, 0.07114); (954.1 K, 0.4536); (954.1 K, 0.4536)

4.19 2.58×10^{13} Hz   **4.20** 0.04540; 0.1707; 0.7241; 0.9207; 0.9638

4.21 $\bar{C}_v/R$: 3.0274; 3.2256; 3.9363; 5.0399   **4.22** (a) 0.229   (b) 1.024×10^{-9}

4.23 0.0831   **4.24** 0.6931; 447 K

Chapter 5

5.1 46 atm   **5.3** 32.2 kJ/mol   **5.4** $p_\infty = 1.450 \times 10^6$ atm; $p_{298} = 0.02819$ atm

5.5 $1/T = (1/T_0) + (M_{\text{air}}gz/Q_{\text{vap}}T_a)$; 94 °C   **5.9** 118.1 kJ/mol; 1177 K

5.10 $a = \alpha_0$; $b = \tfrac{1}{2}(\alpha' + \alpha_0^2)$; $c = \tfrac{1}{6}(\alpha'' + 3\alpha'\alpha_0 + \alpha_0^3)$

Chapter 6

6.1 400 kJ   **6.2** 0.098 J   **6.3** 12 kJ   **6.4** (a) 31%   (b) None

6.5 (a) (t, t'): (0, 0); (25, 2.52); (50, 11.7); (75, 37.6); (100, 100)

(b) $(p/\text{mmHg}, t')$: (40, 0.40); (100, 2.6); (400, 19.7); (760, 46.5)

6.6 $t' = t[1 + b(t - 100)/(a + 100b)]$ **6.7** 409.83

Chapter 7

7.1 (a) -30.3 K   (b) 0 K   (c) 10.1 K   **7.2** 12.6 J/K   **7.3** $Q = W = 4$ kJ; $\Delta U = \Delta H = 0$

7.4 (a) $Q = W = 8.22$ kJ; $\Delta U = \Delta H = 0$ (b) $Q = W = -8.22$ kJ; $\Delta U = \Delta H = 0$

7.5 $Q = W = -20.3$ kJ; $\Delta U = \Delta H = 0$

7.6 $W_{\text{rev}} = nRT \ln(V_2/V_1) + n^2(RTb - a)[(1/V_1) - (1/V_2)]$

7.7 $Q = 2746$ J/mol; $W = 2727$ J/mol; $\Delta U = 18.5$ J/mol; $\Delta H = 31.7$ J/mol

7.8 $Q_p = \Delta H = -1559$ J/mol; $W = -624$ J/mol; $\Delta U = -935$ J/mol

7.9 $Q_V = \Delta U = 1560$ J/mol; $W = 0$; $\Delta H = 2180$ J/mol

7.10 $\Delta U = 9.4$ kJ/mol; $\Delta H = 11.9$ kJ/mol

7.11 (a) $W = 830$ J/mol   (b) $Q = -1250$ J/mol; $\Delta U = -2080$ J/mol; $\Delta H = -2910$ J/mol

7.12 (a) $W = 105$ mJ   (b) $Q_p = \Delta H = 20.90$ kJ; $\Delta U = 20.90$ kJ

7.13 *Case 1*: $T_2 = 1380$ K; $Q = 0$; $\Delta U = -W = 13.5$ kJ/mol; $\Delta H = 22.4$ kJ/mol; *Case 2*: $T_2 = 1071$ K; $Q = 0$; $\Delta U = -W = 16.0$ kJ/mol; $\Delta H = 22.4$ kJ/mol. For n moles, T_2 is the same; W, ΔU, and ΔH are n times larger.

7.14 *Case 1*: $T_2 = 754$ K; $Q = 0$; $\Delta U = -W = 5.66$ kJ/mol; $\Delta H = 9.44$ kJ/mol. *Case 2*: $T_2 = 579$ K; $Q = 0$; $\Delta U = -W = 5.80$ kJ/mol; $\Delta H = 8.12$ kJ/mol

7.15 *Case 1*: $T_2 = 192$ K; $Q = 0$; $\Delta U = -W = -1.35$ kJ/mol; $\Delta H = -2.24$ kJ/mol. *Case 2*: $T_2 = 223$ K; $Q = 0$; $\Delta U = -W = -1.60$ kJ/mol; $\Delta H = -2.24$ kJ/mol

7.16 *Case 1*: $T_2 = 119$ K; $Q = 0$; $\Delta U = -W = -2.26$ kJ/mol; $\Delta H = -3.76$ kJ/mol.
   Case 2: $T_2 = 155$ K; $Q = 0$; $\Delta U = -W = -3.01$ kJ/mol; $\Delta H = -4.22$ kJ/mol

7.17 $T_2 = 202$ K; $Q = 0$; $\Delta U = -W = -1.20$ kJ/mol; $\Delta H = -2.00$ kJ/mol

7.18 $Q = 0$; $\Delta U = -W = -208$ J/mol; $\Delta H = -291$ J/mol

7.19 $T_2 = 235$ K; $Q = 0$; $\Delta U = -W = -1.21$ kJ/mol; $\Delta H = -1.69$ J/mol

7.20 (a) 110.5 kPa   (b) 107.9 kPa   **7.21** 1.66

7.22 $Q = 0$; $\Delta U = -W = 624$ J/mol; $\Delta H = 873$ J/mol

7.23 $p_2 = 452$ kPa; $Q = 0$; $\Delta U = -W = 6.24$ kJ/mol; $\Delta H = 8.73$ kJ/mol

7.24 (a) $\Delta U = \Delta H = 0$; $Q = W = 1.69$ kJ/mol
   (b) $W = 0$; $Q_V = \Delta U = 1.00$ kJ/mol; $\Delta H = 1.66$ kJ/mol
   (a) + (b): $Q = 2.69$ kJ/mol; $W = 1.69$ kJ/mol; $\Delta U = 1.00$ kJ/mol; $\Delta H = 1.66$ kJ/mol

7.25 (a) $\Delta U = \Delta H = 0$; $Q = W = 0.50$ kJ/mol
   (b) $W = 0$; $Q_V = \Delta U = -1.04$ kJ/mol; $\Delta H = -1.46$ kJ/mol
   (a) + (b): $Q_V = -0.54$ kJ/mol; $W = 0.50$ kJ/mol; $\Delta U = -1.04$ kJ/mol; $\Delta H = -1.46$ kJ/mol

7.26 (a) $M = (nRT/gh)(1 - p_2/p_1)$ (b) $M' = (nRT/gh)[(p_1/p_2) - 1]$
   (c) $M' - M = (nRT/gh)(p_1 - p_2)^2/p_1 p_2$
   (d) $M = 1.27$ Mg; $M' = 2.54$ Mg; $M' - M = 1.27$ Mg

7.27 (a) $W = RT[2 - (P'/p_1) - (p_2/P')]$ (b) $P' = (p_1 p_2)^{1/2}$ (c) $W_{max} = 2RT[1 - (p_2/p_1)^{1/2}]$

7.28 -9004 J/mol

7.29 (a) $Q_p = \Delta H = 6195.3$ J/mol; $W = 1662.9$ J/mol; $\Delta U = 4532.4$ J/mol
   (b) $Q_V = \Delta U = 4532.4$ J/mol; $W = 0$; $\Delta H = 6195.3$ J/mol

7.30 -3.54 kJ/mol   **7.31** 490 atm   **7.32** 60 atm

7.33 $Q = 0$; $W = 2400$ J/mol; $\Delta U = -2400$ J/mol; $\Delta H = -2900$ J/mol

7.34 3.47 kJ/mol

7.35 (a) -285.4 kJ/mol   (b) -562.0 kJ/mol   (c) 142 kJ/mol   (d) 172.45 kJ/mol
   (e) -128.2 kJ/mol   (f) -851.5 kJ/mol   (g) -179.06 kJ/mol   (h) -128 kJ/mol
   (i) 178.3 kJ/mol

7.36 (a) -287.9 kJ/mol   (b) -558.3 kJ/mol   (c) 144 kJ/mol   (d) 169.97 kJ/mol
   (e) -120.8 kJ/mol   (f) -851.5 kJ/mol   (g) -176.58 kJ/mol   (h) -130 kJ/mol
   (i) 175.8 kJ/mol

7.37 (a) 49.07 kJ/mol   (b) -631.12 kJ/mol   **7.38** -59.8 kJ/mol

7.39 (a) -5635 kJ/mol   (b) -2232 kJ/mol   (c) 1195 J/K

7.40 (a) -1366.9 kJ/mol   (b) -277.6 kJ/mol

7.41 FeO: -266.3 kJ/mol; Fe_2O_3: -824.2 kJ/mol

7.42 (a) -937 kJ/mol   (b) -933 kJ/mol   **7.43** H_2S: -20.6 kJ/mol; FeS_2: -178 kJ/mol

7.44 -180 kJ/mol

7.45 (a) 44.016 kJ/mol   (b) 2.479 kJ/mol   (c) 41.537 kJ/mol   (d) 40.887 kJ/mol

7.46 -45.98 kJ/mol   **7.47** 132.86 kJ/mol   **7.48** -223.91 kJ/mol   **7.49** -53.87 kJ/mol

7.50 298 K: -1255.5 kJ/mol; 1000 K: -1259.8 kJ/mol   **7.51** -812.2 kJ/mol

7.52 (a) -73 kJ/mol   (b) -804 kJ/mol   **7.53** -57.18 kJ/mol

7.54 -61.9 kJ/mol; -68.3 kJ/mol

7.55 $[n\text{Aq}; \Delta H_S/(\text{kJ/mol})]$: (1; -27.80); (2; -41.45); (4; -53.89); (10; -66.54); (20; -70.93); (100; -73.65); (∞; -95.28)

7.56 $[\Delta H/(\text{kJ/mol}); \Delta U/(\text{kJ/mol})]$ (a) (428.22; 425.74)   (b) (926.98; 922.02)
   (c) (498.76; 496.28)

7.57 SiF: 596 kJ/mol; SiCl: 398 kJ/mol; CF: 490 kJ/mol; NF: 279 kJ/mol; OF: 215 kJ/mol; HF: 568 kJ/mol

7.58 (a) 415.9 kJ/mol   (b) 330.6 kJ/mol   (c) 589.3 kJ/mol   (d) 810.8 kJ/mol

7.59 302.4 kJ/mol   **7.60** (a) 7500 K   (b) 2900 K   (c) 5100 K   **7.61** 27 units

7.62 6.9 min   **7.63** For $\Delta p = 10$ atm, $\Delta H = 18.2$ J/mol; For $\Delta T = 10$ K, $\Delta H = 753$ J/mol

7.64 3.78 °C

7.65 (a) 1.667   (b) 1.286   (c) 1.167   (d) $\gamma_{Ar} = 1.667$; $\gamma_{N_2} = 1.400$; $\gamma_{I_2} = 1.292$; $\gamma_{H_2O} = 1.329$
   (e) $\gamma = 1$

Chapter 8

8.1 (a) Reverse Carnot engine; make $W_{comp} = 0$
(b) Forward Carnot engine; make $Q_{2,\,comp} = 0$
8.2 0.251
8.3 (a) 62.7% (b) 41.9% less 5% other losses = 37% (c) 119 Mg/hr
(d) 36 200 MJ/min (e) 9.2 °C
8.4 (a) 80% (b) 1500 K   **8.5** 6.2 m^2 **8.6** 640 W   **8.7** 0.24 hp   **8.8** 457 g/min
8.9 0.52 hp   **8.10** 255 K   **8.11** 2.79 m^2
8.12 (a) 9.9 (b) 0.69
(c) Case (a): Furnace supplies 0.081 of energy supplied by heat pump per unit of fossil fuel consumed. Heat pump is more economical. Case (b): Furnace supplies 1.16 of energy supplied by heat pump. Furnace is more economical.
8.13 (a) 2.2 (b) 7.5%   **8.14** High temp: 23.0; Low temp: 10.0
8.15 $\eta = 36.0$; EER = 128
8.16 (a) $t = 373.15(1 - T/273.15)$ (b) $t = T - 273.15$
8.17 (a) $-R \ln 2 = -5.76$ J/K mol (b) $-R \ln 2$ (c) $+R \ln 2$
(d) $R \ln 2 \neq 0$; note that $\Delta S_1 > Q_1/T$.

Chapter 9

9.1 (a) 13.7 J/K mol (b) 22.8 J/K mol (c) three times larger in each case
9.2 (a) 47.948 J/K mol (b) 178.540 J/K mol   **9.3** 13.2 J/K mol
9.4 (a) 11.71 J/K mol (b) 40.06 J/K mol   **9.5** 25.00 J/K mol   **9.6** 81.5 J/K mol
9.7 (a) 1.03 J/K mol S (b) 3.14 J/K mol S (c) 8.2 J/K mol S_8; 25.1 J/K mol S_8
9.8 (a) 23.488 J/K mol (b) 154.443 J/K mol   **9.9** 216.127 J/K mol
9.10 (a) 99.89 J/K mol (b) 18.47 kJ/mol   **9.11** 33.77 J/K mol
9.12 $\Delta H = 2849.5$ J/mol; $\Delta S = 8.8934$ J/K mol
9.13 (a) 5.763 J/K mol (b) 28.82 J/K mol
9.14 16.021 J/K mol   **9.15** 10.1 J/K mol
9.16 $Q = 487$ J/mol; $\Delta U = 187$ J/mol; $\Delta H = 312$ J/mol; $\Delta S = 6.78$ J/K mol
9.17 $Q = 2747$ J/mol; $W = 2728$ J/mol; $\Delta U = 18.5$ J/mol; $\Delta H = 31.7$ J/mol; $\Delta S = 9.152$ J/K mol
9.18

	$Q/$(J/mol)	$W/$(J/mol)	$\Delta U/$(J/mol)	ΔH(J/mol)	$\Delta S/$(J/K mol)	$(Q/T)/$(J/K mol)
(a)	1250	0	1250	2080	2.39	—
(b)	2080	830	1250	2080	5.98	—
(c)	1730	1730	0	0	5.77	5.77
(d)	1250	1250	0	0	5.77	4.17
(e)	0	0	0	0	5.77	0
(f)	0	748	−748	−1250	1.12	0
(g)	0	910	−910	−1520	0	0

9.19 16.49 J/K mol   **9.20** $\gamma = 0.00063276$ J/K^2 mol; $a = 0.00007222$ J/K^4 mol
9.21 26.80 J/K mol   **9.22** (a) -0.377 J/K mol (b) -0.369 J/K mol
9.23 -0.0355 J/K mol; -0.0355 J/K mol   **9.24** $(\partial S/\partial p)_T = -V\alpha$ **9.25** $\Delta T = -1.49$ K
9.27 Final state: 11.2 g ice and 38.8 g liquid H_2O at 0 °C; $\Delta H = 0$; $\Delta S = 0.50$ J/K
9.28 (a) 0.28 g; 0.01 J/K mol (b) 0.77 g; 0.02 J/K (c) 34 g; 0.6 J/K (d) 123 g; 1.6 J/K
9.29 (a) All liquid at 64.0 °C (b) 23 J/K   **9.30** (a) 108 g (b) 144 J/K
9.31 (a) 15 (b) 15 (c) $\frac{3}{5}$ **9.32** (a) 10 (b) 1 (c) 2; $\Delta S = k \ln 2$
9.33 (a) N_c^N (b) $N_c(N_c - 1)(N_c - 2) \ldots [N_c - (N - 1)] = N_c!/(N_c - N)!$ (c) 0.4927
9.34 18.27 J/K mol
9.35 $[x_a, S_{mix}/(\text{J/K mol})]$: (0, 0); (0.2, 4.16); (0.4, 5.60); (0.5, 5.76); (0.6, 5.60); (0.8, 4.16); (1.0, 0)

Chapter 10

10.1 $a/\overline{V}^2$ **10.3** 5.29 J/mol; 120 J/mol
10.4 (a) $(\partial S/\partial V)_T = R/(\overline{V} - b)$ (b) $\Delta S = R \ln[(\overline{V}_2 - b)/(\overline{V}_1 - b)]$ (c) $\Delta S_{vdW} > \Delta S_{id}$
10.5 $(\partial U/\partial V)_T = 2a/T\overline{V}^2$; $(\partial U/\partial V)_T = [a/\overline{V}(\overline{V} - b)]e^{-a/RTV}$

10.6 $p = T(\partial p/\partial T)_V$; $p = Tf(V)$

10.9 (a) $\Delta H = -2.48$ kJ/mol; $\Delta S = -38.9$ J/K mol

(b) $\Delta H = -1.75$ kJ/mol; $\Delta S = -38.6$ J/K mol

(c) For both cases: $\Delta H_{id} = 0$; $\Delta S_{id} = -38.3$ J/K mol

10.10 $\Delta H = -4.120$ kJ/mol; $\Delta S = -55.946$ J/K mol   **10.12** $\bar{C}_p \mu_{JT} = (2a/RT) - b$

10.14 -3.44 kJ/mol

10.15 (a) $\bar{A} = \bar{A}°(T) - RT\ln(V/V°)$ (b) $A = \bar{A}° - RT\ln[(\bar{V} - b)/(\bar{V}° - b)] - [a/\bar{V}) - a/\bar{V}°)]$

10.16 -8.03 kJ/mol

10.17 $\Delta G = RT\ln(p/p°) + (b - a/RT)(p - p°)$ where $p° = 1$ atm   **10.18** -7.92 kJ/mol

10.19 (a) 5.74 kJ/mol   (b) 16 J/mol   (c) 6.4 J/mol   (d) 24 J/mol

10.20 $\ln f = \ln p + (b - a/RT)(p/RT)$ **10.23** Set cross derivatives equal.

10.26 $p\kappa \ll T\alpha$; -3.55 J/atm

10.29 (a) $\bar{S} = \bar{S}°(T) - R\ln p$; $\bar{V} = RT/p$; $\bar{H} = \mu°(T) + T\bar{S}°(T) = \bar{H}°(T)$; $U(T) = \bar{H}°(T) - RT = U°(T)$

(b) $\bar{S} = \bar{S}°(T) - R(\ln p) - ap/RT^2$; $\bar{V} = (RT/p) + b - (a/RT)$; $\bar{H} = \bar{H}°(T) + [b - (2a/RT)]p$ where $\bar{H}°(T) = \mu°(T) + T\bar{S}°(T)$; $\bar{U} = \bar{H}° - RT - (ap/RT) = \bar{U}°(T) - ap/RT$

Chapter 11

11.2 $[p/\text{atm}, \mu/(\text{kJ/mol})]$: $(\frac{1}{2}, -18.2)$; $(2, -14.8)$; $(10, -10.8)$; $(100, -5.1)$

11.3 (a) -34.4 kJ   (b) -47.3 kJ   (c) -12.9 kJ

11.4 (a) 18.7 J/K   (b) -5.58 kJ

(c) and (d) $(\xi/\text{mol}, \Delta G_{mix}/\text{kJ}, G/\text{kJ})$; $(0, -5.58, -5.58)$; $(0.2, -7.57, -14.17)$; $(0.4, -7.81, -21.0)$; $(0.6, -6.97, -26.8)$; $(0.8, -4.90, -31.3)$; $(1.0, 0, -33.0)$

(e) $\xi_e = 0.939$ mol; $G = -33.2$ kJ

11.5 (a) $\Delta G_{mix} = 12RT\{\frac{1}{3}\ln\frac{1}{3} + [(8-n)/12]\ln[(8-n)/12] + (n/12)\ln(n/12)\}$ (b) $n = 4$ mol

(c) -2.74 kJ/mol

11.7 (a) $G = \mu°_{H_2(g)} + \mu°_{I_2(g)} + \xi\Delta G° + 2RT[\ln p + (1-\xi)\ln\frac{1}{2}(1-\xi) + \xi\ln\xi]$

(b) $G = \mu°_{H_2(g)} + \mu°_{I_2(s)} + \xi\Delta G° + RT[(1-\xi)\ln(1-\xi) + 2\xi\ln 2\xi - (1+\xi)\ln(1+\xi) + (1+\xi)\ln p]$

11.8 $K_p = 112.9$; $x_{HI} = 0.842$; the same at 10 atm

11.9 (a) 1.6×10^{-5}; 1.6×10^{-4} (b) 1 atm: 0.999969; 10 atm: 0.99969

(c) $K_x = 6.2 \times 10^{-4}$; $K_c = 1.5 \times 10^{-3}$

11.10 (a) 6.6×10^{-58} (c) $K_x = 3.3 \times 10^{-57}$; $K_c = 1.6 \times 10^{-56}$

11.11 5.09×10^{-3}; 2.36×10^{-3} **11.12** (a) 0.186 (b) 0.378 (c) 0.186 **11.13** 1.3×10^{-6}

11.14 (a) $\Delta G° = 37$ kJ/mol; $\Delta H° = 88$ kJ/mol   (b) 19 (c) 1 atm: 0.975; 5 atm: 0.890

11.15 (a) 1.906×10^{-25} (b) 0.06667 (c) 1300 K   **11.16** (a) 6.89×10^{-15} (b) 1350 K

11.17 -11.1 kJ/mol

11.18 (a) 0.982 (b) $\Delta H° = 88.9$ kJ/mol; $\Delta G° = -2.49$ kJ/mol; $\Delta S° = 175.7$ J/K mol

11.19 (a) 0.64 (b) 3.0 kJ/mol   **11.20** (a) 0.14 (b) 2.0×10^{-18} (c) 101 kJ/mol

11.21 (a) 0.379; 1.28 (b) $\Delta G°_{700} = 5.65$ kJ/mol; $\Delta G°_{800} = -1.64$ kJ/mol; $\Delta H° = 56.7$ kJ/mol

11.22 (a) 1.40, -2.80 kJ/mol   (b) -29.72 kJ/mol   **11.23** 5.7 kJ/mol

11.24 (a) 8.6×10^{-6} (b) 3.2×10^{-6} (c) $(K_x)_{5\,\text{atm}} = 5(K_x)_{1\,\text{atm}}$

11.25 40.888 kJ/mol   **11.26** 3.23 kJ/mol

11.27 (a) -19.0 kJ/mol; 0.765 kJ/mol; -22.6 J/K mol   (b) 0.474

11.28 (a) 0.6841; 0.5198 (b) -12.78 kJ/mol; -21.42 J/K mol   (c) -6.40 kJ/mol

11.29 $MgCO_3$: 570 K; $CaCO_3$: 1110 K; $SrCO_3$: 1400 K; $BaCO_3$: 1600 K

11.30 (a) 52.20 kJ/mol   (b) 555 K   (c) 0.160 Torr   (d) 0.0421 Torr   (e) 24.29 kJ/mol

11.31 (298.15 K $-$ 548 K): $\Delta G°/(\text{kJ/mol}) = -369.43 + 0.1530\,(T/\text{K} - 298.15)$

(548 K $-$ 693 K): $\Delta G°/(\text{kJ/mol}) = -331.2 + 0.111(T/\text{K} - 548)$

(693 K $-$ 1029 K): $\Delta G°/(\text{kJ/mol}) = -315.1 + 0.122(T/\text{K} - 693)$

(1029 K $-$ 1180 K): $\Delta G°/(\text{kJ/mol}) = -274.1 - 0.0039(T/\text{K} - 1029)$

(1180 K $-$ T): $\Delta G°/(\text{kJ/mol}) = -274.7 + 0.00934(T/\text{K} - 1180)$

11.32 (a) 2.30×10^{-5} atm

(b) $\ln K_p = 10\,950.1/T - 0.185 \ln T + 1.242 \times 10^{-3} T + 0.051 \times 10^5/T^2 - 12.486$;
$\Delta H^\circ/(\text{J/mol}) = -91\,044 - 1.54 \ln T + 10.33 \times 10^{-3} T^2 - 0.84 \times 10^5/T$;
$\Delta S^\circ/(\text{J/K mol}) = -105.35 - 1.54 \ln T + 20.66 \times 10^{-3} T - 0.42 \times 10^5/T^2$

11.33 (a) 460.3 K

(b) $\log_{10} K_p = -1691.5/T - 0.9047 \log_{10} T + 6.084$;
$\Delta H^\circ/(\text{J/mol}) = 32\,384 - 7.522\,T$;
$\Delta S^\circ/(\text{J/K mol}) = 108.96 - 17.32 \log_{10} T$

11.34 The ratio, O_2/CO_2, is constant; there is relatively less CO at higher pressures.

		O_2	CO	CO_2
(a)	600 K	3.92×10^{-33}	0.121	99.88
	1000 K	6.25×10^{-20}	68.4	31.6
(b)	600 K	3.99×10^{-33}	0.130	99.87
	1000 K	6.16×10^{-20}	71.2	28.8
(c)	1000 K	1.41×10^{-19}	34.0	66.0

11.35

	A/G	B/G	C/G	D/G	E/G	F/G
(a)	9.86×10^{-8}	1.50×10^{-6}	4.12×10^{-6}	2.49×10^{-5}	4.60×10^{-7}	2.03×10^{-4}

(b) No.

	A	B	C	D	E	F	G
(c) mol %:	9.86×10^{-6}	1.50×10^{-4}	4.12×10^{-4}	2.49×10^{-3}	4.60×10^{-5}	0.0203	99.98
(d) mol %:	0.0870	0.414	0.623	1.79	0.110	5.28	91.70

11.36 (a) $(T/K; p/\text{atm})$: $(900; 1.2 \times 10^{-9})$; $(1200; 3.8 \times 10^{-5})$

(b) $(T/K; p/\text{atm})$: $(900; 2.1 \times 10^{-13})$; $(1200; 5.1 \times 10^{-9})$

(c) At no temperature.

11.37 (a) 201.2 kJ/mol; 489.0 kJ/mol

(b) 900 K: 2.27×10^{-13}; 0.00271; 0.9973; 1200 K: 5.25×10^{-9}; 0.117; 0.883

11.38 (c) The entropy is independent of z; $\bar{H}_i = \bar{H}_i^\circ(T) + M_i gz$

11.40 (c) $(T/K; \bar{C}_p/R)$: $(200; 9.68)$; $(240; 14.08)$; $(280; 21.56)$; $(320; 85.23)$; $(330; 90.75)$; $(360; 59.68)$; $(400; 19.13)$; $(440; 10.87)$; $(480; 9.39)(500; 9.17)$

11.41 (a) 1.0×10^{-16}; 1.6×10^{-7} (b) 1.9×10^{25}; 7.8×10^{13}

(c) For (a): 2.0×10^{-8}; 8.0×10^{-4}; For (b): $1 - 1.1 \times 10^{-13}$; $1 - 5.6 \times 10^{-8}$

11.42 ΔG_{mix} is greater for reaction 2.   **11.43** 300

Chapter 12

12.1 76 kJ/mol   **12.2** 60.8 °C

12.3 (a) 29.8 kJ/mol (b) 34 °C; 29 °C (c) 97.4 J/K mol (d) 861 J/mol

12.4 (a) 94.3 °C (b) 134.1 °C   **12.5** 0.03128 atm = 3169 Pa

12.6 1162 K; 101.4 kJ/mol; 87.3 J/K mol

12.7 (a) 48.5 kJ/mol; 489 K; 99.2 J/K mol (b) 7.65 mmHg = 1020 Pa

(c) $\Delta H_{\text{sub}} = 71.0$ kJ/mol; $\Delta H_{\text{fus}} = 22.5$ kJ/mol (d) $T < 226.3$ K

12.8 (a) 384 K; 10.8 kPa (b) 45.1 kJ/mol; 98.9 J/K mol (c) 19.11 kJ/mol

12.9 22.8 kJ/mol; 239 K

12.10 (a) $1/T = (1/T_0) + M_{\text{air}}gh/T_a\Delta H_{\text{vap}}$, where T_0 is b.p. at 1 atm (b) 86 °C (c) 25 °C

12.11 (a) $\ln p = 10.8(1 - T_b/T)$ (b) 72 kPa   **12.12** S_8: 117 J/K mol; P_4: 90.0 J/K mol

12.14 $d \ln \tilde{c}/dT = (\Delta H_{\text{vap}} - RT)/RT^2 = \Delta U_{\text{vap}}/RT^2$

12.15 (a) p_0/RT_b, $p_0 = 1$ atm (b) $\ln(T_H/T_0) = (\Delta H_{\text{vap}}/R)[(1/T_b) - (1/T_H)]$

(c) $(T_H/K; T_b/K)$: $(50; 59.0)$; $(100; 109.9)$; $(200; 205.8)$; $(300; 297.5)$; $(400; 386.7)$

(d) [Substance: T_H/K; $\Delta S_H/(\text{J/K mol})$; $\Delta S_T/(\text{J/K mol})$]; (Ar: 77.4; 84.2, 74.7); (O_2: 80.3; 85.0; 75.6); (CH_4: 101.8; 80.4; 73.3); (Kr: 110.1; 82.0; 75.3); (Xe: 156.9; 80.6; 76.6); (CS_2: 324.3; 82.6; 83.8) Note: $\langle \Delta S_H \rangle = 82.5 \pm 1.9$ J/K mol; $\langle \Delta S_T \rangle = 76.6 \pm 3.7$ J/K mol

12.16 1.50 GPa = 14800 atm   **12.17** (a) 0.36 °C (b) 3400 atm (c) -24 °C

12.18 119 °C; Possible range from 83 °C to 277 °C   **12.19** 13 °C   **12.20** Rhombic
12.21 0.017 mmHg = 2.3 Pa

Chapter 13

13.1 (a) 60 g/mol   (b) 333 g   **13.2** 428 g   **13.3** 0.0099; 0.0050; 0.00010   **13.4** (d) M
13.5 9.986 kPa   **13.6** 242 g/mol; about twice the expected value   **13.7** 3.577 K kg/mol
13.8 $(x, T/K)$; (1.0, 273); (0.8, 252); (0.6, 229); (0.4, 203); (0.2, 170)
13.9 (vol %; T/K); (0; 273); (20; 265); (40; 254); (60; 238); (80; 208)
13.10 $m > 0.59$ mol/kg   **13.11** $a = K_f$; $b = -\frac{1}{2}MK_f[1 + 2(K_f/MT_0) - (\Delta C_p/R)(K_f/MT_0)^2]$
13.12 3.8 K; 0.018; 470 kPa; 250 g/mol   **13.13** K_b/(K kg/mol); 1.730; 2.631; 3.77; 0.188; 2.391
13.14 At 1 atm let $K_b = K_b^\circ$; $T_b = T_0$; Then, $K_b(p) = K_b^\circ/[1 - (RT_0/\Delta H_{vap})\ln p]^2$;
     [p/mmHg; K_b/(K kg/mol)]: (760; 0.5130); (750; 0.5120); (740; 0.5109)
13.15 (a) 0.250   (b) 0.534   **13.16** (a) 0.236   (b) 90.8 g I_2/100 g hexane
13.17 19.1 kJ/mol; 80.0 °C   **13.18** 250 kPa   **13.19** 3.75 m; 36.7 kPa   **13.20** 62.0 kg/mol
13.21 (a) $x = (1 - \tilde{c}\tilde{V}_2^\circ)/(1 + \tilde{c}\tilde{V}^\circ - \tilde{c}\tilde{V}_2^\circ)$

Chapter 14

14.1 (a) 60.44 mmHg   (b) $y_b = 0.6817$ (c) 44.25 mmHg   (d) $x_b = 0.1718$
     (e) 56.42 mmHg; $x_b = 0.3433$; $y_b = 0.6268$
14.2 27.3 mol %   **14.3** (a) 0.560   (b) 0.884
14.4 (a) 25 mmHg; $x_{EtCl} = 0.50$ (b) $x_{EtCl} = 0.61$ **14.5** $p_A^\circ = 2$ atm; $p_B^\circ = 0.5$ atm
14.6 $x_1 = [(p_1^\circ p_2^\circ)^{1/2} - p_2^\circ]/(p_1^\circ - p_2^\circ)$; $p = (p_1^\circ p_2^\circ)^{1/2}$
14.9 If X_1 is overall mole fraction of 1, then $p_{upper} = X_1 p_1^\circ + (1 - X_1)p_2^\circ$; $1/p_{lower} = (X_1/p_1^\circ) + (1 - X_1)/p_2^\circ$.
14.10 b = benzene; t = toluene;
     (a) $\exp(-\Delta S_{vap}^\circ/R) = x_b \exp(-T_{0b}\Delta S_{vap}^\circ/RT) + (1 - x_b)\exp(-T_{0t}\Delta S_{vap}^\circ/RT)$
     (b) $x_b = 0.401$
14.12 (a) $p = (1 - x_2)p_1^\circ + K_h x_2$ (b) $1/p = (y_1/p_1^\circ) + (1 - y_1)/K_h$
14.13 1.71 cm³; 17.1 cm³; $N_2/O_2 = 2.02$ **14.14** -13.9 kJ/mol   **14.15** 380 cm³
14.16 (a) m/(mol/kg): 0.0346; 0.0265   (b) 0.776; 0.594
14.17 (gas; α); (He; 0.0097); (Ne; 0.0097); (Ar; 0.0313); (Kr; 0.0507); (Xe; 0.101)
14.18 (a) 0.036   (b) (1 atm; 0.0373); (4 atm; 0.0776)   **14.19** 0.33
14.20 (a) 10.6 cm³   (b) 5.09 cm³   (c) As $n \to \infty$, 2.71 cm³   **14.21** -608.44 kJ/mol
14.22 -9.957 kJ/mol   **14.23** -17.124 kJ/mol

Chapter 15

15.1 (a) 730 mmHg   (b) $\sim$92 °C   (c) 4.13 g; 4.14 g
15.2 (a) 129.7 g; 50.3 g   (b) 3.99 mol; 2.41 mol   **15.3** 0.858; 249 °C
15.4 (b) [mass % Cu; $t/$°C (ideal)]; (0; 660); (20; 597); (40; 517); (60; 474); (80; 696); (100; 1083)
15.5 ~ -13 °C   **15.6** 15%   **15.7** 62%
15.8 $x_{Ni} = 0.90$; $x_{Cu}' = 0.079$; $x_{Ni}' = 0.921$ $T = 1694$ K
15.9 In Liq, or α, or β: $F = 2$. In (Liq + α), or (Liq + β), or ($\alpha + \beta$): $F = 1$. On abc: $F = 0$.
15.10 In $aCBe$: $F = 2$; in Aab, or bcd, or Ade: $F = 1$; in Abd: $F = 0$.
15.11 (a) $K_2CO_3(s)$ + soln on de; then $K_2CO_3(s)$ + soln d + soln b; then soln bc + soln dc; then one soln.
     (b) $K_2CO_3(s)$ + soln on ab; then $K_2CO_3(s)$ + soln b + soln d; $K_2CO_3(s)$ + soln on de; one soln in region between e and B.
15.12 (a) In sln, $F = 2$; In sln + Na_2SO_4, or sln + hydrate, or Na_2SO_4 + hydrate, or ice + hydrate: $F = 1$; at e and along bc, $F = 0$.
     (b) At 25 °C: solid hydrate precipitates, when sln disappears, Na_2SO_4 appears, hydrate slowly decomposes; finally we are left with Na_2SO_4 only. At 35 °C: anhydrous salt precipitates; liquid evaporates, finally we have Na_2SO_4 only.

15.13 At b, $Fe_2Cl_6 \cdot 12H_2O$ precipitates; at c system looks dry. Between c and d liquid forms in equilibrium with $Fe_2Cl_6 \cdot 12H_2O$; between d and e system is entirely liquid. $Fe_2Cl_6 \cdot 7H_2O$ precipitates at e; at f system looks dry; liquid appears between f and g; between g and h system is entirely liquid; $Fe_2Cl_6 \cdot 5H_2O$ precipitates at h; system goes dry at i; $Fe_2Cl_6 \cdot 5H_2O$ and $Fe_2Cl_6 \cdot 4H_2O$ are present to vertical line, then anhydrous Fe_2Cl_6 appears; at j we have a mixture of $Fe_2Cl_6 \cdot 4H_2O$ and Fe_2Cl_6.

Chapter 16

16.1 (a) 0.99818; 0.99633; 0.99055; 0.9802; 0.9690; 0.9564
(b) 0.99998; 0.99992; 0.99947; 0.9979; 0.9951; 0.9909   (d) 1.24

16.2 $(a; \gamma)$; (0.061; 1.03); (0.135; 1.10); (0.211; 1.14)

16.3 $(a; \gamma)$; (0.9382; 0.997); (0.8688; 0.991); (0.7994; 0.981)

16.4 (a) 1.000; 0.959; 0.898   (b) 1.000; 1.038; 1.074   (c) 177 J/mol

16.5 (a) μ_i° is μ of pure i (b) $RT \ln \gamma_i = w(1 - x_i)^2$ (c) 1.140; 1.0763; 1.0332; 1.00820; 1.000

16.6 $(a_2; \gamma_2)$: (0.0986; 0.986); (0.196; 0.981)

16.7 0.0149; 0.0209; 0.0322; 0.0437; 0.0583; 0.0832; 0.1077

16.8 $(a_\pm; a)$:   (a) 0.0769; 0.00591   (b) 0.0421; 7.44×10^{-5} (c) 0.016; 2.6×10^{-4}
(d) 0.075; 3.2×10^{-5} (e) 0.0089; 5.7×10^{-11}

16.9 (a) $[m/(mol/kg)]$: 0.0794; 0.05; 0.05; 0.114   (b) $[I_c/(mol/kg)]$: 0.15; 0.05; 0.20; 0.30

16.10 HCl: 0.988; 0.964; $CaCl_2$: 0.960; 0.879; $ZnSO_4$: 0.910, 0.743   **16.11** 3.0 nm; 0.30 nm

16.12 (a) 0.736   (b) $1.68/\varkappa$

16.13 $(m/mol/kg)$; 100α; $100\alpha_0)$; (0.01; 4.18; 4.09); (0.10; 1.37; 1.31); (1.0; 4.51; 4.18)

16.14 0.0202; 0.0346; 0.149   **16.15** 10^5s: 1.29; 1.38; 1.56; 1.84

16.16 (a) 2.5×10^{-5} (b) 1.6×10^{-5}

Chapter 17

17.1 (a) -1.473 V; not spontaneous   (b) -0.312 V; not spontaneous   (c) 1.344 V; spontaneous

17.2 (a) 1.56×10^{-53} (b) 5.25×10^{-10} (c) 2.64×10^{44}

17.3 (a) 1.54×10^{37} (b) 8.0×10^{16} (c) 1×10^{-3} (d) 8.7×10^{40} (e) 5×10^{46}
(f) 1.7×10^{-8}

17.4 (a) $Ni_2O_3(s) + Fe(s) \rightarrow 2NiO(s) + FeO(s)$ (b) Independent of a_{KOH} (c) 1100 kJ/kg

17.5 (a) 0.38   (b) $PbO_2(s) + Pb(s) + 4H^+ + 2SO_4^{2-} \rightarrow 2PbSO_4(s) + 2H_2O(l)$; yes
(c) 415 kJ/mol PbO_2 (d) $\mathscr{E} = 2.041 + 0.05916 \log_{10} a$ (e) 605.4 kJ/kg

17.6 (a) $Fe^{2+} + 2Hg(l) + SO_4^{2-} \rightarrow Fe(s) + Hg_2SO_4(s)$
(b) -1.114 V; 2.1×10^{-36}; 2.036 kJ/mol

17.7 (a) 1.8×10^{-4} (b) 0.029

17.8 (a) 10   (b) 0.10   (c) 8.1×10^{-5}; 4.0×10^{-3}; 0.16; 0.50; 0.91; 0.998; 0.99996

17.9 (a) $K = 2.8 \times 10^6$ (b) -37 kJ/mol

17.10 (a) 0.799 V; 0.740 V; 0.681 V; 0.622 V   (b) 0.324 V   (c) -0.151 V

17.11 ϕ/V: 0.298; 0.339; 0.399; 0.458; 0.510; 0.562; 0.621; 0.681; 0.722

17.12 (a) $\phi^\circ < 0$ (b) $\phi^\circ < -0.414$ V   (c) Basic solution

17.13 (a) $\phi^\circ > 0.401$ V   (b) $\phi^\circ > 1.229$ V   (c) $\phi^\circ > 0.815$ (d) Acid Soln.

17.14 Na^+: -261.9 kJ/mol; Pb^{2+}: -24.3 kJ/mol; Ag^+: 77.10 kJ/mol   **17.15** -10.5 kJ/mol

17.16 -131.1 kJ/mol

17.17 $\mathscr{E}^\circ = 0.22238$ V; $[m/(mol/kg); \gamma_\pm]$: (0.001, 0.965); (0.01; 0.905); (0.1; 0.796); (1.0; 0.809); (3; 1.316)

17.18 0.075 V; 0.156 V; 0.190 V

17.19 (a) $[t/°C; \Delta G/(kJ/mol); \Delta S/(J/K \text{ mol}); \Delta H/(kJ/mol)]$: (0; -369.993; 10.83; -367.036); (25; -370.394; 21.25; -364.060)   (b) 0.131

17.20 0.171   **17.21** 0.78

17.22 $2AgCl(s) + H_2(f = 1) \rightarrow 2Ag(s) + 2HCl(aq, m = 0.1)$; $\Delta G = -66.785$ kJ/mol; $\Delta S = -59.886$ J/K mol; $\Delta H = -86.137$ kJ/mol

17.23 (a) $H_2(p = 1 \text{ atm}) \rightarrow H_2(p = 0.5 \text{ atm})$; $\mathscr{E} = 8.90 \text{ mV}$
(b) $Zn^{2+}(a = 0.1) \rightarrow Zn^{2+}(a = 0.01)$; $\mathscr{E} = 29.6 \text{ mV}$

17.24 0.8261; 11.1 mV   **17.25** $\approx 2 \times 10^{-12}$

17.26 (a) and (c) $[\xi/0.1 \text{ mol}; \mathscr{E}/V; \Delta G/\Delta G_{total}]$; (0; 1.100; 0); (0.5; 1.086; 0.505); (0.9; 1.062; 0.903); (0.99; 1.032; 0.9906); (0.999; 1.002; 0.9991); (0.9999; 0.973; 0.9999)

17.27 (a) $(f; \mathscr{E}/V)$: (0.01; 0.653); (0.1; 0.714); (0.3; 0.749); (0.5; 0.771); (0.7; 0.793); (0.9; 0.827); (0.99; 0.889)
(b) $(v/\text{mL}; \mathscr{E}/V)$: (40; 0.735); (49.0; 0.671); (49.9; 0.611); (49.99; 0.552); (50.00; 0.36); (50.01; 0.26); (50.1; 0.23); (51.0; 0.20); (60; 0.17)

Chapter 18

18.1 2.18 J   **18.2** 2.11 J; -315 J   **18.3** (a) $r_0 = 3\gamma/\Delta H_{vap}(1 - T/T_0)$; same   (b) 0.44 nm

18.4 0.108 N/m   **18.5** 1.46 cm   **18.6** 1.49 mm   **18.7** 5×10^{-5} cm

18.8 55.50 mN/m; 48.90 mN/m; 41.10 mN/m   **18.9** 288 Pa   **18.10** 0.0273 N/m

18.11 r_1; $\Delta p = 12$ Pa; r_2; $\Delta p = 6$ Pa; film radius = 2 cm; centered in smaller bubble

18.12 Smaller bubble gets smaller, larger gets larger, until smaller bubble has radius equal to that of the larger bubble.

18.13 (a) 67 mJ/m^2 (b) 57.70 mJ/m^2 for benzene; 145.50 mJ/m^2 for water   (c) 9 mN/m

18.14 (a) -23.9 mN/m   (b) 77.6 mJ/m^2

18.15 $[\delta/\mu m; (T_0 - T)/K]$: (10; 0.013); (1; 0.13); (0.1; 1.3); (0.01; 13); (0.001; 130)

18.16 $(\delta/\mu m; x/x_0)$: (1; 1.066); (0.1; 1.9); (0.01; 590)   **18.17** p/kPa; 11.75; 14.24; 97.5

18.18 $[t/°C; g^\sigma/(\text{mJ/mol}); s^\sigma/(\mu J/K \text{ mol}); u^\sigma/(\text{mJ/mol})$: (0; 75.5; 246; 143); (30; 68.2; 242; 142); (60; 61.0; 238; 140); (90; 53.9; 233; 138); (120; 47.0; 228; 136); (150; 40.3; 222; 134); (180; 33.7; 215; 131); (210; 27.4; 208; 128); (240; 21.3; 199; 124); (270; 15.4; 189; 118); (300; 9.95; 176; 111); (330; 4.95; 156; 99); (360; 0.763; 114; 73.2); (368; 0; 0; 0)

18.19 2.7 mm   **18.20** 81 g

18.21 (a) $k = 0.717$ cm^3; $1/n = 0.567$ (b) 0.292; 0.453; 0.554; 0.623; 81 m^2/g

18.22 (a) $(p/\text{mmHg}; \theta)$; (20; 0.604); (50; 0.792); (100; 0.884); (200; 0.938); (300; 0.958)
(b) 12,000 m^2

18.23 (a) 27.66 cm^3/g   (b) 331 m^2/g
(c) 10 kPa; 0.562; 0.054; 0.0053; 0.378; 20 kPa; 0.634; 0.123; 0.024; 0.213
(d) 0.436; 0.607; 530 m^2/g

18.25 (a) 2.75 μmol/m^2 (b) 3.65 μmol/m^2 **18.26** Water: no change; Hg: forms a ball.

Chapter 19

19.1 $u/(\text{J/m}^3)$: 7.57×10^{-8}; 6.13×10^{-6}; 7.57×10^{-4}

19.2 (a) 5.29×10^{-6} (b) 7.54×10^{-3} **19.3** (a) 1.64×10^{-5} (b) 0.107   (c) 0.458

19.4 (a) 9.660×10^{-6} m   (b) 5.796×10^{-6} m   **19.5** 4830 K

19.6 (a) 4.59 J/s   (b) 219 K   **19.7** 4.52×10^{26} J/s; 5.03×10^6 Mg

19.9 (a) 1.21×10^{15} s^{-1} (b) 650 km/s

19.10 (a) 656.11 nm   (b) 1.05×10^{-34} kg m^2/s   (c) 2188 km/s

19.11 (a) 150 V   (b) 7.27×10^6 m/s   **19.12** 0.0819 V   **19** 6.63×10^{-31} m

19.15 2.41×10^{-21} J

Chapter 20

20.4 $x^2(d^2/dx^2) - (d^2/dx^2)(x^2) = -2 - 4x(d/dx)$

20.5 $\mathbf{M}_z\mathbf{M}_x - \mathbf{M}_x\mathbf{M}_z = i\hbar\mathbf{M}_y$; $\mathbf{M}_y\mathbf{M}_z - \mathbf{M}_z\mathbf{M}_y = i\hbar\mathbf{M}_x$; $\mathbf{M}^2\mathbf{M}_x = \mathbf{M}_x\mathbf{M}^2$; $\mathbf{M}^2\mathbf{M}_y = \mathbf{M}_y\mathbf{M}^2$

20.6 6   **20.7** 9   **20.8** $-\frac{1}{9}$ **20.9** $P_0(x) = \sqrt{\frac{1}{2}}$; $P_1(x) = \sqrt{\frac{3}{2}}x$; $P_2(x) = \sqrt{\frac{5}{2}}(\frac{1}{2} - \frac{3}{2}x^2)$

Chapter 21

21.1 $\frac{1}{2}$ if n is even; $\frac{1}{2} + (-1)^{(n-1)/2}/n\pi$ if n is odd.   **21.2** 6.0×10^{-32} J; golf balls; $n = 1.6 \times 10^7$

21.3 54.8 μm   **21.4** $\approx 2.0 \times 10^{-30}$ m   **21.5** $\langle x^2 \rangle = \beta^2(n + \frac{1}{2})$: $\Delta x = \beta(n + \frac{1}{2})^{1/2}$; $\beta^2 = h\nu/k$

21.6 $\langle p_x^2 \rangle = (n + \frac{1}{2})(\hbar^2/\beta^2)$; $\langle p_x \rangle = 0$; $\Delta p_x = (n + \frac{1}{2})^{1/2}(\hbar/\beta)$ **21.7** $\Delta p_x \Delta x = (2n + 1)(h/4\pi)$

21.8 $\langle E_{\text{kin}} \rangle = \frac{1}{2}(n + \frac{1}{2})h\nu$; $\langle V \rangle = \frac{1}{2}(n + \frac{1}{2})h\nu$; $(n; \langle E_{\text{kin}} \rangle; \langle V_{\text{kin}} \rangle)(0; \frac{1}{4}h\nu; \frac{1}{4}h\nu)$; $(1; \frac{3}{4}h\nu; \frac{3}{4}h\nu)$

21.9 $\Delta E_{\text{kin}} = \frac{1}{2}h\nu[\frac{1}{2}(n^2 + n + 1)]^{1/2}$; $n = 0$; $\Delta E_{\text{kin}} = \frac{1}{2}(\frac{1}{2})^{1/2}h\nu$; $n = 1$, $\Delta E_{\text{kin}} = \frac{1}{2}h\nu(\frac{3}{2})^{1/2}$

21.12 (Energy, degeneracy): (3, 1); (6, 3); (9, 3); (11, 3); (12, 1); (14, 6); (17, 3); (19, 3)

21.13 (a) 4.657×10^{-48} kg m^2; 1.491×10^{-34} kg m^2/s; 2.388×10^{-21} J

      (b) 1.938×10^{-46} kg m^2; 1.491×10^{-34} kg m^2/s; 5.739×10^{-23} J

Chapter 22

22.1 (a) $(k; \lambda/\text{nm})$; Lyman: (2; 121.502); (3; 102.518); (4; 97.202); Balmer: (3; 656.112); (4; 486.009); (5; 433.937); Paschen: (4; 1874.61); (5; 1281.47); (6; 1093.52)

      (b) $(\infty; 91.127)$; $(\infty; 364.507)$; $(\infty; 820.140)$

22.3 (a) $2.66 a_0$ (b) $9.13 a_0$ (c) $7.99 a_0$ (d) $19.44 a_0$

22.5

	$1s$	$2s$	$2p$	$3s$
(a) $\langle r \rangle$	$\frac{3}{2}a_0$	$6 a_0$	$5 a_0$	$\frac{27}{2}a_0$
(b) $\langle (r - \langle r \rangle)^2 \rangle$	$\frac{3}{4}a_0^2$	$6 a_0^2$	$5 a_0^2$	$\frac{99}{4}a_0^2$

22.6 (a) $(\langle p_r \rangle; \langle p_r^2 \rangle; \Delta p_r)$: $1s$: $(0; \hbar^2/a_0^2; \hbar/a_0)$; $2s$: $(0; \hbar^2/4a_0^2; \hbar/2a_0)$; $2p$: $(0; \hbar^2/12a_0^2; \hbar/2\sqrt{3}a_0)$; $3s$: $(0; \hbar^2/9a_0^2; \hbar/3a_0)$

      (b) (state; $\Delta p_r \cdot \Delta r$): $[1s; \sqrt{3}(h/4\pi)]$; $[2s; \sqrt{6}(h/4\pi)]$; $[2p; \sqrt{5/3}(h/4\pi)]$; $[3s; \sqrt{11}(h/4\pi)]$

22.7 (state; $\langle V \rangle/E_h$; $\langle E_{\text{kin}} \rangle/E_h$): $(1s; -1; \frac{1}{2})$; $(2s; -\frac{1}{4}; \frac{1}{8})$; $(2p; -\frac{1}{4}; \frac{1}{8})$; $(3s; -\frac{1}{9}; \frac{1}{18})$

22.8 [state; $\langle F \rangle/(E_h/a_0)$]; $(1s; -2)$; $(2s; -\frac{1}{4})$; $(2p; -\frac{1}{12})$; $(3s; -\frac{2}{27})$

22.9 (a) (state; $\langle E_{\text{rot}} \rangle/E_h$; $\langle E_{\text{kin}}/E_h \rangle$; $\langle E_{\text{rot}} \rangle/\langle E_{\text{kin}} \rangle$): $(2p; \frac{1}{12}; \frac{1}{8}; \frac{2}{3})$; $(3p; \frac{2}{81}; \frac{1}{18}; \frac{4}{9})$; $(3d; \frac{2}{45}; \frac{1}{18}; \frac{4}{5})$

      (b) (state; m; fraction): $(p; 0; 0)$; $(p; 1; \frac{1}{2})$; $(d; 0; 0)$; $(d; 1; \frac{1}{6})$; $(d; 2; \frac{2}{3})$

22.10 [state; $\langle I \rangle/\mu a_0^2$]; $(1s; 3)$; $(2s; 42)$; $(2p; 30)$; $(3s; 207)$; $(3p; 180)$; $(3d; 126)$

22.11 (electron: μ_z/μ_B): $(s: \pm 1)$; $(p: 0, \pm 1, \pm 2)$; $(d: 0, \pm 1, \pm 2, \pm 3)$; $(f: 0, \pm 1, \pm 2, \pm 3, \pm 4)$

Chapter 23

23.2 σ_1, σ_4, and $\sigma_2 + \sigma_3$ are symmetric; $\sigma_2 - \sigma_3$ is antisymmetric.   **23.3** (b); (c)

23.4 tetrahedral

23.8

	E	C_3	$\bar{C}_3$	$\sigma_v^{(1)}$	$\sigma_v^{(2)}$	$\sigma_v^{(3)}$
E	E	C_3	$\bar{C}_3$	$\sigma_v^{(1)}$	$\sigma_v^{(2)}$	$\sigma_v^{(3)}$
C_3	C_3	$\bar{C}_3$	E	$\sigma_v^{(2)}$	$\sigma_v^{(3)}$	$\sigma_v^{(1)}$
$\bar{C}_3$	$\bar{C}_3$	E	C_3	$\sigma_v^{(3)}$	$\sigma_v^{(1)}$	$\sigma_v^{(2)}$
$\sigma_v^{(1)}$	$\sigma_v^{(1)}$	$\sigma_v^{(2)}$	$\sigma_v^{(3)}$	E	$\bar{C}_3$	C_3
$\sigma_v^{(2)}$	$\sigma_v^{(2)}$	$\sigma_v^{(3)}$	$\sigma_v^{(1)}$	C_3	E	$\bar{C}_3$
$\sigma_v^{(3)}$	$\sigma_v^{(3)}$	$\sigma_v^{(1)}$	$\sigma_v^{(2)}$	$\bar{C}_3$	C_3	E

23.9 $a_1: \chi_1; \chi_2 + \chi_3$;

      $b_2: \chi_2 - \chi_3$

23.10 $a_1: \chi_1 + \chi_2; \chi_3; \chi_6; \chi_7; \chi_{10}$ $b_1: \chi_4; \chi_8$;

      $b_2: \chi_1 - \chi_2; \chi_5; \chi_9$

23.11 $a_1: \chi_1; \chi_4; \chi_5 + \chi_6; \chi_9 - \chi_{10}; \chi_{11} + \chi_{12}$

      $a_2: \chi_7 - \chi_8$ $b_1: \chi_2; \chi_7 + \chi_8$ $b_2: \chi_3; \chi_5 - \chi_6; \chi_9 + \chi_{10}; \chi_{11} - \chi_{12}$

23.12 $a_g: \chi_1 + \chi_2 + \chi_3 + \chi_4; \chi_5 + \chi_6; \chi_{11} - \chi_{12}$ $b_{2g}: \chi_7 - \chi_8$

      $b_{3g}: \chi_1 - \chi_2 - \chi_3 + \chi_4; \chi_9 - \chi_{10}$ $b_{1u}: \chi_1 + \chi_2 - \chi_3 - \chi_4; \chi_5 - \chi_6; \chi_{11} + \chi_{12}$

      $b_{2u}: \chi_1 - \chi_2 + \chi_3 - \chi_4; \chi_9 + \chi_{10}$ $b_{3u}: \chi_7 + \chi_8$

23.13 $a_g: \chi_1 + \chi_2 + \chi_3 + \chi_4; \chi_5 + \chi_6; \chi_7 + \chi_8; \chi_{13} - \chi_{14}$ $b_{2g}: \chi_9 - \chi_{10}$

      $b_{3g}: \chi_1 - \chi_2 - \chi_3 + \chi_4; \chi_{11} - \chi_{12}$ $b_{1u}: \chi_1 + \chi_2 - \chi_3 - \chi_4; \chi_7 - \chi_8; \chi_{13} + \chi_{14}$

      $b_{2u}: \chi_1 - \chi_2 + \chi_3 - \chi_4; \chi_{11} + \chi_{12}$ $b_{3u}: \chi_5 - \chi_6; \chi_9 + \chi_{10}$

23.14 $a_g: \chi_1 + \chi_2; \chi_3 + \chi_4; \chi_5 - \chi_6; \chi_7 - \chi_8$ $b_g: \chi_9 - \chi_{10}$

      $a_u: \chi_9 + \chi_{10}$ $b_u: \chi_1 - \chi_2; \chi_3 - \chi_4; \chi_5 + \chi_6; \chi_7 + \chi_8$

23.15 The first ten wave functions are the same as in Problem 23.14 except that the oxygen orbitals are replaced by carbon orbitals. The remaining eight functions are:

$a_g: \chi_{11} + \chi_{12}; \chi_{13} - \chi_{14}; \chi_{15} - \chi_{16}$ $b_g: \chi_{17} - \chi_{18}$

$a_u: \chi_{17} + \chi_{18}$ $b_u: \chi_{11} - \chi_{12}; \chi_{13} + \chi_{14}; \chi_{15} + \chi_{16}$

23.16 $\Gamma = 5a_1 + a_2 + 5e$ **23.17** $\Gamma = 5a_1 + a_2 + 2b_1 + 4b_2$

Chapter 24

24.1 (a) 43.0%   (b) 105 m^{-1} (c) 0.0915   **24.2** 76.2%   **24.3** 6350 m^2/mol   **24.4** 4.36

24.5 0.0269 mm   **24.6** 9.47 μmol/L   **24.7** 4.82 × 10^4 m^2/mol

24.8 (a) s^2: $^3S_1$, $^1S_0$; sp: $^3P_{2,1,0}$; $^1P_1$; sd; $^3D_{3,2,1}$; $^1D_2$; pd: $^3F_{4,3,2}$; $^1F_3$; $^3D_{3,2,1}$; $^1D_2$; $^3P_{2,1,0}$;
   $^1P_1$; d^2: $^3G_{5,4,3}$; $^1G_4$; $^3F_{4,3,2}$; $^1F_3$; $^3D_{3,2,1}$; $^1D_2$; $^3P_{2,1,0}$; $^1P_1$; $^3S_1$; $^1S_0$

  (b) s^2: $^1S_0$; sp: same as (a); sd: same as (a); pd: same as (a); d^2: $^1G_4$; $^3F_{4,3,2}$; $^1D_2$;
   $^3P_{2,1,0}$; $^1S_0$

24.9 (b); (c); (e); (j)

24.10 [term; $\mu_z/(-\mu_B)$]: ($^1D_2$; 0, ±1; ±2); ($^3P_2$; 0; ±$\frac{3}{2}$; ±3); ($^3P_1$; 0; ±$\frac{3}{2}$); ($^3P_0$; 0); ($^1S_0$; 0)

24.11 [term; $\mu_z/(-\mu_B)$]: ($^3D_3$; 0; ±$\frac{4}{3}$; ±$\frac{8}{3}$; ±4); ($^3D_2$; 0; ±$\frac{7}{6}$; ±$\frac{7}{3}$); ($^3D_1$; 0; ±$\frac{1}{2}$); ($^1D_2$; 0, ±1; ±2);
   ($^3P_2$; 0; ±$\frac{3}{2}$; ±3); ($^3P_1$; 0; ±$\frac{3}{2}$); ($^3P_0$; 0); ($^1P_1$; 0, ±1); ($^3S_1$; 0; ±2); ($^1S_0$; 0)

24.12 In $v = v_0 + kv_L$, k is given.

  (a) $^3S_1 \leftrightarrow {}^3P_2$: 0; ±$\frac{1}{2}$; ±1; ±$\frac{3}{2}$; ±2. $^3S_1 \leftrightarrow {}^3P_1$: ±$\frac{1}{2}$; ±$\frac{3}{2}$; ±2. $^3S_1 \leftrightarrow {}^3P_0$; 0; ±2

  (b) $^3P_2 \leftrightarrow {}^3D_3$: 0; ±$\frac{1}{6}$; ±$\frac{1}{3}$; ±1; ±$\frac{7}{6}$; ±$\frac{4}{3}$; ±$\frac{3}{2}$; ±$\frac{5}{3}$. $^3P_2 \leftrightarrow {}^3D_2$: ±$\frac{1}{3}$; ±$\frac{2}{3}$; ±$\frac{5}{6}$; ±$\frac{7}{6}$; ±$\frac{3}{2}$; ±$\frac{11}{6}$.
   $^3P_2 \leftrightarrow {}^3D_1$: 0; ±$\frac{1}{2}$; ±1; ±$\frac{3}{2}$ $^3P_1 \leftrightarrow {}^3D_2$: 0; ±$\frac{1}{3}$; ±$\frac{5}{6}$; ±$\frac{7}{6}$; ±$\frac{3}{2}$.
   $^3P_1 \leftrightarrow {}^3D_1$: ±$\frac{1}{2}$; ±1; ±$\frac{3}{2}$. $^3P_0 \leftrightarrow {}^3D_1$: 0; ±$\frac{1}{2}$

  (c) $^2P_{3/2} \leftrightarrow {}^2D_{5/2}$: ±$\frac{1}{15}$; ±$\frac{1}{5}$; ±1; ±$\frac{17}{15}$; ±$\frac{19}{15}$; ±$\frac{7}{5}$.
   $^2P_{3/2} \leftrightarrow {}^2D_{3/2}$: ±$\frac{4}{15}$; ±$\frac{8}{15}$; ±$\frac{4}{5}$; ±$\frac{16}{15}$; ±$\frac{8}{5}$

24.13 (a) (Element; v_L/MHz): (^{11}B; 13.660); (^{13}C; 10.704); (^{19}F; 40.053); (^{31}P; 17.234)

  (b) (Element; 60 MHz; 100 MHz); (^{11}B; 4.3924 T; 7.3206 T); (^{13}C; 5.6054 T; 9.3423 T); (^{19}F;
   1.4980 T; 2.4967 T); (^{31}P; 3.4815 T; 5.8025 T)

24.14 7.4501 MHz   **24.16** 41.3284 pm

24.17 slope = 0.49748 × 10^8 s$^{-1/2}$; intercept = −0.49673 × 10^8 s$^{-1/2}$

24.18 0.675; 0.647; 1.76 × 10^{-4}; 9.90 × 10^{-7}; 2.56 × 10^{-40}; 8.98 × 10^{-25}

24.19 0.0832 cm; 0.533 cm; 11.70 cm   **24.20** 5.9326 kV   **24.21** 52.40 MJ/mol

24.22 1.76 MJ/mol; 1.40 MJ/mol; 1.22 MJ/mol

Chapter 25

25.1 $I = 8.4585 × 10^{-47}$ kg m^2; $r = 228.03$ pm; $\tilde{v}_0 = 3310$ cm^{-1}

25.2 $I = 4.272 × 10^{-47}$ kg m^2; $2B = 13.11$ cm^{-1}

25.3 (X; I/kg m^2; $2B$/cm^{-1})

  (a) 120.1 pm from terminal N atom; 6.675 × 10^{-46} kg m^2, 0.8387 cm^{-1}

  (b) on C atom; 7.173 × 10^{-46} kg m^2; 0.7806 cm^{-1}

  (c) 295.4 pm from H atom; 1.845 × 10^{-45} kg m^2; 0.3035 cm^{-1}

25.4 5.5594 × 10^{-21} J; 1.6678 × 10^{-20} J; 2.7797 × 10^{-20} J; $k = 322.74$ N/m

25.5 [$\tilde{v}_0$/cm^{-1}; E_0/(J/mol)]: ^{79}Br$_2$: 325.321; 1945.85; ^{79}Br^{81}Br: 323.306; 1933.80; ^{81}Br$_2$: 321.379;
   1921.68

25.6 $\tilde{v}_0 = 2989.14$ cm^{-1}; $x_e = 0.0017288$; $\mu = 0.979593$ g/mol; $k = 515.70$ N/m

25.7 $\tilde{v}_1$: λ/nm = 424.6, 573.7; $\tilde{v}_2$: λ/nm = 465.5, 512.8

25.8 (a) Forbidden: all $g \leftrightarrow g$; all $u \leftrightarrow u$; $B_{ig} \leftrightarrow B_{iu}$ ($i = 1, 2, 3$)

      Allowed: $A_g \leftrightarrow B_{1u}$ $B_{1g} \leftrightarrow A_u$ $B_{2g} \leftrightarrow B_{1u}$ $B_{3g} \leftrightarrow B_{1u}$

           $A_g \leftrightarrow B_{2u}$ $B_{1g} \leftrightarrow B_{2u}$ $B_{2g} \leftrightarrow A_u$ $B_{3g} \leftrightarrow B_{2u}$

           $A_g \leftrightarrow B_{3u}$ $B_{1g} \leftrightarrow B_{3u}$ $B_{2g} \leftrightarrow B_{3u}$ $B_{3g} \leftrightarrow A_u$

  (b) Forbidden: $A_1 \leftrightarrow A_2$. Allowed: $A_1 \leftrightarrow A_1$; $A_1 \leftrightarrow E$; $A_2 \leftrightarrow A_2$; $A_2 \leftrightarrow E$; $E \leftrightarrow E$

  (c) Allowed: $A'_1 \leftrightarrow A'_2$ $A'_2 \leftrightarrow A''_1$ $E' \leftrightarrow E'$ $A''_1 \leftrightarrow E''$ $E'' \leftrightarrow E''$

           $A'_1 \leftrightarrow E'$ $A'_2 \leftrightarrow E'$ $E' \leftrightarrow E''$ $A''_2 \leftrightarrow E''$

25.9 (a) $[\mu/(\text{g/mol}); \tilde{\nu}/\text{cm}^{-1}]$: ($H^{35}Cl$: 0.979593; 2990.946) ($H^{37}Cl$: 0.981077; 2988.682)
($D^{35}Cl$: 1.90441; 2145.12)($D^{37}Cl$: 1.91003; 2141.96)

(b) $(I/10^{-47} \text{ kg m}^2; 2B/\text{cm}^{-1})$: ($H^{35}Cl$: 2.6923; 20.795)($H^{37}Cl$: 2.6964; 20.764);
($D^{35}Cl$: 5.2340; 10.697)($D^{37}Cl$: 5.2495; 10.665)

25.10 (a) Let $(\mu_x)_{mn} = (2/L)\int_0^L \sin(m\pi x/L)(qx)\sin(n\pi x/L)\,dx$ and $\mu = qL$. If $n + m = $ even, $(\mu_x)_{mn} = 0$;
if $n + m = $ odd, $(\mu_x)_{mn} = -8mn\mu/\pi^2(m^2 - n^2)^2$

(b) Even to odd allowed, even to even and odd to odd are forbidden.

Chapter 26

26.1 $\Delta U = a[(1/b) - (p/RT)] = a/b$ **26.2** 1.46×10^{-40} C m^2/V; 13.21×10^{-30} C m

26.3 $R/(\text{cm}^3/\text{mol})$: 20.6; 15.56; 16.04   **26.4** 1.66×10^{-40} C m^2/V

26.5 $\langle U_i\rangle_{\text{liq}}/\langle U_i\rangle_{\text{gas}} = 1000$

26.6 $A = n\epsilon_m r_0^6/(6 - n)$; $B = 6\epsilon_m r_0^n/(6 - n)$; $\epsilon/\epsilon_m = [nr_0^6/(n - 6)r^6] - [6r_0^n/(n - 6)r^n]$; $(\sigma/r_0)^{n-6} = 6/n$

26.7 $\epsilon/\epsilon_m = 2(r_0/r)^6 - (r_0/r)^{12}$; $\epsilon/4\epsilon_m = (\sigma/r)^6 - (\sigma/r)^{12}$

26.8 (J/mol): (a) -15.3 (b) -193 (c) -393 (d) -902

26.9 Dipole–Induced Dipole: -2.46×10^{-21} J; dipole–dipole: -43.6×10^{-21} J;
dispersion: -8.67×10^{-21} J

26.10 (Ion; r/pm): (O^{2-}; 140); (F^-; 100); (Ne; 73); (Na^+; 58); (Mg^{2+}; 49); (Al^{3+}; 41); (Si^{4+}; 34)

Chapter 27

27.1 Co: 1.622; Mg: 1.623; Ti: 1.586; Zn: 1.861   **27.2** fcc: 4 atoms; bcc: 2 atoms

27.3 6 atoms   **27.4** fcc: 26.0% empty; bcc: 32.0% empty   **27.5** $1 Cs^+$; $1 Cl^-$

27.6 $4 Na^+$; $4 Cl^-$ **27.7** fcc: 1 hole/atom; bcc: 1.5 hole/atom   **27.8** $r_h/r_a = 2^{1/2} - 1 = 0.414$

27.9 (a) 8 (b) 4 pairs (c) 2 pairs (d) $4 CaF_2$ units (e) $2 TiO_2$ units

27.10 $\sqrt{3} - 1 = 0.732$ **27.11** $\sqrt{2} - 1 = 0.414$

27.12 1 fourfold axis; 4 twofold axes; 5 planes of symmetry; center of symmetry

27.13 (a) $100; 010; 001; \bar{1}00; 0\bar{1}0; 00\bar{1}$ (b) $110; 1\bar{1}0; 101; \ldots$ **27.14** 111 is close-packed

27.15 316.2 pm   **27.16** 154.4 pm   **27.17** $\theta_{111} = 21.68°$; $\theta_{200} = 25.25°$

Chapter 28

28.1 (a) 1 (b) 1.1667 (c) 1.233 (d) 1.293   **28.2** 880 kJ/mol; 770 kJ/mol

28.3 $U_c(\text{NaCl}) = 738$ kJ/mol; $U_c(\text{CsCl}) = 744$ kJ/mol

28.4 (a) CsF; RbF; KF; NaF; LiF (b) KI; KBr; KCl; KF   **28.5** $\sim 1:4$

28.6 281.9 pm; $\kappa = 4.0 \times 10^{-11}$ Pa^{-1} = 4.1×10^{-6} atm^{-1}

Chapter 29

29.2 $S/Nk = 73.3597 + \ln(V/N \text{ m}^3) + \frac{3}{2}\ln(M \text{ mol/kg}) + \frac{3}{2}\ln(T/\text{K})$; $A/NkT = 1.5 - S/Nk$;
$G = A + NkT$; for Ar: $[S/(\text{J/K mol}); A/(\text{kJ/mol}); G/(\text{kJ/mol})]$: 1 atm, 298.15 K:
(154.735; -42.416; -39.937); 1 atm, 1000 K: (179.890; -167.418; -159.104)

29.3

	U/(kJ/mol)		H/(kJ/mol)		S(J/K mol)		A/(kJ/mol)		G/(kJ/mol)	
	298.15	1000	298.15	1000	298.15	1000	298.15	1000	298.15	1000
Trans.	3.718	12.472	6.197	20.786	150.309	175.464	-41.096	-162.992	-38.617	-154.678
Rot.	2.479	8.314	2.479	8.314	41.186	51.248	-9.801	-42.934	-9.801	-42.934
Vib.	13.885	14.905	13.885	14.905	0.001	1.320	13.885	13.585	13.885	13.585
Total	20.082	35.691	22.561	44.005	191.496	228.032	-37.012	-192.341	-34.533	-184.027

Note: $U_0 = H_0 = N_A\frac{1}{2}h\nu = 13.885$ kJ/mol is included.

29.4 (a) $C_v/Nk = (\theta/T)^2 \exp(-\theta/T)[1 - \exp(-\theta/T)]^2$; $C_p(\infty) = R$;

(c) $[\theta/T; C_v/C_v(\infty)]$: (0; 1); (0.5; 0.9794); (1.0; 0.9207); (1.5; 0.8318); (2; 0.7241);
(3; 0.4963); (4; 0.3041); (5; 0.1707); (6; 0.0897)

29.5 (a) 960.3 K (b) 3.770 J/K mol (c) Doubled; 7.540 J/K mol (d) 13.5 mJ/K mol

29.6

	U/(kJ/mol)	H/(kJ/mol)	S/(J/K mol)	A/(kJ/mol)	G/(kJ/mol)
Trans.	3.718	6.197	135.157	-36.579	-34.100
Rot.	2.479	2.479	54.724	-13.837	-13.837
Vib. θ_1	8.324	8.324	0.079	8.301	8.301
$2\theta_2$	8.647	8.647	2.906	7.781	7.781
θ_3	14.052	14.052	0.001	14.052	14.052
Total	37.220	39.699	192.867	-20.282	-17.803

Note: $U_0 = H_0 = N_A\frac{1}{2}h(v_1 + 2v_2 + v_3) = 30.339$ kJ/mol is included.

29.7 $(T/K; N_J/N, J = 0, 2, 4)$: $(10, 1.00; 2.8 \times 10^{-22}; 5.6 \times 10^{-74})$;
$(50; 0.9998; 1.77 \times 10^{-4}; 1.3 \times 10^{-14})$; $(100; 0.9711; 2.89 \times 10^{-2}; 3.3 \times 10^{-7})$

29.8 $(T/K; S/Nk; C_v/Nk)$:
(a) o-H_2: $(100; 2.2016; 0.0332)$; $(150; 2.2495; 0.2509)$; $(200; 2.3645; 0.5610)$
(b) p-H_2: $(100; 0.1774; 0.7370)$; $(150; 0.6349; 1.4241)$; $(200; 1.0506; 1.3945)$
(c) $(T/K; K)$: $(100; 1.586)$; $(150; 2.494)$; $(200; 2.852)$

29.9 (a) 861.7; 19.62   (b) 2.65×10^{33}

29.10 $(T/K; K_c)$: $(800; 0.245)$; $(1000; 0.720)$; $(1200; 1.76)$

29.11 (a) 87   (b) 2.10×10^9 (c) 263   (d) 5.15×10^{23} **29.12** 7.62×10^{-9}

29.13 Lowers the value of μ.

29.14 (a) $p = v^3(2\pi m)^{3/2}(kT)^{-1/2}\exp(W + \frac{3}{2}Nhv)/NkT$
(b) $\Delta H_{vap} = -(W + \frac{3}{2}N_Ahv + \frac{3}{2}RT)$
(c) Diatomic

Chapter 30

30.1 (a) 1.64×10^{10}/s   (b) 1.27×10^{10}/s   (c) 1.64×10^6/s
(d) 2.01×10^{29}/cm^3 s; 9.35×10^{28}/cm^3 s; 2.01×10^{21}/cm^3 s

30.2 (a) $(p/\text{atm}; \lambda/\text{cm})$: $(1; 6.79 \times 10^{-6})$; $(0.1; 6.79 \times 10^{-5})$; $(0.01; 6.79 \times 10^{-4})$
(b) 67.9 m   (c) 1360

30.3 (a) 1.05×10^{-5} cm   (b) 1.59

30.4 (a) doubled   (b) quadrupled   (c) halved   (d) None

30.5 $(\tilde{N} = \text{number/area}; A = \text{area})$. $\lambda = 1/2\sqrt{2}\sigma\tilde{N}$; $Z_1 = 2\sqrt{2}\sigma\langle c\rangle\tilde{N}$;
$Z_{total} = \frac{1}{2}Z_1\tilde{N}A = \sqrt{2}\sigma\langle c\rangle\tilde{N}^2A$; $\lambda = 22$ ft; $Z_1 = 5.4$/min; $Z_{total} = 54$/min

30.6 $\kappa_{H_2}/\kappa_{O_2} = 4$ **30.7** 3.23×10^{-3} W/cm^2

30.8 C_2H_6 has more internal degrees of freedom; hence $\bar{C}_v$ is much larger.

30.9 (a) 1.226   (b) 279 pm   **30.10** 340 pm   **30.11** 862 W   **30.12** 3.1 W/m^2

30.13 (a) 235 W   (b) 19.75 °C; 1.59 °C   **30.14** 1.7 mm   **30.15** 2.7 s

30.16 (a) 2.8 L/min   (b) 6.5 L/min   (c) 1.4 L/min   **30.17** (a) 4.76 atm   (b) 47 L/min

30.18 3.0 min

30.19 $\dot{V} = (\pi/8\eta l)(p_1 - p_2)(b^2 - a^2)[b^2 + a^2 - (b^2 - a^2)/\ln(b/a)]$;
Note: If $b = a(1 + \Delta)$ where $\Delta \ll 1$, then $\dot{V} = (\pi/6\eta l)(p_1 - p_2)a^4\Delta^3$

30.20 0.3266 m Pa s   **30.21** 6.855 kJ/mol   **30.22** 1.55×10^{-3} mol/cm^2 day

30.23 1.10×10^{-10} m^2/s

Chapter 31

31.1 (a) 4.0Ω; 0.25 S   (b) 50 V/m   (c) 1.27×10^8 A/m^2 (d) $3.93 \times 10^{-7} \Omega$ m; 2.55×10^6 S/m

31.2 6.24×10^{18}/s   **31.3** 61 mA   **31.4** 79 μV   **31.5** $9.73 \times 10^{-8} \Omega$ m

31.6 (a) 1.89×10^{-4} (m/s)/(V/m)   (b) 4.71   (c) 24.0 nV   **31.7** 1.12×10^{-10} m^3/C

31.8 188 cm^3 **31.9** 10.9 hr   **31.10** 6.68 g   **31.11** 26.70 mA   **31.12** 30.60 mA

31.13 720 hr   **31.14** (a) 0.59389/m   (b) 0.57514 S/m

31.15 (a) 2.58476×10^5 S/m   (b) 6.517 S/m   (c) 0.2492 m

31.16 (a)(b) [Ion: $u_i/(10^{-8}$ m^2/V s); $v_i/(\mu$m/s)]: (H$^+$: 36.256; 14.5); (Na$^+$: 5.193; 2.08); (Ca^{2+}: 6.167; 2.47); (La^{3+}: 7.224; 2.89); (OH$^-$: 20.55; 8.22); (Br$^-$: 8.099; 3.24); (SO$_4^{2-}$: 8.294; 3.32); (P$_2$O$_7^{4-}$: 9.94; 3.98)

31.17 (a) 73.55×10^{-4} S m^2/mol   (b) 4.574 μm/s   (c) $t_+ = 0.6426$

31.18 (salt: $\kappa/(S/m)$; $R/k\Omega$): (AgNO$_3$: 0.133; 3.75); (HCl: 0.426; 1.17); (CaCl$_2$: 0.272; 1.84); (MgSO$_4$: 0.266; 1.88); [La$_2$(SO$_4$)$_3$: 0.898; 0.557]

31.19 (a) $\Delta n_c = -\Delta n_a = t_+ Q/F$ (b) $\Delta n_c = -t_- Q/F$; $\Delta n_a = -t_+ Q/F$
(c) $\Delta n_c = t_+ Q/F$; $\Delta n_a = t_- Q/F$

31.20 0.83   **31.21** $(t_+)_{\text{ave}} = 0.4888$ **31.22** 0.33   **31.23** 0.61

31.24 0.1792; 0.6038; 0.5095; 0.5620; 0.5228

31.25 $t(\text{H}^+) = 0.111$; $t(\text{Ca}^{2+}) = 0.379$; $t(\text{Cl}^-) = 0.510$ **31.26** (HCl/NaCl) = 0.462

31.27 (a) 4.01×10^{-8} m^2/V s   (b) 6.0 μm/s   **31.28** 126.3×10^{-4} S m^2/mol

31.29 425.92×10^{-4} S m^2/mol

31.30 (Soln: $\Lambda/10^{-4}$ S m^2/mol): (HCl: 424.58; 421.17; 410.38); (KCl: 148.90; 146.86; 140.40); (LiCl: 114.16; 112.29; 106.37)

31.31 Na$^+$: 97×10^{-4} S m^2/mol; Cl$^-$: 50.9×10^{-4} S m^2/mol   **31.32** 382.19×10^{-4} S m^2/mol

31.33 (a) 388.1×10^{-4} S m^2/mol   (b) 1.840×10^{-5} (c) 0.128; 0.0919; 0.0660

31.34 (a) 92.1/m   (b) 3.65×10^{-4} S m^2/mol   (c) 0.0134   (d) 1.83×10^5

31.35 1.08×10^{-10} **31.36** 6.371×10^{-9}

31.37 (c) [v/cm^3; $\kappa/(S/m)$]: (0; 4.26); (10; 3.05); (25; 1.84); (40; 1.04); (45; 0.823); (50; 0.632); (55; 0.720); (60; 8.01); (75; 1.00); (90; 1.16); (100; 1.25)

31.38 [v/cm^3; $\kappa/(S/m)$]: (0; 0.0052); (10; 0.154); (40; 0.405); (45; 0.431); (50; 0.455); (55; 0.552); (60; 0.639); (90; 1.03); (100; 1.13)

31.39 (a) [Ion: $D_i^\infty/(10^{-9}$ m^2/s)]: (H$^+$: 9.315); (OH$^-$: 5.280); (Na$^+$: 1.334); (Ag$^+$: 1.648); (Ca^{2+}: 0.7922); (Cu^{2+}: 0.7137); (NO$_3^-$: 1.903); (SO$_4^{2-}$: 1.0654); [Co(NH$_3$)$_6^{3+}$: 0.9036]
(b) [Ion: $u_i/(m/sN)$; $u_i/(m^2/s V)$: [Ag$^+$: 4.004(10^{11}); 6.416(10^{-8})]; [Cu^{2+}: 1.734(10^{11}); 5.555(10^{-8})]

31.40 $D^\infty/(10^{-9}$ m^2/s): 3.355; 1.247; 0.8495; 0.8266

31.41 (Cpd: $\Delta\phi$/mV): (CuSO$_4$: 3.46); (HCl: -35.9); (K$_2$SO$_4$: -10.1); [La(NO$_3$)$_3$: 15.4]

Chapter 32

32.1 (a) $d(\xi/V)/dt = (1/RT)\,dp/dt$ (b) $p = p^\infty(1 - \frac{1}{2}e^{-kt})$; Plot $\ln(p^\infty - p)$ vs. t
(c) 2790 s; 15,800 s   (d) 0.916

32.2 (a) $d(\xi/V)/dt = -(1/\epsilon l)\,dA/dt$ (b) $(1/A) = (1/A_0) + (k/\epsilon l)t$; Plot $1/A$ vs. t.

32.3 (a) $c^{1/2} = c_0^{1/2} - \frac{1}{2}kt$; Plot $c^{1/2}$ vs t.
(b) $c^{-1/2} = c_0^{-1/2} + \frac{1}{2}kt$; Plot $c^{-1/2}$ vs. t. $c^{1-n} = c_0^{1-n} + (n-1)kt$; Plot c^{1-n} vs. t.
(c) $\tau = (2^{n-1} - 1)/(n-1)kc_0^{n-1}$

32.4 (a) 2.08×10^{-3}/s   (b) 138 s   **32.5** (a) 0.0231/min   (b) 0.198

32.6 (a) First order   (b) 47600 s   **32.7** (a) 1200 s   (b) 3970 s   **32.8** 72.2 hr

32.9 (a) 75.4%   (b) 1060 days   **32.10** 7400/min   **32.11** 1.5×10^9 yr   **32.12** 6000 yr

32.13 17 min

32.14 (t/min; N/N_0): (30; 1.32); (60; 1.74); (75; 2.00); (90; 2.30); (150; 4.00)

32.15 (a) 7.18%   (b) 6.99%   (c) 6.93%   **32.16** (a) 29000 s   (b) 150000 s

32.17 (a) $k = 0.278$ L/mol-min; $\tau = 180$ min   (b) 24 min

32.18 (a) NO: second; H$_2$: first   (b) $d(\xi/V)/dt = -(RT)^{-1}(dp/dt)$
(c) $-dp/dt = [k/(RT)^2][2p - (2 - x_0)p_0]^2[2p - (1 + x_0)p_0]$

32.19 A: second; B: first; $k = 3.2 \times 10^8$ L^2/mol^2 s

32.20 (a) 1.735   (b) 0.0830 (L/mol)$^{0.735}$/s

32.21 (a) 464000 s   (b) 232000 s

32.22 $\ln[(y_1 - y)/y_1] - \ln[(y_2 - y)/y_2] = 2k_1 K^{-1/2}at$, where $y = \xi/V$; $k_{-1} = k_1/K$;
$y_1 = aK^{1/2}/(2K^{1/2} - 1)$; $y_2 = aK^{1/2}/(2K^{1/2} + 1)$

32.23 $\ln[(y_e - y)/y_e] = -k_1(1 + 1/K)t$; $y_e = aK/(1 + K)$

32.24 $\ln[(y_1 - y)/y_1] - \ln[(y_2 - y)/y_2] = -k_1(1 + 16a/K)^{1/2}t;$
$y_1 = (K/8)[-1 + (1 + 16a/K)^{1/2}]; y_2 = (K/8)[-1 - (1 + 16a/K)^{1/2}]$

32.25 53.6 kJ/mol   **32.26** (a) 150 kJ/mol; 9.93×10^9 L/mol s   (b) 0.0235 L/mol s

32.27 $A = 4.42 \times 10^{15}$/min; $k = 0.00231$/min   **32.28** 233   **32.29** 219 kJ/mol

32.30 1.11×10^{-6} mol/L

32.31 (a) $d[HBr]/dt = 2k_1[Br_2]$
(b) $d[HBr]/dt = 2k_1[Br_2]\{k_2[H_2] - k_3[HBr]\}/\{k_2[H_2] + k_3[HBr]\}$

32.32 $d[CH_4]/dt = k_2(k_1/k_4)^{1/2}[CH_3CHO]^{3/2}$ **32.33** 200 kJ/mol   **32.34** 30 times larger

32.35 (a) CH_4 and CH_2CO (c) 260 kJ/mol

32.36 (a) $-d[O_3]/dt = 2k_1k_2[O_3]^2/\{k_{-1}[O_2] + k_2[O_3]\}$
(b) When $k_{-1}[O_2] \ll k_2[O_3]$, then $-d[O_3]/dt = 2k_1[O_3]$

32.37 $-d[N_2O_5]/dt = [2k_1k_3/(k_{-1} + 2k_3)][N_2O_5]$

32.38 (a) $\tau = 1/(k_f + k_r)$ (b) $\tau = 1/(k_f + 4k_r\bar{c}_A)$

32.39 $c_A/c_0 = \exp(-k_1t);$ $c_B/c_0 = [k_1/(k_2 - k_1)][\exp(-k_1t) - \exp(-k_2t)];$
$c_C/c_0 = 1 - [k_1k_2/(k_2 - k_1)][(1/k_1)\exp(-k_1t) - (1/k_2)\exp(-k_2t)]$

32.40 $k_{H^+} = 4.58 \times 10^{-4}$ L/mol min; $k_{ClCH_2COOH} = 2.35 \times 10^{-5}$ L/mol min

32.41 $v_{max} = 0.0377$ mol/L s; $K_m = 0.0263$ mol/L; $k_2 = 9.41 \times 10^6$/s   **32.43** 2.5×10^{-3} mol/L s

32.44 $v = k_2[E]_0\{[S]_0 - [E]_0[P]/K_2 - [P]/K_1K_2\}/\{[S]_0 + K_m + (k_{-2}/k_1)[P]\}$

Chapter 33

33.1 (a) 0.273 L/mol s   (b) 380 pm   **33.2** 175 kJ/mol   **33.3** 0.00753 L/mol s

33.4 $H_2 + I_2$: $k = 0.0517$ L/mol s; $HI + HI$: $k = 3.26 \times 10^{-4}$ L/mol s

33.5 300 K: $(p/\text{atm}; Z_3/Z_2)$: $(0.1; 2.93 \times 10^{-4})$; $(1; 2.93 \times 10^{-3})$; $(10; 0.0293)$; $(100; 0.293)$.
600 K, values are $\frac{1}{2}$ as large

33.6 0.296; 0.0370

33.8 If (ABC) is linear, $A \approx 4 \times 10^9$ L/mol s; if (ABC) is nonlinear, $A \approx 4 \times 10^{10}$ L/mol s

33.9 $\Delta S^{\ddagger}/(\text{J/K mol}) = -180; -150$ **33.10** $\Delta S^{\ddagger}/(\text{J/K mol})$: 120; 94; -74

33.11 (a) Decreases   (b) Increases   (c) No effect   **33.12** $A = N_A\pi r_{AB}^2(8kT/\pi\mu)^{1/2}$

33.13 7×10^5 m³/mol s; observed: 4×10^6 m³/mol s

Chapter 34

34.1 10^{-3} mol/m² s   **34.2** k_{500}/k_{400}:   (a) 64 000   (b) 35

34.3 (a) HI weakly adsorbed   (b) HI strongly adsorbed   (c) SO_3 very strongly adsorbed
(d) CO_2 more strongly adsorbed than H_2

34.4 (a) Pt: $\pm 2.5 \mu V$ (b) ± 2.5 MV   **34.5** $i_a = 0.41$ A/cm²; $i_c = 1.02$ A/cm²

34.6 (a) 6.2×10^{16}/cm² s   (b) 62/s   **34.7** (a) Decreases by 9.65 kJ/mol   (b) $i_+/i_0 = 49$

34.8 $[-i/(\text{A/cm}^2); \eta]$: $(10^{-3}; -5.61 \mu V)$; $(10^{-2}; -56.1 \mu V)$; $(10^{-1}; -564 \mu V)$; $(1; -5.93$ mV$)$

34.9 (a) -2.5×10^{-9} A/cm²   (b) 2.9×10^{-6}

34.10 $[A_{Pt}/A_{Cd}; \phi_{corr}/V; i_{corr}/(\text{mA/cm}^2)]$: $(0.01; -0.447; 1.86)$; $(0.1; -0.413; 9.71)$;
$(1; -0.374; 45.4)$; $(10; -0.335; 211)$; $(100; -0.296; 978)$

34.11 $[A_c/A_a; \phi_{corr}/V; i_{corr}/(\text{mA/cm}^2)]$: $(1; -0.462; 2.5 \times 10^{-6})$; $(10^3; -0.462; 2.50 \times 10^{-3})$;
$(10^6; -0.447; 1.86)$; $(10^9; -0.335; 211)$

34.12 4.66 m²/mol   **34.13** 3.0×10^{19}/s   **34.14** 0.0393

34.15 $(\lambda/\text{nm}; T)$: $(340; 0.986)$; $(320; 0.813)$; $(310; 0.437)$; $(300; 0.0479)$; $(290; 2.5 \times 10^{-4})$;
$(280; 4.0 \times 10^{-15})$; $(260; 1.0 \times 10^{-39})$; $(240; 2.5 \times 10^{-28})$; $(220; 1.0 \times 10^{-6})$

34.16 2.20×10^{18}; $I_a = 1.83 \times 10^{15}$ s^{-1} **34.17** 4.55×10^{18}; 2.41 mJ/s

34.18 $\phi_F = 0.33$; $\tau_F = 1.7 \times 10^{-7}$ s   **34.19** $k_{ISC}^S = 3 \times 10^6$/s; $\phi_F = 0.25$

34.20 0.044/s; 0.36/s; 0.82   **34.21** 0.045/s; 0.26/s; 7.3

34.22 10^6/s; 0.016/s; 0.42/s; 2.4×10^6/s

34.23 (a) $\phi = k_2[A]/\{k_2[A] + k_3\}$
(b) Fluorescence is weak; most A* are deactivated by collision before fluorescence can occur.

34.24 $d[CO]/dt = k_2(I_a/2k_4)^{1/2}[CH_3CHO]$; $\phi_{CO} = [k_2/(2k_4I_a)^{1/2}][CH_3CHO]$

34.25 (a) $-d[O_3]/dt = 2\phi_1 I_a/\{1 + (k_3/k_2)[O_2][M]/[O_3]\}$
(b) $\phi_0 = 2\phi_1/\{1 + (k_3/k_2)[O_2][M]/[O_3]\}$
(c) $\phi_1 = 1$
34.26 $d[H_2]/dt = \phi_1 I_a + \phi_2 I_a + (2\phi_2 I_a k_4/k_5)[M]$; $\phi_{H_2} = \phi_1 + \phi_2 + (2\phi_2 k_4/k_5)[M]$

Chapter 35

35.1 4.1×10^{-3} **35.2** (a) 2.5 kg/mol   (b) 3.0 kg/mol
35.3 $[\phi; x; \Delta S_{mix}/(J/K \text{ mol})]$: (0; 0; 0); (0.2, 0.00048; 1.86); (0.4; 0.0013; 4.26); (0.6; 0.0029; 7.63); (0.8; 0.0076; 13.4); (1; 1; 0)
35.4 112 Pa   **35.5** $\langle M \rangle_N = 300$ kg/mol; $w = 1.91 \times 10^{-21}$ J $= 1.15$ kJ/mol
35.7 (Raoult; Flory; mod. Flory): (0.9999631; 0.999820; 0.9999551); (0.99808; 0.824; 0.907);
35.8 10.08 kg/mol   **35.9** 9.874 kg/mol   **35.10** 78 kg/mol   **35.11** 18 kg/mol
35.12 $K = 3.00 \times 10^{-3}$ m^3/kg; $a = 0.640$
35.13 $[\theta; (I_\theta/I_0)/(I_\theta/I_0)_{max}]$: (0; 1); $(\frac{1}{8}\pi; 0.927)$; $(\frac{1}{4}\pi; 0.75)$; $(\frac{3}{8}\pi; 0.573)$; $(\frac{1}{2}\pi; 0.5)$
35.14 $I_\theta(540 \text{ nm})/I_\theta(750 \text{ nm}) = 3.72$ **35.15** $K = 3.38 \times 10^{-5}$ m^2 mol/kg^2; $M = 78.2$ kg/mol
35.16 $K = 1.00 \times 10^{-7}$ m^2 mol/kg^2; $R_{90} = 0.02$/m   **35.17** 0.934/m; 123 kg/mol
35.18 (a) $K = 9.18 \times 10^{-5}$ m^2 mol/kg^2; 2.46 times larger   (b) 2.30/m
35.19 (a) $-\Delta I = 0.0156$ W/m^2;   (b) 47.8 kg/mol   **35.20** (a) 150 kg/mol   (b) 4.46 nm
35.21 (a) 576 nm/s   (b) 268 min   (c) 5.47×10^{-11} kg/s   (d) 7.40×10^{-11} m^2/s
35.22 5.22×10^{-14} s   **35.23** 69.9 kg/mol   **35.24** $c_7/c_6 = e^{409}$ **35.25** $c_7/c_6 = e^{-0.00684}$

Index